Tenebrionoidea

Catalogue of Palaearctic Coleoptera

VOLUME 5

The titles published in this series are listed at *brill.com/cpc*

Leptoderis collaris (Linnaeus, 1767) © David Schimrosczyk

Tenebrionoidea

Revised and Updated Second Edition

VOLUME 5

Edited by

Dariusz Iwan
Ivan Löbl

TUTA SUB AEGIDE PALLAS · 1683 ·

BRILL

LEIDEN | BOSTON

Cover illustration: *Leptoderis collaris* (Linnaeus, 1767) © David Schimrosczyk.

Date of publication: 17 September 2020.

The index of the species-group names is available on site of the Koninklijke Brill NV, Leiden (10.6084/m9.figshare.12063672).

The publication of the work was supported by the "Muséum de la Ville de Geneve", Geneva, Switzerland.

VILLE DE GENÈVE

muséum genève

The Library of Congress Cataloging-in-Publication Data is available online at http://catalog.loc.gov
LC record available at http://lccn.loc.gov/2015018977

Typeface for the Latin, Greek, and Cyrillic scripts: "Brill". See and download: brill.com/brill-typeface.

ISBN 978-90-04-43498-1 (hardback)
ISBN 978-90-04-43499-8 (e-book)

Contents

Preface

Assessing the diversity of life has become an important challenge of the Anthropocene, as the massive impact of human activity on natural habitats is associated with the mass extinction of plants and animals. The beetle superfamily Tenebrionoidea is one of the largest groups of animals in terms of species and ecological diversity, and is thus also considerably impacted by the changing landscapes and biotic crisis. Members of Tenebrionoidea are found in a wide array of terrestrial habitats, ranging from tropical forest canopy to extremely depauperate desert soils. Numerous species are associated with fungi or flowering plants, many live in decaying wood and forest floor litter, while others are found on banks of streams and some thrive as synanthropic pests. Adults of many species possess unusual adaptive characters attracting the interest of experts and the public. The larvae are usually inconspicuous, though they are often predators or parasites of other insects. Hence, the study of Tenebrionoidea addresses a variety of evolutionary, ecological and conservation issues, and is recognized by many institutions and individuals as an important study subject.

The present work is based on the first edition of the Catalogue of Palaearctic Coleoptera and follows the same structure: all taxa (including synonyms) are listed with their bibliographic data, primary references are cross-checked, all type species and primary references are verified, and the distributional data compiled using known sources and unpublished information available to the individual authors. Nevertheless, to avoid confusion users of the works should pay adequate attention to the information given in the introductory chapters.

Since the first edition was issued, a great, world-wide effort has been made to fill remaining gaps. In particular, the phylogeny of many taxa has been supported by more robust data, leading to changes in classification, many newly discovered species occurring in previously poorly explored areas have been discovered and described, the ambiguous nomenclature has been fixed for many taxa, and the knowledge of species and subspecies distribution has been considerably expanded.

Taxonomic Information

The present Catalogue includes all available names, both valid and invalid, of extant beetle taxa described before January 1, 2020 and known to occur in the Palaearctic Region, as it is defined below. The classification is subservient to phylogeny, but in taxonomic praxis cannot show all uncertainties resulting from phylogenetic analyses. The higher classification largely follows that in Bouchard & *al.* (2011) and Ślipiński & *al.* (2011). All taxa below subfamily rank are arranged alphabetically within the higher taxon and the synonyms follow the respective valid name alphabetically.

With respect to present facilities in access to information, the editors adopted the policy that the gap between publication date of the Catalogue and the date of the corresponding deadline entry should be as short as possible. Therefore, also data published in 2019 are included although some sources may have remained unknown to the authors. The present volume covers 00,000 names of taxa, including synonyms of the genus- and species-groups.

Extinct taxa, names rejected by the ICZN (*International Code of Zoological Nomenclature, Fourth Edition.* London: International Trust for Zoological Nomenclature, 1999), misspellings, misidentifications and other nomina nuda are not included in the Catalogue. However, concepts that are important for nomenclatural purposes, may be included. Similarly, all infrasubspecific names, such as those established as "morpha", "natio", or "race", "subvariety" and "aberration", or proposed as variety and form of a subspecies or another variety, and names published, e.g., as *A-us b-us c-us*, but specified in the text that they are actually proposed for a "natio" or "race", etc., are not considered subspecific and are therefore excluded from the Catalogue. Names proposed as varieties and forms before 1961 are included, if deemed subspecific under the provisions of the ICZN, Article 45.6. Unjustified emendations may be included.

The currently valid names of the family-group taxa include the name of the author and the year of the publication. Their synonyms are not listed.

The names of the genus-group taxa are given with the name of the author, and the year and page of publication. The page given is the page where the name and the actual description of the taxon is printed. The type species of all genus-group names are given in their original combination. If the type species is currently regarded as a junior synonym, the valid senior synonym is given in brackets in its original combination.

The names of the species-group taxa are given with the name of the author, and the year and page of publication. The page given is the page where the name and the actual description of the taxon is printed. In bi-languaged Chinese/English and Japanese/English publications both respective pages of the actual description may be given. For species-group taxa subsequently transferred to another genus, the name of the original genus is given in parentheses, following the page of publication.

Some authors (e.g., V. Apfelbeck, H. John) published the same description twice, or even more times, in separate papers. Such publications produce, de facto, primary homonyms and objective synonyms. The first publication in such cases is referred to as indicated above, followed by the mark = , the year and first page of the subsequent description/s in square brackets. This is particularly important for taxa that are erroneously associated with their junior description.

The following symbols, all given in square brackets following the page of publication, or the original combination when applicable, are used for taxonomic information : UE for unjustified emendations, HN for homonyms, RN for replacement names, URN for unnecessary replacement names, NO for nomina oblita, NP for nomina protecta, DA for doubtful assignment, and EA for erroneous assignment.

Taxa considered incertae sedis and nomina dubia are listed separately at the end of the nearest applicable taxon.

Distributional Information

The limits of the Palaearctic region, as those of other biogeographical regions, are arbitrarily defined (Map 1). For practical reasons, the boundaries of the Palaearctic Region, as they were established for the Catalogue (see above), usually follow national boundaries. The region includes Europe, Africa north of the Sahara, and Asia except for the part that is arbitrarily defined as belonging to the Oriental Region.

For each species and subspecies an outline of its present distribution is given. Fossil records are not considered. The information is given by means of symbols, presented at three levels.

The first level is the subdivision of the Palaearctic Region into three main parts, Europe (letter **E**, bold), North Africa (letter **N**, bold) and Asia (letter **A**, bold).

Europe includes the Azores, Iceland and Turkey west of the Bosporus. The eastern boundaries are a matter of controversy. In the Catalogue, Europe includes Russia west of the main ridge of the Ural Mountains, the Permsk Oblast, Bashkortostan Republic and Orenburskaya Oblast, and the small part of Kazakhstan west of the Ural River. It includes the Caucasian republics of Georgia, Armenia and Azerbaijan. The south-eastern boundaries are the political boundaries of the Asian part of Turkey, Iran, Kazakhstan, and the Caspian and Black seas. Taxa known from "Russia" are in absence of detailed information given under the symbol RU. The limits of Ukraine are as in the previous volumes of the Catalogue, and Abkhazia and Nahorny-Karabagh are for practical reasons given under Georgia and Armenia, respectively.

Former Yugoslavia (Serbia and Montenegro from 2003 to 2006) has been split into Serbia and Montenegro, and Kosovo. The term Yugoslavia, listed in the body of the Catalogue under the symbol YU, is still used in absence of more exact knowledge of the distribution of the respective taxa.

North Africa includes Morocco (incl. Western Sahara), Algeria, Tunisia, Libya and Egypt west of the Suez Canal, and the Canary and Madeira islands.

Asia includes Sinai and the Arabian Peninsula (including Suqutra), Turkey east of the Bosporus, the Middle East and Central Asian countries, Russia east of the main ridge of the Ural mountains, Korea, Japan (including Ryukyu [= Nansei] Islands and the Japanese Pacific Islands), the entire People's Republic of China, Taiwan, Bhutan, Nepal, North India along the base of the Himalaya (Arunachal Pradesh, Uttarakhand (= Uttaranchal), north-western area of former Uttar Pradesh, Himachal Pradesh, Jammu & Kashmir) and all of Pakistan. Thus, India is the only state for which the strict political boundaries are not respected. Large parts of Uttar Pradesh south of Nepal are overpopulated plains. Information on Coleoptera from this North Indian state is based almost exclusively on its Himalayan districts lying west of Nepal, which are in the present state Uttarakhand.

The second level of the geographic information is provided by two-letter symbols for countries, major areas of Russia and North Indian states, and by three-letter symbols for provinces of mainland China and for Taiwan (see Table 1, Map 3).

The symbols are arranged in alphabetical order within the first-level subdivision. Some, usually older, distributional records cannot be accommodated within the structure of the Catalogue (e.g., Arabia, Caucasus, North India, Siberia). Such information is given in quotation marks (e.g. "Caucasus") behind the last symbol of the respective first-level symbol. Russia is subdivided into six major sub-regions (Map 2), each of which has its own two-letter symbol. These, as well as the symbol RU for Russia, are used only when more detailed information is not available. Similarly the symbol CH for the People's Republic of China, as well as the seven two-letter symbols for China's major regions, are used only in the absence of more detailed geographical information.

In general, the published distributional information is based on both identified material in collections and on published records, scattered in an enormous number of taxonomic and faunistic papers that are virtually impossible to review in their entirety. Revisions of collections reveal a high proportion of misidentifications, attaining 40% of specimens in some large museums. The degree of identification reliability and of the records derived from identifications is correlated to faunal diversity and quality of systematic revisions. Thus, the reliability in general increases from south to north and from poorly studied groups to "popular" groups. At present, a number of modern catalogues or check-lists, covering the beetle faunas of many European and some extra-European countries or archipelagos, are available. The use of data contained in these and other faunistic works is left to the discretion of the authors, who may also add unpublished information available to them. The second-level geographic information is not necessarily exhaustive, it should rather be considered as a base for future faunal research. The third-level geographic information concerns species and subspecies with restricted distribution. Taxa of this category may be strict endemic, or taxa comparatively widely distributed in one area but restricted in another area. For example, the distributional record of a species widely distributed in North Africa with isolated occurrence on Pantelleria would appear as follows: E: IT (Pantelleria) N: AG MO TU. The third-level information is facultative. It is given in parentheses after the respective second-level symbol. The official language of the respective state is used for records in languages using the Latin alphabet, or it is transliterated from the Cyrillic alphabet. Records in languages using non-Latin or Cyrillic characters (e.g., Chinese or Japanese pictographs) are translated into English, and the translated geographical terms are spelled as closely as possible to those used in the Times Atlas® or in other well-known sources. Detailed geographical information may refer to natural geographical features such as islands, mountains, lakes, valleys, caves, or to administrative entities, such as districts.

The extralimital distribution of some Palaearctic species is indicated by three letter symbols in bold, cosmopolitan species by the symbol COS, all located at the end of the respective geographical information (see Table 1). Introductions are indicated by the letter "i" (e.g., Ei: GB). The extralimital regions for the needs of the Catalogue are defined as follows (see Map 1): Nearctic (**NAR**) north of Mexico; Neotropical (**NTR**): south of the United States; Afrotropical (**AFR**): south of the North African states included in the Palaearctic Region; Oriental (**ORR**): areas south of the People's Republic of China and Taiwan, areas south of the Himalaya in India, the Philippines Malaysia and Indonesia south to the Lydekker line; Australian (**AUS**): south of the Lydekker line and Pacific (**PAC**).

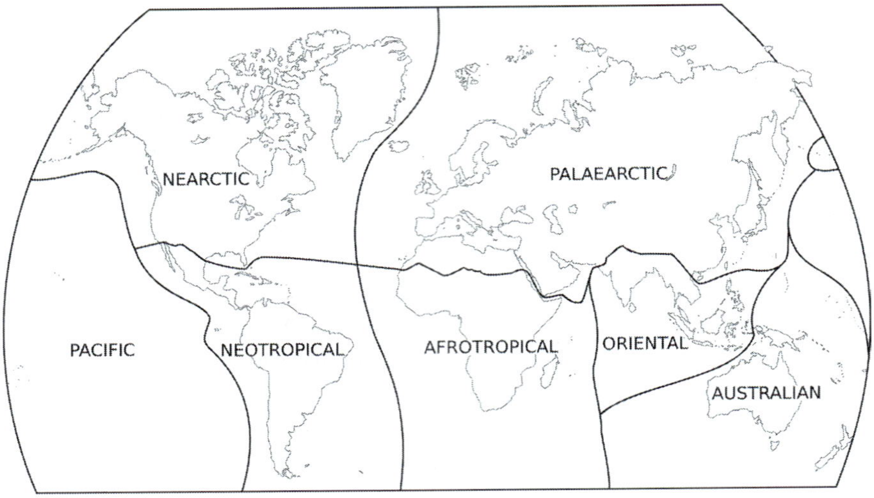

MAP 1 *The limits of the geographic Regions as defined for the purpose of this Catalogue*

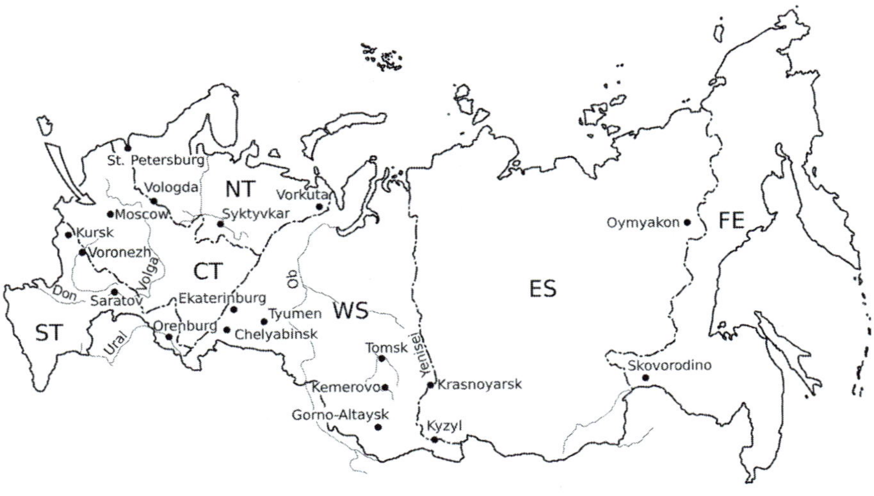

MAP 2 *Subdivisions of Russia*

MAP 3 *Subdivisions and provinces of the People's Republic of China*

TABLE 1 *Geographical symbols*

E	**Europe**	KZ	Kazakhstan
AB	Azerbaijan	LA	Latvia
AL	Albania	LS	Liechtenstein
AN	Andorra	LT	Lithuania
AR	Armenia	LU	Luxembourg
AU	Austria	MA	Malta
AZ	Azores	MC	North Macedonia
BE	Belgium	MD	Moldova (Moldavia)
BH	Bosnia Herzegovina	ME	Montenegro
BU	Bulgaria	NL	The Netherlands
BY	Belarus	NR	Norway
CR	Croatia	NT	Russia: North European Territory
CT	Russia: Central European Territory	PL	Poland
CZ	Czech Republic	PT	Portugal
DE	Denmark	RO	Romania

TABLE 1 *Geographical symbols* (cont.)

EN	Estonia	RU	Russia
FA	Faeroe Islands	SB	Serbia
FI	Finland	SK	Slovakia
FR	France (incl. Corsica, Monaco)	SL	Slovenia
GB	Great Britain (incl. Channel Islands)	SP	Spain (incl. Gibraltar)
GE	Germany	SR	Svalbard (Spitzbergen)
GG	Georgia	ST	Russia: South European Territory
GR	Greece (incl. Crete)	SV	Sweden
HU	Hungary	SZ	Switzerland
IC	Iceland	TR	Turkey
IR	Ireland	UK	Ukraine
IT	Italy (incl. Sardinia, Sicilia, San Marino)	YU	Serbia and Montenegro
KO	Kosovo		
N	**North Africa**	CI	Canary Islands
AG	Algeria	EG	Egypt
LB	Libya	TU	Tunisia
MO	Morocco (incl. Western Sahara)	SG	Salvage Islands
MR	Madeira Archipelago		
A	**Asia**	NO	China: Northern Territory
AE	Arab Emirates	NP	Nepal
AF	Afghanistan	NW	China: Northwest Territory
AP	India: Arunachal Pradesh	OM	Oman
BA	Bahrain	PA	Pakistan
BT	Bhutan	QA	Qatar
CE	China: Central Territory	RU	Russia
CH	China	SA	Saudi Arabia
CY	Cyprus	SC	South Korea
ES	Russia: East Siberia	SD	India: Sikkim, Darjeeling District
FE	Russia: Far East	SE	China: Southeastern Territory (incl. Macao, Hong Kong)
HP	India: Himachal Pradesh		
IN	Iran	SI	Egypt: Sinai
IQ	Iraq	SW	China: Southwestern Territory
IS	Israel	SY	Syria
JA	Japan	TD	Tajikistan
JO	Jordan	TM	Turkmenistan

KA	India: Kashmir
KI	Kyrgyzstan
KU	Kuwait
KZ	Kazakhstan
LE	Lebanon
MG	Mongolia
NC	North Korea
NE	China: Northeast Territory
TR	Turkey
UP	India: Uttarakhand (Uttaranchal, former part of Uttar Pradesh)
UT	Uttar Pradesh
UZ	Uzbekistan
WP	China: Western Plateau
WS	Russia: West Siberia
YE	Yemen (incl. Suqutra)

China: Provinces, Autonomous Regions or Municipalities, and Taiwan

ANH	Anhui (Anhwei)
BEI	Beijing (Peking or Peiping)
CHQ	Chongqing
FUJ	Fujian (Fukien)
GAN	Gansu (Kansu)
GUA	Guangdong (Kwantung)
GUI	Guizhou (Kweichow)
GUX	Guangxi (Kwangsi)
HAI	Hainan
HEB	Hebei (Hopeh)
HEI	Heilongjiang (Heilungkiang)
HEN	Henan (Honan)
HKG	Hong Kong
HUB	Hubei (Hupeh)
HUN	Hunan
JIA	Jiangsu (Kiangsu)
JIL	Jilin (Kirin)
JIX	Jiangxi (Kiangsi)
LIA	Liaoning
MAC	Macao
NIN	Ningxia (Ningsia)
NMO	Nei Mongol (Inner Mongolia)
QIN	Qinghai (Tsinghai)
SCH	Sichuan (Szechwan)
SHA	Shaanxi (Shensi)
SHG	Shanghai
SHN	Shandong (Shantung)
SHX	Shanxi (Shansi)
TAI	Taiwan (Formosa)
TIA	Tianjin (Tsientsin)
XIN	Xinjiang (Sinkiang)
XIZ	Xizang (Tibet)
YUN	Yunnan
ZHE	Zhejiang (Chekiang)

World Zoogeographical Regions

AFR	Afrotropical Region
AUR	Australian Region
NAR	Nearctic Region
NTR	Neotropical Region
ORR	Oriental Region
PAC	Pacific Region

Bibliographic Information

The total number of references in the present volume is 5556. The references are given to primary sources of the genus-group and species-group names included in the Catalogue, and to works supporting the New Acts and other statements published in the present volume. No bibliographical reference is given to family-group names, to designation of type-species, to secondary nomenclatural and taxonomic acts, or to sources of any other information, these are in Bouchard *et al.* 2011. Names of all authors, if applicable, are given for each reference. An effort was put forth to unify the spellings of the names. For names that appeared in publications in two or more different orthographies, the one used in the Catalogue is that which was considered to be correct after thorough verification (e.g., the spelling Josef Müller for the author who also published as Giuseppe Müller). In such cases the alternative spelling follows the original one in brackets. The original name is used for authors that subsequently adopted additional names, or honorary epithets (e.g., A. Semenov, and not A. Semenov-Tian-Shanskyi). Initials are given for all authors, even if they are missing in the original publication. Names published in two or more alternative spellings, that seem to be equally correct, are spelled following Schmitt, Hübner & Gaedike (1998): *Nomina Auctorum. Auflösung von Abkürzungen taxonomischer Autoren-Namen. Nova Supplementa Entomologica* **11**: 3–189. An exception is made for F.L. Castelnau de Laporte, who is referred to as Laporte in recent major works. Therefore, the name Laporte is arbitrarily used in the Catalogue.

Names in Cyrillic are transliterated as for any other Cyrillic text, unless an alternative spelling was preferred by the author, or the alternative name is generally used in the literature (e.g., Jakobson, Tschitschérine). It should be noted that lowercase prepositions such as French "de" and German "von" are not given with the author's name in the list of the taxa.

Titles of references are given in full, i.e. not abbreviated, with subtitles, as originally published. Text written in Cyrillic is transliterated. Titles in languages using non-Latin or non-Cyrillic characters are translated into English and are presented in square brackets, followed by the indication of the original language in parentheses.

Titles of periodicals are given in full. Series numbers are given in parentheses, and always precede the boldface volume number. The numbers of separate issues are only provided when they are paginated separately, in which case they are in parentheses following the volume number. Unnumbered pages are indicated in square brackets. The titles of some periodicals vary over the years. For these periodicals the titles are given as they appeared for the respective items referred to (e.g., *Coleopterologische Rundschau* / *Koleopterologische Rundschau*). Contrarily, the journal *Entomologische Blätter* had several subtitles during its existence, the last of them being *Entomologische Blätter für Biologie und Systematik der Käfer*. Since there is no danger of confusion with any other periodical, the journal is cited simply as *Entomologische Blätter*. If the year of the volume differs from the actual year of publication, the former follows the volume number and is given is square brackets. If periodicals have no volume number, the year of the volume is given.

The *Bulletin de la Société Entomologique de France* is one of the more important periodicals in the field of systematics. The title changed almost continuously for more than 60 years, appearing usually as *Bulletin des séances*, *Bulletin trimestriel* or *Bulletin entomologique*. In addition, the first numbered volume, 37, was published 100 years after the creation of the *Annales* and the *Bulletin* of the French Entomological Society in 1832. To avoid confusion, this periodical is referred to as *Bulletin de la Société Entomologique de France,* and, from 1832 to 1931, is identified by the year in which it was published. From the year 1932, the volume number is given.

The *Bulletin de la Société Entomologique de France* was published from 1873 to 1894, in addition to the well known quarterly version in a bi-mensuel version "*Bulletin Bimensuel*". The pagination of the two versions are not identical, the publication of the bi-mensuel version preceded the quarterly one. The *Bulletin Bimensuel* is ignored by almost all coleopterists, and is not respected in the Catalogue.

The publication data of taxa described by P.H. Lucas in his *Exploration scientifique de l'Algérie* are as given in Evenhuis, 2012; for taxa of doubtful year of publication the later year given in that work is accepted.

For ease of access, references of the same author and year were placed in order based on the first page number of the paper, with Roman numerals being placed last, with respect to the number of the respective issue. Books were placed first for a given year, unless a page range was cited, in which case the first-page rule was followed. Papers of the same year and with the same first page numbers were placed in order of their last page number. Thus the references are not arranged chronologically within each year.

Acknowledgements

The Editors and Authors of the present volume are indebted to following individuals who commented on different parts of the Catalogue, pointed to nomenclatural problems, added distributional information, completed references, and helped in various ways: Miguel ALONSO-ZARAZAGA, Madrid, Spain; Ming BAI, Beijing, China; Stanislav BEČVÁŘ, Prague, Czech Republic; Jonathan COOPER, Hereford, UK; Augusto DEGIOVANNI, Bubano, Italy; Roland GRIMM, Neuenbürg, Germany; Boris M. KATAEV, St. Petersburg, Russia; Richard A.B. LESCHEN, Auckland, New Zealand; Yue-Chen LI, Chongqing, China; Ryan LUMEN, West Lafayette, USA; Yiping NIU, Hebei, China; Guo-Dong REN, Hebei, China; Alexander B. RUCHIN, Saransk, Russia; Laurent SOLDATI, Montferrier-sur-Lez, France; Igor A. SOLODOVNIKOV, Vitebsk, Belarus; Mikhail TYAGLOV, Shanghai, China; Caixia YUAN, Yan'an, China.

Piotr TYKARSKI, Warszawa, Poland, developed the software tools needed to generate the Indexes, and was helpful in solving a number of computer problems.

David SCHIMROSCZYK, Warszawa, Poland, provided the illustration of *Leptoderis collaris* (Linnaeus, 1767).

Marcin RAŚ, Warszawa, Poland, provided the distributional maps.

The work of M.J. KAMIŃSKI was partly supported by the NSF ARTS Program (DEB #1754630).

The part on the tribe Helopini was funded by the Russian Foundation for Basic Research and RPF (project no. 19-54-25001) for M.V. NABOZHENKO.

Editors and Authors

The Editors

Dariusz Iwan
Museum and Institute of Zoology PAS
Warsaw, Poland

Ivan Löbl
Muséum d'histoire naturelle
Genève, Switzerland

Authors

family MYCETOPHAGIDAE Leach, 1815	N.B. NIKITSKY
family ARCHEOCRYPTICIDAE Kaszab, 1964	O. MERKL
family PTEROGENIIDAE Crowson, 1953	I. LÖBL
family CIIDAE Leach, 1819	R. KRÓLIK
family TETRATOMIDAE Billberg, 1820	N.B. NIKITSKY
family MELANDRYIDAE Leach, 1815	N.B. NIKITSKY
family MORDELLIDAE Latreille, 1802	J. HORÁK
family RIPIPHORIDAE Laporte, 1840	M. BARCLEY
family ZOPHERIDAE Solier, 1834	R. SCHUH
family TENEBRIONIDAE Latreille, 1802	D. IWAN, I. LÖBL, F. BOUCHARD, Y. BOUSQUET, M. KAMIŃSKI, O. MERKL, K. ANDO, W. SCHAWALLER
tribe Lagriini Latreille, 1825	O. MERKL
tribe Asidini Fleming, 1821	F. SOLDATI
tribe Dendarini Mulsant & Rey, 1854	D. IWAN, I. LÖBL
tribe Blaptini Leach, 1815	M. NABOZHENKO, I. CHIGRAY
tribe Melanimonini Seidlitz, 1894 (1854)	D. IWAN, I. LÖBL
tribe Opatrini Brullé, 1832	D. IWAN, I. LÖBL
tribe Pedinini Eschscholtz, 1829	D. IWAN, I. LÖBL
tribe Platyscelidini Lacordaire, 1859	L.V. EGOROV
tribe Helopini Latreille, 1802	M. NABOZHENKO
subfamily Alleculinae Laporte, 1840	V. NOVÁK
family PROSTOMIDAE C.G. Thomson, 1859	W. SCHAWALLER
family SYNCHROIDAE Lacordaire, 1859	I. LÖBL
family STENOTRACHELIDAE Thomson, 1859	I. LÖBL
family OEDEMERIDAE Latreille, 1810	D. KUBISZ, D. IWAN
family MELOIDAE Gyllenhal, 1810	M.A. BOLOGNA
family MYCTERIDAE Oken, 1843	I. LÖBL, D. IWAN, D.A. POLLOCK

family BORIDAE C.G. Thomson, 1859	D. TELNOV
family TRICTENOTOMIDAE Blanchard, 1845	D. TELNOV, A. DRUMONT
family PYTHIDAE Solier, 1834	D.A. POLLOCK, D. IWAN
family PYROCHROIDAE Latreille, 1806	D.K. YOUNG, D. TELNOV, D. POLLOCK
family SALPINGIDAE Leach, 1815	D.A. POLLOCK, I. LÖBL, D. IWAN
family ANTHICIDAE Latreille, 1819	D. TELNOV
family ISCHALIIDAE Blair, 1920	D. TELNOV
family ADERIDAE Winkler, 1927	G. NARDI
family SCRAPTIIDAE Mulsant, 1856	D. IWAN, D. KUBISZ

Adresses of the Authors

ANDO Kiyoshi, Entomological Laboratory, Faculty of Agriculture, Ehime University, Tarumi 3-5-7, Matsuyama, 790-8566 Japan. E-mail: tenebrio@entom.biz-web.jp

BARCLAY Maxwell V.L., Natural History Museum, London SW7 5BD, United Kingdom. E-mail: m.barclay@nhm.ac.uk

BOLOGNA A. Marco A., Dipartimento di Scienze, Università Roma Tre, Viale G. Marconi 446, 00146 Roma, Italy. E-mail: bologna@uniroma3.it

BOUCHARD Patrice, Canadian National Collection of Insects, Arachnids and Nematodes, Agriculture and Agri-Food Canada, 960 Carling Avenue, Ottawa, Ontario, K1A 0C6, Canada. E-mail: Patrice.Bouchard@canada.ca.

BOUSQUET Yves, Gatineau, Quebec, Canada.

CHIGRAY Ivan A., Zoological Institute of the Russian Academy of Sciences, Universitetskaya naberezhnaya 1, 199034 St Petersburg, Russia. E-mail: chigray93@bk.ru

DRUMONT Alain, O.D. Taxonomy and Phylogeny-Entomology, Royal Belgian Institute of Natural Sciences, Vautier Street 29, B-1000 Brussels, Belgium. E-mail: alain.drumont@naturalsciences.be

EGOROV V. Leonid, Kadykova st. 11/525, Cheboksary, Chuvashia, 428037 Russia. E-mail: platyscelis@mail.ru

HORÁK Jan, K hádku 567, Dubeček, CZ-107 00 Praha 10- Dubeč, Czech Republic. E-mail: janho_cz@yahoo.com

IWAN Dariusz, Museum and Institute of Zoology, Polish Academy of Sciences, Wilcza 64, 00-679 Warszawa, Poland. E-mail: darek@miiz.waw.pl

KAMIŃSKI Marcin J., Museum and Institute of Zoology, Polish Academy of Sciences, Wilcza 64, 00-679 Warszawa, Poland. E-mail: kaminskientomo@gmail.com

KRÓLIK Roman, Mickiewicza 8, 46–200 Kluczbork, Poland. E-mail: agrilus@poczta.onet.pl

KUBISZ Daniel, Institute of Systematics and Evolution of Animals, Polish Academy of Sciences, Sebastiana 9, 31–049 Kraków, Poland. E-mail: kubisz@muzeum.pan.krakow.pl

LÖBL Ivan, Muséum d'histoire naturelle, Case postale 6434, CH-1211 Genève, Switzerland. E-mail: ivan.lobl@bluewin.ch

MERKL Otto, Hungarian Natural History Museum, Department of Zoology, H-1088 Baross u. 13, Budapest, Hungary. E-mail: merkl.otto@nhmus.hu

NABOZHENKO V. Maxim, Precaspian Institute of Biological Resources of the Daghestan Federal Research Centre of the Russian Academy of Sciences; Dagestan State University, M. Gadzhiev str. 45 and 43a 367000 Makhachkala, Russia. E-mail: nalassus@mail.ru

NARDI Gianluca, Centro Nazionale Biodiversità Carabinieri Bosco Fontana, Strada Mantova 29, I-46045 Marmirolo (Mantova). E-mail: l_nardi@hotmail.com

NIKITSKY Nikolay B., Zoological Museum of Moscow Lomonosov State University, Bolshaya Nikitskaya 2, Moscow 125009, Russia. E-mail: NNikitsky@mail.tu

NOVÁK Vladimír, Nepasické náměstí 796, Prague 9 - Klánovice, Czech Republic. E-mail: alleculinae.vn@centrum.cz

POLLOCK Darren A., Department of Biology, Station #33, Eastern New Mexico University, Portales, NM 88130, USA. E-mail: darren.pollock@enmu.edu

SCHAWALLER Wolfgang, Staatliches Museum für Naturkunde, Rosenstein 1 D-70191 Stuttgart, Germany. E-mail: schawaller.ehrenamt@smns-bw.de

SCHUH Rudolf, Wielandgasse 3/6, A-2700 Wiener Neustadt, Austria. E-mail: rudolf.schuh@at

SOLDATI Fabien, Office National des Forêts, Laboratoire National d'Entomologie Forestière, 2 rue Charles Péguy, 11500 Quillan, France. E-mail: fabien.soldati@onf.fr

TELNOV Dmitry, Natural History Museum London SW7 5BD, United Kingdom. E-mail: anthicus@google-mail.com

YOUNG Daniel K., UW Insect Research Collection, 445 Russell Laboratories, Department of Entomology, 1630 Linden Drive, University of Wisconsin, Madison, WI 53706, USA. E-mail: dkyoung@wisc.edu

The present work is based upon data provided by the following authors of the first edition of the Catalogue:

family MYCETOPHAGIDAE Leach, 1815	N.B. NIKITSKY
family ARCHEOCRYPTICIDAE Kaszab, 1964	O. MERKL
family PTEROGENIIDAE Crowson, 1953	I. LÖBL
family CIIDAE Leach, 1819	J. JELÍNEK
family TETRATOMIDAE Billberg, 1820	N.B. NIKITSKY
family MELANDRYIDAE Leach, 1815	N.B. NIKITSKY, D.A. POLLOCK
family MORDELLIDAE Latreille, 1802	J. HORÁK
family RIPIPHORIDAE Laporte, 1840	J. BATELKA
family ZOPHERIDAE Solier, 1834	S.A. ŚLIPIŃSKI, R. SCHUH
family TENEBRIONIDAE Latreille, 1802	I. LÖBL, O. MERKL, K. ANDO, P. BOUCHARD, M. LILLIG, K. MASOMUTO, W. SCHAWALLER
tribe Lagriini Latreille, 1825	O. MERKL
tribe Asidini Solier, 1834	F. SOLDATI
tribe Blaptini Leach, 1815	I. LÖBL, M. NABOZHENKO, O. MERKL

tribe Melanimini Seidlitz, 1894	D. IWAN, I. LÖBL
tribe Opatrini Brullé, 1832	D. IWAN, I. LÖBL
tribe Pedinini Eschscholtz, 1829	D. IWAN, I. LÖBL
tribe Platyscelidini Lacordaire, 1859	L.V. EGOROV
tribe Helopini Latreille, 1802	M. NABOZHENKO, I. LÖBL
subfamily Alleculinae Laporte, 1840	V. NOVÁK, R. PETTERSSON
family PROSTOMIDAE C.G. Thomson, 1859	W. SCHAWALLER
family SYNCHROIDAE Lacordaire, 1859	I. LÖBL
family STENOTRACHELIDAE Latreille, 1810	I. LÖBL
family MELOIDAE Gyllenhal, 1810	M.A. BOLOGNA
family MYCTERIDAE LeConte, 1862	I. LÖBL
family BORIDAE C.G. Thomson, 1859	D.A. POLLOCK
family TRICTENOTOMIDAE Blanchard, 1845	D.A. POLLOCK
family PYTHIDAE Solier, 1834	D.A. POLLOCK
family PYROCHROIDAE Latreille, 1807	D.A. POLLOCK, D.K. YOUNG
family SALPINGIDAE Leach, 1915	D.A. POLLOCK, I. LÖBL
family ANTHICIDAE Latreille, 1819	D.S. CHANDLER, G. UHMANN, G. NARDI, D. TELNOV
family ADERIDAE Winkler, 1927	G. NARDI
family SCRAPTIIDAE Mulsant, 1856	P. LEBLANC, B. LEVEY, J. HORÁK

New Nomenclatural and Taxonomic Acts, and Comments

Ciidae

R. Królik

New Replacement Names

Cis michiochujoi Królik, **nom. nov.** for *Cis polypori* M. Chûjô, 1939 (nec *Cis polypori* Gistel, 1857)

Tenebrionidae

I. Löbl, P. Bouchard, O. Merkl & Y. Bousquet

Bibliographic Note

The first number of the *Nouveaux Mémoires de la Société Impériale des Naturalistes de Moscou* is dated 1829 on its title page and the permit for publication, 15 February 1827 (recto of title page), a Julian date corresponding to the 27 February 1827 in the Gregorian calendar. The volume, however, was issued by October 1828 as it was lengthily reviewed in the tenth *cahier* (dated October 1828 on the cover) of the first volume of the journal *Bulletin du Nord* (Anonymous 1828). All new taxa made available in the paper by Steven on the "Tentyriae" and "Opatra" are consequently dated 1828, not 1829.

Newly Established Taxa

Genus *Adesmia* Fischer von Waldheim, 1822

Macradesmia Löbl & Merkl, **new subgenus.** Type species *Pimelia cancellata* Klug, 1830, **present designation.**

Macropodesmia Löbl & Merkl, **new subgenus.** Type species *Pimelia reticulata* Klug, 1830, **present designation.**

Oteroscelopsis Löbl & Merkl, **new subgenus.** Type species *Pimelia dilatata* Klug, 1830, **present designation.**

Koch (1944) provided descriptions of six new subgenera of *Adesmia*, but unfortunately failed to designate type species, except *Somaladesmia* – the subgenus is available (type species by monotypy). The subgenera *Macradesmia, Macropodesmia, Oteroscelopsis, Spongesmia* and *Zambesmia* included two or more nominal species each and are therefore unavailable (ICZN 1999, Art 13.3.). They are currently considered as valid. Löbl & Merkl (2003: 244) designated *Pimelia cancellata* Klug, 1830 as the type species of *Macradesmia, Pimelia reticulata* Klug, 1830 as the type of *Macropodesma*, and *Pimelia dilatata* Klug, 1830 as the type species of *Oteroscelopsis*. As they failed to indicate explicitly that the taxa are new (ICZN 1999, Art. 16.1) the three subgeneric names remain *nomina nuda*. In order to make these names available, the following new taxa are established.

Macradesmia Löbl & Merkl, with *Pimelia cancellata* Klug, 1830 designated as the type species

 DOI:10.1163/9789004434998_002

Macropodesmia Löbl & Merkl, with *Pimelia reticulata* Klug, 1830 designated as the type species
Oteroscelopsis Löbl & Merkl, with *Pimelia reticulata* Klug, 1830 designated as the type species.
The description given by Koch, 1944: 146 characterises and differentiates *Macradesmia*, and the descriptions given by Koch, 1944: 147 characterise *Macropodesmia* and *Oteroscelopsis*.

Genus *Arthrodibius* Lesne, 1915

Erodibius Löbl, Bouchard, Merkl & Bousquet, **new subgenus.** Type species *Arthrodeis cyphonotus* Fairmaire, 1887, **present designation.**

Erodibius was introduced by Koch (1960: 354) as a new subgenus of *Arthrodibius* Lesne, 1915. He included in *Erodibius* seven species, *A. plicatus* (Fairmmare, 1883), *A. cicatrix* (Fairmaire, 1879), *A. bernadellii* Koch, 1959, *A. nitidiventris* (Fairmaire, 1884), *A. cyphonotus* (Fairmaire, 1887), and as new taxa, *A. scorteccii* and *A. hemmingi.* The subgenus is currently considered valid. A type species was not designated and therefore this genus-group name is unavailable (ICZN 1999, Art. 13.3.). Löbl, Bouchard, Merkl & Iwan (2008: 40) designated *Arthrodeis cyphonotus* Fairmaire, 1887 as the type species of *Erodibius* but failed to indicate explicitly that the taxon is new (ICZN 1999, Art. 16.1). In order to make the name available, we establish here the new taxon *Erodibius* Löbl, Bouchard, Merkl & Bousquet, with *Arthrodeis cyphonotus* Fairmaire, 1887 designated as the type species. The description given by Koch (1960: 354) characterises the group.

Genus *Catomus* Allard, 1876

Catomodontus Löbl & Merkl, **new subgenus.** Type species *Catomus coronatus* Koch, 1935, **present designation.**

Catomodontus was introduced by Koch (1935: 108) as a subgenus of *Catomus,* for three new species, *C. coronatus* Koch, 1935, *C. ovatus* Koch, 1935 and C. *torretassoi* Koch, 1935, without designation of a type species. Therefore, the name *Catomodontus* Koch, 1935 is unavailable (ICZN 1999, Art. 13.3.). It is currently considered as valid. Löbl & Merkl (2003: 251) designated *Catomus coronatus* Koch, 1935 as the type species of *Catomodontus.* As they have not explicitly indicated that the taxon is new, this name remains unavailable (ICZN 1999, Art. 16.1.). In order to make the name available, the new taxon *Catomodontus* Löbl & Merkl is established, with *Catomus coronatus* Koch, 1935 designated as the type species. Koch (1935: 108) provided the character states that differentiate the taxon.

Genus *Cheirodes* Gené, 1839

Spinanemia Löbl, Bouchard, Merkl & Bousquet, **new subgenus.** Type species *Anemia cornuta* Pic, 1898, **present designation.**

Ardoin (1971: 362, 378) introduced the new subgenus name *Spinanemia* for 16 nominal species, but unfortunately failed to designate a type species. The fact that *Anemia cornuta* Pic, 1898 was listed as the "type" of *Spinanemia* in the Zoological Record for the year 1971 (Anonymous in Commonwealth Institute of Entomology 1975) does not represent a valid type species designation since the nomenclatural act is anonymous (ICZN 1999, Art. 14). The subgenus *Spinanemia*, which has been treated as valid since 1971, is therefore unavailable (ICZN 1999, Art. 13.3). We hereby make the name available by selecting *Anemia cornuta* Pic, 1898 as type species and referring to Ardoin (1971: 362) for the character states that characterise and differentiate *Spinanemia*.

Genus *Cyphostethe* Marseul, 1867

Cyphostethoides Löbl & Merkl, **new subgenus.** Type species *Cyphostethe brunnea* Kaszab, 1962, **present designation.**

Kaszab (1979: 273) introduced a new subgenus, *Cyphostethoides*, for seven species but failed to designate a type species. Therefore, the name is unavailable (ICZN 1999, Art. 13.3.). It is currently considered as valid. Löbl & Merkl (2003: 244) designated *Cyphostethe brunnea* Kaszab, 1962, a species originally included by Kaszab (1962), as the type species of *Cyphostethoides*. As they failed to explicitly indicate that the taxon is new (ICZN 1999, Art. 16.1), this subgenus name remains unavailable. In order to make this name available, the new taxon *Cyphostethoides* Löbl & Merkl is established, with *Cyphostethe brunnea* Kaszab, 1962 designated as the type species. The description given by Kaszab (1979: 273) characterises and differentiates *Cyphostethoides*.

Genus *Hyperops* Eschscholtz, 1831

Belutschistanops Löbl, Bouchard, Merkl & Bousquet, **new subgenus.** Type species *Hyperops schusteri* Koch, 1940, **present designation.**

Belutschistanops was introduced by Koch (1943: 552). It was proposed as a subgenus in *Hyperops* Eschscholtz, 1831 for *Hyperops pygmaea* L. Redtenbacher, 1850, *H. schusteri* Koch, 1940, *H. nitidula* Kraatz, 1865, *H. doriae doriae* Baudi di Selve, 1874 and *H. doriae arabica* Koch, 1940 and is currently considered as valid. A type species was not designated and therefore this genus-group name is unavailable (ICZN 1999, Art. 13.3.). Löbl, Bouchard, Merkl & Iwan (2008: 40) designated *Hyperops schusteri* Koch, 1940 as the type species of *Belutschistanops*. They have not explicitly indicated that the taxon is new. Therefore, this name remains unavailable (ICZN 1999, Art. 16.1.). In order to make the name available, we establish here the new taxon *Belutschistanops* Löbl, Bouchard, Merkl & Bousquet, with *Hyperops schusteri* Koch, 1940 designated as the type species.
The description given by Koch (1943: 524) differentiates the taxon.

Genus *Mesostena* Eschscholtz, 1831

Saxistena Löbl & Merkl, **new subgenus**. Type species *Mesostena longicornis* Kraatz, 1865, **present designation.**

Saxistena was introduced by Koch (1940: 65) as a subgenus of *Mesostena*, for *M. longicornis* Kraatz, 1865, its subspecies *cavatica* Andres, 1926 and two additional new species, but he failed to designate a type species. Therefore, the name *Saxistena* Koch, 1940 is unavailable (ICZN 1999, Art. 13.3.). It is currently considered as valid. Löbl & Merkl (2003: 251) designated *Mesostena longicornis* Kraatz, 1865 as the type species of *Saxistena*. As they have not explicitly indicated that the taxon is new, this name remains unavailable (ICZN 1999, Art. 16.1.). In order to make the name available, the new taxon *Saxistena* Löbl & Merkl is established, with *Mesostena longicornis* Kraatz, 1865 designated as the type species. Koch (1940: 65) provided the character states that differentiate the taxon.

Genus *Neoplamius* Löbl, Bouchard, Merkl & Bousquet

Neoplamius Löbl, Bouchard, Merkl & Bousquet, **new genus**. Type species *Neoplamius zoltani* Masumoto, 1981, **present designation.**

Masumoto (1981: 15) introduced the new genus name *Neoplamius* for two species: *Neoplamius zoltani* Masumoto, 1981 and *N. endoi* Masumoto, 1981, but unfortunately failed to designate a type species. The genus *Neoplamius*, which has been treated as valid since 1981, is therefore unavailable (ICZN 1999, Art. 13.3). We hereby make the name available by selecting *Neoplamius zoltani* Masumoto, 1981 as type species and referring to Masumoto (1981: 17) for the character states that characterise and differentiate *Neoplamius*.

Genus *Paramisolampidius* Merkl & Masumoto

Paramisolampidius Merkl & Masumoto, **new genus.** Type species *Paramisolampidius kagoshimensis* Nakane, 1968, **present designation.**

Nakane (1968: 82) provided a description of a new genus *Paramisolampidius*. He included three species, *Misolampidius shirozui* Mt. Chûjô, 1967, *Misolampidius tokarensis* Nakane, 1963, and *Paramisolampidius kagoshimensis* Nakane, 1968 but failed to designate a type species. The name *Paramisolampidius* Nakane, 1968 was therefore unavailable (ICZN 1999, Art. 13.3.). It is currently considered as valid. Merkl & Masumoto (2007: 1) designated *Paramisolampidius kagoshimensis* Nakane, 1968 as the type species of *Paramisolampidius*. This name remains unavailable because they have not explicitly indicated that the taxon is new (ICZN 1999, Art. 16.1). In order to make the name available, the new taxon *Paramisolampidius* Merkl & Masumoto is here established, with *Paramisolampidius kagoshimensis* Nakane, 1968 designated as the type species.
The description given by Nakane (1968: 82) characterises and differentiates *Paramisolampidius.*

Genus *Paraplatyope* Löbl, Bouchard, Merkl & Bousquet

Paraplatyope Löbl, Bouchard, Merkl & Bousquet, **new genus**. Type species *Leucolaephus arabicus* Blair, 1931, **present designation.**

Gridelli (1953: 45) introduced the new genus name *Paraplatyope* for two species: *Leucolaephus arabicus* Blair, 1931 and *Paraplatyope thesigeri* Gridelli, 1953, but unfortunately failed to designate a type species. The fact that *Leucolaephus arabicus* Blair, 1931 was listed as the "type" of *Paraplatyope* in the Zoological Record for the year 1953 (Anonymous in Commonwealth Institute of Entomology 1955) does not represent a valid type species designation since the nomenclatural act is anonymous (ICZN 1999, Art. 14). The genus *Paraplatyope*, which has been treated as valid since 1953, is therefore unavailable (ICZN 1999, Art. 13.3). We hereby make the name available by selecting

Leucolaephus arabicus Blair, 1931 as type species and referring to Pierre (1964b: 873) for the character states that characterise and differentiate *Paraplatyope*.

Genus *Prochoma* Solier, 1835

Oxypistoma Löbl, Bouchard, Merkl & Bousquet, **new subgenus.** Type species *Prochoma bucculenta* Koch, 1940, **present designation.**

Oxypistoma was introduced by Koch (1940: 258). It was proposed as a new subgenus of *Prochoma* Solier, 1835 for two new species, *P. bucculenta* and *P. blairi*, and for *P. clypealis* Blair, 1923. *Oxypistoma* is currently considered as valid. A type species was not designated and therefore this genus-group name was unavailable (ICZN 1999, Art. 13.3.). Löbl, Bouchard, Merkl & Iwan (2008: 40) designated *Prochoma bucculenta* Koch, 1940 as the type species of *Oxypistoma* but failed to indicate explicitly that the taxon is new (ICZN 1999, Art. 16.1). In order to make the name available, we establish here the new taxon *Oxypistoma* Löbl, Bouchard, Merkl & Bousquet, with *Prochoma bucculenta* Koch, 1940 designated as the type species. The description given by in Koch (1940: 258) characterises the group.

Genus *Thraustocolus* Kraatz, 1865

Leptoderops Löbl, Bouchard, Merkl & Bousquet, **new subgenus.** Type species *Thraustocolus priesneri* Koch, 1934, **present designation.**

Leptoderops was introduced as a new subgenus of *Thraustocolus* Kraatz, 1865 by Koch, 1934: 96 who included two species, *T. montandoni* Schuster, 1915 and his new species *T. priesneri*. This subgenus is currently considered valid although a type species was not designated and therefore this genus-group name is unavailable (ICZN 1999, Art. 13.3.). Subsequently, *T. montandoni* was placed in the subgenus *Prothraustocola* Kaszab, 1957, and *T. afghanicus* Kaszab, 1979 was added to *Leptoderops*. Löbl, Bouchard, Merkl & Iwan (2008: 40) designated *Thraustocolus priesneri* Koch, 1934 as the type species of *Leptoderops* but failed to indicate explicitly that the taxon is new (ICZN 1999, Art. 16.1). In order to make the name available, we establish here the new taxon *Leptoderops* Löbl, Bouchard, Merkl & Bousquet, with *Thraustocolus priesneri* Koch, 1934 designated as the type species. The description given by Koch, 1934: 96 characterises the group.

Comment

Cerogria was described by Borchmann (1909a: 210) and originally included 20 species. The currently accepted type species – *Lagria anisocera* Wiedemann, 1823, by subsequent designation in Borchmann (1937: 119) – is threatened by the discovery of an older designation – *Cerogria dohrni* Borchmann, 1909, by subsequent designation in Lucas (1920: 174). *Cerogria dohrni* Borchmann, 1909 is currently the type species of the valid genus *Acerogria* Borchmann, 1936. Acceptance of the type species designation of *Cerogria* by Lucas (1920) would lead to nomenclatural instability and we therefore recommend that an application be submitted to the International Commission on Zoological Nomenclature to maintain the type species designation proposed by Borchmann (1937: 119).

Tenebrionidae

P. Bouchard & Y. Bousquet

Reversal of Precedence

The previously overlooked genus *Hemeralopius* Gistel, 1848 was proposed as a replacement name for the junior homonym *Calcar* Dejean, 1821. *Belopus* Gebien, 1911 was also proposed as a replacement name for *Calcar* Dejean, 1821 and is currently used as a valid subgenus of *Centorus* Mulsant, 1854. Based on the Principle of Priority (ICZN 1999, Art. 23.1) the well-established name *Belopus* Gebien, 1911 is threatened by the discovery of *Hemeralopius* Gistel, 1848. We hereby provide supporting references for the conservation of *Belopus* Gebien, 1911 ***nomen protectum*** over *Hemeralopius* Gistel, 1848 ***nomen oblitum*** using Reversal of Precedence (ICZN 1999, Art. 23.9.2). The taxon name *Hemeralopius* Gistel, 1848 has not been used as valid after 1899 to our knowledge. Supporting references: Al-Houty 1984, 1997; Aliquò & Soldati 2014; Allsopp 1980; Bujalance & Ferreras 1987; Chikatunov et al. 1997; Dalmon 1993; Doyen 1988; Doyen & Tschinkel 1982; Erber 1990; Iwan et al. 2012; Juan & Petitpierre 1989; Labrique & Chavanon 2002; Leo 1984; Löbl et al. 2008; Mordkovich 2006, 2012; Pavlícek et al. 2005; Ren & Yu 1994; Ren & Zeng 1993; Romero 2015; Skopin 1974; Soldati 1995; Soldati F. & Soldati L. 2003; Tschinkel & Doyen 1980; Watt 1974; Zinovyev 2011.

The genus *Adelphus* Dejean, 1834 originally included two nominal species: *Helops beniniensis* Palisot de Beauvois, 1811 and *Helops marginatus* Fabricius, 1792, and is therefore an available genus-group name. Bousquet and Bouchard (2013: 58) designated *Helops marginatus* Fabricius, 1792 as its type species. Based on the Principle of Priority (ICZN 1999, Art. 23.1) the well-established name *Praeugena* Laporte, 1840 is threatened by the recent discovery that *Adelphus* Dejean, 1834 is an available genus-group name. We hereby provide supporting references for the conservation of *Praeugena* Laporte, 1840 ***nomen protectum*** over *Adelphus* Dejean, 1834 ***nomen oblitum*** using Reversal of Precedence (ICZN 1999, Art. 23.9.2). The taxon name *Adelphus* Dejean, 1834 has not been used as valid after 1899 to our knowledge. Supporting references: Abdel-Dayem et al. 2017; Bouchardet al. 2005, 2011; Datinon et al. 2013; Doyen & Tschinkel 1982; Ferrer 2001, 2004, 2006, 2013, 2016; Ferrer et al. 2016; Gillon & Gillon 1974; Grimm 2002; Kaszab 1982; Kenya & Njoroge 2006; Kouassi et al. 2005; Löbl et al. 2008; MacFadyen & Compion 2009; Merkl et al. 2015; Monks et al. 2019; de Moor 1970; Penny 2009; Robiche 2001; Robiche et al. 2002; Schawaller 2010.

The genus *Euphron* Dejean, 1834 originally included one nominal species: *Helops coerulescens* Guérin-Méneville, 1831, and is therefore an available genus-group name (see Bousquet and Bouchard 2013: 55). Based on the Principle of Priority (ICZN 1999, Art. 23.1) the well-established name *Derosphaerus* J. Thomson, 1858 is threatened by the recent discovery that *Euphron* Dejean, 1834 is an available genus-group name. We hereby provide supporting references for the conservation of *Derosphaerus* J. Thomson, 1858 ***nomen protectum*** over *Euphron* Dejean, 1834 ***nomen oblitum*** using Reversal of Precedence (ICZN 1999, Art. 23.9.2). The taxon name *Euphron* Dejean, 1834 has not been used as valid after 1899 to our knowledge. Supporting references: Ando 2003; Bouchard 2002; Bouchard & Yeates 2001;Bouchard & Steiner 2004; Ferrer 1995, 1996; Grimm 2014, 2015; Hegde 2019; Hegde & Vasanthakumar 2018a, 2018b; Hisamatsu 1970; Kaszab 1987; Löbl et al. 2008; Masumoto & Makihara 1997; Matthews 2019; Matthews & Bouchard 2008; Merkl 1992; Poussereau & Ferrer 2007; Robiche 2005, 2010; Schawaller 2005, 2006, 2007, 2011.

New Synonymy and Designation of a Type Species

Acantophorus Billberg, 1820: 32 **syn. nov.** of *Erodius* Fabricius, 1775, with type species *Erodius gibbus* Fabricius, 1775, **present designation**. The previously overlooked genus *Acantophorus* was introduced by Billberg (1820: 32) for *Erodius gibbus* Fabricius, 1775, *Erodius bilineatus* Cuvier, 1791 and *Pimelia costata* Fabricius, 1801.

Coelometopus clypeatus Solier, 1848 is included here as an available species name. Solier (1848: 280) does not refer to *Blaps clypeata* Germar, 1813, the species listed as the type species of *Coelopmetopus* in the first edition of this Catalogue (Löbl et al. 2008: 340). In addition, *C. clypeatus* Solier, 1848 was previously recorded as a separate species in the *Index Animalium* (link to the online version here https://www.sil.si.edu/DigitalCollections/indexanimalium/TaxonomicNames/IA_Namesearch.cfm). *Coelometopus clypeatus* Solier, 1848 is a **new synonym** of *Blaps clypeata* Germar, 1813 and the type species of *Coelometopus* Solier, 1848 by monotypy.

Comments

Dejean (1821: 64) proposed the generic name *Laena* for one species, *Helops pimelia* Fabricius, 1787 from Austria, and listed *Scaurus viennensis* Sturm, 1807 as a synonym of Fabricius' species. Under Article 68.3 (ICZN 1999), *Helops pimelia* Fabricius is the type species by monotypy. Although described from "Anglia", *Helops pimelia* is currently listed as a valid species in the genus *Penthe* Newman, 1838 (family Tetratomidae) from North America (e.g., Young & Pollock 2002: 415). Obviously, Dejean misidentified Fabricius' species. Therefore the type species of *Laena* is *Helops pimelia* Fabricius sensu Dejean, 1821 (= *Scaurus viennensis* J. Sturm, 1807).

Ocnera was described by Fischer von Waldheim (1822: 169) and originally included four species: *Tenebrio cephalotes* Pallas, 1781, *Tenebrio muricatus* Linnaeus, 1758, *Pimelia imbricata* Fischer von Waldheim, 1820 and *Pimelia nodosa* Fischer von Waldheim, 1820. The currently accepted type species – *Pimelia imbricata* Fischer von Waldheim, 1820, by subsequent designation in Gebien (1937: 814) – is threatened by the discovery of an older designation – *Tenebrio cephalotes* Pallas, 1781, by subsequent designation in Crotch (1870: 241). *Tenebrio cephalotes* Pallas, 1781 is currently the type species of the valid subgenus *Chaetotoma* Motschulsky, 1860 in *Pimelia* Fabricius, 1775. Acceptance of the type species designation of *Ocnera* by Crotch (1870) would lead to nomenclatural instability and we therefore recommend that an application be submitted to the International Commission on Zoological Nomenclature to maintain the type species designation proposed by Gebien (1937: 814).

The genus *Scelace* Marseul (1887: 325) originally included four nominal species: *Pimelia tuberculifera* Lucas, 1858, *Pimelia dayensis* Mulsant & Godart, 1866, *Pimelia serieperlata* Fairmaire, 1876 and *Pimelia semiasperula* Fairmaire, 1876. We could not find a valid species designation for this genus. We hereby select *Pimelia tuberculifera* Lucas, 1858 as the type species of *Scelace* Marseul, 1887. The synonymy between *Scelace* Marseul, 1887 and *Pachyscelodes* Sénac, 1887 was previously established by Reitter (1915: 2). *Scelace* was published by Marseul on June 9, 1877 (wrapper's date for issue 318 in volume 24 of *L'Abeille, Journal d'Entomologie*) while its synonym *Pachyscelodes* was published by Sénac on August 15, 1887 (wrapper's date for the second *cahier* of volume 7, 6th series, of *Annales de la Société Entomologique de France*), therefore *Scelace* is the oldest and valid name.

Bolitophagus was proposed by Illiger (1798: 100) and originally included five nominal species: *Silpha reticulata* Linnaeus, 1767, *Opatrum crenatum* Fabricius, 1792, *Opatrum agricola* Herbst, 1783, *Opatrum gibbum* Fabricius, 1775 and *Hispa cornuta* Thunberg, 1784. The currently accepted type species (*Silpha reticulata* Linnaeus, 1767, by subsequent designation in C.G. Thomson (1859: 115)) is threatened by the discovery of an older designation (*Opatrum agricola* Herbst, 1783, by subsequent designation in Curtis (1836: pl. 586)). *Opatrum agricola* Herbst, 1783 is currently the type species of the valid genus *Eledona* Latreille, 1796. Acceptance of the type species designation of *Bolitophagus* by Curtis (1836) would lead to nomenclatural instability and we therefore recommend that an application be submitted to the International Commission on Zoological Nomenclature to maintain the type species designation proposed by C.G. Thomson (1859: 115).

The spelling *Caedius* Blanchard, 1845, introduced by Lacordaire (1859: 261), is an incorrect spelling of the original spelling *Coedius* in prevailing usage and is deemed to be the correct original spelling (ICZN 1999, Art. 33.3.1).

Brachypophlaeus Fairmaire, 1897 is sometimes misspelled as *Brachypophloeus* in the literature. Since the incorrect subsequent spelling *Brachypophloeus* is not in prevailing usage, we prefer to use the original spelling *Brachypophlaeus*.

Tenebrionidae
K. Ando

New Replacement Names

Diachoriops Ando, **nom. nov.** is proposed for *Schizomma* Gebien, 1920: 392, type species *Schizomma cucumericola* Gebien, 1920, gender masculine (nec *Schizomma* Ehrenberg, 1861).

Tenebrionidae
M.J. Kamiński

Comments

The subgeneric classification of *Tentyria* Latreille, 1802 is not sustained in the present catalogue due to recently introduced taxonomic decisions and errors. Löbl et al. (2008) listed two subgenera within *Tentyria* – *Subtentyria* Löbl & Merkl, 2003 and *Tentyria*. Taking into consideration Article 16.1 of the International Code of Zoological Nomenclature (ICZN 1999), *Subtentyria* Löbl & Merkl, 2003 is not an available name. Furthermore, *Tentyria ligurica* Solier, 1835 – formerly included in *Subtentyria* – was designated the type species of the genus *Tentyria* (ICZN 2010). A subgeneric classification of *Tentyria* requires new studies.

The genus *Pedinus* was originally described by Latreille (1797: 20) without included nominal species; Latreille (1802: 175), by listing *Tenebrio femoralis* Linnaeus, 1767 and *Tenebrio dermestoides* Fabricius, 1787 in association with this genus name, was the first author to subsequently and expressly include nominal species in Pedinus (ICZN, Art. 67.2.2). The type species is *Tenebrio femoralis* Linnaeus, 1767, by subsequent designation (Westwood 1838: 33).

Changes in Rank

Iphthiminus bellardi Truqui, 1857 raised to valid species, from subspecies of *Iphthiminus italicus* Truqui, 1857.

Iphthiminus croaticus Truqui, 1857 raised to valid species, from subspecies of *Iphthiminus italicus* Truqui, 1857.

Tenebrionidae: Alleculinae

V. Novák

New Replacement Name

Loriculoides Novák, **nom. nov.** for *Loricula* Novák, 2016: 45 type species *Allecula subaeneipennis* Pic, 1922, gender masculine (nec *Loricula* Curtis, 1833).

Tenebrionidae: Akidini

L. Soldati

New Replacement Name

Blaps nitens glebmedvedevi L. Soldati, **nom. nov.** for *Blaps nitens medvedevi* L. Soldati, Kergoat & Condamine, 2009 (nec *Blaps medvedevi* Bogatchev, 1961).

Priority

In the first edition of the Catalogue, p. 126, *Akis obtusangula* Thèry, 1932 is listed as a synonym of *Akis heydeni heydeni* Haag-Rutenberg, 1876 while Antoine (1957) had already correctly established its synonymy with *Akis antoinei* Escalera, 1934.

Comments

Akis costitubera Marseul, 1883 was transferred to the genus *Morica* Dejean, 1834 by Kwieton (1980b), based on the presence of four rows of elytral granules, but ignoring the pronotal characters. Schawaller (1987) restored this species to the genus *Akis* Herbst, 1799.

Akis dorsigera Reitter, 1904 was described as a variety of *Akis acuminata* (Fabricius, 1787) because of the presence of sublateral costa on the false epipleuron, near the lateral edge of the elytron. Ferrer et al. (2008) raised *A. dorsigera* to species rank based on irrelevant (and incorrectly illustrated) character states. Therefore, *A. dorsigera* is considered synonymous with *A. acuminata* (Fabricius).

Akis tuberculata Kraatz, 1865 was raised to species rank in the same work (loc. cit. p. 164). Leo & Fanchello (2019) have recently placed it in synonymy with *Akis bacarozzo* Schrank von Paula, 1786. Note that *A. tuberculata* may be an ecotype of *A. bacarozzo* associated with granitic soil.

Tenebrionidae: Blaptini

M. Nabozhenko & I.A. Chigray

New Synonymy

Kolbe (1928) erected several "Artengruppen" in the genus *Blaps* Fabricius, 1775 that are available and to be treated as subgenera (ICZN Art. 10.4 and 12). All were based on characters not relevant for grouping species of *Blaps* (the sculpture of elytra, the form and size of the elytral caudal extension, etc.), and the respective species are currently placed within the nominotypical subgenus. In addition, Kolbe (1928) failed to designate type species. Thus, the following designations of type species and synonyms are proposed:

Hypoblaps Kolbe, 1928: 200 with type species *Blaps rotundata* Solier, 1848, by original designation, **syn. nov.** of *Blaps* Fabricius, 1775.

Mesoblaps Kolbe, 1928: 200 with type species *Blaps acuminata* Fischer von Waldheim, 1820, designated here, **syn. nov.** of *Blaps* Fabricius, 1775. Kolbe included in this subgenus two species, *Blaps acuminata* and *B. pulvinata* Ménétriés, 1849 (a junior synonym of *B. lethifera pterotapha* Fischer von Waldheim, 1832).

Blapidium Kolbe, 1928: 200 **syn. nov.** of *Blaps* Fabricius, 1775 with type species *Blaps ocreata* Allard, 1880, designated here. Kolbe included in this subgenus four species, *B. ocreata*, *B. abbreviata* Ménétriés, 1836, *B. luctuosa* Ménétriés, 1832 (junior synonyms of *B. lethifera pterotapha* Fischer von Waldheim, 1832) and *B. tenuicollis* Solier, 1848.

Notoblaps Kolbe, 1928: 201 **syn. nov.** of *Blaps* Fabricius, 1775 with type species *Blaps juliae* Allard, 1880, designated here. Kolbe included in this subgenus two species, *B. juliae* and *B. punctostriata* Solier, 1848.

Opisthoblaps Kolbe, 1928: 201 **syn. nov.** of *Blaps* Fabricius, 1775 with type species *Blaps sulcifera* Seidlitz, 1893, designated here. Kolbe included in this subgenus five species, *B. sulcifera*, *B. sulcata* Fabricius, 1775 (a junior synonym of *B. polychresta* (Forskål, 1775)), *B. nitens nitens* Laporte, 1840, *B. nitens brachyura* Küster, 1848, *B. nitens requieni* Solier, 1848.

The following subgeneric names proposed by Kolbe are ***nomina nuda***, because he did not provide a separate description or indication for each of them: *Heteroblaps* Kolbe, 1928: 202, *Genoblaps* Kolbe, 1928: 202, *Tracheloblaps* Kolbe, 1928: 203, *Sceloblaps* Kolbe, 1928: 203, *Caraboblaps* Kolbe, 1928: 203.

The following two subgenera of *Prosodes* Eschscholtz, 1829 have the same type species and the following synonymy is proposed: *Peltarium* Fischer von Waldheim, 1844: 106 (type species *Peltarium marginatum* Fischer von Waldheim, 1844 (=*Blaps baerii* Fischer von Waldheim, 1842)) = *Blaposodes* Skopin, 1960: 47 (type species *Prosodes baerii* (Fischer von Waldheim, 1842) by original designation), **syn. nov.**

Subgeneric Placement

Genus *Gnaptorina* Reitter, 1887

Gnaptorina longicornis Z. Li & Ren, 2004 and *Gnaptorina quadratus* Z. Li & Ren, 2004 have remained unassigned to any current subgenus. Therefore, they are listed as incertae sedis.

Gnaptorina curvatibia Ren & Zhu in Ren, Niu, Li & Shi, 2016 was not placed in any subgenus. The authors (Ren, Niu, Li & Shi, 2016: 341) compared it with *Gnaptorina australis* G.S. Medvedev, 2008 (placed in the nominotypical subgenus), but compared it with *G. cordicollis* G.S. Medvedev, 1998 (the subgenus *Boreoptorina* G.S. Medvedev, 2008) in the English abstract. Therefore, the species is listed as incertae sedis.

Gnaptorina rufipes Ren & Zhu in Ren, Niu, Li & Shi, 2016 was compared with *G.* (*Boreoptorina*) *cordicollis* G.S. Medvedev, 1998, though it has a plantar surface on the male protarsomeres 1 to 3 and the protarsomere 2 has a hair-brush and therefore belongs to the nominotypical subgenus (unlike *Boreoptorina* Medvedev, 2009 and *Hesperoptorina* Medvedev, 2009 in which only the 1st segment of the male protarsi has a flat hairy brush at the apical margin of plantar surface).

Designation of a Type Species

Leptocolena Allard, 1880 was erected by Allard (Allard E. 1880: *Tableau des Blapsides. Entomologische Monatsblätter* 15: 71–74) as a subgenus for the following 10 available taxa: *Blaps mucronata* Latreille, 1804, = *Blaps chevrolatii* Solier, 1848: 298, = *Blaps mortisaga* (Linnaeus, 1758) sensu Olivier, 1795, = *Blaps obtusa* J. Sturm, 1807, = *Blaps stricticollis* A. Villa & J.B. Villa, 1835 (this taxon was mentioned by Allard twice, as "*stricticollis* Mén". and as "*striaticollis* Villa".), = *Blaps dilatata* Laporte, 1840, = *Blaps reflexa* Solier, 1848, = *Blaps sublineata* Brullé, 1832, = *Blaps striolata* Füster, 1846, *Blaps japonensis* Marseul, 1879. Two taxa, listed by Allard (1880) in *Leptocolena* are unavailable: *Blaps emoda* Allard, 1880 (this name is unavailable according to the Article 12.1 of ICZN, the species was described by Allard only in 1881; available name is *Blaps emoda* Allard, 1881), *Blaps foveicollis* Allard, 1880 (this name is unavailable according to the Article 12.1 of ICZN, because the species was described by Allard only in 1881; available name is *Blaps foveicollis* Allard, 1881). The type species for *Leptocolena* was not designated in the original publication of Allard (1880). *Blaps emoda* Allard, 1881 was incorrectly designated by Nabozhenko (2008c: 36) as the type species of *Leptocolena* Allard, 1881. As a result we correctly designate here *Blaps mucronata* Latreille, 1804 as the type species of *Leptocolena* Allard, 1880.

Correction of Authorship

Blaps holcus Ren in Ren, Niu, Li & Shi, 2016 was described under the authorship of Ren (Ren et al. 2016) in the Chinese text (p. 124) and under the autorship of Ren & Shi in the English abstract (p. 464). The correct authorship is considered to be Ren in Ren, Niu, Li & Shi, 2016 as in the original description on page 124.

Resurrection

Blap ledereri Fairmaire, 1866, **sp. resurr.** has been listed as a junior synonym of *Blaps graeca* Solier, 1848 in recent literature. We compared the holotype of *Blaps ledereri* (Muséum national d'Histoire naturelle, Paris: https://science.mnhn.fr/institution/mnhn/collection/ec/item/ec2024?listIndex=24) and many other specimens from the type locality of Bozdag, Izmir, Turkey (collection of Zoological department of Ege University) with the holotype of *B. graeca* (Muséum national d'Histoire naturelle, Paris: https://science.mnhn.fr/institution/mnhn/collection/ec/item/ec2023?lang=en_US) and found that *B. ledereri* is a distinct species (i.e., not a synonym of *B. graeca*), endemic for Bozdag, Izmir Province, Turkey.

Tenebrionidae: Helopini

M. Nabozhenko

New Synonymy and New Combinations

Nalassus mediatlantis (Antoine, 1949), **comb. nov.** and *Nalassus baudoni* (Antoine, 1954), **comb. nov.** Two Moroccan species, *mediatlantis* Antoine, 1949 and *baudoni* Antoine, 1954 were originally described in *Stenomax* Allard, 1876, and subsequently erroneously transferred to *Odocnemis* Allard, 1876 (Nabozhenko, 2008c). According to their descriptions (especially the strongly thickened middle antennomeres of "nalassoid" type) these species clearly belong to the nominotypical subgenus *Nalassus* Mulsant, 1854.

Odocnemis aurichalcea (Adams, 1817), **comb. nov.** (from *Blaps*) = *Helops gloriosa* Faldermann, 1837, **syn. nov.** = *Helops pulcher* Allard, 1877, **syn. nov.** The type material of the three taxa have not been found in museum collections and are probably lost. The species is unique among the Transcaucasian Helopini by its large body-size (length up to 18 mm) and copper green metallic shine. Adams provided unambiguous description of *Blaps aurichalcea*, giving "atra laevis *supra* viridi – cuprea nitida" and other characters which fit also *Helops gloriosa*. In addition, large material from the type locality "Sion" (Sioni, Georgia) and from the locality "Kobi" (Georgia) (Nabozhenko, 2001) were examined. The proposed synonymies are supported by the fact that only a single tenebrionid species possessing characters given by Adams is known from these two places. *Blaps aurichalcea* was mentioned subsequently to its description by Löbl et al. (2008) for the first time and listed under "subtribe Blaptina, species incertae sedis". The name is used as valid because the conditions of the ICZN Art. 23.9. are not met.

Euboeus nitidipennis (Küster, 1850), **comb. nov.** and *Euboeus rufipes* (Allard, 1876), **comb. nov.** (both from *Probaticus* Seidlitz, 1896). Reitter (1922) placed these two species in *Probaticus* with a doubtful position. They were not transferred to *Euboeus* Boieldieu, 1865 subsequently to the synonymy *Euboeus* and *Probaticus* (Nabozhenko et al. 2017), and are listed here as nomina dubia within *Euboeus*.

Hedyphanes lutosus Allard, 1877 = *Apolites angustus* Marseul, 1879, **syn. nov.** = *Pseudoprobaticus roznerorum* Nabozhenko, 2008, **syn. nov.** Seidlitz (1896) and subsequent authors identified the sole Western Anatolian species possessing granulate and setose elytra as *Hedyphanes upioides* Faldermann, 1837 (or as its junior synonym *Apolites angustus*). Abdurakhmanov & Nabozhenko (2011) resurrected the synonymy *Hedyphanes tagenioides* Falderman in Ménétriés, 1832 = *Hedyphanes upioides* Faldermann, 1837. Later, Nabozhenko (2013b) resurrected *Hedyphanes* *lutosus* from synonymy of *upioides*. *Hedyphanes tageniodes* occurs only in Transcaucasia (Armenia, Azerbaijan, Georgia) and Iran and was not found in Western Turkey ("Asie mineure" is the type locality of *Apolites angustus*). According to the original description, *Apolites angustus* has convex and granulate elytral interstriae. These characters are also presented in *Hedyphanes* (*Granulophanes*) *lutosus* Allard, 1877 described from Western Turkey. As result, the new synonymy is proposed. The species *roznerorum* Nabozhenko, 2008 was described in *Pseudoprobaticus* Nabozhenko, 2001, later transferred to *Hedyphanes* Fischer von Waldheim 1820 and is found to be also a synonym of *H. lutosus*.

New status

Nalassus kartvelius Nabozhenko, 2013 was described from Georgia (Oni District). The caracters of the type series of *Nalassus svaneticus* Nabozhenko, 2001 (Georgia: Svanetia) exhibit subspecific differences between these two taxa. Therefore, *N. kartvelius* to is downgraded to *N. svaneticus kartvelius* Nabozhenko, 2013, **stat. nov.**

Meloidae

M.A. Bologna

New Replacement Names

Epicauta miroslavi Bologna, **nom. nov.** for *Epicauta suturalis* Dvořák, 2008 (nec *Epicauta suturalis* Germar, 1821).

New Synonymy

Hycleus amrishi Makhan, 2012 **syn. nov.** of *Hycleus cingulatus* (Falderman, 1837) (Serri et al., in press).

Hycleus golnazae Makhan & Ezzatpanah, 2011 **syn. nov.** *Hycleus scabiosae* A.G. Olivier, 1811.

Hycleus javeti (Marseul, 1870) and *H. javeti umbilicatus* Kaszab, 1958 **syn. nov.** of *Hyclues colligatus* L. Redtenbacher, 1850 (Serri et al., in press).

Hycleus ringenbachi Bologna, 2009 **syn. nov.** of *Hycleus rotroui* (Pic, 1930).

Mylabris aschnae Makhan & Ezzatpanah, 2011 **syn. nov.** of *Mylabris* (*Eumylabris*) *calida* (Pallas, 1782).

Mylabris (*Micrabris*) *maculosopunctata mendizabali* de la Rosa, 2008 **syn. nov.** of *Mylabris* (*Micrabris*) *maculosopunctata* Graells, 1858. Moreover, the collection locality of this taxon (Valle d'Aosta, NW Italy) is extremely doubtful.

Mylabris minae Makhan & Ezzatpanah, 2011 **syn. nov.** *Mylabris* (*Eumylabris*) *cincta* A.G. Olivier, 1811.

Mylabris rishwani Makhan 201, **syn. nov.** of *Mylabris* (*Mylabris*) *quadripunctata* (Linnaeus, 1767).

According to unpublished molecular analyses, *Meloe* (*Coelomeloe*) Reitter 1911 is **syn. nov.** of *Meloe* (*Eurymeloe*) Reitter, 1911. This synonymy was already suspected by morphological study of the first instar larva (Di Giulio et al. 2013).

Comments

Records of some species, especially from European countries, probably refer to extinct populations (no more recorded in the last 50 years or more), but we prefer to maintain these data, lacking sure evidences of extinction.

According to unpublished morphological studies, *Lytta vesicatoria freudei* Kaszab, 1958 is elevated to specific rank as *Lytta* (*Lytta*) *freudei*.

According to molecular analyses (Riccieri et al., in press) the validity of *Hycleus humerosus* (Escherich, 1899) and *Hycleus zebraeus* (Marseul, 1870) is doubtful and possibly they are synonyms respectively of *Hycleus polymorphus* (Pallas, 1771) and *Hycleus atratus* (Pallas, 1773).

According to Serri et al. (in press) *Zonabris sialunus* Pic, 1929 probably is a synonym *Hycleus apicipennis* (Reiche, 1866).

According to unpublished molecular evidences, the subgeneric position of *Mylabris* (*Mylabris*) *apiceguttata* Pan & Bologna, 2014 is doubtful, and this species could be referred to the subgenus *Calydabris* Kaszab, 1960.

According to unpublished molecular evidences *Mylabris* (*Mylabris*) *barezensis* Serri, Pan & Bologna, 2012 must be referred to the subgenus *Calydabris* Kaszab, 1960.

Mylabris mimosae Olivier, 1811 is referred to the genus *Hycleus* after morphological and molecular studies (Riccieri et al., in press).

According to unpublished molecular data, *Meloe* (*Meloe*) *aegyptius* Brandt & Erichson, 1832 could be a synonym of *Meloe* (*Meloe*) *proscarabaeus* Linnaeus, 1758. Waiting for a revision of this widespread species, we prefer to maintain *aegyptius* as distinct species.

Coriologiton Marseul, 1879, from southern Africa, probably represents a distinct genus more than a synonym of *Apalus* Fabricius, 1775 (Bologna et al., 2018).

The revision of the genus *Zonitoschema* is under preparation and several new species, especially from eastern Asia, will be described. Moreover, the taxonomy of the genus is greatly confused, and the present arrangement of the species is provisional.

Anthicidae

D. Telnov

New Synonymy

Anthicini Latreille, 1819 = Amblyderini Desbrochers des Loges, 1899 **syn. nov.**

Anthicus elegans Steven, 1806 [= 1811] **syn. nov.** of *Stricticomus transversalis transversalis* (A. Villa & J.B. Villa, 1833). Based on study of type material.

Anthicus floreus LaFerté-Sénectère, 1849k [= 1849n] **syn. nov.** of *Anthicus crinitus* LaFerté-Sénectère, 1849k [= 1849n]. Based on study of type material.

Anthicus globipennis quercicola Sahlberg, 1913a **syn. nov.** of *Microhoria globipennis* (Pic, 1897j). Based on study of type material.

Anthicus merkli Pic, 1897b **syn. nov.** of *Tenuicomus ottomanus* (LaFerté-Sénectère, 1849l). Based on study of type material.

Anthicus pietschmi Pic, 1938 **syn. nov.** of *Tenuicomus luristanicus* (Pic, 1911e). Based on study of type material.

Anthicus simplex Marseul, 1879a **syn. nov.** of *Anthicus ater* Thunberg, 1787. Based on study of type material.

Anthicus testaceus Pic, 1941a **syn. nov.** of *Microhoria pinicola* (Reitter, 1889e), not of *M. nectarina* (Panzer, 1794). Based on study of type material.

Cyclodinus coniger (Krekich-Strassoldo, 1864) **syn. nov.** of *Cyclodinus debilis* (LaFerté-Sénectère, 1849i) [= 1849n]. Based on study of type material.

Cyclodinus dimidiatus (Wollaston, 1864) **syn. nov.** of *Cyclodinus debilis* (LaFerté-Sénectère, 1849i) [= 1849n]. Based on study of type material.

Microhoria feroni Bonadona, 1960 **syn. nov.** of *Microhoria pinicola* (Reitter, 1889e). Based on study of type material.

Notoxus haagi hokkaidensis Kôno, 1931 **syn. nov.** of *Notoxus haagi* Marseul, 1879. Based on study of type material.

Sapintus subgen. *Barbigerosapintus* Telnov, 1998 **syn. nov.** of *Alloeoceras* Krekich-Strassoldo, 1914. *Alloeoceras* was originally established for two Sri Lankan species, primarily based on the shape of the antenane (which are sexually dimorphic) and presence of the antebasal pronotal sulcus of pronotum bearing long, dense setation. Subgenus *Barbigerosapintus* was established for *Sapintus* (*B.*) *confertopunctatus* (the type species) and six other Oriental taxa possessing double elytral setation (like in most *Sapintus* Casey, 1895), in combination with the mentioned pronotal setation. Examination of available material led to conclusion that the unique features of *Alloeoceras* justify a genus rank, and that *Barbigerosapintus* is a juniour synonym of *Alloeoceras*. Thus, only the nominotypical subgenus remains in *Sapintus* Casey. Therefore, a list of ten new combinations (all Oriental taxa) is provided below.

New Combinations

Alloeoceras barbifer (Uhmann, 1990a) **comb. nov.** from *Sapintus* subgen. *Barbigerosapintus* Telnov (Oriental species). Based on study of type material.

Alloeoceras confertopunctatus (Telnov, 1998b) **comb. nov.** (Oriental species). Based on study of type material.

Alloeoceras indicus (Pic, 1896o) **comb. nov.** from *Anthicus* Paykull (Oriental species). Based on study of type material.

Alloeoceras argentatus (Pic, 1903s) **comb. nov.** from *Anthicus* Paykull (Oriental species). Based on study of type material.

Alloeoceras luteitarsis (Pic, 1912s) **comb. nov.** from *Anthicus* Paykull (Oriental species). Based on study of type material.

Alloeoceras parvicollis (Heberdey, 1934) **comb. nov.** from *Sapintus* subgen. *Barbigerosapintus* Telnov (Oriental species). Based on study of type material.

Alloeoceras santali antennalis (Heberdey, 1934) **comb. nov.** from *Sapintus* subgen. *Barbigerosapintus* Telnov (Oriental species). Based on study of type material.

Alloeoceras santali santali (Heberdey, 1934) **comb. nov.** from *Sapintus* subgen. *Barbigerosapintus* Telnov (Oriental species). Based on study of type material.

Alloeoceras satanas (Heberdey, 1934) **comb. nov.** from *Sapintus* subgen. *Barbigerosapintus* Telnov (Oriental species). Based on study of type material.

Alloeoceras punctulaticeps (Heberdey, 1934) **comb. nov.** from *Sapintus* subgen. *Barbigerosapintus* Telnov (Oriental species). Based on study of type material.

Anthicus baicalicus Mulsant & Rey, 1866a **comb. nov.** from *Cordicomus* Pic. Based on study of type material.

Anthicus desertus Marseul, 1879a **comb. nov.** from *Clavicomus* Pic. Based on study of type material.

Anthicus medius Pic, 1905 **comb. nov.** from *Cordicomus* Pic. Based on study of type material.

Anthicus scapularis LaFerté-Sénectère, 1849k **comb. nov.** from *Cordicomus* Pic. Based on study of type material.

Leptaleus aemulus (Krekich-Strassoldo, 1914a) **comb. nov.** from *Stricticomus* Pic. Based on study of type material.

Microhoria mesopotamica (Pic, 1912c) **comb. nov.** from *Tenuicomus* Pic. Based on study of type material.

Microhoria yemenita (Nardi, 2004) **comb. nov.** from *Anthicus* Paykull. Based on study of type material.

Changes in Rank

Anthelephila badia (Krekich-Strassoldo, 1931) **stat. nov.**, originally described as subspecies of *Formicomus braminus* (LaFerté-Sénectère, 1849g) [= 1849n], now *Anthelephila*.

Microhoria inobscura (Pic, 1908i) **stat. nov.**, originally described as variety of *Anthicus tauricus*, now *Microhoria* (Pic, 1904f).

Comments

Marseul (1879) established several informal names for Anthicidae with the suffix '-colles'. Pic (1894g) formalized them as available scientific names of subgenera of *Anthicus* Paykull, 1798, changing their endings from "-colles" to "-comus". ICZN (2016), based on Alonso-Zarazaga (2013), has ruled the unavailability of 26 incorrectly formed generic names in the Anthicidae. Several of them were used as available by Chandler et al. (2008). These names are here listed but marked as supressed [SN].

Spelling

Anthicus pygmoeus LaFerté-Sénectère, 1849j (originally *pygmœus*) is the original, correct spelling and the subsequently used *A. pygmaeus* of various authors is incorrect.

New Distributional Data and Comments

Ciidae

R. Królik

New Records

Atlantocis canariensis Israelson, 1985: Madeira, Funchal, Jardim Botânico, N 32°39.738', W 16°53.680', 325 m, 18.V.2016, leg. & det. R. Królik, (2 specimens, collection of R. Królik); Madeira, Funchal, Santa Catarina Park, N 32° 39.0', W 16° 55.0', 32 m, 18.V.2016, leg. & det. R. Królik (7 specimens reared from fruiting bodies of fungi 15.VI-10.VII.2016, collection of R. Królik). New to Madeira.

Cis boleti (Scopoli, 1763): Bulgaria, 16 km S Berkowica, N 43°11.0', E 23°09.0', 566 m, 4.V.2008, leg. & det. R. Królik, (6 specimens, collection of R. Królik); Turkey, 7 km S İğneada, Kırklareli Prov., N 41°49.0', E 27°57.0', 25 m, 4–5.VI.2001, leg. & det. R. Królik, (8 specimens, collection of R. Królik). New to Bulgaria, Turkey.

Cis castaneus (Herbst, 1793): Bulgaria, 16 km S Berkowica, N 43°11.0', E 23°09.0', 566 m, 4.V.2008, leg. & det. R. Królik, (1 specimen, collection of R. Królik); North Macedonia, Velivar Mt., N Debar, N 41°33.0', E 20°31.0', 22.VI.2016, leg. P. Białooki, det. R. Królik, (38 specimens, collection of R. Królik); Turkey, 7 km S İğneada, Kırklareli Prov., N 41°49.0', E 27°57.0', 25 m, 22–25.VI.2000, leg. & det. R. Królik, (12 specimens, collection of R. Królik); Turkey, 10 km S Bolu, Bolu Prov., N 40°39.0', E 31° 37.0', 30.V.2002, leg. & det. R. Królik, (2 specimens, collection of R. Królik). New to Bulgaria, North Macedonia, Turkey.

Cis comptus Gyllenhal, 1827: Kyrgyzstan, Toktogul, Jalal-Abad Prov., N 41°52.0', E 72°56.0', 870 m, 26.VI.2003, leg. & det. R. Królik, (3 specimens, collection of R. Królik); Madeira, 1 km NE Ribeiro Frio, N 32°44.343', W 16°05.268', 800 m, 19.V.2016, leg. & det. R. Królik, (3 specimens, collection of R. Królik); Turkey, 7 km S İğneada, Kırklareli Prov., N 41°49.0', E 27°57.0', 25 m, 4–5.VI.2001, leg. & det. R. Królik, (25 specimens, collection of R. Królik). New to Kyrgyzstan, Madeira, Turkey.

Cis fusciclavis Nyholm, 1953: Bulgaria, Belogradchik env., N 43°38.0', E 22°41.0', 507 m, 30.IV.2010, leg. & det. R. Królik, (1 specimen, collection of R. Królik); Greece, Katara Pass, N 39°48.0', E 21°14.0', 1690 m, 12–13.VI.2007, leg. & det. R. Królik, (1 specimen, collection of R. Królik); Romania, 25 km NE Băile Herculane, N 45°04.0', E 22°37.0', 830 m, 27.V.2002, leg. & det. R. Królik, (4 specimens, collection of R. Królik); Turkey, 7 km S İğneada, Kırklareli Prov., N 41°49.0', E 27°57.0', 25 m, 22–25.VI.2000, leg. & det. R. Królik, (2 specimens, collection of R. Królik); Ukraine, Gurzuf env., Crimea, south coast, N 44°33.166', E 34°17.250', 170 m, 27–31.V.2005, leg. J. Borowski, det. R. Królik, (1 specimen, collection of R. Królik). New to Bulgaria, Greece, Romania, Turkey, Ukraine.

Cis glabratus Mellié, 1849: Bosnia Herzegovina, Klekovača Mts., Lom Mt., N 44°24.0', E 16°26.0', 2.IX.1966, leg. B. Burakowski, det. R. Królik, (3 specimens reared from fruiting bodies of fungi 22.IX.1966, collection of Museum and Institute of Zoology of the Polish Academy of Sciences,

 | DOI:10.1163/9789004434998_003

Warsaw – MiIZ); Romania, 25 km NE Băile Herculane, N 45°04.0', E 22°37.0', 830 m, 27.V.2002, leg. & det. R. Królik, (16 specimens, collection of R. Królik); Ukraine, Brzuchowice, N 49°55.0', E 23°57.0', 10.VI.1941, ex coll. M. Kłapacz, det. R. Królik, (1 specimen, collection of MiIZ). New to Bosnia Herzegovina, Romania, Ukraine.

Cis jacquemartii Mellié, 1849: Bulgaria, 16 km S Berkowica, N 43°11.0', E 23°09.0', 566 m, 4.V.2008, leg. & det. R. Królik, (1 specimen, collection of R. Królik). New to Bulgaria.

Cis laminatus Mellié, 1849: Turkey, 24 km N Fethiye, Muğla Prov., N 36°50.0', E 29° 07.0', 25.V.2011, leg. P. Białooki, det. R. Królik, (45 specimens, collection of R. Królik). New to Turkey.

Cis micans (Fabricius, 1792): Kyrgyzstan, Ala-Archa valley, 30 km S Bishkek, Chüy Prov., N 42°36.0', E 74°29.0', 1300 m, 15–16.VI.2003, leg. & det. R. Królik, (5 specimens from fruiting bodies of fungi on *Salix* sp., collection of R. Królik); Turkey, 7 km S İğneada, Kırklareli Prov., N 41°49.0', E 27°57.0', 8 m, 22.V.2005, leg. & det. R. Królik, (2 specimens from fruiting bodies of fungi on *Fraxinus* sp., collection of R. Królik). New to Kyrgyzstan, European Turkey.

Cis punctulatus Gyllenhal, 1827: Turkey, 3 km S Aydınlar, Mersin Prov., N 36°44.0', E 34°08.0', 1400 m., 28–30.V.2001, leg. & det. R. Królik, (4 specimens under bark *Abies cilicica*, collection of R. Królik). New to Turkey.

Cis pygmaeus pygmaeus (Marsham, 1802): Greece, Sapka Mts., 35 km N Alexandroupoli, N 41°09.0', E 25°45.0', 583 m, 10.V.2008, leg. & det. R. Królik, (2 specimens, collection of R. Królik); Turkey, 7 km S İğneada, Kırklareli Prov., N 41°49.0', E 27°57.0', 25 m, 4–5.VI.2001, leg. & det. R. Królik, (1 specimen, collection of R. Królik). New to Greece, Turkey.

Cis rugulosus Mellié, 1849: Azerbaijan, Talysh Mts., 2 km NW Lerik, N 38°48.035', E 48°25.348', 940 m, 17.V.2012, leg. & det. R. Królik, (18 specimens, collection of R. Królik); Bulgaria, Belogradchik env., N 43°38.0', E 22°41.0', 507 m, 30.IV.2010, leg. & det. R. Królik, (4 specimens, collection of R. Królik); Greece, Sapka Mts., 35 km N Alexandroupoli, N 41°09.0', E 25°45.0', 583 m, 19.V.2009, leg. & det. R. Królik, (11 specimens from fruiting bodies on *Quercus* sp., collection of R. Królik); Hungary, Mecsek-hegység, N 46°10.0', E 18°19.0', 5.VI.1957, leg. B. Burakowski, det. R. Królik, (1 specimen, collection of MiIZ); Kyrgyzstan, Ala-Archa valley, 30 km S Bishkek, Chüy Prov., N 42°36.0', E 74°29.0', 1300 m, 15–16.VI.2003, leg. & det. R. Królik, (16 specimens from fruiting bodies of fungi on *Salix* sp., collection of R. Królik); Madeira, E Funchal, Palheiro Gardens, N 32°39.743', W 16°52.000', 570 m, 22.V.2016, leg. & det. R. Królik, (3 specimens and 31 specimens reared from small fruiting bodies of fungi untill 30.VI.2016, collection of R. Królik); Romania, Odvoş env., N 46°06.0', E 21°49.0', 148 m, 28.IV.2009, leg. & det. R. Królik, (2 specimens from fruiting bodies of fungi on *Crataegus* sp., collection of R. Królik); Turkey, 7 km S İğneada, Kırklareli Prov., N 41°49.0', E 27°57.0', 25 m, 4–5.VI.2001, leg. & det. R. Królik, (6 specimens, collection of R. Królik); Turkey, Cankurtaran Geçidi, Artvin Prov., N 41°23.0', E 41° 34.0', 15.VI.2002, leg. & det. R. Królik, (18 specimens from fruiting bodies of polyporous fungi on *Alnus* sp. collection of R. Królik). New to Azerbaijan, Bulgaria, Greece, Hungary, Kyrgyzstan, Madeira, Romania, Turkey.

Cis seriatocribratus Reitter, 1913: Kyrgyzstan, Toktogul, Jalal-Abad Prov., N 41°52.0', E 72°56.0', 870 m, 26.VI.2003, leg. & det. R. Królik, (5 specimens, collection of R. Królik). New to Kyrgyzstan.

Cis striatulus Mellié, 1849: Morocco, Azrou, N 33°26.0', W 5°13.0', 3.V.1930, ex coll. W. Eichler, det. R. Królik, (1 specimen, collection of MIIZ). New to Morocco.

Cis submicans Abeille de Perrin, 1874: Romania, 24 km S Sibiu, N 45°35.0', E 24°16.0', 358 m, 26.V.2008, leg. & det. R. Królik, (24 specimens, collection of R. Królik); Turkey, 7 km S İğneada, Kırklareli Prov., N 41°49.0', E 27°57.0', 25 m, 4–5.VI.2001, leg. & det. R. Królik, (7 specimens, collection of R. Królik). New to Romania, European Turkey.

Cis tauriensis Królik, 2002: Turkey, by the road from Pülümur to Tunceli, Tunceli Prov., N 39°20.0', E 39°47.0', 1140 m., 17–19.V.2002, leg. & det. R. Królik, (23 specimens from fruiting bodies of fungi on *Populus* sp., collection of R. Królik). Turkey (second record).

Cis tomentosus Mellié, 1849: Turkey, 7 km S İğneada, Kırklareli Prov., N 41°49.0', E 27°57.0', 25 m, 4–5.VI.2001, leg. & det. R. Królik, (14 specimens, collection of R. Królik); Ukraine, Gurzuf env., Crimea, south coast, N 44°33.166', E 34°17.250', 170 m, 27–31.V.2005, leg. J. Borowski, det. R. Królik, (3 specimens, collection of R. Królik). New to Turkey, Ukraine.

Cis villosulus (Marsham, 1802): Romania, Brezoi, N 45°21.0', E 24°15.0', 27.VI.2002, leg. R. Dobosz, det. R. Królik, (6 specimens, collection of R. Królik). New to Romania.

Cisarthron laevicolle Reitter, 1885: Romania, Băile Herculane env., N 44°50.0', E 22°24.0', 435 m, 2.V.2010, leg. & det. R. Królik, (numerous specimens from fruiting bodies of polyporous fungi on *Carpinus betulus*, collection of R. Królik); Turkey, 7 km S İğneada, Kırklareli Prov., N 41°49.0', E 27°57.0', 25 m, 24–25.VI.2002, leg. & det. R. Królik, (numerous specimens from fruiting bodies of polyporous fungi on *Alnus* sp., collection of R. Królik). New to Romania, Turkey.

Dolichocis laricinus (Mellié, 1849): Ukraine, Gorgany, Chomiak, N 48°22.30', E 24°23.30', 27.VII.1937, ex coll. A. Bartoszyński, det. R. Królik, (2 specimens, collection of MIIZ). New to Ukraine.

Ennearthron cornutum (Gyllenhal, 1827): North Macedonia, Struga env., N 41°11.0', E 20°41.0', 9.V.2012, leg. J. Borowski, det. R. Królik, (14 specimens reared from fruiting bodies on *Salix* sp. 03.IV.2014, collection of R. Królik); Turkey, 7 km S İğneada, Kırklareli Prov., N 41°49.0', E 27°57.0', 25 m, 24–25.VI.2002, leg. & det. R. Królik, (5 specimens, collection of R. Królik); Turkey, Topaktaş vill. env., Hatay Prov., N 36°49.0', E 36°20.0', 1170 m., 5–8.VIII.2006, leg. & det. R. Królik, (2 specimens, collection of R. Królik). New to North Macedonia, Turkey

Octotemnus glabriculus (Gyllenhal, 1827): Greece, Kokkino Nero env., N 39°48.928', E 22°41.480', 1400 m, 10.VI.2013, leg. R. Dobosz, det. R. Królik, (18 specimens, collection of R. Królik); Turkey, 7 km S İğneada, Kırklareli Prov., N 41°49.0', E 27°57.0', 8 m, 22.V.2005, leg. & det. R. Królik, (11 specimens from fruiting bodies of fungi on *Fraxinus* sp., collection of R. Królik). New to Greece, Turkey.

Octotemnus mandibularis (Gyllenhal, 1813): Azerbaijan, Qəmərvan, N 41°03.0', E 47°47.0', 1019 m, 4–9.VI.2010, leg. & det. R. Królik, (2 specimens, collection of R. Królik). New to Azerbaijan.

Octotemnus opacus Mellié, 1849: Canary Islands, Tenerife, Aguamansa, N 28°21.647', W 16°29.767', 1093 m, 19.III.2013, leg. & det. R. Królik, (7 specimens, collection of R. Królik). New to Canary Islands.

Octotemnus rugosopunctatus Drogvalenko, 2002: Azerbaijan, Talysh Mts., 2 km SE Peştətük, N 38°46.287', E 48°33.851', 372 m, 13.VI.2012, leg. A. Lasoń, det. R. Królik, (2 specimens, collection of R. Królik). New to Azerbaijan.

Orthocis alni (Gyllenhal, 1813): Bulgaria, Petrič, N 41°24.0', E 23°13.0', 23–25.IV.1958, leg. B. Pisarski, det. R. Królik, (1 specimen, collection of MiIZ); Romania, Sinaia env., 50 km S Brasov, N 45°21.0', E 25°32.0', 30.V.2003, leg. P. Białooki, det. R. Królik, (1 specimen, collection of R. Królik). New to Bulgaria, Romania.

Orthocis coluber (Abeille de Perrin, 1874): Greece, Sapka Mts., 35 km N Alexandroupoli, N 41°09.0', E 25°45.0', 583 m, 10.V.2008, leg. & det. R. Królik, (1 specimen, collection of R. Królik). New to Greece

Orthocis reflexicollis (Abeille de Perrin, 1874): Azerbaijan, Talysh Mts., 2 km SE Peştətük, N 38°46.396', E 48°33.550', 375 m, 5–12.V.2014, leg. & det. R. Królik, (2 specimens shaken from dead twigs of *Pterocarya* sp., collection of R. Królik); Mongolia, 50 km NE Dzüüncharaa, N 49°05.233', E 107°17.470', 3.VIII.2015, leg. W. Szczepański, det. R. Królik, (1 specimen, collection of R. Królik); Turkey, Hasanbeyli pass, Osmaniye Prov., N 37°07.0', E 36°34.0', 19–25.V.2001, leg. & det. R. Królik, (2 specimens, collection of R. Królik). New to Azerbaijan, Mongolia, Turkey.

Ropalodontus novorossicus Reitter, 1901: Greece, Igoumenitsa env., N 39°29.0', E 20°16.0', 6.V.2012, leg. J. Borowski, det. R. Królik, (35 specimens reared from fruiting bodies of fungi untill 4.IV.2014, collection of R. Królik); Turkey, 7 km S İğneada, Kırklareli Prov., N 41°49.0', E 27°57.0', 25 m, 24–25.VI.2002, leg. & det. R. Królik, (1 specimen, collection of R. Królik); Ukraine, Gurzuf env., Crimea, south coast, N 44°33.166', E 34°17.250', 170 m, 27–31.V.2005, leg. J. Borowski, det. R. Królik, (4 specimens, collection of R. Królik). New to Greece, Turkey, Ukraine.

Ropalodontus perforatus (Gyllenhal, 1813): Bulgaria, Rila plan., Rilski man., N 42°08.0', E 23°20.4', 10.VIII.1959, leg. B. Burakowski, det. R. Królik, (8 specimens reared from fruiting bodies of fungi on *Fagus* sp. 16.IX.1959, collection of MiIZ). New to Bulgaria.

Ropalodontus strandi Lohse, 1969: Romania, Brezoi, N 45°21.0', E 24°15.0', 27.VI.2002, leg. & det. R. Królik, (9 specimens from fruiting bodies of polyporous fungi on *Fagus sylvatica*, collection of R. Królik); Ukraine, Magura nr. Worochta, N 48°17.083', E 24°34.066', 28. VIII.1937, ex coll. A. Bartoszyński, det. R. Królik, (2 specimens, collection of MiIZ). New to Romania, Ukraine.

Strigocis bicornis (Mellié, 1849): Greece, Sapka Mts., 35 km N Alexandroupoli, N 41°09.0', E 25°45.0', 583 m, 19.V.2009, leg. & det. R. Królik, (1 specimen, collection of R. Królik); Turkey, 7 km S İğneada, Kırklareli Prov., N 41°49.0', E 27°57.0', 8 m, 22.V.2005, leg. & det. R. Królik, (5 specimens from fruiting bodies of fungi on *Fraxinus* sp., collection of R. Królik); Turkey, Topaktaş vill. env., Hatay Prov., N 36°49.0', E 36°20.0', 1170 m., 5–8.VIII.2006, leg. & det. R. Królik, (17 specimens, collection of R. Królik). New to Ukraine, Turkey.

Sulcacis bidentulus (Rosenhauer, 1847): Spain, Torrente de Cinca, N 41°28.0', E 0°20.0', 30.VI-15.VIII.2002, leg. K. Rudziński, det. R. Królik, (14 specimens, collection of R. Królik); Kazakhstan, Putintsevo, N 49°52.216', E 84°21.766', 472 m, 20.VI.2017, leg. M. Walczak, det. R. Królik, (9 specimens from fruiting bodies of fungi on *Salix* sp., collection of R. Królik). New to Spain, Kazakhstan.

Sulcacis fronticornis (Panzer, 1805): Bulgaria, 1 km S Byala, NW Sliven, N 42°43.0', E 26°12.0', 709 m, 24.V.2008, leg. & det. R. Królik, (18 specimens, collection of R. Królik); Kazakhstan, Putintsevo,

N 49°52.216', E 84°21.766', 472 m, 20.VI.2017, leg. M. Walczak, det. R. Królik, (16 specimens from fruiting bodies of fungi on *Salix* sp., collection of R. Królik); Kyrgyzstan, Toktogul, Jalal-Abad Prov., N 41°52.0', E 72°56.0', 870 m, 26.VI.2003, leg. & det. R. Królik, (20 specimens, collection of R. Królik); Turkey, 7 km S İğneada, Kırklareli Prov., N 41°49.0', E 27°57.0', 25 m, 4–5.VI.20[illegible]1, leg. & det. R. Królik, (2 specimens, collection of R. Królik); Turkey, Aralık, Iğdır Prov., N 39°54.0', E 44°30.0', 11–12.VI.2002, leg. & det. R. Królik, (42 specimens from fruiting bodies of fungi on *Populus* sp., collection of R. Królik). New to Bulgaria, Kazakhstan, Kyrgyzstan, Turkey.

Sulcacis nitidus (Fabricius, 1792): Greece, Pangeon Mts., 21 km W Kavala, N 40°55.0', E 24°09.[illegible]', 1215 m, 20.V.2009, leg. & det. R. Królik, (26 specimens, collection of R. Królik); Turkey, 7 km S İğneada, Kırklareli Prov., N 41°49.0', E 27°57.0', 25 m, 24–25.VI.2002, leg. & det. R. Królik, (22 specimens, collection of R. Królik); Turkey, Cankurtaran Geçidi, Artvin Prov., N 41°23.0', E 41° 34.0', 15.VI.2002, leg. & det. R. Królik, (1 specimen from fruiting bodies of polyporous fungi on *Alnus* sp., collection of R. Królik). New to Greece, Turkey.

Xylographus bostrichoides (Dufour, 1843): Bulgaria, Obzor, N 42°49.0', E 27°53.0', 22.VII.1971, leg. A. Warchałowski, det. R. Królik, (7 specimens reared from fruiting bodies of fungi 20.07.-14.08.1972 by M. Kak, collection of R. Królik); Kyrgyzstan, Kara-Köl, Jalal-Abad Prov., N 41°38.0', E 72°39.0', 915 m, 19–20.VI.2003, leg. & det. R. Królik, (54 specimens from fruiting bodies of fungi *Fomes* sp. on trunk of *Salix* sp., collection of R. Królik). New to Bulgaria, Kyrgyzstan.

Anthicidae

D. Telnov

New Records

Anthelephila angustiformis (Fairmaire, 1893): Israel, Eilat S, Snuba Diving Center N env., 20.iii.2012, leg. R. Puplosis (2 specimens, collection of D. Telnov). New to Israel.

Anthelephila bispilifasciata bispilifasciata (Pic, 1897): Oman: Dhofar, Wadi Darbaat. 13.x.19[illegible]0. J.C. Deeming (1 specimen, collection of National Museum Wales). New to Oman.

Anthelephila mandarina (Boheman, 1858): China Foochow 1936–37 M. S Yang / Pres.By Imp. Inst. Ent. B.M.1938–720. / Formicomus mandarinus Boh. Dr. R.F. Heberdey (2 specimens, collection of Natural History Museum, London). New to Fujian (China).

Anthelephila panelii (van Hille 1977): Saudi Arabia: Aseer, Maraba. 25.ii.2006. H.A. Dawah (5 specimens, collection of National Museum Wales). New to Saudi Arabia.

Anthelephila parallela (Marseul, 1879a): Persia: Kerman. H.E.J. Briggs. B.M.1933–246. (4 specimens, collection of Natural History Museum, London); Persia / Brit. Mus. 1977–126 (2 specimens, collection of Natural History Museum, London). New to Iran.

Anthicus armatus Truqui, 1855: Kuwait, Al Jahra Block 6, Kuwait Bay shore, 19.V.2016, leg. M. Kozlov (1 specimen, collection of D. Telnov). New to Kuwait.

Anthicus commaculatus Krekich-Strassoldo, 1931: Nepal, Narayani Prov. Sauraha, Chitwan Nat.-P. 27°34'80"N,84°29'49"E 180m 18.IV.2000 Rapti-Ufer LF leg. A. Weigel (12 specimens, collection of Naturkundemuseum Erfurt). New to Nepal.

Anthicus dentaticornis Pic, 1898: Azerbaijan, Gr. Balachan Dschebell F. Hauser 1898. / ♂ (1 specimen, collection of Natural History Museum, London). New to Azerbaijan.

Anthicus dilatatus Krekich-Strassoldo, 1914: China W, Kashgar, at light, viii.2013, leg. S.A. Bukov (1 specimen, collection of D. Telnov). New to Xinjiang (China).

Anthicus gratiosus Pic, 1896: Turkey, Besika Bay (5 specimens, collection of Natural History Museum, London). New to European Turkey.

Anthicus inconspicuus inconspicuus Krekich-Strassoldo, 1931: Bhutan (Mongar Distr.) Mongar City, Thrumshingla (at light) 20.-27.VI.2010 local collector (via Li Jingke) (1 specimen, collection of Naturkundemuseum Erfurt). New to Bhutan.

Anthicus mongolicus Pic, 1896: China: Yung-ting-Shan Mts., 90 m. W. of Tai-yuan-fu, Shensi, N. China. 8.10.08. H.E.M.Douglas. 1910–332. (1 specimen, collection of Natural History Museum, London). New to Shaanxi (China).

Anthicus pindos Kejval, 2015: Armenia, Arkas valley, 28.VII-3.VIII.1999 leg. A. Rubienjan (1 specimen, collection of D. Telnov). New to Armenia.

Anthicus rubiginosus Krekich-Strassoldo, 1931: China: Yunnan, Lincang Pref., Nanding He river bank, 17 km N Lincang, 1288 m, washed from gravel and sand, from roots of vegetation, 24°02'11.9"N, 100°04'56.8"E, 10.IX.2009, leg. M. Schülke (CH09-41) (2 specimens, collection of Museum für Naturkunde – Leibniz-Institut für Evolutions- und Biodiversitätsforschung, Berlin); China: Yunnan, Lincang Pref., river bank, 5 km NE Yunxian, 24°27'45"N, 100°10'37.8"E, 1053 m, washed from gravel and sand, from roots of vegetation, 12.IX.2009, leg. M. Schülke (CH09-46) Museum für Naturkunde – Leibniz-Institut für Evolutions- und Biodiversitätsforschung). New to Yunnan (China).

Anthicus sollicitus LaFerté-Sénectère, 1849: Lebanon, Marjayoun env. Litani river, ~280 m 33°21'27"N, 35°32'48"E 18.XI.-11.XII.2017 pitfall trap, leg. Reuter (2 specimens, collection of Ch. Reuter); LEBANON, Rayfoun ca. 33°59'N, 35°42'E, mix. oak.for., ~900m 11.-28.11.2018, pitfall trap, leg. Reuter (1 specimen, collection of Ch. Reuter). New to Lebanon.

Doubtful Records

Anthelephila arabica (Pic, 1899n), record from Yemen.
Anthelephila bramina bramina (LaFerté-Sénectère, 1849g), record from Pakistan.
Anthelephila pegasus (Reitter, 1892a), record from Israel.
Anthelephila simoni (Pic, 1893u), record from Iran.
Anthicus antherinus antherinus Linnaeus, 1760, record from Saudi Arabia.
Anthicus ater Thunberg, 1787, record from Algeria.
Anthicus axillaris W.L.E. Schmidt, 1842c, record from East Siberia.
Anthicus bimaculatus Illiger, 1801, record from Slovakia.
Anthicus biguttatus LaFerté-Sénectère, 1849k, record from Portugal.
Anthicus fuscicornis LaFerté-Sénectère, 1849k, record from Turkey.
Anthicus latens Bonadona, 1977b, record from Turkey.
Anthicus quadrioculatus LaFerté-Sénectère, 1849k, record from Slovakia.

Anthicus sabulosus Marseul, 1879a, records from Iran and Moldova.
Cordicomus margelanicus (Pic, 1893r), record from Mongolia.
Cyclodinus brunnescens (Pic, 1893o), record from Iran.
Cyclodinus humilis (Germar, 1824a), records from Morocco, Tunisia, Turkmenistan and Uzbekistan.
Cyclodinus lotus (Marseul, 1879a), records from Iran and Uzbekistan.
Endomia lefebvrei (LaFerté-Sénectère, 1849m), record from Turkmenistan.
Leptaleus rodriguesi (Latreille, 1804a), record from Malta.
Microhoria cyrtopyga Bonadona, 1952, record from Southern Russia.
Microhoria leptostemma (Kolenati, 1846), records from Afghanistan, Turkmenistan and Uzbekistan.
Microhoria mollis (Desbrochers des Loges, 1875), record from Iran.
Microhoria pallidula (Pic, 1892e), record from Morocco.
Nitorus unifasciatus unifasciatus (Desbrochers des Loges, 1875), record from Iran.
Notoxus angustulus Krekich-Strassoldo, 1919a, record from Iran.
Notoxus transcaspicus Pic, 1899b, record from Afghanistan.
Omonadus cinctomus (Marseul, 1879a), record from Ningxia Province of China.
Omonadus phoenicius (Truqui, 1855), records from Kazakhstan and Southern Russia.
Stenidius vittatus vittatus (Lucas, 1843), record from Iran.
Steropes pici Abdullah, 1966, records from Bulgaria and former Yugoslavia.
Stricticomus herzi (Pic, 1905m), record from Turkmenistan.
Stricticomus longulus (Pic, 1905m), record from East Siberia.
Tenuicomus depressus (LaFerté-Sénectère, 1849l), record from Iran.
Tenuicomus ocreatus ocreatus (LaFerté-Sénectère, 1847), record from Italy.
Tenuicomus pallicrus (Dufour, 1849) record from Saudi Arabia.
Tomoderus bosnicus Pic, 1892m, record from Greece.

Tenebrionidae: Lagriini

O. Merkl

New Records

Cerogria decorata Borchmann, 1911: China, Sichuan, Mount Qingcheng, 725 m, N 0°53.770', E 103°34.690', 11–16.V.2017, leg. R. Ambrus, det. O. Merkl (1 specimen, collection of R. Ambrus); China, Sichuan, Ya'an, Bifeng, 1140 m, N 30°04.112', E 102°59.427', 17–24.V.2017, leg. R. Ambrus, det. O. Merkl (1 specimen, collection of R. Ambrus). New to China (Sichuan).

Cerogria klapperichi Borchmann, 1941: China, Guangxi A.R., 11–14.IV.2013, Longsheng Hot Spring (forested river valley, wet rock), 25°53.6'N, 110°12.4'E, 360 m, leg. M. Fikáček, J. Hájek & J. Růžička, det. O. Merkl (1 specimen, Mus. Prague); China, Shaanxi Prov., 11.V.2011, ca. 30 km SSE of Xian, Dayu village, Chanha river, 34°01.0'N, 109°06.9'E, 720 m, leg. M. Balke & J. Hájek, det. O. Merkl (2 specimens, Mus. Prague, 1 specimen, Mus. Budapest). New to China (Guanxi, Shaanxi).

Cerogria pachycera (Fairmaire, 1886): China, Guangxi A.R., 5–9.IV.2013, Shiwandashan National Forest Park (forested river valley, at light), 21°54.4'N, 107°54.2'E, 290–360 m, leg. M. Fikáček, J. Hájek & J. Růžička, det. O. Merkl (1 specimen, Mus. Prague). New to China (Guanxi).

Lagria lameyi Fairmaire, 1893: China, Guangdong Prov., 30 km NE Shaoguan, Duanshi vill., Danxia Shan NP (stream, pool) 25°02.7'N, 113°43.8'E, 125 m, 4–5.V.2011, leg. Jiří Hájek, det. O. Merkl (3 specimens, Mus. Prague, 1 specimen, Mus. Budapest); China, Guangdong Prov., Nanling National Nature Reserve, Dadongshan, 18–21.IV.2013, (border of mixed forest; at light), 24°56.0'N, 112°42.9'E, 690 m, leg. J. Hájek & Růžička, det. O. Merkl (1 specimen, Mus. Prague); China, Guangdong Prov., Danxia Shan NP, Wo Long Gang Forest Walkway, 23–26.IV.2013 (border of mixed forest, at light), 25°01.3'N, 113°44.5'E, 100 m, leg. J. Hájek & J. Růžička, det. O. Merkl (8 specimens, Mus. Prague, 2 specimens, Mus. Budapest). New to China (Guangdong).

Lagria lata Fabricius, 1801: Tunisia, 5 km E of Tabarka, 36°57'N, 85 49'E, alt. 20 m, 7–17.IX.2007, leg. R. & H. Fouquè, det. O. Merkl (1 specimen, Mus. Prague). New to Tunisia.

Lagria tenenbaumi Pic, 1929: Lebanon, Northern gov., Tannourine env., 2 km N Harissa, Tannourine Cedars Nat. Reserve, swept & singled, 34°12'34"N, 35°55'45"E, 1750 m, 16.VI.2016, leg. A. Kotán, P. Nemes & T. Németh (5 specimens, Mus. Budapest). New to Lebanon.

Xanthalia nigrovittata (Pic, 1910): China, Hainan Isl., Jianfengling Mts., Tiachi Lake env., Bishu Villa, (at light) 18°44'40"N, 108°50'41"E, 950 m, 9–11.V.2011, leg. M. Fikáček, v. Kubeček & L. Li, det. O. Merkl (1 specimen, Mus. Prague); China, Hainan Isl., Limushan Mts., first administr. centre (at light) 19°10'30"N, 106°44'33"E, 630 m, 4–6.V.2011, leg. M. Fikáček, v. Kubeček & L. Li, det. O. Merkl (1 specimen, Mus. Prague). New to China (Hainan).

Donaciolagria malgorzatae Merkl, 2011: China, S-Yunnan (Xishuangbanna) 45km SW Jinghong vic. Bangzhang vill. N21°44'37", E100°27'02" 16–1700m 3–5.V.2009, blüh. Cast., leg. A. Weigel, det. O. Merkl (3 specimens, Mus. Erfurt, 1 specimen, Mus. Budapest). New for China (Yunnan).

Taiwanolagria merkli Masumoto, 1988: China, Sichuan, Ya'an, Bifeng, 1140 m, N 30°04.112', E 102°59.427', 17–24.V.2017, leg. R. Ambrus, det. O. Merkl (1 specimen, collection of R. Ambrus); CHINA, Sichuan, Mount Qingcheng, 725 m, N 30°53.770', E 103°34.690', 11–16.V.2017, leg. R. Ambrus, det. O. Merkl (1 specimen, collection of R. Ambrus). New to China (Sichuan).

Tenebrionidae: Pimeliinae and Platyscelidini

L. Egorov

New Records

Bioramix (*Bioramix*) *oblongula* Fairmaire, 1898: N. Pakistan, Diamir District, Nanga Parbat m. Raikhot v. Faire Meadows, 3300–3400 m, 19–21.VII.1999, leg. M. Šlachta, det. L. Egorov (1 male, 1 female, collection of S. Bečvar, Prague, Czech Republic). New to Pakistan.

Bioramix (*Leipopleura*) *darbukensis* Kaszab, 1940: China, Xizang, Bange to Shenzha road 4586 m, N 31°31.092', E 089°12.573', 4.IX.2014, leg. Yun-Chun Li, det. L. Egorov (1 male, collection of Museum of China West Normal University, Nanchong, China). New to Xizang (China).

Bioramix (*Leipopleura*) *reinigi* Kaszab, 1940: China, Sichuan, Qionglai Shan, Ruoergai, 3510 m, N 30°35.068', E 102°57.791', 23.VII.2009, leg. Ai-Min Shi and Yong-Fu Wang, det. L. Egorov (2 males and 2 females, collection of Zoological Institute of the Russian Academy of Sciences, St. Petersburg, Russia). New to Sichuan (China).

Bioramix (*Leipopleura*) *rufipalpis* Reitter, 1887: China, Xizang, Nagqu-Tanggula (Dangla) Shan, Naqu, 4580 m, N 31°27.201', E 092°03.078', 7.VIII.2010, leg. Yun-Chun Li and Yong-Sheng Pan, det. L. Egorov (1 male, collection of Zoological Institute of the Russian Academy of Sciences, St. Petersburg, Russia). New to Xizang (China).

Bioramix (*Planoplatyscelis*) *pamirensis* Bates, 1879: Afghanistan, Badakhshan, Shiwa, 2800 m, 23.VII.2008, leg.C. Reuter, det. L. Egorov, (5 specimens: 1 male, collection of Zoological Institute of the Russian Academy of Sciences, St. Petersburg, Russia; 4 specimens, collection of Staatliches Museum für Naturkunde, Stuttgart, Germany). China, Xinjiang, Wuqia county (Xinjiang), N 39.65784°, E 74.05535°, 3220 m, 27.VII.2014, leg. Jin-Cheng Liu, det. L. Egorov (1 male, collection of Zoological Institute of the Russian Academy of Sciences, St. Petersburg, Russia); China, Xinjiang, Wuqia county (Xinjiang), N 39.22418°, E 75.02392°, 2850 m, 2.VII.2014, leg. Jin-Cheng Liu, det. L. Egorov, (1 male, collection of Zoological Institute of the Russian Academy of Sciences, St. Petersburg, Russia). New to Afghanistan, Xinjiang (China).

Diaclina fagi (Panzer, 1799): Russia, Republic of Mordovia, Temnikovsky district, Pushta village, 15.VIII.2019, to light, leg. G.B. Semishin, det. L. Egorov (1 specimens, collection of Zoological Institute of the Russian Academy of Sciences, St. Petersburg, Russia). New to Central European Territory of Russia.

Oodescelis (*Spinoodescelis*) *punctolineata* Kaszab, 1940: Kazakhstan, Pskem Range, slope in the gorge of the Maidantal river, 11.VI.1993, zone of *Juniperus* sp., leg. B.M. Kataev, det. L. Egorov (1 female, collection of Zoological Institute of the Russian Academy of Sciences, St. Petersburg, Russia). Uzbekistan, Pskem Range, neighborhood of Barkarak and Akkapchagay, 20.VI.1993, leg. B.M. Kataev, det. L. Egorov (14 males and 6 females, collection of Zoological Institute of the Russian Academy of Sciences. St. Petersburg, Russia). Pskem Range, 78 KM SSW of Jambul, Tyuzoshu pass and descent to the north (east of Mount Manas), 21.VI.1993, leg. B.M. Kataev, det. L. Egorov (4 males and 5 females, collection of Zoological Institute of the Russian Academy of Sciences, St. Petersburg, Russia). New to Kazakhstan, Uzbekistan.

Platyscelis (*Platyscelis*) *rugifrons* Fischer von Waldheim, 1823: China, Xinjiang, Heishantou (Karjaw Town, Jeminay County, ca. N 47.1°, E 86.6°), Burqin County, 1200 m, 3.IX.1960, leg. Fa-Cai Zhang, IOZ(E) 753037, det. L. Egorov (1 female, collection of Institute of Zoology, Chinese Academy of Sciences, Beijing, China). New to Xinjiang (China).

Comments

Bioramix (*Leipopleura*) *politicollis* Kaszab, 1940. The species is based on collection of R. J. H. Kaulbakh (Kaszab, 1940b). The type locality "Zentral Tibet, Salween Vallay, Bunthang Gompa" was erroneously assumed to be in Nepal (Egorov, 2008). According to the description of the expedition route (Kaulback, 1938), it is in Xizang.

Bioramix (*Leipopleura*) *rubripes* Reitter, 1889. Based on Egorov (1990), the species occurs only in Qinghai, the records in Egorov (2008) are erroneous.

Platyscelis (*Platyscelis*) *subcordata* Seidlitz, 1893. The species was reported from Taiwan and Japan (Bai & Ren, 2019a). However, according to Ando et al. (2016) and personal information from Dr. K. Ando, this species is absent from Japan and Taiwan.

Platyscelis (*Pleiopleura*) *ovata* Ballion, 1878. Bai & Ren (2019a) reported the species from the following provinces of China: Gansu, Qinghai, Inner Mongolia and Guangxi. This information seems dubious.

Catalogue

family MYCETOPHAGIDAE Leach, 1815
subfamily **Mycetophaginae** Leach, 1815
tribe Mycetophagini Leach, 1815

genus *Eulagius* Motschulsky, 1845a: 92 type species *Eulagius acernus* Motschulsky, 1845
Atritomoides Nikitsky, 1988b: 1847 type species *Eulagius ussuriensis* Nikitsky, 1988
Parabaptistes Ganglbauer, 1899: 838 type species *Tritoma univestris* Reitter, 1878
acernus Motschulsky, 1845a: 92 **E:** AB AR GG ST (Kavkaz)
univestris Reitter, 1878a: 166 (*Tritoma*)
chinensis Nikitsky, 1996a: 1372 **A:** SCH
dentatus Nikitsky, 1988b: 1847 **A:** FE
filicornis Reitter, 1887c: 288 (*Atritomus*) **E:** FR GBi SP **N:** AG TU
boissyi Caillol, 1925: 101 (*Atritomus*)
hyrcanicus Esser, Varandi & Farashian, 2018: 240 **A:** IN
irregularis Reitter, 1888a: 174 (*Atritomus*) **E:** AB AR GG ST (Kavkaz)
lewisi Reitter, 1889d: 249 (*Atritomus*) **A:** JA TAI
reitteri Lewis, 1895a: 331 (*Atritomus*) **A:** FE JA
ussuriensis Nikitsky, 1988b: 1845 **A:** FE JA NC SC
irregularis Miyatake, 1985: 288 (*Parabaptistes*) [HN]

genus *Litargops* Reitter, 1880a: 89 type species *Litargus pictus* Wollaston, 1854
maculosus Reitter, 1889d: 247 **A:** FE JA
pictus Wollaston, 1854: 200 (*Litargus*) **N:** MR

genus *Litargus* Erichson, 1846: 415 type species *Ips bifasciatus* Fabricius, 1787
subgenus *Alitargus* Casey, 1900: 136 type species *Litargus balteatus* LeConte, 1856
balteatus LeConte, 1856: 14 **E:** AU AR CZ FR GB GE GG GR IR HU IR IT MA NLi PL SK SL SP ST SVi SZ **N:** CI EG MO MR **A:** JA **AUR NAR NTR ORR**
antennatus Miyatake, 1957: 41
disjunctus Sharp, 1902: 639
ferrantei Reitter, 1908c: 44
infulatus LeConte, 1856: 14
pilosus Wollaston, 1857a: 71
tranisversus LeConte, 1856: 14
subgenus *Litargosomus* Motschulsky, 1858b: 52 type species *Litargosomus maculatus* Motschulsky, 1858
coloratus Rosenhauer, 1856: 105 **E:** FR GR IT MA PT SP SVi SZ **N:** AG AZ CI MO MR TU **A:** IS SY TR **AFR**
obscuripennis Pic, 1901g: 66
trifasciatus Wollaston, 1864: 154

japonicus Reitter, 1877c: 27 **A:** JA MG NC SC
kyushuensis Miyatake, 1985: 287 **A:** FE JA
lewisi Reitter, 1889d: 246 **A:** CH JA
sexsignatus Miyatake, 1957: 40 **A:** JA
unifasciatus Reitter, 1889d: 247 **A:** FE JA
vestitus Sharp, 1879: 88 **A:** JA **Pacific**
subgenus *Litargus* Erichson, 1846: 415 type species *Ips bifasciatus* Fabricius, 1787
caucasicus Pic, 1894h: 71 **E:** "Caucasus"
connexus Geoffroy, 1785: 138 (*Antribus*) **E:** AB AL AR AU BE BH BU BY CR CT CZ DE EN FI FR GB GE GG GR HU IR IT KZ LA LS LT MA MC MD NL NR NT PL PT RO SK SL SP ST SV SZ TR UK YU **N:** CI MO TU **A:** CH CY ES FE IN JA JO KI KZ MG NC NE SY TR UZ
bifasciatus Fabricius, 1787a: 47 (*Ips*)
lunatus Fabricius, 1792b: 514 (*Ips*)
marginalis Panzer, 1792a: 24 (*Ips*)
mediojunctus Pic, 1903: 191
signatus Panzer, 1798: 20 (*Mycetophagus*)
leprieuri Pic, 1901g: 65 **N:** AG TU

genus *Mycetophagus* Fabricius, 1792b: 497 type species *Silpha quadrimaculata* Schaller, 1783
subgenus *Arnoldiellus* Nikitsky, 1989b: 60 type species *Tritoma tschitscherini* Reitter, 1897
tschitscherini Reitter, 1897e: 228 (*Tritoma*) **E:** CT LA NT ST SVi **A:** ES FE MG NC WS
tschitscherini Semenov, 1898: 549 (*Tritoma*) [HN]
subgenus *Ilendus* Casey, 1900: 131 type species *Mycetophagus pluripunctatus* LeConte, 1856
irroratus Reitter, 1879: 224 (*Tritoma*) **A:** ES FE JA NC NE SC
lederi Reitter, 1897e: 228 (*Tritoma*) **A:** ES MG
multipunctatus Fabricius, 1792b: 498 **E:** AB AL AR AU BE BH BU BY CR CT CZ DE EN FI FR GB GE GG HU IN IR IT KZ LA LT MC MD NL NR NT PL RO SK SP ST SV SZ TR UK YU **A:** KZ ES WS
similis Marsham, 1802: 140 (*Boletaria*)
sulcatulus Roubal, 1929a: 47
subgenus *Kovalevius* Nikitsky, 1992b: 415 type species *Tritoma grandis* Reitter, 1889
grandis Reitter, 1889d: 248 (*Tritoma*) **A:** FE (S. Kuril Is., Sakhalin) JA
simplex Reitter, 1889d: 248
obsoletesignatus Miyatake, 1968: 31 **A:** JA
subgenus *Micromycetophagus* Nikitsky, 2007: 58 type species *Tritoma undulata* Reitter, 1889 (= *Mycetophagus reitteri* Nikitsky, 2008a)
reitteri Nikitsky, 2008a **A:** FE JA
undulatus Reitter, 1889d: 249 (*Tritoma*) [HN]
subgenus *Mycetophagoides* Nikitsky, 1988b: 1849 type species *Mycetophagus livshitzi* Nikitsky, 1988
alni Champion, 1917: 53 **A:** UP
hillerianus Reitter, 1877d: 116 **A:** FE JA NC NE SC

livshitzi Nikitsky, 1988b: 1848 **A:** FE
narukawai Saito, 2006: 201 **A:** JA
subgenus *Mycetophagus* Fabricius, 1792b: 497 type species *Silpha quadrimaculata* Schaller, 1783
Boletaria Marsham, 1802: 138 type species *Chrysomela quadripustulata* Linnaeus, 1760
Silphoides Herbst, 1783: 41 type species *Silphoides boleti* Herbst, 1783
ancora Reitter, 1884a: 6 (*Tritoma*) **E:** AB (Talysh) **A:** IN
ater Reitter, 1879: 224 (*Tritoma*) **E:** AU BH BY CR CT CZ EN FI FR GE HU LA LT MD NT PL RO SK SL ST SZ UK YU **A:** CH ES FE JA KZ MG NC NE SC WS
jaroslawensis Semenov, 1898: 547 (*Tritoma*)
quadriornatus Reitter, 1901c: 71 (*Tritoma*) **E:** AB (Talysh) **A:** IN
quadripustulatus Linnaeus, 1760: 168 (*Chrysomela*) **E:** AB AL AR AU BE BH BU BY CR CT CZ DE EN FI FR GB GE GG GR HU IT KZ LA LT MC MD NL NR NT PL PT RO SK SL SP ST SV SZ TR UK YU **N:** MO **A:** IN KZ MG TR WS
antemacularis Dalla Torre, 1879: 98 (*Tritoma*)
bipustulatus Schilsky, 1888: 184 (*Tritoma*)
boleti Herbst, 1784a: 41 pl. 21 (*Silphoides*)
impustulatus Schilsky, 1888: 184 (*Tritoma*)
quadrimaculatus Schaller, 1783: 256 (*Silpha*)
ruficollis Schilsky, 1888: 184 (*Tritoma*)
winteri Reitter, 1911d: 291
subgenus *Mycetoxides* Motschulsky, 1858c: 144 type species *Mycetophagus fulvicollis* Fabricius, 1792
Calilendus Reitter, 1911a: 94 type species *Mycetophagus fulvicollis* Fabricius, 1792
elongatus Reitter, 1889d: 247 (*Tritoma*) **A:** FE JA
fulvicollis Fabricius, 1792b: 499 **E:** AB AL AR AU BE BH BU BY CR CT CZ DE EN FI FR GB GE GG GR HU IT KZ LA LT MD NR NT PL PT RO SK SL SP ST SV SZ UK YU **A:** ES FE KZ MG TR WS
deubeli Méhely, 1890: 272 (*Tritoma*)
massanae Dajoz, 1960: 195
sexmaculatus Ragusa, 1892a: 142
subgenus *Paramycetophagus* Nikitsky, 2007: 73 type species *Mycetophagus sulcicollis* Champion, 1917
sulcicollis Champion, 1917: 53 **A:** UP
subgenus *Parilendus* Casey, 1900: 131 type species *Mycetophagus bipustulatus* Melsheimer, 1844 (= *Mycetophagus quadriguttatus* P.W.J. Müller, 1821)
quadriguttatus P.W.J. Müller, 1821: 198 **E:** AB AL AR AU BE BH BU BY CR CT CZ DE EN FI FR GB GE GG GR HU IT KZ LA LS LT MC MD NL NT PL RO SK SL SP ST SV SZ TR UK YU **N:** AG MO **A:** IS KI KZ SA SY TD TM TR UZ WS **AUR NAR NTR CRR**
bipustulatus Melsheimer, 1844: 114
disjunctemaculatus Roubal, 1936a: 194
pubescens Stephens, 1830: 86
variegatus C.R. Sahlberg, 1837: 168

subgenus *Philomyces* Ganglbauer, 1899: 832 type species *Mycetophagus populi* Fabricius, 1798

populi Fabricius, 1798: 175 **E:** AU BE BY CR CT CZ DE EN FI FR GB GE HU IT LA LT NL NR NT PL RO SK SP ST SV SZ UK YU **A:** WS

subgenus *Ulolendus* Reitter, 1911a: 93 type species *Ips piceus* Fabricius, 1777

akitai M. Saito, 2013: 97 **A:** JA

amamianus Nakane, 1989: 6 **A:** JA

antennatus Reitter, 1879: 225 (*Tritoma*) **A:** CH ES FE JA MG NC NE TAI

jureceki Pic, 1921a: 1 (*Tritoma*)

sauteri Grouvelle, 1914: 75

atomarius Fabricius, 1787a: 46 (*Ips*) **E:** AL AU BE BH BU BY CR CT CZ DE EN FI FR GB GE GR HU IT LA LT MC MD NL NR NT PL PT RO SK SL SP ST SV SZ UK YU **A:** TR

bifasciatus Champion, 1919: 242 **A:** UP

ciscaucasicus Semenov, 1899b: 596 (*Tritoma*) **E:** AB AR GG ST **A:** TR

decempunctatus decempunctatus Fabricius, 1801b: 568 **E:** AB AR AU BH BU BY CR CT CZ EN FI FR GE GG HU IT LA LT MD NR NT PL RO SK SL SP ST SV SZ UK YU **A:** KZ TR

rossicus Semenov, 1898a: 86

decempunctatus syriacus Reitter, 1904c: 165 **A:** SY

hiranoi M. Saito, 2013: 99 **A:** JA

hyrcanicus Nikitsky, 1989b: 59 **E:** AB (Talysh)

incognitus Nikitsky, 1988b: 1850 **A:** ES FE

infulatus Reitter, 1884a: 6 (*Tritoma*) **E:** AB (Talysh) **A:** IN

intermedius Nikitsky, 1992b: 419 **E:** FE

piceus Fabricius, 1777: 214 (*Ips*) **E:** AB AL AR AU BE BH BU BY CR CT CZ DE EN FI FR GB GE GG GR HU IT KZ LA LT LU MA MD NL NR NT PL PT RO SK SL SP ST SV SZ TR UK YU **N:** EG **A:** KZ MG TR UZ WS

bosnicus Apfelbeck, 1911: 220

brunneus Panzer, 1798: 21

decipiens Apfelbeck, 1911: 226

feliciae Ragusa, 1892a: 142

flavotinctus Roubal, 1931: 132

histrio C.R. Sahlberg, 1837: 167

humeralis Schilsky, 1888: 184 (*Tritoma*)

hungaricus Papp, 1946: 38

lunaris Fabricius, 1801b: 568

punctulatus Schilsky, 1888: 184 (*Tritoma*)

salicis C. Brisout de Barneville, 1862: 24

undulatus Marsham, 1802: 140 (*Boletaria*)

variabilis Hellwig, 1792: 397

varius Marsham, 1802: 140 (*Boletaria*)

pustulosus Reitter, 1889d: 248 (*Tritoma*) **A:** FE JA NC SC

funebris Reitter, 1889d: 248 (*Tritoma*)

ramosus Reitter, 1904c: 166 **E:** AB
tauricus Pliginskij, 1923a: 78 **E:** UK (Krym)

genus *Mycetophagus*, species incertae sedis
fraternus Grouvelle, 1911: 315 **A:** SD

genus *Pseudotriphyllopsis* Nikitsky, 1996a: 1370 type species *Pseudotriphyllopsis inseriata* Nikitsky 1996
inseriata Nikitsky, 1996a: 1370 **A:** SCH

genus *Pseudotriphyllus* Reitter, 1880a: 38 type species *Dermestes suturalis* Fabricius, 1801
colchicus Reitter, 1876b: 293 (*Triphyllus*) **E:** AB AR GG ST
lewisianus Wollaston, 1874: 170 (*Triphyllus*) **A:** FE JA
nepalensis Nikitsky, 2003: 49 **A:** NP
ohbayashii Miyatake, 1968: 29 **A:** JA
suturalis Fabricius, 1801a: 317 (*Dermestes*) **E:** AU BH BU CR CZ FR GB GE GR HU IT RO SK SZ UK
humeralis Stephens, 1830: 74 (*Cryptophagus*)
vicarius Peyerimhoff, 1913a: 254 **N:** AG

genus *Triphyllina* Reitter, 1877h: 384 type species *Triphyllina lederi* Reitter, 1878
lederi Reitter, 1878a: 168 **E:** AB AR GG ST (Kavkaz)

genus *Triphylloides* Miyatake, 1959a: 27 type species *Triphyllus seriatus* Reitter, 1889
seriatus Reitter, 1889d: 246 (*Triphyllus*) **A:** FE JA NE

genus *Triphyllopsis* Nikitsky, 1989b: 56 type species *Triphyllopsis altaica* Nikitsky, 1989
altaica Nikitsky, 1989b: 58 **A:** KZ WS

genus *Triphyllus* Dejean, 1821: 102 type species *Mycetophagus punctatus* Fabricius, 1792
bicolor Fabricius, 1777: 216 (*Nitidula*) **E:** AB AL AR AU BE BH BU BY CR CT CZ DE EN FI FR GB GE GG GR HU IT LA LT MD NL NR NT PL PT RO SK SL SP ST SV SZ UK YU **A:** TR
humeralis Marsham, 1802: 123 (*Silpha*)
immaculatus Roubal, 1936a: 192
pilosus Herbst, 1792: 177, pl. 42 (*Kryptophagus*)
punctatus Fabricius, 1792b: 499 (*Mycetophagus*)

tribe Typhaeini C.G. Thomson, 1863

genus *Typhaea* Stephens, 1829a: 8 type species *Mycetophagus testaceus* Fabricius, 1792 (= *Dermestes stercorea* Linnaeus, 1758) *angusta* Rosenhauer, 1856: 106 [EA] **E:** IT SP **N:** AG TU

haagi Reitter, 1874: 527 **E:** AU CT CZ DE FI FR GB GE GG HU LA NL PL SK ST SV UK **A:** CH FE JA NC SC **AFR NAR NTR ORR**

decipiens Lohse, 1989: 145

pallidula Reitter, 1874: 527

stercorea Linnaeus, 1758: 357 (*Dermestes*) **E:** AB AL AN AR AU AZ BE BH BU BY CR CT CZ DE EN FA FI FR GB GE GG GR HU IC IR IT KZ LA LS LT LU MA MC MD NL NR NT PL PT RO RU SK SL SP ST SV SZ TR UK YU **N:** AG CI EG LB MO MR TU **A:** AE AF BT CH CY ES FE IN IQ IS JA JO KI KZ MG NC NP PA SA SC SI SY TD TM TR UZ WS YE **COS**

crenata Melsheimer, 1844: 114 (*Atomaria*)

curvipes Broun, 1880: 238

fumata Linnaeus, 1767: 564 (*Dermestes*)

hirta Broun, 1880: 239

testacea Fabricius, 1792b: 499 (*Mycetophagus*)

tomentosa Stephens, 1830: 71

variabilis Herbst, 1792: 141 (*Dermestes*)

genus *Typhaeola* Ganglbauer, 1899: 836 type species *Typhaea maculata* Perris, 1865

brachati Esser, 2016: 171 **A:** IS TR

curticollis Fairmaire, 1879b: 167 (*Triphyllus*) **N:** AG TU

maculata Perris, 1865: 507 (*Typhaea*) **E:** CR FR MA PT SP **N:** MO

umbrata Baudi di Selve, 1870: 62 (*Typhaea*) **E:** IT MA

subfamily **Berginінae** Leng, 1920

genus *Berginus* Erichson, 1846: 405 type species *Berginus tamarisci* Wollaston, 1854

tamarisci Wollaston, 1854: 195 **E:** AU BE BH CR CZ FR GB GE GG GR HU IT MA PLi RO SK SL SP ST SVi SZ TR **N:** AG EG MO MR TU **A:** IS TR

subfamily **Esarcinae** Reitter, 1882

genus *Esarcus* Reiche, 1864: 238 type species *Esarcus leprieuri* Reiche, 1864

subgenus *Entoxylon* Ancey, 1869: 46 type species *Esarcus abeillei* Ancey, 1870

Atritomus Reitter, 1877h: 384 type species *Atritomus cribratus* Reitter, 1877 (= *Entoxylon abeillei* Ancey, 1870)

abeillei Ancey, 1870: 85 (*Entoxylon*) **E:** FR IT

cribratus Reitter, 1877h: 384 (*Atritomus*)

baudii Seidlitz, 1889: 149 **E:** FR IT

besucheti Dajoz, 1964: 13 **E:** SP

franzi Dajoz, 1964: 14 **E:** SP

inexpectatus Dajoz, 1964: 14 **E:** SP

martini Reitter, 1887c: 285 **N:** AG

subgenus *Esarcus* Reiche, 1864: 238 type species *Esarcus leprieuri* Reiche, 1864

fiorii Reitter, 1887c: 285 **E:** IT
iolensis Peyerimhoff, 1917a: 140 **N:** AG
leprieuri Reiche, 1864: 239 **N:** AG TU
letourneuxi Raffray, 1873: 368 (*Entoxylon*) **N:** AG
parallelus Pic, 1901a: 3

family ARCHEOCRYPTICIDAE Kaszab, 1964

genus *Sivacrypticus* Kaszab, 1964f: 384 type species *Sivacrypticus taiwanicus* Kaszab, 1964
bengalicus Kaszab, 1975a: 34 **A:** NP SD **ORR**
dilliensis Kaszab, 1979b: 200 **A:** PA
indicus Kaszab, 1964f: 386 **A:** NP PA **ORR**
kashmirensis Kaszab, 1975a: 34 **A:** KA
loebli Kaszab, 1979b: 196 **A:** NP **ORR**
sericans Fairmaire, 1896a: 24 (*Platydema*) **A:** NP **ORR**
spinifer Champion, 1920a: 73 (*Enneboeus*)
taiwanicus Kaszab, 1964f: 385 **A:** TAI
uenoi Masumoto & Yin, 1993: 241 **A:** YUN

family PTEROGENIIDAE Crowson, 1953

genus *Laenagenius* Löbl, 2005: 153 type species *Laenagenius apterus* Löbl, 2005
apterus Löbl, 2005: 155 **A:** YUN

family CIIDAE Leach, 1819
subfamily **Ciinae** Leach, 1819
tribe Ciini Leach, 1819

genus *Acanthocis* Miyatake, 1954: 47 type species *Cis bisetosus* Miyatake, 1954 (= *Acanthocis inonoti* Miyatake, 1955)
inonoti Miyatake, 1955b: 95 [RN] **A:** JA (Shikoku)
bisetosus Miyatake, 1954: 48 (*Cis*) [HN]
quadridentatus Nobuchi & Wada, 1959: 52 **A:** JA

genus *Aliocis* Sandoval-Gómez & Lopes-Andrade, 2015: 599 [RN] type species *Ennearthron poriae* Nakane & Nobuchi, 1955
Anoplocis Kawanabe, 1996b: 331 [HN] type species *Ennearthron poriae* Nakane & Nobuchi, 1955
poriae Nakane & Nobuchi, 1955: 49 (*Ennearthron*) **A:** JA
ryukyuensis Kawanabe, 1996b: 333 (*Anoplocis*) **A:** JA (Okinawa)

genus *Atlantocis* Israelson, 1985: 80 type species *Atlantocis gillerforsi* Israelson, 1985
gillerforsi Israelson, 1985: 82 **E:** AZ
canariensis Israelson, 1985: 84 **N:** CI (Tenerife) MR
lauri Wollaston, 1854: 282 (*Cis*) **N:** MR

genus *Ceracis* Mellié, 1849a: 375 type species *Ennearthron sallei* Mellié, 1849a
cucullatus Mellié, 1849a: 372 (*Ennearthron*) Ei: FR **AFR NTR**i
bilamellatus Pic, 1916c: 20 (*Ennearthron*)
lamellatus Pic, 1939a: 8 (*Ennearthron*)
tabelliferus Mellié, 1849a: 373 (*Ennearthron*)
japonus Reitter, 1878c: 36 (*Ennearthron*) **A:** JA
laminicollis Miyatake, 1982b: 77 **A:** TAI
shikokuensis Miyatake, 1954: 56 (*Ennearthron*) **A:** JA (Shikoku)

genus *Cis* Latreille, 1796: 50 type species *Dermestes boleti* Scopoli, 1763
Cisdygma Reitter, 1885b: 209 type species *Cis clavicornis* Baudi di Selve, 1874
Dimerocis Peyerimhoff, 1919b: 277 type species *Cis aurasiacus* Peyerimhoff, 1919
Eridaulus C.G. Thomson, 1859: 91 type species *Cis comptus* Gyllenhal, 1827
Macrocis Reitter, 1878c: 34 type species *Macrocis taurus* Reitter, 1878
Xestocis Casey, 1898: 85 type species *Xestocis levettei* Casey, 1898
aurasiacus Peyerimhoff, 1919b: 277 **N:** AG
aurosericeus Reitter, 1887e: 515 **A:** FE
bidentatus G.-A. Olivier, 1790: 16: 11 (*Anobium*) **E:** AU BE BU BY CR CZ DE EN FI FR GB GE HU IR IT LA LT NL NR NT PL RO SK SV SZ UK **N:** AG
inermis Marsham, 1802: 87 (*Ptinus*)
bifasciatus Reitter, 1877g: 381 (*Cis*) **A:** JA
bilamellatus Wood, 1884: 130 **Ei:** FR GB IR **AUS**
munitus Blackburn, 1888: 268
bisetiger M. Chûjô, 1939: 3 **A:** FE (Sachalin)
bituberculosus Roubal, 1937a: 39 [RN] **E:** ST (Kavkaz)
bituberculatus Roubal, 1912a: 25 [HN]
boleti Scopoli, 1763: 17 (*Dermestes*) **E:** AB AU BE BH BU BY CR CT CZ DE EN FI FR GB GE GG GR HU IR IT LA LS LT NL NR NT PL RO SK SL SP ST SV SZ TR UK **N:** AG **A:** FE IN JA (Hokkaido) MG SCH "Transcaspia"
armicollis Roubal, 1937b: 67
boletorum Marsham, 1802: 85 (*Ptinus*)
latior Pic, 1934f: 86
minor Mellié, 1849a: 241
obliteratus Mellié, 1849a: 241
picipes Fabricius, 1787a: 35 (*Dermestes*)
pruinosus Motschulsky, 1861: 17
substriatus Mellié, 1849a: 241
brevipennis Nakane & Nobuchi, 1955: 48 **A:** JA (Honshu)
capricornis Kawanabe, 1997b: 326 **A:** JA (Ryukyus)

castaneus Herbst, 1793: 17 (*Kateretes*) **E:** AB AU BE BH BU BY CR CT CZ DE EN FI FR GB GE GR HU IR IT LA LS LT MC NL NR NT PL RO SK SL SP ST SV SZ TR UK YU **A:** FE "Buchara" IN TR

caucasicus Ménétriés, 1832: 224 **E:** "Caucasus"

cedri Peyerimhoff, 1915: 25 **N:** AG

chinensis Lawrence, 1991: 288 **E:** FR GE HU IT MA SP **A:** CH IN

chujoi Miyatake, 1982b: 73 **A:** TAI

clavicornis Baudi di Selve, 1874c: 338 **E:** GR UK (Krym) **A:** CY LE TR

comptus Gyllenhal, 1827: 625 **E:** AB AR AU BE BH BU BY CR CT CZ DE EN FI FR GE GG GR HU IT LA MC NL NR NT PL RO SK SL SP ST SV SZ TR UK **N:** MO MR TU **A:** ES FE KI KZ TM TR WS

cognatus Rey, 1892: 17

substriatus Munster, 1927: 185 [= 1928: 295] [HN]

coriaceus Baudi di Selve, 1874c: 337 **A:** CY SY

corioli Peyerimhoff, 1915: 26 (*Cisdygma*) **N:** AG

crenatus C.R. Sahlberg, 1836: 159 **E:** FI

cucullatus Wollaston, 1865: Appendix: 39 **N:** CI (Gomera)

dentatus Mellié, 1849a: 324 **E:** AU BH CR CT CZ FI FR GB GE IT LA NR PL RO SK SV SZ

microgonus C.G. Thomson, 1868: 46

eminenticollis Nobuchi, 1955a: 56 **A:** JA (Honshu)

erinaceus Roubal, 1933a: 72 **A:** FE

fagi Waltl, 1839: 224 **E:** AU BH BY CR CT CZ DE FI FR GB GE GR HU IR IT LT NL NR NT PL RO SK SP SV SZ

fuscatus Mellié, 1849a: 352

fasciculosus Champion, 1922b: 151 **A:** UP

festivus Panzer, 1793a: 7 (*Anobium*) **E:** AU BE BH CT CZ DE FI FR GB GE GG HU IR IT LA LT NL NR NT PL RO SK SL ST SV SZ UK **A:** ES

fissicollis Mellié, 1849a: 247 **E:** AB AU BH CR CZ FR GG HU IT PL RO SK ST UK

fissicornis Mellié, 1849a: 317 **E:** AU BY CT CZ EN FR HU IT LT PL RO ST SZ

germanicus Abeille de Perrin, 1874a: 39

sublaminatus Wankowicz, 1869: 422

flavipes Motschulsky, 1845b: 371 **A:** FE (Kamchatka)

friebi Roubal, 1933a: 72 **A:** FE

fukudai M. Chûjô, 1940b: 135 **A:** JA (Honshu)

fusciclavis Nyholm, 1953: 190 **E:** AU BH BU BY CT CZ DE EN FI FR GE GR IT LA NL NR PL RO SK SV SZ TR UK

castaneus Mellié, 1849a: 351 [HN]

fuscipes Mellié, 1849a: 271 **Ni:** MR **NAR NTR**

gardneri Pic, 1937g: 123 **A:** UP
giganteus Pic, 1934a: 15 **A:** CH
glabratus Mellié, 1849a: 249 **E:** AU BH BY CT CZ EN FI FR GE IT LA LT NL NR PL RO SK SV ST SZ UK
gladiator Flach, 1882: 250 **E:** GE
graecus Schilsky, 1900: 68 **E:** GR **N:** MO **A:** SY TR
hieroglyphicus Reitter, 1877g: 380 **A:** FE JA
incanus Rey, 1892: 17 **E:** FR (Lyon)
jacquemartii Mellié, 1849a: 328 **E:** AB AU BH BU BY CR CT CZ DE FI FR GB GE GG GR HU IT LA LT NR NT PL RO SK SP ST SV SZ UK **A:** TR
japonicus Nobuchi, 1955a: 57 **A:** JA (Honshu, Kyushu)
jezoensis Nobuchi, 1960a: 38 **A:** JA (Hokkaido)
judaeus Reitter, 1901a: 55 **A:** LE
konoi M. Chûjô, 1940a: 132 **A:** FE (Sachalin) JA (Hokkaido, Honshu) JIL
laevigatus Kawanabe, 1997b: 319 **A:** JA
laminatus Mellié, 1849a: 318 **E:** AU CR CT (Kaliningrad) CZ FR GE IT PL SZ **A:** TR "Transcaucasus"
lasoni Królik, 2016: 495 **E:** AB **A:** IN
lederi Reitter, 1880c: 477 **E:** GG ST (Kavkaz)
libanicus Abeille de Perrin, 1874a: 58 **A:** LE
lineatocribratus Mellié, 1849a: 336 **E:** AB AU BH BU BY CR CT CZ DE EN FI FR GB GE GG HU IT LA LS NR NT PL RO SK ST SV SZ UK YU **A:** IN
lineatulus Reitter, 1901a: 51 **A:** UZ
lugowoji Królik, 2016: 497 **E:** AB
maculatus M. Chûjô, 1939: 2 **A:** FE (Sachalin) JA (Hokkaido) LIA
manchuricus M. Chûjô, 1940b: 136 **A:** NE
maurus Peyerimhoff, 1915: 25 **N:** AG
matchanus Reitter, 1915c: 66 **E:** AB BH CZ DE EN FI FR GE GG IT LA PL RO NR NT ST SV SZ UK "Caucasus" **A:** IN TR
  hanseni A. Strand, 1965: 61
micans Fabricius, 1792a: 238 (*Anobium*) **E:** AL AU BE BH BU BY CR CT CZ DE EN FI FR GB GE GR HU IR IT LA LS LT NL NR NT PL RO SK SL SP ST SV SZ TR UK **A:** ES FE KI MG TR
  albohispidulus Reitter, 1901a: 50
  hispidus Paykull, 1798: 310 (*Anobium*)
  nitidicollis Abeille de Perrin, 1874a: 32
  ruficornis Marsham, 1802: 87 (*Ptinus*)
  savilli Donisthorpe, 1936: 220
michiochujoi Królik, 2020 [see New Acts] **A:** JA (Honshu)
  polypori M. Chûjô, 1939: 4 [HN]
mikagensis Nobuchi & Wada, 1955b: 106 **A:** JA (Honshu) LIA

morikawai Miyatake, 1954: 50 **A:** JA (Honshu, Shikoku)
multidentatus Pic, 1917f: 4 (*Ennearthron*) **A:** IQ YUN
nigrorugosus Schilsky, 1900: 66 **A:** FE
nikkoensis Nobuchi, 1960b: 65 **A:** JA
nipponicus M. Chûjô, 1940b: 134 **A:** JA (Hokkaido, Honshu) LIA
onyosi Viñolas & Muñoz-Batet, 2015: 125 **E:** SP (Catalonia)
orius Kompantsev, 1996: 587 **A:** AF KI TD UP UZ
polysticti M. Chûjô, 1939: 5 **A:** TAI
pumilio Baudi di Selve, 1874c: 338 (*Cis*) **A:** CY
puncticollis Wollaston, 1860a: 360 **N:** MR
punctifer Mellié, 1849a: 344 **E:** CR FR IT RO SZ **A:** CY
punctulatus Gyllenhal, 1827: 625 **E:** AU BE BH BY CR CT CZ DE EN FI FR GB GE HU IT LA LT NL NR NT PL RO SK SL SV SZ **A:** TR
pygmaeus glutinosus Peyerimhoff, 1919a: 186 **N:** AG
pygmaeus pygmaeus Marsham, 1802: 86 (*Ptinus*) **E:** AU BE CZ DE EN FR GB GE GR HU IT LA NL NR PL RO SK SV SZ TR
nigricornis Marsham, 1802: 87 (*Ptinus*)
oblongus Mellié, 1849a: 341
pandellei Abeille de Perrin, 1874a: 73
rhododactylus Marsham, 1802: 87 (*Ptinus*)
quadridens Mellié, 1849a: 270 **E:** AU BH BY CR CZ FI FR GE IT LA LT NR NT PL RO SK ST SV SZ UK FR SZ
quadridentulus Perris, 1874: liii **E:** CR FR IT
ragusai Roubal, 1916: 89 **E:** IT (Sicilia)
reitteri Lopes-Andrade, 2002: 54 [RN] **E:** AB (Lenkoran)
setifer Reitter, 1884a: 8 [HN]
robustithorax Pic, 1915k: 22 **N:** AG
rufocastaneus Nakane & Nobuchi, 1955: 48 **A:** JA (Honshu, Shikoku)
rugulosus Mellié, 1849a: 242 **E:** AB AU BE BU CT CZ EN FI FR GE GE HU IT LA LT NL NT PL RO SK ST SV SZ TR **N:** MR **A:** KI TR
pyrrhocephalus Mellié, 1849a: 244 [HN]
rubiginosus Mellié, 1849a: 243
sasajii Kawanabe, 2001: 267 **A:** JA (Hokkaido)
sasakawai Nobuchi, 1960a: 37 **A:** JA (Hokkaido)
seriatocribratus Reitter, 1913b: 124 **A:** KI TD UZ
seriatopilosus Motschulsky, 1861: 17 **A:** FE JA
seriatulus Kiesenwetter, 1879: 320 **A:** JA
sibiricus Schilsky, 1900: 75 **A:** FE
sinensis Pic, 1917f: 4 (*Ennearthron*) **A:** YUN
sinicus Pic, 1954a: 58 **A:** FUJ

striatulus Mellié, 1849a: 264 **E:** AB AU BH CR FR GE GG GR HU IT PL SK SP ST SZ **N:** AG MO **A:** TR
flavipes Lucas, 1846: pl. 40 [HN]
peyronis Abeille de Perrin, 1874a: 65
submicans Abeille de Perrin, 1874a: 28 **E:** AB AR AU BE BH BY CR CT CZ DE EN FI FR GB GE GG GR HU IT LA LT NL NR NT PL RO SK ST SV SZ TR UK **A:** FE TR
subrobustus Miyatake, 1955b: 95 [RN] **A:** JA (Honshu, Kyushu)
robustus Miyatake, 1954: 46 [HN]
taiwanus M. Chûjô, 1939: 1 **A:** JA (Amami-Ôshima) TAI
tauriensis Królik, 2002: 198 **A:** TR
tomentosus Mellié, 1849a: 253 **E:** AB AR BH CR FR GG GR IT (Sicilia) RO SP (Gibraltar) ST TR UK YU **N:** MO
sericeus Mellié, 1849a: 346
tomentoides Reitter, 1901a: 56
vestitus Mellié, 1849a: 354 **E:** AU BE BH BY GB CZ DE EN FI FR GE HU IT LA NL NR PL RO SK SV SZ
villosulus Marsham, 1802: 87 (*Ptinus*) **E:** AU BE BY CR CT CZ DE EN FR GB GE GG GR HU IR IT LA LS LT MD NL NR NT PL PT RO SK SP SV ST SZ UK **A:** ES FE IN TR
flavus Stephens, 1830: 345
petropolitanus Jakobson, 1896: 520
plagiatus C.G. Thomson, 1863: 184
pyrrhocephalus Marsham, 1802: 86 (*Ptinus*)
quadricollis Schilsky, 1900: 67
setiger Mellié, 1849a: 244
striatulus Mellié, 1849a: 246 [HN]
yamamotoi Kawanabe, 1997b: 321 **A:** JA

genus *Cisarthron* Reitter, 1885b: 208 type species *Cisarthron laevicolle* Reitter, 1885
laevicolle Reitter, 1885b: 209 **E:** BH FR RO ST TR (Kavkaz)

genus *Dichodontocis* Kawanabe, 1994a: 186 type species *Dichodontocis uncinatus* Kawanabe, 1994
uncinatus Kawanabe, 1994a: 187 **A:** JA (Shikoku, Yakushima)

genus *Diphyllocis* Reitter, 1885b: 209 type species *Ennearthron opaculum* Reitter, 1878
opaculus Reitter, 1878b: 25 (*Ennearthron*) **E:** BH CR CZ HU IT RO SK UK

genus *Dolichocis* Dury, 1919: 158 type species *Dolichocis manitoba* Dury, 1919
laricinus Mellié, 1849a: 355 (*Cis*) **E:** CT FI LA LT NR NT PL RO UK "Silesia" **A:** FE WS
yuasai M. Chûjô, 1941: 85 (*Ennearthron*) **A:** JA (Hokkaido)

genus *Ennearthron* Mellié, 1847: 110 type species *Cis cornutus* Gyllenhal, 1827
abeillei Caillol, 1914: 160 **E:** FR
amamense Miyatake, 1959b: 121 **A:** JA (Amami-Ôshima)
chujoi Nakane & Nobuchi, 1955: 49 **A:** JA (Honshu, Hokkaido)
cornutum Gyllenhal, 1827: 626 (*Cis*) **E:** AB AU BE BH BY CR CT CZ DE EN FI FR GB GE GG HU IR IT LA LS LT MC NL NR NT PL RO SK SL SP ST SV SZ TR UK **A:** JA (Hokkaido) JIL TR
concinnum Marsham, 1802: 87 (*Ptinus*) [NO]
filum Abeille de Perrin, 1874c: liii (*Cis*) **E:** AU FR IT (Sicilia)
hayashii Nobuchi, 1955b: 108 **A:** JA
ishiharai Miyatake, 1954: 57 **A:** JA (Hokkaido, Shikoku)
mohrii Miyatake, 1954: 56 **A:** JA (Honshu)
ondreji Roubal, 1919: 63 **E:** ST (Kavkaz)
palmi Lohse, 1966: 28 **E:** AU BY CT CZ FI GE HU PL SV
pruinosulum Perris, 1864: 291 (*Cis*) **E:** CZ FR GE HU IT PL SK SZ
reichei Abeille de Perrin, 1874a: 89 **N:** EG
reitteri Flach, 1882: 249 (*Cis*) **E:** CZ GE (Bayern)
robusticorne Kawanabe, 1996a: 105 **A:** JA (Hokkaido)

genus *Euxestocis* Miyatake, 1954: 53 type species *Euxestocis bicornutus* Miyatake, 1954
bicornutus Miyatake, 1954: 54 **A:** JA
formosanus Miyatake, 1982b: 76 **A:** TAI

genus *Hadraule* C.G. Thomson, 1859: 91 type species *Cis elongatulus* Gyllenhal, 1827
Hadreule C.G. Thomson, 1859: 91 [unused original spelling]
Knablia Roubal, 1936b: 54 type species *Cis elongatulus* Gyllenhal, 1827
Maphoca Casey, 1900: 165 type species *Maphoca blaisdelli* Casey, 1900
Pityocis Peyerimhoff, 1918b: 141 type species *Pityocis coarctatus* Peyerimhoff, 1918 (= *Cis elongatulus* Gyllenhal, 1827)
elongatula Gyllenhal, 1827: 627 (*Cis*) **E:** AL AU CT CZ FI FR GE GR HU IT LA LT NR PL SK SV **N:** AG NARi
coarctata Peyerimhoff, 1918b: 142 (*Pityocis*)
striata J.R. Sahlberg, 1901: 11 (*Ennearthron*)

genus *Lipopterocis* Miyatake, 1954: 44 type species *Lipopterocis simplex* Miyatake, 1954
simplex Miyatake, 1954: 45 **A:** JA (Shikoku)

genus *Neoennearthron* Miyatake, 1954: 58 type species *Neoennearthron bicarinatum* Miyatake, 1954
bicarinatum Miyatake, 1954: 59 **A:** JA
hisamatsui Miyatake, 1959b: 119 **A:** JA (Amami-Ôshima)

genus *Nipponapterocis* Miyatake, 1954: 42 type species *Nipponapterocis brevis* Miyatake, 1954
brevis Miyatake, 1954: 43 **A:** JA (Honshu, Kyushu, Shikoku)
hirsutus Kawanabe, 1995: 172 **A:** JA (Yakushima)
inermis Kawanabe, 1995: 170 **A:** JA (Honshu)

genus *Nipponocis* Nobuchi & Wada, 1955a: 54 type species *Nipponocis longisetosus* Nobuchi, 1955
ashuensis Nobuchi, 1959: 51 **A:** JA (Honshu)
longisetosus Nobuchi, 1955a: 55 **A:** JA (Honshu) LIA
magnus Nobuchi, 1955a: 55 **A:** JA (Honshu)
unipunctatus Nakane & Nobuchi, 1956: 107 **A:** JA (Tokara)

genus *Odontocis* Nakane & Nobuchi, 1955: 50 type species *Odontocis denticollis* Nakane & Nobuchi, 1955
denticollis Nakane & Nobuchi, 1955: 50 **A:** JA (Hokkaido, Honshu)

genus *Orthocis* Casey, 1898: 84 type species *Orthocis aterrima* Casey, 1898 (= *Cis punctatus* Mellié, 1849)
Mellieicis Lohse, 1964: 122 type species *Cis alni* Gyllenhal, 1813
alni Gyllenhal, 1813: 386 (*Cis*) **E:** AU BH BU BY CT CZ DE EN FI FR GB GE GG GR HU IR IT LA LT NL NR NT PL RO SK SP ST SV SZ UK **A:** ES MG
betulae Zetterstedt, 1828: 352 (*Cis*)
mitfordi Donisthorpe, 1916: 156
recticollis Abeille de Perrin, 1874a: 60 (*Cis*)
alnoides Reitter, 1884c: 120 (*Cis*) **E:** AB CR GR (Korfu) UK
volhynicus Roubal, 1912b: 71 (*Cis*)
auriculariae Lawrence, 1991: 289 **A:** CH
coluber Abeille de Perrin, 1874c: lii (*Cis*) **E:** FR GB GR HU IT NL SZ
latifrons Pool, 1917: 89 (*Cis*)
ishiharai Kawanabe, 1994b: 187 **A:** JA (Kyushu, Shikoku, Tsushima)
juglandis Reitter, 1885b: 208 (*Cis*) **E:** AU BH FR GE HU YU SZ
linearis J.R. Sahlberg, 1901: 10 (*Cis*) **E:** DE EN FI LA LT NR NT PL SV SZ UK
nigrosplendidus Nobuchi, 1955b: 105 (*Cis*) **A:** JA (Hokkaido)
ornatus Reitter, 1877g: 381 (*Cis*) **A:** JA
perrisi Abeille de Perrin, 1874c: liii (*Cis*) **E:** FR IT
pseudolinearis Lohse, 1965: 179 (*Cis*) **E:** AU GE HU PL
reflexicollis Abeille de Perrin, 1874c: lii (Cis) **E:** AB AU BH BY CR CZ EN FR GE GG HU IT LA PL SK SL SP ST SZ UK YU **A:** FE MG TR **N:** AG
lucasi Abeille de Perrin, 1874a: 62 (Cis)
punctulatus Lucas, 1846: pl. 40 (Cis) [HN]
sagittiferus Israelson, 1980: 189 (*Cis*) **N:** CI (La Palma)

schizophylli Nakane & Nobuchi, 1955: 47 (*Cis*) **A:** JA (Hokkaido, Honshu)
wollastonii Mellié, 1849b: lxxxvi **N:** MR
zoufali Reitter, 1902a: 6 (*Cis*) **E:** BH

genus *Paraxestocis* Miyatake, 1954: 51 type species *Paraxestocis unicornis* Miyatake, 1954
unicornis Miyatake, 1954: 52 **A:** JA (Shikoku)
nakanei Nobuchi, 1955a: 53 (*Xylographus*)

genus *Strigocis* Dury, 1917: 19 type species *Strigocis opacicollis* Dury, 1917
bicornis Mellié, 1849a: 356 (*Cis*) **E:** AB BH CR CZ FR GB GE GR HU IT NL PL RO SK SP SZ TR UK "Caucasus" **A:** IN TR **N:** AG
tokunagai Nobuchi, 1960a: 39 (*Rhopalodontus*) **A:** JA

genus *Sulcacis* Dury, 1917: 20 type species *Sulcacis lengi* Dury, 1917
Entypocis Lohse, 1964: 121 type species *Cis bidentulus* Rosenhauer, 1847
Entypus L. Redtenbacher, 1849: 350 [HN] type species *Cis affinis* Gyllenhal, 1827 (= *Anobium nitidum* Fabricius, 1792a)
bidentulus Rosenhauer, 1847: 58 (*Cis*) **E:** AU BU BY CR CT CZ FR GE HU IT NL PL SK SP SZ UK **A:** ES KZ
alpinus Mellié, 1849a: 347 (*Cis*)
fronticornis Panzer, 1805b: 7 (*Apate*) **E:** AB AU BE BH BU BY CR CT CZ DE FI FR GE GR HU IT LA NL NR NT PL RO SK SP ST SV SZ TR UK YU **A:** KI KZ TR
japonicus Nobuchi, 1960a: 39 (*Rhopalodontus*) **A:** JA (Hokkaido)
nitidus Fabricius, 1792a: 238 (*Anobium*) **E:** AB AU BE BH BU BY CR CT CZ DE EN FI FR GB GE GG GR HU IT LA LT MD NL NR NT PL PT RO SK SL SP ST SV SZ TR UK YU **A:** FE IN JA (Hokkaido, Honshu) TR
affinis Gyllenhal, 1827: 628 (*Cis*)
nobuchii Kawanabe, 1997a: 74 **A:** JA (Hokkaido)

genus *Wagaicis* Lohse, 1964: 121 type species *Ennearthron wagae* Wankowicz, 1869
wagae Wankowicz, 1869: 421 (*Ennearthron*) **E:** AU BY CR CT FI FR GE LT NT PL RO SZ UK

tribe Orophiini C.G. Thomson, 1863

genus *Hyalocis* Kawanabe, 1993: 763 type species *Hyalocis yakushimensis* Kawanabe, 1993
yakushimensis Kawanabe, 1993: 765 **A:** JA (Yakushima)

genus *Octotemnus* Mellié, 1847: 110 type species *Cis glabriculus* Gyllenhal, 1827
Orophius L. Redtenbacher, 1849: 350 type species *Cis mandibularis* Gyllenhal, 1813
aculeatus Kawanabe, 2002: 263 **A:** JA (Iriomote Is.)
assimilis Kobayashi & Sota, 2019: 184 **A:** JA

chinensis J.-K. Li, 1992: 143 **A:** JIL
crassus Kobayashi & Sota, 2019: 184 **A:** JA
glabriculus Gyllenhal, 1827: 629 (*Cis*) **E:** AB AU BE BH BU BY CR CT CZ DE FI FR GB GE GG GR HU IR IT LA LS LT LU MC NL NR NT PL PT RO SK SL SP ST SV SZ TR UK **A:** JA (Hokkaido, Honshu) LIA TR
japonicus Miyatake, 1954: 64 **A:** JA
kawanabei Kobayashi & Sota, 2019: 185 **A:** JA
laminifrons Motschulsky, 1861: 17 (*Orophius*) **A:** JA LIA TAI
mandibularis Gyllenhal, 1813: 717 (*Cis*) **E:** AB AU BH BU BY CR FI FR GE GG IT LA LT NR PL RO SK SL SP ST SV SZ UK **A:** ES MG
inaequidens Chevrolat, 1835a: pl. 40 (*Cis*)
michiochujoi Kawanabe, 2005: 47 **A:** TAI
omogensis Miyatake, 1954: 61 **A:** JA (Honshu, Kyushu, Shikoku)
opacus Mellié, 1849a: 386 **N:** CI (Tenerife) MR
parvulus Miyatake, 1954: 62 **A:** JA
pilosoceps Kawanabe, 2002: 249 **A:** JA (Yakushima)
punctidorsum Miyatake, 1954: 63 **A:** JA (Honshu, Shikoku)
robustus Kawanabe, 2002: 258 **A:** JA (Tsushima)
rugosopunctatus Drogvalenko, 2002: 9 **E:** AB UK (Krym)

genus *Ropalodontus* Mellié, 1847: 109 type species *Cis perforatus* Gyllenhal, 1813
Cedrinus Abeille de Perrin, 1876: 313 type species *Rhopalodontus camelus* Abeille de Perrin, 1876
Rhopalodontus Abeille de Perrin, 1876: 312
armifrons Reitter, 1913d: 655 **N:** AG
baudueri Abeille de Perrin, 1874c: lii **E:** AU DE FR GE GR HU IT PL RO SK ST
camelus berberus Peyerimhoff, 1945: 2 **N:** AG
camelus camelus Abeille de Perrin, 1876: 312 (*Rhopalodontus*) **A:** LE
harmandi Lesne, 1917: 191 **A:** JA (Hokkaido, Honshu)
novorossicus Reitter, 1901a: 58 **E:** CZ FR GE GR IT PL ST SZ TR UK
perforatus Gyllenhal, 1813: 385 (*Cis*) **E:** AB AU BH BU BY CR CZ DE FR GB GE GG GR HU IT LA LT NL NR PL RO SK SP ST SV SZ UK **A:** JA (Hokkaido) JIL
punctiger Waltl, 1839: 224 (*Cis*)
gyllenhali Gistel, 1857: 69 (*Rhopalodontus*)
perrini Reitter, 1878a: 221 **E:** AB GG
populi C. & H. Brisout de Barneville, 1877: cvii **E:** FR IT SP
strandi Lohse, 1969: 50 **E:** CT FI LA NR NT PL RO SV UK **A:** ES FE WS

genus *Xylographus* Mellié, 1847: 109 type species *Cis bostrichoides* Dufour, 1843
bostrichoides Dufour, 1843: 93 (*Cis*) **E:** AB AU BH BU BY CR CZ DE FR GR HU IT (Sardinia) PL RO SK SP ST SZ TR UK **N:** AG MO **A:** CY KI TD TR "Siberia"

aubei Mellié, 1849a: 232

cribratus Lucas, 1846: pl. 40 (*Cis*)

scheerpeltzi Nobuchi & Wada, 1956: 35 **A:** JA (Hokkaido)

tomicoides Reitter, 1901a: 47 **A:** FE

tribe Xylographellini Kawanabe & Miyatake, 1996

subtribe Syncosmetina Lopes-Andrade, 2008

genus *Syncosmetus* Sharp, 1891: 6 type species *Syncosmetus japonicus* Sharp, 1891

Omogocis Miyatake, 1954: 40 type species *Omogocis tuberculifrons* Miyatake, 1954

euryale Lopes-Andrade & Grebennikov, 2015: 467 **A:** YUN

japonicus Sharp, 1891: 7 **A:** JA (Kyushu, Shikoku)

tuberculifrons Miyatake, 1954: 41 (*Omogocis*)

medusa Lopes-Andrade & Grebennikov, 2015: 470 **A:** SCH

perseus Lopes-Andrade & Grebennikov, 2015: 472 **A:** SCH

reticulatus Miyatake, 1985: 280 **A:** JA

stheno Lopes-Andrade & Grebennikov, 2015: 473 **A:** YUN

subtribe Xylographellina Kawanabe & Miyatake, 1996

genus *Scolytocis* Blair, 1928: 95 type species *Scolytocis samoensis* Blair, 1928

danae Lopes-Andrade & Grebennikov, 2015: 465 **A:** YUN

genus *Xylographella* Miyatake, 1985: 279 type species *Xylographella punctata* Miyatake, 1985

punctata Miyatake, 1985: 280 **A:** JA (Honshu, Shikoku)

family Ciidae, nomina dubia

Cis agaricinus Drapiez, 1819: 200 **E:** IT

Cis familiaris Gistel, 1857: 57 **E:** GE

Cis polypori Gistel, 1857: 75 **E:** GE

Cis septembris Gistel, 1857: 8 **E:** AU (Styria)

family TETRATOMIDAE Billberg, 1820

subfamily **Tetratominae** Billberg, 1820

genus *Tetratoma* Fabricius, 1790: 217 type species *Tetratoma fungorum* Fabricius, 1790

subgenus *Abstrulia* Casey, 1900: 167 type species *Tetratoma tessellata* Melsheimer, 1844

ainu Nakane, 1963a: 30 (*Abstrulia*) **A:** FE (S. Sakhalin, S. Kuril Is.) JA

ancora Fabricius, 1790: 217 **E:** AU BE BH BY CT CZ DE EN FI FR GE GB HU IR IT LA LT MD NL NR NT PL RO SK SL SP ST SV CZ UK YU **A:** ES FE KZ MG WS

conjuncta Pic, 1901f: 58

obscurior Pic, 1901f: 58

perrisi Houlbert, 1934: 13

pfefferi Roubal, 1930: 75

picta Newman, 1835: 202

baudueri Perris, 1864: 292 **E:** FR SP PT **N:** MO

flavopicta Fairmaire, 1884f: cxviii (*Phloeophilus*)

crenicollis Baudi di Selve, 1877a: 54 **A:** CY

japonica Miyatake, 1955a: 93 **A:** FE JA

pictipennis Reitter, 1896i: 237 **A:** ES FE NC

tedaldi Reitter, 1887b: 84 **E:** IT

virgo Motschulsky, 1845b: 366 **A:** FE (Kamchatka)

subgenus *Falsoxanthalia* Pic, 1934f: 84 type species *Falsoxanthalia bicoloripes* Pic, 1934

bicoloripes Pic, 1934f: 84 (*Falsoxanthalia*) **A:** SCH

desmarestii Latrielle, 1807: 180 **E:** AU BE BH CZ DE FR HU GB GE GR IT NL SK SP ST SV **N:** AG

pallida Curtis, 1826: pl. 123

subgenus *Paratetratoma* Nikitsky, 1998a: 16 type species *Tetratoma sakagutii* Nakane, 1955

sakagutii Nakane, 1955: 35 **A:** JA

subgenus *Tetratoma* Fabricius, 1790: 217 type species *Tetratoma fungorum* Fabricius, 1790

cyanoptera Champion, 1924c: 263 **A:** SD

fungorum Fabricius, 1790: 217 **E:** AB AR AU BE BY CR CT CZ DE EN FI FR GB GE GG HU IR IT LA LS LT MC MD ND NL NR PL RO SK SP ST SV SZ TR UK YU **A:** TR

bicolor Marsham, 1802: 141 (*Boletaria*)

collaris Quensel, 1790: 8 (*Dermestes*)

dermestoides Herbst, 1791: 88, pl. 38

fuscoguttata Nikitsky, 1998a: 21 **A:** NP

nepalensis Nikitsky, 1998a: 23 **A:** NP

hobuchii Nakane, 1955: 36 **A:** JA

talyshensis Nikitsky, 1989b: 62 **E:** AB (Talysh)

wittmeri Nikitsky, 1998a: 25 **A:** BT

yunnanensis Nikitsky, 2016: 346 **A:** YUN

subfamily **Piseninae** Miyatake, 1960

genus *Pisenus* Casey, 1900: 167 type species *Cryptophagus humeralis* Kirby, 1837

chujoi Miyatake, 1960: 129 **A:** FE JA NC SC

formosanus Miyatake, 1974: 60 **A:** TAI

insignis Reitter, 1889d: 245 (*Pseudotriphyllus*) **A:** FE JA NC SC

rufitarsis Reitter, 1889d: 245 (*Pseudotriphyllus*) **A:** FE JA

genus *Triphyllia* Reitter, 1898f: 359 type species *Triphyllia koenigi*, 1898

koenigi Reitter, 1898f: 359 **E:** GG ST (Kavkaz)

subfamily **Penthinae** Lacordaire, 1859

genus *Cyanopenthe* Nikitsky, 1998a: 29 type species *Penthe metallica* Champion, 1916
metallica Champion, 1916b: 79 (*Penthe*) **A:** BT **ORR**
taiwana Hsiao, Pollock & Barclay, 2015: 581 **A:** TAI
leei Hsiao, Pollock & Barclay, 2015: 584 **A:** TAI

genus *Penthe* Newman, 1838: 373 types species *Helops obliquatus* Fabricius, 1798
  Pyrrocis Laporte, 1840: 236 type species *Helops obliquatus* Fabricius, 1798
almorensis Champion, 1917: 54 **A:** NP UP
japana Marseul, 1876d: 335 **A:** JA
kochi Mařan, 1940b: 87 **A:** FUJ JIA TAI
reitteri Nikitsky, 1998a: 32 **A:** SCH TAI YUN **ORR**
rufopubens Marseul, 1876d: 336 **A:** NP SD "Indes Orientales"
similis Nikitsky, 2005: 24 **A:** "Assam" **ORR**
simplicicornis Gusakov, 2009: 6 **A:** GAN

subfamily **Eustrophinae** Gistel, 1848

tribe Eustrophini Gistel, 1848

genus *Eustrophus* Illiger, 1802b: 301 type species *Mycetophagus dermestoides* Fabricius, 1792
dermestoides Fabricius, 1792b: 498 (*Mycetophagus*) **E:** AU BH BU BY CR CZ EN FI FR GE GR HU IT LA MC PL RO SK SL SP SZ TR UK
  siculus Baudi di Selve, 1892: 121 (*Tritoma*)
niponicus Lewis, 1895a: 259 **A:** FE JA NC NE SC
yunnanensis Nikitsky, 1998a: 54 **A:** SW

genus *Synstrophus* Seidlitz, 1898b: 442 types species *Eustrophus macrophthalmus* Reitter, 1887
klapperichi Pic, 1955c: 29 **A:** FUJ
macrophthalmus Reitter, 1887d: 383 (*Eustrophus*) **A:** JA NE
rollei Pic, 1910q: 75 (*Eustrophus*) **A:** TAI

tribe Holostrophini Nikitsky, 1998

genus *Holostrophus* Horn, 1888: 36 type species *Eustrophus bifasciatus* Say, 1824
subgenus *Holostrophus* Horn, 1888: 36 type species *Eustrophus bifasciatus* Say, 1824
aureofasciatus Pic, 1954b: 63 (*Eustrophinus*) **A:** FUJ
diversefasciatus Pic, 1921a: 1: **A:** FE JA NE NC SC
  katoi Nomura, 1959: 43
dux Lewis, 1895a: 259 **A:** JA
lewisi Csiki, 1924: 10 [RN] **A:** FE (Kunashir) JA
  quadrimaculatus Lewis, 1895a: 260 [HN]

morimotoi Sasaji, 1974: 1 **A:** JA
unicolor Lewis, 1895a: 260 **A:** JA
subgenus *Paraholostrophus* Nikitsky, 2007: 19 type species *Holostrophus orientalis* Lewis, 1895
orientalis Lewis, 1895a: 259 **A:** JA CH NC SC TAI **ORR**
multinotatus Pic, 1911p: 188
toyoshimai Saitô & Konvička, 2017: 2 **A:** JA

genus *Pseudoholostrophus* Nikitsky, 1983: 37 type species *Hallomenus klapperichi* Pic, 1954
chinensis Nikitsky, 1998a: 44 **A:** SCH
klapperichi Pic, 1954a: 53 **A:** FUJ

subfamily **Hallomeninae** Gistel, 1848

genus *Hallomenus* Panzer, 1793e: 17 type species *Hallomenus humeralis* Panzer, 1793 (= *Chrysomela binotata* Quensel, 1790)
subgenus *Hallomenus* Panzer, 1793e: 17 type species *Hallomenus humeralis* Panzer, 1793 (= *Chrysomela binotata* Quensel, 1790)
Dinophorus Illiger, 1807: 335 type species *Dinophorus axillaris* Illiger, 1807
Dryala Mulsant, 1856f: 250 [= 1856c: 58] type species *Hallomenus fuscus* Gyllenhal, 1810
Hallominus Paykull, 1799: 179 type species *Hallominus bipunctatus* Paykull, 1799
axillaris Illiger, 1807: 335 (*Dinophorus*) **E:** AU BE BY CR CT CZ DE EN FI FR GE GG HU IT LA LT MD NR NT PL RO SK SP ST SV SZ UK **A:** CH WS
dimidiatus Stephens, 1835: 426
dorsalis Beck, 1817: 18 pl. 5 (*Dircaea*)
fuscus Gyllenhal, 1810: 528
piceus Marsham, 1802: 494 (*Mordella*)
scapulatus Fairmaire, 1881c: xii
binotatus Quensel, 1790: 14 (*Chrysomela*) **E:** AU BE BH BY CT CZ DE EN FI FR GB GE GG HU IR IT LA LT LU MD NL NR NT PL RO SK SL SP ST SV SZ UK YU **A:** ES WS
bipunctatus Paykull, 1799: 179 (*Hallominus*)
boleti Kugelann, 1792: 239 (*Elateroides*)
elateroides Kugelann, 1792: 239 (*Ips*)
humeralis Panzer, 1793e: 17
chinensis Nikitsky, 1998a: 35 **A:** SCH
nipponicus Nomura & Katô, 1958: 2 **A:** FE JA
orientalis Nikitsky, 1998a: 36 **A:** KA
subgenus *Xeuxes* Champion, 1889: 85 type species *Xeuxes brevicollis* Champion, 1889
Parahallomenus Nomura & Katô, 1958: 1 type species *Hallomenus tokejii* Nomura & Katô, 1958
arimotoi Toyoshima & Y. Ishikawa, 2000: 174 **A:** TAI
tokejii Nomura & Katô, 1958: 1 **A:** FE JA

genus *Mycetoma* Dejean, 1834: 201 type species *Dryops suturalis* Panzer, 1797
affine Nikitsky, 1985c: 267 **A:** FE JA
sapporense Nakane, 1991: 7
ussuriense Nikitsky, 1985c: 266 **A:** FE
suturale Panzer, 1797: 12 (*Dryops*) **E:** AU BH CR CZ FR GE GG HU IT PL RO SK SL SP ST SZ UK YU
richteri Roubal, 1933c: 128

family MELANDRYIDAE Leach, 1815
subfamily **Melandryinae** Leach, 1815
tribe Anisoxiellini Nikitsky, 2007

genus *Anisoxiella* Nikitsky, 1989a: 42 type species *Anisoxya ocularis* Nomura & Katô, 1959
laticollis Nakane, 1963b: 222 (*Anisoxya*) **A:** JA
ocularis Nomura & Katô, 1959: 7 (*Anisoxya*) **A:** JA

tribe Dircaeini Kirby, 1837

genus *Abdera* Stephens, 1832: 37 type species *Mordella bifasciata* Marsham, 1802
subgenus *Abdera* Stephens, 1832: 37 type species *Mordella bifasciata* Marsham, 1802
biflexuosa Curtis, 1829: 255 (*Hypulus*) **E:** AU BU CZ DE EN FR GB GE IT SK SP **N:** AG TU
bifasciata Marsham, 1802: 493 (*Mordella*) [HN]
bifasciata Stephens, 1832: 38
flexuosa G.-A. Olivier, 1790: 31: 53 (*Elater*)
griseoguttata Fairmaire, 1850b: 423 (*Dircaea*)
undata Perris, 1852: 44 [HN]
quadrifasciata Curtis, 1829: 255 (*Hypulus*) **E:** AU BE BU CR FR GE GB GR HU IT LU RO SK SL SP SZ UK YU **N:** TU **A:** TR
subgenus *Caridua* E. Strand, 1929: 24 [RN] type species *Hallomenus affinis* Paykull, 1799
Carida Mulsant, 1856c: 61 [HN]
affinis Paykull, 1799: 181 (*Hallominus*) **E:** AU BE BY CR CT CZ DE EN FI FR GB GE HU IT LA LT NL NR NT PL RO SK SL SP ST SV SZ UK **A:** ES FE KZ WS
picea Walker, 1837: 83
flexuosa Paykull, 1799: 182 (*Hallominus*) **E:** AB AL AR AU BE BH BU BY CR CT CZ DE EN FI FR GB GE GG HU IR IT LA LS LT MD NL NR NT PL RO SK SL ST SV UK **A:** TR
undata Panzer, 1799c: 23 (*Hallomenus*)
scriptipennis Lewis, 1895a: 269 **A:** FE JA NE

genus *Anisoxya* Mulsant, 1856c: 45 type species *Dircaea tenuis* Rosenhauer, 1847
conicicollis Champion, 1916b: 34 **A:** FE JA
fuscula Illiger, 1798: 132 (*Serropalpus*) **E:** AL AU BE BH BU CR CT CZ DE EN FR GB GE GG GR HU IT LT MC MD NL NR PL RO SK SL SP ST SV SZ UK YU
mustela Abeille de Perrin, 1869: 42
tenuis Rosenhauer, 1847: 34 (*Dircaea*)

genus *Dircaea* Fabricius, 1798: 6 type species *Dircaea quadriguttata* Fabricius, 1798
australis Fairmaire, 1856: 546 **E:** AU BE BH BY CT CR CZ FR GE HU PL RO SK SL SV SZ UK YU
imitata Olexa, 1955: 38
quadriguttata Duftschmid, 1812: 264 [HN]
quadrisignata Mulsant, 1856c: 71
erotyloides Lewis, 1895a: 267 **A:** JA SC
quadriguttata Paykull, 1798: 251 (*Hypulus*) **E:** BY CR CT EN FI FR IT LA NT PL SK ST SV UK **A:** ES FE JA KZ MG NE WS
guttata Stierlin, 1898: 166 [DA]
quadriguttata Fabricius, 1798: 122
shibatai Hayashi, 1960: 42
reducta Pic, 1954a: 53 **A:** FUJ
ussuriensis Nikitsky, 1985b: 57 **A:** FE

genus *Dircaea*, nomen dubium
quadrimaculata Illiger, 1798: 130 (*Serropalpus*)

genus *Dircaeomorpha* Fairmaire, 1896b: 122 type species *Dircaeomorpha clavicornis* Fairmaire, 1896
elegans Sasaji, 1974: 4 **A:** JA SC
satoi Ishikawa, Toyoshima & Lee, 2007: 178 **A:** TAI
validicornis Lewis, 1895a: 266 (*Dircaea*) **A:** JA

genus *Paradircaea* Nikitsky, 1998b: 44 type species *Dircaea dentatomaculata* Lewis, 1895
dentatomaculata Lewis, 1895a: 267 (*Dircaea*) **A:** FE JA SC

genus *Phloiotrya* Stephens, 1832: 35 type species *Phloiotrya rufipes* Stephens, 1832
subgenus *Phloiotrya* Stephens, 1832: 35 type species *Phloiotrya rufipes* Stephens, 1832
Phloeotrya Agassiz, 1846b: 286 [UE]
bellicosa Lewis, 1895a: 266 **A:** FE HUN JA NC NE SC YUN
flavitarsis Lewis, 1895a: 266 (*Dircaea*) **A:** JA NE
granicollis Seidlitz, 1898b: 556 **E:** IT **N:** TU
ishikawai Toyoshima, 1999: 98 **A:** JA
obscura Lewis, 1895a: 268 (*Dircaea*) **A:** ES FE JA SC
rufomarginella Hayashi, 1960: 43
subcostulata Pic, 1953c: 335
planiuscula Nomura & Katô, 1959: 8 **A:** JA
quercicola Champion, 1922a: 74 **A:** UP
rufipes Gyllenhal, 1810: 519 (*Dircaea*) **E:** AU BE BH BY CR CZ DE FI FR GB GE GR IT LA MD NL NR PL RO SK SL SV SZ UK YU
testaceipennis Pic, 1900g: 62 (*Dircaea*)

rufoventris Nomura, 1958: 9 **A:** JA
rugicollis Marseul, 1876d: 334 **A:** FE JA SC
shimomurai Toyoshima & Y. Ishikawa, 2000: 177 **A:** JA
subtilis Reitter, 1897g: 246 (*Dircaea*) **E:** AU BH CR CT CZ EN LT PL RO SK SL UK YU **A:** ES FE WS
tenuis Hampe, 1850: 355 (*Serropalpus*) **E:** AB AR AU BE BU CR CZ FR GB GE GG GR HU IT LU NL RO SK SP ST SZ UK
rufipes Stephens, 1832: 35 [HN]
scabra Petri, 1912: 228
stephensi Jacquelin du Val, 1863: 393
vaudoueri Mulsant, 1856c: 79
trisignata Nomura, 1958: 7 **A:** JA SCH

genus *Stenoxylita* Nomura, 1959: 43 type species *Dircaeomorpha trialbofasciata* Hayashi & Katô, 1956
trialbofasciata Hayashi & Katô, 1956: 37 (*Dircaeomorpha*) **A:** JA

genus *Wanachia* Schulze, 1912: 11 type species *Dircaea triguttata* Gyllenhal, 1810
Caridina Seidlitz, 1898b: 520 [HN] type species *Dircaea triguttata* Gyllenhal, 1810
similis Nikitsky, 1984: 679 **A:** FE
triguttata Gyllenhal, 1810: 524 (*Dircaea*) **E:** AU BE BY CT CZ DE EN FI FR GB GG GE HU IT LA LT NL NR NT PL SK SP ST SV SZ UK **A:** ES FE WS
scutellaris Mulsant, 1856c: 50 (*Abdera*)
suturalis Baudi di Selve, 1877a: 62 (*Abdera*)
trisignata Champion, 1916b: 35 (*Abdera*) **A:** FE (Kunashir) JA SHA
viberti Pic, 1905b: 98 (*Abdera*) **N:** AG MO

tribe Dircaeini, species incertae sedis
Hallomenus variegatus Motschulsky, 1872: 42 **A:** ST

tribe Hypulini Gistel, 1848

genus *Hypulus* Paykull, 1798: 251 type species *Elater quercinus* Quensel, 1790
Mystaxus Semenov, 1900e: 685 type species *Mystaxus relictus* Semenov, 1900
acutangulus Lewis, 1895a: 270 **A:** JA SCH
bifasciatus Fabricius, 1792a: 210 (*Notoxus*) **E:** AU BH CR CT CZ DE EN FI FR GE GG GR HU IT MC MD PL RO SK SL ST SV SZ UK YU
alni D.H. Schneider, 1792: 242 (*Ptilinus*)
fasciatus Panzer, 1793a: 3 (*Helops*)
bodemeyeri Reitter, 1909b: 51 **A:**TR
cingulatus Lewis, 1895a: 269 **A:** JA TAI
higonius Lewis, 1895a: 269 **A:** JA
maroliiformis Reitter, 1911c: 234 **A:** KZ

quercinus Quensel, 1790: 17 (*Elater*) **E:** AU BE BH BU BY CR CT CZ DE EN FR GB GE HU IT LA LT NL NR PL RO SK SL SP ST SV SZ UK YU

blekingensis D.H. Schneider, 1792: 241 (*Elater*)

dubius Fabricius, 1792a: 211 (*Notoxus*)

undulatus Pic, 1925d: 11

relictus Semenov, 1900e: 685 (*Mystaxus*) **A:** TD UZ

genus *Marolia* Mulsant, 1856c: 92 type species *Serropalpus variegatus* Bosc, 1791

grandis Peyerimhoff, 1917c: 330 **N:** AG MO

leseigneuri Nicolas, 1977: 282 **E:** FR IT SZ

purkynei Mařan, 1933: 112 **E:** BH BU MC RO YU

variegata Bosc, 1791: 329 (*Serropalpus*) **E:** FR GBi IT SP SZ

obscurithorax Pic, 1925a: 1

genus *Neogonus* Hampe, 1873: 165 type species *Neogonus plasoni* Hampe, 1873

subgenus *Metax* Semenov & Arnoldi, 1937: 221 type species *Neogonus hiemalis* Semenov & Arnoldi, 1937

fausti Reitter, 1880d: 546 **E:** AB

hiemalis Semenov & Arnoldi, 1937: 223 **A:** TM

subgenus *Neogonus* Hampe, 1873: 165 type species *Neogonus plasoni* Hampe, 1873

Opsigonus Baudi di Selve, 1877c: 409 type species *Opsigonus kruperi* Baudi di Selve, 1877

emgei Reitter, 1884b: 91 **E:** GR

hampei Reitter, 1897e: 221 **E:** AR

minos Reitter, 1897e: 222 **E:** GR

plasoni Hampe, 1873: 165 **E:** GR

kruperi Baudi di Selve, 1877c: 409 (*Opsigonus*)

genus *Nipponomarolia* Miyatake, 1982a: 29 type species *Nipponomarolia kobensis* Miyatake, 1982

kidoi Nakane, 1989: 9 **A:** JA

kobensis Miyatake, 1982a: 30 **A:** JA

matsuyamensis Miyatake, 1982a: 33 **A:** JA

nagaii Nakane, 1989: 9 **A:** JA

oharai Nakane, 1989: 9 **A:** JA

tribe Melandryini Leach, 1815

genus *Bonzicus* Lewis, 1895a: 270 type species *Bonzicus hypocrita* Lewis, 1895

hypocrita Lewis, 1895a: 271 **A:** JA NE

notatus Pic, 1934e: 28

incostata Fairmaire, 1889b: 46 (*Melandrya*) **A:** CH

genus *Ivania* Lewis, 1895a: 271 type species *Ivania coccinea* Lewis, 1895
coccinea Lewis, 1895a: 271 **A:** CH FE (Lozovskyi reserve, Kunashir, Sakhalin) JA TAI

genus *Melandrya* Fabricius, 1801a: 163 type species *Helops serratus* Fabricius, 1775
subgenus *Apteromelandrya* Gusakov, 2005: 117 type species *Melandrya minshanensis* Gusakov, 2005
minshanensis Gusakov, 2005: 117 **A:** SCH
monstrum Gusakov, 2009: 4 **A:** YUN
subgenus *Emmesa* Newman, 1838: 376 type species *Emmesa connectens* Newman, 1838
atricolor Lewis, 1895a: 273 **A:** JA
flavipennis Nakane & Hayashi, 1955: 81 **A:** JA
jaromiri Konvička, 2015: 97 **A:** HUB
karafutona Kôno, 1930d: 50 **A:** FE JA NC
ordinaria Lewis, 1895a: 274 **A:** JA
pictipennis Lewis, 1895a: 275 **A:** JA
rausuana Nakane, 1963b: 224 **A:** ES FE JA
rufonotata Nomura, 1958: 13 **A:** JA
utakoae Sasaji, 1988: 43 (*Prothalpia*) **A:** JA
subgenus *Eumelandrya* Nomura, 1959: 45 type species *Melandrya obsoletomaculata* Nomura, 1958
duodecimmaculata Nakane & Hayashi, 1955: 81 **A:** JA
obsoletomaculata Nomura, 1958: 12 **A:** JA
subgenus *Melandrya* Fabricius, 1801a: 163 type species *Helops serratus* Fabricius, 1775
aeneola Gusakov, 2009: 3 **A:** SCH
barbata Fabricius, 1787a: 213 (*Helops*) **E:** AU BE BH CR CT CZ DE FI FR GB GE HU IT LU NR NT PL RO SK SL ST SV SZ UK YU
flavicornis Duftschmid, 1812: 262
caraboides Linnaeus, 1760: 173 (*Chrysomela*) **E:** AB AR AU BE BH BU CR CZ DE FR GB GE GG GR HU IN IR IT LA LS LU MC MD NL NR PL RO SK SL SP ST SV SZ TR UK YU **A:** TR
coerulescens Petri, 1912: 230
luridipes Pic, 1907e: 138
rufibarbis Schaller, 1783: 324 (*Tenebrio*)
serrata Fabricius, 1775: 257 (*Helops*)
tibialis Viturat, 1895: 110
fasciata Nikitsky, 1974: 61 **A:** FE
gloriosa Lewis, 1895a: 272 **A:** FE JA NE
modesta Lewis, 1895a: 273 **A:** FE JA
quadricostata Hayashi, 1960: 46
mongolica Solsky, 1871: 378 **A:** CH ES FE JA MG NC NE SC WS
parallela Nomura & Katô, 1968: 1 **A:** FE JA

rufipes Gebler, 1829a: 130 **A:** ES FE MG WS
fulgida Motschulsky, 1845a: 81
inequalis Pic, 1927a: 2
kieberi Laporte, 1840: 249
splendida Faldermann, 1833: 58
shimoyamai Hayashi, 1960: 45 **A:** FE JA
ussuriensis Nikitsky, 1972: 342 **A:** FE
subgenus *Paramelandrya* Nikitsky, 2002: 24 type species *Tenebrio dubius* Schaller, 1783
dubia Schaller, 1783: 326 (*Tenebrio*) **E:** AU BE BU BY CT CZ DE EN FI FR GE HU IT LA LT NR NT PL RO SK ST SV SZ UK **A:** ES FE JA MG NC SC WS
alternans Motschulsky, 1872: 42
canaliculata Fabricius, 1787a: 213 (*Helops*)
caraboides Latreille, 1818a: 42
goryi Laporte, 1840: 249
niponica Lewis, 1895a: 273
rufipes Chevrolat, 1833a: pl. 33
subgenus *Pseudomelandrya* Nikitsky, 2002: 24 type species *Melandrya flavonotata* Pic, 1938
flavonotata Pic, 1938c: 11 **A:** FE JA NC SC
quadrisignata Nakane & Hayashi, 1955: 81

genus *Phryganophilus* C.R. Sahlberg, 1833: 454 type species *Dircaea ruficollis* Fabricius, 1798
subgenus *Phryganophilus* C.R. Sahlberg, 1833: 454 type species *Dircaea ruficollis* Fabricius, 1798
Longemelandrya Pic, 1953c: 335 type species *Melandrya angustata* Pic, 1953
Phryganophiloides Morishima, 1988: 41 type species *Phryganophiloides elegans* Morishima, 1988
angustatus Pic, 1953c: 335 **E:** CT **A:** ES FE
auritus Motschulsky, 1845a: 81 **E:** AU BY CT CZ EN GE LT NT PL SK UK **A:** ES FE JA KZ MG WS
elegans Morishima, 1988: 42 (*Phryganophiloides*)
nigriventris Hampe, 1850: 355
pseudoauritus Nikitsky, 1988a: 1429 **E:** CT **A:** ES FE
ruficollis Fabricius, 1798: 122 (*Dircaea*) [NP] **E:** AL AU BH BY CR CT CZ EN FI FR GE LA NR NT PT PL RO SK SV SP UK YU **A:** ES FE JA MG NE SCH WS
analis Fabricius, 1792a: 214 (*Cantharis*) [NO]
paradoxus Paykull, 1799: 162 (*Lymexylon*)
rosti Hubenthal, 1905: 57
ruficollis Lewis, 1895a: 274 (*Melandrya*)
subgenus *Pseudophryganophilus* Nikitsky, 2002: 27 type species *Melandrya affinis* Nikitsky, 1985
affinis Nikitsky, 1985d: 757 (*Melandrya*) **A:** FE NC SC

genus *Sallumia* Fairmaire, 1889b: 46 type species *Sallumia davidis* Fairmaire, 1889
Bonzicomorpha Pic, 1927c: 13 type species *Bonzicomorpha unifasciata* Pic, 1927

atra Pic, 1934e: 28 (*Bonzicomorpha*) **A:** CH
davidis Fairmaire, 1889b: 47 **A:** NO SCH

tribe Orchesiini Mulsant, 1856

genus *Eucinetomorphus* Perris, 1875: 7 type species *Eucinetomorphus leprieurii* Perris, 1875
asturiensis Reitter, 1881c: 232 (*Lederia*) **E:** SP
ehlersi Heyden, 1884b: 360 (*Lederia*) **E:** SP
leprieurii Perris, 1875: 8 **N:** AG TU
lucidus Peyerimhoff, 1917a: 143 (*Orchesia*) [DA] **N:** AG
trabuti Peyerimhoff, 1918a: 371 **N:** AG

genus *Lederia* Reitter, 1880c: 478 type species *Lederia suramensis* Reitter, 1880
suramensis Reitter, 1880c: 479 **E:** GG ST (Kavkaz) **A:** TR

genus *Lederina* Nikitsky & Belov, 1982: 113 type species *Lederia kaszabi* Nikitsky & Belov, 1982
angusticanalis Sasaji, 1987: 42 (*Lederia*) **A:** JA
foenilis Lewis, 1895a: 262 (*Microscapha*) **A:** JA
japonica Reitter, 1891a: 30 (*Lederia*) **A:** JA
lata Lewis, 1895a: 262 (*Microscapha*) **A:** JA
imdecorata Ishikawa & Sakai, 2001: 333 **A:** JA
indica Nikitsky & Belov, 1982: 119 (*Lederia*) **A:** SD
kidoi Sasaji, 1995: 426 (*Lederia*) **A:** JA
martensi Nikitsky, 1994b: 346 **A:** NP
mozolevskayae Nikitsky, 2001: 23 **A:** YUN
narukawai Ishikawa & Sakai, 2007: 56 **A:** JA
obscuripennis Nikitsky, 1994b: 347 **A:** NP
prion Sasaji, 1987: 44 (*Lederia*) **A:** JA
schawalleri Nikitsky, 1994b: 348 **A:** NP
similis Nikitsky & Belov, 1982: 119 (*Lederia*) **A:** NP

genus *Microscapha* LeConte, 1866: 152 type species *Microscapha clavicornis* LeConte, 1866
akitai Ishikawa & Sakai, 2004: 128 **A:** JA
isensis Sasaji, 1987: 35 **A:** JA
nepalensis Nikitsky, 1992a: 83 **A:** NP
yamato Ishikawa & Sakai, 2004: 131 **A:** JA

genus *Orchesia* Latreille, 1807: 194 type species *Hallomenus micans* Panzer, 1793
subgenus *Clinocara* C.G. Thomson, 1859: 120 type species *Serropalpus fasciatus* Illiger, 1798
blandula Brancsik, 1874: 229 **E:** AU BH CR CZ IT PL RO SK SZ UK
  carpathica Reitter, 1878d: 60
  transsylvanica Reitter, 1878d: 59

diversenotata Pic, 1932a: 3 **A:** JA
duplicata Nikitsky, 1985c: 269 **A:** CH ES FE JA NC SC
elegantula Lewis, 1895a: 260 **A:** FE JA NE
fasciata Illiger, 1798: 135 (*Serropalpus*) **E:** AU BY CR CT CZ DE EN FI FR GE HU IT LA LT NR NT PL RO SK SL ST SV SZ UK **A:** ES FE KZ WS
laterufa Pic, 1927h: 10
trifasciata Zetterstedt, 1838: 163 (*Dircaea*)
grandicollis Rosenhauer, 1847: 33 **E:** AU BH CR CZ GE HU IT PL RO SL YU
laticollis L. Redtenbacher, 1849: 611
imitans Lewis, 1895a: 261 **A:** FE JA NE SC
kamberskyi Reitter, 1888b: 213 **E:** GG ST (Kavkaz) **A:** TR
fuscofasciata Reitter, 1888b: 213
keili Roubal, 1933c: 128 **E:** RO
maculata Mulsant & Godart, 1856: 108 **E:** IT (Sicilia) **N:** AG
minor Walker, 1837: 83 **E:** AU BE BH BY CT CZ DE EN FI FR GB GE HU IR IT LA LS LT LU NL NR NT PL RO SK SL SP ST SV SZ UK **A:** KZ WS
sepicola Rosenhauer, 1847: 32
tetratoma C.G. Thomson, 1864: 309
undulata Kraatz, 1853: 256 **E:** AB AL AR AU BE BH BU BY CR CT CZ DE EN FI FR GB GE GG HU IR IT LA LT LU MC MD NL NR NT PL RO SK SL SP ST SV SZ UK YU **N:** AG
fasciata Gyllenhal, 1810: 531 (*Hallomenus*) [HN]
vorontsovi Nikitsky, 2001: 19 **A:** YUN
subgenus *Orchesia* Latreille, 1807: 194 type species *Hallomenus micans* Panzer, 1793
fusiformis Solsky, 1871: 377 **E:** AU BY CT CZ LA LT NT PL RO SK SL UK YU **A:** CH ES FE KZ NC MG WS SC
acicularis Reitter, 1886e: 349
nadeshdae Semenov, 1898e: 289
obscuricolor Pic, 1954a: 53
marseuli Lewis, 1895a: 262 **A:** CH JA
micans Panzer, 1793e: 18 (*Hallomenus*) [NP] **E:** AB AL AR AU BE BH BU BY CR CT CZ DE EN FI FR GB GE GG GR HU IR IT LA LT LU MD NL NR NT RO SK SL SP ST SV SZ TR UK YU **N:** AG TU **A:** WS **AFR**
abeillei Guillebeau, 1888: 14
australis Guillebeau, 1887: 3
clavicornis Latreille, 1804a: 417 (*Anaspis*)
corsica Guillebeau, 1888: 13
picea Herbst, 1792: 97 (*Megatoma*) [NO]
reyi Guillebeau, 1887: 3
subimpressa Guillebeau, 1887: 4
ocularis Lewis, 1895a: 261 **A:** FE (S. Kuril Is., Sakhalin) JA NC NE
subgenus *Orchestera* Guillebeau, 1887: 5 type species *Orchesia luteipalpis* Mulsant & Guillebeau, 1857

arnoldii Nikitsky, 1985c: 272 **A:** ES FE JA
luteipalpis Mulsant & Guillebeau, 1857: 414 **E:** AU BE BH CR CZ DE FR GE GG IT LA MD NR PL RO SK SL ST SV YU **A:** TR

genus *Orchesia*, species incertae sedis
anaspioides Motschulsky, 1845a: 81 (*Haliomenus*) **A:** FE

genus *Paralederia* Nikitsky & Belov, 1982: 113 type species *Lederia anatolica* J. Frivaldszky, 1880
anatolica J. Frivaldszky, 1880: 262 (*Lederia*) **E:** BU GR **A:** TR
seidlitzi Reitter, 1916f: 290 (*Lederia*) **A:** TR

tribe Serropalpini Latreille, 1829

genus *Amblyctis* LeConte, 1879: 3 type species *Amblyctis praeses* LeConte, 1879
Dapsiloderus Fairmaire, 1887c: 157 type species *Dapsiloderus costipennis* Fairmaire, 1887
Hikohohodemia Nakane & Hayashi, 1955: 80 type species *Hikohohodemia nomurai* Nakane & Hayashi, 1955
nomurai Nakane & Hayashi, 1955: 81 (*Hikohohodemia*) **A:** JA

genus *Enchodes* LeConte, 1866: 148 type species *Dircaea sericea* Haldeman, 1848
Paramikadonius Nomura, 1959: 44 type species *Synchroa crepuscula* Lewis 1895
crepusculus Lewis, 1895a: 263 (*Synchroa*) **A:** FE (Kunashir) JA
costulatus Pic, 1932a: 3 (*Mikadonius*)
orientalis Nikitsky, 1973b: 1729 **A:** FE SC

genus *Mikadonius* Lewis, 1895a: 264 type species *Mikadonius gracilis* Lewis, 1895
gracilis Lewis, 1895a: 264 **A:** CH JA SC
japonicus Hayashi, 1960: 43 **A:** JA

genus *Mimoserropalpus* Pic, 1922b: 17 type species *Serropalpus binhanus* Pic, 1922
formosanus Nomura, 1958: 10 (*Cuphosis*) **A:** TAI
indicus Pic, 1945: 4 (*Serropalpus*) **A:** HP
striatus Blair, 1931a: 200 (*Talayra*) **A:** UP

genus *Nagakuchikius* Toyoshima & Y. Ishikawa, 2000: 183 type species *Phloeotrinus tenuipes* Toyoshima & Y. Ishikawa, 2000
tenuipes Toyoshima & Y. Ishikawa, 2000: 184 (*Phloeotrinus*) **A:** JA

genus *Perakianus* Pic, 1914k: 61 type species *Perakianus atriceps* Pic, 1914
bicolor Blair, 1931a: 201 **A:** SD
hisamatsui Nakane, 1963b: 223 **A:** JA

genus *Phloeotrinus* Nikitsky, 1989a: 45 type species *Serropalpus filiformis* Marseul, 1876
subgenus *Phloeotrinops* Nikitsky, 1989a: 46 type species *Dircaea femoralis* Lewis, 1895

femoralis Lewis, 1895a: 268 (*Dircaea*) **A:** JA

subgenus *Phloeotrinus* Nikitsky, 1989a: 45 type species *Serropalpus filiformis* Marseul, 1876

filiformis Marseul, 1876d: 333 (*Serropalpus*) **A:** JA

minusculus Nomura, 1962b: 45 (*Phloeotrya*) **A:** JA TAI

genus *Serropalpus* Hellenius, 1786: 310 type species *Serropalpus striatus* Hellenius, 1786 (= *Mordella barbata* Schaller, 1783)

barbatus Schaller, 1783: 322, pl. 1 (*Mordella*) **E:** AU BE BH BY CR CT CZ DE EN FI FR GB GE GR HU IT LA LT MC NL NR NT PL RO SK SL SP ST SV SZ TR UK YU **N:** MOi **A:** ES FE JA KZ MG NC SC TR WS **NAR**

biguttatus Schellenberg, 1802: 10, pl. 3 (*Lymexylon*)

niponicus Lewis, 1895a: 263

striatus Hellenius, 1786: 318

marseuli Nikitsky, 1992d: 659 **E:** CT ST (Kavkaz) **A:** ES FE JA TR

genus *Serropalpus*, species incertae sedis

iriei Toyoshima & Y. Ishikawa, 2000: 187 **A:** JA TAI

tribe Xylitini C.G. Thomson, 1864

genus *Dolotarsus* Jacquelin du Val, 1863: 391 type species *Dolotarsus rufipes* Jacquelin du Val, 1863

lividus C.R. Sahlberg, 1833: 448 (*Dircaea*) **E:** AU BH BU BY CR CT CZ EN FI FR GE HU IT LA LT LS MC NT PL RO SK SL SP SV SZ UK YU **A:** ES FE JA NE WS **NAR**

ephippius Schaum, 1859: 50 (*Dircaea*)

suturus Gredler, 1866: 278 (*Phryganophilus*)

rufipes Jacquelin du Val, 1863: 391

genus *Rushia* Forel, 1893: 167 type species *Carebara longula* LeConte, 1866

Carebara LeConte, 1866: 251 [HN] type species *Carebara longula* LeConte, 1866

Xylitella Reitter, 1911a: 363 type species *Xylita parreyssi* Mulsant, 1856

parreyssii Mulsant, 1856c: 74 (*Xylita*) **E:** AU BH CR DE FR GG GR HU IT LA MC RO SK SL SP ST SV SZ UK YU **N:** MO **A:** CY TR

ferruginea Gredler, 1866: 279 (*Phryganophilus*)

mollis Graells, 1858: 73 pl. 12 (*Dircaea*)

paganettii Obenberger, 1917: 40 (*Xylitella*)

revelierei Mulsant, 1859: 161 (*Dircaea*)

nagaii Nakane, 1975: 168 (*Xylita*) **A:** JA

genus *Spilotus* LeConte, 1862: 251 type species *Hallomenus quadripustulatus* Melsheimer, 1846

Micromarolia Pic, 1937f: 35 type species *Micromarolia uninotata* Pic, 1937

Pseudozilora Nomura, 1959: 44 type species *Pseudozilora quadrimaculata* Nomura, 1959

quadrimaculatus Nomura, 1959: 44 (*Pseudozilora*) **A:** JA

uninotatus Pic, 1937f: 35 (*Micromarolia*) **A:** JA

genus *Xylita* Paykull, 1798: 249 type species *Elater buprestoides* Fabricius, 1792

laevigata Hellenius, 1786: 318 (*Serropalpus*) **E:** AU BE BH BU BY CR CT CZ DE EN FI FR GB GE GR HU IT JA LA LT MC NL NR NT PL RO SK SL SV SZ UK YU **A:** ES FE KZ MG WS **NAR**

buprestoides Fabricius, 1792b: 234 (*Elater*)

decolorata Randall, 1838: 22 (*Dircaea*)

discolor Fabricius, 1798: 121 (*Dircaea*)

producta Hatch, 1965: 70

robusta Motschulsky, 1872: 43

tribe Zilorini Desbrochers des Loges, 1900

genus *Euryzilora* Lewis, 1895a: 265 type species *Euryzilora lividipennis* Lewis, 1895

lividipennis Lewis, 1895a: 265 **A:** JA

ussurica Mamaev, 1978: 47 **A:** FE

genus *Zilora* Mulsant, 1856c: 84 type species *Xylita ferruginea* Paykull, 1798

akitai Nikitsky & Saito, 2014: 206 **A:** JA

anatolica Korge, 1971: 9 **A:** TR

elongata J.R. Sahlberg, 1881: 132 **E:** BY FI CT EN NT **A:** ES FE WS

ferruginea Paykull, 1798: 250 (*Xylita*) **E:** BY EN FI GB LA LT NR NT PL SV

obscura Fabricius, 1794: 445 (*Parnus*) **E:** AU BH BU BY CR CZ DE ES FR GE HU IT LA LS LT PL RO SK SL SP ST SZ UK YU

bifoveolata Dufour, 1851: 337 (*Mycetochara*)

eugeniae Ganglbauer, 1891: 132

ferruginea Mulsant, 1856c: 85

sericea J. Sturm, 1807: 261 (*Helops*)

subfamily Melandryinae, genera incertae sedis

genus *Adobia* Mulsant, 1856c: 56 type species *Abdera bifasciata* Mulsant, 1856

mulsanti Nikitsky, 2008b **E:** FR

bifasciata Mulsant, 1856c: 56 (*Abdera*) [HN]

genus *Hira* Hayashi, 1960: 45 type species *Hira humerosignata* Hayashi, 1960

humerosignata Hayashi, 1960: 45 **A:** JA

suturalis Nomura, 1962b: 42 **A:** JA

genus *Microtonus* LeConte, 1862: 259 type species *Microtonus sericans* LeConte, 1962

dimidiatus Marseul, 1876d: 457 (*Scraptia*) **A:** JA

takaosanus Nomura, 1962b: 44 **A:** JA

genus *Symphora* LeConte, 1866: 150 type species *Scraptia flavicollis* Haldeman, 1848
atra Nomura, 1959: 45 **A:** JA
brunnea amamiana Nomura, 1962b: 44 **A:** JA
brunnea brunnea Marseul, 1876d: 457 (*Scraptia*) **A:** FE (Kunashir) JA
miyakei Nomura & Hayashi, 1960: 44

subfamily **Osphyinae** Mulsant, 1856 (1840)

genus *Conopalpus* Gyllenhal, 1810: 547 type species *Conopalpus flavicollis* Gyllenhal, 1810
brevicollis Kraatz, 1855: 374 **E:** BH CR GE FR IT NL PT SP SZ
testaceus G.-A. Olivier, 1790: 21: 11 (*Melyris*) **E:** AL AU BE BH BU CR CZ CT DE FR GB GE GR HU IT LA LU NL NR PL RO SK SP ST SV SZ UK
flavicollis Gyllenhal, 1810: 547
nigricornis Germar, 1821: nr. 4: 4
obscuriceps Pic, 1925d: 11
ruficollis Newman, 1833: 185
thoracicus Baudi di Selve, 1877a: 67
vigorsi Stephens, 1832: 60

genus *Osphya* Illiger, 1807: 370 [RN] type species *Cantharis bipunctata* Fabricius, 1775
Nothus G.-A. Olivier, 1811: 384 type species *Nothus praeustus* G.-A. Olivier, 1811
Pelecina Illiger, 1807: 333 [HN] type species *Cantharis bipunctata* Fabricius, 1775
aeneipennis Kriechbaumer, 1848: 163 (*Osphya*) **E:** AU FR IT SZ
immaculata Pic, 1897b: 3
maculicollis Pic, 1897b: 3
picticollis Pic, 1897b: 3
simplicipes Seidlitz, 1898b: 674
aerata Seidlitz, 1898b: 666 **E:** AB AR **A:** IN
bilineata Pic, 1922d: 18
discoidalis Pic, 1937a: 1
multiinterrupta Pic, 1937a: 1
seidlitzi Pic, 1908a: 32
subapicalis Pic, 1912g: 34
talyschensis Pic, 1902b: 9
bipunctata Fabricius, 1775: 206 (*Cantharis*) **E:** AU BH BE BU BY CT CR CZ DE FR GB GE GG GR HU IT KZ MC MD NR PL RO SK SL SP ST SV SZ UK YU **A:** KZ WS
apicenotata Pic, 1908a: 31
anceps Stephens, 1839: 339 (*Nothus*)
bimaculata Stephens, 1832: 59 (*Nothus*)
clavipes Olivier, 1811: 385 (*Nothus*)

clavipes Gyllenhal, 1817: 7 (*Dryops*)
impunctata Bouskell, 1899: 267
innotata Pic, 1898g: 98
maculata Bouskell, 1899: 267
obscuripennis Pic, 1897b: 2
praeusta G.-A. Olivier, 1811: 385 (*Nothus*)
quadipunctata Bouskell, 1899: 267
testaceithorax Pic, 1909i: 121
uralensis Motschulsky, 1845a: 84 (*Nothus*)
vittipennis Seidlitz, 1898b: 668
brusteli Konvička, 2016: 272 **E:** GR
cylindromorpha Abeille de Perrin, 1896: 283 **A:** SY TR
delagrangei Pic, 1902h: 64
dissimilis Champion, 1922a: 75 **A:** UP
formosana Pic, 1927i: 49 **A:** TAI
griseofasciata Pic, 1921e: 13 **A:** TR
interrupta Pic, 1921e: 13
rufithorax Pic, 1921e: 13
harmandi Pic, 1926m: 354 **A:** SD
lehnertae Konvička, 2014: 162 **E:** GR
nigriventris Champion, 1920a: 74 **A:** UP
nigroapicalis Pic, 1921a: 10 **A:** "India"
orientalis Lewis, 1895a: 276 (*Nothus*) **A:** JA
testaceipes Pic, 1914k: 61
trilineata Pic, 1910q: 75 **A:** TAI
uniformis Pic, 1921e: 13 **A:** TR
vandalitiae Kraatz, 1868b: 335 (*Nothus*) **E:** PT SP

family RIPIPHORIDAE Laporte, 1840
subfamily **Hemirhipidiinae** Heller, 1921

genus *Heteromeroxylon* Pic, 1939h: 375 type species *Heteromeroxylon ruficeps* Pic, 1939
Ohananomia Tôyama, 1986: 43 type species *Ohananomia malayana* Tôyama, 1986
yagii Tôyama, 1987: 33 (*Ohananomia*) **A:** JA (Yonagunijima Is., Ryukyus)

genus *Nephrites* Shuckard, 1838: 512 type species *Nephrites nitidus* Schuckard, 1838
apicalis Tôyama & Hatayama, 1985: 20 **A:** SE (Lan Hsu Is.)
japonicus Tôyama & Hatayama, 1985: 19 **A:** JA (Miyako Is., Ryukyus)
kurosawai Tôyama & Hatayama, 1985: 19 **A:** JA (Ishigakijima Is., Ryukyus)
obscurus Tôyama & Hatayama, 1985: 20 **A:** JA (Nakanoshima Is., Tokara)

subfamily **Pelecotominae** Seidlitz, 1875

genus *Clinops* Gerstaecker, 1855a: 16 [= 1855b: 9] type species *Clinops badius* Gerstaecker, 1855

Scotoscopus Brenske & Reitter, 1884b: 92 type species *Scotoscopus carbonarius* Reitter, 1884

spectabilis L.W. Schaufuss, 1872: 276 **E:** GR **A:** TR (Antakya)

carbonarius Reitter, 1884b: 93 (*Scotoscopus*)

genus *Micropelecotoides* Pic, 1910g: 21 type species *Pelecotomoides sumatrensis* Pic, 1910

Micropelecotomoides Csiki, 1913: 6 [UE]

aurosericeus Gressitt, 1941: 526 (*Pelecotoides*) **A:** HAI

japonicus Pic, 1910g: 21 (*Pelecotomoides*) **A:** JA TAI

genus *Pelecotoma* Fischer von Waldheim, 1809: 293 type species *Pelecotoma mosquense* Fischer von Waldheim, 1809

fennica Paykull, 1799: 178 (*Ripiphorus*) **E:** AU BH BY CZ FI GE HU LA LT PL RO RU SK UK

latreillei Fischer von Waldheim, 1823: pl. 38

mosquense Fischer von Waldheim, 1809: pl. 18

zoufali Fleischer, 1916: 120

septentrionalis Kôno, 1936b: 87 **A:** JA

genus *Trigonodera* Dejean, 1834: 217 type species *Pelecotoma leachii* Latreille, 1818

Caspyria Fairmaire, 1901a: 194 type species *Caspyria perrieri* Fairmaire, 1901

Hylacter Gistel, 1848: xi [URN]

Pelecotoides Laporte, 1840: 263 type species *Pelecotoma leachii* Latreille, 1818

Pelecotomoides Gemminger, 1870b: 2118 [UE]

tokejii Nomura & Nakane, 1959: 115 (*Pelecotomoides*) **A:** JA

subfamily **Ptilophorinae** Gerstaecker, 1855

genus *Ptilophorus* Dejean, 1834: 218 type species *Pelecotoma dufourii* Latreille, 1818

Evaniocera Guérin-Méneville, 1835: pl. 2 type species *Pelecotoma dufourii* Latreille, 1818

afghanicus Kaszab, 1957a: 52 (*Evaniocera*) **A:** AF

dufourii Latreille, 1818e: 135 (*Pelecotoma*) **E:** AR BU CR FR GR HU IT PT RO SP ST TR "Caucasus" **N:** AG MO TU **A:** IN IS JO TR

boryi Lucas, 1847: 382 (*Evaniocera*)

frivaldjskii Guérin-Méneville, 1834a: pl. 34 (*Pelecotoma*)

stevenii Fischer von Waldheim, 1829b: 369 [= 1836: 69] (*Pelecotoma*)

striolatus Nowicki, 1873: 5 (*Evaniocera*)

fallax Iablokoff-Khnzorian, 1973: 308 **A:** TD

fischeri Ménétriés, 1848: 29 (*Evaniocera*) **A:** KI KZ TD TM UZ

fischeri Ménétriés, 1849: 245 (*Evaniocera*)

novemflabellatus J.R. Sahlberg, 1903b: 34 (*Evaniocera*)

hauseri Reitter, 1903g: 238 (*Evaniocera*) **A:** IN

plumicornis Reitter, 1898f: 349 (*Evaniocera*) **A**: TR
purcharti Batelka, 2012: 272 **A**: YE

subfamily **Ripidiinae** Gerstaecker, 1855
tribe Eorhipidiini Iablokoff-Khnzorian, 1986

genus *Pterydrias* Reitter, 1895e: 202 type species *Pterydrias debilis* Reitter, 1895
Eorhipidius Iablokoff-Khnzorian, 1986: 89 type species *Eorhipidius januschevi* Iablokoff-Khnzorian, 1986
debilis Reitter, 1895e: 202 **A**: IS
januschevi Iablokoff-Khnzorian, 1986: 89 (*Eorhipidius*) **A**: TD
jelineki Batelka & Hájek, 2009: 772 (*Eorhipidius*) **A**: IN
ruzickai Batelka & Hájek, 2009: 776 (*Eorhipidius*) **A**: PA

tribe Ripidiini Gerstaecker, 1855

genus *Blattivorus* Chobaut, 1891: 237 type species *Rhipidius lusitanicus* Gerstaecker, 1855
Rhyzostylops Silvestri, 1906: 316 type species *Rhyzostylops inquirendus* Silvestri, 1906
inquirendus Silvestri, 1906: 317 **E**: IT
lusitanicus Gerstaecker, 1855a: 25 [= 1855b: 15] (*Rhipidius*) **E**: PT
margaritae Richter, 1949: 360 **E**: AR

genus *Ripidius* Thunberg, 1806: 4 type species *Ripidius pectinicornis* Thunberg, 1806
Pararhipidius Coiffait, 1947: 15 type species *Pararhipidius burdigalensis* Coiffait, 1947
Rhipidius Agassiz, 1846b: 324 [UE]
Symbius Sundevall, 1831: 1223 type species *Symbius blattarum* Sundevall, 1831
abeillei abeillei Chobaut, 1891: 235 (*Rhipidius*) **E**: FR **N**: AG MO **A**: IS
hervei Mollandin de Boissy, 1940: 94 (*Rhipidius*)
guignoti Chobaut, 1904a: 228 (*Rhipidius*)
vaulogeri Chobaut, 1893: 99 (*Rhipidius*)
abeillei bolivari Viedma, 1962: 48 (*Rhipidius*) **E**: SP
longicollis Schilder, 1923b: 203 (*Rhipidius*) **A**: TAI
pectinicornis Thunberg, 1806: 5 **Ei**: AU FI FR GB GE NL SL **Ni**: EG **Ai**: CH COS
anceps Stephens, 1835: 427
blattarum Sundevall, 1831: 1223 (*Symbius*)
fairmairei Chobaut, 1894b: ci (*Rhipidius*)
quadriceps Abeille de Perrin, 1872: 35 (*Rhipidius*) **E**: AR AU CZ EN FI FR GE IT LA PL PT SK SF SV SZ UK **N**: AG MO
amori Bolívar y Pieltain, 1917: 249 (*Rhipidius*)
apicipennis Kraatz, 1891: 358 (*Rhipidius*)
boissyi Abeille de Perrin, 1909: 854 (*Rhipidius*)
burdigalensis Coiffait, 1947: 16 (*Pararhipidius*)
denisi Chobaut, 1919: 200 (*Rhipidius*)
kabylianus Chobaut, 1902: 89 (*Rhipidius*)

maroccanus Chobaut, 1922: 127 (*Rhipidius*)
parisiensis Lesne, 1902a: 182 (*Rhipidius*)
tigrani Iablokoff-Khnzorian, 1957b: 231 (*Rhipidius*)
vilarrubiai Español, 1942: 344 (*Rhipidius*)

subfamily **Ripiphorinae** Laporte, 1840
tribe Macrosiagonini Reitter, 1911

genus *Macrosiagon* Hentz, 1830: 462 type species *Mordella dimidiata* Fabricius, 1781
Emenadia Laporte, 1840: 261 type species *Mordella bimaculata* Fabricius, 1787
Siagonadia Reitter, 1910c: 131 type species *Macrosiagon pallidipennis* Reitter, 1898
armata Waterhouse, 1883: 280 (*Emenadia*) **A:** SD
bifasciata Marseul, 1877: 478 (*Emenadia*) **A:** BEI FUJ JA HP NP NC SC SCH SD UP YUN
coreana Pic, 1955a: 2
donceeli Pic, 1908g: 60
medana Pic, 1910s: 86
reducta Pic, 1909m: 147
tschungseni Pic, 1955a: 2
bimaculata Fabricius, 1787a: 218 (*Mordella*) **E:** BU FR GR HU IT KZ PT RO RU SP UK YU **N:** AG EG MO TU **A:** IN IS KZ TR "Turkestan"
angustula Motschulsky, 1849: 128 (*Ripiphorus*)
apicalis Fischer von Waldheim, 1832b: 429 [= 1836: 158] (*Rhipiphorus*)
bicolor G.-A. Olivier, 1795: 65: 5 (*Ripiphorus*)
binotata Fischer von Waldheim, 1829a: 77 (*Rhipiphorus*)
binotata Fischer von Waldheim, 1832b: 427 [= 1836: 157] (*Rhipiphorus*)
carinthiaca Panzer, 1794a: 7 (*Ripiphorus*)
dubius Motschulsky, 1849: 128 (*Ripiphorus*)
larvata Schrank von Paula, 1789: 90 (*Mordella*)
liturata Fabricius, 1798: 126 (*Ripiphorus*)
nigrithorax Fischer von Waldheim, 1832b: 428 [= 1836: 158] (*Rhipiphorus*)
pallidipennis Reitter, 1898f: 350
quadrimaculata Gyllenhal, 1817: 36 (*Ripiphorus*)
tricuspidata Gmelin, 1790: 2024 (*Mordella*)
bipunctata Fabricius, 1801b: 120 (*Ripiphorus*) **A:** KA NP PA UP YE **AFR ORR**
apicalis Westwood, 1839: lvii, pl. 9 (*Ripiphorus*) [HN]
basinotata Pic, 1949b: 12
imbasalis Pic, 1949b: 12
kaschmirensis Pic, 1949b: 12
reductenotata Pic, 1949b: 12
semiobliterata Pic, 1952b: 7
tropica Nietner, 1856: 392 (*Rhipiphorus*)
elegans Marseul, 1876a: 27 (*Rhipiphorus*) **A:** SA YE **AFR**

ferruginea Fabricius, 1775: 262 (*Mordella*) **E:** FR GR IT MA SP **N:** EG MO TU **A:** AE BT CH CY IS KA NP OM TAI TR **AFR ORR**
flabellata Fabricius, 1781: 501 (*Mordella*)
formosicolis Kôno, 1927: 178
indicus Hope, 1831: 32 (*Ripiphorus*)
pectinata Villers, 1789: 408 (*Mordella*)
subattenuata Pic, 1950b: 8
uniformis Motschulsky, 1849: 127 (*Ripiphorus*)
fortieri Chobaut, 1893: 97 (*Emenadia*) **N:** EG MO
grombczewskii Semenov, 1891a: 377 (*Emenadia*) **A:** TD
medvedevi Iablokoff-Khnzorian, 1973: 307 **E:** AB **A:** KZ MG
meridionalis A. Costa, 1859: 24 (*Emenadia*) [RN] **E:** IT **N:** AG EG TU
bipunctata A. Costa, 1853: 5 (*Emenadia*) [HN]
italica Bedel, 1895: 184 [RN]
raffrayi Fairmaire, 1873a: 342 (*Rhipiphorus*)
nasuta Thunberg, 1784: 66 (*Mordella*) **A:** JA NC SC TAI XIZ **ORR**
discoidalis Pic, 1950b: 8
iwatai Kôno, 1936b: 88
obscuricolor Pic, 1912m: 69
semiobscura Pic, 1923c: 15
oberthurii Fairmaire, 1879c: 200 (*Rhipiphorus*) **N:** AG TU **A:** BA "Arabia"
melanura Fairmaire, 1887e: 303 (*Rhipiphorus*)
praeusta Gebler, 1829a: 133 [= 1830: 133] (*Rhipiphorus*) **E:** AR BU CR FR GR IT SP ST "Caucasus" **N:** AG **A:** IN IS KZ SY TM TR
apicalis Küster, 1848a: 79 (*Rhipiphorus*) [HN]
fulvipennis Küster, 1846b: 71 (*Rhipiphorus*) [HN]
rufipennis Chevrolat, 1834: pl. 34 (*Rhipiphorus*)
sulcata Ménétriés, 1832: 205 (*Rhipiphorus*) [HN]
sulcata Fischer von Waldheim, 1832b: 429 [= 1836: 158] (*Rhipiphorus*)
pusilla Gerstaecker, 1855b: 32 (*Rhipiphorus*) **A:** BT CE FE FUJ GUA GUI HEB HP HUN JA NC NO NP SC SCH SD SE SW UP WP XIZ YUN **ORR**
acutipennis Gressitt, 1941: 528 [HN]
atriceps Pic, 1950b: 8
atricolor Pic, 1953a: 15
atronitida Gressitt, 1941: 529
bicoloriceps Pic, 1953a: 15
bicoloripes Pic, 1953a: 15
coomani Pic, 1953a: 15
corporaali Pic, 1950a: 12
cyanivestis Marseul, 1877: 479 (*Rhipiphorus*)
fukienensis Pic, 1955a: 2

gerstaeckeri Harold, 1878a: 82 (*Emenadia*)
pieli Pic, 1953a: 15
thibetana Pic, 1953a: 15
tonkinea Pic, 1906f: 44
variicollis Fairmaire, 1894a: 34 (*Rhipiphorus*)
spinicollis Fairmaire, 1893b: 38 (*Rhipiphorus*) **A:** GAN GUA HAI NP SD TAI UP **ORR**
bipartita Fairmaire, 1894a: 35 (*Rhipiphorus*)
diversipennis Pic, 1951c: 16
inapicalis Pic, 1922c: 22
leseleuci Pic, 1906f: 44
uninotaticollis M. Chûjô, 1935b: 133
vitalisi Pic, 1922c: 21
terminata Laporte, 1840: 262 (*Emenadia*) **N:** AG EG **A:** AE YE **AFR ORR**
bicolor Laporte, 1840: 262 (*Emenadia*) [HN]
gibbifera Abeille de Perrin, 1880: 234 (*Emenadia*)
lateapicalis Pic, 1950a: 11
nigripennis Gerstaecker, 1855b: 24 (*Rhipiphorus*)
notata Pic, 1947a: 8
perroudi Pic, 1947a: 8
scutellaris Gerstaecker, 1855b: 24 (*Rhipiphorus*)

genus *Macrosiagon,* nomen dubium
bifasciata prescutellaris Pic, 1949b: 12 **A:** CH

genus *Metoecus* Dejean, 1834: 218 type species *Mordella paradoxa* Linnaeus, 1760
Cyttaroecus Schilder, 1923a: 169 type species *Mordella paradoxa* Linnaeus, 1760
Ripidastes Gistel, 1848: x type species *Mordella paradoxa* Linnaeus, 1760
morawitzi Semenov, 1890: 376 (*Rhipiphorus*) **A:** AF XIN
paradoxus Linnaeus, 1760: 228 (*Mordella*) **E:** AB AU BE BH CT CZ DE EN FI FR GB GE GR HU IR IT LA LT NL NT PL PT SK SP ST SV SZ "Caucasus" **A:** IS SY TR
abdominalis Gradl, 1882: 326
affinis Fischer von Waldheim, 1832b: 427 [=1836: 157] (*Rhipiphorus*)
antoniae Reitter, 1895d: 157 (*Myiodes*)
angulatus Panzer, 1803a: 3 (*Ripiphorus*)
apicalis Gradl, 1882: 326
atripennis Pic, 1916h: 22
basalis Schilder, 1924: 241
dicipiens Schilder, 1924: 244
erythrogaster Frölich, 1792: 163 (*Mordella*)
flavoniger Gradl, 1882: 326
gradli Schilsky, 1900: 35

humeralis Schilder, 1924: 241
luteonotatus Pic, 1916h: 21
macularis Gradl, 1882: 326
multinotatus Pic, 1916h: 21
nigrescens Gradl, 1882: 326
nigriventris Gradl, 1882: 326
nigrolineatus Pic, 1916h: 21
notiventris Gradl, 1882: 326
obscuripennis Schilder, 1924: 241
pictus Gistel, 1857: 48
semiflavus Gradl, 1882: 326
theresae Pic, 1916h: 22
tricolor Gistel, 1857: 48
trinotatus Pic, 1916h: 21
viturati Pic, 1916h: 21
satanas Schilder, 1924: 238 **A:** CH FE JA
abdominalis Matsumura, 1906: 133 (*Motoecus* [sic]) [HN]
vespae Kôno, 1927: 181 **A:** JA

tribe Ripiphorini Laporte, 1840

genus *Ripiphorus* Bosc, 1791: 327 type species *Ripiphorus subdipterus* Bosc, 1792
Dorthesia Say, 1824a: 274 type species *Dorthesia fasciata* Say, 1824
Myiodes Heyden, Reitter & Weise, 1891: 520 [UE]
Myiodites Fabre, 1886: 221 [UE]
Myodes Latreille, 1818b: 130 [HN] type species *Myodes dorthesii* Latreille, 1818
Myodites Latreille, 1819b: 303 [RN] type species *Myodes dorthesii* Latreille, 1818
Polychronia Gistel, 1848: x type species *Ripiphorus subdipterus* Bosc, 1792
Rhipidophorus Gemminger, 1870b: 2118 type species *Ripiphorus subdipterus* Bosc, 1792
Rhipiphorus Agassiz, 1846b: 324 [UE]
Rhipiphorus Fabricius, 1792a: xiv [HN] type species *Rhipiphorus subdipterus* Fabricus, 1792
arabiafelix Batelka, 2009a: 156 **A:** OM YE
chalcidoides Gressitt, 1941: 532 (*Rhipiphorus*) **A:** GUA HAI
davidis Fairmaire, 1878a: 124 (*Myodites*) **A:** CE HAI
minor Gressitt, 1941: 533 (*Rhipiphorus*) **A:** GUA HAI
spalatensis Obenberger, 1917: 42 (*Rhipiphorus*) **E:** CR
subdipterus Bosc, 1792: 293 **E:** AB AL AR BU CR FR GR IT PT SP YU **N:** AG MO TU **A:** CY IS KZ UZ
ambigua Giorna, 1804: 216 (*Mordella*)
antoniae Pic, 1905h: 165 (*Myiodes*) [HN]
babadjanidesi Reitter, 1912a: 43 (*Myiodes*)
caucasicus Reitter, 1906a: 453 (*Rhipiphorus*) [RN]

clermonti Chobaut, 1906: 224 (*Myiodes*) [RN]
dorthesii Latreille, 1818b: 131 (*Myodes*)
pallescens Solsky, 1881b: 57 (*Rhipidophorus*)
subdipterus Fabricius, 1792b: 109 [HN]
syriacus Pic, 1904h: 90 (*Myiodes*) **A:** TR
tenthredinoides Gressitt, 1941: 534 **A:** GUA HAI
turcicus Pic, 1914g: 42 (*Myodites*) **A:** TR

family Ripiphoridae, nomina dubia
Ripiphorus fulvipennis Fonscolombe, 1845: 504 **E:** FR
Ripiphorus maurus Betta, 1847: 28 **E:** IT
Ripiphorus suturalis Betta, 1847: 28 **E:** IT
Rhipiphorus populneus Fabricius, 1792b: 112 **E:** "Germaniae"

family ZOPHERIDAE Solier, 1834
subfamily **Zopherinae** Solier, 1834
tribe Monommatini Blanchard, 1845

genus *Inscutomonomma* Pic, 1951a: 2 type species *Monomma syriacum* J. Thomson, 1860
mycotretoides Ancey, 1881b: 413 (*Monomma*) **A:** YE
syriacum J. Thomson, 1860b: 12 (*Monomma*) **A:** SY
yemense Freude, 1958: 36 **A:** YE

genus *Monomma* Klug, 1833: 182 type species *Monomma irroratum* Klug, 1833
brittoni Freude, 1955: 20 **A:** PA SD
brunneum J. Thomson, 1860b: 23 **A:** PA SD
fulvum dehradunense Freude, 1955: 24 **A:** UP
glyphysternum Marseul, 1876d: 330 **A:** JA HAI HKG SHG TAI XIN YUN
chinense Freude, 1955: 46
formosanum Pic, 1916e: 1
gressitti Freude, 1978: 221 **A:** HAI
japonicum Motschulsky, 1861: 18 [DA] **A:** JA
similifulvum Freude, 1955: 24 **A:** PA UP

tribe Phellopsini Ślipiński & Lawrence, 1999

genus *Phellopsis* LeConte, 1862: 216 type species *Boletophagus obcordatus* Kirby, 1837
Pseudonosoderma Heyden, 1885b: 305 type species *Pseudonosoderma amurense* Heyden, 1885
amurensis Heyden, 1885b: 306 (*Pseudonosoderma*) **A:** FE SC
imurai Masumoto, 1990b: 87
chinensis Semenov, 1893g: 499 (*Pseudonosoderma*) **A:** GAN SCH SHA
suberea Lewis, 1887c: 219 **A:** JA
yulongensis Foley & Ivie, 2008: 23. **A:** YUN

tribe Pycnomerini Erichson, 1845

genus *Pycnomerus* Erichson, 1842: 214 type species *Ips terebrans* G.-A. Olivier, 1790

Dechomus Jacquelin du Val, 1858: 180 type species *Cerylon sulcicolle* Germar, 1823

Penthelispa Pascoe, 1860b: 111 type species *Penthelispa porosa* Pascoe, 1860

Pycnomeroplesius Ganglbauer, 1899: 885 type species *Pycnomerus inexspectus* Jacquelin du Val, 1858

boninensis Aoki, 2017: 176 **A:** JA

fuliginosus Erichson, 1842: 215 **E:** GB IR

inexspectus Jacquelin du Val, 1858: 179 **E:** AU BE CZ FR GE GB IT SP

italicus Ganglbauer, 1899: 887 (*Dechomus*) **E:** IT

lucidus Dajoz, 1975: 308 (*Penthelispa*) **A:** BT SW

nishii Aoki, 2017: 179 **A:** JA

nitidicollis Reitter, 1877f: 350 (*Penthelispa*) **A:** CH SD

sculpturatus Sharp, 1885a: 77 **A:** JA TAI

strumiger Aoki, 2017: 177 **A:** JA

sulcicollis Germar, 1823: 466 (*Cerylon*) **E:** AB AR BU GG GR RO ST UK **A:** IN TR

coniunctus Říha, 1958: 32 (*Dechomus*)

fagivorus Říha, 1958: 32 (*Dechomus*)

terebrans G.-A. Olivier, 1790: 18: 5 (*Ips*) **E:** AB AU BH BU CR CT CZ FR GE GR HU IT PL PT SK SL ST SZ TR UK YU **A:** IN TR

vilis Sharp, 1885a: 77 **A:** JA

japonicus Reitter, 1899e: 216 (*Penthelispa*)

yoshidai Aoki, 2011b: 13 **A:** JA

tribe Usechini Horn, 1867

genus *Usechus* Motschulsky, 1845a: 79 type species *Usechus lacerta* Motschulsky, 1845

chujoi Kulzer, 1960: 304 **A:** JA

ohdaiensis Sasaji, 1987: 52 **A:** JA

sasajii Saitô, 1999: 106 **A:** JA

tsushimensis Kamiya, 1963: 21 **A:** JA

subfamily **Colydiinae** Billberg, 1820

tribe Colydiini Billberg, 1820

genus *Aulonium* Erichson, 1845: 275 type species *Colydium bidentatum* Fabricius, 1801

Anoectochilus L. Redtenbacher, 1845: 124 type species *Colydium bicolor* Herbst, 1797

ruficorne G.-A. Olivier, 1790: 18: 13 (*Ips*) **E:** AB AR BU CR FR GE GR HU IT MC RO SP ST TR UK YU **N:** AG MO TU **A:** IN IS LE TM TR

bicolor Herbst, 1797: 285 (*Colydium*)

sulcicolle Wollaston, 1864: 127 **N:** CI

trisulcum Geoffroy, 1785: 23 (*Dermestes*) **E:** AR AU BE BH CT CZ FR GB GE HU IT MC NL NR NT PL SK ST SV SZ UK **A:** IN

bicolor Fabricius, 1801a: 555 (*Colydium*) [HN]

sulcatum G.-A. Olivier, 1790: 18: 4 (*Ips*)

genus *Colydium* Fabricius, 1792b: 495 type species *Bostrichus elongatus* Fabricius, 1787

Paschabium Gozis, 1886: 11 type species *Bostrichus elongatus* Fabricius, 1787

elongatum Fabricius, 1787a: 36 (*Bostrichus*) **E:** AB AL AR AU BE BH BU BY CR CT CZ DE FI FR GB GE GR HU IT LT MC NL PL RO SK SL SP ST SW SZ TR UK YU **N:** AG MO TU **A:** CY IN RU SY TR

linearis Rossi, 1790: 50 (*Ips*)

filiforme Fabricius, 1792b: 496 **E:** AB AU BE BH BU BY CR CZ FR GE GG GR HU IT LA LS NR PL SK SP ST SV SZ UK **A:** IN RU

tribe Gempylodini Sharp, 1893

genus *Gempylodes* Pascoe, 1863c: 132 type species *Gempylodes macer* Pascoe, 1863

Mecedanops Reitter, 1878e: 120 type species *Mecedanops ornamentalis* Reitter, 1878

ornamentalis Reitter, 1878e: 121 (*Mecedanops*) **A:** JA TAI **ORR**

lewisii Sharp, 1885a: 72

genus *Pseudendestes* Lawrence, 1980: 298 type species *Pseudendestes australis* Lawrence, 1980

andrewesi Grouvelle, 1908b: 414 (*Endestes*) **A:** JA SD

lawrencei Pal, 1984: 32

namdaphaensis Pal, 1984: 33 **A:** AP

tribe Nematidiini Horn, 1878

genus *Nematidium* Erichson, 1845: 275 type species *Colydium cylindricum* Fabricius, 1801

elongatum Pal & Ślipiński, 1984: 539 **A:** UP

tribe Orthocerini Blanchard, 1845 (1820)

genus *Diplagia* Reitter, 1882a: 117 type species *Diplagia hellenica* Reitter, 1882

hellenica Reitter, 1882a: 118 **E:** GR

genus *Helioctamenus* Schaufuss, 1882b: 531 type species *Helioctamenus hippopotamus* Schaufuss, 1882

Acotulus Reitter, 1891f: 246 type species *Acotulus oranensis* Reitter, 1891

bedeli Dajoz, 1971: 83 **E:** PT SP

curticornis Pic, 1922e: 21 **N:** EG

espanoli Cobos, 1950: 589 **E:** SP

fernandezi Cobos, 1965b: 59 **N:** Cl

fumicornis Bedel, 1891b: cxxxv (*Orthocerus*) **N:** AG MO

aestivus Escalera, 1914: 121 (*Acotulus*)

oranensis Reitter, 189la: 247 (*Acotulus*)

gineri Español, 1948: 311 (*Acotulus*) **N:** MO

hippopotamus Schaufuss, 1882b: 531 **E:** SP (Baleares)

lusitanicus Reitter, 1903d: 174 **E:** PT

occidentalis Oromí, 1984a: 226 **N:** CI
pardoi Español, 1957c: 147 **N:** MO

genus *Orthocerus* Latreille, 1796: 16 type species *Tenebrio hirticornis* DeGeer, 1775
Sarrotrium Illiger, 1798: 339 type species *Hispa mutica* Linnaeus, 1767
clavicornis Linnaeus, 1758: 355 (*Dermestes*) **E:** AU BE BH BY CR CT CZ DE FI FR GE GB HU LA LS LU NL NR IT IR PL RO RU SK SP SV SZ UK YU "Caucasus" **A:** ES FE MG TR
crenulatus Motschulsky, 1845a: 80 (*Sarrotrium*)
hirticornis DeGeer, 1775: 47 (*Tenebrio*)
muticus Linnaeus, 1767: 604 (*Hispa*)
crassicornis Erichson, 1845: 260 (*Sarrotrium*) **E:** AR AU BH BY CZ FR GE HU LA IT PL PT RO SK SZ UK
tereticornis Erichson, 1845: 261 (*Sarrotrium*)

tribe Rhopalocerini Reitter, 1911

genus *Rhopalocerus* W. Redtenbacher, 1842: 21 type species *Rhopalocerus setosus* W. Redtenbacher, 1842 (= *Monotoma rondanii* A. Villa & J.B. Villa, 1833)
nakasatoi Aoki & Narukawa, 2019: 165 **A:** TAI
rondanii A. Villa & J.B. Villa, 1833: 36 (*Monotoma*) **E:** AU BU CR CZ GE GR HU IT PT RO SK UK YU
setosus W. Redtenbacher, 1842: 22

tribe Synchitini Erichson, 1845

genus *Acolophus* Sharp, 1885a: 66 type species *Acolophus debilis* Sharp, 1885
debilis Sharp, 1885a: 66 **A:** JA

genus *Bhutania* Dajoz, 1975: 300 type species *Bhutania elongata* Dajoz, 1975
elongata Dajoz, 1975: 300 **A:** BT NP

genus *Bitoma* Herbst, 1793: 25 type species *Tritoma crenata* Fabricius, 1775
Ditoma Illiger, 1807: 320 type species *Tritoma crenata* Fabricius, 1775
Euditomum Gistel, 1856: 128 type species *Tritoma crenata* Fabricius, 1775
Synchytodes Crotch, 1873a: 45 type species *Bitoma quadriguttata* Say, 1827
Xuthia Pascoe, 1863c: 128 type species *Xuthia siccana* Pascoe, 1863
crenata Fabricius, 1775: 69 (*Tritoma*) **E:** AB AL AR AU BE BH BU BY CR CT CZ DE EN FI FR GB GE GG GR HU IT LA LS KZ MA MC MD NL NR NT PL RO RU SK SL SP ST SV SZ UK YU **N:** AG TU **A:** ES FE KZ MG WS NAR
castanea Dalla Torre, 1879: 93 (*Synchitodes*)
ferruginea Dalla Torre, 1879: 93 (*Synchitodes*)
picipes G.-A. Olivier, 1790: 18: 7 (*Ips*)
rufipennis Fabricius, 1801b: 562 (*Lyctus*)
rufithorax Pic, 1925a: 1 (*Ditoma*)

iranica Mařan, 1954a: 132 **E:** AB **A:** IN
siccana Pascoe, 1863c: 128 (*Xuthia*) **A:** BT CH HP HKG JA NP SA SD TAI YE **N:** AG EG **AFR NTR ORR**
lyctiformis Wollaston, 1867: 63 (*Ditoma*)
parallela Sharp, 1885a: 70
turcica Dajoz, 1973b: 80 **A:** CY TR

genus *Bolcocius* Dajoz, 1975: 297 type species *Colobicus granulosus* Sharp, 1886
bhutanensis Dajoz, 1975: 297 **A:** BT SW
formosanus Sasaji, 1984: 42 **A:** TAI
granulosus Sharp, 1885a: 65 (*Colobicus*) **A:** JA SW
shibatai Sasaji, 1984: 39 **A:** JA
yaeyamensis Sasaji, 1984: 41 **A:** JA (Ryukyus)

genus *Cerchanotus* Erichson, 1845: 257 type species *Syntarsus asperulus* Fairmaire, 1869
Asprecodes Nakane, 1990: 65 type species *Asprecodes ogasawarensis* Nakane, 1990
Syntarsus Fairmaire, 1869b: 205 type species *Syntarsus asperulus* Fairmaire, 1869
jelineki Ślipiński, 1985: 182 (*Syntarsus*) **A:** SW
orientalis Ślipiński, 1985: 183 (*Syntarsus*) **A:** BT JA NP SD
ogasawarensis Nakane, 1990: 66 (*Asprecodes*)

genus *Colobicones* Grouvelle, 1918: 15 type species *Colobicones singularis* Grouvelle, 1918
Wittmeria Dajoz, 1975: 294 type species *Wittmeria pubscens* Dajoz, 1975
pubescens Dajoz, 1975: 295 (*Wittmeria*) **A:** BT
sakaii Okada, 2005: 428 **A:** JA (Amami-Ôshima)
tokarensis Okada, 2005: 426 **A:** JA (Tokara Is.)

genus *Colobicus* Latreille, 1807: 9 type species *Colobicus marginatus* Latreille, 1807
changquingi Liang & Ren, 2013: 110 **A:** TAI
gigas Grouvelle, 1908a: 119 **A:** NP
hirtus Rossi, 1790: 59 (*Nitidula*) **E:** AU BH BU CR CT CZ FR GB GE GR HU IT LS NT PL RU SK SC SL SP ST SV SZ UK YU **N:** MO **A:** AF FE JA NO TR WS
axillaris Duftschmid, 1825: 155 (*Monotoma*)
emarginatus Erichson, 1845: 268
hirtus Brullé, 1835: 410
marginatus Latreille, 1807: 10
parilis Pascoe, 1860b: 102 **E:** FR **A:** AF CH HKG JA NP SD TAI UP

genus *Corticus* Germar, 1823: 146 type species *Sarrotrium celtis* Germar, 1823
Ceropachys A. Costa, 1847: 150 type species *Ceropachys foveicollis* A. Costa, 1847
Horrimantus Gozis, 1881: 113 [RN]

brevipennis Reitter, 1882a: 119 **E:** AB ST **A:** IQ IN TR
celtis Germar, 1823: 146 (*Sarrotrium*) **E:** CR IT **N:** AG TU
foveolatus Fairmaire, 1848: 171
foveicollis A. Costa, 1847: 150

genus *Coxelus* Dejean, 1821: 67 type species *Bolitophagus pictus* J. Sturm, 1807
alinae Dajoz, 1973b: 82 **E:** GR
bituberculatus J. Frivaldszky, 1894: 85 (*Diodesma*) **E:** TR
humeridens Reitter, 1885d: 392 **A:** LE TR
luteopilosus Pic, 1901a: 3 **A:** TR
pictus J. Sturm, 1807: 161 (*Bolitophagus*) **E:** AU BH CR CZ FR GE GR HU IT LS PL RO SK SL SP SZ UK YU
yeti Ślipiński, 1985: 189 **A:** NP

genus *Diodesma* Latreille, 1829a: 97 type species *Diodema subterranea* Latreille, 1829
besucheti Dajoz, 1977: 156 **N:** MO
denticincta Abeille de Perrin, 1899a: 408 **E:** FR IT
parallela Reitter, 1922a: 21 **N:** AG MO
subterranea Latreille, 1829a: 97 **E:** AU BH CZ FR GE HU IT PL SK SL SZ UK YU
picea J.H.C.F. Sturm, 1849: 15

genus *Endeitoma* Sharp, 1894: 450 type species *Endeitoma mexicana* Sharp, 1894
Pabula Carter & Zeck, 1937: 193 type species *Ditoma perforata* Blackburn, 1888
bonina Nakane, 1990: 67 (*Pabula*) **A:** JA

genus *Endophloeus* Dejean, 1834: 195 type species *Eledona spinulosa* Latreille, 1807
exsculptus Germar, 1847: nr. 2 **E:** AB AR GG HU ST UK **A:** IN TR
markovichianus Piller & Mitterpacher, 1783: 105 (*Sylpha* [sic]) **E:** AU BH BU CR FR GB GR HU IT MC PL PT RO RU SK SL SP ST TR UK YU **N:** AG MO TU
spinulosus Latreille, 1807: 179 (*Eledona*)
serratus Sharp, 1885a: 61 **A:** JA TAI
squarrosus Germar, 1847: nr 3 **E:** GR TR YU

genus *Franzorphius* Schuh, 1998a: 228 type species *Franzorphius franzi* Schuh, 1998
franzi Schuh, 1998a: 229 **A:** SW

genus *Glyphocryptus* Sharp, 1885a: 63 type species *Glyphocryptus brevicollis* Sharp, 1885
brevicollis Sharp, 1885a: 64 **A:** JA
grandis Aoki & Okada, 2011: 245 **A:** JA
masumotoi Aoki & Okada, 2011: 247 **A:** TAI
toyoshimai Aoki & Okada, 2011: 246 **A:** JA

genus *Hyberis* Pascoe, 1860b: 112 type species *Hyberis araneiformis* Pascoe, 1860
mizuhoae Makita, 2017b: 10 **A:** TAI
similis Grouvelle, 1913: 106 **A:** HP SD

genus *Langelandia* Aubé, 1842: 227 type species *Langelandia anophthalma* Aubé, 1842
subgenus *Agelandia* Reitter, 1882b: 163 type species *Langelandia grandis* Reitter, 1877
gigantea Dajoz, 1968a: 986 **A:** TR
grandis Reitter, 1877e: 296 **A:** AB ST TR UK "Caucasus"
hummleri Obenberger, 1918: 58 (*Agelandia*) **A:** IT
macedonia Franz, 1980: 110 **A:** GR
merkliana Reitter, 1890f: 210 (*Agelandia*) **E:** TR **A:** IN TR
terricola Reitter, 1912c: 280 **E:** GR
wittmeri Dajoz, 1973b: 80 **A:** IN
subgenus *Fleischerella* Reitter, 1911a: 113 type species *Langelandia incostata* Perris, 1869
exigua Perris, 1869: 9 **E:** FR IT
incostata Perris, 1869: 11 **E:** FR (Corse)
caucasica Nikitsky, 1996b: 1434 **E:** ST
hypogea Normand, 1936a: 93 **N:** AG
khaustovi Drogvalenko 2005: 90 **E:** UK
leonhardi Reitter, 1912c: 285 **E:** IT
niticosta Schuh & Mifsud, 2000: 262 **E:** MA
nitidicollis Reitter, 1910b: 115 **E:** IT
subgenus *Langelandia* Aubé, 1842: 227 type species *Langelandia anophthalma* Aubé, 1842
aglena Reitter, 1908a: 20 **E:** BH MC
anophthalma Aubé, 1842: 228 **E:** AU CZ BE BH FR GB GE GR HU IT NL PL PT RO SK SP SZ UK YU
  insularis Dajoz, 1969: 712
  media Rey, 1889: 55
ausonica Obenberger, 1914: 109 **E:** IT
brachati Daffner, 1983: 22 **N:** CI
excavata Reitter, 1912c: 282 **E:** FR
gjonovici Reitter, 1912c: 281 **E:** BH CR
karamani Reitter, 1912c: 284 **E:** CR YU
mauli Franz, 1970: 3 **N:** MR
reitteri Belon, 1882: cxxv **E:** FR IT SP **N:** AG MO
  galitensis Franz, 1982: 238
  ibizensis Dajoz, 1969: 717
  meridionalis Dajoz, 1969: 717
  parallela Marquet, 1897: 107
  torressalai Español, 1955d: 103
portosantoi Franz, 1970: 1 **N:** MR (Porto Santo)

thessalica Daffner, 1985: 36 **E:** GR
viennensis Reitter, 1912c: 283 **E:** AL AU GR IT LS YU
paganettii Obenberger, 1914: 109
subgenus *Normandella* Binaghi, 1937: 77 type species *Langelandia tunisina* Binaghi, 1937
antennaria Binaghi, 1937: 78 **E:** IT
montalbica Fancello & Magrini, 2013: 284 **E:** IT
tunisina Binaghi, 1937: 77 **N:** TU
zacynthia Heinze, 1943b: 127 **E:** GR
subgenus *Paganettia* Ganglbauer, 1899: 882 type species *Langelandia callosipennis* Reitter, 1881
callosipennis Reitter, 1881a: 217 **E:** BH CR
ozimeci Schuh & Hlaváč, 2015: 244 **E:** CR YU
reflexipennis Reitter, 1912c: 281 **E:** CR YU

genus *Lasconotus* Erichson, 1845: 258 type species *Lasconotus complex* LeConte, 1859
Ithris Pascoe, 1863c: 134 type species *Ithris decisa* Pascoe, 1863
Lado Wankowiez [sic!], 1867: 251 type species *Bitoma jelskii* Wankowicz, 1867
Othismopteryx J.R. Sahlberg, 1871: 441 type species *Othismopteryx carinatus* J.R. Sahlberg, 1871
akitai Aoki, 2018: 72 **A:** JA
hajeki Schuh, 2012: 290 **A:** YE (Suqutra)
jelskii Wankowicz, 1867: 249 (*Bitoma*) **E:** BY CT FI LT PL RU SV **A:** KI
carinatus J.R. Sahlberg, 1871: 441 (*Othismopteryx*)
montanus Nikitsky & Ślipiński, 2003: 43 **A:** KZ KI
nepalensis Ślipiński, 1985: 184 **A:** NE
niponius Lewis, 1879: 462 (*Xuthia*) **A:** JA
okadai Aoki, 2011a: 99 **A:** JA, SC
saudicus Ślipiński, 1984: 253 **A:** SA
sculpturatus Sharp, 1885a: 72 (*Ithris*) **A:** JA

genus *Lastrema* Reitter, 1882a: 129 type species *Pycnomerus verrucicollis* Reitter, 1880
verrucicollis Reitter, 1880d: 543 (*Pycnomerus*) **E:** AB ST (Kavkaz) **A:** IN TR

genus *Lyreus* Aubé, 1861: 196 type species *Lyreus subterraneus* Aubé, 1861
caneparii Magrini, Fancello & Hernando, 2003: 39 **E:** SP (Girona)
septemstriatus Fancello & Leo, 1991: 129 **E:** IT (Sardinia)
subterraneus Aubé, 1861: 196 **E:** FR

genus *Microprius* Fairmaire, 1869a: 779 type species *Microprius terrenus* Fairmaire, 1868
Trionus Sharp, 1885a: 70 type species *Trionus opacus* Sharp, 1885
demissus Pascoe, 1863a: 33 (*Phormesa*) **A:** SD **ORR**
opacus Sharp, 1885a: 70 (*Trionus*) **A:** HKG JA NP TAI UP

rufulus Motschulsky, 1863: 502 (*Bitoma*) **Ei:** FR GE MA SP **N:** AG EG MO TU **A:** AE IS JO LE SY YE

frivaldszkyi Reitter, 1877f: 326 (*Synchitodes*)
linearis Wollaston, 1867: 64 (*Ditoma*)
rufus Reitter, 1882a: 131 (*Synchitodes*)

genus *Namunaria* Reitter, 1882a: 114 type species *Coxelus guttulatus* LeConte, 1863
Sympanotus Sharp, 1885a: 62 type species *Sympanotus pictus* Sharp, 1885
bhutanensis Ślipiński, 1981a: 425 (*Tarphius*) **A:** BT
chinensis Schuh, 1999: 143 **A:** SW
mammillaris Schuh, 1999: 148 **A:** NE
montana Aoki & Narukawa, 2016: 103 **A:** JA
picta Sharp, 1885a: 62 (*Sympanotus*) **A:** JA NP SE SW

genus *Neotrichus* Sharp, 1885a: 60 type species *Neotrichus hispidus* Sharp, 1885
afoveicollis Pal, 2003: 173 **A:** SD
cavatus Aoki, 2009: 143 **A:** JA (Bonin Is.)
hispidus Sharp, 1885a: 61 **A:** JA
lanyuensis Sasaji, 1986: 247 **A:** TAI
serraticollis Sasaji, 1986: 246 **A:** JA (Ryukyus)

genus *Niphopelta* Reitter, 1882a: 129 type species *Niphopelta imperialis* Reitter, 1882
imperialis Reitter, 1882a: 129 **E:** AB GG UK ST "Caucasus" **A:** IN

genus *Nosodomodes* Reitter, 1922a: 16 type species *Corticus rotundicollis* Pic, 1897
diabolicus Schaufuss, 1862b: 309 (*Corticus*) **E:** AR BU GR MC RO UK YU **A:** TR
rotundicollis Pic, 1897h: 78 (*Corticus*) **N:** AG
syriacus Fairmaire, 1892b: 144 (*Corticus*) **A:** SY TR
tauricus Germar, 1832: 4 (*Sarrotrium*) **E:** AB ST UK **A:** TR
tuberculatus Germar, 1832: 4 (*Sarrotrium*) **E:** BU GR MC RO ST UK

genus *Prosteca* Wollaston, 1860a: 254 type species *Prostheca aspera* Wollaston, 1860
Doderonymus Binaghi, 1937: 70 type species *Doderonymus lusitanicus* Binaghi, 1937
aspera Wollaston, 1860a: 255 **E:** AZ PT **N:** MR
lusitanica Binaghi, 1937: 71 (*Doderonymus*)
orientalis Vít, 1986: 453 **A:** PA

genus *Pseudotarphius* Wollaston, 1873: 1 type species *Pseudotarphius lewisii* Wollaston, 1873
bhutanensis Dajoz, 1975: 303 **A:** BT
lewisii Wollaston, 1873: 4 **A:** JA TAI

genus *Sprecodes* Pope, 1961: 39 type species *Caprodes ater* Grouvelle, 1904
socotrensis Schuh, 2012: 288 **A**: YE (Suqutra)

genus *Synchita* Hellwig, 1792: 401 type species *Synchita juglandis* Hellwig, 1792
Cicones Curtis, 1827: 149 type species *Cicones carpini* Curtis, 1827
Microsicus Sharp, 1894: 456 type species *Microsicus setosus* Sharp, 1894
Pseudocicones Fursov, 1939: 88 type species *Pseudocicones montanus* Fursov, 1939
Pseudosynchita Pic, 1922e: 21 type species *Pseudosynchita hirsuta* Pic, 1922
africana Grouvelle, 1905: 246 (Cicones) **A**: YE
amurensis Mamaev, 1977: 28 **A**: FE
angustissima Nakane, 1968a: 75 (*Cicones*) **A**: JA
bitomoides Sharp, 1885a: 69 (*Cicones*) **A**: JA
bonina Nakane, 1991: 2 (*Cicones*) **A**: JA
constricta Aoki, 2012:217 (*Microsicus*) **A**: JA
crenicollis Wollaston, 1867: 62 (*Synchyta*) **N**: EG **A**: AE YE
hirsuta Pic, 1922e: 21 (*Pseudosynchita*)
fallax Schuh, 1998b: 321 **E**: FR GR IT SP **A**: TR **N**: AG MO
hayashii Sasaji, 1971: 43 (*Cicones*) **A**: JA
hirsuta Aoki, 2008: 275 **A**: JA
humeralis Fabricius, 1792a: 205 (*Elophorus*) **E**: AU BE BH BU CR CT CZ DE FI FR GB GE HU IT LA MC MD NR NT PL RO SK SL SP ST SV SZ UK YU **A**: ES FE
juglandis Hellwig, 1792: 403
obscura L. Redtenbacher, 1857: 345
striata Herbst, 1793: 23 (*Monotoma*)
iranica Dajoz, 1973b: 79 (*Cicones*) **E**: AR **A**: IN
mediolanensis A. Villa & J.B. Villa, 1833: 36 **E**: AB AU BU CZ FR GE GR HU IT PT RO SK SP ST SZ UK **N**: AG **A**: CY IN TR
berolinensis Reitter, 1922a: 34
minima Sharp, 1885a: 69 (*Cicones*) **A**: JA
montana Fursov, 1939: 88 (*Pseudocicones*) **A**: TD
nivea Sharp, 1885a: 68 (*Cicones*) **A**: JA
oculata Sharp, 1885a: 67 (*Cicones*) **A**: JA
oblonga Sharp, 1885a: 68 (*Cicones*)
separanda Reitter, 1882a: 126 (*Ditoma*) **E**: AR AU BH BU CR CT CZ FI FR GB GE HU IT LA NR NT PL RO SK SP ST SV SZ UK **A**: IN TR
angularis Abeille de Perrin, 1901a: 61
rufosignata Sasaji, 1984: 34 (*Cicones*) **A**: JA
tokarensis Nakane, 1968a: 75 (*Cicones*) **A**: JA
undata Guérin-Méneville, 1844: 189 **E**: AU BH BU CZ FR GB GE HU IT RO PT SL ST SZ UK
picta Erichson, 1845: 273 (*Cicones*)

ussuriensis Iablokoff-Khnzorian, 1978: 252 (*Cicones*) **A:** FE
uralensis Iablokoff-Khnzorian, 1970: 66 **E:** KZ
variegata Hellwig, 1792: 403 **E:** AU BE BH BU CR CZ FR GB GE GR HU IT LS NL PT RO RU SL SP SV SZ TR UK YU **N:** AG **A:** TR **NARi**
carpini Curtis, 1827: 149 (*Cicones*)

genus *Tarphiosoma* Wollaston, 1862b: 373 type species *Tarphiodes indicus* Wollaston, 1862
Neoplatus Sharp, 1885b: 119 type species *Tarphisoma fasciata* Motschulsky, 1863
Tarphisoma Motschulsky, 1863: 504 type species *Tarphisoma fasciata* Motschulsky, 1863
bouvieri Grouvelle, 1903: 119 **A:** SD
globosa Ślipiński, 1985: 187 (*Tarphius*)
kempi Grouvelle, 1913: 104 **A:** AP

genus *Tarphius* Erichson, 1845: 256 type species *Tarphius gibbulus* Erichson, 1845
Atlantotarphius Franz, 1967: 65 type species *Tarphius canariensis* Wollaston, 1862
Caudatotarphius Franz, 1967: 64 type species*Tarphius deformis* Wollaston, 1862
Glabrotarphius Franz, 1967: 63 type species *Tarphius lutulentus* Wollaston, 1871
abbreviatus Wollaston, 1865: 20 (App.) **N:** CI
acuminatus Gillerfors, 1986a: 4 **E:** AZ
affinis Wollaston, 1862b: 383 **N:** CI
angusticollis Wollaston, 1860a: 252 **N:** MR
angustulus Wollaston, 1862d: 289 **N:** MR
azoricus Gillerfors, 1986a: 2 **E:** AZ
barbarae Gillerfors, 1991: 22 **N:** CI
besucheti Español, 1961b: 83 **E:** SP
brevicollis Wollaston, 1854: 144 **N:** MR
camelus Wollaston, 1862b: 383 **N:** CI
canariensis Wollaston, 1862b: 383 **N:** CI
caudatus Wollaston, 1862b: 386 **N:** CI
cicatricosus Wollaston, 1854: 141 **N:** MR
compactus Wollaston, 1854: 139 **N:** MR
congestus Wollaston, 1862b: 385 **N:** CI
deformis Wollaston, 1862b: 387 **N:** CI
depressus Gillerfors, 1985: 4 **E:** AZ
desertaensis Machado 2012: 637 **N:** MR
echinatus Wollaston, 1854: 143 **N:** MR
elongatus Dajoz, 1971: 86 **N:** CI
epinae Gillerfors, 1986b: 4 **N:** CI
ericae Gillerfors, 1986b: 4 **N:** CI
erosus Wollaston, 1862b: 384 **N:** CI
excisus Wollaston, 1857a: 47 **N:** MR

explicatus Wollaston, 1857a: 47 **N:** MR
fairmairei Marseul, 1872b: 134 **N:** AG
    humerosus Fairmaire, 1866a: 20 [HN]
    wollastonii Fairmaire, 1871a: 372 [HN]
fernandezlopezi Franz, 1967: 74 **N:** CI
floresensis Borges & Serrano 2017: 423 **E:** AZ
formosus Wollaston, 1857a: 44 **N:** MR
furtadoi Borges & Serrano 2017: 418 **E:** AZ
gabrielae Borges & Serrano 2017: 421 **E:** AZ
gibbulus Erichson, 1845: 256 **E:** IT
gigas Wollaston, 1862b: 386 **N:** CI
gomerae Franz, 1967: 76 [RN] **N:** CI
    affinis Wollaston, 1865: 19 (App.) [HN]
huggerti Gillerfors, 1991: 23 **N:** CI
humerosus Wollaston, 1865: 19 (App.) **N:** CI
inornatus Wollaston, 1854: 135 **N:** MR
    spinipes Wollaston, 1854: 136
isabelae Serrano, Amorim & Borges, 2013: 494 **N:** MO
jandiensis Machado 2012: 639 **N:** CI
kiesenwetteri Heyden, 1870: 105 **E:** PT SP
lauri Wollaston, 1854: 138 **N:** MR
liliputanus Lucas, 1846: pl. 30 (*Opatrum*) **N:** AG TU
    oblongulus Fairmaire, 1866a: 21
lowei Wollaston, 1854: 134 **N:** MR
lutulentus Wollaston, 1871: 240 **N:** MR
maroccanus Franz, 1967: 65 **N:** MO
monstrosus Wollaston, 1865: 20 (App.) **N:** CI
moyanus Franz, 1967: 72 **N:** CI
nodosus Wollaston, 1854: 140 **N:** MR
oromii Franz, 1984b: 78 **N:** CI
oulmesensis Franz, 1967: 67 **N:** MO
palmensis Gillerfors, 1990: 149 **N:** CI
    fraudulentus Franz, 1990a: 30
parallelus Wollaston, 1854: 134 **N:** MR
piniphilus Franz, 1967: 89 **N:** CI
pomboi Borges, 1991: 2 **E:** AZ
postcostatus Uyttenboogaart, 1937: 86 **N:** CL
quadratus Wollaston, 1862b: 384 **N:** CI
relictus Borges & Serrano 2017: 414 **E:** AZ
rotundatus Wollaston, 1854: 137 **N:** MR
rufonodulosus Israelson, 1984: 149 **E:** AZ

rugosus Wollaston, 1854: 144 **N:** MR
wolffi Wollaston, 1865: 21 (App.)
sculptipennis Wollaston, 1857a: 46 **N:** MR
serranoi Borges, 1991: 4 **E:** AZ
setosus Wollaston, 1865: 17 (App.) **N:** CI
simplex Wollaston, 1862b: 382 **N:** CI
constricticollis Franz, 1990a: 29
rotundicollis Franz, 1990a: 29
stagosus Gillerfors & Oromí, 1991: 85 **N:** CI
supranubis Franz, 1984b: 77 **N:** CI
sylvicola Wollaston, 1854: 137 **N:** MR
tacorontinus Franz, 1967: 71 **N:** CI
tenerifae Franz, 1967: 68 **N:** CI
testudinalis Wollaston, 1854: 141 **N:** MR
tornvalli Gillerfors, 1985: 5 **E:** AZ
truncatus Wollaston, 1854: 142 **N:** MR
wollastoni Crotch, 1867a: 386 **E:** AZ
zerchei Gillerfors, 1997: 2 **N:** MR

genus *Trachypholis* Erichson, 1845: 257 type species *Opatrum hispidum* Weber, 1801
Labromimus Sharp, 1885a: 64 type species *Labromimus variegatus* Sharp, 1885
Microvonus Sharp, 1885b: 118 type species *Microvonus squalidus* Sharp, 1885
Tarphiodes Wollaston, 1862b: 372 type species *Tarphiodes bowringii* Wollaston, 1862
chinensis Ślipiński, 1985: 186 **A:** SW
cinerea Dajoz, 1975: 307 **A:** BT
decorata Grouvelle, 1911: 126 **A:** HP
harmandi Grouvelle, 1903: 120 (*Labromimus*) **A:** NP SD
hispida Weber, 1801: 38 (*Opatrum*) **A:** AP NP SD
bowringii Wollaston, 1862b: 372 (*Tarphiodes*)
erichsoni Reitter, 1877f: 328
lanyuensis Makita, 2017a: 369 **A:** TAI
luteonigra Dajoz, 1975: 304 **A:** BT NP
montana Grouvelle, 1903: 121 (*Microvonus*) **A:** SD
okinawensis Nakane, 1991: 3 **A:** JA
ornata Grouvelle, 1911: 132 **A:** TAI
proxima Dajoz, 1975: 307 **A:** BT
sauteri Grouvelle, 1914: 47 (*Microvonus*) **A:** TAI
signata Grouvelle, 1911: 128 **A:** AP
variegata Sharp, 1885a: 65 (*Labromimus*) **A:** FE (Sakhalin) JA TAI

genus *Xylolaemus* Dejean, 1835: 313 type species *Lyctus fasciculosus* Gyllenhal, 1827
Canaricerus Oromí & García, 1986: 304 type species *Canaricerus aeonii* Oromí & García, 1986

abnormis Ślipiński, 1984: 251 **A:** AE SA
aeonii Oromí & García, 1986: 304 (*Canaricerus*) **N:** CI
fasciculosus Gyllenhal, 1827: 632 (*Lyctus*) **E:** AU FR IT RO ST SV YU **N:** MO TU
indicus Grouvelle, 1903: 119 [DA] **A:** SD

family Zopheridae, nomen dubium
Ips bifasciatus Schrank, 1798: 513 **E:** GE (Bavaria)

family MORDELLIDAE Latreille, 1802
subfamily **Mordellinae** Latreille, 1802
tribe Conaliini Ermisch, 1956

genus *Conalia* Mulsant & Rey, 1858c: 313 type species *Conalia baudii* Mulsant & Rey, 1858
baudii Mulsant & Rey, 1858c: 313 **E:** AU BY CR CT CZ FR GG GR HU NT SK ST

tribe Mordellini Latreille, 1802

genus *Calycina* Blair, 1922b: 250 [RN] type species *Calycella palpalis* Blair, 1922
Calycella Blair, 1922a: 222 [HN] type species *Calycella palpalis* Blair, 1922
horaki Ruzzier & Kovalev, 2016: **A:** FE
major Nomura, 1967a: 6 **A:** TAI
nigroapicalis Nomura, 1967a: 5 **A:** TAI

genus *Congomorda* Ermisch, 1955: 26 type species *Congomorda atra* Ermisch, 1955
orientalis Horák, 2009: 82 **A:** SD

genus *Curtimorda* Méquignon, 1946: 58 type species *Mordella bisignata* L. Redtenbacher, 1849
bisignata L. Redtenbacher, 1849: 613 (*Mordella*) **E:** AU BY CT CZ FR GE HU NL NT PL ST SK SL SZ TR UK
albosignata Mulsant, 1856f: 333 (*Mordella*)
maculosa Neazen, 1794: 273 (*Mordella*) **E:** AU BY CT CZ FR GE HU IT NL NT PL RO SK ST SZ UK **A:** ES FE GAN JA QIN SCH WS YUN
atomaria Fabricius, 1801b: 123 (*Mordella*)
guttata Paykull, 1798: 186 (*Mordella*)
guttatipennis Pic, 1927l: 183 (*Mordella*)
irrorata Trost, 1801: 27 (*Mordella*)

genus *Glipa* LeConte, 1857: 17 type species *Mordella hilaris* Say, 1835
Neoglipa Franciscolo, 1952: 332 type species *Glipa favareli* Pic, 1917
subgenus *Stenoglipa* Franciscolo, 1952: 344 type species *Mordella annulata* L. Redtenbacher, 1868
azumai Nakane, 1950: 21 **A:** CH JA
hatayamai hatayamai Takakuwa, 2000: 79 **A:** JA
hatayamai izuinsulana Takakuwa, 2000: 80 **A:** GUX HAI JA TAI YUN

hatayamai sejuncta Takakuwa, 2000: 82 **A:** SD
ishigakiana ishigakiana Kôno, 1932b: 152 **A:** JA
ishigakiana kotoensis Nomura, 1967a: 7 **A:** TAI
kurosawai Takakuwa, 1985: 9 **A:** JA
nipponica Nomura, 1957: 40 **A:** CH JA
watanabei Takakuwa, 2000: 71 **A:** JA TAI
subgenus *Macroglipa* Franciscolo, 1952: 335 type species *Mordella durvillei* Boisduval, 1835
alboscutellata Kôno, 1934a: 116 **A:** TAI
angustilineata Fan & J. Yang, 1993: 54 **A:** GUX
asahinai Nakane & Nomura, 1950: 3 **A:** JA
cinereonigra Fairmaire, 1893f: 302 (*Mordella*) **A:** CH JA
  apicalis Pic, 1922c: 21
  schibatai Nakane, 1960: 17
cladoda Fan & J. Yang, 1993: 51 **A:** GUX
curtopyga Fan & J. Yang, 1993: 56 **A:** YUN
fanjingana C. Yang & Fan, 1988: 109 **A:** GUI
fasciata Kôno, 1928a: 32 **A:** HUB FUJ GUA GUX HAI HUN JA JIX SCH TAI YUN ZHE
flava Fan & J. Yang, 1993: 49 **A:** YUN
formosana formosana Pic, 1911q: 191 **A:** CH JA TAI
  apicepubens Pic, 1941c: 11
  laterufa Pic, 1941c: 4
  obscuripennis Pic, 1912i: 46
  satoi Nakane & Nomura, 1950: 1
  subpubens Pic, 1941c: 11
  takashii Nomura, 1967a: 8
  uniformis Pic, 1932d: 29
formosana yanma Takakuwa, 1986: 260 **A:** JA
fukiensis Ermisch, 1940: 166 **A:** FUJ
laosensis Pic, 1922c: 21 **A:** NP
karubei Takakuwa, 1993: 92 **A:** JA
klapperichi Ermisch, 1940: 170 **A:** FUJ HUB SCH
kusamai Takakuwa, 1999: 65 **A:** JA
masatakai M. Chûjô, 1960: 1 **A:** JA
obliquivittata Fan & J. Yang, 1993: 50 **A:** YUN TAI
obscuripennis Pic, 1912i: 46 **A:** FUJ TAI
ogasawarensis Kôno, 1928a: 32 **A:** JA
  salvazai Pic, 1929c: 23
ohgushii M. Chûjô, 1957a: 2 (*Neoglipa*) **A:** JA
oshimana Nomura, 1966: 41 **A:** JA
  ohgushi Nomura, 1963a: 247 [HN]

pici Ermisch, 1940: 163 **A:** FUJ GUA GUX HAI HUB HUN JA JIX SHA SCH TAI YUN ZHE
pseudofasciata Fan & J. Yang, 1993: 57 **A:** HEN
sauteri Pic, 1911p: 189 **A:** JA TAI
separata Pic, 1941c: 4 **A:** CH JA TAI
iriei Takakuwa, 1977: 9
shirozui Nakane, 1949: 29 **A:** HUB HUN JA SC TAI YUN ZHE
thoracica Horák, 1994: 102 **A:** NP
tripartita Pic, 1934b: 30 **A:** HKG NP SD
uenoi Takakuwa, 1986: 261 **A:** CH JA TAI
watanabeorum Takakuwa, 2002: 357 **A:** TAI
zhangi Fan & J. Yang, 1993: 53 **A:** YUN

genus *Glipidiomorpha* Franciscolo, 1952: 347 type species *Glipidiomorpha leucozona* Franciscolo, 1952
curticauda Ermisch, 1968a: 289 **A:** FUJ
kuantunensis Ermisch, 1968a: 289 **A:** FUJ
rufobrunnipennis Ermisch, 1968a: 291 **A:** FUJ

genus *Hoshihananomia* Kôno, 1935b: 124 type species *Mordella perlata* Sulzer, 1776
Machairorophora Franciscolo, 1943a: 34 type species *Mordella tibialis* Brown, 1880
auromaculata amamiana Nomura, 1962b: 39 **A:** JA
auromaculata auromaculata M. Chûjô, 1935a: 72 (*Mordella*) **A:** FE TAI
auromaculata nipponica Nomura, 1962b: 39 **A:** JA
auromaculata nishimurai Takakuwa, 1986a: 243 **A:** JA
borealis Nomura, 1957: 40 **A:** HAI HUB JA JIX TAI ZHE
composita composita Walker, 1858: 286 (*Mordella*) **A:** JA TAI
innotata Pic, 1926i: 67 (*Mordella*)
composita sakishimana Nomura, 1966: 42 **A:** JA
formosana Nakane & Nomura, 1950: 9 **A:** TAI
gacognei Mulsant, 1852: 49 (*Mordella*) **E:** AU BU CZ FR HU RO SL SK SP **N:** AG
brevefasciata Pic, 1925d: 11 (*Mordella*)
godarti Méquignon, 1946: 57
vauclusiana Méquignon, 1946: 57
hananomi Kôno, 1928b: 149 (*Mordella*) **A:** JA SC TAI
katoi katoi Nakane & Nomura, 1957: 3 **A:** JA TAI
katoi boninensis Nomura, 1975: 30 **A:** JA
kirai Nakane & Nomura, 1950: 6 **A:** JA TAI
kuatunensis Ermisch, 1968a: 287 **A:** FUJ
kurosai M. Chûjô & Nakane, 1955: 28 **A:** JA
kusuii Nomura, 1975: 32 **A:** JA

libanica Méquignon, 1941: 147 (*Mordella*) **A:** LE
longecaudata Fairmaire, 1891f: ccxviii (*Mordella*) **A:** SCH
michaelae Horák, 1986: 161 **A:** YE
minuscula Nomura, 1967a: 11 **A:** TAI
mitsuoi Nakane & Nomura, 1950: 7 **A:** JA
nakanei nakanei Takakuwa, 1986a: 240 **A:** JA
nakanei kuraruana Kiyoyama, 1993: 75 **A:** TAI
ochrothorax Nomura, 1975: 33 **A:** JA
oshimae Nomura, 1967b: 59 **A:** JA
perlata Sulzer, 1776: 67 (*Mordella*) **E:** AU BH BY CT CZ ES FE FR GE HU IT PL SK RO SK ST SZ UK **A:** FE GUA HEI JA KZ MG SHA WS
laticornis Stchegoleva-Barovskaya, 1927: 161 (*Tomoxia*)
multipunctata Trost, 1801: 27 (*Mordella*)
octopunctata Schrank von Paula, 1786: 80 (*Mordella*) [HN]
duodecimpunctata Rossi, 1790: 243 (*Mordella*)
sexpunctata Herbst, 1784c: 148 (*Mordella*)
pirika Kôno, 1935b: 125 **A:** JA SC
pseudoauromaculata Kiyoyama, 1993: 76 **A:** TAI
pseudohananomi Kiyoyama, 1993: 71 **A:** JA
splendens Miwa, 1933: 10 (*Mordella*) **A:** JA
transsylvanica Ermisch, 1977: 176 **E:** RO
trichopalpis Nomura, 1975: 30 **A:** JA
ussuriensis Ermisch, 1969a: 102 **A:** FE

genus *Iberomorda* Méquignon, 1946: 59 type species *Mordella sulcicauda* Mulsant, 1856
sulcicauda Mulsant, 1856f: 354 (*Mordella*) **E:** IT SP
viridipennis Mulsant, 1856f: 355 (*Mordella*) **E:** SP

genus *Klapperichimorda* Ermisch, 1968a: 279 type species *Klapperichimorda quadrimaculata* Ermisch, 1968
quadrimaculata Ermisch, 1968a: 280 **A:** FUJ
lutevittata Fan & J. Yang, 1995: 99 **A:** ZHE

genus *Macrotomoxia* Pic, 1922i: 208 type species *Macrotomoxia castanea* Pic, 1928
Higehananomia Kôno, 1935b: 123 type species *Higehananomia palpalis* Kôno, 1935
castanea Pic, 1922i: 208 **A:** FE JA SE TAI
palpalis Kôno, 1935b: 123 (*Higehananomia*)
gardneri Blair, 1931a: 203 (*Calycina*) **A:** SD

genus *Mediimorda* Méquignon, 1946: 62 type species *Mordella bipunctata* Germar, 1827
angeliquae Leblanc, 2002: 318 **E:** FR IT

attalica Schilsky, 1895: 24 (*Mordella*) **E:** AL BU CR GR MC UK **A:** CY IN IS TR SA

basicornis Schilsky, 1895: 24 (*Mordella*)

obtecta Schilsky, 1895: 24 (*Mordella*)

batteni Plaza, 1986a: 267 **E:** SP

bipunctata Germar, 1827: 12 (*Mordella*) **E:** AL AU BU CR FR GE GR IT MC PT SP SZ TR UK **N:** AG MO TU **A:** CY IS SA TR

abbreviata Schilsky, 1899: G [HN] (*Mordella*)

castellanii Franciscolo, 1949c: 84 (*Mordella*)

decora Chevrolat, 1840a: 16 (*Mordella*)

perpicilata A. Costa, 1854: 15 (*Mordella*)

brusteli Leblanc, 2002: 319 **E:** FR SP

fallaciosa Stchegoleva-Barovskaya, 1930b: 750 (*Mordella*) **A:** AF IN KI TD

horaki Leblanc, 2007: 243 **N:** AG LB MO TU

maceki Horák, 1985: 23 **A:** IN IQ

serrahimai Leblanc, 2007: 245 **N:** AG MO TU

socotrensis Horák, Farkač & Nakládal, 2012: 254 **A:** YE (Suqutra)

genus *Mordella* Linnaeus, 1758: 420 type species *Mordella aculeata* Linnaeus, 1758

aculeata Linnaeus, 1758: 420 E: AL AU BE BH BU BY CR CT CZ DE FR GE GG GR HU IT LU MC PL PT RO SK SL SP ST SV SZ TR UK YU **A:** BEI ES FE GAN IN JA KZ NMO SC SHA SHG SY TAI TR WS

communis Matsumura, 1915: 111

aradasiana Patti, 1840: 291 **E:** IT

argenteoguttata Champion, 1922a: 75 **A:** UP

brachyura brachyura Mulsant, 1856f: 354 **E:** AB AL AU BH BU BY CR CT CZ FI FR GE GR GG HU IT MC PL PT RO SK SL SP ST SZ UK YU **A:** IN SC TR

brevicauda A. Costa, 1854: 15

brunneicornis Schilsky, 1895: 28

fleischeri Emery, 1876: 69

brachyura akage Nomura, 1958: 18 **A:** JA

brachyura fuscopurpurea Nomura, 1958: 18 **A:** JIL

brachyura obscurata Ermisch, 1968b: 18 **A:** FE MG

brancuccii Horák, 1995c: 168 **A:** NP SD

brevicornis Méquignon, 1946: 67 **N:** MO

cinerea Gebler, 1829a: 134 **A:** WS

cinereoplagiata Blair, 1931a: 203 **A:** UP

comptei Plaza, 1977: 173 **N:** CI

cuneiformis Ermisch, 1968a: 282 **A:** FUJ

curticornis Ermisch, 1968a: 286 **A:** FUJ

curvipalpis Stchegoleva-Barovskaya, 1930b: 751 **A:** IN PA

duplicata Schilsky, 1895: 27 **E:** AB CR TR **A:** CY IN JO KI KZ SY TR

korgei Ermisch, 1969b: 107

rhodiensis Ermisch, 1963c: 51

rufipalpis Schilsky, 1895: 27

funesta Faldermann, 1838: 108 **A:** UZ

gratiosa Plaza & Compte, 1980: 205 **E:** SP

himalayana Horák, 1995c: 170 **A:** NP SD

hoberlandti Horák, 1985: 19 **A:** IN

hofferi Horák, 1979: 104 **E:** BU

holomelaena holomelaena Apfelbeck, 1914: 614 **E:** AB AL AU BE BH BU CR CT CZ FR GB GE GG GR HU NL NT PL RO SK SL SP ST SZ UK YU **A:** KZ

holomelaena sibirica Apfelbeck, 1914: 615 **A:** ES FE MG

huetheri Ermisch, 1956: 278 **E:** AL AU BH BU CR CT CZ BH FR GE GG GR HU IT MC NT PL SP SK ST SV SZ **A:** KZ WS

inouei Nomura, 1958: 12 **A:** TAI

insidiosa Lucas, 1847: 384 **N:** AG

izraelsoni Plaza, 1977: 175 **N:** CI

kuantunensis Ermisch, 1968a: 283 **A:** FUJ

konpira Takakuwa, 1985a: 71 **A:** JA

latipalpis Nomura, 1958: 15 **A:** JA SC

legionensis Plaza & Compte, 1980: 202 **E:** SP

leucaspis bicoloripilosa Horák, 1985: 18 **E:** AB GR **A:** IN SY TR UZ

leucaspis leucaspis Küster, 1849a: 80 **E:** AL AU BH BU CR CZ FR GB GE GR FR HU IT MC PL RO SK SL SP SZ TR UK YU **A:** IS TR

adnexa Ermisch, 1969b: 104

persica Apfelbeck, 1914: 619

longicauda Roubal, 1921: 248 **E:** SL

meridionalis Méquignon, 1946: 66 **E:** AL AU FR GE HU IT MC SP SZ

micacea Pic, 1941c: 7 **A:** CH

mixta Fabricius, 1801b: 122 **A:** TAI

mongolica Ermisch, 1964b: 341 **A:** MG NMO

multilineata Champion, 1927: 51 **A:** NP SD UP

niveoscutellata Nakane & Nomura, 1950: 15 **A:** JA TAI

ochrotrichia Nomura, 1958: 8 **A:** TAI

onaga Nomura, 1958: 10 **A:** JA

oxyptera Nomura, 1958: 10 **A.** JA SC

quadriguttulata Motschulsky, 1872: 46 **A:** WS

palmae Emery, 1876: 68 **E:** FR IT SP

plagiata Mannerheim, 1849: 240 **A:** ES

purpurascens Apfelbeck, 1914: 620 **E:** AL AU BH BU CR FR GR HU IT MC PT SP TR YU

pygidialis Apfelbeck, 1914: 612 **E:** AL AU BH GE SZ UK

quelpaertensis Pic, 1941c: 5 **A:** SC
sasajii Takakuwa, 2001: 293 **A:** JA
shirozui Nomura, 1967a: 14 **A:** TAI
splendidula Faldermann, 1838: 110 **A:** UZ
strigipennis Faldermann, 1838: 109 **A:** UZ
tadjikistanica Odnosum, 2002: 21 **A:** TD UZ
tenuicauda Nomura, 1958: 15 **A:** TAI
tokejii Nomura, 1958: 13 **A:** JA JIL SC
tunisiaca Ermisch, 1969b: 105 **N:** AG TU
velutina panonica Horák, 1985: 22 **E:** CZ HU SK
velutina velutina Emery, 1876: 69 **E:** AB AL AR IT BH BU CR GR HU IT KI ME NT RO ST UK YU **A:** IN KZ TR UZ
vestita Emery, 1876: 69 **E:** AL AR BH BU CR FR GR IT MC ME SL UK YU
horvathi Apfelbeck, 1914: 616
viridescens A. Costa, 1854: 18 **E:** AU BH BY CR CT CZ FR GE GR HU IT PL UK
yami Nomura, 1967a: 13 **A:** TAI

genus *Mordellaria* Ermisch, 1950b: 69 type species *Mordella scripta* Fairmaire & Germain, 1863
arakii Nakane & Nomura, 1950: 13 (*Mordella*) **A:** JA
aurata Kôno, 1928a: 36 (*Mordella*) **A:** FE JA
aurofasciata Comolli, 1837: 29 (*Mordella*) **E:** AU CR CT CZ GE GG HU IT NT PL RO SK ST YU **N:** AG **A:** IN
conjuncta Schilsky, 1895: 23 (*Mordella*)
sacheri I. Frivaldszky, 1865: 202 (*Mordella*)
vittata Gemminger, 1851: 52 (*Mordella*)
hananoi Nakane & Nomura, 1950: 12 (*Mordella*) **A:** JA TAI
humeralis Nomura, 1961b: 76 **A:** JA
latior Nomura, 1967a: 9 **A:** JA TAI
kanoi Kôno, 1932b: 153 (*Mordella*) **A:** TAI
triguttata Nomura, 1961b: 75 **A:** JA
zenchii Tokeji, 1953: 4 **A:** JA

genus *Neocurtimorda* Franciscolo, 1949a: 1 type species *Neocurtimorda convexa* Franciscolo, 1949
hoanensis Pic, 1941c: 4 (*Mordella*) **A:** NP SD
touzalini Pic, 1941c: 4 (*Mordella*) **A:** FUJ HUN JIX NP SD YUN

genus *Paratomoxia* Ermisch, 1950: 5 type species *Tomoxia pulchella* Ermisch, 1948
subgenus *Metatomoxia* M. Chûjô, 1957a: 1 type species *Tomoxia scutellata* Kôno, 1928
crux Kôno, 1928b: 148 (*Tomoxia*) **A:** JA
nipponica Kôno, 1928b: 148 (*Tomoxia*) **A:** FE JA
scutellata Kôno, 1928b: 147 (*Tomoxia*) **A:** FE JA

genus *Paratomoxioda* Ermisch, 1954a: 173 type species *Paratomoxioda leopardina* Ermisch, 1954
evanescens Normand, 1950a: 82 (*Mordella*) **N:** AG MO TU **AFR**
fallaciosa Stchegoleva-Barovskaya, 1930b: 750 (*Mordella*) **A:** AF IN KI PA TD
harteni Horák, 2009a: 207 **A:** AE

genus *Stenomordella* Ermisch, 1941a: 115 type species *Stenomordella longeantennalis* Ermisch, 1941
longeantennalis Ermisch, 1941a: 116 **A:** FUJ HUB
ochii Kiyoyama, 1975: 46 **A:** JA JIX

genus *Tomoxia* A. Costa, 1854: 8 type species *Tomoxia bucephala* A. Costa, 1854
bucephala bucephala A. Costa, 1854: 8 **E:** AU BH BU BY CR CT CZ DE GB GE FI FR HU IT LU NL NR NT PL PT RO SL SK SP ST SV SZ TR UK YU **N:** AG **A:** FE KZ WS
fasciata Paykull, 1800: 455 (*Mordella*)
sericea Mulsant, 1856f: 324
formosana M. Chûjô, 1935a: 75 **A:** JA TAI
relicta Takakuwa, 1985: 8 **A:** JA
ryukyuana Takakuwa, 1985a: 69 **A:** JA
similaris Nomura, 1967b: 57 **A:** JA

genus *Tomoxioda* Ermisch, 1950: 52 type species *Tomoxioda auropubescens* Ermisch, 1950
truncatoptera Nomura, 1958: 8 (*Mordella*) **A:** CH JA TAI

genus *Variimorda* Méquignon, 1946: 62 type species *Mordella villosa* Schrank von Paula, 1781
Sulcatimorda Méquignon, 1946: 63 type species *Mordella ragusai* Emery, 1876
subgenus *Variimorda* Méquignon, 1946: 62 type species *Mordella villosa* Schrank von Paula, 1781
ainu Nomura, 1958: 6 (*Mordella*) **A:** JA
argyropleura Franciscolo, 1942a: 8 (*Mordella*) **E:** IT
basalis A. Costa, 1854: 13 (*Mordella*) **E:** AL AU BH BU BY CR CT CZ FR GR HU IT PL SK SL SP ST SZ TR UK
pseudobrachyura Franciscolo, 1949c: 81 (*Mordella*)
bistrinotata Pic, 1928: 12 (*Mordella*) **A:** NP
briantea Comolli, 1837: 28 (*Mordella*) **E:** AU BH BU BY CR CT CZ EN FR GE GR HU IT MD PL SK SL SP ST SZ UK **N:** AG **A:** TR
flavimana Marseul, 1877: 472 (*Mordella*) **A:** FUJ GUA GUI HUB JA SCH TAI SHA
gerardi Horák, 2009b: 69 **A:** SD
holzschuhi Horák, 1985: 13 **A:** IN IQ TR
ihai boninensis Nomura, 1975: 34 **A:** JA
ihai ihai M. Chûjô, 1959: 9 **A:** JA
inomatai Takakuwa, 1985: 10 **A:** JA
ishiharai Kiyoyama, 1994: 193 **A:** TAI
kurosawai Takakuwa, 2001a: 310 **A:** JA

kuyrasha Nomura, 1962b: 41 **A**: JA
mendax Méquignon, 1946: 71 **E**: AB AL AU BH BU CR CT CZ FR GE GG GR HU IT PL SK SZ
chobauti Méquignon, 1946: 71
devillei Méquignon, 1946: 71
miyarabi miyarabi Nomura, 1962b: 40 **A**: JA
miyarabi chujoi Tsuru & Takakuwa, 2005: 80 **A**: JA
miyarabi michioi Tsuru & Takakuwa, 2005: 81 **A**: JA
persica Horák, 1985: 8 **A**: IN
quomoi Franciscolo, 1942a: 7 (*Mordella*) **E**: IT
ragusai Emery, 1876: 60 (*Mordella*) **E**: FR IT **N**: MO
nigrata Méquignon, 1946: 73 (*Mordella*)
obsoleta Baudi di Selve, 1878b: 345 (*Mordella*)
oertzeni Schilsky, 1895: 26 (*Mordella*)
sinensis Pic, 1917f: 13 (*Mordella*) **A**: SD YUN
takakuwai Horák, 2009b: 73 **A**: HEB SHA
truncatopyga Pic, 1938e: 16 (*Mordella*) **A**: SE TAI
taiwana Nakane & Nomura, 1950: 14 (*Mordella*)
villosa Schrank von Paula, 1781: 228 (*Mordella*) **E**: AB AL AR AU BE BH BU BY CR CT CZ DE FR GB GE GG GR HU IT LA MC MD NL PL RO SK SL SP ST SZ TR UK YU **A**: FE IN IQ JA KZ TR
conjucta Stchegoleva-Barovskaya, 1931: 56 (*Mordella*)
coronata A. Costa, 1854: 11 (*Mordella*)
fasciata Fabricius, 1775: 263 (*Mordella*) [HN]
fasciolata Rossi, 1792: 103 (*Mordella*)
habelmanni Emery, 1876: 66 (*Mordella*)
interrupta A. Costa, 1854: 13 (*Mordella*)
nigricornis Stchegoleva-Barovskaya, 1931: 57 (*Mordella*)
nigrosuturalis Stchegoleva-Barovskaya, 1931: 56 (*Mordella*)
ruficornis Stchegoleva-Barovskaya, 1931: 57 (*Mordella*)
seriatoguttata Mulsant, 1856f: 345 (*Mordella*)
subbasalis Stchegoleva-Barovskaya, 1931: 56 (*Mordella*)
subcoeca Mulsant, 1856f: 345 (*Mordella*)
subgenus *Galeimorda* Horák, 1985: 15 type species *Mordella fagniezi* Méquignon, 1946
caprai Franciscolo, 1951: 7 (*Mordella*) **E**: AL
fagniezi Méquignon, 1946: 74 (*Mordella*) **E**: FR PT SP
therondi Méquignon, 1946: 74 (*Mordella*)
hladili Horák, 1985: 15 **E**: BU GR MC ST TR UK **A**: KZ
krikkeni Batten, 1977: 167 **E**: CR GR
theryi Méquignon, 1946: 74 (*Mordella*) **E**: PT SP **N**: AG MO
maroccana Méquignon, 1946: 75 (*Mordella*)
nigrescens Méquignon, 1946: 75 (*Mordella*)

genus *Yukahananomia* Kôno, 1935b: 124 type species *Tomoxia yakui* Kôno, 1930
tsuyukii Takakuwa, 1978: 5 **A:** JA
uenoi Takakuwa, 1995: 433 **A:** TAI
yakui Kôno, 1930c: 104 (*Tomoxia*) **A:** CH FE JA

tribe Mordellistenini Ermisch, 1941

genus *Asiatolida* Shiyake, 2000: 26 type species *Asiatolida miyatakei* Shiyake, 2000
melana Fan & J. Yang, 1995: 95 (*Mordellistenoda*) **A:** ZHE
miyatakei Shiyake, 2000: 27 **A:** JA

genus *Dellamora* Normand, 1916: 284 type species *Dellamora palposa* Normand, 1916
palposa Normand, 1916: 285 **E:** FR GR PT SP **N:** LB MO TU **A:** CY TM TR

genus *Ermischiella* Franciscolo, 1930: 2 type species *Ermischiella papuana* Franciscolo, 1950
bejceki Horák, Farkač & Nakládal, 2012: 264 **A:** YE (Suqutra)
castanea Boheman, 1858: 108 (*Mordella*) **A:** JA
  dodonaeae Montrouzier, 1860: 306 (*Mordella*)
chichijimana Nomura, 1975: 37 **A:** JA
fukiensis Ermisch, 1952b: 154 (*Glipostenoda*) **A:** FUJ
hahajimana Nomura, 1975: 38 **A:** JA
hasagawai Nomura, 1951: 69 (*Glipostenoda*) **A:** JA
nigriceps Nomura, 1975: 35 **A:** JA

genus *Falsomordellina* Nomura, 1966: 44 type species *Glipostenoda luteoloides* Nomura, 1961
amamiana Nomura, 1961b: 85 (*Glipostenoda*) **A:** JA
luteoloides Nomura, 1961b: 83 (*Glipostenoda*) **A:** FUJ JA TAI ZHE
  luteola Nomura, 1951: 67 (*Glipostenopa*) [HN]
  luteora Kôno, 1932b: 155 (*Mordellistena*)
  luteorubra Ermisch, 1965c: 201 (*Mordellistena*)
nigripennis Nomura, 1967a: 31 **A:** TAI
ohsumiana Nakane, 1957: 51 (*Glipostenoda*) **A:** JA
takaosana madara Nomura, 1966: 45 **A:** JA
takaosana takaosana Kôno, 1932: 154 (*Mordellistena*) **A:** JA ZHE
vagevittata Nakane, 1957: 50 (*Falsomordellistena*) **A:** JA

genus *Falsomordellistena* Ermisch, 1941c: 724 type species *Mordellistena formosana* Pic, 1911
alpigena Tokeji, 1953: 15 **A:** JA
altestrigata Marseul, 1877: 474 (*Mordellistena*) **A:** HUB HUN JA JIX SE SHA TAI YUN
  auriguttata Nomura, 1951: 64
  sauteri Pic, 1926i: 67 (*Mordellistena*)
aurofasciata Nakane, 1949: 40 (*Mordellistena*) **A:** JA

auromaculata Kôno, 1928a: 38 (*Mordellistena*) **A:** JA SC
aurosuturalis Nomura, 1967a: 25 **A:** TAI
ayahime Nomura, 1967a: 21 **A:** TAI
baishanzuna Fan, 1995b: 223 **A:** ZHE
formosana boninensis Nomura, 1975: 40 **A:** JA
formosana formosana Pic, 1911q: 190 (*Mordellistena*) **A:** TAI
chrysotrichia Nomura, 1951: 49 (*Mordellistena*) **A:** JA SC
hananoi Nomura, 1951: 48 (*Mordellistena*) **A:** JA SC
hime Kôno, 1928a: 41 (*Mordellistena*) **A:** TAI
altestrigata Ermisch, 1950b: 85 [HN]
hiranoi hiranoi Nomura, 1961b: 83 **A:** JA
hiranoi takasagoana Nomura, 1967a: 26 **A:** TAI
hirasana Shiyake, 1996: 13 **A:** JA
humerosignata Nomura, 1967a: 22 **A:** TAI
inouei Nomura, 1951: 50 (*Mordellistena*) **A:** TAI
arisana Nomura, 1951: 53
capillata Nomura, 1951: 54
katoi Nomura, 1961b: 81 **A:** JA
konoi konoi Pic, 1936h: 300 (*Mordellistena*) [RN] **A:** TAI
horni Kôno, 1934a: 118 [HN]
konoi kotoensis Nomura, 1967a: 23 **A:** TAI
konoi yakushimaensis Nomura, 1951: 65 **A:** JA
kuri Nomura, 1951: 51 (*Mordellistena*) **A:** JA
loochooana Nomura, 1964: 51 **A:** JA
okamotoi Kôno, 1935b: 127 (*Mordellistena*) **A:** JA
parca Tokeji, 1953: 14 **A:** JA
pseudoalpigena Nomura, 1975: 42 **A:** JA
satoi Nomura, 1951: 51 (*Mordellistena*) **A:** JA
rosseoloides Nomura, 1975: 41 **A:** JA
shinanoensis Tokeji, 1953: 18 **A:** JA
superfusca Nomura, 1951: 48 (*Mordellistena*) **A:** JA
tokarana Nakane, 1956a: 9 **A:** JA
watanabei Nomura, 1975: 43 **A:** JA
wui Fan & J. Yang, 1995: 96 **A:** ZHE
yoshidai Nomura, 1951: 54 (*Mordellistena*) **A:** JA

genus *Glipostena* Ermisch, 1941c: 723 type species *Mordellistena pelecotomoidea* Pic, 1911
pelecotomoidea Pic, 1911p: 188 (*Mordellistena*) **A:** JA SE TAI

genus *Glipostenoda* Ermisch, 1950b: 81 type species *Glipostenoda castaneicolor* Ermisch, 1950
excisa Nomura, 1967b: 59 **A:** JA

fukiensis Ermisch, 1952b: 154 **A:** FUJ
fusciceps Nomura, 1967b: 61 **A:** JA
higashinoi Nomura, 1967b: 67 **A:** JA TAI
kaihuana Fan & J. Yang, 1995: 98 **A:** ZHE
kawasakii Nomura, 1951: 27 (*Mordellistena*) **A:** TAI
kimotoi M. Chûjô, 1957a: 17 **A:** JA
klapperichi Ermisch, 1952b: 153 **A:** FUJ
lineatisuturalis Nomura, 1967a: 26 **A:** TAI
matsumurai Kôno, 1932b: 154 (*Mordellistena*) **A:** FUJ JA TAI
melanocephala Ermisch, 1952b: 156 **A:** FUJ
pseudoexcisa Nomura, 1975: 39 **A:** JA
rarasana Nomura, 1951: 69 **A:** JA TAI
retusa Nomura, 1967b: 60 **A:** JA
rimogana Nomura, 1967a: 30 **A:** TAI
rosseola Marseul, 1877: 473 (*Mordellistena*) **A:** JA TAI SE
  hisamatsui M. Chûjô, 1956: 175
  phengotrichia Nomura, 1951: 66
rufobrunnea Champion, 1927a: 52 (*Mordellistena*) **A:** UP
sasajii Schiyake, 2001: 333 **A:** JA
shibatai Nomura, 1961b: 84 **A:** JA
shizuokana Kôno, 1935b: 128 (*Mordellistena*) **A:** FUJ JA TAI
taiwana Kôno, 1934a: 117 (*Mordellistena*) **A:** TAI
takashii Nomura, 1967a: 29 **A:** TAI
trichophora Nomura, 1951: 55 (*Mordellistena*) **A:** HEI JA LIA TAI

genus *Mordellina* Schilsky, 1908b: 137 type species *Mordellina gracilis* Schilsky, 1908
subgenus *Mordellina* Schilsky, 1908b: 137 type species *Mordellina gracilis* Schilsky, 1908
atrofusca atrofusca Nomura, 1951: 60 (*Mordellistena*) **A:** JA TAI
atrofusca signatelloides Nakane, 1956a: 7 (*Mordellistena*) **A:** JA
atrofusca reductemaculata Nakane, 1956a: 8 (*Mordellistena*) **A:** JA
brunneotincta Marseul, 1877: 474 (*Mordellistena*) **A:** JA ZHE
chibi Kôno, 1932b: 156 (*Mordellistena*) **A:** JA
gutianshana Fan & J. Yang, 1995: 95 **A:** ZHE
hidakai Nomura, 1963b: 41 (*Pseudomordellistena*) **A:** JA
hirayamai Kôno, 1933: 29 (*Mordellistena*) **A:** JA
koikei Tokeji, 1953: 8 (*Mordellistena*) **A:** JA
kubotai Nomura, 1951: 57 (*Mordellistena*) **A:** JA SC
longula Kôno, 1928a: 39 (*Mordellistena*) **A:** JA
marginalis Nomura, 1967a: 20 **A:** TAI
mentiens Tokeji, 1953: 10 (*Mordellistena*) **A:** JA
mikado Tokeji, 1953: 8 (*Mordellistena*) **A:** JA

nigrofusca Nomura, 1951: 58 (*Mordellistena*) **A:** JA
otohime Nomura & Katô, 1957: 84 (*Mordellistena*) **A:** JA
palleola Nomura, 1966: 43 **A:** JA
palliata Kôno, 1932b: 156 (*Mordellistena*) **A:** JA
quinquenotata Champion, 1927a: 53 (*Mordellistena*) **A:** NP UP
shimoyamai M. Chûjô, 1957a: 6 (*Mordellistena*) **A:** JA
signatella Marseul, 1877: 473 (*Mordellistena*) **A:** JA TAI
nigrofasciata M. Chûjô, 1935a: 74 (*Mordellistena*)
uruma Nomura, 1966: 42 **A:** JA
watanabei T. Tsuru, 2002: 351 **A:** JA
yamamotoi Nomura, 1951: 56 (*Mordellistena*) **A:** JA
yezoensis M. Chûjô, 1957b: 21 (*Mordellistena*) **A:** JA
subgenus *Pseudomordellistena* Ermisch, 1952b: 145 type species *Pseudomordellistena rufobrunnea* Ermisch, 1952
amamiensis Nomura, 1951: 60 (*Mordellistena*) **A:** JA TAI
callichroa callichroa Tokeji, 1953: 11 (*Mordellistena*) **A:** JA
callichroa ami Nomura, 1967a: 18 **A:** TAI
curteapicalis Pic, 1926i: 67 (*Mordellistena*) **A:** TAI
gina Nomura, 1967a: 15 **A:** TAI
janae Horák, Farkač & Nakládal, 2012: 257 **A:** YE (Suqutra)
kaguyahime Nomura & Katô, 1957: 85 (*Mordellistena*) **A:** JA TAI
macrophthalma Ermisch, 1952b: 147 (*Pseudomordellistena*) **A:** FUJ
paiwana Nomura, 1967a: 17 **A:** TAI
pilosovittata Nakane, 1956a: 5 (*Mordellistena*) **A:** JA TAI
pseudohumeralis Nakane, 1956a: 6 (*Mordellistena*) **A:** JA
purcharti Horák, Farkač & Nakládal, 2012: 259 **A:** YE (Suqutra)
rufobrunnea Ermisch, 1952b: 145 (*Pseudomordellistena*) **A:** FUJ
rufohumeralis Nomura, 1967a: 16 **A:** TAI
stastnyi Horák, Farkač & Nakládal, 2012: 261 **A:** YE (Suqutra)
tsutsuii Nakane, 1956a: 4 (*Mordellistena*) **A:** JA TAI
vidua Nakane, 1956a: 7 (*Mordellistena*) **A:** JA

genus *Mordellistena* A. Costa, 1854: 16 type species *Mordellistena confinis* Costa, 1854
subgenus *Mordellokoides* Franciscolo, 1942d: 133 type species *Mordellokoides grandii* Franciscolo, 1942
grandii Franciscolo, 1942d: 134 **E:** IT
subgenus *Mordellistena* A. Costa, 1854: 16 type species *Mordella pumila* Gyllenhal, 1810
Natirrica A. Costa, 1854: 19 type species *Natirrica meridionalis* A. Costa, 1854
aegea Franciscolo, 1949c: 90 **E:** GR
aequalica Ermisch, 1977: 156 **E:** HU
aertsi Ermisch, 1963a: 22 **E:** FR GE

ainonia M. Chûjô, 1957b: 18 **A:** JA
albocapillata Ermisch, 1965a: 70 **A:** MG
algeriensis Ermisch, 1966b: 42 **N:** AG TU
alpicola Ermisch, 1963a: 28 **E:** IT SZ
altestriatoides Horák, 1995b: 171 **A:** NP
altifrons Stchegoleva-Barovskaya, 1927: 159 **E:** ST
amurensis Horák, 1982: 48 **A:** FE
angustula Ermisch, 1977: 168 **E:** HU
anomala Ermisch, 1968b: 18 **A:** MG
apicerufa Ermisch, 1977: 158 **E:** HU
arabissa Franciscolo, 1956c: 467 **A:** YE
atrocincta lincenti Pic, 1938e: 16 **A:** GAN
atrogemellata Ermisch, 1965d: 267 **E:** BU GR MC
aureolopilosa Stchegoleva-Barovskaya, 1932: 222 **A:** KI
aureomicans Ermisch, 1965b: 70 **N:** CI
aureotomentosa Ermisch, 1966b: 40 **N:** MG
austriaca Schilsky, 1899: 77b **E:** AL AU BH CR CZ FR GE GR HU IT PL RO SK SZ UK **A:** KZ TR
micantoides Ermisch, 1954b: 176
austriacensis Ermisch, 1956: 285 **E:** AU CZ HU SK
balcanica Ermisch, 1967b: 110 **E:** BU MC
balearica Compte, 1985: 63 **E:** SP
baliani Franciscolo, 1942c: 77 **E:** IT
bavarica Ermisch, 1963a: 11 **E:** GE
batteni Horák, 1980: 281 **A:** IN KI TD
berbera Horák, 1983: 8 **N:** AG
bicoloripilosa Ermisch, 1967b: 113 **E:** AU BE BU BY CT CZ FR GE GG HU PL RO SK UK **A:** KZ TM UZ WS
biroi Ermisch, 1977: 166 **E:** HU
bodemeyeri Ermisch, 1956: 283 **A:** TR
boldi Ermisch, 1965a: 68 **A:** MG
bolognai Horák, 1990: 136 **A:** LE IN IQ TR SY
borogolensis Ermisch, 1964b: 348 **A:** MG
breddini Ermisch, 1963a: 26 **E:** GE PL
brevicauda Bohemann, 1849: 223 (*Mordella*) **E:** AB AR AU BH BU BY CR CZ FR GB GE GG GR HU IT MC PL PT RO SK SL SP ST SZ TR YU **A:** IN IS JO KI KZ LE SY TD TM TR UZ
brevicornis Schilsky, 1895: 45
obtusa C. Brisout de Barneville, 1859: 233
subtruncata Mulsant, 1856f: 377
brevilineata Nomura, 1961b: 77 **A:** JA
brunneispinosa Ermisch, 1963: 6 **E:** BY GE HU PL TR UK
bulgarica Ermisch, 1977: 166 **E:** BU

canariensis Ermisch, 1965b: 73 **N:** CI
caprai Franciscolo, 1942a: 8 **E:** IT
carinthiaca Ermisch, 1966a: 33 **E:** GE HU
charagolensis Ermisch, 1964b: 343 **A:** MG
chujoi Hatayama, 1985: 389 **A:** JA
  shirozui M. Chûjô, 1957b: 16 (*Falsomordellistena*) [HN]
columbretensis Compte, 1970: 110 **E:** SP
comes Marseul, 1877: 473 **A:** JA SC
confinis A. Costa, 1854: 18 **E:** CR FR IT SL SP SZ **N:** AG MO
  africana Roubal, 1911b: 272
connata Ermisch, 1969b: 108 **E:** AL AU BY CZ FI FR HU NT PL SK SV UK
consonbrina Ermisch, 1977: 157 **E:** HU SZ
cuneigera Champion, 1922a: 76 **A:** NP UP
curtelineata Pic, 1927k: 153 **A:** CH
curticornis Ermisch, 1977: 164 **E:** HU
cymbalistria Peyerimhoff, 1925: 8 **N:** TU
cypria Ermisch, 1963c: 62 **A:** CY IS
dalmatica Ermisch, 1956: 289 **E:** AU BH BU CR CZ GG GR HU IT MC RO SK UK **A:** IN TD TM TR
  mostarensis Ermisch, 1977: 171
daurica Motschulsky, 1860c: 141 **A:** WS
dentata Batten, 1978: 79 **E:** GR
desertorum Horák, 2009a: 213 **A:** AE
dieckmanni Ermisch, 1963a: 26 **E:** AU BU CR CZ FI FR GE GR HU PL SL SK SZ
difficilis Ermisch, 1963c: 60 **A:** CY
dives Emery, 1876: 95 **E:** AR CT GG HU ST
dlabolai Ermisch, 1967d: 126 **A:** MG
dolini Odnosum, 2005: 53 **E:** AB
dvoraki Ermisch, 1956: 286 **E:** BU CZ FR GE HU MC PL SK TR
edashigei M. Chûjô, 1956: 174 **A:** JA
  tokaraensis Nakane, 1956a: 1
echingolensis Ermisch, 1969c: 172 **A:** MG
elbrusicola Ermisch, 1969b: 111 **E:** ST
eludens Allen, 1999: 205 **E:** GB
emeryi Schilsky, 1895: 53 **E:** AL CR FR GR IT SP **N:** AG MO
episternalis Mulsant, 1856f: 371 **E:** AU BH BU CR FR GE GR HU IT PT SK SP SZ **N:** AG MO **A** IN IS SY TM
  extensa Rosenhauer, 1856: 227 (*Mordella*)
episternaloides Ermisch, 1963a: 7 **E:** FR GE
erdoesi Ermisch, 1977: 158 **E:** HU UK
ermischi Compte, 1966: 252 **E:** SP

eversi Ermisch, 1965b: 71 **N:** CI
exisa Stchegoleva-Barovskaya, 1927: 158 **A:** IN TD TM
exclusa Ermisch, 1977: 155 **E:** HU
fageli Ermisch, 1969b: 112 **E:** FR PT SP **N:** AG MO
falsoparvula Ermisch, 1956: 281 **E:** AU BU BY CZ EN FR GE GG HU NL IT PL PT SL SZ UK YU
falsoparvuliformis Ermisch, 1963a: 4 **E:** AL AU BY CZ GE GG HU RO SK ST UK **A:** IN TD TM TR
feigei Ermisch, 1956: 287 **E:** AU CR GE PL
fernandezi Palm, 1976: 31 **N:** CI
festiva Ermisch, 1977: 154 **E:** HU
flavella Ermisch, 1963c: 58 **A:** CY
flavospinosa Hubenthal, 1911: 186 **E:** AU GE
flavospinulosa Ermisch, 1977: 163 **E:** HU
fujiyamai Nomura, 1951: 57 **A:** JA
fulvipennis Stchegoleva-Barovskaya, 1932: 219 **A:** KZ
fusciapicalis Nomura, 1961b: 79 **A:** JA
fuscogemellata Ermisch, 1963b: 17 **E:** FR IT NL PT SP **N:** AG MO TU
fuscogemellatoides Ermisch, 1977: 165 **E:** BU GR HU **A:** KI KZ TM
fuscosuturalis Nakane, 1961: 78 **A:** JA
gallagheri Horák, 2009a: 210 **A:** AE
gallica Ermisch, 1956: 283 **E:** FR
gardneri Blair, 1930b: 180 **A:** BT NP SD
gemellata Schilsky, 1899: 79 **E:** EN GR PT SP **A:** CY
geronensis Ermisch, 1977: 169 **E:** FR SP
gfelleri Horák, 1990: 130 **E:** FR SP
ghanii Franciscolo, 1974: 71 **A:** PA
gianassoi Horák, 1992: 137 **N:** MO
goetzi Ermisch, 1969b: 114 **E:** GG ST
grisea Mulsant, 1856f: 376 **E:** CR FR SP
gutianshana Fan & C. Yang, 1995: 95 **A:** ZHE
hatorii Tokeji, 1953: 7 **A:** JA
helvetica Ermisch, 1967b: 110 **E:** SZ
hirtipes Schilsky, 1895: 46 **E:** CT FR GR MC TR UK **A:** CY IN IS JO LE TM TR SY
hollandica Ermisch, 1966a: 36 **E:** NL
hoberlandti Horák, 1983: 3 **E:** AR GG **A:** IN IQ IS SY TM TR UZ
horaki Piňo, 1985: 103 [RN] **A:** IN IQ KI TM
    bicolor Horák, 1982: 46 [HN]
horioni Ermisch, 1956: 285 **E:** AU CZ GE HU SK SZ
horvathi Ermisch, 1977: 173 **E:** BU CZ HU MC SK
humeralis Fabricius, 1758: 420 (*Mordella*) **E:** AB AU BY CR CT CZ ES FE GB GE GG HU IT MC NL PL SL SK SZ YU **A:** FE JA MG
    nigricollis Schilsky, 1895: 35

humeronotata Champion, 1922a: 76 **A:** NP SP UP
humeropicta Ermisch, 1963c: 63 **E:** AB AR **N:** AG **A:** CY IN TR
hungarica Ermisch, 1977: 163 **E:** CR HU
hypopygialis Ermisch, 1968b: 20 **A:** MO
inaequalis Mulsant, 1856f: 365 **E:** FR
incerta Ermisch, 1965a: 72 **A:** MG
inexpectata Ermisch, 1967b: 111 **E:** AU BU CZ GE SK ST SZ
inornata Nomura, 1961b: 80 **A:** JA
insignata Ermisch, 1965c: 200 **A:** JA
  inornata Nomura, 1961b: 80 [HN]
intersecta Emery, 1876: 93 **E:** AB AR BU CR GE GG GR FR MC GG GR HU IT SP ST TR UK **A:** KZ LE SY TD TM TR
irritans Franciscolo, 1991: 168 **E:** IT
istrica Ermisch, 1977: 169 **E:** FR SL
jelineki Horák, 1980: 283 **A:** TD
kaszabi Ermisch, 1965a: 65 **A:** MG
kirai Nomura, 1951: 52 **A:** JA SC
kirghizica Odnosum, 2003: 47 **A:** KI TD
kochi Ermisch, 1956: 284 **E:** SP **A:** MO
kolleri Ermisch, 1956: 289 **E:** AU BH BU CZ DE GE HU MC PL SK SP SV SZ YU
koma Nomura, 1951: 56 **A:** SC
korschefskyana Ermisch, 1963a: 5 [RN] **E:** GE
  korschefskyi Ermisch, 1956: 283 [HN]
kotenkoi Odnosum, 1990: 54 **E:** UK
kraatzi kraatzi Emery, 1876: 91 **E:** AB AL AR BH BU CR CT CZ FR GE GG GR HU PL PT RO SK ST SZ UK **A:** CY IN IQ KI KZ TD TM TR
kraatzi kretica Czetó, 1990: 28 **E:** GR
krauseri Plaza, 1986b: 286 **E:** SP
krujanensis Ermisch, 1963a: 14 **E:** AL BH
kuro Nomura, 1951: 51 **A:** JA
kurosai M. Chûjô, 1957b: 20 **A:** JA
labialis Motschulsky, 1860c: 142 **A:** WS
latitarsis Batten, 1983a: 88 **A:** TR
lenensis Ermisch, 1977: 173 **A:** ES
lesbia Ermisch, 1965d: 266 **E:** GR
leveyi Batten, 1989: 105 **A:** KA
lichtneckerti Ermisch, 1977: 160 **E:** AU HU RO SK TR
lindbergi Ermisch, 1963c: 59 **E:** BU GR TR **A:** CY TR
lonai Franciscolo, 1949c: 92 **E:** AL MC YU
longicornis Mulsant, 1856f: 373 **E:** GR FR
longicornoides Ermisch, 1965d: 258 **E:** BU SP TR

longipalpis Emery, 1891: 39 **N:** TU
aypassai Batten, 1978: 78
lusitanica Ermisch, 1963b: 16 **E:** PT
luteipalpis Schilsky, 1895: 51 **E:** AU BU CZ GE GG PL SK
perparvula Ermisch, 1966a: 30
maedai Nakane, 1956a: 2 **A:** JA
majuscula Ermisch, 1977: 171 **E:** IT
manteroi Franciscolo, 1942c: 76 **E:** IT
maroccana Ermisch, 1966b: 42 **N:** MO TU
mazedonica Ermisch, 1965d: 262 **E:** BU FR GR MC
mediogemellata Ermisch, 1977: 163 **E:** HU TR
meuseli Ermisch, 1956: 283 **E:** AU CZ FR GE HU SK UK **A:** KZ UZ
minutuloides Ermisch, 1966a: 31
micans Germar, 1817: 212 (*Mordella*) **E:** AU BH CR CT CZ HU IT SL SP TR **A:** IS
infima Emery, 1876: 117
microgemellata Ermisch, 1965d: 266 **E:** BU GR HU **A:** CY
michalki Ermisch, 1956: 288 **E:** AU BH CT CZ GE MC SK **A:** KZ TD
mihoki Ermisch, 1977: 158 **E:** CR GE HU RO UK
minima A. Costa, 1854: 18 **E:** AB BU CR FR GR HU MC IT SK SP TR YU **A:** CY IS KI TR
pseudorhenana Ermisch, 1977: 164
minutula Ermisch, 1956: 285 **E:** AU HU
miyamotoi Nakane, 1956a: 3 **A:** JA TAI
mongolica Ermisch, 1964b: 344 **A:** MG
muchei Ermisch, 1965a: 75 **A:** MG
multicicatrix Kangas, 1986: 139 **E:** FI
neglecta Ermisch, 1977: 156 **E:** HU
nepalensis Horák, 1995b: 168 **A:** NP
nessebaricus Batten, 1980: 43 **E:** BU
neuwaldeggiana Panzer, 1796d: 8 (*Mordella*) **E:** AU BE BH BU BY CR CZ GB GR FR HU IT LU NL PL SK SL SP SV SZ UK YU **A:** IN
brunnea Fabricius, 1801b: 125 (*Mordella*)
ferruginea Marsham, 1802: 490 (*Mordella*)
meridionalis A. Costa, 1854: 19 (*Nattirica*)
nigritarsis Horák, 1996: 172 **E:** AU CZ GE GG RO SK SL
nigrogemellata Ermisch, 1964b: 343 **A:** MG
nomurai Tokeji, 1953: 6 **A:** JA
obscurosuturalis Ermisch, 1964b: 345 **A:** MG
olympica Ermisch, 1965d: 268 **E:** BU GR MC YU **A:** TR
okinawana Nomura, 1963b: 40 **A:** JA
oranensis Pic, 1900l: 124 **E:** IT (Lampione Is.) **N:** AG
ozeana Nakane, 1957: 58 (*Pseudomordellina*) **A:** JA

parabrevicauda Ermisch, 1965a: 71 **A:** IN KI KZ MG UZ
paradohumeralis Ermisch, 1963c: 58 **A:** CY TR
paraepisternalis Ermisch, 1965d: 260 **E:** BU CR GR RO SP **A:** IQ IS SP UZ TR
paraintersecta Ermisch, 1956: 284 **E:** AR GR HU ST **A:** IN IQ SY TR
paraobscurosuturalis Ermisch, 1965a: 66 **A:** FE MG
parapentas Ermisch, 1977: 171 **E:** IT
pararhenana Ermisch, 1977: 164 **E:** CR HU
paraweisei Ermisch, 1977: 160 **E:** RO
parvicauda Ermisch, 1967b: 116 **E:** AB AL BH BU CR CT GG GR MC RO TR **A:** TR
parvula Gyllenhal, 1827: 519 (*Mordella*) **E:** AB AU BH BU BY CT CZ GB HU IT MC NL PL RO SK SL ST SZ **A:** HUB GAN IN KZ NMO
liliputana Mulsant, 1856f: 374
pusilla L. Redtenbacher, 1849: 614 (*Mordella*)
rectangula Roubal, 1911b: 272 [HN]
troglodytes Mannerheim, 1844a: 198 (*Mordella*)
parvuliformis Stchegoleva-Barovskaya, 1930: 57 **E:** AB BY CZ GE HU SK ST UK **A:** IN UZ WS
paucistrigosa Ermisch, 1965a: 64 **A:** MG
peloponnesensis Batten, 1980: 42 **E:** GR **A:** CY
pentas Mulsant, 1856f: 380 **E:** AL AU CT CZ FR GG GR HU IT MC NL PL PT RO SK SP SZ UK YU **N:** AG MO **A:** CY
perroudi Mulsant, 1856f: 382 **E:** AU CR FR GR HU IT MC PL SK SL SP TR **A:** IS
persica Horák, 1983: 7 **A:** IN
permira Franciscolo, 1949c: 91 **E:** GR
phaea Nomura, 1951: 46 **A:** JA
podlussanyi Czetó, 1990: 26 **E:** GR
pontica Ermisch, 1977: 169 **E:** HU
porphyria Nomura, 1951: 55 **A:** TAI
psammophila Peyerimhoff, 1943: 16 **N:** AG
praesagita Kangas, 1988: 89 **E:** FI SV
pseudobrevicauda Ermisch, 1963a: 12 **E:** AB AR AU CR CZ BH BU EN FR GE GG HU MC PL RO SK SL SP ST TR **A:** IN KZ MG TD TM TR UZ
pseudohirtipes krotosensis Czetó, 1990: 28 **E:** GR
pseudohirtipes pseudohirtipes Ermisch, 1965d: 268 **E:** AB BU CR FR GG GR MC SP TR UK YU **N:** AG **A:** CY TR
pseudomicans Ermisch, 1977: 169 **E:** CR MC
pseudoparvula Ermisch, 1956: 281 **E:** AU BY CZ GB GE HU IT PL SZ
eludens Allen, 1999: 205
parvuloides Ermisch, 1956: 281
pseudopumila Ermisch, 1963a: 28 **E:** BY DE FI FR GB GE NL NT PL PT SP **N:** MO **A:** FE KZ ST TD WS
pseudoreichei Ermisch, 1977: 158 **E:** IT

pseudorugipennis Ermisch, 1963c: 60 **E:** BU GR MC **A:** CY IN IQ IS JO LE SY TR
pseudotarsata Ermisch, 1964b: 345 **A:** FUJ JA MG NE
pumila Gyllenhal, 1810: 605 (*Mordella*) **E:** AL AR AU BE BH BU BY CR CT CZ FI FR GB GE GG GR HU IT LU MC NL PL RO SK SL SP ST SV SZ TR UK YU **N:** AG **A:** ES GAN HUB IN IS KZ MG NC NMO SC SY TAI TD TM TR
deficiens Mulsant, 1856f: 380
elongata Kraatz, 1868a: 295 (*Mordella*)
stricta A. Costa, 1854: 16
purpurascens A. Costa, 1854: 17 **E:** CR FR IT SL SP **N:** AG MO TU **A:** KZ TD TM WS
purpureonigrans Ermisch, 1963a: 24 **E:** AL AU BU BY CZ DE EN FI GE HU IT MC NL PL RO SK SL SV SZ **A:** CY TR
pygmaeola Ermisch, 1956: 288 **E:** AU BU CR CZ FR GB GE GG HU LU NL NR PL SK ST SV SZ **A:** TM
pyrenaea Ermisch, 1966a: 35 **E:** FR SP
quadrisulcata Pic, 1938g: 220 **A:** JIA
reichei Emery, 1876: 91 **E:** BU CR CZ FR GE HU MC SK ST UK
mulsanti Ermisch, 1956: 283
reitteri Schilsky, 1894a: 94 **E:** AL AU BU CR FI GE GG GR HU IT RO SK SP TR YU **A:** CY
rhenana Ermisch, 1956: 285 **E:** BE EN FR GE HU **A:** KZ TM
rufifrons Schilsky, 1894a: 92 **E:** AL AU BY CR GE MG NT PL RU **A:** ES FE KZ MG
rufospina Ermisch, 1977: 159 **E:** CR
rugipennis Schilsky, 1895: 57 **E:** CT GR UK **A:** IN TD TM TR
sajoi Ermisch, 1977: 165 **E:** HU
saxonica Ermisch, 1967b: 115 **E:** GE PL
savioi Pic, 1924a: 23 **A:** JIA
schatzmayri Franciscolo, 1949c: 91 **E:** BU MC GR
pfefferi Ermisch, 1963a: 14
secreta Horák, 1983: 10 **E:** AB AR AU BE BU CR CT CZ FR GB GG GR HU IT MC NR PL RO SK SL SP ST SZ UK YU **A:** IS TR
sejugata Ermisch, 1965a: 73 **A:** MG
semiferruginea Reitter, 1911a: 376 **E:** AU BU CZ FR GE HU RO SK SZ UK **A:** TR
balazuci Ermisch, 1966a: 37
taurica Csiki, 1949: 45
semipygmaeola Ermisch, 1964b: 349 **A:** MG
semirubra Champion, 1927: 52 **A:** NP UP
sericata Wollaston, 1864: 515 **N:** CI
serraticornis Horák, 1991: 135 **N:** MO
shimoyamai M. Chûjô, 1957a: 6 **A:** JA
shirozui Nomura, 1951: 53 **A:** JA
sibirica Ermisch, 1977: 156 **A:** WS
signicollis Schilsky, 1894a: 93 **A:** SY

solarii Franciscolo, 1942c: 79 **E:** IT
squamipilosa Ermisch, 1967a: 45 **A:** MG
stenidea Mulsant, 1856f: 381 **E:** AL AU BH BU BY CR CT FR GE GR HU IT PL SL SP SZ
  flexipes Rey, 1857: 411
steppensis Ermisch, 1965a: 68 **A:** MG
stoeckleini Ermisch, 1956: 282 **E:** CZ GE HU PL SK SZ UK **A:** FE
subepisternalis Ermisch, 1965d: 262 **E:** AL BU FR HU YU **N:** MO **A:** IS TR
subincana Ermisch, 1967a: 49 **A:** MG
subsquamosa Schilsky, 1899: 80 **A:** IN IQ MG WS TM
syrites Ermisch, 1977: 162 **A:** SY
tadjikistanica Horák, 1980: 280 **A:** TD UZ
takizawai Kôno, 1932b: 157 **A:** JA
tamana Tokeji, 1953: 17 (*Falsomordellistena*) **A:** JA
taorminensis Ermisch, 1965a: 71 **E:** GR IT
tarsata Mulsant, 1856f: 381 **E:** AL AU BH BU CR CT CZ FR GE GG GR HU IT RO SK SP ST SZ UK YU **A:** KI TM KZ TR WS
teneriffensis Ermisch, 1965: 69 **N:** CI
tenuicornis Schilsky, 1899: T [RN] **E:** GR
  gracilicornis Schilsky, 1895: 47 [HN]
tenuimanus Champion, 1927a: 52 **A:** NP UP
thurepalmi Ermisch, 1965c: 201 [RN] **E:** FI HU NR PL SV
  palmi Ermisch, 1963a: 10 [HN]
thuringiaca Ermisch, 1963a: 23 **E:** AU BY CT CZ FR GE HU PL SK **A:** FE KZ TM WS
tibiella Ermisch, 1965a: 67 **A:** MG
tokeji Nomura, 1951: 52 **A:** JA SC
tondui Horák, 1996: 176 **A:** AF
tosgoniensis Ermisch, 1969c: 174 **A:** MG
trilineata Mulsant, 1856f: 384 **E:** FR
turknenica Odnosum, 2003: 43 **A:** TM
uvsensis Ermisch, 1968b: 17 **A:** MG
variegata Fabricius, 1798: 127 (*Mordella*) **E:** AU BU BY CR CT CZ FR GB GE HU IT LU NL PL SK SL SP ST SZ UK YU **A:** ES FE KZ
  bicolor Marsham, 1802: 490 (*Mordella*)
  lateralis G.-A. Olivier, 1795: 64: 8 (*Mordella*) [HN]
visai Horák, 1983: 5 **A:** FE
vogti Ermisch, 1963a: 8 **E:** GE
wankai Ermisch, 1966a: 33 **E:** GE HU
weisei Schilsky, 1895: 41 **E:** AU BU BY CR CZ GE HU NL PL RO SK ST SZ **N:** AG **A:** FE KZ TD WS
wiebesi Batten, 1977: 170 **E:** CR GR IT SP **N:** AG
yangi Fan, 1995a: 23 **A:** HEB

zaisanensis Ermisch, 1967d: 125 **A:** MG
zoltani Ermisch, 1977: 164 **E:** CR HU
zosijniensis Ermisch, 1969c: 173 **A:** MG
subgenus *Pseudomordellina* Ermisch, 1952a: 47 type species *Mordellistena nana* Motschulsky, 1860
acuticollis Schilsky, 1895: 40 **E:** AU BU BY CZ GB GE HU IT LU NL PL RO SK SZ UK **A:** KI
imitatrix Allen, 1995: 85
aliena Ermisch, 1967d: 128 **A:** MG
arabica Chobaut, 1904b: 244 **A:** SA
aritai Nomura, 1964: 50 **A:** JA
csiki Ermisch, 1977: 175 **E:** HU SZ
ferruginipes Ermisch, 1966a: 39 **E:** NL
flavospinulosa Ermisch, 1977: 163 **E:** HU
klapperichi Ermisch, 1956: 291 **E:** AU BH CZ FR GE HU SK SZ
lebedevi Roubal, 1929a: 48 **A:** KI TM UZ
kubani Horák, 1982: 50
luteispina Ermisch, 1977: 175 **E:** RO TR
magyarica Ermisch, 1977: 174 **E:** AU HU
mediana Ermisch, 1977: 175 **E:** CR GR HU
microscopica Ermisch, 1977: 175 **E:** CR HU
nana Motschulsky, 1860c: 142 **E:** ST
nanula Ermisch, 1967b: 118 **E:** AL AU BU BY GE PL RO SK ST UK **A:** FE KZ TM TR WS
nanuloides Ermisch, 1967b: 118 **E:** GB GE NL
nigrocapillata Ermisch, 1965d: 270 **E:** BU
paranana Ermisch, 1977: 174 **E:** HU
pseudonana Ermisch, 1956: 291 **E:** AU CR CZ GE HU MC PL SK SL SP SZ
separanda Ermisch, 1965a: 76 **A:** MG
ulanbatorensis Ermisch, 1967a: 50 **A:** MG
umbra Franciscolo, 1949b: 54 **E:** IT
zavchanensis Ermisch, 1970: 174 **A:** MG
zuuncharaensis Ermisch, 1965d: 75 **A:** MG

genus *Mordellistenochroa* Horák, 1982: 52 type species *Mordellistenochroa strejceki* Horák, 1982
fallaciosa Ermisch, 1969b: 110 (*Mordellistena*) **E:** HU IT SZ
strejceki Horák, 1982: 52 **E:** GG UK **A:** TD TM

genus *Mordellistenoda* Ermisch, 1941b: 590 type species *Mordellistenoda fukiensis* Ermisch, 1941
aka Kôno, 1928a: 43 (*Mordellistena*) **A:** JA TAI SE
donan Tsuru, 2004: 408 **A:** JA

fukiensis Ermisch, 1941b: 591 **A**: FUJ
ohsumiana Nakane, 1957: 51 (*Glipostenoda*) **A**: JA

genus *Mordellistenula* Stchegoleva-Barovskaya, 1930a: 57 type species *Mordellistena planifrons* Stchegoleva-Barovskaya, 1930
anomala Ermisch, 1957: 47 **E**: FR SP
lacinicollis Czetó, 1990: 25 **E**: GR
longipalpis Ermisch, 1965d: 256 **E**: AB AR BU GR MC TR YU **A**: KZ
perrisi Mulsant, 1857b: 412 (*Mordellistena*) **E**: AB AN AR AU BE BU CR CZ DE FR GE GG GR HU IT KZ MC NL PL PT SK SP SV SZ UK YU **N**: MO **A**: IN KZ TR
engelharti Schilsky, 1910: 255 (*Mordellistena*)
rectangula C.G. Thomson, 1868: 134 (*Mordellistena*)
planifrons Stchegoleva-Barovskaya, 1930a: 56 **E**: AB AL AU BU CR GE GG GR HU IT MC RO RU SK UK **N**: AG MO
plutonica Compte, 1969: 114 **E**: SP

genus *Mordellochroa* Emery, 1876: 83 type species *Mordella abdominalis* Fabricius, 1775
abdominalis Fabricius, 1775: 264 (*Mordella*) **E**: AB AU BY BE BH BU CR CT CZ FR GB GE GG GR HU IT LT LU NL NT PL RO SK SL ST SZ UK **A**: FE IN TM TR
bicolor Sulzer, 1776: 67 (*Mordella*)
nigra Marsham,: 1802: 490 (*Mordella*)
ventralis Fabricius, 1792b: 114 (*Mordella*)
hasegawai Nomura & Katô, 1959: 5 **A**: JA
humerosa Rosenhauer, 1847: 38 (*Mordella*) **E**: AL BH BU FR GR HU MC PT RU SP ST SZ UK **A**: CY IS KZ MG SY TR
freyi Ermisch, 1944: 211 (*Tolida*)
milleri Emery, 1876: 83 (*Mordellistena*) **E**: AB AU CZ FR HU PL SK SL UK
pulchella Mulsant & Rey, 1859f: 232 (*Mordella*) **E**: FR IT **N**: AG MO TU
pygidialis Nomura, 1961b: 85 **A**: JA
shibatai Kiyoyama, 1987: 31 **A**: TAI
tournieri Emery, 1876: 102 (*Tolida*) **E**: AL AU BU BY CR CT CZ ES FI FE FR GE GG GR HU IT MC NT PL RU SK SP ST SZ UK **A**: FE TR SV SZ
graeca Schilsky, 1895: 56 (*Mordellistena*)
schusteri Schilsky, 1895: 55 (*Mordellistena*)
similis Stschegoleva-Barovskaya, 1930b: 752 (*Mordellistena*)
yanoi Nomura, 1951: 56 (*Mordellistena*) **A**: TAI

genus *Pseudodellamora* Ermisch, 1941c: 720 type species *Mordellistena brevicollis* Emery, 1876
brevicollis Emery, 1876: 102 (*Mordellistena*) **N**: AG
championi Schilsky, 1899: 82 (*Mordellistena*) **E**: SP **M**: MO

distinguenda Ermisch, 1963b: 15 **E:** CR FR PT SP **A:** TR
grossepalpalis Ermisch, 1942a: 87 **E:** AB

genus *Pseudotolida* Ermisch, 1950b: 86 type species *Mordellistena ephippiata* Champion, 1891
awana Kôno, 1932b: 157 (*Mordellistena*) **A:** HUB JA TAI
mitsuhashii Nomura, 1951: 68 (*Glipostenoda*)
tokyoensis Nomura & Katô, 1959: 6
menoko Kôno, 1932b: 158 (*Mordellistena*) **A:** JA
morimotoi Nomura, 1967a: 32 **A:** TAI
multisulcata Nomura, 1966: 45 **A:** JA
sinica Fan & J. Yang, 1995: 96 **A:** ZHE

genus *Pulchrimorda* Ermisch, 1968: 37 type species *Mordellistena grelaki* Pic, 1923
savioi Pic, 1924a: 23 (*Mordellistena*) **A:** JIA

genus *Sinopalpus* Horák, 2007: 79 type species *Sinopalpus maculatus* Horák, 2007
maculatus Horák, 2007: 79 **A:** SHX

genus *Tolida* Mulsant, 1856f: 387 type species *Mordellistena artemisiae* Mulsant, 1856
artemisiae Mulsant, 1856f: 385 (*Mordellistena*) **E:** BE FR GE IT SP

genus *Tolidopalpus* Ermisch, 1952b: 144 type species *Tolidopalpus castaneicolor* Ermisch, 1952
castaneicolor Ermisch, 1952b: 149 **A:** FUJ JA
galloisi Kôno, 1932b: 158 (*Mordellistena*) **A:** FE JA TAI

genus *Tolidostena* Ermisch, 1942b: 674 type species *Tolidostena tarsalis* Ermisch, 1942
subgenus *Tolidostena* Ermisch, 1942b: 674 type species *Tolidostena tarsalis* Ermisch, 1942
atripennis Nakane, 1956a: 8 **A:** FE JA TAI
ermischi Nakane, 1956a: 9 **A:** JA
fusei Tokeji, 1953: 14 (*Mordellochroa*) **A:** JA
japonica Tokeji, 1953: 13 (*Mordellochroa*) **A:** JA
montana Kiyoyama, 1991: 54 **A:** TAI
similator Kiyoyama, 1991: 56 **A:** TAI
tarsalis Ermisch, 1942b: 675 **A:** FUJ GUI SHX
subgenus *Neotolidostena* Kiyoyama, 1991: 53 type species *Mordellochroa taiwana* Kiyoyama, 1987
hayashii Kiyoyama, 1991: 59 **A:** TAI
taiwana Kiyoyama, 1987: 34 (*Mordellochroa*) **A:** TAI

genus *Uhligia* Horák, 1990: 139 type species *Mordellistena schilskyi* Csiki, 1915
schilskyi Csiki, 1915: 50 (*Mordellistena*) **E:** AR **A:** TM UZ
deserticola Horák, 1981: 337 (*Mordellistenula*)
murina Schilsky, 1903: 100 (*Mordellistena*) [HN]

tribe Stenaliini Franciscolo, 1955

genus *Brodskyella* Horák, 1989: 37 type species *Stenalia angustata* Pic, 1923

holzschuhi Horák, 1989: 42 **A:** NP SD

genus *Stenalia* Mulsant, 1856f: 387 type species *Mordella testacea* Fabricius, 1787

araxicola Khnzorian, 1957a: 163 **E:** AB AR GG GR TR **N:** TU **A:** IN IQ IS KZ TD TM TR UZ

muchei Ermisch, 1965d: 251

ascaniaenovae Lazorko, 1975: 110 **E:** AR CT UK **A:** KZ UZ

atra Perris, 1875: 8 (*Mordella*) **N:** AG EG TU

argentata Schilsky, 1899: L

balcanica Franciscolo, 1949c: 88 **E:** AL

bilyi Horák, 1977: 400 **A:** KI TD UZ

bisecta Baudi di Selve, 1883a: 5 **E:** IT

biskrensis Schilsky, 1903: 99 **N:** AG TU

brunneipennis brunneipennis Mulsant, 1856f: 389 **E:** AR GR **N:** EG **A:** IN IQ IS JO LE SY TR

brunneipennis unistrigosa Chobaut, 1924d: 182 **A:** SY

dolini Lazorko, 1974: 112 **E:** UK

dzhulfae Odnosum, 2001: 22 **E:** AB

ermolenkoi Odnosum, 2000: 46: **E:** AB

escherichi Schilsky, 1899: 76 **E:** AR HU EO TR SK **N:** MO **A:** IN KZ TD TR UZ WS

uralensis Csiki, 1915: 45 (*Mordellistena*)

singularis Reitter, 1911a: 377 (*Mordellistena*) [HN]

gracilicornis Baudi di Selve, 1878b: 348 **E:** AB AR **A:** CY IN

graeca Ermisch, 1965d: 254 **E:** GR

gridellii Franciscolo, 1949c: 87 **E:** GR

hispana Schilsky, 1895: 32 **E:** SP

indica Horák, 1995a: 97 **A:** SD

iranica Horák, 1981: 335 **A:** IN

jordanensis Ermisch, 1965d: 254 **A:** IS JO

lindbergi Ermisch, 1963c: 53 **E:** GR **A:** CY IS

merkli Schilsky, 1895: 33 **E:** BU TR **A:** IN IQ IS JO LE SY TR

rufohumeralis Pic, 1926a: 2 **A:** FE

sefrensis peyerimhofferi Chobaut, 1924d: 183 **N:** AG

sefrensis sefrensis Chobaut, 1924d: 183 **N:** AG

stipae Chobaut, 1924d: 183 **E:** SP **N:** AG

testacea testacea Fabricius, 1787a: 218 (*Mordella*) **E:** AR BU CR FR GE IT PT SL SP ST SZ UK **N:** AG TU MO

chiragra Dufour, 1843: 72 (*Mordella*)

dolini Lazorko, 1974: 112

luteicornis Schilsky, 1895: 30

testacea tristrigosa Chobaut, 1924d: 181 **N:** AG

vladimiri Odnosum, 2001: 21 E: AB

family Mordellidae, nomen dubium
Mordella perlata Roemer, 1789: 8 **E:** SZ

family TENEBRIONIDAE Latreille, 1802
subfamily **Kuhitangiinae** G.S. Medvedev, 1962
tribe Foranotini Nabozhenko & Sadeghi, 2017

genus *Foranotum* Nabozhenko & Sadeghi, 2017: 165 type species *Foranotum perforatum* Nabozhenko & Sadeghi, 2017
perforatum Nabozhenko & Sadeghi, 2017: 135 **A:** IN

tribe Kuhitangiini G.S. Medvedev, 1962

genus *Kuhitangia* G.S. Medvedev, 1962: 1184 type species *Kuhitangia kryzhanovskii* G.S. Medvedev, 1962
kryzhanovskii G.S. Medvedev, 1962: 1185 **A:** TM

subfamily **Lagriinae** Latreille, 1825
tribe Belopini Reitter, 1917

genus *Centorus* Mulsant, 1854: 272 type species *Calcar procerus* Mulsant, 1854
subgenus *Belopomerus* Reitter, 1920a: 3 type species *Calcar zoufali* Reitter, 1915
lucasi Mulsant & Godart, 1859b: 243 **N:** AG LB
zoufali Reitter, 1915e: 89 (*Calcar*) **N:** LB TU
subgenus *Belopus* Gebien, 1911: 459 [RN, NP] type species *Tenebrio elongatus* Herbst, 1797
Calcar Dejean, 1821: 67 [HN] type species *Tenebrio elongatus* Herbst, 1797
Hemeralopius Gistel, 1848: viii, 125 [RN, NO] type species *Tenebrio elongatus* Herbst, 1797
aegyptiacus Zoufal, 1893: 116 (*Calcar*) **N:** EG LB **AFR**
elongatus ecalcaratus Seidlitz, 1896: 655 (*Calcar*) **E:** IT (Sicilia) MA **N:** AG TU **A:** CY **AFR**
siculus Canzoneri, 1977: 279 (*Belopus*)
elongatus elongatus Herbst, 1797: 259 (*Tenebrio*) **E:** FR IT (Sardinia) PT SP **N:** CI MO MR TU
calcar Fabricius, 1801a: 153 (*Trogosita*)
variabilis Gebien, 1906: 222 (*Calcar*)
variolosus Fabricius, 1801a: 146 (*Tenebrio*)
heydeni Zoufal, 1893: 116 (*Calcar*) **N:** EG **A:** CY IN IQ IS SA SY TR
humeralis Champion, 1891d: 387 (*Calcar*) **N:** AG MO
truncaticollis Zoufal, 1893: 117 (*Calcar*)
syriacus Zoufal, 1893: 117 (*Calcar*) **A:** IS "Syria"
subgenus *Centorus* Mulsant, 1854: 272 type species *Calcar procerus* Mulsant, 1854
alashanicus Skopin, 1974a: 90 (*Belopus*) **A:** NMO
bucharicus Obenberger, 1914: 141 **A:** KZ TD
porrecticollis Reitter, 1915e: 90
calcaroides calcaroides Reitter, 1920a: 6 (*Belopus*) **E:** ST **A:** AF KI KZ TM UZ
calcaroides gobiensis Kaszab, 1964g: 399 (*Belopus*) **A:** ES (Tuva) MG
calcaroides intermedius Skopin, 1974a: 93 (*Belopus*) **A:** KZ

calcaroides tsharynensis Skopin, 1974a: 95 (*Belopus*) **A**: KZ
chappuisi Peyerimhoff, 1933: 346 (*Belopus*) **N**: AG
crassipes borealis Skopin, 1974a: 93 (*Belopus*) **A**: KZ
crassipes crassipes Fischer von Waldheim, 1844: 123 (*Calcar*) **E**: GR IT MC ST UK **A**: KZ TR
reitteri Zoufal, 1893: 118
sulcatus Fischer von Waldheim, 1844: 124 (*Calcar*)
csikii bagdadensis Reitter, 1920a: 9 (*Belopus*) **A**: CY IQ IS JO SA YE
csikii csikii Reitter, 1920a: 6 (*Belopus*) **E**: ST **A**: AF IN KZ TM
ferganensis Skopin, 1974a: 92 (*Belopus*) **A**: UZ (Fergana)
filiformis Motschulsky, 1872: 40 **E**: AB AR ST **A**: AF GAN IN KZ TM XIN
araxidis Reitter, 1920a: 10 (*Belopus*)
americaus [sic] Motschulsky, 1872: 40
tenuicornis Motschulsky, 1872: 40
helanensis Ren & Wu, 1994: 351 (*Belopus*) **A**: NIN
lineatus Zhang & Ren, 2009: 213 **A**: XIN
luculentus Ren, 1999: 177 (*Belopus*) **A**: XIN
lucalentus Ren, 1999: 393 (*Belopus*)
medvedevi Zhang & Ren, 2009: 213 **A**: NIN
microceps Motschulsky, 1872: 39 **N**: EG
ocularis Reitter, 1920a: 7 (*Belopus*) **A**: "Turkestan"
pakistanus Kaszab, 1961d: 226 (*Belopus*) **A**: PA
parvulus G.S. Medvedev, 1964a: 55 (*Belopus*) **A**: TM (Kugitang)
pilosus Skopin, 1974a: 92 (*Belopus*) **A**: IQ
proceroides Leo, 1984a: 165 (*Belopus*) **E**: IT (Sardinia)
procerus moldaviensis Reitter, 1920a: 7 (*Belopus*) **E**: MD RO ST UK **A**: KZ UZ
procerus procerus Mulsant, 1854: 272 (*Calcar*) **E**: FR SP ST **N**: TU
procerus Rosenhauer, 1856: 217 (*Calcar*) [HN]
raffrayi Fairmaire, 1873a: 340 (*Calcar*) **N**: AG TU
reymondi Antoine, 1951a: 76 (*Belopus*) **N**: MO
rufipes Gebler, 1833: 288 (*Calcar*) **E**: ST UK **A**: KZ WS
saccardyi Peyerimhoff, 1939a: 25 (*Belopus*) **N**: AG
sefiranus Reitter, 1920a: 11 (*Belopus*) **A**: AF KZ
sublaevis Reitter, 1920a: 11 **A**: AF KZ TM
steppensis Kaszab, 1964g: 401 (*Belopus*) **A**: ES MG
tibialis Zoufal, 1893: 118 **E**: AB UK (Krym) **A**: KZ
trogosita Motschulsky, 1872: 39 **E**: AR ST **A**: AF IN KZ TM UZ
tumotensis Ren & Zheng, 1993d: 52 (*Belopus*) **A**: NMO
turcicus Kaszab, 1959d: 81 (*Belopus*) **A**: TR
tuvensis Knor, 1976: 106 (*Belopus*) **A**: ES (Tuva)
subgenus *Nanocalcar* Skopin, 1974a: 102 type species *Nanocalcar nanus* Skopin, 1974
nanus Skopin, 1974a: 95 (*Belopus*) **A**: KZ (Almaatinskaya Oblast)

tribe Cossyphini Latreille, 1802

genus *Cossyphus* G.-A. Olivier, 1791: 119 type species is *Lampyris depressa* Fabricius, 1781

subgenus *Acontodactylus* Desbrochers des Loges, 1894: 5 type species *Cossyphus minutissimus* Laporte, 1840

Paracossyphus Viñolas & Cartagena, 2005: 29 [URN]

minutissimus Laporte, 1840: 228 **E:** PT SP **N:** MO

pygmaeus Brême, 1846: 29

tuberculatus Brême, 1846: 27

punctatostriatus Kocher, 1953: 67 **N:** MO

pusillus Pascoe, 1887: 14 **A:** NP **ORR**

subgenus *Cossyphus* G.-A. Olivier, 1791: 119 type species *Lampyris depressa* Fabricius, 1781

algiricus Laporte, 1840: 228 **N:** AG MO TU

algyricus Laporte, 1840: 228 [incorrect original spelling]

ovatus Brême, 1846: 22

dejeani Brême, 1846: 21 **E:** PT SP **N:** AG MO

depressus Fabricius, 1781: 254 (*Lampyris*) **A:** NP SD YUN **ORR**

planus Fabricius, 1801b: 98

subgranulatus Reitter, 1917b: 138

hoffmanseggii Herbst, 1797: 229 **E:** PT SP **N:** AG MO

bletoni Antoine, 1953b: 211

rifensis Kocher, 1963c: 198

sublaevis Kocher, 1953: 66

laevis Laporte, 1840: 228 **E:** SP **N:** AG MO

barbarus Brême, 1846: 24

incostatus Brême, 1846: 25

substriatus Laporte, 1840: 229

moniliatus Reitter, 1917b: 135 **E:** GR **A:** SY

moniliferus moniliferus Chevrolat, 1833a: pl. 31 **E:** GR IT MA SP (Baleares) **N:** AG CI EG LB MO TU **A:** SY **AFR**

insularis Laporte, 1840: 229

ruffoi indicus Scupola, 2000: 241 **A:** NP **ORR**

rugosulus Peyron, 1854: 224 **A:** CY IN IS SY TR

subgranulatus Reitter, 1917b: 138 **A:** SD **ORR**

tauricus Fischer von Waldheim, 1832b: 425 **E:** AL AR BU GR IT ST TR UK **A:** AF CY IN IQ IS SY TD TM TR UZ

tribe Goniaderini Lacordaire, 1859

genus *Anaedus* Blanchard, 1842: pl. 14 type species *Anaedus punctatissimus* Blanchard, 1842

Anaedes Agassiz, 1846: 36 [UE]

Aspidosoma Agassiz, 1846: 36 [UE]

Aspisoma Duponchel & Chevrolat, 1841: 210 [HN] type species *Aspisoma fulvipenne* Duponchel & Chevrolat, 1841

buricus Schawaller, 1994: 266 **A:** NP PA
nepalicus Schawaller, 1994: 267 **A:** NP SD
mroczkowskii Kaszab, 1968a: 10 **A:** NP SC
spinicornis Kaszab, 1973a: 39 **A:** NP TAI **ORR**
substriatus Pic, 1938a: 17 **A:** SCH

tribe Laenini Seidlitz, 1895

genus *Hypolaenopsis* Masumoto, 2001a: 45 type species *Hypolaenopsis uenoi* Masumoto, 2001 (= *Laena nanpingica* Schawaller, 2001)
benesi Schawaller, 2001a: 7 (*Laena*) **A:** SCH
honggyuanica Schawaller, 2001a: 21 (*Laena*) **A:** SCH
hualongica Schawaller, 2001a: 22 (*Laena*) **A:** SCH
nanpingica Schawaller, 2001a: 29 (*Laena*) **A:** SCH
uenoi Masumoto, 2001a: 46
nomurai Schawaller, 2001a: 29 (*Laena*) **A:** SCH

genus *Laena* Dejean, 1821: 64 type species *Helops pimelia* sensu Dejean (1821) (= *Scaurus viennensis* J. Sturm, 1807)
Catolaena Reitter, 1900h: 282 type species *Laena turkestanica* Reitter, 1897
Ebertius Jedlička, 1965: 98 type species *Ebertius nepalensis* Jedlička, 1965 (= *Laena longipilis* Schuster, 1926)
Psilolaena Heller, 1923: 70 type species *Psilolaena schusteri* Heller, 1923
acuticollis Kaszab, 1978a: 175 **A:** SD
aenea Schuster, 1926a: 49 **A:** NP SD
darjeelingiana Kaszab, 1938e: 226
affinis Schuster, 1935d: 451 **A:** SD
alaiensis Reitter, 1906h: 446 **A:** KI UZ (Fergana)
edda Reitter, 1906h: 447
alesi Schawaller, 2008a: 389 **A:** YUN
almorensis Schuster, 1926a: 44 **A:** UP
alpina Kaszab, 1977b: 264 **A:** NP
alticola Blair, 1923b: 284 **A:** XIZ
amplofemura Wei & Ren, 2019b: 524 **A:** XIZ
amrishi Makhan, 2019: 1 **A:** CHQ
anglofemura Wei & Ren, 2019b: 526 **A:** SCH
angorensis Reitter, 1900h: 284 **A:** TR
angulifemoralis Musumoto, 1996: 169 **A:** YUN
angusta Weise, 1878a: 233 **E:** UK (Krym) "Caucasus"
annapurna Schawaller, 2002: 41 **A:** NP
apfelbecki Schuster, 1915c: 358 **E:** GR **A:** TR
armidenta Wei & Ren, 2019b: 527 **A:** ANH
arun Schawaller, 2002: 41 **A:** NP

atripennis Pic, 1919c: 6 **E**: GR (Taygetos)
badakschanica Kaszab, 1960a: 170 **A**: AF
badrinathica Schawaller, 2009e: 349 **A**: UP
baiorum Schawaller, 2008a: 389 **A**: YUN
baishuica Schawaller, 2001a: 5 **A**: YUN
baoshanica Schawaller, 2008a: 390 **A**: YUN
barclayi Schawaller, 2009e: 349 **A**: HP
barkamica Schawaller, 2008a: 390 **A**: SCH
barypithoides Schuster, 1916: 623 **A**: KA
basumtsoica Schawaller, 2008a: 390 **A**: XIZ
baudii Weise, 1878a: 239 **E**: GG
becvari Schawaller, 2001a: 6 **A**: YUN
beesoni Schuster, 1935d: 459 **A**: HP
bhatiai Schuster, 1935d: 447 **A**: HP UP
bhutanensis Kaszab, 1975e: 329 **A**: BT
bidentata Zhao & Ren, 2012b: 43 **A**: SCH
bifoveolata Reitter, 1889k: 709 **A**: GAN SHA SCH
  imurai Masumoto, 1996d: 179
  shaanxiica Masumoto, 1996d: 178
blairi Schuster, 1926a: 47 **A**: UP
bogatschevi Iablokoff-Khnzorian, 1984: 43 **E**: AR
bohrni Schawaller, 2008a: 391 **A**: SCH
bomdilaica Schawaller, 2018: 73 **A**: AP
bowaica Schawaller, 2001a: 8 **A**: SCH
brahmae Masumoto, 1990f: 190 **A**: SD
  gomcheyensis Masumoto, 1990f: 194
breiti Schuster, 1916: 588 **E**: GR (Taygetos)
brendelli Schawaller, 2001a: 9 **A**: YUN
brevipennis Reitter, 1901b: 63 **A**: KZ
broscosomoides Kaszab, 1977b: 268 **A**: NP
businskyorum Schawaller, 2001a: 9 **A**: YUN
byzantina Apfelbeck, 1901: 466 **E**: TR **A**: TR
  bodemeyeri Reitter, 1903e: 224
cameroni Schuster, 1926a: 36 **A**: UP
cardiothorax Kaszab, 1978a: 175 **A**: SD
carinata Schuster, 1926a: 38 **A**: UP
carinipennis Schuster, 1935d: 448 **A**: HP
carolinae Schuster, 1935d: 450 **A**: KA
caucasica Motschulsky, 1845a: 75 **E**: "Caucasus"
championi Schuster, 1926a: 41 **A**: UP

chatkalica Schawaller, 1995a: 69 **A:** KI
chatterjeei Schuster, 1935d: 452 **A:** HP
chaukiensis Masumoto, 1990f: 192 **A:** NP
chejuana Mt. Chûjô, 1992: 41 **A:** SC (Chejudo Is.)
chetri Schawaller, 2002: 42 **A:** NP
chileyla Schawaller, 2012b: 265 **A:** BT
chiloriluxa Zhao & Ren, 2012a: 29 **A:** YUN
chinensis Kaszab, 1965c: 284 **A:** SCH YUN
  youcida Masumoto, 1996d: 171
chitralica Schawaller, 2014b: 200 **A:** PA
cholanica Schawaller, 2001a: 11 **A:** XIZ
chomolungma Schawaller, 2002: 43 **A:** NP
clivinoides Baudi di Selve, 1876c: 243 **E:** BH **A:** CY SY
clypealis Fairmaire, 1896b: 98 **A:** HP PA
confusa Löbl & Schawaller, 2008: 30 **A:** ANH
  chinensis J.-K. Li & Chen, 1993: 166 [HN]
coniceps Kaszab, 1973a: 56 **A:** NP
consimilis Kaszab, 1973a: 45 **A:** NP
  mandli Kaszab, 1975b: 125
constricta Iablokoff-Khnzorian, 1957a: 166 **E:** AR
convexicollis Reitter, 1908k: 275 **A:** HP KA PA
cooteri Schawaller, 2008a: 391 **A:** JIX FUJ
corallipes Reitter, 1908k: 280 **A:** KA
crenulicollis Kaszab, 1977b: 265 **A:** NP
cribrella Reitter, 1908k: 279 **A:** HP KA PA
curvipes Desbrochers des Loges, 1881: 134 **A:** CY
cylindrica Schuster, 1926a: 52 **A:** XIZ
dabashanica Schawaller, 2008a: 392 **A:** HUB
daliensis Masumoto & Yin, 1994: 168 **A:** YUN
  tumidoculata Masumoto, 1996d: 173
  xiaoi Masumoto & Yin, 1994: 169
  yunnanensis Masumoto & Yin, 1994: 167
dampaensis Kaszab, 1975b: 125 **A:** NP
davidi Schawaller, 2008a: 392 **A:** SCH
daxueica Schawaller, 2001a: 13 **A:** SCH
dentata Zhao & Ren, 2012a: 30 **A:** YUN
dentatocrassa Zhao & Ren, 2012a: 34 **A:** HAI
denticollis Schuster, 1926a: 35 **A:** KA UP
denticrus Fairmaire, 1896b: 97 **A:** HP
  dentipennis Reitter, 1908k: 279

dentipes Schuster, 1935d: 458 **A:** NP SD
adriani Schawaller, 2001a: 34 [RN]
flavicincta Kaszab, 1938e: 228
schusteri Kaszab, 1938e: 228 [HN]
dentitibia Reitter, 1901b: 62 **A:** UZ (Bukhara)
denudata Kaszab, 1975e: 330 **A:** BT
deplanata Weise, 1878a: 240 **A:** TR
deqenica Schawaller, 2001a: 13 **A:** YUN
dhorpatanica Kaszab, 1977b: 267 **A:** NP
diancangica Schawaller, 2001a: 14 **A:** YUN
dickorei Schawaller, 2001a: 14 **A:** XIZ
dilutella Solsky, 1881b: 55 **A:** AF IN KI KZ TD TM UZ
dumialensis Nakane, 1966: 244 **A:** PA
durga Schawaller, 2015: 457 **A:** NP
edmundi Schuster, 1915c: 359 **A:** KA PA
nitida Schuster, 1935d: 453
elburzica Schawaller, 2009d: 122 **A:** IN
etalinica Schawaller, 2018: 73 **A:** AP
emeishana Masumoto, 1996d: 175 **A:** SCH
espagnoli Kaszab, 1965a: 128 **A:** SD
exjuncta Zhao & Ren, 2012b: 43 **A:** SCH
ezimai Masumoto & Akita, 2001a: 15 **A:** JA
fangjingshanana Ren & Hua, 2006: 267 **A:** GUI
farkaci Schawaller, 2008a: 392 **A:** YUN
fengileana Masumoto, 1996d: 180 **A:** SCH SHA
ferruginea Küster, 1846b: 68 **E:** AL BH BU CR GR YU **A:** IS LE TR
krueperi Reitter, 1881a: 222
marthae Reitter, 1881a: 224
minima Motschulsky, 1858b: 112
parvula Reitter, 1885e: 402
villosula Motschulsky, 1851a: 594
fikaceki Schawaller, 2012b: 266 **A:** AP
flatiforeheada Wei & Ren, 2019b: 528 **A:** XIZ
flectotibia Zhao & Ren, 2012b: 43 **A:** YUN
formaneki Schuster, 1916: 561 **A:** XIZ
tibetana Schuster, 1916: 563
fouquei Schawaller, 2008a: 393 **A:** YUN
franzi Kaszab, 1973a: 44 **A:** NP
franziana Kaszab, 1973a: 61 **A:** NP
freudei Kaszab, 1961d: 228 **A:** NP
fulunga Kaszab, 1973a: 63 **A:** NP

strigosa Kaszab, 1973a: 62
gairibasensis Masumoto, 1990f: 177 **A:** NP SD
galyatica Schawaller, 2014b: 200 **A:** PA
ganesha Schawaller, 2015: 458 **A:** NP
ganglbaueri Reitter, 1891b: 33 **E:** BU TR **A:** TR
turcica Reitter, 1900h: 282
ganzica Schawaller, 2001a: 17 **A:** SCH
gaoligongica Schawaller, 2008a: 393 **A:** YUN
gardneri Schuster, 1935d: 454 **A:** UP
gebieni Reitter, 1906h: 443 **A:** HP KA
kaschmirensis Reitter, 1906h: 443
gigantea Schuster, 1940b: 864 **A:** SCH
glabriuscula J.R. Sahlberg, 1908: 87 **A:** LE
goetzi Kaszab, 1970c: 427 **A:** NP
gracilis Schuster, 1926a: 46 **A:** XIZ
graeca Weise, 1878a: 231 **E:** BU GR TR
grandis Schuster, 1935d: 465 **A:** UP
guangxica Schawaller, 2008a: 394 **A:** GUX
guizhouica Schawaller, 2008a: 394 **A:** GUI
gyalthangica Schawaller, 2008a: 394 **A:** YUN
gyamdaica Schawaller, 2001a: 18 **A:** XIZ
gyatsoi Schawaller, 2018: 73 **A:** AP
habashanica Schawaller, 2001a: 19 **A:** YUN
haigouica Schawaller, 2001a: 19 **A:** SCH
hanuman Schawaller, 2015: 458 **A:** NP
hanzhongana Wei & Ren, 2019b: 530 **A:** SHA
hartmanni Schawaller, 2002: 43 **A:** NP
hauseri Reitter, 1906h: 445 **A:** TD (Hissar Alai)
hazaraca Schawaller, 2014: 201 **A:** PA
heinzi Schawaller, 2001a: 20 **A:** XIZ
hengduanica Schawaller, 2001a: 20 **A:** SCH YUN
herbertfranzi Kaszab, 1973a: 60 **A:** NP
augur Kaszab, 1973a: 59
magarkensis Masumoto, 1990f: 188
schusteriana Kaszab, 1973a: 58
heydeni Weise, 1880a: 102 **E:** BU GR
himalayana Schuster, 1915c: 357 **A:** HP
hingstoni Schuster, 1926a: 42 **A:** XIZ
hiranoi Masumoto & Akita, 2001a: 15 **A:** JA (Ryukyus)
hirtella Solsky, 1881b: 54 **A:** KI
auliensis Reitter, 1902c: 189

hirtipes Reitter, 1881a: 222 **E**: AB **A**: IN
hlavaci Schawaller, 2008a: 395 **A**: JIX
hopffgarteni Weise, 1878a: 230 **E**: BH BU CR MC RO YU
holdhausi Schuster, 1916: 622 **A**: KZ (Transili Alatau)
holzschuhi Schawaller, 2012b: 266 **A**: BT
hongkongica Schawaller & Aston, 2017: 170 **A**: HKG
hongqiao Zhou & Song, 2019: 91 **A**: HKG
houzhenzica Schawaller, 2001a: 22 **A**: SHA
hubeica Schawaller, 2001a: 23 **A**: HUB
incomperta Kaszab, 1973a: 50 **A**: NP
indica Fairmaire, 1896b: 96 **A**: HP
insularis Kaszab, 1964a: 5 **A**: JA (Ryukyus)
interrupta Schuster, 1935d: 460 **A**: SD
irregularis Schuster, 1935d: 464 **A**: UP
jalaorana Reitter, 1908k: 278 **A**: HP KA PA
janatai Schawaller, 2008a: 395 **A**: SCH
jiangxica Schawaller, 2008a: 395 **A**: JIX
jinpingica Schawaller, 2008a: 398 **A**: SCH
jizushana Masumoto, 1996d: 168 **A**: YUN
jumlana Kaszab, 1975b: 123 **A**: NP
justinae Reitter, 1887e: 523 **E**: GG
kabakovi Schawaller, 2010: 131 **A**: AF PA
kaghanica Schawaller, 2014b: 201 **A**: PA
kalabi Schawaller, 2008a: 399 **A**: SCH
kali Schawaller, 2015: 458 **A**: NP
kaliensis Schuster, 1935d: 445 **A**: UP
kangchendzongensis Masumoto, 1990f: 182 **A**: NP SD
kangdingica Schawaller, 2001a: 24 **A**: SCH
karakorumensis Kaszab, 1961f: 354 **A**: PA (Karakoram)
kaszabi Masumoto, 1990f: 181 **A**: NP SD
kaufmanni Reitter, 1881a: 222 **E**: AL BH CR YU **A**: TR
khumbuana Kaszab, 1977b: 266 **A**: NP
kopetzi Schawaller, 2015: 459 **A**: NP
korbi Heyden, 1890d: 374 **A**: TR
kraatzii Weise, 1878a: 236 **E**: TR
    alternata Reitter, 1885e: 401
krishna Schawaller, 2015: 459 **A**: NP
kubani Schawaller, 2001a: 25 **A**: YUN
kuluana Reitter, 1908k: 278 **A**: HP KA PA
kunarica Schawaller, 2010: 131 **A**: AF
lacordairei Marseul, 1876c: 133 **A**: UP

ladakhica Schawaller, 2014b: 201 **A**: KA (Ladakh)
laevigata Schuster, 1926a: 48 **A**: UP
laevipennis Schuster, 1926a: 45 **A**: SD
sandakphuensis Masumoto, 1990f: 189
lakshmi Schawaller, 2015: 460 **A**: NP
lamjurensis Masumoto, 1990f: 180 **A**: NP
langmusica Schawaller, 2001a: 25 **A**: SCH SHA
lawaraica Schawaller, 2014b: 201 **A**: PA
lebedevi Roubal, 1929c: 193 **A**: KZ
leonhardi Schuster, 1916: 612 **A**: KI XIN
spaethi Schuster, 1916: 613
lederi Weise, 1878a: 229 **E**: GG
liangi Zhao & Ren, 2012a: 32 **A**: YUN
libanotica Reitter, 1885e: 401 **A**: LE
lilliputana Kaszab, 1968e: 462 **A**: TR
limbu Schawaller, 2002: 44 **A**: NP
lindbergi Kaszab, 1959c: 55 **A**: TR
lisuorum Schawaller, 2008a: 399 **A**: YUN
longichaeta Wei & Ren, 2019b: 532 **A**: XIZ
longipilis Schuster, 1926a: 52 **A**: NP SD
nepalensis Jedlička, 1965: 98 (*Ebertius*)
longula Marseul, 1876c: 133 **A**: CY IS LE SY
loricera Kaszab, 1973a: 54 **A**: NP
ludingica Schawaller, 2001a: 26 **A**: SCH
luguica Schawaller, 2001a: 27 **A**: SCH
luhuoica Schawaller, 2001a: 27 **A**: SCH
luprops Kaszab, 1973a: 57 **A**: NP
magar Schawaller, 2002: 45 **A**: NP
manang Schawaller, 2002: 46 **A**: NP
maowenica Schawaller, 2008a: 400 **A**: SCH
martensi Kaszab, 1973a: 47 **A**: NP
dedita Kaszab, 1973a: 48
zurstrasseni Kaszab, 1977b: 263
martinhauseri Schawaller, 2002: 46 **A**: NP
medvedevi Wei & Ren, 2019a: 1058 **A**: JIA
merditana Schuster, 1916: 505 **E**: AL
merkli Weise, 1883: 313 **E**: TR **A**: TR
merklottoi Masumoto, 1990f: 184 **A**: NP SD
miandamica Schawaller, 2014b: 202 **A**: PA
michaeli Schawaller, 2008a: 400 **A**: YUN
minuta Fairmaire, 1896b: 97 **A**: HP

mirabilis Kaszab, 1970b: 121 **A**: SCH
monpa Schawaller, 2018: 75 **A**: AP
motogana Zhao & Ren, 2012a: 27 **A**: XIZ
moxica Schawaller, 2008a: 400 **A**: SCH
mussoorica Schawaller, 2009e: 358 **A**: UP
mulica Schawaller, 2001a: 28 **A**: SCH
naxiorum Schawaller, 2008a: 401 **A**: YUN
newar Schawaller, 2002: 47 **A**: NP
nigritissima Reitter, 1906h: 444 **A**: KA
nishikawai Masumoto, 1990f: 176 **A**: SD
nujiangica Schawaller, 2008a: 401 **A**: YUN
nyingchica Schawaller, 2001a: 30 **A**: XIZ
oblonga Wei & Ren, 2019a: 1057 **A**: JIA
obscuricornis Reitter, 1900h: 283 **A**: TR
ocys Kaszab, 1973a: 54 **A**: NP
oedipus Kaszab, 1977b: 261 **A**: NP
oertzeni Reitter, 1885e: 401 **E**: GR (Crete)
opaca Kaszab, 1970b: 120 **A**: NP SD
bikhebhanjangensis Masumoto, 1990f: 178
opacicollis Kaszab, 1970c: 425 **A**: NP
bembidion Kaszab, 1973a: 50
orbicollis Schuster, 1926a: 43 **A**: UP
osmanlis Reitter, 1906h: 447 **E**: TR **A**: TR
ovipennis Schuster, 1926a: 41 **A**: ZHE
pahakhola Schawaller, 2002: 47 **A**: NP
pakistanica Kaszab, 1962c: 316 **A**: PA
paomaica Schawaller, 2001a: 30 **A**: SCH
parallelocollis Schuster, 1926a: 40 **A**: XIZ
parateneta Kaszab, 1973a: 57 **A**: NP
parkeri Schuster, 1935d: 446 **A**: UT
parvati Schawaller, 2015: 460 **A**: NP
parvidens Reitter, 1890i: 373 **A**: TR
phithangensis Masumoto, 1990f: 175 **A**: SD
shunichii Masumoto, 1990f: 187
pici Reitter, 1900c: 87 **A**: TR
piligera Weise, 1878a: 235 **E**: AR
longicollis Weise, 1878a: 237
pilosissima Reitter, 1906h: 448 **E**: RO
planipennis Schuster, 1926a: 50 **A**: UP
pokharana Kaszab, 1973a: 64 **A**: NP
silvicola Kaszab, 1973a: 63

prehimalayaica Kaszab, 1977b: 269 **A**: NP
    moodlungensis Masumoto, 1990f: 193
pseudofranzi Kaszab, 1975b: 124 **A**: NP
punctulata Wei & Ren, 2019b: 535 **A**: XIZ
puetzi Schawaller, 2001a: 31 **A**: SCH
pulchella Fischer von Waldheim, 1823: 201 **E**: BU
    pubella Solier, 1848: 190
    rubiginosa Solier, 1848: 190
punctatissima Schuster, 1926a: 44 **A**: UP
puncticollis Schuster, 1935d: 463 **A**: UP
punctiventris Schuster, 1926a: 51 **A**: KA
qinlingica Schawaller, 2001a: 32 **A**: SCH SHA
quadrata Zhao & Ren, 2012a: 25 **A**: XIZ
quadricollis Weise, 1878a: 232 **E**: "Caucasus"
quadrifoveolata Wei & Ren, 2019b: 536 **A**: SCH
quinquagesima Schawaller, 2008a: 401 **A**: YUN
rai Schawaller, 2002: 48 **A**: NP
rama Schawaller, 2015: 460 **A**: NP
reitteri Weise, 1877: 27 **E**: AU BH BU CR HU PL RO SK SL UK
    ormayi Reitter, 1887e: 523
rhododendri Kaszab, 1977b: 269 **A**: NP
robusta Reitter, 1885e: 401 **A**: UZ (Samarkand)
rosti Reitter, 1906h: 444 **A**: HP KA
rotundicollis Marseul, 1876c: 134 **A**: JA SC
rubripes Reitter, 1908k: 279 **A**: KA
rugosa Schuster, 1916: 553 **A**: HP KA
rupalica Schawaller, 2014b: 202 **A**: PA
safraneki Schawaller, 2001a: 33 **A**: XIZ
sagarmatha Schawaller, 2002: 49 **A**: NP
sakaii Masumoto, 1990f: 188 **A**: SD
schmidti Schawaller, 2002: 50 **A**: NP
schulkei Schawaller, 2001a: 33 **A**: SCH
schusteri Heller, 1923: 70 (*Psilolaena*) **A**: SCH
schwarzi Reitter, 1885e: 402 **E**: AL GR YU
sehnali Schawaller, 2008a: 402 **A**: SCH
septuagesima Schawaller, 2008a: 402 **A**: YUN
shaluica Schawaller, 2001a: 35 **A**: SCH
sherpa Schawaller, 2002: 50 **A**: NP
shiva Schawaller, 2015: 460 **A**: NP
simillima Schuster, 1935d: 462 **A**: KA PA
similis Schuster, 1926a: 37 **A**: UP
    transversicollis Schuster, 1926a: 37

simlaica Schawaller, 2009e: 360 **A**: HP
singalilensis Masumoto, 1990f: 180 **A**: SD
smetanai Schawaller, 2001a: 35 **A**: SCH
sparsepunctata Kaszab, 1970c: 428 **A**: NP
starcki Reitter, 1887e: 523 **E**: ST (Kavkaz)
striolata Reitter, 1900h: 285 **A**: TR
studiosa Kaszab, 1973a: 46 **A**: NP
subalpina Kaszab, 1977b: 262 **A**: NP
subcoeca Kaszab, 1973a: 52 **A**: NP
sulcata Schuster, 1915c: 360 **A**: KA
swatica Schawaller, 2014b: 202 **A**: PA
syriaca Baudi di Selve, 1876c: 243 **A**: CY SY TR
tabanai Masumoto, 1998b: 316 **A**: SCH
tachysoides Kaszab, 1973a: 55 **A**: NP
takara Mt. Chûjô, 1996: 136 **A**: JA (Ryukyus)
takolana Kaszab, 1973a: 50 **A**: NP
tamang Schawaller, 2002: 51 **A**: NP
tamur Schawaller, 2002: 57 **A**: NP
tawang Schawaller, 2012b: 267 **A**: AP
tenga Schawaller, 2012b: 267 **A**: AP
thameogensis Masumoto, 1990f: 183 **A**: NP
theana Reitter, 1900h: 283 **A**: TR
thimphuica Schawaller, 2012b: 265 **A**: BT
thodunga Kaszab, 1973a: 60 **A**: NP
tibialis Schuster, 1926a: 39 **A**: "India"
tryznai Schawaller, 2001a: 36 **A**: SCH XIZ
tuntalica Schawaller, 2001a: 37 **A**: SCH XIZ
turkestanica Reitter, 1897e: 221 **A**: KI UZ (Margilan)
turnai Schawaller, 2001a: 38 **A**: SCH
viennensis J. Sturm, 1807: 180 (*Scaurus*) **E**: AU BH CR CZ HU IT SK SL
villosa Schuster, 1915c: 357 **A**: KA UT
vishnua Masumoto, 1990f: 186 **A**: SD
  kephakensis Masumoto, 1990f: 193
walkeri Schawaller & Aston, 2017: 171 **A**: HKG
wanensis Schuster, 1940b: 863 **A**: TR (Van)
watanabei Masumoto & Yin, 1993: 240 **A**: YUN
waygalica Schawaller, 2010: 133 **A**: AF
weigeli Schawaller, 2015: 461 **A**: NP
weiperti Schawaller, 2015: 461 **A**: NP
weisei Reitter, 1878a: 238 **E**: GR TR
wittmeri Kaszab, 1973a: 72 **E**: GR (Crete)

wolongica Schawaller, 2001a: 39 **A:** SCH
wrasei Schawaller, 2001a: 39 **A:** SCH
xuerenensis Masumoto, 1996d: 172 **A:** YUN
xueshanica Schawaller, 2008a: 403 **A:** YUN
yajiangica Schawaller, 2001a: 40 **A:** SCH
yasuakii Masumoto, 1996d: 166 **A:** YUN
yodai Kaszab, 1970c: 430 **A:** NP
yufengsi Masumoto, 1996d: 167 **A:** YUN
yulongica Schawaller, 2001a: 42 **A:** YUN
yusmargica Schawaller, 2014b: 203 **A:** KA
yuzhuensis Masumoto & Yin, 1994: 165 **A:** YUN
zhengi Zhao & Ren, 2011: 568 **A:** XIZ
zogqenica Schawaller, 2001a: 43 **A:** SCH

genus *Nepalolaena* Schawaller, 2001d: 277 type species *Nepalolaena kira* Schawaller, 2001
kira Schawaller, 2001d: 278 **A:** NP

tribe Lagriini Latreille, 1825
subtribe Lagriina Latreille, 1825

genus *Adynata* Fåhraeus, 1870: 330 type species *Adynata tricolor* Fåhraeus, 1870
Microlagria Seidlitz, 1898b: 336 type species *Lagria poupillieri* Reiche, 1864
poupillieri Reiche, 1864: 242 (*Lagria*) **N:** AG MO TU

genus *Aulonogria* Borchmann, 1929: 9 type species *Lagria rugosa* Fabricius, 1801
chinensis Borchmann, 1936: 149 **A:** HAI
concolor Blanchard, 1853: 184 (*Lagria*) **A:** NP PA **ORR**

genus *Bothynogria* Borchmann, 1915a: 128 type species *Bothynogria calcarata* Borchmann, 1915
bicolor L. Redtenbacher, 1844: 534 (*Lagria*) **A:** NP SD UP
bhutanica Merkl, 1990: 292 **A:** BT
calcarata Borchmann, 1915a: 129 **A:** CHQ FUJ GUA GUI HAI HUB HUN JIX SCH TAI YUN ZHE **ORR**
himalayana Borchmann, 1936: 116 **A:** NP UP
ruficollis Hope, 1831: 32 (*Lagria*) **A:** NP
simillima Merkl, 2019a: 72 **A:** BT NP SD UP **ORR**

genus *Cerogria* Borchmann, 1909a: 210 type species *Lagria anisocera* Wiedemann, 1823
subgenus *Cerogria* Borchmann, 1909a: 210 type species *Lagria anisocera* Wiedemann, 1823
Aeschrocera Chen & Chou, 1996: 265 type species *Cerogria brunneocollis* Chen & Chou, 1996 (= *Cerogria diversicornis* Pic, 1933)
Cerogriodes Borchmann, 1941: 25 type species *Cerogria klapperichi* Borchmann, 1941

abnormicornis Borchmann, 1942a: 8 **A:** SD YUN **ORR**

anisocera Wiedemann, 1823: 81 (*Lagria*) **A:** SCH TAI YUN **ORR**

corporaali Borchmann, 1929: 8

basalis Hope, 1831: 32 (*Lagria*) **A:** BT NP SD YUN **ORR**

dichroa Reitter, 1880b: 257 (*Lagria*)

birmana Borchmann, 1942a: 7 **A:** YUN **ORR**

castaneipennis Borchmann, 1936: 124 **A:** SCH YUN **ORR**

chinensis Fairmaire, 1886c: 349 (*Lagria*) **A:** HUB FUJ SCH TAI XIZ YUN

crassa Borchmann, 1911: 221 **A:** NP SD **ORR**

decorata Borchmann, 1911: 215 **A:** SCH **ORR**

distincticornis Reitter, 1880b: 256 (*Lagria*) **A:** SD

diversicornis Pic, 1933b: 9 **A:** GAN SCH

brunneocollis Chen & Chou, 1996: 266

harpacorna Chen & Chou, 1996: 267

flavicornis Borchmann, 1911: 214 **A:** YUN **ORR**

glabricollis Chen & Xia, 2001: 5 (*Schevodera*)

forteimpressa Pic, 1955c: 31 **A:** FUJ

hajimei Masumoto, 1988: 37 **A:** TAI

hauseri Borchmann, 1936: 125 **A:** YUN

janthinipennis Fairmaire, 1886c: 349 (*Lagria*) **A:** FUJ GUI JIX HUN HUB SC SCH SHA ZHE

antennata Borchmann, 1909b: 714 (*Lagria*) [RN]

distincticornis Heyden, 1887a: 269 (*Lagria*) [HN]

kikuchii Kôno, 1929a: 28 (*Lagria*) **A:** FUJ SCH TAI

praecox Borchmann, 1941: 24

klapperichi Borchmann, 1941: 25 **A:** FUJ GUX JIX SHA TAI ZHE

miwai Kôno, 1930a: 18 **A:** FUJ JIX SCH TAI YUN

gardneri Borchmann, 1936: 127

montana Merkl, 1991a: 10 **A:** BH NP SD

nepalensis Hope, 1831: 32 (*Lagria*) **A:** BT HP NP SD UP

aenea L. Redtenbacher, 1844: 553 (*Lagria*)

nigrita Borchmann, 1915b: 180 (*Lagria*) **A:** NP

nigrosparsa Pic, 1928g: 248 (*Lagria*) **A:** YUN **ORR**

odontocera Fairmaire, 1886c: 348 (*Lagria*) **A:** SCH SHA YUN

pachycera Fairmaire, 1886c: 347 (*Lagria*) **A:** GUX JIX SCH

popularis Borchmann, 1936: 121 **A:** FUJ GUI GUX HEN SCH SHN YUN

quadrimaculata Hope, 1831: 32 (*Lagria*) **A:** BT FUJ GAN GUX HP HUB NP PA SCH SD UP YUN ZHE **ORR**

cardonii Fairmaire, 1894a: 31 (*Lagria*)

fukienensis Borchmann, 1941: 24

variabilis L. Redtenbacher, 1844: 534 (*Lagria*)

sikkimensis Borchmann, 1936: 128 **A:** HP NP PA SD UP

genus *Chrysolagria* Seidlitz, 1898b: 339 type species *Lagria viridipennis* Fabricius, 1798
viridipennis Fabricius, 1798: 118 (*Lagria*) **E:** SP **N:** AG MO TU

genus *Lagria* Fabricius, 1775: 124 type species *Chrysomela hirta* Linnaeus, 1758
subgenus *Ammocera* Borchmann, 1942a: 5 type species *Lagria unicolor* Borchmann, 1942
bhutanicola Merkl, 2019a: 67 **A:** BT
spinulicornis Merkl, 2019a: 70 **A:** BT
tibetana Blair, 1927: 249 **A:** NP XIZ
wangduensis Merkl, 2019a: 71 **A:** BT
subgenus *Apteronympha* Seidlitz, 1898b: 336 type species *Lagria rubida* Graells, 1858
rubida Graells, 1858: 74 **E:** SP
elliptica Fairmaire, 1880c: 242
parvula Perris, 1864: 291
rugosula Rosenhauer, 1856: 223 **E:** FR IT PT SP SZ **N:** MO
glabrata G.-A. Olivier, 1797: 446 [HN]
melichari Reitter, 1890e: 173
pubescens Latreille, 1804a: 351 [HN]
tristis Bonelli, 1812: 172 [HN]
tenenbaumi Pic, 1929a: 2 **A:** IS LE
subgenus *Lagria* Fabricius, 1775: 124 type species *Chrysomela hirta* Linnaeus, 1758
Lachna Billberg, 1820: 35 type species *Chrysomela hirta* Linnaeus, 1758
atripes Mulsant & Guillebeau, 1855a: 74 **E:** AB AR AU BE BH BU CR CT CZ FR GE GG GR HU IT MC MD PL PT RO SK SL SP ST SZ TR UK YU **A:** IN TM TR
puncticollis Seidlitz, 1898b: 337
tenuicollis Seidlitz, 1898b: 338 [HN]
brevipilis Desbrochers des Loges, 1881: 143 **N:** AG
carinulata Fairmaire, 1891f: ccxvii **A:** HUB
comosella Fairmaire, 1894a: 30 **A:** NP SD **ORR**
conspersa Reitter, 1880b: 256 **A:** GUX NP SD XIZ YUN
multipunctata Fairmaire, 1894a: 31
formosensis Borchmann, 1912a: 6 **A:** JA TAI
grenieri C. Brisout de Barneville, 1867: 181 **E:** FR PT SP
hirta Linnaeus, 1758: 377 (*Chrysomela*) **E:** AB AL AN AR AU BE BH BU BY CR CT CZ DE EN ES FI FR GB GE GG GR HU IR IS IT KZ LA LS LT LU MC MD NL NR NT PL PT RO SK SL SP ST SV SZ TR UK YU **N:** MO **A:** ES CY IN IQ KZ TD TM TR UZ WS
caucasica Motschulsky, 1860c: 144
depilis Mulsant, 1856: 21
flava Geoffroy, 1758: 155 (*Cantharis*)
fuscata Motschulsky, 1860: 144
glabrata Fabricius, 1775: 125
limbata Desbrochers des Loges, 1881: 142

lurida Krynicki, 1832: 137
nudipennis Mulsant, 1856a: 20
pontica Motschulsky, 1860c: 144
seminuda Reitter, 1889e: 259
spadicea Scopoli, 1763: 43 (*Cantharis*)
villosa DeGeer, 1775: 44 (*Tenebrio*)
ignota Borchmann, 1941: 22 (*Xenocera*) **A:** FUJ **ORR**
indicola Bates, 1879: 483 **A:** HP NP PA
costatula Borchmann, 1909b: 714 [RN]
subcostata Fairmaire, 1896a: 59 [HN]
klapperichi Pic, 1955c: 32 **A:** FUJ
semiobscurior Pic, 1955c: 32
kondoi Masumoto, 1988: 35 **A:** SE TAI
lameyi Fairmaire, 1893g: 325 **A:** FUJ GUA GUI HUB HUN JIX NMO SCH **ORR**
annulipes PIC, 1955c: 32 [HN]
geniculata Seidlitz, 1898b: 339
lata Fabricius, 1801b: 70 **E:** SP **N:** AG MO TU
rugata Desbrochers des Loges, 1881: 145
laticollis Motschulsky, 1860c: 143 **E:** CT **A:** ES MO WS FE
moutoni Pic, 1928d: 4 **A:** SHG
notabilis Lewis, 1896: 341 **A:** JA (Ryukyus)
nigricollis Hope, 1843: 63 **A:** FE FUJ HEN HUB HUN JA JIX LIA NC SC SCH XIN ZHE
babai Nakane, 1994b: 243
picea Brancsik, 1914: 58
subtilipunctata Seidlitz, 1898b: 340
oharai Masumoto, 1988: 34 **A:** TAI
okinawana M. Chûjô, 1972: 13 **A:** JA (Ryukyus)
okinawana M. Chûjô, 1985: 88 (*Cerogria*) [HN]
ophthalmica Fairmaire, 1891f: ccxvi **A:** HUB
pallidipennis Borchmann, 1936: 33 **A:** GUI FUJ SHN
paracomosella Merkl, 1991a: 6 **A:** NP
pici Normand, 1936a: 161 **N:** TU
picta Borchmann, 1911: 203 **A:** HKG YUN **ORR**
coadunata Borchmann, 1932: 96
rubella Borchmann, 1932: 95 **A:** SCH XIZ YUN
rubiginea Fairmaire, 1886c: 347 **A:** CH
rufipennis Marseul, 1876d: 337 **A:** BEI FE HUB GAN JA JIX NC NIN SC SCH SHA
vervex Marseul, 1876d: 338
rufiventris Pic, 1928d: 4 **A:** XIZ
schawalleri Merkl, 1991a: 8 **A:** NP SD UP
scutellaris Pic, 1910v: 259 **A:** TAI
sakaii Masumoto, 1988: 36

theryi Pic, 1938d: 14 **N:** MO
tigrina Fairmaire, 1893g: 325 **A:** GUI GUX YUN **ORR**
tristicula Fairmaire, 1886c: 346 **A:** CH
ventralis Reitter, 1880b: 255 **A:** GUI HP SD NP UP YUN **ORR**
villosa Fabricius, 1781: 160 **A:** YE **AFR NTR**
confusa Reiche, 1850: 374
obscura Fabricius, 1792b: 79

genus *Neogria* Borchmann, 1911: 222 type species *Neogria sulcipennis* Borchmann, 1911
cyanipennis Borchmann, 1911: 226 **A:** NP **ORR**

genus *Schevodera* Borchmann, 1936: 59 type species *Lagria hirticollis* Borchmann, 1911
gracilicornis Borchmann, 1911: 204 (*Lagria*) **A:** YUN **ORR**
inflata Borchmann, 1925: 348 (*Lagria*) **A:** YUN **ORR**

genus *Xanthalia* Fairmaire, 1894e: 395 [RN] type species *Xanthia curticollis* Fairmaire, 1893
Diversogria Pic, 1954c: 229 type species *Heterogria lepersonnei* Pic, 1954
Heterogria Fairmaire, 1896a: 42 type species *Heterogria punctatissima* Fairmaire, 1896
Lagriocera Fairmaire, 1896a: 41 type species *Lagriocera cavicornis* Fairmaire, 1896
Lagriodes Borchmann, 1930d: 432 type species *Heterogria armigera* Borchmann, 1930
Pachylagria Borchmann, 1912b: 17 type species *Pachylagria ovata* Borchmann, 1912
Wallardilagria Pic, 1910q: 74 type species *Heterogria pallidicolor* Pic, 1910
Xanthia Fairmaire, 1893b: 31 [HN] type species *Xanthia curticollis* Fairmaire, 1893
atricolor Pic, 1920f: 15 (*Heterogria*) **A:** SD
cavicornis Fairmaire, 1896a: 41 (*Lagriocera*) **A:** NP **ORR**
clavata Merkl, 1991a: 13 **A:** NP
franzi Kaszab, 1973a: 40 **A:** NP
nepalica Kaszab, 1973a: 42 **A:** NP
nigrovittata Pic, 1910q: 75 (*Lagriocera*) **A:** HAI NP SCH SHA YUN **ORR**
brunneovittata Pic, 1918d: 15 (*Heterogria*)
martensi Merkl, 1991a: 12
reducta Pic, 1955a: 2 (*Heterogria*) **A:** FUJ
obliterata Pic, 1955a: 3 (*Heterogria*)
serrifera Borchmann, 1930d: 440 (*Heterogria*) **A:** FUJ NC SCH
sinensis Pic, 1918d: 15 (*Heterogria*) **A:** CH
spinosa Kaszab, 1965d: 289 **A:** FUJ **ORR**

genus *Xenocerogria* Merkl, 2007: 269 [RN] type species *Lagriocera feai* Borchmann, 1911
Xenocera Borchmann, 1936: 116 [HN] type species *Lagriocera feai* Borchmann, 1911
feai Borchmann, 1911: 209 (*Lagriocera*) **A:** YUN **ORR**
ruficollis Borchmann, 1912a: 7 (*Lagriocera*) **A:** CHQ FUJ GUA GUI HUB HUN JIA TAI ZHE
xanthisma Chen 2002: 178 (*Xenocera*)

subtribe Statirina Blanchard, 1845

genus *Anisostira* Borchmann, 1915c: 296 type species *Anisostira varicolor* Borchmann, 1915 (= *Macrolagria rugipennis* Lewis, 1896)

elegans M. Chûjô, 1959: 8 **A:** JA (Ryukyus) TAI

rugipennis Lewis, 1896: 341 (*Macrolagria*) **A:** FUJ GUA GUI HUB JA (Ryukyus) SC SCH TAI ZHE

abdominalis Kôno, 1929a: 33

abnormipes Borchmann, 1912a: 10 (*Nemostira*)

cognata Borchmann, 1912a: 11 (*Nemostira*)

lucidicollis Borchmann, 1915c: 298

ferriei Pic, 1914r: 304 (*Nemostira*)

flavipes Kôno, 1929a: 33

kikuchii Kôno, 1929a: 34

nigripes Pic, 1911q: 190 (*Nemostira*)

rufoscutellaris Pic, 1914n: 76 (*Nemostira*)

similaris Borchmann, 1915c: 297

sinuatipes Pic, 1911b: 7 (*Nemostira*)

testaceithorax Pic, 1914r: 305 (*Nemostira*)

varicolor Borchmann, 1915c: 298

genus *Arthromacra* Kirby, 1837: 238 type species *Arthromacra donacioides* Kirby, 1837 (= *Lagria aenea* Say, 1824)

Macrarthra Agassiz, 1846b: 219 [UE]

amamiana Nakane, 1963a: 30 **A:** JA (Ryukyus)

arimotoi Akita & Masumoto, 2009b: 248 **A:** JA

bhutanica Merkl, 2011a: 302 **A:** BT

bicolor Saitô, 1995: 178 **A:** TAI

chifengi Merkl, 2004: 300 **A:** TAI

cuprina Borchmann, 1942a: 11 **A:** YUN **ORR**

decora Marseul, 1876d: 339 (*Lagria*) **A:** JA

apicalis Kôno, 1929a: 31

bicoloripes Pic, 1917f: 16

obscuripes Pic, 1917f: 16

donckieri Pic, 1910s: 86 **A:** YUN

entypodera Chen & X. Yang, 1997: 2 **A:** FUJ

flavipes Nakane, 1994a: 85 **A:** JA

formosana Sasaji, 1986: 11 **A:** TAI

gracilis Borchmann, 1915a: 133 **A:** FUJ

higoniae Lewis, 1895a: 278 **A:** JA

imasakai Saitô, 1995: 180 **A:** TAI

ishizuchiensis Akita, 2012: 216 **A:** JA [RN]

imasakai Akita, 2011: 271 [HN]

kasuga Ando, 2010: 241 **A:** JA
kimioi Merkl, 2011a: 304 **A:** NP
kinodai Nakane, 1997: 5 **A:** JA
kyushuensis Nakane, 1997: 4 **A:** JA
majuscula Nakane, 1994a: 84 **A:** JA
masumotoi Merkl, 2011a: 307 **A:** BT
minuta Masumoto, 1988: 39 **A:** TAI
narukawai Imasaka, 2005: 91 **A:** JA (Honshu)
okiana Imasaka, 2005: 100 **A:** JA (Oki Is.)
oodaigahara Imasaka, 2005: 80 **A:** JA (Honshu)
pretiosa Reiche, 1862c: 544 **E:** GG ST (Kavkaz) **A:** TR
rubidorsalis Chen & X. Yang, 1997: 3 **A:** ZHE
schawalleri Merkl, 2011a: 307 **A:** NP
shiraishii Imasaka, 2005: 97 **A:** JA (Shikoku)
sikkimensis Borchmann, 1915a: 132 **A:** SD
subopaca Pic, 1910s: 86 (*Chlorophila*) **A:** YUN
sumptuosa Lewis, 1895a: 277 **A:** JA
takahashii Imasaka, 2005: 81 **A:** JA (Honshu)
takedai Saitô & Ando, 2016: 97 **A:** TAI
tana Saitô, 1995: 182 **A:** TAI
tibetana Chen & X. Yang, 1997: 3 **A:** XIZ
tsurimakii Akita & Masumoto, 2019: 90 **A:** JA (Honshu)
tsuyukii Masumoto, 1988: 38 **A:** TAI
viridissima Lewis, 1895a: 277 **A:** JA SC (Chejudo)
yakushimana Imasaka, 2005: 93 **A:** JA (Yakushima)

genus *Arunogria* Merkl, 1991a: 14 type species *Arunogria pubescens* Merkl, 1991
pubescens Merkl, 1991a: 17 **A:** NP

genus *Casnonidea* Fairmaire, 1882c: 264 type species *Casnonidea holomelaena* Fairmaire, 1882
subgenus *Casnonidea* Fairmaire, 1882c: 264 type species *Casnonidea holomelaena* Fairmaire, 1882
Falsocasnonidea Pic, 1934: 31 type species *Nemostira diversipennis* Pic, 1923
Synatractus W.J. Macleay, 1887: 312 type species *Synatractus variabilis* W.J. Macleay, 1887
nigripennis Fairmaire, 1894: 32 **A:** BT NP SD
occipitalis Borchmann, 1912a: 8 (*Nemostira*) **A:** JA (Ryukyus) TAI
patruelis Borchmann, 1936: 384 **A:** SD
rufipennis Borchmann, 1912a: 9 (*Nemostira*) **A:** FUJ TAI
tibialis Fairmaire, 1893f: 301 **A:** YUN
viridimetallica Pic, 1911b: 7 (*Nemostira*) **A:** TAI
subgenus *Pilosocasnonidea* Pic, 1934e: 31 type species *Nemostira bipartita* Pic, 1911
terminata Fairmaire, 1889a: 365 **A:** HAI ORR

genus *Chlorophila* Semenov, 1890: 374 type species *Lagria portschinskii* Semenov, 1890
campestris Fairmaire, 1894b: 221 **A:** SCH YUN
cyanea Pic, 1915g: 15 **A:** FUJ
davidi Fairmaire, 1898d: 399 **A:** SCH
portschinskii Semenov, 1890: 374 **A:** FUJ GAN NIN SCH SD SHA YUN
basipennis Fairmaire, 1897j: 229
carolina Fairmaire, 1899: 632
immarginata Fairmaire, 1897j: 231
ingeniculata Pic, 1917f: 15
nitidicollis Fairmaire, 1897j: 230
obscuripennis Fairmaire, 1894b: 220
semenowi Fairmaire, 1899: 633 **A:** FUJ HUB

genus *Donaciolagria* Pic, 1914b: 14 type species *Donaciolagria impressipennis* Pic, 1914
admirabilis Sasaji, 1986: 9 (*Chlorophila*) **A:** TAI
anthracina Merkl, 2019b: 1017 **A:** SCH
densicornis Merkl, 2019a: 79 **A:** BT
femoralis Borchmann, 1915a: 135 (*Arthromacra*) **A:** GUI FUJ SCH
flavitibialis Chen, 1997a: 125 **A:** SCH
kurosawai Masumoto, 1988: 40 (*Chlorophila*) **A:** TAI
malgorzatae Merkl, 2011b: 362 **A:** YUN **ORR**
pici Borchmann, 1936: 179 **A:** NP SD
viridimetallica Chen, 1997a: 124 **A:** ANH SCH

genus *Exostira* Borchmann, 1925: 353 type species *Exostira sellata* Borchmann, 1925
punctator Borchmann, 1930d: 523 **A:** TAI
schroederi Borchmann, 1936: 421 **A:** FUJ GUA GUI JIX TAI

genus *Hosohamudama* Masumoto, 1988: 43 type species *Hosohamudama sasajii* Masumoto, 1988
abnormalis Kôno, 1929a: 31 (*Arthromacra*) **A:** TAI
hanatanii Masumoto & Akita, 2008: 38 **A:** JA (Ryukyus) TAI
miyakei Masumoto, 1988: 45 **A:** TAI
sasajii Masumoto, 1988: 44 **A:** TAI

genus *Lagriogonia* Fairmaire, 1891: ccxvii type species *Lagriogonia humerosa* Fairmaire, 1891
humerosa Fairmaire, 1891: ccxvii **A:** HUB

genus *Macrolagria* Lewis, 1895a: 422 type species *Statyra rufobrunnea* Marseul, 1876
denticollis Fairmaire, 1891: ccxviii (*Casnonidea*) **A:** FUJ HUB ZHE
hirsuta Lewis, 1895a: 423 **A:** FE (Kuril Is.) JA
robusticeps Lewis, 1895a: 276 (*Eutrapela*) **A:** FE (Kuril Is.) JA

rufobrunnea Marseul, 1876d: 340 (*Statyra*) **A:** JA
fujisana Lewis, 1895a: 422
shibatai Masumoto, 1987a: 51 **A:** JA

genus *Merklia* Chen, 1997b: 307 type species *Merklia bimaculata* Chen, 1997
bimaculata Chen, 1997b: 307 **A:** YUN **ORR**

genus *Mimoborchmania* Pic, 1934e: 31 type species *Nemostira coloripes* Pic, 1922
yangi Merkl & Chen, 1997: 113 **A:** YUN **ORR**

genus *Odontocerostira* Merkl, 2007: 269 [RN] type species *Odontocera qinlingensis* Chen & Yuan, 1996
Odontocera Chen & Yuan, 1996: 183 [HN] type species *Odontocera qinlingensis* Chen & Yuan, 1996
qinlingensis Chen & Yuan, 1996: 184 (*Odontocera*) **A:** SHA TAI
hexamaculata Chen & Yuan, 1996: 185 (*Odontocera*) **A:** SCH SHA

genus *Pachystira* Chen, 1997b: 308 type species *Pachystira impressipennis* Chen, 1997
impressipennis Chen, 1997b: 309 **A:** YUN

genus *Sora* Walker, 1859a: 259 type species *Sora marginata* Walker, 1859
subgenus *Sora* Walker, 1859a: 259 type species *Sora marginata* Walker, 1859
Nemostira Fairmaire, 1869a: 815 type species *Nemostira coquerelii* Fairmaire, 1869
aeneipennis Pic, 1927f: 5 (*Nemostira*) **A:** CH
angulicollis Pic, 1920f: 15 (*Nemostira*) **A:** CH
barapanica Merkl, 2019a: 76 **A:** BT
cavaleriei Pic, 1937h: 148 **A:** GUI
crenulipennis Fairmaire, 1896b: 120 (*Nemostira*) **A:** HP
inopaca Pic, 1923c: 16 (*Nemostira*) **A:** CH
humeralis Pic, 1907m: 182 (*Nemostira*) **A:** SCH
klapperichi Pic, 1955c: 31 **A:** FUJ
longissima Pic, 1920f: 15 (*Nemostira*) **A:** CH
marmoreipennis Merkl, 2019a: 78 **A:** BT NP
mimica Pic, 1912f: 31 (*Nemostira*) **A:** YUN **ORR**
rugulicollis Fairmaire, 1899: 633 (*Nemostira*) **A:** CH
sinensis Pic, 1912f: 30 (*Nemostira*) **A:** YUN
subcordicollis Pic, 1937f: 35 **A:** CH
subrecticollis Pic, 1914h: 44 (*Nemostira*) **A:** HUB
thibetana Pic, 1914r: 304 (*Nemostira*) **A:** XIZ

genus *Splichalia* Reitter, 1913d: 664 type species *Splichalia tigrinella* Reitter, 1913
andrewesi Borchmann, 1936: 340 **A:** KA

nigrofasciata Pic, 1912l: 63 (*Nemostira*) **A:** KA PA UP
tigrinella Reitter, 1913d: 665 **A:** KA

genus *Strongylagria* Pic, 1915c: 4 type species *Strongylagria metallica* Pic, 1915
metallica Pic, 1915c: 5 **A:** XIZ

genus *Taiwanolagria* Masumoto, 1988: 41 type species *Taiwanolagria merkli* Masumoto, 1988
merkli Masumoto, 1988: 41 **A:** FUJ GUI SCH TAI

tribe Lupropini Lesne, 1926

genus *Bolitrium* Gebien, 1914d: 390 type species *Bolitrium crenulicolle* Gebien, 1914
crenulicolle Gebien, 1914d: 391 **A:** TAI **ORR**
 chinensis Kaszab, 1940c: 157 (*Lorelus*)
sasajii Masumoto & Akita, 2001c: 248 (*Lorelus*) **A:** JA (Ryukyus)

genus *Falsotithassa* Pic, 1934a: 18 type species *Falsotithassa sumatrana* Pic, 1934
 Derispiolina Kaszab, 1979: 279 type species *Derispiolina pterolomoides* Kaszab, 1979
sumatrana Pic, 1934a: 19 **A:** BT GUX NP **ORR**

genus *Indenicmosoma* Ardoin, 1964: 688 type species *Enicmosoma indochinensis* Kaszab, 1940
ardoini Kaszab, 1970a: 176 **A:** AF
indochinensis Kaszab, 1940c: 158 (*Enicmosoma*) **A:** JA NP TAI **ORR**
nathani Ardoin, 1964: 689 **A:** BT NP **ORR**
pocsi Kaszab, 1965d: 291 **A:** NP **ORR**

genus *Luprops* Hope, 1833: 63 type species *Luprops chrysophthalmus* Hope, 1833
 Etazeta Fairmaire, 1889a: 358 type species *Etazeta aeneicolor* Fairmaire, 1889
 Oligorus Dejean, 1834: 206 type species *Tagenia indica* Wiedemann, 1823
 Syggona Fåhraeus, 1870: 330 type species *Syggona concinna* Fåhraeus, 1870
 Syngona Rye, 1873: 293 [UE]
aeneicolor Fairmaire, 1889a: 359 (*Etazeta*) **A:** FUJ **ORR**
afghanicus Kaszab, 1973b: 57 **A:** AF
bruyanti Pic, 1917d: 18 (*Lyprops*) **A:** YE **AFR**
cribratellus Fairmaire, 1896b: 102 (*Lyprops*) **A:** SD
cribrifrons Marseul, 1876c: 125 (*Lyprops*) **A:** JA SC TAI **ORR**
curticollis Fairmaire, 1896a: 28 (*Lyprops*) **A:** PA SD **ORR**
gracilior Fairmaire, 1896a: 28 (*Lyprops*) **A:** AP BT PA SD **ORR**
horni Gebien, 1914b: 34 (*Lyprops*) **A:** TAI
indicus Wiedemann, 1823: 42 (*Tagenia*) **A:** NP
irregularis Gebien, 1914b: 35 (*Lyprops*) **A:** TAI
kaszabi Schawaller, 1997: 296 **A:** NP

luzonicus Gebien, 1914d: 405 (*Lyprops*) **A:** TAI **ORR**
orientalis Motschulsky, 1868: 195 (*Anaedus*) **A:** BT FE FUJ GAN HAI HEB HEI HEN HUB JA JIA JIL JIX LIA MG NC NIN NMO NP SC SCH SHA SHX TAI ZHE **ORR**
sinensis Marseul, 1876c: 126 (*Lyprops*)
tristis Fabricius, 1801b: 70 (*Lagria*) **A:** SD **ORR**
piceus Fairmaire, 1894a: 24 (*Lyprops*)
yunnanus Fairmaire, 1887b: 128 (*Lyprops*) **A:** BT GUA HAI NP YUN

genus *Micropedinus* Lewis, 1894: 379 type species *Micropedinus algae* Lewis, 1894 (= *Heterophaga pullulus* Boheman, 1858)
Noteprataeus Carter, 1924a: 37 type species *Notoprataeus litoralis* Carter, 1924 (= *Mesotertis inconstans* Lea, 1917)
pallidipennis Lewis, 1894: 380 **A:** FE JA SC TAI "East China" **ORR**
maderi Kaszab, 1940c: 154 (*Microlyprops*)
rotundicollis Mt. Chûjô, 1996: 135
pullulus Boheman, 1858: 94 (*Heterophaga*) **A:** FE HKG JA NC SC TAI
algae Lewis, 1894: 379

genus *Sphingocorse* Gebien, 1921: 110 type species *Sphingocorse angulicollis* Gebien, 1921
nepalica Kaszab, 1975b: 122 **A:** NP UP

genus *Spinolyprops* Pic, 1917c: 12 type species *Spinolyprops rufithorax* Pic, 1917
cribricollis Schawaller 2012a: 86 **A:** GUX YUN **ORR**
himalayicus Kaszab, 1965a: 124 **A:** BT SD UP **ORR**
pakistanus Schawaller, 1996c: 692 **A:** PA

subfamily **Phrenapatinae** Solier, 1834

tribe Archaeoglenini Watt, 1974

genus *Archaeoglenes* Broun, 1893a: 188 type species *Archaeoglenes costipennis* Broun, 1893
orientalis Sasaji, 1983: 47 **A:** JA

tribe Penetini Lacordaire, 1859

genus *Clamoris* Des Gozis, 1886: 25 [RN] type species *Phtora crenata* Mulsant, 1854
Phthora Champion, 1893a: 531 [UE]
Phtora Mulsant, 1854: 228 [HN] type species *Phtora crenata* Mulsant, 1854
canalicollis Lewis, 1894: 466 (*Phthora*) **A:** FE JA
crenata Mulsant, 1854: 229 (*Phtora*) **E:** FR IT MA PT SP **N:** AG MO TU
insurgens Des Gozis, 1886: 25
formosana Masumoto, 1982a: 143 (*Phthora*) **A:** TAI
narana Nakane, 1988b: 82 (*Phthora*) **A:** JA (Honshu)
salinae L. Soldati, 2009: 79 (*Phthora*) **A:** QA

genus *Dioedus* J.F. LeConte, 1862: 238 type species *Dioedus punctatus* J.F. LeConte, 1862

Arrhabaeus Champion, 1886a: 144 type species *Arrhabaeus convexus* Champion, 1886

Brachycilibe Carter, 1911: 207 type species *Brachycilibe antennata* Carter, 1911

Tagalus Gebien, 1914d: 388 type species *Tagalus impressicollis* Gebien, 1914

miyakensis Nakane, 1963a: 27 (*Tagalus*) **A:** JA (Izu Is.)

tokaranus Nakane, 1963a: 28 (*Tagalus*)

genus *Taiwanotagalus* Masumoto, 1982a: 143 type species *Taiwanotagalus klapperichi* Masumoto, 1982

klapperichi Masumoto, 1982a: 143 **A:** TAI

nepalicus Schawaller, 1999: 425 **A:** NP

subfamily **Pimeliinae** Latreille, 1802

tribe Adelostomini Solier, 1834

genus *Adelostoma* Duponchel, 1827: 338 type species *Adelostoma sulcata* Duponchel, 1827

subgenus *Adelostoma* Duponchel, 1827: 338 type species *Adelostoma sulcata* Duponchel, 1827

Polyscopus Waltl, 1853: 73 type species *Polyscopus costatus* Waltl, 1835 (= *Adelostoma sulcatum* Duponchel, 1827)

bedeli Reitter, 1885c: 317 **N:** AG

bicarinata Waterhouse, 1881: 475 **A:** YE (Suqutra)

cordata Solier, 1837: 169 **N:** EG LB **A:** IS SI

granti Gahan, 1900: 10 **A:** YE (Abd al-Kuri, Suqutra)

granulithorax Escalera, 1914: 292 **N:** MO

hauseri Reitter, 1895e: 203 **A:** LE

longiceps Reitter, 1895e: 203 **N:** TU

metatarsale Carl, 1991a: 25 **A:** IQ

ovalipenne Chevrolat, 1877a: 117 **A:** TR

schawalleri Purchart, 2009: 24 **A:** YE (Suqutra)

subtile arabica Kaszab, 1981c: 366 **A:** SA

subtile sabulosa Kaszab, 1981c: 366 **A:** SA

subtile subtile Reitter, 1900d: 193 **A:** IN IS JO SY TR

sulcata crassicorne Peyerimhoff, 1931b: 97 **N:** AG MO

sulcata sulcata Duponchel, 1827: 338 **E:** CR SP **N:** AG EG LB MO **A:** CY IQ IS JO SI SY TR

carinata Solier, 1837: 168

costata Waltl, 1835: 74 (*Polyscopus*)

cristata Eschscholtz, 1831: 11 (*Nyctoporis*)

deplanata Haag-Rutenberg, 1875d: 422

grandiforme Koch, 1935a: 39

nitida Haag-Rutenberg, 1875d: 421

parallela Haag-Rutenberg, 1875d: 421

parallela Chevrolat, 1877a: 117 [HN]

scabra Chevrolat, 1877a: 118

waterhousei Schawaller, 2004c: 447 **A:** YE (Samha, Suqutra)
subgenus *Omandelostoma* Purchart, 2017: 282 type species *Omandelostoma muqalensis* Purchart, 2017
muqalensis Purchart, 2017: 283 **A:** OM
subgenus *Zarudnionymus* Semenov & Bogatchev, 1947: 175 type species *Zarudnionymus persis* Semenov & Bogatchev, 1947 (= *Adelostoma grande* Haag-Rutenberg, 1879)
Adelostomoides Carl, 1991a: 24 type species *Adelostoma grande* Haag-Rutenberg, 1879
Falsaspila Koch, 1952: 28 type species *Adelostoma batesi* Haag-Rutenberg, 1872
abyssinicum hirsutum Koch, 1935a: 39 **N:** EG **AFR**
abyssinicum obockensis Purchart, 2012b: 24 **A:** OM
batesi Haag-Rutenberg, 1875d: 426 **A:** OM SA YE
grandicolle Fairmaire, 1890: 554
grande Haag-Rutenberg, 1879a: 294 **A:** IN
intermedium J.R. Sahlberg, 1913b: 164
perissinottoi Canzoneri, 1965: 168 (*Falsaspila*)
persis Semenov & Bogatchev, 1947: 176 (*Zarudnionymus*)

genus *Machlopsis* Pomel, 1871: 236 type species *Eurychora levaillantii* Lucas, 1850
subgenus *Hidrosella* Koch, 1952: 32 type species *Hidrosis incostata* Haag-Rutenberg, 1875
incostata Haag-Rutenberg, 1875d: 425 (*Hidrosis*) **N:** MO (Western Sahara) **AFR**
subgenus *Machlopsis* Pomel, 1871: 236 type species *Eurychora levaillantii* Lucas, 1850
Hidrosis Haag-Rutenberg, 1875c: 120 type species *Steira crenatocostata* L. Redtenbacher, 1868
alata Fairmaire, 1875a: 523 (*Eurychora*) **N:** TU
crenatocostata L. Redtenbacher, 1868: 120 (*Steira*) **N:** EG LB TU **A:** IS SI SY **AFR**
aegyptiaca Kirsch, 1871: 389 (*Steira*)
squalida Baudi di Selve, 1875c: 65 (*Eurychora*)
doderoi Gridelli, 1930: 235 **E:** IT (Lampedusa Is.) **N:** LB
elongatula G. Quedenfeldt, 1890: 395 (*Hidrosis*) **N:** LB TU
levaillantii Lucas, 1850b: vii (*Eurychora*) **N:** AG
pseudoalata Gridelli, 1930: 236 **N:** EG LB
subfusca Pic, 1910m: 42 **N:** AG

genus *Pogonobasis* Solier, 1837: 161 type species *Pogonobasis opatroides* Solier, 1837 (= *Eurychora rugosula* Guérin-Méneville, 1831)
Peristeptus Haag-Rutenberg, 1875d: 382 type species *Pogonobasis laevigata* Gerstaecker, 1854
opaca Haag-Rutenberg, 1875d: 388 **A:** SA YE
ornata Solier, 1837: 163 **N:** EG **AFR**

genus *Serrichora* Koch, 1952: 15 type species *Eurychora crenata* Solier, 1837
arabica Kaszab, 1981c: 364 **A:** OM

tribe Adesmiini Lacordaire, 1859

genus *Adesmia* Fischer von Waldheim, 1822: 153 type species *Adesmia longipes* Fischer von Waldheim, 1822 (= *Pimelia anomala* Fischer von Waldheim, 1820)

subgenus *Adesmia* Fischer von Waldheim, 1822: 153 type species *Adesmia longipes* Fischer von Waldheim, 1822 (= *Pimelia anomala* Fischer von Waldheim, 1820)

Sarachus Gistel, 1848: viii [URN]

anomala anomala Fischer von Waldheim, 1820: pl. 14 (*Pimelia*) **A:** UZ

longipes Fischer von Waldheim, 1822: 153 [HN]

anomala dejeanii Gebler, 1841b: 589 **A:** ES KZ MG XIN

aweinsis Ren & Wang, 1993a: 3 **A:** XIN

beduina Reitter, 1916a: 21 **N:** TU

biskreensis asperrima Peyerimhoff, 1931b: 94 **N:** MO

biskreensis biskreensis Lucas, 1844: 264 **N:** AG TU

immargo Reitter, 1916a: 25

biskreensis recticollis Peyerimhoff, 1931b: 92 **N:** AG **AFR**

fischeri caspia Bogatchev, 1964b: 177 **E:** AB

fischeri fischeri Faldermann, 1837: 22 **A:** IN TR

ehrenbergi Faldermann, 1837: 24

tatjanae Bogatchev, 1941a: 97

fischeri interposita Bogatchev, 1964b: 176 **A:** IN (Azarbayjan)

jugalis conveniens Bogatchev, 1941a: 97 **A:** AF

kabuliensis Kaszab, 1959f: 400

jugalis gracilenta Reitter, 1916a: 29 **A:** TD UZ (Fergana)

jugalis gridellii Kaszab, 1959e: 243 [= 1959f: 399] **A:** AF

jugalis jugalis Reitter, 1916a: 29 **A:** PA XIN

jugalis kasyi Kaszab, 1970a: 10 **A:** AF

jugalis kulzeri Kaszab, 1959f: 401 **A:** AF

jugalis septentrionalis Kaszab, 1973b: 40 **A:** AF

karelini karelini Fischer von Waldheim, 1835: 312 **E:** AB **A:** IN KZ TM UZ XIN

karelini qalatensis Kaszab, 1970a: 12 **A:** AF

lehmanni Ménétriés, 1849: 224 **A:** KZ

biseriata Reitter, 1916a: 26

intersticta Reitter, 1916a: 26

septemcostata Reitter, 1916a: 27

maillei Solier, 1835c: 530 **E:** AB AR **A:** IQ IN TR

pulcherrima Fischer von Waldheim, 1835: 315

strophium Fischer von Waldheim, 1838: 316

montana acervata Klug, 1830: nr. 26, pl. 12 **N:** AG EG LB TU

dubia Solier, 1835c: 527

klugii Solier, 1835c: 539

montana hamadae Koch, 1937: 393 **N:** LB (Hamada)

montana minor Gridelli, 1930: 232 **N:** EG LB
montana montana Klug, 1830: nr. 25, pl. 12 (*Pimelia*) **A:** IQ IS JO SA SI
perpolita Reitter, 1916a: 24
montana parallela Miller, 1861: 171 **N:** EG **AFR**
muricata Linnaeus, 1758: 418 (*Tenebrio*) **N:** EG **A:** YE
austera Baudi di Selve, 1881: 275
tuberculosa Champion, 1896b: 250 [RN]
tuberculifera Gahan, 1896a: 289
panderi iranica Bogatchev, 1947a: 15 **A:** IN
panderi lindbergi Kaszab, 1970a: 11 **A:** AF
panderi panderi Fischer von Waldheim, 1835: 313 **A:** AF KZ UZ XIN
laevicollis Kraatz, 1882c: 107
panderi pupillata Kaszab, 1970a: 11 **A:** AF
planidorsis Reitter, 1916a: 28 **A:** TD TM UZ
qapqalensis Ren & Wang, 1993a: 4 **A:** XIN
tenebrosa Solier, 1835c: 536 **A:** IN
punctipennis Haag-Rutenberg, 1875a: 18
subgenus *Adesmina* Reitter, 1916a: 29 type species *Adesmia arabica* Reitter, 1916
arabica arabica Reitter, 1916a: 30 **A:** OM
arabica guichardi Kaszab, 1981c: 362 **A:** OM
arabica thesigeri Gridelli, 1953: 32 **A:** OM YE
arabica wittmeri Kaszab, 1981c: 362 **A:** OM
persiana Reitter, 1916a: 30 **A:** IN PA
subgenus *Macradesmia* Löbl & Merkl, 2020 [see New Acts] type species *Pimelia cancellata* Klug, 1830
aberrans Kwieton, 1978a: 8 **N:** AG (Tassili-n'Ajjer)
anatolica Ganglbauer, 1905: 275 **A:** TR SY
anthracina Klug, 1830: nr. 28, pl. 13 (*Pimelia*) **A:** IN IS LE SY TR
libana Baudi di Selve, 1875c: 30
olivieri Reiche & Saulcy, 1857: 224
procera Miller, 1861: 170
ramburi Solier, 1835c: 529
antiqua Klug, 1830: nr. 35, pl. 13 (*Pimelia*) **N:** EG **A:** IQ **AFR**
aethiopia Allard, 1885a: 182
asperata Baudi di Selve, 1875c: 29 **A:** YE
assimilis Gahan, 1896a: 290 **A:** YE
cancellata abbreviata Klug, 1830: nr. 27, pl. 13 (*Pimelia*) **A:** SI
aspera Reitter, 1916a: 20
hieroglyphica Baudi di Selve, 1875c: 30
latreillei Solier, 1835c: 540
sinaitica Crotch, 1872: 267

cancellata cancellata Klug, 1830: nr. 37, pl. 13 (*Pimelia*) **A:** BA IS JO KU OM PA QA SA SY YE
cancellata clathrata Solier, 1835c: 541 **A:** IN IQ OM SA SY
babylonica Reitter, 1916a: 14
cancellata hadramautina Kaszab, 1981c: 360 **A:** YE
conjungens Koch, 1951: 130 **A:** IQ TR
gibbula Reitter, 1916a: 17 **A:** TR
interrupta interrupta Klug, 1830: nr. 36, pl. 13 (*Pimelia*) **A:** SA YE
interrupta pyriformis Schuster, 1938a: 57 **A:** SA YE
mesopotamica Chevrolat, 1877a: 113 **A:** IN IQ
mimica Reitter, 1916a: 15 **A:** IQ
monilis fougieri Koch, 1937: 390 **N:** LB
monilis monilis Klug, 1830: nr. 38, pl. 13 (*Pimelia*) **N:** EG LB
monilis pluriseriata Schuster, 1928e: 375 **N:** LB
monilis reducta Schuster, 1928e: 375 **N:** LB
sculptilis Reitter, 1916a: 17 **N:** LB
servillei chaldaea Koch, 1951: 124 **A:** IN
servillei fagergreenii Baudi di Selve, 1874a: 92 **A:** IN
servillei holdhausi Reitter, 1916a: 13 **A:** IN
servillei parisii Koch, 1940d: 287 **A:** AF
servillei provincialis Kaszab, 1959f: 399 **A:** AF PA
servillei schatzmayri Koch, 1940d: 286 **E:** ST **A:** AF IN KZ TM UZ
servillei schusteri Reitter, 1916a: 13 **A:** IN PA
servillei servillei Solier, 1835c: 538 **A:** IN
immarginicollis Koch, 1951: 124
ulcerosa arca Reiche & Saulcy, 1857: 222 **A:** IS LE
ulcerosa confusa Ardoin, 1978b: 310 **A:** IS
ulcerosa ulcerosa Klug, 1830: nr. 34, pl. 13 (*Pimelia*) **A:** IS JO SY
perplexa Schaum, 1864: 671 [RN]
reticulata Solier, 1835c: 528 [HN]
wahabita Koch, 1951: 129 **A:** YE
elateralis Koch, 1951: 129
subgenus *Oteroscelis* Solier, 1835c: 546 type species *Adesmia pulcherrima* Solier, 1835 (= *Adesmia audouini* Solier, 1835)
Heteroscelis Agassiz, 1846: 181, 266 [UE]
aenescens Kühnelt, 1951: 245 **A:** IN KU
allardi allardi Gebien, 1910a: 83 [RN] **A:** SA
fossulata Allard, 1885a: 165 [HN]
allardi interpunctata Schuster, 1938a: 56 **A:** YE
audouini aegrota Faldermann, 1837: 25 **A:** IN

audouini audouini Solier, 1835c: 550 **A:** IN
faldermanni Fischer von Waldheim, 1835: 314
nodulosa Faldermann, 1837: 26
pulcherrima Solier, 1835c: 546
audouini kandaharica Kaszab, 1959f: 397 **A:** AF
carinata carinata Solier, 1835c: 547 **A:** IN IQ JO SA SY TR
aenea L. Redtenbacher, 1850: 48
elevata Solier, 1835c: 549
carinata persica Koch, 1940d: 284 **A:** IN
cothurnata cothurnata Forskål, 1775: 80 (*Tenebrio*) **N:** EG **A:** AE BA IQ IS JO OM SA SI YE **AFR**
bicarinata Klug, 1830: nr. 31, pl. 13 (*Pimelia*)
cursor Chevrolat, 1877a: 113
extensa Klug, 1830: nr. 29, pl. 13 (*Pimelia*)
glabrior Schatzmayr & Koch, 1934b: 23
lefebvrii Fischer von Waldheim, 1835: 315
omanensis Kaszab, 1981c: 353
pariseti Solier, 1835c: 552
subserrata Chevrolat, 1877a: 113
villae Osculati, 1844: 72
cothurnata montisatri Peyerimhoff, 1929b: 156 **N:** AG (Hoggar) LB
cothurnata rathjensi Schuster, 1938a: 56 **A:** AE SA YE
coucyi Lucas, 1850c: xi **N:** AG
devecchii Osculati, 1844: 72 **E:** AR **A:** IN SY
filippovi Bogatchev, 1941a: 98 **A:** YE
gebleri elliptica Reitter, 1916a: 6 **A:** AF TM
gebleri gebleri Gebler, 1844: 101 **A:** AF KZ TD TM
oschana Reitter, 1916a: 6
khaliensis Blair, 1931b: 270 **A:** OM QA SA
lacunosa fossulata Allard, 1885a: 191 **A:** SA YE
lacunosa lacunosa Klug, 1830: nr. 33, pl. 13 (*Pimelia*) **A:** SA YE
polita Baudi di Selve, 1874a: 95
semiglabra Chevrolat, 1877a: 113
legatoria Bogatchev, 1941a: 99 **A:** AF
lii Ren & Wang, 1993a: 1 **A:** XIZ
metallica atlantica Escalera, 1935a: 296 **N:** MO (Ifni)
metallica besnardi Kaszab, 1981c: 356 **A:** SA
metallica brozai Ardoin, 1978b: 300 **A:** IS SI
metallica laevior Ardoin, 1978b: 299 **A:** IS SI
metallica mediocostata Koch, 1940d: 273 **N:** AG MO

metallica metallica Klug, 1830: nr. 23, pl. 12 (*Pimelia*) **N:** AG EG LB MO TU
andresi Böhm, 1908: 165
chakouri Reitter, 1916a: 6
faremontii Lucas, 1844: 264
ganglbaueri Reitter, 1916a: 10
metallica syriaca Baudi di Selve, 1874a: 95 **A:** IN IS JO
carmelitana Reitter, 1916a: 8
fortesculpta Reitter, 1916a: 8
micrognanulosa Carl, 1992b: 332 [RN] **A:** IQ
microgranulata Carl, 1990a: 73 [HN]
scorteccii Koch, 1965a: 117 **A:** YE
sodalis elegantissima Koch, 1940d: 282 **A:** IN (Beluzhistan)
sodalis ghazniana Kaszab, 1974: 194 **A:** AF
sodalis laghmanica Kaszab, 1959f: 396 **A:** AF
sodalis sodalis Waterhouse, 1889: 125 (*Adesmia*) **A:** AF PA (Balochistan)
afghanica Koch, 1940d: 280
belutschistana Schuster, 1930b: 237
stoeckleini stoeckleini Koch, 1940d: 271 **A:** JO SA
stoeckleini rasalkhaymana Kaszab, 1981c: 355 **A:** AE
subgenus *Oteroscelopsis* Löbl & Merkl, 2020 [see New Acts] type species *Pimelia dilatata* Klug, 1830
dilatata affinis Solier, 1835c: 535 **N:** AG TU
duponchelii Allard, 1885a: 179
reitteri Koch, 1940d: 297
solieri Lucas, 1844: 266
tibialis Reitter, 1916a: 21
dilatata dilatata Klug, 1830: nr. 32, pl. 13 (*Pimelia*) **N:** EG **A:** IS JO SI
convergens Walker, 1871: 14
drakii Crotch, 1872: 267
insignis Miller, 1861: 172
macropus Solier, 1835c: 542
dilatata douei Lucas, 1844: 265 **N:** AG TU
dilatata getula Peyerimhoff, 1935b: 24 **N:** AG MO
dilatata tripolitana Reitter, 1916a: 22 **N:** LB TU
dilatata tunisia Reitter, 1916a: 22 **N:** TU
dilatata vermiculata Koch, 1940d: 300 **N:** AG MO
longipes Fabricius, 1775: 251 (*Pimelia*) **N:** EG
concisa Walker, 1871: 15
maroccana Haag-Rutenberg, 1877b: 283 **N:** MO
microcephala Solier, 1835c: 533 **N:** AG
zanoni Schuster, 1922b: 16 **N:** LB

genus *Adesmia*, species incertae sedis
baudii Kraatz, 1882c: 106 **A:** IN

tribe Akidini Billberg, 1820

genus *Akis* Herbst, 1799: 124 type species *Pimelia reflexa* Fabricius, 1775
Acis Billberg, 1820: 32 [UE]
acuminata Fabricius, 1787a: 209 (*Pimelia*) **E:** FRi PT SP **N:** MOi
dorsigera Reitter, 1904a: 45
andromeda Bogatchev, 1947a: 15 **A:** IN
bacarozzo Schrank von Paula, 1786: 23 (*Carabus*) **E:** FR IT SP (Baleares)
plicata Latreille, 1816: 280
punctata Thunberg, 1787: 49 (*Pimelia*)
reflexa G.-A. Olivier, 1795: 25
tuberculata Kraatz, 1865a: 249
bayardi Solier, 1836d: 670 **E:** SP PT
carinata Drapiez, 1821: 276
lusitanica Solier, 1836d: 670
bernhaueri Schuster, 1919a: 29 **N:** LB
biskrensis Reitter, 1904a: 43 **N:** AG TU
bremeri Ardoin, 1979: 155 **E:** SP (Formentera)
costitubera Marseul, 1883b: 182 **N:** EG LB
obstituber Champion, 1893: 43
schweinfurthi Quedenfeldt, 1890: 139
pharao Reitter, 1904a: 38
cyrenaica Schuster,1928e: 376 **N:** EG LB
discoidea discoidea Quensel,1806: 138 **E:** SP
discoidea nevadensis Español, 1954c: 114 **E:** SP (Granada)
duplicata Reitter,1904a: 41 **E:** SP
elegans Charpentier, 1825: 216 (*Acis*) **E:** SP **N:** AG MO
carinata Solier, 1836d: 674
kobelti Heyden, 1883b: 236
pachecoi Escalera, 1934b: 303
elevata Solier,1837a: 671 **N:** EG **A:** IQ JO SA SI
elongata elongata Brullé, 1832: 194 **E:** GR
deplanata Waltl, 1836: 461
taygetana Koch, 1948a: 326
elongata opaca Heyden, 1897: 31 **A:** CY TR **E:** AL BU MC
anatolica Reitter, 1904a: 40
otoes Fischer von Waldheim, 1837b: 14
genei Solier, 1836d: 668 **E:** SP
gestroi Schuster, 1922b: 18 **N:** LB (Cyrenaica)

goryi goryi Solier, 1836d: 676 **N:** AG LB MO TU
goryi tatae Théry, 1932a: 191 **N:** MO
granulifera granulifera C.R. Sahlberg, 1823: 11 **E:** PT SP
triseriata zur Strassen, 1957: 54
granulifera ilonka zur Strassen, 1957: 54 **E:** SP
granulifera saltesiana López-Pérez, 2010: 507 **E:** SP
heydeni Haag-Rutenberg, 1876c: 76 **N:** MO
araneipes Antoine, 1957: 53
longipennis Escalera, 1934b: 309
lozanoi Escalera, 1934b: 307
penoti Antoine, 1957: 51
rhatica Antoine, 1957: 47
susica Escalera, 1934b: 309
hispanica Solier, 1836d: 667 **E:** SP
salzei Solier, 1836d: 662
italica Solier, 1836d: 674 **E:** IT **N:** TU, LB
romana Luigioni, 1921: 61
speluncaria Koch, 1939: 266
latreillei Solier, 1836d: 675 **E:** GR (Morea) **A:** CY IQ IS SY TR
maresi Ardoin, 1971a: 49 **N:** TU (Galite)
obtusangula obtusangula Théry, 1932a: 190 **N:** MO
antoinei Escalera, 1934b: 306
obtusangula mideltensis Antoine, 1957: 50 **N:** MO
obtusangula sabasi Antoine, 1957: 50 **N:** MO
perissinottoi Canzoneri, 1971: 155 **A:** IN
reflexa Fabricius, 1775: 252 (*Pimelia*) **A:** IS SI SY **N:** EG
hispida Herbst, 1799: 131
pterygodes Pallas, 1781: 41 (*Tenebrio*)
richteri G. Quedenfeldt, 1889: 354 **N:** MO
rotundicollis Escalera, 1934b: 301
sansi Solier, 1836d: 660 **E:** SP
spinosa Linnaeus, 1764: 101 (*Tenebrio*) **N:** AG (Hoggar) EG LB **A:** IQ IS JO KU QA SA SI SY
sculptior Koch, 1935a: 51
stricticollis Théry, 1931a: 26 **N:** MO
subterranea Solier, 1836d: 656 **E:** IT MA
melitana Reitter, 1904a: 40
subtricostata Redtenbacher, 1850: 48 **A:** IN IQ SY
tingitana tingitana Lucas, 1859b: cxiv **N:** MO
tingitana cordicolis Fairmaire, 1884c: 446 **N:** MO
venerata Antoine, 1926: 256

trilineata trilineata Herbst, 1799: 130 **E:** IT **N:** AG LB TU **A:** SY TR
angusticollis Baudi di Selve, 1875c: 76
olivieri Solier, 1836d: 665
ragusae Reitter, 1904b: 97
sardea Solier, 1836d: 669
terricola Ménétriés, 1840: 36
trilineata algeriana Solier, 1836d: 663 **N:** AG MO
nitida Solier, 1836d: 666
planicollis Solier, 1836d: 664
trilineata barbara Solier, 1836d: 673 **E:** IT (Sardinia) **N:** AG TU

genus *Cyphogenia* Solier, 1836d: 677 type species *Tenebrio auritus* Pallas, 1781
subgenus *Cyphogenia* Solier, 1836d: 677 type species *Tenebrio auritus* Pallas, 1781
Eocyphogenia G.S. Medvedev, 1968c: 897 type species *Akis rugipennis* Faldermann, 1835
aurita aurita Pallas, 1781: 40 (*Tenebrio*) **E:** ST **A:** AF IN KZ TD TM UZ XIN
brevicostata Kraatz, 1882b: 83
truncata Gebler, 1844: 101 (*Akis*)
unicostata Kraatz, 1882b: 84
aurita cratii A.F. Morawitz, 1865: 40 **A:** IN TM
cassiopea Bogatchev, 1952b: 284
chinensis Faldermann, 1835b: 392 (*Akis*) **A:** BEI GAN HEB KZ LIA MG NIN NMO SHA XIN
intermedia Bogatchev, 1962: 45
gibba gibba Fischer von Waldheim, 1820: pl. 15 (*Akis*) **A:** AF IN KI KZ PA TM
angustata Zubkov, 1833: 329 (*Akis*)
zablotzkii Zubkov, 1837: 67 (*Akis*)
gibba persica Baudi di Selve, 1874a: 108 [= 1875c: 78] (*Akis*) **A:** AF IN
humeralis Bates, 1879: 471 **A:** GAN KA NMO XIN XIZ
semicarinata Reitter, 1887d: 360
rugipennis Faldermann, 1835b: 389 (*Akis*) **A:** BEI HEB HEN GAN LIA NMO SHA SHN SHX TIA
subgenus *Lechriomus* A.F. Morawitz, 1865: 21 type species *Akis lucifuga* Adams, 1817
bienerti A.F. Morawitz, 1865: 27 (*Akis*) **A:** AF IN TD TM
bogatshevi G.S. Medvedev, 2004b: 560 **A:** KZ
funesta Faldermann, 1835b: 390 (*Akis*) **A:** "China borealis" "Mongolia"
sepulchralis Faldermann, 1835b: 391 (*Akis*)
limbata Fischer von Waldheim, 1820: pl. 15 (*Akis*) **A:** KZ TM UZ
lucifuga Adams, 1817: 304 (*Akis*) **E:** AB AR GG ST UK **A:** IN IQ KZ SY TM TR
depressa Küster, 1848b: 74 (*Akis*)

genus *Morica* Dejean, 1834: 182 type species *Akis planata* Fabricius, 1801
favieri Lucas, 1859b: cxiii **E:** SP **N:** MO

grossa Linnaeus, 1767: 676 (*Tenebrio*) **N:** EG LB TU

octocostata Solier, 1836d: 649

hybrida Charpentier, 1825: 215 (*Acis*) **E:** SP

obtusa Solier, 1836d: 650

jevini Lucas, 1850a: iv (*Acis*) **N:** AG MO

planata Fabricius, 1801a: 134 (*Akis*) **E:** SP **N:** MO

tingitana Baudi di Selve, 1875c: 75

genus *Sarothropus* Kraatz, 1865a: 268 type species *Akis depressa* Zubkov, 1837

depressus Zubkov, 1837: 68 (*Akis*) **A:** AF KZ TM (Karakumy)

fallax G.S. Medvedev, 1961: 365 **A:** KZ

kiritshenkoi G.S. Medvedev, 1961: 366 **A:** TD TM UZ

medvedevi Bogatchev, 1962: 45 (*Sarathropus*) **A:** TD

genus *Solskyia* Solsky, 1881b: 48 type species *Solskyia peregrina* Solsky, 1881

afghanica Kaszab, 1970a: 48 (*Solskya*) **A:** AF

caporiaccoi Gridelli, 1934: 53 **A:** CH KA

concavidorsis Reitter, 1909f: 296 (*Solskia*) **A:** TD UZ

depressiuscula Fairmaire, 1891d: xcii (*Cyphogenia*) **A:** KA

angusticollis Della Beffa, 1931: 185

inermis Della Beffa, 1931: 185

grombczewskii Semenov, 1889a: 221 (*Akis*) **A:** AF KA PA TD

kaszabi Español, 1961c: 125 **A:** KA

lhasana Ren & Yu, 2000: 325 **A:** XIZ

lopatini G.S. Medvedev, 1968c: 893 **A:** TD

parvicollis Kraatz, 1865a: 251 (*Akis*) **A:** KA XIN XIZ "Himalaya"

kuenluna Kaszab, 1965c: 282

morawitzi Semenov, 1890: 363

plana Bates, 1879: 471 (*Cyphogenia*)

peregrina Solsky, 1881b: 49 **A:** TD UZ (Samarkand)

schmidi Español, 1961c: 128 **A:** HP KA

subperegrina Reitter, 1904a: 49 **A:** "Turkestan"

tribe Asidini Fleming, 1821

genus *Alphasida* Escalera, 1905a: 380 type species *Asida gaditana* Escalera, 1905 (= *Alphasida typica* Gebien 1937)

subgenus *Alphasida* Escalera, 1905a: 380 type species *Asida gaditana* Escalera, 1905 (= *Alphasida typica* Gebien 1937)

almeriensis Escalera, 1901: 174 (*Asida*) **E:** SP

antoniocobosi Martínez Fernández & F. Soldati, 2017: 119 **E:** SP

becerrae Escalera, 1905a: 387 (*Asida*) **E:** SP

bolivari Escalera, 1906e: 382 **E**: SP
calmonti Martínez Fernández & F. Soldati, 2014: 39 **E**: SP
clementei Pérez Arcas, 1865b: 432 (*Asida*) **E**: SP
solieri Rambur, 1838: pl. 19 (*Asida*) [HN]
depilata Escalera, 1906e: 384 **E**: SP
escalerae Oberthür, 1903: 74 (*Asida*) **E**: SP
alpujarrensis Escalera, 1905a: 386 (*Asida*)
lazaroi Escalera, 1906c: 235
leonhardi Reitter, 1917a: 12
fuscopepla Koch, 1940d: 302 **E**: SP
holosericea Germar, 1824: 139 (*Asida*) **E**: SP
bicostata Escalera, 1905a: 386 (*Asida*)
ramburi Solier, 1836c: 438 (*Asida*)
lopezi Escalera, 1906e: 383 **E**: SP
lorcana Pérez Arcas, 1865: 45 (*Asida*) **E**: SP
martinezi Escalera, 1901: 175 (*Asida*) **E**: SP
martini Escalera, 1903: 75 (*Asida*) **E**: SP
mendizabali Koch, 1944b: 234 **E**: SP
merceti Bolívar & Pieltain, 1914: 237 (*Asida*) **E**: SP
oberthueri Escalera, 1901: 175 (*Asida*) **E**: SP
perezverai F. Soldati & Martínez Fernández, 2019: 642 **E**: SP
rufopubescens Escalera, 1905a: 388 (*Asida*) **E**: SP
calva Escalera, 1905a: 388 (*Asida*) [HN]
sanchezgomezi Escalera, 1901: 173 (*Asida*) **E**: SP
subbaetica Obregón y Verdugo, 2012: 54 **E**: SP
typica Gebien, 1937: 717 [RN] **E**: SP
gaditana Escalera, 1905a: 385 (*Asida*) [HN]
volxemi Escalera, 1905a: 387 (*Asida*) **E**: PT
ziziphi Cobos, 1988a: 9 **E**: SP
subgenus *Betasida* Reitter, 1917a: 11 type species *Asida argenteolimbata* Escalera, 1901
Subalphasida Escalera, 1927a: 136 type species *Asida luctuosa* Boisduval, 1835
argenteolimbata Escalera, 1901: 172 (*Asida*) **E**: SP
luctuosa Rambur, 1838: pl. 19 [HN]
espanoli Cobos, 1962: 39 **E**: SP
ferreri Cobos, 1988b: 54 **E**: SP
luctuosa Boisduval, 1835: 250 (*Asida*) **E**: SP
luctuosa Rosenhauer, 1856: 193 (*Asida*) [HN]
minor Rosenhauer, 1856: 194 (*Asida*) [HN]
subgenus *Elongasida* Escalera, 1906d: 306 type species *Asida grandipalpis* Allard, 1869
Cribrasida Reitter, 1917a: 38 type species *Asida grandipalpis* Allard, 1869
Pseudoelongasida Escalera, 1922b: 173 type species *Asida silvestrei* Escalera, 1922

alonensis Martinez y Sáez, 1873b: 409 (*Asida*) **E**: SP
calumniata Escalera, 1906d: 311 (*Asida*) **E**: SP
grandipalpis Allard, 1869a: 288 (*Asida*) **E**: SP
granosa Escalera, 1921a: 118 (*Asida*) **E**: SP
hesperica Rambur, 1838: pl. 20 (*Asida*) **E**: SP
hispalensis Escalera, 1906d: 309 (*Asida*) **E**: SP
levantina Pérez Vera, Avila & Martínez Fernández, 2017: 15 **E**: SP
moroderi Pérez Vera, Avila & Martínez Fernández, 2017: 17 **E**: SP
rectipennis Escalera, 1906d: 312 (*Asida*) **E**: SP
rufomarginalis Escalera, 1906d: 310 (*Asida*) **E**: SP
silvestrei Escalera, 1922b: 174 (*Asida*) **N**: MO
tenuecostata Fairmaire, 1880b: 190 (*Asida*) **N**: MO
subgenus *Glabrasida* Escalera, 1910c: 408 type species *Glabrasida conspuata* Escalera, 1910
Aplanasida Reitter, 1917a: 11 type species *Alphasida brevicosta* Solier, 1836
Aulonasida Reitter, 1917a: 19 type species *Asida chauveneti* Solier, 1836
Durasida Reitter, 1917a: 14 type species *Tenebrio silphoides* Linnaeus, 1767
Gymnetasida Reitter, 1917a: 22 type species *Asida tricostata* Allard, 1869
Melambasida Reitter, 1917a: 21 type species *Alphasida interjecta* Reitter, 1917
Mimelasida Reitter, 1917a: 21 type species *Asida puncticollis* Solier, 1836
Pedarasida Reitter, 1917a: 11 type species *Asida cariosicollis* Solier, 1836
alcirensis Reitter, 1917a: 38 **E**: SP
alcirensis Escalera, 1923a: 466 (*Asida*) [HN]
alluaudi alluaudi Escalera, 1925a: 331 (*Glabrasida*) **N**: MO
alluaudi maritima Antoine, 1934: 117 (*Gymnetasida*) **N**: MO
altomirana Escalera, 1923b: 38 (*Asida*) **E**: SP
amorii Pérez Arcas, 1868: 510 (*Asida*) **E**: SP
gaditana Reitter, 1917a: 37 [HN]
annamariae annamariae Antoine, 1934: 118 (*Gymnetasida*) **N**: MO
annamariae hantzi Antoine, 1937: 186 **N**: MO
annina Reitter, 1917a: 37 **E**: SP
ardoisi Reitter, 1917a: 31 **E**: SP
ardoisi Escalera, 1922e: 391 (*Asida*) [HN]
asperata Solier, 1836c: 450 (*Asida*) **E**: SP
rugosula Rambur, 1838: pl. 19 (*Asida*)
squalida Allard, 1869a: 289 (*Asida*)
asturica Schuster, 1928f: 946 [RN] **E**: SP
heydeni Escalera, 1923c: 87 (*Asida*) [HN]
auriculata Solier, 1836c: 451 (*Asida*) **N**: EG LB
bengasiana Wilke, 1921: 254
mystica Wilke, 1921: 255
azruensis azruensis Escalera, 1925a: 326 (*Glabrasida*) **N**: MO

azruensis mrirtensis Antoine, 1933a: 202 (*Gymnetasida*) **N:** MO
baezensis Escalera, 1923a: 467 (*Asida*) **E:** SP
balboana Koch, 1937: 411 **N:** LB
barbara Allard, 1869a: 262 (*Asida*) **N:** MO
bifoveata Allard, 1869a: 294 (*Asida*) **N:** MO
bonacherai Español, 1958a: 16 **E:** SP
boscai Escalera, 1922e: 389 (*Asida*) **E:** SP
bounaceurensis F. Soldati & François, 2015: 524 **N:** MO
breuiliana Escalera, 1923c: 86 (*Asida*) **E:** SP
brevicosta Solier, 1836c: 449 (*Asida*) **E:** SP (Baleares)
  crenata Schaufuss, 1882c: 623 (*Asida*)
  depressa Solier, 1836c: 466 (*Asida*) [HN]
  laevigata Rambur, 1838: pl. 19 (*Asida*) [HN]
brevipubens Reitter, 1917a: 32 **E:** SP
  calagurriensis Escalera, 1943a: 203 (*Asida*)
cariosicollis Solier, 1836c: 446 (*Asida*) **N:** AG
candeledana Pérez Vera & Avila, 2016: 10 **E:** SP
castellana bejarana Pérez Vera & Avila, 2016: 7 **E:** SP
castellana castellana Graells, 1858: 69 (*Asida*) **E:** SP
castellana piedralavesana Pérez Vera & Avila, 2016: 7 **E:** SP
celtica Pérez Vera & Avila, 2015c: 28 **E:** PT SP
centralis centralis Escalera, 1925a: 330 (*Glabrasida*) **N:** MO
centralis ouezzania Kocher, 1968: 109 **N:** MO
centralis volubilitana Escalera, 1929: 108 (*Glabrasida*) **N:** MO
cerebellum cerebellum Antoine, 1933a: 211 (*Gymnetasida*) **N:** MO
cerebellum gundafica Antoine, 1933a: 213 (*Gymnetasida*) **N:** MO
cereocosta Kwieton, 1986a: 8 **N:** AG
chauveneti chauveneti Solier, 1836c: 440 (*Asida*) **N:** AG TU
  lapidaria Lucas, 1847: pl. 29 (*Asida*)
  quadricarinata Reitter, 1917a: 19
chauveneti quadricostata Allard, 1869a: 233 (*Asida*) **N:** AG TU
  appulsa Reitter, 1917a: 20
cobosi Pérez Vera & Avila, 2012c: 19 **N:** MO
conjuncta Escalera, 1923b: 40 (*Asida*) **E:** SP
conspuata Escalera, 1910c: 409 (*Glabrasida*) **N:** MO
cordubensis Escalera, 1923c: 82 (*Asida*) **E:** SP
cortesensis cortesensis Reitter, 1917a: 33 **E:** SP
  grossepunctata Escalera, 1923a: 473 (*Asida*)
  millarensis Escalera, 1943a: 205 (*Asida*)
cortesensis sacanetensis Escalera, 1943a: 205 (*Asida*) **E:** SP
cossyrensis Reitter, 1917a: 30 **E:** IT (Sicilia)

costulata Solier, 1836c: 455 (*Asida*) **E**: PT SP
grossa Thunberg, 1821b: 51 (*Opatrum*) [NO]
scabrosa Allard, 1869a: 270 (*Asida*)
crassicollis Fairmaire, 1868: 487 (*Asida*) **N**: AG
felicitana Reitter, 1917a: 18
silphiformis Reitter, 1917a: 17
cylindrica cylindrica Reitter, 1917a: 27 **N**: MO
cylindrica rehamnensis Antoine, 1937: 184 **N**: MO
dantini Escalera, 1923a: 472 (*Asida*) **E**: SP
dendriticosta Pérez Vera & Avila, 2012c: 20 **N**: MO
deyrollei Reitter, 1917a: 34 **E**: SP
terolensis Escalera, 1923a: 473 (*Asida*)
discostriata Escalera, 1923a: 468 (*Asida*) **E**: SP
djebiletensis Antoine, 1937: 185 **N**: MO
dubia Rambur, 1838: pl. 19 (*Asida*) **E**: SP
edithae Reitter, 1917a: 16 **N**: AG
tuberculosa Reitter, 1917a: 19
elongata Solier, 1836c: 463 (*Asida*) **E**: SP
embergeri embergeri Antoine, 1936b: 71 (*Glabrasida*) **N**: MO
embergeri infrasedens Antoine, 1937: 241 (*Machlasida*) **N**: MO
embergeri supersedens Antoine, 1937: 241 (*Machlasida*) **N**: MO
exilata Escalera, 1925b: 153 (*Glabrasida*) **E**: SP
flens Antoine, 1933a: 200 (*Gymnetasida*) **N**: MO
frigidissima Escalera, 1923a: 470 (*Asida*) **E**: FR SP
gaditana Rambur, 1838: pl. 20 (*Asida*) **E**: SP
atrata Baudi di Selve, 1875c: 107 (*Asida*)
gigas Dufour, 1820: 309 **E**: SP
globipennis Escalera, 1910c: 412 (*Glabrasida*) **N**: MO
goudoti Solier, 1836c: 457 (*Asida*) **E**: PT SP
perezi Chevrolat, 1865: 390 (*Asida*)
ventricosa Solier, 1836c: 458 (*Asida*)
gracilis Allard, 1869a: 295 (*Asida*) **E**: SP
granadina Escalera, 1922e: 391 (*Asida*) **E**: SP
granifera Solier, 1836c: 460 (*Asida*) **E**: SP
granipennis breiti Antoine, 1937: 188 **N**: MO
granipennis granipennis Escalera, 1925a: 328 (*Glabrasida*) **N**: MO
granipennis punctibasis Escalera, 1925a: 330 (*Glabrasida*) **N**: MO
granipennis schusteri Antoine, 1937: 188 **N**: MO
granipunctipennis Escalera, 1923c: 86 (*Asida*) **E**: PT SP
granulata Fabricius, 1787a: 49 (*Silpha*) **N**: TU

lethierryi Allard, 1869a: 236 (*Asida*)
variolosa Fabricius, 1792a: 108 (*Blaps*)
grassellyi Antoine, 1937: 189 **N**: MO
gredosana Pérez Vera & Avila, 2016: 13 **E**: SP
gregoirei Antoine, 1933a: 213 (*Pedarasida*) **N**: MO
grossa Solier, 1836c: 453 (*Asida*) **E**: IT
cadamuroi Canzoneri, 1963b: 36
otrantina Leoni, 1929: 151 (*Asida*)
halophila halophila Koch, 1937: 406 **N**: LB
halophila syrtica Koch, 1939: 265 **N**: LB
henonii Fairmaire, 1868: 486 (*Asida*) **N**: AG TU
dissimilis Allard, 1869a: 241 (*Asida*)
herminae Reitter, 1917a: 17 **N**: AG
heydeni Reitter, 1917a: 36 **E**: SP
himerera Reitter, 1917a: 18 **E**: IT (Sicilia)
hispanica Solier, 1836c: 461 (*Asida*) **E**: SP
sibirica Solier, 1838c: 462 (*Asida*)
ibicensis Pérez Arcas, 1868: 507 (*Asida*) **E**: SP (Baleares)
iblanensis iblanensis Antoine, 1939: 128 **N**: MO
ighrezranensis Koch, 1940d: 304
iblanensis reymondi Antoine, 1952: 113 **N**: MO
indecisa indecisa Escalera, 1923c: 80 (*Asida*) **E**: SP
indecisa vivesi Viñolas & Cartagena, 2005: 35 (*Glabrasida*) [RN] **E**: SP
calva Escalera, 1923c: 80 (*Asida*) [HN]
inesperata Escalera, 1921a: 117 (*Asida*) **E**: SP
interjecta Reitter, 1917a: 22 **N**: AG MO
brachycera Antoine, 1937: 191
integra Reitter, 1917a: 22
interstrata Reitter, 1917a: 22
ithana Reitter, 1917a: 30 **E**: SP
javalambrensis Escalera, 1943a: 206 (*Asida*) **E**: SP
jumillensis Escalera, 1923a: 467 (*Asida*) **E**: SP
kocheri Antoine, 1934: 120 (*Gymnetasida*) **N**: MO
koltzei Reitter, 1917a: 36 **E**: SP
insularis Escalera, 1922e: 391 (*Asida*)
lacunosa Escalera, 1923c: 83 (*Asida*) **E**: SP
laevis approximata Escalera, 1923a: 464 (*Asida*) **E**: SP
laevis insulicola Pérez Vera & Avila, 2015d: 512 **E**: SP
laevis laevis Solier, 1836c: 464 (*Asida*) **E**: SP
laevis salinatrix Koch, 1944b: 238 **E**: SP

lapini Kocher, 1966a: 45 **N:** MO
babberretiana Cobos, 1968: 913
lassallei J. Ferrer, 1983: 137 **E:** SP
lecerfi lecerfi Koch, 1940d: 302 **N:** MO
lecerfi timeliltensis Koch, 1940d: 304 **N:** MO
leonensis gallaeciana Escalera, 1923c: 84 (*Asida*) **E:** SP
leonensis leonensis Reitter, 1917a: 32 **E:** PT SP
leonensis Escalera, 1923c: 84 (*Asida*) [HN]
leperina Reitter, 1917a: 21 **N:** AG
loroi Escalera, 1923a: 471 (*Asida*) **E:** SP
louati Antoine, 1933a: 198 (*Gymnetasida*) **N:** MO
lucens Escalera, 1923c: 87 (*Asida*) **E:** SP
lusitana Escalera, 1923c: 82 (*Asida*) **E:** PT SP
magnifica Escalera, 1923b: 41 (*Asida*) **E:** SP
mariae Cobos, 1988a: 9 **E:** SP
maroccana Allard, 1869a: 260 (*Asida*) **N:** MO
blattiformis Reitter, 1917a: 27
marseuli Allard, 1869a: 274 (*Asida*) **E:** PT SP
mazaganica baudoni Antoine, 1957: 43 **N:** MO
mazaganica mazaganica Escalera, 1910c: 413 (*Glabrasida*) **N:** MO
mazaganica micrigranifera Escalera, 1922b: 173 **N:** MO
medae Español, 1940: 11 **E:** SP (Baleares)
melillensis Escalera, 1910c: 411 (*Glabrasida*) **N:** MO
meridionalis ilurquensis Pérez Vera & Avila, 2014: 330 **E:** SP
meridionalis meridionalis Escalera, 1923b: 40 (*Asida*) **E:** SP
meridionalis plana Escalera, 1923b: 41 (*Asida*) **E:** SP
multicostata Escalera, 1923c: 84 (*Asida*) **E:** PT
multigranulosa Escalera, 1923c: 83 (*Asida*) **E:** SP
nemethi ganayi Antoine, 1942b: 46 (*Glabrasida*) **N:** MO
nemethi nemethi Théry, 1932a: 189 **N:** MO
nigerrima Allard, 1869a: 242 (*Asida*) **N:** AG
nigroopaca G. Quedenfeldt, 1890: 398 (*Asida*) **N:** LB TU
nitida Bedel, 1918: 363 (*Asida*) **N:** MO
nitidicollis Reitter, 1917a: 35 **N:** TU
obsoleta obsoleta Fairmaire, 1867: 405 (*Asida*) **N:** AG
obsoleta tuberculifera Allard, 1869a: 259 (*Asida*) **N:** AG
algeriana Gebien, 1910a: 126 (*Asida*) [RN]
tournieri Allard, 1869a: 264 (*Asida*)
tuberculata Allard, 1869a: 263 (*Asida*)
occulta Cobos, 1988a: 9 **E:** SP

olmedensis Escalera, 1923a: 470 (*Asida*) **E**: SP
opatroides Allard, 1869a: 249 (*Asida*) **N**: AG
  sulcipennis Fairmaire, 1868: 485 (*Asida*)
ovalaris Escalera, 1925e: 409 (*Asida*) **E**: SP (Baleares)
ovalis Escalera, 1923b: 38 (*Asida*) **E**: SP
orientalis Escalera, 1929: 108 (*Glabrasida*) **N**: MO
parallela Solier, 1836c: 459 (*Asida*) **E**: SP
  montana Rambur, 1838: pl. 19 (*Asida*)
  morenoi Koch, 1944b: 239
  vuillefroyi Pérez Arcas, 1868: 505 (*Asida*)
perezarcasi J. Ferrer, 2008: 64 **E**: SP
pernitida Antoine, 1942b: 45 (*Glabrasida*) **N**: MO
pilosipennis Escalera, 1923c: 85 (*Asida*) **E**: SP
politicollis gracilitarsis Antoine, 1957: 43 **N**: MO
politicollis politicollis Fairmaire, 1868: 489 (*Asida*) **N**: AG
porcata Solier, 1836c: 448 (*Asida*) **E**: SP
pozuelensis Escalera, 1923b: 41 (*Asida*) **E**: SP
  calva Escalera, 1923b: 42 (*Asida*) [HN]
primoti Antoine, 1933a: 204 (*Gymnetasida*) **N**: MO
proconsularis Kwieton, 1986a: 20 **N**: TU
pseudogoudoti Escalera, 1923b: 39 (*Asida*) **E**: SP
puncticollis moltonii Canzoneri, 1972b: 291 **E**: IT (Sicilia)
puncticollis puncticollis Solier, 1836c: 439 (*Asida*) **N**: AG TU
  vagecostata Fairmaire, 1875a: 528 (*Asida*)
puncticollis tirellii Leoni, 1929: 149 (*Asida*) **E**: IT (Sicilia)
  ragusae Leoni, 1929: 151 (*Asida*)
punctigranosa Escalera, 1923b: 39 (*Asida*) **E**: SP
punctipennis Pérez Arcas, 1865b: 435 **E**: SP
querensis Escalera, 1923b: 39 (*Asida*) **E**: SP
rabatica Escalera, 1910c: 414 **N**: MO
raymondae Antoine, 1937: 190 **N**: MO
renata Koch, 1937: 416 **N**: LB
riffensis Antoine, 1933a: 210 (*Gymnetasida*) **N**: MO
robledana Pérez Vera & Avila, 2016: 10 **E**: SP
robusta Escalera, 1923a: 471 (*Asida*) **E**: SP
rolphii Fairmaire, 1867: 404 (*Asida*) **N**: MO
rotroui Escalera, 1929: 108 (*Glabrasida*) **N**: MO
rotundicollis Escalera, 1913: 40 (*Glabrasida*) **N**: MO
rugosa Fabricius, 1775: 252 (*Pimelia*) **N**: MO
  fabricii Allard, 1869a: 245 (*Asida*)

ruiderensis Escalera, 1923a: 472 (*Asida*) **E:** SP
sagrensis Escalera, 1923a: 471 (*Asida*) **E:** SP
saharensis Koch, 1937: 408 **N:** LB
saintpierrei Allard, 1869a: 239 (*Asida*) **N:** AG
salmantina Escalera, 1923b: 38 (*Asida*) **E:** SP
scabrata Fairmaire, 1868: 485 (*Asida*) **N:** AG
allardi Baudi di Selve, 1875c: 112 (*Asida*)
opaca Allard, 1869a: 235 (*Asida*)
tumida Reitter, 1917a: 22
schatzmayri Koch, 1940d: 305 **N:** MO
scobina otini Antoine, 1937: 188 **N:** MO
scobina scobina Antoine, 1933a: 206 (*Gymnetasida*) **N:** MO
scobina semiresecta Antoine, 1933a: 208 (*Gymnetasida*) **N:** MO
msunensis Pérez Vera & Avila, 2012d: 396
sefrouensis Pérez Vera & Avila, 2012c: 22 **N:** MO
serpiginosa dermatodes Fairmaire, 1868: 488 (*Asida*) **N:** AG
clypeata Allard, 1869a: 256 (*Asida*)
serpiginosa laevicollis Allard, 1869a: 243 (*Asida*) **N:** AG MO
serpiginosa serpiginosa Erichson, 1841: 180 (*Asida*) **N:** AG
servillei Solier, 1836c: 443 (*Asida*) **N:** AG
affinis Lucas, 1847: 323 (*Asida*)
depressa Solier, 1836c: 444 (*Asida*) [HN]
dufourii Baudi di Selve, 1875c: 110 (*Asida*)
miliaris Erichson, 1841: 179 (*Asida*)
minor Solier, 1836c: 444 (*Asida*) [HN]
pseudotuberculifera Reitter, 1917a: 24
subcostata Solier, 1836c: 446 (*Asida*)
sicula melitana Reitter, 1894b: 82 (*Asida*) **E:** MA
sicula sicula Solier, 1836c: 454 (*Asida*) **E:** IT (Sicilia)
silphoides Linnaeus, 1767: 677 (*Tenebrio*) **N:** AG MO
laevigata Fabricius, 1792a: 108 (*Blaps*) [HN]
stenomarginicollis Pérez Vera & Avila, 2014: 333 **E:** SP
strangulata Escalera, 1923a: 465 (*Asida*) **E:** SP
subgracilis Escalera, 1914: 304 (*Glabrasida*) **N:** MO
sulcata Allard, 1869a: 271 (*Asida*) **E:** PT SP
syriaca Allard, 1869a: 287 (*Asida*) **A:** SY
tangeriana Solier, 1836c: 456 (*Asida*) **N:** MO
tazekensis Antoine, 1939: 149 **N:** MO
timaditensis hebriensis Antoine, 1933a: 202 (*Gymnetasida*) **N:** MO
pulvisculosa Antoine, 1953b: 211
timaditensis timaditensis Escalera, 1925a: 324 (*Glabrasida*) **N:** MO

tijolensis almagrerensis Cobos, 1988a: 10 **E:** SP
tijolensis tijolensis Escalera, 1923a: 468 (*Asida*) **E:** SP
toletana Escalera, 1923a: 472 (*Asida*) **E:** SP
tricostata Allard, 1869a: 238 (*Asida*) **N:** AG MO
crassipes Allard, 1869a: 256 (*Asida*)
tuberculipennis sarroensis Antoine, 1951b: 314 (*Glabrasida*) **N:** MO
tuberculipennis tuberculipennis Escalera, 1910c: 410 (*Glabrasida*) **N:** MO
turrillensis Escalera, 1923a: 466 (*Asida*) **E:** SP
uhagoni Escalera, 1912: 166 (*Asida*) **E:** SP
vetusta Antoine, 1937: 187 **N:** MO
vidali Antoine, 1937: 192 **N:** MO
villososulcata Allard, 1869a: 247 (*Asida*) **N:** AG
vinalopensis Viñolas & Cartagena, 2005: 23 (*Glabrasida*) **E:** SP
zapaterii granulosa Escalera, 1923a: 470 (*Asida*) **E:** SP
zapaterii zapaterii Pérez Arcas, 1872: 103 (*Asida*) **E:** SP
minor Escalera, 1923a: 470 (*Asida*) [HN]
subgenus *Granasida* Reitter, 1917a: 14 type species *Asida granulifera* Chevrolat, 1840
granulifera Chevrolat, 1840a: 16 (*Asida*) **E:** PT SP
subgenus *Machlasida* Escalera, 1907: 338 type species *Asida muleyhafidi* Escalera, 1907
Protomachlasida Escalera, 1927a: 137 type species *Machlasida liouvillei* Escalera, 1925
acuticosta acuticosta Fairmaire, 1880b: 190 (*Asida*) **N:** MO
acuticosta lucusi Escalera, 1922b: 171 (*Machlasida*) **N:** MO
antoinei antoinei Escalera, 1922b: 172 (*Machlasida*) **N:** MO
latior Antoine, 1937: 184
antoinei bremondi Antoine, 1934: 108 (*Machlasida*) **N:** MO
azilalensis Escalera, 1925c: 315 (*Machlasida*) **N:** MO
bletoni Antoine, 1937: 182 **N:** MO
boissangeri Antoine, 1942b: 44 (*Machlasida*) **N:** MO
bordei Théry, 1931b: 122 **N:** MO
corvus Antoine, 1934: 117 (*Machlasida*) **N:** MO
crassimargo Antoine, 1937: 184 **N:** MO
finei Kocher, 1964: 43 **N:** MO
gattefossei Antoine, 1934: 115 (*Machlasida*) **N:** MO
granipuncticollis Escalera, 1927a: 146 (*Machlasida*) **N:** MO
hachtamii Escalera, 1910a: 283 (*Machlasida*) **N:** MO
joliveti Cobos, 1968: 911 **N:** MO
kraatzi kraatzi Allard, 1869a: 293 (*Asida*) **N:** MO
kraatzi occidentalis Escalera, 1922b: 170 (*Machlasida*) **N:** MO
kraatzi parallelissima Escalera, 1927a: 138 (*Machlasida*) **N:** MO
liouvillei Escalera, 1925c: 313 (*Machlasida*) **N:** MO
merkli Pérez Vera & Avila, 2012a: 79 **N:** MO

muleyhafidi Escalera, 1907: 336 (*Machlasida*) **N:** MO
nadaii Pérez Vera & Avila, 2015b: 220 **N:** MO
nitidicosta Antoine, 1934: 114 (*Machlasida*) **N:** MO
olcesii Fairmaire, 1867: 405 (*Asida*) **N:** MO
safiana Antoine, 1937: 184 **N:** MO
segonzaci Escalera, 1914: 302 (*Machlasida*) **N:** MO
secsaui Escalera, 1913: 39 (*Machlasida*) **N:** MO
singularis Reitter, 1917a: 14 **N:** MO
surcoufi Antoine, 1934: 106 (*Machlasida*) **N:** MO
telueti Escalera, 1910a: 283 (*Machlasida*) **N:** MO
theryi Escalera, 1925c: 314 (*Machlasida*) **N:** MO
variesculpta tazultensis Kocher, 1953: 63 (*Machlasida*) **N:** MO
variesculpta variesculpta Antoine, 1951b: 314 (*Machlasida*) **N:** MO
zemensis meskina Antoine, 1934: 109 (*Machlasida*) **N:** MO
zemensis zemensis Escalera, 1925c: 316 (*Machlasida*) **N:** MO

genus *Asida* Latreille, 1802: 167 type species *Opatrum griseum* Fabricius, 1781
subgenus *Asida* Latreille, 1802: 167 type species *Opatrum griseum* Fabricius, 1781
Dolichasida Reitter, 1917a: 40 type species *Asida moraguezi* L.W. Schaufuss, 1879
Euryasida Reitter, 1917a: 40 type species *Asida barceloi* Pérez Arcas, 1868
Insulasida Escalera, 1922a: 64 type species *Asida moraguezi* L.W. Schaufuss, 1879
Leptasida Reitter, 1917a: 60 [HN] type species *Asida diecki* Allard, 1869
Rugasida Escalera, 1922a: 65 type species *Asida reticulata* Solier, 1836
abrupta Fairmaire, 1867: 406 **N:** AG
acuticollis acuticollis Allard, 1869a: 205 **E:** AL BH CR
acuticollis ganglbaueri J. Müller, 1917: 15 **E:** BH CR
alcarazensis Reitter, 1917a: 44 **E:** SP
anachoreta Leo, 2009: 404 **E:** IT (Sardinia)
androgyna Leo, 2012: 97 **E:** IT (Sardinia)
argentierae Leo, 1980c: 189 **E:** IT (Sardinia)
ascoensis F. Soldati & L. Soldati, 2001b: 180 **E:** FR (Corse)
australis Baudi di Selve, 1875b: 27 **E:** IT (Sardinia)
baguenai Escalera, 1943a: 201 **E:** SP
banatica I. Frivaldszky, 1865: 200 **E:** BU RO
barceloi Pérez Arcas, 1868: 494 **E:** SP (Baleares)
bayardi bayardi Solier, 1836c: 423 **E:** IT
bayardi blaptoides Leoni, 1909: 222 **E:** IT
bayardi leosinii Leoni, 1909: 226 **E:** IT
bejarensis Escalera, 1922a: 68 **E:** SP
calabra Leoni, 1909: 222 **E:** IT
cardonae Pérez Arcas, 1868: 496 **E:** SP (Baleares)

horrens L.W. Schaufuss, 1869: 24
reichei Allard, 1869a: 211
carinata carinata Solier, 1836c: 426 **E**: FR (Corse)
carinata lepidoptera Allard, 1869a: 194 **E**: FR (Corse)
squamigera Reitter, 1917a: 58
cazorlensis Escalera, 1922a: 68 **E**: SP
cephalonica Reitter, 1917a: 48 **E**: GR
christianperezi F. Soldati, 2007: 271 **E**: FR
christinae F. Soldati & Leo, 2005: 342 **E**: FR (Corse)
coachei F. Soldati & Leo, 2005: 343 **E**: FR (Corse)
cocquempoti F. Soldati & L. Soldati, 2001c: 186 **E**: AL MC GR
coimbrensis Escalera, 1922a: 67 **E**: PT
combae Gené, 1839: 71 **E**: IT (Sardinia)
proxima Leoni, 1910: 74
corsica corsica Laporte, 1833: 33 **E**: FR (Corse) IT (Sardinia)
corsica genei Solier, 1836c: 435 **E**: IT (Sardinia)
dominula Reitter, 1917a: 53
ignorata Reitter, 1917a: 53
curta Allard, 1869a: 220 **N**: AG TU
defricta Peyerimhoff, 1920c: 245 **N**: AG
dejeanii Solier, 1836c: 420 **E**: FR IT
crispata Rey, 1892: 42
detrita Rey, 1892: 42
sinuans Rey, 1892: 42
devillei Leoni, 1910: 54 **E**: FR (Corse)
diecki Allard, 1870: 50 **E**: SP
meridionalis Escalera, 1922a: 69 [HN]
doderoi Leoni, 1910: 49 **E**: IT (Sardinia)
dorgaliensis dorgaliensis Leoni, 1911: 3 **E**: IT (Sardinia)
dorgaliensis montalbica Reitter, 1917a: 56 **E**: IT (Sardinia)
dryas Leo, 2009: 401 **E**: IT (Sardinia)
fairmairei fairmairei Boieldieu, 1865: 8 **E**: GR
fairmairei graeca Allard, 1869a: 290 **E**: GR
fairmairei moriasica F. Soldati & L. Soldati, 2001c: 193 **E**: GR
fairmairei parnassiana F. Soldati & L. Soldati, 2001c: 192 **E**: GR
fascicularis brattiensis J. Müller, 1917: 9 **E**: CR
fascicularis fascicularis Germar, 1817: 190 (*Platynotus*) **E**: CR IT SL
fascicularis fiorii Leoni, 1909: 224 **E**: IT
fascicularis lineatocollis Küster, 1849a: 26 **E**: AL BH CR GR
setulifera Küster, 1849a: 25
terricola Küster, 1849a: 27

fascicularis meledana J. Müller, 1917: 10 **E**: CR
fascicularis pubipennis J. Müller, 1917: 12 **E**: CR
fodori Pérez Vera & Avila, 2015a: 103 **E**: SP
gestroi capraiensis Gridelli, 1972: 196 **E**: IT
gestroi gardinii Lo Cascio, 2000: 159 **E**: IT
gestroi gestroi Leoni, 1910: 45 **E**: IT
obliterata Leoni, 1910: 46
gestroi lanzai Leo, 1998: 74 **E**: IT
gestroi thyrrena Leoni, 1910: 46 **E**: IT
gibbicollis Pérez Arcas, 1865b: 423 **E**: PT
angusta Allard, 1869a: 217
fuentei Fairmaire, 1896e: 345
glacialis barbaricina Leoni, 1911: 2 **E**: IT (Sardinia)
glacialis glacialis Gené, 1839: 69 **E**: IT (Sardinia)
solarii Leoni, 1910: 77
glacialis rustica Gené, 1839: 70 **E**: IT (Sardinia)
exculpta Baudi di Selve, 1875b: 30
piriensis Leoni, 1910: 79
undulata Leoni, 1910: 81
goryi Solier, 1836c: 424 **E**: IT
angustata Leoni, 1910: 15
atriventris Reitter, 1917a: 57
cribricollis Leoni, 1910: 15
grisea Fabricius, 1781: 89 (*Opatrum*) **E**: IT
heikertingeri Schuster, 1928f: 949 [RN] **E**: SP
segurensis Escalera, 1922a: 68 [HN]
inaequalis carthagina Kwieton, 1986b: 11 **N**: TU
inaequalis inaequalis Solier, 1836c: 428 **N**: AG TU
foveicollis Allard, 1883: 27
rhytirrhina Reitter, 1917a: 50
kaminskii F. Soldati, 2016a: 572 **E**: FR (Corse)
ligurica Baudi di Selve, 1875b: 27 **E**: FR IT
longicollis Solier, 1836c: 425 **E**: FR (Corse) IT
squamulata Leoni, 1910: 88
lostiae Allard, 1888: 333 **E**: IT (Sardinia)
luigionii doriae Leoni, 1910: 43 **E**: IT
luigionii insularis Leoni, 1910: 94 **E**: IT
luigionii luigionii Leoni, 1910: 40 **E**: IT
lulensis Reitter, 1917a: 57 **E**: IT (Sardinia)
lutosa Solier, 1836c: 422 **E**: BU MD RO UK ST
maltinii Ardoin, 1971a: 47 **N**: TU
melisellensis melisellensis J. Müller, 1946: 115 **E**: CR

melisellensis pomoensis J. Müller, 1950: 13 **E**: CR
minima Reitter, 1917a: 46 **E**: IT (Sicilia)
minuta Rambur, 1838: pl. 20 **E**: SP
  hebes Rosenhauer, 1856: 200
moraguezi L.W. Schaufuss, 1879: 534 **E**: SP (Baleares)
novasiriensis Grimm, 1985: 10 **E**: IT
nuragica Leo, 1984b: 73 **E**: IT (Sardinia)
nurrae Leo, 2012: 103 **E**: IT (Sardinia)
ochsi Ardoin, 1958: 37 **E**: FR
paulae Leo, 2012: 107 **E**: IT (Sardinia)
piligera Leoni, 1909: 221 **E**: IT
pirazzolii pirazzolii Allard, 1869a: 200 **E**: IT
  intermedia Leoni, 1909: 184
pirazzolii sardiniensis Allard, 1869a: 291 **E**: IT
  baudii Leoni, 1909: 186
planipennis cabrerensis Viñolas & Cartagena, 2005: 24 **E**: SP (Baleares)
planipennis minoricensis Español, 1954a: 37 **E**: SP (Baleares)
planipennis planipennis L.W. Schaufuss, 1869: 23 **E**: SP (Baleares)
  balearica Escalera, 1922a: 65
  anodonta Wilke, 1921: 258
pseudoreticulata Escalera, 1922a: 67 **E**: PT
punica Kwieton, 1986b: 10 **N**: TU
sabulosa reticulata Solier, 1836c: 419 **E**: FR PT SP
  convexicollis Allard, 1883: 27
  reitteri Gebien, 1906b: 248 [RN]
sabulosa sabulosa Fuessly, 1775: 20 (*Tenebrio*) **E**: BE FR GE IT LU SZ
  catenulata Mulsant, 1854: 92
  consanguinea Allard, 1869a: 292
  glabricosta Solier, 1836c: 416
  helvetica Solier, 1836c: 418
  insidiosa Mulsant, 1854: 92
  morbillosa Fabricius, 1801a: 140 (*Platynotus*)
  obesa Allard, 1869a: 179
  rugosa Geoffroy, 1785: 157 (*Tenebrio*)
  undata Fabricius, 1801a: 139 (*Platynotus*)
  vicina Solier, 1836c: 417
sardoa Leoni, 1910: 82 **E**: IT (Sardinia)
schawalleri F. Soldati & L. Soldati, 2003b: 43 **E**: GR
schusteri Reitter, 1917a: 52 **E**: FR (Corse) IT (Sardinia)
serripes Chevrolat, 1874: 157 **E**: SP
solieri caroli Leo, 2009: 407 **E**: IT (Sardinia)
solieri fancelloi Leo, 1984b: 76 **E**: IT (Sardinia)

solieri solieri Gené, 1836b: iv [= 1836a: 197] **E**: IT (Sardinia)
spaethi Schuster, 1928f: 947 [RN] **E**: SP
  lanceocollis Reitter, 1917a: 44 [HN]
stierlini Reitter, 1917a: 56 **E**: FR (Corse)
transmontana Escalera, 1922a: 68 **E**: PT
subgenus *Globasida* Escalera, 1905a: 430 type species *Asida oblonga* Rambur, 1838
almeriana almeriana Escalera, 1905a: 437 **E**: SP
almeriana zarcoi Koch, 1944b: 241 **E**: SP
asturiana Pérez Vera & Avila, 2015a: 107 **E**: SP
barrosi Escalera, 1921a: 119 **E**: PT
cartagenica Escalera, 1905a: 433 **E**: SP
cincta Rambur, 1838: pl. 20 **E**: SP
  cincta Rosenhauer, 1856: 196 [HN]
curvatipennis curvatipennis Escalera, 1905a: 440 **E**: SP
curvatipennis dubiosa Escalera, 1905a: 440 **E**: SP
curvatipennis mateui Viñolas & Cartagena, 2005: 25 **E**: SP
  setosa Escalera, 1905a: 441 [HN]
curvatipennis quadrata Escalera, 1905a: 439 **E**: SP
curvatipennis schrammi Escalera, 1906b: 193 **E**: SP
deformis Escalera, 1905a: 437 **E**: SP
gambeyi Allard, 1869a: 184 **N**: AG
ludovici Pérez Arcas, 1874: 136 **E**: SP (Baleares)
  minorata Escalera, 1925e: 408
monchiquensis Escalera, 1921a: 118 **E**: PT
mollicoma Reitter, 1917a: 70 **E**: SP
nerjensis Escalera, 1905a: 438 **E**: SP
oblonga bacaresensis Escalera, 1905a: 435 **E**: SP
oblonga frigida Escalera, 1905a: 436 **E**: SP
oblonga intermedia Escalera, 1905a: 435 **E**: SP
oblonga novissima Escalera, 1905a: 436 **E**: SP
oblonga oblonga Rambur, 1838: pl. 20 **E**: SP
  bonvouloiri Allard, 1869a: 210
  elongata Rambur, 1838: pl. 20
oblonga rotunda Escalera, 1905a: 436 **E**: SP
segurensis Escalera, 1905a: 439 **E**: SP
setosa Pérez Arcas, 1874: 139 **E**: SP
  setigera Gebien, 1910a: 137 [RN]
sinuaticollis cristata Cobos, 1963a: 19 **N**: MO
sinuaticollis sinuaticollis Solier, 1836c: 433 **N**: AG MO
sinuaticollis mauritana Escalera, 1909a: 135 **N**: MO
subgenus *Gracilasida* Escalera, 1905: 445 type species *Asida ariasi* Escalera, 1909
  Granulasida Escalera, 1921c: 427 type species *Asida setipennis* Allard, 1869

Planasida Escalera, 1907: 337 type species *Asida paulinoi* Pérez Arcas, 1868
Pseudoplanasida Escalera, 1921b: 360 type species *Asida pygmaea* Rosenhauer, 1856
Trachasida Reitter, 1917a: 62 type species *Asida ruficornis* Solier, 1836
akisoides Reitter, 1917a: 62 **E**: SP
akisoides Escalera, 1921b: 358 [HN]
alcoyana Escalera, 1921c: 436 **E**: SP
amplicollis Escalera, 1921b: 356 **E**: SP
basiplicata Heyden, 1884b: 360 **E**: SP
bereai africana Escalera, 1925a: 333 (*Pseudoplanasida*) **N**: MO
bereai bereai Escalera, 1907: 337 **N**: MO
bereai espanoli Pérez Vera & Avila, 2012c: 26 **N**: MO
bereai silicicola Pérez Vera & Avila, 2012d: 397 **N**: MO
bereai vaucheri Escalera, 1907: 339 **N**: MO
breiti Schuster, 1928f: 949 [RN] **E**: SP
acuticollis Escalera, 1921b: 355 [HN]
breuili Escalera, 1921b: 352 **E**: SP
brucki Allard, 1869a: 292 **E**: SP
approximata Escalera, 1921c: 429
garruchensis Escalera, 1921c: 429
ghilianii Baudi di Selve, 1875b: 25
larrinuae Escalera, 1921c: 430
nuda Escalera, 1921c: 430
candidoi Escalera, 1912: 167 **E**: SP
confusa Kraatz, 1874: 114 **E**: SP
cossyphoides Antoine, 1951b: 315 **N**: MO
curvicollis Escalera, 1921b: 353 **E**: PT
esteparia Reitter, 1917a: 59 **E**: SP
fallax Escalera, 1921c: 431 **E**: SP
granularis Escalera, 1921c: 431
granatensis Escalera, 1921c: 428 **E**: SP
hispidula Pic, 1903g: 114 **N**: TU
pilosula Gebien, 1910a: 134 [RN]
humeralis Escalera, 1943a: 199 **E**: SP
inquinata Rambur, 1838: pl. 20 **E**: SP **N**: MO
inquinata Rosenhauer, 1856: 195 [HN]
inquinatoides Escalera, 1921b: 356 **E**: PT
ithae Reitter, 1917a: 62 **E**: SP
lanceocollis Escalera, 1914: 300 (*Planasida*) **N**: MO
latissima Wilke, 1921: 259 **E**: PT
lauzannei arribasi Cobos, 1961: 87 **N**: MO
lauzannei freyi Koch, 1940d: 308 **N**: MO
lauzannei lauzannei Rotrou, 1938: 119 **N**: MO

marginicollis Rambur, 1838: pl. 20 **E:** SP
marginicollis Rosenhauer, 1856: 197 [HN]
mater cunicularia Escalera, 1925e: 406 **E:** SP (Baleares)
mater gasulli Español, 1951a: 16 **E:** SP (Baleares)
mater inmarginata Escalera, 1925e: 407 **E:** SP (Baleares)
mater josefinae Español, 1951a: 17 **E:** SP (Baleares)
mater mater Escalera, 1925e: 404 **E:** SP (Baleares)
morae Pérez Arcas, 1868: 502 (*Asida*) **E:** SP
moczarskii Schuster, 1928f: 950 [RN] **E:** SP
minima Escalera, 1921b: 363 [HN]
moroderi Reitter, 1917a: 65 **E:** SP
moroderi Escalera, 1921c: 435 [HN]
normandi Koch, 1940d: 309 **N:** TU
oertzeni Escalera, 1921c: 434 **E:** SP
paludosa Escalera, 1921c: 436 **E:** SP
paulinoi Pérez Arcas, 1868: 499 **E:** PT
planidorsis Escalera, 1921b: 357 **E:** SP
pseudosetipennis Escalera, 1921c: 431 **E:** SP
pusillima Kraatz, 1874: 112 **E:** SP
pygmaea filabrica Cobos, 1988a: 8 **E:** SP
pygmaea lencinai Viñolas & Cartagena, 2005: 26 **E:** SP
pygmaea lindaraja Escalera, 1921b: 361 **E:** SP
pygmaea moraima Escalera, 1921b: 362 **E:** SP
pygmaea pygmaea Rosenhauer, 1856: 199 **E:** SP
pygmaea suarezi Cobos, 1988a: 7 **E:** SP
radulicollis Kwieton, 1986b: 4 **N:** TU
ricoi cobosi Viñolas, 1989: 33 **E:** SP
ricoi laufferi Escalera, 1921c: 434 **E:** SP
ricoi murciana Escalera, 1921c: 433 **E:** SP
ricoi ricoi Martinez y Sáez, 1873b: 412 **E:** SP
ricoi salinosa Escalera, 1921c: 433 **E:** SP
rufa Escalera, 1921b: 351 **E:** PT
ruficornis antoinei Cobos, 1963a: 35 **N:** MO
ruficornis ariasi Escalera, 1909a: 135 **N:** MO
ruficornis escalerae Pérez Vera & Avila, 2012b: 186 **N:** AG MO
ruficornis kocheri Cobos, 1965b: 55 **N:** MO
ruficornis lefrancei Reitter, 1917a: 66 **N:** AG MO
ruficornis pardoi Cobos, 1963a: 35 **N:** MO
ruficornis rotroui Cobos, 1963a: 35 **N:** MO
ruficornis ruficornis Solier, 1836c: 421 **N:** AG MO TU
complanata Lucas, 1847: pl. 29
gebieni Reitter, 1917a: 66

sacculipennis Kwieton, 1986b: 8 **N:** TU
septemsis Pérez Vera, Ruiz & Avila, 2012: 386 **N:** MO
setipennis Allard, 1869a: 218 **E:** SP
timida Escalera, 1921b: 350 **E:** SP
valentina Reitter, 1917a: 59 **E:** SP
valentina Escalera, 1921c: 432 [HN]
wankai Reitter, 1917a: 65 **E:** SP
zaida Reitter, 1917a: 61 **E:** SP
zaida Escalera, 1921c: 437 [HN]
zoraida elegans Cobos, 1988a: 7 **E:** SP
zoraida zoraida Escalera, 1921c: 437 **E:** SP
subgenus *Peltasida* Reitter, 1917a: 39 type species *Asida favieri* Fairmaire, 1880
favieri confluens Antoine, 1933a: 195 **N:** MO
favieri favieri Fairmaire, 1880b: 190 (*Asida*) **N:** MO
subgenus *Polasida* Reitter, 1917a: 41 type species *Opatrum sericeum* G.-A. Olivier, 1795
Opatrasida Escalera, 1922a: 69 type species *Asida jurinei* Solier, 1836
jurinei jurinei Solier, 1836c: 429 **E:** FR SP
bigorrensis Solier, 1836c: 430
mahonis Boieldieu, 1865: 9
pazii Pérez Arcas, 1865b: 425
subcylindrica Leoni, 1909: 178
jurinei pyrenaea Baudi di Selve, 1875c: 117 **E:** FR
levantina Escalera, 1922a: 70 **E:** SP
marmottani C. Brisout de Barneville, 1863: 88 **E:** AN FR SP
poneli F. Soldati & L. Soldati, 2001a: 103 **E:** FR IT
saguntina Escalera, 1922a: 71 **E:** SP
sericea G.-A. Olivier, 1795: 56: 5 (*Opatrum*) **E:** FR SP
difficilis Allard, 1869a: 200

tribe Asidini, species incertae sedis
Alphasida beduina Reitter, 1917a: 26 **N:** MO
Asida anceyi Allard, 1870: 50 **A:** "Syria"
Asida incerta Leoni, 1910: 87 [DA] **E:** IT
Asida massiliensis Baudi di Selve, 1875c: 115 [DA] **E:** FR

tribe Boromorphini Skopin, 1978

genus *Boromorphus* Wollaston, 1854: 492 type species *Boromorphus maderae* Wollaston, 1854 (= *Boros tagenioides* Lucas, 1846)
aegyptiacus Reitter, 1908h: 217 **N:** EG
armeniacus Reitter, 1889a: 33 **E:** AB AR
italicus Gardini, 2010: 2 **E:** IT
libanicus Baudi di Selve, 1881: 286 **A:** IS LE "Syria"

opaculus Reitter, 1887e: 521 **E:** AB AR **A:** AF IQ KZ TM UZ
parvus Wollaston, 1864: 502 **N:** CI
saudicus Schawaller, Al Dhafer & Fadl, 2013: 149 **A:** SA
tagenioides Lucas, 1846: pl. 30 (*Boros*) **E:** SP **N:** AG MO MR TU
maderae Wollaston, 1854: 493

tribe Ceratanisini Gebien, 1937

genus *Ceratanisus* Gemminger, 1870b: 1818 [RN] type species *Anisocerus tristis* Faldermann, 1837
Anisocerus Faldermann, 1837: 39 [HN] type species *Anisocerus tristis* Faldermann, 1837
Apolites Jacquelin du Val, 1861: 324 [HN] type species *Helops mucoreus* Waltl, 1838
Haemerophygus Baudi di Selve, 1876c: 266 type species *Helops mucoreus* Waltl, 1838
Idastrandiella E. Strand, 1929: 23 [RN] type species *Helops mucoreus* Waltl, 1838
Seidlitzellus Reitter, 1920b: 221 [RN] type species *Anisocerus tristis* Faldermann, 1837
allardi Reitter, 1884b: 88 (*Apolites*) **E:** BU GR **A:** TR
costipennis Nabozhenko, J. Ferrer, Kalashian & Abdurakhmanov, 2016: 613 **A:** TR
funebris Reitter, 1898d: 106 (*Anisocerus*) **A:** TR
graecus Kraatz, 1877: 304 (*Apolites*) **E:** GR
guerroumii F. Soldati & L. Soldati, 2002b: 177 (*Idastrandiella*) **E:** GR
keskini Nabozhenko, J. Ferrer, Kalashian & Abdurakhmanov, 2016: 615 **A:** TR
khnzoriani Nabozhenko, J. Ferrer, Kalashian & Abdurakhmanov, 2016: 612 **E:** AR
mucoreus Waltl, 1838: 464 (*Helops*) **E:** GR TR **A:** IN TR
osellai Scupola, 1984: 69 (*Idastrandiella*) **E:** GR
purcharti Nabozhenko, J. Ferrer, Kalashian & Abdurakhmanov, 2016: 616 **A:** TR
talyshensis Nabozhenko, J. Ferrer, Kalashian & Abdurakhmanov, 2016: 609 **E:** AB
taygetanus Reitter, 1898d: 105 (*Apolites*) **E:** GR
transcaucasicus Nabozhenko, J. Ferrer, Kalashian & Abdurakhmanov, 2016: 609 **E:** AB AR
tristis Faldermann, 1837: 40 (*Anisocerus*) **E:** GR
turkestanicus Reitter, 1898d: 106 (*Anisocerus*) **A:** UZ (Tashkent)

genus *Tenebriocephalon* Pic, 1925b: 10 type species *Tenebriocephalon piceum* Pic, 1925
Klapperichia Kaszab, 1954: 249 type species *Klapperichia mirabilis* Kaszab, 1954 (= *Tenebriocephalon thoracicum* Pic, 1925)
piceum Pic, 1925b: 10 **A:** YUN
rufum Pic, 1925b: 10
thoracicum Pic, 1925b: 10 **A:** FUJ
mirabilis Kaszab, 1954: 250 (*Klapperichia*)

tribe Cnemeplatiini Jacquelin du Val, 1861

genus *Cnemeplatia* A. Costa, 1847: 146 type species *Cnemeplatia atropos* A. Costa, 1847
Autocera Wollaston, 1857a: 154 type species *Autocera laticeps* Wollaston, 1857
Cnemoplatia Wollaston, 1865: 411 [UE]
atropos africana Kaszab, 1938c: 81 **E:** PT SP **N:** AG MO TU

atropos atropos A. Costa, 1847: 148 **E:** AB AR CR FR GR HU IT MA PT RO SP TR **N:** EG MO TU **A:** AF IQ IN IS KZ SA SY TD TM TR UZ YE **AFR**
anticipes Wollaston, 1857a: 155 (*Autocera*)
rufa Tournier, 1874: lxxxvi
indica indica Fairmaire, 1896a: 16 **A:** AF NP **ORR**
laticeps Wollaston, 1857a: 155 (*Autocera*) **N:** CI MR
oxiana G.S. Medvedev, 1982: 40 **A:** UZ (Karakalpakistan)
theryi Kaszab, 1938c: 82 **E:** SP (Algecirás)
zarudnyi G.S. Medvedev, 1991: 568 **A:** IN

genus *Lepidocnemeplatia* Bousquet & Bouchard, 2018: 20 type species *Cnemeplatia sericea* Horn, 1870
szekessyi Kaszab, 1942a: 41 **A:** BT **ORR**

genus *Philhammus* Fairmaire, 1871a: 393 type species *Philhammus sericans* Fairmaire, 1871
subgenus *Philhamellus* Kaszab, 1962a: 83 type species *Philhammus myrmecophilus* Kaszab, 1960
leei Kaszab, 1962a: 81 **A:** XIN
myrmecophilus Kaszab, 1960a: 137 **A:** AF IN
tadzhikistanicus G.S. Medvedev, 1991: 569 **A:** TD
subgenus *Philhammus* Fairmaire, 1871a: 393 type species *Philhammus sericans* Fairmaire, 1871
Canariella Uyttenboogaart, 1929: 342 type species *Canariella arenapta* Uyttenboogaart, 1929
Psilachnopus Reitter, 1901e: 161 type species *Psilachnopus cribratellus* Reitter, 1901
aharonii Reitter, 1910d: 197 (*Psilachnopus*) **N:** EG MO **A:** IN IS JO SA **AFR**
andresi Koch, 1935a: 89 (*Philhammus*)
cribratellus Reitter, 1901e: 162 (*Psilachnopus*) **E:** AB AR
sericans Fairmaire, 1871a: 393 **E:** SP **N:** CI EG MO
arenapta Uyttenboogaart, 1929: 343 (*Canariella*)
parallelus Théry, 1931b: 124 (*Cnemoplatia*)
zaitzevi G.S. Medvedev, 1979b: 43 **E:** RU **A:** KZ TM

tribe Cossyphodini Wasmann, 1899

genus *Cossyphodes* Westwood, 1851: 168 type species *Cossyphodes wollastonii* Westwood, 1851
asiricus Schawaller, 2011b **A:** SA
raffrayi Gestro, 1874: 443 **A:** YE **AFR**
wollastonii Westwood, 1851: 170 **N:** CI MR **AFR**

genus *Mimocossyphus* Pic, 1923f: 5 type species *Mimocossyphus minor* Pic, 1923
minor Pic, 1923f: 5 **N:** EG

genus *Paramellon* Waterhouse, 1882: iv type species *Paramellon sociale* Waterhouse, 1882
sociale Waterhouse, 1882: iv **A:** AE IN
socotraense Schawaller, 2006b: 248 **A:** YE (Suqutra)

tribe Edrotini Lacordaire, 1859

genus *Ascelosodis* L. Redtenbacher, 1868: 117 type species *Ascelosodis serripes* L. Redtenbacher, 1868

abdurakhmanovi Chigray, 2018: 167 **A:** TD

assimilis Bates, 1879: 468 **A:** KA

punctatissima Reitter, 1915b: 62 (*Gnathosia*)

ciliatis Bates, 1879: 468 **A:** KA

concinna Bates, 1879: 468 **A:** AF TD XIN

semenowi Reitter, 1897h: 303 (*Orocina*)

emarginata Ren & Shi, 2006a: 4 **A:** XIN

emarginiventris Koch, 1949: 76 **A:** KA XIZ

everestina Blair, 1923b: 278 **A:** SD XIZ

forsteri Kaszab, 1961d: 214 **A:** NP

grandis Bates, 1879: 469 **A:** KA

ovoidea Fairmaire, 1891d: xcii

haagi Bates, 1879: 469 **A:** KA UZ XIZ

inermis Fairmaire, 1891d: xciii **A:** KA

intermedia Bates, 1879: 469 **A:** HP KA

kochi Kaszab, 1959f: 357 **A:** AF

lindbergi Kaszab, 1968b: 52 **A:** AF

longstaffi Blair, 1923b: 279 **A:** XIZ

minor Kaszab, 1978b: 219 **A:** KA

nitida Blair, 1923b: 281 **A:** SD XIZ

punctulata Ren & Shi, 2006a: 4 **A:** XIN

reinigi Koch, 1949: 78 **A:** HP XIZ

rugulosa Ren & Shi, 2006a: 4 **A:** XIN

schmidi Kaszab, 1965a: 108 **A:** SD XIZ

serripes L. Redtenbacher, 1868: 118 **A:** KA XIZ

thibetana Blair, 1923b: 281 **A:** XIZ

waltoni Blair, 1923b: 280 **A:** XIZ

tribe Elenophorini Solier, 1837

genus *Leptoderis* Billberg, 1820: 31 type species *Tenebrio collaris* Linnaeus, 1767

Elenophorus Dejean, 1821: 64 type species *Tenebrio collaris* Linnaeus, 1767

collaris Linnaeus, 1767: 677 (*Tenebrio*) **E:** CR FR IT MA SP **N:** MO

tribe Erodiini Billberg, 1820

genus *Ammodoides* Lesne, 1915: 233 type species *Arthrodeis lateripunctatus* Fairmaire, 1890

franchettii Gridelli, 1931: 220 **A:** "Arabia" **AFR**

lateripunctatus lateripunctatus Fairmaire, 1891a: 553 (*Arthrodeis*) **A:** YE

lateripunctatus yemenesis Gebien, 1938: 53 (*Spyrathus*) **A:** YE
yemenensis Gebien, 1938: 53 (*Spyrathus*) **A:** YE

genus *Ammozoides* Kaszab, 1979a: 91 type species *Ammozoum validicorne* Reitter, 1900
bulla Semenov, 1896c: 353 (*Ammozoum*) **A:** AF TM UZ
hauseri Reitter, 1894a: 48 (*Arthrodeis*) **A:** TM (Karakumy)
validicorne Reitter, 1900k: 301 (*Ammozoum*) **A:** AF TM UZ

genus *Ammozoum* Semenov, 1890: 352 type species *Ammozoum hyalinum* Semenov, 1890
hyalinum Semenov, 1890: 353 **A:** AF TM

genus *Amnodeis* Miller, 1858: 117 type species *Amnodeis grandis* Miller, 1858
asiaticus Miller, 1858: 122 **A:** SY TR
  wagneri Chevrolat, 1877a: 113 [DA]
confluens confluens Miller, 1858: 123 **A:** IQ IS JO SI
  arabicus Reitter, 1914b: 83
  diffluens Reitter, 1914b: 83
  nitidus Chevrolat, 1877a: 113
confluens freyi Schuster, 1937a: 46 **A:** SY
gebieni Reitter, 1914b: 82 **A:** IS JO
giganteus Reiche & Saulcy, 1857: 187 (*Anodesis*) **A:** IQ SY
grandis Miller, 1858: 120 **A:** IN IQ SY TM TR
intermedius Reitter, 1914b: 82 **A:** SY TR
milleri Reitter, 1914b: 81 **A:** SY TR

genus *Apentanodes* Reitter, 1914b: 53 type species *Arthrodeis globosus* Reiche & Saulcy, 1857
subgenus *Apentanodes* Reitter, 1914b: 53 type species *Arthrodeis globosus* Reiche & Saulcy, 1857
arabicus arabicus Kirchsberg, 1877b: 204 (*Arthrodeis*) **A:** AE OM SA
arabicus edomitus Koch, 1940c: 87 **A:** IS JO
buettikeri Kaszab, 1979c: 270 **A:** OM SA
globosus globosus Reiche & Saulcy, 1857: 186 (*Arthrodeis*) **E:** GR (Crete) **A:** CY LE
  syriacus Kraatz, 1865a: 11 (*Arthrodeis*) [RN]
globosus reductepleuralis Koch, 1935e: 309 **E:** GR (Rhodes)
longulus Kulzer, 1956: 635 **A:** IN
morgani Kulzer, 1956: 636 **A:** IN
mesopotamicus Blair, 1923a: 119 **A:** IQ
pakistanicus Kwieton, 1981a: 2 **A:** PA (Balochistan)
persicus Koch, 1940c: 87 **A:** IN
philbyi Blair, 1933: 4 **A:** SA
subgenus *Rasphytus* Kulzer, 1956: 639 type species *Rasphytus freyi* Kulzer, 1956

freyi Kulzer, 1956: 640 **A:** PA **ORR**
globatus Kulzer, 1956: 642 **A:** PA **ORR**

genus *Arthrodeis* Solier, 1834: 513 type species *Arthrodeis rotundatus* Solier, 1834
subgenus *Apentanes* Reitter, 1914b: 45 type species *Arthrodeis occidentalis* Fairmaire, 1868
bolivari Español, 1951b: 364 **N:** MO (Tan-Tan)
murati Koch, 1943a: 500 **N:** MO
gattefossei Antoine, 1946a: 26
occidentalis erodioides Fairmaire, 1875c: 543 **N:** MO
occidentalis occidentalis Fairmaire, 1868: 484 (*Arthrodeis*) **N:** MO
opaculus Fairmaire, 1884f: cxix
oblongior Escalera, 1914: 266 **N:** MO
subgenus *Arthrodeis* Solier, 1834: 513 type species *Arthrodeis rotundatus* Solier, 1834
Arthrodes Agassiz, 1846: 35 [UE]
costifrons Wollaston, 1864: 445 (*Arthrodes*) **N:** CI (Fuerteventura, Isla de Lobos, Lanzarote)
emarginatus Wollaston, 1864: 447 (*Athrodes*) **N:** CI (Fuerteventura)
geotrupoides Wollaston, 1864: 447 (*Arthrodes*) **N:** CI (Fuerteventura)
glomeratus Fairmaire, 1871a: 384 **N:** MO
fairmairei Reitter, 1914b: 52
globulosus Escalera, 1922d: 375
prosternalis Reitter, 1914b: 53
subasperatus Reitter, 1914b: 52
uamarensis Escalera, 1922d: 376
hartungii Wollaston, 1864: 442 (*Arthrodes*) **N:** CI (Fuerteventura, Lanzarote)
heydeni Reitter, 1914b: 53 **N:** MO
densepunctatus Escalera, 1914: 268
impressifrons Baudi di Selve, 1881: 274 **A:** SA YE
kocheri Español, 1963e: 379 **N:** MO
longicornis Ecalera, 1914: 269 **N:** MO
malleatus Wollaston, 1864: 446 (*Arthrodes*) **N:** CI (Fuerteventura, Lanzarote, Graciosa)
montanus atlanticus Escalera, 1922d: 373 **N:** MO
montanus montanus Escalera, 1914: 265 **N:** MO
montanus rehhamensis Antoine, 1946a: 32 **N:** MO
parcepunctatus Wollaston, 1864: 443 (*Arthrodes*) **N:** CI (Gran Canaria)
punctator Reitter, 1914b: 53 **N:** MO
punctatulus Wollaston, 1864: 443 (*Arthrodes*) **N:** CI (Fuerteventura, Lanzarote, Isla de Lobos, Graciosa)
rotundatus confalonierii Gridelli, 1930: 196 **N:** LB
rotundatus rotundatus Solier, 1834: 515 **N:** EG **A:** SY YE
cruciatus Solier, 1834: 517
gebieni Reitter, 1914b: 53
obliteratus Solier, 1834: 516

subgenus *Arthrodinus* Reitter, 1900j: 299 type species *Erodius obesus* Brullé, 1839
byrrhoides Wollaston, 1864: 441 (*Arthrodes*) **N:** CI (Fuerteventura)
curtus Brullé, 1839: 63 (*Erodius*) **N:** CI (Gran Canaria)
inflatus Wollaston, 1864: 439 (*Arthrodes*) **N:** CI (Graciosa, Lanzarote, Montaña Clara)
laticollis Brullé, 1839: 63 (*Erodius*) **N:** CI (Fuerteventura, Isla de Lobos, Graciosa, Lanzarote)
obesus crassus Wollaston, 1865: 389 (*Arthrodes*) **N:** CI (Gran Canaria)
  perraudieri Wollaston, 1865: 58 [Appendix] (*Arthrodes*)
obesus gomerensis Oromí, 1977: 15 **N:** CI (La Gomera)
obesus obesus Brullé, 1839: 63 (*Erodius*) **N:** CI (Tenerife)
obesus simillimus Wollaston, 1864: 440 (*Arthrodes*) **N:** CI (El Hierro, La Palma)
subciliatus Wollaston, 1864: 444 (*Arthrodes*) **N:** CI
subcostatus Brullé, 1839: 64 (*Erodius*) **N:** CI (Gran Canaria)
subgenus *Kocheria* Antoine, 1946a: 25 type species *Arthrodeis rungsi* Español, 1943
rungsi Español, 1943c: 121 **N:** MO (Western Sahara) **AFR**
  bulloni Español, 1943c: 122
  mauritanicus Koch, 1943a: 501

genus *Arthrodeis*, nomen dubium
crypticoides Fairmaire, 1879a: 4 **A:** SA **N:** EG

genus *Arthrodibius* Lesne, 1915: 235 type species *Arthrodeis laxepunctatus* Fairmaire, 1884
subgenus *Erodibius* Löbl, Bouchard, Merkl & Bousquet, 2020 [see New Acts] type species *Arthrodeis cyphonotus* Fairmaire, 1887
cicatrix Fairmaire, 1879a: 4 (*Arthrodeis*) **A:** SA YE
  wissmanni Gebien, 1938: 52

genus *Arthrodosis* Reitter, 1900j: 299 type species *Erodius globosus* Faldermann, 1837
afghanicus Kaszab, 1959f: 332 **A:** AF
bogatshevi G.S. Medvedev, 1976: 898 **A:** IN
bucharicus Reitter, 1915i: 131 **A:** KZ UZ
carmanicus G.S. Medvedev, 1976: 899 **A:** IN
castaneus Vauloger de Beaupré, 1897: 294 (*Arthrodes*) **A:** AF TM UZ
costatus Kaszab, 1959f: 329 **A:** AF
globosus Faldermann, 1837: 3 (*Erodius*) **E:** AB AR **A:** IN
intermedius Reitter, 1889a: 27 (*Arthrodeis*) **A:** TM
kawirensis Schuster, 1935b: 22 **A:** IN
klapperichi Kaszab, 1959f: 332 **A:** AF
lindbergi Kaszab, 1968b: 36 **A:** AF
lucidus Kaszab, 1968b: 37 **A:** AF
orientalis Faust, 1875: 165 (*Arthrodeis*) **A:** KZ TM
  lobicollis Reitter, 1914b: 57
pappi Kaszab, 1979a: 111 **A:** AF

planosternum Reitter, 1915i: 130 **A:** KZ UZ
pleuralis Kaszab, 1979a: 112 **A:** AF
politus Waterhouse, 1889: 124 (*Spirathus*) **A:** AF PA
povolnyi Kaszab, 1968b: 40 **A:** AF
puncticollis Reitter, 1902f: 214 **A:** KZ UZ
pusztae Kaszab, 1979a: 108 **A:** IN
rhomboidalis Kwieton, 1981a: 6 **A:** PA (Balochistan)
schusteri Reitter, 1914b: 57 **A:** AF KZ TM UZ
fuscocastaneus Reitter, 1914b: 57
taravatii Lillig, 2015: 167 **A:** IN
tuxeni shvrtshinorum Kwieton, 1981a: 3 **A:** PA (Quetta)
tuxeni tuxeni Kaszab, 1959e: 238 [= 1959f: 331] **A:** AF
waziristanicus Kaszab, 1979a: 115 **A:** PA

genus *Arthrohyalosis* Kaszab, 1979a: 75 type species *Arthrodosis mostofii* Pierre, 1975
mostofii Pierre, 1975: 299 (*Arthrodosis*) **A:** IN

genus *Bulbulus* Lesne, 1915: 240 type species *Arthrodeis byrrhiformis* Fairmaire, 1892
bidentulus Fairmaire, 1879a: 4 (*Arthrodeis*) **A:** SA YE
immarginatus Gebien, 1938: 51 (*Arthrodibius*)
pinguis Ancey, 1881b: 509 (*Arthrodes*)

genus *Capricephalius* Koch, 1943a: 508 type species *Arthrodibius bazmanicus* Schuster, 1938
bazmanicus Schuster, 1938b: 79 (*Arthrodibius*) **A:** AE IN OM QA
zarudnyi Bogatchev, 1950: 235 (*Spyrathus*) **A:** IN

genus *Diaphanidus* Reitter, 1900j: 299 type species *Arthrodeis antennatus* Reitter, 1894
subgenus *Diaphanidus* Reitter, 1900j: 299 type species *Arthrodeis antennatus* Reitter, 1894
Globularthrodosis Kaszab, 1979a: 95 type species *Diaphanidus semenowi* Reitter, 1900
antennatus Reitter, 1894a: 48 (*Arthrodeis*) **E:** ST **A:** TM (Karakumy)
ferrugineus Fischer von Waldheim, 1821: 12 (*Erodius*) **E:** ST **A:** KZ TM TR UZ
volganus Semenov & Bogatchev, 1940: 208 (*Diaphanidius*)
globosus Skopin, 1961b: 381 **A:** TD UZ
granulatus Bogatchev, 1950: 233 **A:** UZ
narynensis G.S. Medvedev & Kaltaev, 1979: 59 **A:** KI
semenowi Reitter, 1900l: 302 **A:** KZ TM
subgenus *Pseudodiaphanidus* Bogatchev, 1950: 234 type species *Diaphanidus robustus* Bogatchev, 1950
khashensis Kaszab, 1979a: 101 **A:** IN
lutensis Kaszab, 1979a: 100 **A:** IN
robustus Bogatchev, 1950: 235 (*Diaphanidus*) **A:** IN
talatabensis G.S. Medvedev, 1976: 900 (*Arthrodosis*) **A:** IN

genus *Erodiontes* Reitter, 1914b: 79 type species *Erodiontes vermiculatus* Reitter, 1914
Iranarthrodosis Kaszab, 1959f: 334 type species *Arthrodosis pfaundleri* Schuster, 1935
aelleni Kaszab, 1968b: 33 (*Arthrodosis*) **A:** IN
pfaundleri Schuster, 1935b: 21 (*Arthrodosis*) **A:** IN
menetriesi Bogatchev, 1950: 232 (*Arthrodosis*)
vermiculatus Reitter, 1914b: 80 **A:** IN

genus *Erodius* Fabricius, 1775: 258 type species *Erodius gibbus* Fabricius, 1775
subgenus *Dimeriseis* Solier, 1834: 530 type species *Erodius laevigatus* Olivier, 1791
bicostatus bicostatus Solier, 1834: 536 **N:** AG EG LB MO
tripolitanus Mader, 1943: 41
bicostatus coccinelloides Koch, 1937: 330 **N:** LB
bicostatus costatulus Koch, 1939: 253 **N:** AG
bicostatus leptis Koch, 1937: 330 **N:** LB
bicostatus vittatus Peyerimhoff, 1943: 22 **N:** AG MO
bicostatus pulvereus Reitter, 1914b: 60 **N:** AG TU
bilineatus bungemensis Koch, 1937: 328 **N:** LB (Tripolitana)
bilineatus mocciai Koch, 1937: 327 **N:** LB (Tripolitana, Fezzan)
bilineatus rohlfi Koch, 1941c: 258 **N:** LB (Cyrenaica)
chudeaui Koch, 1941: 256 **N:** AG **AFR**
exilipes exilipes Lucas, 1858a: clxxx **N:** AG LB MO
exilipes thiebaultii Fairmaire, 1877: 142 **N:** AG (Grand Erg Oriental)
glabratus Solier, 1834: 544 **A:** SA
graniventris Peyerimhoff, 1921c: 233 **N:** AG **AFR**
kneuckeri kneuckeri Andres, 1920: 61 **N:** EG (Ismailia) **A:** IS SI
kneuckeri semisculptus Koch, 1935a: 12 **N:** EG (Gebel Asfar)
latreillei Solier, 1834: 537 **N:** AG LB **AFR**
mateui Español, 1946: 108 **N:** MO
omanensis Lillig, 2009a: 75 **A:** OM
opacus Kraatz, 1865a: 25 **N:** EG **A:** IS SI
peyerimhoffi Koch, 1939: 258 **N:** AG (In Salah)
piestognathoides Koch, 1941c: 257 **N:** AG TU
subgenus *Dirosis* Miller, 1858: 115 type species *Dirosis nervosus* Miller, 1858 (= *Erodius servillei* Solier, 1834)
antonei Kocher, 1951a: 214 **N:** MO
barbarus agedabicus Koch, 1939: 255 **N:** LB (Cyrenaica)
barbarus barbarus Solier, 1834: 538 **N:** LB TU
barbarus syrticus Koch, 1937: 334 **N:** LB
bicarinatus bicarinatus Erichson, 1841: 175 **N:** AG TU
bicarinatus oranensis Koch, 1941c: 260 **N:** AG (Oran)
bicarinatus tunesius Koch, 1941c: 263 **N:** TU
brevicollis Kraatz, 1865a: 64 **N:** AG MO

chauveneti Solier, 1834: 568 **N:** AG MO
subnitidus Solier, 1834: 579 [DA]
costatus costatus Solier, 1834: 539 **N:** EG LB (Cyrenaica)
costatus iforax Peyerimhoff, 1931b: 92 **N:** AG LB (Fezzan) **AFR**
dejeanii Solier, 1834: 543 **A:** IS SI SY
duponchelii Allard, 1865a: 388 **A:** IS SI
lineolatus Reitter, 1914b: 73 (*Derosis*)
octolineatus Reitter, 1914b: 73 (*Derosis*)
edomitus Koch, 1940c: 94 **A:** IS (Negev) JO
elegans elegans Kraatz, 1865a: 51 **N:** AG TU
semicostatus Allard, 1873: 160
elegans hadjouensis Koch, 1941c: 263 **N:** AG (Oran)
elegans oujdensis Koch, 1941c: 264 **N:** MO (Oujda)
escalerai Blair, 1935a: 103 [RN] **E:** SP **N:** MO
rabaticus Koch, 1941c: 266
reticulatus Escalera, 1914: 270 [HN]
rugosus Kraatz, 1865a: 32 [HN]
festae Schuster, 1922a: 4 **N:** LB (Tripolitania, Cyrenaica)
goryi goryi Solier, 1834: 564 **E:** PT SP (Cadiz, Andalusia)
faroensis Reitter, 1914b: 68 (*Derosis*)
inconstatus Reitter, 1914b: 68 (*Derosis*)
subobtusu Koch, 1944b: 219
goryi obtusus Allard, 1873: 221 **E:** PT SP
bolivari Koch, 1944b: 220
sevillanus Koch, 1944b: 220
granipennis fauconnieri Kocher, 1957: 372 **N:** MO (Meknès)
granipennis granipennis Fairmaire, 1871a: 384 **N:** MO
allardi Kocher, 1951a: 208
deplanatus Kocher, 1951a: 210
maximus Allard, 1873: 206
minimus Kocher, 1951a: 211
reymondi Kocher, 1951a: 211
testudo Reitter, 1914b: 69 (*Derosis*)
granipennis reitteri Kocher, 1951a: 213 **N:** MO
granipennis validus Baudi di Selve, 1874b: 192 **N:** MO
hebraicus Lillig, 1997: 55 **A:** IS SI
henoni Allard, 1873: 147 [RN] **N:** AG MO
antennarius Vauloger de Beaupré, 1897: 307
deserticola Vauloger de Beaupré, 1897: 307
granulosus Allard, 1865b: 391 [HN]
impressicollis impressicollis Vauloger de Beaupré, 1897: 307 **N:** AG TU

impressicollis zavattarii Koch, 1939: 254 **N:** LB
klugii Allard, 1865a: 388 **A:** IS LE
kruegeri Schuster, 1928e: 374 **N:** EG LB
laticollis laticollis Solier, 1834: 558 **N:** AG
laticollis lucescens Peyerimhoff, 1949b: 277 **N:** MO
lefrancii Kraatz, 1865a: 26 **N:** AG LB TU
maroccanus Koch, 1941c: 264 **N:** MO
multicostis Reitter, 1914b: 73 (*Derosis*) **A:** "Palestina"
nanus Vauloger de Beaupré, 1897: 308 **N:** LB TU
opaciventris garamas Peyerimhoff, 1931b: 92 **N:** AG (Hoggar)
opaciventris opaciventris Peyerimhoff, 1931b: 91 **N:** AG (Hoggar)
parvus Solier, 1834: 557 **E:** SP
  pseudogoryi Koch, 1944b: 221
pesvalgus Lillig, 2009a: 78 **A:** SY
puncticollis maillei Solier, 1834: 546 [DA] **A:** TR
puncticollis puncticollis Solier, 1834: 556 **N:** EG LB **AFR**
  contractus Kraatz, 1865a: 31
  ecostatus Crotch, 1872: 267
puncticollis sinaiticus Crotch, 1872: 267 **A:** IS SI
reticulatus Reitter, 1914b: 69 (*Derosis*) **N:** MO
retrangulus Reitter, 1914b: 66 (*Derosis*) **N:** AG
rugosus Allard, 1865a: 387 **N:** AG TU
  rugicollis Allard, 1873: 145
semenovi Bogatchev, 1950: 231 **A:** IQ TR
servillei Solier, 1834: 540 **A:** AF IN IQ SA SY
  nervosus Miller, 1858: 116 (*Dirosis*)
  scaber Solier, 1834: 542
singularis Kocher, 1951a: 227 **N:** AG MO
theryi Koch, 1940c: 97 **A:** SY
wagneri Erichson, 1841: 175 **N:** AG
subgenus *Eodirosis* Kwieton, 1980b: 25 type species *Erodius quadrilineatus* Kraatz, 1865
acutangulus Bogatchev, 1957: 530 **A:** IN
fabricii Solier, 1834: 552 **A:** CY
freyi Schuster, 1934b: 181 **A:** JO (Dead Sea)
  subpuncticollis Schuster, 1934b: 183
matthiasi Lillig, 1994: 153 **A:** SA (Hail Prov.)
octocostatus octocostatus Peyerimhoff, 1907a: 28 **A:** IQ JO SA
octocostatus verruculiferus Kaszab, 1981c: 308 **A:** SA (Hail Prov.)
orientalis boyeri Solier, 1834: 553 **E:** GR (Crete)
orientalis brevicostatus Solier, 1834: 551 **E:** GR
orientalis oblongus Solier, 1834: 555 **E:** GR (Dodecanese Is., Crete, Lesvos) **A:** TR

orientalis orientalis Brullé, 1832: 192 **E:** GR IN
duponti Solier, 1834: 562
longus Solier, 1834: 553
orientalis Solier, 1834: 553 [HN]
pavliceki Lillig, 1994: 152 **A:** IS (Negev)
quadrilineatus Kraatz, 1865a: 31 **N:** EG **A:** IS
gibbus Solier, 1834: 547 [HN]
reichei besnardi Kaszab, 1981c: 313 **A:** SA (Eastern Prov.)
reichei curvipes Kaszab, 1981c: 313 **A:** SA (Riyad Prov.)
reichei reichei Allard, 1875: 164 **A:** IS (Negev) JO
rothi Koch, 1940c: 90 **A:** IS (Negev) JO
wohlberedti Koch, 1940c: 93
rubalkhalianus Kaszab, 1981c: 314 **A:** OM SA
sauditus boreocentralis Kaszab, 1981c: 312 **A:** SA (Hail Prov.)
sauditus costatissimus Kaszab, 1981c: 312 **A:** AE OM SA YE
sauditus granularis Kaszab, 1981c: 311 **A:** SA (Gasim Prov.)
sauditus insularis Kaszab, 1981c: 312 **A:** BA
sauditus occidentalis Kaszab, 1981c: 311 **A:** SA (Tabuk Prov.)
sauditus sauditus Kaszab, 1981c: 310 **A:** KU SA
sauditus simplex Kaszab, 1981c: 311 **A:** SA
subgenus *Erodius* Fabricius, 1775: 258 type species *Erodius gibbus* Fabricius, 1775
Acantophorus Billberg, 1820: 32 type species *Erodius gibbus* Fabricius, 1775
Cephacerus Rafinesque, 1815: 113 [URN]
Herodius Agassiz, 1846a: 143 [UE]
audouini audouini Solier, 1834: 569 **N:** MO
audouini destefanii Failla-Tebaldi, 1887: 102 **E:** IT (Lampedusa Is.)
audouini peiroleri Solier, 1834: 590 **E:** IT (Sardinia)
belutschistanicus Kaszab, 1957d: 63 **A:** IN
carinatus carinatus Solier, 1834: 549 **E:** SP **N:** MO
latus Solier, 1834: 574
carinatus mendizabali Koch, 1944b: 221 **E:** SP (Almería)
chouberti Kocher, 1951a: 199 **N:** MO
convexus Grimm & Lillig, 2000: 32 **N:** AG MO
dentifer Peyerimhoff, 1920b: 21 **N:** AG (Grand Erg Occidental)
dilatatus Reitter, 1914b: 79 **N:** AG
emondi emondi Solier, 1834: 585 **N:** AG
ambiguus Solier, 1834: 586
marginicollis Solier, 1834: 587
emondi laevis Solier, 1834: 588 **E:** SP **N:** AG MO TU
affinis Solier, 1834: 589
africanus Solier, 1834: 578 [DA]

imberbitus Reitter, 1914b: 79
mittrei Solier, 1834: 591
nitidicostis Allard, 1873: 220
externus duplex Escalera, 1914: 271 **N:** MO
externus externus Fairmaire, 1875c: 543 **N:** MO
gibbus cyrenaicus Schuster, 1926b: 130 **E:** LB (Cyrenaica)
gibbus gibbus Fabricius, 1775: 258 **N:** EG **A:** IS SI
syriacus Solier, 1834: 592
lusitanicus Solier, 1834: 565 **E:** PT SP **N:** MO
nitidicollis Solier, 1834: 583 **N:** AG
pulverulentus Allard, 1873: 208
solieri Allard, 1873: 212
proximus Solier, 1834: 575 **E:** SP
siculus dalmatinus Kraatz, 1865a: 44 **E:** AL CR GR IT
siculus melitensis Reitter, 1914b: 77 **E:** MA
siculus neapolitanus Solier, 1834: 571 **E:** IT
siculus siculus Solier, 1834: 570 **E:** IT (Sicilia, Favignana)
pyriformis Allard, 1873: 203
rotundatus Kraatz, 1865a: 43 [DA]
vicinus Solier, 1834: 582
subparallelus Solier, 1834: 584 **N:** AG
tangerianus Solier, 1834: 581 **N:** MO
nitidiventris Solier, 1834: 576
pseudocarinatus Kocher, 1951a: 218
tibialis Linnaeus, 1766: 678 (*Tenebrio*) **E:** PT SP **N:** MO (Tanger, Ceuta)
bilineatus Herbst, 1799: 174
curvipes Solier, 1834: 567
distinctus Allard, 1865b: 394
europaeus Solier, 1834: 560
subgenus *Zophoserodius* Reitter, 1914b: 58 type species *Erodius zophosoides* Allard, 1865
planatus Reitter, 1914b: 61 **N:** AG
melillensis Kocher, 1951a: 222
zophosoides marrakensis Kocher, 1951a: 223 **N:** MO (Marrakesh)
zophosoides nunicus Kocher, 1956b: 248 **N:** MO
zophosoides schatzmayri Koch, 1937: 331 **N:** LB (Tripolitania)
zophosoides subbicostatus Koch, 1937: 333 **N:** LB (Agedabia)
zophosoides zophosoides Allard, 1865a: 387 **N:** AG EG LB TU **A:** SI
dimidiatipennis Kraatz, 1865a: 30

genus *Erodius*, nomen dubium
kraatzi Reitter, 1914b: 66 (*Derosis*) **E:** SP

genus *Farsarthrosis* Kaszab, 1979a: 86 type species *Farsarthrosis benardi* Kaszab, 1979
benardi Kaszab, 1979a: 87 **A:** IN

genus *Foleya* Peyerimhoff, 1916a: 71 type species *Foleya brevicornis* Peyerimhoff, 1916
brevicornis Peyerimhoff, 1916a: 71 **N:** AG

genus *Histeromimus* Gahan, 1896a: 288 type species *Histeromimus arabicus* Gahan, 1896
arabicus Gahan, 1896a: 288 **A:** YE

genus *Histeromorphus* Kraatz, 1865a: 11 type species *Histeromorphus plicatus* Kraatz, 1865
carinatus Purchart, 2014a: 215 **A:** YE (Suqutra)
convexicostatus Purchart, 2014a: 216 **A:** YE (Suqutra)
nogedensis Purchart, 2014a: 219 **A:** YE (Suqutra)
plicatipennis Waterhouse, 1881: 473 **A:** YE (Suqutra)
plicatus Kraatz, 1865a: 12 **A:** YE (Suqutra)
  socotranus Lesne, 1915: 229
undatus Gahan, 1900: 8 **A:** YE (Suqutra)

genus *Hyalarthrodosis* Kaszab, 1979a: 74 type species *Arthrodosis monodi* Pierre, 1974
monodi Pierre, 1974: 230 (*Arthrodosis*) **A:** IN

genus *Hyalerodius* Kaszab, 1979a: 80 type species *Hyalerodius jirofti* Kaszab, 1979
jirofti Kaszab, 1979a: 80 **A:** IN

genus *Iranerodius* Kaszab, 1959f: 334 type species *Arthrodosis richteri* Kaszab, 1957
iranicus Kaszab, 1979a: 84 **A:** IN
menetriesi Bogatchev, 1950: 232 (*Arthrodosis*) **A:** IN
richteri Kaszab, 1957d: 62 (*Arthrodosis*) **A:** IN
  beludzhistanicus G.S. Medvedev, 1976: 896 (*Arthrodosis*)
semenovi Bogatchev, 1949a: 40 [= 1950: 233] (*Arthrodosis*) **A:** IN

genus *Leptonychoides* Schawaller, 1990b: 51 type species *Leptonychoides juengeri* Schawaller, 1990
juengeri Schawaller, 1990b: 52 **A:** SA

genus *Leptonychus* Chevrolat, 1833b: 26 type species *Leptonychus erodioides* Chevrolat, 1833
curvicornis Peyerimhoff, 1916a: 73 **N:** AG
laevisternus Fairmaire, 1883l: clxxviii **N:** AG LB TU
reymondi Pierre, 1954: 24 **N:** MO
rufipennis Guérin-Méneville, 1859: clxxxvii **N:** AG
pellucidus iguidiense Pierre, 1954: 28 **N:** AG
pellucidus pellucidus Mulsant & Godart, 1866: 450 (*Erodius*) **N:** AG
  convexiventris Fairmaire, 1875a: 517

pellucidus tunisiana Dajoz, 1983: 65 **N:** TU
sabulicola Chobaut, 1898a: 81 **N:** AG

genus *Piestognathoides* Kaszab, 1981c: 305 type species *Piestognathoides bahrainicus* Kaszab, 1981
bahrainicus Kaszab, 1981c: 306 **A:** BA

genus *Piestognathus* Lucas, 1858b: clxxxviii type species *Piestognathus douei* Lucas, 1858
douei Lucas, 1858b: clxxxviii **N:** AG TU
asperipennis Fairmaire, 1875a: 516

genus *Spyrathus* Kraatz, 1865a: 9 type species *Spyrathus indicus* Kraatz, 1865
assimilis Kaszab, 1957f: 290 **A:** PA
fabrii Fairmaire, 1883f: lv **A:** "Hindoustan"
gebieni Reitter, 1914b: 55 **A:** PA **ORR**
indicus Kraatz, 1865a: 10 **A:** "India orientale anteriore"
maindroni Vauloger de Beaupré, 1897: 294 (*Arthrodes*) **A:** PA **ORR**

tribe Idisiini G.S. Medvedev, 1973

genus *Idisia* Pascoe, 1866: 452 type species *Idisia ornata* Pascoe, 1866
ornata Pascoe, 1866: 452 **A:** FE JA NE NMO SC "Northeast China"
vestita Marseul, 1876c: 100 **A:** JA

tribe Klewariini Gebien, 1910

genus *Klewaria* Reitter, 1910a: 20 type species *Klewaria colydiiformis* Reitter, 1910
colydiiformis Reitter, 1910a: 21 **A:** TD TM

tribe Lachnogyini Seidlitz, 1894

subtribe Lachnodactylina Reitter, 1904

genus *Lachnodactylus* Seidlitz, 1898a: 838 [RN] type species *Lachnopus digitalis* Seidlitz, 1894
Lachnopus Seidlitz, 1894: 476 [HN] type species *Lachnopus digitalis* Seidlitz, 1894
digitalis Seidlitz, 1894: 476 (*Lachnopus*) **A:** KZ TM

subtribe Lachnogyina Seidlitz, 1894

genus *Lachnogya* Ménétriés, 1849: 228 type species *Lachnogya squamosa* Ménétriés, 1849
orientalis Kaszab, 1982a: 57 **A:** "Bengal" **ORR**
squamosa Ménétriés, 1849: 229 **E:** AR ST **A:** AF KZ PA TM UZ
skopini J. Ferrer & Yvinec, 2004: 45

subtribe Netuschiliina J. Ferrer & Yvinec, 2004

genus *Netuschilia* Reitter, 1904a: 35 type species *Lachnopus hauseri* Reitter, 1897
hauseri Reitter, 1897e: 217 (*Lachnopus*) **E:** GG ST **A:** AF GAN HEB KZ LIA NMO PA TD TM UZ ZIN

tribe Leptodini Lacordaire, 1859

genus *Leptodes* Dejean, 1834: 181 type species *Sepidium boisduvalii* Zubkov, 1833

subgenus *Leptodes* Dejean, 1834: 181 type species *Sepidium boisduvalii* Zubkov, 1833

Leptodinopsis Kaszab, 1959g: 352 type species *Sepidium boisduvali* Zubkov, 1833

boisduvalii Zubkov, 1833: 328 (*Sepidium*) **A:** KZ (Ustyurt) TM

cavicola Kaszab, 1959g: 355 **A:** AF

chakchakensis Tahami, Merkl & Sadeghi, 2016: 592 **A:** IN

farashahi Tahami, Merkl & Sadeghi, 2016: 595 **A:** IN

heydeni Reitter, 1892b: 95 **A:** TM

kabakovi G.S. Medvedev, 1997a: 176 **A:** AF (Herat)

olegi G.S. Medvedev, 1997a: 178 **A:** AF (Herat)

rimicola Bogatchev & Kryzhanovskiy, 1960: 265 **A:** TM (Kopet Dag)

sagarensis G.S. Medvedev, 1997a: 179 **A:** AF (Ghor)

transcaspicus Kaszab, 1959g: 352 **A:** UZ

zubkovi Semenov, 1909: 300 **E:** ST **A:** TM

subgenus *Leptodopsis* Haag-Rutenberg, 1879b: 408 type species *Leptodopsis insignis* Haag-Rutenberg, 1879

andrejevae G.S. Medvedev, 1990a: 110 **A:** TD

asanovitshae asanovitshae G.S. Medvedev, 1990a: 107 **A:** UZ (Uramskiy Khrebet)

asanovitshae kassansaicus G.S. Medvedev, 1990a: 108 **A:** KI (Chatkalsiy Khrebet)

belousovi G.S. Medvedev, 1997a: 180 **A:** KI

brevicarina Ren, 1999: 48 **A:** XIN

chinensis Kaszab, 1962a: 78 **A:** GAN NIN NMO XIN

fedtschenkoi Semenov, 1893b: 233 **A:** TD TM UZ

quadricostatus Kaszab, 1959g: 358

gurjevae G.S. Medvedev, 1990a: 105 **A:** KI

insignis Haag-Rutenberg, 1879b: 409 (*Leptodopsis*) **A:** KI XIN

karatavicus G.S. Medvedev, 1990a: 109 **A:** KZ (Karatau)

kaszabi G.S. Medvedev, 1972a: 309 **A:** TD

medvedevi Kaszab, 1970a: 21 **A:** KZ (Altyn-Emel Mts)

mongolicus G.S. Medvedev, 1990b: 136 **A:** MG

panfilovi G.S. Medvedev, 1967: 356 **A:** KI

reimovi G.S. Medvedev & B.P. Pirnazarov, 1972: 125 **A:** UZ (Ustyurt)

schusteri Reitter, 1916d: 134 **A:** UZ

shurabensis Bogatchev, 1967c: 359 **A:** TD (Shurab)

stackelbergi G.S. Medvedev, 1967: 357 **A:** KI

susamyrica G.S. Medvedev, 1997a: 179 **A:** KI

suworowi Reitter, 1907f: 205 (*Leptodopsis*) **A:** KZ (Almaatinskaya Oblast)

szekessyi Kaszab, 1962a: 79 **A:** NIN NMO SHA SHX

tadzhikistanicus G.S. Medvedev, 1979c: 71 **A:** TD

tjanschanicus egorovi G.S. Medvedev, 1990a: 113 **A:** KI
tjanschanicus tjanschanicus Semenov, 1893b: 232 **A:** KI KZ
turkestanicus Semenov, 1889a: 219 (*Leptodopsis*) **A:** TD UZ
subgenus *Mesoleptodes* Medvedev & Iljina, 2007 type species *Leptodes semenowi* Reitter, 1892
daghestanicus G. Medvedev & Iljina, 2007: 1036 **E:** ST (Dagestan)
karmaniae Tahami, Merkl & Sadeghi, 2016: 596 **A:** IN
khanensis Tahami, Merkl & Sadeghi, 2016: 598 **A:** IN
kulzeri Kaszab, 1959g: 356 **A:** IN
lederi Reitter, 1889a: 32 **E:** AB AR
ovtshinnikovi G.S. Medvedev, 1990a: 106 **A:** KI
persiae Tahami, Merkl & Sadeghi, 2016: 599 **A:** IN
semenowi Reitter, 1892b: 95 **E:** AB AR GG
shapouri Tahami, Merkl & Sadeghi, 2016: 600 **A:** IN
solieri Semenov, 1909: 302 [RN] **A:** IN TM UZ
boisduvali Solier, 1838b: 193 [HN]
tschitscherini Semenov, 1895a: 139 **A:** IN
subgenus *Paraleptodes* G.S. Medvedev, 1967: 354 type species *Leptodes lindbergi* Kaszab, 1959
badakhshanicus G.S. Medvedev, 1990a: 104 **A:** TD
kryzhanovskiji Kaszab, 1970a: 24 **A:** AF
lindbergi Kaszab, 1959g: 354 **A:** AF TM UZ
tadzhibaevi G.S. Medvedev, 1979c: 69 **A:** TD
terminassianae G.S. Medvedev, 1970: 404 **A:** TM
subgenus *Proleptodes* G.S. Medvedev, 1967: 354 type species *Leptodes sulcicollis* Reitter, 1889
reitteri Semenov, 1891: 271 **A:** XIN
sulcicollis Reitter, 1889k: 687 **A:** GAN XIZ

genus *Tapenopsis* Solier, 1843: 217 type species *Tapenopsis costata* Solier, 1843
Tapinopsis Agassiz, 1846: 361 [UE]
costata Solier, 1843: 218 **A:** AF IN

tribe Pimeliini Latreille, 1802

genus *Afghanopachys* Kwieton, 1978b: 30 type species *Pachyscelis haarlovi* Gridelli, 1954
freyi freyi Kaszab, 1960a: 35 (*Pachyscelis*) **A:** AF
freyi kabuliensis Kaszab, 1960a: 37 (*Pachyscelis*) **A:** AF
haarlovi Gridelli, 1954: 230 [= 1955: 41] (*Pachyscelis*) **A:** AF
zhenzhuristi Bogatchev, 1939a: 124 (*Pachyscelis*) **A:** AF
afghanica Gridelli, 1954: 228 [= 1955: 40] (*Pachyscelis*)

genus *Allotadzhikistania* Bogatchev, 1960a: 43 type species *Allotadzhikistania comata* Bogatchev, 1960

comata Bogatchev, 1960a: 44 **A**: TD
setosa G.S. Medvedev, 1970: 403 **A**: TD (Zaalaskij Ridge)

genus *Apatopsis* Semenov, 1890: 368 type species *Apatopsis grombczewskii* Semenov, 1890
conradti Semenov, 1890: 370 **A**: XIN
grombczewskii Semenov, 1890: 368 **A**: XIN

genus *Argyradelpha* G.S. Medvedev, 2005b: 306 type species *Argyradelpha lopatini* G.S. Medvedev, 2005
lopatini G.S. Medvedev, 2005b: 307 **A**: AF (Registan)

genus *Argyrophana* Semenov, 1889a: 222 type species *Argyrophana deserti* Semenov, 1889
caspia Semenov, 1910c: 43 **A**: TM
deserti Semenov, 1889a: 224 **A**: TM
diaphana Semenov, 1910c: 44 **A**: UZ (Kyzylkum)
vicaria Semenov, 1910c: 43 **A**: UZ (Kyzylkum)

genus *Astorthocnemis* Lillig & Pavlíček, 2002: 98 type species *Storthocnemis saudita* Koch, 1965
becvarorum Lillig & Pavlíček, 2002: 99 **A**: JO SI
saudita Koch, 1965a: 130 (*Storthocnemis*) **A**: SA YE

genus *Balachowskya* Peyerimhoff, 1928: 61 type species *Balachowskya portentosa* Peyerimhoff, 1928
portentosa Peyerimhoff, 1928: 62 **N**: AG

genus *Bogatshevia* G.S. Medvedev & Iwan, 2006: 618 [RN] type species *Achaemenes villosus* Bogatchev, 1949 (= *Thriptera bogatchevi* Kwieton, 1982)
Achaemenes Bogatchev, 1949a: 39 [HN] type species *Achaemenes villosus* Bogatchev, 1949 (= *Thriptera bogatchevi* Kwieton, 1982)
bogatchevi Kwieton, 1982b: 23 (*Thriptera*) [RN] **A**: IN PA
villosa Bogatchev, 1949a: 40 [HN]

genus *Cyclocnera* Leo, 2018: 50 type species *Cyclocnera azarovi* Leo, 2018
azarovi Leo, 2018: 53 **A**: OM

genus *Diesia* Fischer von Waldheim, 1820: pl. 14 type species *Diesia quadridentata* Fischer von Waldheim, 1820
Diesiola Skopin, 1961b: 397 type species *Diesia quadridentata* Fischer von Waldheim, 1820
aralensis Skopin, 1960a: 301 (*Idiesa*) **A**: KZ
quadridentata iliensis Skopin, 1961b: 401 **A**: KZ (Alamtinskaya oblast)
quadridentata quadridentata Fischer von Waldheim, 1820: pl. 14 **A**: AF KI KZ TM

sexdentata auliensis Reitter, 1901c: 78 (*Trigonoscelis*) **A:** KZ
sexdentata sexdentata Fischer von Waldheim, 1820: pl. 14 **A:** KZ TM UZ
sexdentata termesiana Skopin, 1960a: 302 **A:** KZ TM
sexdentata vachshiana Skopin, 1960a: 302 **A:** TD

genus *Dietomorpha* Reymond, 1938: 143 type species *Dietomorpha pardalis* Reymond, 1938
pardalis Reymond, 1938: 143 **A:** IN PA
szekessyi Kaszab, 1957f: 295 (*Kawiria*)

genus *Earophanta* Semenov, 1903b: 172 [RN] type species *Platyope serrata* Semenov, 1893
Earophila Semenov, 1903a: 9 [HN] type species *Platyope serrata* Semenov, 1893
Earophilina E. Strand, 1917: 99 [RN] type species *Platyope serrata* Semenov, 1893
beludzhistana Bogatchev, 1957: 531 **A:** PA (Balochistan)
collaris Fischer von Waldheim, 1844: 62 (*Platyope*) **A:** KZ XIN
regeli Ballion, 1878: 318 (*Platyope*)
holdhausi Reymond, 1937: 247 **A:** KZ (Dzungaria)
pilosissima Reitter, 1895d: 155 (*Przewalskia*) **A:** AF TM
loudoni Semenov, 1903a: 11
planidorsis Reitter, 1889h: 285 (*Platyope*) **A:** AF TM
autumnalis Semenov, 1903a: 9 (*Platyope*)
pubescens Skopin, 1960a: 306 **A:** KZ (Zhambylskaya oblast)
serrata Semenov, 1893c: 254 (*Platyope*) **A:** TM (Karakumy)
tomentosa Semenov, 1893c: 256 (*Platyope*) **A:** UZ

genus *Euryostola* Reitter, 1893b: 207 type species *Pachyscelis minor* Baudi di Selve, 1875
minor Baudi di Selve, 1875d: 689 (*Pachyscelis*) **A:** IN
zophosioides Baudi di Selve, 1875d: 691 (*Leucolaephus*)

genus *Euthriptera* Reitter, 1893b: 229 type species *Thryptera grisescens* Fairmaire, 1875
delicata Bedel, 1893: 102 (*Thriptera*) **N:** TU **AFR**
grisescens Fairmaire, 1875a: 528 (*Thryptera*) **N:** AG LB TU
piceola Desbrochers des Loges, 1881: 86 [= 1898: 81] (*Ocnera*)

genus *Gedeon* Reiche & Saulcy, 1857: 219 type species *Gedeon hierichonticus* Reiche & Saulcy, 1857
hierichonticus Reiche & Saulcy, 1857: 221 **A:** IS JO SA SY
arabicus Solier, 1836a: 126 (*Pimelia*) [HN]
magniceps Reitter, 1915a: 60 **A:** AF UZ
oblongoovatus Schuster, 1937a: 48 (*Pimelia*) **A:** IN IQ
parallelus Solier, 1836a: 125 (*Pimelia*) **A:** IN IQ SY TR
borrei Haag-Rutenberg, 1878: 91

persicus Baudi di Selve, 1875d: 694 **A:** IN SY

baudii Ganglbauer, 1905: 277 (*Pimelia*)

remaudierei Pierre, 1964c: 1045

werneri Ganglbauer, 1905: 277 (*Pimelia*) **A:** TR

genus *Habrobates* Semenov, 1903a: 11 type species *Habrobates vernalis* Semenov, 1903

agnesae Schuster, 1938b: 86 **A:** IN

vejisovi Kelejnikova, 1977: 653 **A:** TM (Karakumy)

vernalis Semenov, 1903a: 12 **A:** TM (Karakumy)

genus *Habrochiton* Semenov, 1907a: 179 type species *Habrochiton vernus* Semenov, 1907

primaeverris Semenov, 1907a: 181 **A:** XIN

vernus Semenov, 1907a: 179 **A:** XIN

genus *Homopsis* Semenov, 1893c: 258 type species *Homopsis grumi* Semenov, 1893

grumi Semenov, 1893c: 258 **A:** XIN

genus *Idiesa* Reitter, 1893b: 245 type species *Diesia fischeri* Ménétriés, 1849

eversmanni afghanica Kaszab, 1970a: 51 **A:** AF

eversmanni eversmanni Kraatz, 1865a: 276 (*Diesia*) **A:** KZ UZ

fischeri Ménétriés, 1849: 225 (*Diesia*) **A:** TM KZ UZ

trigonosceloides Lebedev, 1931: 78 **A:** TM (Karakum)

genus *Iranolasiostola* Pierre, 1968: 1020 type species *Iranolasiostola davatchii* Pierre, 1968

davatchii Pierre, 1968: 1020 **A:** IN

genus *Iranopachyscelis* Pierre, 1968: 1027 type species *Iranopachyscelis eghbali* Pierre, 1968 (= *Thriptera persica* L. Redtenbacher, 1850)

persica lahlezarica Kwieton, 1982a: 37 **A:** IN

persica persica L. Redtenbacher, 1850: 48 (*Thriptera*) **A:** IN

eghbali Pierre, 1968: 1027

ispahansis Pierre, 1968: 1028

genus *Kawiria* Schuster, 1935b: 26 type species *Kawiria gabrieli* Schuster, 1935

gabrieli Schuster, 1935b: 28 **A:** IN

genus *Lasiostola* Dejean, 1834: 179 type species *Tenebrio pubescens* Pallas, 1781

subgenus *Centrocnemita* E. Strand, 1935: 284 [RN] type species *Centrocnemis mollis* Kraatz, 1882

Centrocnemis Kraatz, 1882d: 330 [HN] type species *Centrocnemis mollis* Kraatz, 1882

bendamira Kaszab, 1973b: 42 **A:** AF

lindbergi Kaszab, 1970a: 59 **A:** AF

linnei Kaszab, 1973b: 43 **A:** AF
lopatini Skopin, 1960a: 304 **A:** TD
mollis Kraatz, 1882d: 330 (*Centrocnemis*) **A:** UZ (Samarkand)
seminuda pimela Kaszab, 1970a: 58 **A:** AF
seminuda seminuda Reitter, 1900g: 277 **A:** AF TD TM
tomentosa Reitter, 1900g: 278 **A:** UZ
subgenus *Lasiocnema* G.S. Medvedev, 1993: 110 type species *Lasiostola heterogena* Fischer von Waldheim, 1844
amseli Kaszab, 1962c: 312 [RN] **A:** AF
afghanica Kaszab, 1961e: 237 [HN]
carinata Kraatz, 1882b: 89 **A:** KZ UZ
costulifera Heyden, 1892b: 107 **A:** UZ (Margilan)
espanoli Skopin, 1960a: 304 **A:** TD
gemmata Reitter, 1889h: 286 **A:** AF IN PA TM
grandis aideriensis G.S. Medvedev, 1993: 117 **A:** TM
grandis elongata Kraatz, 1883b: 359 **A:** TM
gracilis Waterhouse, 1889: 127 (*Ocnera*)
grandis grandis Kraatz, 1883b: 358 **A:** IN TM
grisescens Kraatz, 1883a: 350 **A:** AF KZ TM
heterogena Fischer von Waldheim, 1844: 58 **A:** IN TM
minuta Kraatz, 1865a: 285
nitens Reitter, 1887f: 28
kabakovi G.S. Medvedev, 1993: 116 **A:** AF
minima Ménétriés, 1849: 223 **A:** KZ UZ (Samarkand)
montana Kaszab, 1970a: 62 **A:** AF
nephelidis Reitter, 1893b: 223 **A:** AF IN KZ TM
pulla Reitter, 1893b: 224 **A:** TM
stackelbergiana Bogatchev, 1949b: 279 **A:** TD
zarudnyi G.S. Medvedev, 1993: 115 **A:** IN
subgenus *Lasiograna* G.S. Medvedev, 1993: 111 type species *Lasiostola interrupta* Reitter, 1901
interrupta Reitter, 1901e: 161 (*Centrocnemis*) **A:** TM
machnovskii Skopin, 1960a: 305 **A:** TD
rigrevana Kaszab, 1970a: 62 **A:** AF
seriegranosa Kaszab, 1970a: 60 **A:** AF
boreoafghanica G.S. Medvedev, 1970: 401
subgenus *Lasiostola* Dejean, 1834: 179 type species *Tenebrio pubescens* Pallas, 1781
affinis Kraatz, 1882b: 89 **A:** UZ (Margilan)
simillima Kraatz, 1882b: 89
afghanica Kaszab, 1960a: 26 **A:** AF
ashkhabadensis Bogatchev & Kryzhanovskiy, 1960: 267 **A:** TM (Kopet Dag)
crassa G.S. Medvedev, 1993: 111 **A:** TD

granulata Kraatz, 1882b: 90 **A:** UZ (Margilan)
hirta alakuliensis Skopin, 1961a: 188 **A:** KZ (Alakol)
hirta hirta Fischer von Waldheim, 1820: pl. 14 (*Pimelia*) **E:** ST **A:** UZ
nigrohirta Reitter, 1893b: 222
hirta robusta G.S. Medvedev, 1964a: 61 **A:** TM
hirtipes Kraatz, 1885: 331 **A:** UZ
laticollis Kraatz, 1882b: 90 **A:** UZ (Margilan)
piligera Kraatz, 1882d: 331 **A:** KZ UZ (Samarkand)
pubescens Pallas, 1781: 53 (*Tenebrio*) **E:** ST **A:** KI KZ TM
plustschevskyi Reitter, 1893a: 112
scabricollis Reitter, 1907f: 206 **A:** KZ (Almaatinskaya Oblast)
setifera Kaszab, 1970a: 64 **A:** AF
variolaris Gebler, 1841a: 373 **A:** KZ
zarjanovi G.S. Medvedev, 1993: 113 **A:** KZ

genus *Leucolaephus* Lucas, 1859a: xxii type species *Leucolaephus perrisii* Lucas, 1859 (= *Pimelia liliputana* Lucas, 1857)
Mecopisthopus Karsch, 1881: 46 type species *Mecopisthopus rohlfsi* Karsch, 1881
liliputanus Lucas, 1857: lvi (*Pimelia*) **N:** AG
perrisii Lucas, 1859a: xxiii
nigropunctatus Lucas, 1858a: clxxx (*Pimelia*) **N:** AG TU
rohlfsi Karsch, 1881: 47 (*Mecopisthopus*) **N:** LB
tripolitanus G. Quedenfeldt, 1889: 353 **N:** LB

genus *Mantichorula* Reitter, 1889k: 695 type species *Mantichorula semenowi* Reitter, 1889
grandis Semenov, 1893c: 263 **A:** GAN NIN NMO
mongolica Schuster, 1940a: 23 **A:** NMO
semenowi Reitter, 1889k: 695 **A:** GAN MG NIN NMO SHA XIN

genus *Meladiesia* Reitter, 1909g: 309 type species *Meladiesia miritarsis* Reitter, 1909
miritarsis Reitter, 1909g: 309 **A:** TM

genus *Ocnera* Fischer von Waldheim, 1822: 169 type species *Pimelia imbricata* Fischer von Waldheim, 1820
imbricata Fischer von Waldheim, 1820: pl. 14 (*Pimelia*) **A:** AF IN TM UZ
lepidacantha Fischer von Waldheim, 1830: 187 **A:** AF KZ TD TM
raddeana Reitter, 1889b: 116
longicollis Solsky, 1881b: 51 **A:** IN
mevedevi Ren & Ba, 2009: 222 **A:** XIN
pilicollis Faldermann, 1836: 378 (*Trachyderma*) **A:** AF IN KA KI KZ SY TM UZ
przewalskyi Reitter, 1887d: 380 **A:** KI XIN

sublaevigata schusteri Gridelli, 1954: 210 **A:** AF
sublaevigata sublaevigata Bates, 1879: 477 **A:** GAN KA XIN

genus *Pachylodera* G. Quedenfeldt, 1890: 399 type species *Pachylodera brevicornis* G. Quedenfeldt, 1890
brevicornis G. Quedenfeldt, 1890: 400 **N:** EG LB TU

genus *Pachyscelina* Kwieton, 1978b: 30 type species *Pachyscelis micros* Kaszab, 1970
micros Kaszab, 1970a: 74 (*Pachyscelis*) **A:** AF

genus *Pachyscelis* Solier, 1836a: 54 [NP] type species *Pimelia musiva* Faldermann, 1832
subgenus *Pachyscelis* Solier, 1836a: 54 [NP] type species *Pimelia musiva* Faldermann, 1832
Brachyscelis Dejean, 1834: 179 [NO] type species *Pimelia mussiva* Ménétriés, 1832
banghaasi Reitter, 1893b: 208 **A:** AF TD TM UZ
euphratica Kraatz, 1865a: 310 **A:** IN IQ TR
gemmans Baudi di Selve, 1875d: 690 [= 1876a: 14] **A:** IN
klapperichi centralis Kaszab, 1970a: 72 **A:** AF (Saleha-lamg)
klapperichi imitata Kaszab, 1973b: 46 **A:** AF (Naurak, Terbolaq)
klapperichi klapperichi Kaszab, 1960a: 33 **A:** AF (Bande-e-Mir)
matilei Pierre, 1968: 1027 **A:** IN
mollis Kraatz, 1882d: 329 (*Stalagmoptera*) **A:** UZ (Samarkand)
musiva achaemenia Bogatchev, 1949b: 278 **A:** IN
descampsi Pierre, 1968: 1025
escalerai Pierre, 1968: 1025
musiva divisa Reitter, 1893b: 210 **E:** AR **A:** IQ TR
musiva gastridula Faldermann, 1837 **E:** AB AR **A:** IN
musiva musiva Ménétriés, 1832: 193 (*Pimelia*) **E:** AB AR ST **A:** IN IQ TM TR
bilineata Baudi di Selve, 1876a: 12
clavaria Faldermann, 1832: 193 (*Pimelia*)
depressa Solier, 1836a: 57
leprosa Faldermann, 1837: 14 (*Brachyscelis*)
mammillata Faldermann, 1837: 17 (*Brachyscelis*)
ordinata Solier, 1836a: 58
narynensis G.S. Medvedev, 1992: 793 **A:** KI
pauxilla Hampe, 1852: 312 (*Pimelia*) **A:** IN
remaudierei Pierre, 1968: 1026 **A:** IN (Elburz)
subgenus *Parapachyscelis* Kwieton, 1978b: 29 type species *Pimelia villosa* Drapiez, 1820
chrysomeloides G.-A. Olivier, 1795: 59 (*Pimelia*) **E:** GR **A:** LE
coriacea Klug, 1830: nr. 1, pl. 11 (*Pimelia*)
kraatzi Baudi di Selve, 1875d: 689
piochardi Desbrochers des Loges, 1881: 95

rotundata Kraatz, 1865a: 313 **A:** IS JO LE SY
villosa Drapiez, 1820b: 118 (*Pimelia*) **E:** GR "Transcaucasus" **N:** EG TR **A:** IN IQ IS SY TR
crinita Solier, 1836a: 61
granulosa Solier, 1836a: 60
hirtella Solier, 1836a: 62
tenebrosa Solier, 1836a: 60
wettsteini Schuster, 1936b: 72

genus *Paraplatyope* Löbl, Bouchard, Merkl & Bousquet, 2020 [see New Acts] type species *Leucolaephus arabicus* Blair, 1931
arabica arabica Blair, 1931b: 270 (*Leucolaephus*) **A:** AE KU SA
arabica gridellii Koch, 1965a: 133 **A:** SA
arabica thesigeri Gridelli, 1953: 47 **A:** OM
arabica tihamana Koch, 1965a: 133 **A:** SA
popovi Koch, 1965a: 134 **A:** SA YE

genus *Pelorocnemis* Solsky, 1876: 283 type species *Pimelia punctigera* Ménétriés, 1849
darwini Faust, 1875: 214 (*Ocnera*) **A:** IN TM
hesperia Semenov & Bogatchev, 1940: 203
kraatzi Solsky, 1876: 285 **A:** KZ UZ
punctigera Ménétriés, 1849: 219 (*Pimelia*) [RN] **A:** AF KZ TD TM UZ
punctata Gebler, 1844: 101 (*Pimelia*) [HN]
spectabilis Kraatz, 1882d: 329 (*Pimelia*)

genus *Phymatiotris* Solier, 1836a: 63 type species *Tentyria quadricollis* Brullé, 1832
Graecopachys Skopin, 1968b: 99 type species *Tentyria quadricollis* Brullé, 1832
parallela Solier, 1836a: 64 (*Pachyscelis*) **E:** GR
cycladica Reitter, 1893b: 212 (*Pachyscelis*)
quadricollis cretica Koch, 1948a: 327 (*Pachyscelis*) **E:** GR (Crete)
quadricollis obscura Solier, 1836a: 65 (*Pachyscelis*) **E:** GR
porphyrea Solier, 1836a: 65 (*Pachyscelis*)
quadricollis quadricollis Brullé, 1832: 199 (*Tentyria*) **E:** GR MC TR **A:** IN TR
quadricollis smyrnensis Kraatz, 1865a: 309 (*Pachyscelis*) **A:** TR
quadraticollis thessalica Reitter, 1893b: 212 (*Pachyscelis*) **E:** GR

genus *Pimelia* Fabricius, 1775: 251 type species *Pimelia angulata* Fabricius, 1775
subgenus *Amblyptera* Solier, 1836a: 195 type species *Pimelia scabrosa* Solier, 1836
fornicatiformis Koch, 1941a: 105 **N:** MO
intermedia Fischer von Waldheim, 1837b: 13 **A:** TM
marginata Fischer von Waldheim, 1844: 53 **A:** "Turkomania"

mogadora Fairmaire, 1871a: 387 **N:** MO

pseudorugosa Koch, 1941a: 113

subglabra Koch, 1941a: 112

obesa Solier, 1836a: 191 **E:** SP

sparsegranulata Koch, 1941a: 110

plinthota Fischer von Waldheim, 1837b: 17 **A:** TM

rotundipennis rotundipennis Kraatz, 1865a: 370 **N:** MO

incostata Escalera, 1914: 321

glabra Koch, 1941a: 112

rifensis Antoine, 1951c: 96 **N:** MO (Rif)

scabrosa costulata Schuster, 1929: 40 **N:** MO

scabrosa scabrosa Solier, 1836a: 188 **E:** SP **N:** MO

ifranensis Antoine, 1949a: 52

simonis Sénac, 1887a: 124

tristis apicesculpta Koch, 1941a: 109 **N:** MO (Mechra ben Abbou)

tristis ifniensis Antoine, 1949a: 54 **N:** MO

tristis rhatica Antoine, 1949a: 54 **N:** MO

tristis tristis Haag-Rutenberg, 1875a: 54 **N:** MO

uniformis Koch, 1941a: 110

zaerensis Antoine, 1949a: 54

ventricosa Faldermann, 1837: 7 **E:** AB "Transcaucasus"

cordata Allard, 1886: (2) (*Trigonoscelis*)

subgenus *Amblypteraca* Mas-Peinado, Buckley, Ruiz & García-París, 2018: 543 type species *Pimelia rugosa* Fabricius, 1792 (= *Pimelia fairmairii* Kraatz, 1865)

chrysomeloides bathyglypta Antoine, 1949a: 56 **N:** MO

chrysomeloides chrysomeloides Pallas, 1781: 52 (*Tenebrio*) **E:** SP **N:** CI

fornicata Herbst, 1799: 79

sculptilis Reitter, 1915a: 57

variolosus Thunberg, 1804: 6

fairmairii Kraatz, 1865a: 369 **N:** MO

curticollis Haag-Rutenberg, 1875a: 53

haagi Koch, 1941a: 105

haroldi Sénac, 1887a: 127

laevisulcata Kraatz, 1887: 363

rugosa Fabricius, 1792a: 103 [HN]

tumidipennis Haag-Rutenberg, 1875a: 54

subris Koch, 1941a: 111 **N:** MO

subgenus *Aphanaspis* Wollaston, 1864: 478 type species *Pimelia granulicollis* Wollaston, 1864

ascendens Wollaston, 1864: 473 **N:** CI (Tenerife)

canariensis Brullé, 1838: 67 **N:** CI (Tenerife)

estevezi Oromí, 1990: 246 **N:** CI (Gran Canaria)

fernandezlopezi Machado, 1979: 123 **N:** CI (Gomera)
granulicollis Wollaston, 1864: 478 **N:** CI (Gran Canaria)
auriculata Wollaston, 1864: 479
hybrida Uyttenboogaart, 1930: 233
laevigata costipennis Wollaston, 1864: 476 **N:** CI (El Hierro)
ambigua Wollaston, 1864: 475
laevigata laevigata Brullé, 1838: 67 **N:** CI (La Palma)
laevigata validipes Wollaston, 1864: 476 **N:** CI (Gomera)
alluaudi Sénac, 1892: 107
lutaria Brullé, 1838: 68 **N:** CI (Lanzarote, Fuerteventura)
lutulenta Wollaston, 1865: 403
radula oromii Viñolas, 1994: 140 **N:** CI (Tenerife)
asperata Har. Lindberg, 1950: 10 [HN]
granulata Wollaston, 1864: 474 [HN]
radula radula Solier, 1836a: 136 **N:** AG
pseudoradula Har. Lindberg, 1950: 10
sparsa albohumeralis Har. Lindberg, 1950: 11 **N:** CI (Gran Canaria)
sparsa serrimargo Wollaston, 1864: 477 **N:** CI (Gran Canaria)
parvula Har. Lindberg, 1950: 10 [HN]
sparsa sparsa Brullé, 1838: 67 **N:** CI (Gran Canaria)
fritschii Heyden, 1875: 141
subgenus *Camphonota* Solier, 1836a: 195 type species *Tenebrio subglobosus* Pallas, 1781
Eurypimelia Reitter, 1915a: 9 type species *Tenebrio subglobosus* Pallas, 1781
akbesiana akbesiana Fairmaire, 1884a: 170 **A:** IN SY TR
akbesiana evanescens Kwieton, 1982a: 3 **A:** TR
anatolica Reitter, 1915a: 51 **A:** TR
latitibia Reitter, 1915a: 51
asperula Solier, 1836a: 182 **E:** GR
exanthematica Solier, 1836a: 183 [HN]
monilifera Solier, 1836a: 184
lendli Kaszab, 1940a: 4 **A:** TR
longula Kwieton, 1981b: 405 **A:** BA SA
monikovae Kwieton, 1982a: 3 **A:** TR
phymatodes Solier, 1836a: 187 **E:** GR (Crete)
minos Lucas, 1853: 575
praetermissa Sénac, 1887a: 62 **E:** GR
puncticollis Reitter, 1915a: 55 **E:** GR
bittneri Reitter, 1915a: 56
repleta Reitter, 1915a: 50 **A:** TR
sericella Solier, 1836a: 185 **E:** GR (Crete)
calculosa Solier, 1836a: 186
trachyderma Solier, 1836a: 186

subglobosa cephalenica Kraatz, 1865a: 348 **E**: GR
subglobosa graeca Brullé, 1832: 192 **E**: GR
subglobosa krueperi Reitter, 1915a: 54 **E**: GR (Morea)
subglobosa mongeneti Solier, 1836a: 177 **E**: GR
balcanica Kirchsberg, 1876: 30
granipenna Solier, 1836a: 178
subglobosa polita Solier, 1836a: 176 **E**: BU GR RO TR **A**: TR
declivis Reitter, 1915a: 52
euboica Boieldieu, 1865: 7
lesbicola J.R. Sahlberg, 1913b: 168
subglobosa salaria Reitter, 1915a: 51 **A**: TR
subglobosa separanda Reitter, 1915a: 54 **E**: GR
subglobosa subglobosa Pallas, 1781: 50 (*Tenebrio*) **E**: BU GR IN RO ST UK **A**: KZ
fulvopuberula Reitter, 1915a: 53
inaequalis Fischer von Waldheim, 1837b: 18
tuberculatus Thunberg, 1787: 16 (*Erodius*)
uralensis Reitter, 1915a: 53
subglobosa verruculifera Solier, 1836a: 180 **E**: BU GR MC TR **A**: TR
coordinata Fischer von Waldheim, 1837b: 17
diffusa Solier, 1836a: 180
phymatoptera Solier, 1836a: 177
varicosa Ménétriés, 1836a: 150
tomentifera Reitter, 1915a: 53
varicellosa J.R. Sahlberg, 1913b: 169
testudo Kraatz, 1865a: 362 **A**: IN TR
timarchoides Ménétriés, 1836a: 150 **A**: IN TR **N**: MO
intersticta Fischer von Waldheim, 1837b: 16
scabrosiformis Reitter, 1915a: 57
subgenus *Chaetotoma* Motschulsky, 1860d: 436 type species *Tenebrio cephalotes* Pallas, 1781
abnormis Reitter, 1915a: 23 **A**: AF TM
capito Krynicki, 1832: 131 **E**: AB AR GG ST **A**: IN KZ TM
anomala Solier, 1836a: 118
metopotapha Fischer von Waldheim, 1832a: 192
neglecta Fischer von Waldheim, 1832a: 194
schonherri Solier, 1836a: 117
cephalotes bactriana Bogatchev, 1964a: 44 **A**: UZ (Surkhandarinskaya oblast)
cephalotes cephalotes Pallas, 1781: 49 (*Tenebrio*) **E**: ST **A**: IN KZ TD TM
kirgisica Motschulsky, 1859c: 472 [= 1860e 533]
menetriesi Sénac, 1884a: 93
oxysterna Solier, 1836a: 121
cursor Ménétriés, 1832: 192 **E**: AB **A**: AF IN
dubia bellica Bogatchev, 1953: 305 **A**: IN (Teheran)

dubia dubia Faldermann, 1837: 8 **E:** AB AR **A:** IN TR

indistincta Reitter, 1915a: 41

gracilipes Sénac, 1886: xlvi **A:** AF IN

matthiesseni Reitter, 1915a: 26 **A:** IN

schoenherri Faldermann, 1837: 11 **E:** AR **A:** IN

stellifera Motschulsky, 1860e: 533 (*Chaetotoma*) **A:** "déserts méridionaux des Kirghises"

torquata Baudi di Selve, 1875d: 692 **A:** IN

atarnites Baudi di Selve, 1875d: 692

areolata Reitter, 1915a: 24

tuberculata Fischer von Waldheim, 1832a: 194 **E:** AB **A:** IN

verrucosa Fischer von Waldheim, 1820: pl. 14 **A:** AF KI TM UZ

verrucosa Brullé, 1838: 67 [HN]

subgenus *Ecphoroma* Solier, 1836a: 195 type species *Pimelia hemisphaerica* Solier, 1836

capillata Solier, 1836a: 194 **N:** AG MO

hemisphaerica Solier, 1836a: 193 **N:** AG MO

insignis Fairmaire, 1867: 407

subgenus *Hispanomelia* Mas-Peinado, Buckley, Ruiz & García-París, 2018: 541 type species *Pimelia manchega* Lauffer, 1905

brevicollis Solier, 1836a: 172 **E:** SP

coelloi Castro Tovar & J. Ferrer, 2012: 235 **E:** SP

cognata Castro Tovar & J. Ferrer, 2012: 235 **E:** SP

crassipes Solier, 1836a: 190 **E:** SP

ingrata Castro Tovar & J. Ferrer, 2012: 234 **E:** SP

punctata Solier, 1836a: 148

integra Rosenhauer, 1856: 190 **E:** PT SP

manchega Lauffer, 1905: 406 **E:** SP

monticola Rosenhauer, 1856: 192 **E:** SP

rotundata Solier, 1836a: 149 **E:** SP

hispanica Solier, 1836a: 150

stroebli Reitter, 1915a: 47 **E:** SP

subgenus *Iberomelia* Mas-Peinado, Buckley, Ruiz & García-París, 2018: 539 type species *Pimelia castellana* Perez Arcas, 1865

baetica Solier, 1836a: 170 **E:** PT SP

castellana Pérez Arcas, 1865a: 176 **E:** SP

costata Waltl, 1835: 69 **E:** PT SP

graphica Baudi di Selve, 1876a: 22

cribra Solier, 1836a: 151 **E:** SP (Baleares)

distincta Solier, 1836a: 172 **E:** PT SP

elevata Sénac, 1887a: 75 **E:** SP (Baleares)

evorensis Reitter, 1915a: 39 **E:** PT SP

ferreri Caro-Pintos, 2015: 167 **E:** SP

furcicosta Reitter, 1915a: 41 **E**: SP
hesperica Solier, 1836a: 167 **E**: SP
  bipunctatoides Reitter, 1915a: 40
impercepta Castro Tovar & J. Ferrer, 2012: 225 **E**: SP
interjecta Solier, 1836a: 152 **E**: FR SP (Baleares)
modesta Herbst, 1799: 96 **E**: PT SP
  incerta Solier, 1836a: 166
  obscura Reitter, 1915a: 37
perezi Sénac, 1887a: 69 **E**: SP
ruida Solier, 1836a: 153 **A**: SP
variolosa Solier, 1836a: 130 **E**: SP
zuzartei Castro Tovar & J. Ferrer, 2012: 229 **E**: PT
subgenus *Italomelia* Mas-Peinado, Buckley, Ruiz & García-París, 2018: 539 type species *Pimelia rugulosa* Germar, 1824
angusticollis angusticollis Solier, 1836a: 163 **E**: FR (Corse)
angusticollis sulcitana Leo & Pisano, 1978: 80 **E**: IT (Sardinia)
cajetana Sénac, 1887a: 96 **E**: IT
goryi cassolai Ardoin, 1973: 305 **E**: IT (Sardinia)
goryi goryi Solier, 1836a: 162 **E**: IT (Sardinia)
  corsica Solier, 1836a: 164
  sardea Solier, 1836a: 164
interstitialis Solier, 1836a: 105 **N**: AG LB MO TU
  dejeanii Solier, 1836a: 132
payraudi payraudi Latreille, 1829: 7 **E**: FR (Corse) IT
  latericosta Reitter, 1915a: 42
  rugatula Solier, 1836a: 159
payraudi punctatorugosa Reitter, 1915a: 42 **E**: IT (Sardinia)
payraudi subalpina Ardoin, 1973: 306 **E**: IT (Sardinia)
rugulosa apula Gridelli, 1950: 162 **E**: IT
rugulosa melitana Reitter, 1915a: 43 **E**: MA
rugulosa pelagosana J. Müller, 1912: 291 **E**: GR (Ionioi)
rugulosa rugulosa Germar, 1824b: 134 **E**: IT
  bifurcata Solier, 1836a: 157
  calabrica Reitter, 1915a: 44
  oblita Reitter, 1915a: 44
rugulosa sublaevigata Solier, 1836a: 154 **E**: IT (Sicilia)
  subscabra Solier, 1836a: 160
subgenus *Magrebmelia* Mas-Peinado, Buckley, Ruiz & García-París, 2018: 540 type species *Pimelia xauenensis* Escalera, 1923
atlantis atlantis Solier, 1836a: 139 **N**: MO (Middle Atlas)
atlantis frigidoides Koch, 1941a: 85 **N**: MO (Middle Atlas)

boyeri boyeri Solier, 1836a: 143 **N:** AG TU

asphaltina Baudi di Selve, 1875c: 100

granifera Solier, 1836a: 147

graniventris Baudi di Selve, 1876a: 25

rugifera Solier, 1836a: 144

boyeri cedrorum Koch, 1941a: 87 **N:** MO (Rif)

boyeri duponti Solier, 1836a: 146 **N:** AG MO

boyeri rifatlanta Koch, 1941a: 86 **N:** MO (Rif)

boyeri septemcostata Antoine, 1949a: 45 **N:** MO

escalerai Castro Tovar & J. Ferrer, 2012: 233 **E:** MO SP

maura Solier, 1836a: 137 E: PT SP **N:** MO

mauritanica breiti Koch, 1941a: 89 **N:** AG MO

rotroui Antoine, 1949a: 41

mauritanica hollandei Pierre, 1985: 290 **N:** MO (High Atlas)

mauritanica mauritanica Solier, 1836a: 139 **N:** AG

mauritanica moltonii Koch, 1941a: 90 **N:** MO (Melilla)

pseudotimarcha Antoine, 1949a: 42

mauritanica parisii Koch, 1941a: 90 **N:** MO (Midelt)

bletoni Antoine, 1949a: 42

lapidicola Antoine, 1949a: 42

mauritanica verecostata Koch, 1941a: 89 **N:** AG

thomsoni brunneipedes Escalera, 1913: 44 [= 1914: 316] **N:** MO (High Atlas)

thomsoni thomsoni Nevinson, 1889: 255 **N:** MO (High Atlas)

tatae Théry, 1934: 47

thomsoni tirsalensis Antoine, 1937: 242 **N:** MO (High Atlas)

thomsoni villiersi Koch, 1941a: 92 **N:** MO (High Atlas)

xauenensis Escalera, 1923d: 130 **N:** MO

subgenus *Massadraamelia* Mas-Peinado, Buckley, Ruiz & García-París, 2018: 541 type species *Pimelia granulithorax* Escalera, 1914

granulithorax Escalera, 1914: 317 **N:** MO

subgenus *Melanostola* Solier, 1836: 123 type species *Pimelia simplex* Solier, 1836

Balius Gistel, 1848: xi, xiv [URN]

echidna densa Koch, 1941a: 78 **N:** MO

echidna echidna Fairmaire, 1875c: 544 **N:** MO

echidna schatzmayri Koch, 1941a: 78 **N:** MO

echidna tafrautensis Antoine, 1952: 113 **N:** MO

laeviuscula brachythorax Escalera, 1914: 311 **N:** MO

parvula Antoine, 1949: 28

laeviuscula laeviuscula Kraatz, 1865a: 367 **N:** MO

externeserrata Fairmaire, 1875c: 544

oblonga mediatlanta Koch, 1941a: 79 **N:** MO (Middle Atlas)

oblonga oblonga Sénac, 1885: lxxxi **N:** MO
simplex simplex Solier, 1836a: 123 **N:** AG LB TU
simplex oasis Koch, 1941a: 75 **N:** AG TU
simplex subtriseriata Koch, 1941a: 76 **N:** MO
subgenus *Pimelia* Fabricius, 1775: 251 type species *Pimelia angulata* Fabricius, 1775
Agelarches Gistel, 1848: x [UE]
Doderoella Schuster, 1926b: 133 type species *Doderoella cyrenaica* Schuster, 1926
Pimidia Rafinesque, 1815: 113 [UE]
adramantina Koch, 1965a: 142 **A:** YE
adriani Reitter, 1915a: 36 **A:** JO
aegyptiaca Andres, 1929: 15 **N:** EG
afghanistanica Gridelli, 1954: 236 [= 1955: 44] **A:** AF
albinea Reitter, 1915a: 11 **A:** YE
alboepisternalis Bogatchev, 1953: 302 **A:** IN
amseli Kaszab, 1961e: 238 **A:** AF
angulata angulata Fabricius, 1775: 352 **N:** EG **A:** SI
aculeata Klug, 1830: nr. 17, pl. 12
alternata Klug, 1830: nr. 16 nota
alternata Sénac, 1884a: 24 [HN]
asperrima Pallas, 1781: 55 (*Tenebrio*)
spinosa Forskål, 1775: 80 (*Tenebrio*)
angulata antiaegypta Koch, 1937: 446 **N:** MO (Western Sahara)
angulata expiata Peyerimhoff, 1923: 319 **N:** AG LB MO **AFR**
angulata nilotica Sénac, 1884a: 27 **N:** EG **AFR**
angulata sinaitica Schatzmayr & Koch, 1934b: 26 **A:** SI
angulata syriaca Sénac, 1884a: 24 **A:** IS
angulosa G.-A. Olivier, 1795: 59: 11 **N:** LB **AFR**
anomaloides Löbl, Bouchard & Merkl, 2008: 40 **N:** AG
anomala Sénac, 1880: xxiv [= 1880: 261] [HN]
arabica arabica Klug, 1830: nr. 18, pl. 12 **A:** AE BA IQ JO OM SA YE
serricosta Solier, 1836a: 102
arabica edomita Koch, 1940c: 102 **A:** IS JO SI
arabica emiri Kaszab, 1982b: 149 **A:** AE BA YE
arabica omanica Kaszab, 1982b: 148 **A:** AE OM
arenacea Solier, 1836a: 114 **N:** AG TU
arenosa Reitter, 1915a: 21 **N:** AG (Oran)
ariasi Escalera, 1914: 311 **N:** MO
bajula bajula Klug, 1830: nr 8, pl. 11 **N:** EG **A:** IN IQ IS JO LE SY
bajula solieri Mulsant & Wachanru, 1853: 8 **A:** CY IN TR
cylindrica Solier, 1836a: 124
mulsanti Reiche, 1861b: 88 [RN]

balachowskyi Pierre, 1968: 1029 **A:** IN
barthelemyi Solier, 1836b: 350 **N:** EG LB **A:** SI
bengasiana bengasiana Schuster, 1922b: 20 **N:** LB
bengasiana plurigranula Koch, 1939: 277 **N**
bipunctata Fabricius, 1781: 316 **E:** FR IT
papii Canzoneri, 1963a: 14
sulcata Geoffroy, 1785: 159 (*Tenebrio*)
boehmi Reitter, 1915a: 34 **N:** EG
bottae Sénac, 1887a: 48 **A:** SA SI YE
carinosa Sénac, 1887a: 48
buqueti Lucas, 1858c: ccxxi **N:** AG
canescens arabicola Schatzmayr & Koch, 1934b: 26 **N:** EG
canescens canescens Klug, 1830: nr. 3, pl. 11 **N:** EG LB **A:** IS SI
depilata Solier, 1836a: 129
claudia Buquet, 1840: 242 **N:** AG
spectabilis Haag-Rutenberg, 1875a: 52
comata Klug, 1830: nr. 4, pl. 11 **N:** EG LB **A:** "Syria"
ornata Miller, 1861: 179
confusa Sénac, 1884c: xxiv **N:** TU
conjuncta Kwieton, 1976: 1 **A:** IQ
consobrina confalonierii Gridelli, 1930: 274 **N:** AG LB
consobrina consobrina Lucas, 1858c: ccxx **N:** AG
densata Reitter, 1915a: 15
lesnei Peyerimhoff, 1911: 347
sparsidens Reitter, 1915a: 13
cordata cordata Kraatz, 1865a: 368 **N:** MO
atheta Antoine, 1949a: 29
maroccana Fairmaire, 1875c: 543
cordata desertorum Escalera, 1914: 313 **N:** MO (Western Sahara)
cordata mackenziei Escalera, 1914: 312 **N:** MO
cordata zarcoi Español, 1943c: 133 **N:** MO (Tiznit, Tan-Tan)
crenata crenata Fabricius, 1798: 44 **N:** MO
acroterica Antoine, 1949: 31
denisi Antoine, 1949: 31
grandicollis Kraatz, 1865a: 368
intermittens Reitter, 1915a: 15 (*Pieterotarsa*)
gracilenta Haag-Rutenberg, 1875a: 52
granulicosta Escalera, 1914: 313
crenata lenta Antoine, 1949a: 32 **N:** MO
cribripennis Solier, 1836a: 117 **N:** AG MO
davletschinae Bogatchev, 1964a: 42 **A:** UZ (Kyzylkum)

damasci Sénac, 1880: xlviii [= 1880: 265] **A:** SY
depressa Solier, 1836a: 116 **N:** AG
  papulosa Solier, 1836a: 113
  subquadrata Solier, 1836a: 113 [HN]
derasa barcana Koch, 1941c: 311 [RN] **N:** LB
  tricostata Koch, 1937: 452 [HN]
derasa derasa Klug, 1830: nr. 7, pl. 11 **N:** EG LB
derasa derasoides Schuster, 1922b: 20 **A:** IS JO SY
discoides Kwieton, 1978a: 10 **A:** IN
discicollis Fairmaire, 1875c: 544 **A:** YE **N:** MO (Taroudant)
  ciliativentris Koch, 1941
  pilosa Escalera, 1914: 314
  taroudantica Koch, 1941a: 62
doderoi deckeli Koch, 1939: 276 **N:** LB
doderoi doderoi Schuster, 1925a: 30 **N:** LB
doderoi doloresina Koch, 1937: 453 **N:** LB
edura Reitter, 1915a: 34 **A:** JO
errans errans Miller, 1861: 181 **A:** SY
errans kraatzi Sénac, 1887a: 17 **A:** IN IQ SA SY
freyi Schuster, 1937a: 46 **A:** IQ
fulvovillosa Schuster, 1935a: 17 **N:** LB
gestroana Gridelli, 1930: 279 **N:** LB
gestroi Sénac, 1884a: 94 **A:** IN
gracilis Schuster, 1935a: 20 **N:** LB
grandis amplipennis Pierre, 1978: 200 **N:** MO (Tizi n'Tazazert)
grandis depressipennis Pierre, 1978: 201 **N:** MO
grandis echidniformis Reitter, 1915a: 18 **N:** AG MO
  politidorsum Koch, 1941a: 73
grandis grandis Klug, 1830: nr. 5, pl. 11 **N:** EG **AFR**
  denticostata Koch, 1935a: 70
  sudanica Fairmaire, 1882b: 66
grandis latastei Sénac, 1884b: x **N:** AG LB MO
  mixta Koch, 1941a: 73
grandis minor Pierre, 1978: 198 **N:** MO (Ouarzazate)
granulata doumeti Sénac, 1882a: xxxi **N:** TU
granulata granulata Solier, 1836a: 103 **N:** AG
  intertuberculata Lucas, 1858c: ccxx
  lucasii Reiche, 1861b: 87
granulata imitata Reitter, 1915a: 20 **N:** TU
granulata papulenta Reiche, 1861b: 88 **N:** AG
granulata profetei Sénac, 1884: x **N:** AG

granulata realii Koch, 1937: 441 **N:** LB (Tripolis)
granulata rectangularis Kwieton, 1987: 13 **N:** AG
granulata sefrensis Pic, 1894p: 247 **N:** AG
granulata tenuicornis Solier, 1836a: 100 **N:** LB
dohrni Reitter, 1915a: 12 (*Piesterotarsa*)
tripolitana Sénac, 1884a: 39
gridellii Schuster, 1928a: 26 **N:** LB
grossa Fabricius, 1792a: 101 **E:** IT **N:** TU
inflata Herbst, 1799: 98
vestita Sénac, 1884a: 54
haarlovi Gridelli, 1954: 240 [= 1955: 46] **A:** AF
hayekae Kaszab, 1957f: 297 **A:** PA
hirtella Sénac, 1887a: 47 **N:** EG **A:** SA SI
horni Schuster, 1936c: 195 **A:** AF
indica Sénac, 1882b: lvi **A:** PA **ORR**
inexspectata Sénac, 1882a: xxxi **A:** PA
convexicollis Sénac, 1882b: lvii
interpunctata interpunctata Klug, 1830: nr. 2, pl. 11 (*Pimelia*) **N:** EG LB
andreinii Gridelli, 1929a: 10 (*Doderoella*)
carinata Solier, 1836a: 97
cyrenaica Schuster, 1925a: 29 (*Doderoella*)
intervallaris Kwieton, 1981b: 405 **A:** OM
iscariota Reitter, 1915a: 35 **A:** IQ
jansseni Reitter, 1915a: 35 **A:** JO
kruegeri Schuster, 1928e: 377 **N:** LB
laevidorsis Reitter, 1915a: 33 **A:** JO
latreillei Solier, 1836a: 93 **N:** EG **AFR**
denticulata Solier, 1836a: 95
permixta Sénac, 1881: xx [RN]
letourneuxi gridelliana Koch, 1939: 278 **N:** LB
letourneuxi letourneuxi Sénac, 1880: xxiv [= 1880: 263] **N:** EG LB
letourneuxi platytubera Koch, 1939: 280 **N:** LB
lineata Solier, 1836a: 169 **E:** SP
gadium Solier, 1836a: 169
marseuli asirensis Koch, 1965a: 140 **A:** SA
marseuli marseuli Sénac, 1887a: 50 **A:** SA
matutinalis Antoine, 1954: 205 **N:** MO
mittrei Solier, 1836a: 134 **A:** IS LE SI
monilis Haag-Rutenberg, 1875a: 55 **N:** MO
fortecarinata Schuster, 1929: 39
seriegranulata Escalera, 1913: 43

moralesi Español, 1943c: 135 **N:** MO (Western Sahara)
muricata cyrnensis Soldati & Rougon, 2017: 404 **E:** FR (Corse)
muricata muricata G.-A. Olivier, 1795: 59: 9 **E:** FR
aspera Germar, 1824b: 134
nazarena Miller, 1861: 178 **A:** IS IQ SY
nazarenoides Kaszab, 1982b: 150 **A:** SA YE
nesterovi Bogatchev, 1953: 305 **A:** IN
obsoleta giorgii Koch, 1937: 448 **N:** LB (Tripolis)
obsoleta insinuata Kwieton, 1977a: 21 **N:** AG TU
obsoleta latipes Solier, 1836a: 109 **N:** TU
amicta Baudi di Selve, 1876a: 20
obsoleta nitida Schuster, 1935a: 16 **N:** LB
obsoleta obsoleta Solier, 1836a: 104 **N:** LB TU
wohlberedti Schuster, 1928d: 154
obsoleta pilifera Sénac, 1884b: xi **N:** AG MO TU
rugosipennis Escherich, 1896d: 276
orientalis Sénac, 1886: xlv **A:** IS JO SY
nabathaea Peyerimhoff, 1921a: 328
otini otini Antoine, 1949a: 48 **N:** MO (central High Atlas)
otini sarroensis Antoine, 1949a: 43 **N:** MO (Jbel Sarro)
pakistana Kaszab, 1961d: 222 **A:** PA
peyerimhoffi Pierre, 1976: 19 **N:** AG (Beni Abbés)
platynota Fairmaire, 1875c: 544 **N:** MO
prolongata Miller, 1861: 180 **A:** IS
persica Faldermann, 1837: 10 **E:** AB AR **A:** IN TR
pseudinterstitialis Pierre, 1982: 2 **N:** AG
pseudopilifera Reitter, 1915a: 20 **N:** TU
puberula Chevrolat, 1873: 203 **A:** LE
remaudieri Pierre, 1964c: 1045 **A:** SY
retrospinosa Lucas, 1858a: clxxix **N:** AG
semihispida Fairmaire, 1874b: 407
robusta Kraatz, 1865a: 362 **E:** AR **A:** TR
roseni Koch, 1940c: 103 **A:** JO
ryssos Herbst, 1799: 90 [RN] **N:** AG
rugosa G.-A. Olivier, 1795: 59: 10 [HN]
salebrosa Solier, 1836a: 142
scabricollis Sénac, 1886: xlv **N:** AG
schusteri Reitter, 1915a: 33 **A:** JO SA
scorteccii Gridelli, 1939d: 439 **N:** AG
semenovi balkhensis Kaszab, 1974: 197 **A:** AF
semenovi klapperichi Kaszab, 1960a: 41 **A:** AF PA

semenovi lindbergi Kaszab, 1970a: 80 **A:** AF
semenovi saroubiana Kaszab, 1973b: 47 **A:** AF
semenovi semenovi Bogatchev, 1953: 305 **A:** AF
afghanica Gridelli, 1954: 235 [= 1955: 42]
seminuda capannii Koch, 1937: 442 **N:** LB (Misurata)
seminuda seminuda Schuster, 1935a: 19 **N:** LB (Benghazi, Derna)
seminuda subpustulata Koch, 1937: 442 **N:** LB
semiopaca Sénac, 1884c: xxiv **N:** AG MO
senegalensis G.-A. Olivier, 1795: 59: 7 **N:** MO **AFR**
cultrimargo Sénac, 1884c: xxv
sericea G.-A. Olivier, 1795: 59: 8 **N:** EG
asperata Klug, 1830: nr. 10
mohamedis Reitter, 1915a: 22
servillei Solier, 1836a: 112 **N:** AG
similis Kwieton, 1978a: 9 **A:** IN
simplicior medialticola Pierre, 1978: 204 **N:** MO
simplicior simplicior Escalera, 1914: 318 **N:** MO
subcanescens Pierre, 1982: 5 **N:** AG LB TU
subquadrata chudeaui Koch, 1940d: 312 **N:** MO (Western Sahara) **AFR**
subquadrata subquadrata J. Sturm, 1826: 68 **N:** EG **A:** IS SI
irrorata Solier, 1836a: 99
subquadrata valdanii Guérin-Méneville, 1859: clxxxvii **N:** AG LB MO TU **AFR**
malzyi Peyerimhoff, 1938: 178 (*Euthriptera*)
superba Schuster, 1928b: 123 **N:** LB
tenuitarsis Reitter, 1915a: 33 **N:** EG
theveneti Sénac, 1880: xlix [= 1880: 267] **N:** EG **A:** SI
thomasi nefudana Koch, 1965a: 138 **A:** SA
thomasi thomasi Blair, 1931b: 270 **A:** SA
titan Bogatchev, 1953: 304 **A:** IN
tizintesti kumshensis Antoine, 1954: 211 **N:** MO (Middle Atlas)
tizintesti multigranifera Escalera, 1913: 46 **N:** MO
tizintesti tizintesti Escalera, 1913: 45 **N:** MO
tricostata kandaharica Kaszab, 1960a: 42 **A:** AF
tricostata mukurensis Kaszab, 1974: 198 **A:** AF
tricostata paktiana Kaszab, 1974: 199 **A:** AF
tricostata tricostata Gebien, 1913: 4 **A:** IN PA (Balochistan)
tricostata vartiani Kaszab, 1970a: 80 **A:** AF
tunisea Fairmaire, 1879c: 191 **N:** TU
undulata Solier, 1836a: 161 **E:** IT (Sardinia)
humericosta Reitter, 1915a: 44

urticata Klug, 1830: nr. 11, pl. 11 **N:** EG LB

exanthematica Klug, 1830: nr. 13, pl. 11

tuberosa Klug, 1830: nr. 12, pl. 11

valida amblyptericollis Koch, 1941a: 94 **N:** MO

valida valida Erichson, 1841: 176 **N:** MO

scabra Fabricius, 1792a: 101

villanovae Sénac, 1887a: 78 **E:** SP (Granada)

zarudnyi Bogatchev, 1953: 304 **A:** IN

zhenzhuristi Bogatchev, 1953: 303 **A:** SA YE

subgenus *Pseudamblyptera* Pierre, 1985: 295 type species *Pimelia brisouti frigida* Escalera, 1925

brisouti brisouti Sénac, 1885: lxxxii **N:** AG MO

brisouti frigida Escalera, 1925f: 498 **N:** MO

antoinei Schuster, 1929: 38

genus *Pimelia*, species incertae sedis

parisi J. Ferrer, 2016: 97 **E:** SP (Baleares)

genus *Pimelia*, nomen dubium

rugosicollis Levrat, 1859b: 41 **E:** "Sicile, Espagne"

genus *Pimeliocnera* Reitter, 1909g: 310 type species *Pimeliocnera gebieni* Reitter, 1909

gebieni Reitter, 1909g: 310 **A:** TD TM UZ

genus *Pimelipachys* Skopin, 1962: 232 type species *Pachyscelis laevicollis* Reitter, 1893

laevicollis laevicollis Reitter, 1893b: 211 (*Pachyscelis*) **A:** KZ TM

laevicollis mesophilus Skopin, 1960a: 303 (*Pachyscelis*) **A:** KZ (Yuzhnyy Kazakhastan)

rickmersi Reitter, 1898a: 13 (*Pachyscelis*) **A:** UZ

genus *Pisterotarsa* Motschulsky, 1860e: 532 [= 1860d: 436] type species *Pimelia gigantea* Fischer von Waldheim, 1820

Piesterotarsa Sénac, 1884b: xi [UE]

Sympiezocnemis Solsky, 1876: 290 type species *Sympiezocnemis kessleri* Solsky, 1876

gigantea gigantea Fischer von Waldheim, 1820: pl. 14 (*Pimelia*) **A:** AF KZ TM UZ

gigantea zoubkoffi Reitter, 1915a: 61 (*Sympiezocnemis*) **A:** TM UZ

kessleri Solsky, 1876: 291 (*Sympiezocnemis*) **A:** AF TD TM

kiritshenkoi Semenov, 1914: 506 (*Sympiezocnemis*) **A:** UZ (Bukhara)

pavlovskii Bogatchev, 1959: 63 **A:** TD

genus *Platyesia* Skopin, 1971b: 335 type species *Diesia karelini* Fischer von Waldheim, 1844

iduna Reitter, 1902b: 179 (*Trigonoscelis*) **A:** KZ (Almaatinskaya Oblast) XIN

karelini Fischer von Waldheim, 1844: 63 (*Diesia*) **A:** KZ TD TM UZ
medvedevi Skopin, 1971b: 344 [= 1972a: 122] **A:** KZ TM
sefirana Reitter, 1894d: 244 (*Diesia*) **A:** AF IN TM
barchanica Sumakov, 1902: 42 (*Diesia*)
sericata Zubkov, 1833: 326 (*Pimelia*) **E:** ST (Astrakhanskaya Oblast) **A:** KZ
argentata Bogatchev, 1961b: 369 (*Diesia*)

genus *Platyope* Fischer von Waldheim, 1820: pl. 15 type species *Tenebrio leucographa* Pallas, 1781 (= *Tenebrio leucogramma* Pallas, 1773)
altaiensis Ren & Wang, 1993b: 46 **A:** XIN
bairinana Ren & Dong, 2001: 299 **A:** NMO
balteiformis Ren & Wang, 1993b: 44 **A:** NIN
granulata Fischer von Waldheim, 1820: pl. 15 **A:** XIN
korgasica Wu & Huang, 2005: 285 **A:** XIN
korlaensis Fan & Huang, 2005: 286 **A:** XIN
leucogramma Pallas, 1773: 719 (*Tenebrio*) **E:** ST **A:** KZ
bassii Solier, 1836a: 14
dichotoma Fischer von Waldheim, 1830: 187
leucographa Pallas, 1781: 54 (*Tenebrio*)
lineata Fabricius, 1781: 319 (*Akis*)
obliterata Fischer von Waldheim, 1844: 60
mongolica Faldermann, 1835b: 388 **A:** ES (Tuva) JIL LIA MG NIN NMO
ordossica Semenov, 1907a: 183 **A:** GAN NIN NMO
pointi Schuster & Reymond, 1937: 234 **A:** NMO
proctoleuca chinensis Kaszab, 1962c: 311 **A:** MG XIN
proctoleuca proctoleuca Fischer von Waldheim, 1820: pl. 15 **A:** KZ (Zaisan)
qitaiensis Wu & Huang, 2005: 282 **A:** XIN
trichophora Ren & Dong, 2001: 297 **A:** NMO
unicolor Zubkov, 1829: 160 **E:** ST **A:** KZ
victor Schuster & Reymond, 1937: 235 **A:** GAN

genus *Podhomala* Solier, 1836a: 72 type species *Podhomala suturalis* Solier, 1836
subgenus *Podhomala* Solier, 1836a: 72 type species *Podhomala suturalis* Solier, 1836
Podomala Agassiz, 1846b: 300 [UE]
Pterocomodes Reitter, 1901e: 159 type species *Pterocomodes acutus* Reitter, 1901
Uriela Reitter, 1887e: 518 type species *Podhomala fausti* Kraatz, 1881
acuta Reitter, 1901e: 160 (*Pterocomodes*) **A:** KZ
bicostata Sénac, 1884a: 3 (*Pimelia*) **A:** "Siberia"
fausti Kraatz, 1881d: 332 **A:** UZ (Margilan) "Kurdistan"
cristata Sénac, 1882a: xxx (*Pimelia*)

heydeni Reitter, 1907h: 414 **A:** UZ
lucidula Krynicki, 1832: 132 (*Pimelia*) **E:** ST **A:** KZ TM
suturalis Solier, 1836a: 74
medvedevi S. Chigray, 2019b: 927 **A:** KZ
serrata Fischer von Waldheim, 1830: 187 (*Pimelia*) **A:** KI KZ
bicarinata Gebler, 1841a: 373 (*Pimelia*)
subgenus *Urielina* Reitter, 1888c: 332 type species *Podhomala nitida* Baudi di Selve, 1876
nitida Baudi di Selve, 1876a: 30 **A:** AF IN TM UZ

genus ***Prionotheca*** **Dejean, 1834: 179** type species *Pimelia coronata* G.-A. Olivier, 1795
coronata coronata G.-A. Olivier, 1795: 59: 4 (*Pimelia*) **N:** AG EG LB MO TU **A:** SI **AFR**
coronata freyi Bytinski-Salz, 1955: 313 **A:** IS (Negev)
coronata ovalis Ancey, 1881a: 397 **A:** BA IN IQ OM SA YE

genus ***Przewalskia*** **Semenov, 1893c: 262** type species *Platyope dilatata* Reitter, 1887
dilatata Reitter, 1887d: 375 (*Platyope*) **A:** QIN XIZ
lineata Reitter, 1887d: 375 (*Platyope*) **A:** XIZ
trinkleri Gebien, 1940a: 1 **A:** XIZ

genus ***Pseudopachyscelis*** **Skopin, 1968: 99** type species *Trigonoscelis pygmaea* Ménétriés, 1849
galinae G.S. Medvedev, 1964c: 659 (*Pachyscelis*) **A:** AF TM
pygmaea Faust, 1875: 218 (*Pachyscelis*) [HN]
bucharensis Sumakov, 1902: 41 (*Pachyscelis*) **A:** UZ
kaszabi Pierre, 1964c: 1048 (*Pachyscelis*) **A:** IN
pygmaea major Kraatz, 1882b: 86 (*Pachyscelis*) **A:** AF UZ (Margilan)
pygmaea pygmaea Ménétriés, 1849: 222 (*Trigonoscelis*) **A:** TD TM UZ
nitidula Kraatz, 1882b: 86 (*Pachyscelis*)
pygmaea tadzika Skopin, 1960a: 303 **A:** TD

genus ***Pseudoplatyope*** **Pierre, 1964b: 867** type species *Storthocnemis antoinei* Español, 1951
niger Pierre, 1964b: 867 **N:** AG

genus ***Pseudopodhomala*** **Schuster, 1938b: 88** type species *Pseudopodhomala gabrieli* Schuster, 1938
Gedrosia Bogatchev, 1961a: 111 [HN] type species *Gedrosia monstrosa* Bogatchev, 1961 (= *Pseudopodhomala gabrieli* Schuster, 1938)
Pseudopodhomalina Kaszab, 1960a: 22 type species *Diesia costifera* Waterhouse, 1889
barclayi G.S. Medvedev, 2004b: 561 **A:** IN
costifera Waterhouse, 1889: 127 (*Diesia*) **A:** PA (Balochistan)

gabrieli eremita Pierre, 1968: 1019 **A:** IN
gabrieli gabrieli Schuster, 1938b: 89 **A:** AF IN
monstrosa Bogatchev, 1961a: 112 (*Gedrosia*)
vanharteni Schawaller, 2014a: 274 **A:** AE

genus *Pseudostorthocnemis* Gridelli, 1952: 83 type species *Storthocnemis patrizii* Gridelli, 1933
patrizii longicornis Koch, 1937: 428 (*Storthocnemis*) **N:** LB (Fezzan)
patrizii patrizii Gridelli, 1933: 232 (*Storthocnemis*) **N:** LB
renatae Koch, 1937: 429 (*Storthocnemis*) **N:** LB

genus *Pterocoma* Dejean, 1834: 178 type species *Pimelia piligera* Gebler, 1830
subgenus *Dzhungaropterocoma* Skopin, 1974b: 150 type species *Pterocoma subnuda* Reitter, 1897
autumnalis Semenov, 1908: 123 **A:** QIN XIN
kashtshenkoi Semenov, 1908: 124 **A:** QIN
pachyscelis Skopin, 1974b: 141 **A:** QIN XIN
submetallica Semenov, 1908: 124 **A:** QIN XIN
subnuda Reitter, 1897e: 218 **A:** QIN XIN
biimpressa Reitter, 1897e: 219
subgenus *Eupterocoma* Skopin, 1974b: 152 type species *Pterocoma ganglbaueri* Reitter, 1890
beicki Skopin, 1974b: 141 **A:** GAN XIN XIZ
ganglbaueri ganglbaueri Reitter, 1890h: 362 **A:** KI UZ
ganglbaueri punctipleuris Skopin, 1974b: 142 **A:** KZ XIN
iliensis Semenov, 1907c: 264 **A:** KZ XIN
grigorievi Semenov, 1908: 122 **A:** XIN
lepechini Semenov, 1908: 121 **A:** XIN
suworowi Reitter, 1907f: 206 **A:** KZ
subgenus *Hemipterocoma* Skopin, 1974b: 159 type species *Pterocoma nikolskii* Semenov, 1910
nikolskii Semenov, 1910d: 151 **A:** KZ
subgenus *Mesopterocoma* Skopin, 1974b: 147 type species *Pterocoma semicarinata* Bates, 1879
amandana amandana Reitter, 1887d: 374 **A:** NIN XIN XIZ
amandana edmundi Skopin, 1974b: 141 **A:** XIN
loczyi J. Frivaldszky, 1890: 209 **A:** XIN
semicarinata Bates, 1879: 477 **A:** KA XIN (Kashgar)
subgenus *Mongolopterocoma* Skopin, 1974b: 146 type species *Pterocoma reitteri* J. Frivaldszky, 1890
parvula J. Frivaldszky, 1890: 209 **A:** GAN MG NMO
reitteri J. Frivaldszky, 1890: 208 **A:** GAN MG NIN NMO
obesa J. Frivaldszky, 1890: 209
subgenus *Neopterocoma* Skopin, 1974b: 158 type species *Pterocoma balchachense* Skopin, 1974
balchashense Skopin, 1974b: 143 **A:** KZ
subgenus *Pachypterocoma* Skopin, 1974b: 157 type species *Pterocoma pallasi* Semenov, 1910
pallasi Semenov, 1910d: 152 **A:** KZ

valida Skopin, 1974b: 143 **A:** KZ

subgenus *Parapterocoma* Skopin, 1974b: 149 type species *Pterocoma vittata* J. Frivaldszky, 1890

vittata J. Frivaldszky, 1890: 208 **A:** GAN NIN NMO XIN

hedini Schuster, 1936a: 5

subgenus *Poopterocoma* Skopin, 1974b: 145 type species *Pterocoma zaidamica* Skopin, 1974

zaidamica Skopin, 1974b: 141 **A:** QIN

subgenus *Propterocoma* Skopin, 1974b: 144 type species *Pterocoma tibialis* Bates, 1879

convexa Bates, 1879: 476 **A:** KI TD XIN

plicicollis Reitter, 1902d: 191 **A:** XIN (Kashgar)

tibialis Bates, 1879: 476 **A:** KA XIN XIZ

alutacea Reitter, 1902d: 190

chan Reitter, 1897e: 218

fuscopilosa Reitter, 1902d: 191

serrimargo Bates, 1879: 476

subgenus *Pseudopimelia* Motschulsky, 1860e: 536 type species *Pterocoma tuberculata* Motschulsky 1845

mongolica Kaszab, 1968c: 371 **A:** MG

tuberculata Motschulsky, 1845a: 62 **A:** KZ NMO XIN

subgenus *Pseudopterocoma* Skopin, 1974b: 145 type species *Pterocoma trapezicollis* Skopin, 1974

trapezicollis Skopin, 1974b: 140 **A:** KI

subgenus *Pterocoma* Dejean, 1834: 178 type species *Pimelia piligera* Gebler, 1830

baeckmanni Semenov, 1910d: 151 **A:** KZ

brevicollis Ménétriés, 1849: 220 **A:** KZ

costata Pallas, 1781: 52 (*Tenebrio*) **E:** ST **A:** KI KZ

aucta Semenov, 1910d: 150

dubianskii Semenov, 1910d: 150

gracilicornis Solier, 1836a: 47

sarpae Solier, 1836a: 46

subarmata Motschulsky, 1860e: 535 [= 1860d: 440]

denticulata denticulata Gebler, 1829a: 114 (*Pimelia*) **A:** KZ XIN

subrugosa Motschulsky, 1860e: 535

denticulata piligera Gebler, 1829a: 115 (*Pimelia*) **A:** KZ

denticulata reducta Skopin, 1974b: **A:** XIN

gebleri Skopin, 1974b: 142 **A:** XIN (Tarbag)

genus *Pterolasia* Solier, 1836a: 66 type species *Pterolasia squalida* Solier, 1836

multicostata Blair, 1931b: 271 **A:** OM

squalida Solier, 1836a: 68 **N:** EG LB (Cyrenaica) MO **A:** IS SI AFR

boehmi Reitter, 1908c: 48 (*Thryptera*)

distincta Solier, 1836a: 69

lanata Peyerimhoff, 1907a: 35

genus *Scelace* Marseul, 1887: 325 type species *Pimelia tuberculifera* Lucas, 1858

Pachyscelodes Sénac, 1887b: 189 type species *Pachyscelis henoni* Sénac, 1887

ferdinandi ferdinandi Escalera, 1911: 300 **N:** MO

orientalis Antoine, 1937: 241

ferdinandi reymondi Antoine, 1957: 55 **N:** MO

henoni Sénac, 1887b: 191 (*Pachyscelis*) **N:** AG TU

malleata Wollaston, 1875: 219 (*Pimelia*) **N:** MO (High Atlas)

occidentalis Escalera, 1914: 309

semiasperula Fairmaire, 1876: 38 (*Pimelia*) **N:** AG

tuberculifera debduensis Antoine, 1957: 54 **N:** MO

tuberculifera kocheri Antoine, 1934: 122 **N:** MO (Middle Atlas)

tuberculifera lyauteyi Escalera, 1925c: 312 **N:** MO (Middle Atlas)

tuberculifera tuberculifera Lucas, 1858c: ccxxi (*Pimelia*) **N:** AG

dayensis Mulsant & Godart, 1866: 451 (*Pimelia*)

serieperlata Fairmaire, 1876: 38 (*Pimelia*)

genus *Spectrocnera* Kwieton, 1981b: 402 type species *Spectrocnera anguliceps* Kwieton, 1981

anguliceps Kwieton, 1981b: 403 **A:** OM YE

genus *Stalagmoptera* Solsky, 1876: 286 type species *Stalagmoptera tuberculatocostata* Solsky, 1876

Arnoldiola Semenov & Bogatchev, 1940: 201 type species *Arnoldiola peculiaris* Semenov & Bogatchev, 1940 (= *Stalagmoptera ruginota* Reitter, 1896)

alaica Skopin, 1972b: 309 **A:** UZ (Fergana)

clypeata Skopin, 1972b: 305 **A:** TD

confusa confusa Solsky, 1876: 289 **A:** IN TD UZ

confusa deserticola Skopin, 1972b: 306 **A:** TD

confusa incostata Kraatz, 1882d: 327 **A:** UZ (Samarkand)

confusa tuberculosa Kraatz, 1882d: 327 **A:** KZ TD UZ

tomentosa Reitter, 1893b: 226

gurievae Skopin, 1972b: 309 **A:** TD

hybrida hybrida Skopin, 1969: 886 **A:** TM (Kughitangay)

hybrida tadzhika Skopin, 1972b: 304 **A:** TD UZ

intermedia Kraatz, 1882d: 328 **A:** TD UZ (Samarkand)

dubia Kraatz, 1882d: 328

modesta Kraatz, 1882d: 328

laticollis fernanensis Skopin, 1972b: 304 **A:** UZ (Fergana)

laticollis laticollis Solsky, 1876: 289 **A:** KI KZ TD TM UZ

striata Kraatz, 1882d: 328

oblonga Skopin, 1972b: 309 **A:** TD

pachyscelis alticola Skopin, 1972b: 303 **A:** TD

pachyscelis pachyscelis Kraatz, 1882b: 85 (*Pimelia*) **A:** UZ (Fergana)

ruginota Reitter, 1896a: 39 **A:** IN TM (Kopet Dag)
peculiaris Semenov & Bogatchev, 1940: 202 (*Arnoldiola*)
seriatogemmata Solsky, 1876: 288 **A:** TD
heydeni Kraatz, 1882d: 326
sordida costipennis Skopin, 1972b: 305 **A:** UZ (Fergana)
sordida sordida Kraatz, 1881d: 331 (*Pimelia*) **A:** UZ (Fergana)
staudingeri badakschanica Kaszab, 1960a: 39 **A:** AF TD
staudingeri humeralis Skopin, 1972b: 308 **A:** TD
staudingeri soluta Skopin, 1972b: 308 **A:** TD
staudingeri staudingeri Reitter, 1893b: 227 **A:** TD
tuberculatocostata pterocomoides Reitter, 1893b: 225 **A:** KI (Kirghiz Range) KZ (Karatau)
tuberculatocostata simulatrix Kraatz, 1882b: 85 (*Pimelia*) **A:** UZ (Margilan)
tuberculatocostata solskyi Semenov & Skopin, 1951: 127 **A:** KZ
tuberculatocostata tienshanica Skopin, 1972b: 302 **A:** KI UZ
tuberculatocostata tuberculatocostata Solsky, 1876: 287 **A:** KZ TD UZ
variabilis Kraatz, 1881d: 331 (*Pimelia*)

genus *Sternocnera* Skopin, 1964: 394 type species *Sternocnera lindti* Skopin, 1964
lindti Skopin, 1964: 396 **A:** TD

genus *Sternodes* Fischer von Waldheim, 1837b: 10 type species *Sternodes karelini* Fischer von Waldheim, 1837 (= *Tenebrio caspicus* Pallas, 1781)
caspicus caspicus Pallas, 1781: 47 (*Tenebrio*) **A:** IN TM UZ
karelini Fischer von Waldheim, 1837: 11 (*Sternodes*)
pallasi Semenov & Bogatchev, 1940: 205
caspicus eous Semenov & Bogatchev, 1940: 204 **A:** AF TM

genus *Sternoplax* J. Frivaldszky, 1890: 207 type species *Trigonoscelis szechenyii* J. Frivaldszky, 1890
subgenus *Mesosternoplax* Skopin, 1973b: 142 type species *Trigonoscelis laeviuscula* Kraatz, 1882
belutschinstanica Skopin, 1973b: 148 **A:** PA (Balochistan)
blairi Skopin, 1974c: 259 **A:** "Baluchistan"
bogatshevi Skopin, 1960a: 300 **A:** TM UZ
laeviuscula Kraatz, 1882b: 87 (*Trigonoscelis*) **A:** UZ
matthiesseni Reitter, 1907b: 87 **A:** IN
persica Schuster, 1935b: 25 (*Trigonoscelis*) **A:** IN
planiuscula Kraatz, 1882b: 88 (*Trigonoscelis*) **A:** AF TM UZ
sexseriata Reitter, 1913d: 659 **A:** IN
souvorowiana Reitter, 1907b: 88 **A:** AF GAN KZ UZ XIN
subgenus *Pachysternoplax* Skopin, 1973b: 154 type species *Trigonoscelis armeniaca* Faldermann, 1837

armeniaca Faldermann, 1837: 21 (*Trigonoscelis*) **E:** AB AR

caucasica Motschulsky, 1860e: 439 (*Trigonoscelis*)

nicomedia Reitter, 1908g: 214 (*Trigonoscelis*) **A:** TR

subgenus *Parasternoplax* Skopin, 1973b: 148 type species *Pimelia deplanata* Krynicki, 1832

affinis affinis Zubkov, 1833: 327 (*Pimelia*) **A:** KZ

abbreviata Motschulsky, 1860e: 533 [= 1860d: 437] (*Ocnera*)

affinis australis Skopin, 1973b: 152 **A:** AF IN TM UZ

ballioni Skopin, 1973b: 153 **A:** GAN XIN

deplanata deplanata Krynicki, 1832: 130 (*Pimelia*) **E:** ST **A:** KZ TM

perevostchikovii Zubkov, 1833: 326 (*Pimelia*)

deplanata kuldzhana Skopin, 1973b: 149 **A:** XIN

fossor Kaszab, 1967b: 320 **A:** MG

subgenus *Pseudosternoplax* Skopin, 1973b: 141 type species *Trigonoscelis lacerta* Bates, 1879

lacerta Bates, 1879: 475 (*Trigonoscelis*) **A:** KA XIN XIZ

pustulosa Reitter, 1887d: 378 (*Diesia*)

subgenus *Sternoplax* J. Frivaldszky, 1890: 207 type species *Trigonoscelis szechenyii* J. Frivaldszky, 1890

Pseudeuthriptera Bogatchev & Kryzhanovskiy, 1955: 240 type species *Trigonoscelis uzboica* Bogatchev & Kryzhanovskiy, 1955

costatissima Reitter, 1899d: 204 (*Trigonoscelis*) **A:** QIN XIN

echinata Fischer von Waldheim, 1844: 57 (*Trigonoscelis*) **E:** AB ST (Dagestan) **A:** KZ TM

apscheronica Bogatchev, 1934: 70 (*Trigonoscelis*)

sinuatocollis Desbrochers des Loges, 1881: 93 (*Trigonoscelis*)

kughitangi G.S. Medvedev, 1979a: 40 **A:** TM

schusteri Günther, 1937: 132 **A:** XIN

steinbergi Bogatchev & Kryzhanovskiy, 1960: 266 (*Trigonoscelis*) **A:** TM UZ

szechenyii J. Frivaldszky, 1890: 207 **A:** GAN XIN

uzboica Bogatchev & Kryzhanovskiy, 1955: 240 (*Trigonoscelis*) **A:** TM

genus *Sternotrigon* Skopin, 1973b: 109 type species *Trigonoscelis setosa* Bates, 1879

albipulis Ba & Ren, 2013b: 570 **A:** CH

boldi Kaszab, 1964b: 17 [= 1964f: 385] **A:** MG

grandis Faldermann, 1835b: 387 (*Platyope*) **A:** GAN MG NMO XIN

mongolica Reitter, 1889k: 696 (*Trigonoscelis*)

hiekei Skopin, 1973b: 176 **A:** XIN

kraatzi J. Frivaldszky, 1890: 206 (*Trigonoscelis*) **A:** GAN MG NMO XIN

opaca Reitter, 1907b: 91 (*Sternoplax*) **A:** NMO

setosa juvencus Reitter, 1893b: 241 (*Trigonoscelis*) **A:** UZ XIN XIZ

kashgarensis Reitter, 1907b: 90 (*Sternoplax*)

seidlitzi Reitter, 1893b: 243 (*Trigonoscelis*)

setosa setosa Bates, 1879: 475 (*Trigonoscelis*) **A:** KA NIN XIN
impressicollis Reitter, 1893b: 242 (*Trigonoscelis*)
niana Reitter, 1887d: 377 (*Diesia*)
zichyi Csiki, 1901a: 110 (*Trigonoscelis*) **A:** CH MG

genus *Storthocnemis* Karsch, 1881: 47 type species *Storthocnemis steckeri* Karsch 1881
antoinei Español, 1953c: 298 **N:** MO
kulzeri Koch, 1969: 484 **A:** YE
steckeri dufouri Peyerimhoff, 1935c: 90 **N:** AG MO **AFR**
steckeri steckeri Karsch, 1881: 48 **N:** LB

genus *Tadzhikistania* Bogatchev, 1960b: 35 type species *Tadzhikistania mystacea* Bogatchev, 1960
mystacea Bogatchev, 1960b: 35 **A:** TD
stoljarovi G.S. Medvedev, 1970: 402 **A:** AF

genus *Thriptera* Solier, 1836a: 48 type species *Thriptera maillei* Solier, 1836 (= *Pimelia crinita* Klug, 1830)
afghanica Kaszab, 1970a: 70 **A:** AF
ballionis Kraatz, 1882b: 87 **A:** TM UZ
crinita crinita Klug, 1830: nr. 22, pl. 12 (*Pimelia*) **N:** EG **A:** IN JO SA SI YE **AFR**
maillei Solier, 1836a: 50
pygmaea Miller, 1861: 181 (*Ocnera*)
dubia Kaszab, 1973b: 44 (*Thryptera*) **A:** AF
foucauldi Peyerimhoff, 1929c: 285 **N:** AG (Hoggar)
freyi freyi Gridelli, 1954: 223 **A:** AF
freyi gracillima Kaszab, 1960a: 32 (*Thryptera*) **A:** AF **ORR**
jaffana Miller, 1861: 182 (*Ocnera*) **A:** IS SY
asphaltidis Reiche & Saulcy, 1875: 218
kashmirensis Kaszab, 1957f: 297 **A:** KA
klapperichi Kaszab, 1960a: 24 (*Lasiostola*) **A:** AF
kraatzi Haag-Rutenberg, 1876b: 75 **A:** IN JO OM PA SA SI YE **ORR**
groschkei Kaszab, 1957d: 65
sorocula Reitter, 1894e: 303
yemenensis Gridelli, 1940: 236
laperrinei laperrinei Peyerimhoff, 1929c: 284 **N:** AG (Hoggar)
longipilis Reitter, 1895d: 156 **A:** AF TD TM
sparsicollis Reitter, 1909g: 311 (*Tryptera*)
malyi Obenberger, 1914: 139 **A:** UZ (Bukhara)
maroccana Théry, 1932a: 191 **N:** AG MO
minuta Pic, 1897m: 124 **N:** LB

murina murina Baudi di Selve, 1881: 280 **A:** IN SA YE
olegi G.S. Medvedev, 2004b: 565 **A:** AF
pastor Reitter, 1893b: 216 **A:** KZ
protensa Fairmaire, 1891d: xciv (*Ocnera*) **A:** KA
varvasi pilipes Kraatz, 1865a: 299 **N:** EG **A:** SI
guyoti Andres, 1920: 70
heydeni Reitter, 1893b: 214
varvasi varvasi Solier, 1836a: 52 **N:** AG LB TU
bedeli Allard, 1886: (3)
debilicornis Kraatz, 1865a: 300
hamadensis Koch, 1937: 436
lybica Koch, 1937: 435
phazanica Koch, 1937: 436
villosa Solier, 1836a: 53 **N:** LB
alternans Baudi di Selve, 1876a: 10 (*Ocnera*)
tunicensis Desbrochers des Loges, 1898a: 82 (*Ocnera*)
zarudnyi G.S. Medvedev, 2004b: 567 **A:** IN

genus *Trachyderma* Latreille, 1828: 576 type species *Pimelia hispida* Fabricius, 1775 (= *Tenebrio hispidus* Forskål, 1775)
subgenus *Atrachyderma* Skopin, 1962: 228 type species *Pimelia setosa* Fischer von Waldheim, 1832
afghanicum afghanicum Kaszab, 1960a: 29 **A:** AF
afghanicum kandaharicum Kaszab, 1970a: 69 **A:** AF
allardi Gridelli, 1928: 17 (*Ocnera*) [RN] **N:** EG
grande adriani Kaszab, 1974: 196 **A:** AF
grande grande Schuster, 1930b: 238 (*Ocnera*) **A:** IN PA (Beluzhistan)
maroccanum Schuster, 1921b: 46 (*Ocnera*) **N:** MO
menetriesi afghanimorphum Skopin, 1968b: 98 (*Brachyscelis*) **A:** TD
menetriesi hafizi Kwieton, 1981a: 14 **A:** PA (Beluchistan)
menetriesi macrogranulatcum Gridelli, 1954: 218 [= 1955: 38] (*Trachyderma*) **A:** AF
menetriesi menetriesi Kraatz, 1865a: 296 (*Ocnera*) **A:** AF KZ TM
asperula Ménétriés, 1849: 221 [RN]
muricata Fischer von Waldheim, 1837b: 15 (*Brachyscelis*)
menetriesi robustus Faust, 1875: 213 (*Ocnera*) **A:** AF IN
perlata Baudi di Selve, 1875d: 685 (*Ocnera*)
setosum Faldermann, 1832: 192 (*Pimelia*) **E:** AB AR GG ST **A:** AF IN TM TR
beckeri Desbrochers des Loges, 1881: 90 (*Ocnera*)
granulata Fischer von Waldheim, 1844: 56 (*Ocnera*)
triangulare australe Medvedev, 1964c: 658 (*Ocnera*) **A:** TM (Badkhyz)
triangulare triangulare Faust, 1875: 212 (*Ocnera*) **A:** AF KZ TM UZ
subgenus *Trachyderma* Latreille, 1828: 7 type species *Pimelia hispida* Fabricius, 1775 (= *Tenebrio hispidus* Forskål, 1775)

andresi Gridelli, 1928: 16 (*Ocnera*) **A:** SI
christophi benardi Pierre, 1968: 1023 **A:** IN
christophi christophi Faust, 1875: 208 (*Ocnera*) **E:** AR GG **A:** AF IN TM
longicollis Baudi di Selve, 1875d: 687 (*Ocnera*)
cyrenaicum Gridelli, 1930: 270 (*Ocnera*) **N:** LB
emondi Solier, 1836a: 38 **N:** EG
longicollis Desbrochers des Loges, 1881: 87 (*Ocnera*) [HN]
genei Solier, 1836a: 38 **N:** EG LB **A:** SI
habelmanni Kraatz, 1865a: 292 (*Ocnera*)
gomorrhanum Reiche & Saulcy, 1857: 215 **A:** IS
hispidum Forskål, 1775: 79 (*Tenebrio*) **N:** AG EG LB MO TU **A:** IN IQ IS SA SI SY YE **AFR**
depressa Baudi di Selve, 1875d: 685 (*Ocnera*)
hispida Fabricius, 1775: 251 [HN] (*Pimelia*)
latreillei Solier, 1836a: 36
leprieuri Allard, 1886: (4) (*Ocnera*)
major Baudi di Selve, 1875d: 684 (*Ocnera*)
minor J.R. Sahlberg, 1913b: 170
sparsispina Böhm, 1916b: 67 (*Ocnera*)
limum L. Petagna, 1819: 29 (*Pimelia*) **E:** GR IT MA **N:** AG EG LB TU **A:** CY IN IS LE SY TR
angustata Solier, 1836a: 37
persea farahense Kaszab, 1974: 196 **A:** AF
persea persea Baudi di Selve, 1875d: 687 (*Ocnera*) **A:** IN PA
persea khashensis Pierre, 1968: 1022 **A:** IN
philistinum Reiche & Saulcy, 1857: 214 **E:** GR (Rhodes) **E:** GR **N:** EG **A:** AE BA CY IN IQ IS JO OM QA SA SI SY TR YE **ORR**
graeca Desbrochers des Loges, 1881: 89 (*Ocnera*)
ovalipennis Baudi di Selve, 1875d: 685 [= 1876: 5] (*Ocnera*)
parvicolle Baudi di Selve, 1875d: 686 (*Ocnera*)
seurati Peyerimhoff, 1929c: 283 (*Ocnera*) **N:** AG (Hoggar) LB
striatogranosum Fairmaire, 1882a: 67 (*Thriptera*) **A:** PA SA YE **AFR**

genus *Trigonocnera* Reitter, 1893b: 202 type species *Trigonoscelis pseudopimela* Reitter, 1889
pseudopimelia pseudopimelia Reitter, 1889k: 697 (*Trigonoscelis*) **A:** GAN
amitina Kolbe, 1908: 90 (*Trigonoscelis*)
pseudopimelia reitteri Csiki, 1901a: 111 (*Trigonoscelis*) **A:** MG NMO

genus *Trigonopachys* Skopin, 1968a: 300 type species *Trigonopachys michailovi* Skopin, 1968
michailovi Skopin, 1968a: 302 **A:** UZ

genus *Trigonoscelis* Dejean, 1834: 179 type species *Pimelia nodosa* Fischer von Waldheim, 1820
subgenus *Chinotrigon* Skopin, 1973b: 170 type species *Trigonoscelis sublaevigata* Reitter, 188-
holdereri Reitter, 1900e: 161 **A:** XIN

sublaevigata granicollis Kaszab, 1964b: 15 **A:** MG
sublaevigata sublaevigata Reitter, 1887e: 519 **A:** NMO XIN
subgenus *Echinotrigon* Skopin, 1973b: 170 type species *Trigonoscelis granulata* Reitter, 1915
granulata Reitter, 1915e: 90 **A:** TD
tadzhikistana Bogatchev, 1963b: 94
subgenus *Trigonoscelis* Dejean, 1834: 179 type species *Pimelia nodosa* Fischer von Waldheim, 1820
apicalis Reitter, 1907b: 83 **A:** TM
borosi Kaszab, 1951a: 2 **A:** TM
ceromatica Bogatchev, 1963b: 95 **A:** UZ (Surkhandarinska oblast)
fasciculitarsis fasciculitarsis Reitter, 1893b: 236 **A:** KI UZ
fasciculitarsis saeptiarius Skopin, 1973b: 168 **A:** UZ (Fergana)
gemmulata callosa Motschulsky, 1860e: 534 **A:** KZ
gemmulata gemmulata Ménétriés, 1849: 221 **A:** KZ UZ
gemmulata intermedia Skopin, 1973b: 160 **A:** KZ
gemmulata kabuliensis Kaszab, 1974: 195 **A:** AF
gemmulata oxiana Skopin, 1973b: 160 **A:** UZ
gemmulata punctipleuris Reitter, 1893b: 235 **A:** TM
gemmulata seriatula Reitter, 1907b: 84 **A:** AF TM
aequalis Reitter, 1907b: 84
hayekae Skopin, 1973b: 165 **A:** PA
muricata abnormis Skopin, 1973b: 157 **A:** KZ
muricata mirabilis Faldermann, 1836: 380 **A:** KZ
muricata muricata Pallas, 1781: 48 (*Tenebrio*) **E:** ST **A:** KZ TM
spinigera Motschulsky, 1860d: 439 [RN]
muricata sparsa Reitter, 1893b: 233 **A:** TM
nodosa gigas Reitter, 1893b: 231 **A:** TM
nodosa grandis Kraatz, 1865a: 279 **A:** TM
contraria Desbrochers des Loges, 1881: 91
corallifera Reitter, 1893b: 234
elegans Allard, 1886: [1]
nodosa mixta Skopin, 1973b: 162 **A:** AF KZ
nodosa nodosa Fischer von Waldheim, 1820: pl. 14 (*Pimelia*) **A:** AF KZ TM
nodosa pustulifera Skopin, 1973b: 164 **A:** TM (Karakumy)
nodosa schrenkii Gebler, 1844: 101 **A:** KZ XIN
seriata Ménétriés, 1849: 222 **A:** TD TM UZ
pecirkana Reitter, 1909g: 308
sublaevicollis Reitter, 1893b: 236 **A:** IN TM UZ
submuricata Kraatz, 1882b: 95 **A:** UZ (Fergana)
zoufali Reitter, 1893b: 237 **A:** KZ UZ

genus *Waterhousia* Skopin, 1973b: 109 type species *Trigonoscelis longipes* Waterhouse, 1889
Heinrichesia Carl, 2000: 258 type species *Heinrichesia schaeferi* Carl, 2000

longipes Waterhouse, 1889: 128 (*Trigonoscelis*) **A:** AF PA (Balochistan)
lopatini G.S. Medvedev, 2005b: 310 **A:** AF (Regestan)
schaeferi Carl, 2000: 258 (*Heinrichesia*) **A:** IN (Sistan)
warei Skopin, 1973b: 179 **A:** PA

tribe Pimeliini, species incertae sedis
Trigonoscelis occidentalis Motschulsky, 1860d: 438 **A:** LE (Beirut)

tribe Sepidiini Eschscholtz, 1829
subtribe Molurina Solier, 1834

genus *Psammophanes* Lesne, 1922: 690 type species *Moluris catenata* Reiche, 1850
subgenus *Somalarabes* Koch, 1953: 155 type species *Psammodes gracilentus* Fairmaire, 1882
arabicus Gebien, 1938: 58 (*Psammodes*) **A:** YE
benardellii Koch, 1965a: 126 **A:** YE

subtribe Sepidiina Eschscholtz, 1829

genus *Sepidiostenus* Fairmaire, 1884e: lxxv type species *Sepidiostenus erinaceus* Fairmaire, 1884
  Sepidiacis Fairmaire, 1884h: cxlvi type species *Sepidiacis compressa* Fairmaire, 1884
pradierii Guérin-Méneville, 1858: 128 (*Sepidium*) **A:** SA

genus *Sepidium* Fabricius, 1775: 250 type species *Sepidium tricuspidatum* Fabricius, 1775
  Espidium Rafinesque, 1815: 113 [URN]
aitagiae Escalera, 1913: 41 **N:** MO
aliferum Erichson, 1841: 178 **E:** SP **N:** AG MO TU
  douei Solier, 1843: 18
bathelemyi Solier, 1843: 236 **N:** EG
barbarum Solier, 1843: 235 **N:** AG TU
  pallens Allard, 1874: 137
  servillei Solier, 1843: 24
  solieri Desbrochers des Loges, 1881: 101
bicaudatum Fairmaire, 1871a: 388 **N:** MO
bidentatum Solier, 1843: 227 **E:** PT SP **N:** MO
capricorne Desbrochers des Loges, 1881: 96 **N:** AG
cristatum Fabricius, 1775: 250 **A:** SA YE
  notoceros Pallas, 1781: 59 (*Tenebrio*)
cyrenaicum Schuster, 1928b: 122 **N:** LB
dathan Crotch, 1872: 268 **N:** EG LB **A:** IS LE SI SY
  abiram Crotch, 1872: 268
  vietaeformis Reitter, 1914h: 385
fusiforme Kwieton, 1980a: 17 **N:** MO
gypsicola Escalera, 1913: 42 **N:** MO

hoseini Escalera, 1911: 301 **N:** MO (Tiznit)
hystrix desertica Español, 1944: 12 **N:** MO (Western Sahara)
hystrix hystrix Antoine, 1932: 185 **N:** MO (Sous)
hystrix ifniensis Escalera, 1940: 5 **N:** MO (Tiznit)
inaequale Reitter, 1914h: 386 **N:** LB
lusitanicum Kaszab & Pinheiro, 1972: 6 **E:** PT SP
mali J. Ferrer & Martínez, 2012: 286 [RN] **E:** PT
elongatum Mal, 1984: 200 [HN]
marraquense Escalera, 1911: 302 **N:** MO
mesopotamicum Reitter, 1914h: 386 **A:** IQ
mskalicum Escalera, 1914: 307 **N:** MO
penicilligerum Karsch, 1881: 49 **N:** LB
perforatum Allard, 1874: 130 **N:** MO
peyerimhoffi Antoine, 1932: 183 **N:** MO (Sous)
reichei Allard, 1870: 49 **N:** EG LB TU
bispinicollis Reitter, 1914h: 389
requieni Solier, 1843: 241 **N:** AG LB TU
siculum siculum Solier, 1843: 231 **E:** IT (Sicilia)
genei Solier, 1843: 230
tricuspidatum korah Crotch, 1872: 268 **N:** EG
tricuspidatum mogadoricum Escalera, 1914: 306 **N:** MO
guisseri Kocher, 1968: 111
immundum Reitter, 1914h: 388
schrammi Kocher, 1968: 110
tricuspidatum multispinosum Solier, 1843: 241 **N:** AG TU
laghoatense Baudi di Selve, 1875d: 695
tricuspidatum tomentosum Erichson, 1841: 178 **E:** MA **N:** EG TU
barthelemyi Solier, 1843: 24
maillei Solier, 1843: 239
remotum J.R. Sahlberg, 1903f: 50
serratum Solier, 1843: 240
tricuspidatum tricuspidatum Fabricius, 1775: 250 **N:** EG LB **A:** IS SI SY
alexandrinus Forskål, 1775: 78 (*Tenebrio*)
cerisyi Solier, 1843: 238
flexuosum Solier, 1843: 237 **N:** EG
tualensis Escalera, 1940: 9 **N:** MO (Ifni)
uncinatum Erichson, 1841: 178 **N:** AG LB MO TU
bicorne Solier, 1843: 229
mittrei Solier, 1843: 228 **N:** AG
sefrense Rotrou, 1943: 235
subcornutum Escalera, 1925d: 376

variegatum Fabricius, 1792a: 112 (*Tenebrio*) **N:** AG MO TU
angustatum Solier, 1843: 233
dispar Desbrochers des Loges, 1881: 101
dufouri Solier, 1843: 234
integrum Desbrochers des Loges, 1881: 100
laterale Allard, 1874: 133
subfurcatum Desbrochers des Loges, 1881: 100
wagneri Erichson, 1841: 179 **N:** AG MO TU
confusum Allard, 1874: 141 [HN]
macrotys Antoine, 1951c: 94

genus *Vieta* Laporte, 1840: 196 type species *Sepidium vestitum* Guérin-Méneville, 1831
Divieta Reitter, 1914h: 390 type species *Vieta costata* Allard, 1874
Dymonus Solier, 1843: 219 type species *Sepidium vestitum* Guérin-Méneville, 1831
algeriana Allard, 1870: 50 [= 1874: 146] **N:** AG
clypeata Gahan, 1896b: 454 **A:** YE
costata Allard, 1874: 149 **N:** EG **AFR**
gracilenta Ancey, 1881a: 397 **A:** YE
luxorii Allard, 1874: 150 **N:** EG
millingenii Kirchsberg, 1877a: 203 **A:** SA YE
schawalleri Mal, 2019: 3 **A:** SA
tuberculata Solier, 1843: 223 (*Dymonus*) **N:** EG **A:** SA SI YE
gibbicollis Solier, 1843: 224 (*Dymonus*)

genus *Vietomorpha* Fairmaire, 1887a: 186 type species *Vietomorpha foveipennis* Fairmaire, 1887
crassipes socotraense Mal, 2019: 3 **A:** YE
foveipennis Fairmaire, 1887a: 186 **A:** YE **AFR**
arabica Gebien, 1938: 59

tribe Stenosini Schaum, 1859 (1834)

genus *Afghanillus* Kaszab, 1960a: 1 type species *Afghanillus klapperichi* Kaszab, 1960
klapperichi Kaszab, 1960a: 2 **A:** AF

genus *Aspidocephalus* Motschulsky, 1839: 63 type species *Aspidocephalus desertus* Motschulsky, 1839
desertus Motschulsky, 1839: 64 **E:** AB AR GG ST **A:** KZ

genus *Dichillus* Jacquelin du Val, 1861: 253 type species *Tagenia minuta* Solier, 1838
subgenus *Dichillesthes* Reitter, 1916d: 156 type species *Dichillus cordicollis* Reitter, 1886
cordicollis Reitter, 1886a: 134 **A:** KZ TD TM UZ

kuhitangi G.S. Medvedev, 1994b: 854 **A:** TM (Kuhitanga)
subgenus *Dichillinus* Reitter, 1916d: 161 type species *Tagenia pusilla* Ménétriés, 1849
afghanicus afghanicus Kaszab, 1960a: 11 **A:** AF
afghanicus badakschanicus Kaszab, 1960a: 12 **A:** AF
afghanicus khinjanensis Kaszab, 1960a: 12 **A:** AF
afghanicus klapperichi Kaszab, 1960a: 12 **A:** AF
buettikeri Kaszab, 1982b: 127 **A:** SA
cylindricus Baudi di Selve, 1875a: 298 **A:** IS LE TR
  unistriatus Desbrochers des Loges, 1881: 80
desioi Gridelli, 1934: 58 **A:** KA
dolosus Reitter, 1886a: 138 **A:** IS LE "Syria"
epipleuralis Reitter, 1916d: 163 **A:** AF TD UZ
explanatus explanatus Reitter, 1900b: 54 **A:** KZ TD UZ
explanatus kataghanicus Kaszab, 1960a: 13 **A:** AF
farsistanicus G.S. Medvedev, 1976: 894 **A:** IN
fusicornis Reitter, 1916d: 166 **A:** KZ TM UZ
  karatavicus Skopin, 1966b: 341
ganquanensis Ren, 1999: 333 **A:** SHA
heydeni Reitter, 1886a: 137 **A:** "Syria"
iranicus Kaszab, 1963a: 74 **A:** IN TR
kryzhanovskii G.S. Medvedev, 1975: 602 **A:** TM (Kopet Dag)
laeviusculus Kraatz, 1862a: 99 **E:** SP **N:** AG MO TU
  distinguendus Fairmaire, 1884d: lxvi
  modestus Fairmaire, 1884d: lxvi
lisae Carl, 1990c: 105 **A:** IQ
nitidulus nitidulus Reitter, 1886a: 137 **A:** IS
nitidulus transjordanicus Koch, 1936: 296 **A:** JO
nitidus Kraatz, 1862a: 99 **A:** IQ SY
ocellaris Kaszab, 1970a: 36 **A:** AF TM
pauxillus filum Reitter, 1916d: 163 **A:** AF TM
pauxillus pauxillus Reitter, 1900b: 54 **A:** AF TD TM UZ
pusillus Ménétriés, 1849: 228 (*Tagenia*) **A:** AF KI KZ TD UZ
  brunneus Kraatz, 1882d: 321
  fulvescens Reitter, 1916d: 162
  nigritulus Reitter, 1916d: 162
reitteri Semenov, 1890: 361 **A:** KZ TM UZ
sanguinipes sanguinipes Fairmaire, 1888b: 126 **A:** BEI HEB
sanguinipes shantungensis G.S. Medvedev, 1994a: 309 **A:** SHN
schmidi Kaszab, 1961a: 116 **A:** AF
semipolitus Reitter, 1916d: 164 **A:** JO
simeki Kaszab, 1970a: 33 **A:** AF TD UZ

simillimus Kaszab, 1960a: 13 **A:** AF
sinensis G.S. Medvedev, 1994a: 308 **A:** SCH YUN
tenebrosus Reitter, 1886a: 136 **A:** KZ TM
subgenus *Dichillocerus* Reitter, 1916d: 167 type species *Hyperops pertusa* Kiesenwetter, 1861
araxidis Reitter, 1889a: 30 **E:** AB AR **A:** TR
crassicornis Reitter, 1916d: 167 **E:** GR **N:** LB **A:** IQ LE IS TR
cyrenaicus Gridelli, 1930: 242 **N:** EG LB
damascenus Reitter, 1916d: 167 **A:** SY
obenbergeri Mařan, 1935b: 147 **E:** GR (Crete)
pertusus Kiesenwetter, 1861a: 224 (*Hyperops*) **E:** GR IT MA **A:** TR
biseriatus Reiche, 1862a: 371 (*Oedenocera*)
subgenus *Dichillodontus* Reitter, 1916: 166 type species *Dichillus adriani* Reitter, 1916
adriani Reitter, 1916d: 167 **A:** TD TM UZ
aruktavicus G.S. Medvedev, 1991: 561 **A:** TD
dentipes G.S. Medvedev, 1964a: 61 **A:** TM (Kuhitang)
subgenus *Dichillomessor* Reitter, 1916d: 155 type species *Dichillus himalayanus* Fairmaire, 1896
himalayanus Fairmaire, 1896a: 57 **A:** HP
indicus Wasmann, 1897: 239 **A:** SD ORR
kashmirensis Kaszab, 1975a: 15 **A:** KA
lindemannae Kaszab, 1961d: 219 **A:** PA
pakistanus Kaszab, 1961d: 221 **A:** PA
piffli Kaszab, 1961f: 344 **A:** PA (Karakoram)
subgenus *Dichillus* Jacquelin du Val, 1861: 253 type species *Tagenia minuta* Solier, 1838
alfierii Koch, 1935a: 46 **A:** SI
algiricus Lucas, 1846: pl. 28 (*Tagenia*) **N:** AG TU
bicarinatus Baudi di Selve, 1875a: 299
castanescens Fairmaire, 1884d: lxvi
henoni Allard, 1869b: 468 (*Stenosis*)
angelicae Reitter, 1897e: 217 **E:** AB AR **A:** IN TR
angusticollis Kaszab, 1959e: 244 [= Kaszab, 1960a: 10] **A:** AF
carinatus Küster, 1848b: 89 (*Tagenia*) **E:** BH BU CR GR MC TR
corsicus Solier, 1838a: 33 (*Tagenia*) **E:** FR IT
pumilus Solier, 1838a: 33 (*Tagenia*)
formicophilus Breit, 1916: 61 **E:** GR RO UK
laevicornis Reitter, 1886a: 132 **N:** MO
libanicus Koch, 1936: 295 **A:** LE
lindbergi Kaszab, 1970a: 31 **A:** AF
michailovi G.S. Medvedev, 1975: 598 **A:** TD
minutus Solier, 1838a: 32 (*Tagenia*) **E:** FR IT SP
myrmecophilus Fairmaire, 1884d: lxvii **N:** AG
palaestinus Baudi di Selve, 1875a: 296 **A:** "Palestina"

rugatus bogatshevi G.S. Medvedev, 1975: 597 **E:** AB (Talish)
rugatus rugatus Baudi di Selve, 1874a: 105 [= 1875a: 298, 1875c: 70] **E:** AR **A:** IN IQ
scheibei Kaszab, 1960a: 7 **A:** AF
schusteri Reitter, 1916d: 160 **A:** IN TM
seminitidus Solsky, 1881b: 47 **A:** AF KZ TD TM UZ
inhumeralis Pic, 1906b: 18
socius Rottenberg, 1871: 247 **E:** IT (Sicilia)
stackelbergi G.S. Medvedev, 1975: 595 **A:** UZ (Tian Shan
stoeckleini Kaszab, 1960a: 9 **A:** AF
strigiceps Reitter, 1886a: 141 **A:** SY
subcostatus Solier, 1838a: 34 (*Tagenia*) **E:** PT SP **N:** MO
andulusicus Rosenhauer, 1856: 202 (*Tagenia*)
elevatus Reitter, 1886a: 142
valentinus Reitter, 1916d: 159
subsetulosus Reitter, 1886a: 140 **E:** GR (Rhodos) **A:** CY TR
subtilis Kraatz, 1862a: 98 **E:** IT (Sicilia)
tyrrhenicus Leo, 2008: 610 **E:** FR IT
weberi J. Müller, 1916: 93 **E:** CR
subgenus *Myrmecodichillus* Kaszab, 1960a: 6 type species *Dichillus reichenspergeri* Kaszab, 1960
reichenspergeri Kaszab, 1960a: 5 **A:** AF
subgenus *Pushtunillus* G.S. Medvedev, 1994b: 855 type species *Dichillus skopini* G.S. Medvedev & Kabakov, 1994
skopini G.S. Medvedev & Kabakov, 1994: 855 **A:** AF

genus *Dichillus*, species incertae sedis
duricornis Reitter, 1886a: 139 **A:** TR

genus *Ethas* Pascoe, 1862: 324 type species *Ethas carbonarius* Pascoe, 1862
anomalus Fairmaire, 1896a: 56 **A:** HP
semicribrosus Fairmaire, 1896a: 56 **A:** "North India"

genus *Eutagenia* Reitter, 1886a: 125 type species *Tagenia smyrnensis* Solier, 1838
aegyptiaca aegyptiaca Reitter, 1889h: 285 **N:** EG LB
aegyptiaca tunisea Normand, 1936b: 359 **E:** IT (Lampedusa Is.) **N:** BL TU
annae Grimm, 1991: 20 **A:** CY
cribricollis Reitter, 1916d: 152 **A:** IS TR
elvirae Marcuzzi & Turchetto, 1982: 235 **E:** IT (Lucania)
minutissima Pic, 1903m: 162 **E:** GR (Crete, Kos, Rhodes, Dodecanese Is.) **A:** TR
smyrnensis Solier, 1838a: 31 (*Tagenia*) **E:** BU GR **A:** IN JO SY TR
hellenica Reitter, 1889h: 285
longicollis Rey, 1892: 41 (*Stenosis*)

syriaca fortepunctata Koch, 1935d: 273 **A:** JO SA
syriaca ovaliceps Koch, 1935d: 272 **A:** IS
syriaca syriaca Reitter, 1889h: 284 **A:** IS SY
turcomania Reitter, 1889h: 285 **A:** AF KZ TD TM UZ

genus *Gebieniella* Koch, 1940e: 736 type species *Ethas stenosides* Pascoe, 1862
hainanensis G.S. Medvedev, 1994a: 307 **A:** HAI

genus *Herbertfranzia* Kaszab, 1973a: 26 type species *Herbertfranzia nepalica* Kaszab, 1973
nepalica Kaszab, 1973a: 28 **A:** NP

genus *Herbertfranziella* Kaszab, 1973a: 26 type species *Herbertfranzia eutagenoides* Kaszab, 1973
almorensis Fouquè, 2013: 187 **A:** UP (Kumaon)
bhimtaleica Fouquè, 2013: 191 **A:** UP (Kumaon)
brancuccii Fouquè, 2013: 192 **A:** HP
eutagenoides Kaszab, 1973a: 29 (*Herbertfranzia*) **A:** NP
himalayica Fouquè, 2013: 193 **A:** NP
janushevi G.S. Medvedev, 1991: 560 **A:** PA TD
kumaona G.S. Medvedev, 1994b: 854 **A:** UP (Kumaon)
loebli Fouquè, 2013: 193 **A:** UP

genus *Indostola* G.S. Medvedev, 1991: 557 type species *Indostola pulchella* G.S. Medvedev, 1991
pulchella G.S. Medvedev, 1991: 558 **A:** SD UP

genus *Microblemma* Semenov, 1889a: 213 type species *Microblemma simplex* Semenov, 1889
afghanica Kaszab, 1960a: 18 **A:** AF
cordicollis Kaszab, 1960a: 19 **A:** AF
kasyi Kaszab, 1970a: 42 **A:** AF
lindbergi Kaszab, 1970a: 40 **A:** AF
quadricollis Kaszab, 1970a: 41 **A:** AF
simpex Semenov, 1889a: 213 **A:** TM

genus *Microtelus* Solier, 1838a: 9 type species *Microtelus asiaticus* Solier, 1838
afghanicus Kaszab, 1970a: 37 **A:** AF
angulatus Carl, 1992b: 334 **A:** IN
asiaticus Solier, 1838a: 10 **E:** GR **A:** TR
cariniceps binodiceps Reitter, 1907c: 115 **N:** EG **A:** SI
  sinaiticus Koch, 1935a: 48
cariniceps cariniceps Reiche & Saulcy, 1857: 227 **N:** EG **A:** IQ IS JO LE SY
  limitis Reitter, 1914a: 14
insolitus Carl, 1992a: 78 **A:** IQ

interstitialis Reitter, 1914a: 14 **A:** IQ IS JO SA SY
    cordicollis Kaszab, 1982b: 128
lethierryi Reiche, 1860: 334 **N:** AG EG IS LB MO TU
    torretassoi Schatzmayr & Koch, 1934: 24
nejdanus Kaszab, 1982b: 128 **A:** SA
persis Baudi di Selve, 1874a: 106 **A:** AE IN IQ KU
schusteri Carl, 1992a: 75 **N:** EG

genus *Mitotagenia* Reitter, 1916d: 153 type species *Stenosis arabs* Baudi di Selve, 1881
aegyptiaca Koch, 1941b: 40 **N:** EG **A:** IS SA SI
afghanica G.S. Medvedev, 2004b: 558 **A:** AF
arabs Baudi di Selve, 1881: 279 (*Stenosis*) **A:** YE **AFR**
biroi Koch, 1941b: 40 **A:** OM
pohli Schawaller, 2004c: 448 **A:** YE (Suqutra)
priesneri Andres, 1931: 56 **N:** EG

genus *Nepalofranziella* Fouquè, 2013: 194 type species *Nepalofranziella kaszabi* Fouquè, 2013
kaszabi Fouquè, 2013: 194 **A:** NP
schawalleri Fouquè, 2013: 195 **A:** NP

genus *Oogaster* Faldermann, 1837: 30 type species *Oogaster menetriesii* Faldermann, 1837 (= *Tagenia picea* Ménétriés, 1832)
lehmanni Ménétriés, 1849: 226 **A:** AF IN KZ TM UZ
piceus Ménétriés, 1832: 196 (*Tagenia*) **E:** AB AR **A:** IN
    doriae Baudi di Selve, 1874a: 106
    menetriesii Faldermann, 1837: 31

genus *Platamodes* Ménétriés, 1849: 233 type species *Platamodes dentipes* Ménétriés, 1849
dentipes dentipes Ménétriés, 1849: 233 **E:** AB AR **A:** AF IN KZ TM UZ
dentipes orientalis Kaszab, 1970a: 43 **A:** AF

genus *Pseudethas* Fairmaire, 1896a: 57 type species *Pseudethas quadraticeps* Fairmaire, 1896
subgenus *Pseudethas* Fairmaire, 1896a: 57 type species *Pseudethas quadraticeps* Fairmaire, 1896
    Dischizillus Wasmann, 1902: 244 [RN] type species *Schizillus rogersi* Wasmann, 1899
    Schizillus Wasmann, 1899: 166 [HN] type species *Schizillus rogersi* Wasmann, 1899
afghanicus Kaszab, 1960a: 3 **A:** AF
antennalis Kaszab, 1981b: 298 **A:** UP (Kumaon)
convexigenus Ren & Shi, 2006b: 182 (*Indochillus*) **A:** XIZ
cordiformis Ba & Ren, 2013a: 699 **A:** XIZ
costatus Kaszab, 1981b: 303 **A:** UP

incostatus Kaszab, 1981b: 302 **A:** PA (Swat, Chitral)
jaegeri Schawaller, 2001c: 143 **A:** XIZ
ladakhensis Kaszab, 1978b: 223 **A:** KA (Ladakh)
longigenus Fouquè, 2008: 361 **A:** NP
nepalicus Kaszab, 1981b: 299 **A:** NP
pakistanus Kaszab, 1981b: 301 **A:** PA (Hazara)
quadraticeps Fairmaire, 1896a: 57 **A:** HP
rogersi Wasmann, 1899: 166 (*Schizillus*) **A:** HP UP **ORR**
schawalleri Fouquè, 2008: 366 **A:** NP
subgenus *Stenillus* Blair, 1927: 245 type species *Stenillus monticola* Blair, 1927
monticola Blair, 1927: 245 (*Stenillus*) **A:** XIN
weigeli Fouquè, 2008: 367 **A:** NP

genus *Reitterella* Semenov, 1890: 362 type species *Reitterella fusiformis* Semenov, 1890
fusiformis Semenov, 1890: 363 **A:** TM
gracillima Semenov, 1903c: 356 **A:** TM
turkestanica Pic, 1906e: 33 **A:** UZ (Fergana)

genus *Stenosis* Herbst, 1799: 160 type species *Stenosis angustata* Herbst, 1799 (= *Stenosis intermedia* Küster, 1848)
subgenus *Stenosis* Herbst, 1799: 160 type species *Stenosis angustata* Herbst, 1799 (= *Stenosis intermedia* Küster, 1848)
Tagenia Latreille, 1802: 170 type species *Tenebrio filiformis* Fabricius, 1792
affinis Solier, 1838a: 26 (*Tagenia*) **N:** EG **A:** SI
algirica Lucas, 1846: pl. 28 (*Tagenia*) **N:** AG
alluaudi Antoine, 1936a: 14 **N:** MO
altiatlantis Antoine, 1936a: 14 **N:** MO (High Atlas)
angustata Fabricius, 1792a: 105 (*Pimelia*) **E:** SP **N:** TU
angusticollis Reiche, 1861c: 209 (*Tagenia*) **E:** FR (Var, Corse) IT (Sardinia, Tuscany)
elongatissima Koch, 1940e: 702
antoinei Koch, 1937: 398 **N:** LB
armeniacus Motschulsky, 1849: 123 (*Tagenia*) **E:** AB AR **A:** TR
austini Lemaire & F. Soldati, 2009: 128 **E:** CY
bavel Castro Tovar, 2013: 144 **E:** SP
bengala Fairmaire, 1894a: 18 **A:** UP **ORR**
brentoides brentoides Bosc d'Antic, 1790: 135 (*Rhinomacer*) **E:** CR IT
italica Kraatz, 1862a: 99
maroccana Desbrochers des Loges, 1881: 73
brentoides carfii Ardoin, 1971a: 46 **N:** TU
brentoides solieri Koch, 1940e: 706 **E:** PT SP

brignonei Koch, 1935e: 314 **E:** IT (Sicilia)
canariensis Franz, 1990b: 93 **N:** CI (Gran Canaria)
cephalonica Koch, 1940e: 726 **E:** GR (Kephalonia)
cobosi Español, 1958b: 46 **E:** SP (Málaga)
colfalonierii Gridelli, 1933: 224 **N:** LB
comata comata Reiche & Saulcy, 1857: 230 (*Tagenia*) **A:** IS SA SY
comata longiceps Koch, 1935d: 271 **A:** IS LE
consiglioi Canzoneri, 1976: 10 **E:** IT (Palmarola)
costulata Baudi di Selve, 1881: 278 **A:** SA YE
ecostata Gridelli, 1953: 42
cretica Koch, 1940e: 726 **E:** GR
crivellarii Koch, 1935e: 313 **E:** GR (Kastellorhizo)
cugiae Gridelli, 1930: 238 **N:** EG LB
dianae J.R. Sahlberg, 1908: 77 **A:** SY TR
dilutipes Reitter, 1887a: 77 **A:** IS LE TR
reitteri Marseul, 1887: 311 [RN]
duplex Antoine, 1948: 284 **N:** MO
elongata Solier, 1838a: 24 (*Tagenia*) **N:** EG
elvirae arganeticola Antoine, 1957: 42 **N:** MO
elvirae elvirae Antoine, 1936a: 14 **N:** MO
elvirae reymondi Kocher, 1965a: 26 **N:** MO
elvirae sarhoensis Kocher, 1965a: 26 **N:** MO
esau J.R. Sahlberg, 1908: 77 **E:** GR (Samos) **A:** TR
espanoli Antoine, 1948: 289 **N:** MO
fallaciosa J. Ferrer & Murria Beltrán, 2009: 512 **E:** SP
fausti Reitter, 1888d: 430 **A:** KZ
figuerasi Español, 1958c: 111 **N:** MO
filiformis Fabricius, 1792a: 114 (*Tenebrio*) **N:** AG MO
foveiventris Desbrochers des Loges, 1881: 77 **E:** UK (Krym)
freyi Koch, 1940e: 700 **E:** IT MA
frioli Solier, 1838a: 19: (*Tagenia*) **N:** AG TU
fulvipennis fulvipennis Pic, 1912o: 90 **N:** AG MO
safiana Antoine, 1948: 283
fulvipennis ramburi Antoine, 1948: 282 **E:** SP
fulvipes Reiche & Saulcy, 1857: 232 (*Tagenia*) **A:** IS JO LE SY
canaliculata Miller, 1861: 176
planicollis Reitter, 1916d: 147
gattefossei Antoine, 1936a: 15 **N:** MO
grandis grandis Solier, 1838a: 23 (*Tagenia*) **N:** TU
grandis ravasinii Gestro, 1880: 417 **N:** TU
hesperica Solier, 1838a: 29 (*Tagenia*) **N:** EG

hispanica abti Antoine, 1949c: 215 **N:** MO (Middle Atlas)
hispanica hispanica Solier, 1838a: 25 (*Tagenia*) **E:** PT SP **N:** EG MO
elongata Solier, 1838a: 24 (*Tagenia*)
rabatica Théry, 1921: 53
tangeriana Desbrochers des Loges, 1881: 72
tazekensis Antoine, 1949c: 215
insignita Antoine, 1949c: 226 [RN] **N:** MO
peyerimhoffi Antoine, 1948: 286 [HN]
intermedia dalmatina Reitter, 1916d: 145 **E:** AL CR GR
intermedia intermedia Solier, 1838a: 17 (*Tagenia*) **E:** FR IT SP
angustata Herbst, 1799: 160 [HN]
coarctata Gemminger, 1870a: 121 [RN]
fiorii Antoine, 1949c: 231
sicula Solier, 1838a: 18 (*Tagenia*)
intermedia muelleri Koch, 1940e: 710 **E:** BH CR
intermedia pelagosana J. Müller, 1912: 290 **E:** GR (Ionioi) **A:** TR
intricata Reitter, 1886a: 120 **E:** SP (Baleares, Catalán)
keosana Koch, 1948a: 322 **E:** GR (Kyklades)
kochi Antoine, 1948a: 288 **N:** MO
kuntzeni Koch, 1941c: 295 **A:** AF TD TM
laevicollis Solier, 1838a: 19 (*Tagenia*) **E:** SP **N:** TU
hipponensis Desbrochers des Loges, 1881: 75
impunctipennis Chevrolat, 1877a: 118
lafanecherei Antoine, 1951c: 94 **N:** MO (Rif)
lateralis bogosensis Reitter, 1886a: 105 **N:** EG **AFR**
lateralis lateralis Reitter, 1886a: 104 **N:** EG **AFR**
leprieuri Antoine, 1949c: 232 **N:** AG
lindbergi Schuster, 1933a: 16 **N:** MO
longipilis Reitter, 1916d: 151 **E:** PT SP
mamorensis Théry, 1924: 149 **N:** MO
maxima Desbrochers des Loges, 1881: 70 **N:** AG MO
alticola Antoine, 1949c: 212
mazetieri ascendens Antoine, 1948: 287 **N:** MO
mazetieri mazetieri Antoine, 1936a: 13 **N:** MO
mazetieri rehamnensis Antoine, 1948: 287 **N:** MO
mediatlantis kocheri Antoine, 1957: 42 **N:** MO
mediatlantis mediatlantis Antoine, 1936a: 15 **N:** MO (Middle Atlas)
mediatlantis robustior Antoine, 1957: 42 **N:** MO
melitana Reitter, 1894b: 82 **E:** IT MA
mendizabali Español, 1960a: 109 **E:** SP (Almería)
milosana Koch, 1948a: 323 **E:** GR (Milos)

mimeuri Antoine, 1936a: 16 **N:** MO
mogadorica Antoine, 1936a: 16 **E:** SP **N:** MO
normandi Koch, 1940e: 696 **N:** AG
  vaulogeri Koch, 1940e: 697
obliterata bathyglypta Antoine, 1949c: 218 **N:** MO
obliterata chobauti Antoine, 1949c: 219 **N:** AG
obliterata obliterata Solier, 1838a: 30 (*Tagenia*) **E:** SP **N:** MO TU
  ferruginea Desbrochers des Loges, 1881: 71
  oblongicollis Desbrochers des Loges, 1881: 73
  obsoleta Desbrochers des Loges, 1881: 71
occidentalis Théry, 1921: 53 **N:** MO
oertzeni Koch, 1948a: 323 **E:** GR (Nikaria)
olcesii jacentipilis Antoine, 1949c: 221 **N:** MO
  errans Antoine, 1949c: 221
olcesii olcesii Fairmaire, 1871a: 388 (*Tagenia*) **N:** MO
opaca Reitter, 1886a: 112 **N:** AG TU
opacoides Normand, 1939: 508 **N:** TU
orientalis jonica Koch, 1940e: 728 **E:** GR (Ionioi)
orientalis orientalis Brullé, 1832: 194 (*Tagenia*) **E:** AL GR TR **A:** TR
oteroi Español, 1981: 58 **E:** SP (Pontevedra)
otini Antoine, 1948: 283 **N:** MO
pardoi Español, 1966: 174 **N:** MO
parvicollis Desbrochers des Loges, 1881: 76 **N:** AG MO
peyerimhoffi Koch, 1940e: 714 **A:** JO
pilifera Solier, 1838a: 21 (*Tagenia*) **E:** GR (Morea)
pubescens Solier, 1838a: 22 (*Tagenia*) **N:** EG
pueli Antoine, 1949c: 219 **N:** AG
punctiventris Eschscholtz, 1831: 12 (*Tagenia*) **E:** ST (Taman' Peninsula) UK **A:** TR
  quadraticollis Desbrochers des Loges, 1881: 79
punica Pic, 1905b: 98 **N:** LB TU
  quedenfeldti Reitter, 1916d: 146
rhodica profundepunctata Koch, 1935d: 269 **A:** IS
rhodica rhodica Koch, 1935e: 311 **E:** GR (Rhodes) **A:** LE
rotroui Antoine, 1936a: 13 **N:** AG
  pauper Antoine, 1948: 286
rungsi Antoine, 1936a: 14 **N:** MO
saintjeani Normand, 1939: 510 **N:** TU
sardoa ardoini Canzoneri, 1970: 63 **E:** IT (Sicilia)
sardoa laeviventris Desbrochers des Loges, 1881: 78 **E:** UK (Krym) TR **A:** TR
  brevicollis Reiche, 1857: 232 [RN]
sardoa sardoa Küster, 1848b: 83 (*Tagenia*) **E:** FR IT SP **N:** AG TU
  duchoni Antoine, 1949c: 231

schembrii Canzoneri, 1979: 20 **E:** MA
schrammi Antoine, 1933b: 225 **N:** MO
silvestrii schusteri Koch, 1940e: 726 **A:** TR
silvestrii silvestrii Koch, 1935e: 312 **E:** GR (Rhodes)
striolata Escalera, 1914: 291 **N:** MO
sulcata Miller, 1861: 177 **A:** CY IN TR
sulcicollis Ménétriés, 1849: 227 (*Tagenia*) **A:** KZ TM UZ
syrensis naxica Koch, 1940e: 724 **E:** GR (Kyklades)
syrensis syrensis Koch, 1936: 293 **E:** GR (Syra)
teluetica Antoine, 1949c: 223 **N:** MO (High Atlas)
temporalis Reitter, 1916d: 142 **N:** AG
tenuicornis Baudi di Selve, 1874a: 104 [= 1875c: 69, 1875a: 295] **A:** IN
theryi macra Antoine, 1936a: 16 **N:** MO
theryi theryi Antoine, 1936a: 15 **N:** MO
torretassoi Koch, 1935a: 42 **N:** EG
villosa andalusica Español, 1958b: 50 **E:** SP
villosa villosa C. Brisout de Barneville, 1866: 382 **E:** SP
pilifera Reitter, 1886a: 115
webbi ramburi Antoine, 1936a: 13 **E:** SP
webbii webbii Guérin-Méneville, 1844: 113 (*Tagenia*) **N:** AG MO
pici Antoine, 1936a: 15
zaiana Antoine, 1936a: 15 **N:** MO
zarcoi Español, 1947b: 186 **E:** SP (Castellón)

genus *Tagenostola* Reitter, 1916d: 151 type species *Stenosis turkestanica* Reitter, 1886
albovillosa Koch, 1940e: 735 **A:** BT NP UP
pilosa Motschulsky, 1839: 65 (*Tagenia*) **E:** AB AR **A:** IN KZ TD TM UZ
seriepilosa muelleri Reitter, 1886a: 123 (*Stenosis*) **N:** EG **A:** SA **AFR**
turkestanica Reitter, 1886a: 123 (*Stenosis*) **E:** AB **A:** AF KI KZ TD TM UZ

genus *Tetranillus* Wasmann, 1899: 167 type species *Tetranillus costatus* Wasmann, 1899
costatus Wasmann, 1899: 167 **A:** IN **ORR**
kobmanni Koch, 1940e: 744 **A:** IN
longicarinatus Ren & Shi, 2006b: 181 **A:** XIZ
nyingchiensis Ba & Ren, 2013: 698 **A:** XIZ

genus *Tetranosis* G.S. Medvedev, 1994b: 858 type species *Tetranosis clypeolobus* Koch, 1940
subgenus *Hypermicrotelopsis* Koch, 1940e: 743 type species *Microtelopsis thibetana* Koch, 1940
thibetanus Koch, 1940e: 743 (*Microtelopsis*) **A:** HP XIZ
subgenus *Microtelopsis* Koch, 1940e: 742 type species *Tetranillus simplicifrons* Gridelli, 1934
simplicifrons Gridelli, 1934: 55 (*Tetranillus*) **A:** KA
topali Kaszab, 1975a: 10 **A:** KA

subgenus *Tetranosis* G.S. Medvedev, 1994b: 858 type species *Tetranosis clypeolobus* Koch, 1940
ciliaticostis Koch, 1940e: 742 **A:** SD
clypeolobus Koch, 1940e: 741 **A:** KA
franzi Kaszab, 1973a: 25 **A:** NP
wittmeri Kaszab, 1981b: 295 **A:** HP NP

tribe Tentyriini Eschscholtz, 1831

genus ***Abigopsis*** **Escalera, 1914: 277** type species *Scelosodis ustus* Fairmaire, 1879
atlantica Escalera, 1914: 279 **N:** MO (Sous)
usta Fairmaire, 1879c: 189 (*Scelosodis*) **N:** MO (Mogador)

genus ***Alcinoeta*** **E. Strand, 1929: 23** [RN] type species *Alcinoe helopioides* Ménétriés, 1849
Alcinoe Ménétriés, 1849: 230 [HN] type species *Alcinoe helopioides* Ménétriés, 1849
Allodengitha Bogatchev, 1963a: 57 type species *Allodengitha deserta* Bogatchev, 1963
deserta Bogatchev, 1963a: 58 (*Allodengitha*) **A:** KZ UZ
microderoides Skopin, 1964: 292 (*Alcinoe*)
helopioides helopioides Ménétriés, 1849: 231 (*Alcinoe*) **A:** AF KZ TM UZ
helopioides spectabilis Kraatz, 1882b: 82 (*Alcinoe*) **A:** TM UZ
iranica G.S. Medvedev, 1988b: 800 **A:** IN

genus ***Amblycarenum*** **Gebien, 1910a: 77** [RN] type species *Amblycara biskrensis* Fairmaire, 1893 (= *Melancrus alutaceum* Fairmaire, 1875)
Amblycara Fairmaire, 1893i: cxlvii [HN] type species *Amblycara biskrensis* Fairmaire, 1893 (= *Melancrus alutaceum* Fairmaire, 1875)
alutaceum Fairmaire, 1875b: 495 (*Melancrus*) **N:** AG TU
biskrensis Fairmaire, 1893i: cxlviii (*Amblycara*)

genus ***Ammogiton*** **Peyerimhoff, 1920a: 326** type species *Ammogiton viberti* Peyerimhoff, 1920
buettikeri buettikeri Kaszab, 1979c: 281 **A:** SA (Prov. Riyadh, Prov. Eastern)
buettikeri major Kaszab, 1981c: 348 **A:** SA (Prov. Riyadh)
buettikeri medinae Kaszab, 1981c: 349 **A:** SA (Prov. Medina)
omanicum Schawaller, 1990a: 44 **A:** OM
peltieri jolyi Español, 1953c: 296 **N:** MO (Baja Daoura)
peltieri peltieri Peyerimhoff, 1924: 75 **N:** AG (Grand Erg Occidental)
schawalleri L. Soldati, 2009: 69 **A:** QA
sonyae Kaszab, 1979c: 279 **A:** SA
viberti Peyerimhoff, 1920a: 326 **N:** TU

genus ***Anatolica*** **Eschscholtz, 1831: 7** type species *Tentyria subquadrata* Tauscher, 1812
subgenus *Anatolica* Eschscholtz, 1831: 7 type species *Tentyria subquadrata* Tauscher, 1812

abbreviata Gebler, 1832: 116 **E:** ST UK **A:** KZ WS
gnathosioides Motschulsky, 1870: 389
amoena Faldermann, 1835b: 397 (*Tentyria*) **A:** ES (Tuva) MG NMO
auctiformis Skopin, 1964: 376
iduna Reitter, 1900d: 122
typonota Ménétriés, 1836b: 181
amoenula Reitter, 1889k: 683 **A:** GAN MG NIN NMO
dschungarica Kaszab, 1967a: 21
angustata Steven, 1828: 85 (*Tentyria*) **E:** ST **A:** KZ
audouini Solier, 1835b: 385
conica Motschulsky, 1845a: 73
elongata Gebler, 1829a: 119 (*Tentyria*)
glaberrima Baudi di Selve, 1875c: 38
mallei Solier, 1835b: 386
arethusa Reitter, 1900d: 119 **A:** UZ
arnoldii Skopin, 1964: 398 **A:** KZ
atshitnura Kaszab, 1969b: 308 **A:** MG
aucta Faldermann, 1835b: 401 (*Tentyria*) **A:** ES (Tuva) MG NMO
boldi Kaszab, 1965e: 303 **A:** MG
cecchiniae Bogdanov-Katjkov, 1915: 4 **A:** MG
cellicola Faldermann, 1835b: 398 (*Tentyria*) **A:** ES (Tuva) HEB LIA MG NMO
chogsonzhavi Kaszab, 1967a: 18 **A:** MG
crassiuscula Bogatchev, 1961a: 117 **A:** MG
dashidorzsi dashidorzsi Kaszab, 1965e: 306 **A:** MG
dashidorzsi fortepunctata Kaszab & Knor, 1976: 295 **A:** ES (Tuva)
dashidorzsi subalpina Kaszab & Knor 1976: 296 **A:** ES (Tuva) WS (Altay)
dashidorzsi temporalis Kaszab, 1965e: 308 **A:** ES (Tuva) MG
deserticola deserticola Skopin, 1960a: 296 **A:** KZ
deserticola orientalis Skopin, 1964: 397 **A:** KZ
elegans Skopin, 1961b: 389 **A:** KZ (Almatynskaya oblast)
eremita Steven, 1828: 87 (*Tentyria*) **E:** ST UK
besseri Krynicki, 1829: 193 (*Tentyria*)
besseri Solier, 1835b: 390 [HN]
ventralis Motschulsky, 1845a: 73
externecostata Fairmaire, 1888b: 125 **A:** BEI HEB
fausti Reitter, 1900d: 118 **A:** KI
gibbosa gibbosa Steven, 1828: 87 (*Tentyria*) **E:** ST **A:** KZ UZ
balassogloi Dohrn, 1882: 245
depressa Besser, 1832: 8 (*Tentyria*)
lineata Steven, 1828: 87 (*Tentyria*)
saisanensis Motschulsky, 1870: 396

gibbosa punctipleuris Bogdanov-Katjkov, 1915: 3 **E:** ST
gibbosa thoracica Besser, 1832: 8 (*Tentyria*) **A:** KZ XIN
gobialtaica altaica Kaszab & G.S. Medvedev, 1972: 217 **A:** MG
gobialtaica gobialtaica Kaszab, 1965e: 314 **A:** MG
gobialtaica ovalis Kaszab & G.S. Medvedev, 1976: 109 **A:** MG
granulipleuris Reitter, 1900d: 124 **A:** MG NMO
gravidula J. Frivaldszky, 1890: 204 **A:** GAN
grebenscikovi Kaszab, 1964b: 6 **A:** MG
guentheri Kaszab, 1967a: 20 **A:** MG
hammarstromi Poppius, 1904: 12 **A:** ES (Tuva) MG NMO
hauseri Reitter, 1900d: 117 **A:** KI (Tian Shan)
hoboksarana Ren & Ba, 2010: 134 **A:** XIN
holdereri Reitter, 1900d: 123 **A:** XIN XIZ
humerangula Kaszab, 1965e: 318 **A:** MG
immarginata Reitter, 1889k: 681 **A:** GAN MG NMO XIN
hummeli Schuster, 1936a: 4
impressa Tauscher, 1812a: 30 (*Tentyria*) **E:** ST **A:** KI KZ XIN
integra Reitter, 1887d: 355 **A:** MG QIN XIZ
knori Kaszab, 1976d: 294 **A:** ES (Tuva)
koltzei Reitter, 1900d: 117 **A:** KZ (Kirghiz Range)
konurolena Skopin, 1964: 400 **A:** KZ (Almaatinskaya Oblast)
kozlovi Bogatchev, 1961a: 117 **A:** GAN
kozloviana Skopin, 1964: 379 **A:** XIN
kulzeri Kaszab, 1965c: 280 **A:** XIN
lacustris Kaszab, 1967a: 291 **A:** MG
lata Steven, 1828: 86 (*Tentyria*) **E:** ST **A:** KZ MG NMO WS XIN
albovittis Motschulsky, 1845a: 73
oblonga Allard, 1883: 19
tristis Solier, 1835b: 388
lepida Faldermann, 1835b: 394 (*Tentyria*) **A:** ES MG NMO
atramentaria Faldermann, 1835b: 395 (*Tentyria*)
damascena Kraatz, 1865a: 98
longithorax Reitter, 1900d: 118 **A:** KZ (Kirghiz Range)
lucidula Bogatchev, 1961a: 116 **A:** GAN MG
malkovskii Skopin, 1961b: 390 **A:** KZ (Muyunkum)
medusa Reitter, 1900d: 126 **A:** XIN
midas Reitter, 1900d: 116 **A:** KZ (Vostochnyy Kazakhstan)
mirabilis Kaszab, 1962c: 307 **A:** XIN
modesta Bogatchev, 1961a: 118 **A:** ES (Tuva) MG
uljasutaja Kaszab, 1965e: 311
muchei Kaszab, 1965e: 309 **A:** MG

mucronata Reitter, 1889k: 682 **A**: GAN MG NIN NMO QIN XIZ
mustacea Kolbe, 1908: 88 **A**: GAN QIN
ningxiana Ren & Ba, 2010: 173 **A**: NIN
nureti Schuster & Reymond, 1937: 237 **A**: GAN MG NIN NMO
chauveti Schuster & Reymond, 1937: 236
omnoensis Skopin, 1964: 374 **A**: MG NIN NMO
pandaroides Reitter, 1889k: 680 **A**: GAN NIN
paphia Reitter, 1900d: 117 **A**: TD (Alai)
paradoxa Reitter, 1900d: 120 **A**: ES (Tuva) MG NMO
planata J. Frivaldszky, 1890: 203 **A**: GAN
polita borealis Kaszab, 1964b: 3 **A**: GAN MG NIN NMO
polita polita J. Frivaldszky, 1890: 204 **A**: GAN NMO
potanini Reitter, 1889k: 683 **A**: GAN MG NIN NMO XIZ
psammophila Skopin, 1964: 377 **A**: MG
montagui Kaszab, 1967c: 308
pseudaucta Reitter, 1900d: 124 **A**: ES MG NMO
pseudiduna Kaszab, 1965c: 281 **A**: XIN
pusilla Kaszab, 1967c: 297 **A**: MG
quadricollis Ballion, 1878: 297 (*Colposcelis*) **A**: KZ XIN
lucina Reitter, 1900d: 116
relicta Kaszab, 1964b: 8 **A**: MG
rugata Ren & Ba, 2010: 172 **A**: GAN
salinicola Kaszab, 1967b: 295 **A**: MG
scythisoides Kaszab, 1968c: 359 **A**: MG
semenowi Reitter, 1889k: 680 **A**: GAN
splendida G.S. Medvedev, 1990b: 134 **A**: MG
sternalis gobiensis Kaszab, 1964g: 371 **A**: MG
sternalis sternalis Reitter, 1889k: 681 **A**: GAN NMO XIN
loczyi J. Frivaldszky, 1890: 205
strigosa redikorzevi Bogdanov-Katjkov, 1916b: xxx **A**: "Altai"
strigosa strigosa Germar, 1823: 138 (*Tentyria*) **A**: ES (Tuva) MG WS (Altay) XIN XIZ
suavis J. Frivaldszky, 1890: 205 **A**: GAN
subpolita Ren & Ba, 2010: 140 **A**: XIN
subquadrata Tauscher, 1812a: 31 (*Tentyria*) **E**: ST **A**: KZ
coriacea Motschulsky, 1845a: 73
subtrapezicollis Kaszab, 1965e: 309 **A**: MG
sulcipennis laevior Kaszab, 1965e: 316 **A**: MG
sulcipennis sulcipennis Reitter, 1900d: 125 **A**: MG NMO
suturalis Reitter, 1889k: 682 **A**: NMO
syrtensis Kaszab, 1962c: 306 **A**: KI
tibialis Reitter, 1900d: 127 **A**: ES (Tuva) MG XIN

tsharynensis Skopin, 1964: 399 **A:** KZ (Almaatinskaya Oblast)
tsendsureni Skopin, 1964: 372 **A:** MG XIN
colposcina Skopin, 1964: 374
undulata Gebler, 1832: 55 (*Tentyria*) **A:** ES MG NMO
implana Faldermann, 1835b: 400 (*Tentyria*)
propinqua Faldermann, 1835b: 396 (*Tentyria*)
torulosa Fischer von Waldheim, 1844: 65
subgenus *Eurepipleura* Bogdanov-Katjkov, 1915: 2 type species *Anatolica minima* Bogdanov-Katjkov, 1915
ebenina Fairmaire, 1887f: 323 **A:** BEI HEB
minima Bogdanov-Katjkov, 1915: 2 **A:** "China borealis"

genus *Calyptopsis* Solier, 1835b: 269 type species *Calyptopsis emondi* Solier, 1835 (= *Hegeter caraboides* Brullé, 1832)
Choristopsis Kraatz, 1865a: 228 type species *Choristopsis caucasica* Kraatz, 1865
amaroides Baudi di Selve, 1874a: 96 [= 1875c: 32] **A:** IN
armeniaca Baudi di Selve, 1874a: 97 [= 1875: 34] **A:** TR
bazmanica Schuster, 1938b: 82 **A:** IN
bogatchoevi Nabozhenko & Abdurakhmanov, 2009: 79 **E:** ST (Dagestan)
capnisiformis capnisiformis Reitter, 1903b: 44 **A:** IN TR
capnisiformis volcanica Chigray, Nabozhenko, Keskin & Abdurakhmanov, 2018: 332 **A:** TR
capnisoides Reitter, 1897i: 307 **E:** GR TR **A:** TR
caraboides Brullé, 1832: 201 (*Hegeter*) **E:** GR
emondi Solier, 1835b: 271
caucasica Kraatz, 1865a: 228 (*Choristopsis*) **E:** AB AR **A:** IN TR
morawitzi Faust, 1877a: 37
clypeata Faust, 1875: 179 **A:** AF IN TM
convexicollis Desbrochers des Loges, 1881: 65 **E:** AB GG
deplanata Faust, 1875: 177 **A:** IN TM TR
egecemi fouquei Chigray, Nabozhenko, Keskin & Abdurakhmanov, 2018: 343 **A:** TR
elbursica Chigray, 2019: 225 **A:** IN
emarginata Reitter, 1889a: 28 **E:** AB AR GG **A:** IN TR
escherichi Reitter, 1900d: 133 **A:** TR
fausti Chigray & Nabozhenko, 2018: 1130 **E:** TM
fouquei Chigray, Nabozhenko, Keskin & Abdurakhmanov, 2018: 340 **A:** TR
gigas Kaszab, 1962c: 309 **A:** IN
goliath Kaszab, 1959e: 242 [= 1959f: 385] **A:** AF
harpaloides Baudi di Selve, 1874a: 96 [= 1875c: 33] **E:** AR **A:** IN
incerta amaratica S. Chigray, 2019a: 236 **A:** IN
incerta incerta Faust, 1875: 179 **A:** IN TM
jeremias Reiche & Saulcy, 1857: 197 **A:** IS SY
kadleci S. Chigray, 2019a: 210 **A:** IN

kaszabi Chigray, Nabozhenko, Keskin & Abdurakhmanov, 2018: 336 **A:** TR
kermanica S. Chigray, 2019a: 235 **A:** IN
lata S. Chigray, 2019a: 218 **A:** IN
lezginica Nabozhenko & Abdurakhmanov, 2009: 82 **E:** ST
lineimargo Reitter, 1897i: 310 **A:** TR
merkli S. Chigray, 2019a: 214 **A:** IN
monstruosa S. Chigray, 2019a: 227 **A:** IN
nitescens Reitter, 1897i: 310 **E:** AR
ottoi S. Chigray, Nabozhenko, Keskin & Abdurakhmanov, 2018: 336 **A:** TR
pandaroides Reitter, 1897i: 308 **E:** BU GR
    equalis Reitter, 1897i: 308
    ovalis Reitter, 1897i: 308
pulchella avarica Nabozhenko & Abdurakhmanov, 2009: 82 **E:** ST (Dagestan)
pulchella pulchella Faldermann, 1837: 36 (*Gnathosia*) **E:** AB AR GG **A:** IN TR
    antoniae Reitter, 1896a: 29
    rugipennis Faldermann, 1837: 37 (*Gnathosia*)
punctiventris Baudi di Selve, 1874a: 100 [= 1875c: 34] (*Tentyria*) **A:** IN KZ TD TM (Kopet Dag), UZ
    laevipennis Reitter, 1897i: 310
rosti Reitter, 1897i: 306 **E:** AB **A:** IN TM
schach Kaszab, 1963a: 70 **A:** IN
sistanica S. Chigray, 2019a: 222 **A:** IN
solieri libanica Koch, 1936: 291 **A:** LE
solieri solieri Reiche & Saulcy, 1857: 199 **A:** IS SY TR
sulcimargo Reitter, 1897i: 306 **A:** IN LE
theodoridesi Kaszab, 1962c: 308 **A:** IN
zarudnyi S. Chigray, 2019a: 234 **A:** IN

genus *Cantopipleurus* Koch, 1943a: 578 type species *Tentyria mesostenoides* Baudi di Selve, 1881
mesostenoides Baudi di Selve, 1881: 276 (*Tentyria*) **A:** YE

genus *Capnisiceps* Chatanay, 1914c: 215 type species *Capnisiceps maindroni* Chatanay, 1914
brunneus Schawaller, 1993a: 103 **A:** OM
lilligi G. Wagner, 2005: 178 **A:** OM
maindroni Chatanay, 1914c: 215 **A:** AE OM
masirahensis G. Wagner, 2005: 179 **A:** OM (Masirah)

genus *Catomulus* Reitter, 1897h: 302 type species *Catomulus subaeneus* Reitter, 1897 (= *Oxycara olcesii* Fairmaire, 1883)
olcesii olcesii Fairmaire, 1883g: cxi (*Oxycara*) **N:** MO
    minimus Escalera, 1914: 280
    subaeneus Reitter, 1897h: 302
olcesii rugosicollis Español, 1944: 9 **N:** MO (Western Sahara)

genus *Cimipsa* Peyerimhoff, 1911: 346 type species *Cimipsa sergenti* Peyerimhoff, 1911
peyerimhoffi Théry, 1931b: 120 **N:** MO
hamadensis Kocher, 1951b: 129
reymondi Kocher, 1951b: 126 **N:** MO
sergenti Peyerimhoff, 1911: 346 **N:** AG
zarcoi rungsi Kocher, 1966b: 136 **N:** MO (Western Sahara)
zarcoi zarcoi Español, 1944: 8 **N:** MO (Western Sahara)

genus *Colposcelis* Dejean, 1834: 185 type species *Tentyria longicollis* Zubkov, 1833
subgenus *Colposcelis* Dejean, 1834: 185 type species *Tentyria longicollis* Zubkov, 1833
afghanica Gridelli, 1954: 192 [= 1955: 32] **A:** AF
clypealis Kaszab, 1968b: 110 **A:** AF
globicollis Kaszab, 1968b: 108 **A:** AF
humerangula Reitter, 1900d: 108 **A:** KZ
hybrida Reitter, 1900d: 109 **A:** KI KZ
lindbergi Kaszab, 1968b: 112 **A:** AF
longicollis Zubkov, 1833: 328 (*Tentyria*) **E:** AB **A:** KZ TM
microderoides microderoides Reitter, 1900d: 106 **A:** MG NMO XIN
mongolica Skopin, 1965: 375
microderoides strigipleuris Kaszab, 1967c: 286 **A:** MG
minor Kaszab, 1968b: 107 **A:** AF
minuta Reitter, 1900d: 108 **A:** TM
oschensis Kaszab, 1962c: 305 **A:** TD
regeli G.S. Medvedev, 1990b: 132 **A:** XIN
simillima Kaszab, 1968b: 111 **A:** AF
skopini Kaszab, 1962c: 304 **A:** KZ
subgenus *Colposceloides* Schuster, 1940a: 20 type species *Colposcelis licenti* Schuster, 1940
licenti Schuster, 1940a: 20 **A:** NMO
subgenus *Colposcythis* Reitter, 1889b: 113 type species *Colposcythis walteri* Reitter, 1889
walteri Reitter, 1889b: 114 (*Colposcythis*) **A:** TM
zarudnyi G.S. Medvedev, 1978c: 857 **A:** IN
subgenus *Scelocolpis* Reitter, 1900d: 105 type species *Colposcelis damone* Reitter, 1900
amanda Reitter, 1900d: 107 **A:** XIN
balchashense Skopin, 1964: 407 **A:** KZ
brunnea G.S. Medvedev, 1989b: 371 **A:** MG
bulganica Kaszab, 1967c: 288 **A:** MG
damone Reitter, 1900d: 107 **A:** MG XIN
elegans Kaszab, 1968c: 352 **A:** MG
foersteri Reitter, 1900d: 107 **A:** XIN
kamalovi G.S. Medvedev, 1989b: 373 **A:** MG
montivaga Bates, 1879: 470 (*Anatolica*) **A:** KA

pavlovskii Bogatchev, 1961a: 115 **A**: XIN (Turfan)
sulcatipleuris Skopin, 1961b: 385 **A**: KZ
tentyrioides Skopin, 1961b: 385 **A**: KZ
trisulcata Reitter, 1900d: 106 **A**: MO XIN
subgenus *Turcmenicola* Bogatchev, 1952a: 44 type species *Turcmenicola jachontovi* Bogatchev, 1952
jachontovi Bogatchev, 1952a: 45 (*Turcmenicola*) **A**: KZ TM UZ
kerzhneri G.S. Medvedev, 1970: 406 **A**: UZ
lopatini G.S. Medvedev, 1970: 406 **A**: TM
selevini Skopin, 1951: 124 **A**: KZ (Betpak-Dala)

genus *Colposphena* Semenov, 1889: 217 type species *Sphenaria brevicollis* Solsky, 1882
brevicollis Solsky, 1881b: 45 (*Sphenaria*) **A**: KZ (Kyzylkum) TM
breviceps Semenov, 1889a: 216 (*Sphenaria*)
obtusa Semenov, 1889a: 214 (*Sphenaria*)

genus *Cyphostethe* Marseul, 1867: xxxix type species *Himatismus ferrugineus* Marseul, 1867
subgenus *Apterocyphostethe* Kaszab, 1962c: 302 type species *Himatismus koltzei* Reitter, 1895
amseli Kaszab, 1961e: 236 **A**: AF
iranica Kaszab, 1962c: 302 **A**: IN
koltzei Reitter, 1895d: 155 (*Himatismus*) **A**: AF XIN
mongolica Kaszab, 1968c: 348 **A**: MG
subgenus *Cyphostethe* Marseul, 1867: xxxix type species *Himatismus ferrugineus* Marseul, 1867
Asphena Semenov, 1889a: 218 type species *Aphena komarowi* Semenov, 1889
antonowi Semenov, 1889b: 310 **A**: KZ TM UZ
belutschistanica Kaszab, 1957d: 64 **A**: IN PA SA
ferruginea Marseul, 1867: xxxviii (*Himatismus*) **N**: AG LB MO TU **A**: IS SA **AFR**
komarowi Semenov, 1889a: 218 (*Asphena*) **A**: TM
seidlitzi Reitter, 1916e: 142 **A**: KZ TD TM UZ
stehliki Kaszab, 1968b: 44 **A**: AF
wittmeri Kaszab, 1979: 270 **A**: SA
subgenus *Cyphostethoides* Löbl & Merkl, 2020 [see New Acts] type species *Cyphostethe brunnea* Kaszab, 1962
angulosicolle Chobaut, 1924c: 166 **N**: AG
brunnea Kaszab, 1962c: 303 **A**: IN
brunneoides Kaszab, 1979c: 274 **A**: PA
grombczewskii Semenov, 1890: 354 (*Himatismus*) **A**: XIN
heydeni Haag-Rutenberg, 1877a: 282 (*Himatismus*) **N**: AG EG LB MO TU **A**: IS SA SI YE **AFR**
saharensis Chobaut, 1897d: 205 (*Himatismus*)
jelineki Merkl, 1991b: 63 **A**: IN
komarowii Reitter, 1889b: 115 (*Sphenaria*) **A**: IN KZ TM
semenovi Bogatchev, 1947b: 80 **E**: AB AR **A**: IN

genus *Dailognatha* Steven, 1828: 88 type species *Tentyria aequalis* Tauscher, 1812

Delognatha Agassiz, 1846b: 118 [UE]

aequalis Tauscher, 1812a: 311 [40] (*Tentyria*) **E:** AB ST (Dagestan)

depressicollis Kraatz, 1865a: 89

levigata Brullé, 1832: 199 (*Tentyria*)

arnoldii Bogatchev, 1947d: 460 **A:** TM UZ

bogatshevi afghanica Kaszab, 1968b: 102 **A:** AF

bogatshevi bogatshevi G.S. Medvedev, 1964a: 63 **A:** TD TM

caraboides Eschscholtz, 1831: 9 (*Hegeter*) **E:** AB AR BU GG GR TR **A:** IN IQ SY TR

audouini Solier, 1835b: 266

caucasica Desbrochers des Loges, 1881: 67 (*Gnathosia*)

depressicornis Faldermann, 1837: 35 (*Gnathosia*)

crenata Reiche & Saulcy, 1857: 196 **A:** CY IN IS JO LE SI SY TR

cylindritarsis cylindritarsis Koch, 1948a: 304 **E:** GR (Crete)

cylindritarsis probsti Picka, 1984: 20 **E:** GR (Crete)

ferghanensis Kaszab, 1962c: 305 (*Colposcelis*) **A:** KI

gracilicornis Reitter, 1897k: 317 **A:** KI KZ XIN

gracilitarsis Kaszab, 1968b: 104 **A:** TM UZ

granimentum Reitter, 1897k: 317 **A:** KZ UZ

hellenica Reitter, 1897k: 316 **E:** GR

humeridens Kaszab, 1968b: 100 **A:** AF

impressicollis Solier, 1835b: 263 **A:** TR

marginicollis Kaszab, 1968b: 97 **A:** AF

minuscula Desbrochers des Loges, 1881: 68 (*Gnathosia*) **A:** TR

nasuta Ménétriés, 1849: 231 (*Anatolica*) **A:** AF KZ TD TM UZ

planata Miller, 1861: 173 (*Gnathosia*) **A:** JO SY TR

pumila Baudi di Selve, 1874b: 197 [= 1875c: 31] (*Gnathosia*) **E:** AR **A:** TR

humeralis Desbrochers des Loges, 1881: 69 (*Gnathosia*)

quadricollis anaphiana Koch, 1948a: 301 **E:** GR (Cyclades)

quadricollis carceli Solier, 1835b: 265 **E:** AR CR GR MC **A:** TR

quadricollis laticollis Besser, 1832: 9 (*Tentyria*) **E:** AL GR (Kerkyra) TR

chevrolati Allard, 1883: 17 (*Gnathosia*)

quadricollis montana Koch, 1948a: 302 **E:** GR (Crete)

quadricollis obtusangula Reitter, 1897k: 314 **E:** GR

quadricollis plicata Koch, 1948a: 303 **E:** GR (Crete)

quadricollis pseudobtusangla Koch, 1948a: 302 **E:** GR

quadricollis quadricollis Brullé, 1832: 193 (*Pimelia*) **E:** BU GR RO **A:** LE TR

crassicornis Desbrochers des Loges, 1881: 66 (*Gnathosia*)

sinuatocollis Desbrochers des Loges, 1881: 66 (*Gnathosia*)

variabilis Solier, 1835b: 262

quadricollis rhodica Koch, 1948a: 301 **E:** GR (Rhodes)

quadricollis rugata Solier, 1835b: 263 **E:** GR (Crete)
quadricollis samosana Koch, 1948a: 302 **E:** GR (Samos)
quadricollis sporadica Koch, 1948a: 301 **E:** GR (Sporades)
rugipleuris Reitter, 1897k: 316 **E:** GR (Kastellorhizo) **A:** TR
vicina Brullé, 1832: 200 (*Tentyria*) **E:** AL GR
opaca Allard, 1883: 16 (*Gnathosia*)

genus *Dengitha* Reitter, 1887d: 516 type species *Dengitha lutea* Reitter, 1887
crystallina Semenov, 1896b: 49 **A:** TM UZ
hyalina Semenov, 1896b: 52 **A:** TM
lutea Reitter, 1887e: 517 **A:** UZ
symmetrica G.S. Medvedev, 1964b: 160 **A:** KZ

genus *Dichomma* Solier, 1835b: 271 type species *Dichomma duponti* Solier, 1835 (= *Tentyria dardana* Steven, 1828)
dardanum Steven, 1828: 89 (*Tentyria*) **E:** GR **A:** SY TR
chevrolatii Reiche & Saulcy, 1857: 200
duponti Solier, 1835b: 274
foraminosum Küster, 1851: 64 (*Helops*)
glabrum Brullé, 1832: 201 (*Hegeter*)
maillei Solier, 1835b: 273

genus *Epitrichia* Gebler, 1859: 475 type species *Helops tomentosus* Gebler, 1843
fusca Ren & Zheng, 1993c: 51 **A:** NIN
kerzhneri Kaszab & G.S. Medvedev, 1972: 215 **A:** MG
intermedia Kaszab & G.S. Medvedev, 1976: 100
knori Kaszab, 1976b: 99 **A:** ES (Tuva)
mongolica Kaszab, 1965e: 299 **A:** ES (Tuva) MG
ningsiana Kaszab, 1965c: 279 **A:** NIN
semenovi semenovi Bogatchev, 1949b: 277 **A:** KZ MG NMO
semenovi tsendsureni Kaszab, 1968c: 343 **A:** MG
tomentosa Gebler, 1843: 39 (*Helops*) **A:** KZ
zaissanica Skopin, 1960a: 295 **A:** KZ

genus *Eschatostena* Kelejnikova, 1977: 654 type species *Eschatostena kuznetzovi* Kelejnikova, 1977
kuznetzovi Kelejnikova, 1977: 655 **A:** TM

genus *Eulipus* Wollaston, 1864: 448 type species *Tentyria elongata* Brullé, 1839
Mogadoria Escalera, 1905b: 467 type species *Tentyria subelegans* Fairmaire 1871
elongatus elongatus Brullé, 1839: 66 (*Tentyria*) **N:** CI
brullaei Wollaston, 1865: 393 (*Tentyria*) [RN]

elongatus punctidorsis Reitter, 1900d: 184 **N:** MO

foveifrons Escalera, 1935a: 302

quirogai Escalera, 1905b: 468

kruegeri Schuster, 1926b: 132 **N:** LB (Cyrenaica)

subelegans Fairmaire, 1871a: 386 (*Tentyria*) **N:** MO

vittembeschii Koch, 1937: 378 **N:** LB (Tripolitania)

genus *Eusyntelia* Waterhouse, 1881: 473 type species *Eusyntelia balfouri* Waterhouse, 1881

balfouri Waterhouse, 1881: 474 **A:** YE (Darsa, Suqutra)

ebenina Waterhouse, 1881: 474 **A:** YE (Suqutra)

exigua Gahan, 1900: 8 (*Rhytidonota*) **A:** YE (Suqutra)

glabra glabra Waterhouse, 1881: 475 **A:** YE (Suqutra)

glabra infraplicata Koch, 1970: 104 **A:** YE (Samha, Suqutra)

opacicollis Gahan, 1900: 9 **A:** YE (Suqutra)

socia Gahan, 1900: 9 (*Rhytidonota*) **A:** YE (Samha, Suqutra)

genus *Falsocatomulus* Pic, 1914e: 10 type species *Falsocatomulus euphraticus* Pic, 1914

Microhionthis Blair, 1923a: 120 type species *Microhionthis patriciae* Blair, 1923 (= *Falsocatomulus euphraticus* Pic, 1914)

euphraticus Pic, 1914e: 11 **A:** IN IQ QA

patriciae Blair, 1923a: 121 (*Microhionthis*)

genus *Freudeia* Kaszab, 1961d: 216 type species *Freudeia nepalica* Kaszab, 1961

granulipleuris Kaszab, 1977b: 245 **A:** NP

martensi Kaszab, 1977b: 245 **A:** NP

nepalica Kaszab, 1961d: 217 **A:** NP

punctipleuris Kaszab, 1977b: 243 **A:** NP

genus *Girardius* L. Soldati, 2009: 73 type species *Micipsa persica* Baudi di Selve, 1874

persica Baudi di Selve, 1874a: 102 [= 1875c: 58] (*Micipsa*) **A:** QA

pubescens Haag-Rutenberg, 1875b: 90

genus *Gnathosia* Fischer von Waldheim, 1821: 13 type species *Gnathosia glabra* Fischer von Waldheim, 1821

Capnisa Dejean, 1836: 197 type species *Bradyus karelini* Faldermann, 1836

afganica Bogatchev, 1947d: 461 **A:** AF

agaboides Kaszab, 1968b: 86 **A:** UZ

andersoni Bogdanov-Katjkov, 1915: 1 **A:** TD UZ (Bukhara)

aphodina Kaszab, 1959f: 370 **A:** TM

apicepilosa Gridelli, 1954: 186 [= 1955: 28] **A:** AF

aphodina Kaszab, 1959f: 370 **A:** AF

balachana Koch, 1949: 80 **A:** TM
barschewskyi Reitter, 1896c: 130 (*Capnisa*) **A:** UZ
interrupta Reitter, 1896c: 133 (*Capnisa*)
biconiger Kaszab, 1968b: 91 **A:** AF
blapoides Kaszab, 1959f: 365 **A:** AF
bogatschevi Kaszab, 1968b: 74 **A:** UZ
brincki Kaszab, 1973b: 36 **A:** AF
compressa Reitter, 1896c: 134 (*Capnisa*) **A:** TM (Kopet Dag)
cribratella Fairmaire, 1892e: cliii (*Capnisa*) **A:** TM
crypticola Reitter, 1896c: 132 (*Capnisa*) **A:** AF KZ TD TM UZ
declivis Reitter, 1896c: 131 (*Capnisa*) **A:** TD UZ
effugens Koch, 1949: 78 **A:** TD UZ (Fergana)
elliptica Ménétriés, 1849: 219 (*Capnisa*) **A:** UZ
elongata Reitter, 1900d: 98 **A:** KZ TD UZ
fausti Reitter, 1889a: 29 **A:** IN
glabra Fischer von Waldheim, 1821: 14 **A:** KZ TD UZ
gridellii Kaszab, 1959e: 241 [= 1959f: 368] **A:** AF
haarlovi Gridelli, 1954: 187 [= 1955: 30] **A:** AF
humerosa Reitter, 1900d: 100 **A:** UZ
hygrobiiformis Reitter, 1896c: 133 (*Capnisa*) **A:** TM UZ
jacobsoni Bogdanov-Katjkov, 1916a: 69 **A:** AF IN
kandaharica Kaszab, 1959f: 373 **A:** AF
karelini Faldermann, 1836: 375 (*Bradyus*) **A:** AF KZ TD TM UZ
depressiuscula Kraatz, 1882b: 81 (*Capnisa*)
kashmirensis Kaszab, 1961f: 343 **A:** PA (Karakoram)
klapperichi Kaszab, 1959f: 367 **A:** AF
kuhitangi G.S. Medvedev, 1964a: 62 **A:** TM (Kuhitang)
kulzeri Kaszab, 1959f: 363 **A:** AF
kushkensis G.S. Medvedev, 1964c: 660 **A:** TM (Kushka)
lindbergi Kaszab, 1968b: 83 **A:** AF
lopatini Kaszab, 1968b: 72 **A:** TD
lucida Kaszab, 1968b: 70 **A:** AF
minutissima G.S. Medvedev, 1979d: 78 **A:** TD
modesta Faldermann, 1837: 28 (*Bradytus*) **E:** AB AR GG **A:** IN TM
montana Bogatchev, 1947d: 462 **A:** UZ
nuristanica Kaszab, 1959f: 371 **A:** AF
orozgana Kaszab, 1968b: 93 **A:** AF
paghmatica Kaszab, 1968b: 81 **A:** AF
piffli Kaszab, 1961f: 342 **A:** PA (Karakoram)
praeoccupata Kaszab, 1968b: 78 [RN] **A:** AF
afghanica Gridelli, 1954: 183 [= 1955: 27] [HN]

proxima Reitter, 1915b: 61 **A:** IN
puncticeps Kaszab, 1968b: 79 **A:** AF
schrenkii Gebler, 1844: 100 (*Capnisa*) **A:** AF KZ TM UZ
caudata Gridelli, 1954: 189 [= 1955: 31]
semenovi Bogatchev, 1946c: 393 **A:** AF
simillima Kaszab, 1968b: 92 **A:** AF
skobelewi Starck, 1889: 53 (*Capnisa*) **A:** AF TM
somocoelioides Kaszab, 1968b: 77 **A:** AF
stehliki Kaszab, 1968b: 89 **A:** AF
sublaevigata Bogatchev & Kryzhanovskiy, 1960: 264 **A:** TM (Kopet Dag)
suturalis Reitter, 1896c: 130 (*Capnisa*) **A:** UZ
tadzhikistanica G.S. Medvedev & Kompantseva, 1989: 98 **A:** TD
trimarginata Kaszab, 1968b: 87 **A:** AF
turcomanica Kaszab, 1968b: 73 **A:** TM

genus *Hegeter* Latreille, 1802: 172 type species *Blaps elongata* G.-A. Olivier, 1795 (= *Blaps tristis* Fabricius, 1792)
subgenus *Hegeter* Latreille, 1802: 172 type species *Blaps elongata* G.-A. Olivier, 1795 (= *Blaps tristis* Fabricius, 1792)
abbreviatus Brullé, 1839: 66 **N:** CI (Gran Canaria)
amaroides Solier, 1835b: 378 **N:** CI
brevicollis Brullé, 1839: 65 **N:** CI (Tenerife)
brevicornis Har. Lindberg, 1953: 9 **N:** CI (Gomera)
costipennis Wollaston, 1864: 457 **N:** CI (Gran Canaria)
excisus Har. Lindberg, 1950: 9 **N:** CI (Tenerife)
glaber Brullé, 1839: 65 **N:** CI (La Palma)
gomerensis Wollaston, 1864: 456 **N:** CI (Gomera)
grancanariensis Har. Lindberg, 1950: 8 **N:** CI
impressus Brullé, 1839: 64 **N:** CI (Gran Canaria)
intercedens Har. Lindberg, 1950: 9 **N:** CI (Tenerife)
latebricola Wollaston, 1854: 510 **N:** SG
lateralis Brullé, 1839: 65 **N:** CI (Tenerife)
proximus Har. Lindberg, 1950: 9 **N:** CI (Tenerife)
subrotundatus Wollaston, 1864: 459 **N:** CI (Gran Canaria)
tenuipunctatus Brullé, 1839: 65 **N:** CI (Tenerife)
transversus Brullé, 1839: 65 **N:** CI (Tenerife)
tristis Fabricius, 1792a: 108 (*Blaps*) **E:** AZ **N:** CI MO MR SG **AFR**
elongatus G.-A. Olivier, 1795: 60: 9 (*Blaps*)
picicornis Fabricius, 1792a: 121 (*Helops*)
striatus Latreille, 1804a: 276
webbianus Heineken, 1835: 40 **N:** CI (Gran Canaria)

subgenus *Homalapipleurus* Español, 1957d: 166 type species *Hegeter gonzalezi* Español, 1957
deyrollei Wollaston, 1864: 462 (*Thalpophila*) **N:** CI
gonzalezi Español, 1957d: 169 **N:** CI (Fuerteventura)
subgenus *Pseudotalpophila* Reitter, 1900d: 95 type species *Thalpophila plicifrons* Wollaston, 1864
fernandezi Español, 1957a: 12 **N:** CI (Fuerteventura)
plicifrons Wollaston, 1864: 461 (*Thalpophila*) **N:** CI (Fuerteventura)

genus *Hegeterocara* Reitter, 1900d: 190 type species *Hegeterocara arabica* Reitter, 1900
Fourtaus Pic, 1921g: 47 type species *Fourtaus brevicornis* Pic, 1921 (= *Hegeterocara arabica* Reitter, 1900)
arabica Reitter, 1900d: 190 **A:** IS JO SI
brevicornis Pic, 1921g: 48 (*Fourtaus*)

genus *Herlesa* Reitter, 1897h: 298 type species *Herlesa globicollis* Reitter 1897 (= *Micipsa cavifrons* Fairmaire, 1863)
cavifrons Fairmaire, 1863: 642 (*Micipsa*) **N:** AG LB MO
gastonis Fairmaire, 1871a: 387 (*Micipsa*)
globicollis Reitter, 1897h: 299
grandis Peyerimhoff, 1933: 348 **N:** AG
semiopaca Peyerimhoff, 1916b: 83 (*Tentyronota*) **N:** MO

genus *Hionthis* Miller, 1861: 174 type species *Hionthis tentyroides* Miller, 1861
angusticollis Schuster, 1920b: 235 **N:** EG
heydeni Reitter, 1900d: 139 **A:** IS
mesopotamica Schuster, 1920b: 234 **A:** IQ
saulcyi Reitter, 1900d: 139 **A:** IS SY
tentyrioides Miller, 1861: 176 **N:** EG LB (Cyrenaica) **A:** SI SY "Palestina"

genus *Homoeonota* Fairmaire, 1882a: 63 type species *Homoeonota subopaca* Fairmaire, 1882
Isonota Fairmaire, 1887a: 171 type species *Isonota opaca* Fairmaire, 1887
dailognathoides Kaszab, 1981c: 350 **A:** OM
omanica Kaszab, 1981c: 351 **A:** OM

genus *Hyonthosoma* Reitter, 1900d: 142 type species *Hionthis occidentalis* Fairmaire, 1897
Hionthisoma Gebien, 1937: 609 [UE]
balboi Koch, 1937: 354 (*Hionthisoma*) **N:** LB (East Tripolitania)
bernardi Pierre, 1958: 136 (*Hionthisoma*) **N:** AG (Hoggar, Tassili n'Ajjer)
coutayari coutayari Chobaut, 1918: 74 (*Hionthisoma*) **N:** LB (West Tripolitania) TU
coutayari lybica Koch, 1937: 354 (*Hionthisoma*) **N:** LB (Tripolitania)
occidentalis grandis Español, 1953c: 288 **N:** AG MO
occidentalis occidentalis Fairmaire, 1897h: 180 (*Hionthis*) **N:** AG
sicardi Peyerimhoff, 1921b: 62 (*Hionthisoma*) **N:** MO

genus *Hyperops* Eschscholtz, 1831: 9 type species *Hyperops tagenoides* Eschscholtz, 1831
subgenus *Belutschistanops* Löbl, Bouchard, Merkl & Bousquet, 2020 [see New Acts] type species *Hyperops schusteri* Koch, 1940
doriae Baudi di Selve, 1874a: 102 [= 1875c: 61] **A:** IN
pygmaea afghanica Kaszab, 1959f: 387 (*Pachycera*) **A:** AF PA
pygmaea arabica Koch, 1940e: 686 **A:** AE BA OM QA SA
pygmaea pygmaea L. Redtenbacher, 1850: 48 **A:** IN PA
schusteri Koch, 1940e: 686 **A:** PA
subgenus *Hyperops* Eschscholtz, 1831: 9 type species *Hyperops tagenoides* Eschscholtz, 1831
Oedenocera Reiche, 1862a: 372 [RN] type species *Akis laevigata* sensu Eschscholtz, 1831 (= *Tenebrio buprestoides* Fabricius, 1781)
Pachycera Eschscholtz, 1831: 7 [HN] type species *Akis laevigata* sensu Eschscholtz, 1831 (= *Tenebrio buprestoides* Fabricius, 1781)
Tetromma Dejean, 1834: 183 type species *Upis unicolor* Herbst, 1797
masirahana Kaszab, 1981c: 328 (*Pachycera*) **A:** OM
nitidula Kraatz, 1865a: 237 **A:** PA **ORR**
pfeifferi pfeifferi Koch, 1940e: 689 **A:** PA
tagenoides aegyptiaca Kraatz, 1865a: 236 **N:** EG **AFR**

genus *Hypsosoma* Ménétriés, 1854: 30 type species *Hypsosoma mongolica* Ménétriés, 1854
juxtalacum Skopin, 1964: 383 (*Scythis*) **A:** MG
mongolicum Ménétriés, 1854: 31 **A:** BEI HEB LIA NMO SHA SHX
extrema Reitter, 1889k: 679 (*Anatolica*)
rotundicolle Fairmaire, 1888b: 125 **A:** BEI HEB SHX
sulciceps Gebler, 1841b: 592 (*Anatolica*) **A:** ES MG NMO
rectangulum Poppius, 1904: 13 (*Scythis*)

genus *Imatismus* Dejean, 1834: 202 type species *Helops fasciculatus* Fabricius, 1798
subgenus *Imatismus* Dejean, 1834: 202 type species *Helops fasciculatus* Fabricius, 1798
Curimosphena Gebien, 1920: 43 type species *Himatismus villosus* Haag-Rutenberg, 1870
arabicus Kaszab, 1981c: 324 (*Himatismus*) **A:** QA SA YE
chinensis Reitter, 1916e: 149 (*Himatismus*) **A:** YUN
fasciculatus Fabricius, 1798: 52 (*Helops*) **A:** PA SA **N:** MR **AFR ORR**
luctuosus Fairmaire, 1895a: 444 (*Himatismus*) **N:** MR **AFR**
ovipennis Koch, 1950b: 191 (*Himatismus*) **A:** YE **AFR**
sublinearis Walker, 1871: 15 (*Himatismus*) **A:** SA
villosus Haag-Rutenberg, 1870b: 90 (*Himatismus*) **E:** GR (Crete) IT (Linosa Is.) SP **N:** EG LB MO TU **A:** CY IN IQ IS LE OM PA SA SI SY YE **AFR ORR**
forticornis Baudi di Selve, 1874a: 103 (*Himatismus*)

genus *Kokeniella* Reitter, 1906c: 41 type species *Kokeniella mesostenoides* Reitter, 1906
afghanistana Kaszab, 1968b: 116 (*Mesostena*) **A:** AF
convergens Koch, 1940a: 87 (*Mesostena*) **A:** IN
freyi Koch, 1940a: 92 (*Mesostena*) **A:** IN (Bushire)
gracilicornis boeckelei Kaszab, 1974: 194 (*Mesostena*) **A:** AF
gracilicornis gracilicornis Reitter, 1915h: 128 **A:** IN
lineatopunctata Kraatz, 1865a: 173 (*Rhytinota*) **A:** IN PA
major interjecta Koch, 1940a: 91 (*Mesostena*) **A:** PA (Beluchistan)
major major Baudi di Selve, 1874a: 101 [= 1875c: 56] (*Mesostenopa*) **A:** IN PA
major persica Schuster, 1938b: 83 (*Mesostenopa*) **A:** IN
major pfaundleri Koch, 1940a: 89 (*Mesostena*) **A:** IN (Bijnabad)
matthiesseni Reitter, 1915h: 127 **A:** IN
mesostenoides afghanica Koch, 1940a: 85 (*Mesostena*) **A:** AF
mesostenoides mesostenoides Reitter, 1906c: 42 **A:** PA
perforalva luristana Koch, 1943a: 570 **A:** IN (Lorestan)

genus *Leptosphena* Semenov, 1890: 356 type species *Sphenaria tomentosa* Semenov, 1889 (= *Sphenaria rubripes* Reitter, 1889)
rubripes Reitter, 1889b: 115 (*Sphenaria*) **A:** TM (Karakumy)
tomentosa Semenov, 1889a: 214 (*Sphenaria*)

genus *Megagenius* Solier, 1835c: 513 type species *Megagenius frioli* Solier, 1835
frioli frioli Solier, 1835c: 514 **N:** AG MO TU
frioli troglophilus Kocher, 1960: 233 **N:** MO

genus *Melanochrus* Wollaston, 1864: 467 type species *Melanochrus lacordairii* Wollaston, 1864
Melasmocara Reitter, 1900d: 95 type species *Melanochrus lacordairii* Wollaston, 1864
blairi Uyttenboogaart, 1929: 347 **N:** CI
lacordairii Wollaston, 1864: 468 **N:** CI

genus *Melaxumia* Reitter, 1895f: 280 type species *Melaxumia acutangula* Reitter, 1895 (= *Tentyria angulosa* Gebler, 1832)
angulosa Gebler, 1832: 53 (*Tentyria*) **A:** ES MG NMO SHX
acutangula Reitter, 1895f: 280

genus *Mesostena* Eschscholtz, 1831: 9 type species *Mesostena punctata* Eschscholtz, 1831 (= *Pimelia angustata* Fabricius, 1775)
subgenus *Mesostena* Eschscholtz, 1831: 9 type species *Mesostena punctata* Eschscholtz, 1831 (= *Pimelia angustata* Fabricius, 1775)

angustata angustata Fabricius, 1775: 253 (*Pimelia*) **N**: EG LB TU **A**: IN IQ IS JO LE SA SI SY **AFR**

elegans Solier, 1835b: 399

laevicollis Solier, 1835b: 402

oblonga Solier, 1835b: 401

politipennis Fairmaire, 1877: 141

punctata Eschscholtz, 1831: 9 (*Mesostena*)

punctipennis Solier, 1835b: 403

angustata deserticola L. Soldati, 2009: 67 **A**: QA

angustata longicollis Lucas, 1858c: ccxxii **N**: AG LB

angustata praesahariana Koch, 1940a: 78 **N**: AG LB MO TU

klugii Solier, 1835b: 404 **N**: EG

parvula Reiche & Saulcy, 1857: 211 **A**: LE

puncticollis Solier, 1835b: 405 **E**: AB TR **A**: IN IQ IS JO OM QA SA SI SY TM YE **AFR**

armeniaca Motschulsky, 1870: 397

avenae J.R. Sahlberg, 1908: 76 (*Mesostenopa*)

bramina Motschulsky, 1870: 397

yemenensis Scupola, 1994: 25 **A**: YE

subgenus *Mesostenopa* Kraatz, 1865a: 179 type species *Mesostenopa picea* Kraatz, 1865

aegyptiaca Koch, 1940a: 44 **N**: EG

arabica arabica Gestro, 1881: 661 (*Mesostenopa*) **A**: YE

arabica titschaki Gebien, 1938: 55 (*Mesostenopa*) **A**: OM YE

caeciliae G. Wagner, 2003: 114 **A**: YE

dentrix Baudi di Selve, 1874a: 102 [= 1875c: 57] (*Micipsa*) **A**: IN

gridellii Koch, 1940a: 53 **A**: SA YE

habbanensis G. Wagner, 2003: 112 **A**: YE

habessinica Kraatz, 1865a: 181 (*Mesostenopa*) **N**: EG **AFR**

kuriamuriaca Schawaller, 1993a: 104 **A**: OM (Is. Kuria Muria)

nabathaea Peyerimhoff, 1907a: 31 (*Mesostenopa*) **A**: JO SY

perimensis Koch, 1943a: 563 (*Mesostena*) **A**: YE (Perim Is.)

peyerimhoffi Gridelli, 1938: 109 [RN] **A**: SI

gracilis Peyerimhoff, 1907a: 31 (*Mesostenopa*) [HN]

picea agilis Gestro, 1881: 661 (*Mesostenopa*) **N**: EG **A**: SI **AFR**

picea caporiaccoi Gridelli, 1938: 115 (*Mesostena*) **N**: LB

picea kulzeri Koch, 1940a: 45 **A**: SA YE

picea picea Kraatz, 1865a: 180 (*Mesostenopa*) **N**: EG

picea sinaitica Schatzmayr & Koch, 1934b: 17 (*Mesostenopa*) **A**: SI

rathjensi rathjensi Gebien, 1938: 54 (*Mesostenopa*) **A**: YE

sericea moltonii Schatzmayr, 1938: 173 **N**: LB

sericea sericea Fairmaire, 1884i: cccxlvi (*Mesostenopa*) **A**: YE **AFR**

socotraensis Koch, 1970: 95 (*Mesostenopa*) **A**: YE (Samha, Suqutra)

transnilotica Koch, 1937: 350 (*Mesostenopa*) **N**: LB

subgenus *Platystena* Koch, 1940a: 95 type species *Mesostena blairi* Koch, 1940

blairi Koch, 1940a: 96 **A:** OM
subgenus *Saxistena* Löbl & Merkl, 2020 [see New Acts] type species *Mesostena longicornis* Kraatz, 1865
cavatica Andres, 1926: 287 (*Mesostenopa*) **N:** EG
darsiana Koch, 1970: 100 **A:** YE (Darsa, Suqutra, Samba)
ibnsaudi Koch, 1940a: 73 **A:** SA
longicornis longicornis Kraatz, 1865a: 181 **A:** IS JO
longicornis rothi Koch, 1940c: 100 **A:** JO
mequignoni Koch, 1940a: 69 **A:** YE
tibialis Gahan, 1900: 9 (*Rhytidonota*) **A:** YE (Abd al-Kuri, Suqutra)

genus *Micipsa* Lucas, 1855b: xxxiv type species *Micipsa rufitarsis* Lucas, 1855 (= *Pimelia mulsanti* Levrat, 1853)
subgenus *Cirsa* Lucas, 1857: lvi type species *Cirsa striaticollis* Lucas, 1857
chobauti Peyerimhoff, 1926b: 194 **N:** AG MO
instriata Pic, 1922g: 29 **N:** MO
striaticollis Lucas, 1857: lvi (*Cirsa*) **N:** AG
subgenus *Micipsa* Lucas, 1855b: xxxiv type species *Micipsa rufitarsis* Lucas, 1855 (= *Pimelia mulsanti* Levrat, 1853)
agnesae Schuster, 1935b: 24 **A:** IN
angustipennis Fairmaire, 1875a: 520 **N:** TU
rufescens Pic, 1922g: 29
substrigosa Reitter, 1900d: 187
arabica Kaszab, 1981c: 343 **A:** SA (Eastern Prov.)
batesi Haag-Rutenberg, 1875b: 92 **A:** "Syria"
burtoni Baudi di Selve, 1881: 277 **A:** SY
douei distincticornis Peyerimhoff, 1933: 347 **N:** AG LB
garamantium Koch, 1937: 381
douei douei Lucas, 1856: xlvi **N:** AG EG **A:** JO LB SI **AFR**
grandis Kraatz, 1865a: 188
longiceps Pic, 1922g: 30
elliptipennis Schuster, 1925a: 25 **N:** LB
gracilipes Fairmaire, 1875a: 522 **N:** AG TU
kerimii Fairmaire, 1875a: 520 **N:** TU
monilifera Carl, 1991c: 141 **A:** IQ
mulsanti Levrat, 1853: 1 (*Pimelia*) **N:** AG TU
cursor Guérin-Méneville, 1859: clxxxviii (*Cirta*)
rufitarsis Lucas, 1855b: xxxiv
velox Guérin-Méneville, 1859: clxxxix (*Cirta*)
ovoidea Fairmaire, 1883d: 460 **N:** LB
philistina Reiche & Saulcy, 1857: 212 **A:** IQ IS JO SI "Syria"
poripennis Fairmaire, 1875a: 523 **N:** AG

princeps occidentalis Español, 1953c: 291 **N:** MO
princeps princeps Peyerimhoff, 1934: 106 **N:** AG LB MO **AFR**
rufitarsis Lucas, 1855b: xxxiv **N:** AG
schaumii Kraatz, 1865a: 111 **N:** EG **A:** SI
similis Haag-Rutenberg, 1875b: 91 **A:** IN
variabilis balbus Koch, 1937: 388 **N:** LB (Fezzan)
variabilis granulipennis Koch, 1937: 388 **N:** LB
variabilis variabilis Koch, 1937: 385 **N:** LB

genus *Microdera* Eschscholtz, 1831: 6 type species *Tentyria deserta* Tauscher, 1812
subgenus *Amicrodera* Kaszab, 1966b: 292 type species *Microdera lindbergi* Kaszab, 1966
lindbergi Kaszab, 1966b: 293 **A:** AF
subgenus *Dordanea* Reitter, 1887d: 357 type species *Dordanea elegans* Reitter, 1887
Adordanea Reitter, 1897f: 229 type species *Microdera subseriata* Reitter, 1889
duplicatipunctata Ren, 1999: 116 **A:** GAN
elegans Reitter, 1887d: 358 (*Dordanea*) **A:** GAN LIA NIN NMO QIN SHX XIN
globata Faldermann, 1835b: 402 (*Tentyria*) **A:** GAN NIN MG NMO SHX
interrupta Reitter, 1897f: 230 **A:** MG NMO
jurganovae Skopin, 1964: 385 **A:** ES (Tuva) MG
kanssuana Kaszab, 1957f: 292 **A:** GAN
kraatzi alashanica Skopin, 1964: 384 **A:** GAN NIN NMO
kraatzi kraatzi Reitter, 1889k: 685 (*Dordanea*) **A:** MG NIN NMO
lampabilis Ren, 1999: 113 **A:** GAN GUA NIN
luoshanica Ren, 1999: 109 **A:** NIN
luoshanana Ren, 1999: 391
promptipuncta Ren & Ba, 2010: 72 **A** GAN
ordossica Schuster, 1940a: 21 **A:** NMO
rotundithorax Ren, 1999: 118 **A:** NIN
shenmuana Ren, 1999: 111 **A:** SHA
shenmana Ren, 1999: 391
subseriata Reitter, 1889k: 686 **A:** ES (Tuva) GAN NMO
subfoveolata Skopin, 1964: 387
subgenus *Falsomicrodera* Kaszab, 1966b: 294 type species *Microdera turkerstanica* Schuster, 1915
turkestanica Schuster, 1915b: 86 (*Microdera*) **A:** XIN
subgenus *Iliodera* Skopin, 1961b: 396 type species *Microdera desertoides* Skopin, 1961
desertoides Skopin, 1961b: 396 **A:** KZ (Alamatinskaya oblast)
subgenus *Microdera* Eschscholtz, 1831: 6 type species *Tentyria deserta* Tauscher, 1812
Rhostax Fischer von Waldheim, 1844: 67 type species *Rhostax menetriesii* Fischer von Waldheim, 1844 (= *Microdera gracilis* Eschscholtz, 1831)
aciculata aciculata Reitter, 1897f: 232 **A:** KZ XIN
aciculata penkinae Kaszab, 1966b: 298 **A:** KZ

afghanica Gridelli, 1954: 198 [= 1955: 34] **A**: AF
aurita Reitter, 1889k: 686 (*Dordanea*) **A**: GAN NMO XIN
badghysi G.S. Medvedev & Nepesova, 1985b: 358 **A**: TM
balchaschensis balchaschensis Skopin, 1960a: 297 **A**: KZ
balchaschensis wernoyensis Kaszab, 1966b: 298 **A**: KZ
campestris campestris Steven, 1828: 93 **A**: KZ TM
margelanica Kraatz, 1882b: 81
campestris karanogaica Abdurakhmanov & Nabozhenko, 2011: 183 **E**: ST
chan Kaszab, 1963a: 71 **A**: IN
charpentieri Kaszab, 1973b: 39 **A**: AF
convexa convexa Tauscher, 1812a: 39 (*Tentyria*) **E**: ST **A**: KZ TM
convexa kaszabi Skopin, 1964: 409 **A**: KZ (Almaatinska Oblast)
deserta deserta Tauscher, 1812a: 38 (*Tentyria*) **E**: ST **A**: KZ
deplanata Gebler, 1841a: 373 (*Tentyria*)
deserta skopini Kaszab, 1966b: 296 **A**: KZ
dzhungarica G.S. Medvedev, 1964b: 161 **A**: KZ
extabesa Skopin, 1961b: 393 **A**: KZ TM UZ
fausti Kraatz, 1888b: 294 **A**: TM
ferganensis Kaszab, 1957f: 290 **A**: KZ UZ
fortepunctata Reitter, 1915f: 94 **A**: KZ
gigas G.S. Medvedev, 1964b: 162 **A**: KZ XIN
globulicollis Ménétriés, 1849: 230 **A**: TM
globithorax Steven, 1828: 93 (*Tentyria*) **A**: KZ "Siberia"
gorochovi G.S. Medvedev, 2004b: 555 **A**: TM
gracilis Eschscholtz, 1831: 6 **A**: KI KZ TM
macrops Reitter, 1915f: 93
menetriesii Fischer von Waldheim, 1844: 68 (*Rostrax*)
grandicollis Bogatchev, 1946a: 211 **A**: AF KZ TD
badakshanica Kaszab, 1959f: 390
grandipunctata Ren, 1999: 122 **A**: XIN
habaensis Ren, 1999: 123 **A**: XIN
heydeni Kraatz, 1882b: 82 **A**: KZ TD TM UZ
hybrida Skopin, 1966c: 69 **A**: KZ
iliensis Skopin, 1961b: 394 **A**: KZ
iranica Kaszab, 1966b: 299 **A**: IN
karelini Fischer von Waldheim, 1844: 67 (*Rhostrax*) **A**: "Songaria"
kaszabi Skopin, 1961a: 179 **A**: KZ (Amatinskaya oblast)
kazenasi Skopin, 1960a: 297 **A**: KZ
keramana Ren, 1999: 124 **A**: XIN
karamiensis Ren, 1999: 390
kermanica Kaszab, 1966b: 294 **A**: IN

laevigatula Reitter, 1915f: 95 **A:** AF IN UZ
laticollis Bates, 1879: 470 **A:** KA KZ XIN
 przewalskyi Reitter, 1887d: 359 (*Tentyria*)
minax Reitter, 1897f: 233 **A:** AF KZ TM UZ
mongolica kozlovi Kaszab, 1966b: 295 **A:** MG NMO
mongolica mongolica Reitter, 1889k: 686 (*Dordanea*) **A:** GAN NMO QIN XIN
obesitas Ren, 1999: 127 **A:** XIN
 obsitas Ren, 1999: 390
ocularis Reitter, 1915f: 95 **E:** AB AR GG
parvicollis Bates, 1879: 470 **A:** KA KZ XIN
 basalis Reitter, 1887d: 360 (*Tentyria*)
 semenowi Reitter, 1897f: 231
pleuralis Reitter, 1887d: 360 (*Tentyria*) **A:** KZ NMO XIN
pulla Reitter, 1897f: 234 **A:** IN
punctipennis Kaszab, 1967c: 311 **A:** KZ MG XIN
reitteri Kaszab, 1966b: 280 [RN] **E:** AB
 parvicollis Reitter, 1897f: 235 [HN]
schusteriana Kaszab, 1957f: 292 **A:** KZ
scyphiforma Ren & Ba, 2010: 84 **A** XIN
scyta Baudi di Selve, 1875c: 55 **A:** TM
semenoviana Bogatchev, 1947d: 463 [RN] **A:** AF TD UZ
 semenovi Bogatchev, 1946a: 210 [HN]
shandanana Ren & Ba, 2010: 79 **A** GAN
shasenema G.S. Medvedev & Nepesova, 1985b: 360 **A:** TM
strigipleuris Reitter, 1897f: 232 **A:** IN
strigiventris Reitter, 1900d: 157 **A:** MG XIN
 halophila Skopin, 1961b: 392
subglobosa Kraatz, 1865a: 187 **A:** KZ
subsulcata Kraatz, 1865a: 186 **A:** KZ
thoracica Kaszab, 1966b: 300 **A:** IN
transversicollis Reitter, 1887e: 517 **E:** AR **A:** AF IN TM UZ
tscharynensis Kaszab, 1966b: 297 **A:** KZ
vachshiana Skopin, 1964: 411 **A:** TD
xinjiangana Ren & Ba, 2010: 96 **A** XIN
subgenus *Tentyrodera* Koch, 1943a: 572 type species *Microdera marginata* Baudi di Selve, 1874
marginata brandti Koch, 1943a: 573 **A:** IN
marginata deserticola Blair, 1933: 5 (*Rhytinota*) **A:** AE OM QA SA YE
 perezi Koch, 1943a: 574
marginata evansi Blair, 1923a: 121 (*Rhytinota*) **A:** IQ SA
marginata marginata Baudi di Selve, 1874a: 97 [= 1874a: 278, 1875c: 53] **A:** IN

genus *Microdera*, species incertae sedis
convexicollis Motschulsky, 1870: 379 (*Rhostax*) **A:** KI
excavata Motschulsky, 1870: 382 **A:** KI

genus *Neognathosia* Kaszab, 1959f: 383 type species *Gnathosia pseudanemia* Reitter, 1915
pseudanemia Reitter, 1915b: 62 (*Gnathosia*) **A:** UZ

genus *Orostegastopsis* Koch, 1962: 255 type species *Orostegastopsis scorteccii* Koch, 1962
planioculata Lillig, 2014: 346 **A:** OM

genus *Oterophloeus* Desbrochers des Loges, 1881: 52 type species *Oterophloeus picipes* Desbrochers des Loges, 1881 (= *Pachychila humerosus* Fairmaire, 1875)
Tynteria Reitter, 1897h: 301 type species *Pachychila humerosa* Fairmaire, 1875
alveatus alveatus Peyerimhoff, 1931b: 95 **N:** AG LB MO **AFR**
alveatus peyerimhoffi Koch, 1935a: 23 **N:** EG **A:** SI
deflexangulus deflexangulus Reitter, 1900d: 143 **N:** LB (Tripolitania) TU
deflexangulus syrticus Koch, 1937: 359 **N:** LB (Cyrenaica, Tripolitania)
densepunctatus Escalera, 1935a: 298 (*Pachychila*) **N:** MO **AFR**
desioi desioi Gridelli, 1930: 210 **N:** LB
desioi divinus Koch, 1935a: 25 **N:** EG (Siwa)
haagii haagii Kraatz, 1865a: 155 (*Tentyria*) **N:** EG
tenuimarginatus Böhm, 1916a: 50 (*Tentyria*)
haagii orientalis Koch, 1935a: 25 **N:** EG **A:** SI
humerosus Fairmaire, 1875a: 518 (*Pachychila*) **N:** AG LB TU **AFR**
picipes Desbrochers des Loges, 1881: 53
scuticollis Fairmaire, 1879c: 188 (*Tentyria*) **N:** AG MO
torretassoi Koch, 1937: 360 **N:** LB (Fezzan)

genus *Oxycara* Solier, 1835b: 254 type species *Oxycara blapsoides* Solier, 1835
subgenus *Oxycara* Solier, 1835b: 254 type species *Oxycara blapsoides* Solier, 1835
Crypticoides Fairmaire, 1898d: 389 type species *Crypticoides mellyi* Fairmaire 1898
Emmenastus Motschulsky, 1845a: 75 type species *Emmenastus rugosus* Motschulsky, 1845
Melancrus Reiche & Saulcy, 1857: 190 type species *Melancrus laevigatum* Reiche & Saulcy, 1857
aethiopium aethiopium Gredler, 1878: 515 **N:** AG (Tassili n'Ajjer) LB (Fezzan, Tibesti) **AFR**
ardoini Kaszab, 1979c: 285 **A:** IS (Negev) JO SI
balachowskyi Pierre, 1958: 138 **N:** AG (Tassili n'Ajjer)
becharense Chobaut, 1923c: 165 **N:** AG
buettikeri Kaszab, 1979c: 285 **A:** AE BA OM QA SA
chudeaui Peyerimhoff, 1942c: 276 **N:** MO (Western Sahara) **AFR**
cyrenaicum Gridelli, 1933: 222 **N:** LB

deronectinum Reitter, 1900d: 190 **N:** EG
dhofarense Lillig, 2009b: 227 **A:** OM
gastonis Reitter, 1903b: 45 **N:** AG LB MO **AFR**
chobauti Gebien, 1910a: 76 [RN]
trapezicolle Chobaut, 1898a: 82 [HN]
laevigatum Reiche & Saulcy, 1857: 192 (*Melancrus*) **A:** IS JO SY
lavocati Español, 1953c: 294 **N:** AG MO
liebegotti Lillig, 2009b: 228 **A:** OM
llorensi Escalera, 1935a: 306 **N:** AG MO **AFR**
mandli Kaszab, 1963a: 72 **A:** IN
mellyi Fairmaire, 1898d: 390 (*Crypticoides*) **A:** AF PA XIZ **ORR**
tharense Kulzer, 1956: 645
pierrei Girard, 1966b: 204 **A:** IQ SY
murati Peyerimhoff, 1942b: 217 **N:** MO (Western Sahara)
productum Peyerimhoff, 1907a: 33 **A:** SI
pygmaeum Reiche & Saulcy, 1857: 194 (*Melancrus*) **N:** EG LB **A:** SI
rihanei Lillig, 2009b: 229 **A:** OM
sanaaense forte Kaszab, 1981c: 346 **A:** SA
sanaaense sanaaense Gridelli, 1953: 30 **A:** SA YE
saudarabicum Kaszab, 1979c: 286 **A:** SA
subgenus *Symphoxycara* Koch, 1943a: 577 type species *Oxycara breviuscula* Fairmaire, 1892
evae Lillig, 2001: 379 **A:** OM
ferreri Lillig, 2001: 372 **N:** EG (Red See: Is. Zabargad)
gallagheri Lillig, 2001: 375 **A:** OM
grimmi Lillig, 2001: 378 **A:** OM
girardi Lillig, 2001: 382 **A:** IN
grande Kaszab, 1981c: 345 **A:** OM YE
hansbremeri Lillig, 2001: 380 **A:** AE BA IN OM QA YE
hegetericum Reiche & Saulcy, 1857: 193 (*Melancrus*) **N:** EG **A:** "Dead Sea" **AFR**
laticolle Lillig, 2001: 383 **A:** IN
malgorzatae Purchart, 2014b: 236 **A:** YE (Suqutra)
minutum Lillig, 2001: 382 **A:** IN
occidentale Lillig, 2001: 369 **N:** EG
peyerimhoffi Lillig, 2001: 369 **A:** IS JO SI
pumilum Reitter, 1900d: 189 **A:** EG (Red See: Is. Homan)
schawalleri Lillig, 2001: 377 **A:** OM
subcostatum Guérin-Méneville, 1862: 375 (*Melancrus*) **A:** YE
ulfbremeri Lillig, 2001: 373 **A:** SA

genus *Oxycarops* Reitter, 1900d: 94 type species *Hegeter fuscipes* Brullé, 1839
fuscipes Brullé, 1839: 66 (*Hegeter*) **N:** CI (Fuerteventura, Lanzarote)
submetallicus Wollaston, 1864: 464 (*Thalpophila*)

genus *Pachychila* Eschscholtz, 1831: 5 type species *Pimelia punctata* Fabricius, 1798
subgenus *Anebacis* Peyerimhoff, 1927a: 4 type species *Tentyria cribricollis* Fairmaire, 1875
cribricollis Fairmaire, 1875a: 519 (*Tentyria*) **N:** AG TU
elegans Escalera, 1935a: 296 **N:** MO (Tiznit)
mairei Peyerimhoff, 1920a: 327 (*Tentyria*) **N:** MO (region of Figuig)
roulleaui Antoine, 1952: 112 **N:** MO
subgenus *Hegeteromorpha* Escalera, 1913: 38 type species *Pachychile externecostata* Haag-Rutenberg, 1875
externecostata Haag-Rutenberg, 1875a: 48 (*Pachychile*) **N:** MO
obtusecostata Reitter, 1900d: 152 (*Pachychile*) **N:** MO
trisulcata Antoine, 1942b: 65 **N:** MO
subgenus *Neocisba* Reitter, 1900d: 154 type species *Pachychile dissecta* Kraatz, 1865
Neacisba Peyerimhoff, 1927a: 53 [UE]
dissecta Kraatz, 1865a: 204 (*Pachychile*) **N:** AG
latevagans Antoine, 1952: 110 **N:** MO
mediatlantis Antoine, 1939: 148 **N:** MO (Middle Atlas)
medioxima Peyerimhoff, 1921b: 64 (*Tentyria*) **N:** AG
muluyensis Antoine, 1942b: 60 **N:** MO (Moyenne Moulouya)
prosternalis ayachiensis Antoine, 1942b: 59 **N:** MO (High Atlas)
prosternalis prosternalis Bedel, 1903: 369 **N:** MO (Middle Atlas)
rotroui perissa Antoine, 1942b: 59 **N:** MO
rotroui rotroui Antoine, 1932: 181 **N:** MO
sicardi interrupta Antoine, 1957: 40 **N:** MO (Moyenne Moulouya)
sicardi reymondi Antoine, 1957: 40 **N:** MO (High Atlas)
sicardi sicardi Peyerimhoff, 1927b: 218 **N:** MO
subgenus *Pachychila* Eschscholtz, 1831: 5 type species *Pimelia punctata* Fabricius, 1798
Acisba Dejean, 1834: 185 type species *Tentyria cribrosa* Besser, 1832 (= *Pimelia punctata* Fabricius, 1798)
Lophoma Solier, 1835b: 285 type species *Pimelia punctata* Fabricius, 1798
Pachychile Lacordaire, 1859: 46 [UE]
acuminata Erichson, 1841: 177 **N:** AG
angulicollis angulicollis Fairmaire, 1875c: 543 **N:** MO
fritschi Haag-Rutenberg, 1875a: 45 (*Pachychile*)
angulicollis haagi Antoine, 1932: 179 **N:** MO (Sous)
angulicollis macra Antoine, 1932: 178 **N:** MO (High Atlas)
angulicollis ventriosa Antoine, 1932: 177 **N:** MO
antiatlantis Antoine, 1952: 111 **N:** MO (Anti Atlas)
aulonoscelis Reitter, 1900d: 148 (*Pachychile*) **E:** SP
bergevini bergevini Peyerimhoff, 1925: 12 **N:** MO (Atlantic)
bergevini theryi Peyerimhoff, 1925: 12 **N:** MO (Sebou)
bifida Rosenhauer, 1865: 184 **E:** SP (Andalucía)
bremondi Koch, 1941c: 286 **N:** MO

brevis Besser, 1832: 11 (*Tentyria*) **N**: AG MO

cognata Haag-Rutenberg, 1875a: 46 (*Pachychile*)

pedinoides Solier, 1835b: 294

breviuscula breviuscula Haag-Rutenberg, 1875a: 47 (*Pachychile*) **N**: MO

breviuscula discrepans Antoine, 1942b: 61 **N**: MO

calida calida Escalera, 1925d: 375 **N**: MO

calida subangulicollis Antoine, 1942b: 61 **N**: MO (Middle Atlas)

catalanica J. Ferrer, 2018: 39 **E**: SP

chobauti Antoine, 1926: 253 **N**: MO

compacta compacta Antoine, 1942b: 41 **N**: MO (High Atlas)

compacta subaspera Antoine, 1942b: 42 **N**: MO (Middle Atlas)

cornuta Peyerimhoff, 1925: 10 **N**: MO

crassicollis cossyrensis Ragusa, 1875: 252 (*Pachychile*) **E**: IT (Pantelleria Is.) **N**: TU

parallela Schuster, 1919b: 42 (*Pachychilina*)

crassicollis crassicollis Kraatz, 1865a: 210 **N**: AG

cursitans Escalera, 1925d: 374 **N**: MO

denticollis Grimm, 1987: 22 **E**: SP (Almería)

desbrochersi Reitter, 1900d: 150 (*Pachychile*) **N**: AG

forticornis Escalera, 1923d: 129 **N**: MO (Rif)

foveipennis Kraatz, 1865a: 201 (*Pachychile*) **N**: MO

soligena Antoine, 1932: 182

freyi Koch, 1944b: 223 **E**: SP (Almería)

frioli Solier, 1835b: 301 **N**: AG LB TU

germari Solier, 1835b: 302 **N**: AG MO TU

glabella Herbst, 1799: 61 (*Pimelia*) **E**: SP

grandis debduensis Antoine, 1942b: 61 **N**: MO

grandis grandis Desbrochers des Loges, 1881: 54 **N**: AG

rugatula Desbrochers des Loges, 1881: 55

grandis gravida Antoine, 1933b: 223 **N**: MO

kocheri Antoine, 1934: 106

grandis montana Peyerimhoff, 1925: 13 **N**: MO (Middle Atlas)

grandis strenua Antoine, 1932: 180 **N**: MO

haroldi Kraatz, 1865a: 211 (*Pachychile*) **N**: AG MO TU

hispanica Solier, 1835b: 292 **E**: SP

impressifrons Solier, 1835b: 299 **N**: AG

breviuscula Desbrochers des Loges, 1881: 61

conformis Desbrochers des Loges, 1881: 62

continua Desbrochers des Loges, 1881: 58

curtula Gebien, 1910a: 58 [RN]

emarginata Desbrochers des Loges, 1881: 59

rugatipennis Desbrochers des Loges, 1881: 59

subcylindrica Solier, 1835b: 300

impunctata Fairmaire, 1860: 632 (*Lophoma*) **N**: MO

ifniensis Escalera, 1935a: 300

intermedia Haag-Rutenberg, 1875: 45 (*Pachychile*)

incrassata Rosenhauer, 1856: 183 **E**: SP

insolita J. Ferrer, 2018: 42 **E**: SP

insculpta Escalera, 1914: 288 **N**: MO

italica J. Ferrer, 2018: 42 **E**: IT

kunzei Solier, 1835b: 298 **N**: AG

discedens Desbrochers des Loges, 1881: 56

lesnei Peyerimhoff, 1925: 13 **N**: AG MO

maroccana Haag-Rutenberg, 1875a: 50 (*Pachychile*) **N**: MO

mskalica Escalera, 1914: 281 **N**: MO (High Atlas)

mscalica Peyerimhoff, 1927a: 42 [UE]

opaca Breit, 1908: 63 **E**: SP (Baleares)

otini otini Antoine, 1939: 146 **N**: MO (Middle Atlas)

otini ranarum Antoine, 1939: 148 **N**: MO (Middle Atlas)

parallelipennis Escalera, 1914: 284 **N**: MO

plasoni beidensis Antoine, 1926: 255 **N**: MO

plasoni plasoni Haag-Rutenberg, 1875a: 50 (*Pachychile*) **N**: MO

prujai Antoine, 1962: 51 **N**: MO

punctata mamorensis Antoine, 1942b: 60 **N**: MO

punctata punctata Fabricius, 1798: 45 (*Pimelia*) **E**: SP **N**: AG MO

cribrosa Besser, 1832: 11 (*Tentyria*)

punctulata Lucas, 1846: pl. 28 **E**: SP **N**: AG

quadricollis Reiche, 1860: 729 **E**: IT (Sicilia)

reflexímargo Desbrochers des Loges, 1881: 59 **N**: AG MO

reini asniensis Antoine, 1942b: 63 **N**: MO (High Atlas)

reini reini Haag-Rutenberg, 1875a: 49 (*Pachychile*) **N**: MO

fairmairei Haag-Rutenberg, 1875a: 49 (*Pachychile*)

sabulosa Lucas, 1846: pl. 28 **N**: AG

sardea Kraatz, 1865a: 209 **E**: IT (Sardinia)

salzmanni jordanai Escalera, 1925d: 373 **N**: MO (Rif)

salzmanni salzmanni Solier, 1835b: 297 **E**: SP **N**: MO

salzmanni simoni Escalera, 1914: 285 **N**: MO

sarroensis Antoine, 1947: 148 **N**: MO (Djebel Sarro)

schrammi Antoine, 1942b: 63 **N**: MO

secsaui imperfecta Peyerimhoff, 1927a: 23 **N**: MO (High Atlas)

secsaui insidiosa Antoine, 1942b: 63 **N**: MO (Azilal)

secsaui secsaui Escalera, 1913: 39 **N**: MO (Western High Atlas)

servillei Solier, 1835b: 309 (*Microdera*) **E**: FR (Corse) IT (Sardinia)

pygmaea Solier, 1835b: 310 (*Microdera*)

simytae Antoine, 1934: 104 **N**: MO

sublunata Solier, 1835b: 308 (*Microdera*) **E:** SP (Baleares)

sulcifrons Solier, 1835b: 295 **N:** MO

tetuanica Escalera, 1914: 287

susiana Escalera, 1914: 289 **N:** MO

susana Peyerimhoff, 1927a: 48 [UE]

tazmaltensis Desbrochers des Loges, 1881: 56 **E:** IT (Lampedusa Is.) **N:** AG TU

lampedusae Kraatz, 1896: 101

transversithorax Escalera, 1914: 284 **N:** MO

schenklingi Schuster, 1919b: 34

tripoliana Solier, 1835b: 303 **N:** AG MO

tumidifrons Kraatz, 1865a: 200 (*Pachychile*) **N:** AG MO

ariasi Escalera, 1914: 286

cardaillaci Peyerimhoff, 1913b: 520 (*Pachychile*)

orientalis Escalera, 1914: 286

subgenus *Pachychilina* Reitter, 1900d: 145 type species *Tentyria dejeani* Besser, 1832

alluaudi alluaudi Peyerimhoff, 1925: 9 **N:** MO (High Atlas)

alluaudi escalerai Peyerimhoff, 1925: 9 **N:** MO (High Atlas)

andreinii andreinii Dodero, 1916: 350 **N:** LB TU

gracilis Schuster, 1919b: 38 (*Pachychilina*)

andreinii djerbensis Normand, 1938a: 302 [= 1938b: 365] (*Pachychilina*) **N:** TU (Djerba)

cedreticola Antoine, 1947: 147 **N:** MO (Middle Atlas)

dejeani dejeani Besser, 1832: 11 (*Tentyria*) **E:** IT **N:** AG TU

steveni Solier, 1835b: 296

subovata Besser, 1832: 11 (*Tentyria*)

wiedemanni Fischer von Waldheim, 1837b: 14 (*Acispa*)

dejeani doderoi Peyerimhoff, 1927a: 17 **E:** IT (Lampedusa Is.)

dejeani infrastriata Escalera, 1914: 283 **N:** MO (Middle Atlas)

bolivari Antoine, 1926: 249

rotunda Escalera, 1925d: 372

dejeani longipennis Kraatz, 1865a: 212 (*Pachychile*) **N:** TU

doriae Haag-Rutenberg, 1875a: 51

leonhardi Schuster, 1919b: 38 (*Pachychilina*)

oxyophthalmus Reitter, 1900d: 146

dejeani neomontana Leo, 2017: 173 [RN] **N:** LB

montana Koch, 1937: 364 [HN]

dejeani nitens Solier, 1835b: 293 **N:** AG

brevis Kraatz, 1865a: 202 (*Pachychile*)

violantis Gestro, 1880: 416

subgenus *Tentyromorpha* Escalera, 1913: 37 type species *Tentyromorpha telueti* Escalera, 1913

Tentyriomorpha Peyerimhoff, 1927a: 53 [UE]

antoinei Peyerimhoff, 1927a: 53 [RN] **N:** MO
mairei Antoine, 1926: 252 [HN]
fallaciosa Antoine, 1933b: 221 **N:** MO
gattefossei Antoine, 1933b: 220 **N:** MO
telueti Escalera, 1913: 38 (*Tentyromorpha*) **N:** MO

genus *Paivaea* Wollaston, 1864: 449 type species *Tentyria hispida* Brullé, 1839
Paivea Gebien, 1937: 609 [UE]
hispida Brullé, 1839: 66 (*Tentyria*) **N:** CI
mogadorica Escalera, 1914: 274 **N:** MO

genus *Parabigopsis* Español, 1946: 111 type species *Parabigopsis peyerimhoffi* Español, 1946
peyerimhoffi Español, 1946: 115 **N:** MO (Tan-Tan)

genus *Paracirta* Schuster, 1930a: 66 type species *Paracirta schatzmayri* Schuster, 1930
peyerimhoffi daourensis Español, 1953c: 293 **N:** AG
peyerimhoffi peyerimhoffi Español, 1944: 10 **N:** MO (Western Sahara)
schatzmayri algirica Koch, 1941c: 295 **N:** AG
schatzmayri schatzmayri Schuster, 1930a: 67 **N:** LB

genus *Phaeotribon* Kraatz, 1865a: 242 type species *Phaeotribon pulchellus* Kraatz, 1865
pulchellus Kraatz, 1865a: 243 **N:** EG **A:** SA YE **AFR**

genus *Prochoma* Solier, 1835b: 393 type species *Prochoma audouini* Solier, 1835
subgenus *Oxypistoma* Löbl, Bouchard, Merkl & Bousquet, 2020 [see New Acts] type species *Prochoma bucculenta* Koch, 1940
blairi Koch, 1940d: 262 **A:** PA **ORR**
bucculenta Koch, 1940d: 260 **A:** AE IN OM QA SA
clypealis Blair, 1923a: 119 **A:** BA IQ OM SA
margolateralis Carl, 1994: 86 **A:** OM
subgenus *Prochoma* Solier, 1835b: 393 type species *Prochoma audouini* Solier, 1835
audouini Solier, 1835b: 395 **A:** IN IQ IS JO SI SY
membranea Carl, 1994: 83 **A:** IN
pierrei Girard, 1966b: 204 **A:** IN IQ
pilosa Carl, 1994: 81 **A:** IN

genus *Psammocryptus* Kraatz, 1865a: 259 type species *Tentyria minuta* Tauscher, 1812
bayeri Koch, 1943a: 583 **A:** TM
minutus bergi Kuzin, 1934: 160 **A:** KZ
minutus minutus Tauscher, 1812a: 33 (*Tentyria*) **E:** AB ST **A:** KZ TM

prosternalis Koch, 1943a: 581 **A:** AF PA
semenovi Bogatchev, 1946c: 393

genus *Psammoica* Solier, 1835b: 307 type species *Microdera lucida* Solier, 1835
Psammoeca Agassiz, 1846b: 311 [UE]
laticollis Schuster, 1926b: 131 **N:** LB
lucida lucida Solier, 1835b: 307 (*Microdera*) **N:** EG
lucida schusteri Gridelli, 1933: 217 **N:** LB

genus *Rhomaleus* Chatanay, 1915: 64 type species *Rhomaleus scauroides* Chatanay, 1915
scauroides Chatanay, 1915: 64 **N:** MR **AFR**

genus *Rhytinota* Eschscholtz, 1831: 7 type species *Rhytinota scabriuscula* Eschscholtz, 1831
subgenus *Sphenariopsis* Kraatz, 1865a: 175 type species *Sphenariopsis tristis* Kraatz, 1865
subfossulata Solier, 1835b: 369 (*Tentyria*) **A:** SD **ORR**
morio Baudi di Selve, 1875c: 52
pilicollis Kraatz, 1865a: 173
pubens Fairmaire, 1896a: 8
rugipennis Kraatz, 1880b: 92
tristis belutschistana Koch, 1943b: 856 **A:** PA (Balochistan)
tristis kokeni Reitter, 1906c: 42 **A:** PA (Punjab)
tristis tristis Kraatz, 1865a: 176 (*Sphenariopsis*) **A:** HP UP

genus *Scelosodis* Solier, 1835b: 283 [RN] type species *Cratopus castaneus* Eschscholtz, 1831
Abiga Guérin-Méneville, 1859: clxxxix type species *Abiga humilis* Guérin-Méneville, 1859
Cratopus Eschscholtz, 1831: 8 [HN] type species *Cratopus castaneus* Eschscholtz, 1831
Sceleodis Gemminger, 1870b: 1836 [UE]
Sceleoides Agassiz, 1846b: 333 [UE]
alluaudi Peyerimhoff, 1933: 345 **N:** MO (Western Sahara) **AFR**
besnardi aindarensis Kaszab, 1981c: 335 **A:** SA (Eastern Prov.)
besnardi besnardi Kaszab, 1981c: 334 **A:** SA
castaneus castaneus Eschscholtz, 1831: 8 (*Cratopus*) **N:** AG EG LB **A:** SI **AFR**
cerisyi Guérin-Méneville, 1859: cxc (*Abiga*)
humilis Guérin-Méneville, 1859: clxxxix (*Abiga*) **N:** AG TU
laticollis Chatanay, 1914a: 19 **N:** AG MO

genus *Schweinfurthia* Andres, 1922: 26 type species *Schweinfurthia sinaitica* Andres, 1922
alfierii Schatzmayr & Koch, 1934b: 20 **A:** SI
nana Fairmaire, 1893a: 28 (*Psammodes*) **A:** YE **AFR**
schusteri Gridelli, 1930: 211 **N:** LB
sinaitica Andres, 1922: 26 **A:** SI

genus *Scythis* Schaum, 1865: 102 type species *Tentyria macrocephala* Tauscher, 1812

Megascythis Kelejnikova, 1963: 622 type species *Megascythis panfilovi* Kelejnikova, 1963

Semenovonymus Bogatchev, 1946c: 391 type species *Semenovonymus tenuis* Bogatchev, 1946

affinis Ballion, 1878: 299 **A:** KZ (Almaatinskaya Oblast) XIN

altaica G.S. Medvedev, 1990b: 135 **A:** XIN

alticola Reinig, 1931: 901 **A:** TD (Karategin)

angusticollis Gebler, 1829a: 117 (*Tentyria*) **A:** KZ XIN

integrina Reitter, 1900d: 165

kraatzi Reitter, 1900d: 164

laticeps Reitter, 1915d: 67

longipes Kraatz, 1865a: 104

ardoini Skopin, 1960a: 299 **A:** KZ (Alamatinskaya oblast)

arenaria arenaria Faldermann, 1835b: 403 (*Tentyria*) **A:** KZ MG XIN

bulganica Kaszab, 1967a: 31

arenaria australis Skopin, 1979: 179 **A:** KZ

athelea Reitter, 1915d: 72 **A:** KI KZ XIN

gorgone Reitter, 1915d: 72

substriatula Reitter, 1915d: 72

ballioni Reitter, 1900d: 166 **A:** KZ TD (Mogol-Tau)

banghaasi Reitter, 1900d: 166 **A:** ES (Tuva) GAN MG NMO XIN

constricta Steven, 1828: 86 (*Tentyria*) **A:** ES UZ

tenebricosa Faldermann, 1835b: 399 (*Tentyria*)

denticulata Gebler, 1841b: 591 (*Anatolica*) **A:** ES (Altai)

dschungarica Kaszab, 1967a: 23 **A:** MG

edda Reitter, 1915d: 69 **A:** KZ (Almaatinskaya Oblast)

gobiensis Kaszab, 1965g: 1 **A:** MG

grossepunctata Kaszab, 1965e: 326 **A:** MG

intermedia intermedia Ballion, 1878: 300 **A:** KZ (Almaatinskaya Oblast) XIN

anthoini Skopin, 1960a: 298

intermedia scythiformis Reitter, 1915f: 93 (*Microdera*) **A:** XIN

ivanovi G.S. Medvedev, 2004b: 553 **A:** ES (Tuva)

latithorax Skopin, 1979: 175 **A:** XIN

lehri Skopin, 1961a: 182 [= 1964c: 404] **A:** KZ

macrocephala Tauscher, 1812a: 32 (*Tentyria*) **E:** ST **A:** KZ TM

panfilovi Kelejnikova, 1963: 623 (*Megascythis*) **A:** KI

piechockii Kaszab, 1964b: 13 **A:** MG

pusilla Skopin, 1964: 381 **A:** ES (Tuva) MG

septentrionalis Kaszab, 1965e: 328

rotundicollis alexandrae Skopin, 1960a: 299 **A:** KZ (Vostochnyy Kazakhstan)

rotundicollis rotundangula Skopin, 1979: 176 **A:** KZ

rotundicollis rotundicollis Reitter, 1915d: 69 **A:** KI KZ (Zhambylskaya oblast)

sculptilis Reitter, 1900e: 163 **A:** BEI HEB NMO SHX
selivanovskii Skopin, 1961b: 391 **A:** TD (Pamir)
sithon Reitter, 1915e: 83 (*Anatolica*) **A:** KZ
skopini Kaszab, 1964b: 12 **A:** MG
tatarica gracilis Ballion, 1878: 298 **A:** KI KZ (Alamtinskaya oblast) XIN
tatarica pseudoscythis Reitter, 1900d: 119 (*Anatolica*) **A:** KI KZ XIN
tatarica suturalis Reitter, 1915d: 70 **A:** KI KZ
tatarica tatarica Gebler, 1844: 102 (*Anatolica*) **A:** KI KZ
aglaja Reitter, 1915d: 71
dominula Reitter, 1915d: 70
prosternalis Reitter, 1915d: 69
thetis Reitter, 1915d: 70
tatarica weichboldi Melichar, 1912: 33 (*Anatolica*) **A:** KI KZ
tenuimarginis Ren & Ba, 2010: 120 **A** XIN
tenuis Bogatchev, 1946c: 391 (*Semenovonymus*) **A:** MG NMO
tuvae Kaszab & Knor, 1976: 296 **A:** ES (Tuva)
vitabunda Skopin, 1961a: 181 [= 1964c: 403] **A:** KZ (Almaatinskaya Oblast)
vtorovianus Skopin, 1979: 177 **A:** KI

genus *Scytosoma* Reitter, 1895f: 281 type species *Scytosoma arcibasis* Reitter, 1895 (= *Tentyria pygmaea* Gebler, 1832)
Scythodonta Reitter, 1897h: 300 type species *Scythis humeridens* Reitter, 1887
dissitimarginis Ren & Ba, 2010: 110 **A:** NIN
fascia Ren & Zheng, 1993b: 38 **A:** NIN
funebre Reitter, 1900d: 162 **A:** KZ (*Dzungaria*)
humeridens Reitter, 1887d: 356 (*Scythis*) **A:** NMO QIN XIN
constrictum Ren & Zheng, 1993b: 40
inhumeridens Reitter, 1915e: 83 (*Scytodonta*) **A:** XIN
obesum Ren & Zheng, 1993b: 35 **A:** NMO
opacum Reitter, 1889k: 684 (*Scytis*) **A:** HEB NMO XIN
ovale Ren & Zheng, 1993b: 37 **A:** GAN
pygmaeum Gebler, 1832: 54 (*Tentyria*) **A:** ES MG NMO NIN
aequale Reitter, 1895f: 281
arcibasis Reitter, 1895f: 281
bellum Faldermann, 1835b: 393 (*Tentyria*)
rufiabdominum Ren & Zheng, 1993b: 39 **A:** NMO
scalare Ren & Zheng, 1993b: 35 **A:** GAN

genus *Sinoecia* Chatanay, 1914c: 220 type species *Sinoecia puncticollis* Chatanay, 1914
puncticollis Chatanay, 1914c: 221 **A:** HEB

genus *Sphenaria* Ménétriés, 1849: 240 type species *Sphenaria elongata* Ménétriés, 1849
barchanica Skopin, 1961b: 384 **A:** TM
elongata Ménétriés, 1849: 241 **A:** IN TM UZ
hauseri Reitter, 1894a: 48 **A:** TM UZ
acuta Semenov, 1894c: 532 (*Himatismus*)
karelini Ménétriés, 1849: 242 **A:** TD TM UZ
lubricula Reitter, 1916e: 144 **A:** TM UZ
menetriesi Semenov, 1890: 360 (*Himatismus*) **A:** TD TM UZ
glabra Semenov, 1889b: 309
longula Reitter, 1916e: 144
reitteri Brancsik, 1899: 99
moritzi Shestoperov, 1930: 101 **A:** IN
nana Skopin, 1961b: 385 **A:** UZ
obtusa Semenov, 1889a: 214 **A:** UZ
pierrei Peyerimhoff, 1949a: 70 **N:** AG
sewertzowi Semenov, 1894c: 530 (*Himatismus*) **A:** UZ

genus *Stegastopsis* Kraatz, 1865a: 176 type species *Stegastopsis babylonica* Kraatz, 1865
Ohyonthis Reitter, 1898f: 347 type species *Ohyonthis microderoides* Reitter, 1898
arabica Allard, 1883: 23 (*Stegatopsis*) **A:** OM
babylonica Kraatz, 1865a: 177 **A:** IN IQ LE
crassicornis Baudi di Selve, 1874a: 101 [= 1875c: 60] **A:** IN
microderoides Reitter, 1898f: 347 (*Ohyonthis*) **A:** IN TR
persica Baudi di Selve, 1874a: 101 [= 1875c: 59] **A:** IN

genus *Stenosida* Solier, 1835b: 281 type species *Stenosida tenuicollis* Solier, 1835 (=*Tagenia striato-punctata* Wiedeman, 1821)
Aprosphaena Reitter, 1916e: 142 type species *Aprosphaena adriani* Reitter, 1916 (= *Tagenia striatopunctata* Wiedemann, 1821)
Notioscythis Fairmaire, 1883b: 31 type species *Notioscypthis punctoseriata* Fairmaire, 1883
indica Haag-Rutenberg, 1877a: 278 (*Himatismus*) **A:** GUA HAI **ORR**
indicola Reitter, 1916e: 143 (*Aprosphaena*)
striatopunctata Wiedemann, 1821: 149 (*Tagenia*) **A:** NP YUN **ORR**
adriani Reitter, 1916e: 143 (*Aprosphaena*)
tenuicollis Solier, 1835b: 282

genus *Syachis* Bates, 1879: 467 [= 1883: 55] type species *Syachis himalaicus* Bates, 1879
Orocina Reitter, 1897h: 303 type species *Orocina capnisiceps* Reitter, 1897
afghanicus Schuster, 1936c: 193 **A:** AF
ajmonis Gridelli, 1934: 44 **A:** KA XIN

badakschanicus Kaszab, 1959f: 352 **A:** AF
capnisiceps Reitter, 1897h: 303 (*Orocina*) **A:** AF IN TD (Pamir)
marginatus Reinig, 1931: 899 (*Ascelosodis*)
cugiae Gridelli, 1934: 45 **A:** KA
darvazianus Skopin, 1964: 408 **A:** TD (Pamir)
globuliformis Della Beffa, 1931: 184 **A:** KA
gnathosoides Kaszab, 1959f: 350 **A:** AF
haafi Kaszab, 1959f: 348 **A:** AF TD
himalaicus Bates, 1879: 467 **A:** KA XIZ
klapperichi Kaszab, 1959f: 351 **A:** AF
marussii Gridelli, 1957: 49 **A:** KA XIN
mediocris Fairmaire, 1891d: xciii (*Capnisa*) **A:** KA TM
osellai Kaszab, 1985: 450 **A:** KA
paludani Kaszab, 1959f: 239 **A:** AF
picicornis Bates, 1879: 467 **A:** KA XIN
proximus Kaszab, 1959f: 346 **A:** AF
rugosipennis Della Beffa, 1931: 184 **A:** KA
xizangana Ba & Ren, 2004: 763 **A:** XIZ

genus *Tamena* Reitter, 1900d: 143 type species *Psammocryptus rugiceps* Reitter, 1897
rugiceps Reitter, 1897j: 311 (*Psammocryptus*) **A:** XIN

genus *Tentyria* Latreille, 1802: 170 type species *Tentyria ligurica* Solier, 1835
Heliodromus Brullé, 1832: 196 type species *Heliodromus rotundatus* Brullé, 1832
acuticollis Reitter, 1900d: 182 **A:** KI KZ
ganglbaueri Reitter, 1900d: 182
arabica arabica Schuster, 1938a: 55 **A:** SY YE
arabica scorteccii Koch, 1965a: 115 **A:** YE
aragonica Koch, 1944b: 231 **E:** SP
atlantis Antoine, 1942b: 43 **N:** MO (High Atlas)
balachowskyi Girard, 1965: 139 **N:** AG
barbara Solier, 1835b: 340 **N:** AG
bassii Solier, 1835b: 356 **E:** PT SP
cantabrica Koch, 1944b: 230
gredosana Koch, 1944b: 230
meridionalis Koch, 1944b: 231
bifida Bujalance, Cárdenas, J. Ferrer & Gallardo, 2016: 350 **E:** SP
bogdanovi Gebien, 1937: 634 [RN] **A:** AF IN TM UZ
castillana Koch, 1944b: 226 **E:** SP
castrogironai Escalera, 1923d: 128 **N:** MO
clavata Kraatz, 1865a: 151 **E:** "Caucasus"

collatina Reiche & Saulcy, 1857: 207 **A:** IS
latiuscula Kraatz, 1865a: 159
confalonierii confalonierii Gridelli, 1930: 222 **N:** LB
confalonierii coniophthalma Koch, 1937: 376 **N:** LB
corrugata Rosenhauer, 1856: 189 **E:** PT SP
andalusiaca Kraatz, 1865a: 142
curculionoides Herbst, 1799: 58 (*Pimelia*) **E:** PT
curta Steven, 1828: 89
cylindrica Solier, 1835b: 353 **A:** CY SY TR
acuminata Reiche & Saulcy, 1857: 209
lata Baudi di Selve, 1875a: 284
cypria Kraatz, 1865a: 160 **A:** CY TR
cyrenaica Schuster, 1919a: 27 **N:** LB
subpunctata Schuster, 1922b: 15
discicollis Reiche & Saulcy, 1857: 207 **A:** IS JO SI SY
donanensis Bujalance, Cárdenas, J. Ferrer & Gallardo, 2016: 351 **E:** SP
duponti Solier, 1835b: 323 **N:** AG EG
elegans Schuster, 1919a: 28 **N:** LB
elongata elongata Waltl, 1835: 70 **E:** SP (Andalucía oriental)
arenaria Kraatz, 1865a: 140
elongata sinuatocollis Rosenhauer, 1856: 185 **E:** SP (Andalucía occidental)
rugosostriata Kraatz, 1865a: 140
elongata subcostata Solier, 1835b: 325 **N:** AG MO **E:** PT SP
beidensis Antoine, 1926: 248
emarginata Kraatz, 1865a: 141
godartiana Lucas, 1855a: 291
eulipoides Koch, 1944b: 231 **E:** SP
euphratica escalerai Girard, 1966b: 204 **A:** SY
euphratica euphratica Girard, 1966a: 84 **A:** IQ
euphratica kursistana Girard, 1966a: 84 **A:** "Kurdistan"
excavata Solier, 1835b: 364 **N:** AG TU
scabriuscula Solier, 1835b: 365
fossulata C. Brisout de Barneville, 1888: cxcii **N:** AG
frivaldskii Kraatz, 1865a: 150 **E:** YU
banatica J. Frivaldszky, 1865: 43
gaditana Rosenhauer, 1856: 186 **E:** SP
geliae Carl, 1989: 25 **A:** IQ
gigas Faldermann, 1836: 376 **A:** KZ TD TM UZ XIN
asiatica Skopin, 1966a: 5
laevicollis Gebler, 1844: 102
robusta Motschulsky, 1870: 383
valida Motschulsky, 1870: 397 (*Mesostena*)

gibbicollis Lucas, 1855a: 291 **N:** AG TU
strauchii Kraatz, 1865a: 166
grossa angustata Kraatz, 1896: 103 **E:** IT (Pantelleria Is.)
grossa basalis L.W. Schaufuss, 1869: 21 **E:** SP
grossa grossa Besser, 1832: 11 **E:** GR IT MA SP **N:** AG MO TU
dejeanii Solier, 1835b: 345
grandis Solier, 1835b: 343
sicula Solier, 1835b: 342
tristis Solier, 1835b: 344
grossa sardiniensis Ardoin, 1973: 305 **E:** IT (Sardinia)
grossa sommieri Baudi di Selve, 1874a: 99 [= 1875c: 47] **E:** IT (Linosa Is., Lampione)
herculeana Reiche & Saulcy, 1857: 204 **A:** IS JO LE SY TR
saulcyi Reiche, 1861d: 212 [RN]
solieri Reiche, 1857: 206
heydeni Haag-Rutenberg, 1870a: 130 **E:** SP
incerta Solier, 1835b: 359 **E:** SP
intermittens Skopin, 1968b: 106 [RN] **A:** KZ UZ
monticola Skopin, 1966a: 4 [HN]
interrupta Latreille, 1807: 155 **E:** FR PT SP
gallica Solier, 1835b: 333
italica Solier, 1835b: 348 **E:** CR GR IT
calabrica Reitter, 1906e: 130
romana Baudi di Selve, 1875a: 283
tarquinia Baudi di Selve, 1875c: 44
jordani Koch, 1944b: 229 **E:** SP
kantarae Fairmaire, 1879c: 187 **N:** AG
kriegi Koch, 1940d: 264 **A:** SA YE
laevigata laevigata Steven, 1828: 92 **E:** IT
laevigata leachii Baudi di Selve, 1875c: 42 **E:** MA
lateritia Reitter, 1900d: 169 **E:** PT SP
laticollis Kraatz, 1865a: 158 **A:** CY IS JO SY
latreillei brachythorax Gridelli, 1930: 219 **N:** LB
latreillei latreillei Solier, 1835b: 339 **E:** IT
sardea Solier, 1835b: 340
latreillei oblongipennis Fairmaire, 1875a: 518 **N:** TU
latreillei pseudogrossa Koch, 1937: 372 **N:** LB
latreillei simplicibasalis Koch, 1937: 372 **N:** LB
latreillei syrtica Koch, 1937: 374 **N:** LB
levis Solier, 1835b: 358 **E:** SP
leptidea Fairmaire, 1879c: 188
sublaevis Kraatz, 1865a: 144

ligurica confusa Ardoin, 1973: 304 **E:** IT (Sardinia)
ligurica ligurica Solier, 1835b: 324 **E:** FR IT
cribrata Gené, 1836b: iii
dentibasis Reitter, 1900d: 168
ottii Lucas, 1855a: 291
ligurica pseudorugosa Ardoin, 1973: 304 **E:** IT (Sardinia)
marocana Solier, 1835b: 333 **N:** MO
maura Erichson, 1841: 177 **N:** AG
acuminipennis Lucas, 1855a: 291
mauritanica Baudi di Selve, 1875c: 45 **N:** EG LB
mesopotamica Holdhaus, 1919: 53 **A:** IQ
morgani Girard, 1968: 873 **A:** IN
mucronata mucronata Steven, 1828: 90 **E:** FR SP
glabra Herbst, 1799: 60 [HN]
oblonga Solier, 1835b: 321
nitida Solier, 1835b: 323
mucronata ophiusae Codina, 1918: 265 **E:** SP (Formentera)
mulsanti Lucas, 1855a: 291 **N:** AG LB TU
nomas Pallas, 1781: 43 (*Tenebrio*) **E:** AB RO ST UK **A:** KZ "Siberia"
coerulescens Bogdanov-Katjkov, 1915: 6
gebleri Besser, 1832: 15 (*Tenebrio*)
kindermanni Fischer von Waldheim, 1844: 64
podolica Steven, 1828: 91
rugulosa Germar, 1823: 138
saisanensis Motschulsky, 1845a: 74 (*Anatolica*)
salzmanni Solier, 1835b: 351
sibirica Gebler, 1829a: 120
valentinae Bogdanov-Katjkov, 1915: 7
occidentalis Peyerimhoff, 1925: 14 **N:** MO (Middle Atlas)
ocularis calcophila Koch, 1937: 368 **N:** LB (Tripolitania)
ocularis gusbatensis Koch, 1937: 369 **N:** LB (Tripolitania)
ocularis ocularis Reitter, 1900d: 174 **N:** LB
parallela Baudi di Selve, 1874a: 98 [= 1875c: 49] **A:** IN
pazi Español, 1958a: 13 **E:** SP (Islas Columbretes)
peiroleri Solier, 1835b: 357 **E:** PT SP
pierrei Girard, 1965: 138 **N:** AG
platyceps Steven, 1828: 92 **E:** PT SP
calcarata Reitter, 1900d: 170
goudoti Solier, 1835b: 360
modesta Rosenhauer, 1856: 188
prolixa Rosenhauer, 1856: 187

pseudogigas Skopin, 1968b: 107 **A:** KI KZ UZ
pseudolaevis Koch, 1944b: 231 **E:** SP
punctatostriata Solier, 1835b: 337 **N:** EG LB **A:** SI
puncticeps puncticeps Miller, 1861: 173 **A:** CY JO SY
puncticeps variabilis Girard, 1966b: 205 **A:** IQ SY
ramburi Solier, 1835b: 327 **E:** FR (Corse) IT (Sardinia)
angusticollis Solier, 1835b: 328
levicollis Solier, 1835b: 329
maillei Solier, 1835b: 330
substriata Solier, 1835b: 332
robustoides Reitter, 1900d: 183 **A:** AF IN TM TR
rotundata angulata Brullé, 1832: 197 (*Heliodromus*) **E:** BU GR MC **A:** IN TR
subdepressa Solier, 1835b: 346
rotundata convexipennis Koch, 1948a: 316 **A:** TR
rotundata daghestris Koch, 1948a: 316 **A:** TR
rotundata ionica Koch, 1948a: 313 **E:** GR (Ionioi)
rotundata mittrei Solier, 1835b: 341 **E:** GR (Chios, Kos, Nisyros, Rhodes, Samos) **A:** TR
subsulcata Koch, 1948a: 316
rotundata orbicollis Solier, 1835b: 362 **E:** GR
rotundata paganettii Schuster, 1915a: 6 **E:** GR (Crete)
rotundata rotundata Brullé, 1832: 197 **E:** GR
grossa Solier, 1835b: 361 [HN]
rotundata sulcatipennis Schuster, 1936b: 69 **E:** GR
rotundata winkleri C. Koch, 1936: 293 **A:** IN TR
rugosa rugosa Gené, 1836b: iii [= 1836a: 193] **E:** IT (Sardinia)
rugosa cassolai Ardoin, 1973: 304 **E:** IT (Sardinia)
rugosa floresii Gené, 1836b: iii [= 1836a: 194] **E:** FR (Corse) IT (Sardinia)
monticola Gené, 1839: 69
schaumi Kraatz, 1865a: 142 **E:** SP (Baleares)
sinaitica Peyerimhoff, 1907a: 32 **A:** SI
solieri Lucas, 1846: pl. 28 **N:** AG TU
sporadica gavdosana Koch, 1948a: 307 **E:** GR (Gavsos)
sporadica sporadica Reitter, 1900d: 179 **E:** GR (Sporades)
sporadica zimmermanni Koch, 1948a: 308 **E:** GR (Kufonisi)
striatopunctata Ménétriés, 1832: 195 **E:** AB AR GG ST **A:** IN TR
olgae Bogdanov-Katjkov, 1915: 5
punctipleuris Reitter, 1900d: 182
subrugosa Solier, 1835b: 352 **E:** SP
syrtana Normand, 1936b: 357 **N:** TU
taurica Tauscher, 1812a: 37 **E:** ST UK (Krym)
fischeri Solier, 1835b: 355

tessulata scabripennis Solier, 1835b: 366 **A:** IN TR
tessulata tessulata Tauscher, 1812a: 316 **E:** AB AR GG ST **A:** IN TR
  alpina L. Redtenbacher, 1850: 48
  incusa Ménétriés, 1832: 195
  olivieri Solier, 1835b: 368
  rugipleuris Bogdanov-Katjkov, 1916: xxix
tessulata villiersi Girard, 1968: 871 **A:** IN
thunbergii Steven, 1828: 90 **N:** AG MO
  affinis Lucas, 1846: pl. 28
  bipuncta Solier, 1835b: 336
transversicollis Fairmaire, 1879c: 187 **N:** AG
varioculata Carl, 1991b: 173 **A:** IQ
velox Chevrolat, 1865: 390 **E:** SP
vieta Faldermann, 1835b: 402 **A:** NMO
wiedemanni Ménétriés, 1840: 38 (*Heliodromus*) **A:** IN TR
  bisinuatibasalis Koch, 1936: 292
  sinuatocollis Kaszab, 1940a: 3 [HN]

genus *Tentyria*, nomina dubia
glabra Fabricius, 1775: 253 (*Pimelia*) distribution unknown
pygmaea Gené, 1836b: iii **E:** IT (Sardinia)

genus *Tentyrina* Reitter, 1900d: 166 type species *Pimelia orbiculata* Fabricius, 1775
  Tentyriina Peyerimhoff, 1907b: 186 [UE]
deserta deserta Kaszab, 1981c: 338 **A:** SA (Eastern Prov.)
deserta hadraumatina Kaszab, 1981c: 339 **A:** YE
deserta sparsepunctata Kaszab, 1981c: 339 **A:** AE
deserta thesigeri Kaszab, 1981c: 339 **A:** OM
duplicata boehmi Reitter, 1909a: 29 **N:** EG LB
duplicata brachythorax Gridelli, 1929b: 55 (*Tentyriina*) **N:** LB (Cyrenaica)
duplicata duplicata Reitter, 1900d: 172 (*Tentyria*) **N:** LB TU
duplicata manzonii Gridelli, 1929b: 53 (*Tentyriina*) **A:** SA YE **AFR**
duplicata schusteri Schatzmayr & Koch, 1934b: 22 (*Tentyriina*) **N:** EG
gracilis Schuster, 1930b: 236 (*Tentyriina*) **A:** IN PA SA
granulicauda Koch, 1940a: 123 **A:** IN
longicollis Lucas, 1855a: 291 (*Tentyria*) **N:** AG LB MO TU
orbiculata orbiculata Fabricius, 1775: 253 (*Pimelia*) **N:** EG
  aegyptiaca Solier, 1835b: 334 (*Tentyria*)
orbiculata subsulcata Reiche & Saulcy, 1857: 203 (*Tentyria*) **N:** EG **A:** CY IS LE SI SY
palmeri palmeri Crotch, 1872: 267 (*Tentyria*) **N:** EG LB **A:** IN IQ JO SA SI SY
  giraffa Allard, 1883: 22 (*Tentyria*)

palmeri persicola Schuster, 1938b: 84 (*Tentyria*) **A:** IN PA
palmeri thomasi Blair, 1931b: 269 (*Tentyria*) **A:** OM QA SA
sculptissima Koch, 1940c: 101 (*Tentyria*) **A:** JO
senegalensis Solier, 1835: 326 (*Tentyria*) **N:** AG LB MO
hircipes Peyerimhoff, 1916a: 75

genus *Tentyrina*, nomen dubium
reichii Besser, 1832: 14 (*Tentyria*)

genus *Tentyronota* Reitter, 1900d: 188 type species *Micipsa rotundicollis* Kraatz, 1865
atrimembris Pic, 1922g: 30 (*Micipsa*) **N:** MO
ifnica Escalera, 1935a: 302 (*Tentyrionota*) **N:** MO
rotundicollis lucida Peyerimhoff, 1931b: 96 **N:** AG
rotundicollis rotundicollis Kraatz, 1865a: 108 (*Micipsa*) **N:** AG LB

genus *Thalpobia* Fairmaire, 1871a: 385 type species *Thalpobia laevipennis* Fairmaire, 1871
Micipsina Reitter, 1900d: 188 type species *Micipsina rolphi* Reitter, 1900
fonti fonti Escalera, 1935a: 305 **N:** MO
fonti korsakovi Español, 1946: 117 **N:** MO (Tan-Tan)
laevipennis Fairmaire, 1871a: 386 **N:** MO
gracilitarsis Antoine, 1933b: 224
rolphi gattefossei Antoine, 1946b: 37 **N:** MO (Tan-Tan)
clermonti Español, 1946: 116
rolphi rolphi Reitter, 1900d: 189 (*Micipsina*) **N:** MO
alluaudi Kocher, 1957: 272
maxima Kocher, 1957: 272
meridionalis Escalera, 1914: 280
reitteri Escalera, 1913: 36 **N:** MO
zemensis Antoine, 1946b: 36 **N:** MO

genus *Thraustocolus* Kraatz, 1865b: 414 [RN] type species *Calobamon leptoderus* Kraatz, 1865
subgenus *Leptoderops* Löbl, Bouchard, Merkl & Bousquet, 2020 [see New Acts] type species *Thraustocolus priesneri* Koch, 1934
afghanicus Kaszab, 1979c: 278 **A:** AF
priesneri Koch, 1934a: 97 **N:** EG **AFR**
subgenus *Prothraustocola* Kaszab, 1957f: 293 type species *Ibnsaudia belutschistanica* Kaszab, 1957
belutschistanicus Kaszab, 1957f: 294 (*Ibnsaudia*) **A:** IN
hormozganus Grimm & Merkl, 2018 **A:** IN
montandoni Schuster, 1915b: 87 **N:** EG
pitcheri Johnson, 1989: 126 **A:** SA (Eastern Prov.)

subgenus *Thraustocolus* Kraatz, 1865b: 414 [RN] type species *Calobamon leptoderus* Kraatz, 1865

Calobamon Kraatz, 1865a: 105 [HN] type species *Calobamon leptoderus* Kraatz, 1865

Ibnsaudia Koch, 1941c: 290 type species *Ibnsaudia propheta* Koch, 1941

arabicus Kaszab, 1979c: 277 **A:** SA

catomoides Kaszab, 1981c: 342 **A:** SA

klapperichi Kaszab, 1959f: 393 **A:** AF

leptoderus Kraatz, 1865a: 106 (*Calobamon*) **N:** EG **A:** IS SI "Dead Sea"

propheta Koch, 1941c: 290 (*Ibnsaudia*) **A:** SA

rugosus Kaszab, 1979c: 276 **A:** AE OM SA

genus *Trichosphaena* Reitter, 1916e: 145 type species *Sphenaria olgae* Semenov, 1889

arabica Kaszab, 1961c: 170 **A:** OM SA YE

chogsonzhavi Kaszab, 1968c: 346 **A:** MG

chotanica Semenov, 1890: 354 (*Himatismus*) **A:** NMO XIN

compactilis Merkl, 1991b: 63 **A:** IN

dunhangensis Ren & Zheng, 1993c: 52 **A:** GAN

gobica Kaszab & G.S. Medvedev, 1972: 215 **A:** MG

iranica Bogatchev, 1967b: 350 (*Epitrichia*) **A:** IN SA

olgae Semenov, 1889b: 309 (*Sphenaria*) **A:** AF TM UZ

perraudieri perraudieri Marseul, 1867: xxxix (*Himatismus*) **N:** AG EG MO TU **A:** IS OM SA SI YE **AFR**

persica Schuster, 1935b: 23 **A:** IN

quadrate [sic!] Ren & Zheng, 1993c: 55 **A:** GAN

reitteri Semenov, 1890: 353 (*Himatismus*) **A:** MG NMO

sahlbergi Reitter, 1916e: 147 **A:** TM

schusteri Reitter, 1916e: 147 **A:** UZ

suturalis Semenov, 1889b: 309 (*Sphenaria*) **A:** TM UZ

ulanbuhensis Ren & Zheng, 1993c: 55 **A:** NMO

vestita Reitter, 1889b: 116 (*Sphenaria*) **A:** MG NMO UZ XIN

turkestanica Semenov, 1890: 357 (*Himatismus*)

vestita Semenov, 1889b: 215 (*Sphenaria*) [HN]

zarudnyana Schuster, 1938b: 80 **A:** IN

zarudnyi Bogatchev, 1949b: 277

genus *Uyttenboogaartia* Koch, 1943a: 595 type species *Hegeter cribricollis* Brullé, 1839

cribricollis Brullé, 1839: 66 (*Hegeter*) **N:** CI (Gran Canaria)

inaequalis Wollaston, 1864: 466 (*Gnophota*) **N:** CI (Gran Canaria)

punctipennis Wollaston, 1864: 467 (*Gnophota*) **N:** CI (Gran Canaria)

tribe Tentyriini, nomina dubia
Dailognatha hispana Solier, 1835b: 261 **E:** PT SP
Tentyria oblonga Tauscher, 1812a: 314 **E:** "Iberia"
Tentyria punctata Tauscher, 1812a: 313 [41] **E:** "regionibus Caucasicis"

tribe Zophosini Solier, 1834

genus *Zophosis* Latreille, 1802: 167 type species *Erodius testudinarius* Fabricius, 1781
subgenus *Cardiosis* Deyrolle, 1867: 81 type species *Cardiosis mouffleti* Deyrolle, 1867
mouffleti Deyrolle, 1867: 81 **N:** TU
subgenus *Cheirosis* Deyrolle, 1867: 220 type species *Zophosis ovata* Faldermann, 1837 (= *Pedinus acuminatus* Fischer von Waldheim, 1832)
Chirosis Gemminger, 1870b: 1806 [UE]
acuminata Fischer von Waldheim, 1832a: 202 (*Pedinus*) **E:** AB **A:** IN
ovata Faldermann, 1837: 5 [HN]
subgenus *Gahanosis* Penrith, 1983b: 379 type species *Zophosis undulata* Gahan, 1900
undulata Gahan, 1900: 8 **A:** YE (Suqutra)
subgenus *Hologenosis* Deyrolle, 1867: 82 type species *Hologenosis laceratus* Deyrolle, 1867
angustula Reitter, 1916c: 88 **N:** TU
atlantica Escalera, 1913: 36 **N:** MO
emiliana Reitter, 1916c: 88 **N:** MO
nigroaenea Deyrolle, 1867: 181 **N:** AG EG LB MO **A:** SI **AFR**
aelanitica Peyerimhoff, 1907a: 26
convergens Koch, 1941c: 269
mozabita Fairmaire, 1897f: 120
theryi Koch, 1941c: 270
picipennis arabica Lesne, 1920: 15 **A:** YE
oxycaroides Koch, 1935a: 6 **N:** EG
subgenus *Oculosis* Penrith, 1977: 126 type species *Zophosis boei* Solier, 1834
abbreviata Solier, 1834: 606 **N:** EG **A:** SA **AFR**
reichei Guérin-Méneville, 1849: 315
semilineata Deyrolle, 1867: 107
aequalis Waterhouse, 1881: 471 **A:** YE (Darsa, Suqutra)
dilatata Deyrolle, 1876: 208 **E:** GR **A:** IN LE SY TR
minuta Fabricius, 1775: 259 (*Erodius*) **N:** EG **A:** IS KU
errans Solier, 1834: 614
minutecostulata Koch, 1945: 393
pygmaea Solier, 1834: 611
unicostulata Koch, 1945: 393
persis Chatanay, 1914d: 379 **A:** IN
punctata alborana Baudi di Selve, 1883c: 760 **E:** SP (Alboran Is.)

punctata punctata Brullé, 1832: 191 **E:** AE AL AR GR IT SP TR **N:** AG LB MO TU **A:** AF CY IN IQ IS JO KZ LE PA SA SI SY TD TM TR UZ XIN

algeriana Solier, 1834: 613

barbara Solier, 1834: 608

curta Solier, 1834: 611

deflexa Reitter, 1916c: 87

faldermanni Deyrolle, 1867: 116

gibbicollis Solier, 1834: 612

kollari Deyrolle, 1867: 112

maillei Solier, 1834: 613

mediocris Deyrolle, 1867: 114

nitida Gebler, 1844: 100 [= 1860: 10]

nitidula Motschulsky, 1860e: 526 [= 1860d: 426] [HN]

orientalis Deyrolle, 1867: 115

osmanlis Deyrolle, 1867: 114

ovata Brullé, 1832: 190

polita Lucas, 1854: 28

steppensis Gemminger, 1870a: 121 [RN]

subinflata Brullé, 1832: 191

truquii Deyrolle, 1867: 116

variepunctata Deyrolle, 1867: 113

vescoi Deyrolle, 1867: 113

punctata pygmaea Solier, 1834: 611 **N:** EG **A:** IS KU

quadrilineata G.-A. Olivier, 1791: 426 (*Erodius*) **N:** AG LB **AFR**

subgenus *Ophthalmosis* Deyrolle, 1867: 81 type species *Ophthalmosis longipes* Deyrolle, 1867

longipes Deyrolle, 1867: 81 **N:** TU

subgenus *Septentriophosis* Penrith, 1982a: 167 type species *Erodius planus* Fabricius, 1775

acuticosta Fairmaire, 1886a: 448 **A:** YE **AFR**

aciculaticollis C. Müller, 1887: 297

bottegoi Chatanay, 1917b: 602

gestroi Chatanay, 1917b: 600

pinardi Théry, 1892: clxxx

nitidicostis Chatanay, 1917b: 598

ogadenica Gestro, 1892: 762

subcostulata Fairmaire, 1887a: 166

aegresinuata Penrith, 1986: 429 **A:** IN

afghanica Kaszab, 1959f: 339 **A:** AF

alticola Peyerimhoff, 1929b: 154 **N:** AG

asiatica Miller, 1861: 169 **A:** IN SY TR

puncticeps Baudi di Selve, 1874a: 90

bicarinata bicarinata Solier, 1834: 617 **N:** CI
clarkii Deyrolle, 1864: 435
ecostata Håk. Lindberg, 1962: 25
gomerae Håk. Lindberg, 1962: 25
nitidula Solier, 1834: 619
quadricostata Deyrolle, 1864: 433
vagans Brullé, 1839: 64
verrucosa Håk. Lindberg, 1962: 25
bicarinata cognata Miller, 1861: 170 **A:** IS LE
orbiculata Deyrolle, 1867: 209
bicarinata dinauxi Peyerimhoff, 1929b: 154 **N:** AG MO
bicarinata plicata Brullé, 1839: 64 **N:** CI (Fuerteventura, Lanzarote)
bicarinata renatae Penrith, 1982a: 197 **N:** MO
bicarinata quadricostata Solier, 1834: 628 **N:** EG **A:** JO SA YE **AFR**
intermedia Chatanay, 1917b: 619
bicarinata supposita Peyerimhoff, 1931b: 90 **N:** AG
bicarinata susica Escalera, 1914: 272 **N:** MO **AFR**
graphita Koch, 1941c: 274
tricarinata Escalera, 1914: 273
clypeosuturata Carl, 1990b: 80 **A:** JO
complanata Solier, 1834: 626 **N:** EG **A:** IN IQ IS JO SA SI
deplanata Böhm, 1909: 112
humberti Deyrolle, 1867: 177
miliaris Deyrolle, 1867: 179
pulverulenta Deyrolle, 1867: 178
sulphurea Deyrolle, 1867: 180
deserticola Kulzer, 1956: 643 **A:** AF PA **ORR**
dimorpha Carl, 1990b: 79 **A:** IQ
farinosa Blair, 1914b: 483 [RN] **A:** IN IQ JO SA SI
latissima Bogatchev, 1957: 529
testudinaria Solier, 1834: 622 [HN]
finzii Koch, 1937: 343 **N:** LB
globulipennis Escalera, 1914: 271 **N:** MO
hydrobiiformis hydrobiiformis Reitter, 1916c: 98 **A:** SY
hydrobiiformis khuzistanica Kaszab, 1959a: 2 **A:** IN
inlustris Penrith, 1984: 411 **A:** IN IQ
kaszabi Penrith, 1984: 407 **A:** PA
klapperichi Kaszab, 1959f: 341 **A:** AF PA **ORR**
leonardii Kaszab, 1972: 370 **A:** SA YE
lethierryi Deyrolle, 1867: 211 **N:** AG LB TU **A:** SI

maindroni Chatanay, 1914d: 381 **A**: PA
circumducta Kwieton, 1881: 6
robusta Kwieton, 1981a: 8
marseuli Deyrolle, 1867: 210 **N**: AG TU
aerifer Antoine, 1933b: 218
migneauxi Deyrolle, 1867: 171 **A**: SA YE
novaki Purchart, 2014b: 233 **A**: YE (Suqutra)
oblonga Solier, 1834: 625 **A**: LE IQ SA SY TR YE
cursor Deyrolle, 1867: 170
persica ardistana Reitter, 1916c: 95 **A**: IN TM
persica gridellii Kaszab, 1959e: 239 [= 1959f: 343] **A**: AF IN
persica persica Kraatz, 1882b: 94 **A**: AF IN TM
personata Erichson, 1841: 176 **N**: AG EG LB TU **A**: SI
curta Deyrolle, 1867: 211
depressipennis Lucas, 1858c: ccxxii
tunisea Reitter, 1916c: 91
wollastoni Deyrolle, 1867: 212
zuberi Deyrolle, 1867: 213
pharaonis pharaonis Reitter, 1916c: 90 [RN] **N**: EG **A**: IS SA SI
rotundata Deyrolle, 1867: 215 [HN]
saudita Koch, 1965a: 112
pharaonis simplex Kaszab, 1981c: 319 **A**: AE
plana Fabricius, 1775: 259 (*Erodius*) **N**: AG EG LB MO **A**: SA SI **AFR**
approximata Deyrolle, 1867: 195
astacura Peyerimhoff, 1931b: 90
beraudi Peyerimhoff, 1920a: 325
carinata Solier, 1834: 630
elongata Koch, 1935a: 8
germari Solier, 1834: 631
phazanica Koch, 1937: 346
praedeserta Koch, 1937: 348
reaumuri Solier, 1834: 630
schoenherri Solier, 1834: 629
posticalis Deyrolle, 1867: 199 **N**: AG EG LB **AFR**
lapruni Chatanay, 1914d: 381
niegeri Peyerimhoff, 1929b: 155
punctatissima Peyerimhoff, 1929b: 153
rotundata Ménétriés, 1849: 218 **A**: IN KZ TM
rugosa Faldermann, 1837: 4 **E**: AB AR **A**: IN
armeniaca Deyrolle, 1867: 172

sabaea Baudi di Selve, 1881: 273 **A:** YE **AFR**
  aenea Chatanay, 1917b: 586
  blairi Chatanay, 1917b: 586
  costulata Chatanay, 1917b: 584
  longa Chatanay, 1917b: 584
  tenuitarsis Chatanay, 1917b: 586
scabriuscula lata Kraatz, 1882b: 94 **A:** AF KZ UZ
  blattoides Reitter, 1916c: 98
scabriuscula scabriuscula Ménétriés, 1849: 218 **A:** KZ TM UZ
  nivosa Dohrn, 1884: 40
scortecciana Kaszab, 1979c: 259 [RN] **A:** SA YE
  scorteccii Kaszab, 1972: 370 [HN]
stilla Penrith, 1986: 431 **A:** IN
suborbicularis Solier, 1834: 616 **E:** PT SP
sulcata sulcata Deyrolle, 1867: 189 **N:** EG **A:** SA YE **AFR**
  alternans Deyrolle, 1867: 190
  arabs Chatanay, 1917b: 606
  aromatum Gestro, 1895: 358
  bicostis Chatanay, 1917b: 607
  bohemanni Deyrolle, 1867: 188
  chalybaea Chatanay, 1917b: 611
  crispata Fairmaire, 1891g: ccxci
  dakhlensis Andres, 1928: 48
  lineata Chatanay, 1917b: 606
  maeklini Deyrolle, 1867: 186
  opacipennis Chatanay, 1917b: 605
  planidorsis Chatanay, 1917b: 606
  plicatipennis Deyrolle, 1867b: 190
  pulchra Chatanay, 1917b: 613
  rodolphi Chatanay, 1917b: 608
  strigipleuris Chatanay, 1917b: 605
  testudo Chatanay, 1917b: 589
  zhenzhuristi Bogatchev, 1957: 529
viridilimbata Chobaut, 1899: 409 **N:** EG LB TU
  cyrenaica Schuster, 1928e: 375
  scorteccii Gridelli, 1939d: 421

subfamily **Tenebrioninae** Latreille, 1802

tribe Alphitobiini Reitter, 1917

genus *Alphitobius* Stephens, 1829a: 19 type species *Helops picipes* Panzer, 1794 (= *Opatrum laevigatum* Fabricius, 1781)

Cryptops Solier, 1851: 235 type species *Cryptops ulomoides* Solier, 1851 (= *Tenebrio diaperinus* Panzer 1797)

Heterophaga Dejean, 1834: 199 type species *Tenebrio mauritanicus* Fabricius, 1792 (= *Opatrum laevigatum* Fabricius, 1781)

Latetribolium Lepesme, 1943: 45 type species *Latetribolium risbeci* Lepesme, 1943 (= *Opatrum laevigatum* Fabricius, 1781)

Microphyes W.J. MacLeay, 1873b: 286 type species *Microphyes rufipes* W.J. MacLeay, 1873 (= *Opatrum laevigatum* Fabricius, 1781)

Proselytus Fåhraeus, 1870: 302 type species *Proselytus caffer* Fåhraeus, 1870 (= *Alphitobius diaperinus* Panzer, 1796)

diaperinus Panzer, 1796e: 16 (*Tenebrio*) **E:** AB AR AU AZ BE BY CR CT CZ DE FI GB GE GG GR HU IR IT LU MA NL PL RO SA SK SL SP ST SZ TR UK **N:** EG CI LB MO MR TU **A:** AF ANH BA BT CY FE FUJ GUA GUX HAI HEB HEI HKG HUB HUN IQ IS JA JIA JIX KZ LIA MG NE NP SA SCH SHA SHX SI TAI TIA TM YE YUN ZHE "Korea" **COS**

caffer Fåhraeus, 1870: 266 (*Proselytus*)

longipennis Walker, 1858: 284 (*Crypticus*)

opatroides Brullé, 1839: 70 (*Uloma*)

ovatus Herbst, 1799: 16 (*Tenebrio*)

rufipes Walker, 1858: 284 (*Phaleria*)

ulomoides Solier, 1851: 236 (*Cryptops*)

laevigatus Fabricius, 1781: 90 (*Opatrum*) **E:** AB AR AZ BE CT GB GE GG GR HU IT LU MA NL PL SK SP ST SZ TR UK **N:** CI EG LB MR TU **A:** AF BA BT CY FE FUJ GUA GUI GUX HAI HEB HEI HEN HUB HUN IQ JA JIA JIL JIX KZ LIA NMO SA SCH SHA TAI YE YUN "Korea" **COS**

granivorus Mulsant & Godart, 1868: 288

mauritanicus Fabricius, 1792a: 113 (*Tenebrio*)

piceus G.-A. Olivier, 1795: 58: 17 (*Helops*)

picipes Panzer, 1794c: 4 (*Helops*)

risbeci Lepesme, 1943 (*Latetribolium*)

ruficolor Pic, 1925d: 11

rufipes W.J. MacLeay, 1873b: 286 (*Microphyes*)

striatulus Fairmaire, 1869b: 231 (*Cataphronetis*)

viator Mulsant & Godart, 1868: 290 **Ei:** FR SP **AFR**

genus *Diaclina* Jacquelin du Val, 1861: 296 type species *Tenebrio chrysomelinus* Herbst, 1799 (= *Tenebrio testudineus* Piller & Mitterpacher, 1783)

bisbinotata Pic, 1926d: 27 **A:** NP **ORR**

ceylonica Kaszab, 1980a: 185 **A:** NP **ORR**

fagi Panzer, 1799a: 3 (*Tenebrio*) **E:** AB AL AU CT CZ FR GE GG GR HU IT RO SK SL ST SZ TR UK YU **A:** CY IS TR

gracilipes Baudi di Selve, 1876c: 231 (*Alphitobius*)

nakatai Akita & Masumoto, 2003: 178 **A:** JA (Ishigaki)

plagiata Marseul, 1876c: 108 (*Alphitophagus*) **A:** SC TAI
quadrimaculata Gebien, 1914d: 394 (*Alphitobius*) **A:** AP JA (Ryukyus) NP TAI **ORR**
testudinea Piller & Mitterpacher, 1783: 83 (*Tenebrio*) **E:** AB AL AR AU BH BU CR CZ FR GE GR HU GG IT MC RO SK SL ST UK YU **A:** IN TR
chrysomelina Rossi, 1790: 233 (*Tenebrio*)
chrysomelina Herbst, 1799: 17 (*Tenebrio*) [HN]
glabrata Fabricius, 1801b: 567 (*Mycetophagus*)
picea L. Redtenbacher, 1874: 113 (*Alphitobius*)
yaeyamaensis Masumoto & Akita, 2002: 150 **A:** JA (Ryukyus)

genus *Hoplopeltis* Fairmaire, 1894a: 22 type species *Hoplopeltis tricornis* Fairmaire, 1894
tricornis Fairmaire, 1894a: 23 **A:** NP **ORR**

genus *Peltoides* Laporte, 1832: 401 type species *Peltoides senegalensis* Laporte, 1832
subgenus *Peltoides* Laporte, 1832: 401 type species *Peltoides senegalensis* Laporte, 1832
Oopiestus Chevrolat, 1833c: 30 type species *Oopiestus ovalis* Chevrolat, 1833
capensis longulus Fairmaire, 1897e: 126 **A:** SA YE **AFR**

tribe Amarygmini Gistel, 1848

genus *Amarygmus* Dalman, 1823: 60 type species *Chrysomela micans* Fabricius, 1794
subgenus *Amarygmus* Dalman, 1823: 60 type species *Chrysomela micans* Fabricius, 1794
Anacycus Fairmaire, 1896a: 33 type species *Dietysus navicularis* Fairmaire, 1894
Apelina Saha, 1988: 429 type species *Apelina keralaensis* Saha, 1988
Aphyllocerus Fairmaire, 1881a: 282 type species *Aphyllocerus decipiens* Fairmaire, 1881
Becvaramarygmus Masumoto, 1999: 369 type species *Dietysus atricolor* Pic, 1922 (= *Dietysus filicornis* Gravely, 1915)
Dictysus Rye, 1873: 281 [UE]
Dietysus Pascoe, 1866: 486 type species *Dietysus confusus* Pascoe, 1866 (= *Cnodalon metallicum* Perty, 1831)
Elixota Pascoe, 1866: 475 type species *Elixota cuprea* Pascoe, 1866 (= *Amarygmus hopei* Bremer, 2001)
Eurypera Pascoe, 1870: 106 type species *Eurypera cuprea* Pascoe, 1870 (= *Amarygmus pascoei* Gebien, 1911)
Platolenes Gebien, 1914d: 420 type species *Platolenes rufipes* Gebien, 1914
Plesiamarygmus Masumoto, 1989: 314 type species *Dietysus ovoideus* Fairmaire, 1882
Pseudamarygmus Pic, 1915f: 9 type species *Pseudamarygmus testaceipes* Pic, 1915
adonis Pic, 1922b: 12 **A:** NE NP **ORR**
antecedens Ando & Bremer, 2014: 9 **A:** TAI
antecursor Ando & Bremer, 2014: 11 **A:** TAI
ardoini Bremer, 2001: 57 [RN] **A:** SCH YUN
viridiscens Pic, 1926j: 75 (*Anacycus*) [HN]
bremeri Masumoto, Akita & Lee, 2005: 248 **A:** TAI
carbonarius Hope, 1843: 63 (*Cnodulon*) **A:** FUJ
cuprarius cuprarius Weber, 1801: 40 (*Helops*) **A:** JA TAI **AUR ORR**
callichromus Fairmaire, 1897b: 70

iridipennis Fairmaire, 1893c: 58
laosensis Pic, 1922j: 304
curvus Marseul, 1876d: 316 **A:** GAN JA SC TAI ZHE
pellegrini Pic, 1922f: 24 (*Anacycus*)
cyaneipennis Pic, 1938a: 10 **A:** TAI **ORR**
diversiceps Pic, 1926b: 21 **A:** TAI **ORR**
bicoloriceps Pic, 1923b: 20 [HN]
sakaii Masumoto, 1981: 29
falsicurva Masumoto, 1982b: 51 (*Elixota*) **A:** TAI
familiaris Masumoto, 1985c: 122 **A:** HKG
filicornis Gravely, 1915: 532 (*Dietysus*) **A:** AP **ORR**
atricolor Pic, 1922j: 304 (*Dietysus*)
hoanus Pic, 1929f: 44 **A:** TAI **ORR**
duporti Pic, 1938a: 9
hsiaohsingus Masumoto, 1984: 16
hopei Bremer, 2001: 57 [RN] **A:** FUJ GUX
cupreus Pascoe, 1866: 475 (*Elixota*) [HN]
incognitus Bremer & Ando, 2009: 14 **A:** JA
iridicollis Nakane, 1968b: 83 (*Elixota*) **A:** JA (Ryukyus)
izumii Masumoto, 1983a: 81 (*Elixota*) **A:** JA (Ryukyus) TAI
latifrons Gravely, 1915: 534 (*Dietysus*) **A:** AP
maliburuensis Masumoto, Akita & Lee, 2015a: 166 **A:** TAI
nakanei Masumoto, 1982b: 53 (*Elixota*) **A:** TAI
nantouensis Masumoto, Akita & Lee, 2015a: 163 **A:** TAI
navicularis Fairmaire, 1894a: 26 (*Dietysus*) **A:** AP NP UP **ORR**
nepalensis Kaszab, 1977b: 281 (*Platolenes*) **A:** NP **ORR**
brancuccii Bremer, 2003: 51
nodicornis Gravely, 1915: 533 (*Dietysus*) **A:** AP
vitalisi Pic, 1922j: 305
parvus Pic, 1926b: 20 **A:** HAI GUA YUN **ORR**
picitarsis Fairmaire, 1882c: 250 (*Dietysus*) **A:** JA (Ryukyus) TAI **ORR**
aeneus Wiedemann, 1821: 154 (*Cnodalon*) [HN]
blaisei Pic, 1923b: 21
inadai Masumoto & Akita, 2001a: 21
rouyeri Pic, 1951a: 18
pilipes Gebien, 1914a: 42 **A:** AP FUJ JIX NP SD TAI YUN **ORR**
formosanus Pic, 1915g: 21
longipillis Pic, 1922j: 303
punctatus Pic, 1922b: 11 (*Anacycus*) **A:** YUN
purpureofossus Fairmaire, 1896a: 32 **A:** AP NP SD **ORR**
darjilingensis Saha, 1988: 434 (*Platolenes*)
rufipennis Pic, 1922b: 11 (*Anacycus*) **A:** AP

sikkimensis Saha, 1988: 435 (*Platolenes*) **A:** SD
sinensis Pic, 1922b: 11 **A:** ZHE
socer Bremer, 2013: 201 **A:** TAI
speciosus Dalman, 1823: 61 **A:** NP YUN **ORR**
taiwanus Masumoto, 1981c: 31 **A:** TAI
trichopus Kaszab, 1941c: 69 **A:** TAI
vicinus Pic, 1922b: 10 (*Anacycus*) **A:** HAI SHG
wangi Masumoto, Akita & Lee, 2015a: 161 **A:** TAI
yangmingensis Masumoto, Akita & Lee, 2015a: 168 **A:** TAI
subgenus *Oogeton* Kaszab, 1941c: 69 type species *Oogeton nigrocoeruleum* Kaszab, 1941 (= *Amarygmus makii* Miwa, 1939)
makii Miwa, 1939: 412 **A:** TAI
nigrocoeruleum Kaszab, 1941c: 70

genus *Azarelius* Fairmaire, 1892c: vii type species *Azarelius sculpticollis* Fairmaire, 1892
tenuicornis Blair, 1914c: 315 **A:** SD **ORR**

genus *Eumolpamarygmus* Pic, 1923h: 11 type species *Eumolpamarygmus bigibbosus* Pic, 1923
becvari Masumoto, 1999c: 367 **A:** YUN
paotao Masumoto, 1989a: 107 **A:** TAI

genus *Gonocnemis* J. Thomson, 1858: 101 type species *Gonocnemis strigipennis* J. Thomson, 1858
Acastus Péringuey, 1896: 177 type species *Acastus agrestis* Péringuey, 1896
senegalensis Fairmaire, 1894f: 673 **A:** YE **AFR**
surcoufi Pic, 1928h: 258 **A:** SA YE **AFR**

genus *Hoplobrachium* Fairmaire, 1886b: 74 type species *Hoplobrachium asperipenne* Fairmaire, 1886 (= *Helops dentipes* Fabricius, 1781)
Cephaloplonyx Pic, 1922b: 12 type species *Cephaloplonyx opacus* Pic, 1922 (= *Helops dentipes* Fabricius, 1781)
dentipes Fabricius, 1781: 326 (*Helops*) **A:** NP YUN **AFR ORR**
asperipenne Fairmaire, 1886b: 74
ebeninus Walker, 1858: 285 (*Helops*)
opacus Pic, 1922b: 12 (*Cephaloplonyx*)

genus *Insolitoplonyx* Bremer, 2014a: 179 type species *Insolitoplonyx seorsus* Bremer, 2014
seorsus Bremer, 2014a: 179 **A:** SD

genus *Javamarygmus* Pic, 1928b: 22 type species *Javamarygmus tristis* Pic, 1928
chingshanus Masumoto, 1982b: 49 (*Eumolpocyriogeton*) **A:** TAI
maculosus Pic, 1927g: 8 (*Eumolpamarygmus*) **A:** HAI **ORR**

genus *Nepaloplonyx* Bremer, 2014a: 175 type species *Nepaloplonyx caelebs* Bremer, 2014
caelebs Bremer, 2014a: 175 **A:** NP
singularis Wasmann, 1896b: 151 (*Azarelius*) **A:** SD

genus *Plesiophthalmus* Motschulsky, 1858a: 34 type species *Plesiophthalmus nigrocyaneus* Motschulsky, 1858
subgenus *Cyriogeton* Pascoe, 1871: 356 type species *Cyriogeton insignis* Pascoe, 1871
convexus Pic, 1914d: 13 (*Cyriogeton*) **A:** GUI GUX HKG YUN **ORR**
auricolle Pic, 1952a: 3 (*Cyriogeton*)
pulchrum Pic, 1926d: 28 (*Cyriogetum*)
lijainus Masumoto, Akita & Lee, 2008b: 15 **A:** TAI
subgenus *Inspinogeton* Pic, 1937i: 174 type species *Inspinogeton impressipennis* Pic, 1937
anmashanus Masumoto, Akita & Lee, 2008: 218 **A:** TAI
impressipennis Pic, 1937i: 174 (*Cyriogeton*) **A:** FUJ ZHE
mizusawai Yamazaki, 1968: 27 (*Plesiophthalmus*) **A:** JA (Amami-Ôshima)
suzukii Masumoto, Akita & Lee, 2014b: 237 (*Amarygmus*) **A:** TAI
subgenus *Opacoplonyx* Bremer, 2014b: 35 type species *Plesiophthalmus davidis* Fairmaire, 1878
davidis Fairmaire, 1878a: 121 **A:** BEI FE HUB HEN NC SC **ORR**
lentus Heyden, 1887a: 268
subgenus *Plesiophthalmus* Motschulsky, 1858a: 34 type species *Plesiophthalmus nigrocyaneus* Motschulsky, 1858
antrax Fairmaire, 1903a: 17 **A:** YUN **ORR**
subnitidipennis Pic, 1934a: 17
ater Pic, 1930b: 34 **A:** FUJ GUI SCH YUN
aurichalceus M. Chûjô, 1978a: 75 **A:** JA (Ryukyus)
borchmanni Kaszab, 1957c: 56 **A:** FUJ
brevipennis brevipennis Lewis, 1896: 340 **A:** JA (Amami-Ôshima)
brevipennis makiharai M. Chûjô, 1977: 16 **A:** JA (Tokunoshima Is.)
caeruleus Pic, 1914d: 14 **A:** "South China" **ORR**
chujoi Masumoto, 2005a: 76 **A:** TAI
colossus Kaszab, 1957c: 59 **A:** FUJ GUA
cruralis Fairmaire, 1903a: 17 **A:** FUJ GUI **ORR**
donckieri Pic, 1914d: 14 (*Cyriogeton*) **A:** YUN
excellens Gebien, 1914b: 41 (*Cyriogeton*) [RN] **A:** BT SD
formosanus Miwa, 1939: 412 **A:** TAI
fujitai Masumoto, 1981c: 24 (*Cyriogeton*) **A:** TAI
fukiensis Masumoto, 1990g: 717 **A:** FUJ GUA
fuscoaenescens Fairmaire, 1898a: 14 **A:** JA (Ryukyus)
fushanus Masumoto, Akita & Lee, 2015: 171 **A:** TAI
gaoligongensis Masumoto, 2000a: 148 **A:** YUN
gibbosulus Fairmaire, 1893c: 41 (*Eucyrtus*) **A:** FUJ **ORR**

guizhouensis Masumoto, 1999c: 355 **A:** GUI
gracilis M. Chûjô, 1979: 45 **A:** JA (Iriomote)
hainanensis Masumoto, 1990d: 268 **A:** HAI
hsinhuiae Masumoto, Akita & Lee, 2014b: 240 **A:** TAI
impressus Pic, 1927j: 132 (*Cyriogeton*) **A:** FUJ
inexpectatus Masumoto, 1989d: 764 **A:** SCH YUN
jessopi Masumoto, 1990d: 269 **A:** BH **ORR**
kanoi Masumoto, 1981c: 26 **A:** TAI
kaszabi Masumoto, 1990a: 40 **A:** FUJ
kentingensis Masumoto, Akita & Lee, 2015a: 171 **A:** TAI
kimurai Masumoto, 1996a: 211 (*Pseudoogeton*) **A:** JA (Ryukyus)
kondoi Masumoto, 1981c: 18 (*Cyriogeton*) **A:** TAI
kulzeri Kaszab, 1954: 263 **A:** FUJ GUA
kumejimensis Masumoto, 1987b: 87 **A:** JA (Ryukyus)
kurosawai Masumoto, 1991a: 30 **A:** SD
laevicollis Harold, 1878a: 79 **A:** JA
lalashanus Masumoto, Akita & Lee, 2015a: 173 **A:** TAI
latus Pic, 1914d: 13 (*Cyriogeton*) **A:** NP **ORR**
semilatus Pic, 1952a: 3 (*Cyriogeton*)
lineipunctatus Fairmaire, 1891c: xx **A:** HUB
longipes Pic, 1938a: 8 **A:** FUJ GUI SCH YUN XIZ
longulus Fairmaire, 1893b: 33 (*Rehumius*) **A:** FUJ
mayumiae Masumoto, 1981c: 21 (*Cyriogeton*) **A:** TAI
michitakai Masumoto, 1990f: 716 **A:** TAI
morio Pic, 1938a: 8 **A:** HUB HUN JIA ZHE
nakamurai Masumoto, 1987b: 89 **A:** JA (Amami-Ôshima)
nanshanchiense Masumoto, 1981c: 25 (*Cyriogeton*) **A:** TAI
nigroaeneus Gebien, 1914b: 40 (*Cyriogeton*) **A:** TAI
nigrocyaneus aeneus Motschulsky, 1862: 19 **A:** JA
kikutii Yamazaki, 2002: 377
yakushimanus Yamazaki, 1964: 24
nigrocyaneus nigrocyaneus Motschulsky, 1858a: 35 **A:** FE JA NE "Korea"
aenescens Marseul, 1876d: 318
mikuraensis Nakane, 1968b: 83
nigratus Motschulsky, 1862: 20
nitidior Pic, 1952a: 3
olivaceus Pic, 1939g: 342
sericeifrons Marseul, 1876d: 317
nigrocyaneus oyamai Nakane, 1956b: 169 **A:** JA
nishikawai Masumoto, 1981c: 22 (*Cyriogeton*) **A:** TAI
oblongus Masumoto, 1991a: 10 **A:** YUN

paiwanus Masumoto, Akita & Lee, 2008a: 220 **A:** TAI
pallidicrus Fairmaire, 1889b: 46 **A:** FUJ GUX HUN SCH YUN
bicoloripes Pic, 1914d: 14
pernitidus Fairmaire, 1897k: 251 (*Cyriogeton*) **A:** GUI SCH
perpulchrus Pic, 1930b: 34 (*Cyriogeton*) **A:** YUN **ORR**
pulchrus Pic, 1927m: 187 (*Cyriogeton*) [HN]
piceus M. Chûjô, 1978a: 76 **A:** JA (Okinawa)
pieli Pic, 1937i: 175 **A:** FUJ GUA ZHE **ORR**
niger Kaszab, 1957c: 58
pseudometallicus Masumoto, 1990e: 478 **A:** YUN
punctatus M. Chûjô, 1979: 50 **A:** JA (Ryukyus)
puncticollis Yamazaki, 1964: 26 **A:** JA
rubrofasciatus Ren, 1998: 233 **A:** ZHE
sameshimai Masumoto, 2001b: 58 **A:** JA (Kyushu)
semipurpureus Pic, 1917c: 9 (*Coriogeton*) **A:** SD
shibatai Masumoto, 1990g: 711 **A:** TAI
shigeoi Masumoto, 1981c: 19 (*Cyriogeton*) **A:** TAI
shoyamai Akita & Masumoto, 2015: 410 **A:** JA
sichuanicus Masumoto, 1999c: 356 **A:** SCH
spectabilis okinawanus Nomura, 1964: 49 **A:** JA (Okinawa)
spectabilis spectabilis Harold, 1875: 293 **A:** BEI HEI JA LIA NC NMO SC TAI ZHE
arciferens Fairmaire, 1878a: 120
obesus Marseul, 1876d: 319
subparallelus Pic, 1916d: 11
splendens Masumoto, 1990e: 494 **A:** BT **ORR**
taiwanus Nomura, 1964: 49 **A:** TAI
takakuwai Masumoto, Akita & Lee, 2015b: 305 **A:** TAI
yunnanus Pic, 1952a: 3 (*Cyriogeton*) **A:** YUN

genus *Pseudoogeton* Masumoto, 1989b: 304 type species *Plesiophthalmus amplipennis* Fairmaire, 1897
amplipennis Fairmaire, 1897k: 252 (*Plesiophthalmus*) **A:** SCH
gebieni Masumoto, 1989b: 312 **A:** SCH
maoxianum Dong & Ren, 2016: 75 **A:** SCH
ovipennis Fairmaire, 1891c: xx (*Plesiophthalmus*) **A:** GUA HUB JIA
uenoi Masumoto, 1981: 93 (*Plesiophthalmus*) [RN] **A:** TAI
taiwanus Masumoto, 1981c: 28 (*Plesiophthalmus*) [HN]

genus *Ziaelas* Fairmaire, 1892d: cx type species *Ziaelas insolitus* Fairmaire, 1892
formosanus Hozawa, 1914: 484 **A:** TAI

tribe Apocryphini Lacordaire, 1859

genus *Diplocyrtus* G. Quedenfeldt, 1887: 257 type species *Diplocyrtus floccosus* G. Quedenfeldt, 1887

baudoni Kocher, 1962a: 200 **N:** MO
beidensis Antoine, 1936c: 141 (*Diplocyrthus*) **N:** MO
floccosus G. Quedenfeldt, 1887: 258 **N:** MO
formicoides Kocher, 1963b: 114 **N:** MO
hiemalis Escalera, 1914: 356 **N:** MO
hirtus Escalera, 1914: 357 **N:** MO
longithorax Escalera, 1914: 356 **N:** MO
marmoratus Kocher, 1962b: 306 **N:** MO
minimus Antoine, 1952: 118 **N:** MO (Anti Atlas)
panousei Kocher, 1962a: 201 **N:** MO
schrammi Antoine, 1936c: 139 (*Diplocyrthus*) **N:** MO
semihispidus Antoine, 1951b: 319 **N:** MO (Sous)
susicus Escalera, 1913: 50 **N:** MO
thamii Kocher, 1962a: 202 **N:** MO
vaulogeri Escalera, 1914: 358 **N:** MO

tribe Blaptini Leach, 1815
subtribe Blaptina Leach, 1815

genus *Ablapsis* Reitter, 1887d: 368 type species *Blaps compressipes* Reitter, 1887
compressipes Reitter, 1887d: 368 (*Blaps*) **A:** GAN QIN

genus *Blaps* Fabricius, 1775: 254 type species *Tenebrio mortisagus* Linnaeus, 1758
subgenus *Arenoblaps* G.S. Medvedev, 1999a: 400 type species *Blaps hiemalis* Semenov & Bogatchev, 1940
hiemalis Semenov & Bogatchev, 1940: 206 **A:** TM UZ
subgenus *Blaps* Fabricius, 1775: 254 type species *Tenebrio mortisagus* Linnaeus, 1758
Acanthoblaps Reitter, 1889k: 687 type species *Blaps dentitibia* Reitter, 1889
Agroblaps Motschulsky, 1860e: 531 type species *Blaps fatidica* J. Sturm, 1807 (= *Blaps lethifera* Marsham, 1802)
Blapidium Kolbe, 1928: 200 type species *Blaps ocreata* Allard, 1881
Blapidurus Fairmaire, 1891d: xcvi type species *Blapidurus crassicornis* Fairmaire, 1891
Blapimorpha Motschulsky, 1860e: 531 type species *Blaps reflexa* Gebler, 1832
Blapisa Motschulsky, 1860e: 530 type species *Blaps jaegeri* Hummel, 1827 (= *Tenebrio mortisagus* Linnaeus, 1758)
Hypoblaps Kolbe, 1928: 200 type species *Blaps rotundata* Solier, 1848
Leptocolena Allard, 1880: 74 type species *Blaps mucronata* Latreille, 1804
Leptomorpha Faldermann, 1835b: 406 type species *Leptomorpha chinensis* Faldermann, 1835
Lithoblaps Motschulsky, 1860e: 532 type species *Tenebrio gigas* Linnaeus, 1767

Mesoblaps Kolbe, 1928: 200 type species *Blaps acuminata* Fischer von Waldheim, 1820

Nanoblaps Semenov & Bogatchev, 1936: 565 type species *Blaps jakovlevi* Semenov & Bogatchev, 1936

Notoblaps Kolbe, 1928: 201 type species *Blaps juliae* Allard, 1881

Opisthoblaps Kolbe, 1928: 201 type species *Blaps sulcifera* Seidlitz, 1893

Platyblaps Motschulsky, 1860e: 531 type species *Blaps holconota* Fischer von Waldheim, 1844

Rhizoblaps Motschulsky, 1860e: 532 type species *Blaps pruinosa* Falderman, 1833

Uroblaps Motschulsky, 1860e: 530 type species *Blaps producta* Brullé, 1832 (= *Blaps lusitanica* Herbst, 1799)

abbreviata abbreviata Ménétriés, 1836a: 150 **E:** AU BU CR GR HU MC RO SK SL YU **A:** IN SY TR

ecaudata Küster, 1845b: 45 [HN]

ecaudata Seidlitz, 1893: 306 [HN]

ovata Solier, 1848: 305

rotundicollis Reiche & Saulcy, 1857: 240

abbreviata indagator Reiche & Saulcy, 1857: 238 **A:** IS

acuminata acuminata Fischer von Waldheim, 1820: pl. 16 **A:** GAN MG NMO QIN XIN "Siberia"

przewalskyi Reitter, 1887: 370

acuminata kulszhana Skopin, 1973a: 872 **A:** XIN

acutangula Ren & Wang, 2001: 19 **A:** NIN SHA

adelae L. Soldati & F. Soldati, 1999: 29 **A:** TR

afghanistanica Gridelli, 1954: 266 [= 1955: 61] **A:** AF

alternans Brullé, 1839: 68 **N:** CI

amurensis Allard, 1881a: 318 **A:** NMO

angusta Seidlitz, 1893: 307 **A:** IN

antennalis Allard, 1881a: 305 **N:** MO

anura Seidlitz, 1893: 306 **A:** AF TD UZ

appendiculata Motschulsky, 1851b: 653 [RN] **N:** AG MO TU

caudata Solier, 1848: 332 [HN]

caudigera Gemminger, 1870a: 121 [RN]

approximans Seidlitz, 1893: 256 **N:** LB TU

araxicola Seidlitz, 1893: 306 **E:** AR

aruktavica G.S. Medvedev, 2004b: 572 **A:** TD UZ

aspericollis Schuster, 1923b: 159 **A:** BEI

badakschanica Kaszab, 1960a: 73 **A:** AF TD

balashovi Bogatchev & G.S. Medvedev, 1974: 645 **A:** TM (Kopet Dag)

ballioni Skopin, 1977: 151 **A:** KZ (Tien Shan) XIN

barrancoi Castro Tovar, 2014: 238 **E:** SP

batesi Allard, 1881a: 304 **A:** IQ SA YE

rathjensi Schuster, 1938a: 61

bedeli Chatanay, 1914b: 75 **N:** AG MO

bengalensis Hope, 1831: 31 **A:** AF IN **ORR**

spathulata Solier, 1848: 328

berezowskii G.S. Medvedev, 1998b: 206 **A:** SCH

bifurcata bifurcata Solier, 1848: 348 **N:** EG LB **A:** IS (Negev) SI
andresi Koch, 1935a: 27
dimidiata Solier, 1848: 349
bifurcata strauchii Reiche, 1861b: 88 **N:** AG LB MO TU
gridellii Pierre, 1961: 1042
binominata Escalera, 1914: 324 [RN] **N:** AG MO TU
caudata Allard, 1881a: 314 [HN]
caudigera Allard, 1881a: 306 [HN]
bogatshevi G.S. Medvedev, 1964a: 60 **A:** TD TM
brevicollis Fairmaire, 1891e: cxxxi (*Leptomorpha*) **A:** KA
brunnea Allard, 1881a: 316 **A:** XIZ "Himalayas"
bucharensis Seidlitz, 1893: 308 **A:** UZ
bushirensis Kaszab, 1959c: 54 **A:** IN
caraboides caraboides Allard, 1882a: 135 **A:** AF GAN KI KZ NIN QIN SHA TD XIN XIZ
aberrans Reinig, 1931: 878
alaiensis Reinig, 1931: 877
chinensis Reinig, 1931: 878 [HN]
emarginata Reinig, 1931: 878
licinoides Seidlitz, 1893: 308
ovata Reinig, 1931: 879
schusteri Reinig, 1934: 162 [RN]
caraboides intermittens Kaszab, 1962c: 313 **A:** KZ
caspica I. Chigray, Abdurakhmanov, Nabozhenko & Shapovalov, 2016: 3 **A:** Kazakhstan (Kulaly Island)
chinensis Faldermann, 1835b: 407 (*Leptomorpha*) **A:** BEI GAN HEB HEN HUB JIA LIA NIN NMO SHN SHX
choui Ren & Wang, 2001: 21 **A:** SHA
clotzeri Fischer von Waldheim, 1844: 105 **A:** IN TM
cognata Solier, 1848: 336 **N:** EG
schweinfurthii Seidlitz, 1893: 263
concii Koch, 1965a: 143 **A:** YE
conradtii Seidlitz, 1893: 308 **A:** KI KZ TD TM UZ
contracta Ren, 2016: 103 **A:** XIZ
cordicollis Solier, 1848: 346 **N:** EG **A:** SY
crassicornis Fairmaire, 1891d: xcvi (*Blapidurus*) **A:** KA
cretensis Koch, 1948a: 332 **E:** GR (Crete)
cribrosa Solier, 1848: 304 **A:** IS SY TR
angulata Reiche & Saulcy, 1857: 247
cychroides Fairmaire, 1887b: 127 **A:** YUN
davidis Deyrolle, 1878: 119 **A:** BEI GAN HEI HUB MG NIN NMO SHA SHN "Korea"
davidea Allard, 1886: [6] (*Blapimorpha*)

debdouensis Obenberger, 1914: 138 **N:** MO
dehaani Baudi di Selve, 1875c: 100 **A:** AF IN TM
dentitibia Reitter, 1889k: 687 **A:** GAN NIN QIN XIN
deplanata Ménétriés, 1832: 199 **E:** AB **A:** IN KZ TM UZ
cordata Seidlitz, 1893: 288
curvipes Baudi di Selve, 1874a: 111 [= 1875c: 91]
muricata Fischer von Waldheim, 1832a: 199
dispar Seidlitz, 1893: 308 **A:** UZ "Turkestan"
divergens Fairmaire, 1875a: 527 **N:** TU
doderoi Schuster, 1922b: 21 **N:** LB
dorsogranata Fairmaire, 1887b: 128 **A:** YUN
effeminata L. Soldati, 2017: 6 **N:** LB
eleodes Kaszab, 1962c: 313 **A:** HEB
elliptica Seidlitz, 1893: 309 **A:** UZ "Turkestan"
emoda Allard, 1881a: 320 **A:** HP SCH XIZ
emondi Solier, 1848: 331 **N:** AG TU
nitidula Solier, 1848: 330
evanida Seidlitz, 1893: 285 **A:** KZ
eximia Semenov & Bogatchev, 1936: 561 **A:** KZ XIN
fascinosa Seidlitz, 1893: 305 **A:** IQ
faustii bactriana Bogatchev, 1959: 65 **A:** TD
faustii faustii Seidlitz, 1893: 305 **A:** AF KZ TD TM UZ
felix Waterhouse, 1889: 126 **A:** AF IN TM
regia Seidlitz, 1893: 308
femoralis femoralis Fischer von Waldheim, 1844: 141 (*Pandarus*) **A:** GAN HEB MG NIN NMO SHA SHX
femoralis rectispina Kaszab, 1968c: 377 **A:** MG
ferganica Bogatchev, 1959: 65 **A:** KI
fortesculpta Gridelli, 1953: 55 **A:** YE
freibergi Skopin, 1969: 881 **A:** TM (Kugitang-Tay)
furcala Ren & Wang, 2001: 16 **A:** NMO
garzica Ren & Wang, 2001: 15 **A:** SCH
gibba Laporte, 1840: 200 **E:** CR FR GR IT SP
australis Solier, 1848: 322
impressicollis Solier, 1848: 322
italica Baudi di Selve, 1875c: 97
planicollis Solier, 1848: 322
rectangularis Solier, 1848: 316
gigas Linnaeus, 1767: 676 (*Tenebrio*) **E:** AL AUi AZ BU CR CZ FR GEi GR IC IT MA PT SL SP SZ UK **N:** AG CI LB MR SG TU **A:** IQ KZ TR **AFR**
avenae J.R. Sahlberg, 1903d: 13

azorica Seidlitz, 1893: 258
gages Linnaeus, 1767: [Errata] (*Tenebrio*) [RN]
gigantea L. Petagna, 1819: 30
obtusangula Rey, 1892: 41
occulta Seidlitz, 1893: 312
producta Brullé, 1833: 302 [HN]
glabrata Seidlitz, 1893: 310 **A:** CH
glazunovi Semenov & Bogatchev, 1936: 562 **A:** IN (Elburs)
gobiensis J. Frivaldszky, 1890: 207 **A:** GAN HEB NMO QIN NIN XIN
mongolica Seidlitz, 1893: 274
graeca Solier, 1848: 312 **E:** GR
granulata altynemelis Skopin, 1966b: 337 **A:** KZ (Almaatinskaya Oblast)
granulata granulata Gebler, 1825: 47 **A:** GAN KI KZ XIN
seidlitzi Schuster, 1936a: 5
granulata granulosa Ménétriés, 1854: 34 **A:** KI XIN
granulata kirgisica Skopin, 1966b: 336 **A:** KI
granulata obliterata Ménétriés, 1849: 236 **A:** KZ
granulata stackelbergi Bogatchev, 1952b: 284 **A:** KZ (Vostochnyy Kazakhstan)
granulata turcomana Fischer von Waldheim, 1843: 16 [= 1844: 88] **A:** KI KZ NMO UZ XIN
caudata Gebler, 1844: 103
granulipennis Skopin, 1966b: 337 **A:** KZ (Almaatinskaya Oblast)
gressoria Reitter, 1889k: 689 **A:** GAN NIN NMO QIN
haarlovi Gridelli, 1954: 264 [= 1955: 59] **A:** AF TD
haberti Peyerimhoff, 1931b: 102 **N:** AG (Hoggar) LB
helopioides Seidlitz, 1893: 309 **A:** XIN
heydeni Allard, 1881a: 305 **N:** MO
taciturna Peyerimhoff, 1949b: 279
hispanica Laporte, 1840: 200 **E:** SP
holconota Fischer von Waldheim, 1844: 71 **A:** AF KZ TM UZ
corrosa Fischer von Waldheim, 1844: 74
quinquecostata Fischer von Waldheim, 1844: 76
stenothorax Fischer von Waldheim, 1844: 73
holcus Ren, 2016: 124 **A:** XIZ
indicola Bates, 1879: 472 **A:** AF IN KA XIN XIZ
lucens Fairmaire, 1891d: xcv
inflata Allard, 1881a: 305 **N:** MO
inflatipennis J.R. Sahlberg, 1908: 75 **A:** LE
inflexa Zubkov, 1833: 331 **A:** KZ TM
carbonaria Kraatz, 1888a: 197
haberhaueri Seidlitz, 1893: 308
intermedia L. Soldati, 2017: 9 **N:** MO

iraquensis Kaszab, 1959c: 53 **A:** IQ
jakovlevi Semenov & Bogatchev, 1936: 565 **A:** AF TM
japonensis Marseul, 1879c: 99 **A:** BEI GUI HEB JA LIA SHA SHX TAI "Korea"
jeannei J. Ferrer & L. Soldati, 1999: 66 **A:** TR
judaeorum Miller, 1861: 178 **N:** EG **A:** IQ IS LE SI SY
judaica Seidlitz, 1893: 265
pharao Seidlitz, 1893: 266
juliae Allard, 1881a: 311 **A:** IS
kabuliensis Kaszab, 1960a: 66 **A:** AF
kadyrbekovi G.S. Medvedev, 2004b: 570 **A:** KZ
kaifensis Seidlitz, 1893: 305 **A:** IQ IS
kasatkini I. Chigray & Nabozhenko, 2016: 268 **A:** IN
kashgarensis Bates, 1879: 473 **A:** XIN
eusoma Kraatz, 1888a: 195
lobnoriana Reitter, 1887d: 371
kiritshenkoi Semenov & Bogatchev, 1936: 555 **A:** GAN MG NIN NMO
klapperichi Kaszab, 1960a: 71 **A:** AF
kolbei Schuster, 1914b: 142 **A:** SCH XIZ
kollari kollari Seidlitz, 1893: 263 **N:** EG **A:** IQ JO OM SA YE
kollari zhenzhuristi Bogatchev, 1939a: 125 **A:** YE
kovali Abdurakhmanov & Nabozhenko, 2011: 213 **E:** AR
kulzeri Pierre, 1964c: 1049 **A:** IN
kulzeriana Pierre, 1964c: 1049 (*Caenoblaps*)
ladakensis Bates, 1879: 472 **A:** KA XIZ
lata Seidlitz, 1893: 309 **A:** TR
laticollis Solier, 1848: 324 **N:** EG
ledereri Fairmaire, 1866b: 263 (*Prosodes*) **A:** TR (Izmir: Bozdag)
lethifera lethifera Marsham, 1802: 479 **E:** AL AU AZ BE BH BU CR CT CZ EN FI FR DE FI GB GE GR HU IR IT LA LS MC NL NR PL PT RO SK SP ST SV SZ UK WS **N:** AG MR CI XIN **A:** ES JO KZ TR WS XIN **NAR**i
abdita Picka, 1978: 100
anthracina Faldermann, 1837: 50
asiatica Solier, 1848: 310
bipunctata Allard, 1881a: 320
brevis Fischer von Waldheim, 1844: 96
carinula Seidlitz, 1893: 306
convexa Fischer von Waldheim, 1844: 92
crassa Reiche & Saulcy, 1857: 245
damascena Fischer von Waldheim, 1844: 101
dorsata Fischer von Waldheim, 1844: 98
fatidica J. Sturm, 1807: 205

laevicollis Rey, 1892: 41
milleri Seidlitz, 1893: 306
orbicollis Motschulsky, 1845a: 68
plicaticollis Ménétriés, 1836a: 151
proxima Solier, 1848: 315
reflexicollis Solier, 1848: 314 [HN]
robusta Motschulsky, 1845a: 68
similis Latreille, 1804a: 279
striatopunctata Allard, 1881a: 314
subquadrata Brullé, 1832: 204
lethifera pterotapha Fischer von Waldheim, 1832a: 201 **E:** AB AR GG ST **A:** IN TM TR
angulicollis Motschulsky, 1860e: 528 (*Agroblaps*)
luctuosa Ménétriés, 1832: 201
menetriesii Kraatz, 1881a: 56
obliterata Ménétriés, 1848: 20 [=1849: 236]
pallasii Seidlitz, 1893: 306
pulvinata Ménétriés, 1849: 235
tarda Motschulsky, 1860e: 530 (*Agroblaps*)
lindbergi Kaszab, 1959c: 52 **A:** TR
lineata Laporte, 1840: 200 **N:** AG
longicornis Kraatz, 1882b: 95 **A:** KA KI UZ
longula Reiche & Saulcy, 1857: 234 **A:** IS TR
lucidula G.S. Medvedev, 1998b: 204 **A:** SCH
lugens Seidlitz, 1893: 304 **E:** PT
lusitanica ceballosi Koch, 1944b: 245 **E:** SP (Madrid)
lusitanica espanoli Koch, 1944b: 246 **E:** SP (Baleares)
lusitanica lusitanica Herbst, 1799: 197 **E:** FR PT SP
kordofana Baudi di Selve, 1875c: 92
mequignoni Koch, 1944b: 246
producta Brullé, 1832: 203
magica Erichson, 1841: 183 **N:** AG
maldesi L. Soldati, 2017: 12 **N:** AG
marginicollis Fairmaire, 1891e: cxxxi (*Blapidurus*) **A:** KA
maroccana Seidlitz, 1893: 257 **N:** MO
martensi Kaszab, 1978b: 227 **A:** KA
medusa Reitter, 1900e: 161 **A:** GAN MG NIN NMO
medusula Skopin, 1964: 372 **A:** ES (Tuva) MG NMO
medvedevi Bogatchev, 1961a: 107 **A:** TD TM
megalatlantica Koch, 1945: 409 **N:** MO (High Atlas)
menetriesiana Bogatchev, 1948: 152 **E:** AB
miliaria Fischer von Waldheim, 1844: 103 **A:** ES (Tuva) GAN MG NIN NMO SHX XIN

moerens Allard, 1881a: 319 **A:** HP SCH UP YUN XIZ

mortisaga Linnaeus, 1758: 418 (*Tenebrio*) **E:** AB AR AU BE BU BY CR CT CZ DE EN FI GB GE GG GR HU IC LA LS LT NL NR NT PL RO SK SL SP ST SV SZ UK **A:** IN IQ TM TR **NAR**

acuminata DeGeer, 1775: 31 (*Tenebrio*)

brevicornis Seidlitz, 1893: 307

canaliculata Faldermann, 1837: 48

carbo Fischer von Waldheim, 1844: 87

confusa Fischer von Waldheim, 1844: 81

dahlii Solier, 1848: 317

elongata Ménétriés, 1832: 199

encifer Motschulsky, 1845a: 67

jaegeri Hummel, 1827: 40

koenigii Seidlitz, 1893: 307

planicollis Motschulsky, 1845a: 65

reflexicollis Fischer von Waldheim, 1844: 82

motschulskiana Bogatchev, 1947e: 515 [RN] **A:** AF TM

gigantea Motschulsky, 1845a: 65 [HN]

mucronata Latreille, 1804a: 278 **E:** AU BE BU CR CZ DEi FIi FR GE GB GR IR IT LS MA NL NRi PL RO SK SP SVi SZ UK **N:** AG **A:** TR **NAR**i

chevrolatii Solier, 1848: 298

dilatata Laporte, 1840: 200

foveicollis Allard, 1881a: 320

obtusa J. Sturm, 1807: 206 [HN]

reflexa Solier, 1848: 299

stricticollis A. Villa & J.B. Villa, 1835: 48

striolata Küster, 1846c: 46

sublineata Brullé, 1832: 204

subovata Solier, 1848: 299

murati Peyerimhoff, 1943: 23 **N:** MO **AFR**

murgabensis Bogatchev, 1961a: 109 **A:** TD TM

mutata Gemminger, 1870a: 122 [RN] **A:** IQ IS LE TR SY

convexa Reiche & Saulcy, 1857: 243 [HN]

nadaii G.S. Medvedev, 2004a: 182 **A:** UP

nanshanica Semenov & Bogatchev, 1936: 556 **A:** GAN QIN

nefzauensis nefzauensis Seidlitz, 1893: 257 **N:** TU

nefzauensis vespertina Koch, 1937: 456 **N:** LB

nitens barbara Solier, 1848: 343 **N:** LB MO

substriata Solier, 1848: 345

nitens brachyura Küster, 1848a: 66 **E:** SP

abbreviata Solier, 1848: 339

nitens cyrenaia Seidlitz, 1893: 264 **N:** LB

nitens glebmedvedevi L. Soldati, 2020 [see New Acts] [RN] **A:** MO
medvedevi L. Soldati, Kergoat & Condamine, 2009: 232 [HN]
nitens laportei Ardoin, 1973: 292 [RN] **N:** EG LB **A:** IS (Negev) SI
sulcata Laporte de Castelnau, 1840: 200 [HN]
nitens mercatii Canzoneri, 1969: 44 **E:** IT (Sardinia)
nitens nitens Laporte, 1840: 200 **N:** AG TU
stygia Erichson, 1841: 182
subteres Solier, 1848: 341
tunisia Seidlitz, 1893: 264
vicina Solier, 1848: 342 [HN]
nitens praedeserta Koch, 1944b: 249 **N:** AG TU
nitens requieni Solier, 1848: 340 **N:** AG MO TU
nitida Fischer von Waldheim, 1844: 96 **A:** KZ WS
amoena Fischer von Waldheim, 1844: 99
nitiduloides L. Soldati, 2017: 15 **N:** TU
nuristanica Kaszab, 1959e: 250 [= 1960a: 75] **A:** AF
oberthuerii Seidlitz, 1893: 299 **A:** XIZ
oblonga Kraatz, 1883a: 349 **A:** GAN TD UZ WS XIN
gracilicollis Kraatz, 1885: 292
major Kraatz, 1883a: 350
ocreata Allard, 1881a: 314 **E:** GR (Crete) **A:** TR
oertzenii Seidlitz, 1893: 309 **E:** GR **A:** TR
oglobini Semenov & Bogatchev, 1936: 558 **A:** UZ (Bukhara)
ominosa Ménétriés, 1832: 198 **E:** AB AR GG **A:** AF IN TR
armeniaca Faldermann, 1837: 42
opaca Reitter, 1889k: 691 **A:** GAN NIN NMO QIN
orientalis Solier, 1848: 327 **A:** PA TAI **ORR**
ovipennis Seidlitz, 1893: 254 **N:** MO
paludani Kaszab, 1959e: 249 [= 1960a: 73] **A:** AF
parvicollis parvicollis Zubkov, 1829: 160 **E:** AB ST **A:** KZ
affinis Seidlitz, 1893: 307
krynickii Krynicki, 1829: 195
scabra Fischer von Waldheim, 1842: 15
parvicollis quadricollis Ballion, 1878: 304 **A:** KZ
subcordata Seidlitz, 1893: 306
pauliani Koch, 1945: 408 **N:** MO
pavlovskii Bogatchev, 1959: 64 **A:** TD
perlonga Bates, 1879: 472 **A:** KA XIZ
prolongata Fairmaire, 1887g: clxxix (*Leptomorpha*)
persica Seidlitz, 1893: 305 **A:** IN
persicola Seidlitz, 1893: 306 **A:** IN

petiti Pierre, 1964a: 408 **A:** LE
petra I. Chigray, Nabozhenko & Keskin, 2015a: 63 **A:** TR
peyerimhoffi Koch, 1945: 410 **N:** MO (Middle Atlas)
pilosa Ren & Wang, 2001: 19 **A:** XIZ
pinguis Allard, 1881a: 309 **N:** MO
plana Solier, 1848: 338 **E:** FR SP SV YU **N:** AG EG TU **A:** TR
planicollis Laporte, 1840: 200 **N:** EG
platythorax Gemminger, 1870a: 121 [RN] **A:** IN
 laticollis L. Redtenbacher, 1850: 49
polychresta Forskål, 1775: 79 (*Tenebrio*) **N:** EG LB (Cyrenaica) **A:** IS SI AUR
 kuesteri Ballion, 1888a: 276
 maura Solier, 1848: 347
 minor Solier, 1848: 347
 sulcata Fabricius, 1775: 254 [HN]
potanini Reitter, 1889k: 690 **A:** GAN NIN QIN XIZ
povolnyi Kaszab, 1970a: 113 **A:** AF
prodigiosa Erichson, 1841: 182 **N:** AG
 multicosta Solier, 1848: 326
propheta fiorii Español, 1967b: 179 **N:** LB (Tripolitania)
propheta propheta Reiche, 1861b: 89 **N:** AG LB MO TU
 algirica Ballion, 1888b: 695
pruinosa Eversmann, 1833: 53 **E:** ST **A:** KZ TM
 fischeri Fischer von Waldheim, 1844: 94
 longipes Zubkov, 1833: 331
 rorulenta Motschulsky, 1845a: 65
pseudocaudata Kaszab, 1960a: 69 **A:** AF
pterosticha Fischer von Waldheim, 1844: 93 **A:** KI KZ MG XIN
 akinina Allard, 1882a: 135 (*Agroblaps*)
 multistriata Ballion, 1878: 302
pubescens Allard, 1881a: 307 **N:** TU "Bou-Saada"
pudica Ballion, 1888b: 704 **E:** AB AR **A:** IN
puella Allard, 1881a: 317 **E:** AB AR GG **A:** TR
punctostriata Solier, 1848: 329 **A:** IN
putrida Motschulsky, 1845a: 68 **A:** ES KZ WS
quedenfeldtii Seidlitz, 1893: 254 **N:** MO
ratalaria Ren & Wang, 2001: 22 **A:** QIN XIZ
reflexa Gebler, 1832: 55 **A:** ES HEB MG NIN NMO SHA
 variolaris Gemminger, 1870a: 122 [RN]
 variolosa Fischer von Waldheim, 1844: 89 [HN]
 vermiculata Allard, 1882a: 93
reichardti Semenov & Bogatchev, 1936: 558 **A:** KI TD UZ (Bukhara)

reitteri Allard, 1885b: 180 (*Uroblaps*) **E:** AB
rhynchoptera Fairmaire, 1886c: 344 **A:** SCH YUN **ORR**
inermis Fairmaire, 1887b: 127 (*Coelocnemodes*)
rimskii Semenov & Bogatchev, 1936: 560 **A:** GAN QIN
rotundata Solier, 1848: 325 **A:** IS SY TR
sodalis Reiche & Saulcy, 1857: 249
rougemonti Kaszab, 1986: 148 **A:** YE
rugosa Gebler, 1825: 48 **A:** BEI ES GAN HEB JIL LIA MG NIN NMO QIN SHA
scabripennis Faldermann, 1835b: 405
variolota Gemminger, 1870a: 122 [RN]
rugulipennis Fairmaire, 1891d: xcvii **A:** KA
ruhmeni Seidlitz, 1893: 260 **N:** LB
rybalovi Tschernyshev & Mordkovitsh, 2002: 385 **A:** WS
scabiosa Baudi di Selve, 1874a: 111 **A:** IN
neoscabiosa Nabozhenko, 2008c: 35
scabiosa Faust, 1875: 229
scabriuscula scabriuscula Ménétriés, 1832: 200 **E:** AB
scabriuscula subalpina Ménétriés, 1832: 200 **E:** AB GG ST
montana Motschulsky, 1839: 62
scutellata Fischer von Waldheim, 1844: 72 **A:** AF KZ TM
semenovi Bogatchev, 1939a: 127 **A:** KI (Arslanbob, Ak-Terek)
semenoviana Bogatchev, 1941b: 109 **A:** YE
seriata Fischer von Waldheim, 1820: pl. 16 **A:** IN TM UZ
acuta Seidlitz, 1893: 307
lajoyei Allard, 1883: 26
seriatimpunctata Fischer von Waldheim, 1844: 86
striola Motschulsky, 1860e: 528
shach Kaszab, 1970a: 121 **A:** AF
simplex Kaszab, 1960a: 72 **A:** AF
sinuatocollis sinuatocollis Solier, 1848: 314 **E:** ST
sinuatocollis suecica J. Ferrer & Picka, 1990: 30 **E:** SV
sonamarga Skopin, 1978: 228 **A:** KA
spinosa Allard, 1881a: 304 **A:** IQ
splichali Gebien, 1913: 1 **A:** CY TR
srinagarica Kaszab, 1975a: 17 **A:** KA
stoetzneri Schuster, 1923b: 158 **A:** SCH XIZ
stoliczkana Bates, 1879: 471 **A:** TD
strandi Semenov, 1936: 557 **A:** GAN NMO XIN
sulcatipennis Schuster, 1920a: 200 **A:** XIN
sulcifera Seidlitz, 1893: 264 **N:** LB
superstitiosa icosiensis Peyerimhoff, 1925: 14 **N:** AG

superstitiosa superstitiosa Erichson, 1841: 183 **N:** AG TU
szetschwana Schuster, 1923b: 156 **A:** AHN FUJ GUI JIA SCH YUN
taeniolata Ménétriés, 1832: 198 **E:** AB AR GG GR TR **N:** EG **A:** CY IN IQ LE SY TM TR
hians Fischer von Waldheim, 1844: 78
taiyuanica Ren & Wang, 2001: 17 **A:** SHX
tatsienlua Schuster, 1923a: 31 **A:** SCH
tenuicauda Seidlitz, 1893: 307 **A:** KZ
tenuicollis Solier, 1848: 301 **E:** GR **A:** IS LE
tenuepunctata Motschulsky, 1860e: 528 (*Agroblaps*) **A:** KZ
vicina Ménétriés, 1849: 235 [HN]
teocchii L. Soldati, 2017: 18 **N:** TU
theodoridis Pierre, 1964a: 408 **A:** IS
thibetana Blair, 1922c: 559 **A:** NP XIZ
thibetanoides Ren, 2004: 69, 72 **A:** XIZ
tianshanica Semenov & Bogatchev, 1936: 559 **A:** XIN (Dzhungaria)
tibialis Reiche & Saulcy, 1857: 241 **E:** BU GR ST TR UK **A:** IS TR
strigicollis Baudi di Selve, 1875b: 20
tibiella Seidlitz, 1893: 306 **A:** TR
tichyi Martínez Fernández, 2010: 182 **E:** SP
tingitana Allard, 1881a: 305 **N:** MO
desertica Escalera, 1913: 47
titanus Ménétriés, 1849: 234 **E:** AB **A:** IN TM
torresallai Español, 1961f: 408 **E:** SP (Menorca)
transversalis Fischer von Waldheim, 1844: 105 **A:** KZ XIN
trapezoidalis Kaszab, 1970a: 119 **A:** AF
tridentata Waterhouse, 1889: 125 **A:** AF
tripolitanica Karsch, 1881: 48 **N:** AG (Hoggar) LB MO
vialattei Peyerimhoff, 1920a: 328
tristiciae Bogatchev, 1949a: 39 **A:** KA
tschitscherini Semenov & Bogatchev, 1936: 564 **A:** SCH
tsharynensis balchashensis Skopin, 1966b: 338 **A:** KZ (Almaatinskaya Oblast)
tsharynensis tsharynensis Skopin, 1961a: 190 **A:** KZ (Almaatinskaya Oblast)
turcomanorum Seidlitz, 1893: 305 **A:** UZ
turkestanica Seidlitz, 1893: 309 **A:** UZ
tuxeni Gridelli, 1954: 262 [= 1955: 57] **A:** AF
umbilicata Seidlitz, 1893: 310 **A:** GAN QIN
urophora Fairmaire, 1891d: xcv **A:** KA
varicosa Seidlitz, 1893: 308 **A:** HEB
variolosa Faldermann, 1835b: 404 **A:** BEI ES (Tuva) GAN MG NIN NMO SCH TM WS (Altay) XIN
tschiliana Wilke, 1921: 256 [RN]
verrucosa Adams, 1817: 305 **E:** GG

virgo Seidlitz, 1893: 307 **A:** KI XIN
sagitta Seidlitz, 1893: 307
waltlii Seidlitz, 1893: 304 **E:** PT SP
waschana Schuster, 1923a: 31 **A:** QIN SCH XIZ
wiedemannii Solier, 1848: 350 **N:** EG LB **A:** IS SA SI YE
wolinskii Bodemeyer, 1927: 24 **A:** IN
yini Ren X.-P. Wang & Y.-Z. Yu, 2000: 27 (*Blaptogonia*) **A:** XIZ
zarudniana chorassanica Semenov & Bogatchev, 1936: 555 **A:** IN (Chorassan)
zarudniana persis Semenov & Bogatchev, 1936: 555 **A:** IN (Kuhistan)
zarudniana zarudniana Semenov & Bogatchev, 1936: 554 **A:** IN
zugmayeri Schuster, 1935c: 239 **A:** IN (Baluchestan) PA (Balochistan)
subgenus *Dineria* Motschulsky, 1860e: 530 type species *Blaps confusa* Ménétriés, 1832 (= *Blaps halophila* Fischer von Waldheim, 1820)
Laraliprosodes Bogatchev, 1947e: 513 type species *Prosodes lar* Bogatchev, 1947.
halophila Fischer von Waldheim, 1820: pl. 16 **E:** AB AU BU CR CT CZ GG HU MC MO PL RO SK SL ST UK YU **A:** KZ TR WS
confusa Ménétriés, 1832: 201
confluens Fischer von Waldheim, 1844: 84
convexicollis Motschulsky, 1845a: 68
coriacea Fischer von Waldheim, 1842: 15
intrusa Fischer von Waldheim, 1844: 95
longicollis Fischer von Waldheim, 1842: 83
solieri Allard, 1881a: 317
songorica Fischer von Waldheim, 1844: 97
lar Bogatchev, 1947e: 513 (*Prosodes*) **A:** UZ (Tashkentskaya oblast)
subgenus *Prosoblapsia* Skopin & Kaszab, 1978: 208 type species *Leptocolena allardiana* Reitter, 1889
allardiana alaiensis Seidlitz, 1893: 309 **A:** UZ (North Alaisky Ridge)
sunpantinga Schuster, 1923a: 30
allardiana allardiana Reitter, 1889k: 691 (*Leptocolena*) **A:** GAN QIN SCH XIZ
znoikoi Semenov & Bogatchev, 1936: 563
apicecostata Blair, 1922c: 559 **A:** BT NP QIN SCH SD XIZ
breiti Reitter, 1913d: 658 **A:** HP XIZ
trapezicollis Gebien, 1913: 3
brevipes Seidlitz, 1893: 310 **A:** XIZ "Himalaya"
brevis Ren & Wang, 2001: 21 **A:** XIN
conica Ren & Shi, 2016: 189 **A:** XIZ
gentilis gentilis Fairmaire, 1887b: 127 **A:** BT GAN QIN SCH YUN
gentilis semistriatomorpha Skopin & Kaszab, 1978: 211 **A:** NP
gentilis transversithorax Skopin & Kaszab, 1978: 211 **A:** NP
helleri Schuster, 1923b: 158 **A:** SCH QIN

himalaica Blair, 1923b: 283 **A:** NP SD XIZ
latericosta Reitter, 1889k: 688 **A:** GAN XIN
maeander Kraatz, 1885: 291 **A:** KI UZ (Tian Shan)
socia socia Seidlitz, 1893: 309 **A:** NP SD XIZ
gentiloides Kaszab, 1977b: 254
punctithorax Kaszab, 1977b: 253
radula Kaszab, 1977b: 254
semistriata Kaszab, 1977b: 255
tentyroides Seidlitz, 1893: 310
tenuicornis Gebler, 1847: 476 **A:** ES WS
altaica Kelejnikova, 1970: 27
depressa Gebler, 1829a: 121 [HN]
transversimsulcata meandroides Skopin, 1977: 151 **A:** KI KZ
transversimsulcata transversimsulcata Ballion, 1878: 301 **A:** KZ XIN
rudesculpta Semenov & Bogatchev, 1936: 560

genus *Coelocnemodes* Bates, 1879: 474 [= 1886: 64] type species *Coelocnemodes stoliczkanus* Bates, 1879
Neoblaps Ren & Li, 2001: 310 type species *Neoblaps huizensis* Ren & Li, 2001
heqingensis Ren, 2016: 437 **A:** YUN
huizensis Ren & Li, 2001: 311 (*Neoblaps*) **A:** YUN
aspericollis Fairmaire, 1886c: 341 **A:** YUN
stoliczkanus Bates, 1879: 475 **A:** PA
tibialis Ren, 2016: 435 **A:** YUN

genus *Dila* Fischer von Waldheim, 1844: 111 type species *Blaps laevicollis* Gebler, 1841
Caenoblaps König, 1906: 24 type species *Caenoblaps difformis* König, 1906
afghanica Kaszab, 1960a: 45 **A:** AF
alaica Semenov & Bogatchev, 1940: 207 **A:** KI (Alai Range)
angustata Baudi di Selve, 1875c: 102 (*Blaps*) **A:** IN
laticollis Baudi di Selve, 1875c: 103 (*Blaps*)
mniszechi Allard, 1881a: 277
armata Blair, 1913: 56 (*Blaps*) **A:** PA
baeckmanni Schuster, 1928c: 125 (*Caenoblaps*) **E:** ST (Dagestan)
daghestanica G.S. Medvedev & Abdurachmanov, 1984: 291
bomiana Ren & Li, 2001: 311 **A:** XIZ
caudata Kaszab, 1970a: 86 **A:** AF
crenatopunctata I. Chigray, Nabozhenko, Abdurakhmanov & Keskin, 2019: 12 **A:** IN
difformis König, 1906: 24 (*Caenoblaps*) **A:** GG TR
hakkarica I. Chigray, Nabozhenko, Abdurakhmanov & Keskin, 2019: 10 **A:** TR (Hakkari)
kulzeri Schuster, 1928c: 123 (*Caenoblaps*) **A:** TR (Van, Hakkari)

kuntzeni Schuster, 1914c: 182 **A:** IN
laevicollis Gebler, 1841a: 373 (*Blaps*) **A:** AF KI KZ TD TM UZ XIN
leptoscelis Reitter, 1900a: 239 **A:** IN
lindbergi Kaszab, 1970a: 84 **A:** AF
nitida Schuster, 1920a: 199 (*Caenoblaps*) **A:** IN
persiana Reitter, 1900i: 295 **A:** IN
rugelytra Ren, 2016: 441 **A:** XIZ
sedecimstriata Reitter, 1900i: 295 **A:** KI UZ
seriata Reitter, 1894d: 242 **A:** TD UZ
bucharica Reitter, 1900i: 296
svetlanae I. Chigray, Nabozhenko, Abdurakhmanov & Keskin, 2019: 14 **A:** TR (Hakkari)
transversecordata Reitter, 1899d: 203 **A:** TR (Mardin)

genus *Dilablaps* Bogatchev, 1976: 98 type species *Dilablaps paradoxa* Bogatchev, 1976
paradoxa Bogatchev, 1976: 99 **A:** TD

genus *Hoplitoblaps* Fairmaire, 1888a: 26 type species *Hoplitoblaps fallaciosa* Fairmaire, 1888
fallaciosa Fairmaire, 1888a: 27 **A:** YUN

genus *Medvedevia* I. Chigray, 2019: 915 type species *Medvedevia glebi* I. Chigray, 2019
glebi I. Chigray, 2019: 915 **A:** AF

genus *Nalepa* Reitter, 1887d: 366 type species *Blaps cylindracea* Reitter, 1887
cylindracea Reitter, 1887d: 366 (*Blaps*) **A:** GAN SCH XIZ

genus *Protoblaps* G.S. Medvedev, 1998b: 200 type species *Protoblaps kashkarovi* G.S. Medvedev, 1998
kashkarovi G.S. Medvedev, 1998b: 202 **A:** SCH

genus *Thaumatoblaps* Kaszab & G.S. Medvedev, 1984: 79 type species *Thaumatoblaps marikovskiji* Kaszab & G.S. Medvedev, 1984
marikovskiji Kaszab & G.S. Medvedev, 1984: 80 **A:** TM
zhengi Ren & Luo, 1995: 28 **A:** XIN

subtribe Blaptina, species incertae sedis
Blaps indica Hope, 1831: 31 **A:** NP

subtribe Gnaptorina G.S. Medvedev, 2001

genus *Gnaptor* Brullé, 1832: 203 type species *Tenebrio spinimanus* Pallas, 1781
subgenus *Gnaptor* Brullé, 1832: 203 type species *Tenebrio spinimanus* Pallas, 1781
Petrobius Brullé, 1832: 202 [HN] type species *Tenebrio spinimanus* Pallas, 1781

boryi Laporte, 1840: 208 (*Petrobius*) **E:** GR
graecus Küster, 1848a: 64
medvedevi I. Chigray, Nabozhenko & Keskin, 2015b: 1278 **E:** TR
spinimanus Pallas, 1781: 55 (*Tenebrio*) **E:** AL AU BH BU HU MC RO SK SL ST TR UK YU
glabratus Fabricius, 1792a: 100 (*Pimelia*)
laevigatus Fabricius, 1792a: 100 (*Pimelia*)
subgenus *Plesiognaptor* I. Chigray, Nabozhenko & Keskin, 2015b: 1277 type species *Gnaptor prolixus* Fairmaire, 1866
prolixus Fairmaire, 1866b: 264 **A:** IN TR

subtribe Gnaptorinina G.S. Medvedev, 2001

genus ***Agnaptoria*** **Reitter, 1887d: 372** type species *Agnaptoria rubripes* Reitter, 1887
abdita G.S. Medvedev, 2002: 861 **A:** GAN
amdoensis G.S. Medvedev, 2006a: 197 **A:** GAN QIN XIZ
anomala G.S. Medvedev, 2007b: 132 **A:** YUN
anthracina anthracina G.S. Medvedev, 2002: 863 **A:** GAN
anthracina labrana G.S. Medvedev, 2002: 865 **A:** GAN
apicopilosa Shi & Ren, 2007a: 283 **A:** XIZ
belousovi G.S. Medvedev, 2002: 847 **A:** SCH
crassa G.S. Medvedev, 2006a: 180 **A:** SCH
danbaensis G.S. Medvedev, 2006a: 195 **A:** SCH
dzaensis G.S. Medvedev, 2002: 853 **A:** SCH
fugonga G.S. Medvedev, 2007b: 136 **A:** YUN
gansuensis G.S. Medvedev, 2002: 849 **A:** GAN
grummi G.S. Medvedev, 2002: 859 **A:** QIN
jinchengi Li & Shi, 2019: 3 **A:** NP
kabaki G.S. Medvedev, 2002: 845 **A:** SCH
kamensis G.S. Medvedev, 2002: 853 **A:** QIN
kozlovi G.S. Medvedev, 2002: 855 **A:** QIN XIZ
kunda G.S. Medvedev, 2006a: 186 **A:** GAN
lauta G.S. Medvedev, 2006a: 184 **A:** SCH
lecta G.S. Medvedev, 2008: 821 **A:** SCH YUN
markana Shi & Ren, 2005: 279 **A:** XIZ
minuta G.S. Medvedev, 2006a: 178 **A:** SCH
miroshnikovi G.S. Medvedev, 1998b: 183 **A:** SCH
nigriceps Shi & Ren, 2007a: 281 **A:** XIZ
nigripes Ren & Liu, 2009: 56 **A:** XIZ
nyingcha Ren, 2016: 238 **A:** XIZ
proba G.S. Medvedev, 2006a: 190 **A:** GAN
rubripes Reitter, 1887d: 372 **A:** QIN SCH
rufipes Z. Li & Ren, 2004: 257 (*Gnaptorina*) **A:** QIN

ruida G.S. Medvedev, 2008: 824 **A:** SCH
seidlitzi Reitter, 1893d: 315 **A:** XIN
shamaevi shamaevi G.S. Medvedev, 1999a: 401 **A:** GAN
shamaevi sitshuana G.S. Medvedev, 1999a: 403 **A:** SCH
shangmengi G.S. Medvedev, 2006a: 192 **A:** SCH
tenuicorpra Z. Li & Ren, 2004: 258 (*Gnaptorina*) **A:** QIN
zamotailovi G.S. Medvedev, 1998b: 180 **A:** SCH

genus *Asidoblaps* Fairmaire, 1886c: 342 type species *Asidoblaps davidis* Fairmaire, 1886
altimontana G.S. Medvedev, 2002: 73 **A:** SCH
angustitarsis G.S. Medvedev, 2005c: 545 **A:** SCH
annae G.S. Medvedev, 2002: 85 **A:** SCH
attigua G.S. Medvedev, 2009b: 833 **A:** YUN
belousovi G.S. Medvedev, 2002: 68 **A:** SCH
berezowskii G.S. Medvedev, 1998b: 196 **A:** SCH
bianera G.S. Medvedev, 2005c: 543 **A:** SCH
bifida G.S. Medvedev, 2005c: 565 **A:** SCH
confinis G.S. Medvedev, 2005c: 534 **A:** SCH
danba G.S. Medvedev, 2005c: 531 **A:** SCH
davidiani G.S. Medvedev, 2002: 70 **A:** SCH
davidis Fairmaire, 1886c: 342 **A:** YUN XIZ
dentipes G.S. Medvedev, 2009b: 825 **A:** YUN SCH
edentata G.S. Medvedev, 2009b: 831 **A:** YUN
elongata G.S. Medvedev, 2008: 819 (*Agnaptoria*) **A:** YUN
expressa G.S. Medvedev, 2005c: 552 **A:** GAN
faceta G.S. Medvedev, 2009b: 828 **A:** YUN
fida G.S. Medvedev, 2007b: 144 **A:** SCH
firma G.S. Medvedev, 2009b: 839 **A:** SCH
friedrichi G.S. Medvedev & Merkl, 2001: 625 **A:** SCH
galinae G.S. Medvedev, 2002: 80 **A:** SCH
gansuensis G.S. Medvedev, 2002: 82 **A:** GAN
glyptoptera Fairmaire, 1886c: 343 **A:** SCH YUN
gorgneri G.S. Medvedev & Merkl, 2001: 623 **A:** SCH
gracilis G.S. Medvedev, 2002: 87 **A:** SCH
ignorata G.S. Medvedev, 2005c: 541 **A:** SCH
igori G.S. Medvedev, 2002: 91 **A:** SCH
inermis G.S. Medvedev, 1998b: 198 **A:** SCH
inopinata G.S. Medvedev, 2005c: 558 **A:** GAN
justa G.S. Medvedev, 2007b: 148 **A:** SCH
kabaki G.S. Medvedev, 2002: 94 **A:** SCH
konstantinovi G.S. Medvedev, 2007b: 142 **A:** YUN
longicornis G.S. Medvedev, 2001: 83 **A:** SCH

miroshnikovi G.S. Medvedev, 1998b: 191 A: SCH YUN
mekonga G.S. Medvedev, 2007b: 139 A: YUN
mesa G.S. Medvedev, 2009b: 819 A: YUN
modica G.S. Medvedev, 2002: 77 A: SCH
multa G.S. Medvedev, 2005c: 548 A: GAN
munda G.S. Medvedev, 2005c: 537 A: GAN
physoptera G.S. Medvedev, 2009b: 823 A: SCH YUN
pedinoides G.S. Medvedev, 2009b: 842 A: SCH
polita Schuster, 1935c: 162 A: SCH
potanini G.S. Medvedev, 1998c: 583 A: XIZ
prasolovi G.S. Medvedev, 2009b: 845 A: YUN
reperta G.S. Medvedev, 2007b: 151 A: SCH
rilonga G.S. Medvedev, 2005c: 539 A: SCH
silvicola G.S. Medvedev, 2005c: 560 A: SCH
sinensis G.S. Medvedev, 1998b: 194 A: SCH
siniaevi G.S. Medvedev, 2002: 65 A: SCH
smetanai G.S. Medvedev, 2001: 81 A: TAI
subnotata G.S. Medvedev, 2007b: 154 A: SCH
subopaca Schuster, 1935c: 162 A: GAN SHA SCH
yaanensis G.S. Medvedev, 2005c: 563 A: SCH
yuana G.S. Medvedev, 2002: 75 A: SCH
zamotailovi G.S. Medvedev, 1998b: 189 A: YUN
zhongdiana G.S. Medvedev, 2009b: 837 A: YUN

genus *Belousovia* G.S. Medvedev, 2007b: 157 type species *Belousovia helenae* G.S. Medvedev, 2007
helenae G.S. Medvedev, 2007b: 159 A: YUN
kabaki intermedia G.S. Medvedev, 2007b: 164 A: YUN
kabaki kabaki G.S. Medvedev, 2007b: 162 A: YUN

genus *Blaptogonia* G.S. Medvedev, 1998b: 186 type species *Tagonoides costulata* Fairmaire, 1901
costulata Fairmaire, 1901b: 267 (*Tagonoides*) A: SD
subcarinata Blair, 1927: 243 (*Blaps*) A: XIZ
tshernjachovskii G.S. Medvedev, 2004a: 177 A: NP
zhentanga Li, Bai & Ren, 2018: 72 A: XIZ
zurstrasseni Kaszab, 1977b: 246 (*Blaptyscelis*) A: NP

genus *Colasia* Koch, 1965b: 131 type species *Colasia akisoides* Koch, 1965
akisoides Koch, 1965b: 131 A: SCH YUN

genus *Gnaptorina* Reitter, 1887d: 364 type species *Gnaptorina felicitana* Reitter, 1887
subgenus *Boreoptorina* Medvedev, 2009a: 417 type species *Gnaptorina cordicollis* G.S. Medvedev, 1998

cordicollis G.S. Medvedev, 1998c: 555 (*Gnaptorina*) **A:** QIN XIZ
subgenus *Gnaptorina* Reitter, 1887d: 364 type species *Gnaptorina felicitana* Reitter, 1887
acutangula Li, Bai & Ren, 2019: 85 **A:** SCH
australis G.S. Medvedev, 2008: 814 **A:** YUN
crassitibia Li, Bai & Ren, 2019: 85 **A:** SCH
cylindricollis Reitter, 1889k: 693 **A:** GAN NIN SCH XIZ
fairmairei Koch, 1965b: 130 **A:** GAN SCH
felicitana Reitter, 1887d: 365 **A:** NMO QIN XIZ
kashkarovi G.S. Medvedev, 1998c: 564 **A:** QIN SCH XIZ
kozlovi G.S. Medvedev, 1998c: 560 **A:** QIN XIZ
lii Z. Li & Ren, 2004: 256 **A:** NIN
media G.S. Medvedev, 1998c: 558 **A:** QIN SCH XIZ
miroshnikovi G.S. Medvedev, 1998c: 562 **A:** QIN XIZ
montana G.S. Medvedev, 1998c: 557 **A:** QIN XIZ
platytarsia Z. Li & Ren, 2004: 259 **A:** XIZ
potanini potanini Reitter, 1889k: 694 **A:** GAN SCH
potanini minxiana G.S. Medvedev, 2008: 816 **A:** GAN
proxima Reitter, 1889k: 694 **A:** GAN SCH
reitteri Koch, 1965b: 129 **A:** SCH
rufipes Ren & Zhu, 2016: 369 **A:** QIN
rugosipennis G.S. Medvedev, 1998c: 565 **A:** QIN XIZ
subgenus *Hesperoptorina* Medvedev, 2009a: 417 type species *Gnaptorina brucei* Blair, 1923
brucei Blair, 1923b: 282 **A:** NP SD XIZ
compressa Shi, Ren & Merkl, 2007: 222 **A:** XIZ
cuonaensis Shi, 2013a: 468 **A:** XIZ
dongdashanensis Shi, 2013a: 466 **A:** XIZ
globithoracalis Shi, Ren & Merkl, 2007: 220 **A:** XIZ
himalayana Shi, Ren & Merkl, 2007: 220 **A:** XIZ
kangmar Shi, Ren & Merkl, 2007: 224 **A:** XIZ
medvedevi Shi, 2013a: 464 **A:** XIZ
nigra Shi, Ren & Merkl, 2007: 220 **A:** XIZ
pilifera Shi, Ren & Merkl, 2007: 220 **A:** XIZ
sikkimensis Kaszab, 1965a: 110 **A:** SD
tishkovi G.S. Medvedev, 1998c: 567 **A:** XIZ

genus *Gnaptorina*, species incertae sedis
curvatibia Ren & Zhu, 2016: 341 **A:** QIN
longicornis Z. Li & Ren, 2004: 261 **A:** XIZ

genus *Itagonia* Reitter, 1887d: 362 type species *Itagonia gnaptorinoides* Reitter, 1887
baxoica Liu & Ren, 2009a: 447 **A:** XIZ
bisetosa G.S. Medvedev, 1998c: 568 **A:** SCH

cordiformis Shi & Ren, 2007b: 34 **A:** XIZ
deplanata Reinig, 1931: 866 **A:** TD UZ (Bukhara)
elegans G.S. Medvedev, 1998c: 574 **A:** SCH
ganglbaueri Schuster, 1914a: 58 **A:** HEB NMO SHA
gnaptorinoides Reitter, 1887d: 362 **A:** TD UZ
kabaki G.S. Medvedev, 1998c: 578 **A:** TD
litangensis Shi, 2013b: 8 **A:** SCH
longicornis Shi & Ren, 2007c: 176 **A:** XIZ
lopatini lopatini G.S. Medvedev, 1998c: 576 **A:** TD
lopatini turkestanica G.S. Medvedev, 1998c: 578 **A:** TD
medvedevi Shi, 2010: 62 **A:** SCH
mera G.S. Medvedev, 1998c: 572 **A:** SCH XIZ
quadrata Z. Li & Ren, 2004: 260 (*Gnaptorina*) **A:** XIZ
provostii Fairmaire, 1889c: cci (*Platyscelis*) **A:** BEI HEB NIN NMO SHA
  ganglbaueri Schuster, 1914a: 58
semenowi Reitter, 1889k: 694 **A:** GAN QIN SCH
shamaevi G.S. Medvedev, 2004a: 168 **A:** SCH
szetschwana Schuster, 1923b: 162 (*Asidoblaps*) **A:** SCH
tibialis Shi, 2013b: 4 **A:** SCH
trisetosa G.S. Medvedev, 1998c: 570 **A:** SCH
tuberculata Shi, 2010: 58 **A:** SCH
xinlongensis Shi, 2010: 61 **A:** SCH
zayica Shi & Ren, 2007b: 36 **A:** XIZ

genus *Montagona* G.S. Medvedev, 1998b: 176 type species *Tagonoides asperula* Fairmaire, 1896
asperula Fairmaire, 1896b: 95 (*Tagonoides*) **A:** HP XIN XIZ
pustulosa Fairmaire, 1886c: 344 (*Tagonoides*) **A:** YUN
sculpturata Gebien, 1913: 2 (*Blaps*) **A:** KA

genus *Nepalindia* G.S. Medvedev, 1998b: 187 type species *Taganoides* [sic] *alpina* Kaszab, 1965
alpina Kaszab, 1965a: 109 (*Tagenoides*) **A:** NP SD
dreuxi Pierre, 1961: 213 (*Blaptyscellis*) **A:** NP
nepalica Kaszab, 1973a: 30 (*Tagenoides*) **A:** NP
volkovitshi infida G.S. Medvedev, 2004a: 173 **A:** NP
volkovitshi volkovitshi G.S. Medvedev, 2004a: 170 **A:** NP

genus *Pseudognaptorina* Kaszab, 1977b: 250 type species *Pseudognaptorina nepalica* Kaszab, 1977
exsertogena Shi, Ren & Merkl, 2005: 164 **A:** XIZ
flata Liu & Ren, 2009b: 46 **A:** SCH
nepalica Kaszab, 1977b: 251 **A:** NP
obtusa Shi, Ren & Merkl, 2005: 166 **A:** XIZ

genus *Sintagona* G.S. Medvedev, 1998c: 585 type species *Sintagona miranda* G.S. Medvedev, 1998
miranda G.S. Medvedev, 1998c: 586 **A:** QIN

genus *Tagonoides* Fairmaire, 1886c: 343 type species *Tagonoides delavayi* Fairmaire, 1886
ampliata Fairmaire, 1887b: 126 **A:** YUN
belousovi G.S. Medvedev, 2008: 808 **A:** YUN
delavayi Fairmaire, 1886c: 343 **A:** YUN
kabaki G.S. Medvedev, 2004a: 164 **A:** SCH
skopini G.S. Medvedev & Merkl, 2001: 620 **A:** YUN
volkovitshi G.S. Medvedev, 2004a: 166 **A:** YUN
yunnana G.S. Medvedev, 2008: 812 **A:** YUN
zamotailovi G.S. Medvedev, 1998b: 173 **A:** YUN

subtribe Prosodina Skopin, 1960

genus *Prosodes* Eschscholtz, 1829: 9 type species *Blaps attenuata* Fischer von Waldheim, 1820 (= *Blaps obtusa* Fabricius, 1798)
subgenus *Dilopersina* Reitter, 1909d: 117 type species *Prosodes jakowlewi* Semenov, 1894
cribrella Baudi di Selve, 1874a: 110 [= 1875c: 104] (*Blaps*) **A:** AB IN TM (Kopet Dag)
pustulata Faust, 1875: 237
consanguinea Allard, 1885b: 180
emiri Sumakov, 1902: 39 **A:** TM
jakowlewi Semenov, 1894a: 192 **A:** IN TM
grouvellei Reitter, 1909d: 132
mithras Reitter, 1904d: 258 **A:** IN
chorassanica G.S. Medvedev, 1996: 603
neopersis Semenov, 1909: 435 [RN] **A:** IN (Khorasan)
jakowlewi Reitter, 1909: 131 [HN]
exilis G.S. Medvedev, 1996: 599
parthenica G.S. Medvedev, 1996: 596 **A:** TM (Kopet Dag)
rishwani Makhan, 2012a: 1 **A:** IN
vestita Allard, 1881a: 289 **A:** IN TM (Kopet Dag)
subgenus *Diprosodes* Reitter, 1909d: 115 type species *Prosodes biformis* Semenov, 1894
biformis Semenov, 1894a: 181 **A:** TD UZ
biformiopaca biformiopaca Reitter, 1909d: 127 **A:** UZ (Bukhara)
biformiopaca occidentalis G.S. Medvedev, 1997b: 571 **A:** UZ
biformiopaca saganiana G.S. Medvedev, 1997b: 574 **A:** UZ
elateroides Semenov, 1894a: 185 **A:** UZ (Bukhara)
callosa Semenov, 1894a: 188
gymnandra Semenov, 1894a: 189 **A:** TD UZ (Bukhara)
juliae Bogatchev, 1960c: 82
kunduzensis Kaszab, 1970a: 104 **A:** AF

lopatini Bogatchev, 1960c: 81 **A:** TD (Vakhshskiy Ridge)
psammophila G.S. Medvedev, 1997b: 579 **A:** TD
sphodroides Semenov, 1894a: 190 **A:** UZ (Bukhara)
stackelbergi G.S. Medvedev, 1997b: 557 **A:** TD
subgenus *Euryprosodes* Reitter, 1909d: 122 type species *Prosodes areolata* Reitter, 1893
areolata Reitter, 1893c: 269 **A:** TD UZ (Bukhara)
  plicata Reitter, 1909d: 166
brevipes Reitter, 1893c: 269 **A:** UZ (Bukhara)
steppensis Kaszab, 1960a: 64 **A:** AF
trifossa Reitter, 1909d: 165 **A:** TD
subgenus *Ferganoprosodes* G.S. Medvedev, 1997b: 595 type species *Prosodes angulicollis* Kraatz, 1883
angulicollis Kraatz, 1883a: 347 **A:** UZ (Fergana)
  nitida Kraatz, 1885: 289
pleuralis Reitter, 1893c: 266 **A:** UZ
subgenus *Gebleria* Motschulsky, 1846: 410 type species *Dila philacoides* Fischer von Waldheim, 1844
  Aulonoscelis Reitter, 1896g: 173 type species *Platynoscelis hauseri* Reitter, 1895 (= *Dila philacoides* Fischer von Waldheim, 1844)
philacoides Fischer von Waldheim, 1844: 118 (*Dila*) **A:** ES KI KZ XIN
  brevis Gebler, 1844: 103 [= 1860: 22] (*Tagona*)
  gnaptorina Reitter, 1909d: 123
  hauseri Reitter, 1895d: 156 (*Platyscelis*) [HN]
regeli Semenov, 1894a: 202 **A:** KZ XIN
subgenus *Hypoprosodes* Reitter, 1909d: 122 type species *Prosodes minuta* Kraatz, 1881
minuta Kraatz, 1881b: 62 **A:** KZ UZ
  angulata Reitter, 1890h: 361
  sulcipubens Reitter, 1909d: 165
subgenus *Indoprosodes* G.S. Medvedev, 2003: 690 type species *Prosodes boorpi* Kaszab, 1956
boorpi Kaszab, 1956d: 420 **A:** UP
subgenus *Iranosodes* G.S. Medvedev, 1996: 605 type species *Prosodes kaszabi* G.S. Medvedev & Kabakov, 1996
diversa Waterhouse, 1889: 126 **A:** AF TM
  quadriimpressa Reitter, 1909d: 149
jakesi Kaszab, 1970a: 105 **A:** AF
kaszabi G.S. Medvedev & Kabakov, 1996: 607 **A:** AF
laticauda Reitter, 1896a: 38 **A:** IN TM (Kopet Dag)
subgenus *Lyprosodes* Reitter, 1909d: 116 type species *Prosodes quadricostata* Reitter, 1893
aciculata Reitter, 1893c: 274 **A:** TD UZ (Bukhara)
fallax Kraatz, 1886: 189 **A:** UZ (Margilan)
pavlovskii Bogatchev, 1960c: 80 **A:** TD

pseudaciculata Reitter, 1909d: 129 **A:** UZ (Bukhara)
quadricostata Reitter, 1893c: 278 **A:** UZ
subgenus *Megaprosodes* Reitter, 1909d: 118 type species *Prosodes striata* Reitter, 1893
Altiprosodes G.S. Medvedev, 1997b: 580 type species *Prosodes kuhistanica* G.S. Medvedev, 1996
janushevi G.S. Medvedev, 1996: 616 **A:** TD
kuhistanica G.S. Medvedev, 1996: 624 **A:** TD
merkli G.S. Medvedev, 1996: 619 **A:** TD
oschanini Semenov, 1890: 366 **A:** UZ (Bukhara)
pseudalaiensis Skopin, 1971a: 496 **A:** UK (Samarkand)
striata Reitter, 1893c: 289 **A:** UZ (Bukhara)
obliterata Semenov, 1894a: 206
svetlanae G.S. Medvedev & Michailov, 1978: 61 **A:** TD
vashgirda G.S. Medvedev, 1996: 618 **A:** TD
subgenus *Meropersina* Reitter, 1909d: 117 type species *Prosodes cordicollis* Allard, 1883
amica G.S. Medvedev, 2001: 86 **A:** TR
banghaasi Reitter, 1893c: 293 **A:** TR
cordicollis Allard, 1883: 25 **A:** IN
vermicularis Reitter, 1909d: 131
fabiani G.S. Medvedev & Merkl, 2005: 172 **A:** IN (Zagros)
kasatkini I. Chigray, Nabozhenko, Merkl & Kovalev, 2018: 472 **A:** IN
laevigata Baudi di Selve, 1874a: 109 [= 1875c: 101] (*Blaps*) **A:** IN
persica Faust, 1875: 236
shokhini I. Chigray, Nabozhenko, Merkl & Kovalev, 2018: 476 **A:** IN
vermiculosa Reitter, 1909d: 131 **A:** IN
vigi G.S. Medvedev & Merkl, 2005: 172 **A:** IN (Zagros)
subgenus *Mesoprosodes* G.S. Medvedev, 1995: 827 type species *Prosodes nuratensis* Semenov, 1894
fausti Semenov, 1894a: 217 **A:** UZ
foveicollis Semenov, 1894a: 213 **A:** TD
karatavica Skopin, 1966b: 339 **A:** KZ (Karatay)
kyzylkumensis G.S. Medvedev, 1995: 828 **A:** UZ
nuratensis Semenov, 1894a: 214 **A:** UZ
turkestanica Semenov, 1894a: 211 **A:** TD UZ
undulata Kraatz, 1882d: 322 **A:** UZ (Samarkand)
subgenus *Montiprosodes* Medvedev, 2001: 83 type species *Prosodes alaiensis* Kraatz, 1885
alaiensis alaiensis Kraatz, 1885: 288 **A:** UZ
sparsa Reitter, 1909d: 135
alaiensis transmontana G.S. Medvedev, 1996: 612 **A:** TD
baratovi G.S. Medvedev, 1996: 613 **A:** TD
jardanus Reitter, 1909d: 135 **A:** UZ (Bukhara)
latitibia Kaszab, 1960a: 62 **A:** AF
praelonga Reitter, 1893c: 286 **A:** UZ (Bukhara)

pubistriata Reitter, 1893c: 307 **A:** UZ (Bukhara)
lateralis Reitter, 1893c: 276
subgenus *Oliprosodes* Reitter, 1909d: 118 type species *Prosodes trisulcata* Bates, 1879
kuluana G.S. Medvedev, 1999b: 857 **A:** HP
laevicostata G.S. Medvedev, 1999b: 853 **A:** AF
rufosulcata Fairmaire, 1891d: xciv **A:** KA
trisulcata Bates, 1879: 473 **A:** KA PA XIZ
vicina Bates, 1879: 474
subgenus *Peltarium* Fischer von Waldheim, 1844: 106 type species *Peltarium marginatum* Fischer von Waldheim, 1844 (=*Blaps baerii* Fischer von Waldheim, 1842)
Blaposodes Skopin, 1960b: 47 type species *Blaps baeri* Fischer von Waldheim, 1842
baerii Fischer von Waldheim, 1842: 16 (*Blaps*) **A:** KZ
marginata Fischer von Waldheim, 1842: 28 (*Blaps*)
marginata Fischer von Waldheim, 1844: 108 (*Peltarium*)
karelini Gebler, 1841b: 593 (*Blaps*) **A:** KZ XIN
altaica Reitter, 1909d: 147
kataevi G.S. Medvedev, 1995: 831 **A:** KI (Alaiyskiy Ridge)
staudingeri Kraatz, 1882d: 305 **A:** UZ (Samarkand)
subgenus *Planoprosodes* G.S. Medvedev, 2005a: 68 type species *Prosodes reitteri* Reitter, 1893
reitteri Reitter, 1893c: 285 **A:** KZ
asperipennis Reitter, 1893c: 311
subgenus *Prosodella* Reitter, 1909d: 120 type species *Prosodes bactriana* Semenov, 1894
Paraprosodes Reitter, 1909d: 119 type species *Prosodes triangulipes* Reitter, 1909
bactriana Semenov, 1894a: 199 **A:** AF TD UZ
bactrianiformis Reitter, 1901c: 77 **A:** UZ (Bukhara)
bogatshevi G.S. Medvedev, 1996: 631 **A:** TD
convergens Reitter, 1893c: 283 **A:** UZ (Bukhara)
divergens Reitter, 1893c: 283 **A:** UZ (Bukhara)
singularis Reitter, 1909d: 145 **A:** UZ (Bukhara)
triangulipes Reitter, 1893c: 282 **A:** "Turkestan" (Tien Shan)
subgenus *Prosodes* Eschscholtz, 1829: 9 type species *Blaps attenuata* Fischer von Waldheim, 1820 (= *Blaps obtusa* Fabricius, 1798)
Blaptoprosodes Reitter, 1909d: 120 type species *Prosodes mucronata* Reitter, 1893
Eupomeca Solier, 1848: 289 type species *Blaps cylindrica* Herbst, 1799 (= *Blaps obtusa* Fabricius, 1798)
Lioprosodes Reitter, 1909d: 121 type species *Prosodes dilaticollis* Motschulsky, 1859
Nyctipates Gebler, 1841a: 373 type species *Nyctipates rugulosa* Gebler, 1841
Platyprosodes Reitter, 1909d: 121 type species *Nyctipates rugulosa* Gebler, 1841
Prosodila Reitter, 1909d: 121 type species *Prosodes strigiventris* Reitter, 1893
Pseudoprosodes Reitter, 1909d: 120 type species *Prosodes transfuga* Reitter, 1893
angustipleuris angustipleuris Reitter, 1893c: 292 **A:** KZ UZ
angustipleuris boroldaica Skopin, 1966b: 341 **A:** KZ (Boroldai Tau)
angustipleuris ugamica Skopin, 1966b: 341 **A:** UZ (Ugamsky Ridge)

blapoides Reitter, 1893c: 292 **A:** KZ UZ (Margilan, Tashkent)
  cribripennis Reitter, 1909d: 155
brevicauda brevicauda Reitter, 1909d: 156 **A:** UZ (Margilan, Tashkent)
brevicauda laevicauda Skopin, 1966b: 341 **A:** UZ (Ugamsky Ridge)
edmundi Semenov, 1894a: 220 **A:** XIN
dilaticollis Motschulsky, 1859a: 302 **A:** KI XIN
fulminans Reitter, 1893c: 303 **A:** KI
galinae G.S. Medvedev, 1995: 830 **A:** UZ (Nuratau)
gracilis Faust, 1875: 241 **A:** KZ UZ XIN
granulipleuris G.S. Medvedev, 1995: 822 **A:** UZ
hiekei G.S. Medvedev, 1995: 819 **A:** TD (Kuramin Mts)
kassansaica G.S. Medvedev, 1995: 821 **A:** KI
kumbeliana G.S. Medvedev, 1995: 826 **A:** UZ (Fergana)
longicornis Kraatz, 1883a: 349 **A:** KI
  tenuicornis G.S. Medvedev, 1995: 825
lucida Ballion, 1878: 316 **A:** KI
  deplanata Reitter, 1909d [HN]
ludmilae G.S. Medvedev, 1995: 816 **A:** TD (Kuramin Mts)
mogoltavica G.S. Medvedev, 1995: 813 **A:** TD (Mogoltay)
mucronata Reitter, 1893c: 291 **A:** UZ
  composita Reitter, 1893c: 291
obtusa Fabricius, 1798: 46 (*Blaps*) **E:** ST UK **A:** KZ
  attenuata Fischer von Waldheim, 1820: pl. 16 (*Blaps*)
  cylindrica Herbst, 1799: 185 (*Blaps*)
  cylindropastica Krynicki, 1832: 133 (*Blaps*)
  pastica Solier, 1848: 290 (*Nyctipates*)
  punctata Fischer von Waldheim, 1844: 109 (*Peltarium*)
pangazica G.S. Medvedev, 1995: 814 **A:** TD (Kuramin Mts)
parallelocollis Kraatz, 1882c: 108 **A:** UZ (Margilan)
  rudesculpta Reitter, 1909d: 158
pecirkana Reitter, 1909d: 157 **A:** KI (Koksuiskiy Ridge)
pekinensis Fairmaire, 1887f: 323 **A:** BEI HEB GAN NIN NMO QIN SHA SHX XIN
  kreitneri J. Frivaldszky, 1890: 206
przewalskii Semenov, 1894a: 195 **A:** XIN
rugosicollis padshaatinus Skopin, 1971a: 500 **A:** KI
rugosicollis rugosicollis Kraatz, 1885: 290 **A:** UZ (Namanganskaya Oblast)
rugosicollis sarytshelekis Skopin, 1971a: 500 **A:** KI
rugulosa granulosa Ballion, 1878: 307 **A:** KZ (Almaatinskaya Oblast)
rugulosa degenerata Semenov, 1894a: 198 **A:** XIN
rugulosa hetschkoi Reitter, 1909d: 163 **A:** KI

rugulosa nitidula Motschulsky, 1859b: 309 [= 1859a: 300] **A**: KI XIN
  costipennis Ballion, 1878: 312
  deplanata Ballion, 1878: 314
  dubia Ballion, 1878: 316
  similis Ballion, 1878: 309
rugulosa rugulosa Gebler, 1841a: 373 (*Nyctipates*) **A**: KI KZ GAN UZ XIN
  bicostata Fischer von Waldheim, 1844: 108 (*Peltarium*)
  costata Fischer von Waldheim, 1844: 120 (*Nyctipates*)
  foveata Fischer von Waldheim, 1842: 15 (*Blaps*)
sobolevae G.S. Medvedev, 1995: 812 **A**: TD (Kuraminsky Ridge)
strigiventris Reitter, 1893c: 277 **A**: KI TD UZ
sulcicauda Reitter, 1893c: 290 **A**: UZ
transfuga cariosa Skopin, 1961a: 193 **A**: KI KZ
transfuga transfuga Reitter, 1893c: 304 **A**: KI KZ UZ
  sulcatissima Reitter, 1901c: 78
tshistjakovae G.S. Medvedev, 1995: 824 **A**: UZ (Chatkalskiy Ridge)
valida Reitter, 1893c: 304 **A**: KI UZ (Margilan)
zarudnyi G.S. Medvedev, 1995: 818 **A**: UZ XIN
subgenus *Prosodestes* Reitter, 1909d: 114 type species *Prosodes grandicollis* Kraatz, 1883
bucharica Reitter, 1915f: 92 **A**: UZ (South West)
dolini Kh. Dadabaev, 1986: 80 **A**: TD (Khodzha-Maston)
grandicollis grandicollis Kraatz, 1883a: 328 **A**: KI (Osh)
grandicolli centralis G.S. Medvedev, 2005a: 68 **A**: TD (Shartuskiyv Rayon)
grandicolli interga G.S. Medvedev, 2005a: 69 **A**: TD (South West)
lindbergi Kaszab, 1970a: 95 **A**: AF
malyi malyi Obenberger, 1914: 139 **A**: TD (South West)
malyi tupalangi G.S. Medvedev, 2005a: 77 **A**: UZ (Tupalangdarya Valley)
subgenus *Prosodinia* Reitter, 1909d: 115 type species *Prosodes calcarata* Reitter, 1893
afghanica Schuster, 1936c: 197 **A**: AF
badakschanica Kaszab, 1960a: 51 **A**: AF
baghlana G.S. Medvedev, 2003a: 374 **A**: AF
boeckeleri Kaszab, 1974: 199 **A**: AF
brunnea Kaszab, 1960a: 50 **A**: AF
calcarata Reitter, 1893c: 309 **A**: IN TM
chazariensis G.S. Medvedev, 2003a: 386 **A**: AF
ferozensis G.S. Medvedev, 2003a: 378 **A**: AF
ghourbandensis Kaszab, 1970a: 99 **A**: AF
haarlovi Gridelli, 1954: 256 [= 1955: 54] **A**: AF
higginsi Kaszab, 1956d: 418 **A**: KA PA
kabakovi G.S. Medvedev, 1999b: 862 **A**: AF

kabulensis G.S. Medvedev, 1999b: 878 **A:** AF
klapperichi Kaszab, 1960a: 55 **A:** AF
kraatzi Reitter, 1893c: 309 **A:** AF
refleximargo Gridelli, 1954: 251 [= 1955: 50]
kunarensis G.S. Medvedev, 2003a: 366 **A:** AF
olegi G.S. Medvedev, 1999b: 874 **A:** AF
oruzgana G.S. Medvedev, 2003a: 394 **A:** AF
provincialis Kaszab, 1959e: 246 [= 1960a: 48] **A:** AF
quadraticollis Kaszab, 1960a: 47 **A:** AF
scheibei Schuster, 1936c: 199 **A:** AF
shahranensis Medvedev, 2003a: 382 **A:** AF
simplicithorax Gridelli, 1954: 255 [= 1955: 52] **A:** AF
speciosa Kaszab, 1970a: 102 **A:** AF
subbrunnea Kaszab, 1970a: 100 **A:** AF
tuxeni Kaszab, 1959e: 247 [= 1960a: 52] **A:** AF
undulatipennis Gridelli, 1954: 257 [= 1955: 55] **A:** AF
zoltani G.S. Medvedev, 1997b: 564 **A:** AF (Hindu Kush)
subgenus *Prosodopria* Reitter, 1909d: 116 type species *Blaps angustata* Zubkov, 1833
Nyctipates Solier, 1848: 285 [HN] type species *Nyctipates coriacea* Solier, 1848 (= *Blaps angustata* Zubkov, 1833)
angustata Zubkov, 1833: 329 (*Blaps*) **A:** KI
asperata Zubkov, 1833: 330 (*Blaps*)
carinata Fischer von Waldheim, 1844: 119 (*Nyctipates*)
coriacea Solier, 1848: 289 (*Nyctipates*)
mammillata Faldermann, 1836: 382 (*Blaps*)
subgenus *Prosodoscelis* Reitter, 1909d: 117 type species *Prosodes solskyi* Faust, 1875
dentimana Reitter, 1909d: 130 **A:** IN
irinae Skopin, 1964: 284 **A:** KZ (Mangistauskaya Oblast)
solskyi Faust, 1875: 239 **A:** IN TM (Kopet Dag)
subgenus *Prosodura* Reitter, 1909d: 118 type species *Prosodes semenowi* Reitter, 1893
peloroides Reitter, 1909d: 134 **A:** UZ (Bukhara)
prosternalis Reitter, 1893c: 284 **A:** UZ (Bukhara)
punctimana Reitter, 1893c: 284 **A:** UZ (Bukhara)
semenowi darvazica G.S. Medvedev, 1996: 611 **A:** TD
semenowi semenowi Reitter, 1893c: 273 **A:** TD UZ
punctatostriata Reitter, 1909d: 133
semenowi sexualis Kaszab, 1960a: 61 **A:** AF
semenowi vachshiana G.S. Medvedev, 1996: 609 **A:** TD
subgenus *Tangiprosodes* G.S. Medvedev, 2005a: 68 type species *Prosodes pygmaea* Kraatz, 1882
kuhitangiana G.S. Medvedev, 1964a: 56 **A:** TM (Kuhitang)
pygmaea pygmaea Kraatz, 1882d: 325 **A:** AF KZ UZ
digitula Reitter, 1915e: 91

kelessiana Skopin, 1971a: 495
minima Kraatz, 1882d: 326
subgenus *Uroprosodes* Reitter, 1909d: 119 type species *Prosodes costifera* Kraatz, 1886
altimontana G.S. Medvedev, 1995: 848 **A**: TD (Zheranshanskiy Ridge)
amathia Reitter, 1893c: 275 **A**: TD
arnoldii G.S. Medvedev, 1995: 852 **A**: UZ (Fergana)
ballionis Kraatz, 1885: 288 [RN] **A**: UZ
belousovi G.S. Medvedev, 1996: 627 **A**: TD
brevitarsis Kraatz, 1882d: 305 **A**: UZ (Samarkand)
georgii Reitter, 1909d: 153
catenulata Kraatz, 1883a: 346 **A**: KI
chodzhaksaris Skopin, 1969: 885 **A**: TM (Kuhitang-Tay)
costifera Kraatz, 1886: 187 **A**: KI (Alai) TD XIN
noethi Reinig, 1931: 872
crassitibia G.S. Medvedev, 1995: 839 **A**: UZ
danijarovi G.S. Medvedev, 1995: 847 **A**: TD
diloides Kraatz, 1882d: 323 **A**: UZ (Samarkand)
dushenkoi G.S. Medvedev, 1996: 629 **A**: TD
fleischeri Reitter, 1894d: 243 **A**: UZ (Samarkand)
glasunowi Semenov, 1894a: 208 **A**: UZ (Bukhara)
semenoviana Bogatchev, 1941b: 110
grombczewskii Semenov, 1894a: 203 **A**: UZ (Bukhara)
hauseri Reitter, 1893c: 300 **A**: UZ
heydeni Kraatz, 1882d: 324 **A**: UZ
gebieni Reitter, 1909d: 152
parumpunctata Kraatz, 1882d: 321
pubibasis Reitter, 1909d: 142
hissarica hissarica G.S. Medvedev, 1995: 837 **A**: UZ
hissarica infida G.S. Medvedev, 1995: 837 **A**: UZ
hissarica inopinata G.S. Medvedev, 1997b: 584 **A**: TD
kondarensis H.R. Dadabaev, 1986: 127 **A**: TD
korotyaevi G.S. Medvedev, 1997b: 588 **A**: TD
kryzhanovskii G.S. Medvedev, 1995: 845 **A**: TD
michailovi G.S. Medvedev, 1995: 846 **A**: TD
monticola G.S. Medvedev, 1964a: 58 **A**: TM (Kuhitang)
nalepoides Reitter, 1893c: 299 **A**: UZ (Bukhara)
novemcostata Semenov, 1890: 367 **A**: UZ
acutimarga Skopin, 1971a: 497
obliquesulcata Kraatz, 1882d: 322 **A**: UZ (Samarkand, Tashkent)
oblonga G.S. Medvedev, 1995: 841 **A**: UZ
opaca Bogatchev, 1960c: 84 **A**: TD
ovtshinnikovi G.S. Medvedev, 1995: 843 **A**: TD

rectilineata Bogatchev, 1960c: 83 **A:** TD
reflexicollis Reitter, 1893c: 290 **A:** UZ (Bukhara)
rugicostata Reitter, 1900b: 55 **A:** UZ (Bukhara)
rugosipennis G.S. Medvedev, 1995: 850 **A:** TD
sogdiana G.S. Medvedev, 1995: 840 **A:** TD (Zerafshanskiy Ridge)
spectabilis Reitter, 1893c: 287 **A:** UZ (Arslanbovskiy Massiv)
inconstans Reitter, 1909d: 139
kiritchenkoi Semenov & Bogatchev, 1939: 125
subpilosa G.S. Medvedev, 1964a: 59 **A:** TM (Kuhitang)
freibergi Skopin, 1969: 882
suturangula Reitter, 1900b: 55 **A:** UZ (Bukhara)
suturidens Reitter, 1909d: 142 **A:** UZ
tachtaensis tachtaensis G.S. Medvedev, 1995: 834 **A:** UZ
tachtaensis umnovi G.S. Medvedev, 1995: 837 **A:** UZ
tshikatunovi Kh. Dadabaev, 1986: 78 **A:** TD
tudaensis G.S. Medvedev, 1995: 832 **A:** UZ
vereshtshaginae G.S. Medvedev, 1997b: 591 **A:** UZ (Kutikangay)
vincens Reitter, 1900b: 55 **A:** UZ (Samarkand)
ypsilon Reitter, 1893c: 296 **A:** UZ (Samarkand)

genus *Prosodes*, species incertae sedis
motschulskyi Motschulsky, 1853: 78 **A:** XIN
transversosulcata Allard, 1881a: 292 **A:** "Inde orientale"

genus *Tagona* Fischer von Waldheim, 1820: pl. 16 type species *Tagona acuminata* Fischer von Waldheim, 1820
acuminata Fischer von Waldheim, 1820: pl. 16 **A:** KZ TM
lata Sumakov, 1908b: 245
macrophthalma macrophthalma Fischer von Waldheim, 1820: pl. 16 **E:** ST (Astrakhanskaya Oblast) **A:** AF KZ TM ZU
longicollis Brancsik, 1899: 100
tinodactyla Fischer von Waldheim, 1844: 121
macrophthalma rugipleuris Reitter, 1901c: 76 **A:** KZ (Muyumkum)
robusta Skopin, 1970: 166 **A:** UZ

subtribe Remipedellina Semenov, 1907

genus *Remipedella* Semenov, 1907b: 257 type species *Remipedella deserti* Semenov, 1907
deserti Semenov, 1907b: 259 **A:** TM (Karakumy)
semenovi G.S. Medvedev, 1989a: 90 **A:** TM

tribe Bolitophagini Kirby, 1837
subtribe Bolitophagina Kirby, 1837

genus *Atasthalomorpha* Miyatake, 1964: 77 type species *Atasthalus dentifrons* Lewis, 1894
dentifrons Lewis, 1894: 385 (*Atasthalus*) **A:** ES FE JA

genus *Boletoxenus* Motschulsky, 1858b: 63 type species *Boletoxenus gibber* Motschulsky, 1858
assamicus Kaszab, 1965a: 115 (*Bolitoxenus*) **A:** AP
bellicosus Lewis, 1894: 386 (*Atasthalus*) **A:** FE JA SC
galloisi Pic, 1932a: 3 (*Bolitonaeus*)
formosanus Masumoto, 1982c: 31 **A:** TAI
incurvatus Lewis, 1894: 386 (*Atasthalus*) **A:** FE JA
kawadai Masumoto & Akita, 2003: 176 **A:** JA (Ryukyus)
sasajii Masumoto & Akita, 2001c: 247 **A:** JA (Amami-Ôshima)
yichingae Masumoto, Akita & Lee, 2008a: 213 **A:** TAI

genus *Bolitonaeus* Lewis, 1894: 387 type species *Bolitonaeus mergae* Lewis, 1894
mergae Lewis, 1894: 387 **A:** JA
quadridentatus Candèze, 1861: 368 (*Bolitotherus*) **A:** NP TAI **ORR**

genus *Bolitophagiella* Miyatake, 1964: 71 type species *Bolitophagus pannosus* Lewis, 1894
pannosa Lewis, 1894: 384 (*Bolitophagus*) **A:** JA SC

genus *Bolitophagus* Illiger, 1798: 100 type species *Silpha reticulata* Linnaeus, 1767
Boletophagus Agassiz, 1846b: 48 [UE]
granulatus Fischer von Waldheim, 1842: 16 [= 1844: 129] **A:** KZ (Dzungaria)
interruptus Illiger, 1800: 112 **E:** AB AU BH CR CZ FR GE HU IT PL RO SK SL **A:** IN
goedeni Panzer, 1803b: 1 (*Bolitophagus*)
reticulatus Linnaeus, 1767: 572 (*Silpha*) **E:** AB AL AR AU BE BH BU BY CR CT CZ DE EN FI FR GB GE GG GR HU IT LS LT NL NR NT PL RO SK SP ST SV SZ UK YU **A:** FE ES JA KZ SC TR WS
boleti Piller & Mitterpacher, 1783: 113 (*Silpha*)
cornutus Thunberg, 1784: 15 (*Hispa*)
crenatus Fabricius, 1792a: 89 (*Opatrum*)
subinteger Reitter, 1896b: 74 **E:** AB (Talysh) **A:** IN

genus *Bolitotrogus* Miyatake, 1964: 80 type species *Bolitotrogus kurosonis* Miyatake, 1964
formosanus Masumoto, 1982c: 30 **A:** TAI
kurosonis Miyatake, 1964: 82 **A:** JA
ohdaiensis Miyatake, 1964: 83 **A:** JA

genus *Byrsax* Pascoe, 1860a: 42 type species *Byrsax coenosus* Pascoe, 1860
fukiensis Kaszab, 1954: 251 **A:** FUJ
hiranoi Akita & Masumotio, 2012: 58 **A:** JA (Ryukyus)
kaszabi Miyatake, 1970: 120 **A:** JA
lewisi Akita & Masumoto, 2012: 64 **A:** JA SC
niponicus amamiensis Akita & Masumoto, 2012: 72 A: JA (Tokara, Amami-Ôshima)
niponicus niponicus Lewis, 1894: 388 **A:** JA
nishinoi Akita & Masumoto, 2012, 67 **A:** JA (Ryukyu)
shibatai Masumoto, 1982c: 26 **A:** JA TAI
kawadai Masumoto, 1982c: 28
spiniceps Lewis, 1894: 388 **A:** FE JA SC TAI
kimurai Miyatake, 1970: 118
takakuwai Akita & Masumoto, 2012: 68 **A:** JA (Mikura Is.)
tsushimensis Miyatake, 1970: 122 **A:** JA (Tsushima)
tuberculatus Gravely, 1915: 522 **A:** AP NP YUN **ORR**

genus *Eledona* Latreille, 1796: 19 type species *Opatrum agricola* Herbst, 1783
Heledona Agassiz, 1846: 135, 174 [UE]
agricola Herbst, 1783: 35 (*Opatrum*) **E:** AB AR AU BE BH BY CR CT CZ DE EN FI FR GB GE GG GR HU IT LA LS LT MA NL NR PL PT SK SL SP ST SV SZ TR UK **N:** TU **A:** TM TR **NAR**
fungicola Horn, 1870: 389
sulcata Thunberg, 1784: 2 (*Dermestes*)
turcica Seidlitz, 1891: 553
hellenica Reitter, 1885d: 381 (*Heledona*) **E:** GR TR

genus *Eledonoprius* Reitter, 1911a: 338 type species *Opatrum armatum* Panzer, 1799
armatus Panzer, 1799a: 2 (*Opatrum*) **E:** AB AU BU CR CZ DE FR GE GR HU IT PL RO SK SL SP SV SZ UK **N:** MO **A:** CY **NAR**
serrifrons Reitter, 1890e: 172 (*Bolitophagus*) **E:** AB AR IT SP

genus *Lanhsia* Shibata, 1980: 64 type species *Lanhsia bucca* Shibata, 1980
bucca Shibata, 1980: 64 **A:** TAI

genus *Parabolitophagus* Miyatake, 1964: 70 type species *Bolitophagus felix* Lewis, 1894
felix Lewis, 1894: 384 (*Bolitophagus*) **A:** JA SC

subtribe Rhipidandrina LeConte, 1862

genus *Rhipidandrus* J.F. LeConte, 1862: 236 type species *Xyletinus flabellicornis* Sturm, 1826 (= *Melolontha paradoxa* Palisot de Beauvois, 1818)
Bolitopertha Gebien, 1910b: 379 type species *Bolitopertha novemcostata* Gebien, 1910
Cherostus Waterhouse, 1894: 68 type species *Cherostus walkeri* Waterhouse, 1894

Eutomus Lacordaire, 1866: 369 type species *Eutomus micrographus* Lacordaire, 1866

Heptaphylla Friedenreich, 1883: 375 type species *Heptaphylla fungicola* Friedenreich, 1883

speculifrons Gebien, 1922a: 288 (*Cherostus*) **A:** JA (Okinawa) TAI **AFR AUR ORR**

dybasi Kulzer, 1957: 204

scolytoides M. Chûjô, 1985: 61

sodalis Kulzer, 1957: 206

tribe Dendarini Mulsant & Rey, 1854
subtribe Dendarina Mulsant & Rey, 1854

genus *Bioplanes* Mulsant, 1854: 144 type species *Bioplanes meridionalis* Mulsant, 1854

meridionalis Mulsant, 1854: 145 **E:** FR IT

genus *Dendarophylan* Español, 1945b: 3[illegible]2 type species *Phylan pardoi* Español, 1945

pardoi Español, 1945b: 349 (*Phylan*) **E:** SP **N:** MO

genus *Dendarus* Dejean, 1821: 65 type species *Helops tristis* sensu Laporte, 1840 (= *Pandarus coarcticollis* Mulsant, 1854)

subgenus *Dendaroscelis* Reitter, 1904a: 79 type species *Dendarus serripes* Reitter, 1904

serripes Reitter, 1904a: 92 (*Dendarus*) **E:** SP **N:** MR

subgenus *Dendarus* Dejean, 1821: 65 type species *Helops tristis* sensu Laporte, 1840 (= *Pandarus coarcticollis* Mulsant, 1854)

Microdendarus Escalera, 1944: 84 type species *Dendarus schusteri* Español, 1937

Pandarus Dejean, 1834: 191 [UE] [HN]

Reitterellus Escalera, 1944: 91 type species *Dendarus depressus* Reitter, 1915

aubei aubei Mulsant & Rey, 1854a: 64 [= 1854b: 208] (*Pandarus*) **E:** SP (Andalusia)

aubei cazorlensis Koch, 1944b: 254 **E:** SP (Sierra de Cazorla)

seguranus Escalera, 1944: 89

calcaratus Baudi di Selve, 1881: 281 **A:** IS LE

carinatus Mulsant & Rey, 1854a: 58 [= 1854b: 202] (*Pandarus*) **E:** FR (Corse) IT (Sardinia)

corsicus Perris, 1875: 5 (*Pandarinus*)

tricostatus Mulsant & Rey, 1854a: 60 = 1854b: 204] (*Pandarus*)

castilianus Piochard de la Brûlerie, 1869: 31 (*Pandarus*) **E:** PT SP

cedrorum F. Soldati & L. Soldati, 2000: 68 **A:** LB

coarcticollis Mulsant, 1854: 142 (*Pandarus*) **E:** AU CR FR IT SZ

depressus Reitter, 1915e: 86 **E:** SP (Baleares)

disparicus Clermont, 1925: 183

hildti Tenenbaum, 1915: 837

emarginatus Brullé, 1832: 211 (*Phylax*) **E:** GR **A:** TR

tibialis Seidlitz, 1893: 398

graecus Brullé, 1832: 207 **E:** GR (Crete)

montanus Koch, 1948a: 343

insidiosus alcojonensis Español, 1961a: 62 **E**: SP (Málaga)
insidiosus bolivari Escalera, 1944: 87 **E**: SP (Cádiz)
insidiosus insidiosus Mulsant & Rey, 1854a: 66 [= 1854b: 210] (*Pandarus*) **E**: SP **N**: MO
mohgrebicus Escalera, 1944: 88
insidiosus riffensis Escalera, 1944: 88 **N**: MO
pardoi Escalera, 1944: 89 **N**: MO
pectoralis bejarensis Español, 1961a: 45 **E**: SP
pectoralis ilerdensis Español, 1937: 49 **E**: SP
pectoralis pectoralis Mulsant & Rey, 1854a: 62 [= 1854b: 206] (*Pandarus*) **E**: PO SP **N**: MO
coriaceus Baudi di Selve, 1876a: 36
piochardi Español, 1937: 51 **E**: PT SP
sinuatus Mulsant & Rey, 1854a: 69 [= 1854b: 213] (Pandarus) **E**: GR (Cyclades)
stygius kochi G.S. Medvedev, 1968a: 100 [RN] **E**: GR (Evvoia)
oertzeni Koch, 1948a: 349 [HN]
stygius stygius Waltl, 1838: 462 **E**: GR **A**: TR
schusteri Español, 1937: 54 **E**: SP
tauricus Koch, 1948a: 350 **A**: SY TR
trichasi Leo, 2017: 174 **E**: GR (Kythnos Is.)
kochi Trichas, 2008: 433 [HN]
werneri Koch, 1948a: 342 **E**: GR (Kyklades)
zariquieyi almeriensis Escalera, 1944: 90 **E**: SP
zariquieyi vivesi Español, 1961a: 58 **E**: SP (Granada)
zariquieyi zariquieyi Español, 1937: 62 **E**: SP **N**: MO
subgenus *Pandarinus* Mulsant & Rey, 1854a: 103 [= 1854b: 247] type species *Pandarinus tenellus* Mulsant & Rey, 1854
alleonis Seidlitz, 1893: 399 **E**: TR **A**: TR
ananensis Chatzimanolis, Engel & Trichas, 2002: 260 **E**: GR (Ananes Is., Astypalaia Is.)
anaphianus Koch, 1948a: 344 **E**: GR (Anaphis Is.)
angulitibia Koch, 1948a: 345 **E**: GR (Tinos Is.)
antikythirensis Chatzimanolis, Engel & Trichas, 2002: 264 **E**: GR (Antythira Is.)
armeniacus Baudi di Selve, 1876c: 266 [= 1976b: 140] **E**: AR **A**: AF IN TM TR
coelatus Brullé, 1832: 208 **E**: AL GR IT **A**: TR
crenulatus Ménétriés, 1832: 202 (*Heliophilus*) **E**: AB AR GG ST **A**: IN TR
dardanus Faldermann, 1837: 51 (*Pandarus*)
cribratus Waltl, 1838: 462
dalmatinus Germar, 1823: 142 (*Pedinus*) **E**: AL BH CR GR IT SL **A**: TR
dentitibia Koch, 1948a: 344 **E**: GR (Andros Is.)
dragonadanus Koch, 1948a: 348 **E**: GR (Crete, Dragonada Is)
extensus Faldermann, 1837: 52 (*Pandarus*) **E**: AR **A**: IN TM
falassamensis Chatzimanolis, Engel & Trichas, 2002: 263 **E**: GR (Crete)
foveolatus Seidlitz, 1893: 399 **E**: AR **A**: TR

grampusanus Koch, 1948a: 347 **E**: GR (Grampusa)
jonicus Koch, 1948a: 352 **E**: GR
leonhardi Schuster, 1940b: 861 **A**: IN TM (Kopet Dag)
maximus Koch, 1948a: 348 **E**: GR (Gavdos Is., Gavdopoula Is.)
messenius Brullé, 1832: 211 (*Phylax*) **E**: AL GR MC TR YU **A**: TR
  gravidus Brullé, 1832: 210 (*Phylax*)
  ottomanus Mulsant & Rey, 1854a: 101 [=1854b: 245] (*Pandarus*)
  tentyrioides Brullé, 1832: 212 (*Phylax*)
moesiacus Mulsant & Rey, 1854a: 87 [= 1854b: 231] (*Pandarus*) **E**: AL BU GR MC **A**: TR
mylonasi Chatzimanolis, Engel & Trichas 2002: 262 **E**: GR (Augo Is.)
opacus Koch, 1948a: 347 **E**: GR (Crete)
orientalis Seidlitz, 1893: 399 **A**: CY SY
paganettii Koch, 1948a: 353 **E**: GR
peslieri F. Soldati, 2012b: 53 **E**: GR
plicatulus Brullé, 1832: 210 (*Phylax*) **E**: GR TR
  simius Mulsant & Rey, 1854a: 76 [= 1854b: 220] (*Pandarus*)
  seidlitzi Reitter, 1904a: 80
punctatus Audinet-Serville, 1825: 26 (*Pedinus*) **E**: AR ST UK **A**: TR
puncticollis Koch, 1948a: 348 **E**: GR (Crete)
schatzmayri Koch, 1948a: 344 **E**: GR (Syra, Kythera Is.)
scoparipes Reitter, 1904a: 86 **E**: GR (Kythera)
scrobiculatus Seidlitz, 1893: 399 **A**: TR
simplex Seidlitz, 1893: 399 **E**: AR **A**: IQ IN SY TR
sporadicus Koch, 1948a: 347 **E**: GR (Sporades)
stampalicus Koch, 1948a: 347 **E**: GR (Sporades)
subopacus Baudi di Selve, 1881: 282 (*Pandarinus*) **A**: LE SY
tenellus Mulsant & Rey, 1854a: 106 [= 1854b: 250] (*Pandarinus*) **E**: GR **A**: IQ TR
torpidus Mulsant & Rey, 1854a: 82 [= 1854b: 226] (*Pandarus*) **A**: TR
transcaspicus afghanicus Kaszab, 1973b: 53 **A**: AF
transcaspicus medvedevi Kaszab, 1973b: 55 **A**: AF
transcaspicus transcaspcius Brancsik, 1899: 101 **A**: TM (Kopet Dag)
  vagabundus Reitter, 1904a: 84
  vagans Reitter, 1902b: 180
victoris Mulsant & Rey, 1854a: 84 [= 1854b: 228] (*Pandarus*) **E**: CR GR
wernerianus Koch, 1948a: 344 **E**: GR (Sikinos)
wettsteini Koch, 1948a: 348 **E**: GR (Crete)
subgenus *Paroderus* Mulsant & Rey, 1854a: 111 [= 1854b: 255] type species *Pandarinus elongatus* Mulsant & Rey, 1854
  Dichromma Mulsant & Rey, 1855c: 194 type species *Pandarinus foraminosus* Mulsant & Rey, 1855
corcyrensis Koch, 1948a: 352 **E**: GR (Kerkyra)
elongatus Mulsant & Rey, 1854a: 111 [= 1854b: 255] (*Pandarinus*) **E**: SP (Andalusia)

creticus Motschulsky, 1858d: 188 (*Bioplanes*)
nevadensis Piochard de la Brûlerie, 1869: 33 (*Pandarinus*)
foraminosus Mulsant & Rey, 1855b: 95 [= 1855c: 195] **E:** GR (Crete)
punctulatus Lucas, 1854: 30 (*Philax*)
lugens Mulsant & Rey, 1854a: 77 [= 1854b: 221] (*Pandarus*) **E:** IT MA
libanicus Desbrochers des Loges, 1881: 116 [HN]
politus Reitter, 1904a: 89 **E:** GR (Crete)
rhodius Baudi di Selve, 1876a: 39 **E:** GR (Rhodes, Sporades)
subgenus *Rhizalemus* Reitter, 1904a: 79 type species *Dendarus reitteri* Seidlitz, 1893
calcaroides Peyerimhoff, 1907a: 36 (*Phylax*) **A:** SI
crassiusculus Mulsant & Rey, 1854a: 246 [= 1855a: 134] (*Bioplanes*) **A:** IS
impressus Reiche & Saulcy, 1857: 255 (*Bioplanes*) **A:** IS "Syria"
oertzenii Seidlitz, 1893: 393 **A:** IS
pauper Mulsant & Rey, 1854a: 114 [= 1854b: 258] (*Pandarinus*) **A:** IS JO
plorans Mulsant & Rey, 1854a: 245 [= 1855a: 133] (*Bioplanes*) **A:** IS
saginatus Baudi di Selve, 1876a: 41
reitteri Seidlitz, 1893: 398 **A:** LE "Syria"
libanicus Baudi di Selve, 1875d: 697
vagans Seidlitz, 1893: 399 **A:** LE
subgenus *Rizalus* Mulsant & Rey, 1854a: 104 [= 1854b: 248] type species *Opatrum piceum* G.-A. Olivier, 1811
piceus G.-A. Olivier, 1811: 501 (*Opatrum*) **N:** EG **A:** CY IS LE SY
tenuicornis Miller, 1861: 201 (*Pandarus*)
syriacus Reiche & Saulcy, 1857: 257 (*Bioplanes*) **N:** EG **A:** IS JO SI
nitidulus Seidlitz, 1893: 394
viduus Reiche, 1857: 256 (*Bioplanes*)

genus *Dendarus*, nomen dubium
obscuripennis Brullé, 1832: 212 (*Phylax*) **E:** GR

genus *Heliopates* Dejean, 1834: 191 type species *Tenebrio lusitanicus* Herbst, 1797
subgenus *Heliocrates* Reitter, 1904a: 98 type species *Heliophilus humerangulus* Reitter, 1904
forcadelli Español, 1958a: 23 **E:** SP (Columbrete)
humerangulus Reitter, 1904a: 99 (*Heliophilus*) **E:** SP
morandi Español, 1969: 81 **E:** SP (Salinas, Torrevieja)
murriai J. Ferrer, 2011: 79 **E:** SP
sculpturatus Reitter, 1904a: 102 (*Heliophilus*) **E:** SP
strigosus Reitter, 1904a: 98 (*Heliophilus*) **E:** SP
transversalis Mulsant & Rey, 1854a: 232 [= 1855a: 120] **E:** SP

subgenus *Heliopates* Dejean, 1834: 191 [RN] type species *Tenebrio lusitanicus* Herbst, 1797
Heliocaes Bedel, 1906: 92 [RN]
Heliophilus Dejean, 1821: 65 [HN] type species *Pedinus hybridus* Latreille, 1804 (= *Tenebrio abbreviatus* G.-A. Olivier, 1795)
agrestis Mulsant & Rey, 1854a: 239 [= 1855a: 127] **E:** SP
angustipennis Español, 1958c: 115 **E:** SP
atlasicus Escalera, 1925f: 501 (*Heliocaes*) **N:** MO (Middle Atlas, High Atlas)
altiatlantis Antoine, 1936b: 72 (*Heliophilus*)
bergmanni Antoine, 1951b: 318 (*Heliophilus*)
avarus avarus Mulsant & Rey, 1854a: 234 [= 1855a: 122] **E:** IT MA **N:** AG
ambiguus Baudi di Selve, 1876b: 164
avarus donatellae Canzoneri, 1970: 72 (*Heliocaes*) **E:** IT (Pantelleria Is.)
avarus dwejrensis Scupola & Mifsud, 2001: 1 **E:** MA
balearicus Español, 1951a: 21 [= 1954a: 56] **E:** SP (Ibiza Is.)
batnensis Mulsant & Godart, 1866: 455 **E:** IT (Sicilia) **N:** AG MO TU
tunisiensis Dajoz, 1985: 193
castrotovari J. Ferrer & Iwan, 2012: 130 **E:** SP
conjuctus Reitter, 1904a: 101 (*Heliophilus*) **E:** SP (Huesca, Zaragoza, Lérida)
cribratostriatus Mulsant & Rey, 1854a: 234 [= 1855a: 122] **E:** PT SP **N:** AG MO
cribratus Chevrolat, 1865: 391 **E:** PT SP
emarginatus Fabricius, 1792a: 108 (*Blaps*) **E:** PT SP **N:** AG MO
subvariolosus Lucas, 1847: 330 (*Dendarus*)
heydeni Reitter, 1904a: 102 (*Heliophilus*) **E:** SP
ibericus Mulsant & Rey, 1854a: 235 [= 1855a: 123] **E:** PT SP
immarginatus Reitter, 1904a: 101 (*Heliophilus*) **E:** SP
interstitialis Mulsant & Rey, 1854a: 231 [= 1855a: 119] **N:** AG MO
strigicollis Baudi di Selve, 1876b: 163
liebei Schuster, 1926c: 243 (*Heliophilus*) **E:** SP (Pontevedra)
littoralis Español, 1958a: 20 **E:** FR SP
luctuosus Le Peletier de Saint-Fargeau & Audinet-Serville, 1825: 27 (*Pedinus*) **E:** FR SP
lusitanicus Herbst, 1797: 244 (*Tenebrio*) **E:** PT SP
martinezi J. Ferrer & Iwan, 2012: 133 **E:** SP
montivagus Mulsant & Rey, 1854a: 233 [= 1855a: 121] **E:** SP (Granada)
neptunius Baudi di Selve, 1875b: 164 **E:** IT (Sicilia: Mti Madone)
obsoletus Marsham, 1802: 475 (*Tenebrio*) **E:** PT
franzi Español, 1955c: 97
parcefoveatus Reitter, 1904: 100 (*Heliophilus*) **E:** PT SP
perroudi Mulsant, 1854: 237 **E:** SP
rex Antoine, 1942c: 48 (*Heliophilus*) **N:** MO (Rif)

rectangularis Antoine, 1951c: 97 (*Heliophilus*) **N:** MO
rotundicollis Lucas, 1847: pl. 29 (*Dendarus*) **E:** PT SP **N:** AG LB TU
rotundicollis Mulsant & Rey, 1854a: 237 [= 1855a: 125] (*Heliopathes*) [HN]
tripolianus Baudi di Selve, 1876a: 50
rupestris Mulsant & Rey, 1854: 240 **E:** SP
simulans Chevrolat, 1865: 391 **E:** SP
sprecherae J. Ferrer & Iwan, 2012: 131 **E:** PT SP
striatoforatus Mulsant & Rey, 1854: 239 **E:** SP
sulcipennis Reitter, 1904a: 99 (*Heliophilus*)
subimpressus Reitter, 1904a: 103 (*Heliophilus*) **E:** SP
tazakensis Antoine, 1942c: 48 (*Heliophilus*) **N:** MO
torresmendezi J. Ferrer & Iwan, 2012: 133 **E:** SP
vivesi Español & Viñolas, 1983b: 216 **E:** SP

genus *Heliopates*, nomen dubium
meridianus Brullé, 1832: 208 (*Heliophilus*) **E:** GR

genus *Litoboriolus* Español, 1945b: 310 type species *Omocrates collaris* Mulsant & Rey, 1854
barclayi J. Ferrer, 2010b: 136 (*Phylan*) **E:** SP (Guadarrama)
bremeri J. Ferrer, 2010b: 136 (*Phylan*) **E:** SP (Galicia Pontevedra)
cobosi J. Ferrer, 2010b: 137 (*Phylan*) **E:** SP (Sierra de Montánchez)
collaris Mulsant & Rey, 1854a: 195 [= 1855a: 83] (*Omocrates*) **E:** SP (Guadarrama)
companyae Viñolas, 1997a: 98 (*Phylan*) **E:** SP (Zamora)
gloriae J. Ferrer, 2010b: 137 (*Phylan*) **E:** SP
mulsanti Piochard de la Brûlerie, 1869: 36 (*Heliopathes*) **E:** SP
gadeai Español & Viñolas, 1981b: 108 (*Phylan*)
reyi Piochard de la Brûlerie, 1869: 37 (*Heliopathes*) **E:** SP

genus *Litororus* Reitter, 1904a: 96 type species *Micrositus semicostatus* Mulsant & Rey, 1854
nitidicollis Perez Arcas, 1872: 107 (*Micrositus*) **E:** SP (Mallorca)
cabrerensis Tenenbaum, 1915: 839 (*Dendarus*)
melas Tenenbaum, 1915: 838 (*Dendarus*)
semicostatus curtulus Breit, 1908: 63 (*Phylan*) **E:** SP (Cabrera)
semicostatus semicostatus Mulsant & Rey, 1854a: 178 [= 1854b: 322] (*Micrositus*) **E:** SP (Baleares)
balearicus Schaufuss, 1882c: 623 (*Phylax*)

genus *Micrositus* Mulsant & Rey, 1854a: 148 [= 1854b: 292] type species *Micrositus orbicularis* Mulsant & Rey, 1854
opacus Reitter, 1904a: 116 **A:** TR
orbicularis Mulsant & Rey, 1854a: 155 [= 1854b: 299] **E:** GR **A:** TR

genus *Neoisocerus* Bouchard, Lawrence, Davies & Newton, 2005: 499 [RN] type species *Tenebrio purpurascens* Herbst, 1799 (= *Helops ferrugineus* Fabricius, 1798)

Isocerus Dejean, 1821: 66 [HN] type species *Tenebrio purpurascens* Herbst, 1799 (= *Helops ferrugineus* Fabricius, 1798)

balearicus L.W. Schaufuss, 1879: 535 (*Isocerus*) **E:** SP (Baleares)

barbieri Ardoin, 1972: 239 (*Isocerus*) **N:** AG

ferrugineus Fabricius, 1798: 53 (*Helops*) **E:** PT SP **N:** MO

cameroni Reitter, 1904a: 78 (*Isocerus*)

lusitanicus Reitter, 1886d: 152 (*Isocerus*)

purpurascens Herbst, 1799: 20 (*Tenebrio*)

tunisiensis E. Strand, 1935: 294 (*Isocerus*) [RN] **N:** AG TU

ferrugineus Reitter, 1886d: 152 (*Isocerus*) [HN]

genus *Phylan* Sturm, 1826: 23 type species *Opatrum gibbum* Fabricius, 1775

subgenus *Eumicrositus* Viñolas, 1990: 62 type species *Pedinus ulissiponensis* Germar, 1823

emmanueli Español, 1958c: 108 (*Micrositus*) **E:** SP (Guadalupe)

laufferi Reitter, 1915e: 87 (*Micrositus*) **E:** SP (Granada)

martinezi Español, 1947a: 34 (*Micrositus*) **E:** SP (Ávila)

miser almeriensis Español, 1947a: 30 (*Micrositus*) **E:** SP (Almeriá)

miser miser Mulsant & Rey, 1854a: 168 [= 1854b: 312] (*Micrositus*) **E:** PT SP **N:** MO

obesus agricola Mulsant & Rey, 1854a: 164 [= 1854b: 308] (*Micrositus*) **E:** SP

obesus obesus Waltl, 1835: 71 (*Heliophilus*) **E:** PT SP

baeticus Mulsant & Rey, 1854b: 310 [= 1854a: 166] (*Micrositus*)

saxicola Mulsant & Rey, 1854a: 193 [= 1855a: 81] (*Omocrates*) **E:** PT SP

saxeticola Graells, 1858: 71 (*Phylan*)

ulissiponensis conveximargo Español, 1947a: 24 (*Micrositus*) **E:** PT SP

ulissiponensis montanus Mulsant & Rey, 1854a: 160 [= 1854b: 304] (*Micrositus*) **E:** SP

ulissiponensis paivae Perez Arcas, 1865b: 239 (*Micrositus*) **E:** SP

ulissiponensis ulissiponensis Germar, 1823 143 (*Pedinus*) **E:** PT SP

obesus Mulsant & Rey, 1854a: 165 [= 1854b: 309] (*Micrositus*) [HN]

ulyssiponensis Le Peletier de Saint-Fargeau & Audinet-Serville, 1825: 26 (*Pedinus*)

ulissiponensis ventralis Marseul, 1868b: 20 (*Micrositus*) [RN] **E:** SP

levis Perez Arcas, 1865a: 178 (*Micrositus*)

vivesi Español, 1963d: 185 (*Micrositus*) **E:** SP

subgenus *Meladeras* Mulsant & Rey, 1854a: 219 [= 1855a: 107] type species *Meladeras quadratulus* Mulsant & Rey, 1854

amoenus Mulsant & Rey, 1854a: 224 [= 1855a: 112] (*Meladeras*) **N:** AG

barbarus Lucas, 1847: 329 (*Dendarus*) **N:** AG

quadratulus Mulsant & Rey, 1854a: 220 [= 1855a: 108] (*Meladeras*)

ferreri Labrique, Chavanon & Janati-Idrissi, 2011: 78 **N:** MO

marocanus Obenberger, 1914: 140 (*Meladeras*) **N:** MO
obscurus Mulsant & Rey, 1854a: 223 [= 1855a: 111] (*Meladeras*) **N:** AG MO
subgenus *Meladocrates* Reitter, 1904a: 96 type species *Omocrates planiusculus* Mulsant & Rey, 1854
costulatus Escalera, 1914: 327 **N:** MO
orientalis Dajoz, 1985: 195 **N:** TU
planiusculus Mulsant & Rey, 1854a: 216 [= 1855a: 104] (*Omocrates*) **E:** SP (Cádiz) **N:** MO
lacertosus Antoine, 1942c: 56
latipennis Champion, 1891d: 384 (*Heliopathes*)
mediatlantis Antoine, 1942c: 57
tenuicornis Antoine, 1942c: 55 **N:** MO
viaticus Mulsant & Rey, 1854a: 218 [= 1855a: 106] (*Omocrates*) **E:** SP **N:** MO
subgenus *Phylan* Sturm, 1826: 23 type species *Opatrum gibbum* Fabricius, 1775
Olocrates Mulsant, 1854: 383 [RN] type species *Tenebrio abbreviatus* G.-A. Olivier, 1795
Phylax Brullé, 1832: 211 [URN] type species *Opatrum gibbum* Fabricius, 1775
abbreviatus abbreviatus G.-A. Olivier, 1795: 57: 17 (*Tenebrio*) **E:** FR SP
dubius Baudi di Selve, 1875b: 161 (*Heliopathes*)
hybridus Latreille, 1804a: 284 (*Pedinus*)
tristis Herbst, 1797: 243 (*Tenebrio*)
abbreviatus chossoni Pic, 1937d: 10 **N:** AG
abbreviatus italicus Reitter, 1904a: 93 (*Olocrates*) **E:** IT
abbreviatus hamilcaris Reitter, 1904a: 94 (*Olocrates*) **E:** SP
abbreviatus wankai Reitter, 1915e: 87 **E:** SP
ceballosi Español, 1945b: 335 **E:** SP (Granada, Murcia)
difficilis Antoine, 1942c: 54 **N:** MO (Middle Atlas)
minutus Antoine, 1942c: 54
foveipennis foveipennis Mulsant & Rey, 1854a: 203 [= 1855a: 91] (*Omocrates*) **E:** SP
dendaroides Baudi di Selve, 1876a: 46 (*Heliopates*)
foveostriatus Baudi di Selve, 1875b: 160 (*Heliopathes*)
foveipennis kochi Español, 1945b: 319 **E:** PT
foveipennis lineatopunctatus Mulsant & Rey, 1854a: 205 [= 1855a: 93] (*Omocrates*) **E:** SP
foveolatus Graells, 1858: 71 (*Heliopates*) **E:** SP
fossulatus Mulsant & Rey, 1854a: 200 [= 1855a: 88] (*Omocrates*)
milleporus Baudi di Selve, 1875b: 160 (*Heliopathes*)
gemellatus Marsham, 1802: 475 (*Tenebrio*) **E:** GB IR
gibbus Fabricius, 1775: 76 (*Opatrum*) **E:** EN FI LA LT LU NR NT PT SV
arenosus Marsham, 1802: 475 (*Tenebrio*)
pilipes Thunberg, 1821: 52 (*Opatrum*) [HN]
indiscretus indiscretus Mulsant & Rey, 1854a: 208 [= 1855a: 96] (*Omocrates*) **E:** SP
indiscretus mendizabali Español, 1945b: 332 **E:** SP (Almería, Jaén)
indiscretus nivalis Baudi di Selve, 1876a: 50 (*Heliopathes*) [= 1876b: 161] **E:** SP (Granada)
intermedius Dajoz, 1965: 131 **E:** FR SP

labriquei J. Ferrer, 2013a: 59 **E**: SP
mahjoubi Antoine, 1942c: 52 **N**: MO (surroundings of Fez)
mediterraneus Piochard de la Brûlerie, 1869: 37 (*Heliopathes*) **E**: SP (Baleares)
nitidior Antoine, 1942c: 53 **N**: MO (Meknès)
paludicola Chevrolat, 1865: 391 (*Olocrates*) **E**: SP (Zamora)
ardoini Español & Viñolas, 1981b: 102
pilipes Herbst, 1797: 260 (*Tenebrio*) **E**: BE DE FR GE NL PL
poggii J. Ferrer, 2013b: 59 **E**: IT
ribesi Español, 1969: 83 **E**: SP
rouxi J. Ferrer, 2013b: 58 **E**: FR
subheliopates Reitter, 1904a: 95 (*Olocrates*) **E**: SP
viriathus J. Ferrer, 2013b: 57 **E**: PT SP
gibbus Fuente, 1935: 114 [HN]
subgenus *Platyolus* Mulsant & Rey, 1854a: 153 [= 1854b: 297] type species *Micrositus longulus* Mulsant & Rey, 1854
furvus filabricus Español, 1963e: 387 (*Micrositus*) **E**: SP (Almería)
furvus furvus Mulsant & Rey, 1854a: 181 [= 1854b: 324] (*Micrositus*) **E**: SP
furvus stenopterus Español, 1947a: 49 (*Micrositus*) **E**: SP (Almería)
gibbulus Motschulsky, 1849: 124 (*Pandarus*) **E**: SP
gustavii Mulsant & Rey, 1854a: 244 (*Micrositus*)
heeri Mulsant & Rey, 1854a: 171 [= 1854b: 315] (*Micrositus*) **E**: SP (Almería)
lapidarius lapidarius Reitter, 1904: 121 (*Micrositus*) **E**: SP (Almería, Granada)
lapidarius strasseni Español, 1954d: 161 (*Micrositus*) **E**: SP (Granada)
longulus Mulsant & Rey, 1854a: 185 [= 1854b: 329] (*Micrositus*) **E**: PT SP
melancholicus Mulsant & Rey, 1854a: 176 [= 1854b: 320] (*Micrositus*) **E**: PT SP **N**: MO
milleri Reitter, 1904a: 120 (*Micrositus*) [DA] **E**: SP (Cádiz)
subcylindricus Motschulsky, 1849: 123 (*Pandarus*) **E**: SP
tenebrioides Reitter, 1904a: 118 (*Micrositus*) **E**: PT SP

genus *Phylanmania* J. Ferrer, 2013a: 38 type species *Phylan ilerdensis* Español & Viñolas, 198
escalerai Español, 1943b: 106 (*Heliophilus*) **E**: SP (Navarra)
ilerdensis Español & Viñolas, 1981b: 105 (*Phylan*) **E**: SP

subtribe Melambiina Mulsant & Rey, 1854

genus *Allophylax* Bedel, 1906c: 177 [RN] type species *Phylax littoralis* Mulsant, 1854 (= *Opatrum picipes* G.-A. Olivier, 1811)
subgenus *Allophylax* Bedel, 1906c: 177 [RN] type species *Phylax littoralis* Mulsant, 1854 (= *Opatrum picipes* G.-A. Olivier, 1811)
Melambiophylax Schuster, 1922c: 48 type species *Phylax sardous* Baudi di Selve, 1876
Neophylax Bedel, 1906b: 92 [HN] type species *Phylax littoralis* Mulsant, 1854 (= *Opatrum picipes* G.-A. Olivier, 1811)
brevicollis Baudi di Selve, 1876a: 55 [= 1876b: 214] (*Phylax*) **E**: IT (Sardinia)

picipes melitensis Baudi di Selve, 1876a: 54 (*Phylax*) **E:** MA
picipes picipes G.-A. Olivier, 1811: 500 (*Opatrum*) **E:** CR FR IT **N:** AG TU
callosum Thunberg, 1821: 52 (*Opatrum*)
dalmatinus Reitter, 1904: 109 (*Phylax*)
fraternus Mulsant & Rey, 1854: 287 (*Phylax*)
littoralis Mulsant, 1854: 148 (*Phylax*)
sardous Baudi di Selve, 1876a: 55 [= 1876b: 214] (*Phylax*) **E:** IT (Sardinia)
subgenus *Litoboromimus* Koch, 1948b: 413 type species *Litoborus parallelus* Schuster, 1919
parallelus Schuster, 1919a: 30 (*Litoborus*) **N:** LB
rotundicollis Chobaut, 1925: 228 **N:** TU
subgenus *Phylaximon* Koch, 1948b: 414 type species *Opatrum variolosus* G.-A. Olivier, 1811
costatipennis costatipennis Lucas, 1847: 328 (*Philax*) **E:** IT (Linosa Is., Lampedusa Is.) **N:** AG TU
sicardi Reitter, 1904a: 111 (*Phylax*)
costatipennis godenigoi Canzoneri, 1970: 75 **E:** IT (Marettimo Is.)
ignavus Mulsant & Rey, 1854a: 147 [= 1854b: 291] (*Phylax*) **N:** AG
ingratus Mulsant & Rey, 1854a: 143 [= 1854b: 287] (*Phylax*) **N:** AG TU
segnis Mulsant & Rey, 1854a: 145 [= 1854b: 289] (*Phylax*) **N:** AG
undulatus Mulsant & Rey, 1854a: 136 [= 1854b: 280] (*Phylax*) **N:** AG TU
ovipennis Fairmaire, 1871a: 391 (*Phylax*) **N:** AG
variolosus G.-A. Olivier, 1811: 497 (*Opatrum*) **N:** AG TU

genus *Allophylax*, species incertae sedis
insignis Reitter, 1904a: 110 (*Phylax*) **A:** "Syria"

genus *Bermejoina* Español, 1944: 12 type species *Bermejoina aiunica* Español, 1944
aiunica Español, 1944: 15 **N:** MO (Western Sahara)

genus *Guildia* Antoine, 1956: 346 type species *Micrositus punctistriatus* Escalera, 1925
punctistriata Escalera, 1925d: 379 (*Micrositus*) **N:** MO (Middle Atlas)

genus *Hoplarion* Mulsant & Rey, 1854a: 150 [= 1854b: 294] type species *Micrositus tumidus* Mulsant & Rey, 1854
subgenus *Atlasion* Koch, 1948b: 428 type species *Micrositus bedeli* Escalera, 1914
Megatlasion Koch, 1948b: 427 type species *Micrositus atlantis* Escalera, 1914
angulatum angulatum Escalera, 1913: 48 (*Micrositus*) **N:** MO (High Atlas)
angulatum gundaficum Antoine, 1942b: 37 (*Melambius*) **N:** MO (High Atlas)
angulatum kochi Antoine, 1946c: 55 (*Hoplariobius*) **N:** MO (High Atlas)
atlantis atlantis Escalera, 1914: 333 (*Micrositus*) **N:** MO (High Atlas)
atlantis gattefossei Antoine, 1936b: 74 (*Micrositus*) **N:** MO (High Atlas)
atlantis subtile Antoine, 1956: 355 **N:** MO (High Atlas)
bedeli baudoni Antoine, 1956: 358 **N:** MO (High Atlas)

bedeli bedeli Escalera, 1914: 334 (*Micrositus*) N: MO (Middle Atlas)
globosum Escalera, 1925a: 374 (*Micrositus*)
lecerfi Antoine, 1956: 358 (*Hoplariobius*)
bedeli escalerai Koch, 1945: 417 (*Micrositus*) N: MO (Middle Atlas)
nitens Antoine, 1956: 359 (*Hoplariobius*)
bedeli insolens Antoine, 1942b: 46 (*Melambius*) N: MO (High Atlas)
bedeli major Antoine, 1942b: 40 (*Melambius*) N: MO (High Atlas)
bedeli remotum Antoine, 1956: 359 N: MO (High Atlas)
bedeli septentrionale Antoine, 1942b: 38 (*Melambius*) N: MO (Middle Atlas)
bedeli subcostatum Koch, 1945: 418 (*Hoplariobius*) N: MO (Middle Atlas)
latissimum Antoine, 1942b: 38 (*Melambius*) N: MO (Middle Atlas)
theryi sarroense Antoine, 1956: 354 N: MO (Sarhro)
theryi theryi Antoine, 1933b: 227 (*Micrositus*) N: MO (High Atlas)
rugosium Antoine, 1956: 354
subgenus *Glyptariobius* Koch, 1948b: 423 type species *Glyptariobius excavatus* Koch, 1948
excavatum Koch, 1948b: 423 N: AG
subgenus *Hoplariobius* Reitter, 1904a: 115 type species *Micrositus decurtatus* Fairmaire, 1884
declive Antoine, 1951b: 317 (*Melambius*) N: MO (High Atlas)
decurtatum Fairmaire, 1884f: cxx (*Micrositus*) N: MO
granulosum Billberg, 1815: 276 (*Opatrum*) N: AG
porcus Thunberg, 1821: 52 (*Opatrum*)
tenuepunctatum Escalera, 1914: 333 (*Micrositus*) N: MO
subgenus *Hoplarion* Mulsant & Rey, 1854a: 150 [= 1854b: 294] type species *Hoplarion tumidus* Mulsant & Rey, 1854
attritum Bedel, 1918: 364 N: MO
humile Antoine, 1942b: 52 (*Melambius*) N: MO
kocheri Antoine, 1934: 122 (*Hoplarion*) N: MO
torretassoae Koch, 1948b: 429 N: AG
tumidum Mulsant & Rey, 1854a: 150 [=1854b: 294] (*Micrositus*) N: AG
subgenus *Mentariobius* Koch, 1948b: 425 type species *Micrositus distinguendus* Mulsant & Rey, 1854
pueli Koch, 1948b: 425 (*Hoplariobius*) N: AG
distinguendum Mulsant & Rey, 1854a: 159 [= 1854b: 303] (*Micrositus*) N: AG
subgenus *Saharoplarion* Koch, 1948b: 431 type species *Micrositus compactus* Fairmaire, 1880
compactum Fairmaire, 1880a: 16 (*Micrositus*) N: AG LB TU

genus ***Litoborus*** **Mulsant & Rey, 1854a: 126** [= 1854b: 270] type species *Philax moreletii* Lucas, 1847
subgenus *Litoborus* Mulsant & Rey, 1854a: 126 [= 1854b: 270] type species *Philax moreletii* Lucas, 1847
clermonti Antoine, 1931: 187 N: AG MO

inmarginatus Thunberg, 1821: 53 (*Opatrum*) **N:** MO (High Atlas)
maroccanus Escalera, 1914: 329
moreletii mediatlantis Antoine, 1942a: 22 **N:** MO (Middle Atlas)
moreletii moreletii Lucas, 1847: pl. 29 (*Philax*) **N:** AG MO
baudii Reitter, 1904a: 113
moreletii peyerimhoffi Antoine, 1931: 185 **N:** MO
subgenus *Paralitoborus* Antoine, 1931: 190 type species *Litoborus sternalis* Fairmaire, 1884
chobauti Antoine, 1931: 190 **N:** MO
escalerai Antoine, 1931: 192 **E:** SP
forticostis forticostis Escalera, 1914: 330 **N:** MO
forticostis saharensis Español, 1946: 120 **N:** MO (Western Sahara)
lindbergi Español, 1967a: 39 **N:** MO
planicollis bolivari Antoine, 1931: 195 **N:** MO
planicollis planicollis Waltl, 1835: 71 (*Phylan*) **E:** PT SP **N:** AG MO
algiricus Baudi di Selve, 1875b: 158
planicollis Mulsant & Rey, 1854a: 128 [= 1854b: 272] [HN]
rugosum atlantis Antoine, 1931: 203 **N:** MO (High Atlas)
rugosum rugosum Thunberg, 1821: 52 (*Opatrum*) **N:** MO
olcesii Fairmaire, 1871a: 392 (*Phylax*)
sternalis Fairmaire, 1884c: 446 **N:** MO
subtilimargo Reitter, 1904a: 113 **N:** AG

genus *Melambius* Mulsant & Rey, 1854a: 124 [= 1854b: 268] type species *Opatrum barbarum* Erichson, 1841
subgenus *Hadromelambius* Koch, 1948b: 432 type species *Melambius telueticus* Escalera, 1914
telueticus Escalera, 1914: 331 **N:** MO
subgenus *Hoplambius* Reitter, 1904a: 114 type species *Hoplarion melambioides* Fairmaire, 1893
melamboides Fairmaire, 1893i: cxlviii (*Hoplarion*) **N:** AG TU
subgenus *Melambatlasus* Koch, 1948b: 433 type species *Micrositus hebes* Antoine, 1933
androgynus Antoine, 1942b: 31 **N:** MO (High Atlas)
antoinei Koch, 1945: 421 (*Micrositus*)
anemophilus Antoine, 1942b: 30 **N:** MO (High Atlas)
asniensis Antoine, 1942b: 28 **N:** MO (High Atlas)
cacuminorum Antoine, 1936b: 73 (*Micrositus*) **N:** MO (High Atlas)
cultriger Antoine, 1942b: 33 **N:** MO (High Atlas)
hassani hassani Antoine, 1942b: 26 **N:** MO (High Atlas)
hassani jejunus Antoine, 1956: 362 **N:** MO (High Atlas)
hebes Antoine, 1933b: 228 (*Micrositus*) **N:** MO (High Atlas)
lasciniosus Antoine, 1956: 364 **N:** MO (High Atlas)
meridionalis Antoine, 1951b: 317 **N:** MO (Sarhro)
mirus Antoine, 1956: 361 **N:** MO (Middle Atlas)

pauliani Koch, 1945: 422 (*Micrositus*) **N:** MO (High Atlas)
rugatithorax Antoine, 1951b: 316 **N:** MO (High Atlas)
venustus Antoine, 1942b: 32 **N:** MO (Middle Atlas)
subgenus *Melambius* Mulsant & Rey, 1854a: 124 [= 1854b: 268] type species *Opatrum barbarum* Erichson, 1841
barbarus Erichson, 1841: 181 (*Opatrum*) **N:** AG
bidens Antoine, 1932: 188 **N:** MO (High Atlas)
breviusculus breviusculus Fairmaire, 1871a: 390 **N:** AG
breviusculus syrtanus Normand, 1938a: 303 [= 1938b: 366] **N:** TU
denticollis Escalera, 1913: 48 **N:** MO (High Atlas)
mideltensis Antoine, 1932: 188 **N:** MO
  sulcatus Antoine, 1956: 366
otini Antoine, 1942b: 42 **N:** MO
  schrammi Antoine, 1956: 365
rotroui Antoine, 1951b: 315 **N:** MO (Rif)
rungsi Antoine, 1956: 364 **N:** MO (High Atlas)
teinturieri Mulsant & Godart, 1866: 454 **N:** AG TU
  asperocostatus Fairmaire, 1871a: 391
tuniseus Levrat, 1859a: 34 (*Philax*) **N:** TU

genus ***Melansis*** **Wollaston, 1864: 491** type species *Phylax costatus* Brullé, 1839
angulata angulata Wollaston, 1864: 492 **N:** CI (La Palma)
angulata hierroensis Español, 1962a: 206 **N:** CI
costata Brullé, 1839: 69 (*Phylax*) **N:** CI (Gran Canaria)
kaszabi J. Ferrer, 1992: 83 **N:** CI (Gran Canaria)
reyesi García & Oromí, 2000: 53 **N:** CI (Tenerife)
tibicena Liberto & Leo, 2002: 20 **N:** CI (Gran Canaria)

genus ***Melasmana*** **E. Strand, 1935: 297** [RN] type species *Phylax lineatus* Brullé, 1839
subgenus *Heliomelasma* Koch, 1948b: 408 type species *Melasmana appenhageni* Koch, 1948
appenhageni Koch, 1948b: 408 **N:** CI (Fuerteventura)
subgenus *Melasmana* E. Strand, 1935: 297 [RN] type species *Phylax lineatus* Brullé, 1839
  Melasma Wollaston, 1864: 484 [HN] type species *Phylax lineatus* Brullé, 1839
lineatum Brullé, 1839: 69 (*Phylax*) **N:** CI

genus ***Orarabion*** **Leo & Liberto, 2011: 157** type species *Orarabion dominici* Leo & Liberto, 2011
dominici Leo & Liberto, 2011: 158 **A:** OM

genus ***Oreomelasma*** **Español, 1975b: 238** type species *Oreomelasma oromii* Español, 1975
oromii Español, 1975b: 241 **N:** CI (Fuerteventura)

genus *Otinia* Antoine, 1942b: 22 type species *Otinia iblanensis* Antoine, 1942
subgenus *Antoineius* Koch, 1948b: 418 type species *Micrositus jeanneli* Koch, 1945 (= *Melambius inermis* Antoine, 1942)
espanoli Antoine, 1956: 350 **N:** MO (Rif)
inermis Antoine, 1942b: 35 (*Melambius*) **N:** MO (Middle Atlas)
jeanneli Koch, 1945: 420 (*Micrositus*)
kocheri Antoine, 1946c: 54 (*Melambius*) **N:** MO (High Atlas)
vidali Antoine, 1942b: 27 (*Melambius*) **N:** MO (Rif)
subgenus *Orophylaxus* Koch, 1948b: 417 type species *Philax incertus* Mulsant & Godart, 1866
incerta disparis Escalera, 1925d: 377 (*Micrositus*) **N:** MO (Middle Atlas)
occidentalis Antoine, 1956: 350
incerta incerta Mulsant & Godart, 1866: 453 (*Philax*) **N:** AG
oxyholma Fairmaire, 1876: 38 (*Phylax*)
serripes Desbrochers des Loges, 1881: 119 (*Micrositus*)
lacouri debduensis Antoine, 1942b: 52 **N:** MO
lacouri lacouri Antoine, 1942b: 41 (*Melambius*) **N:** MO
subgenus *Otinia* Antoine, 1942b: 22 type species *Otinia iblanensis* Antoine, 1942
iblanensis Antoine, 1942b: 23 **N:** MO (Middle Atlas)
embergeri Koch, 1945: 419 (*Micrositus*)

genus *Peyerimhoffius* Koch, 1948b: 420 type species *Philax plicatus* Lucas, 1847
plicatus Lucas, 1847: 326 (*Philax*) **N:** AG
plicatus Mulsant & Rey, 1854a: 157 [= 1854b: 301] (*Micrositus*) [HN]

genus *Psammoardoinellus* Leo, 1980a: 34 type species *Isocerus sardiniensis* Ardoin, 1972
sardiniensis Ardoin, 1972: 240 (*Isocerus*) **E:** IT (Sardinia)

tribe Dissonomini G.S. Medvedev, 1968

genus *Bradyus* Dejean, 1834: 190 type species *Erodius pygmaeus* Fischer von Waldheim, 1821
Aphaleria Reitter, 1896i: 235 type species *Aphaleria capnisoides* Reitter, 1896 (= *Erodius pygmaeus* Fischer von Waldheim, 1821)
brodskyi Picka, 1987: 33 (*Aphaleria*) **A:** TM (Kurtlinsk)
oblongus G.S. Medvedev, 1964c: 657 (*Aphaleria*) **A:** TM UZ
pygmaeus Fischer von Waldheim, 1821: 12 (*Erodius*) **A:** AF KZ TM UZ
capnisoides Reitter, 1896i: 236 (*Aphaleria*)

genus *Dissonomus* Jacquelin du Val, 1861: 280 [RN] type species *Heliopates picipes* Faldermann, 1837
subgenus *Dissonomus* Jacquelin du Val, 1861: 280 [RN] type species *Heliopates picipes* Faldermann, 1837
Heterophylus Mulsant & Rey, 1859b: 69 [= 1859e: 7] [HN] type species *Helipathes picipes* Faldermann, 1837

cavicolus Kaszab, 1970: 137 **A**: AF
ellipticus Desbrochers des Loges, 1881: 120 (*Heterophylus*) **E**: "Russia mer".
iranicus G.S. Medvedev, 1968a: 226 **A**: IN
kermanicus G.S. Medvedev, 1968a: 228 **A**: IN
picipes Faldermann, 1837: 55 (*Helipathes*) **E**: AB AR ST **A**: IN TR
parvulus Reitter, 1896e: 159 (*Heterophylus*)
tibialis Reitter, 1904a: 70 (*Heterophylus*) **A**: KZ TM
subgenus *Eudissonomus* G.S. Medvedev, 1968a: 218 type species *Heterophylus substriatus* Reitter, 1898
luristanicus G.S. Medvedev, 1968a: 219 **A**: IN
ovatus G.S. Medvedev, 1968a: 222 **A**: IN
politus Kaszab, 1968e: 460 **A**: IN TR
substriatus Reitter, 1898f: 348 (*Heterophylus*) **A**: IQ TR
zarudnyi G.S. Medvedev, 1968a: 220 **A**: IN
subgenus *Neodissonomus* G.S. Medvedev, 1968a: 235 type species *Heterophylus angustitarsis* Reitter, 1896
angustitarsis Reitter, 1896e: 160 (*Heterophylus*) **A**: TM
amplicollis Reitter, 1896e: 160 (*Heterophylus*)
desertus Skopin, 1966b: 332 **A**: KZ TD UZ
latiusculus Mulsant & Rey, 1859b: 72 (*Heterophylus*) **A**: AF KZ TD TM UZ
subgenus *Paradissonomus* G.S. Medvedev, 1968a: 229 type species *Dissonomus longulus* Bogatchev & Kryzhanovskiy, 1960
badghysi G.S. Medvedev, 1964c: 656 **A**: IN TM
franzi Kaszab, 1960a: 131 **A**: AF
gracilis G.S. Medvedev, 1968a: 232 **A**: TD
kazachstanicus Skopin, 1968b: 80 **A**: KZ (Karatau)
longulus Bogatchev & Kryzhanovskiy, 1960: 268 **A**: TM
rosae Schuster, 1915b: 89 **A**: KZ
validior G.S. Medvedev, 1968a: 234 **A**: IN

genus *Dissonomus*, species incertae sedis
seriepunctatus Reitter, 1904a: 69 (*Heterophylus*) **N**: AG

tribe Heleini Fleming, 1821
subtribe Cyphaleina Lacordaire, 1859

genus *Pachycoelia* Boisduval, 1835: 248 type species *Pachycoelia sulcicollis* Boisduval, 1835
Lepispilus Westwood, 1841: 44 type species *Pachycoelia sulcicollis* Boisduval, 1835
Lepidospilus Agassiz, 1846b: 205 [UE]
Tyndarisus Pascoe, 1869: 294 type species *Tyndarisus longitarsis* Pascoe, 1869 (= *Lepispilus stygianus* Pascoe, 1896)
punctatolineata Allard, 1896: 318 [EA] **A**: SD

tribe Helopini Latreille, 1802

subtribe Cylindrinotina Español, 1956

genus *Armenohelops* Nabozhenko, 2002a: 42 type species *Armenohelops armeniacus* Nabozhenko, 2002

amasiae Seidlitz, 1895: 732 (*Helops*) **A:** TR

strangulatus Reitter, 1922b: 130 (*Cylindronotus*)

armeniacus Nabozhenko, 2002a: 42 **E:** AR **A:** TR

bosphoranus Allard, 1876a: 56 (*Stenomax*) **E:** TR **A:** TR

fundator Reitter, 1908g: 215 (*Stenomax*)

fouquei Nabozhenko, Keskin & Alpagut Keskin, 2016: 264 **A:** TR

kagyzmanicus Nabozhenko, 2002a: 44 **A:** TR (Kars)

kasatkini Nabozhenko, 2011a: 135 **A:** TR

medvedevi Nabozhenko, 2009a: 186 **A:** IN

parvocularis Nabozhenko, Keskin & Alpagut Keskin, 2016: 263 **A:** TR

scutellatus Reitter, 1902g: 221 (*Stenomax*) **A:** TR

genus *Asialassus* Nabozhenko & Ando 2018: 311 type species *Helops cordicollis* Marseul, 1876

adachii Akita & Masumoto, 2009c: 77 (*Tarpela*) **A:** JA (Oshima)

amamiensis Kaszab, 1964a: 7 (Tarpela) **A:** JA (Ryukyus)

cordicollis Marseul, 1876c: 141 (*Helops*) **A:** JA SC

kimurai Masumoto, 1996a: 213 (*Tarpela*) **A:** JA (Ryukyus)

tokunoshimana Masumoto & Akita, 2001a: 20 (*Tarpela*)

todai Ando, 2006: 29 (*Tarpela*) **A:** JA (Ishigaki Is.)

tsushimanus Nakane, 1979: 114 (*Tarpela*) **A:** JA (Tsushima)

genus *Ceratopelius* Antoine, 1962: 52 type species *Ceratopelius mussardi* Antoine, 1962

mussardi Antoine, 1962: 52 **N:** MO (Rif)

genus *Cylindrinotus* Faldermann, 1837: 73 type species *Helops femoratus* Faldermann, 1837

Cylindronotus Agassiz, 1846b: 111 [UE]

Stenomacidius Seidlitz, 1895: 791 type species *Hedyphanes acutangulus* Seidlitz, 1895

acutangulus Seidlitz, 1895: 797 (*Hedyphanes*) **E:** GG

bellator Reitter, 1902g: 218 (*Helops*)

batesi Allard, 1876a: 38 **A:** TR

charlesi Nabozhenko, 2015a: 112 **A:** TR

constrictus Seidlitz, 1895: 722 (*Helops*) E: TR

funestoides Reitter, 1900a: 159 (*Helops*)

erivanus Reitter, 1902g: 219 (*Helops*) **E:** AR

femoratus Faldermann, 1837: 71 (*Helops*) **E:** AB AR IN TR
funestus Faldermann, 1837: 76
lugubris Faldermann, 1837: 74
gibbicollis Faldermann, 1837: 78 **E:** AR GG **A:** TR
corallipes Reitter, 1902g: 219 (*Helops*)
flavipes Allard, 1877a: 219
umbrinus Faldermann, 1837: 77
gibbosus Seidlitz, 1895: 721 (*Helops*) **E:** TR
nitidus Seidlitz, 1895: 721 (*Helops*) **A:** SY TR
tchorokhicus Nabozhenko 2011c: 336 **A:** TR

genus *Ectromopsis* Antoine, 1949b: 145 type species *Catomus politicollis* Allard, 1876
bogatschevi Iablokoff-Khnzorian, 1957a: 169 (*Helops*) **E:** AR
bulgarica G.S. Medvedev & Angelov, 1981: 312 (*Gunarus*) **E:** BU GR
mendizabali Cobos, 1953: 127 **E:** SP (Almería)
ovipennis Allard, 1877a: 123 (*Xanthomus*) **E:** GR
politicollis Allard, 1876a: 31 (*Catomus*) **N:** AG MO
tantilla Ménétriés, 1848: 28 (*Helops*) **E:** ST **A:** KZ
monilicornis Baudi di Selve, 1876d: 273 (*Helops*)

genus *Eustenomacidius* Nabozhenko, 2006b: 807 type species *Helops luridus* Ménétriés, 1848 (= *Stenomax laevicollis* Kraatz, 1882)
subgenus *Caucasohelops* Nabozhenko, 2006b: 816 type species *Eustenomacidius svetlanae* Nabozhenko, 2006
egeuniversitatis Keskin, Nabozhenko & Alpagut-Keskin, 2017a: 239 **A:** TR
glebi Nabozhenko, 2019b: 1030 **A:** IN
svetlanae araxi Nabozhenko, 2006b: 819 **E:** AB (Nakhichevan) AR **A:** IN
svetlanae svetlanae Nabozhenko, 2006b: 817 **E:** AB (Talysh) **A:** IN
subgenus *Eustenomacidius* Nabozhenko, 2006b: 807 type species *Helops luridus* Ménétriés, 1848 (= *Stenomax laevicollis* Kraatz, 1882)
hirtipennis Seidlitz, 1895: 792 (*Hedyphanes*) **A:** KI KZ
laevicollis Kraatz, 1882d: 333 (*Stenomax*) **A:** KI KZ TD TM UZ
lucidicollis Kraatz, 1882d: 333 (*Stenomax*)
luridus Ménétriés, 1848: 27 (*Helops*) [HN]
provocator Reitter, 1922b: 9 (*Catomus*)
mongolicus Kaszab, 1968c: 395 (*Catomus*) **A:** MG
turcmenicus G.S. Medvedev, 1964c: 652 (*Catomus*) **A:** TM (Kopet Dag)
wagnae Ren, 1999: 312 (*Catomus*) **A:** GAN

genus *Gunarus* Des Gozis, 1886: 25 type species *Helops hirtulus* Reiche, 1862
hirtulus Reiche, 1862a: 374 (*Helops*) **E:** SP **N:** AG MO
minutus Mulsant & Godart, 1870: 51 (*Helops*)
kaszabi Grimm, 1981: 10 **E:** GR (Santorin)
lapidicola Küster, 1850b: 62 (*Helops*) **E:** BU CR GR TR YU
rufescens Küster, 1851: 59 (*Helops*)
splendidulus L.W. Schaufuss, 1862a: 92 (*Helops*)
laeviusculus Kraatz, 1883c: 395 (Nalassus) **E:** SP
parvulus Lucas, 1846: pl. 31 (*Helops*) **E:** GR IT MA SP **N:** AG MO TU
heydeni Allard, 1876a: 42 (*Nalassus*)
intersparsus Küster, 1850b: 58 (*Helops*)
nanus Küster, 1850b: 60 (*Helops*)
velai Castro Tovar, 2016: 1 **E:** SP

genus *Idahelops* Keskin & Nabozhenko, 2012: 63 type species *Idahelops alpagutae* Keskin & Nabozhenko, 2012
alpagutae Keskin & Nabozhenko, 2012: 64 **A:** TR

genus *Microdocnemis* Nabozhenko & Keskin, 2010: 841 type species *Microdocnemis xerophilicus* Nabozhenko & Keskin, 2010
xerophilicus Nabozhenko & Keskin, 2010: 841 **A:** TR

genus *Nalassus* Mulsant, 1854: 323 type species *Helops dryadophilus* Mulsant, 1854
subgenus *Caucasonotus* Nabozhenko, 2000: 107 type species *Cylindrinotus dombaicus* Nabozhenko, 2000
adriani Reitter, 1922b: 143 (*Cylindronotus*) **E:** GG (Abkhazia)
alanicus Nabozhenko, 2000: 111 (*Cylindronotus*) **E:** ST (Severnaya Osetiya)
avaricus Nabozhenko & Abdurakhmanov, 2007: 190 **E:** ST (Dagestan)
colchicus colchicus Nabozhenko, 2001b: 651 **E:** GG
colchicus madlenae Nabozhenko, 2013a: 263 **E:** GG
diteras Allard, 1876a: 66 (*Helops*) **E:** AB AR GG ST
dombaicus Nabozhenko, 2000: 108 (*Cylindronotus*) **E:** GG ST
ludmilae Nabozhenko, 2001b: 648 **E:** GG (Abkhazia)
pharnaces Allard, 1876a: 41 **E:** GG ST
aeneipennis Reitter, 1888b: 212 (*Stenomax*) [HN]
glorificus Seidlitz, 1895: 729 (*Helops*) [RN]
svaneticus kartvelius Nabozhenko, 2013a: 261 **E:** GG
svaneticus svaneticus Nabozhenko & Dzhambazishvili, 2001: 90 **E:** GG
zakatalensis Nabozhenko, 2001b: 646 **E:** AB
subgenus *Helopondrus* Reitter, 1922b: 153 type species *Stenomax sareptanus* Allard, 1876
abkhasicus Nabozhenko, 2001b: 644 **E:** GG (Abkhazia)

adzharicus Nabozhenko & Dzhambazishvili, 2001: 92 **E:** GG **A:** TR
ahngeri G.S. Medvedev, 1998a: 138 (*Cylindronotus*) **A:** TM
arcanus Nabozhenko, 2001b: 657 **E:** GG (Adzharia)
assimilis Küster, 1850b: 36 (*Helops*) **E:** FR IT SP
aeneoniger Küster, 1850b: 35 (*Helops*)
convexicollis Küster, 1850b: 34 (*Helops*)
cambyses Seidlitz, 1895: 738 (*Helops*) **A:** IN
clavicornis Allard, 1876a: 20 (*Nalassus*) **A:** TR
cunctator Reitter, 1922b: 155 (*Cylindronotus*) **A:** IN
heres Reitter, 1922b: 154 (*Cylindronotus*) **A:** IN
kabakovi Nabozhenko, 2010: 53 **A:** IN
lineatus Allard, 1877a: 259 (*Stenomax*) **E:** AB GG ST **A:** IN
propheta Seidlitz, 1895: 731 (*Helops*)
vexator Reitter, 1887e: 524 (*Stenomax*)
planivittis Allard, 1876a: 39 (*Stenomax*) **A:** TR
ponticus Seidlitz, 1895: 737 (*Helops*) **A:** IN TR
rejseki Nabozhenko, 2014: 237 **A:** IN
sareptanus Allard, 1876a: 54 (*Stenomax*) **E:** BU GG (Abkhazia) MD ST UK **A:** KZ (Uralsk)
schmalfussi Nabozhenko, 2011b: 266 **A:** TR
striatissimus Nabozhenko & Grimm, 2018: 312 **A:** IN
szalokii Nabozhenko, 2011b: 264 **A:** TR
subgenus *Nalassus* Mulsant, 1854: 323 type species *Helops dryadophilus* Mulsant, 1854
Helopocerodes Reitter, 1922b: 144 type species *Helops faldermanni* Faldermann, 1837
abeillei Seidlitz, 1895: 736 (*Helops*) **E:** FR (Corse)
akitai Masumoto, 1998b: 317 (*Tarpela*) **A:** JA (Ryukyus)
aemulus aemulus Küster, 1850b: 61 (*Helops*) **E:** IT MA SP **N:** LB MO TU
fusculus Allard, 1876a: 40 (*Omalus*)
aemulus calaritanus Leo, 1985: 144 **E:** IT (Sardinia)
alpigradus Fairmaire, 1883k: clxix (*Helops*) **E:** FR IT
balchashensis G.S. Medvedev, 1970: 396 **A:** KZ
baudoni Antoine, 1954: 213 (*Stenomax*) **N:** MO (Middle Atlas)
brevicollis Krynicki, 1832: 137 (*Helops*) **E:** CT GG ST UK
brevicollis Küster, 1850b: 80 (*Helops*) [HN]
gilvipes Ménétriés, 1848: 27 (*Helops*)
brunneus brunneus Marseul, 1876c: 140 (*Helops*) **A:** JA
japonica Allard, 1876a: 46 (*Lamperos*)
brunneus konoi Nakane, 1963a: 28 (*Tarpela*) **A:** JA (Ryukyus)
calpensis Champion, 1891d: 389 (*Helops*) **E:** SP
hispanus Seidlitz, 1895: 753 (*Helops*)
colasi cobosi Español, 1974: 215 **E:** SP (Sierra Filabres)
colasi colasi Español, 1954c: 119 (*Cylindronotus*) **E:** SP (Sierra Nevada)

convexus Comolli, 1837: 26 (*Helops*) **E:** AU GE HU IT LS SZ
convexus Küster, 1850b: 74 (*Helops*) [HN]
laevigatus Küster, 1850b: 79 (*Helops*)
cyrensis Bogatchev, 1946b: 246 (*Cylindronotus*) **E:** AB GG
dilaticornis Reitter, 1922b: 146 (*Cylindronotus*) **A:** TR
dissonus Nabozhenko, 2001b: 635 **E:** ST (Severnaya Osetiya)
dermestoides Illiger, 1798: 120 (*Helops*) **E:** AL AU BH BY CR CT CZ EN GE HU IT LA LT MD PL RO SK SV SZ UK YU
brevis Allard, 1876a: 22 (*Cylindronotus*) [HN]
fabricii Gemminger, 1870a: 123 (*Helops*) [RN]
picipes Küster, 1850b: 78 (*Helops*)
quisquilius Panzer, 1794c: 3 (*Helops*) [HN]
dongurii Masumoto, Akita & Lee, 2017: 6 (*Tarpela*) **A:** TAI
dryadophilus Mulsant, 1854: 337 (*Helops*) **E:** AL BU FR GR IT RO SP SZ TR **A:** IS TR
curticollis Reitter, 1922b: 149 (*Cylindronotus*)
phaeacus J.R. Sahlberg, 1903c: 33 (*Helops*)
ecoffeti Küster, 1850: 70 (*Helops*) **E:** FR IT SZ
picinus Küster, 1850: 81 (*Helops*)
rufocastaneus Pic, 1919c: 5 (*Helops*)
schaeferi Ardoin, 1959: 41
temperei Ardoin, 1959: 41
elegantulus Lewis, 1894: 478 (*Lamperos*) **A:** JA
estrellensis Kraatz, 1870: 133 (*Helops*) **E:** PT SP
faldermanni Faldermann, 1837: 70 (*Helops*) **E:** AB AR GG ST UK **A:** IN TM TR
amaroides Küster, 1850b: 69 (*Helops*)
ambrosius Reitter, 1922b: 146 (*Cylindronotus*)
confluens Seidlitz, 1895: 740 (*Helops*)
eligius Reitter, 1922b: 145 (*Cylindronotus*)
impressicollis Faldermann, 1837: 69 (*Helops*)
semiopacus Reitter, 1902g: 222 (*Xanthomus*)
zabroides Küster, 1850b: 67 (*Helops*)
farsistanus Nabozhenko & Grimm, 2018: 310 **A:** IN
formosanus Masumoto, 1981c: 35 (*Tarpela*) **A:** TAI
genei genei Gené, 1839: 74 (*Helops*) **E:** FR (Corse)
genei melonii Leo, 1982a: 1 **E:** IT (Sardinia)
graecus Seidlitz, 1895: 745 (*Helops*) **E:** BU GR TR
halophilus G.S. Medvedev, 1987b: 101 (*Cylindronotus*) **A:** KZ
harpaloides Küster, 1850b: 73 (*Helops*) **E:** FR
kadleci Nabozhenko, 2019b: 1026 **A:** IN
kalashiani Nabozhenko, 2001b: 640 **E:** ST (Daghestan, Chechnya)
kaszabi Nabozhenko, 2001b: 643 **A:** TR
kataevi Nabozhenko, 2019b: 1028 **A:** IN

kawabatai Akita & Masumoto, 2012a: 276 (*Tarpela*) **A:** JA
kazachstanicus G.S. Medvedev, 1964b: 159 (*Cylindronotus*) **A:** KZ
laevioctostriatus Goeze, 1777: 683 (*Tenebrio*) **E:** BE CZ FR GE GB IR IT NL PL PT SK SP SZ **A:** TR
caraboides Panzer, 1794c: 3 (*Helops*)
latiusculus Kraatz, 1880d: 300
lusitanus Kraatz, 1870: 134 (*Helops*)
ruficollis Fabricius, 1787a: 214 (*Helops*)
striatus Geoffroy, 1785: 157 (*Tenebrio*)
longipennis Küster, 1850b: 68 (*Helops*) **E:** SP
lutshniki Nabozhenko, 2001b: 636 **E:** ST
magyari Kaszab, 1968a: 12 (*Tarpela*) **A:** NC SC
marteni Español, 1953a: 73 (*Stenomax*) **N:** MO
mediatlantis Antoine, 1949b: 129 (*Stenomax*) **N:** MO (Middle Atlas)
medvedevi Nabozhenko, 2006a: 248 **A:** IN
melchiades Reitter, 1922b: 147 (*Cylindronotus*) **A:** KZ
merkli Masumoto, Akita & Lee, 2017: 9 (*Tarpela*) **A:** TAI
mojynkumensis G.S. Medvedev, 1987b: 100 (*Cylindronotus*) **A:** KZ
olgae Nabozhenko & Ivanov, 2015: 138 **A:** FE
pastai Aliquò, Leo & Lo Cascio, 2006: 303 **E:** IT (Vulcano)
pekinensis Fairmaire, 1888b: 130 (*Helops*) **A:** BEI
pilushenmuus Masumoto, Akita & Lee, 2017: 2 (*Tarpela*) **A:** TAI
planipennis Küster, 1850b: 83 (*Helops*) **E:** IT
ruficornis Baudi di Selve, 1876c: 259 (*Helops*)
plebejus Küster, 1850b: 84 (*Helops*) **E:** AL BG CR GR IT TR **A:** TR
distinguendus Küster, 1850b: 82 (*Helops*)
rhaticus Antoine, 1937: 75 (*Helops*) **N:** MO
romashovi Nabozhenko, 2001b: 639 **E:** UK (Krym)
tengchihensis Masumoto, Akita & Lee, 2019: 124 **A:** TAI
xiaoxueshanus Masumoto, Akita & Lee, 2017: 6 (*Tarpela*) **A:** TAI
yuanfengus Masumoto, Akita & Lee, 2017: 4 (*Tarpela*) **A:** TAI
zaratustrai Nabozhenko, 2006a: 246 **A:** IN
zoltani Masumoto, 1981c: 34 (*Tarpela*) **A:** TAI
subgenus *Nipponalassus* Nabozhenko & Ando, 2018: 306 type species *Tarpela andoi* Masumoto, 1993
andoi Masumoto, 1993d: 134 (*Tarpela*) **A:** JA (Yakushima)
lewisi Masumoto, 1993d: 132 (*Tarpela*) **A:** JA SC

genus *Odocnemis* Allard, 1876a: 4 type species *Odocnemis caudata* Allard, 1876 (= *Helops praelongus* Baudi di Selve, 1876)
subgenus *Heloponotus* Reitter, 1922b: 126 type species *Helops perplexus* Ménétriés, 1848 (= *Helops gracilis* Fischer von Waldheim, 1823)

arborea Fischer von Waldheim, 1823: 200 (*Helops*) **E:** UK (Krym)
douei Allard, 1876a: 39 (*Stenomax*)
excavatus Seidlitz, 1895: 752 (*Helops*)
gracilis Fischer von Waldheim, 1823: pl. 22 (*Helops*) **E:** CT (Lipetsk) ST UK **A:** KZ (Uralsk)
grandicollis Küster, 1851: 62 (*Helops*)
perplexus Ménétriés, 1848: 26 (*Helops*)
saturninus Reitter, 1922: 155 (*Cylindronotus*)
sturmi Küster, 1851: 60 (*Helops*)
tauricus Mulsant & Godart, 1870: 49 (*Helops*)
subgenus *Odocnemis* Allard, 1876a: 4 type species *Odocnemis caudata* Allard, 1876 (= *Helops praelongus* Baudi di Selve, 1876)
Homaleis Rye, 1879: 62 [UE]
Odontocnemis Rye, 1878: 69 [URN]
Omaleis Allard, 1876: 4 type species *Helops congener* Reiche, 1861
aegaeica Nabozhenko & Keskin, 2016: 53 **A:** TR
alcida Reitter, 1922b: 140 (*Cylindronotus*) **E:** GR
allardi Nabozhenko & Keskin, 2016: 62 **A:** AR AB IR TR
altimontana Nabozhenko & Keskin, 2016: 39 **A:** TR
amanda Reitter, 1922b: 132 (*Cylindronotum*) **E:** "Pyrenees"
amanosica Nabozhenko et Keskin, 2016: 21 **A:** TR SY
anatolica Pic, 1899p: 411 (*Helops*) **A:** TR
korbi Reitter, 1902g: 219 (*Stenomax*)
aurichalcea Adams, 1817:307 (*Blaps*) **E:** AB AR GG ST (Nalchik, Adler) **A:** TR
gloriosa Faldermann, 1837: 67 (*Helops*)
pulcher Allard, 1877a: 256 (*Helops*)
badia L. Redtenbacher, 1849: 601 (*Helops*) **E:** BU GR RO
asphaltina Küster, 1850b: 55 (*Helops*)
bekiri M. Nabozhenko & S. Nabozhenko, 2015: 367 **E:** GR
clara Allard, 1876a: 37 **N:** EG
clypeata Küster, 1851: 68 (*Helops*) **E:** IT (Sicilia)
colasi Español, 1954c: 119 (*Nalassus*) **E:** SP (Sierra Nevada)
congener cedrorum Antoine, 1962: 55 (*Stenomax*) **N:** MO (Rif)
congener congener Reiche, 1862a: 372 (*Helops*) **N:** MO
cordiformis Nabozhenko & Keskin, 2016: 24 **A:** TR
crenatostriata Allard, 1877a: 146 (*Stenomax*) **E:** GR
pindica Apfelbeck, 1901: 465
cretensis Seidlitz, 1895: 733 (*Helops*) **E:** GR (Crete)
dasypa Seidlitz, 1895: 725 (*Helops*) **A:** TR
denisi Nabozhenko, 2019b: 1029 **A:** IN
dichroa Nabozhenko & Keskin, 2016: 66 **A:** TR
erseni Nabozhenko & Keskin, 2016: 13 **A:** TR

euritopica Nabozhenko & Keskin, 2016: 55 **A**: TR
evestigata Nabozhenko & Keskin, 2016: 28 **A**: TR
exarata Germar, 1817: 192 (*Helops*) **E**: AL BH CR GR IT SL
brunnitarsis Küster, 1851: 69 (*Helops*)
fleischeri Reitter, 1922b: 130 (*Cylindronotum*) **E**: GR (Heraklea)
hakkariensis Nabozhenko & Keskin, 2016: 43 **A**: TR (Hakkary)
inornata Nabozhenko & Keskin, 2016 36 **A**: TR
intricata Allard, 1877a: 257 (*Stenomax*) **E**: BU
intrusicollis Seidlitz, 1895: 733 (*Helops*) **A**: CY
kakunini Nabozhenko & Keskin, 2016: 24 **A**: IN
kochi Español, 1955: 99 **E**: SP
malgorzatae Keskin & Nabozhenko, 2011: 344 **A**: TR
merkli Nabozhenko & Keskin, 2016: 68 **A**: TR
moabitica J.R. Sahlberg, 1908: 88 (*Helops*) **A**: JO
molecularica Nabozhenko & Keskin, 2016: 16 **A**: TR
nasreddini Keskin & Nabozhenko, 2011: 343 **A**: TR
nigropicea Küster, 1850b: 56 (*Helops*) **E**: BU GR
attica Allard, 1881b: ciii (*Omalois*)
operta Reitter, 1922b: 132 (*Cylindronotus*) **A**: TR
ophonoides Lucas, 1847: 352 (*Helops*) **E**: SP **N**: TU
pardoi Español, 1953: 75 (*Cylindronotus*) **N**: MO
perarmata Nabozhenko & Tichý, 2006: 183 **A**: TR
poeciloides Seidlitz, 1895: 739 (*Helops*) **N**: AG TU
praelonga Baudi di Selve, 1876c: 250 [= 1876b: 101] (*Helops*) **A**: IS JO LE SY
caudata Allard, 1876a: 36
protina Reitter, 1900a: 167 (*Helops*) **A**: TR IN
punctata Allard, 1876a: 38 **A**: IQ IS LE SY TR
pseudoclara Reitter, 1922b: 134 (*Cylindronotus*)
rhatica Antoine, 1936: 75 (*Helops*) **N**: MO (High Atlas)
recticollis Allard, 1876a: 38 (*Stenomax*) **A**: IN TR
kurdistana Reitter, 1902g: 221 (*Stenomax*)
rufffoi osellai Gardini, 1979: 75 (*Cylindronotus*) **E**: IT (Montecristo Is.)
ruffoi ruffoi Canzoneri, 1970: 81 (*Cylindronotus*) **E**: IT (Sicilia, Marettimo Is.)
schusteri Reitter, 1922b: 133 (*Cylindronotus*) **E**: AL GR
seducta Nabozhenko & Keskin, 2016 58 **A**: TR
serbica J. Frivaldszky, 1894: 86 (*Helops*) **E**: SB
seriegranata Seidlitz, 1895: 733 (*Helops*) **E**: TR **A**: TR
shokhini Nabozhenko & Keskin, 2016: 39 **A**: TR
skopini Español, 1961d: 152 (*Nalassus*) **E**: SP
steindachneri Apfelbeck, 1907: 1671 (*Helops*) **E**: AL
subtuberculigera subtuberculigera Nabozhenko & Keskin, 2016: 70 **A**: TR

subtuberculigera thracica Nabozhenko & Keskin, 2016: 73 **E:** TR
svetlanae beydagensis Keskin & Nabozhenko, 2011: 343 **A:** TR
svetlanae svetlanae Keskin & Nabozhenko, 2011: 342 **A:** TR
teluetica Escalera, 1914: 350 (*Helops*) **N:** MO
tenebrioides Germar, 1813: 123 (*Helops*) **E:** PT SP
terminassianae Nabozhenko, 2011a: 138 (*Armenohelops*) **A:** TR
torosica subangulata Nabozhenko & Keskin, 2016: 33 **A:** TR
torosica torosica Nabozhenko & Keskin, 2016: 30 **A:** TR
tuberculata Küster, 1851: 67 (*Helops*) **E:** AL GR
tuberculigera Reiche & Saulcy, 1857: 265 (*Helops*) **E:** BU GR MC
valga Baudi di Selve, 1881: 291 (*Helops*) **A:** IS JO
zoltani Nabozhenko, 2008d: 36 **A:** TR
  anatolica Kaszab, 1961c: 173 [HN]

genus *Pseudoprobaticus* Nabozhenko, 2001a: 513 type species *Helops granipennis* Allard, 1876
granipennis Allard, 1876a: 51 (*Helops*) **E:** TR **A:** TR

genus *Reitterohelops* Skopin, 1960a: 308 type species *Zophohelops lazarus* Reitter, 1922
ahngeri G.S. Medvedev, 1964c: 655 (*Zophohelops*) **A:** IN TM (Kopet Dag)
badghysi G.S. Medvedev, 1964c: 654 (*Zophohelops*) **A:** TM
kocheri Skopin, 1960a: 309 (*Zophohelops*) **A:** KI KZ UZ
kulzeri Skopin, 1960a: 310 (*Zophohelops*) **A:** TD UZ
lazarus Reitter, 1922b: 118 (*Zophohelops*) **A:** AF TM
steinbergi G.S. Medvedev, 1964c: 653: (*Zophohelops*) **A:** TM (Kopet Dag)
striatipennis G.S. Medvedev, 1970: 395 (*Zophohelops*) **A:** TD

genus *Stenomax* Allard, 1876a: 4 type species *Tenebrio lanipes* Linnaeus, 1771 (= *Tenebrio aeneus* Scopoli, 1763)
subgenus *Asyrmatus* Canzoneri, 1959b: 149 type species *Helops piceus* J. Sturm, 1826
foudrasii Mulsant, 1854: 361 (*Helops*) **E:** FR IT
meridianus Mulsant, 1854: 370 (*Helops*) **E:** FR
piceus J. Sturm, 1826: 70 (*Helops*) **E:** FR IT
  cordatus Küster, 1850b: 53 (*Helops*)
  subaeneipennis Pic, 1919c: 5 (*Helops*)
subgenus *Stenomax* Allard, 1876a: 4 type species *Tenebrio lanipes* Linnaeus, 1771 (= *Tenebrio aeneus* Scopoli, 1763)
aeneus Scopoli, 1763: 82 (*Tenebrio*) **E:** AU BH BU BY CR CZ FR GE HU IT LA LS PL RO RU SK SL ST SZ TR UK YU **A:** KZ
  arboreus Schrank von Paula, 1781: 219 (*Tenebrio*)
  cupreus Geoffroy, 1785: 158 (*Tenebrio*)
  incurvus Küster, 1850b: 54 (*Helops*)
  lanipes Linnaeus, 1771: 533 (*Tenebrio*)

genus *Stenomax*, species incertae sedis
litigiosus Vauloger de Beaupré, 1900: 701 (*Helops*) **N:** AG
martini Vauloger de Beaupré, 1900: 700 (*Helops*) **N:** AG

genus *Stygohelops* Leo & Liberto, 2003: 299 type species *Probaticus kalavriticus* Schawaller, 2001
kalavriticus Schawaller, 2001b: 44 (*Probaticus*) **E:** GR (Peloponnese)

genus *Taurohelops* Keskin & Nabozhenko, 2015: 84 type species *Stenomax incultus* Allard, 1877
incultus Allard, 1877a: 132 (*Stenomax*) **A:** TR
lordiscelis Reitter, 1900a: 157 (*Helops*)
triplehorni Keskin & Nabozhenko 2015: 88 **A:** TR

genus *Turkmenohelops* G.S. Medvedev, 1987a: 102 type species *Zophohelops balchanicus* G.S. Medvedev & Nepesova, 1985
arvatensis G.S. Medvedev, 1964c: 654 (*Zophohelops*) **A:** TM (Kopet Dag)
balchanicus G.S. Medvedev & Nepesova, 1985b: 362 (*Zophohelops*) **A:** IN TM (Kopet Dag)

genus *Turkonalassus* Keskin, Nabozhenko & Alpagut-Keskin 2017b: 727 type species *Helops adimonius* Allard, 1876
adimonius Allard, 1876: 35 (*Helops*) **A:** TR
bozdagus Keskin & Nabozhenko, 2010: 24 (*Nalassus*) **A:** TR
macedonicus Keskin, Nabozhenko & Alpagut-Keskin, 2017b: 743 **E:** BU GR
pentheri Reitter, 1905e: 278 (*Helops*) **A:** TR
hoberlandti Kaszab, 1959d: 81 (*Cylindronotus*)
petrophilus Keskin, Nabozhenko & Alpagut-Keskin, 2017b: 740 **A:** TR
pineus Keskin, Nabozhenko & Alpagut-Keskin, 2017b: 737 **A:** TR
quercanus Keskin, Nabozhenko & Alpagut-Keskin, 2017b: 739 **A:** TR

genus *Xanthohelops* Nabozhenko, 2006b: 822 type species *Xanthohelops karakumicus* Nabozhenko & G.S. Medvedev, 2006
karakumicus Nabozhenko & G.S. Medvedev, 2006: 823 **A:** TM

genus *Xanthomus* Mulsant, 1854: 302 type species *Helops pallidus* Curtis, 1830
cyprius Grimm, 1991: 41 **E:** GR (Rhodes) **N:** EG **A:** CY TR
graecus Dajoz, 1984: 89 **E:** GR
grimmi J. Ferrer & Whitehead, 2002: 393 **E:** IT
interstitialis Grimm, 1991: 42 **A:** CY
israelsoni J. Ferrer & Whitehead, 2002: 388 **N:** CI (Gran Canaria)
ovulus Seidlitz, 1895: 739 (*Helops*) **E:** GR (Crete)
pallidus Curtis, 1830: 298 (*Helops*) **E:** BE FR GB GE IR IT MA NL PT SP SZi **N:** AG CI EG MO MR **A:** CY

ghidinii Canzoneri, 1959a: 146
residuus Canzoneri, 1959a: 145
testaceus Küster, 1850b: 63 (*Helops*)
pellucidus Mulsant & Rey, 1856: 15 (*Helops*) **E:** FR GR IT SP **N:** AG
subsulcatus Rey, 1892: 65 (*Helops*)
viklundi J. Ferrer & Whitehead, 2002: 393 **N:** TU

genus *Zophohelops* Reitter, 1902h: 221 [RN] type species *Helops tiro* Reitter, 1902
subgenus *Zophohelops* Reitter, 1902h: 221 [RN] type species *Helops tiro* Reitter, 1902
Euryhelops Reitter, 1902g: 214 [HN] type species *Helops tiro* Reitter, 1902
bozbutavicus G.S. Medvedev, 1987a: 115 **A:** KI (Mts Bozbutau)
cylindronotoides Reitter, 1902g: 216 (*Helops*) **A:** KI KZ
gebieni Reitter, 1922b: 120
davletshinae G.S. Medvedev, 1987a: 106 **A:** KZ (Ugamskiy Ridge)
demetrius Reitter, 1922b: 119 **A:** UZ (Ferganskiy Ridge)
dolini Nabozhenko, 2008a: 298 **A:** KI
ferganensis G.S. Medvedev, 1987a: 119 **A:** UZ (Ferganskiy Ridge)
gurjevae G.S. Medvedev, 1987a: 106 **A:** KI (Mts Bozbutau)
hirtus G.S. Medvedev, 1987a: 121 **A:** KI UZ
humeridens Reitter, 1902g: 223 (*Xanthomus*) **E:** AB (Nakhichevan) AR
pullus Seidlitz, 1895: 705: 750 (*Helops*) [HN]
karatavicus G.S. Medvedev, 1987a: 108 **A:** KZ (Karatau Ridge)
kaszabi G.S. Medvedev, 1987a: 108 **A:** UZ (Ferganskiy Ridge)
kerzhneri Nabozhenko, 2008a: 295 **A:** KI
kuramensis G.S. Medvedev, 1987a: 104 **A:** UZ
michajlovi Nabozhenko, 2001a: 512 **A:** TD
montanatolicus Nabozhenko & Keskin, 2014: 245 **A:** TU
nadari Reitter, 1902g: 215 (*Helops*) **A:** KI (Alai)
pangazensis G.S. Medvedev, 1987a: 105 **A:** TD
pilosus G.S. Medvedev, 1987a: 120 **A:** KZ
protzenkoi Skopin, 1964: 389 **A:** KI
pyriformis G.S. Medvedev, 1987a: 110 **A:** KZ (Ugamskiy Ridge)
sergei G.S. Medvedev, 1987a: 112 **A:** KZ (Ugamskiy Ridge)
schawalleri Nabozhenko, 2008a: 300 **A:** KI
skopini G.S. Medvedev, 1987a: 113 **A:** KZ (Karatau Ridge)
spiridion Reitter, 1922b: 121 **A:** KI KZ UZ
staveni Nabozhenko & Grimm, 2019: 113 **A:** TR
stephanus Reitter, 1922b: 121 **A:** KI (Osch)
tar G.S. Medvedev, 1987a: 118 **A:** UZ (Ferganskiy Ridge)
tiro Reitter, 1902g: 215 (*Helops*) **A:** KI UZ
ugamicus Skopin, 1966b: 325 **A:** KZ (Ugamskiy Ridge)

zarudnyi G.S. Medvedev, 1987a: 114 **A:** KI (Chatkalskiy Ridge)
zolotarewi Reitter, 1902c: 190 (*Helops*) **A:** UZ (Tashkent)
Subgenus *Zophondrus* Nabozhenko, 2014: 240 type species *Zophohelops iranensis* Nabozhenko, 2014
iranensis Nabozhenko, 2014: 240 **A:** IN

subtribe Enoplopodina Reitter, 1917

genus ***Accanthopus*** **Dejean, 1821: 71** type species *Tenebrio caraboides* V. Petagna, 1786 (= *Tenebrio velikensis* Piller & Mitterpacher, 1783)
Enoplopus Solier, 1848: 158 [URN] type species *Tenebrio caraboides* V. Petagna, 1786 (= *Tenebrio velikensis* Piller & Mitterpacher, 1783)
reitteri Brenske, 1884: 88 (*Acanthopus*) **E:** AL GR MC
velikensis Piller & Mitterpacher, 1783: 83 (*Tenebrio*) **E:** AL BH BU CR FR GR HU IT MC RO SL SZ TR YU
caraboides V. Petagna, 1786: 15 [= 1787: 15] (*Tenebrio*) [HN]
dentipes Rossi, 1790: 236 (*Helops*)
latissimus Stierlin, 1902: 426 (*Dendarus*)
rufescens Pic, 1919c: 5

subtribe Helopina Latreille, 1802

genus ***Adelphinus*** **Fairmaire & Coquerel, 1866: 44** type species *Eutrapela suturalis* Lucas, 1847
subgenus *Adelphinops* Reitter, 1922b: 168 type species *Adelphinus ordubadensis* Reitter, 1890
afghanicus Kaszab, 1960a: 174 **A:** AF
dmitrii Nabozhenko, 2015c: 363 **A:** IN
ordubadensis Reitter, 1890d: 150 **E:** AB AR GG
fulvipennis Reitter, 1906f: 242
fulvovittatus Reitter, 1890d: 150
subgenus *Adelphinus* Fairmaire & Coquerel, 1866: 44 type species *Eutrapela suturalis* Lucas, 1847
baehri Nabozhenko & Grimm, 2019: 108 **N:** MO
ciliatus Seidlitz, 1898a: 853 **N:** AG MO
rotundicollis Kocher, 1958a: 124 **N:** MO
suturalis Lucas, 1847: 363 (*Eutrapela*) **N:** AG TU
maculatus Pic, 1899h: 115
pallidipennis Pic, 1899h: 114
rasus Seidlitz, 1898: 799 (*Nephodes*)

genus ***Allardius*** **Ragusa, 1898: 130** type species *Parablops oculatus* Baudi di Selve, 1876
Pseudoparablops Heyden, 1908: 132 type species *Paraplops sardiniensis* Allard, 1877
oculatus Baudi di Selve, 1876c: 264 [= 1876b: 135] (*Paraplops*) **E:** IT (Sicilia)
sardiniensis Allard, 1877a: 261 (*Paraplops*) **E:** IT (Sardinia)

genus *Apterotarpela* Kaszab, 1954: 262 type species *Apterotarpela klapperichi* Kaszab, 1954
clypealis Kaszab, 1954: 260 (*Tarpela*) **A:** FUJ
klapperichi Kaszab, 1954: 262 **A:** FUJ
subasperipennis Kaszab, 1954: 261 (*Tarpela*) **A:** FUJ

genus *Catomus* Allard, 1876a: 4 type species *Catomus persicus* Allard, 1876
subgenus *Catomodontus* Löbl & Merkl, 2020 [see New Acts] type species *Catomus coronatus* Koch, 1935
coronatus Koch, 1935a: 108 **N:** LB
subgenus *Catomus* Allard, 1876a: 4 type species *Catomus persicus* Allard, 1876
Catomidius Seidlitz, 1895: 792 type species *Hedyphanes rhynchophorus* Seidlitz, 1895
acutipennis Reiche & Saulcy, 1857: 269 (*Helops*) **A:** IS JO
laenoides Reitter, 1922b: 11
amabilis Vauloger de Beaupré, 1900: 716 (*Helops*) **N:** AG TU
antennatus Bogatchev, 1963b: 100 **A:** TD
anthicoides Vauloger de Beaupré, 1900: 710 (*Helops*) **N:** AG
antoniae Reitter, 1890e: 172 **E:** AB AR IN
arabicus Kaszab, 1982b: 183 **A:** OM
batnensis Fairmaire, 1883e: xliv (*Helops*) **N:** AG
biskrensis Vauloger de Beaupré, 1900: 692 (*Helops*)
soricinus Vauloger de Beaupré, 1900: 692 (*Helops*)
bengasianus Schuster, 1925b: 88 **N:** LB
capilatus Vauloger de Beaupré, 1900: 693 **N:** AG TU
ciliatus Vauloger de Beaupré, 1900: 710 (*Helops*) **N:** AG
comatus Vauloger de Beaupré, 1900: 710 (*Helops*) **N:** AG
consentaneus Küster, 1851: 63 (*Helops*) **E:** AL FR GR IT MC SP **N:** AG EG TU **A:** CY SY TR
convexifrons Fairmaire, 1877: 141 (*Hedyphanes*) [sic] **N:** AG TU
dlusskyi Nabozhenko, 2015b: 33 **N:** MO
echinatus Vauloger de Beaupré, 1900: 692 (*Helops*) [RN] **N:** AG
fallax Vauloger de Beaupré, 1900: 694 (*Helops*) **N:** LB TU
flavus Allard, 1876a: 28 **E:** AL
fragilis Ménétriés, 1848: 27 (*Helops*) **A:** IN TM UZ
subniger Reitter, 1902b: 181
fulvipes Reiche & Saulcy, 1857: 267 (*Helops*) **A:** IS SY
gossypiatus Reiche, 1862a: 373 (*Helops*) **N:** AG TU
grosclaudei Normand, 1936b: 367 **N:** AG TU
henoni Allard, 1876a: 49 **N:** AG
hesperides Reiche, 1861a: 7 (*Helops*) **A:** CY IS JO LE SY TR
berytensis Kraatz, 1880d: 299 (*Isopedes*)
lepidus Reitter, 1922b: 10
hirtus Mulsant & Godart, 1870: 52 (*Hedyphanes*) **N:** AG

improbulus Normand, 1950a: 85 **N:** TU
indubitatus Nabozhenko, 2006b: 838 **A:** TD UZ
karakalensis G.S. Medvedev, 1964c: 652 **A:** IN TM
dolini G.S. Medvedev, 2004b: 575
kuesteri Weise, 1878b: 193 (*Helops*) **A:** TR
dichrous Reitter, 1902g: 224
frivaldszkyi Seidlitz, 1895: 754 (*Helops*)
gracilis Küster, 1850b: 43 (*Helops*) [HN]
liliputanus Reitter, 1906g: 375 (*Helops*) **E:** SP
longiceps longiceps Chobaut, 1918: 75 (*Helops*) **N:** TU
longiceps minutepunctatus Chobaut, 1918: 76 (*Helops*) **N:** TU
lopezi Castro Tovar, 2015: 185 **E:** SP
niger Kraatz, 1882d: 332 (*Hedyphanes*) **A:** KZ TD UZ
minor Kraatz, 1882d: 332 (*Hedyphanes*)
nitidior Antoine, 1949b: 143 **N:** MO (High Atlas)
noctivagus Nabozhenko, 2006b: 840 **A:** UZ
obsoletus biasiolii Koch, 1937: 495 **N:** LB
obsoletus obsoletus Allard, 1876a: 30 **N:** AG LB TU
ovatus Koch, 1935a: 109 **N:** LB
persicus Allard, 1876a: 30 **A:** IN
pilosus Allard, 1876a: 49 **N:** AG (Oran)
puber Allard, 1876a: 50 **N:** AG
pullatus Peyerimhoff, 1917b: 318 (*Helops*) **N:** AG (Djurdjura)
rhynchophorus Seidlitz, 1895: 792 (*Hedyphanes*) **N:** AG
rotundicollis Guérin-Méneville, 1825b: 112 (*Helops*) **E:** FR IT MA SP **N:** TU
agonus Mulsant, 1854: 377 (*Helops*)
angustatus Lucas, 1847: 354 (*Helops*)
curvipes Reitter, 1922b: 12
juncorum Küster, 1850b: 41 (*Helops*)
obtusangulus Küster, 1850b: 37 (*Helops*)
pygmaeus Küster, 1850b: 38 (*Helops*)
rotundicollis Küster, 1850b: 40 (*Helops*) [HN]
siculus Küster, 1850b: 39 (*Helops*)
tagenioides Küster, 1850b: 42 (*Helops*)
rungsi Antoine, 1949b: 143 **N:** MO
schusteri Normand, 1936b: 366 **N:** LB TU
seidlitzi Gebien, 1911: 552 (*Helops*) [RN] **A:** IS "Syria"
hierochonticus J.R. Sahlberg, 1913b: 182 [HN]
semiruber Allard, 1876a: 29 **A:** IN
senussianus Koch, 1937: 495 **N:** LB
sicardi Vauloger de Beaupré, 1900: 690 (*Helops*) **N:** AG TU

submetallicus Vauloger de Beaupré, 1900: 713 (*Helops*) **N:** AG
subparallelus Koch, 1935a: 109 **N:** EG
sulcatus G.S. Medvedev, 1964c: 652 **A:** UZ
testaceipes Fairmaire, 1884d: lxvii (*Helops*) **N:** AG MO TU
thamii Antoine, 1955: 364 **N:** MO
torretassoi Koch, 1935a: 109 **N:** EG
ursus Antoine, 1949b: 143 **N:** MO
vaulogeri Théry, 1931b: 126 **N:** MO (Haute Moulouya)
virgilius Reitter, 1922b: 11 **N:** AG (Dra el Mizan)
subgenus *Montanocatomus* Nabozhenko, 2006b: 842 type species *Catomus grandis* G.S. Medvedev, 1978
badachshanicus G.S. Medvedev, 1970: 393 **A:** TD
fabiani Nabozhenko, 2006b: 849 **A:** KZ
grandis G.S. Medvedev, 1978b: 50 **A:** KI (Alayskiy ridge)
kabakovi Nabozhenko, 2007: 41 **A:** AF
pilosulus Kraatz, 1886: 190 **A:** UZ (Fergana)
gracilicollis Kraatz, 1886: 189 (*Stenomax*)
reinigi Shuster, 1931: 903 (*Stenomacidius*) **A:** KI TD
alaensis G.S. Medvedev, 1970: 394
subgenus *Sinocatomus* Nabozhenko, 2006b: 852 type species *Catomus solitarius* Nabozhenko, 2006
interruptus Liu et Ren, 2010: 71 **A:** GAN
solitarius Nabozhenko, 2006b: 852 **A:** GAN SCH
stanislavi Nabozhenko & Ando, 2018: 288 **A:** YUN

genus *Deretus* Gahan, 1900: 10 type species *Deretus denticollis* Gahan, 1900
bezdeki Purchart 2012a: 58 **A:** YE (Suqutra)
denticollis Gahan, 1900: 10 **A:** YE (Suqutra)
wraniki Schawaller, 2004c: 457
hajeki Purchart, 2013: 80 **A:** YE (Suqutra)
hulai Purchart, 2012a: 61 **A:** YE (Suqutra)
maderai Purchart, 2012a: 62 **A:** YE (Suqutra)
necopinatus Purchart, 2012a: 63 **A:** YE (Suqutra)
spinicollis Schawaller, 2004c: 456 **A:** YE (Suqutra)

genus *Entomogonus* Solier, 1848: 155 type species *Entomogonus barthelemyi* Solier, 1848
subgenus *Delonurops* Reitter, 1922b: 24 type species *Entomogonus clavimanus* Reitter, 1903
Macrophanes Iablokoff-Khnzorian, 1957a: 165 type species *Hedyphanes corax* Iablokoff-Khnzorian, 1957
amandanus Reitter, 1902g: 211 (*Helops*) **E:** AB AR
corax Iablokoff-Khnzorian, 1957a: 163
amri Nabozhenko & Tichý, 2019: 3 **A:** JO SY

bialookii Nabozhenko & Tichý, 2011: 45 **A:** TR
clavimanus Reitter, 1903a: 18 **E:** AR TR
doguanatolicus Nabozhenko, 2018b: 551 **A:** TR
duchoni Reitter, 1903a: 20 **A:** IQ SY TR
egregius Seidlitz, 1895: 703 (*Helops*) **A:** TR
elongatus Allard, 1876a: 42 (*Entomogonus*) **A:** TR
makovskyi Nabozhenko & Tichý, 2011: 48 **A:** TR
obtusus Seidlitz, 1895: 704 (*Helops*) **A:** CY
peyronis Reiche, 1861a: 5 (*Helops*) **A:** TR "Syria"
  fusiformis Allard, 1877a: 76 (*Helops*)
saphyrinus Allard, 1876a: 43 [= 1877b: 69] (*Helops*) **A:** IQ TR
  fausti Allard, 1877a: 255 (*Helops*)
subgenus *Entomogonus* Solier, 1848: 155 type species *Entomogonus barthelemyi* Solier, 1848
angulicollis aleppensis Reitter, 1922b: 28 **A:** SY TR
angulicollis angulicollis Mulsant & Wachanru, 1853: 11 (*Hediphanes*) **A:** SY TR
  gassneri Reitter, 1908b: 30
  helopinus Reitter, 1922b: 28
angulicollis gebieni Reitter, 1922b: 28 **A:** TR
barthelemyi Solier, 1848: 157 **A:** TR "Syria"
subgenus *Eutelogonus* Reitter, 1922b: 24 type species *Helops davidis* Fairmaire, 1884
davidis Fairmaire, 1884a: 171 (*Helops*) **A:** TR
nylanderi J. Ferrer & L. Soldati, 1999: 74 **A:** TR (Antalya)

genus *Erionura* Reitter, 1903a: 18 type species *Helops giganteus* Kraatz, 1862
gigantea Kraatz, 1862b: 124 (*Helops*) **E:** GR (Taygetos)

genus *Euboeus* Boieldieu, 1865: 10 type species *Euboeus mimonti* Boieldieu, 1865
subgenus *Euboeus* Boieldieu, 1865: 10 type species *Euboeus mimonti* Boieldieu, 1865
  Probaticus Seidlitz, 1895: 764 type species *Helops mori* Brullé, 1832
mimonti Boieldieu, 1865: 11 **E:** BU CR GR **A:** TR
mori Brullé, 1832: 222 (*Helops*) **E:** GR **A:** TR
  angulicollis Fairmaire, 1883k: clxx (*Helops*)
  terrenii I. Frivaldszky, 1845: 183 (*Helops*)
parvostriatus Nabozhenko, Nikitsky & Keskin, 2017: 499 **A:** TR
tentyrioides Küster, 1851: 71 (*Helops*) **E:** GR **A:** TR
  bysantica Motschulsky, 1858d: 189 (*Micipsa*)
  pyraei Motschulsky, 1858d: 189 (*Micipsa*)
  graeca Motschulsky, 1858d: 189 (*Micipsa*)
  parvicollis Baudi di Selve, 1876d: 280 (*Helops*) [NO]
subgenus *Helopidoxus* Reitter, 1922b: 44 type species *Helops superbus* Mulsant & Godart, 1855
superbus Mulsant & Godart, 1855b: 53 [= 1855a: 83] (*Helops*) **E:** FR (Corse)

subgenus *Helopostygnus* Antoine, 1949b: 133 type species *Helops atlantis* Antoine, 1926
atlantis atlantis Antoine, 1926: 258 (*Helops*) **N:** MO (High Atlas)
atlantis secedens Antoine, 1933b: 230 (*Helops*) **N:** MO
confinalis confinalis Antoine, 1949b: 133 (*Probaticus*) **N:** MO
confinalis kertensis Español, 1953a: 78 (*Probaticus*) **N:** MO
longulus Reiche, 1862a: 373 (*Helops*) **N:** AG
subgenus *Helopotrichus* Reitter, 1922b: 44 type species *Helops villosipennis* Lucas, 1846
gibbithorax Gemminger, 1870a: 123 (*Helops*) [RN] **E:** IT (Sardinia)
  gibbicollis Küster, 1850b: 86 (*Helops*) [HN]
sphaericollis Küster, 1850b: 87 (*Helops*) **E:** IT
tomentosus Reitter, 1906g: 376 (*Helops*) **E:** IT (Sicilia)
  siculus Canzoneri, 1960b: 50 (*Nephodinus*)
villosipennis Lucas, 1846: pl. 31 (*Helops*) **N:** AG TU
  normandi Vauloger de Beaupré, 1900: 681 (*Helops*)
subgenus *Pelorinus* Vauloger de Beaupré, 1900: 675 type species *Helops anthracinus* Germar, 1813
anthracinus Germar, 1813: 124 (*Helops*) **E:** FR PT SP **N:** AG MO
  coriaceus Küster, 1851: 66 (*Helops*)
anthrax Seidlitz, 1895: 711 (*Helops*) **E:** IT
balearicus Español, 1980: 108 (*Probaticus*) **E:** SP (Baleares)
banaszkiewiczae Nabozhenko, 2011c: 337 **A:** TR
bodemeyeri Reitter, 1900a: 156 (*Helops*) **A:** TR
chorasanicus G.S. Medvedev, 1976: 893 (*Probaticus*) **A:** IN
corrugatus Seidlitz, 1895: 707 (*Helops*) **A:** TR
corrugosus Seidlitz, 1895: 707 (*Helops*) **E:** GR **A:** IN TR
cossyrensis Sparacio, 2007: 256 **E:** IT (Pantelleria Is.)
dieckii Kraatz, 1870: 132 (*Helops*) **E:** SP
  castilianus Reitter, 1922b: 40 (*Probaticus*)
dorsalis dorsalis Allard, 1877a: 97 (*Helops*) **A:** TR
dorsalis violaceus Pic, 1899p: 411 **A:** TR
ebeninus cassolai Ardoin, 1973: 306 (*Probaticus*) **E:** IT (Sardinia)
ebeninus ebeninus A. Villa & J.B. Villa, 1838: 63 (*Helops*) **E:** FR IT
  germari Küster, 1850b: 20 (*Helops*)
  puncticollis Rey, 1892: 65 (*Helops*)
  robustus Mulsant, 1854: 320 (*Helops*)
euboeuicus Reitter, 1885d: 382 (*Helops*) **E:** GR
foveolatus Seidlitz, 1895: 710 (*Helops*) **E:** SP
  obesus Allard, 1876a: 32 (*Helops*) [HN]
  tardus Vauloger de Beaupré, 1900: 682 (*Helops*) [RN]
freyi Español, 1956a: 109 (*Probaticus*) **E:** SP
fulei Küster, 1850b: 21 (*Helops*) **E:** BU GR

globicollis Seidlitz, 1895: 707 (*Helops*) **A**: CY
glyptus Antoine, 1949b: 132 (*Probaticus*) **N**: MO
gorganicus G.S. Medvedev, 1976: 892 (*Probaticus*) **A**: IN
grajus Allard, 1876a: 34 (*Helops*) **E**: GR
granicollis Seidlitz, 1895: 716 (*Helops*) **E**: TR
granosus Seidlitz, 1895: 716 (*Helops*) **A**: TR
granulatus Allard, 1876a: 32 (*Helops*) **E**: PT SP
cordicollis Baudi di Selve, 1876c: 245 (*Helops*)
ehlersi Kraatz, 1882a: 47 (*Helops*)
granulifer Seidlitz, 1895: 711 (*Helops*) **E**: PT SP
huedepohli Kulzer, 1964: 228 (*Probaticus*) **A**: IN
immarginatus Reitter, 1902g: 212 (*Helops*) **A**: TR
interstitialis Küster, 1850b: 25 (*Helops*) **E**: SP
elongatus Español, 1956a: 100 (*Probaticus*)
recticollis Español, 1956a: 100 (*Probaticus*)
kobeltii Seidlitz, 1895: 717 (*Helops*) **N**: TU
kopetzi Nabozhenko & Grimm, 2019: 110 **A**: TR
lacertosus Küster, 1845a: 30 (*Helops*) **E**: BU GR MC **A**: TR
laticollis Küster, 1850b: 76 (*Helops*) **E**: FR (Pyrénées) SP
cerberus Mulsant, 1854: 316 (*Helops*)
ponferradanus Reitter, 1922b: 38 (*Probaticus*)
linearis Vauloger de Beaupré, 1900: 681 (*Helops*) [RN] **N**: AG
maurus Waltl, 1838: 464 (*Helops*) **E**: BU GR MC MD RO TR UK **A**: TR
gibbus Küster, 1851: 60 (*Helops*)
medvedevi Abdurakhmanov & Nabozhenko, 2011: 306 (*Probaticus*) **E**: AR GG
melas Küster, 1850b: 22 (*Helops*) **E**: SP
micantipennis Allard, 1876a: 55 (*Stenomax*) **E**: PT
myops Allard, 1876a: 50 (*Helops*) **A**: TR
obesus I. Frivaldszky, 1835: 262 (*Helops*) **E**: AL BU GR MC RO TR **A**: TR
oliveirae Seidlitz, 1895: 709 (*Helops*) **E**: PT SP
parthorum G.S. Medvedev, 1976: 891 (*Probaticus*) **A**: IN
peropacus Reitter, 1902g: 214 (*Helops*) **A**: TR (Erzurum)
peslieri F. Soldati, 2012a: 222 (*Probaticus*) **E**: GR (Lesvos)
picianus Reitter, 1907a: 30 (*Helops*) [RN] **A**: TR
carinatus Pic, 1899p: 411 (*Helops*) [HN]
priapus Seidlitz, 1895: 751 (*Helops*) **A**: TR
prometheus Reitter, 1902g: 213 (*Helops*) **E**: AB **A**: IN (Talysh)
quadricollis Ménétriés, 1832: 196 (*Hedyphanes*) **E**: AB (Talysh)
subrugosus Duftschmid, 1812: 284 (*Blaps*) **E**: BU CR GR HU KZ MD MC RO SK ST UK
arboreus Germar, 1824b: 160 (*Helops*) [HN]

damascenus Küster, 1850b: 28 (*Helops*)
fovoelatostriatus Motschulsky, 1845a: 82 (*Helops*)
rugosus Küster, 1850b: 26 (*Helops*)
schmidti Ahrens, 1814: 3 (*Helops*)
strigicollis Seidlitz, 1895: 712 (*Helops*) [HN]
strigithorax Reitter, 1906a: 497 (*Helops*) [RN]
sulcatus Fischer von Waldheim, 1844: 124 (*Helops*)
tarsatus Küster, 1850b: 29 (*Helops*) **E:** BU MD RO
tazekensis Antoine, 1952: 116 (*Probaticus*) **N:** MO
tenebricosus Brullé, 1832: 223 (*Helops*) **E:** GR (Peloponessos, Naxos Is.)
theanus Reitter, 1902g: 213 (*Helops*) **A:** TR
tripolitanus Seidlitz, 1895: 717 **N:** LB
valdanii Guérin-Méneville, 1859: cxc (*Helops*) **N:** AG
vicinus Allard, 1876a: 34 (*Helops*) **E:** AB (Nakhichevan) AR
zoroaster Seidlitz, 1895: 714 (*Helops*) **A:** TM (Kopet Dag, Ashkhabat)
diabolinus Brancsik, 1899: 101 (*Helops*)

genus *Euboeus*, nomina dubia
nitidipennis Küster, 1850b: 23 (*Helops*) **E:** IT (Sicilia)
rufipes Allard, 1876a: 33 (*Helops*) **A:** TR

genus *Hedyphanes* Fischer von Waldheim, 1820: pl. 15 type species *Hedyphanes coerulescens* Fischer von Waldheim, 1820
subgenus *Granulophanes* Nabozhenko, 2013b: 2 type species *Hedyphanes lutosus* Allard, 1877a: 229 [= 1877b: 97]
lutosus Allard, 1877a: 229 [= 1877b: 97] **A:** TR
angustus Marseul, 1879c: 75 (*Apolites*)
roznerorum Nabozhenko, 2008b: 721 (*Pseudoprobaticus*)
subgenus *Hedyphanes* Fischer von Waldheim, 1820: pl. 15 type species *Hedyphanes coerulescens* Fischer von Waldheim, 1820
Coelophanes Iablokoff-Khnzorian, 1964: 304 type species *Hedyphanes impressicollis* Faldermann, 1837 (= *Hedyphanes laticollis* Fischer von Waldheim, 1832)
aalbui Nabozhenko, 2018a: 621 **A:** IN
afghanicus Nabozhenko, 2018a: 623 **A:** AF
besseri Faldermann, 1837: 95 **A:** KZ TM
parvicollis Seidlitz, 1895: 797
bodemeyeri Reitter, 1914e: 185 **A:** IN
gebieni Reitter, 1914e: 186
chalybaeus Faldermann, 1837: 96 **E:** AR
coerulescens Fischer von Waldheim, 1820: pl. 15 **E:** ST **A:** AF KZ TD TM UZ
cordicollis Seidlitz, 1895: 797 **A:** TR

cribripennis Lucas, 1854: 32 **E**: GR (Crete)
europs Reitter, 1914e: 187 **A**: IN
helopinus Gemminger, 1870a: 123 **E**: GR (Crete)
helopioides Lucas, 1854: 33 [HN]
helopioides Faldermann, 1837: 83 **E**: AR
igori Nabozhenko, 2018a: 620 **A**: IN
iranicus G.S. Medvedev, 1976: 890 **A**: IN
kadleci Nabozhenko, 2018a: 611 **A**: IN
khachikovi Nabozhenko, 2013b: 5 **A**: TR
kiritshenkoi G.S. Medvedev, 1978b: 51 **A**: TD UZ
koltzei Heyden, 1892b: 108 **A**: KI KZ
laticollis Fischer von Waldheim, 1832a: 195 **E**: AB AR GG IN
desertus Motschulsky, 1845a: 82
hegeteroides Faldermann, 1837: 92
impressicollis Faldermann, 1837: 94
mannerheimi Faldermann, 1837: 91 **E**: AR
matthiesseni Reitter, 1914e: 186 **A**: IN
mesopotamicus Nabozhenko, 2005: 351 **A**: IQ
muminovi Bogatchev, 1963b: 98 (*Catomus*) **A**: UZ (Surkhandarya Province)
nycterinoides Faldermann, 1837: 93 **E**: ST (Dagestan)
nycteroides Reitter, 1914e: 185
ocularis Reitter, 1914e: 187 **A**: IN
seidlitzi seidlitzi Reitter, 1914e: 184 **A**: IN TM
albertus Reitter, 1922b: 16
seidlitzi kuschkensis Kaszab, 1960a: 172 **A**: AF TM
semnanicus Nabozhenko, 2005: 352 **A**: IN
tagenoides Faldermann, 1832: 197 **E**: AB AR GG **A** IN
cruralis Fischer von Waldheim, 1837a: 153
dejeanii Faldermann, 1837: 88
menetriesii Fischer von Waldheim, 1832a: 197
pubiventris Reitter, 1922b: 18
tentyrioides Faldermann, 1837: 85
upioides Faldermann, 1837: 87
tuxeni Kaszab, 1959e: 253 [= 1960a: 172] **A**: AF IN
zarudnyi G.S. Medvedev, 1976: 890 **A**: IN
subgenus *Microhedyphanes* Nabozhenko & Lillig, 2013: 189 type species *Microhedyphanes chikatunovi* Nabozhenko & Lillig, 2013
chikatunovi Nabozhenko & Lillig, 2013: 188 **A**: IS **N**: EG

genus *Helopelius* Reitter, 1922b: 152 type species *Stenomax aeneipennis* Allard, 1876
aeneipennis Allard, 1876a: 18 (*Stenomax*) **E**: GR (Rhodes) **A**: TR "Syria"

disgregus Reitter, 1922b: 152 (*Cylindronotus*) **E:** GR (Rhodes)
nodifer Kraatz, 1880d: 299 (*Isopedes*) **E:** SP
otini Antoine, 1949b: 141 **N:** MO (Middle Atlas)
subsinuatus Antoine, 1951b: 318 **N:** MO (Middle Atlas)
verrucosus Vauloger de Beaupré, 1900: 683 (*Helops*) **E:** PT **N:** AG MO
zaianus Antoine, 1949b: 142 **N:** MO

genus *Helops* Fabricius, 1775: 257 type species *Tenebrio caeruleus* Linnaeus, 1758
Anteros Laporte, 1840: 235 [HN] type species *Helops chalibeus* Rossi, 1790 (= *Tenebrio caeruleus* Linnaeus, 1758)
Biomorphus Motschulsky, 1872: 38 type species *Biomorphus tuberculatus* Motschulsky, 1872 (= *Amphidora attenuata* LeConte, 1851)
Coscinoptilix Allard, 1876: 15 type species *Coscinoptilix gracilicornis* Allard, 1876
Hypulus Rafinesque, 1815: 114 [URN, HN] type species *Tenebrio caeruleus* Linnaeus, 1758
Mesohelops Reitter, 1922b: 31 type species *Helops cyanipes* Allard, 1877
Stenotrichus LeConte, 1862a: 239 type species *Amphidora rufipes* LeConte, 1851
caeruleus caeruleus Linnaeus, 1758: 418 (*Tenebrio*) **E:** AB AL BH BU CR FR GB GE GR IT MC NL RO SL SP ST SZ UK
chalibeus Rossi, 1790: 236
violaceus Marsham, 1802: 480 (*Blaps*)
caeruleus stevenii Krynicki, 1834: 167 **E:** GG ST UK (Krym) **A:** TR
caeruleus talyshensis Bogatchev, 1949a: 38 **A:** AB IN
cyanipes Allard, 1877a: 256 **A:** IS LE TR
carinimentum Reitter, 1885d: 383
glabriventris glabriventris Reitter, 1885d: 382 **E:** GR **A:** CY TR
glabriventris jelineki Picka, 1984: 27 **E:** GR (Crete)
insignis insignis Lucas, 1846: pl. 31 **N:** AG TU
insignis maroccanus Fairmaire, 1873a: 341 **N:** MO
magnificus Escalera, 1925a: 334
punctatissimus Nabozhenko & Keskin, 2017: 46 **A:** TR
rossii Germar, 1817: 191 [NP] **E:** AL BH CR FR GR IT LS MA RO SL TR **A:** TR
marloysi Laporte, 1840: 235
violaceus Thunberg, 1817: 43 (*Tenebrio*) [NO]
thoracicus Grimm, 1991: 41 **A:** CY

genus *Helops*, species incertae sedis
Helops caucasicus Allard, 1877b: 97 **E:** "Caucasus"

genus *Italohelops* Español, 1961e: 295 type species *Parablops subchalybaeus* Reitter, 1907
subchalybaeus Reitter, 1907e: 204 (*Parablops*) **E:** IT

genus *Mamorina* Antoine, 1951c: 98 type species *Mamorina sulcaticeps* Antoine, 1951
sulcaticeps Antoine, 1951c: 99 **N:** MO

genus *Nephodinus* Gebien, 1943: 900 [RN] type species *Nephodes villiger* Rosenhauer, 1856
subgenus *Nephodinus* Gebien, 1943: 900 [RN] type species *Nephodes villiger* Rosenhauer, 1856
Nephodes Blanchard, 1845: 34 [HN] type species *Nephodes villiger* Rosenhauer, 1856
gracilior Fairmaire, 1884d: lxvii (*Nephodes*) **N:** AG
metallescens Küster, 1846c: 47 (*Helops*) **E:** FR (Corse) IT (Sardinia)
corsicus Allard, 1877a: 222 (*Nephodes*)
pubescens modestus Kraatz, 1870: 142 (*Nephodes*) **E:** SP
infoveicollis Pic, 1910c: 10 (*Nephodes*)
pubescens pubescens Küster, 1850b: 88 (*Helops*) **E:** SP
foveicollis Küster, 1851: 72 (*Helops*)
villiger barbarus Kraatz, 1872: 182 (*Nephodes*) **N:** AG MO TU
villiger villiger Rosenhauer, 1856: 218 (*Nephodes*) **E:** SP **N:** MO
incanus Fairmaire, 1879c: 199 (*Nephodes*)
subdepressus Fairmaire, 1868: 492 (*Nephodes*) **N:** AG
subgenus *Paranephodes* Antoine, 1954: 208 type species *Paranephodes camusi* Antoine, 1954
camusi Antoine, 1954: 208 (*Nephodes*) **N:** MO

genus *Nesotes* Allard, 1876a: 5 type species *Helops asper* Küster, 1850
subgenus *Helopogonus* Reitter, 1922b: 122 type species *Helops viridicollis* L.W. Schaufuss, 1869
viridicollis ibicensis Español, 1946: 105 **E:** SP (Ibiza Is.)
viridicollis viridicollis L.W. Schaufuss, 1869: 25 (*Helops*) **E:** SP (Baleares)
rugipennis Tenenbaum, 1915: 841 (*Helops*)
subgenus *Nesotes* Allard, 1876a: 5 type species *Helops asper* Küster, 1850
Diastixus Allard, 1876a: 5 type species *Helops punctipennis* Lucas, 1847 (= *Helops heteromorphus* Lucas, 1846)
aenescens Reiche, 1861b: 90 (*Helops*) **N:** AG MO
acutangulus Fairmaire, 1879c: 199 (*Diastixus*)
lucens Escalera, 1914: 352 (*Helops*)
aethiops Wollaston, 1864: 509 (*Helops*) **N:** CI
afer Erichson, 1841: 184 (*Helops*) **N:** AG
altivagans Wollaston, 1864: 503 (*Helops*) **N:** CI (Tenerife)
andreinii Schuster, 1925b: 89 (*Diastixus*) **N:** LB
arboricola Wollaston, 1862d: 338 (*Helops*) **N:** MR
asper asper Küster, 1850b: 65 (*Helops*) **N:** MR
vulcanus Wollaston, 1854: 513 (*Helops*)
asper obliteratus Wollaston, 1871: 280 (*Helops*) **N:** MR
aterrimus Wollaston, 1864: 506 (*Helops*) **N:** CI (Gomera)

azoricus Crotch, 1867a: 390 (*Helops*) **N:** AZ
carbo Küster, 1850b: 33 (*Helops*) **E:** SP
carbunculus Wollaston, 1854: 519 (*Helops*) **N:** CI
catomoides bibersoni Antoine, 1954: 208 **N:** MO
catomoides catomoides Español, 1952a: 74 **N:** MO (Western Sahara)
cinnamomeus Wollaston, 1854: 520 (*Helops*) **N:** MR
confertus colasi Ardoin, 1960a: 118 **N:** MR
confertus confertus Wollaston, 1854: 515 (*Helops*) **N:** MR
conformis conformis Gemminger, 1870a: 123 [RN] **N:** CI
 congener Wollaston, 1864: 504 (*Helops*) [HN]
conformis grancanariensis Español, 1962b: 339 **N:** CI (Gran Canaria)
conformis turgiticollis Wollaston, 1864: 504 (*Helops*) **N:** CI
congestus Wollaston, 1864: 505 **N:** CI
congregatus Wollaston, 1854: 518 (*Helops*) **N:** MR
crassicollis Küster, 1850b: 32 (*Helops*) **E:** SP
 ghilianii Baudi di Selve, 1876c: 253 [= 1877b: 48] (*Helops*)
 ibericus Allard, 1877a: 182 (*Diastixus*)
effeminatus Antoine, 1949b: 135 **N:** MO
elliptipennis Wollaston, 1864: 503 (*Helops*) **N:** CI (Tenerife)
fusculus Wollaston, 1864: 511 (*Helops*) **N:** CI
 rufipes Har. Lindberg, 1953: 10
futilis Wollaston, 1854: 520 (*Helops*) **N:** MR
gagatinus Küster, 1850b: 66 (*Helops*) **N:** MR
 pluto Wollaston, 1854: 516 (*Helops*)
gomerensis Wollaston, 1965: 64 [Appendix] (*Helops*) **N:** CI (Gomera)
graniger Küster, 1850b: 64 (*Helops*) **N:** MR
 validus Reitter, 1922b: 162 (*Cylindronotus*)
helleri Reitter, 1922b: 163 (*Cylindronotus*) **N:** CI (Tenerife)
heteromorphus Lucas, 1846: pl. 31 (*Helops*) **N:** AG
 punctipennis Lucas, 1847: 352 (*Helops*)
infernus infernus Wollaston, 1854: 517 (*Helops*) **N:** MR (Porto Santo)
infernus wollastoni Ardoin, 1960a: 121 **N:** MR (Porto Santo)
leacockianus Wollaston, 1854: 517 (*Helops*) **N:** SG
lindbergi Español, 1962b: 342 **N:** CI (Gran Canaria)
lucasi Méquignon, 1937: 194 [RN] **N:** AG MO TU
 rotundicollis Lucas, 1846: 350 (*Helops*) [HN]
lucifugulus Reitter, 1922b: 519 (*Cylindronotus*) **N:** CI
lucifugus Wollaston, 1854: 518 (*Helops*) **N:** MR (Porto Santo)
marseulii Wollaston, 1865: 63 [Appendix] (*Helops*) **N:** CI
mogadoricus Escalera, 1914: 351 (*Helops*) **N:** MO
monodi Alluaud, 1935: 41 (*Helops*) **N:** SG

nigroaeneus Küster, 1850b: 31 (*Helops*) E: FR SP
alicantinus Kraatz, 1883c: 395 (*Nalassus*)
subaeneus Baudi di Selve, 1876c: 258 [= 1877b: 49] (*Helops*)
nitens Wollaston, 1864: 506 (*Helops*) N: CI (Tenerife)
nitidicollis Lucas, 1846: pl. 31 (*Helops*) N: AG MO
nigricans Vauloger de Beaupré, 1900: 688 (*Helops*)
thalassinus Allard, 1877a: 181 (*Diastixus*)
obtusatus Fairmaire, 1883e: xliv (*Helops*) N: AG TU
deserticola Vauloger de Beaupré, 1900: 688 (*Helops*)
picescens Wollaston, 1864: 509 (*Helops*) N: CI
piger Vauloger de Beaupré, 1900: 705 (*Helops*) N: TU
piliger Har. Lindberg, 1953: 10 (*Cylindronotus*) N: CI (Gran Canaria)
porrectus Wollaston, 1864: 508 (*Helops*) N: CI
portosanctanus Wollaston, 1854: 521 (*Helops*) N: MR (Porto Santo)
pulvinatus Vauloger de Beaupré, 1900: 704 (*Helops*) N: AG
puncticollis Lucas, 1846: 349 (*Helops*) N: AG TU
quadratus Brullé, 1839: 70 (*Helops*) N: CI (Gran Canaria)
rimosus Wollaston, 1864: 508 (*Helops*) N: CI (Fuerteventura)
rufulus Vauloger de Beaupré, 1900: 708 N: AG
sabulicola Israelson, 1980: 192 N: CI (Lanzarote)
sefruensis Antoine, 1949b: 137 N: MO (Middle Atlas)
subdepressus Wollaston, 1857a: 158 (*Helops*) N: MR
tenuipunctatus Har. Lindberg, 1950: 12 (*Cylindronotus*) N: CI (Tenerife)
transversus Brullé, 1839: 70 (*Helops*) N: CI
tuberculipennis ramirezi Español, 1961e: 304 E: SP (Cádiz, Gibraltar)
tuberculipennis schrammi Antoine, 1949b: 135 N: MO
tuberculipennis tuberculipennis Lucas, 1846: pl. 31 (*Helops*) E: SP (Gibraltar) N: AG MO
tuberculipennis vilarrubiai Español, 1943c: 144 (*Cylindronotus*) N: MO (Tiznit)

genus *Nipponohelops* Masumoto, Ando & Akita, 2006: 33 type species *Nipponohelops ishikawai* Masumoto, Ando & Akita, 2006
ishikawai Masumoto, Ando & Akita, 2006: 34 A: JA (Honshu)

genus *Physohelops* Schuster, 1937a: 50 type species *Physohelops freyi* Schuster, 1937
freyi Schuster, 1937a: 51 A: IQ (Bacuba) SY

genus *Raiboscelis* Allard, 1876a: 5 [NP] type species *Helops corvinus* Küster, 1850
Hipponome Laporte, 1840: 235 [NO] type species *Helops azureus* Brullé, 1832
Rhaebosceles Rye, 1878: 69 [UE]
angustitarsis Reitter, 1902g: 210 (*Helops*) A: TR

azureus azureus Brullé, 1832: 224 (*Helops*) **E:** GR IT (Sicilia)
tumidicollis Küster, 1850b: 45 (*Helops*)
azureus stichi Mader, 1943: 42 (*Raiboscelis*) **E:** GR (Is. Thera)
coelestinus coelestinus Waltl, 1838: 464 (*Helops*) **E:** GR **A:** TR
quadraticollis Küster, 1850b: 48 (*Helops*)
coelestinus rugipleuris Reitter, 1922b: 30 **E:** GR (Rhodes)
corvinus bilyi Picka, 1984: 26 **E:** GR (Gaudos Is.)
corvinus brodskyi Picka, 1984: 24 **E:** GR (Crete: Matala)
corvinus corvinus Küster, 1850b: 44 (*Helops*) **E:** GR (Crete) **A:** TR
corvinus slamai Picka, 1984: 25 **E:** GR (Crete: Nida-Ida)
cyprius Allard, 1876a: 44 **A:** CY
eleodinus Reiche, 1861a: 5 (*Helops*) **A:** IS SY TR (Antakya)
latimargo Seidlitz, 1895: 705 (*Helops*) **A:** TR
obliteratus Allard, 1876a: 56 **E:** GR
syriacus immargo Reitter, 1922b: 30 **A:** TR
syriacus syriacus Reiche, 1861a: 6 (*Helops*) **A:** LE SY TR
iconiensis Pic, 1900d: 31 (*Helops*)

genus *Sabularius* Escalera, 1914: 353 type species *Helops fossor* Escalera, 1914
fossor Escalera, 1914: 354 (*Helops*) **N:** MO
hoplites Antoine, 1937b: 177 **N:** MO
kocheri Antoine, 1953b: 215 **N:** MO
cephalotes Antoine, 1953b: 217
mimeuri Antoine, 1937b: 179 **N:** MO
punctatostriatus Antoine, 1953b: 214 **N:** MO
serrulifer Antoine, 1937b: 178 **N:** MO
talpa Antoine, 1937b: 175 **N:** MO

genus *Socotraphanes* Nabozhenko, 2019a: 150 type species *Socotraphanes krali* Nabozhenko, 2019
dementor Nabozhenko, 2019a: 152 **A:** YE (Suqutra)
krali Nabozhenko, 2019a: 152 **A:** YE (Suqutra)

genus *Stenohelops* Reitter, 1922b: 113 type species *Isopedus plicatulus* Kraatz, 1880
subgenus *Gunarellus* Reitter, 1922b: 114 type species *Helops gratus* J. Frivaldszky, 1894
calvus Seidlitz, 1895: 719 (*Helops*) **N:** MO TU
carlofortinus Leo, 1980b: 36 **E:** IT (Sardinia)
gratus J. Frivaldszky, 1894: 85 (*Helops*) **E:** RO YU
macellus Kraatz, 1870: 136 (*Helops*) **E:** SP
mauritanicus Seidlitz, 1895: 719 (*Helops*) **N:** MO
piligerus Kraatz, 1870: 135 (*Helops*) **E:** SP
silvestrei Escalera, 1914: 349 (*Helops*) **N:** MO
tronqueti F. Soldati & L. Soldati, 2002a: 139 **E:** SP

villosus L.W. Schaufuss, 1861: 92 (*Helops*) **E**: SP
walkeri Champion, 1891d: 390 (*Helops*)
subgenus *Stenohelops* Reitter, 1922b: 113 type species *Isopedus plicatulus* Kraatz, 1880
berettai Marcuzzi, 2001: 1 **N**: TU (Djerba)
filum Reitter, 1922b: 114 **E**: SP
montanus Kraatz, 1870: 137 (*Helops*) **E**: PT
plicatulus Kraatz, 1880d: 299 (*Isopedus*) **E**: SP
prolixus Seidlitz, 1895: 748 (*Helops*) **E**: SP
sublinearis Kraatz, 1870: 139 (*Helops*) **E**: SP
subgenus *Stenomaleis* Español, 1957b: 22 type species *Stenohelops ardoini* Español, 1957 (= *Helops protensulus* Seidlitz, 1896)
barcinensis Lagar & Fresneda, 2014: 96 **E**: SP
chinensis Nabozhenko & Ando, 2018: 291 **A**: SCH
effeminatus Español, 1963f: 392 **N**: MO (Rif)
gayirbegi Nabozhenko & Keskin, 2009: 54 (*Gunarus*) **A**: TR
korkutelensis Nabozhenko & Keskin, 2009: 58 (*Gunarus*) **A**: TR
protensulus Seidlitz, 1895: 735 (*Helops*) **E**: SP
ardoini Español, 1957b: 22
pyrenaeus Mulsant, 1854: 374 (*Helops*) **E**: FR SP

genus *Stenohelops*, species incertae sedis
carinatus Seidlitz, 1895: 719 (*Helops*) **E**: "Caucasus"
narcissus Reitter, 1922b: 114 **E**: IT (Sicilia)

subtribe Helopina, species incertae sedis
Gunarus arenicola Antoine, 1949b: 140 **N**: MO
Gunarus bremondi Antoine, 1949b: 140 **N**: MO

tribe Helopini, species incertae sedis
Gunarus nodicornis Reitter, 1922b: 117 **N**: AG
Helops seminitens Fairmaire, 1897k: 226 **A**: JIX
Xanthomus tingitanus Allard, 1877a: 116 **N**: MO

tribe Melanimonini Seidlitz, 1894 (1854)

genus *Cheirodes* Gené, 1839: 73 type species *Cheirodes sardous* Gené, 1839
subgenus *Cheirodes* Gené, 1839: 73 type species *Cheirodes sardous* Gené, 1839
Anemia Laporte, 1840: 218 type species *Anemia granulata* Laporte, 1840
Eremonomus Wollaston, 1861: 199 type species *Eremonomus huttoni* Wollaston, 1861
Chirodes Agassiz, 1846b: 81 [UE]
aethiopia Ardoin, 1971b: 376 (*Anemia*) **A**: YE **AFR**
chobauti Reitter, 1898f: 349 (*Anemia*) **N**: AG LB MO TU **A**: IS SA YE **AFR**
reitteri Pic, 1899n: 278 (*Anemia*) [RN]

dentipes Ballion, 1878: 332 (*Microzoum*) **E:** AB ST **A:** AF IN KZ MG SA TM UZ XIN
reichardti Kontkanen, 1956: 57 (*Anemia*)
drurei Pic, 1923f: 5 (*Anemia*) **A:** IQ
granulatus Laporte, 1840: 218 (*Anemia*) **A:** SA **AFR**
sardous denticulatus Wollaston, 1867: 197 (*Anemia*) **A:** OM YE **AFR**
sardous sardous Gené, 1839: 73 **E:** AB AR FR GR IT PT SP **N:** CI EG LB MO TU **A:** CY IN IQ IS JO SI TR **AUR**
caulobioides Carter, 1920: 224 (*Anemia*)
sinuatifrons Reitter, 1887d: 389 (*Anemia*) **A:** XIN
subgenus *Histiaea* Fairmaire, 1892a: 107 type species *Histiaea bidentula* Fairmaire, 1892
ardoini Kaszab, 1972: 378 (*Anemia*) **A:** YE
asperulus Reitter, 1884f: 260 (*Anemia*) **N:** EG LB TU **A:** IN IS JO SA SI SY YE **AFR**
aegyptiacus Pic, 1936h: 300 (*Anemia*) [RN]
denticulatus Pic, 1923f: 5 (*Anemia*) [HN]
seriesetosus Baudi di Selve, 1894: 8 (*Anemia*)
bidentulus Fairmaire, 1892a: 108 (*Histiaea*) **A:** SA **AFR**
subgenus *Pseudanemia* Wollaston, 1864: 492 type species *Pseudanemia brevicollis* Wollaston, 1864
Ammidanemia Reitter, 1904a: 132 type species *Anemia fausti* Solsky, 1882
brevicollis Wollaston, 1864: 493 **E:** FR IT (South, Sicilia) MA SP **N:** AG CI EG LB MO TU **A:** AF BA IN IS IQ JO KZ MG OM PA SA SI TM UZ XIN YE **AFR**
aphodioides Walker, 1871: 16 (*Anemia*)
curticollis Pic, 1923f: 5 (*Anemia*)
fausti Solsky, 1881b: 52 (*Anemia*)
hauseri Reitter, 1894a: 50 (*Anemia*)
pharao Reitter, 1897b: 43 (*Anemia*)
rotundicollis Desbrochers des Loges, 1881: 126 (*Anemia*)
striolatus Fairmaire, 1894c: 311 (*Anemia*)
scalarithoracus Ren, Ning & Jia, 2011: 565 **A:** NMO
submetallicus Raffray, 1873: 377 (*Anemia*) **E:** SP (Gibraltar) **N:** AG EG LB MO TU **A:** JO SA **AFR**
palaestinus Pic, 1899n: 278 (*Anemia*)
zhengi Ren & Yu, 1994: 88 **A:** NIX
subgenus *Spinanemia* Löbl, Bouchard, Merkl & Bousquet, 2020 [see New Acts] type species *Anemia cornuta* Pic, 1898
cornutus cornutus Pic, 1898h: 121 (*Anemia*) **A:** AF IN SA
cornutus panelii Ardoin, 1971b: 379 (*Anemia*) **A:** SA **AFR**
pilosus Tournier, 1868: 146 (*Anemia*) **N:** AG CI EG LB MO TU **A:** IS (Negev) OM SA SI YE **AFR**
fenyesi Reitter, 1897b: 42 (*Anemia*)
fissidens Reitter, 1898f: 348
thoracicus Fairmaire, 1891b: 250 (*Anemia*) **A:** SA **AFR**

genus *Dolamara* Reichardt, 1935: 251 type species *Melanimon cupreomicans* Reitter, 1915
cupreomicans Reitter, 1915e: 88 (*Melanimon*) **A:** KI MG TD UZ

genus *Melanimon* Steven, 1828: 98 type species *Opatrum tibiale* Fabricius, 1781

Fungulus Gistel, 1848: 125 [RN]

Microzoon Hope, 1841: 112 [UE]

Microzoum Dejean, 1834: 193 type species *Opatrum tibiale* Fabricius, 1781

amalitae J. Ferrer & Castro Tovar, 2010: 41 **E:** SP

inermus Picka, 1983b: 265 **E:** BU

kiritshenkoi Reichardt, 1936a: 68 **E:** AB (Talysh) GG **A:** TR

tibialis subaereus Reitter, 1904a: 129 **A:** KI KZ

tibialis tibialis Fabricius, 1781: 90 (*Opatrum*) **E:** AU BE BH BU BY CR CZ DE EN FI FR GB GE GR HU IR IT LA LT NL NR NT PL RO SK SL SP ST SV SZ UK **N:** MO **A:** KZ TM UZ WS

tribe Metaclisini Steiner, 2016

genus *Metaclisa* Jacquelin du Val, 1861: 296 [NP] type species *Platydema parallela* Fairmaire 1855 (= *Diaperis azurea* Waltl, 1838)

Amarantha Motschulsky, 1860: 141 [NO] type species *Amarantha viridis* Motschulsky, 1860

Tharsus LeConte, 1862: 233 type species *Tharsus seditiosus* LeConte, 1862

andoi Schawaller, 2016b: 552 **A:** YUN

atrocyanea Lewis, 1891a: 70 (*Amarantha*) **A:** GAN JA SC SCH YU

cyanescens Pic, 1934f: 85 (*Androsus*)

azurea Waltl, 1838: 463 (*Diaperis*) **E:** AB BU CR GG HU IT MC RO SL ST (Kavkaz) TR **A:** IS SY TR

parallela Fairmaire, 1855: 316 (*Platydema*)

gracilis M. Chûjô, 1968b: 395 **A:** TAI

hamai Shibata, 1978: 24 **A:** JA (Ryukyus)

nagaii Nakane, 1968b: 79 **A:** JA (Ryukyus)

ornata Shibata, 1978: 26 **A:** TAI

viridis Motschulsky, 1860b: 141 (*Amarantha*) **E:** AB **A:** IN

chlorodia Baudi di Selve, 1876c: 247 (*Platydema*)

tribe Opatrini Brullé, 1832

subtribe Ammobiina Desbrochers des Loges, 1902

genus *Adavius* Mulsant & Rey, 1859e: 138 [= 1860: 46] type species *Adavius clavipes* Mulsant & Rey, 1859

Udebra Reitter, 1896i: 236 type species *Udebra hauseri* Reitter, 1896 (= *Erodius fimbratus* Ménétriés, 1849)

clavipes Mulsant & Rey, 1859e: 138 [= 1860: 46] **A:** NP XIZ **ORR**

fimbriatus Ménétriés, 1849: 217 (*Erodius*) **E:** ST **A:** AF IN KZ TM UZ

hauseri Reitter, 1896i: 236 (*Udebra*)

genus *Ammobius* Guérin-Méneville, 1844: 121 type species *Trachyscelis rufus* Guérin-Méneville, 1844

Ammophtorus Lacordaire, 1859: 284 type species *Trachyscelis rufus* Guérin-Méneville, 1844

asperatus Champion, 1894b: 448 (*Ammophthorus*) **A:** GUA (Nam'ao Is.)

cyprius Grimm, 1991: 29 **A:** CY TR
rufus rufus Guérin-Méneville, 1844: 121 (*Trachyscelis*) **E:** AL BU CR FR GG GR IT MA MC RO SP ST UK **N:** AG LB MO TU **A:** CY IS LE SI SY TR
insularis Reitter, 1893e: 315
jakovlevi Semenov, 1905: 123
minotaurus Obenberger, 1922: 41
rugosus Rosenhauer, 1856: 214
tarsalis Grimm, 2012: 37 **A:** AE

genus *Amphithrixoides* Bouchard & Löbl, 2008: 39 type species *Amphithrix peyerimhoffi* Español, 1953
Amphithrix Español, 1953c: 306 [HN] type species *Amphithrix peyerimhoffi* Español, 1953
peyerimhoffi Español, 1953c: 309 **N:** AG

genus *Asiocaedius* G.S. Medvedev & Nepesova, 1985a: 138 [RN] type species *Pseudocaedius kiseritzkii* Medvedev, 1966
Pseudocaedius G.S. Medvedev, 1966: 98 [HN] type species *Pseudocaedius kiseritzkii* Medvedev, 1966
kiseritzkii G.S. Medvedev, 1966: 98 (*Pseudocaedius*) **A:** TM
kuznetzovi G.S. Medvedev, 1970: 400 (*Pseudocaedius*) **A:** TM UZ

genus *Brachyidium* Fairmaire, 1883b: 33 type species *Brachyidium breviusculum* Fairmaire, 1883
Cnemodasus Gebien, 1914d: 374 type species *Cnemodasus rectangulus* Gebien, 1914
edentatum Ando, 2017: 300 **A:** JP
iwojimanum Masumoto, 1991b: 27 **A:** JA (Iwojima Is., Volcano Is.)
palauense Kulzer, 1957: 196 (*Caedius*) **A:** JA (Volcano Is.) **AUR**

genus *Caediexis* Lebedev, 1932: 125 type species *Caediexis arenicola* Lebedev, 1932
arenicola Lebedev, 1932: 125 **A:** TD TM UZ

genus *Caedius* Blanchard, 1845: 13 type species *Opatrum sphaeroides* Hope, 1841
Coedius Blanchard, 1845: 13 [unused original spelling]
Isarida Pascoe, 1866: 456 type species *Isarida testacea* Pascoe, 1866 (= *Caedius fulvus* Mulsant & Rey, 1859)
cassidoides Fairmaire, 1894f: 661 **N:** EG
chinensis Kaszab, 1942a: 34 **A:** FUJ **ORR**
ciliger Mulsant & Rey, 1859e: 135 [= 1860: 43] **A:** XIZ
eberti Kaszab, 1980a: 145 **A:** AF **ORR**
fluviatilis Nakane & Masumoto, 1979: 112 **A:** FE JA
formosanus Kaszab, 1942a: 39 **A:** TAI
fulvus Mulsant & Rey, 1859e: 133 [= 1860: 41] **A:** IN NP PA UP XIZ
asper Fairmaire, 1896a: 18
testaceus Pascoe, 1866: 456
himalayensis Kaszab, 1941d: 119 **A:** BT HP NP

hiraii Akita & Masumoto, 2012b: 211 **A:** JA (Ryukyus)
indicus Fairmaire, 1896a: 18 **A:** NP **ORR**
maderi Kaszab, 1942a: 35 **A:** JA "Korea" **ORR**
minor Nakane, 1963a: 26 [HN]
nakanei J. Ferrer, 2003: 303 [RN]
marinus Marseul, 1876c: 96 **A:** JA NC
obesus Kulzer, 1964: 266 **A:** "S. China" **AUR ORR**
orientalis Fairmaire, 1893g: 316 (*Cyptus*) **A:** TAI **ORR**
shoitii Mt. Chûjô, 1992: 34 **A:** SC
yangi Ren, 1995a: 24 **A:** HEB

genus *Clitobius* Mulsant & Rey, 1859e: 141 [= 1860: 49] type species *Clitobius sabulicola* Mulsant & Rey, 1859 (= *Opatrum ovatum* Erichson, 1843)
Apithesis Waterhouse, 1881: 476 type species *Apithesis obesa* Waterhouse, 1881
Apteroclitobius Koch, 1960: 391 type species *Clitobius grandis* Fairmaire, 1896
Halonomus Wollaston, 1861a: 201 type species *Halonomus grayii* Wollaston, 1861 (= *Opatrum ovatum* Erichson, 1843)
cribricollis Allard, 1882b: lxxxvi (*Halonomus*) **A:** YE **AFR**
obesa Waterhouse, 1881: 477 (*Apithesis*) **A:** YE (Suqutra)
oblongiusculus Fairmaire, 1875b: 495 (*Halonomus*) **E:** AB TR **N:** AG EG LB MO TU **A:** CY IN IQ IS JO QA SA SI SY TD TM UZ **AFR**
borkouensis Pierre, 1961
lineicollis Fairmaire, 1879a: 12 (*Halonomus*)
schneideri Allard, 1882b: lxxxvi (*Halonomus*)
omanicus Purchart & Kamiński, 2017: 149 **A:** OM
ovatus Erichson, 1843: 249 (*Opatrum*) **E:** IT MA **N:** AG CI EG LB MO TU **A:** TR **AFR ORR**
grayii Wollaston, 1861a: 203 (*Halonomus*)
immarginatus Fairmaire, 1888c: 195
opacus Har. Lindberg, 1950: 12
sabulicola Mulsant & Rey, 1859e: 141 [= 1860: 49]
salinicola Wollaston, 1861a: 203 (*Halonomus*)
subplumbea Fairmaire, 1856: 533 (*Platydema*)

genus *Cyptus* Gerstaecker, 1871: 61 type species *Cyptus scabrosus* Gerstaecker, 1871
Neocaedius Pierre, 1972: 960 type species *Caedius halli* Kaszab, 1949
aegyptiacus Mulsant & Rey, 1859e: 136 [= 1860: 44] (*Caedius*) **N:** EG **A:** AF IS **AFR**
alternatus Fairmaire, 1894d: 321 (*Coedius*) **N:** AG TU **A:** SA YE **AFR**
auriculatus Chobaut, 1924b: 154 (*Caedius*) **N:** AG MO (south) **AFR**

genus *Dilamus* Jacquelin du Val, 1861: 279 type species *Boros rufipes* Lucas, 1846
subgenus *Choresmolamus* G.S. Medvedev, 1978a: 150 type species *Dilamus oxianus* G.S. Medvedev, 1978

oxianus G.S. Medvedev, 1978a: 150 **A:** UZ
subgenus *Dilamus* Jacquelin du Val, 1861: 279 type species *Boros rufipes* Lucas, 1846
arabicus Kaszab, 1979c: 286 **A:** SA
aralensis G.S. Medvedev, 1978a: 149 **A:** UZ (Karapalkistan)
caspicus G.S. Medvedev, 1978a: 147 **A:** TM
gnom Skopin, 1961b: 405 **A:** KZ MG
grandiceps Bogatchev & Kryzhanovskiy, 1960: 269 **A:** TM UZ
gurjevae G.S. Medvedev, 1972b: 491 **A:** MG
iranicus G.S. Medvedev, 1968a: 259 **A:** IN
laticollis Baudi di Selve, 1876c: 241 **A:** "Palestina"
mandli Kaszab, 1963a: 75 **A:** IN
mongolicus Kaszab, 1968c: 380 **A:** MG
myrmecophilus Skopin, 1966b: 331 **A:** KZ
obsoletus obsoletus Baudi di Selve, 1876c: 240 **A:** IS
obsoletus transjordanicus Koch, 1936: 300 **A:** JO
pardilloi Español, 1956b: 277 **N:** MO (Western Sahara)
planicollis laticeps Peyerimhoff, 1931b: 103 **N:** AG (Hoggar)
planicollis planicollis Fairmaire, 1883h: xliv **E:** IT **N:** AG EG LB MO TU
rufipes Lucas, 1846: pl. 30 (*Boros*) **E:** IT SP **N:** AG TU
  andalusiae Baudi di Selve, 1876c: 240
  congener Rottenberg, 1871: 250
rungsi Antoine, 1955: 363 **N:** MO
scoparius Antoine, 1952: 114 **N:** MO (Anti Atlas)
tangerianus Desbrochers des Loges, 1881: 132 **N:** MO
volkonskyi battareli Peyerimhoff, 1943: 24 **N:** AG
volkonskyi volkonskyi Peyerimhoff, 1943: 24 **N:** AG
zarudnyi G.S. Medvedev, 1968a: 262 **A:** IN
subgenus *Ochrolamus* Reitter, 1904a: 73 type species *Dilamus pictus* Baudi di Selve, 1881
pictus Baudi di Selve, 1881: 285 **N:** EG **AFR**

genus *Falsammidium* Koch, 1960: 395 type species *Clitobius laevipennis* Fairmaire, 1892
medvedevi Schawaller, 2009b: 201 **A:** OM
rugulipennis Fairmaire, 1896c: 224 (*Clitobius*) **A:** SA YE **AFR**

genus *Falsocaedius* Español, 1943c: 140 type species *Clitobius fossulatus* Escalera, 1914
fossulatus Escalera, 1914: 340 (*Clitobius*) **N:** CI MO

genus *Freyula* Koch, 1959: 593 type species *Freyula psammarina* Koch, 1959
psammarina psammarina Koch, 1959: 593 **A:** YE (Suqutra) **AFR**

genus *Mateuina* Español, 1944: 22 type species *Mateuina kaszabi* Español, 1944
kaszabi Español, 1944: 24 **N:** MO (Western Sahara)

genus *Moragacinella* Español, 1954d: 162 [RN] type species *Moralesia longepilosa* Kaszab, 1944
Moralesia Kaszab, 1944a: 18 [HN] type species *Moralesia longepilosa* Kaszab 1944
lindbergi G.S. Medvedev, 2004b: 574 **N:** MO (south)
longepilosa Kaszab, 1944a: 19 (*Moralesia*) **N:** MO (south, Western Sahara)

genus *Nesocaedius* Kolbe, 1915: 262 type species *Nesocaedius schultzei* Kolbe, 1915
minimus M. Chûjô, 1966: 9 (*Caedius*) **A:** JA (Kyushu, Ryukyus) TAI **AUR** (Palau)
pardoi Español, 1956b: 280 **N:** MO (Western Sahara)
taiwanus Shibata, 1979: 72 **A:** TAI
vermiculus Shibata, 1979: 71 **A:** TAI

genus *Perithrix* Fairmaire, 1879c: 193 type species *Perithrix granidorsum* Fairmaire, 1879
granidorsis Fairmaire, 1879c: 193 **N:** AG TU
tunisiensis Dajoz, 1983: 65 **N:** AG TU

genus *Prodilamus* Ardoin, 1969: 258 type species *Dilamus brevicollis* Fairmaire, 1894
boehmi Reitter, 1904a: 71 (*Dilamus*) **N:** EG **AFR**
fausti fausti Reitter, 1890d: 150 (*Dilamus*) **E:** AR **A:** CY IN IQ TM
fausti major Kaszab, 1982b: 162 **A:** SA
ferrantei Reitter, 1908c: 49 (*Dilamus*) **N:** EG **AFR**

genus *Proscheimus* Desbrochers des Loges, 1881: 127 type species *Proscheimus arabicus* Desbrochers des Loges, 1881
arabicus Desbrochers des Loges, 1881: 128 **A:** IS JO SA SI **N:** EG
fulvipes Ménétriés, 1849: 240 (*Microzoum*) **A:** TM

genus *Psammestus* Reichardt, 1936a: 194 type species *Ammobius dilatatus* Reitter, 1893
dilatatus Reitter, 1893e: 315 (*Ammobius*) **A:** KZ MG NMO TM UZ XIN
caspicus Semenov, 1905: 124 (*Ammobius*)
kozlovi Semenov, 1905: 124 (*Ammobius*)
panfilovi G.S. Medvedev, 1970: 398 **A:** TM
pseudocaedius Reitter, 1895d: 156 (*Ammobius*) **A:** AF TM

genus *Trigonopoda* Gebien, 1914b: 2 type species *Trigonopoda crassipes* Gebien, 1914
carmanicus Reichardt, 1936a: 199 (*Caedius*) **A:** IN
crassipes Gebien, 1914b: 3 **A:** TAI
ovalipennis Shibata, 1979: 68 **A:** TAI
piffli Kaszab, 1961f: 348 **A:** AF PA

genus *Weisea* Semenov, 1890: 370 type species *Weisea sabulicola* Semenov, 1890
sabulicola Semenov, 1890: 371 **A:** TM (Karakumy)

subtribe Heterotarsina Blanchard, 1845

genus *Diphyrrhynchus* Fairmaire, 1849: 445 type species *Diphyrrhynchus chalceus* Fairmaire, 1849

Abantiades Fairmaire, 1894e: 395 [RN] [HN] type species *Abantis aenescens* Fairmaire, 1892

Abantis Fairmaire, 1892a: 109 [HN] type species *Abantis aenescens* Fairmaire, 1892

Acanthosternus Montrouzier, 1860: 289 type species *Acanthosternus halorageos* Montrouzier, 1860

Neoabantis Gebien, 1910a: 341 [RN] type species *Abantis aenescens* Fairmaire, 1892

aenescens Fairmaire, 1892a: 110 (*Abantis*) **A:** SA SI YE **AFR**

dilatipennis Schawaller, 1991b: 290 **A:** IN OM YE (Suqutra)

iriomotensis M. Chûjô, 1966: 4 (*Diphyrrhinchus*) **A:** JA TAI **ORR**

oharensis botelensis Nakane, 1973b: 105 **A:** TAI

oharensis ohariensis Nakane, 1973b: 104 **A:** JA (Ryukyus) **ORR**

shibatai Kaszab, 1964c: 43 (*Diphyrrhinchus*) **A:** JA (Amami, Asani) TAI **ORR**

genus *Heterotarsus* Latreille, 1829: 26 type species *Heterotarsus tenebrioides* Guérin-Méneville, 1831

Helopimorphus Desbrochers des Loges, 1881: 141 type species *Helopimorphus angulipennis* Desbrochers des Loges, 1881

Hopatropteron Reitter, 1889k: 701 type species *Hopatropteron subcostatum* Reitter, 1889 (= *Heterotarsus carinula* Marseul, 1876)

Oubanghinum Pic, 1933d: 4 type species *Oubanghinum atrum* Pic, 1933

aphodioides Fairmaire, 1886c: 346 **A:** CH

auricularis Marseul, 1876c: 128 **A:** CH TAI **ORR**

subcrenatus Fairmaire, 1886c: 345

carinula Marseul, 1876c: 127 **A:** ANH FE FUJ GAN GUI HAI HUB JA JIA NC SC SCH TAI ZHE **ORR**

subcostatus Reitter, 1889k: 701 (*Hopatropteron*)

crenulifer Kaszab, 1941c: 58 **A:** JA (Ryukyus) TAI

indicus indicus Marseul, 1876c: 128 **A:** NP SD YUN **ORR**

inflatus Lacordaire, 1859: 399 **A:** NP YUN **ORR**

subinflatus Pic, 1923j: 75

metallifer Kaszab, 1976a: 50 **A:** YUN

pustulifer Fairmaire, 1889a: 361 **A:** ANH FUJ GAN JIA **ORR**

takahashii Akita & Masumoto, 2012b: 207 **A:** JA (Ryukyus)

topali Kaszab, 1976a: 48 **A:** NP **ORR**

urbani Kaszab, 1941c: 59 **A:** TAI

yemeniticus Schawaller, 2007: 180 **A:** YE

subtribe Heterotarsina, species incertae sedis

Diphyrrhynchus geminatus Allard, 1896: 320 **A:** SD

subtribe Neopachypterina Bouchard, Löbl & Merkl, 2007

genus *Neopachypterus* Bouchard, Löbl & Merkl, 2007: 386 [RN] type species *Pachypterus mauritanicus* Lucas, 1847
Pachypterus Lucas, 1847: pl. 29 [HN] type species *Pachypterus mauritanicus* Lucas, 1847
gebieni Kaszab, 1941b: 18 (*Pachypterus*) **A:** SD **ORR**
mauritanicus Lucas, 1847: pl. 29 (*Pachypterus*) **E:** GR SP **N:** AG MO TU
niloticus niloticus Miller, 1861: 205 (*Pachypterus*) **N:** EG
serrulatus Reitter, 1904a: 75 (*Pachypterus*) **E:** AB AR **A:** AF KZ TD TM UZ
zarudnyi G.S. Medvedev, 1968a: 251 (*Pachypterus*) **A:** IN

genus *Pseudolamus* Fairmaire, 1874a: 388 type species *Pseudolamus seriatoporus* Fairmaire, 1874
schatzmayri Koch, 1937: 459 **N:** LB
seriatoporus Fairmaire, 1874a: 389 **E:** SP **N:** MO
pusillus Baudi di Selve, 1876b: 219 (*Pachypterus*)

subtribe Opatrina Brullé, 1832

genus *Anatrum* Reichardt, 1936a: 84 type species *Anatrum songoricum* Reichardt, 1936
shandanicum Ren, 1999: 226 **A:** GAN
songoricum Reichardt, 1936a: 85 **A:** MG XIN

genus *Brachyesthes* Fairmaire, 1868: 490 type species *Melanesthes pilosella* Marseul, 1867
brevior Fairmaire, 1883h: cxiii [RN] **N:** AG LB TU
chrysomelina A. Costa, 1878: 23 (*Caedius*) **N:** EG **A:** TR **ORR**
indica Kaszab, 1938d: 230
gastonis gastonis Fairmaire, 1873b: 392 **N:** AG EG LB
gastonis kebiliensis Picka, 1988: 111 (*Brachyestes*) **N:** TU
olexai Picka, 1988: 111 **N:** AG
peyerimhoffi Koch, 1937: 473 **N:** LB
pilosella pilosella Marseul, 1867: xxxvi (*Melanesthes*) **N:** AG MO TU
approximans Fairmaire, 1873b: 391
indentata Pic, 1929d: 6
pilosella salina Koch, 1937: 473 **N:** LB

genus *Eumylada* Reitter, 1904a: 170 type species *Myladina punctifera* Reitter, 1889
glandilosa X.-J. Yang & Ren, 2004: 305 **A:** NMO
obenbergeri Schuster, 1933b: 97 (*Myladina*) **A:** NMO
ordosana Reitter, 1889k: 708 (*Myladina*) **A:** NMO
potanini Reitter, 1889k: 708 (*Myladina*) **A:** GAN NIN NMO
punctifera amaroides Reichardt, 1936a: 171 **A:** MG
punctifera punctifera Reitter, 1889k: 707 (*Myladina*) **A:** GAN MG NMO

genus *Falsolobodera* Kaszab, 1967a: 24 type species *Falsolobodera skopini* Kaszab, 1967
skopini Kaszab, 1967a: 25 **A:** MG

genus *Gonocephalum* Solier, 1834: 498 type species *Opatrum fuscum* Herbst, 1793 (= *Opatrum rusticum* G.-A. Olivier, 1811)
subgenus *Gonocephalum* Solier, 1834: 498 type species *Opatrum fuscum* Herbst, 1793 (= *Opatrum rusticum* G.-A. Olivier, 1811)
Dasus Motschulsky, 1845a: 78 type species *Opatrum fuscum* Herbst, 1793 (= *Opatrum rusticum* G.-A. Olivier, 1811)
Hasticollinum Kaszab, 1939: 96 type species *Hasticollinum podagrarium* Kaszab, 1939
Megadasus Reitter, 1904a: 146 type species *Opatrum lefranci* Fairmaire, 1863
abnormale Kaszab, 1952d: 634 **A:** UP
acoriaceum M. Chûjô, 1975: 15 **A:** JA (Ryukyus)
adpressiforme Kaszab, 1951c: 182 **A:** GUA JA (Volcano Is.) TAI **AUR ORR**
aequatoriale Blanchard, 1853: 152 (*Opatrum*) **A:** HAI UP **ORR**
alaticolle Fairmaire, 1893b: 23 (*Bradymerus*) **A:** SD **ORR**
andrewesi Kaszab, 1952d: 649 **A:** TAI **AUR ORR**
annamita Chatanay, 1917a: 236 **A:** ANH AP FUJ GUA HAI HEN HKG JA NMO UP YUN **AUR ORR**
pubens Gravely, 1915: 520 [HN]
arisi Reitter, 1905a: 92 **A:** AF IN TM
assimile Küster, 1849a: 52 (*Opatrum*) **E:** IT FR
belli Kaszab, 1952d: 513 **A:** NP **ORR**
besnardi besnardi Kaszab, 1982b: 163 **A:** AE AF OM SA
besnardi muehlei J. Ferrer, 1995: 58 [= 1993a: 91] **A:** SA YE
besnardi pierrei J. Ferrer, 2004: 87 **A:** IN
bilineatum Walker, 1858: 284 (*Opatrum*) **A:** BT FE FUJ GUA GUX HAI HKG HUN JA NP SC SCH SD UP YUN ZHE **AUR NAR ORR**
kamtschaticum Motschulsky, 1860c: 139
latifrons LeConte, 1874a: 70 (*Blapstinus*)
orarium Lewis, 1894: 380 (*Opatrum*)
birmanicum Kaszab, 1952d: 484 **A:** NP TAI UP **ORR**
biseriatum Kaszab, 1975b: 121 **A:** NP
brachelytra Kaszab, 1952d: 632 **A:** AF PA UP **ORR**
calcaripes Karsch, 1881: 59 (*Opatrum*) **E:** FR PT SP **N:** AG MO **AFR**
yelamosi Español & Viñolas, 1983a: 37
catenulatum Fairmaire, 1896a: 19 (*Hopatrum*) **A:** SD **ORR**
chinense Gebien, 1910a: 322 [RN] **A:** HAI YUN **ORR**
subsetosum Reitter, 1904a: 147 [HN]
civicum Kaszab, 1952d: 612 **A:** NP SD UP **ORR**

coarcticolle Kaszab, 1952d: 500 **A:** NP 'Indus" **ORR**
coenosum Kaszab, 1952d: 643 **A:** JA FUJ GUA HKG HUB JIA SC SCH TAI XIN ZHE
japanum Marseul, 1876: 96 [HN]
depressum Miwa, 1931: 152 HN]
recticolle Kaszab, 1941: 51 HN]
consobrinum Blair, 1923a: 122 **A:** AF IN IQ PA QA SA UP YE **AFR ORR**
controversum Gridelli, 1948: 11 [RN] **N:** EG **A:** BA JO SA SY YE **AFR**
coriaceum Motschulsky, 1858a: 34 **A:** FE FUJ GAN GUA GUI GUX HEN JA NMO NP SC SCH SHN SHX TAI ZHE
costatum costatum Brullé, 1832: 218 (*Opatrum*) **E:** AB AR CR GR IT SP TR **A:** CY IN IS SY TR
lineare Küster, 1849a: 54 (*Opatrum*)
costatum rugulosum Küster, 1849a: 55 (*Opatrum*) **E:** AR GR (Lesvos) **A:** CY IN IS SY TR
crassepunctatum Kaszab, 1952d: 526 **A:** BT NP **ORR**
curiosum Kaszab, 1952d: 460 **A:** NP **ORR**
curvicolle Reitter, 1889k: 705 **A:** GAN HP KA MG NMO SCH XIN
deliense Kaszab, 1952d: 505 **A:** GUX **ORR**
dentipes Kaszab, 1952d: 655 **A:** TAI **ORR**
depressum Fabricius, 1801a: 116 (*Opatrum*) **A:** AF AP BT GUA GUX HAI HP KA NP PA SD TAI UP YUN **AUR ORR**
contrahens Walker, 1858: 284 (*Opatrum*)
cinereum Thunberg, 1821: 54 (*Opatrum*)
dilatatum Wollaston, 1854: 501 (*Opatrum*) **N:** SG
dorsogranosum Fairmaire, 1896a: 19 (*Hopatrum*) **A:** AF PA SD UP **ORR**
foveolatus Kaszab, 1963b: 343
endroedii Kaszab, 1952d: 480 **A:** NP **ORR**
ermischi Kaszab, 1960a: 139 **A:** AF PA
espagnoli Kaszab, 1965a: 111 **A:** AP
ficifolium Mt. Chûjô, 1995a: 113 **A:** JA
formosanum Gebien, 1914b: 4 **A:** TAI
freudei Kaszab, 1960a: 143 **A:** AF IN PA
gallagheri J. Ferrer, 1997b: 214 **A:** OM
gebienianum Kaszab, 1952d: 470 **A:** SD **ORR**
geneirotundum Ren, 1998: 111 **A:** ZHE
gracile Bates, 1879: 482 (*Penthicus*) **A:** HP KA NP SD XIN XIZ
curvicolle Gridelli, 1934: 66 [HN]
ruficorne Gravely, 1915: 250
granulatipenne Kaszab, 1978a: 173 **A:** BT SD
granulatum afghanicum Gridelli, 1954: 283 [= 1955: 72] **A:** AF TM TR
granulatum granulatum Fabricius, 1792a: 90 (*Opatrum*) **N:** MO
granulatum meridionale Küster, 1849a: 57 (*Opatrum*) **E:** FR IT SP **N:** AG TU **A:** SY

granulatum nigrum Küster, 1849a: 58 (*Opatrum*) **E:** BU FR GR IT SP TU **A:** TU
granulatum pusillum Fabricius, 1792a: 91 (*Opatrum*) **E:** AB AL AR AU BU CR CZ HU GR IT MC RO SK SP ST UK YU **N:** AG MO **A:** AF IN KZ MG NIN SY TM TR XIZ
costatum Thunberg, 1821: 51 (*Opatrum*)
interstitiale Küster, 1849a: 59 (*Opatrum*)
muricatum Brullé, 1832: 215 (*Opatrum*)
nitidulum Küster, 1849a: 56 (*Opatrum*)
parvulum Lucas, 1846: 334 (*Opatrum*)
pruinosum Fischer von Waldheim, 1844: 128 (*Opatrum*)
ragusanum Küster, 1849c: 52 (*Opatrum*)
subnodosum Rey, 1892: 42
viennense Duftschmid, 1812: 293
griseovittatum griseovittatum Gridelli, 1940 [1939]: 241 **A:** OM SA **AFR**
guerryi Chatanay, 1917a: 244 **A:** AF HKG HUB PA SCH SD YUN **ORR**
potanini Reichardt, 1936a: 108
guignoti Kaszab, 1960a: 144 **A:** AF PA
gustafssoni J. Ferrer, 1995: 73 **A:** UP
hadroide Fairmaire, 1888b: 129 (*Hopatrum*) **A:** BEI SCH
reichardti Gebien, 1939: 447 [RN]
hauschildi Kaszab, 1952d: 498 **A:** GUX NP SD UP "Thibet" **ORR**
helopioide Fairmaire, 1894a: 19 (*Hopatrum*) **A:** NP SD UP **ORR**
himalayense Kaszab, 1952d: 528 **A:** BT NP SD **ORR**
hingstoni Kaszab, 1952d: 524 **A:** SD **ORR**
hintoni Kaszab, 1952d: 668 **A:** JA XIZ
hiranoi Akita & Masumoto, 2003: 173 **A:** JA (Honshu)
hispidulum Kaszab, 1952d: 657 **A:** FUG HKG XIZ ZHE
hoffmannseggii Steven, 1828: 96 (*Opatrum*) **A:** BT NP SD **ORR**
japanum Motschulsky, 1861: 16 **A:** JA SC "East China"
expansicolle Lewis, 1894: 380 (*Opatrum*)
kandaharicum Kaszab, 1960a: 148 **A:** AF
karakorumense Kaszab, 1961f: 353 **A:** KA
klapperichi Kaszab, 1952d: 550 **A:** FUJ JA SHA
kochi Kaszab, 1952d: 578 **A:** FUJ GUI
kondoi Masumoto, 1983b: 173 **A:** JA (Ryukyus) TAI
konoi Kaszab, 1952d: 482 **A:** FUJ TAI
koreanum Kaszab, 1952d: 548 **A:** SC
kuenelti Kaszab, 1961f: 351 **A:** KA
kuluanum Kaszab, 1952d: 659 **A:** AF HP NP PA
kuntzeni Kaszab, 1952d: 580 **A:** HAI
labriquei J. Ferrer, 2010a: 234 **A:** YUN
lefranci Fairmaire, 1863: 643 (*Opatrum*) **E:** IT (Sardinia) **N:** AG TU

longitarse Kaszab, 1952d: 465 **A:** YUN **ORR**
macrophthalmum Kaszab, 1952d: 582 **A:** JA (Shikoku) **ORR**
malayanum Gebien, 1935: 53 **A:** HAI HEN SC YUN **ORR**
merensi estevezi J. Ferrer, 2000b: 71 [=1993a: 97] **N:** CI (Gran Canaria)
merensi merensi Uyttenboogaart, 1929: 348 **N:** CI (Gran Canaria)
mesopotamicum Blair, 1923a: 123 **A:** IQ
minusculum Fairmaire, 1894a: 20 (*Hopatrum*) **A:** UP **ORR**
bidentulus Fairmaire, 1896: 23 (*Bradymerus*)
miyakense Nakane, 1963a: 26 **A:** JA (Izu Isls.) ZHE
modestum Reiche, 1850: 368 (*Opatrum*) **A:** IN SA YE **AFR**
sulcipenne J. Thomson, 1858: 84 (*Opatrum*)
moluccanum Blanchard, 1853: 153 (*Opatrum*) **A:** FUJ GUA HAI JA TAI **AUR ORR**
coriaceum Miwa, 1931: 152
muelleri muelleri Gridelli, 1945: 23 **A:** SA YE **AFR**
nepalicum Kaszab, 1973a: 32 **A:** BT NP SD
niloticum Ardoin, 1965a: 73 **A:** SA **AFR**
oblitum Wollaston, 1864: 489 (*Opatrum*) **N:** CI
oblongum Fabricius, 1801a: 117 (*Opatrum*) **A:** NP SD **ORR**
obscurum aethiops J. Ferrer, 1995: 13 [=1993a: 78] **A:** OM SA **AFR**
obscurum obscurum Küster, 1849a: 46 (*Opatrum*) **E:** FR (Alpes-Maritimes, Corse) IT SP **N:** AG EG LB MO **AFR**
parvulum Lucas, 1846: 334 (*Opatrum*)
okinawanum M. Chûjô, 1963b: 149 **A:** JA (Ryukyus.)
outreyi Chatanay, 1917a: 240 **A:** GUA HKG SC TAI YUN **ORR**
papulosum Fairmaire, 1896a: 19 (*Hopatrum*) **A:** PA **ORR**
parcesetosum Kaszab, 1952d: 516 **A:** NP "North India" **ORR**
patricium Kaszab, 1952d: 604 **A:** PA SD UP **ORR**
patruele patruele Erichson, 1843: 248 (*Opatrum*) **N:** AG CI EG LB MO TU **A:** AE AF IN IQ IS OM SA SI SY YE **AFR**
lutosum Wollaston, 1864: 486 (*Opatrum*)
patruele turchestanicum Gridelli, 1948: 9 **N:** EG **A:** AF KI KZ TD TM UZ XIN
pengi Masumoto, Akita & Lee, 2013a: 113 **A:** TAI
perplexum Lucas, 1846: pl. 29 (*Opatrum*) [NP] **E:** CR GR IT (Lampedusa Is., Pantelleria Is.) SP **N:** AG EG LB MO TU **A:** CY IS SA SI SY YE **AFR**
famelicum G.-A. Olivier, 1811: 497 (*Opatrum*) [NO]
persimile Lewis, 1894: 381 (*Opatrum*) **A:** FE HEI JA MG NC NMO SC
platipenne Kaszab, 1960a: 149 **A:** AF
pottsi Kulzer, 1957: 194 **A:** JA (Bonin Is.) **AUR**
prolixum Erichson, 1843: 248 (*Opatrum*) **E:** IT SP **N:** AG CI EG LB MO TU **A:** CY IN SA SY YE **AFR**
clavipes Wollaston, 1867: 188 (*Opatrum*)

hespericum Motschulsky, 1849: 126 (*Dasus*)
inornatum Schuster, 1938a: 61
lugens Küster, 1849a: 50 (*Opatrum*)
pseudopubens Kaszab, 1952d: 592 **A:** HAI HKG TAI ZHE **ORR**
pubens Marseul, 1876c: 97 (*Opatrum*) **A:** AP HAI JA NE SC SHN TAI
obtusicolle Kolbe, 1886: 95 (*Opatrum*)
pubiferum Reitter, 1904a: 142 **E:** AB AR ST **A:** AE AF IN IQ KZ MG OM PA SA TD TM UZ XIN
pygmaeum Steven, 1828: 98 (*Opatrum*) **E:** AU BU CR FR GE GG HU IT SK SL ST SZ UK **A:** KZ TM TR XIN
pedestre Rosenhauer, 1847: 31 (*Opatrum*)
pesthiense Besser, 1832: 16 (*Opatrum*)
pusillum Küster, 1849a: 60 [HN] (*Opatrum*)
sturmii Küster, 1849a: 62 (*Opatrum*)
viennense Küster, 1849a: 61 [HN] (*Opatrum*)
quadrinodosum Reitter, 1904a: 146 **A:** FUJ
recticolle Motschulsky, 1866: 173 **A:** FE GAN HEB HEI JA JIA JIL LIA MG NIN NMO QIN SC SHN SHX TAI
reticulatum Motschulsky, 1854: 47 **A:** BEI ES GAN HEB HEI HEN JIA JIL MG NC NIN NMO QIN SHA SHN SHX TIA
mongolicum Reitter, 1889k: 706
rileyi Kaszab, 1952d: 473 **A:** NP NP **ORR**
roseni Kaszab, 1952d: 546 **A:** AP BT NP SD UP **ORR**
rusticum G.-A. Olivier, 1811: 498 (*Opatrum*) [NP] **E:** AB AL AR AZ CR FR GR IT MA MC SP ST UK **N:** CI EG LB MO MR TU **A:** AF CY ES HEN IN IQ IS JO KZ LIA MG NMO OM SA SHN SI TD TM TR UZ XIN
errans Wollaston, 1854: 501 (*Opatrum*)
famelicum Küster, 1849a: 48 (*Opatrum*)
fuscum Herbst, 1793: 225 (*Opatrum*) [NO]
modestum Küster, 1849a: 53 (*Opatrum*) [HN]
obductum Gebler, 1860: 38 (*Opatrum*)
patruele Küster, 1849a: 51 (*Opatrum*) [HN]
pulverulentum G.-A. Olivier, 1811: 500 (*Opatrum*)
rusticum Brullé, 1832: 218 (*Opatrum*) [HN]
setulosum Küster, 1849a: 47 (*Opatrum*) [HN]
substriatum Besser, 1832: 18 (*Opatrum*)
terrosum Küster, 1845a: 28 (*Opatrum*)
trichopterum Gemminger, 1870a: 122 (*Hopatrum*) [RN]
tricostatum Fischer von Waldheim, 1844: 128 (*Bolitophagus*)
vestitum Küster, 1849a: 45 (*Opatrum*)
vicinum Gemminger, 1870a: 122 (*Hopatrum*)
villosum Steven, 1828: 97 (*Opatrum*)

sauteri Kaszab, 1952d: 529 **A:** TAI
sawadai Masumoto, 1985b: 28 **A:** JA (Honshu)
schneideri Reitter, 1898a: 14 **E:** AB **A:** KZ TM UZ
semipatruele Kaszab, 1952d: 608 **A:** IN PA UP YE **ORR**
senkakuense M. Chûjô, 1973: 13 **A:** JA (Ryukyus) TAI
sericeum Baudi di Selve, 1875d: 701 [= 1876b: 228] (*Opatrum*) **N:** EG LB **A:** IS SA **AFR NAR**
setulosum angustum Har. Lindberg, 1950: 11 **E:** IT (Sicilia) **N:** AG CI EG MO TU
lindbergi J. Ferrer, 1993a: 75 [RN]
setulosum kashgarense Bates, 1879: 481 (*Opatrum*) **A:** IN IQ KA PA UZ XIZ
setulosum setulosum Faldermann, 1837: 60 (*Opatrum*) **E:** AB AR GG GR IT MA SP ST **N:** CI EG LB MO TU **A:** AE AF CY IN IQ IS KZ LE SA SI SY TD TM TR UZ XIN YE **AFR**
demaisonis Allard, 1883: 32
minutum Ménétriés, 1849: 239 (*Opatrum*)
parallelum Normand, 1936b: 363
sexuale Marseul, 1876c: 98 (*Opatrum*) **A:** BEI FE JA NC SC SCH
sikkimense Kaszab, 1952d: 517 **A:** NP SD
simplex yemenicum J. Ferrer, 1993a: 103 [=2000b: 91] **A:** YE
simulatrix Fairmaire, 1891d: c (*Pseudoblaps*) **A:** AF GAN HP KA NP PA UP **ORR**
infimum Fairmaire, 1896a: 57 (*Scleropatrum*)
martensi Kaszab, 1977b: 258
sinicum Reichardt, 1936a: 109 **A:** BT GAN SCH **ORR**
soederbomi J. Ferrer, 2010a: 232 **A:** MG
somalicum Gridelli, 1945: 19 **A:** YE (Suqutra) **AFR**
soricinum insidiosum Fairmaire, 1880a: 16 (*Opatrum*) **N:** AG MO
soricinum soricinum Reiche & Saulcy, 1857: 259 (*Opatrum*) **A:** AE AF IN IS JO PA SA SI SY YE **AFR**
spinicolle Fairmaire, 1896a: 22 (*Bradymerus*) **A:** HP **ORR**
stoeckleini Kaszab, 1952d: 662 **A:** KA PA UP **ORR**
strangulatum Fairmaire, 1888b: 128 (*Hopatrum*) **A:** BEI HAI HEI HUB JA YUN
strigosum Reiche, 1850: 369 **N:** AG EG TU **A:** IS **AFR**
subrugulosum Reitter, 1887d: 388 **A:** KI MG NMO XIN XIZ
subspinosum Fairmaire, 1894a: 19 (*Hopatrum*) **A:** AP BT FUJ GAN GUA GUI GUX HUB HUN JIA JIX NP SCH SHA SD TAI XIZ YUN **ORR**
takara Nakane, 1963a: 26 **A:** JA (Ryukyus)
tellinii Gridelli, 1945: 20 **A:** AE OM SA YE **AFR**
tenuicorne Kaszab, 1952d: 590 **A:** SD **ORR**
tenuipes Kaszab, 1952d: 587 **A:** NP **ORR**
terminale Reichardt, 1936a: 110 **A:** FE JA NC SC
tibetanum Kaszab, 1952d: 511 **A:** XIZ
titschacki Kaszab, 1952d: 509 **A:** FUJ JA YUN
tonkinense Kaszab, 1952d: 532 **A:** NP **ORR**

tschilianum Kaszab, 1952d: 630 **A:** FUJ JA SHX
tuberculatum Hope, 1831: 31 (*Opatrum*) **A:** AF HAI HKG HP HUB KA NP PA SCH SD UP TAI YUN **ORR**
elongatum Guérin-Méneville, 1834: 32 (*Opatrum*)
quadrinodosum Reitter, 1904: 146
vagum Steven, 1828: 96 (*Opatrum*) **A:** NP SD YUN **ORR**
arenarium Fairmaire, 1894: 17 [HN] (*Hopatrum*)
planatum Walker, 1858: 284 (*Opatrum*)
wittmeri Kaszab, 1975e: 320 **A:** BT NP
woynarovichi Kaszab, 1973a: 33 **A:** NP
wui Ren, 1995b: 234 **A:** ZHE
zoltani Iwan, J. Ferrer & Raś, 2010: 302 [RN] **A:** AF IQ KA PA **ORR**
parallelum Kaszab, 1952d: 606 [HN]
subgenus *Myladanesthes* Skopin, 1961a: 202 type species *Myladina fortidens* Reitter, 1915
fortidens Reitter, 1915e: 88 (*Myladina*) **A:** KZ (Almaatinskaya Oblast) XIN
kalidii Skopin, 1964: 281 **A:** KZ
karatalica Skopin, 1966b: 330 [= 1967: 209] **A:** KZ
kazenasi Skopin, 1961a: 203 **A:** KZ
subgenus *Opatropis* Reitter, 1904a: 159 type species *Opatrum hispidum* Brullé, 1839 (= *Opatrum affine* Billberg, 1815)
affine Billberg, 1815: 275 (*Opatrum*) **E:** GR SP **N:** CI EG MO MR **A:** IS YE **AFR**
hispidum Brullé, 1839: 68 (*Opatrum*)
rufipes Thunberg, 1821: 54 (*Opatrum*)
virgatum Erichson, 1843: 249 (*Opatrum*)
blairi Gebien, 1922a: 276 **A:** YE (Suqutra) **AFR**
oculare Kaszab, 1952d: 622 (*Opatropis*) **A:** NP PA UP **ORR**
pilosum J. Ferrer, 2000b: 124 **N:** EG

genus *Gonocephalum*, species incertae sedis
tomentosum Walker, 1871: 16 (*Hopatrum*) **A:** "Arabia"

genus *Gonocephalum*, nomen dubium
spoliatum Rey, 1892: 42 **A:** TR

genus *Jintaium* Ren, 1999: 228 type species *Jintaium sulcatum* Ren, 1999
sulcatum Ren, 1999: 228 **A:** GAN NIN

genus *Melanesthes* Dejean, 1834: 191 type species *Opatrum sibiricum* Faldermann, 1833
subgenus *Lesbidana* Reitter, 1904a: 173 type species *Melanesthes simplex* Reitter, 1897
coriaria Reitter, 1904a: 173 **A:** UZ
chinganica Reichardt, 1936a: 182 **A:** GAN MG NMO

kasachstanica Kaszab, 1962c: 315 **A**: KZ
simplex Reitter, 1897e: 220 **A**: KI XIN
subcoriaria Reitter, 1897e: 220 **A**: KI XIN
subgenus *Melanesthes* Dejean, 1834: 191 type species *Opatrum sibiricum* Faldermann, 1833
Hemitrichestes Reitter, 1904a: 169 type species *Penthicus hirsutus* Reitter, 1896
Miglica Reitter, 1904a: 171 type species *Pedinus laticollis* Gebler, 1829
ciliata Reitter, 1889k: 703 **A**: MG NIN NMO XIN
psammophila Kaszab, 1964g: 397
conica Kaszab, 1965c: 283 **A**: XIN
davadshamsi Kaszab, 1964g: 394 **A**: MG NIN
desertora Ren, 1993b: 487 [486] **A**: NIN
exilidentata Ren, 1993a: 30 **A**: NIN
faldermanni Mulsant & Rey, 1859b: 98 [= 1859e: 36] **A**: ES (Tuva) MG NMO XIN
laticollis Faldermann, 1835a: 167 [HN]
hirsuta balchashensis Skopin, 1966b: 328 [= 1967: 207] **A**: KZ
hirsuta hirsuta Reitter, 1896f: 170 (*Penthicus*) **A**: KZ TM
hirsuta seminuda Skopin, 1966b: 329 [= 1967: 207] **A**: KZ
iliensis Skopin, 1966b: 329 [= 1967: 208] **A**: KZ (Almaatinskaya Oblast)
jenseni jenseni Schuster, 1922d: 164 **A**: ES MG WS XIN
furva Kontkanen, 1956: 58
jenseni lacustris G.S. Medvedev, 1989b: 385 **A**: MG
jenseni meridionalis Kaszab, 1968c: 390 **A**: MG NMO
jintaiensis Ren, 1992: 330 **A**: NIN
kaszabi Bogatchev, 1972: 629 **A**: KZ (Karagiye)
laticollis Gebler, 1829a: 123 (*Pedinus*) **E**: ST **A**: KI KZ
lingwuensis Ren, 1993b: 487 **A**: NIN
maowusuensis Ren, 1992: 329 **A**: NMO
maxima borealis Kaszab, 1968c: 391 **A**: MG
maxima maxima Ménétriés, 1854: 33 **A**: MG NIN NMO
bielawskii Kaszab, 1964g: 391
medvedevi Kaszab, 1973c: 103 **A**: MG
mongolica Csiki, 1901a: 112 **A**: MG NIN NMO
opaca Reitter, 1889k: 703 **A**: NMO
parvula Kaszab, 1967b: 348 **A**: MG
sibirica Faldermann, 1833: 55 (*Opatrum*) **A**: ES MG NMO
alutacea Ménétriés, 1836b: 181
undentata Ren, 1993b: 488 **A**: NIN
subgenus *Mongolesthes* Reitter, 1904a: 174 type species *Melanesthes heydeni* Csiki, 1901
altaica altaica Kaszab, 1967a: 26 [= 1967c: 344] **A**: MG
altaica dschungarica Kaszab, 1967b: 345 **A**: MG
csikii Kaszab, 1965e: 343 **A**: GAN MG NIN NMO

heydeni australis G.S. Medvedev, 1989b: 383 **A:** MG
heydeni heydeni Csiki, 1901a: 112 **A:** MG NMO
multidentata Ren, 1993a: 29 **A:** XIN
tridentata Ren & Meng, 1995: 55 **A:** XIN
subgenus *Opatronesthes* Reitter, 1904a: 174 type species *Melanesthes punctipennis* Reitter, 1889
monatroides Reichardt, 1936a: 185 **A:** GAN
ningxianensis Ren, 1993a: 31 **A:** NIN
punctipennis Reitter, 1889k: 704 **A:** GAN
rugipennis Reitter, 1889k: 704 **A:** GAN NIN
tuberculosa Reitter, 1889k: 704 **A:** GAN

genus *Mesomorphus* Miedel, 1880: 140 type species *Opatrum murinum* Baudi di Selve, 1876 (= *Opatrinus setosus* Mulsant & Rey, 1853)
Hopatromorpha Blackburn, 1907: 286 [RN]
Pentholasius Reitter, 1904a: 178 type species *Halonomus variolatus* Allard, 1884
annamitus Kaszab, 1963b: 348 **A:** BT **ORR**
apterus Kaszab, 1963c: 364 **A:** SA **AFR**
belutschistanicus Kaszab, 1963b: 341 **A:** AF NP PA **ORR**
birmanicus Kaszab, 1963b: 343 **A:** BT NP YUN **ORR**
bremeri J. Ferrer, 1997a: 75 **A:** PA **ORR**
brevis Kaszab, 1963b: 344 **A:** YUN **ORR**
feai Kaszab, 1963b: 351 **A:** BT YUN **ORR**
fulvoscutatus Fairmaire, 1898c: 233 **N:** MR
gabrielae Schawaller, 2000a: 42 **A:** NP
hispidus Español, 1956b: 279 **N:** MO (Western Sahara) **AFR**
kulzeri Kaszab, 1963b: 349 **A:** NP **ORR**
latiusculus Chatanay, 1917a: 234 **A:** NP YUN **ORR**
longulus Reiche & Saulcy, 1857: 263 (*Crypticus*) **A:** CY IN IS JO LE SY TR
cyprius Kaszab, 1963b: 338
pellitus maroccanus Español, 1963e: 381 **N:** MO
rugulosus Chatanay, 1917a: 234 **A:** HAI NP UP **ORR**
maquilingius Gebien, 1921: 440
siamicus Kaszab, 1963b: 350 **A:** YUN **ORR**
setosus Mulsant & Rey, 1853c: 88 [= 1853d: 313] (*Opatrinus*) **N:** EG **A:** IS SI **AFR**
murinus Baudi di Selve, 1876b: 227 (*Opatrum*)
striolatus Fairmaire, 1896a: 21 (*Penthicus*) **A:** AF KA PA **ORR**
variolatus Allard, 1884: 248 (*Halonomus*) **A:** IS SY
subpubescens Reitter, 1896f: 163 (*Penthicus*)
villiger Blanchard, 1853: 154 (*Opatrum*) **A:** AF ANH FE FUJ GUA GUI GUX HAI HEB HEI HEN HKG HUB HUN JA JIA KA LIA NIN NMO NP SC SCH SD SHA SHN SHX TAI YUN **AFR AUR ORR**
asperulus Fairmaire, 1898c: 234

dermestoides Reitter, 1904a: 74
dispersus Champion, 1894a: 361 (*Opatrum*)
mustelinus Fairmaire, 1882c: 221 (*Opatrum*)
puberulus Fauvel, 1867: 187 (*Gonocephalum*)
ussuriensis Solsky, 1871: 374 (*Gonocephalum*)
wittmeri Kaszab, 1960a: 133 **A:** AF

genus *Myladina* Reitter, 1889k: 706 type species *Myladina unguiculina* Reitter, 1889
unguiculina Reitter, 1889k: 707 **A:** NIN NMO SHA

genus *Opatroides* Brullé, 1832: 219 type species *Opatroides punctulatus* Brullé, 1832
Hopatroides Agassiz, 1846: 185, 260 [UE]
curtulus Fairmaire, 1892b: 149 **A:** IS LE SY TR
asiaticus Desbrochers des Loges, 1881: 122 (*Penthicus*)
judaicus Baudi di Selve, 1881: 283 **A:** IS SY TR
punctulatus parvulus Faldermann, 1837: 62 (*Penthicus*) **E:** AB AR **A:** AF IN OM TD TM TR UZ
subcylindricus Ménétriés, 1849: 238 (*Penthicus*)
punctulatus punctulatus Brullé, 1832: 220 **E:** GR IT MA **N:** AG EG LB MO TU **A:** BA CY ES IQ IS JO KA KZ LE PA SA SI SY TR UZ WS YE **AFR NAR**
minutus Mulsant & Rey, 1859b: 78 [= 1859e: 16]
thoracicus Rosenhauer, 1856: 205 **E:** SP **N:** AG MO
vicinus angulatus Baudi di Selve, 1876b: 215 **A:** IQ
vicinus vicinus Fairmaire, 1896a: 21 (*Penthicus*) **A:** AE AF BA IN KU NP OM PA QA SA UP YE **ORR**

genus *Opatrum* Fabricius, 1775: 76 type species *Silpha sabulosa* Linnaeus, 1760
subgenus *Colpopatrum* Reitter, 1904a: 148 type species *Opatrum asperipenne* Reitter, 1897
asperipenne Reitter, 1897e: 219 **A:** GAN MG NIN NMO
subgenus *Colpophorus* Mulsant & Rey, 1859d: 106 [= 1859e: 44] type species *Opatrum emarginatum* Lucas, 1847
Colpophorinus Escalera, 1914: 336 type species *Opatrum gonocephaloides* Escalera, 1914
arrundarum Koch, 1945: 427 **N:** MO (High Atlas)
baeticum almeriense Español, 1954e: 323 **E:** SP (Andalucía)
baeticum baeticum Rosenhauer, 1856: 206 **E:** SP **N:** AG MO
granatum Fairmaire, 1871a: 392
baeticum gregarium Rosenhauer, 1856: 208 **E:** SP (Andalucía)
baeticum orientale Español, 1954e: 322 **E:** SP (Andalucía)
brevepilosum Antoine, 1942a: 22 **N:** MO (High Atlas)
emarginatum Lucas, 1847: pl. 29 **E:** IT MA **N:** AG TU
inaequale Reitter, 1904a: 149
freudei Español, 1963b: 73 **N:** AG
gonocephaloides Escalera, 1914: 336 **N:** MO

hookeri hookeri Wollaston, 1875: 219 (*Hopatrum*) **N:** MO
hookeri malandriosum Antoine, 1942a: 23 **N:** MO (High Atlas)
lanzai Ardoin, 1971a: 51 **N:** TU
moralesi Español, 1959: 247 **N:** MO
nivale Gené, 1839: 72 (*Philax*) **E:** IT (Sardinia)
otini mediatlantis Antoine, 1942a: 27 **N:** MO
otini otini Antoine, 1942a: 25 **N:** MO
porcatum Fabricius, 1792a: 89 **N:** AG MO TU
granuliferum Lucas, 1847: 332
rhaticum Antoine, 1942a: 24 **N:** MO
ualdunum Koch, 1945: 426
rifense Koch, 1945: 426 **N:** MO (Rif)
rotundatum ifrense Español, 1956b: 280 **N:** MO (Ifra)
rotundatum rotundatum Escalera, 1914: 338 **N:** MO
simplicior antoinei Koch, 1945: 426 **N:** MO (High Atlas)
simplicior megalatlanticum Koch, 1945: 427 **N:** MO (High Atlas)
simplicior rugultense Koch, 1945: 428 **N:** MO (Tizi n'Rougoult)
simplicior simplicior Escalera, 1914: 337 **N:** MO
simplicior simplicissimum Antoine, 1942a: 24 **N:** MO
talboti Koch, 1945: 430
tangerianum Koch, 1945: 426 **N:** MO (Tanger)
thalense Mayet, 1890: civ **N:** TU
trisinuatum Koch, 1945: 429 **N:** MO
validum marcuzzii Canzoneri, 1972a: 190 **E:** IT (Ustica)
validum obscurum Thunberg, 1821: 51 **E:** IT (Pantelleria Is.) **N:** AG MO TU
schickli Gebien, 1906a: 213 [RN]
validum rottenbergi Canzoneri, 1972a: 188 **E:** IT (Lapione)
validum validum Rottenberg, 1871: 250 **E:** IT (Sicilia) **N:** AG
venustulum Antoine, 1942a: 24 **N:** MO
subgenus *Opatrum* Fabricius, 1775: 76 type species *Silpha sabulosa* Linnaeus, 1758
Hopatrum Agassiz, 1846: 185, 260 [UE]
Thoracon Gistel, 1848: xi [URN]
alternatum Küster, 1849a: 39 **E:** GR RO TR **A:** TR
asperidorsum Fairmaire, 1878b: lvi **E:** FR IT (Sardinia)
cypraeum Reitter, 1904a: 155 **A:** CY
dahli Küster, 1849a: 34 **E:** FR (Corse) IT (Sardinia)
excisum Seidlitz, 1894: 429 **N:** AG
geminatum Brullé, 1832: 213 **E:** AB AR GR **A:** CY LE SY TR
asiaticum Reitter, 1904a: 153
granulosum Fischer von Waldheim, 1842: 16 **A:** KZ
grenieri Perris, 1870: 19 **E:** FR (Corse)

italicum Reitter, 1904a: 156 **E:** IT
iwani F. Soldati, 2016b: 5 **E:** FR (Corse)
libanii Baudi di Selve, 1876a: 64 [= 1876b: 222] **A:** IS LE SY
melitense Küster, 1849a: 38 **E:** MA
obesum G.-A. Olivier, 1811: 497 **E:** GR IT TR **A:** IQ SY TR
curtum Miller, 1861: 205
dardanum Steven, 1828: 94
elevatum Brullé, 1832: 215
gibbum Besser, 1832: 17 [HN]
messeniacum Rottenberg, 1871: 251
sulcatum Küster, 1849c: 49
perlatum Germar, 1823: 144 **E:** PT SP
riparium W. Scriba, 1865: 50 **E:** AU BY CR CT CZ DE EN FI FR HU GE HU LT NR NT PL RO SK ST SZ UK **A:** WS
sabulosum amauropterum Reichardt, 1936a: 131 **E:** ST (Kavkaz)
sabulosum guttifer Seidlitz, 1898a: 830 **E:** SP
sabulosum lucifugum Küster, 1849a: 41 **E:** AU CR GR IT
sabulosum pyrenaeum Español, 1963c: 480 **E:** SP (Pirineos centrales)
sabulosum sabulosum Linnaeus, 1758: 361 (*Silpha*) **E:** AL AB AR AU BE BH BU BY CR CT CZ DE EN CZ FI FR GB GE GG GR HU IR IT LA LS LT NL NR NT PL RO SK SL ST SV SZ TR UK YU **A:** ES GAN HEN IN KZ NMO TD TM TR WS XIZ
distinctum A. Villa & J.B. Villa, 1835: 49
intermedium Fischer von Waldheim, 1844: 127
reitteri Schuster, 1915b: 90
rugosum DeGeer, 1775: 43 (*Tenebrio*)
tricarinatum Motschulsky, 1859b: 307
sabulosum sculptum Rey, 1892: 42 **E:** FR IT
subdilatatum Reitter, 1908a: 23
sculpturatum caprariense Razzauti, 1919: 5 **E:** IT (Capraia)
sculpturatum igiliense Razzauti, 1919: 6 **E:** IT (Giglio)
sculpturatum ilvense Razzauti, 1919: 7 **E:** IT (Elba)
sculpturatum sculpturatum Fairmaire, 1860: 630 **E:** FR IT
sculpturatum urgonense Razzauti, 1919: 6 **E:** IT (Gorgona)
subaratum Faldermann, 1835b: 413 (*Opatrum*) **A:** ANH ES FE GAN GUI GUX HEI HEB HEN HUB HUN JIL JIX HEI JA LIA MG NC NIN NMO QIN SC SCH SHA SHN SHX TAI
tebessanum Reitter, 1904a: 153 **N:** AG TU
triste Steven, 1828: 94 **E:** UK (Krym)
areolatum Küster, 1849a: 40
granulatostriatum Küster, 1849a: 43
politum Besser, 1832: 17

verrucosum Germar, 1817: 188 **E:** AL AR BH CR GR IT TR YU **A:** TR
graniger Brullé, 1832: 214
granigerum Besser, 1832: 16
hispidosum Brullé, 1832: 217
trilobatum Baudi di Selve, 1875b: 221
verrucosum Waltl, 1835: 72 [HN]

genus *Penthicinus* Reitter, 1896f: 161 type species *Penthicinus koltzei* Reitter, 1896
ghilarovi G.S. Medvedev, 1988a: 55 **A:** KZ
koltzei Reitter, 1896f: 171 **A:** UZ XIN
netshaevae G.S. Medvedev, 1970: 397 **A:** TM
pedinoides Reitter, 1896f: 172 **A:** UZ

genus *Penthicus* Faldermann, 1836: 386 type species *Penthicus pinguis* Faldermann, 1836
subgenus *Allomyladion* Bogatchev, 1972: 626 type species *Penthicus poricollis* Reitter, 1896
poricollis Reitter, 1896f: 165 **A:** AF TD UZ
zoltani G.S. Medvedev & Iwan, 2006: 617 [RN] **A:** TD
kaszabi Bogatchev, 1972: 626 [HN]
subgenus *Aulonolcus* Reitter, 1904a: 168 type species *Pedinus altaicus* Gebler, 1830
altaicus Gebler, 1829a: 123 (*Pedinus*) **A:** ES (Tuva) MG NMO WS (Altay) XIN
sulcibasis Reitter, 1904a: 169 (*Lobothorax*)
tenebrioides Faldermann, 1835b: 411 (*Heliophilus*)
subgenus *Discotus* Reitter, 1904a: 161 type species *Opatrum dilectans* Faldermann, 1836
dilectans Faldermann, 1836: 389 (*Opatrum*) **E:** AB AR ST **A:** AF IN KI KZ MG TM UZ
echingolensis Kaszab & G.S. Medvedev, 1984: 82 (*Lobodera*) **A:** MG
kaszabi Skopin, 1960a: 307 (*Lobodera*) **A:** KZ
semenovi Reichardt, 1936a: 141 (*Lobodera*) **A:** KZ MG TD TM UZ XIN
viberti Reitter, 1906f: 240 (*Lobothorax*) **N:** AG EG **A:** IS **AFR**
subgenus *Myladion* Reitter, 1887d: 385 type species *Myladion acuticolle* Reitter, 1887
Penthicomelanesthes Bogatchev, 1972: 631 type species *Heliophilus gibbulus* Faldermann, 1836
acuticollis acuticollis Reitter, 1887d: 386 (*Myladion*) **A:** XIN XIZ
acuticollis pevtzovi Bogatchev, 1972: 627 **A:** TD
alashanicus Reichardt, 1936a: 163 (*Lobodera*) **A:** GAN MG NIN NMO
beicki beicki Reichardt, 1936a: 164 (*Lobodera*) **A:** MG XIN
beicki cerberus Reichardt, 1936a: 165 (*Lobodera*) **A:** MG
brutus Reichardt, 1936a: 160 (*Lobodera*) **A:** MG
bulganicus G.S. Medvedev, 1990b: 137 **A:** MG QIN XIN
davadshamsi davadshamsi Kaszab, 1965e: 336 (*Lobodera*) **A:** MG XIN
davadshamsi kobdoensis Kaszab, 1977c: 278 (*Lobodera*) **A:** MG
davadshamsi vulgaris Kaszab, 1967b: 334 (*Lobodera*) **A:** MG
dschungaricus Kaszab, 1967b: 339 (*Lobodera*) **A:** MG

explanatus explanatus Reitter, 1896f: 164 **A**: ES (Tuva) WS (Altay)
explanatus laesus Reichardt, 1936a: 159 (*Lobodera*) **A**: MG
explanatus reichardti Kaszab, 1969b: 325 (*Lobodera*) **A**: MG
frater Kaszab, 1967b: 336 (*Lobodera*) **A**: MG NMO
ferganensis Schuster, 1919a: 32 (*Lobothorax*) **A**: KI (Alai Range)
gibbulus Faldermann, 1835b: 412 (*Heliophilus*) **A**: "Altayskiy Kray"
hypolithus Gebler, 1836: 341 (*Heliophilus*)
gobiensis Kaszab, 1964b: 24 (*Lobodera*) **A**: MG
heydeni Reitter, 1896f: 168 **A**: UZ (Fergana)
humeridens Reitter, 1896f: 164 **A**: MG XIN "Altay"
kerzhneri G.S. Medvedev, 1989b: 380 **A**: MG
kiritshenkoi Reichardt, 1936a: 160 (*Lobodera*) **A**: MG NIN
kochi kochi Skopin, 1960a: 308 **A**: KZ (Vostochnyy Kazakhstan)
kochi tarbagataicus Skopin, 1964: 393 **A**: KZ (Vostochnyy Kazakhstan)
obsoletus G.S. Medvedev, 1964b: 163
korotyaevi G.S. Medvedev, 1984: 298 **A**: ES (Tuva)
laelaps Reichardt, 1936a: 161 (*Lobodera*) **A**: MG
lycaon Reichardt, 1936a: 161 (*Lobodera*) **A**: NMO
marginalis Reichardt, 1936a: 160 (*Lobodera*) **A**: MG NMO
nanshanicus Reichardt, 1936a: 164 (*Lobodera*) **A**: GAN MG
nojonicus Kaszab, 1968c: 385 (*Lobodera*) **A**: GAN MG NIN NMO
oblongipennis Bogatchev, 1961a: 113 (*Lobodera*) **A**: TD (Turkestanskiy Ridge)
obtusangulus bajanus G.S. Medvedev, 1989b: 379 **A**: MG
obtusangulus obtusangulus Reitter, 1889k: 709 (*Opatroides*) **A**: ES MG NMO
parvulus Kaszab, 1967b: 340 (*Lobodera*) **A**: MG
pavlovskyi Reichardt, 1935: 253 (*Lobodera*) **A**: AF TD UZ
reitteri Csiki, 1901a: 111 (*Gonocephalum*) **A**: MG NMO
serpens Reitter, 1896f: 168 **A**: KI
schusteri Reichardt, 1936a: 157 (*Lobodera*) **A**: GAN
sequensi Reitter, 1901c: 79 **A**: MG
tannuolensis G.S. Medvedev & Mordkovich, 1970: 30 **A**: ES (Tuva)
teter major Kaszab, 1969b: 322 (*Lobodera*) **A**: MG
teter teter Reichardt, 1936a: 158 (*Lobodera*) **A**: MG
truncatus Reitter, 1896f: 167 **A**: UZ (Bukhara)
tuberculosus Heyden, 1887c: 319 **A**: KI
tzaidamicus Bogatchev, 1951: 72 (*Lobodera*) **A**: QIN
vulneratus tenebrioides Reitter, 1896f: 163 **A**: KI
vulneratus vulneratus Kraatz, 1885: 293 **A**: KI UZ
waziristanica Kaszab, 1957f: 298 (*Lobodera*) **A**: PA
subgenus *Penthicus* Faldermann, 1836: 384 type species *Penthicus pinguis* Faldermann, 1836
Lobodera Mulsant & Rey, 1859b: 80 [= 1859e: 18] type species *Lobodera rufescens* Mulsant & Rey, 1859

Lobothorax Gemminger, 1870a: 124 [RN] type species *Lobodera rufescens* Mulsant & Rey, 1859
Penthomegus Reitter, 1904a: 161 type species *Penticus corpulentus* Reitter, 1896
Stonavus Reitter, 1904a: 161 type species *Penticus alaiensis* Reitter, 1896
adustus Bogatchev, 1951: 71 (*Lobodera*) **A:** AF
alaiensis Reitter, 1896f: 171 **A:** UZ (Alai Mts)
antoinei Español, 1945a: 228 (*Lobodera*) **N:** MO (Western Sahara)
ardoini Kaszab, 1969d: 454 (*Lobodera*) **A:** IN
auliensis Reitter, 1904a: 163 (*Lobothorax*) **A:** KZ UZ
belutschistanicus Kaszab, 1961a: 119 (*Lobodera*) **A:** PA
bogatshevi Kaszab, 1970a: 164 **A:** AF
corpulentus Reitter, 1896f: 163 **A:** IN
curtus Kaszab, 1970a: 163 (*Lobodera*) **A:** AF
fartilis Reitter, 1899d: 205 **A:** TM (Kopet Dag)
freyi Schuster, 1937b: 79 (*Lobodera*) **A:** IN
granulifera Reichardt, 1936a: 145 (*Lobodera*) **A:** UZ
granulosus Ménétriés, 1849: 239 **A:** KI KZ TD TM UZ XIN
semigranosus Reitter, 1904a: 162 (*Lobothorax*)
horni Schuster, 1922e: 212 (*Lobothorax*) **A:** IN
iners Ménétriés, 1832: 202 (*Heliophilus*) **E:** AB **A:** IN TR
molestus Faldermann, 1836: 388
saginatus Ménétriés, 1832: 203 (*Opatrum*)
kirgizicus Bogatchev, 1961a: 114 (*Lobodera*) **A:** KZ (Bednak-Dala)
lenczyi Kaszab, 1968c: 383 (*Lobodera*) **A:** GAN MG
lindbergi Kaszab, 1970a: 161 (*Lobodera*) **A:** AF
medvedevorum Bogatchev, 1972: 628 **A:** TD
netuschili Reitter, 1904a: 162 (*Lobothorax*) **A:** XIN
obesus Kulzer, 1964: 223 (*Lobodera*) **A:** IN
oblongopunctatus Reitter, 1904a: 164 (*Lobothorax*) **A:** AF IN IQ JO PA QA SA TM
persicus Schuster, 1919a: 31 (*Lobothorax*) **A:** IN
pinguis arnoldianus Bogatchev, 1951: 70 (*Lobodera*) **A:** UZ
pinguis kughitangi G.S. Medvedev & Nepesova, 1985b: 361 **A:** TM (Kugitangtay)
pinguis medvedevi Iwan & Löbl, 2007: 734 [RN] **A:** TM (Badkhyz)
granulifera G.S. Medvedev, 1964c: 658 (*Lobodera*) [HN]
pinguis nitescens Reichardt, 1935: 255 **A:** UZ
pinguis pinguis Faldermann, 1836: 386 **E:** AB AR **A:** AF TM
remotus Reitter, 1896f: 170 **A:** AF IN TM
rufescens rufescens Mulsant & Rey, 1859b: 80 [= 1859e: 18] (*Lobodera*) **E:** AB AR GG ST **A:** KZ TM UZ
rufescens tenuepunctatus Bogatchev, 1951: 72 (*Lobodera*) **A:** KZ
rugosicollis Bogatchev, 1951: 69 (*Lobodera*) **A:** UZ
samgarensis Reitter, 1896f: 168 **A:** UZ

similis Schuster, 1937b: 80 (*Lobodera*) **A:** IN
turcomanicus G.S. Medvedev, 1964c: 658 (*Lobodera*) **A:** TM
villiersi Kaszab, 1969d: 452 (*Lobodera*) **A:** IN
winkleri Schuster, 1915b: 92 (*Lobodera*) **A:** KZ XIN
zhenzhuristi afghana Gridelli, 1954: 289 (*Lobodera*) **A:** AF PA
zhenzhuristi zhenzhuristi Bogatchev, 1951: 70 (*Lobodera*) **A:** AF
subgenus *Pseudopenthicinus* Bogatchev, 1972: 628 type species *Penthicus medvedevi* Bogatchev, 1972
medvedevi Bogatchev, 1972: 628 **A:** TD

genus *Penthicus*, species incertae sedis
cribellatus Fairmaire, 1888b: 127 (*Opatroides*) **A:** BEI
gracilis Bates, 1879: 482 **A:** KA XIN
granulatus Desbrochers des Loges, 1881: 121 **E:** "South Russia"

genus *Phelopatrum* Marseul, 1876c: 100 type species *Hadrus scaphoides* Marseul, 1876
Pseudadrus Fairmaire, 1897a: 39 type species *Hadrus scaphoides* Marseul, 1876
scaphoides Marseul, 1876c: 99 (*Hadrus*) **A:** JA

genus *Polycoelogastridion* Reichardt, 1936a: 85 type species *Scleron sexcostatum* Motschulsky, 1858
sexcostatum Motschulsky, 1858d: 188 (*Scleron*) **E:** AB GR **A:** SY TR
scleroide Baudi di Selve, 1876b: 234 (*Opatrum*)
vinciguerrae Gridelli, 1930: 297 (*Scleropatrum*) **N:** LB

genus *Reichardtiellina* Kaszab, 1982a: 79 [RN] type species *Reichardtiella armata* Kaszab, 1942
Reichardtiella Kaszab, 1942a: 18 [HN] type species *Reichardtiella armata* Kaszab, 1942
armata Kaszab, 1942a: 19 (*Reichardtiella*) **A:** HP
tibetana Kaszab, 1942a: 21 (*Reichardtiella*) **A:** XIN

genus *Scleropatroides* Löbl & Merkl, 2003: 250 type species *Opatrum hirtulum* Baudi di Selve, 1876
Scleropatrum Reitter, 1890d: 149 [HN] type species *Opatrum hirtulum* Baudi di Selve, 1876
breviusculus Reitter, 1889a: 32 (*Scleropatrum*) **E:** AB AR **A:** AF KZ TM UZ
gallagheri J. Ferrer, 2000a: 37 (*Scleropatrum*) **A:** OM
hirtulus Baudi di Selve, 1876a: 69 (*Opatrum*) **E:** AB AR ST UK **A:** AF IN IQ KZ SA SI SY TD TM UZ
seidlitzi Reitter, 1898b: 39 (*Scleropatrum*) **E:** AB AR ST **A:** AF GAN IN KZ NMO TM UZ
strigatus Fabricius, 1798: 41 (*Opatrum*) **A:** OM **AFR ORR**
turanicus Reitter, 1890d: 149 (*Scleropatrum*) **A:** KZ TM

genus *Scleropatrum* Reitter, 1887d: 388 type species *Scleropatrum tuberculatum* Reitter, 1887
Monatrum Reichardt, 1936a: 81 type species *Opatrum carinatum* Gebler, 1830
carinatum carinatum Gebler, 1829a: 126 (*Opatrum*) **A:** KI
carinatum relictum Skopin, 1964: 394 (*Monatrum*) **A:** KZ
csikii Kaszab, 1967c: 332 (*Monatrum*) **A:** GAN XIN
elongatum Guérin-Méneville, 1834b: 32 (*Opatrum*) **A:** HAI HP KA **ORR**
horridum horridum Reitter, 1898b: 37 **A:** GAN NIN NMO SHX XIN
horridum humeralis Kaszab, 1964g: 390 (*Monatrum*) **A:** MG
mongolicum Kaszab, 1967b: 329 (*Monatrum*) **A:** MG
prescottii Faldermann, 1833: 54 (*Opatrum*) **A:** ES (Tuva) MG NMO XIN
carinatum Csiki, 1901a: 91
tuberculatum Reitter, 1887d: 388 **A:** GAN NMO QIN XIN XIN
tuberculiferum Reitter, 1890d: 148 **A:** GAN NIN NMO
striatogranulatum Reitter, 1890d: 149

genus *Sinorus* Mulsant & Revelière, 1860: 153 type species *Sinorus ciliaris* Mulsant & Revelière, 1860 (= *Opatrum colliardi* Fairmaire, 1860)
colliardi Fairmaire, 1860: 630 (*Opatrum*) **E:** FR (Corse) IT (Sardinia)
ciliaris Mulsant & Revelière, 1860: 154

genus *Socotropatrum* Koch, 1970: 109 type species *Opatrum costiferum* Waterhouse, 1881
costiferum Waterhouse, 1881: 476 (*Opatrum*) **A:** YE (Samha, Suqutra)
guichardi Koch, 1970: 113 **A:** YE (Suqutra)

genus *Tidiguinia* Español, 1959: 249 type species *Tidiguinia bolivari* Español, 1959
bolivari Español, 1959: 252 **N:** MO (Rif)

genus *Wolladrus* Iwan & Kamiński, 2016: 483 [RN] type species *Hadrus alpinus* Wollaston, 1854
Hadrus Wollaston, 1854: 502 [HN] type species *Hadrus alpinus* Wollaston, 1854
alpinus Wollaston, 1854: 502 (*Hadrus*) **N:** MR
carbonarius carbonarius Quensel, 1806: 142 (*Platynotus*) **N:** MR
cinerascens Wollaston, 1854: 503 (*Hadrus*)
verrucatus Desbrochers des Loges, 1881: 124 (*Hadrus*)
carbonarius paivae Wollaston, 1860b: 50 (*Hadrus*) **N:** MR
carbonarius sousai Ardoin, 1960a: 108 (*Hadrus*) **N:** MR
illotus Wollaston, 1854: 503 (*Hadrus*) **N:** MR
europaeus Motschulsky, 1858d: 189 (*Hadrus*)
subellipticus Desbrochers des Loges, 1881: 124 (*Hadrus*)
subplumbeus Fairmaire, 1856: 533 (*Platydema*)

tribe Opatrini, species incertae sedis
Gonocephalum angustatum Chevrolat, 1873: 204 **A:** "Syria"
Hopatrum opacum Walker, 1871: 15 **N:** EG
Heliopates damascenus Mannerheim, 1825: 30 **E:** ST
Heliopates rugipennis Faldermann, 1837: 56 **A:** IN

subtribe Sclerina Lacordaire, 1859

genus *Eurycaulus* Fairmaire, 1868: 492 type species *Eurycaulus marmottani* Fairmaire, 1868
subgenus *Eurycaulinus* Koch, 1937: 468 type species *Eurycalus quedenfeldti* Heyden, 1890
quedenfeldti Heyden, 1890b: 75 (*Eurycalus*) **N:** LB TU
subgenus *Eurycaulus* Fairmaire, 1868: 492 type species *Eurycaulus marmottani* Fairmaire, 1868
Ammotrypes Fairmaire, 1879c: 194 type species *Ammotrypes crenulicollis* Fairmaire, 1879
Scleronimon Reitter, 1904a: 127 type species *Eurycaulus granulatus* Reitter, 1904
Scleronopsis Koch, 1935a: 87 type species *Scleron hirsutus* Miller, 1861
buettikeri Schawaller, 1993a: 106 **A:** OM
crenulicollis Fairmaire, 1879c: 195 (*Ammotrypes*) **N:** AG TU
koestlini Español & Viñolas, 1981a: 2
dubius dubius Gridelli, 1930: 295 (*Scleron*) **N:** EG LB
dubius scleronoides Koch, 1937: 465 **N:** LB
granulatus Reitter, 1904a: 127 **N:** EG
hellmanni Grimm, 2000: 90 **N:** AG
henoni Fairmaire, 1897f: 121 **N:** EG **A:** IS SI
peyerimhoffi Reitter, 1904a: 127
hirsutus heliopolos Koch, 1935a: 86 **N:** EG
hirsutus hirsutus Miller, 1861: 204 (*Scleron*) **N:** EG
marmottani Fairmaire, 1868: 492 **N:** AG LB TU
kabakovi G.S. Medvedev, 1974: 946 **A:** AF
lilligi Grimm, 2005: 7 **A:** KU
nasreddinovi G.S. Medvedev, 1970: 947 **A:** AF TD
pachecoi Escalera, 1943b: 221 **N:** MO (Western Sahara)
reymondi Español, 1953c: 303 **N:** AG MO
schawalleri Grimm, 2005: 5 **A:** OM SA

genus *Platynosum* Mulsant & Rey, 1859b: 143 [= 1859e: 81] type species *Platynosum paulinae* Mulsant & Rey, 1859
Melanimon Motschulsky, 1845: 78 [HN] type species *Microzoum collare* Motschulsky, 1839
collare Motschulsky, 1839: 60 (*Microzoum*) E: AR **A:** IQ SA SY
paulinae Mulsant & Rey, 1859b: 143 [= 1859e: 81] **E:** GR (Rhodes) **A:** CY IS JO KU SA SY
sabulorum Chobaut, 1900a: 31 (*Melanimon*) **N:** LB TU
zacheus J.R. Sahlberg, 1908: 84 **A:** IS

genus *Sclerum* Dejean, 1834: 193 type species *Opatrum orientale* Fabricius, 1775
Anticlia Gistel, 1848: x [RN]
Chlamydion Gistel, 1848: x [RN]
Scleron Hope, 1841: 111 type species *Opatrum orientale* Fabricius, 1775
aequale Reitter, 1904a: 124 **A:** SA
andreinii Gridelli, 1930: 295 **N:** LB
armatum Waltl, 1835: 72 (*Opatrum*) **E:** FR IT SP **N:** AG CI EG LB MO TU **A:** SY
algiricum Lucas, 1849: pl. 29 (*Opatrum*)
damascenum Baudi di Selve, 1876a: 60 (*Sclerum*)
fernandezi Har. Lindberg, 1950: 11
asperulum Wollaston, 1864: 486 (*Sclerum*) **N:** CI (Gran Canaria)
bengalensis Kaszab, 1942a: 10 **A:** AF **ORR**
bodenheimeri Koch, 1935d: 282 **A:** IS
bremeri Grimm, 2013: 81 **A:** AE OM
carinatum Baudi di Selve, 1875d: 700 [= 1876a: 60] **E:** AB **A:** AF CY IN IQ SA SY TD TM TR UZ
dubium Gridelli, 1930: 295 **N:** EG LB
evansi Blair, 1923a: 122 **A:** IN IQ SA
ferrugineum Fabricius, 1801a: 118 (*Opatrum*) **A:** AP NP TAI "S. China" **ORR**
fossulatum Mulsant & Wachanru, 1852: 9 (*Sclerum*) **A:** CY IQ IS LE SY TR
angustum Miller, 1861: 203 (*Scleron*)
humerosum Miller, 1861: 202 (*Scleron*) **A:** CY SY TR
mariae Mulsant & Rey, 1859b: 140 [= 1859e: 78] (*Sclerum*)
kandaharicum Kaszab, 1960a: 135 **A:** AF
multistriatum Forskål, 1775: 77 (*Silpha*) **E:** GR IT MA ST TR **N:** EG LB **A:** IS SY TR
abbreviatum Reiche & Saulcy, 1857: 260
foveolatum G.-A. Olivier, 1811: 500 (*Opatrum*)
orientale orientale Fabricius, 1775: 76 (*Opatrum*) **N:** EG **A:** IS JO SA SI YE **AFR**
subclathratum Reitter, 1904a: 125 [RN]
orientale yemenense Kaszab, 1972: 376 **A:** YE
reitteri Gebien, 1906a: 234 [RN] **A:** NP UP **ORR**
socotranum Schawaller, 2004c: 449 **A:** YE (Suqutra)
sudanicum Koch, 1935a: 84 (*Scleron*) **N:** EG **AFR**
sulcatum lomii Gridelli, 1953: 59 **A:** SA **AFR**
sulcatum saharense Peyerimhoff, 1931b: 103 **N:** AG LB MO
sulcatum scotti Gridelli, 1953: 58 **A:** YE
sulcatum sulcatum Baudi di Selve, 1876a: 59 (*Sclerum*) **A:** SA

tribe Palorini Matthews, 2003

genus *Palorinus* Blair, 1930a: 135 type species *Palorus humeralis* Gebien, 1914
bicolor Blair, 1930a: 143 **A:** NP **ORR**
nepalicus Grimm, 2003: 4 **A:** NP

genus *Paloropsis* Masumoto & Grimm, 2004: 127 type species *Paloropsis iriei* Masumoto & Grimm, 2004

iriei Masumoto & Grimm, 2004: 127 **A:** JA TAI

genus *Palorus* Mulsant, 1854: 250 type species *Hypophloeus depressus* Fabricius, 1790

Caenocorse C.G. Thomson, 1859: 117 type species *Hypophlaeus depressus* Fabricius, 1790

Circomus Fleischer, 1900: 236 type species *Hypophloeus subdepressus* Wollaston, 1864

Eba Pascoe, 1863c: 129 type species *Eba cerylonoides* Pascoe, 1863

Platyotus Gerstaecker, 1871: 62 type species *Platyotus glabratus* Gerstaecker, 1871

Stenopalorus Blair, 1930a: 136 type species *Palorus hypophloeoides* Blair, 1930

beesoni Blair, 1930a: 8 **A:** NP ORR

cerylonoides Pascoe, 1863c: 129 (*Eba*) **A:** FUJ GUA GUI GUX HAI HEB HEN HUB HUN IN JA JIA JIX NMO NP SCH SHA SHX TAI ZHE **AFR AUR ORR**

exilis Marseul, 1876c: 116 (*Hypophloeus*)

minor Waterhouse, 1894: 71

praslinensis Gebien, 1922a: 304

papuanus Kaszab, 1939b: 218

zimmermanni Kaszab, 1955c: 657

depressus Fabricius, 1790: 223 (*Hypophlaeus*) **E:** AL AU BH BY CR CT CZ EN FI FR GE GR HU IT LA NL NR NT PL RO SK SL SP SV SZ UK **N:** EG **A:** JA KZ TAI TD TR **ORR**

formiceticola Munster, 1928: 296 (*Caenocorse*)

unicolor G.-A. Olivier, 1790: 18: 12 (*Ips*)

euphorbiae Wollaston, 1862a: 183 (*Hypophloeus*) **N:** CI

ficicola Wollaston, 1867: 205 (*Hypophloeus*) **N:** AG EG LB MO **A:** PA SA **AFR ORR**

deserticola J.R. Sahlberg, 1913a: 51 (*Caenocorse*)

subfilum A. Fleischer, 1900: 237

hypophloeoides Blair, 1930a: 136 **A:** NP UP **ORR**

longifoliae Blair, 1930a: 138 **A:** UP **ORR**

orientalis Fleischer, 1900: 237 **E:** AB **A:** IN

ratzeburgii Wissmann, 1848: 77 (*Hypophloeus*) **E:** AR AU AZ BE BY CR CT CZ DE FR GB GE GR HU IR IT LU NL PL SK SL SP SV SZ UK **N:** CI EG MR **A:** GAN GUI GUX HEB HEI HEN HUB HUN FE IQ IS JA JIA JIL JIX LIA NMO SCH SHA SHN SY TR YUN **COS**

ambiguus Wollaston, 1857a: 152 (*Hypophloeus*)

floricola Marseul, 1876c: 115 (*Hypophloeus*)

galilaeus J.R. Sahlberg, 1913a: 49 (*Caenocorse*)

shoreae Blair, 1930a: 138 **A:** NP TAI **ORR**

sinuaticollis Blair, 1930a: 137 **A:** TAI **ORR**

fuhoshoanus Kaszab, 1941c: 56

subdepressus Wollaston, 1864: 499 (*Hypophloeus*) **E:** AZ BH CR CT CZ FR GB GE GR HU IR IT MA PL SK SP SZ UK **N:** EG CI LB MO MR TU **A:** AF ANH FE FUJ GUI GUX HEB HEN HUB IS JA JIA JIX KZ MG NE NMO SA SCH SHA SHN TAI TM TR YE YUN ZHE **COS**

bifoveolatus Baudi di Selve, 1876c: 235

genus *Platycotylus* Olliff, 1883b: 182 type species *Platycotylus inusitatus* Olliff, 1883 (= *Ipsaphes nitidulus* MacLeay, 1873)

Thurea J. Ferrer, 1998: 151 type species *Thurea palmi* J. Ferrer, 1998

ferrugineus Kaszab, 1939a: 105 (*Doliema*) **A:** NP TAI **ORR**

nitidulus W.J. MacLeay, 1873a: 168 (*Ipsaphes*) **A:** NP TAI **AUR ORR**

coloratus Blackburn, 1903: 147

inusitatus Olliff, 1883b: 183

laticollis Pic, 1923d: 25 (*Doliema*)

nigripes Olliff, 1883a: 60

spinicollis Fairmaire, 1893b: 27 (*Sitophagus*)

suturalis Fairmaire, 1893b: 28 (*Sitophagus*)

genus *Ulomina* Baudi di Selve, 1876c: 235 type species *Ulomina carinata* Baudi di Selve, 1876

Coelopalorus Blair, 1930a: 135 type species *Palorus foveicollis* Blair, 1930 (= *Ulomina carinata* Baudi di Selve, 1876)

carinata Baudi di Selve, 1876c: 236 **Ei:** GB IT **A:** HAI HEN GUA GUX JA NP SD TAI YUN **AFR AUR NAR NTR ORR**

foveicollis Blair, 1930a: 136 (*Palorus*)

martinae Scupola, 2002: 188 [RN] **A:** HAI **ORR**

carinata Blair, 1930a: 143 (*Palorus*) [HN]

tribe Pedinini Eschscholtz, 1829
subtribe Helopinina Lacordaire, 1859

genus *Drosochrus* Erichson, 1843: 243 type species *Drosochrus brunnipes* Erichson, 1843

subgenus *Helopinus* Solier, 1848: 197 type species *Helopinus costatus* Solier, 1848

Emyon Gerstaecker, 1854: 532 type species *Emyon caelatus* Gerstaecker, 1854

costatus aegyptiacus Gridelli, 1939b: 198 (*Helopinus*) **N:** EG **A:** SI **AFR**

costatus costatus Solier, 1848: 199 (*Helopinus*) **N:** EG **A:** JO SA **AFR**

misolampoides Lacordaire, 1876: pl. 58 (*Helopinus*)

pilosus Gebien, 1921: 122 (*Helopinus*) **A:** YE

psalidiformis Ancey, 1881a: 397 (*Helopinus*) **A:** YE

genus *Micrantereus* Solier, 1848: 175 type species *Acanthomerus anomalus* Guérin-Méneville, 1834

Solenomerus Fåhraeus, 1870: 306 type species *Solenomerus longipes* Fåhraeus, 1870

arabicus Blair, 1933: 6 **A:** SA

desertus Koch, 1965a: 146 **A:** SA

nitidus nitidus Gahan, 1896b: 455 **A:** SA YE

nitidus yemensis Koch, 1965a: 145 **A:** SA YE

szalaymarzsoi Kaszab, 1972: 382 **A:** YE

tentyrioides Pascoe, 1882: 29 **A:** YE

subtribe Leichenina Mulsant, 1854

genus *Apsheronellus* Bogatchev, 1967a: 157 type species *Apsheronellus arenarius* Bogatchev, 1967

Microleichenum G.S. Medvedev, 1973: 648 type species *Microleichenum choresmensis* G.S. Medvedev, 1973

arabicus L. Soldati, 2009: 75 **A:** QA

arenarius Bogatchev, 1967a: 159 **E:** AB

choresmensis G.S. Medvedev, 1973: 648 (*Microleichenum*) **A:** KZ TM UZ

genus *Leichenum* Dejean, 1834: 194 type species *Opatrum pictum* Fabricius, 1801

Endothina Carter, 1924b: 536 type species *Endothina squamosa* Carter, 1924 (= *Opatrum canaliculatum* Fabricius, 1798)

Lichenum Agassiz, 1846b: 209 [UE]

canaliculatum canaliculatum Fabricius, 1798: 42 (*Opatrum*) **E:** GG GR **A:** FUJ IQ JA NP TA TM UP YE **AUR ORR**

argillaceum Motschulsky, 1863: 475

foveistrium Marseul, 1876c: 102 (*Lichenum*)

seriehispidum Marseul, 1876c: 101 (*Lichenum*)

squamosum Carter, 1924b: 537 (*Endothina*)

variegatum Klug, 1833: 88 (*Opatrum*)

mucronatum Küster, 1849a: 67 **E:** AB GR **A:** AF IN IQ LE KZ SY TR

caucasicum Reitter, 1899a: 84

caudatum Reitter, 1899a: 85

muelleri muelleri Gridelli, 1939c: 222 **A:** SA YE **AFR**

pictum Fabricius, 1801a: 117 (*Opatrum*) **E:** AL AU BH BU CR HU GR IT RO SK ST SZ UK

heyrovskyi Obenberger, 1922: 41 (*Lichenum*)

pulchellum pulchellum Lucas, 1846: pl. 30 (*Opatrum*) **E:** FR IT SP **N:** AG MO

gebieni Reitter, 1906a: 489 (*Lichenum*) [RN]

pulchellum Küster, 1849a: 65 [HN]

variegatum Küster, 1849a: 66 [HN]

pulchellum pumilum Baudi di Selve, 1876a: 73 **E:** AB GR **N:** EG LB TU **A:** BA CY IQ IN IS LE PA SA SI SY TR YE **AFR**

incisum Reitter, 1899a: 85

subtribe Pedinina Eschscholtz, 1829

genus *Cabirutus* E. Strand, 1929: 24 [RN] type species *Cabirus minutissimus* Mulsant & Rey, 1853

subgenus *Cabirutus* E. Strand, 1929: 24 [RN] type species *Cabirus minutissimus* Mulsant & Rey, 1853

Asiobirus G.S. Medvedev, 1968a: 179 type species *Cabirus validipes* Reitter, 1891

Cabirus Mulsant & Rey, 1853c: 223 [= 1853e: 139] [HN] type species *Cabirus minutissimus* Mulsant & Rey, 1853

Dentibirus G.S. Medvedev, 1968a: 204 type species *Heliopathes pusillus* Ménétriés, 1849

ampliatus Desbrochers des Loges, 1881: 109 (*Cabirus*) **A:** SY
castaneus Reitter, 1904a: 67 (*Cabirus*) **A:** IS
cavimanus J.R. Sahlberg, 1908: 81 (*Cabirus*) **A:** LE TR
cilicicus J.R. Sahlberg, 1908: 82 (*Cabirus*) **A:** TR
convexiusculus Desbrochers des Loges, 1881: 110 (*Cabirus*) **A:** "Syria"
cordicollis Kaszab, 1970b: 117 **A:** IN
cribricollis Baudi di Selve, 1876b: 156 (*Cabirus*) **E:** GR (Rhodes) **A:** CY TR
  cribricollis Desbrochers des Loges, 1881: 115 [HN]
cyrenaicus Gridelli, 1930: 290 (*Cabirus*) **N:** EG LB
desbrochersi Reitter, 1904a: 68 **A:** IS LE
elenae G.S. Medvedev, 1978c: 855 **A:** IN
elongatus J.R. Sahlberg, 1908: 79 (*Cabirus*) **A:** SY
ferganensis G.S. Medvedev, 1968a: 201 **A:** KI
glazunovi G.S. Medvedev, 1968a: 193 **A:** UZ
gracilis Reitter, 1904a: 66 (*Cabirus*) **A:** SY TR
gurjevae G.S. Medvedev, 1968a: 199 **A:** KI
iranicus G.S. Medvedev, 1968a: 176 **A:** IN
kirghizicus G.S. Medvedev & Kaltaev, 1979: 57 **A:** KI
kuhitangi G.S. Medvedev, 1968a: 195 **A:** TM
kuramensis G.S. Medvedev, 1968a: 191 **A:** TD
libanicus Desbrochers des Loges, 1881: 110 (*Cabirus*) **A:** LE
minutissimus Mulsant & Rey, 1853c: 224 [= 1853e: 140] (*Cabirus*) **A:** IS LE
mulsanti Desbrochers des Loges, 1881: 111 (*Cabirus*) **A:** IS LE
obsoletus Baudi di Selve, 1881: 281 (*Cabirus*) **A:** IS LE
  latiusculus Desbrochers des Loges, 1881: 111 (*Cabirus*)
obtusicollis Reitter, 1891d: 224 (*Cabirus*) **A:** KI KZ UZ
oxianus G.S. Medvedev, 1968a: 203 **A:** TM
persis Baudi di Selve, 1875d: 697 (*Cabirus*) **A:** IN
  persicus Baudi di Selve, 1876b: 156 (*Cabirus*)
procerulus Reitter, 1904a: 65 (*Cabirus*) **A:** KI KZ
puncticollis Reitter, 1904a: 66 (*Cabirus*) **A:** TD
pusillus Ménétriés, 1849: 327 (*Heliopathes*) **A:** AF KZ TD TM UZ
  dentimanus Reitter, 1904a: 64 (*Cabirus*)
  femoralis Reitter, 1915e: 85 (*Cabirus*)
  pusillus Mulsant & Rey, 1853c: 225 [= 1853e: 141] (*Cabirus*) [HN
rotundicollis Miller, 1861: 202 (*Cabirus*) **A:** IS LE
  pubescens Desbrochers des Loges, 1881: 108 (*Cabirus*)
ruficornis G.S. Medvedev, 1968a: 185 **A:** KI TD UZ
simonis Reitter, 1904a: 68 (*Cabirus*) **A:** IN IS SY
strictulus Desbrochers des Loges, 1881: 110 (*Cabirus*) **A:** IS
subpubescens Reitter, 1904a: 67 (*Cabirus*) **A:** LE

thoracicus J.R. Sahlberg, 1908: 83 (*Cabirus*) **A:** LE
tibialis Reitter, 1891d: 224 (*Cabirus*) **A:** UZ
turcmenicus G.S. Medvedev, 1968a: 178 **A:** TM (Kopet Dag)
validipes Reitter, 1891d: 224 (*Cabirus*) **A:** KI KZ TD UZ
vicinus Desbrochers des Loges, 1881: 112 (*Cabirus*) **A:** "Syria"
zaslavskii G.S. Medvedev, 1968a: 189 **A:** "Tian-Shan"
subgenus *Neocabirutus* Kulzer, 1964: 221 type species *Neocabirutus indicus* Kulzer, 1964
indicus Kulzer, 1964: 221 (*Neocabirutus*) **A:** NP **ORR**

genus ***Colpotinus*** **Fairmaire, 1891c: xvii** type species *Colpotinus simulator* Fairmaire, 1891
simulator Fairmaire, 1891c: xvii **A:** HUB SCH

genus ***Pedinus*** **Latreille, 1796: 20** type species *Tenebrio femoralis* Linnaeus, 1767
subgenus *Blindus* Mulsant & Rey, 1853c: 206 [= 1853e: 122] type species *Pedinus strigosus* Faldermann, 1835
contractus Liu & Ren, 2013: 199 (*Blindus*) **A:** SCH
curvotibius Ren & Zhang, 2010: 64 (*Blindus*) **A:** HEN
discolor Liu & Ren, 2013: 202 (*Blindus*) **A:** SCH
fulvicornis Reitter, 1889k: 700 **A:** GAN NMO SCH SHX
potanini nudiventris G.S. Medvedev, 1968a: 162 (*Blindus*) **A:** SCH
potanini potanini G.S. Medvedev, 1968a: 160 (*Blindus*) **A:** SCH
reichardti G.S. Medvedev, 1968a: 163 (*Blindus*) **A:** SCH
strigosus Faldermann, 1835b: 410 **A:** ANH BEI FE GUI HEB HEN HUB JA JIA JIX LIA MG NC NMO SC SCH SHN TAI TIA
  faldermanni Baudi di Selve, 1876a: 46 (*Colpotus*)
  japonicus Seidlitz, 1893: 376 (*Blindus*)
thibetanus Fairmaire, 1897i: 217 **A:** SCH XIZ
subgenus *Colpotus* Mulsant & Rey, 1853c: 208 [= 1853e: 124] type species *Pandarus strigosus* A. Costa, 1847 (= *Colpotus strigicollis* Mulsant & Rey, 1853)
  Pandarus A. Costa, 1847: 144 [HN] type species *Pandarus strigosus* A. Costa 1847
adebratti J. Ferrer & L. Soldati, 1999: 70 (*Colpotus*) **A:** TR
byzantinicus Waltl, 1838: 462 **E:** BU GG GR SY TR **A:** TR
  angustulus Reitter, 1887e: 520 (*Colpotus*)
cobosi Español, 1953b: 139 (*Colpotus*) **E:** SP
dentatus Grimm & Schawaller, 2000: 6 (*Colpotus*) **A:** TR
godarti Mulsant & Rey, 1853c: 214 [= 1853e: 130] (*Colpotus*) **E:** FR (Corse) IT (Sardinia)
magnicollis Fairmaire, 1892b: 148 [DA] (*Colpotus*) **A:** TR
ovoideus Grimm & Schawaller, 2000: 8 (*Colpotus*) **E:** GR
pectoralis Mulsant & Rey, 1853c: 220 [= 1853e: 136] (*Colpotus*) **E:** AL GR
  corcyrensis Koch, 1948a: 358 (*Colpotus*)
  damorei Koch, 1948a: 358 (*Colpotus*)
  ruficornis Reitter, 1884b: 94 (*Pandarinus*)

punctipleuris Reitter, 1904a: 63 (*Colpotus*) **A:** TR (Boz Dagh)
reitteri Zürcher, 1911: 292 (*Colpotus*) **A:** CY
schusteri Koch, 1936: 298 (*Colpotus*) **A:** TR
lasallei J. Ferrer & L. Soldati, 1999: 71 (*Colpotus*)
similaris nevadensis Español, 1953b: 138 (*Colpotus*) **E:** SP
similaris similaris Mulsant & Rey, 1853c: 212 [= 1853e: 128] (*Colpotus*) **E:** PT SP
skalei Grimm, 2018: 307 (*Colpotus*) **A:** TR
strigicollis ganglbaueri D'Amore-Fracassi, 1907: 191 (*Colpotus*) **E:** IT
strigicollis oglasensis Gardini, 1975: 44 (*Colpotus*) **E:** IT (Montecristo Is.)
strigicollis ragusai D'Amore-Fracassi, 1907: 190 (*Colpotus*) **E:** IT (Sicilia)
strigicollis strigicollis Mulsant & Rey, 1853c: 210 [= 1853e: 126] (*Colpotus*) **E:** IT
strigosus A. Costa, 1847: 144 (*Pandarus*) [HN]
sulcatus Ménétriés, 1840: 39 **E:** GR **A:** TR
asiaticus Koch, 1948a: 356 (*Colpotus*)
rhodius Koch, 1948a: 355 (*Colpotus*)
sulcatus Mulsant, 1853c: 218 [= 1853e: 134] (*Colpotus*) [HN]
vogti Koch, 1948a: 357 (*Colpotus*) **E:** GR (Samos) **A:** TR
subgenus *Pedinus* Latreille, 1796: 20 type species *Tenebrio femoralis* Linnaeus, 1767
Pedinulus Seidlitz, 1893: 373 type species *Pedinus ragusae* Baudi di Selve, 1875
Vadalus Mulsant & Rey, 1854e: 66 [= 1854c: 150] type species *Pedinus punctulatus* Mulsant & Rey, 1853
aetolicus Apfelbeck, 1901: 464 **E:** GR
affinis affinis Brullé, 1832: 206 **E:** GR
gibbosus Mulsant & Rey, 1853c: 165 [=1853e: 81] [HN]
affinis alziari Ardoin, 1978a: 134 **E:** GR (Peloponnisos)
affinis krueperi Seidlitz, 1893: 375 **E:** GR
affinis reitteri Seidlitz, 1893: 375 **E:** GR TR
aliquoi F. Soldati & L. Soldati, 2002b: 178 **E:** GR
balcanicus Apfelbeck, 1901: 465 **E:** AL BH GR YU
bodemeyeri Reitter, 1904a: 56 **A:** TR
borysthenicus Reichardt, 1936b: 676 **E:** ST UK
knechteli Panin, 1943: 179
cimmerius caucasicus G.S. Medvedev, 1968a: 149 **E:** ST
cimmerius cimmerius G.S. Medvedev, 1968a: 147 **E:** UK (Krym)
cimmerius znoikoi G.S. Medvedev, 1968a: 150 **E:** ST UK
circassicus Reitter, 1887e: 520 (*Micrositus*) **E:** GG ST **A:** TR
clementis Seidlitz, 1893: 373 **E:** GR
curvipes Mulsant & Rey, 1853c: 189 [= 1853e: 105] **E:** BU GR RO TR **A:** TR
tibipinnus Koch, 1936: 296
dilaticollis J.R. Sahlberg, 1908: 78 **A:** SY
fallax fallax Mulsant & Rey, 1853c: 168 [=1853e: 84] **E:** AL BH BU CR IT SK SL ST UK YU
fallax gracilis Mulsant & Rey, 1853c: 171 [=1853e: 87] **E:** CR HU RO
ulrichii Seidlitz, 1893: 371

fatuus Mulsant & Rey, 1853c: 178 [= 1853e: 94] **E:** GR
debilis Kiesenwetter, 1880: 68
femoralis femoralis Linnaeus, 1767: 679 (*Tenebrio*) **E:** AR AU BU BY CT CZ FR GE GR HU IT MD PL RO SK ST SZ TR UK **A:** ES KZ WS XIN
aequalis Faldermann, 1837: 57
curtulus Mulsant & Rey, 1853c: 196 [= 1853e: 112]
dermestoides Fabricius, 1787a: 212 (*Tenebrio*)
fornicatus Faldermann, 1837: 58
gemellatus Geoffroy, 1785: 157 (*Tenebrio*)
ovatus Faldermann, 1837: 59
striopunctatus Goeze, 1777: 687 (*Tenebrio*)
fodori Kaszab, 1941e: 161 **E:** GR
helopioides Germar, 1814: 2 **E:** AL BH BU CR GR IT RO SK SL TR YU **A** TR
gibbosus Brullé, 1832: 206
inflatus Reiche & Saulcy, 1857: 262 (*Crypticus*)
laticollis Herbst, 1799: 198 (*Blaps*)
hungaricus Seidlitz, 1893: 376 **E:** BH HU YU
infortunatus Kamiński & Iwan, 2017: 601 [RN] **E:** UK (Krym)
strigicollis Reitter, 1904a: 61 [HN]
intermedius Seidlitz, 1893: 375 **A:** "Syria"
ionicus Kiesenwetter, 1880: 66 **E:** GR
kiesenwetterii Seidlitz, 1893: 375 **E:** TR **A:** TR
longulus Rottenberg, 1871: 249 **E:** IT (Sicilia)
marani Kaszab, 1959d: 80 **A:** TR
meridianus Mulsant & Rey, 1853c: 175 [= 1853e: 91] **E:** CR FR IT
androgyne Reitter, 1904: 57
natolicus Mulsant & Rey, 1853c: 187 [= 1853e: 103] **A:** TR
oblongus Mulsant & Rey, 1853c: 179 [= 1853e: 95] **E:** AL BU GR
olivieri Mulsant & Rey, 1853c: 157 [= 1853e: 73] **E:** GR
punctulatus Kiesenwetter, 1880: 70
olympicus Kiesenwetter, 1880: 67 **E:** GR
paradoxus Reitter, 1904a: 59 **A:** SY
peristericus Apfelbeck, 1901: 464 **E:** AL GR
podager Seidlitz, 1893: 375 **A:** TR
punctatostriatus Mulsant & Rey, 1853c: 172 [= 1853e: 88] **E:** IT (Sicilia)
punctulatus Mulsant & Rey, 1853c: 150 [= 1853e: 66] **A:** TR
quadratus Brullé, 1832: 205 **E:** AL CR GR TR
ragusae Baudi di Selve, 1876a: 44 [= 1876b: 151] **E:** GR IT
schaumii Mulsant & Rey, 1853c: 181 [= 1853e: 97] **E:** GR **A:** "Syria Turc"
sicanus Canzoneri, 1984: 3 **E:** IT (Sicilia)
siculus Seidlitz, 1893: 367 **E:** IT (Sicilia)
simplex Seidlitz, 1893: 376 **E:** GR

strabonis Seidlitz, 1893: 375 **E:** AB AR **A:** IN TR
subdepressus Brullé, 1832: 207 **E:** GR
cylindricus Waltl, 1838: 462
szekessyi Kaszab, 1941e: 163 **E:** GR (Zakynthos)
tauricus Mulsant & Rey, 1853c: 197 [= 1853e: 113] **E:** RO UK (Krym) **A:** TR
problematicus Reichardt, 1936b: 674
taygetanus Kiesenwetter, 1880: 67 **E:** GR
tibipinnus Koch, 1936: 296 **A:** TR
tieffenbachi Reitter, 1915e: 84 **E:** GR
volgensis Mulsant & Rey, 1853c: 203 [= 1853e: 119] **E:** ST UK

genus *Pedinus*, nomen dubium
sibiricus Gebler, 1833: 287 **A:** "Montis Altaicis"

tribe Platynotini Mulsant & Rey, 1853
subtribe Platynotina Mulsant & Rey, 1853

genus *Adamus* Iwan, 1997: 255 type species *Platynotus micrositoides* Kaszab, 1975
sikkimensis Kaszab, 1975d: 315 (*Platydendarus*) **A:** SD **ORR**

genus *Notocorax* Dejean, 1834: 191 type species *Opatrum javanum* Wiedemann, 1819
Platydendarus Kaszab, 1975d: 312 [HN] type species *Opatrum javanum* Wiedemann, 1819
javanus Wiedemann, 1819: 163 (*Opatrum*) **A:** HAI YUN **ORR**
westermanni Mannerheim, 1844b: 862 (*Notocorax*)

genus *Penthicoides* Fairmaire, 1896a: 20 type species *Penthicoides seriatoporus* Fairmaire, 1896
seriatoporus Fairmaire, 1896a: 20 **A:** SD **ORR**

genus *Platynotus* Fabricius, 1801a: 138 type species *Blaps excavata* Fabricius, 1775
perforatus Mulsant & Rey, 1853c: 44 [= 1853d: 269] **A:** UP XIZ **ORR**

genus *Pseudoblaps* Guérin-Méneville, 1834b: 28 type species *Pseudoblaps substriata* Guérin-Méneville, 1834
Nyctalops Gistel, 1848: 125 [HN] [UE]
becvari Iwan, 2001: 149 **A:** PA
deserticola Kaszab, 1975d: 329 **A:** PA
indica Kaszab, 1975d: 328 **A:** KA
iranica Kaszab, 1975d: 327 **A:** IN
kabuliensis kabuliensis Kaszab, 1970a: 134 **A:** AF
kabuliensis lindemannae Kaszab, 1975d: 327 **A:** PA
kazakhstaniensis Iwan, 2001: 151 **A:** KZ
medvedevi Kaszab, 1975d: 325 **A:** KA

nuristanica Kaszab, 1960a: 128 **A:** AF KA
pakistana Kaszab, 1961d: 223 **A:** IN PA
substriata Guérin-Méneville, 1834b: 29 **A:** KA KZ **ORR**
ambigua Mulsant & Rey, 1853c: 55 [= 1853d: 280] (*Notocorax*)
mellyi Mulsant & Rey, 1853c: 54 [= 1853d: 279] (*Notocorax*)
parallela Mulsant & Rey, 1853c: 56 [= 1853d: 281] (*Notocorax*)
strigipennis Mulsant & Rey, 1853c: 61 [= 1853d: 286] (*Notocorax*)
zarudnyi G.S. Medvedev, 1968b: 178 **A:** IN

genus *Zidalus* Mulsant & Rey, 1853c: 71 [= 1853d: 296] type species *Opatrinus corvinus* Mulsant & Rey, 1853
Apterozidalus Ardoin, 1965b: 1315 type species *Apterozidalus royi* Ardoin, 1965
Zodinus Mulsant & Rey, 1853c: 90 [= 1853d: 315] type species *Opatrinus servus* Mulsant & Rey, 1853
corvinus Mulsant & Rey, 1853: 71 [= 1853: 296] (*Opatrinus*) **N:** EG **AFR**
pinkey Koch, 1956: 96 (*Opatrinus*)
erythraeus Gridelli, 1940: 239 (*Opatrinus*) **N:** EG **AFR**
niloticus Mulsant & Rey, 1853c: 87 [= 1853d: 312] (*Opatrinus*) **N:** CI (Fuerteventura) LB **A:** YE **AFR**
angusticollis Fairmaire, 1887e: 283 (*Opatrinus*)
edentatus Koch, 1956: 99 (*Opatrinus*)
olivensis Wollaston, 1864: 501 (*Tenebrio*)
zolotarevskyi Español, 1943c: 138 (*Opatrinus*)
tribe Platynotini, nomen dubium
Platynotus pekinensis Fairmaire, 1888b: 127 **A:** "Montagnes de Pékin"

tribe Platyscelidini Lacordaire, 1859

genus *Bioramix* Bates, 1879: 478 [= 1890: 69] type species *Bioramix ovalis* Bates, 1879
subgenus *Bioramix* Bates, 1879: 478 [= 1890: 69] type species *Bioramix ovalis* Bates, 1879
Botiras Fairmaire, 1891d: xcviii type species *Botiras striatellus* Fairmaire, 1891
Euryhelops Reitter, 1902g: 209 type species *Helops championi* Reitter, 1891
espanoli Kaszab, 1961a: 118 (*Platynoscelis*) **A:** PA
freudei Kaszab, 1970a: 126 (*Platynoscelis*) **A:** AF
glacialis Kaszab, 1975a: 19 (*Platynoscelis*) **A:** KA
kashmirensis Kaszab, 1940b: 178 (*Platynoscelis*) **A:** HP KA
klapperichi Kaszab, 1960a: 85 (*Platynoscelis*) **A:** AF
laeviuscula Fairmaire, 1891d: xcvii (*Faustia*) **A:** KA
oblongula Fairmaire, 1898d: 389 (*Botiras*) **A:** HP KA NP PA XIZ
himalajensis Kaszab, 1940b: 179 (*Platynoscelis*)
ovalis Bates, 1879: 478 **A:** HP NP KA XIZ
paghmanica Kaszab, 1960a: 86 (*Platynoscelis*) **A:** AF
puncticeps Bates, 1879: 479 **A:** HP KA NP SD
princeps Bogatchev, 1952b: 286 (*Platynoscelis*)

rotundicollis Kaszab, 1940b: 175 (*Platynoscelis*) **A:** NP UP XIZ
stoeckleini Kaszab, 1960a: 87 (*Platynoscelis*) **A:** AF
subgenus *Cardiobioramix* Kaszab, 1940b: 183 type species *Bioramix asidioides* Bates, 1879
asidioides Bates, 1879: 479 **A:** KA PA XIN
punctatella Fairmaire, 1891d: xcix (*Botiras*)
bechynei Kaszab, 1960a: 89 (*Platynoscelis*) **A:** AF
blairi Kaszab, 1940b: 183 (*Platynoscelis*) **A:** KA
championi Reitter, 1891e: 226 (*Helops*) [RN] **A:** GAN
subaenea Reitter, 1889k: 710 (*Helops*) [HN]
chinensis Kaszab, 1940b: 197 (*Platynoscelis*) **A:** SCH
dubiosa Kaszab, 1940b: 207 (*Platynoscelis*) **A:** KA
fairmairei Kaszab, 1940b: 184 (*Platynoscelis*) **A:** HP KA
globipunctata Bai & Ren, 2016: 186 **A:** NIN
gridellii Kaszab, 1940b: 186 (*Platynoscelis*) **A:** KA
haafi Kaszab, 1960a: 88 (*Platynoscelis*) **A:** AF
jinchuanensis Li, L.V. Egorov & Shi, 2016: 585 **A:** SCH
kabaki L.V. Egorov, 2006: 794 **A:** GAN
korschefskyi Kaszab, 1940b: 196 (*Platynoscelis*) **A:** SCH
kulzeri Kaszab, 1960a: 91 (*Platynoscelis*) **A:** SCH
liupanshana Bai & Ren, 2016: 188 **A:** NIN
maoxianensis Li, L.V. Egorov & Shi, 2016: 587 **A:** SCH
ovata Kaszab, 1940b: 189 (*Platynoscelis*) **A:** KA
punjabensis L.V. Egorov, 2004: 597 [RN] **A:** HP KA
championi Kaszab, 1940b: 194 (*Platynoscelis*) [HN]
schusteri Kaszab, 1940b: 188 (*Platynoscelis*) **A:** KA
sikkimensis Kaszab, 1940b: 191 (*Platynoscelis*) **A:** SD
splendida L.V. Egorov, 2006: 789 **A:** SCH
subaenescens Schuster, 1923b: 160 (*Platyscelis*) **A:** SCH
szetschuana Kaszab, 1940b: 192 (*Platynoscelis*) **A:** SCH
tshartymensis L.V. Egorov, 1990a: 120 **A:** TD
wittmeri L.V. Egorov, 1990a: 121 **A:** PA
zabriformis Fairmaire, 1896a: 55 (*Tagonoides*) **A:** KA
subgenus *Cardiochianalus* Kaszab, 1940b: 202 type species *Botiras sculptipennis* Fairmaire, 1891
batesi Kaszab, 1940b: 203 (*Platynoscelis*) **A:** HP KA
cordicollis Kaszab, 1940b: 202 (*Platynoscelis*) **A:** HP KA
schawalleri L.V. Egorov, 2004: 599 [RN] **A:** HP KA
kaszabi Koch, 1965b: 133 (*Platynoscelis*) [RN] [HN]
sculptipennis Fairmaire, 1891d: xcix (*Botiras*) **A:** KA
sommerssmithi Fairmaire, 1896a: 54 (*Tagonoides*)
subgenus *Chianalus* Bates, 1879: 479 [= 1890: 72] type species *Chianalus costipennis* Bates, 1879
costipennis Bates, 1879: 480 (*Chianalus*) **A:** KA PA

falsa Kaszab, 1961f: 347 (*Platynoscelis*) **A**: KA (Karakoram)
striatella Fairmaire, 1891d: xcviii (*Botiras*) **A**: KA PA
subcostipennis Gridelli, 1934: 64 (*Chiaralus*)
subgenus *Faustia* Kraatz, 1882b: 92 type species *Faustia modesta* Kraatz, 1882
lopatini L.V. Egorov, 1990a: 117 **A**: TD
modesta Kraatz, 1882b: 93 (*Faustia*) **A**: KI TD
subgenus *Leipopleura* Seidlitz, 1893: 354 type species *Faustia integra* Reitter, 1887
aenescens Blair, 1923b: 284 (*Platyscelis*) **A**: NP SD XIZ
banbarensis Li & L.V. Egorov, 2016: 35 **A**: XIZ
baqenensis Li & L.V. Egorov, 2016: 31 **A**: XIZ
crypticoides Reitter, 1887d: 384 (*Faustia*) **A**: GAN QIN SCH XIN XIZ
darbukensis Kaszab, 1940b: 160 (*Platynoscelis*) **A**: KA XIZ
frivaldszkyi Kaszab, 1940b: 167 (*Platynoscelis*) **A**: GAN QIN
hongyuanensis Li & L.V. Egorov, 2013: 90 **A**: SCH
igori Li & L.V. Egorov, 2013: 90 **A**: XIZ
integra Reitter, 1887d: 382 (*Faustia*) **A**: GAN QIN SCH
kochi Kaszab, 1940b: 169 (*Platynoscelis*) **A**: NP SD XIZ
medvedevi Bai, Merkl & Ren, 2019: 900 **A**: NP
micans Reitter, 1889k: 699 (*Faustia*) **A**: GAN QIN XIZ
brevipilosa Meng & Ren, 2005: 106 (*Myatis*)
siningensis J. Frivaldszky, 1890: 210 (*Faustia*)
nagquana Meng & Ren, 2005: 107 (*Myatis*) **A**: XIZ
nyainrongensis Li & L.V. Egorov, 2016: 33 **A**: XIZ
picipes Gebler, 1833: 287 (*Platyscelis*) **A**: ES MG WS
angustata Faldermann, 1835b: 409 (*Platyscelis*)
reitteri Seidlitz, 1893: 354 (*Platyscelis*)
politicollis Kaszab, 1940b: 171 (*Platynoscelis*) **A**: XIZ
reinigi Kaszab, 1940b: 163 (*Platynoscelis*) **A**: QIN SCH XIZ
tibetana Kaszab, 1940b: 162 (*Platynoscelis*)
rubripes Reitter, 1889k: 700 (*Faustia*) **A**: QIN
rufipalpis Reitter, 1887d: 383 (*Faustia*) **A**: QIN XIZ
subgenus *Nudoplatyscelis* Kaszab, 1940b: 222 type species *Platynoscelis turanica* Reitter, 1896
kaszabi Gridelli, 1954: 270 [= 1955: 64] (*Platynoscelis*) **A**: AF
ghorana Kaszab, 1974: 200 (*Platynoscelis*)
turanica Reitter, 1896g: 176 (*Platynoscelis*) **A**: IN TM
subgenus *Ovalobioramix* L.V. Egorov, 2004: 603 type species *Platynoscelis molesta* Bogatchev, 1947
molesta Bogatchev, 1947c: 87 (*Platynoscelis*) **A**: TD UZ
hissarica L.V. Egorov, 1992: 796 **A**: KI TD
subgenus *Planoplatyscelis* Kaszab, 1940b: 206 type species *Platyscelis margelanica* Kraatz, 1882 (= *Bioramix pamirensis* Bates, 1879)
Pleioplatyscelis Kaszab, 1940b: 220 type species *Platynoscelis lederi* Seidlitz, 1893

andreevae L.V. Egorov, 1990b: 891 **A:** KI (Talasskiy Ridge)
aruktavica L.V. Egorov, 1990a: 119 **A:** TD (Mts. Aruktau)
caraboides Kaszab, 1960a: 92 (*Platynoscelis*) **A:** AF
conradti Seidlitz, 1893: 354 (*Platyscelis*) **A:** TD
  pseudohelops Bogatchev, 1947c: 84 (*Platynoscelis*)
faldermanni Seidlitz, 1893: 354 (*Platyscelis*) **A:** TD UZ
  auliensis Kaszab, 1940b: 212 (*Platynoscelis*)
  blaptiformis Bogatchev, 1946d: 437 (*Platynoscelis*)
  duplicata Kaszab, 1940b: 218 (*Platynoscelis*)
haberhaueri Seidlitz, 1893: 354 (*Platyscelis*) **A:** KI
  humeralis Reitter, 1896g: 175 (*Platynoscelis*)
latipennis Kaszab, 1940b: 217 (*Platynoscelis*) **A:** TD
lederi Seidlitz, 1893: 355 (*Platyscelis*) **A:** KI TD
pamirensis Bates, 1879: 478 **A:** AF KI TD XIN
  caroli Kaszab, 1940b: 221 (*Platynoscelis*)
  difficilis Kraatz, 1883a: 351 (*Platyscelis*)
  latipes Kraatz, 1886: 189 (*Platyscelis*)
  margelanica Kraatz, 1882b: 84 (*Platyscelis*)
  rotundangula Kraatz, 1883a: 351 (*Platyscelis*)
  simplex Kraatz, 1883a: 352 (*Platyscelis*)
subgenus *Platynoscelis* Kraatz, 1882b: 91 type species *Platynoscelis helopioides* Kraatz, 1882
afghanica Schuster, 1936c: 200 **A:** AF
afghanistana Gridelli, 1954: 275 [= 1955: 65] (*Platynoscelis*) **A:** AF
bogatshevi Kaszab, 1970a: 128 (*Platynoscelis*) [RN] **A:** AF
  rotundicollis Kaszab, 1960a: 93 (*Platynoscelis*) [HN]
cylindrica Kaszab, 1960a: 104 (*Trichomyatis*) **A:** AF
cylindricollis L.V. Egorov, 1990a: 130 **A:** AF
delerei Kaszab, 1960a: 94 (*Platynoscelis*) **A:** AF
gazni L.V. Egorov, 1990a: 127 **A:** AF
gigantea Kaszab, 1974: 202 (*Platynoscelis*) **A:** AF
gracilioides Kaszab, 1970a: 130 (*Platynoscelis*) **A:** AF
gracilipenis Kaszab, 1960a: 98 (*Platynoscelis*) **A:** AF
gracilis Seidlitz, 1893: 355 (*Platyscelis*) **A:** TD
granulipennis Kaszab, 1960a: 95 (*Platynoscelis*) **A:** AF
gurjevae L.V. Egorov, 1990a: 123 **A:** TD
haarlovi Kaszab, 1959e: 251 [= 1960a: 99] (*Platynoscelis*) **A:** AF
helopioides Kraatz, 1882b: 92 (*Platynoscelis*) **A:** KI TD
hirtipennis Kaszab, 1960a: 103 (*Trichomyatis*) **A:** AF
kabakovi L.V. Egorov, 1990a: 131 **A:** AF
kapisensis L.V. Egorov, 1990a: 133 **A:** AF
lindbergi Kaszab, 1973b: 49 (*Platynoscelis*) **A:** AF

lucida Gridelli, 1954: 278 [= 1955: 68] (*Platynoscelis*) **A**: AF
merkli L.V. Egorov, 1990a: 129 **A**: AF
montana Kaszab, 1960a: 97 (*Platynoscelis*) **A**: AF
subalpestris Kaszab, 1973b: 48 (*Platynoscelis*)
oruzganensis L.V. Egorov, 1990a: 132 **A**: AF
ovipennis Kaszab, 1960a: 106 (*Trichomyatis*) **A**: AF
parvula L.V. Egorov, 1990a: 127 **A**: TD
rectangularis Kaszab, 1960a: 107 (*Trichomyatis*) **A**: AF
rufipes Kaszab, 1940b: 230 (*Platynoscelis*) **A**: AF TD
scheerpeltzi Kaszab, 1960a: 100 (*Platynoscelis*) **A**: AF
setosa L.V. Egorov, 1990a: 126 **A**: "Ost Bukhara"
shugnanensis L.V. Egorov, 1990a: 125 **A**: TD
sistensis L.V. Egorov, 1990a: 122 **A**: TD
tuxeni Gridelli, 1954: 280 [= 1955: 70] (*Platynoscelis*) **A**: AF
waziristanica Kaszab, 1940b: 233 (*Platynoscelis*) **A**: AF PA
paktiana Kaszab, 1974: 201
subgenus *Trichochianalus* Kaszab, 1940b: 201 type species *Platynoscelis monticola* Kaszab, 1940
monticola Kaszab, 1940b: 201 (*Platynoscelis*) **A**: KA
subgenus *Tricholeipopleura* Kaszab, 1940b: 223 type species *Platynoscelis lucidicollis* Kraatz, 1882
constricta Seidlitz, 1893: 355 (*Platyscelis*) **A**: TD
horni Kaszab, 1940b: 227 (*Platynoscelis*) **A**: TD
lucidicollis Kraatz, 1882b: 92 (*Platynoscelis*) **A**: KI XIN
psalidium Seidlitz, 1893: 355 (*Platyscelis*) **A**: TD
sinuatocollis Reitter, 1902b: 180 (*Platyscelis*) **A**: TD UZ
szekessyi Kaszab, 1938a: 1 (*Platyscelis*) **A**: TD
subgenus *Trichoplatyscelis* Reinig, 1931: 895 type species *Trichoplatyscelis pamirensis* Reinig, 1931 (= *Bioramix lapidicola* Kaszab, 1940)
Pseudotrichoplatyscelis Kaszab, 1960a: 82 type species *Platynoscelis badakschanica* Kaszab, 1960 (= *Bioramix lapidicola* Kaszab, 1940)
lapidicola Kaszab, 1940b: 166 (*Platynoscelis*) [RN] **A**: AF (Badakhshan) TD (West Pamir)
badakschanica Kaszab, 1960a: 83 (*Platynoscelis*)
pamirensis Reinig, 1931: 895 (*Trichoplatyscelis*) [HN]
tadzhika Bogatchev, 1947c: 85 (*Platynoscelis*)
paludani Gridelli, 1954: 269 [= 1955: 63] (*Platynoscelis*) **A**: AF (Kunar)

genus *Bioramix*, species incertae sedis
gebieni Kaszab, 1940b: 211 (*Platynoscelis*) **A**: KZ ("Karabutak")

genus *Microplatyscelis* Kaszab, 1940b: 144 type species *Faustia seriepunctata* Reitter, 1890
seriepunctata Reitter, 1890h: 362 (*Faustia*) **A**: AF KZ TD TM UZ

genus *Myatis* Bates, 1879: 480 [= 1890: 73] type species *Myatis humeralis* Bates, 1879
humeralis Bates, 1879: 480 **A:** KA TD XIN XIZ
quadraticollis Bates, 1879: 481
tenuissima Reitter, 1896g: 174 (*Leipopleura*)
variabilis Bates, 1879: 481
schaeferi Kaszab, 1940d: 900 **A:** XIZ

genus *Oodescelis* Motschulsky, 1845a: 76 type species *Blaps polita* J. Sturm, 1807
subgenus *Acutoodescelis* Kaszab, 1940d: 951 type species *Platyscelis punctatissima* Fairmaire, 1886
punctatissima emmerichi Kaszab, 1940d: 953 **A:** GAN SHA SHX
punctatissima henanana Bai & Ren, 2019b: 408 **A:** HEN
punctatissima punctatissima Fairmaire, 1886c: 345 (*Platyscelis*) **A:** BEI GAN HEB HEN NIN NMO SCH SHA SHN TAI TIA
pyripenis Ren, 1999: 293 [393]
subgenus *Clavatoodescelis* Kaszab, 1940d: 974 type species *Platyscelis melas* Fischer von Waldheim, 1823
Oblongoodescelis Kaszab, 1940d: 958 type species *Platyscelis oblonga* Ballion, 1878
acutanguloides Kaszab, 1940d: 976 **A:** XIN
melas Fischer von Waldheim, 1823: pl. 20 [= 1824: 194] (*Platyscelis*) **E:** CT HU KZ PL RO ST UK **A:** KZ
oblonga Ballion, 1878: 327 (*Platyscelis*) **A:** KZ XIN
gebieni Kaszab, 1940d: 978
kuntzeni Kaszab, 1940d: 977
regeli Ballion, 1878: 326 (*Platyscelis*)
subgenus *Convexoodescelis* L.V. Egorov, 2004: 593 type species *Platyscelis brevipennis* Kaszab, 1938
brevipennis brevipennis Kaszab, 1938d: 92 (*Platyscelis*) **A:** KI KZ XIN
brevipennis wernoyensis Kaszab, 1940d: 950 **A:** KI KZ
subgenus *Longuloodescelis* Kaszab, 1940d: 957 type species *Platyscelis hirta* Seidlitz, 1893
Trichoodescelis Kaszab, 1940d: 954 type species *Platyscelis acutangula* Kraatz, 1884
acutangula Kraatz, 1884: 224 (*Platyscelis*) **A:** KI
hirta Seidlitz, 1893: 355 (*Platyscelis*) **A:** KI
medvedevi L.V. Egorov, 1989a: 102 **A:** KZ
pilosa L.V. Egorov, 1987: 800 **A:** KZ UZ
songariensis Kaszab, 1940d: 960 **A:** KI
ballioni Skopin, 1973a: 873
subgenus *Montanoodescelis* L.V. Egorov, 2004: 592 type species *Platyscelis sahlbergi* Reitter, 1900
femoralis Kaszab, 1938d: 91 (*Platyscelis*) **A:** KZ (Zailiyskiy Alatau)
ketmeniana Skopin, 1961a: 198 [= 1965: 66] **A:** KZ (Ridge Ketmen) XIN
sahlbergi Reitter, 1900f: 220 (*Platyscelis*) **A:** KZ (Zailiyskiy Alatau)
transiliensis L.V. Egorov, 1987: 801 **A:** KZ (Zailiyskiy Alatau)

subgenus *Oodescelis* Motschulsky, 1845a: 76 type species *Blaps polita* J. Sturm, 1807

Oodeoscelis Agassiz, 1846: 260 [UE]

depilata Skopin, 1966b: 333 **A:** KZ

polita J. Sturm, 1807: 208 (*Blaps*) **E:** AB AU BU CT CZ GG HU KZ LT RO SK SL ST UK **A:** ES KZ WS

arnoldii Skopin, 1964: 390

gages Fischer von Waldheim, 1823: pl. 20, fig. 5 [= 1824: 195] (*Platyscelis*)

globosa Seidlitz, 1893: 355 (*Platyscelis*)

minima Motschulsky, 1860e: 527

subgenus *Ovaloodescelis* Kaszab, 1940d: 947 type species *Platyscelis affinis* Seidlitz, 1893

adriani Kaszab, 1940d: 948 **A:** KZ

affinis Seidlitz, 1893: 355 (*Platyscelis*) **A:** KI KZ XIN

heydeni Seidlitz, 1893: 355 (*Platyscelis*) **A:** KZ XIN

xerophila Skopin, 1965: 68

karaganae Skopin, 1965: 67 **A:** KZ

ovalis Ballion, 1878: 329 (*Platyscelis*) **A:** XIN

chinensis Kaszab, 1962a: 84

przewalskii Bogatchev, 1946d: 436

pseudotibialis Skopin, 1973a: 874 **A:** KI KZ XIN

similis similis Kaszab, 1938b: 55 (*Platyscelis*) **A:** KI KZ XIN

clavatipes Kaszab, 1938b: 55 (*Platyscelis*)

iliensis Skopin, 1958: 146

similis blattiformis Kaszab, 1938b: 54 (*Platyscelis*) **A:** KI KZ UZ

subattenuata L.V. Egorov, 1991: 146 [RN] **A:** KZ XIN

attenuata Kaszab, 1940d: 982 [HN]

tibialis Ballion, 1878: 331 (*Platyscelis*) **A:** KI KZ XIN

curvipes Skopin, 1966b: 335

heptapotamica Skopin, 1966b: 335

sachtlebeni Kaszab, 1940d: 980

subgenus *Planoodescelis* L.V. Egorov, 2004: 591 type species *Oodescelis kansouensis* Kaszab, 1940

abdurakhmanovi L.V. Egorov, 2018: 187 **A:** GAN

kansouensis Kaszab, 1940d: 954 **A:** GAN

lii Bai & Ren, 2019b: 417 **A:** HUB

subgenus *Spinoodescelis* Kaszab, 1940d: 966 type species *Platyscelis somocoeloides* Seidlitz, 1893

acuta Kaszab, 1940d: 968 **A:** KI KZ UZ

dispar Skopin, 1965: 65

semenoviana Bogatchev, 1946d: 435

hirtipennis Kaszab, 1940d: 972 **A:** UZ TD

longisterna Kaszab, 1940d: 971 **A:** "Turkestan"

necopinata Skopin, 1965: 66 **A:** KZ

ovulum Seidlitz, 1893: 355 (*Platyscelis*) **A:** KI TD UZ
latipleura Kaszab, 1940d: 969
punctolineata Kaszab, 1940d: 973 **A:** KI KZ UZ
somocoeloides grandis Skopin, 1965: 69 **A:** KZ
alticola Skopin, 1965: 69
somocoeloides somocoeloides Seidlitz, 1893: 356 (*Platyscelis*) **A:** KI KZ UZ
blapoides Reitter, 1896g: 175 (*Platynoscelis*)
transcaspica Kaszab, 1940d: 960 **A:** KZ
subgenus *Splenoodescelis* L.V. Egorov, 2004: 593 type species *Platyscelis turkestanica* Seidlitz, 1893
turkestanica Seidlitz, 1893: 355 (*Platyscelis*) **A:** KI
hirsuta Reitter, 1896g: 176
turul Kaszab, 1940d: 966
subgenus *Truncatoodescelis* Kaszab, 1940d: 962 type species *Platyscelis longicollis* Kraatz, 1884
glebi L.V. Egorov, 2009: 217 **A:** KI
kirghizica ferghanensis L.V. Egorov, 1989a: 105 **A:** KI (Ferganskiy Ridge)
kirghizica kirghizica L.V. Egorov, 1989a: 104 **A:** KI (Talasskiy Alatay)
longicollis Kraatz, 1884: 224 (*Platyscelis*) **A:** KI
kiritchenkoi Bogatchev, 1939a: 123 (*Platyscelis*)
schusteri schusteri Kaszab, 1940d: 961 **A:** KI (Kirgizskiy Ridge)
schusteri truncatoides Skopin, 1968b: 83 **A:** KI KZ

genus *Oodescelis*, species incertae sedis
attenuata Motschulsky, 1860e: 527 **A:** KZ

genus *Platyscelis* Latreille, 1818c: 23 type species *Tenebrio hypolithus* Pallas, 1781
subgenus *Oblongoplatyscelis* Kaszab, 1940d: 916 type species *Platyscelis ganglbaueri* Seidlitz, 1893
ganglbaueri Seidlitz, 1893: 354 **A:** KI KZ
seidlitzi Reitter, 1896g: 173
subgenus *Paraplatyscelis* Kaszab, 1940d: 936 type species *Platyscelis sinuata* Seidlitz, 1893
sinuata Seidlitz, 1893: 354 **A:** KI
subgenus *Platyscelis* Latreille, 1818c: 23 type species *Tenebrio hypolithus* Pallas, 1781
Kaszaboscelis Löbl & Merkl, 2003: 246 type species *Tenebrio hypolithus* Pallas, 1781
acutipenis Bai & Ren, 2019a: 103 **A:** GAN NIN
amdoensis L.V. Egorov, 1989b: 347 **A:** GAN
angusticollis angusticollis Kaszab, 1940d: 931 **A:** KI KZ XIN
skopini Kaszab, 1962c: 314
angusticollis platytarsis Kaszab, 1940d: 932 **A:** KZ XIN
bogatshevi L.V. Egorov, 1989b: 341 **A:** GAN SCH
brevis Baudi di Selve, 1876a: 35 **A:** ES HEB HEI JIL LIA NMO MG SHN SHX
ballioni Reitter, 1889k: 698
rugifrons Germar, 1823: 139 (*Blaps*) [HN]

confusa Schuster, 1934a: 75 [RN] **A:** GAN NIN QIN
hauseri Reitter, 1899d: 205 [HN]
densipunctata Bai & Ren, 2019a: 113 **A:** GAN
freyi Kaszab, 1940d: 925 **A:** GAN
gebieni Schuster, 1915b: 88 **A:** BEI HEB NIN NMO SHA SHX
helanensis Bai & Ren, 2019a: 117 **A:** NIN NMO
hungarica I. Frivaldszky, 1865: 200 **E:** HU
hypolitha Pallas, 1781: 44 (*Tenebrio*) [NP] **E:** CT GG KZ ST UK **A:** KZ WS
laevis Forskål, 1775: 78 (*Tenebrio*) [NO]
spinolae Solier, 1848: 211
ketmeniana L.V. Egorov, 1989b: 343 **A:** KZ (Mts. Ketmen)
kirghisica Reitter, 1896g: 174 **E:** ST KZ **A:** KZ
aralensis Reitter, 1896g: 174
licenti Kaszab, 1940d: 926 **A:** HEB NMO
rugifrons Fischer von Waldheim, 1823: pl. 20 (*Trachyscelis*) [= 1824: 194 (*Platyscelis*)] **A:** KI KZ WS XIN
intermedia Motschulsky, 1859c: 481
potanini Bogatchev, 1961a: 110
subcordata Seidlitz, 1893: 354 **A:** BEI HEB HEI LIA SHX SHN
strigicollis Lewis, 1895a: 447
suiyuana Kaszab, 1940d: 928 **A:** GAN HEB HEN NIN NMO SHA SHX
subgenus *Pleiopleura* Seidlitz, 1893: 343 type species *Platyscelis striatus* Motschulsky, 1859
ovata Ballion, 1878: 324 **A:** KI KZ XIN
striata Motschulsky, 1859b: 307 [= 1859a: 299] **A:** KZ XIN
przewalskii Bogatchev, 1961a: 109
sulcata Ballion, 1878: 322

genus *Platyscelis*, species incertae sedis
obovata Ren, 1999: 307 (393) [DA] **A:** QIN

genus *Somocoelia* Heyden & Kraatz, 1882: 331 type species *Somocoelia pinguis* Heyden & Kraatz, 1882
arnoldii L.V. Egorov, 1992: 797 **A:** TM
gracilipes Kaszab, 1940d: 907 **A:** TD
pinguis kuhitangi L.V. Egorov, 1998: 213 **A:** TM (Mts. Kugitangtau)
pinguis pinguis Heyden & Kraatz, 1882: 332 **A:** TD UZ
globulus Seidlitz, 1893: 354 (*Platyscelis*)
triplehorni Merkl & L.V. Egorov, 2015: 73 **A:** IN (Mts. Eastern Alborz)

genus *Somocoeloplatys* Skopin, 1968b: 82 type species *Platynoscelis boroldaica* Skopin, 1965
boroldaica Skopin, 1965: 64 (*Platynoscelis*) **A:** KZ

genus *Trichomyatis* Schuster, 1931: 893 type species *Trichomyatis conradti* Schuster, 1931

Trichoplatynoscelis Kaszab, 1940d: 896 type species *Trichoplatynoscelis pamirensis* Kaszab, 1940

conradti Schuster, 1931: 894 **A:** TD

michailovi L.V. Egorov, 1992: 798 **A:** TD

nuristanensis L.V. Egorov, 1992: 798 **A:** AF

pamirensis Kaszab, 1940d: 897 (*Trichoplatynoscelis*) **A:** TD

rugicollis Kaszab, 1960a: 108 (*Trichoplatynoscelis*) **A:** AF TD

tadzhika Bogatchev, 1952b: 285 (*Myatis*) **A:** TD

tribe Praeugenini De Moor, 1970

genus *Miltoprepes* Gerstaecker, 1871: 65 type species *Miltoprepes laetus* Gerstaecker, 1871

Anephyctus Fairmaire, 1891: 257 type species *Anephyctus hirtulus* Fairmaire, 1891

erythraeus Gridelli, 1939a: 78 **A:** YE **AFR**

genus *Praeugena* Laporte, 1840: 241 [NP] type species *Helops marginatus* Fabricius, 1792

Adelphus Dejean, 1834: 208 [NO] type species *Helops marginatus* Fabricius, 1792

Ergenna Fairmaire, 1897e: 139 type species *Ergenna caerulescens* Fairmaire, 1897

Lamprobothris Fairmaire, 1887e: 302 type species *Lamprobothris attenuata* Fairmaire, 1887 (= *Praogena fossulata* C. Müller, 1887)

Praogena Agassiz, 1846: 306 [UE] type species *Helops marginatus* Fabricius, 1792

Tactoderus Fairmaire, 1892a: 112 type species *Tactoderus subopacus* Fairmaire, 1892 (= *Praogena gagatina* Mäklin, 1863)

gagatina Mäklin, 1863: 579 (*Praogena*) **A:** SA YE **AFR**

subopaca Fairmaire, 1892a: 112 (*Tactoderus*)

tenuicollis Fairmaire, 1894d: 328 (*Dysgena*) **A:** SA **AFR**

tribe Rhysopaussini Wasmann, 1896

genus *Rhysopaussus* Wasmann, 1896a: 617 type species *Rhysopaussus dohertii* Wasmann, 1896

dohertii Wasmann, 1896a: 617 **A:** NP **ORR**

tribe Scaurini Billberg, 1820

genus *Cephalostenus* Solier, 1838b: 184 type species *Cephalostenus dejeanii* Solier, 1838 (= *Scaurus elegans* Brullé, 1832)

Stenocephalus Agassiz, 1846: 351 [UE] [HN]

alziari Grimm, 1991: 24 **A:** CY

elegans Brullé, 1832: 195 (*Scaurus*) **E:** GR **A:** TR

dejeanii Solier, 1838b: 186

demaisoni Reitter, 1903c: 132 **A:** TR

orbicollis Ménétriés, 1836a: 151 **E:** BU GR TR **A:** SY TR

genus *Scaurus* Fabricius, 1775: 253 type species *Scaurus atratus* Fabricius, 1775

Scauris Rafinesque, 1815: 113 [UE]

aegyptiacus Solier, 1838b: 170 **E:** GR (Castellorizo) IT (Sardinia, Sicilia, Stromboli) MA **N:** AG EG LB MO TU **A:** CY IQ LE SI SY TR
giganteus Küster, 1848a: 54
alticola Escalera, 1914: 296 **N:** MO
angustus Reiche, 1864: 242 **N:** AG TU
araxinus Richter, 1945: 103 **E:** AB AR TR
atratus Fabricius, 1775: 253 **E:** FR IT (Sicilia, Sardinia) **N:** AG TU
breviatus Desbrochers des Loges, 1881: 81
morio Fabricius, 1777: 241 (*Helops*)
planidorsis Fairmaire, 1879c: 190
quadraticollis Fairmaire, 1875a: 526
vagecostatus Fairmaire, 1879c: 189
barbarus barbarus Solier, 1838b: 165 **N:** LB TU
gracilicornis Fairmaire, 1875a: 524
interruptus Baudi di Selve, 1875b: 11
pellegrinii Koch, 1937: 424
bougonii bougonii Fairmaire, 1883i: cxxv **N:** EG LB TU
bougonii schusteri Gridelli, 1930: 260 **N:** LB
calcaratus Fabricius, 1798: 42 **N:** MO
distinctus Solier, 1838b: 173
latipennis Baudi di Selve, 1875b: 13
sublaevis Desbrochers des Loges, 1881: 84
camelus Kocher, 1958b: 195 **N:** MO
carinatus carinatus Solier, 1838b: 176 **N:** EG LB **A:** SI
carinatus vicinoides Schuster, 1935a: 15 **N:** LB
contractus Fairmaire, 1880a: 13 **N:** MO
incostulatus Escalera, 1935a: 308
dubius Solier, 1838b: 181 **N:** AG MO TU
amplicollis Fairmaire, 1875a: 526
barbarus Erichson, 1841: 181 (*Scaurus*) [HN]
elongatus Mulsant & Godart, 1866: 452 **N:** AG TU
gigas Waltl, 1835: 70 **E:** PT SP **N:** AG MO
ferreri Español, 1943c: 130
hespericus Solier, 1838b: 178
maroccanus Fairmaire, 1875a: 525
girardi Labrique, 2009: 9 **N:** MO
hypogeus Antoine, 1954: 203 **N:** MO (Tleta des Akhsas)
jahandiezi Théry, 1931b: 123 **N:** MO
kocheri Peyerimhoff, 1948b: 171 **N:** MO (High Atlas)
kraatzi Haag-Rutenberg, 1876a: 10 **N:** MO (Atlantic)
mairei Peyerimhoff, 1929c: 282 **N:** AG (Hoggar)
manueli Labrique, 1992: 123 **N:** MO

mesatlanticus horstensis Labrique & Chavanon, 2005: 41 **N:** MO (Middle Atlas: Horst)
mesatlanticus mesatlanticus Peyerimhoff, 1948b: 170 **N:** MO (Middle Atlas: Taza)
harbuloti Reymond, 1956b: 266 [= 1956a: 40]
microcephalus Escalera, 1914: 297 **N:** MO
lopezi Español, 1943c: 131
pierrei Antoine, 1954: 204
quirogai Escalera, 1914: 299
puncticollis dlabolai Kaszab, 1959d: 79 **A:** TR
puncticollis getula Peyerimhoff, 1946: 97 **N:** MO
puncticollis macricollis Allard, 1882b: lxxxvii [= 1883: 24] **A:** IN IQ TR
puncticollis puncticollis Solier, 1838b: 172 **N:** AG EG LB TU **A:** IS SA SI SY **AFR**
parvicollis Fairmaire, 1875a: 525
rugicollis Reitter, 1914g: 374
puncticollis syriacus Reitter, 1914g: 374 **A:** CY SY TR
rugulosus Solier, 1838b: 174 **E:** SP
eleonorae Tenenbaum, 1915: 840
gracilis Küster, 1848a: 61
sanctiamandi Solier, 1838b: 177 **N:** LB MO TU
lucidulus Desbrochers des Loges, 1881: 82
ovipennis Fairmaire, 1875: 525
striatus Fabricius, 1792a: 93 **E:** FR GR IT MA SP
maremmanus Solier, 1838b: 180
sardous A. Costa, 1884: 13
tingitanus antarensis Labrique, 1995: 124 **N:** AG (Dj. Antar)
tingitanus gimeli Antoine, 1953a: 59 **N:** MO
tingitanus tingitanus Peyerimhoff, 1948b: 169 **E:** SP **N:** MO
sabasi Kocher, 1965b: 34
tristis G.-A. Olivier, 1795: 62: 4 **E:** FR GR IT MA SP SZ **N:** AG MO TU
abbreviatus Reitter, 1914g: 371
algericus Küster, 1848a: 56
gracilipes Solier, 1838b: 167
lugens Küster, 1848a: 57
uncinus Forster, 1771: 60 (*Tenebrio*) **E:** FR GR IT (Sardinia) PT SP **N:** AG CI MO **AFR**
asperulus Fairmaire, 1875a: 525
punctatus Fabricius, 1798: 43
mendax Reitter, 1914g: 379
punctatus Herbst, 1799: 138 [HN]
sticticus Gemminger, 1870a: 121 [RN]
variolosus Wollaston, 1867: 178
varvasi Solier, 1838b: 175 **N:** AG MO
porcatus Erichson, 1841: 182

vicinus Solier, 1838b: 171 **E:** SP **N:** MO
balearicus Reitter, 1914g: 372
balearicus Tenenbaum, 1915: 840 [HN]

genus *Scaurus*, nomina dubia
carinatus V. Petagna, 1787: 26 (*Pimelia*) **E:** IT
dentipes Villers, 1789: 396 (*Tenebrio*) **E:** "Galia"
sulcatus Fabricius, 1792a: 93 **E:** FR

tribe Tenebrionini Latreille, 1802

genus *Ariarathus* Fairmaire, 1891f: ccxi type species *Ariarathus ulomoides* Fairmaire, 1891 (= *Tenebrio atronitens* Fairmaire, 1891)
Teneatopus Reitter, 1920a: 23 type species *Tenebrio atronitens* Fairmaire, 1891
atronitens Fairmaire, 1891c: xvii (*Tenebrio*) **A:** BEI FUJ GUI HUB JA SCH SHG
ulomoides Fairmaire, 1891f: ccxi (*Ariarathus*)

genus *Bius* Dejean, 1834: 205 type species *Trogosita thoracicus* Fabricius, 1792
Bia Hope, 1841: 132 [UE]
Dendroscopius Gistel, 1848: 125 [URN]
thoracicus Fabricius, 1792a: 116 (*Trogosita*) **E:** AU BY CR CT ES FI FR GE NR NT RO SK SV SZ **A:** ES FE JA KZ MG WS
laevis Stephens, 1835: 426 (*Tenebrio*)

genus *Neatus* J.L. LeConte, 1862: 233 type species *Helops tenebrioides* Palisot de Beauvois, 1812
noctivagus Mulsant & Rey, 1853a: 7 [= 1853b: 9] (*Tenebrio*) **E:** AL CR IT
picipes Herbst, 1797: 245 (*Tenebrio*) **E:** AU BU BY CR CT CZ DE ES FR GB GE GR HU IT LA PL RO SK ST SZ UK **A:** KZ NE SC WS **NAR**
badius Say, 1824a: 265 (*Tenebrio*)
intermedius Fiori, 1903: 226 (*Tenebrio*)
laticollis Stephens, 1832: 8 (*Tenebrio*)
loripes J. Sturm, 1807: 223 (*Tenebrio*)
transversalis Duftschmid, 1812: 300 (*Tenebrio*)
subaequalis Reitter, 1920a: 22 **E:** AB AL AR GG GR MC ST **A:** IN TR
ventralis Marseul, 1876c: 123 (*Tenebrio*) **A:** ANH BEI ES FE GAO HUB JA NC SHA SHX

genus *Tenebrio* Linnaeus, 1758: 417 type species *Tenebrio molitor* Linnaeus, 1758
subgenus *Tenebrio* Linnaeus, 1758: 417 type species *Tenebrio molitor* Linnaeus, 1758
Menedrio Motschulsky, 1872: 27 type species *Tenebrio obscurus* Fabricius, 1792
Tenebrionellus Crotch, 1873a: 105 [URN]
alternicostis Marseul, 1876c: 123 **A:** JA (Kyushu)
angustus Zoufal, 1892: 284 **E:** AB **A:** AF IN KZ TD TM UZ

cedrorum Pic, 1925c: 6 **N:** AG
fuliginosus Fairmaire, 1874a: 388 **N:** AG MO
medius Desbrochers des Loges, 1881: 131 **N:** AG
molitor Linnaeus, 1758: 417 **E:** AB AL AU BE BH BU BY CR CT CZ DE EN FI FR GB GE GR HU IC IR IT LA LS LT LU MA NL NR NT PL RO SK SP ST SV SZ UK YU **N:** CI EG MR **A:** FE FUJ GAN GUA GUX HAI HEB HEI HEN HKG HUB JA JIA JIL JIX LIA NIN NMO SC SCH SHN SHX TAI TD TM YUN **AFR AUR NAR NTR**
obscurus Fabricius, 1792a: 111 **E:** AB AL AU AZ BE BH BU BY CR CT CZ DE EN FI FR GB GE GR HU IR IT LT LU MA NL NT PL RO SK SL SP ST SV SZ UK YU **N:** CI EG MO MR TU **A:** AF ANH CY FE FUJ GUA GUI GUX HAI HEB HEI HEN HUN IN IQ JA JIA JIL JIX KZ LIA NMO QIN SC SCH SHA SHN SHX TAI TD TM TR UZ XIN ZHE **COS**
codinae Reitter, 1912d: 310
longicollis Motschulsky, 1872: 37 (*Menedrio*)
pollens Casey, 1924: 321
opacus Duftschmid, 1812: 299 **E:** AU BU CR CT CZ DE FR GE GG GR HU IT MC PL RO SK SL SV SZ UK
punctipennis Seidlitz, 1895: 633 **E:** BU GR IT **N:** TU
punctatissimus Denisova, 1940: 243 **A:** AF TD UZ
syriacus Zoufal, 1892: 284 **N:** EG **A:** SY TR

tribe Toxicini Oken, 1843

genus *Cryphaeus* Klug, 1833: 177 type species *Cryphaeus aries* Klug, 1833
Anthracias Dejean, 1834: 205 type species *Uloma cornuta* Fischer von Waldheim, 1823
Ardelio Gistel, 1848: xi [URN]
albopilosus Mt. Chûjô & Lee, 1993: 116 **A:** FE SC SHA TAI
amurensis Kulzer, 1950: 41 **A:** FE HAI JA SC TAI
duellicus Lewis, 1894: 470 (*Anthracias*)
bicornutus Pic, 1921c: 28 (*Toxicum*) **A:** FUJ YUN
boleti Lewis, 1894: 471 (*Anthracias*) **A:** JA
cavifrons Kulzer, 1950: 38 **A:** FUJ YUN
chengi Masumoto, Akita & Lee, 2005: 252 **A:** TAI
cornutus Fischer von Waldheim, 1823: 199 (*Uloma*) **E:** AB BH BU FR GG GR HU RO SL ST UK **N:** CI **A:** CY SY TR
bicornis L. Redtenbacher, 1857: 617 (*Anthracias*)
dissolutus Kulzer, 1950: 39 **A:** SCH
formosanus Masumoto, 1982d: 62 **A:** TAI
inomatai Masumoto, 1982d: 65 **A:** TAI
kuantoushanus Masumoto, Akita & Lee, 2012b: 194 **A:** TAI
laticeps Lillig, 2006: 49 **A:** IS LE
punctatulus Lewis, 1894: 470 (*Anthracias*) **A:** JA SC
fagi Lewis, 1894: 471 (*Anthracias*)

punctipennis Gravely, 1915: 529 **A:** AF AP BT NP PA SD **ORR**
rotundicollis Mt. Chûjô & Lee, 1993: 117 **A:** SC
satoi Kaszab, 1964a: 1 **A:** JA (Ryukyus) TAI
incurvatus Masumoto, 1982d: 63
iriomotensis M. Chûjô, 1979: 42
taiwanus Masumoto, 1982d: 64 **A:** TAI
taurus Fabricius, 1801a: 153 (*Trogosita*) **A:** YE **AFR**
tenuis Fairmaire, 1896a: 26 (*Toxicum*) **A:** NP **ORR**

genus *Epitoxicum* Bates, 1873a: 46 type species *Epitoxicum haplandroides* Bates, 1873
haplandroides Bates, 1873a: 46 **A:** NP **ORR**

genus *Taiwanocryphaeus* Masumoto, 1996c: 67 type species *Taiwanocryphaeus rhinoceros* Masumoto, 1996
rhinoceros Masumoto, 1996c: 68 **A:** TAI

genus *Toxicum* Latreille, 1802: 174 type species *Toxicum richesianum* Latreille, 1804
subgenus *Mutiloxicum* Nabozhenko & Ivanov, 2018: 893 type species *Toxicum elvirae* Nabozhenko & Ivanov, 2018
elvirae Nabozhenko & Ivanov, 2018: 893 **A:** FE
subgenus *Toxicum* Latreille, 1802: 174 type species *Toxicum richesianum* Latreille, 1804
Trestonia Rafinesque, 1815: 113 [URN]
angustatum Pic, 1921c: 27 **A:** AP BT SCH
assamense Pic, 1913c: 18 **A:** AP NP SD XIZ YUN **ORR**
cavifrons Pic, 1913c: 17 **A:** TAI
fangjingshananum Ren & Hua, 2006: 271 **A:** GUI
formosanum Kulzer, 1950: 37 **A:** TAI
funginum Lewis, 1894: 469 **A:** JA TAI
morii Masumoto & Akita, 2008: 35 (*Cryphaeus*) **A:** JA (Honshu)
subtricornutum M. Chûjô, 1978a: 70 **A:** JA (Ryukyus)
tricornutum Waterhouse, 1874: 126 **A:** JA SC
tuberculifrons Lewis, 1894: 469
umbrosum Harold, 1880: 165

tribe Triboliini Gistel, 1848

genus *Latheticus* Waterhouse, 1880: 147 type species *Latheticus oryzae* Waterhouse, 1880
oryzae Waterhouse, 1880: 148 **Ei:** AU BE CT CZ DE FI FR GB GE HU IT LA LU NL NR PL SK SP ST SV SZ UK **Ni:** CI LB MO TU **A:** FE HEB HEN HUB GUA GUX IN IQ IS JA JIA JX KZ MG NE NMO SA SCH SHA SHX TAI TM YE "Korea" **COS**
striolatus Fairmaire, 1892a: 111 (*Lyphia*)

genus *Lyphia* Mulsant & Rey, 1859d: 166 type species *Lyphia ficicola* Mulsant & Rey, 1859 (= *Bius tetraphyllus* Fairmaire, 1856)

Lindia Blackburn, 1888: 275 [HN] type species *Lindia angusta* Blackburn, 1888

abyssinica Fairmaire, 1893a: 29 **A:** YE **AFR**

angusta Lucas, 1846: pl. 30 (*Hypophloeus*) **E:** FR (Corse) SP (Mallorca) **N:** AG CI

exigua Marseul, 1876c: 113 **A:** JA (Honshu)

formosana Masumoto, 1982d: 67 **A:** TAI YUN **ORR**

indicola Gebien, 1920: 281 **A:** NP UP YUN **ORR**

subopaca Blair, 1930a: 18

instriata Pic, 1924: 20 **A:** TAI YUN **ORR AUR**

elongata Pic, 1924: 20

longiceps Pic, 1924: 20

orientalis Blair, 1930a: 18

punctaticeps Pic, 1924: 20 **A:** NP YUN **ORR**

subopaca Blair, 1930a: 150 **A:** NP UP **ORR**

tetraphylla Fairmaire, 1856: 534 (*Bius*) **Ei:** CR CZ FR GR IT SP **Ai:** CY IS TR **NAR**

ficicola Mulsant & Rey, 1859d: 167

rugosa Dury, 1902: 171 (*Hypophloeus*)

genus *Platybolium* Blair, 1938: 222 type species *Platybolium alvearium* Blair, 1938

alvearium Blair, 1938: 223 **A:** FUJ **ORR**

genus *Tribolium* W.S. MacLeay, 1825: 47 type species *Colydium castaneum* Herbst, 1797

subgenus *Tribolium* W.S. MacLeay, 1825: 47 type species *Colydium castaneum* Herbst, 1797

Leanum Uyttenbenboogaart, 1934: 29 type species *Tribolium myrmecophilum* Lea, 1904

Margus Dejean, 1834: 200 type species *Tenebrio ferrugineus* Fabricius, 1787a (= *Colydium castaneum* Herbst, 1797)

Stene Stephens, 1829: 19 type species *Tenebrio ferrugineus* Fabricius, 1787a (= *Colydium castaneum* Herbst, 1797)

Tenebrioloma Gebien, 1910b: 386 type species *Tenebrioloma semicostata* Gebien, 1910

castaneum Herbst, 1797: 282 (*Colydium*) **E:** AL AU AZ BE BU BY CR CT CZ EN DE FI FR GB GE GR HU IC IR IT LA LS LT LU MA NL NR NT PL RO SK SL SP SV SZ UK **N:** CI EG LB MO MR TU **A:** AF ANH BT CY FE FUJ GUA GUI GUX HAI HEN HKG HUB HUN IQ IS JA JIA JIX KZ MG NE NIN SA SCH SHX SI TAI TM TR YE YUN ZHE **COS**

bifoveolatum Duftschmid, 1812: 304 (*Tenebrio*)

ferrugineum Fabricius, 1787a: 212 (*Tenebrio*) [HN]

japonicum Thunberg, 1822 (*Hypophlaeus*)

testaceum Fabricius, 1798: 179 (*Ips*)

confusum Jacquelin du Val, 1861: 181 **E:** AL AU AZ BE BU BY CR CT CZ DE EN FI FR GB GE GR HU IC IR IT LT LU NL NR NT PL SK SP ST SV SZ UK **N:** AG CI EG LB MO TU **A:** ANH CY ES FE FUJ GUI GUX HAI HEI HEN HUB HUN IS JA JIA JIL JIX KZ LIA MG NIN SA SCH SHA SHN SHX SI TAI TM YUN ZHE "Korea" **COS** *destructor* Uyttenboogaart, 1933: xlii [= 1934: 26] **E:** BE BY CT CZ DE EN FI FR GB GE HU IC LT LU NL NR PL SK SV SZ UK **N:** CI **A:** FE MG TM YE **COS**

freemani Hinton, 1948: 49 **A:** JA GAN KA NIN SHX XIN

indicum Blair, 1930a: 148 **A:** IN SA YE **AFR ORR**

ares Hinton, 1948: 44

seres Hinton, 1948: 44

madens Charpentier, 1825: 218 (*Tenebrio*) **E:** AR AU CT CZ FR GE GR HU IC IT NT PL RO SK ST SV SZ UK **N:** EG **A:** NIN TD TM WS **NAR**

obscurum W. Redtenbacher, 1842: 17 (*Margus*)

uezumii Nakane, 1963a: 28 **A:** JA (Honshu)

tribe Ulomini Blanchard, 1845

genus *Basanopsis* Gebien, 1914b: 21 type species *Basanopsis curvipes* Gebien, 1914

curvipes Gebien, 1914b: 22 **A:** TAI

genus *Brachypophlaeus* Fairmaire, 1897d: 113 type species *Hypophloeus dimidiatipennis* Fairmaire, 1880

Leptoscapha Fairmaire, 1886b: 73 [RN, HN] type species *Stenoscapha spissicornis* Fairmaire, 1884

Stenoscapha Fairmaire, 1884b: 234 [HN] type species *Stenoscapha spissicornis* Fairmaire, 1884

lignicola Gravely, 1915: 531 (*Leptoscapha*) **A:** AP NP **ORR**

pulcher Gravely, 1915: 530 (*Leptoscapha*) **A:** AP BT

unifasciatus Kaszab, 1964c: 47 (*Leptoscapha*) **A:** JA (Ryukyus)

genus *Cenoscelis* Wollaston, 1867: 179 type species *Cenoscelis tibialis* Wollaston, 1867 (= *Uloma pulla* Erichson, 1843)

atra Pic, 1926c: 29 (*Eutochia*) **A:** JA TAI **AUR ORR**

lateralis Boheman, 1858: 94 (*Heterophaga*) **A:** HAI HKG JA TAI "Korea" **AUR ORR**

picescens Fairmaire, 1885b: cix (*Holaniara*)

pulla Erichson, 1843: 253 (*Uloma*) **A:** SA YE **AFR ORR**

distinguendus Fairmaire, 1869b: 230 (*Alphitobius*)

tibialis Wollaston, 1867: 180

genus *Cneocnemis* Gebien, 1914a: 32 type species *Uloma haemorrhoa* Fairmaire, 1893

angustula Fairmaire, 1893b: 26 (*Uloma*) **A:** HKG NP TAI **ORR**

chetri Schawaller, 1998: 464

neglecta Grimm, 2017: 21

arabica Schawaller, El Torkey & Al Dhafer, 2013: 342 **A:** SA

indica Kaszab, 1939a: 104 **A:** TAI **ORR**

newar Schawaller, 1998: 465 **A:** NP

genus *Curtopeltoides* Pic, 1916e: 14 type species *Curtopeltoides rufescens* Pic, 1916

rufonotatus Pic, 1916e: 14 **A:** NP **ORR**

genus *Uloma* Dejean, 1821: 67 type species *Tenebrio culinaris* Linnaeus, 1758

subgenus *Uloma* Dejean, 1821: 67 type species *Tenebrio culinaris* Linnaeus, 1758

Melasia Perroud & Mulsant, 1856: 160 type species *Melasia gagatina* Perroud & Mulsant, 1856 (= *Tenebrio culinaris* Linnaeus, 1758)

Prioscelida White, 1846: 11 type species *Prionoscelida tenebrionoides* White, 1846

bonzica Marseul, 1876c: 114 **A:** JA NC

robustior Nakane, 1956b: 167

bhutanensis Kaszab, 1975e: 327 **A:** BT

castanea Ren & Liu, 2004: 299 **A:** GUX YUN

contracta Fairmaire, 1882c: 226 **A:** GUX HAI YUN **ORR**

culinaris Linnaeus, 1758: 417 (*Tenebrio*) **E:** AB AL AU BE BH BU BY CR CT CZ DE EN FR GB GE GG GR HU IT LA LS LT NR PL RO SK SL SP ST SV SZ TR UK YU **A:** IN WS

cucullata Fischer von Waldheim, 1832a: 204

ferruginea Panzer, 1793c: 2 (*Tenebrio*)

gagatina Perroud & Mulsant, 1856: 161 (*Melasia*)

rubens Laporte, 1840: 220

cypraea Kraatz, 1873: 196 **E:** GR **A:** CY TR

crenulata Baudi di Selve, 1876c: 230

cypria Allard, 1876b: 21

cypriotes Marseul, 1876b: 40

euphorbiae Blair, 1930b: 177 **A:** NP SD

sherpa Schawaller, 1996a: 121

excisa excisa Gebien, 1914b: 24 **A:** FUJ GUX HAI TAI **ORR**

tschungseni Kaszab, 1954: 254

excisa nanseiensis Masumoto & Nishikawa, 1986: 27 **A:** JA (Ryukyus)

formosana Kaszab, 1941c: 54 **A:** TAI

fukiensis Kaszab, 1954: 254 **A:** FUJ TAI

gongshanica Ren & Liu, 2004: 300 **A:** YUN

hikosana Nakane, 1956b: 167 **A:** JA

hiranoi Masumoto & Akito, 2002: 148 **A:** JA (Ryukyus)

ichoi Nakane, 1956b: 168 **A:** JA (Ryukyus)

ishigakiensis M. Chûjô, 1983: 49 **A:** JA (Ryukyus)

kondoi Nakane, 1968b: 77 **A:** JA (Yakushima)

latimanus Kolbe, 1886: 202 **A:** JA NC SC

lewisi Nakane, 1956b: 167 **A:** JA

liangi Ren & Liu, 2004: 298 **A:** YUN

marseuli fujitai Masumoto, 1985a: 21 **A:** JA (Ryukyus)

marseuli marseuli Nakane, 1956b: 168 **A:** JA SC

meifengensis Masumoto, 1982c: 25 **A:** TAI

metogana Ren, 2004: 69, 71 **A:** XIZ

miyakei Masumoto & Nishikawa, 1986: 23 **A:** TAI

mulidenta Ren & Liu, 2004: 296 **A:** YUN

nakanei Masumoto & Nishikawa, 1986: 25 **A:** TAI
nanshanchica Masumoto & Nishikawa, 1986: 30 **A:** TAI
nepalica Kaszab, 1977b: 259 **A:** NP
nomurai Masumoto, 1982c: 25 **A:** TAI
picea Küster, 1846a: 77 **E:** CR
picicornis Fairmaire, 1882: 224 **A:** JA **ORR**
polita Wiedemann, 1821: 149 (*Phaleria*) **A:** BT GUX JA NP SD TAI **AFR ORR**
anomala Herbst, 1799: 23 (*Tenebrio*) [HN]
latipes Fairmaire, 1903b: 208 (*Eutochia*)
retenta Walker, 1858: 283 (*Tenebrio*)
reitteri Kaszab, 1941d: 123 **A:** SCH
rubripes minor Gebien, 1914b: 28 **A:** AP **ORR**
rubripes rubripes Hope, 1831: 31 (*Tenebrio*) **A:** AP BT NP TAI **AUR ORR**
prehimalayana Kaszab, 1975e: 325
rufa Piller & Mitterpacher, 1783: 39 (*Tenebrio*) **E:** AB AU BH BY CR CT CZ EN FI FR GE GR HU IT LA LT NR NT PL RO SK SP ST SV SZ UK **N:** MO **A:** ES FE KZ
perroudi Mulsant & Guillebeau, 1855b: 421
ruficornis Allard, 1896: 319 **A:** NP SD
kulzeri Ardoin, 1960b: 273
rufilabris Fairmaire, 1882c: 226 **A:** NP SD UP **ORR**
sakuraii Masumoto, 1983a: 89 **A:** JA (Ryukyus)
sauteri Kaszab, 1941c: 55 **A:** TAI
scita Walker, 1858: 284 **A:** BT NP PA SD **ORR**
splendida Ren & Liu, 2004: 301 **A:** YUN
spinipes Kaszab, 1975e: 328 **A:** BT NP
tadafumii Masumoto & Akita, 2003: 179 **A:** JA (Ryukyus)
takagii Masumoto & Nishikawa, 1986: 28 **A:** TAI
takarai M. Chûjô, 1983: 51 **A:** JA (Iriomote)
tsugeae Masumoto, 1982c: 23 **A:** TAI
versicolor Ren & Liu, 2004: 297 **A:** YUN

genus *Ulomimus* Bates, 1873: 201 type species *Ulomimus indicus* Bates, 1873
Pseuduloma Fairmaire, 1893: 27 type species *Pseuduloma cribricollis* Fairmaire, 1893 (= *Ulomimus indicus* Bates, 1873)
indicus Bates, 1873 **A:** GUX HAI **ORR**
cribricollis Fairmaire, 1893: 27 (*Pseuduloma*)
laminipes Gebien, 1914b: 26 (*Cneocnemis*) **A:** JA TAI ZHE (Putuo Shan Is.) "Korea"

subfamily Tenebrioninae, genus incertae sedis
genus *Anophthalmolamus* J. Ferrer, 1993b: 122 type species *Anophthalmolamus fuerteventurae* J. Ferrer, 1993
fuerteventurae J. Ferrer, 1993b: 122 **N:** CI (Fuerteventura)

subfamily **Diaperinae** Latreille, 1802

tribe Crypticini Brullé, 1832

genus *Cechenosternum* Gebien, 1921: 9 type species *Cechenosternum nigromaculatum* Gebien, 1921

bengalense Kaszab, 1960a: 165 **A:** PA "Bengale"

klapperichi Kaszab, 1960a: 164 **A:** AF NP

rufulum Motschulsky, 1873: 480 (*Neomida*) **N:** EG **AFR**

genus *Crypticus* Latreille, 1817: 298 [= 1817b: 525] type species *Helops glaber* Fabricius, 1775 (= *Tenebrio quisquilius* Linnaeus, 1760)

subgenus *Crypticopsis* Antoine, 1945: 270 type species *Crypticus corticeus* Fairmaire, 1871

corticeus Fairmaire, 1871a: 389 **N:** MO

proximus Escalera, 1914: 342

meridionalis meridionalis Escalera, 1914: 343 **N:** MO

meridionalis validus Español, 1963e: 382 **N:** MO

subgenus *Crypticus* Latreille, 1817: 298 [= 1817b: 525] type species *Helops glaber* Fabricius, 1775 (= *Tenebrio quisquilius* Linnaeus, 1760)

aksuensis Español, 1955a: 14 **E:** AB **A:** AF KZ XIN

antoinei Español, 1950a: 144 **E:** SP

arandai arandai Escalera, 1914: 344 **N:** MO (High Atlas)

gundaficus Antoine, 1945: 266

major Antoine, 1945: 266

arandai faucium Antoine, 1933b: 229 **N:** MO

atlasicus atlasicus Théry, 1931b: 125 **N:** MO (Middle Atlas)

atlasicus culminicola Antoine, 1936b: 74 **N:** MO (High Atlas)

azruensis Escalera, 1925a: 335 **N:** MO

beitunensis Ren & Zheng, 1993a: 14 **A:** XIN

burqinensis Ren & Zheng, 1993a: 17 **A:** XIN

capsianus capsianus Antoine, 1945: 268 **N:** MO

capsianus littoralis Español, 1950a: 141 **N:** LB TU

cyrenaicus Koch, 1937: 477 **N:** LB

dactylispinus Marseul, 1870a: 381 **N:** AG LB MO TU

melancroides Fairmaire, 1879c: 196

fairmairei Desbrochers des Loges, 1881: 101 **N:** AG

fernandezi Español, 1954b: 105 **N:** CI

gibbulus Quensel, 1806: 163 (*Helops*) **E:** FR IT MA PT SP **N:** AG MO TU **A:** CY JO SY

bicolor Thunberg, 1821b: 52 (*Opatrum*)

pubescens Dufour, 1820: 311 (*Pedinus*)

gomerensis Oromí, 1978: 105 **N:** CI (Gomera)

iblanensis Antoine, 1945: 265 **N:** MO

impunctatus Théry, 1932a: 193 **N:** MO

kochi Español, 1950a: 146 **E:** SP
latiusculus Ménétriés, 1849: 327 (*Heliopathes*) **A:** GAN HUB JIA KZ TM XIN
substriatus Desbrochers des Loges, 1881: 166
leptodentatus Ren & Zheng, 1993a: 13 **A:** XIN
lecerfi Théry, 1932a: 194 **N:** MO (High Atlas)
minutus Brullé, 1839: 69 **N:** CI (Gran Canaria)
navicularis amaroides Har. Lindberg, 1953: 9 **N:** CI (Tenerife)
navicularis calvus Wollaston, 1865: 59 [Appendix] **N:** CI (El Hierro)
navicularis canariensis Wollaston, 1864: 481 **N:** CI (Tenerife)
navicularis latihumeralis Har. Lindberg, 1953: 9 **N:** CI (Tenerife)
navicularis navicularis Brullé, 1839: 69 **N:** CI (Tenerife)
navicularis punctatissimus Wollaston, 1864: 480 **N:** CI (La Palma)
nitidulus Wollaston, 1865: 60 [Appendix] **N:** CI (Gomera)
nondentatus Ren & Zheng, 1993a: 12 **A:** XIN
obesus Lucas, 1847: pl. 29 **N:** AG
oblongus oblongus Wollaston, 1864: 482 **N:** CI (Tenerife)
oblongus subnitens Har. Lindberg, 1950: 12 **N:** CI (Tenerife)
obtusus Reitter, 1896b: 74 **A:** KZ (Aral sea)
pentheri Reitter, 1905e: 278 **A:** TR
petraeus Escalera, 1913: 49 **N:** MO
pettersoni J. Ferrer & L. Soldati, 1999: 73 **A:** TR
platystaturalis Ren & Zheng, 1993a: 16 **A:** XIN
platytibialis Ren & Zheng, 1993a: 10 **A:** XIN
pubens balearicus Español, 1950a: 144 **E:** SP (Baleares)
pubens pubens Fairmaire, 1881b: cxxxii **E:** SP **N:** MO
quisquilius aprutianus Gridelli, 1949: 272 **E:** IT
quisquilius pyrenaeus Baudi di Selve, 1876a: 33 **E:** FR SP
quisquilius quisquilius Linnaeus, 1760: 226 (*Tenebrio*) **E:** AL AR AU BE EH BU BY CR CT CZ DE EN FI FR GB GE GR HU IR IT LA LT NL NR NT PL RO SK SP ST SV SZ TR UK YU **A:** ES FE KI KZ MG TR WS XIN
glaber Fabricius, 1775: 258 (*Helops*)
laevis Gmelin, 1790: 2010 (*Pimelia*)
laticollis Panzer, 1796d: 1 (*Helops*)
luctuosus Schrank von Paula, 1781: 219 (*Tenebrio*)
niger Geoffroy, 1785: 158 (*Tenebrio*)
pusillus Herbst, 1799: 199 (*Blaps*)
opacus Brancsik, 1899: 100
strauchi Roubal, 1911a: 133
rhaticus Antoine, 1945: 263 **N:** MO
schrammi Antoine, 1945: 269 **N:** MO
sibiricus Solsky, 1871: 373 **A:** ES

subpunctatus Walker, 1871: 15 **N:** EG
tachdirtensis Antoine, 1945: 263 **N:** MO
tauri Pic, 1907c: 123 **A:** TR
tenietensis Desbrochers des Loges, 1890: clxxxv **N:** AG
thessalicus Reitter, 1896d: 150 **E:** GR
yosiii Nakane, 1966: 243 **A:** AF
zuberi Marseul, 1870a: 382 **E:** AB ST UK **A:** AF KZ MG UZ XIN
ovalis Ballion, 1878: 321
subgenus *Platycrypticus* Español, 1952b: 117 [RN] type species *Crypticus viaticus* Fairmaire, 1851
Ulomoides Escalera, 1927b: 501 [HN] type species *Crypticus viaticus* Fairmaire, 1851
brusteli F. Soldati & Kakiopoulos, 2010: 226 **E:** GR (Kos, Lesvos, Samos)
castaneus Baudi di Selve, 1876a: 34 **E:** GR **A:** TR
frigidus Escalera, 1927b: 503 **E:** SP (Andaluzía)
pardoi Español, 1952b: 118 **E:** SP (Ciudad Real)
punctatolineatus Fairmaire, 1880d: 251 **N:** MO
vaucheri Bedel, 1906a: 272 **N:** MO
viaticus Fairmaire, 1851: 528 **E:** SP
ulomoides Fairmaire, 1852: 85
subgenus *Seriscius* Motschulsky, 1845a: 77 type species *Seriscius pubescens* Motschulsky, 1845 (= *Crypticus rufipes* Gebler, 1830)
paradoxus Reitter, 1896d: 149 **N:** MO
rufipes Gebler, 1829a: 125 **E:** ST **A:** BEI ES FE GAN MG NIN NMO SCH WS XIN "Korea"
asiaticus Reitter, 1889k: 698
fuscovariegatus Reitter, 1889k: 698
ovatulus Reitter, 1896d: 150
pubescens Motschulsky, 1845a: 77 (*Seriscius*)
soricinus Fairmaire, 1887f: 324

genus *Ellipsodes* Wollaston, 1854: 485 type species *Sphaeridium glabratum* Fabricius, 1781
subgenus *Anthrenopsis* Koch, 1950a: 74 type species *Platydema scriptipenne* Fairmaire, 1875 (= *Basites ziczac* Motschulsky, 1873)
hieroglyphicus Fairmaire, 1896a: 25 (*Platydema*) **A:** NP **ORR**
scriptus Lewis, 1894: 396 (*Platydema*) **A:** ANH FUJ GUA GUI GUX HEB HEN HUB HUN JA JIA JIX NMO NP SCH SD SHA SHN SHX TAI UP YUN ZHE **AUR NAR ORR**
minutissimus Gebien, 1914b: 6 (*Crypticus*)
submaculatus Champion, 1894a: 362 (*Crypticus*)
ziczac nuristanicus Kaszab, 1960a: 163 **A:** AF
ziczac ziczac Motschulsky, 1873: 475 (*Basides*) **A:** HEN HUB JIX NP QIN SD **AFR AUR ORR**
bifasciatus Chevrolat, 1878a: 222 (*Platydema*)
scriptipennis Fairmaire, 1875d: xxxiii (*Platydema*)

subgenus *Ellipsodes* Wollaston, 1854: 485 type species *Sphaeridium glabratum* Fabricius, 1792
glabratus Fabricius, 1781: 78 (*Sphaeridium*) **N:** MR
oblongior Wollaston, 1857a: 150 **N:** MR

genus *Lamprocrypticus* Español, 1950b: 127 type species *Crypticus alpinus* Comolli, 1837
subgenus *Lamprocrypticus* Español, 1950b: 127 type species *Crypticus alpinus* Comolli, 1837
alpinus Comolli, 1837: 26 (*Crypticus*) **E:** IT SZ (Valais)
baenai Castro Tovar, 2012: 99 **E:** SP
baguenai Español, 1948b: 488 (*Crypticus*) **E:** SP
kraatzii C. Brisout de Barneville, 1866: 383 (*Crypticus*) **E:** PT SP
minutissimus Escalera, 1934a: 211 (*Crypticus*) **E:** SP (Sierra de Segura)
pusillus Rosenhauer, 1856: 212 (*Crypticus*) **E:** SP
tonsilis Bedel, 1893: 103 (*Crypticus*) **N:** AG MO
zophosioides Heyden, 1870: 131 (*Crypticus*) **E:** PT SP

genus *Oochrotus* Lucas, 1852: xxix type species *Oochrotus unicolor* Lucas, 1852
glaber glaber Demaison, 1905: 287 **E:** GR (Rhodes) **A:** CY SY TR
  boyadjiani Lokay, 1907: 91
glaber rhodicus Koch, 1935e: 319 **E:** GR (Rhodes)
unicolor ardoini Canzoneri, 1961a: 21 **E:** IT
unicolor lusitanicus Canzoneri, 1961a: 16 **E:** PT
unicolor moltonii Canzoneri, 1961a: 20 **E:** IT (Sicilia)
unicolor unicolor Lucas, 1852: xxix **E:** FR IT (Sardinia) SP **N:** AG MO TU
  chilivanii Rallo, 1974: 62
  espagnoli Canzoneri, 1961a: 20
  hispanus Canzoneri, 1961a: 17
  laurae Canzoneri, 1961: 23
  meridionalis Canzoneri, 1961a: 19
  sardous Canzoneri, 1961a: 23

genus *Pseudoseriscius* Español, 1950b: 125 type species *Pedinus pruinosus* Dufour, 1820
subgenus *Pseudoseriscius* Español, 1950b: 125 type species *Pedinus pruinosus* Dufour, 1820
adspersus adspersus Küster, 1851: 51 (*Crypticus*) **E:** SP
adspersus nebulosus Fairmaire, 1871a: 389 (*Crypticus*) **N:** AG LB MO TU
  uniformis Reitter, 1896d: 150 (*Crypticus*)
adspersus timaditensis Théry, 1932a: 194 (*Crypticus*) **N:** MO (Middle Atlas)
cameroni Reitter, 1902e: 194 (*Crypticus*) **E:** MA
fonti Escalera, 1923d: 132 (*Crypticus*) **N:** CI MO (Western Sahara)
  alluaudi Peyerimhoff, 1942a: 11 (*Crypticus*)
gridellii Español, 1949: 211 (*Crypticus*) **E:** IT **N:** LB

griseovestis Fairmaire, 1879c: 192 (*Crypticus*) **E:** IT (Linosa Is.) **N:** AG EG LB TU **A:** CY JO KU
pelitus Reitter, 1896d: 150 (*Crypticus*)
helvolus adriaticus Español, 1949: 208 (*Crypticus*) **E:** AL GR IT
helvolus helvolus Küster, 1852a: 83 (*Crypticus*) **E:** IT **N:** TU
ibicensis Español, 1955b: 27 **E:** SP (Ibiza Is.)
maculosus maculosus Fairmaire, 1871a: 390 (*Crypticus*) **A:** IN IS LE SA SY
maculosus murinus Allard, 1882b: lxxxvii [= 1883: 30] (*Crypticus*) **N:** EG **A:** SI
mus Reitter, 1896d: 149 (*Crypticus*)
mollis Reitter, 1896d: 150 (*Crypticus*) **N:** AG LB TU
munyozi Viñolas, 1997b: 97 **E:** SP (Almería)
normandi normandi Español, 1949: 209 (*Crypticus*) **N:** TU
normandi pacificii Leo, 1982b: 104 **E:** IT
olivierii olivierii Desbrochers des Loges, 1881: 103 (*Crypticus*) **N:** AG LB MO TU
olivierii sardiniensis Leo, 1982b: 102 **E:** IT (Sardinia)
pruinosus Dufour, 1820: 310 (*Pedinus*) **E:** FR PT SP **N:** MO
richteri Kaszab, 1959a: 3 **A:** IN

tribe Diaperini Latreille, 1802

genus *Adelina* Dejean, 1835: 315 type species *Cucujus planus* Fabricius, 1801
Doliema Pascoe, 1860a: 50 type species *Doliema platisoides* Pascoe, 1860
Schedarosus Reitter, 1876a: 42 type species *Schedarosus cucujiformis* Reitter, 1876 (= *Pytho pallida* Say, 1823)
platysoides Pascoe, 1860a: 50 (*Doliema*) **A:** YUN **ORR**
turcica Reitter, 1877b: 8 (*Sitophagus*) **E:** GR RO TR **A:** IS

genus *Alphitophagus* Stephens, 1832: 12 type species *Alphitophagus quadripustulatus* Stephens, 1832 (= *Diaperis bifasciatus* Say, 1824)
Phyletes L. Redtenbacher, 1845: 128 type species *Phylethus populi* L. Redtenbacher, 1849 (= *Diaperis bifasciatus* Say, 1824)
bifasciatus Say, 1824a: 268 (*Diaperis*) **E:** AB AL AR AU BE BY CR CT CZ DE EN FI FR GB GE GR HU IT LT MC NL NR PL SK SP ST SV UK **N:** EG TU **A:** AF CY GAN GUA HAI HEB HEN HUB IS JA JIA JIL JIX KZ LIA NIN NMO SCH SHA SHN SHX SY TAI TM TR XIN "Korea" **COS**
iudaeus Roubal, 1929b: 97
lilliputanus Carter, 1937: 130 (*Platydema*)
pictus Ménétriés, 1832: 203 (*Diaperis*)
populi L. Redtenbacher, 1848: 589 (*Phylethus*)
quadripustulatus Stephens, 1832: 12
carteianus Castro Tovar, Torres & Baena, 2012: 75 **E:** SP
convexicollis Reitter, 1911d: 268 **A:** TR
nigrotrabeatus Fairmaire, 1879c: 197 **N:** MO (West Sahara)

obtusangulus J. Müller, 1904: 209 **E:** CR (Meleda) GR FR SP
socotranus Schawaller, 2004c: 454 **A:** YE (Suqutra)
xaxarsi Reitter, 1914b: 82 (*Alphitobius*) **E:** SP (Baleares)

genus *Ceropria* Laporte & Brullé, 1831: 396 type species *Helops indutus* Wiedemann, 1819
Dictysomorphus Pic, 1921c: 24 type species *Dictysomorphus dentaticornis* Pic, 1921
Epilamprus Gistel, 1848: xiv [UE]
Epilampus Dejean 1834: 198 [URN] type species *Helops indutus* Wiedemann, 1819
chinensis Masumoto, 1995a: 3 **A:** FUJ
formosana Gebien, 1914b: 20 **A:** TAI
induta induta Wiedemann, 1819: 164 (*Helops*) **A:** AP BT FUJ HAI JA NP TAI YUN XIZ "Korea" **ORR**
kinugasai Masumoto, 1982a: 151
subocellata Laporte & Brullé, 1831: 398
induta purpurina Gebien, 1925a: 277 **A:** AP HP NP SD **ORR**
krausei Masumoto, 1994b: 772 **A:** YUN
laticollis Fairmaire, 1903a: 13 **A:** AP FUJ JA SC SD SW YUN **ORR**
opacipennis Pic, 1921c: 26
lii Ando & Ren, 2006: 82 **A:** YUN
schenklingi Gebien, 1914b: 19 **A:** TAI
serripes Gebien, 1925a: 284 **A:** NP YUN **ORR**
striata Lewis, 1894: 399 **A:** JA SC
subocellata Laporte & Brullé, 1831: 398 **A:** AP NP SD **ORR**
sulcifrons Harold, 1878c: 353 **A:** JA SC (Jejudo Is.)
superba Wiedemann, 1823: 43 (*Cnodalon*) **A:** "South China" **ORR**
chrysosticta Hope, 1843: 65 [= 1845: 52] (*Epilampus*)
festiva Laporte & Brullé, 1831: 400
taiwana Masumoto, 1995a: 1 **A:** TAI
thailandica Masumoto, 1995b: 724 **A:** YUN **ORR**
umbrata Marseul, 1876c: 107 (*Platydema*) **A:** JA (Kyushu)
versicolor Laporte & Brulle, 1831: 401 **A:** YUN **ORR**
concolor Pic, 1923c: 20
impressifrons Fairmaire, 1882c: 222
yaoi Ren, 2004: 68, 71 **A:** XIZ

genus *Cynaeus* J.L. LeConte, 1862: 233 type species *Platydema angustum* J.L. LeConte, 1851
angustus J.L. LeConte, 1851: 149 (*Platydema*) **Ei:** FI FR GE HU LA PL ST SV UK **Ai:** JA SC NAF **ORR**
opacus Champion, 1886a: 156 **Ei:** FI SV **NTR**

genus *Diaperis* Geoffroy, 1762: 337 type species *Chrysomela boleti* Linnaeus, 1758
Allophasia Pascoe, 1871: 351 type species *Allophasia fryi* Pascoe, 1871

boleti Linnaeus, 1758: 371 (*Chrysomela*) **E:** AR AU BE BU BY CT CZ DE EN FI FR GB GE GR HU IT LA LS LT NL NR NT PL PT SK SP ST SV SZ UK **N:** AG MO TU **A:** CY FE ES IS KI KZ SY TR WS

bipustulata Laporte & Brullé, 1831: 337

cypria Baudi di Selve, 1876c: 229

fasciata Scopoli, 1763: 79 (*Coccinella*)

fagniezi Méquignon, 1947: 60

fungi Motschulsky, 1873: 466

interrupta Heyden, 1890b: 78

morio Baudi di Selve, 1876c: 228

posticalis Heyden, 1890b: 78

quercana Baudi di Selve, 1876c: 228

lewisi fujitai Akita & Masumoto, 2015c: 434 **A:** JA (Izu Is.)

lewisi intersecta Gebien, 1914b: 15 **A:** HKG JA TAI **ORR**

lewisi lewisi Bates, 1873a: 14 **A:** ES FE GUA HKG HUB JA NC SC TAI "North China" **ORR**

rubrofasciata Reitter, 1879: 226

sinensis Gebien, 1925a: 155

niponensis Lewis, 1887b: 217 **A:** FE JA

quadrimaculata Pic, 1926c: 22 **A:** XIZ

sanguinipennis Bates, 1873a: 14 **A:** YUN **ORR**

genus *Espagnolina* Kaszab, 1965a: 117 type species *Espagnolina assamica* Kaszab, 1965

assamica Kaszab, 1965a: 117 **A:** AP

genus *Falsocosmonota* Kaszab, 1962a: 75 type species *Falsocosmonota cheni* Kaszab, 1962

cheni Kaszab, 1962a: 77 **A:** GUA HAI TAI

genus *Gnatocerus* Thunberg, 1814: 47 type species *Gnatocerus ruber* Thunberg, 1814 (= *Trogosita cornuta* Fabricius, 1798)

subgenus *Echocerus* Horn, 1870: 366 type species *Trogossita maxillosa* Fabricius, 1801

maxillosus Fabricius, 1801a: 155 (*Trogossita*) **E:** BE FR GB IT LU NL **N:** CI MR **A:** CH JA **AFR NAR NTR ORR**

subgenus *Gnatocerus* Thunberg, 1814: 47 type species *Gnatocerus ruber* Thunberg, 1814 (= *Trogosita cornuta* Fabricius, 1798)

Cerandria Dejean, 1834: 200 type species *Trogosita cornuta* Fabricius, 1798

Gnathocerus Agassiz, 1846a: 73 [UE]

Sicinus Champion, 1886a: 146 type species *Sicinus guatemalensis* Champion, 1886

cornutus Fabricius, 1798: 51 (*Trogosita*) **E:** AU AZ BE CR CT CZ DE EN FI FR GB GE HU IC IR IT LA LS LT LU MA NL NR NT PL SK SP ST SV SZ UK **N:** CI LB MO MR TU **A:** FE FUJ GUA GUI GUX HEN HKG JA JIA KZ MG NMO SCH TAI TM YUN **COS**

cucullatus Montrouzier, 1860: 294 (*Neomida*)

laeviusculus Stephens, 1832: 10 (*Uloma*)
maxillosus Palisot de Beauvois, 1805: 125 (*Trogossita*) [HN]
ruber Thunberg, 1814: 47

genus *Ischnodactylus* Chevrolat, 1877d: 173 type species *Ischnodactylus quadridentatus* Chevrolat, 1877
bisbifasciatus Gebien, 1925b: 446 **A:** HKG TAI
ruficornis Pic, 1925g: 16 [Pic, 1925j: 437] (*Basides*)
formosanus Gebien, 1925b: 445 **A:** FUJ TAI
iriomotensis Ando, 1978: 81 **A:** JA (Iriomote)
parallelicornis Nakane, 1956b: 164 (*Platydema*) **A:** JA
rubromarginatus yunnanus Kaszab, 1965c: 284 **A:** YUN

genus *Neomida* Latreille, 1829: 29 type species *Ips haemorrhoidalis* Fabricius, 1787
Arrhenoplita Kirby, 1837: 235 type species *Ips haemorrhoidalis* Fabricius, 1787
Evoplus J.L. LeConte, 1866: 128 type species *Evoplus ferruginea* J.L. LeConte, 1866
Hoplocephala Agassiz, 1846: 185 [UE]
Oplocephala Laporte & Brullé, 1831: 338 type species *Ips haemorrhoidalis* Fabricius, 1787
haemorrhoidalis Fabricius, 1787b: 378 (*Ips*) **E:** AL AU BH BU BY CR CT CZ DE EN FI FR HU GE GR IT LA LT NR NT PL RO SK SL SP ST SV SZ UK **A:** AF ES CY FE IN JA KI TR WS
asiatica Lewis, 1894: 392 (*Arrhenoplita*)
cornifrons Rossi, 1792: 92 (*Tenebrio*)
quadricornis Motschulsky, 1873: 466 (*Oplocephala*) **E:** AB GG **A:** IN
quadricornis Reitter, 1884a: 8 (*Hoplocephala*) [HN]

genus *Pelleas* Bates, 1872: 98 type species *Tenebrio crotchii* Wollaston, 1865
crotchii Wollaston, 1865: 62 (*Tenebrio*) **N:** CI

genus *Pentaphyllus* Dejean, 1821: 68 type species *Mycetophagus testaceus* Hellwig, 1792
Iphicorynus Jacquelin du Val, 1861: 299 type species *Pentaphyllus melanophthalmus* Mulsant 1854 (= *Nitidula chrysomeloides* Rossi, 1792)
biconiger Gebien, 1914: 23 **A:** JA (Kyushu, Ryukyus) **ORR**
lewisi Kaszab, 1982a: 63
chrysomeloides Rossi, 1792: 24 (*Nitidula*) **E:** AB AL AU BU CR DE EN FI FR GE GG GR HU IT LA RO SK SL SP ST SV SZ UK **N:** MO TU **A:** AF CY IN TD TM TR
melanophthalmus Mulsant, 1854: 197
ensifera Fauvel, 1904: 172 (*Arrhenoplita*) **A:** JA (Ryukyus) TAI **AUS ORR**
dilatipes Shibata, 1978: 22
punctatus Kulzer, 1957: 210
izuanus Nakane, 1978: 161 **A:** JA (Izu Is.)
kampiraensis Akita & Masumoto, 2015c: 435 **A:** JA (Ryukyus)

nitidulus Reitter, 1884a: 8 **E**: AB
quadricornis Gebien, 1914: 23 **A**: JA (Ryukyus) TAI **ORR**
philippinensis Kaszab, 1956a: 94
tokarensis Nakane, 196a3: 27
reibnitzi Schawaller & Merkl, 2012: 723 **A**: CY TR
testaceus Hellwig, 1792: 400 (*Mycetophagus*) **E**: AL AU BE BY CR CT CZ DE FI FR GB GE GG GR HU IT MA NL PL RO SK SP ST SZ UK **N**: AG TU **A**: TD
barbarus Motschulsky, 1873: 481

genus *Platydema* Laporte & Brullé, 1831: 350 type species *Diaperis violacea* Fabricius, 1790
Anisochara Gebien, 1925d: 101 type species *Anisochara gynadromorpha* Gebien, 1925
Histeropsis Chevrolat, 1878a: 221 type species *Platydema americanum* Laporte & Brullé, 1831
Typhobia Pascoe, 1869: 279 type species *Typhobia fuliginea* Pascoe, 1869
alticornis Gravely, 1915: 524 **A**: AP NP TAI **ORR**
gyandromorpha Gebien, 1925d: 102 (*Anisochara*)
aurimaculata Gravely, 1915: 523 **A**: AP TAI **ORR**
cederholmi Kaszab, 1980a: 167
monoceratoides Masumoto, 1982a: 145
brahma Schawaller, 2003b: 271 **A**: NP
caesifrons Marseul, 1876b: 38 **N**: EG
capreola Chevrolat, 1877d: 170 (*Hoplocephala*) **A**: NP **ORR**
celata Nakane, 1983b: 146 **A**: JA
coerulea Gebien, 1925c: 570 **A**: TAI
dejeanii Laporte & Brullé, 1831: 351 **E**: AU BH BY CR CT CZ FR HU GE GR PL RO SK SL ST SZ UK YU **A**: ES FE WS
detersa Walker, 1858: pl. 4 **A**: AP FE NP TAI YUN **AUR ORR**
endoi Masumoto, 1984: 22 **A**: TAI
europaea Laporte & Brullé, 1831: 365 **E**: CR FR GR IT SP ST SZ UK **N**: AG TU **A**: TR
atra Motschulsky, 1873: 477 (*Neomida*)
subsericea Motschulsky, 1873: 478 (*Neomida*)
flavopicta Gebien, 1914b: 17 **A**: TAI **ORR**
fumosa akaashi Akita & Masumoto, 2015c: 439 **A**: JA (Ryukyus)
fumosa formosana Gebien, 1925c: 588 **A**: TAI
morimotoi Nakane, 1968a: 76
fumosa fumosa Lewis, 1894: 395 **A**: FUJ HUB JA SC ZHE
fumosa kawaii Nakane, 1968a: 76 **A**: JA (Ryukyus)
ganesha Schawaller, 2003b: 271 **A**: NP
guangxica Schawaller, 2004b: 24 **A**: GUX
haemorrhoidalis Gebien, 1914b: 16 **A**: FUJ NP TAI **ORR**
higonia Lewis, 1894: 394 **A**: JA
imamurai Akita & Masumoto, 2015c: 436 **A**: JA (Ryukyus)

kochi Schawaller, 2004c: 452 **A:** YE (Suqutra)
koreana Mt. Chûjô, 1992: 37 **A:** SC
krishna Schawaller, 2003b: 271 **A:** NP
kulzeriana Nakane, 1977: 158 **A:** JA (Bonin Is.)
kurama Nakane, 1963a: 26 **A:** JA SC TAI
lyncea Lewis, 1894: 395 **A:** FE JA SC
ussuriana Kaszab, 1977a: 118
marseuli Lewis, 1894: 393 **A:** HKG JA TAI ZHE **ORR**
benakatensis Masumoto & Makihara, 1997: 119
monoceros Gebien, 1925c: 556 **A:** TAI **ORR**
quadrispilotus Chevrolat, 1878c: cxlix (*Histeropsis*)
nigroaenea Motschulsky, 1861: 18 **A:** JA SC TAI "Korea"
musiva Harold, 1878a: 78
nigropicta Nakane, 1963a: 27 **A:** FUJ JA TAI **ORR**
pallidicollis Lewis, 1894: 398 (*Alphitophagus*) **A:** CH JA TAI **AUR ORR**
sodalis Waterhouse, 1894: 70
parachalcea Masumoto, 1982a: 147 **A:** FUJ GUI HEN HUB SHA TAI
zoltani Masumoto, 1985c: 121
pentaphylloides Kaszab, 1980b: 186 **A:** YUN **ORR**
planicornis M. Chûjô, 1978: 3 **A:** JA (Ryukyus)
recticornis amami Akita & Masumoto, 2015c: 440 **A:** JA (Ryukyus)
recticornis recticornis Lewis, 1894: 394 **A:** JA SC TAI **ORR**
reitteri Pic, 1934f: 85 (*Ceropria*) **A:** SCH
sakishimensis Nakane, 1973b: 106 **A:** JA TAI
satoi Kaszab, 1964c: 45 **A:** JA (Tsushima Is.)
sauteri omoto Nakane, 1973b: 107 **A:** JA (Ryukyus)
sauteri sauteri Gebien, 1914b: 15 **A:** TAI
sawadai Masumoto, 1991b: 25 **A:** JA (Ryukyus) TAI
semimetallica Blair, 1930b: 177 **A:** NP SD XIZ
martensi Schawaller, 1994: 262
subfascia kumejimaensis Akita & Masumoto, 2015c: 442 **A:** JA (Ryukyus)
subfascia nishii Akita & Masumoto, 2015c: 441 **A:** JA (Ryukyus)
subfascia subfascia Walker, 1858: 284 (*Alphitophagus*) **A:** FUJ JA NP SC TAI **AFR AUR ORR**
celeba Chevrolat, 1877d: 177 (*Hoplocephala*)
diversidens Fairmaire, 1893b: 24 (*Hoplocephala*)
houana Pic, 1929c: 33
japana Marseul, 1876c: 109 (*Alphitophagus*)
picicollis Motschulsky, 1873: 474 (*Basides*)
subfascia yaeyama Nakane, 1973b: 105 **A:** JA (Ryukyus)
subfascia yoroi Akita & Masumoto, 2015c: 442 **A:** JA (Ryukyus)
sylvestris Lewis, 1894: 394 **A:** JA

takeii Nakane, 1956b: 165 **A:** JA TAI
terusane Masumoto, 1984: 21 **A:** TAI
tricuspis Motschulsky, 1873: 479 (*Neomida*) **A:** TAI **AUS ORR**
reflexum Chevrolat, 1878a: 222
timorensis Marseul, 1876a: 108
tristis Laporte & Brullé, 1831: 359 **E:** AR GG ST UK **A:** IN TM
armeniaca Desbrochers des Loges, 1881: 129 (*Scaphidema*)
cophosioides Faldermann, 1837: 64 (*Neomida*)
tuchinlongi Masumoto, 1982a: 149 **A:** TAI
violacea Fabricius, 1790: 219 (*Diaperis*) **E:** AL AR AU BE BH BU BY CR CT CZ DE EN FI FR GB GE GR HU IT LA LS LT NL NT PL RO SK SL SP ST SV SZ UK
yangmingensis Masumoto, 1982a: 148 **A:** TAI
yunnanica Schawaller, 2004b: 46 **A:** YUN

genus *Sitophagus* Mulsant, 1854: 264 type species *Sitophagus solieri* Mulsant, 1854 (= *Uloma hololeptoides* Laporte, 1840)
hololeptoides Laporte, 1840: 220 (*Uloma*) **N:** MR **A:** TR
solieri Mulsant, 1854: 265

genus *Ulomoides* Blackburn, 1888: 274 type species *Ulomoides humeralis* Blackburn, 1888
Martianus Fairmaire, 1893h: 540 type species *Martianus castaneus* Fairmaire, 1893 (= *Palembus ocularis* Casey, 1891)
Palembus Casey, 1891: 65 type species *Palembus ocularis* Casey, 1891
Phayllidius Gebien, 1922: 451 type species: *Phayllidius dispar* Gebien, 1922
Tenebriomimus Kolbe, 1901: 342 type species *Tenebriomimus adansoniarum* Kolbe, 1901 (= *Palembus ocularis* Casey, 1891)
dermestoides Chevrolat, 1878a: 242 (*Histeropsis*) **E:** SVi **A:** FUJ GUA GUX HAI HUB JA JIA TAI "Korea" **AUR ORR**

genus *Yamatotakeru* Ando, 2015: 385 type species *Ischnodactylus loripes* Lewis, 1894
loripes Lewis, 1894: 392 (*Ischnodactylus*) **A:** HEI JA SC

tribe Gnathidiini Gebien, 1921

genus *Enanea* Lewis, 1894: 467 type species *Enanea testacea* Lewis, 1894
chujoi Ando & Mt. Chûjô, 2005: 72 **A:** JA (Amami-Ôshima)
shuneii Masumoto, Akita & Lee, 2019b: 160 **A:** TAI
testacea Lewis, 1894: 467 **A:** JA (Kyushu)
yakushimana Nakane, 1984: 597 **A:** JA (Yakushima)

genus *Menimus* Sharp, 1876: 73 type species *Menimus batesi* Sharp, 1876
subgenus *Menimus* Sharp, 1876: 73 type species *Menimus batesi* Sharp, 1876
Ceramba Fauvel, 1904: 206 type species *Ceramba hydrovatina* Fauvel, 1904

Microcilibe Carter, 1919: 147 type species *Microcilibe castaneum* Carter, 1919
Neomenimus Kaszab, 1939b: 190 type species *Neomenimus clavatus* Kasazab, 1939
Paita Fauvel, 1904: 173 type species *Paita setosella* Fauvel, 1904
Tjikoraia Pic, 1921c: 18 type species *Tjikoraia javana* Pic, 1921
belousovi Medvedev, 2007a: 868 **A:** YUN ORR
formosanus Ando, 2018: 129 **A:** TAI
gairibansicus Schawaller, 2016a: 192 **A:** SD
hunlicus Schawaller, 2016a: 192 **A:** AP
kabaki Medvedev, 2007a: 873 **A:** YUN
medvedevi Schawaller, 2009c: 364 **A:** YUN
niponicus Lewis, 1894: 398 **A:** JA
puetzi Schawaller, 2009c: 364 **A:** YUN
wittmeri Kaszab, 1982a: 64 **A:** SD
yunnannus Medvedev, 2007a: 870 **A:** YUN

genus *Micropeneta* Pic, 1921: 19 type species *Micropeneta testacea* Pic, 1921
qilin Li & Jiang, 2019: 196 **A:** GUI

genus *Nanocaecus* Schawaller & Purchart, 2012: 305 type species *Nanocaecus hlavaci* Schawaller & Purchart, 2012
hlavaci Schawaller & Purchart, 2012: 305 **A:** YE (Suqutra)

genus *Sakaiomenimus* Ando, 2003c: 135 type species *Sakaiomenimus todai* Ando, 2003
oblongus Lewis, 1894: 398 (*Pentaphyllus*) **A:** JA
todai Ando, 2003c: 139

genus *Taiwanomenimus* Masumoto, Akita & Lee, 2019b: 161 type species *Taiwanomenimus nakasatoi* Masumoto, Akita & Lee, 2019
nakasatoi Masumoto, Akita & Lee, 2019b: 161 **A:** TAI

tribe Hypophlaeini Billberg, 1820

genus *Corticeus* Piller & Mitterpacher, 1783: 87 type species *Corticeus unicolor* Piller & Mitterpacher, 1783
subgenus *Corticeus* Piller & Mitterpacher, 1783: 87 type species *Corticeus unicolor* Piller & Mitterpacher, 1783
Hypophlaeus Fabricius, 1790: 222 type species *Hypophlaeus castaneus* Fabricius, 1790 (= *Corticeus unicolor* Piller & Mitterpacher, 1783)
Paraphloeus Seidlitz, 1894: 553 type species *Hypophloeus longulus* Gyllenhal, 1827
Syncolydium Kolbe, 1898: 110 type species *Syncolydium glabratum* Kolbe, 1897
amamiensis Kaszab, 1964c: 48 (*Hypophloeus*) **A:** JA (Honshu, Ryukyus)
baehri Bremer & Grimm, 2019: 52 **A:** YUN
basalis Reitter, 1884a: 9 **E:** AB IN

becvari Bremer, 1999: 29 **A:** YUN
bicolor G.-A. Olivier, 1790: 18: 12 (*Ips*) **E:** AL AR AU BE BH BU BY CR CT CZ DE EN FI FR GB GE GR HU IT LA LT NL NR NT PL RO SK SP ST SV SZ UK **N:** AG **A:** ES FE WS
bicoloroides Roubal, 1933b: 78 (*Hypophloeus*) **E:** AU CZ FR GE HU IT RO SK SZ
colydioides Lewis, 1894: 468 **A:** JA SC
engelhardtiae Blair, 1937: 36 **A:** SD TAI **ORR**
zoltani Masumoto, 1981e: 92 (*Hypophloeus*)
fasciatus Fabricius, 1790: 223 (*Hypophlaeus*) **A:** TR **E:** AL AU BY CR CT CZ DE EN FR GE GR HU IT LA PL RO SK SP ST SV SZ UK **N:** AG MO TU
pusillus Rossi, 1790a: 50 (*Ips*)
rufobasalis Pic, 1917g: 5 (*Hypophloeus*)
unicolor Ragusa, 1892b: 31 [HN]
fraxini Kugelann, 1794: 527 (*Hypophloeus*) **E:** AU BE BU BY CR CT CZ EN FI GB GE GR HU IT LA LT NL NR NT PL RO SK ST SV SZ UK **A:** ES FE KZ MG TR WS
ferrugineus Mulsant, 1854: 254 (*Hypophloeus*)
fusciventris Reitter, 1884e: 256 **A:** CY **E:** BH CR GR
haoi J.-K. Li, 1992: 165 (*Hypophloues*) **A:** LIA
krueperi Obenberger, 1922: 42 (*Hypophloues*) **E:** GR
leonhardi Reitter, 1906f: 241 (*Hypophloeus*) **E:** SP
linearis Fabricius, 1790: 223 (*Hypophlaeus*) **E:** AU BE BY CT CZ DE EN FI FR GB GE GR HU IT LA LT NL NR NT PL RO SK SP ST SV SZ **N:** TU **A:** ES FE WS
perrisi Pic, 1917g: 5 (*Hypophloeus*)
longicollis Wollaston, 1867: 206 **A:** YE **AFR**
longulus Gyllenhal, 1827: 517 (*Hypophloeus*) **E:** AL AU BY CR CT CZ EN FI FR GE GR HU LA LT NR NT PL RO SK SL SP ST SV SZ UK YU **A:** ES FE MG WS
maehleri maehleri Kulzer, 1957: 233 (*Hypophloeus*) **A:** TAI **ORR PAC**
maehleri mayumiae Masumoto, 1983b: 175 (*Hypophloeus*) **A:** JA (Ryukyus)
nepalicus Bremer, 1999: 27 **A:** BT NP
okuezonis Kôno, 1939: 85 (*Hypophloeus*) **A:** FE JA
pini Panzer, 1799b: 19 (*Hypophloeus*) **E:** AU BE BY CR CT CZ EN FR GE GR HU IT LS PL PT RO SK SP ST SZ UK **N:** CI MO TU **A:** CY ES FE IS LE MG SY TR
ferrugineus Creutzer, 1799: 126 (*Hypophloeus*)
leonardi Reitter, 1906f: 241 (*Hypophlaeus*)
nocivus Wollaston, 1862: 442 (*Hypophloeus*)
praenuntius Bremer, 1999: 23 **A:** HP NP PA UP
rufithorax Pic, 1903m: 162 **E:** GR SP
socotranus Purchart & Schawaller 2012: 317 **A:** YE (Suqutra)
stebbingi Blair, 1921: 5 (*Hypophloeus*) **A:** HP UP
suberis Lucas, 1846: pl. 30 (*Hypophloeus*) **E:** AU CR CZ FR GE HU IT RO SK ST **N:** AG TU **A:** CY IS TR
insidiosus Mulsant, 1854: 384 (*Hypophloeus*)
rufulus Rosenhauer, 1847: 32 (*Hypophloeus*)

suturalis Paykull, 1800: 323 (*Hypophlaeus*) **E:** BY CT CZ EN FI LA LT NR NT PL RO SK ST SV **A:** ES FE JA
bivittatus Reitter, 1875: 362 (*Hypophloeus*)
turcicus F. Soldati, 2019: 167 **A:** TR
ulomoides Blair, 1921: 4 (*Hypophloeus*) **A:** KA PA
unicolor Piller & Mitterpacher, 1783: 87 **E:** AR AU BE BU BY CR CT CZ DE EN FR GB GE GR HU IT LA LS LT MA MC NL NR PL RO SK SP ST SV SZ UK **A:** IN IS TR
castaneus Fabricius, 1790: 223 (*Hypophlaeus*)
nitidus Melsheimer, 1846: 62 (*Hypophloeus*)
pseudocastaneus Reitter, 1906f: 242 (*Hypophlaeus*)
purkynei Roubal, 1932b: 130 (*Hypophloeus*)
taxicornis Rossi, 1790a: 49 (*Ips*)
vanmeeri F. Soldati & L. Soldati, 2014: 114 **E:** FR
versipellis Baudi di Selve, 1876c: 234 (*Hypophloeus*) **E:** AU CZ GE HU IT RO SK SZ
hopffgarteni Reitter, 1877a: 26
subgenus *Stenophloeus* Blair, 1921: 1 type species *Hypophloeus filum* Fairmaire, 1893
Cnemophloeus Bremer, 1998: 10 type species *Hypophloeus filum* Fairmaire, 1893
analis Gebien, 1914d: 395 (*Hypophloeus*) **A:** JA (Ryukyus) TAI **ORR**
andamanus Pic, 1945: 7 (*Hypophloeus*)
attenuicollis Pic, 1946: 3 (*Hypophlaeus*)
borneensis Pic, 1914d: 16 (*Hypophloeus*)
javanus Pic, 1946: 3 (*Hypophlaeus*)
obscurior Pic, 1946: 3 (*Hypophlaeus*)
cephalotes Gebien, 1914b: 28 (*Hypophloeus*) **A:** AP UP TAI YUN **ORR**
andaiensis Pic, 1946: 3 (*Hypophlaeus*)
cornutus Pic, 1914d: 16 (*Hypophloeus*)
palawanus Pic, 1945: 4 (*Hypophloeus*)
subcastaneus Pic, 1945: 8 (*Hypophloeus*)
filum Fairmaire, 1893: 29 (*Hypophloeus*) **A:** CH NP UP TAI **AFR AUR ORR**
niasensis Pic, 1914d: 16 (*Hypophloeus*) **A:** TAI **ORR**
similis Bremer, 1999: 45 **A:** TAI **ORR**
subgenus *Pogonophloeus* Bremer, 1998: 9 type species *Hypophloeus thoracicus* Melsheimer, 1846.
exiguus Bremer, 1998: 13 (*Corticeus*) **A:** JA (Kyushu) TAI **ORR**
subgenus *Tylophloeus* Bremer, 1998: 10 type species *Hypophloeus flavipennis* Motschulsky, 1860
curtithorax Pic, 1924b: 14 (*Hypophlaeus*) **A:** GUA HAI NP SD YUN **ORR**
beesoni Blair, 1930a: 146 (*Hypophloeus*)
dorsalis Blair, 1930a: 145 (*Hypophloeus*) **A:** UP
flavipennis Motschulsky, 1860a: 99 (*Hypophlaeus*) **A:** AP HP NP PA UP YUN **ORR**
brunneicollis Pic, 1924a: 29 (*Hypophloeus*)
gentilis Lewis, 1894: 468 **A:** HKG JA NP TAI YUN **ORR**
robustus Gebien, 1914b: 29 (*Hypophloeus*)

tribe Leiochrinini Lewis, 1894

genus *Ades* Guérin-Méneville, 1857: 277 type species *Ades hemisphericus* Guérin-Méneville, 1857

Leiochrodes Westwood, 1883: 69 type species *Leiochrodes discoidalis* Westwood, 1883

assimilis Kaszab, 1961i: 461 (*Leiochrodes*) **A:** AP NP SD

bengalicus Kaszab, 1975a: 36 (*Leiochrodes*) **A:** SD

convexus Lewis, 1894: 391 (*Leiochrodes*) **A:** HKG JA SD TAI **ORR**

subovatus Pic, 1921b: 8 (*Leiochrodes*)

coomani Pic, 1927b: 14 (*Leichrodes*) **A:** SCH SD **ORR**

diaphanus Fabricius, 1798: 76 (*Coccinella*) **A:** GUA NP UP **ORR**

discoidalis Westwood, 1883: 71 (*Leiochrodes*) **A:** YUN **ORR**

emeicus Schawaller, 1993b: 357 (*Leiochrodes*) **A:** SCH

formosanus Kaszab, 1946: 157 (*Leiochrodes*) **A:** GUI SCH TAI

glabratus Walker, 1859b: 219 (*Lycoperdina*) **A:** BT FUJ GUX SD TAI **ORR**

obscuricollis Fairmaire, 1893b: 25 (*Liochrinus*)

tibialis Gebien, 1914b: 23 (*Leiochrodes*)

tonkineus Pic, 1922h: 100 (*Leiochrodes*)

himalayensis Kaszab, 1946: 158 (*Leiochrodes*) **A:** BT GUA HP **ORR**

lanceolatus Kaszab, 1961i: 456 (*Leiochrodes*) **A:** AP NP SCH YUN ZHE

masidai Nakane, 1963a: 26 (*Leiochrodes*) **A:** JA

nigronotatus Pic, 1934f: 84 (*Leiochrodes*) **A:** SCH SHX YUN ZHE

politus Kaszab, 1961i: 455 (*Leiochrodes*) **A:** YUN

rufofulvus Westwood, 1883: 71 (*Leiochrinus*) **A:** SD **ORR**

sheduwicus Schawaller, 1992: 9 (*Leiochrodes*) **A:** NP

sikkimensis Kaszab, 1961i: 464 (*Leiochrodes*) **A:** AP NP SD

subaeneus Pic, 1918a: 14 (*Leiochrodes*) **A:** SD **ORR**

taiwanus Masumoto, 1981d: 25 (*Leiochrodes*) **A:** TAI

testaceicollis Kaszab, 1946: 155 (*Leiochrodes*) **A:** HUN TAI

genus *Crypsis* Waterhouse, 1877: 73 type species *Crypsis violaceipennis* Waterhouse, 1877

Leiochrotina Westwood, 1883: 76 type species *Leiochrinus indicus* Westwood, 1883

bimaculatus Kaszab, 1946: 191 **A:** AP NP SD

blairi Kaszab, 1946: 192 **A:** NP SD

chinensis Kaszab, 1946: 196 **A:** FUJ GUA GUX SCH ZHE

gebieni Kaszab, 1946: 195 **A:** TAI

malgosiae Schawaller, 2011a: 355 **A:** SCH

minutus Kaszab, 1946: 197 **A:** NP SD **ORR**

purpuripennis Neervoort van de Poll, 1886: 34 (*Leiochrotina*) **A:** SD **ORR**

ruficollis Kaszab, 1946: 193 **A:** SD **ORR**

rufomarginatus Kaszab, 1946: 195 **A:** BT NP SD

migliaccioi J. Ferrer, 1983: 140

scotti Kaszab, 1946: 191 **A:** YUN **ORR**

shangrongae Masumoto, Akita & Lee, 2013c: 259 **A:** TAI
szekessyi Kaszab, 1946: 194 **A:** SCH SHN YUN
violaceipennis Waterhouse, 1877: 74 **A:** NP SD UP **ORR**
braeti Pic, 1918a: 15 (*Leiochrotina*)
vitalisi Kaszab, 1961h: 375 **A:** YUN **ORR**
wrasei Schawaller, 2005: 222 **A:** HUB SCH SHA
yunnanus Kaszab, 1961h: 376 **A:** YUN **ORR**

genus *Derispia* Lewis, 1894: 389 type species *Diaperis maculipennis* Marseul, 1876
aeneonigra Kaszab, 1946: 65 **A:** SD **ORR**
akitai Masumoto, 1994a: 42 **A:** JA (Ryukyus)
akiyamai Masumoto, Akita & Lee, 2014b: 242 **A:** TAI
amamiana Miyatake, 1961: 37 **A:** JA (Amami-Ôshima)
annamita Pic, 1921b: 8 (*Leiochrodes*) **A:** TAI **ORR**
arunachala Schawaller, 2016c: 198 **A:** AP
bisexnotata Kaszab, 1961b: 177 **A:** GUX HKG **ORR**
bisquadrimaculata Kaszab, 1961b: 179 **A:** GUX **ORR**
bisquinquemaculata Blair, 1937: 36 **A:** SD
bistrimaculata bistrimaculata Blair, 1937: 36 **A:** GUX NP SD
bistrimaculata championi Kaszab, 1961b: 178 **A:** BT UP
bhutanensis Kaszab, 1975e: 323 **A:** BT
bomdila Schawaller, 2016c: 198 **A:** AP
chinensis Kaszab, 1946: 108 **A:** GUI HKG YUN
confluens Kaszab, 1946: 112 **A:** NP SD **ORR**
cooteri Schawaller, 2005: 216 **A:** JIX
diaperoides Kaszab, 1946: 105 **A:** NP SD
flavicornis Kaszab, 1946: 79 **A:** SD
formosana Masumoto, 1981d: 20 **A:** TAI
fukiensis Kaszab, 1961b: 181 **A:** FUJ GUA ZHE
hajeki Schawaller, 2016c: 199 **A:** GUA GUX
heishidinga Schawaller, 2016c: 199 **A:** GUA
indica Kaszab, 1946: 79 **A:** BT NP SD **ORR**
janetscheki Kaszab, 1970c: 432 **A:** NP
japonica japonica Kaszab, 1961b: 181 **A** JA (Okinawa)
japonica yaeyamana Mt. Chûjô, 1997b: 200 **A:** JA (Ryukyus)
japonicola Miyatake, 1961: 36 **A:** JA
jizushanica Schawaller, 2005: 218 **A:** YUN
klapperichi Kaszab, 1942c: 216 **A:** FUJ SCH TAI
kondoi Masumoto, 1981d: 23 **A:** TAI
kraatzi Kaszab, 1946: 93 **A:** GUX **ORR**
kryzhanovskii Kaszab, 1961b: 182 **A:** YUN

lineolata Pic, 1922f: 23 (*Leiochrodes*) **A:** FUJ GUA GUX HUN JIX SCH **ORR**
maculipennis Marseul, 1876c: 105 (*Diaperis*) **A:** JA FUJ GUA GUI GUX HUN SCH SHA ZHE
klapperichiana Kaszab, 1954: 253
melli Kaszab, 1946: 106 **A:** GUA
miyatakei Masumoto, 1994a: 42 **A:** JA (Okinawa)
nanshanchiensis Masumoto, 1981d: 21 **A:** TAI
notata Kaszab, 1946: 81 **A:** BT NP **ORR**
ocellata Kaszab, 1946: 109 **A:** SD "Himalaya" **ORR**
osawai Masumoto, Akita & Lee, 2012b: 197 **A:** TAI
sauteri Kaszab, 1946: 108 **A:** TAI
lalashana Masumoto, Akita & Lee, 2013b: 126
liyuanensis Masumoto, Akita & Lee, 2014b: 244
tatachiaensis Masumoto, Akita & Lee, 2013b: 130
tengchiensis Masumoto, Akita & Lee, 2013b: 123
tsaoi Masumoto, Akita & Lee, 2013b: 132
wangi Masumoto, Akita & Lee, 2013b: 128
zoltani Masumoto, 1981d: 22
shibatai M. Chûjô & Miyatake, 1961: 38 **A:** JA (Amami-Ôshima)
sichuanensis Schawaller, 1993b: 354 **A:** GUI SCH
sikkimensis Kaszab, 1946: 103 **A:** SD
similis Kaszab, 1961b: 176 **A:** GUA YUN
takahashii Masumoto, Akita & Lee, 2013b: 134 **A:** TAI
tenuipunctata Kaszab, 1946: 105 **A:** SD
titschacki Kaszab, 1946: 101 **A:** GUA HKG JIA SHG **ORR**
tricolor Kaszab, 1942c: 216 **A:** FUJ GUA JIX
truncata Kaszab, 1961b: 152 **A:** BT NP SD
undulata Pic, 1922h: 101 (*Leiochrodes*) **A:** YUN
walkeri Kaszab, 1961b: 180 **A:** HKG
wittmeri Kaszab, 1975e: 321 **A:** BT NP

genus *Derispiella* Kaszab, 1961h: 364 type species *Derispiella hingstoni* Kaszab, 1961
bhutanensis Kaszab, 1975e: 324 **A:** BT
hingstoni Kaszab, 1961h: 364 **A:** AP BT SD

genus *Derispiola* Kaszab, 1946: 115 type species *Derispiola fruhstorferi* Kaszab, 1946
blairi Kaszab, 1946: 119 **A:** AP NP SD YUN **ORR**
darjeelingiana Kaszab, 1946: 117 **A:** NP SD
fruhstorferi Kaszab, 1946: 118 **A:** SCH **ORR**
unicornis Kaszab, 1946: 116 **A:** FUJ GUA GUX HUB HUN JIX SCH ZHE **ORR**

genus *Leiochrinus* Westwood, 1883: 68 type species *Leiochrinus fulvicollis* Westwood, 1883
bifurcatus Kaszab, 1946: 132 **A:** SHG "North India" **ORR**

lutescens Westwood, 1883: 71 **A:** NC **ORR**
metallicus Schawaller, 2005: 221 **A:** SD
satzumae Lewis, 1894: 390 **A:** JA **ORR**
sauteri minor Kaszab, 1961h: 371 **A:** HP UP
sauteri sauteri Kaszab, 1946: 132 **A:** GUI NP SD TAI "South China" **ORR**

genus *Leiochrodontes* Kaszab, 1946: 200 type species *Leiochrodes madurensis* Pic, 1918
madurensis Pic, 1918a: 15 (*Leiochrodes*) **A:** SD **ORR**

genus *Leiochrota* Westwood, 1883: 70 type species *Leiochrinus uniformis* Westwood, 1883
varicolor Westwood, 1883: 76 **A:** SD YUN **ORR**

genus *Pimplema* Pascoe, 1887: 16 type species *Pimplema ampliata* Pascoe, 1887 (= *Platydema hemisphaerica* Laporte & Brullé, 1831)
Hades J. Thomson, 1860: 13 [HN] type species *Hades tenebrosus* Thomson, 1860 (= *Platydema hemisphaerica* Laporte & Brullé, 1831)
hemisphaerica Laporte & Brullé, 1831: 367 (*Platydema*) **A:** NP SD **ORR**
ampliata Pascoe, 1887: 17
rufolimbatus Fairmaire, 1882c: 257 (*Hades*)
tenebrosus J. Thomson, 1860: 13 (*Hades*)

genus *Stethotrypes* Gebien, 1914: 26 type species *Stethotrypes bicornutus* Gebien, 1914
Leichrodomorphus Pic, 1921b: 6 type species not designated *Leichrodomorphus brevicornis* Pic, 1921
korschefskyana Kaszab, 1942c: 215 (*Derispia*) **A:** FUJ

tribe Myrmechixenini Jacquelin du Val, 1858

genus *Myrmechixenus* Chevrolat, 1835b: 267 type species *Myrmechixenus subterraneus* Chevrolat, 1835
Myrmecoxenus Agassiz, 1846b: 243 [UE]
picinus Aubé, 1850: 330 (*Myrmecoxenus*) **E:** CI FR IT MA SK SP **N:** AG EG MR **AFR ORR**
calvus Reitter, 1877: 184 (*Myrmecoxenus*)
sordidus Wollaston, 1864: 152 (*Myrmecoxenus*)
subterraneus Chevrolat, 1835b: 268 **E:** AU BY CT CZ DE FI FR GE HU IT LS NR NT PL SK SV SZ
espulo Mäklin, 1846: 184 (*Myrmecoxenus*)
vaporariorum Guérin-Méneville, 1843a: 70 **E:** AU CT CZ DE FI FR GE HU IT PL SK SV SZ **A:** IS **AFR AUR NAR ORR**
beturiensis Reitter, 1880c: 93 (*Myrmecoxenus*)
lathridioides Crotch, 1873b: 363 (*Myrmecoxenus*)

tribe Phaleriini Blanchard, 1845

genus *Emypsara* Pascoe, 1866: 460 type species *Emypsara adamsii* Pascoe, 1866 (= *Diaperis riederii* Faldermann, 1833)

Callicomus Motschulsky, 1860c: 138 [HN] type species *Diaperis riederii* Faldermann, 1833

Obenbergeria E. Strand, 1929: 24 [RN]

riederii Faldermann, 1833: 57 (*Diaperis*) **A:** FE JA NE SC

adamsii Pascoe, 1866: 461 (*Emypsara*)

flexuosus Pascoe, 1866: 461 (*Emypsara*)

hilgendorfi Harold, 1878a: 76 (*Phaleria*)

genus *Halammobia* Semenov, 1901a: 91 type species *Tenebrio pellucidus* Herbst, 1799

pellucida Herbst, 1799: 19 (*Tenebrio*) **E:** FR IT

hemisphaerica Küster, 1852b: 67 (*Phaleria*)

pallens Latreille, 1802: 162 (*Phaleria*)

genus *Paranemia* Heyden, 1892a: 103 type species *Paranemia schroederi* Heyden, 1892

Taclamacanius J. Ferrer & Yvinec, 2005: 121 [RN] type species *Taklamakania lepetzi* J. Ferrer & Yvinec, 2004

Taklamakania J. Ferrer & Yvinec, 2004: 42 [HN] type species *Taklamakania lepetzi* J. Ferrer & Yvinec, 2004

bicolor Reitter, 1895d: 157 **A:** ES (Tuva) MG NMO XIN

lepetzi J. Ferrer & Yvinec, 2004: 42 (*Taklamakania*)

schroederi Heyden, 1892a: 103 **E:** AB ST **A:** KZ TM UZ

argiropuloi Bogatchev, 1965: 372

argyropuloi Bogatchev, 1967a: 159

genus *Phaleria* Latreille, 1802: 162 type species *Tenebrio cadaverinus* Fabricius, 1792

subgenus *Epiphaleria* Lewis, 1894: 382 type species *Epiphaleria atriceps* Lewis, 1894

anjae Lillig, 2012: 89 **A:** OM

atriceps Lewis, 1894: 384 (*Epiphaleria*) **A:** JA TAI **ORR**

prolixa Fairmaire, 1869a: 797 **N:** EG **A:** AE BA IN KU OM SA SI YE **AFR**

aegyptiaca Seidlitz, 1894: 482

fuscata Fairmaire, 1886a: 437

munda Walker, 1871: 16

pusilla Boheman, 1858: 93 **A:** HKG

subgenus *Eremophaleria* Español, 1951a: 32 type species *Phaleria bedeli* Chobaut, 1900

bedeli Chobaut, 1900b: 340 **N:** AG

subgenus *Phaleria* Latreille, 1802: 162 type species *Tenebrio cadaverinus* Fabricius, 1792

Atahualpina Español, 1960b: 113 type species *Atahualpina peruviana* Español, 1960

Halophalerus Crotch, 1874: 107 type species *Phaleria rotundata* LeConte, 1851

Phalerisida Kulzer, 1959: 563 type species *Phalerisida maculata* Kulzer, 1959

Sepedonastes Gistel, 1856: 382 type species *Tenebrio bimaculatus* Herbst, 1799 (= *Dytiscus bimaculatus* Linnaeus, 1767)

Uria Gistel, 1848: x [URN]

acuminata acuminata Küster, 1852b: 59 **E**: CR FR GR IT MA SP **N**: AG EG LB MO TU **A**: IS LE TR

incerta Rey, 1891a: 86

limbata Baudi di Selve, 1876c: 226

maculosa Seidlitz, 1894: 481

seductor Rey, 1891a: 85

submaculata Rey, 1891a: 85

tunisia Reitter, 1916b: 8

acuminata nigriceps Mulsant & Wachanru, 1853: 10 **A**: TR

acuminata syriaca Reitter, 1916b: 7 **N**: EG **A**: IS LE

atlantica Fauvel, 1899: 77 [RN] **E**: FR PT SP **N**: MR SG

divisa Pic, 1923a: 3

bimaculata adriatica Rey, 1891a: 86 **E**: CR GR (Dodecanese) IT **A**: TR

bimaculata bimaculata Linnaeus, 1767: 667 (*Dytiscus*) **E**: AL BU FR GR IT MA RO SP UK **N**: EG LB TU **A**: TR

abdominalis Seidlitz, 1894: 484

bimaculatus Herbst, 1799: 16 (*Tenebrio*)

bisignata Rey, 1892: 53

calabrica Canzoneri, 1961b: 25

concii Canzoneri, 1961b: 25

conicicollis Reitter, 1918: 154

cordata O. Schneider, 1902: 10

delata Rey, 1891a: 85

dorsalis Fabricius, 1792a: 85 (*Byrrhus*)

dorsigera Fabricius, 1787a: 201 (*Carabus*)

giordanii Canzoneri, 1960a: 44

irregularis Pic, 1923a: 3

italica Reitter, 1916b: 8

kochi Canzoneri, 1961b: 26

ligurica O. Schneider, 1902: 10

longicollis Rey, 1892: 53

nigricollis J.R. Sahlberg, 1903c: 3

nigricordis O. Schneider, 1902: 11

obscuriceps O. Schneider, 1902: 11

ornata O. Schneider, 1902: 11

schatzmayri Canzoneri, 1961b: 25

signata O. Schneider, 1902: 11

bimaculata jonica Canzoneri, 1968: 155 **E**: IT (Calabria)

bimaculata marcuzzii Aliquò, 1993: 113 **E**: IT (Isole Eolie: Vulcano)

cadaverina cadaverina Fabricius, 1792a: 113 (*Tenebrio*) **E**: AZ BE DE FR GB GE IR NL PT SP SV **N**: CI

apicalis O. Schneider, 1902: 12

cava Herbst, 1799: 15 (*Tenebrio*)
fuscata O. Schneider, 1902: 12
leegei O. Schneider, 1902: 12
cadaverina maroccana Pic, 1923a: 3 **N:** CI MO
ciliata Wollaston, 1854: 488 **N:** MR
insulana Rey, 1890: 329 **E:** FR IT (Sardinia)
ornata Wollaston, 1864: 494 **N:** CI
biplagiata Heyden, 1909: 486
boeckeri Heyden, 1909: 486
giebeleri Heyden, 1909: 486
nigrothoracica Heyden, 1909: 486
picta Wollaston, 1861a: 246 [HN]
perissinottoi Canzoneri, 1973: 41 **A:** TR
pontica Semenov, 1901a: 90 **E:** BU GG RO ST UK
provincialis cypria Grimm, 1991: 30 **A:** CY
provincialis ghidinii Canzoneri, 1961c: 61 **E:** IT
provincialis intermedia Schuster, 1930a: 65 **E:** IT
provincialis provincialis Fauvel, 1901: 253 **E:** FR SP
pueli Pic, 1902h: 64
pujeti Español, 1951a: 27 [= 1954a: 65] **E:** SP (Baleares)
revellierii Mulsant & Rey, 1858b: 310 [= 1859a: 52] **E:** FR IT SP **N:** AG TU
beloni Pic, 1923a: 3
cinctella Rey, 1891a: 84
circumducta Rey, 1891b: 238
heydenii Seidlitz, 1894: 482
lineolata Rey, 1891b: 238
oblonga Küster, 1852b: 70
reyi Seidlitz, 1894: 482
seidlitzi O. Schneider, 1902: 11
sublaevicollis Rey, 1891b: 236

genus *Phaleromela* Reitter, 1916b: 4 type species *Phaleria subhumeralis* Marseul, 1876
humeralis Laporte, 1840: 219 (*Phaleria*) **A:** NMO **NAR**
subhumeralis Marseul, 1876c: 102 (*Phaleria*) **A:** FE JA SC

genus *Phtora* Germar, 1836c: pl. 11 type species *Phtora crenata* Germar, 1836
subgenus *Clypeophtora* F. Soldati & L. Soldati, 2003a: 4 type species *Phtora tronqueti* F. Soldati & L. Soldati, 2003
tronqueti F. Soldati & L. Soldati, 2003a: 4 **E:** SP
subgenus *Phtora* Germar, 1836c: 11 type species *Phtora crenata* Germar, 1836
Cataphronetis Lucas, 1846: pl. 30 type species *Cataphronetis levaillantii* Lucas, 1846 (= *Phtora crenata* Germar, 1836)

Phthora Gemminger, 1870b: 1959 [UE]
Pseudostene Wollaston, 1861a: 247 type species *Pseudostene angusta* Wollaston, 1861
aegyptiaca Reitter, 1899c: 159 (*Phthora*) **N:** EG LB
alashanensis Ren & Zheng, 1993e: 77 **A:** NMO
angusta Wollaston, 1861a: 249 (*Pseudostene*) **E:** SP **N:** AG CI MO SG **AFR**
fossoria Wollaston, 1861a: 250 (*Pseudostene*)
apicilaevis Marseul, 1876b: 36 (*Cataphronetis*) **N:** EG LB TU **A:** SI
bicolor Reitter, 1899c: 159 (*Phthora*) **N:** AG TU
confluens Reitter, 1894c: 104 **E:** PT SP **N:** LB TU
crenata Germar, 1836c: 11 **E:** FR IT MA SP **N:** AG LB MO TU
attenuata Motschulsky, 1849: 121 (*Cataphronetes*)
brunnea Mulsant, 1854: 241 (*Cataphronetis*)
levaillanti Lucas, 1846: pl. 30 (*Cataphronetis*)
hauseriana Reitter, 1895c: 149 (*Cataphronetis*) **E:** AB AR **A:** TM UZ
kandaharica Kaszab, 1960a: 168 **A:** AF
millingeni Reitter, 1899c: 159 (*Phthora*) **A:** IQ
plagiocnema Iablokoff-Khnzorian, 1956b: 41 (*Cataphronetis*) **E:** AR
priva G.S. Medvedev, 2004b: 577 **A:** TM
prolixa Fairmaire, 1879c: 198 (*Cataphronetis*) **N:** AG TU
quadricollis Reitter, 1895c: 148 (*Cataphronetis*) **E:** ST **A:** AF KI KZ MG TM UZ XIN
reitteri cypria Grimm, 1991: 34 **A:** CY
reitteri reitteri Seidlitz, 1894: 572 (*Cataphronetis*) **E:** GR UK **A:** TR
hellenica Reitter, 1895c: 148 (*Cataphronetis*)
subclavata Wollaston, 1861a: 250 (*Pseudostene*) **N:** EG **A:** IS SI SA
tenuis Marseul, 1876b: 36 (*Cataphronetis*)
tenuicornis Reitter, 1895c: 148 (*Cataphronetis*) **E:** ST **A:** AF IN KZ TM

tribe Scaphidemini Reitter, 1922

genus *Basanus* Lacordaire, 1859: 306 type species *Basanus forticornis* Lacordaire, 1859
amamianus M. Chûjô, 1966: 13 **A:** JA (Ryukyu, Yakushima) SC (Cheju-do)
annamitus Gebien, 1925a: 147 **A:** BT FUJ SHA ZHE **ORR**
erotyloides Lewis, 1891a: 71 **A:** JA
flaviventris Blair, 1937: 35 **A:** SD **ORR**
formosanus Masumoto, 1982d: 68 **A:** TAI
fukudai Nakane, 1968b: 77 **A:** JA (Ryukyus)
himalayanus Kaszab, 1965a: 119 **A:** AP
lui Masumoto & Merkl, 2003: 169 **A:** TAI
manangensis Schawaller, 1995b: 332 **A:** NP
soppongensis Schawaller, 1995b: 332 **A:** HUB **ORR**
tsushimensis kompancevi Kaszab & G.S. Medvedev, 1984: 84 **A:** FE "Korea"
tsushimensis tsushimensis M. Chûjô, 1963a: 17 **A:** JA SC

genus *Pseudoscaphidema* Pic, 1926a: 2 type species *Pseudoscaphidema rufonotata* Pic, 1926
rufonotata Pic, 1926a: 3 **A:** FE (Kamtchatka)

genus *Scaphidema* L. Redtenbacher, 1849: 591 type species *Scaphidium bicolor* Fabricius, 1798
Microbasanus Pic, 1921a: 1 type species *Microbasanum jureceki* Pic, 1921
Nelites J.L. LeConte, 1850: 232 type species *Nelites aeneolus* J.L. LeConte, 1850
aikoae M. Chûjô, 1983: 47 **A:** JA (Yakushima)
angustata Pic, 1935a: 25 **A:** CH
emeishana Schawaller, 2008b: 382 **A:** SCH
formosana Masumoto, 1982a: 150 **A:** TAI
haruhiae M. Chûjô, 1985: 63 **A:** JA (Tsushima)
hiranoi Masumoto & Akita, 2001b: 75 **A:** JA (Okinawa)
insularis G.S. Medvedev & Kompantseva, 1989: 100 **A:** FE (Sachalin)
jureceki Pic, 1921a: 1 (*Microbasanus*) **A:** FE HEI
khnzoriani Kaszab, 1975c: 127
kayokoae Mt. Chûjô, 1992: 36 **A:** SC (Chejodo Is.)
kondoi Nakane, 1968a: 76 **A:** JA (Amami-Ôshima)
metallica bicolor Fabricius, 1798: 179 (*Scaphidium*) **E:** HU UK
metallica metallica Fabricius, 1792b: 499 (*Mycetophagus*) **E:** AB AR AU BE BH BY CT CZ DE EN FI FR GB GE GG HU IT LA LS LT NL NR NT PL SK SP ST SV SZ UK YU **A:** WS
aenea Panzer, 1793b: 2 (*Scolytus*)
ahenea Marsham, 1802: 176 (*Chrysomela*)
sulcipennis Petri, 1926: 194
michihidei Mt. Chûjô & Lee, 1993: 111 **A:** SC
ornatella Lewis, 1894: 396 **A:** JA
pictipennis Lewis, 1894: 397 **A:** JA
discalis Lewis, 1894: 397
nigricornis Lewis, 1894: 397
rousi Picka, 1983a: 262 **E:** GG
satoi Masumoto, Akita & Lee, 2007: 154 **A:** TAI
shaanxica Schawaller, 2008b: 382 **A:** SHX
shinichii Akita & Masumoto, 2015c: 430 **A:** JA
schmidti Schawaller, 2003a: 2 **A:** NP
sichuana Schawaller, 2003a: 3 **A:** SCH
trimaculata M. Chûjô, 1968b: 393 **A:** TAI
turnai Schawaller, 2008b: 382 **A:** HUB

genus *Spiloscapha* Bates, 1873b: 202 type species *Spiloscapha crassicornis* Bates, 1873 (= *Platydema thallioides* Pascoe, 1869)
bipunctata Schawaller, 1997: 4 **A:** YUN **ORR**
cooteri Schawaller, 2004a: 3 **A:** JIX
hiraii Masumoto & Akita, 2001b: 76 **A:** JA (Okinawa)

ichihashii Nakane, 1956b: 166 (*Platydema*) **A:** JA
elegans M. Chûjô, 1968c: 398 (*Basanus*)
kobayashii Shibata, 1978: 19 **A:** TAI
nepalica Schawaller, 1997: 7 **A:** NP **ORR**
taiwana Masumoto & Merkl, 2003: 169 **A:** TAI
unicolor Blair, 1937: 35 **A:** SD

tribe Trachyscelini Blanchard, 1845

genus *Taiwanotrachyscelis* Masumoto, Akita & Lee, 2012a: 29 type species *Taiwanotrachyscelis chengi* Masumoto, Akita & Lee, 2012
chengi Masumoto, Akita & Lee, 2012a: 29 **A:** TAI

genus *Trachyscelis* Latreille, 1809: 379 type species *Trachyscelis aphodioides* Latreille, 1809
aphodioides Latreille, 1809: 379 **E:** AL AZ BU CR FR GG GR IT MA MC PT SP ST UK **N:** CI EG LB MO SG TU **A:** CY KZ TR **AFR NAR**
flavipes Melsheimer, 1846: 61
hopei Costa, 1844: 19 (*Hybosorus*)
lopadusae Koch, 1935a: 98
subopacus Pic, 1923a: 1 (*Tachyscelis*)
chinensis Champion, 1894b: 448 **A:** GUA (Nam'ao Is.) JA (Ryukyus)
kochi Nabozhenko & Purchart, 2017: 567 **A:** YE (Suqutra)
leoi Lo Cascio & Grita, 2011: 85 **A:** YE (Samha, Suqutra)
litoralis L. Soldati, 2009: 76 **A:** IN QA YE (Suqutra)
sabuleti Lewis, 1894: 383 **A:** JA SC
shokhini Nabozhenko & Purchart, 2017: 569 **N:** MO
tenuestriatus Fairmaire, 1886a: 449 **A:** IN JO SA SI YE **AFR**

subfamily **Alleculinae** Laporte, 1840
tribe Alleculini Laporte, 1840
subtribe Alleculina Laporte, 1840

genus *Allecula* Fabricius, 1801b: 21 type species *Cistela morio* Fabricius, 1787
subgenus *Allecula* Fabricius, 1801b: 21 type species *Cistela morio* Fabricius, 1787
amamiensis amamiensis Maeda & Nakane, 1988: 1 **A:** JA (Amami-Ôshima)
amamiensis okinawaensis Maeda & Nakane, 1988: 2 **A:** JA (Okinawa)
angusta Borchmann, 1941: 26 **A:** FUJ
arthritica Fairmaire, 1896a: 36 **A:** SD **ORR**
arunachalica Novák, 2017a: 21 **A:** AP BT
balthasari Mařan, 1940c: 170 **A:** NO
bicoloripes Pic, 1920a: 23 **A:** FUJ
cincta Pic, 1929b: 22 [RN] **A:** SW YUN
cinctipennis Pic, 1915b: 15 [HN]

coreana Kolbe, 1886: 211 **A:** HEB NC SC SHA
dahanshana Masumoto, Novák, Lee & Akita, 2017: 3 **A:** TAI
densaticollis Fairmaire, 1891f: ccxv **A:** HEB HUB
divisa Reitter, 1883a: 115 **E:** AR "Caucasus" **A:** TM UZ
endemica Muche, 1979b: 269 **A:** PA
estriata Seidlitz, 1896b: 37 **E:** MC **A:** TR
femorata Borchmann, 1915b: 183 **A:** SD **ORR**
fenghuangshana Masumoto, Novák, Lee & Akita, 2017: 6 **A:** TAI
formosana Pic, 1910u: 94 **A:** TAI
guangdongica Novák, 2017a: 24 **A:** GUA
guerryi Pic, 1934b: 26 **A:** YUN
himalaica Blair, 1927: 248 **A:** XIZ
impunctaticollis Borchmann, 1932: 91 **A:** SCH
indica Borchmann, 1909b: 714 [RN] **A:** SD **ORR**
brachydera Fairmaire, 1896a: 38 [HN]
ishigakiensis Maeda & Nakane, 1988: 3 **A:** JA
janssoni Novák, 2011a: 238 **E:** TR
jaroslavi Novák, 2017a: 27 **A:** FUJ JIX
klapperichi Pic, 1955c: 31 **A:** FUJ
kuatunensis Pic, 1955c: 31 **A:** FUJ
lijiangica Novák, 2017a: 29 **A:** YUN
longeimpressa Pic, 1955c: 30 **A:** FUJ
matsudai Masumoto, Novák, Akita & Lee, 2019b: 25 **A:** TAI
maxima Pic, 1910u: 94 **A:** FUJ SW TAI YUN
mazandaranica Novák, 2017a: 431 **A:** IN
melanaria Mäklin, 1875b: 669 **A:** FUJ HEN JA SC TAI
rufipes Marseul, 1876d: 322
metallicipennis Pic, 1926d: 29 **A:** YUN
mongolica Muche, 1972b: 223 **A:** MG NMO
morio Fabricius, 1787a: 86 (*Cistela*) **E:** AU BE BH BU BY CR CZ DE FI FR GE GR HU IT LA LT NL PL RO RU SK SL SP SV SZ UK
opaca Illiger, 1794: 610 (*Cistela*)
semilivida Pic, 1891a: 51
moupinea Fairmaire, 1891f: ccxv **A:** SCH
nanlingica Novák, 2018a: 58 **A:** GUA
nitidipennis Borchmann, 1932: 93 **A:** SCH
noctivaga Lewis, 1895a: 251 **A:** JA SC
akitai Hanatsuka, Masumoto & Kon, 2006: 179
olexai Novák, 2016a: 124 **E:** ST
oronthea Baudi di Selve, 1881: 292 **A:** LE SY TR
paulonotata Pic, 1936a: 29 **A:** JA

platicera Dubrovin, 1992b: 513 **A:** FE NW
rhenana Bach, 1856: 228 **E:** AU BE BH CR CZ DE FR GE GG GR HU IT NL NR PL RO RU SB SK SL SP ST SV SZ TR UK YU
loevendali Reitter, 1886c: 140
ryukyuensis Akita & Masumoto, 2015a: 363 **A:** JA
satsumae Fairmaire, 1896b: 116 **A:** JA
savioi Pic, 1926l: 302 **A:** CH
sculpturata Borchmann, 1932: 92 **A:** SCH
semeti Pic, 1909l: 142 **A:** SW YUN
seminitens Blair, 1927: 249 **A:** XIZ
shibatai Nomura, 1964: 49 **A:** JA
sichuanica Novák, 2017a: 31 **A:** SCH SHA
sobrina Borchmann, 1915b: 186 **A:** SD
striatipennis Pic, 1924b: 7 **A:** CE
subbinotata Pic, 1934c: 23 **A:** CH
suberina Novák, 2012a: 33 **E:** IT
tatsunosukei Akita & Masumoto, 2015a: 367 **A:** JA
tenuis Marseul, 1876d: 326 **A:** JA
touzalini Pic, 1936a: 30 **A:** YUN
toyoshimai Akita & Masumoto, 2015a: 365 **A:** JA
tschungseni Pic, 1955c: 31 **A:** FUJ
tsudai Akita & Masumoto, 2012a: 280 **A:** JA
turcica Novák, 2011a: 342 **E:** TR
ussuriensis Borchmann, 1937: 210 **A:** FE ES NE NC SC XIN
mandshurica Mařan, 1940c: 169 (*Allecula*)
yasakai Akita & Masumoto, 2012a: 282 **A:** JA
yipinglangica Novák, 2017a: 38 **A:** YUN
subgenus *Dietopsis* Solier, 1835: 236 type species *Allecula solieri* Laporte, 1840
costipennis Bates, 1879: 482 **A:** PA

genus *Alogista* Fåhraeus, 1870: 318 type species *Alogista abnormis* Fåhraeus, 1870
kadleci Novák, 2017b: 125 **A:** YE
serricornis Kolbe, 1897: 250 (*Synallecula*) **A:** YE (Suqutra) **AFR**

genus *Anthracula* Fairmaire, 1897d: 236 type species *Anthracula latifrons* Fairmaire, 1897
fouquei Novák, 2017c: 43 **A:** NP
humlaica Novák, 2017c: 45 **A:** NP
latifrons Fairmaire, 1897d: 236 **A:** AP BT UP
lawaraiica Novák, 2017c: 48 **A:** PA
renei Novák, 2017c: 51 **A:** NP
sanamica Novák, 2017c: 53 **A:** NP

genus *Bobina* Novák, 2015a: 125 type species *Bobina jendeki* Novák, 2015
cuspidenta Li & Ren, 2019: 164 **A:** YUN
emeiica Novák, 2015a: 129 **A:** SCH
fikaceki Novák, 2015a: 132 **A:** GUA TAI **ORR**
jendeki Novák, 2015a: 135 **A:** HUN JIX SCH **ORR**

genus *Bolbostetha* Fairmaire, 1896b: 117 type species *Bolbostetha soleata* Fairmaire, 1896
Alleculodes Borchmann, 1925: 335 type species *Alleculodes discrepans* Borchmann, 1925
kimioi Akita, 2011: 278 **A:** JA
opaca Borchmann, 1925: 342 (*Alleculodes*) **A:** HKG
proavia Borchmann, 1925: 343 (*Alleculodes*) **A:** HKG
oshimana Nakane, 1968c: 85 (*Alleculodes*) **A:** JA (Ryukyus)
sauteri Borchmann, 1925: 341 (*Alleculodes*) **A:** TAI
yoshitakei Masumoto, Novák, Lee & Akita, 2017: 8 **A:** TAI

genus *Borbonalia* Novák, 2014a: 136 type species *Borbonalia brancuccii* Novák, 2014
akiyamai Masumoto, Novák, Lee & Akita, 2017: 10 **A:** TAI
barclayi Masumoto, Novák, Akita & Lee, 2019b: 27 **A:** TAI
becvari Novák, 2019j: 166 **A:** YUN
brancuccii Novák, 2014a: 138 **A:** YUN
beinanica Masumoto, Novák, Lee & Akita, 2017: 13 **A:** TAI
chanmeilingae Masumoto, Novák, Akita & Lee, 2019a: 94 **A:** TAI
diaolinica Novák, 2019j: 168 **A:** YUN
ewersi Masumoto, Novák, Akita & Lee, 2019b: 28 **A:** TAI
gongashanica Novák, 2019j: 170 **A:** SCH
jizuica Novák, 2014a: 141 **A:** YUN
murzini Novák, 2014a: 144 **A:** YUN SCH XIZ
nepalica Novák, 2014a: 147 **A:** NP
schneideri Novák, 2014a: 150 **A:** SCH
tibetica Novák, 2014a: 153 **A:** XIZ
tienchihica Masumoto, Novák, Akita & Lee, 2019a: 92 **A:** TAI
wangtaichuani Masumoto, Novák, Akita & Lee, 2019a: 90 **A:** TAI
wrasei Novák, 2014a: 155 **A:** YUN
xueshana Masumoto, Novák, Lee & Akita, 2017: 16 **A:** TAI
yunfengica Novák, 2019j: 173 **A:** YUN

genus *Borboresthes* Fairmaire, 1897k: 253 type species *Allecula cruralis* Marseul, 1876
acicularis Marseul, 1876d: 325 (*Allecula*) **A:** JA SC
aeneipennis Harold, 1878a: 80 (*Allecula*) **A:** JA
aglia Novák, 2018b: 17 **A:** GUI
amamiana Maeda & Nakane, 1988: 5 **A:** JA

andreasi Novák, 2016b: 26 **A:** YUN
aokii Akita & Masumoto, 2015a: 371 **A:** JA
arimotoi Akita & Masumoto, 2015a: 368 **A:** JA
arctia Novák, 2018b: 20 **A:** ZHE
baxienshana Masumoto, Novák, Akita & Lee, 2018: 78 **A:** TAI
bertrandi Pic, 1934b: 20 **A:** "Himalaya"
bicoloricornis Novák, 2018a: 62 **A:** NP
bilamellata Marseul, 1876d: 323 (*Allecula*) **A:** CH FE JA TAI
borchmanni Novák, 2008d: 38 [RN] **A:** FUJ
 tibialis Borchmann, 1941: 27 [HN]
castanea Pic, 1928c: 15 **A:** YUN
catocala Novák, 2018b: 22 **A:** GUI
ciena Masumoto, Novák, Akita & Lee, 2019a: 103 **A:** TAI
cinctipennis Pic, 1909a: 19 (*Allecula*) **A:** SW TAI YUN
cruralis Marseul, 1876d: 324 (*Allecula*) **A:** JA
dahanshana Masumoto, Novák, Akita & Lee, 2018: 81 **A:** TAI
daxueensis Novák, 2016b: 35 **A:** SCH
doncunensis Novák, 2018b: 24 **A:** HUB
elliptica Fairmaire, 1891f: ccxiv (*Cistela*) **A:** HUB SCH SHN
fainanensis diversicollis Pic, 1937i: 173 **A:** CH
fainanensis fainanensis Pic, 1922h: 102 **A:** FUJ TAI ZHE
flavicornis Borchmann, 1941: 28 **A:** FUJ GAN NE
fokienensis Pic, 1922f: 24 **A:** FUJ GUI
formosensis Pic, 1934b: 21 **A:** TAI
fuliginosa Fairmaire, 1897k: 253 **A:** SCH TAI WP
fushana Masumoto, Novák, Lee & Akita, 2017: 16 **A:** TAI
hainanensis Novák, 2018b: 28 **A:** HAI
hemaris Novák, 2018b: 30 **A:** GUA
hiraii Akita & Masumoto, 2008: 110 **A:** JA
hovactyliformis hovactyliformis Pic, 1934d: 22 **A:** CH
hovactyliformis lemoulti Pic, 1934d: 23 **A:** CH
howanga Masumoto, Novák, Akita & Lee, 2018: 76 **A:** TAI
hsiehi Masumoto, Novák, Lee & Akita, 2017: 19 **A:** TAI
huangfuchengi Masumoto, Novák, Akita & Lee, 2019a: 110 **A:** TAI
hyles Novák, 2018b: 33 **A:** GUA
impressithorax Pic, 1922f: 24 **A:** YUN SCH
insulcata Pic, 1922f: 24 **A:** FUJ SW YUN
jaegeri Novák, 2005: 126 **A:** NP
jenisi Novák, 2016b: 37 **A:** YUN
jindrai Novák, 2012b: 241 **A:** HUB SHA
jizuensis Novák, 2012b: 244 **A:** YUN

kabirensis Maeda & Nakane, 1988: 6 **A:** JA
keiichii Masumoto, Novák, Akita & Lee, 2019a: 106 **A:** TAI
kimurai Akita & Masumoto, 2008: 112 **A:** JA
klapperichi klapperichi Pic, 1955c: 30 **A:** FUJ SHA TAI ZHE
klapperichi obscurior Pic, 1955c: 30 **A:** FUJ
klapperichi rufescens Pic, 1955c: 30 **A:** FUJ
kuanwuensis Masumoto, Novák, Lee & Akita, 2018: 91 **A:** TAI
kuluensis Pic, 1910u: 94 (*Allecula*) **A:** HP NP
lanyenchiuae Masumoto, Novák, Akita & Lee, 2019a: 108 **A:** TAI
laothoe Novák, 2018b: 36 **A:** FUJ
lijiana Masumoto, Novák, Akita & Lee, 2018: 86 **A:** TAI
litungshana Masumoto, Novák, Lee & Akita, 2017: 19 **A:** TAI
lushanensis Masumoto, Novák, Akita & Lee, 2019b: 30 **A:** TAI
maguanensis Novák, 2012b: 249 **A:** YUN **ORR**
major Pic, 1934b: 21 **A:** FUJ HUB SCH SHA
mimas Novák, 2018b: 39 **A:** HAI
minor Pic, 1930b: 29 **A:** FUJ SHA YUN
minutula Mařan, 1944b: 76 **A:** CE FUJ
mufuensis Novák, 2018b: 41 **A:** HUB
nakanei Akita & Masumoto, 2008: 118 **A:** JA
nanxiensis Masumoto, Novák, Akita & Lee, 2019a: 96 **A:** TAI
niisatoi Masumoto, Novák, Akita & Lee, 2018: 96 **A:** TAI
nuceipennis Fairmaire, 1893: 322 (*Allecula*) **A:** YUN
obliquefasciata Pic, 1926d: 29 (*Allecula*) **A:** YUN
obscurithorax Pic, 1922f: 24 **A:** TAI
ohmomoi Akita & Masumoto, 2008: 116 **A:** JA
okinawana Akita & Masumoto, 2015a: 359 (*Allecula*) **A:** JA
ongi Masumoto, Novák, Akita & Lee, 2019b: 32 **A:** TAI
oshimana Maeda & Nakane, 1988: 4 (*Allecula*) **A:** JA
pahsienshana Masumoto, Novák, Akita & Lee, 2019a: 101 **A:** TAI
picea Borchmann, 1941: 27 **A:** FUJ TAI
pieli Pic, 1937i: 174 (*Allecula*) **A:** CE JIX
ruficollis Borchmann, 1940a: 155 **A:** TAI
rufina piceicollis Borchmann, 1940a: 156 **A:** TAI
rufina rufina Borchmann, 1940a: 156 **A:** TAI
rufosuturalis Pic, 1934b: 21 **A:** YUN
sanensis Pic, 1934b: 20 **A:** CH
satoi Akita & Masumoto, 2015a: 370 **A:** JA
saturnia Novák, 2018b: 44 **A:** HUB
shedingshana Masumoto, Novák, Akita & Lee, 2018: 83 **A:** TAI

simiola Lewis, 1895a: 251 (*Allecula*) **A:** JA
kurinoensis Maeda & Nakane, 1988: 5
siniaevi Novák, 2018a: 64 **A:** HEN HEB HUB JIX SHA
sphinx Novák, 2018b: 46 **A:** GUI
spilosoma Novák, 2018b: 49 **A:** HUB
subapicalis Pic, 1934b: 22 **A:** CH
submetallica Pic, 1934b: 20 **A:** CH
subsulcata Pic, 1922f: 24 **A:** FUJ SW YUN
suni Masumoto, Novák, Lee & Akita, 2017: 22 **A:** TAI
suturalis Borchmann, 1915b: 186 **A:** SD
suzukii Masumoto, Novák, Akita & Lee, 2018: 88 **A:** TAI
tengchihensis Masumoto, Novák, Akita & Lee, 2018: 98 **A:** TAI
thibetana Pic, 1934b: 20 **A:** FUJ XIZ
tomokunii Masumoto, Novák, Akita & Lee, 2019b: 34 **A:** TAI
tsaijingfui Masumoto, Novák, Akita & Lee, 2019a: 99 **A:** TAI
tsinensis Pic, 1934b: 21 **A:** CH
tulanshana Masumoto, Novák, Akita & Lee, 2019a: 113 **A:** TAI
tyria Novák, 2018b: 51 **A:** GUA
umbilicata Seidlitz, 1896: 38 (*Allecula*) **A:** CE JA TAI
ussiriana Dubrovin, 1992b: 513 **A:** FE
vaclavhaveli Novák, 2015b: 89 **A:** YUN **ORR**
vicina Pic, 1930b: 28 **A:** YUN
walamiensis Masumoto, Novák, Akita & Lee, 2018: 93 **A:** TAI
wangi Masumoto, Novák, Akita & Lee, 2019b: 36 **A:** TAI
weigeli Novák, 2015b: 96 **A:** YUN
yakushimana Akita & Masumoto, 2015a: 358 (*Allecula*) **A:** JA
yunnanensis Novák, 2012b: 264 **A:** YUN
zeuzera Novák, 2018b: 54 **A:** GUA

genus *Chitwania* Novák, 2015c: 91 type species *Chitwania kejvali* Novák, 2015
kejvali Novák, 2015c: 93 **A:** NP UP

genus *Cistelopsis* Fairmaire, 1896a: 39 type species *Cistelopsis rufina* Fairmaire, 1896
aborensis Borchmann, 1915b: 187 **A:** AP
kejvali Novák, 2019b: 64 **A:** UP
klapperichi Mařan, 1944b: 77 **A:** CE FUJ
rufimembris Pic, 1930e: 76 **A:** NMO **ORR**
sinensis Pic, 1955c: 30 **A:** FUJ
tryznai Novák, 2019b: 66 **A:** UP

genus *Fifina* Novák, 2018c: 470 type species *Fifina romani* Novák, 2018
stanislavi Novák, 2018c: 474 **A:** YUN

genus *Gerdacula* Novák, 2015d: 145 type species *Gerdacula fujianica* Novák, 2015
brancuccii Novák, 2015d: 146 **A:** NP
fujianica Novák, 2015d: 150 **A:** FUJ
hubeica Novák, 2015d: 152 **A:** HUB
nepalica Novák, 2015d: 155 **A:** NP
taiwana Masumoto, Novák, Lee & Akita, 2017: 23 **A:** TAI

genus *Hymenalia* Mulsant, 1856b: 48 type species *Cistela fusca* Illiger, 1794 (= *Cistela rufipes* Fabricius, 1792)
subgenus *Hymenalia* Mulsant, 1856b: 48 type species *Cistela fusca* Illiger, 1794 (= *Cistela rufipes* Fabricius, 1792)
afghanica Novák, 2015f: 72 **A:** AF
alenae Novák, 2007a: 152 **A:** YE
asahiensis Maeda & Nakane, 1991: 6 **A:** JA
atronitens Fairmaire, 1892b: 151 (*Gonodera*) **E:** SB **A:** IN IS SY TR
badia Kiesenwetter, 1861a: 234 (*Cistela*) **E:** AB GR TR
basalis Faust, 1877b: 320 (*Allecula*) **E:** AB AR "Caucasus" **A:** IN TM
becvari Novák, 2010a: 194 **A:** YUN
bocaki Novák, 2010a: 196 **A:** SCH
brignolii Muche, 1974: 219 (*Allecula*) **A:** TR
castaneipennis Fairmaire, 1884a: 172 **A:** TR
crassicollis crassicollis Fairmaire, 1866a: 47 **N:** AG MO
bispilosa Desbrochers des Loges, 1884: 169 (*Isomira*)
crassicollis obscurimembris Pic, 1938d: 14 **N:** MO
darjeelingica Novák, 2015e: 373 **A:** SD
denticulata Muche, 1982: 122 (*Prionychus*) **A:** AE OM SA
ehdenica Novák, 2017d: 434 **A:** LE
elongata Pic, 1925a: 2 **E:** GR
genuensis Novák, 2007a: 156 **A:** IN
graeca Seidlitz, 1896: 80 **E:** AL BH BU CR GR MC ME RO SB SL TR YU **A:** TR
delagrangei Fairmaire, 1892b: 150 (*Gonodera*)
gravida Küster, 1850a: 77 (*Cistela*) **E:** BH CR GR **A:** TR
habashanica Novák, 2010a: 198 **A:** YUN
holzschuhi Novák, 2010a: 200 **A:** SHA SCH
horaki Novák, 2010a: 202 **A:** YUN **ORR**
iranica Novák, 2007a: 159 **A:** IN
jakli Novák, 2007a: 163 **A:** OM YE
jaroslavi Novák, 2010a: 204 **A:** HUB HUN JIX

kadleci Novák, 2015f: 76 **A**: JO
kakadu Novák, 2010a: 206 **A**: HUB
klapperichi Pic, 1955c: 30 **A**: FUJ SCH TAI
 tschungseni Pic, 1955c: 30
lalai Novák, 2007a: 166 **A**: IN
leigongshanica Novák, 2015e: 380 **A**: GUI
merkli Novák, 2010a: 210 **A**: TAI
minuta Pic, 1910d: 14 **A**: YUN
minyops Peyerimhoff, 1943: 17 (*Mycetocharina*) **N**: AG MO TU
morio L. Redtenbacher, 1849: 602 (*Cistela*) **E**: AU GR HU RU SK UK **A**: TR
 amplicollis Linder, 1864: 251 (*Cistela*)
murzini Novák, 2008a: 208 **A**: YUN
obscuriceps Pic, 1925a: 2 **E**: GR MC
obscuripennis Pic, 1905g: 162 **E**: GR **A**: TR
orszuliki Novák, 2015f: 80 **A**: IN
pallidipennis Pic, 1926d: 31 **A**: YUN
picheyrei Peyerimhoff, 1943: 20 (*Mycetocharina*) **N**: AG
pseudojakli Novák, 2015f: 82 **A**: YE
pseudominuta Novák, 2015e: 383 **A**: YUN
puetzi Novák, 2010a: 214 **A**: SCH (Ganzi)
purkynei Obenberger, 1917: 38 **E**: BU RO
reticulata Seidlitz, 1896: 77 **A**: IN IQ
rufipennis Marseul, 1876d: 328 (*Cistela*) **A**: CH ES FE JA SC TAI
rufipes Fabricius, 1792b: 44 (*Cistela*) **E**: AL AR AU BE BH BU CR CT CZ DE FR GE GR HU IT LA MC MD ME NL NR PL PT RO RU SB SK SL SP ST SV SZ TR UK YU "Caucasus" **A**: IN TR
 fusca Illiger, 1794: 610 (*Cistela*)
 pauliani Pic, 1931e: 13 (*Gonodera*)
 sericans Rey, 1892: 65
rungsi Peyerimhoff, 1943: 19 (*Mycetocharina*) **N**: MO
ryoi Akita & Masumoto, 2016: 6 **A**: JA
sasajii Saitô, 2001: 347 **A**: JA
smirnovi Dubrovina, 1978: 54 **E**: UK
unicolor Nakane, 1963a: 30 **A**: FE JA SC
wrasei Novák, 2008a: 211 **A**: YUN
wuliangica Novák, 2010a: 218 **A**: YUN
yunnanica Novák, 2010a: 220 **A**: YUN
zabidica Novák, 2015f: 85 **A**: YE
subgenus *Nikomenalia* Dubrovina, 1975: 166 type species *Hymenalia kaszabi* Muche 1972
impunctaticollis Dubrovina, 1975: 171 **A**: GAN NMO
jinshanica Novák, 2015e: 377 **A**: YUN

kaszabi Muche, 1972b: 224 (*Hymenalia*) **A:** MG NMO
medvedevi Dubrovina, 1975: 170 **A:** MG NMO
schawalleri Novák, 2010a: 216 **A:** YUN

genus *Hymenorus* Mulsant, 1852: 68 type species *Hymenorus doublieri* Mulsant, 1852
baudii Seidlitz, 1896: 52 **A:** CY
candeli Pardo Alcaide, 1978: 90 **N:** MO
doublieri Mulsant, 1852: 68 **E:** AU BH BU BY CR CT CZ FI FR GE GG GR HU IT NT PL RO RU SK SL SP ST SV SZ UK "Caucasus" **A:** ES FE MG RU SHA TR WS
andalusiacus Cobos, 1954: 36
avajewi Semenov, 1901b: 168
rugicollis Mulsant, 1856e: 20
scutellatus Pic, 1901o: 254
zoufali Mařan, 1935a: 141 (*Hymenalia*)
evae Novák, 2006c: 318 (*Hymenophorus*) **A:** IN
gerdae Novák, 2006c: 322 (*Hymenophorus*) **A:** IN
halebensis Novák, 2017d: 438 **A:** SY
hartmanni Novák, 2007b: 234 (*Hymenophorus*) **A:** NP
indicus Fairmaire, 1896a: 58 **A:** HP IN
veterator Lewis, 1895a: 252 **A:** JA FE

genus *Liodocistela* Pic, 1930: 28 type species *Cistelopsis rufomarginata* Pic, 1930
brancuccii Novák, 2014b: 535 **A:** NP
raii Novák, 2014b: 547 **A:** SD
tryznai Novák, 2014b: 549 **A:** UP
weigeli Novák, 2007b: 237 (*Cistelopsis*) **A:** NP

genus *Loriculoides* Novák, 2020 [see New Acts] type species *Allecula subaeneipennis* Pic, 1922
Loricula Novák, 2016c: 45 [HN] type species *Allecula subaeneipennis* Pic, 1922
flosmanni Novák, 2016c: 47 (*Loricula*) **A:** YUN
unidentata Li & Ren, 2019: 167 (*Loricula*) **A:** YUN
yunnanica Novák, 2016c: 53 (*Loricula*) **A:** YUN

genus *Makicula* Novák, 2012c: 275 type species *Makicula phoupaneica* Novák, 2012
andreasi Novák, 2012c: 277 **A:** YUN
mengi Novák, 2012c: 285 **A:** YUN

genus *Microsthes* Novák, 2011b: 320 type species *Microsthes barborae* Novák, 2011
holzschuhi Novák, 2014c: 560 **A:** NP
taiwanus Masumoto, Novák, Lee & Akita, 2017: 25 **A:** TAI
tryznai Novák, 2014c: 571 **A:** UP

genus *Mycetocharina* Seidlitz, 1890: 136 type species *Allecula orientalis* Faust, 1877
subgenus *Alleculopsis* Semenov, 1894: 361 type species *Alleculopsis deserticola* Semenov, 1894
bahukalatensis Novák, 2008b: 74 **A:** IN SA
deserticola Semenov, 1893c: 362 [= 1894d: 610] **A:** KZ TM UZ
gafsaica Novák, 2016d: 118 **N:** TU
macrophthalma Gebler, 1859: 341 (*Allecula*) **A:** KI KZ MG TM
puncticollis Reitter, 1898c: 65 **A:** UZ
ruficeps Pic, 1923g: 10 **A:** YE
subgenus *Mycetocharina* Seidlitz, 1890: 136 type species *Allecula orientalis* Faust, 1877
Caristela Fairmaire, 1894c: 311 type species *Caristela megalops* Fairmaire, 1894
adusta Reitter, 1898c: 65 **A:** TR
afghanica Muche, 1986: 181 **A:** AF IN
arabica Muche, 1982: 119 **A:** SA
braaschi Muche, 1982: 117 **A:** AE SA
castanea Faust, 1877b: 322 (*Allecula*) **E:** AR CT ST **A:** WS
cypria Novák, 2018a: 67 **A:** CY
jelineki Novák, 2006a: 87 **A:** AE IN
kabateki Novák, 2006a: 90 **A:** IN
konardanica Novák, 2006a: 94 **A:** IN
marocana Pic, 1951e: 75 **N:** MO
megalops Fairmaire, 1894c: 311 (*Caristela*) **N:** AG EG MO SI TU **A:** IS
longipennis Pic, 1923g: 10
megalops pallidior Pic, 1938d: 14 **N:** MO
mogadorica Escalera, 1914: 360 **N:** MO
orientalis Faust, 1877b: 318 (*Allecula*) **E:** AB AR "Caucasus" **A:** IS JO TM TR
beckeri Kiesenwetter, 1878: 245 (*Hymenalia*)
riabovi Iablokoff-Khnzorian, 1959c: 141 **E:** AB AR
rufofusca Muche, 1982: 119 **A:** SA
rufotestacea Reitter, 1898c: 65 **E:** GR **A:** IN SY TR
seminigra Pic, 1925i: 228 **N:** EG
syriaca Baudi di Selve, 1881: 293 (*Cistela*) **A:** IS SY
vanharteni Novák, 2008c: 259 **A:** AE
volkonskyi Peyerimhoff, 1943: 21 **N:** "Sahara centralis"
wittmeri Muche, 1982: 116 **A:** SA

genus *Mycetocula* Novák, 2015g: 78 type species *Mycetocharina subcruciata* Pic, 1922
hainanica Novák, 2015g: 82 **A:** HAI
subcruciata Pic, 1922: 102 (*Mycetocharina*) **A:** NP YUN **ORR**

genus *Netopha* Fairmaire, 1893f: 299 type species *Netopha pallidipes* Fairmaire, 1893
pallidipes Fairmaire, 1893f: 300 **A:** CE NO TAI YUN **ORR**

genus *Oracula* Novák, 2019c: 54 type species *Oracula bicolor* Novák, 2019
subgenus *Duocula* Novák, 2019c: 56 type species *Oracula clara* Novák, 2019
amica Novák, 2019c: 57 **A:** NP
magnifica Novák, 2019c: 63 **A:** NP
tenebrosa Novák, 2019c: 68 **A:** YUN
subgenus *Oracula* Novák, 2019c: 71 type species *Oracula bicolor* Novák, 2019
bicolor Novák, 2019c: 72 **A:** YUN **ORR**

genus *Prionychus* Solier, 1835a: 237 type species *Helops ater* Fabricius, 1775
Eryx Stephens, 1832: 27 [HN] type species *Pyrochroa nigra* DeGeer, 1775 (= *Helops ater* Fabricius, 1775)
Pelops Gistel, 1834: 22 [NO] type species *Helops ater* Fabricius, 1775
alfonsoblancoi Blanco Villero & Sáez Bolaño, 2011: 177 **N:** MO
anthracinus Mulsant, 1856e: 23 (*Eryx*) **N:** AG TU
asiaticus Fairmaire, 1892b: 151 (*Eryx*) **E:** AB **A:** IN TR
ater Fabricius, 1775: 258 (*Helops*) **E:** AL AU BU BY CR CT CZ DE EN FI FR GB GE GR HU IT LA LT MD NL NR NT PL PT RO RU SK SP ST SV SZ UK **A:** WS
ater Gmelin, 1790: 2011 (*Pimelia*)
niger DeGeer, 1775: 25 (*Pyrochroa*)
subsulcatus Fairmaire, 1862: 583 (*Eryx*)
cisteloides Seidlitz, 1896: 60 **E:** GR (Rhodes) **A:** CY IS IN LE SY TR
fairmairii Reiche, 1860: 731 (*Eryx*) **E:** FR IT PT SP **N:** MO
lugens Küster, 1850a: 70 (*Cistela*) **E:** FR (Corse) IT (Sardinia, Sicilia) **N:** AG MO
bellieri Reiche, 1860: 730 (*Eryx*)
mauritanica Mulsant, 1856e: 23 (*Eryx*)
melanarius Germar, 1813: 123 (*Helops*) **E:** AL AU BE BH BU CR CT CZ DE FI FR GB GE GR HU IT ME NL NR NT PL RO RU SK SL SP ST SV SZ UK YU
ater L. Redtenbacher, 1849: 602 [HN]
ater Baudi di Selve, 1877c: 385 (*Cistela*) [HN]
laevis Rosenhauer, 1847: 122 (*Cistela*)
nitidulus C.G. Thomson, 1864: 286 (*Cistela*)
nitidissimus Pic, 1905l: 288 **A:** SY TR
ottoi Novák, 2017d: 440 **A:** LE
sardegnaensis Novák, 2017d: 443 **E:** IT (Sardinia)
striatipennis Pic, 1909f: 114 **E:** BU RO
montandoni Reitter, 1913a: 26

genus *Psis* Novák, 2019d: 71 type species *Psis nanensis* Novák, 2019
bheriensis Novák, 2019d: 72 **A:** NP PA **ORR**
mechiensis Novák, 2019d: 75 **A:** NP

genus *Socotralia* Novák, 2007: 321 type species *Socotralia major* Novák, 2007
brunnea Novák, 2007c: 323 **A:** YE (Suqutra)
intermedia Novák & Purchart, 2012: 325 **A:** YE (Suqutra)
major Novák, 2007c: 326 **A:** YE (Suqutra)
minor Novák, 2007c: 330 **A:** YE (Suqutra)
montana Novák & Purchart, 2012: 327 **A:** YE (Suqutra)
reitteri Novák & Purchart, 2012: 331 **A:** YE (Suqutra)
vybirali Novák & Purchart, 2012: 333 **A:** YE (Suqutra)

genus *Spinecula* Novák, 2019e: 437 type species *Spinecula houaphanica* Novák, 2019
angustatissima Pic, 1930: 29 (*Allecula*) **A:** YUN
elongatissima Pic, 1926: 30 (*Allecula*) **A:** YUN
weigeli Novák, 2019e: 453 **A:** YUN

genus *Synallecula* Kolbe, 1883: 25 type species *Allecula livida* Sahlberg, 1823,
borchmanni Novák, 2017b: 130 **A:** IN OM SA YE

genus *Upinella* Mulsant, 1856b: 17 type species *Allecula aterrima* Rosenhauer, 1847
subgenus *Thornella* Novák, 2019f: 83 type species *Allecula holomelaena* Fairmaire, 1894
holomelaena Fairmaire, 1894: 27 (*Allecula*) **A:** NP PA **ORR**
subgenus *Tibinella* Novák, 2019f: 90 type species *Upinella pahangica* Novak, 2019
cryptomeriae Lewis, 1895a: 250 (*Allecula*) **A:** JA
lanrenxiensis Masumoto, Akita & Lee 2015b: 307 (*Allecula*) **A:** TAI **ORR**
meifengensis Masumoto, Novák, Akita & Lee, 2018: 101 **A:** TAI
taiwana Masumoto, Akita & Lee, 2015b: 306 (*Allecula*) **A:** TAI
subgenus *Upinella* Mulsant, 1856b: 17 type species *Allecula aterrima* Rosenhauer, 1847
akiyamai Akita & Masumoto, 2012a: 288 (*Allecula*) **A:** JA
amamiensis Maeda & Nakane, 1988: 1 (*Allecula*) **A:** JA
aterrima Rosenhauer, 1847: 122 (*Allecula*) **E:** AR AU BH BU CR GE GR HU IT ME RO RU SB ST UK
frankenbergeri Mařan, 1940c: 168 (*Allecula*) **A:** FUJ TAI
fuliginosa Mäklin, 1875b: 666 (*Allecula*) **A:** GAN JA (Ryukyus) SC
obscura Harold, 1878b: 132 (*Allecula*)
velutina Marseul, 1876d: 322 (*Allecula*)
hirokii Akita & Masumoto, 2012a: 286 (*Allecula*) **A:** JA
jiangxiica Novák, 2015h: 98 **A:** JIX
jingfui Masumoto, Novák, Akita & Lee, 2019a: 115 **A:** TAI
kubani Novák, 2015h: 101 **A:** YUN
mendeli Masumoto, Novák, Akita & Lee, 2019b: 21 **A:** TAI

nipponica Miyatake, 1985: 347 (*Allecula*) **A:** JA
ontakensis Maeda & Nakane, 1988: 2 (*Allecula*)
okinawaensis Maeda & Nakane, 1988: 2 (*Allecula*) **A:** JA
ruzickai Novák, 2015h: 104 **A:** YUN
takaii Akita & Masumoto, 2016: 6 **A:** JA
turnai Novák, 2015h: 107 **A:** HEN
ukenensis Akita & Masumoto, 2012a: 284 (*Allecula*) **A:** JA
yaeyamaensis Akita & Masumoto, 2012a: 290 (*Allecula*) **A:** JA

genus *Zizu* Novák, 2019g: 186 type species *Zizu kejvali* Novák, 2019
nepalensis Novák, 2019g: 192 **A:** NP
tenebris Novák, 2019g: 197 **A:** PA

subtribe Gonoderina Seidlitz, 1896

genus *Asiomira* Dubrovina, 1973b: 367 type species *Isomira ophthalmica* Seidlitz, 1896
Kirgisomira Weise, 1974: 71 type species *Isomira ophthalmica* Seidlitz, 1896
firjusana Dubrovina, 1973b: 371 (*Isomira*) **A:** TD TM
kabadiana Dubrovina, 1973b: 372 (*Isomira*) **A:** TD
kadleci Novák, 2016e: 182 **A:** TD
keleinikovae Dubrovina, 1973b: 373 (*Isomira*) **A:** TD
ophthalmica Seidlitz, 1896: 107 (*Isomira*) **A:** IN KI TM UZ
rufescens Dubrovina, 1982: 142 (*Isomira*) **A:** TD

genus *Copistethus* Seidlitz, 1890: 524 type species *Cistela spadix* Kiesenwetter, 1861
spadix abdominalis Pic, 1934d: 26 **E:** GR
spadix spadix Kiesenwetter, 1861a: 235 (*Cistela*) **E:** BU GR MC

genus *Cornucistela* Campbell, 1980: 133 type species *Cornucistela serrata* Campbell, 1980
anichtchenkoi Novák & Nabozhenko, 2019: 268 **A:** IN
serrata Campbell, 1980: 135 **A:** SA

genus *Gerandryus* Rottenberg, 1873: 217 [RN] type species *Parablops aetnensis* Rottenberg, 1871
Parablops Rottenberg, 1871: 254 [HN] type species *Parablops aetnensis* Rottenberg, 1871
aetnensis Rottenberg, 1871: 256 (*Parablops*) **E:** FR IT (Sicilia)

genus *Gonodera* Mulsant, 1856b: 41 type species *Cistela fulvipes* Fabricius, 1792 (= *Cistela luperus* Herbst, 1783)
bicolor Reitter, 1884b: 89 **E:** GR
obscuripennis Pic, 1902g: 57
obscuripes Pic, 1902g: 56

corinthia Fairmaire, 1884a: 172 **A:** SY TR
fulvipennis Reitter, 1895b: 84
luperoides Reitter, 1895b: 83
flavida Fairmaire, 1892b: 150 (*Cistela*) **A:** TR
forticornis Fairmaire, 1892b: 150 (*Cistela*) **A:** TR
gilanica Novák, 2011c: 186 **A:** IN
kasatkini Nabozhenko, 2018c: 174 **A:** LE
luperus densepunctata Roubal, 1935c: 10 **E:** BU
luperus hellenica Mařan, 1944d: 191 **E:** GR
luperus luperus Herbst, 1783: 65 (*Cistela*) **E:** AL AU BH BU BY CR CZ DE EN FR GB GE GR HU IR IT MC MD ME PL PT RO SK SL SP ST SV SZ YU
castanea Marsham, 1802: 223 (*Crioceris*)
erythropa Marsham, 1802: 223 (*Crioceris*)
ferruginea Fabricius, 1792b: 45 (*Cistela*)
fulvipes Fabricius, 1792b: 44 (*Cistela*)
obscuriceps Pic, 1926f: 9
macrophthalma Reitter, 1884b: 89 **E:** AR ST "Caucasus" **A:** TR
jureceki Mařan, 1944d: 193
latior Pic, 1903m: 163
metallica Küster, 1850a: 73 (*Cistela*) **E:** IT (Sicilia) SZ
pulcherrima Faldermann, 1837: 103 (*Cistela*) **E:** AR GG "Caucasus"
viridis Allard, 1877a: 234 (*Euboeus*)
rufoaenea Reitter, 1900a: 159 **A:** SY TR
subaenea Küster, 1850a: 72 (*Cistela*) **E:** BU GR MC

genus *Isomira* Mulsant, 1856b: 52 type species *Chrysomela murina* Linnaeus, 1758
subgenus *Apteromira* Weise, 1974: 123 type species *Cistela ovulum* Kiesenwetter, 1863
ovulum Kiesenwetter, 1863: 427 (*Cistela*) **E:** SP **N:** MO
acuminata Fairmaire, 1866a: 46 (*Isomera*)
subgenus *Danielomira* Weise, 1974: 119 type species *Isomira cantabrica* Weise, 1974
cantabrica Weise, 1974: 120 **E:** SP
schedli Muche, 1984: 207 **E:** SP
scutellaris Baudi di Selve, 1877c: 388 (*Cistela*) **E:** IT
subgenus *Heteromira* Hölzel, 1958: 19 type species *Isomira moroi* Hölzel, 1958
costessii Bertolini, 1868: 119 (*Cistela*) **E:** AU IT SZ
moroi Hölzel, 1958: 19 **E:** AU IT
subgenus *Isomira* Mulsant, 1856b: 52 type species *Chrysomela murina* Linnaeus, 1758
Tedinus Casey, 1891: 153 type species *Tedinus angustus* Casey, 1891
ajlounica Novák, 2019h: 88 **A:** JO
alleuculoides Faldermann, 1837: 102 (*Cistela*) **E:** ST

aliquoi Soldati & Lemaire, 2010: 306 **A:** CY
anaspiformis Weise, 1974: 102 **E:** IT
antennalis Reitter, 1884a: 9 **E:** AB AR GG "Caucasus" **A:** IN TR
  antennalis Reitter, 1890j: 394 [HN]
antennata Panzer, 1798: 8 (*Cistela*) **E:** AL AU BU CZ FR GR HU IT MC MD ME PT RO SB SK SP UK YU
  budensis Csiki, 1940b: 919
  euboica Pic, 1903l: 146
  tristicula Reitter, 1890j: 393
armena Iablokoff-Khnzorian, 1976: 324 **E:** AR
atlantis Escalera, 1914: 359 **N:** MO
avula Seidlitz, 1896: 107 **E:** AU GE
bodemeyeri Reitter, 1903b: 45 **A:** TR
caucasica Reitter, 1890j: 394 **E:** AB AR GG ST **A:** IN
  atriceps Reitter, 1896b: 75
convexior Weise, 1974: 100 **E:** SP
dhanaica Novák, 2019h: 92 **A:** JO
discithorax Pic, 1917b: 20 **A:** SD
estrellana Kiesenwetter, 1870: 144 **E:** PT SP
eustrophoides Pic, 1930b: 30 **A:** CH
excavata Dubrovina, 1982: 138 **E:** AB GG ST
formosana Pic, 1917b: 20 **A:** TAI
funerea Kiesenwetter, 1861a: 236 (*Cistela*) **E:** GR
genistae Rottenberg, 1871: 256 (*Cistela*) **E:** IT
georgica Iablokoff-Khnzorian, 1976: 321 **E:** GG
gracilicornis Roubal, 1919: 63 **E:** AR GG ST
  fallax Iablokoff-Khnzorian, 1976: 323
granifera Kiesenwetter, 1878: 245 **E:** AR GG "Caucasus"
hispanica Kiesenwetter, 1870: 143 **E:** FR PT SP **N:** AG MO
hypocrita Mulsant, 1856b: 60 **E:** AU FR GE IT SL SZ
  icteropa H. Wagner, 1917: 346
  oblonga Küster, 1850a: 76 (*Cistela*)
icteropa Küster, 1852b: 72 (*Cistela*) **E:** AL AU BH BU CR CT EN FR GE IT (Sicilia) MC ME RO RU SP ST SZ UK **A:** WS
intrusa Seidlitz, 1896: 107 **E:** AB GG "Caucasus"
  murina Kiesenwetter, 1878: 246
ishigakiensis Hanatsuka, Masumoto & Kon, 2005: 414 **A:** JA (Ishigaki Is.)
kandelouica Novák, 2019h: 95 **A:** IN
khnzoriani Dubrovina, 1982: 139 **E:** AR
kompantzevae Dubrovin, 1992a: 14 **A:** TM

konoi Pic, 1931b: 171 **A:** TAI
  formosana Kôno, 1930b: 96
maltaica Novák, 2019h: 97 **E:** MA
marcida Kiesenwetter, 1863: 425 **E:** AU IT SL SZ
matsumurai Kôno, 1930b: 95 (*Isomia*) **A:** TAI
melanophthalma Lucas, 1846: pl. 31 (*Cistela*) **E:** BH CR FR IT MA PT SP **N:** AG MO TU
  corsica Mulsant, 1856e: 22
  ferruginea Küster, 1850a: 78 (*Cistela*)
meybohmi Weise, 1974: 110 **E:** SP
minuta Pic, 1934b: 24 **A:** CH
mitis Bonadona, 1979: 2 **E:** FR (Corse)
muchei Dubrovina, 1982: 133 **E:** GG ST
  monticola Muche, 1972a: 127
murina ciscaucasica Dubrovina, 1982: 135 **A:** FE WS
murina murina Linnaeus, 1758: 377 (*Chrysomela*) **E:** AL AR AU BH BU BY CR CT CZ FR DE EN FI GB GE GR HU IT LA LT MD NR NT PL RO RU SK SL SP ST SV SZ UK **A:** FE WS
  aemiliae Bedel, 1904: 211
  arenaria Gerhardt, 1904: 79
  ecchelii Bertolini, 1892: 350 (*Gonodera*)
  evonymi Fabricius, 1792b: 45 (*Cistela*)
  fulva Geoffroy, 1785: 162 (*Mordella*)
  fusca Marsham, 1802: 223 (*Crioceris*)
  galii Brahm, 1790: 100 (*Chrysomela*)
  reppensis Herbst, 1783: 65 (*Cistela*)
  semiflava Küster, 1852b: 71 (*Cistela*)
  subnitida Rey, 1892: 65
nitida Reitter, 1890j: 394 **E:** GR **A:** TR
  impressithorax Pic, 1903l: 146
nitidula Kiesenwetter, 1861a: 237 (*Cistela*) **E:** GR ST TR **A:** CY IN LE SY TR
  pallidior Pic, 1901a: 3
  rhodius Pic, 1901a: 3
ochropus Küster, 1850b: 92 (*Cistela*) **E:** BH CR IT
  murina Kiesenwetter, 1863: 424 (*Cistela*) [HN]
  parvula Rottenberg, 1871: 257 (*Cistela*)
oertzeni Reitter, 1889e: 257 **E:** GR
parvuloides Weise, 1974: 103 **E:** IT
pilosicollis Escalera, 1913: 51 **N:** AG MO
radula Weise, 1974: 106 **E:** SP
sanguinicollis Reitter, 1911a: 354 **A:** IS SY TR
sericea Drapiez, 1820a: 351 (*Cistela*) **E:** GE

striatipennis Dubrovina, 1982: 141 **E:** GR **A:** TR
stricta Iablokoff-Khnzorian, 1976: 319 **E:** AR
subelongata Pic, 1917c: 20 **A:** TAI
testacea Seidlitz, 1896: 121 **E:** AU BH CR FR HU IT SZ
  paupercula Baudi di Selve, 1883a: 3 (*Cistela*)
testaceicornis obscuricolor Pic, 1904d: 27 **A:** FE
testaceicornis testaceicornis Pic, 1904d: 26 **A:** ES FE
thibetana Pic, 1917c: 20 **A:** XIZ
thoracica Fabricius, 1792b: 45 (*Cistela*) **E:** CZ GE HU RO SK
  hungarica Csiki, 1901b: 488
  rubricollis Panzer, 1796b: 9 (*Cistela*)
tonkinea Pic, 1917c: 19 **A:** TAI **ORR**
umbellatarum Kiesenwetter, 1863: 426 (*Cistela*) **E:** AU BH CR IT
unicolor Iablokoff-Khnzorian, 1976: 325 **E:** AR GG ST
viridimetallicus Pic, 1909l: 142 **A:** PA
yunnana Pic, 1930b: 30 **A:** YUN
subgenus *Mucheimira* Novák, 2016e: 188 type species *Isomira stoetzneri* Muche, 1981
bicolorata Muche, 1982: 121 **A:** SA YE
farkaci Novák, 2014d: 154 **A:** NP
martensi Novák, 2014d: 155 **A:** NP
murzini Novák, 2009a: 364 **A:** YUN
nepalica Novák, 2009a: 367 **A:** NP
pizurae Novák, 2009a: 369 **A:** NP
schawalleri Novák, 2014d: 157 **A:** NP
sichuanica Novák, 2009a: 372 **A:** SCH
stoetzneri Muche, 1981: 157 **A:** GUX HUB JIX SCH YUN ZHE
zhongdianica Novák, 2014d: 158 **A:** YUN
subgenus *Paraisomira* Dubrovina, 1982: 138 type species *Cistela oculata* Marseul, 1876
kasugaensis Akita & Masumoto, 2016: 7 **A:** JA
oculata Marseul, 1876d: 327 (*Cistela*) **A:** JA RU

genus *Kralia* Novák, 2013a: 500 type species *Kralia minshanica* Novák, 2013
minshanica Novák, 2013a: 502 **A:** GAN

genus *Microcistela* Pic, 1904d: 26 type species *Microcistela rosinae* Pic, 1904
haagi Harold, 1878a: 80 (*Pseudocistela*) **A:** FE JA
rosinae Pic, 1904d: 26 **A:** ES FE SHA
  infernalis Pic, 1904d: 26
  muhldorfi Borchmann, 1930: 244 (*Pseudocistela*)

genus *Paracistela* Borchmann, 1941: 30 type species *Paracistela variabilis* Borchmann, 1941
bengali Novák, 2011d: 353 **A:** SD

rufithorax Pic, 1913e: 19 (*Pseudocistela*) **A:** JIX ORR
sinensis Pic, 1934b: 24 (*Isomira*) **A:** CE
smetanai Novák, 2011d: 371 **A:** JIX
weigeli Novák, 2011d: 376 **A:** YUN

genus *Pseudocistela* Crotch, 1873a: 108 type species *Cistela brevis* Say, 1824
ceramboides Linnaeus, 1758: 377 (*Chrysomela*) **E:** AB AL AU BE BU BY CR CT CZ DE EN FI FR GB GE GR HU IR IT LA LT MD NL NR NT PL RO RU SK SP ST SV SZ UK "Caucasus" **A:** TR
rufa DeGeer, 1775: 23 (*Pyrochroa*)
saperdoides Küster, 1850a: 71 (*Cistela*)
serrata Chevrolat, 1844: 125 (*Cistela*)
striata Geoffroy, 1785: 161 (*Mordella*)
hajeki Novák, 2013b: 296 **A:** IN
impressicollis Chevrolat, 1873: 205 (*Cistela*) **A:** SY
pamirica Plavilshchikov, 1927: 133 **A:** TD
rubroflava Fairmaire, 1878a: 123 (*Cistela*) **A:** CE
semirubra Pic, 1910q: 75 **A:** TAI
watanabei Kôno, 1934b: 37 **A:** FE
znojkoi Iablokoff-Khnzorian, 1959a: 60 (*Gonodera*) **E:** AB

genus *Pseudohymenalia* Novák, 2008a: 213 type species *Pseudohymenalia yunnanica* Novák, 2008
andreasi Novák, 2016f: 195 **A:** ZHE
guizhouica Novák, 2016f: 197 **A:** GUI
saliica Masumoto, Novák, Akita & Lee, 2019b: 41 **A:** TAI
sichuanica Novák, 2016f: 207 **A:** SCH
taiwana Masumoto, Novák, Akita & Lee, 2019b: 38 **A:** TAI
turnai Novák, 2008a: 214 **A:** HUB YUN
viktorai Novák, 2016f: 212 **A:** YUN
xihouica Novák, 2016f: 215 **A:** YUN
yunnanica Novák, 2008a: 216 **A:** YUN

subtribe Mycetocharina Gistel, 1848

genus *Mycetochara* Guérin-Méneville, 1827: 346 [RN] type species *Cistela scapularis* Illiger, 1805 (= *Cistela humeralis* Fabricius, 1787)
subgenus *Ernocharis* C.G. Thomson, 1859: 118 type species *Cistela brevis* Illiger, 1794 (= *Cistela maura* Fabricius, 1792)
Stigmatoma LeConte, 1862: 244 type species *Cistela fraterna* Say, 1823
abschasica Pic, 1925c: 6 **E:** "Caucasus"
amamiensis Akita & Masumoto, 2014: 214 **A:** JA
angustifrons Reitter, 1899b: 155 **E:** ST
brenskei Seidlitz, 1896: 136 **E:** GR

chihiroae Akita & Masumoto, 2014: 219 **A**: JA
collina Lewis, 1895a: 253 (*Mycetochares*) **A**: JA
cretica Mařan, 1954b: 169 **E**: GR (Crete)
deguchii Akita & Masumoto, 2016: 8 **A**: JA
elongata Miyatake, 1985: 348 **A**: JA
erbilensis Faraj, 2018: 97 **A**: IQ
excelsa Reitter, 1884d: 246 **E**: AB GG ST "Caucasus" **A**: IN
flavicornis Miller, 1883: 265 (*Mycetochares*) **E**: GR
flavipennis Reitter, 1908e: 129 **E**: IT
gracilicornis Roubal, 1935a: 24 **E**: ST "Caucasus"
graciliformis Reitter, 1899b: 156 **E**: BU CR GR RO
gracilis Faldermann, 1837: 98 (*Mycetochares*) **E**: AR GG MD ST "Caucasus"
croceipes Weise, 1880b: 478 (*Mycetochares*)
hiranoi Akita & Masumoto, 2014: 220 **A**: JA
hirta J.R. Sahlberg, 1908: 89 **A**: TR
humeralis Fabricius, 1787a: 85 (*Cistela*) **E**: AU BE BH BU CR CT CZ DE EN FI FR GB GE GG GR HU IT LA LT MD ME NL NR PL RO RU SB SK SL ST SV SZ UK YU
bipustulata Illiger, 1794: 606 (*Cistela*)
scapularis Illiger, 1805: 90 (*Cistela*)
impressa Pic, 1925a: 2 **N**: AG
jonica Obenberger, 1917: 38 **E**: GR
kimotoi Hanatsuka, Masumoto & Kon, 2006: 181 **A**: JA
koltzei Reitter, 1896b: 75 **A**: ES FE JA (Hokkaido) MG
satanula Reitter, 1899b: 155
laticornis Reitter, 1884d: 249 **A**: IS LE
longior Fairmaire, 1892b: 149 **A**: IS TR
maculipes Reitter, 1899b: 158 **A**: TR
maura Fabricius, 1792b: 45 (*Cistela*) **E**: AB AR AU BE BU BY CR CT CZ DE EN FI FR GE GG GR HU IT LA LT ME NL NR PL PT SK SL SP SV SZ UK "Caucasus" **N**: MO **A**: IN TR
barbata Laporte, 1840: 244
barbatus Latreille, 1804a: 348 (*Helops*)
brevis Illiger, 1794: 608 (*Cistela*)
dalmatina Baudi di Selve, 1877d: 588
hirsuta Pic, 1925a: 1
linearis Illiger, 1794: 607 (*Cistela*)
rufipennis Pic, 1925a: 1
melandryina Roubal, 1934b: 35 (*Mycetochares*) **E**: SK
mersinica Novák, 2013c: 54 **A**: TR
metallica Pic, 1934c: 22 **A**: XIZ
mimica Lewis, 1895a: 253 (*Mycetochares*) **A**: JA
aomoriensis Nakane, 1991: 6
kawasei Saitô, 2015: 382

myrmecophila Obenberger, 1917: 37 **E:** GR
nakamurai Akita & Masumoto, 2014: 221 **A:** JA
netolitzkyi Penecke, 1912: 242 **E:** BU
obscura Zetterstedt, 1840: 162 (*Mycetocharis*) **E:** CT CZ EN FI LA LT NR NT PL RU SK SV
obtusicollis Reitter, 1899b: 157 **E:** AR
ocularis Reitter, 1884d: 245 **E:** AB IN "Caucasus"
okinawaensis Akita & Masumoto, 2014: 218 **A:** JA
ontakensis Akita & Masumoto, 2014: 228 **A:** JA
oodaigaharaensis Akita & Masumoto, 2014: 227 **A:** JA
orientalis Dubrovin, 1992b: 516 **A:** FE
pygmaea L. Redtenbacher, 1874: 133 (*Mycetochares*) **E:** AU BH CR CZ GE HU IT ME RO RU SB SK SL SP YU "Caucasus"
quadrimaculata Latreille, 1804a: 349 (*Helops*) **E:** AL AU BH BU CR CZ FR GE GR HU IT ME RO RU SB SK SP SZ UK YU **A:** TR
schwarzi Reitter, 1888d: 431
quadrispilota Mařan, 1954b: 170 **A:** NC FE SC
roubali Mařan, 1935a: 141 **E:** CZ HU RO SK UK
rudis Küster, 1850b: 97 (*Mycetochares*) **E:** AL GR TR **A:** CY
ruficollis Baudi di Selve, 1877d: 589 **A:** CY LE SY
ruficornis Pic, 1925a: 1 **A:** TR
sakaii Akita & Masumoto, 2014: 223 **A:** JA
scutellaris Lewis, 1895a: 253 (*Mycetochares*) **A:** JA
seidlitzi Reitter, 1899b: 156 **E:** ST
straussii Seidlitz, 1896: 166 **E:** AU BH CR IT
striatipennis Pic, 1904f: 74 **A:** TR
tanakai Akita & Masumoto, 2016: 8 **A:** JA
thoracica Gredler, 1854: 15 (*Mycetochares*) **E:** AU FR IT
fasciata E. Mulsant & V. Mulsant, 1855: 155 (*Mycetochares*)
tsuyukii Akita & Masumoto, 2014: 230 **A:** JA
zolotareffi Reitter, 1896b: 76 **E:** GG ST "Caucasus" **A:** TR
subgenus *Mycetochara* Guérin-Méneville, 1827: 346 [RN] type species *Cistela scapularis* Illiger, 1805 (= *Cistela humeralis* Fabricius, 1787)
Bolitophilus Gistel, 1832: 137 [RN] type species *Cistela scapularis* Illiger, 1805 (= *Cistela humeralis* Fabricius, 1787)
Mycetochares Latreille, 1829: 42 [RN] type species *Cistela scapularis* Illiger, 1805 (= *Cistela humeralis* Fabricius, 1787)
Mycetocharis Gyllenhal, 1827: 510 [RN] type species *Cistela scapularis* Illiger, 1805 (= *Cistela humeralis* Fabricius, 1787)
Mycetophila Gyllenhal, 1810: 541 [HN] type species *Cistela scapularis* Illiger, 1805 (= *Cistela humeralis* Fabricius, 1787)
abdurakhmanovi Nabozhenko, 2018c: 174 **A:** LE

axillaris Paykull, 1799: 123 (*Cistela*) **E:** AL AU BE BU BY CT CZ DE EN FI FR GE GR HU IT LA LT MC MD ME NL NR NT PL RO RU SK SP ST SV SZ UK **A:** "Siberia"
linearis Gyllenhal, 1810: 541 (*Mycetophila*)
maurina Mulsant, 1856b: 33 (*Mycetochares*)
morio L. Redtenbacher, 1849: 605
flavipes Fabricius, 1792b: 45 (*Cistela*) **E:** AU BY CT CZ DE EN FI FR GE GG HU IT KZ LA LT MD NR NT PL RO RU SB SK SL ST SV SZ UK YU "Caucasus" **A:** FE JA MG "South Siberia"
bimaculata Mannerheim, 1844a: 197 (*Mycetochares*)
bipustulata Thunberg, 1784: 17 (*Leptura*)
grandicollis Fairmaire, 1892b: 149 **A:** TR
kazdagiica Novák, 2015i: 138 **A:** TR
nigripes Petri, 1926: 193 **E:** AU RO
sulcipennis Reitter, 1896b: 76 **E:** CZ RO
subgenus *Pterna* Reitter, 1884: 249 type species *Ernocharis auricoma* Reitter, 1884
arnoldii Dubrovina, 1976: 1408 **A:** TD
auricoma Reitter, 1884d: 249 (*Ernocharis*) **E:** CT MD RO ST UK
retowskyi Reitter, 1889j: 373 (*Ernocharis*)

tribe Cteniopodini Solier, 1835

genus *Balassogloa* Semenov, 1890: 372 type species *Balassogloa sphenarioides* Semenov, 1890
minor Semenov, 1890: 373 **A:** TM
sphenarioides Semenov, 1890: 372 **A:** TM

genus *Cistelina* Seidlitz, 1896: 195 type species *Cistela davidis* Fairmaire, 1878
crassicornis Borchmann, 1916: 103 **A:** FUJ TAI
davidis Fairmaire, 1878a: 123 (*Cistela*) **A:** CE FUJ HUB HUN JIX MG SHA SCH YUN **ORR**
iriomotensis Nakane, 1968b: 84 (*Cteniopinus*) **A:** JA
rufa Borchmann, 1930a: 118 **A:** SCH
sulphurea Seidlitz, 1896: 196 **A:** CE SCH
tokaraensis Nomura, 1961a: 38 (*Cteniopinus*) **A:** JA
tricolor Borchmann, 1916: 101 **A:** TAI **ORR**

genus *Cistelomorpha* L. Redtenbacher, 1868: 134 type species *Cistelomorpha straminea* L. Redtenbacher, 1868
alternans Fairmaire, 1894a: 40 **A:** SD
apicipalpis Fairmaire, 1889b: 47 (*Cistela*) **A:** FUJ GUI HUB SCH SHA
atripes Fairmaire, 1899: 631 **A:** FUJ
bina Fairmaire, 1899: 631 **A:** FUJ TAI
bisbinotata bisbinotata Pic, 1909k: 134 **A:** "Himalaya"
bisbinotata subobliterata Pic, 1909k: 134 **A:** "Himalaya"
costatipennis Pic, 1908c: 38 **A:** YUN

costulata Pic, 1908h: 62 **A:** YUN
densestriata Fairmaire, 1896b: 119 **A:** SD "Himalaya"
diversestriata Pic, 1908e: 47 **A:** CH
donckieri Pic, 1908e: 48 **A:** YUN
grandis Pic, 1908e: 47 **A:** XIN
holoxanta Pic, 1908e: 48 **A:** JIX YUN
hummeli Pic, 1933b: 8 **A:** GAN NW
infranigra Fairmaire, 1899: 632 **A:** CH
lameyi Pic, 1908c: 39 **A:** SCH **ORR**
luteicolor Pic, 1915c: 7 **A:** XIZ
maculicornis Fairmaire, 1878a: 121 (*Cistela*) **A:** CE **ORR**
mausonensis Borchmann, 1938a: 296 **A:** CH TAI "Kanshirei" **ORR**
 flavipilis Borchmann, 1938a: 297
melanopyga Fairmaire, 1893g: 322 **A:** CE TAI **ORR**
nigripilis Borchmann, 1940a: 154 **A:** TAI
nigrotibialis Fairmaire, 1893g: 301 **A:** CH TAI **ORR**
obscuriceps Pic, 1909a: 19 **A:** YUN
rufina Fairmaire, 1893b: 322 **A:** TAI **ORR**
savioi Pic, 1924a: 31 **A:** SHG
simillima Fairmaire, 1891c: xxi (*Cteniopus*) **A:** HEB HUB
spurcaticornis Fairmaire, 1891c: xxi (*Cteniopus*) **A:** CE JIX HUB
straminea L. Redtenbacher, 1868: 135 **A:** CE
thibetana Pic, 1915c: 7 **A:** XIZ
ustiventris Fairmaire, 1878a: 122 (*Cistela*) **A:** CE
verbascum Novák, 2018d: 74 **A:** NP

genus *Cnecosochara* Reitter, 1913d: 660 type species *Cnecosochara petriiformis* Reitter, 1913
filia Znojko, 1936: 74 **A:** KZ UZ
petriiformis Reitter, 1913d: 661 **A:** TD

genus *Cteniopinus* Seidlitz, 1896: 200 type species *Cistela altaica* Gebler, 1830
subgenus *Cteniopinus* Seidlitz, 1896: 200 type species *Cistela altaica* Gebler, 1830
altaicus altaicus Gebler, 1829a: 128 (*Cistela*) **E:** CT ST **A:** CE ES FE MG NMO WS
 longicornis Borchmann, 1930b: 155
altaicus kaszabi Muche, 1965a: 207 **A:** MG
andrewesii Fairmaire, 1896a: 58 (*Cistelomorpha*) **A:** ES HP MG NP WS XIZ **ORR**
andrewesii annuligera Fairmaire, 1896b: 118 (*Cistelomorpha*) **A:** HP
andrewesii confluens Borchmann, 1930b: 151 **A:** KA **ORR**
antennatus Ren & Bai, 2005: 385 **A:** GAN
ater Borchmann, 1930b: 160 **A:** TAI
borchmanni Znojko, 1950: 43 **A:** XIZ

brevithoracus Bai & Ren, 2003: 682 **A:** YUN
brunneicapilus Yu & Ren, 1997: 9 **A:** NIN SHA
diversipes Pic, 1937i: 173 **A:** CE ZHE
diversipunctatus Yu & Ren, 1997: 10 **A:** NMO
elegans Nomura, 1961a: 39 **A:** GUX TAI
fangchenghensis Yang & Ren, 2011: 575 [RN] **A:** GUX
flavifemur Bai & Ren, 2004: 320 [HN]
flavifemur Borchmann, 1930b: 161 **A:** CH SD
flavipalpulus Bai & Ren, 2004: 321 **A:** GUX
fouquei Novák, 2019a: 107 **A:** NP
foveicollis Borchmann, 1930b: 154 **A:** TAI
glabratus Yu & Ren, 1997: 8 **A:** GAN NIN
hypocrita Marseul, 1876d: 329 (*Cteniopus*) **A:** BEI FUJ GAN GUA GUX HUN JA JIA NE SC SCH SHA SHG XIZ **ORR**
potanini Heyden, 1889b: 677 (*Cteniopus*)
iyonis Kôno, 1930b: 97
impressithorax Pic, 1910w: 272 **A:** TAI
jindrai Novák, 2019a: 109 **A:** SCH
koltzei Heyden, 1884a: 295 (*Cteniopus*) **A:** FE NE NO SC
koreanus atricornis Pic, 1904d: 26 **A:** "Sibérie"
koreanus koreanus Seidlitz, 1896: 203 **A:** ES FE NE NO SC
koreanus subelongata Pic, 1908k: 75 **A:** "Corée"
kubani Novák, 2018e: 84 **A:** YUN
kwanhsienensis Borchmann, 1930b: 159 **A:** CE GUX NE NO SCH WP XIZ
mikyskai Novák, 2019a: 112 **A:** YUN
moxiensis Novák, 2019a: 114 **A:** SCH
nigricornis Borchmann, 1930b: 153 **A:** ANH FUJ HEN JIX
nigriepipleuron Yang & Ren, 2010: 59 **A:** XIZ
nigritarsis Borchmann, 1930b: 156 **A:** NC
nigrocapitis Bai & Ren, 2003: 682 **A:** YUN
nigromaculatus Bai & Ren, 2004: 322 **A:** GUX
nigrosparsa Fairmaire, 1899: 630 **A:** GAN QIN SICH XIZ YUN
separata Pic, 1908c: 40
nigroventrus Bai & Ren, 2003: 684 **A:** SCH (Jiangjin)
parvus Yu & Ren, 1997: 10 **A:** NIN
protuberans Yu & Ren, 1997: 8 **A:** GAN SHA
ruber Pic, 1923j: 76 **A:** YUN GUX
rubipes Bai & Ren, 2003: 686 **A:** SCH
rubithoracus Yang & Ren, 2011: 574 **A:** GUI HUN
scapulatus Borchmann, 1930b: 162 **A:** CE SCH XIZ YUN
schneideri Novák, 2019a: 117 **A:** SCH

semicirculoris Yang & Ren, 2010: 57 **A**: XIZ
semirufus Pic, 1923j: 76 **A**: XIZ YUN
similis Borchmann, 1930b: 158 **A**: CE JIL LIA MG NE
siniaevi Novák, 2019a: 119 **A**: SHA
subobscurus Pic, 1910w: 271 **A**: TAI
sulcatus Pic, 1925a: 2 **E**: "Caucasus"
taiwanus Kôno, 1930b: 98 **A**: TAI
tatsienluensis Borchmann, 1932: 94 (*Cistelina*) **A**: HEN SCH
tenuitarsis Borchmann, 1930b: 157 **A**: NC
tschiliensis Borchmann, 1930b: 157 **A**: HEB JIL MG NE NMO
unicolor Kôno, 1930b: 98 **A**: TAI
varicolor Heyden, 1889b: 676 (*Cteniopus*) **A**: GAN NO SCH SHA SHX "Korea"
varicornis Ren & Bai, 2005: 383 **A**: SHA GAN
yunlongensis Novák, 2019a: 124 **A**: YUN
zayicus Yang & Ren, 2010: 58 **A**: XIZ
subgenus *Lechinius* Blair, 1922c: 561 type species *Lechinius catenulatus* Blair, 1922 (= *Cistelomorpha fossulata* Pic, 1913)
Lechinius Borchmann, 1930b: 151 [HN] type species *Cistelomorpha fossulata* Pic, 1913
fossulatus Pic, 1913m: 142 (*Cistelomorpha*) **A**: NP SD XIZ **ORR**
catenulatus Blair, 1922c: 561
holzschuhi Novák, 2009b: 250 **A**: BT
semicoccineus Blair, 1922c: 560 **A**: NP XIZ

genus *Cteniopus* Solier, 1835a: 246 type species *Chrysomela sulphurea* Linnaeus, 1758
subgenus *Cteniopsomus* Reitter, 1906d: 131 type species *Cteniopus frater* Reitter, 1903
bodemeyeri Reitter, 1903h: 257 (*Cteniopus*) **A**: TR
drurei Pic, 1915: 14 **A**: IQ (Mossoul)
frater Reitter, 1903h: 258 (*Cteniopus*) **A**: TR
gibbosus Baudi di Selve, 1877c: 394 (*Cteniopus*) **A**: IS SY **N**: EG
infraniger Pic, 1915: 14 **A**: TR
intermedius Fairmaire, 1895b: cx (*Cteniopus*) **A**: TR
notatipennis Pic, 1915: 14 **A**: SY (Alep)
persicus Ogloblin, 1950: 72 **A**: AF IN IQ SY TR
priesneri luteiceps Pic, 1939f: 146 **A**: EG SI
priesneri priesneri Wittmer, 1936: 17 **N**: EG
testaceitarsis Pic, 1915: 14 **A**: TR
turcicus Pic, 1913h: 121 **A**: TR
subgenus *Cteniopus* Solier, 1835a: 246 type species *Chrysomela sulphurea* Linnaeus, 1758
Cistela Fabricius, 1775: 116 [HN] type species *Chrysomela sulphurea* Linnaeus, 1758
Cistella Gistel, 1848b: xi
Sarandonyx Des Gozis, 1881: cxiii [RN]
Telacis Poey, 1854: 322 [RN]

anatolicus Pic, 1903i: 123 **E:** TR **A:** TR
angustatus Pic, 1905j: 177 **A:** IN IQ
erbilensis Nabeel, Gazang & Hazim, 2016: 216 **A:** IQ
expulsus Roubal, 1917b: 47 **E:** GG
impressicollis Fairmaire, 1892b: 152 **A:** IN TR
magnus Novák, 2018e: 88 **A:** SY
neapolitanus Baudi di Selve, 1877c: 393 **E:** IT
trifossus Pic, 1903i: 123
nigrifrons Fairmaire, 1895b: cxi **A:** TR
pallidus Küster, 1850a: 82 (*Cistela*) **A:** CY IN IQ SA SY TR
intrusus Seidlitz, 1896: 210
persimilis Reitter, 1890g: 258 **E:** AB AR **A:** IN TR
purkynei Mařan, 1936: 205 **E:** AL MC
sulphureus Linnaeus, 1758: 376 (*Chrysomela*) **E:** AL AU BH BU BY CR CT CZ DE FI FR GB GE GR HU IT KZ LA LT MD ME NL NR PL RO SK SL SP ST SV SZ UK YU "Caucasus" **A:** ES FE MG TR WS
altaicus Reitter, 1906d: 133
analis Seidlitz, 1896: 212
bicolor Fabricius, 1794: 447 (*Cistela*)
flavus Scopoli, 1763: 84 (*Tenebrio*)
gilvus Seidlitz, 1896: 212
luteus Geoffroy, 1785: 159 (*Tenebrio*)
murinus Herbst, 1783: 64 (*Cistela*)
palpalis Seidlitz, 1896: 212
sulphuratus Gmelin, 1790: 1717 (*Cryptocephalus*)
varians Fabricius, 1787a: 85 (*Cistela*)
tokatensis Pic, 1904d: 26 **A:** TR
subgenus *Rhinobarus* Reitter, 1906d: 131 type species *Cistela sulphuripes* Germar, 1823
elegans Faldermann, 1837: 104 (*Cistela*) **E:** AR GG "Caucasus"
kalambakaensis Novák, 2017e: 137 **E:** GR
punctatissimus Baudi di Selve, 1877c: 395 **E:** GR
graecus Heyden, 1883c: 312
sulphuripes Germar, 1823: 162 (*Cistela*) **E:** AB AU BU CR CZ FR GR HU IT MC MD RO RU SK ST UK YU **A:** ES TR
collaris Küster, 1850a: 75 (*Cistela*)
notatus Pic, 1925a: 2
sulfuripes W. Redtenbacher, 1842: 18 (*Cistela*)
sveci Novák, 2017e: 142 **E:** BU

genus *Falsomophlus* Pic, 1925e: 11 type species *Falsomophlus niger* Pic, 1925
niger Pic, 1925e: 12 **A:** "Himalaya"

genus *Gastrhaema* Jacquelin du Val, 1863: 353 type species *Cistela rufiventris* Waltl, 1835
abdominalis Laporte, 1840: 247 (*Omophlus*) **E:** SP **A:** AL MO
boiteli Peyerimhoff, 1931a: 5 (*Heliotaurus*) **N:** MO
gharbensis Théry, 1932b: 133 **N:** MO
hoemorroidale Fairmaire, 1866a: 50 **N:** AG
martini Bedel, 1894: 164 **N:** AG TU
nigripennis discipennis Reitter, 1906d: 136 **N:** AG
nigripennis erythrocephala Solier, 1835a: 248 (*Megischia*) **N:** AG
nigripennis nigripennis Fabricius, 1792b: 44 (*Cistela*) **N:** AG TU
crassicornis Desbrochers des Loges, 1881: 155 (*Omophlus*)
nigripennis nigrocephala Reitter, 1906d: 137 **N:** AG
nigripennis subrubrica Reitter, 1906d: 137 **N:** AG
ovalis Laporte, 1840: 247 (*Omophlus*) **N:** AG MO TU
rufiventris anceps Mulsant, 1856b: 41 **N:** MO
rufiventris rufiventris Waltl, 1835: 75 (*Cistela*) **E:** SP **N:** AG MO
tingitana Théry, 1932b: 135 (*Heliotaurus*) **N:** MO

genus *Heliomophlus* Reitter, 1906d: 147 type species *Heliotaurus scabriusculus* Fairmaire, 1866
albopilosus Kocher, 1956a: 240 **N:** MO
dasytoides Fairmaire, 1871a: 395 (*Heliotaurus*) **N:** AG
epipleuricus Reitter, 1908a: 22 **N:** MO
mimeuri Kocher, 1956a: 236 **N:** MO
scabriusculus Fairmaire, 1866a: 54 (*Heliotaurus*) **N:** AG MO TU
gracilior Fairmaire, 1871a: 394 (*Heliotaurus*)
kirschii Reitter, 1872: 174 (*Omophlus*)
mulsanti Kirsch, 1869: 103 (*Omophlus*)
theryi mairei Kocher, 1956a: 239 **N:** MO
theryi occipitalis Kocher, 1956a: 239 **N:** MO
theryi theryi Escalera, 1926: 184 **N:** MO

genus *Heliosthraema* Reitter, 1890b: 34 type species *Heliotaurus rolphii* Fairmaire, 1867
griseolineatus Reitter, 1890b: 36 **N:** MO
alterneseriata Escalera, 1911: 303
poweli Théry, 1932b: 136 **N:** MO
prudhommei Bouyon, 2015a: 219 **N:** MO
reymondi Pic, 1953b: 62 **N:** MO
rolphii Fairmaire, 1867: 408 (*Heliotaurus*) **N:** MO
susica Escalera, 1913: 52 (*Heliotaurus*) **N:** MO
vegepunctata Escalera, 1913: 53 (*Heliotaurus*) **N:** MO

genus *Heliotaurus* Mulsant, 1856b: 73 type species *Cistela distincta* Laporte, 1840
subgenus *Atlasotaurus* Bouyon, 2011: 464 type species *Omophlus maroccanus* Lucas, 1847

brisouti Bedel, 1891a: 794 (*Omophlus*) **N:** AG
casseti Bouyon, 2011: 466 **N:** MO
chobauti Bedel, 1894: 169 **N:** AG
doriae Bedel, 1891a: 795 (*Omophlus*) **N:** AG LB TU
obscuripes Pic, 1936c: 22
gestroi Peyerimhoff, 1918c: 257 **N:** AG
maroccanus Lucas, 1847: 357 (*Omophlus*) **N:** AG MO
janthinus Raffray, 1873: 378
splendidus Bedel, 1905: 272 (*Omophlus*) **N:** MO
pulchellus Reitter, 1908a: 21
subgenus *Heliotaurus* Mulsant, 1856b: 73 type species *Cistela distincta* Laporte, 1840
aestivalis Escalera, 1914: 363 **N:** MO
analis analis Desbrochers des Loges, 1881: 137 (*Heliotaurus*) **N:** AG MO TU
analis brevicrinitus Muche, 1967: 23 **N:** TU
analis henoni Pic, 1936c: 22 **N:** AG
angusticollis Mulsant, 1856b: 48 **N:** AG EG LB MO TU **AFR**
oberthuri Reitter, 1890b: 38
chalybaeus Kirsch, 1869: 103 **N:** EG LB
coeruleus Fabricius, 1787a: 85 (*Cistela*) **N:** AG LB MO TU
coerulescens G.-A. Olivier, 1791: 6 (*Cistela*)
confusus Reitter, 1890b: 40 **N:** AG MO TU
corallinus Reitter, 1890b: 51 **N:** LB
crassidactylus crassidactylus Seidlitz, 1896: 228 **E:** PT SP
crassidactylus stenodactylus Reitter, 1906d: 140 **E:** SP
distinctus distinctus Laporte, 1840: 246 (*Cistela*) **E:** IT (Sicilia) SP **N:** AG MO LB TU
distinctus kabylianus Pic, 1896h: 93 **N:** AG
distinctus perroudi Mulsant, 1856b: 45 **N:** AG
distinctus plenifrons Fairmaire, 1866a: 51 **N:** AG
distinctus riffensis Escalera, 1922c: 360 **N:** MO
distinctus seidlitzi Reitter, 1906d: 141 **N:** AG
distinctus variventris Desbrochers des Loges, 1881: 136 **N:** AG TU
erythrogaster Lucas, 1846: pl. 31 (*Omophlus*) **N:** AG
testacea Fabricius, 1787a: 85 (*Cistela*)
gasthraemoides Reitter, 1890b: 52 **N:** MO
goedeli Reitter, 1890b: 38 **A:** SY **N:** EG **AFR**
larbii Escalera, 1911: 304 **N:** MO
longitarsis Escalera, 1914: 365 **N:** MO
menticornis ciliatus Reitter, 1890b: 37 **N:** AG
menticornis menticornis Reitter, 1872: 172 (*Omophlus*) **N:** AG MO TU
anthracinus Fairmaire, 1875a: 530 (*Omophlus*)

moderatus Peyerimhoff, 1939b: 120 **N:** AG
monticola Normand, 1955: 85 **N:** TU
parvicollis Reitter, 1872: 144 **N:** EG LB MO TU
    parvicollis Reitter, 1890b: 52 [HN]
productus Rosenhauer, 1856: 220 (*Omophlus*) **E:** SP **N:** MO
rasalis Normand, 1955: 86 **N:** TU
ruficollis incertus Mulsant, 1856e: 48 **E:** SP
ruficollis punctosulcatus Fairmaire, 1879d: 242 **E:** SP
ruficollis ruficollis Fabricius, 1781: 147 (*Cistela*) **E:** PT SP **N:** AG MO
    maculicollis Desbrochers des Loges, 1881: 92 (*Omophlus*)
ruficollis tangerianus Escalera, 1922c: 359 **N:** MO
rufithorax cincticollis Escalera, 1933: 130 **N:** MO
rufithorax nigritulus Escalera, 1914: 365 **N:** MO
rufithorax rufithorax Reitter, 1890b: 36 **E:** SP **N:** MO
sanchezperezi nigricolor Escalera, 1933: 131 **N:** MO
sanchezperezi sanchezperezi Escalera, 1933: 131 **N:** MO
sanguinicollis nigricollis Reitter, 1906d: 143 **E:** SP
sanguinicollis rubronotatus Fuente, 1907: 319 **E:** SP
sanguinicollis sanguinicollis Reitter, 1906d: 143 **E:** SP
subpilosus Seidlitz, 1896: 232 **N:** TU
tenuipes Seidlitz, 1896: 229 **E:** SP
tournieri scultipennis Escalera, 1933: 130 **N:** MO
tournieri tournieri Pic, 1896f: 61 **N:** MO
tuniseus Fairmaire, 1875a: 529 **N:** AG LB TU
subgenus *Julogenius* Reitter, 1906d: 138 type species *Heliotaurus reichei* Mulsant, 1856
longipilus Fairmaire, 1866a: 54 **N:** AG MC TU
quedenfeldti Reitter, 1890b: 52 (*Heliotaurus*) **N:** MO
reichii Mulsant, 1856b: 50 (*Heliotaurus*) **N:** AG LB MO
    grilati Mulsant & Godart, 1876b: 255 (*Heliotaurus*)
    oranensis Reitter, 1872: 173 (*Omophlus*)

genus *Holdhausia* Reitter, 1906d: 125 type species *Cteniopus crassus* Fairmaire, 1892
crassus Fairmaire, 1892b: 152 (*Cteniopus*) **A:** IS SY TR

genus *Hypocistela* Bates, 1879: 482 type species *Hypocistela tenuipes* Bates, 1879
tenuipes Bates, 1879: 483 **A:** NW

genus *Megischia* Solier, 1835a: 247 type species *Cistela curvipes* Brullé, 1832
curvimana Reitter, 1890b: 40 **A:** CY SY
curvipes corcyrea Mařan, 1936: 207 **E:** GR (Corfu)

curvipes curvipes Brullé, 1832: 226 (*Cistela*) **E:** FR GR IT SP

prosternalis Reitter, 1890b: 40

galbanata Kiesenwetter, 1861a: 238 (*Omophlus*) **E:** AL BH BU CR GR MC **A:** TR

curvipes Reitter, 1890b: 40

laticollis Mařan, 1936: 207

schmalfussi Muche, 1987: 135 **E:** GR

genus *Megischina* Reitter, 1906d: 171 type species *Cistela armillata* Brullé, 1832

adaliae Reitter, 1890b: 41 (*Omophlus*) **A:** TR

armillata Brullé, 1832: 225 (*Cistela*) **E:** AL BH BU CR GR IT ME RO SB TR YU **A:** TR

epipleuralis Seidlitz, 1896: 241 (*Odontomophlus*)

glamocensis Obenberger, 1917: 39 (*Odontomophlus*)

interstitialis Reitter, 1906d: 172

armillata merthae Reitter, 1898f: 349 (*Omophlus*) **E:** RO

bozdagleriensis Novák, 2006b: 100 **E:** GR (Samos) **A:** TR

cypria Mařan, 1936: 206 **A:** CY

renei Novák, 2013d: 266 **E:** GR (Rhodes)

rosinae Seidlitz, 1896: 263 (*Odontomophlus*) **E:** FR MA SP

genus *Omophlina* Reitter, 1890b: 34 type species *Omophlina podontoides* Reitter, 1890

alpina Mulsant, 1856e: 30 (*Podonta*) **A:** TR UZ

arcuata Gebler, 1829a: 129 (*Cistela*) **A:** AF IN KI KZ MG TD TM UZ WS

heydeni Reitter, 1890b: 35

corva Solsky, 1881a: 253 (*Omophlus*) **A:** KI KZ TD TM UZ

matthiesseni Reitter, 1907f: 207

diversipilosa Muche, 1973: 39 **A:** UZ

hirtipennis Solsky, 1881a: 251 (*Podonta*) **A:** KZ TD TM UZ

pubifer Reitter, 1890b: 35

rubricollis Reitter, 1898a: 14

tenuis Kraatz, 1882c: 114 (*Podonta*)

kirgisica Muche, 1973: 37 **A:** KI

muminovi Muche, 1980: 291 **A:** TD

pentheri Ganglbauer, 1905: 279 **A:** TR

podontoides podontoides Reitter, 1890b: 35 **A:** IN SY

podontoides syriaca Muche, 1963b: 63 **A:** SY

reitteri Ogloblin, 1950: 34 **A:** TD

seidlitzi Reitter, 1898a: 15 **A:** KI UZ

turcmenica Iablokoff-Khnzorian, 1970: 64 **A:** TM

willbergi Reitter, 1892c: 135 (*Omophlus*) **A:** TM UZ

bicolorata Reitter, 1906d: 128

genus *Omophlus* Dejean, 1834: 213 type species *Cistela lepturoides* Fabricius, 1787
subgenus *Euomophlus* Iablokoff-Khnzorian, 1983: 134 type species *Cistela rugosicollis* Brullé, 1832
caucasicus caucasicus Kirsch, 1869: 123 **E:** AB AR GG ST TR "Caucasus" **A:** IN TR
somcheticus Seidlitz, 1896: 265
caucasicus svaneticus Znojko, 1950: 107 **E:** GG
caucasicus talyschensis Reitter, 1906d: 160 **E:** AR TR
dilatatus Faldermann, 1837: 99 **E:** AB AR "Caucasus" **A:** IN TR
pallitarsis Reitter, 1890b: 50
dubitatus Reitter, 1906d: 164 **E:** AB AR
laevigatus Seidlitz, 1896: 265 **E:** AR **A:** IN SY TR
politipennis Reitter, 1906d: 161
nitidicollis Seidlitz, 1896: 265 **E:** AR **A:** IN IQ TR
oblongus Znojko, 1950: 117 **E:** AR GG **A:** TM
pilicollis filitarsis Reitter, 1890b: 45 **A:** IN TR
pilicollis fortipes Reitter, 1906d: 168 **A:** UZ
pilicollis longipilis Reitter, 1906d: 168 **A:** UZ XIZ
pilicollis marginatus Kirsch, 1869: 110 **E:** "Caucasus"
pilicollis pilicollis Faldermann, 1832: 204 (*Cistela*) **E:** AB AR CT GG GR KZ RU ST UK **A:** IN KI KZ TM TR UZ WS XIN
mitis Reitter, 1906d: 165
volgensis Kirsch, 1869: 115
pilicollis zolotarewi Reitter, 1906d: 168 **A:** UZ
pruinosus Reitter, 1890b: 45 **E:** AR GG ST "Caucasus" **A:** XIN
rugosicollis Brullé, 1832: 225 (*Cistela*) **E:** AL AN AU BH BU CR CZ FR GR HU IT MC ME RO SK SZ TR YU **A:** TR
brevicollis Mulsant, 1856b: 91
rugicollis Küster, 1849c: 60
subgenus *Micromophlus* Znojko, 1950: 124 type species *Omophlus subtilis* Solsky, 1881
subtilis Solsky, 1881a: 255 (*Omophlus*) **A:** IN TM UZ
subgenus *Odontomophlus* Seidlitz, 1896: 240 type species *Omophlus ochraceipennis* Faldermann, 1837
Pleuromophlus Reitter, 1906: 146 type species *Omophlus baudueri* Baudi di Selve, 1877
afghanus Muche, 1965b: 96 **A:** AF IN PA
anatolicus Reitter, 1906d: 155 **A:** TR
atticus Reitter, 1906d: 153 **E:** GR TR
baudueri Baudi di Selve, 1877c: 401 **A:** SY TR
blumenthali Muche, 1979a: 173 **A:** IN
bodemeyeri Reitter, 1906d: 154 **A:** TR
compressus Seidlitz, 1896: 263 **E:** GR
crinifer Seidlitz, 1896: 263 **A:** CE TM

cyrenaicus Gridelli, 1930: 190 **N:** LB
dispar A. Costa, 1847: 154 **E:** IT
emgei Reitter, 1891c: 199 **E:** BU GR
fallaciosus Rottenberg, 1871: 258 **E:** IT (Sicilia)
falsarius Kirsch, 1869: 109 **E:** AR BU GR RO TR **A:** IN IQ SY TR
pilifer Seidlitz, 1896: 248
flavipennis Küster, 1849c: 57 **E:** AB AR BU CR CT GG GR IT MD RO ST TR UK **A:** IN IQ SY TR
atripes Küster, 1850a: 61
basicornis Reitter, 1890b: 48
dalmatinus Kirsch, 1869: 116
quadricollis Brullé, 1832: 224 (*Cistela*)
forficula Seidlitz, 1896: 263 **A:** IS JO SY TR
foveicollis Fairmaire, 1892b: 153 **A:** LE SY TR
furca Seidlitz, 1896: 264 **A:** IS LE SY
furcula Seidlitz, 1896: 264 **A:** TR
hirsutus Seidlitz, 1896: 264 **A:** IS SY TM TR UZ
infirmus Kirsch, 1869: 119 **E:** GR IT
irbidensis Novák, 2019i: 460 **A:** JO
kalalae Mařan, 1935a: 145 **A:** IQ
kruegeri Gridelli, 1930: 193 **N:** LB
kurda Znojko, 1950: 97 **A:** IQ
laciniatus Seidlitz, 1896: 263 **E:** AB AR GG **A:** IN IQ KI SY TR
latipleuris Reitter, 1890b: 47 **E:** AR **A:** TM TR
lepturoides Fabricius, 1787a: 85 (*Cistela*) **E:** AL AU BH BU CR CZ FR GE GR IT MC MD ME RO SK SL SP ST SZ TR UK **A:** IN TR
alpinus Miller, 1851: 112
elongatus Küster, 1850a: 59
libanonensis Muche, 1964b: 592 **A:** LE
longipleuris Znojko, 1950: 89 **E:** AR **A:** TR
lucidus lucidus Kirsch, 1869: 125 **E:** GR **N:** EG **A:** IS JO SY
lucidus validiusculus Muche, 1965b: 93 **A:** JO
luciolus Seidlitz, 1896: 264 **E:** ST **A:** TR
medius Muche, 1964b: 591 **A:** IN TR
melitensis Baudi di Selve, 1877c: 400 **E:** MA
championi Reitter, 1891g: 260
nadayi Muche, 1965b: 85 **A:** TR
obscurus Reitter, 1890b: 46 **E:** AB AR TR
ochraceipennis Faldermann, 1837: 100 **E:** AB AR GG **A:** IN TR
kuesteri Kirsch, 1869: 105
latitarsis Reitter, 1890b: 41
tarsalis Kirsch, 1869: 115

pilosellus Kirsch, 1869: 120 **E:** BU CR GR IT **A:** IQ TR
pseudolucidus Muche, 1965b: 95 **A:** AF
sandneri Reitter, 1906d: 157 **A:** IN IQ SY TR
schmidi Muche, 1965b: 89 **A:** IN PA
silifkeensis Muche, 1964b: 593 **A:** TR
sulcipleuris Seidlitz, 1896: 263 **E:** GR
syriacus Mulsant, 1856e: 57 **A:** CY IS JO LE SY TR
  varicolor Kirsch, 1869: 112
terminatus Fairmaire, 1884a: 171 **A:** TR
tibialis Reitter, 1890b: 46 **A:** IS JO SY
tumidipes Kirsch, 1878: 247 **E:** AB AR **A:** TM TR
turcicus Kirsch, 1869: 117 **E:** GR MC TR
  foveola Seidlitz, 1896: 264
subgenus *Omophlus* Dejean, 1834: 213 type species *Cistela lepturoides* Fabricius, 1787
anthracinus Znojko, 1950: 102 **E:** ST
brulléi Kirsch, 1869: 109 **E:** GR TR
  hirtellus Kirsch, 1869: 111
candiota Obenberger, 1917: 40 **E:** GR
caspius Znojko, 1950: 105 **E:** AB **A:** IN TM TR
coriaceus Seidlitz, 1896: 264 **E:** AB GG
cribricollis Fairmaire, 1892b: 153 **A:** TR
deserticola Kirsch, 1869: 125 **E:** AB CT KZ ST **A:** KI KZ TM WS XIN
eckerleini Muche, 1964b: 594 **A:** LE TR
ganglbaueri Reitter, 1890b: 51 **A:** IN
gracilipes Kirsch, 1869: 108 **N:** EG **A:** IS JO SY TR
hirtipennis Seidlitz, 1896: 265 **A:** TR
iranicus Muche, 1964a: 238 **A:** IN
kaszabi Muche, 1965b: 88 **A:** IN
klapperichi Muche, 1965b: 91 **A:** JO
kuznetzovi Znojko, 1950: 120 **A:** IN
lepidus Reitter, 1906d: 166 **A:** TR
lindti Muche, 1978: 358 **A:** TM
longicornis Bertolini, 1868: 119 **E:** AU CR CZ BH GR IT ME RO SK SL "Caucasus" **A:** TR
luristanicus Muche, 1965b: 94 **A:** IN
nanulus Muche, 1964a: 235 **E:** RO
nasreddini Reitter, 1890b: 51 **A:** IN IQ SY TR
  conicicollis Reitter, 1906d: 164
nigrinus Reitter, 1889e: 257 **E:** GR
nigropilosellus Muche, 1964a: 239 **A:** TR
ocularis Kirsch, 1869: 128 **A:** IS JO
orestias Muche, 1964b: 596 **A:** TR

orientalis Mulsant, 1856b: 53 **E:** AB AR GR ST TR UK **A:** TR
pollinosus Znojko, 1950: 104 **E:** AB **A:** IN
propagatus Kirsch, 1869: 111 **A:** CY
proteus proteus Kirsch, 1869: 123 **E:** AU CZ BU GR HU IT MD RO SK ST TR UK YU "Caucasus" **A:** IN SY TM TR
longicornis Reitter, 1890b: 50
proteus megista Muche, 1964b: 597 **A:** TR
proteus prometheus Reitter, 1906d: 160 **A:** TM
reitteri König, 1906: 25 **E:** AR **A:** TR
rugipennis Seidlitz, 1896: 265 **E:** AB AR "Caucasus"
salbachi Reitter, 1906d: 163 **A:** TR
scutellaris Mulsant, 1856b: 56 **N:** EG
subcostatus Reitter, 1908d: 56 **A:** TR
tenellus tenellus Kirsch, 1869: 126 **N:** EG LB **A:** IS
tenellus tripolitanus Muche, 1965b: 87 **N:** LB
transitus Muche, 1965b: 98 **A:** AF IN
turanicus Reitter, 1906d: 167 **E:** AR **A:** TM UZ
turcomanicus Znojko, 1950: 118 **A:** TM
subgenus *Paurodontomophlus* Muche, 1979a: 171 type species *Omophlus wittmeri* Muche, 1979
wittmeri Muche, 1979a: 171 **A:** TR
subgenus *Phibalus* Gistel, 1856: 384 type species *Chrysomela pubescens* Linnaeus, 1767
Paromophlus Iablokoff-Khnzorian, 1983: 134 type species *Cistela picipes* Fabricius, 1792
agrapha Reitter, 1890b: 42 **E:** AB AR GR MC
armeniacus Znojko, 1950: 126 **E:** AR
curtellus Seidlitz, 1896: 265 **A:** SY
curtulus Kirsch, 1878: 246 **E:** AB AR GG RU ST "Caucasus" **A:** IN TR
curtus Küster, 1850a: 60 **E:** GG GR **A:** TR
densepunctatus Fairmaire, 1892b: 153 **A:** TR
emmae Iablokoff-Khnzorian, 1959b: 191 (*Heliotaurus*) **E:** AR
excavatus Seidlitz, 1896: 265 **A:** IN
gracilis Znojko, 1950: 123 **E:** AB TR
hirtus Seidlitz, 1896: 265 **E:** GR IT (Sicilia)
irrasus Seidlitz, 1896: 265 **A:** TR
irregularis Reitter, 1906d: 170 **A:** TR
lividipes Mulsant, 1856b: 87 **E:** AU BH BU CT CZ FR GE GR HU IT KZ MD ME RU SK ST UK **A:** WS XIN
nigricolor Reitter, 1913c: 290 **A:** IN
nigripes Küster, 1850a: 65 **E:** GR **A:** IN IQ SY TR
picipes Fabricius, 1792b: 43 (*Cistela*) **E:** AU BU CZ FR GE GR HU IT SK SP
sericeicollis Küster, 1850a: 64
tibialis A. Costa, 1847: 156 (*Omophlus*)

pubescens Linnaeus, 1767: 303 (*Chrysomela*) **E:** AB AL AU BE BH BU BY CZ EN FR GB GE GR HU IT LA LT MC NT PL RU SK SP SV SZ UK **A:** IN
armeriae Curtis, 1836: 622
betulae Herbst, 1783: 65 (*Cistela*)
frigidus Mulsant, 1856b: 81
pinicola L. Redtenbacher, 1849: 604
pubescens Mulsant, 1856b: 84 [HN]
rufitarsis Leske, 1785: 15 (*Chrysomela*)
subalpinus Ménétriés, 1832: 204 (*Cistela*) E: ST "Caucasus"

genus *Petria* Semenov, 1894: 363 [= **1894c: 611**] type species *Petria tachyptera* Semenov, 1894
ales Znojko, 1936: 77 **A:** UZ
antennata Semenov, 1894: 366 [= 1894c: 614] **A:** TM
nitidicollis Semenov, 1896a: 29 **A:** TM
tachyptera Semenov, 1894: 365 [= 1894c: 613] **A:** TM

genus *Podonta* Solier, 1835a: 247 type species *Cistela nigrita* Fabricius, 1794
Pododonta Agassiz, 1846: 300 [UE]
allemandi Bouyon, 2014: 250 **E:** GR
ambigua Kiesenwetter, 1873: 11 **A:** SY TR
anatolica Muche, 1965c: 86 **A:** TR
angelinii Bouyon, 2014: 250 **E:** GR
antennata Muche, 1965c: 83 **E:** AL GR
atrata Kiesenwetter, 1873: 15 **A:** CY IS SY TR
dentella Muche, 1962b: 145
aubei Mulsant, 1856e: 29 **E:** GR
babadagiica Novák, 2016g: 453 **A:** TR
biformis Reitter, 1889j: 374 **E:** AR **A:** IN IQ SY TR
ruficollis Seidlitz, 1896: 189
bodemeyeri Reitter, 1906d: 123 **A:** TR
brevicornis Seidlitz, 1896: 189 **E:** BU TR
carbonaria Kiesenwetter, 1873: 16 **A:** IN IQ IS SY
cassetorum Bouyon, 2015b: 321 **A:** TR
convexicollis Küster, 1850a: 81 (*Cistela*) **E:** TR
corvina Kiesenwetter, 1873: 21 **E:** GR RO TR
daghestanica daghestanica Reitter, 1885d: 383 **E:** BU MC MD RO RU ST UK YU **A:** IN
daghestanica macedonica Mařan, 1935a: 143 **E:** BU MC
dalmatina Baudi di Selve, 1877c: 395 **E:** AL BH CR MC MD RO RU ST TR UK YU
dogueti Bouyon, 2014: 253 **E:** GR
eckerleini Muche, 1963b: 61 **A:** TR
elongata Ménétriés, 1832: 205 (*Cistela*) **E:** AB AR GG ST **A:** IN SY TR

flecki Reitter, 1906d: 122 **E:** BU RO
frater Seidlitz, 1896: 190 **E:** GR TR
graeca Seidlitz, 1896: 190 **E:** GR MC
heydeni Kiesenwetter, 1873: 19 **E:** BU GR RO TR **A:** TR
italica Baudi di Selve, 1877c: 397 **E:** IT
kesanensis Muche, 1965c: 85 **A:** TR
lugubris Küster, 1850a: 80 (*Cistela*) **E:** BU CR RO **A:** TR
marmarensis Bouyon, 2014: 254 **E:** TR
milleri Kiesenwetter, 1873: 19 **E:** GR MC
montreuili Bouyon, 2014: 251 **A:** IN
morio Kiesenwetter, 1873: 21 **E:** AB BU GG GR MC TR
murzini Novák, 2016g: 460 **A:** IN TR
nigrita Fabricius, 1794: 447 (*Cistela*) **E:** AL AU BH BU CR CZ GR HU IT MD RO RU SK ST SZ TR UK YU "Caucasus" **A:** TR
paradoxa Schilsky, 1889: 354
ottomanica Bouyon, 2014: 252 **A:** TR
patagiata Seidlitz, 1896: 190 **A:** TR
pseudoatrata Novák, 2016g: 464 **A:** SY
purkynei Mařan, 1935a: 143 **E:** MC
rambouseki Mařan, 1935a: 144 **E:** MC
rotundangula Reitter, 1906d: 120 **A:** TR
simplex Seidlitz, 1896: 190 **E:** GR
soror Seidlitz, 1896: 189 **A:** SY TR
turcica Kiesenwetter, 1873: 14 **E:** BG RO **A:** TR
korbii Seidlitz, 1896: 190

genus *Podontinus* Seidlitz, 1896: 197 type species *Cteniopus punctatissimus* Kiesenwetter, 1861
farsensis Novák, 2019i: 465 **A:** IN
mujibensis Novák, 2019i: 468 **A:** JO
punctatissimus Kiesenwetter, 1861a: 238 (*Cteniopus*) **A:** IS JO SY

genus *Proctenius* Reitter, 1890g: 256 type species *Cistela granatensis* Rosenhauer, 1856
chamaeleon chamaeleon Reitter, 1894e: 305 (*Cteniopus*) **E:** SP
chamaeleon ictericus Viedma, 1959: 262 **E:** SP
chamaeleon sanguinosus Viedma, 1959: 262 **E:** SP
gaditanus canescens Viedma, 1959: 265 **E:** SP
gaditanus gaditanus Viedma, 1959: 264 **E:** SP
granatensis granatensis Rosenhauer, 1856: 221 (*Cistela*) **E:** SP
granatensis incanus Viedma, 1959: 263 **E:** SP
luteus Küster, 1847c: 82 (*Cistela*) **E:** SP
obscurithorax Pic, 1899o: 300

genus *Stenerophlina* Reitter, 1906d: 130 type species *Omophlina hauseri* Reitter, 1894
hauseri Reitter, 1894a: 50 (*Omophlina*) **A:** TM UZ
levipes Dubrovina, 1973a: 284 **A:** UZ

genus *Steneryx* Reitter, 1890g: 256 type species *Cistela dejeanii* Faldermann, 1836
angustatus Pic, 1938g: 221 **A:** BEI CE
dejeanii Faldermann, 1836: 391 (*Cistela*) **A:** KZ MG TM UZ
gussakovskyi Znojko, 1936: 73 **A:** UZ
hauseri Seidlitz, 1896: 205 **A:** KZ NW TM UZ XIN
semenovi Znojko, 1936: 73 **A:** TM

genus *Tripolicryptus* Strand, 1929: 24 [RN] type species *Brachycryptus tripolitanus* Quedenfeldt, 1891
Brachycryptus Quedenfeldt, 1891: 129 [HN] type species *Brachycryptus tripolitanus* Quedenfeldt, 1891
tripolitanus G. Quedenfeldt, 1891: 130 **N:** LB

subfamily **Stenochiinae** Kirby, 1837
tribe Cnodalonini Oken, 1843

genus *Ainu* Lewis, 1894: 479 type species *Ainu tenuicornis* Lewis, 1894
subgenus *Ainu* Lewis, 1894: 479 type species *Ainu tenuicornis* Lewis, 1894
angustus Ren & Yuan, 2005: 98 **A:** GUX SCH
basifemoratum Nabozhenko & Ren, 2018: 1294 **A:** YUN
fukudai Masumoto, 1981a: 41 **A:** TAI
linwenxini Nabozhenko & Ren, 2018: 1296 **A:** TAI
masumotoi Nakane, 1968b: 83 **A:** JA (Ryukyus)
medoganus Ren & Yuan, 2005: 100 **A:** XIZ
multicolor Pic, 1927c: 18 (*Erulipus*) **A:** FUJ **ORR**
sichuanus Ren & Yuan, 2015: 131 **A:** SCH
tenuicornis Lewis, 1894: 480 **A:** JA "Korea"
subgenus *Erulipus* Fairmaire, 1903a: 14 type species *Erulipus fruhstorferi* Fairmaire, 1903
andoi Ruzzier, 2014: 659 **A:** YUN **ORR**
fruhstorferi Fairmaire, 1903: 14 **A:** GUX **ORR**
grandis Ren & Yuan, 2005: 98 (*Ainu*)

genus *Andocamaria* Masumoto, 1993b: 230 type species *Campsiomorpha formosana* Pic, 1930
andoi Masumoto, 1992: 89 (*Campsiomorpha*) **A:** TAI
formosana Pic, 1930b: 31 (*Campsiomorpha*) **A:** TAI
ryukyuensis Masumoto, 1985a: 25 (*Campsiomorpha*) **A:** JA (Ryukyus)

genus *Androsus* Gebien, 1920: 386 type species *Chariotheca violacea* Pascoe, 1887
adonis Pic, 1927c: 18 (*Chariotheca*) **A:** NP **ORR**
spectabilis Kaszab, 1980b: 212

sinensis Pic, 1934b: 33 (*Chariotheca*) **A:** SCH
taiwanus Masumoto, Akita & Lee, 2012a: 30 **A:** TAI

genus *Aptereucyrtus* Gebien, 1922b: 477 type species *Aptereucyrtus hemichalceus* Gebien, 1922
hemichalceus Gebien, 1922b: 477 **A:** TAI **ORR**
shibatai Ando, 1984: 69

genus *Artactes* Pascoe, 1868: xii type species *Artactes nigritarsis* Pascoe, 1868
cyaneolimbatus Fairmaire, 1893c: 30 **A:** HKG **ORR**

genus *Asbolodomimus* Pic, 1921c: 20 type species *Asbolodomimus subcarinatus* Pic, 1921
Allopezus Gebien, 1922b: 504 type species *Allopezus miritarsis* Gebien, 1922
bhutanicus Masumoto, 2002: 373 (*Allopezus*) **A:** BT

genus *Augolesthus* Motschulsky, 1872: 34 type species *Augolesthus purpureofasciatus* Motschulsky, 1872
Chrysomaia Kulzer, 1952: 755 type species *Eucyrtus carbunculus* Fairmaire, 1885
dentatus Xu & Ren, 2012: 773 **A:** YUN

genus *Bradymerus* Perroud & Montrouzier , 1864: 110 type species *Bradymerus tuberculatus* Perroud & Montrouzier, 1864
Bradynocerus Fairmaire, 1883b: 36 type species *Bradynocerus aulacopterus* Fairmaire, 1883
Isostira Pascoe, 1870: 97 type species *Isostira crenata* Pascoe, 1870
Planibates Kaszab, 1939: 221 type species *Planibates papuanus* Kaszab, 1939
Pseudobradymerus Pic, 1926g: 16 type species *Bradymerus simplicithorax* Pic, 1926
clathratus L.W. Schaufuss, 1887: 134 **A:** JA (Ryukyus) TAI **ORR**
aequecostatus Fairmaire, 1893c: 20
junctus Shibata, 1980: 66
fukiensis Kaszab, 1954: 257 (*Planibates*) **A:** FUJ **ORR**
kabakovi Kaszab, 1980b: 184 **A:** YUN NP SD **ORR**
kanchenjungicus Schawaller, 2006a: 47 **A:** NP
kondoi Kulzer, 1957: 199 **A:** JA (Bonin Is.)
masumotoi Schawaller, 2006a: 53 **A:** TAI (Lanyu Is.)
serricollis Walker, 1858: 284 (*Opatrum*) **A:** YUN **ORR**

genus *Camarimena* Motschulsky, 1863: 473 type species *Camarimena ovicauda* Motschulsky, 1863
Espitomorphus Pic, 1921c: 24 type species *Espitomorphus multicolor* Pic, 1921
Sinopium Pascoe, 1866: 487 type species *Strongylium variabile* Walker, 1858
viridans Kulzer, 1954: 56 **A:** SD

genus *Campsiomorpha* Pic, 1917f: 19 type species *Camaria lata* Pic, 1915
lata Pic, 1915g: 17 (*Camaria*) **A:** FUJ JIX SCH
corinthia Heller, 1923: 71 (*Eucamaria*)
siemsseni Gebien, 1919b: 150 (*Eucamaria*)
subangulata Pic, 1920b: 2
napoana Ren & Yang, 2004: 313 **A:** GUX
regia regia Fairmaire, 1903a: 16 (*Camaria*) **A:** GUX HUN CHQ **ORR**

genus *Catapiestus* Perty, 1831: xxxviii type species *Catapiestus piceus* Perty, 1831
Plateia Laporte, 1840: 215 [HN] type species *Plateia orientalis* Laporte, 1840
bispinosus Yang & Guo, 2018: 50 **A:** GUI
clavipes Lang & Ren, 2009: 196 **A:** HAI
crenulicollis Fairmaire, 1889a: 357 **A:** FUJ HAI YUN **ORR**
indicus Fairmaire, 1896a: 28 **A:** AP NP SD **ORR**
medvedevi Lang & Ren, 2009: 197 **A:** YUN
rugipennis M. Chûjô, 1984: 1 **A:** JA (Ryukyus)
subrufescens Pic, 1911i: 134 **A:** TAI
tonkineus Pic, 1912b: 17 **A:** YUN **ORR**

genus *Cleomis* Fairmaire, 1893c: 54 type species *Cleomis violaceipes* Fairmaire, 1893
Oyanus Pic, 1921c: 23 type species *Oyanus curticornis* Pic, 1921
Pseudeucyrtus Pic, 1916f: 14 type species *Pseudeucyrtus niasensis* Pic, 1916 (= *Cleomis violaceipes* Fairmaire, 1893)
marginicollis Gebien, 1914b: 38 (*Psydus*) **A:** TAI YUN **ORR**
substriatus Pic, 1930b: 30 (*Psydus*)

genus *Coelometopus* Solier, 1848: 278 type species *Coelometopus clypeatus* Solier, 1848 (= *Blaps clypeata* Germar, 1813)
clypeatus Germar, 1813: 122 (*Blaps*) **E:** PT SP
clypeatus Solier, 1848: 280
cobosi Español, 1963a: 52 **E:** SP (Granada)

genus *Cybopiestes* Reitter, 1917b: 150 type species *Cybopiestes csikii* Reitter, 1917
csikii Reitter, 1917b: 150 **N:** MO

genus *Diachoriops* Ando, 2020 [see New Acts] type species *Schizomma cucumericola* Gebien, 1920
Schizomma Gebien, 1920: 392 [HN] type species *Schizomma cucumericola* Gebien, 1920
kondoi Kulzer, 1957: 245 **A:** JA (Bonin Is.)

genus *Dioscoridemus* Koch, 1970: 115 type species *Dioscoridemus vonhayekae* Koch, 1970
vonhayekae Koch, 1970: 116 **A:** YE (Suqutra)

genus *Eucyrtus* Pascoe, 1866: 473 type species *Scotaeus pretiosus* Lacordaire, 1859

Microeucyrtus Pic, 1926: 47 type species *Microeucyrtus coomani* Pic, 1926

annulipes Kraatz, 1880c: 103 **A:** GUX HAI YUN **ORR**

anthracinus anthracinus Kraatz, 1880c: 102 **A:** HAI YUN **ORR**

deyrollei deyrollei Kraatz, 1880c: 100 **A:** YUN ORR

opacus Kraatz, 1880c: 102 **A:** "Himalaya" **ORR**

rondoni Ando, 2003a: 96 **A:** YUN **ORR**

septentrionalis Ando, 2003a: 91 **A:** BT YUN **ORR**

splendens elatus Ando, 2003a: 54 **A:** BT NP SD **ORR**

splendens splendens Lacordaire, 1859: 417 (*Scotaeus*) **A:** YUN AP BT SD **ORR**

semiviolacea Pic, 1923c: 19 (*Hemicera*)

coeruleicollis Pic, 1924e: 27 (*Hemicera*)

genus *Euhemicera* Ando, 1996: 189 type species *Epilampus pulcher* Hope, 1842

aeneovirgulata Ando, 2003a: 200 **A:** FUJ GUI

alternata Gebien, 1914b: 37 (*Hemicera*) **A:** TAI **ORR**

chapaensis Pic, 1923c: 19 (*Hemicera*)

chinensis Kulzer, 1954: 42 (*Hemicera*) **A:** HUB

chujoi Ando, 2003a: 233 **A:** FUJ

conspicuua Ando, 2003a: 164 **A:** YUN **ORR**

formosana Ando, 1997: 110 **A:** TAI

foveoseriata Kaszab, 1941c: 63 (*Hemicera*) **A:** TAI

parafoveoseriata Masumoto, 1981c: 37 (*Hemicera*)

gebieni Kaszab, 1941c: 61 (*Hemicera*) **A:** FUJ TAI **ORR**

fukiensis Kaszab, 1954: 258 (*Hemicera*)

hajimei hajimei Masumoto, 1983a: 86 (*Hemicera*) **A:** JA (Ryukyus)

hajimei kumejimana Akita & Masumoto, 2015b: 414 **A:** JA (Kumejima, Okinawa)

himalaica Blair, 1930b: 178 (*Hemicera*) **A:** BT NP SD **ORR**

inexpecta Ando, 2003: 307 **A:** NP YUN **ORR**

japonica Masumoto, 1983a: 83 (*Hemicera*) **A:** JA (Ryukyus)

lanae Masumoto, Akita & Lee, 2013a: 117 **A:** TAI

nanrenensis Masumoto, Akita & Lee, 2014a: 83 **A:** TAI

nodokai Nakane, 1963a: 28 (*Hemicera*) **A:** JA (Ryukyus)

pingtita Masumoto, 1981e: 84 (*Hemicera*) **A:** ANH TAI

pulchra Hope, 1842: 52 [= 1845: 16] (*Epilampus*) **A:** GUA SHG JA TAI

zigzaga Marseul, 1876c: 111 (*Hemicera*)

sakaii Masumoto, 1981c: 38 (*Hemicera*) **A:** TAI

sakishimensis M. Chûjô, 1978b: 8 (*Hemicera*) **A:** JA (Ryukyus) TAI

shihchiehi Masumoto, Akita & Lee, 2014a: 87

shantita Masumoto, 1981e: 86 (*Hemicera*) **A:** TAI

shibatai Ando, 1997: 107 **A:** TAI
taichuai Masumoto, Akita & Lee, 2014a: 85 **A:** TAI
theresae Pic, 1921c: 30 (*Hemicera*) **A:** YUN **ORR**
undulata Pic, 1923k: 271 (*Hemicera*) **A:** FUJ GUA **ORR**
viridiapicalis Ando, 1994: 111 (*Hemicera*) **A:** NP **ORR**
vittatipennis Pic, 1923c: 19 (*Hemicera*) **A:** TAI **ORR**
tenuestriata Kaszab, 1941c: 61 (*Hemicera*)

genus *Euphloeus* Pascoe, 1887: 15 type species *Euphloeus verrucosus* Pascoe, 1887
verrucosus Pascoe, 1887: 16 **A:** HP **ORR**
bigranosus Fairmaire, 1897i: 221
granosus Allard, 1894: 104 (*Osdara*)

genus *Derosphaerus* J. Thomson, 1858: 99 [NP] type species *Derosphaerus globicollis* J. Thomson, 1858
Cholipus Pascoe, 1866: 471 type species *Cholipus brevicornis* Pascoe, 1866
Encyalesthus Motschulsky, 1860c: 139 type species *Encyalesthus subviolaceus* Motschulsky, 1860
Euphron Dejean, 1834: 206 [NO] type species *Helops coerulescens* Guérin-Méneville, 1831
Falsoencyalesthus Pic, 1923b: 29 type species *Falsoencyalesthus latipennis* Pic, 1923
Neandrosus Pic, 1921e: 12 type species *Neandrosus singularipes* Pic, 1921
Notiolesthus Motschulsky, 1872: 25 type species *Notiolesthus natalensis* Motschulsky, 1872
Pachyurgus J.L. LeConte, 1862: 230 type species *Iphthinus aereus* Melsheimer, 1846
aeruginosus Fabricius, 1787a: 213 (*Helops*) **A:** NP **ORR**
asperipennis Allard, 1896: 320 (*Derosphaerus*) **A:** BT NP SD UP XIZ
angustatus Pic, 1922c: 24 (*Leprocaulus*)
brevicornis Mäklin, 1863: 553 (*Encyalesthus*) **A:** SD YUN **ORR**
crassicrus Fairmaire, 1903a: 14 (*Derosphaerus*) **A:** FUJ **ORR**
cribripennis Fairmaire, 1891f: ccxi (*Derosphaerus*) **A:** CH
exularis Gebien, 1914b: 31 (*Encyalesthus*) **A:** AP JA (Ryukyus) NP SD TAI **ORR**
foveoseriatus Fairmaire, 1889a: 358 (*Derosphaerus*) **A:** CH **ORR**
grandipes Fairmaire, 1893f: 297 (*Derosphaerus*) **A:** CH **ORR**
coomani Pic, 1926d: 26 (*Derosphaerus*)
leseleuci Pic, 1923b: 32 (*Derosphaerus*)
instriatus Pic, 1922c: 27 (*Derosphaerus*) **A:** XIN
niger Wiedemann, 1819: 164 (*Helops*) **A:** "Bengal"
nitidus Pic, 1922c: 27 (*Derosphaerus*) **A:** FUJ TAI
rotundicollis Laporte, 1840: 213 (*Upis*) **A:** TAI **ORR**
rugosus Gravely, 1915: 528 (*Derosphaerus*) **A:** AP NP SD UP **ORR**
sinensis Hope, 1842: 52 (*Upis*) **A:** GUA SHG HKG JA (Kyushu) **ORR**
foveolatus Marseul, 1876c: 119 (*Upis*)
foveolatus Pic, 1923b: 31 (*Encyalesthus*) [HN]
inaequalis Fairmaire, 1889a: 357 (*Nyctobates*)

stevensi Gravely, 1915: 527 (*Encyalesthus*) **A:** UP
subviolaceus Motschulsky, 1860c: 139 (*Encyalesthus*) **A:** FE HEI JA JIL NC NE SC
violaceipennis Marseul, 1876c: 118 (*Upis*)
thibetanus Pic, 1922c: 26 (*Derosphaerus*) **A:** XIZ

genus *Falsocamaria* Pic, 1917f: 19 type species *Falsocamaria obscura* Pic, 1917
Eucamaria Gebien, 1919b: 149 type species *Camaria spectabilis* Pascoe, 1860
distinctestriata Pic, 1917f: 18 (*Camaria*) **A:** HAI **ORR**
fruhstorferi Fairmaire, 1903a: 15 (*Camaria*) **A:** YUN GUX **ORR**
imperialis Fairmaire, 1903a: 18 (*Camaria*) **A:** ZHE JIX FUJ GUA GUX HUN HUB GUI **ORR**
microdera Fairmaire, 1899: 630 (*Camaria*) **A:** FUJ SCH GUX GUA SCH GUI
obscura Pic, 1917f: 19
subopaca Pic, 1920b: 2 (*Camaria*)
muelleri Kaszab, 1954: 259 (*Campsiomorpha*) **A:** FUJ
obscurovientia Wang, Ren & Liu, 2012: 321 **A:** ZHE JIX HAI
rufisutura Wang, Ren & Liu, 2012: 322 **A:** GUX
spectabilis Pascoe, 1860a: 52 (*Camaria*) **A:** GUX ZHE "North China"

genus *Foochounus* Pic, 1921c: 22 type species *Foochounus convexipennis* Pic, 1921
Anobriomaia Kaszab, 1941c: 67 type species *Anobriomaia sulcata* Kaszab, 1941
Chariophenus Blair, 1929: 239 type species *Chariophenus wasmanni* Blair, 1929
Microcameria Ren, 1998: 108 type species *Microcameria pygmaea* Ren, 1998
confusus Schawaller & Ando, 2009: 263 **A:** SD BT
convexipennis Pic, 1921c: 22 **A:** FUJ GUX HUN SCH ZHE
kentingensis Masumoto, Akita & Lee, 2015b: 308 **A:** TAI
manmiaoae Masumoto, Akita & Lee, 2008: 216 **A:** TAI
medvedevi Schawaller, 2019: 1042 **A:** YUN
pygmaeus Ren, 1998: 108 (*Microcameria*) **A:** GUX SHA ZHE
reni Ando & Schawaller, 2018: 138 **A:** YUN **ORR**
sulcatus Kaszab, 1941c: 68 (*Anobriomaia*) **A:** TAI YUN **ORR**
tibialis Blair, 1930b: 178 (*Chariophenus*) **A:** BH SD NP
schmidti Schawaller, 2000b: 354 (*Chariophenus*)
thoracicus Kaszab, 1965d: 292 (*Anobriomaia*) **A:** YUN **ORR**
wasmanni Blair, 1929: 240 (*Chariophenus*) **A:** AP SD NP **ORR**
assamicus Kaszab, 1965a: 127 (*Anobriomaia*)
yamasakoi Schawaller & Ando, 2009: 283 **A:** YUN **ORR**

genus *Gauromaia* Pascoe, 1866: 473 type species *Gauromaia dives* Pascoe, 1866
subgenus *Gauromaia* Pascoe, 1866: 473 type species *Gauromaia dives* Pascoe, 1866
Cephaleucyrtus Pic, 1923b: 22 type species *Cephaleucyrtus viridicollis* Pic, 1923

formosana Masumoto, 1981c: 41 **A:** TAI
himalaica Schawaller, 2006c: 479 **A:** NP SD
tenuestriata Fairmaire, 1891e: cxxxii **A:** KA

genus *Gnesis* Pascoe, 1866: 477 type species *Gnesis helopioides* Pascoe, 1866
Tromosternus Harold, 1878b: 130 type species *Tromosternus haagi* Harold, 1878 (=*Gnesis helopioides* Pascoe, 1866)
dulanensis Masumoto, Akita & Lee, 2014a: 76 **A:** TAI
formosanus M. Chûjô, 1973: 20 (*Phaedis*) **A:** TAI
fujitai Akita & Masumoto, 2013: 250 **A:** JA
guoi Masumoto, Akita & Lee, 2019a: 123 **A:** TAI
helopioides helopioides Pascoe, 1866: 478 **A:** JA TAI NE "Korea" "Mantchuria"
haagi Harold, 1878b: 131 (*Tromosternus*)
kentingensis Masumoto, 1981e: 87 **A:** TAI
lini Masumoto, Akita & Lee, 2014a: 79 **A:** TAI
liukueiensis Masumoto, 1981c: 45 **A:** TAI
magnipunctatus M. Chûjô, 1966: 23 (*Phaedis*) **A:** JA (Ryukyus)
meilingae Masumoto, Akita & Lee, 2014a: 74 **A:** TAI
naitoi Akita & Masumoto, 2013: 246 **A:** JA (Ryukyus)
onoderai Akita & Masumoto, 2013: 249 **A:** JA (Ryukyus)
purpurascens Nakane, 1968b: 79 (*Phaedis*) **A:** JA (Ryukyus)
amamianus M. Chûjô, 1977: 12 (*Phaedis*)
okinawanus Nakane, 1968b: 80 (*Phaedis*)
purpurascens adachii Akita & Masumoto, 2013: 253 **A:** JA (Danjo-Guntô)
purpurascens tokaranus Akita & Masumoto, 2013: 253 **A:** JA (Ryukyus)
purpurascens yakuanus Nakane, 1968b: 80 (*Phaedis*) **A:** JA (Yakushima)
senkakuensis M. Chûjô, 1973: 19 (*Phaedis*) **A:** JA (Senkaku Is.)

genus *Hemicera* Laporte & Brullé, 1831: 393 type species *Cnodalon splendens* Wiedemann, 1823
subgenus *Hemicera* Laporte & Brullé, 1831: 393 type species *Cnodalon splendens* Wiedemann, 1823
Chrysolinoides Jolivet, 1951: 2 type species *Chrysolinoides philippinensis* Jolivet, 1951 (= *Hemicera iridicolor* Gebien, 1921)
consonbrina Ando, 2003a: 388 **A:** SCH
gilvipes gilvipes Ando, 2003a: 422 **A:** NP **ORR**
kurosawai Masumoto, 1981c: 40 **A:** SCH TAI
masumotoi dilatilineata Ando, 2003a: 383 **A:** GUI **ORR**
masumotoi masumotoi Ando, 2003a: 383 **A:** GUX **ORR**
nakamurai Masumoto, 1982d: 56 **A:** TAI
nakanei Ando, 2003a: 467 **A:** NP **ORR**
oblita Ando, 2003a: 394 **A:** NP SD YUN **ORR**

pyrozonia Fairmaire, 1896a: 29 **A:** YUN **ORR**
tsuyukii Masumoto, 1981e: 83 **A:** TAI

genus *Hexarhopalus* Fairmaire, 1891c: xix type species *Hexarhopalus sculpticollis* Fairmaire, 1891
subgenus *Hexarhopalus* Fairmaire, 1891c: xix type species *Hexarhopalus sculpticollis* Fairmaire, 1891
Apteroleprocaulus Kaszab, 1983b: 182 type species *Leprocaulus montanus* Kaszab, 1983
Hexaroptrum Fairmaire, 1894a: 38 [URN] type species *Hexarhopalus sculpticollis* Fairmaire, 1891
Laosocryptobates Pic, 1928b: 25 type species *Laosocryptobates tuberculatus* Pic, 1928
attenuatus Pic, 1922c: 24 (*Leprocaulus*) **A:** YUN
mangshanicus Ren & Xu, 2011: 25 **A:** HUN
sculpticollis Fairmaire, 1891c: xix **A:** HUB **ORR**
sculptilis Kaszab, 1960e: 288 **A:** YUN **ORR**
subgenus *Leprocaulus* Fairmaire, 1896b: 95 type species *Leprocaulus clavipes* Fairmaire, 1896
Pseudocaelophus Pic, 1922c: 28 type species *Strongylium difforme* Pic, 1922
Pseudoderosphaerus Pic, 1922c: 24 type species *Leprocaulus rotundicollis* Pic, 1922
bisinuatus Ren & Xu, 2011: 28 **A:** GUI
xui Ren & Xu, 2011: 30 **A:** YUN

genus *Iphthiminus* Spilman, 1973: 42 type species *Iphthimus italicus* Truqui, 1857
bellardi Truqui, 1857: 94 (*Iphthimus*) **E:** GR **A:** CY IS SY TR
croaticus Truqui, 1857: 93 (*Iphthimus*) **E:** AL BH CR GR
italicus Truqui, 1857: 93 (*Iphthimus*) **E:** FR IT
reitteri Gebien, 1913: 7 (*Iphthimus*)

genus *Iphthimulus* Reitter, 1920a: 17 type species *Iphthimus truquii* Marseul, 1869
truquii Marseul, 1869: 274 (*Iphthimus*) **E:** GR **A:** SY TR

genus *Macrostethus* Wollaston, 1854: 504 type species *Macrostethus tuberculatus* Wollaston, 1854
tuberculatus Wollaston, 1854: 505 **N:** MR

genus *Menephilus* Mulsant, 1854: 291 type species *Tenebrio curvipes* Fabricius, 1792 (= *Tenebrio cylindricus* Herbst, 1784)
aborensis Gravely, 1915: 529 **A:** AP
arciscelis Marseul, 1876c: 119 **A:** JA
cylindricus cylindricus Herbst, 1784c: 144 (*Tenebrio*) **E:** AL AR AU BH BU CR CZ FR GE GG GR HU IT PT RO SK SL SP ST SV SZ UK **A:** CY SY TR **NAR**
affinis Rossi, 1790a: 232 (*Tenebrio*)
curvipes Fabricius, 1792a: 111 (*Tenebrio*)
loripes Illiger, 1798: 113 (*Tenebrio*)

cylindricus maroccanus Théry, 1931b: 127 **N:** AG MO
formosanus Masumoto, 1981e: 91 **A:** TAI
lucens Marseul, 1876c: 121 **A:** JA ES FE SC
medius atronitens Kulzer, 1957: 235 (*Setenis*) **A:** JA (Bonin)
medius medius Marseul, 1876c: 120 **A:** FUJ JA **ORR**
nodai Akita & Masumoto, 2016: 10 **A:** JA (Ryukyus)
striatifrons Fairmaire, 1896b: 101 (*Nyctobates*) **A:** JIX
taiwanus Masumoto, 1981e: 92 **A:** TAI

genus *Misolampidius* Solsky, 1876: 292 type species *Misolampidius tentyrioides* Solsky, 1876
Ptilonix Allard, 1877a: 62 type species *Helops clavicrus* Marseul, 1876
adachii Mt. Chûjô & Imasaka, 1982: 128 **A:** JA (Tsushima)
chagangensis Li & Gao, 2015: 137 **A:** NC
changbaiensis J.-K. Li, 1992: 166 **A:** JIL
chejudoensis Mt. Chûjô & Imasaka, 1982: 124 **A:** SC (Chejudo)
clavicrus Marseul, 1876c: 139 (*Helops*) **A:** JA (Kyushu)
entomogonoides Allard, 1896: 319 (*Ptilonix*) **A:** NP SD HP
foveipennis Fairmaire, 1894a: 38 (*Hexaroptrum*) **A:** SD
gotoensis M. Chûjô & Imasaka, 1982: 130 **A:** JA (Goto)
hanshungliengensis J.-K. Li, 1992: 166 **A:** JIL
hisamatsui Ando, 2009: 187 **A:** JA (Shikoku)
imasakai Akita & Masumoto, 2016: 12 **A:** JA
indicus Allard, 1877a: 63 (*Ptilonix*) **A:** "Himalaya"
iriei M. Chûjô, 1975: 21 **A:** JA (Yakushima)
kawadai Nakane, 1983b: 147 **A:** JA (Izu Is., Mikura Is.)
kazuoi Akita & Masumoto, 2016: 12 **A:** JA (Danjo)
koreanus Mt. Chûjô & Imasaka, 1982: 123 **A:** SC
laevicollis Allard, 1877a: 62 (*Ptilonix*) **A:** "Himalaya"
mantapicus Li & Gao, 2015: 141 **A:** NC
maoershanensis J.-K. Li, 1992: 165 **A:** HEI LIA
molytopsis Marseul, 1876c: 135 (*Heliophygus*) **A:** JA "Korea"
hikosanus Nakane, 1963a: 28
tsurugisanus Nakane, 1963a: 28
morii Akita & Masumoto, 2016: 11 **A:** JA (Yakushima)
ohminesanus Masumoto & Akita, 2001a: 19 **A:** JA (Honshu)
okiensis Nakane, 1983b: 147 **A:** JA (Oki Is.)
okumurai Nakane, 1968b: 81 **A:** JA
paratentyrioides J.-K. Li, 1992: 165 **A:** HEI
qianshanensis J.-K. Li, 1992: 166 **A:** LIA
liaoningensis J.-K. Li, 1992: 166

rugipennis rugipennis Lewis, 1894: 476 **A:** JA

subnitidus Nakane, 1968b: 81

sankuaishiensis Li & Lin, 2015: 137 **A:** LIA

sanoi Akita & Masumoto, 2016: 14 **A:** JA

satoi Masumoto, Akita & Utsunomiya, 2003: 327 **A:** SCH

sobosanus Mt. Chûjô & Imasaka, 1982: 132 **A:** JA (Kyushu)

takakuwai Ando, 2017: 298 **A:** JA (Honshu)

tentyrioides Solsky, 1876: 293 **A:** FE HEI JIL NC SC

morio Kolbe, 1886: 209

tsushimanus Nakane, 1968b: 80 **A:** JA (Tsushima)

wulongshanus Li & Lin, 2015: 141 **A:** LIA

yakushimanus Nakane, 1968b: 81 **A:** JA (Yakushima)

genus *Misolampomorphus* Kaszab, 1941a: 6 type species *Misolampomorphus kochi* Kaszab, 1941 (= *Lyprocaulus reitteri* Pic, 1934)

reitteri Pic, 1934f: 84 (*Lyprocaulus*) **A:** SCH

kochi Kaszab, 1941a: 7

genus *Misolampus* Latreille, 1807: 160 type species *Misolampus hoffmannseggii* Latreille, 1807 (= *Pimelia gibbula* Herbst, 1799)

gibbulus Herbst, 1799: 51 (*Pimelia*) **E:** PT SP

hoffmannseggii Latreille, 1807: 161

goudotii erichsoni Vauloger de Beaupré, 1900: 674 **E:** SP **N:** AG

goudotii Erichson, 1841: 184 [HN]

goudotii goudotii Guérin-Méneville, 1834b: 28 **N:** MO

goudotii peyerimhoffi Antoine, 1926: 257 **N:** MO (High Atlas)

lusitanicus Brème, 1842: 83 **E:** PT SP

ramburii Brème, 1842: 83 **E:** PT SP

scabricollis Graells, 1849: 621 **E:** PT SP

subglaber Rosenhauer, 1856: 204 **E:** SP

genus *Morphostenophanes* Pic, 1925b: 7 type species *Morphostenophanes aenescens* Pic, 1925

Promorphostenophanes Kaszab, 1960e: 277 type species *Promorphostenophanes atavus* Kaszab, 1960

aenescens Pic, 1925b: 7 **A:** YUN

diversus Pic, 1925b: 7

subparallelus Pic, 1925b: 7

atavus Kaszab, 1960e: 278 (*Promorphostenophanes*) **A:** YUN

cuproviridis Gao & Ren, 2009: 313 **A:** GUI

jendeki jendeki Masumoto, 1998b: 307 **A:** YUN

jendeki similis Masumoto, 1998b: 308 **A:** YUN

papillatus Kaszab, 1941a: 11 **A:** GUI SCH
tanikadoi Masumoto, 1998b: 305 **A:** SCH
tuberculatus Gao & Ren, 2009: 315 **A:** YUN

genus *Necrobioides* Fairmaire, 1882c: 234 type species *Necrobioides coeruleatus* Fairmaire, 1882
gibbicollis Fleutiaux, 1887: 65 (*Derosphaerus*) **A:** YUN **ORR**
kabakovi Kaszab, 1980b: 199 **A:** TAI **ORR**

genus *Neoplamius* Löbl, Bouchard, Merkl & Bousquet, 2020 [see New Acts] type species *Neoplamius zoltani* Masumoto, 1981
akiyamai Masumoto, Akita & Lee, 2012b: 199 **A:** TAI
endoi Masumoto, 1981d: 18 **A:** TAI
kusamai Masumoto, Akita & Lee, 2014a: 90 **A:** TAI
oharai Masumoto, Akita & Lee, 2013c: 261 **A:** TAI
yamasakoi Ando, 2013: 187 **A:** TAI
zoltani Masumoto, 1981d: 15 **A:** TAI

genus *Oedemutes* Pascoe, 1860a: 51 type species *Oedemutes tumidus* Pascoe, 1860
subgenus *Tamdaous* Pic, 1923h: 11 type species *Oedemutes impressus* Pic, 1923
chinensis Ando & Ren, 2006: 85 **A:** YUN
formosanus Masumoto, 1981c: 42 **A:** TAI
hirashimai hirashimai M. Chûjô, 1973: 17 (*Oedemutes*) **A:** JA (Ryukyus)
hirashimai inadai Ando, 2000: 309 **A:** JA (Ryukyus)
itoi Ando, 1989: 108 **A:** TAI
lutaoensis Masumoto, Akita & Lee, 2014a: 81 **A:** TAI

genus *Paramisolampidius* Merkl & Masumoto, 2020 [see New Acts] type species *Paramisolampidius kagoshimensis* Nakane, 1968
csorbai Merkl & Masumoto, 2008: 3 **A:** TAI
formosanus Masumoto, 1981c: 32 **A:** TAI
kagoshimensis Nakane, 1968b: 82 **A:** JA
kentingensis Masumoto, Akita & Lee, 2012a: 33 **A:** TAI
kinugasai Masumoto, 1984: 17 **A:** TAI
lichiaensis Masumoto, Akita & Lee, 2019a: 126 **A:** TAI
shirozui M. Chûjô, 1967: 379 (*Misolampidius*) **A:** TAI
  alishanus Masumoto, 1981e: 79
  wufengus Masumoto, 1981e: 80
tenghsiensis Masumoto, 1984: 19 **A:** TAI
tokarensis Nakane, 1963a: 29 (*Misolampidius*) **A:** JA (Tokara Is.)
yehi Masumoto, Akita & Lee, 2015b: 310 **A:** TAI

genus *Phaedis* Pascoe, 1866: 474 type species *Phaedis elysius* Pascoe, 1866

Microgauromaia Pic, 1921c: 23 type species *Microgauromaia minuta* Pic, 1921

Phaedeucyrtus Pic, 1916f: 14 type species *Pseudeumolpus obscuripes* Pic, 1916

Pseudeumolpus Kraatz, 1880c: 111 type species *Pseudeumolpus bicolor* Kraatz, 1880

marmoratus M. Chûjô, 1978b: 11 **A:** JA (Ryukyus)

mushanus Masumoto, 1981c: 44 **A:** TAI

oshimensis iriei M. Chûjô, 1980: 5 **A:** JA (Ryukyus)

oshimensis oshimensis Nakane, 1979: 113 **A:** JA (Ryukyus)

oshimensis rinae Akita & Masumoto, 2016: 11 **A:** JA (Iheya Is.)

quadricollis Ando & Schawaller, 2015: 403 **A:** TAI

sinensis Ando & Schawaller, 2015: 405 **A:** GUA

taiwanus Masumoto, 1981c: 42 **A:** TAI

genus *Pigeus* Gebien, 1919b: 153 type species *Camarimena nitidipes* Fairmaire, 1893

Pseudocamarimena Pic, 1923d: 21 type species *Pseudocamarimena striata* Pic, 1923

striatus Pic, 1923d: 22 (*Pseudocamarimena*) **A:** SHG

genus *Plamius* Fairmaire, 1896a: 30 type species *Plamius tenuestriatus* Fairmaire, 1896

Cnemandrosus Gebien, 1927: 42 type species *Eucyrtus semipurpureus* Fairmaire, 1896

convexus M. Chûjô, 1973: 15 **A:** HKG

fukiensis Picka, 1990: 110 **A:** FUJ

kaszabi Picka, 1990: 111 **A:** ZHE

quadrimaculatus Kaszab, 1941c: 65 (*Cnemandrosus*) **A:** TAI

quadrinotatus Pic, 1925b: 11 (*Rehumius*) **A:** FUJ TAI **ORR**

yaeyamensis Nakane, 1968b: 78 **A:** JA (Ryukyus)

genus *Platycrepis* Lacordaire, 1859: 418 type species *Platycrepis violaceus* Lacordaire, 1859

schmidi Kaszab, 1965a: 125 **A:** UP (Kumaon)

violaceus Lacordaire, 1859: 418 **A:** BT FUJ TAI **ORR**

impressithorax Pic, 1924a: 26 (*Hemicera*)

latitarsis Fairmaire, 1882c: 239 (*Eucyrtus*)

tonkineus Pic, 1922a: 23

yangi Masumoto, 1986b: 64 **A:** TAI

genus *Promethis* Pascoe, 1869: 148 type species *Upis angulata* Erichson, 1842

Mederis Motschulsky, 1872: 24 type species *Upis angulata* Erichson, 1842

Pediris Motschulsky, 1872: 24 type species *Pediris longipes* Motschulsky, 1872

Pseudobates Fairmaire, 1882c: 231 type species *Nyctobates coracina* Fairmaire, 1882 (= *Nyctobates subrobusta* Motschulsky, 1872)

Setenis Motschulsky, 1872: 24 type species *Tenebrio valgus* Wiedemann, 1823

aequatorialis Blanchard, 1853: 30 (*Nyctobates*) **A:** YUN ORR
almorensis Kaszab, 1988: 101 **A:** UP (Almora)
amanoi Masumoto, 1987b: 92 **A:** JA
angulicollis Kaszab, 1988: 113 **A:** YUN
barbereti Fairmaire, 1888a: 27 (*Nyctobates*) **A:** ZHE (Zhoushan)
chinensis Kaszab, 1988: 106 **A:** YUN **ORR**
confusa Fairmaire, 1896a: 27 (*Nyctobates*) **A:** NP **ORR**
semivalga Blair, 1913: 57 (*Setenis*)
cordicollis Kaszab, 1988: 95 **A:** FUJ GUA YUN **ORR**
crassihornis Ba & Ren, 2009: 28 **A:** HAI
crenatostriata Motschulsky, 1872: 30 (*Setenis*) **A:** AP BT HAI NP SCH SD YUN **ORR**
foveicollis Fairmaire, 1893b: 29 (*Nyctobates*)
sinuatipes Pic, 1928i: 377 (*Setenis*)
evanescens Gebien, 1918: 124 [= 1919a: 9] (*Setenis*) **A:** FUJ HAI SHN GUI GUX JIX **ORR**
quadraticollis Fairmaire, 1893f: 297 (*Nyctobates*) [HN]
exigua M. Chûjô, 1980: 1 (*Setenis*) **A:** JA (Yakushima)
formosana Masumoto, 1981e: 88 (*Setenis*) **A:** TAI
nitidula M. Chûjô, 1981: 133 (*Setenis*)
frontosulcatis Ren & Yang, 2004: 315 **A:** GUX
gebieni Kaszab, 1941d: 124 (*Setenis*) **A:** SCH
glabra Hope, 1831: 31 (*Tenebrio*) **A:** BT NP SD **ORR**
glabricula Motschulsky, 1872: 31 (*Nyctobates*) **A:** NP UP XIZ
cribrifrons Fairmaire, 1894a: 38 (*Nyctobates*)
semiopaca Blair, 1913: 57 (*Setenis*)
guangxiana Ren & Yan, 2004: 314 **A:** CUX GUX HAI YUN
harmandi Allard, 1896: 318 **A:** BT NP PA SD YUN XIZ **ORR**
heros Gebien, 1918: 125 (*Setenis*) **A:** YUN **ORR**
impressa Fabricius, 1801a: 146 (*Setenis*) **A:** NP ORR
insomnis Lewis, 1894: 472 (*Setenis*) **A:** JA "Korea"
higonia Lewis, 1894: 472 (*Setenis*)
iriomotensis M. Chûjô, 1979: 40 (*Setenis*) **A:** JA (Ryukyus)
kaoshana Masumoto, 1982d: 60 (*Setenis*) **A:** TAI
kempi kempi Gravely, 1915: 526 (*Setenis*) **A:** AP YUN **ORR**
kurosawai Masumoto, 1982d: 58 (*Setenis*) **A:** TAI
microcephala Fairmaire, 1899: 629 (*Nyctobates*) **A:** SCH
microthorax Kaszab, 1988: 101 **A:** SCH YUN
nitouana Kaszab, 1988: 108 **A:** SCH
noctivigila Lewis, 1894: 473 (*Setenis*) **A:** JA
okinawana M. Chûjô, 1978a: 66 (*Setenis*) **A:** JA (Ryukyus)
oshimana Miwa, 1935: 21 (*Setenis*) **A** HAI JA (Ryukyus)

parallela cheni Kaszab, 1988: 95 **A:** SCH XIZ YUN
parallela parallela Fairmaire, 1897k: 251 (*Nyctobates*) **A:** SCH GUA FUJ JIX GUI YUN ZHE XIZ **ORR**
pauperula Gebien, 1918: 124 (*Setenis*) **A:** HAI YUN **ORR**
persimilis M. Chûjô, 1980: 3 (*Setenis*) **A:** JA (Yakushima)
piluensis Masumoto, Akita & Lee, 2005: 249 **A:** TAI
punctatostriata Motschulsky, 1872: 29 (*Setenis*) **A:** AP BT HAI NP SD ZHE YUN "Korea" **ORR**
foveiceps Pic, 1929b: 17 (*Setenis*)
indosinica Fairmaire, 1893f: 296 (*Nyctobates*)
pilipes Pic, 1929b: 17 (*Setenis*)
puncticollis Motschulsky, 1872: 29 (*Setenis*) **A:** CH **ORR**
rectangula Motschulsky, 1872: 28 (*Setenis*) **A:** HAI GUA GUX FUJ JIX ZHE SC YUN **ORR**
biangulata Gebien, 1918: 122 (*Setenis*)
semisulcata Fairmaire, 1882c: 228 (*Nyctobates*)
sinuatocollis Kaszab, 1988: 103 **A:** HAI SCH YUN GUX **ORR**
striatipennis Lewis, 1894: 473 (*Setenis*) **A:** JA SC
subbiangulata subbiangulata M. Chûjô, 1978a: 68 (*Setenis*) **A:** JA (Ryukyus)
subrobusta indochinensis Kaszab, 1988: 109 **A:** HAI SCH YUN **ORR**
subrobusta subrobusta Motschulsky, 1872: 32 (*Nyctobates*) **A:** "Korea" **ORR**
coracina Fairmaire, 1882c: 230 (*Nyctobates*) [HN]
fairmairei Kolbe, 1900: 74 (*Setenis*) [RN]
sulcicollis Kaszab, 1988: 119 **A:** YUN **ORR**
sulcigera Boisduval, 1835: 256 (*Upis*) **A:** TAI **AUR ORR**
longipes Motschulsky, 1872: 28 (*Pediris*)
subopacus Waterhouse, 1884: 216 (*Pediris*)
sulcitibia Ren & Hua, 2006: 268 **A:** GUI GUX
szetchuanica Kaszab, 1988: 104 **A:** SCH YUN **ORR**
taiwana Masumoto, 1981e: 89 (*Setenis*) **A:** TAI
similis M. Chûjô, 1981: 136 (*Setenis*)
tatsienlua Kaszab, 1988: 101 **A:** SCH YUN
tokunoshimana Mt. Chûjô, 1995b: 119 **A:** JA (Ryukyus)
tonkinensis Gebien, 1918: 124 [= 1919a: 8] (*Setenis*) **A:** TAI YUN GUX **ORR**
aritai M. Chûjô, 1981: 131 (*Setenis*)
hamatilis Ren & Yang, 2004: 314
transversicollis Motschulsky, 1872: 29 (*Setenis*) **A:** AP BT NP SD YUN **ORR**
laevis Fairmaire, 1896d: 230 (*Nyctobates*)
valgipes subvalgipes M. Chûjô, 1981: 138 (*Setenis*) **A:** CH TAI JA (Shimojishima)
valgipes valgipes Marseul, 1876c: 117 (*Nyctobates*) **A:** GUI GUA HAI YUN ZHE FUJ HEN GUX HUB HUN JA JIX SC **ORR**
davidis Fairmaire, 1878a: 119 (*Nyctobates*)
villosipes Marseul, 1876c: 118 (*Nyctobates*)

vietnamica Kaszab, 1980b: 198 (*Setenis*) **A:** HAI YUN FUJ **ORR**
wenxiana Ren & Bai, 2005: 383 **A:** GAN
yunnanica Kaszab, 1988: 96 **A:** YUN

genus *Pseudonautes* Fairmaire, 1893c: 52 type species *Pseudonautes vagevittatus* Fairmaire, 1893
Thydemus Lewis, 1894: 475 type species *Scotaeus purpurivittatus* Marseul, 1876
purpurivittatus Marseul, 1876c: 130 (*Scotaeus*) **A:** JA TAI **ORR**
unicolor Pic, 1923b: 24 (*Thydemus*) **A:** FUJ NP **ORR**
viridiaeneus Allard, 1896: 321 (*Strongylium*) **A:** NP SD

genus *Psydomorphus* Pic, 1921c: 24 type species *Psydomorphus diversipes* Pic, 1921
diversipes Pic, 1921c: 24 **A:** YUN

genus *Rhopalobates* Fairmaire, 1897j: 230 type species *Rhopalobates villardi* Fairmaire, 1897
inexpectatus Schawaller, 2009b: 310 **A:** BT **ORR**
villardi Fairmaire, 1897j: 231 **A:** FUJ GUA SD SCH YUN HP **ORR**
klapperichi Kaszab, 1954: 255 (*Encyalesthus*)

genus *Scotaeus* Hope, 1834a: 14 type species *Scotaeus corallipes* Hope, 1834
focalis Gebien, 1935: 70 **A:** FUJ JIX

genus *Simalura* Gebien, 1914c: 71 type species *Simalura jacobsoni* Gebien, 1914
Microhemicera Pic, 1921c: 30 type species *Hemicera humeralis* Pic, 1921
atra Kaszab, 1980b: 208 **A:** YUN ORR
coerulea Lewis, 1894: 475 (*Eucyrtus*) **A:** JA CH "Korea"
godavarica Schawaller, 2006c: 480 **A:** NP
splendens Ando & Ren, 2006: 87 **A:** GUX

genus *Steneucyrtus* Fairmaire, 1896a: 31 type species *Steneucyrtus pexicollis* Fairmaire, 1896
Pseudochariotheca Pic, 1934b: 32 type species *Pseudochariotheca minutissima* Pic, 1934
major Pic, 1934b: 33 (*Pseudochariotheca*) **A:** CH **ORR**

genus *Stenochinus* Motschulsky, 1860a: 102 type species *Stenochinus reticulatus* Motschulsky, 1860
Brachypilium Fairmaire, 1896a: 23 type species *Brachypilium sculpturatum* Fairmaire, 1896
Dicraeosis Gebien, 1911: 355 [RN] type species *Dicraeus bacillus* Marseul, 1876
Dicraeus Marseul, 1876c: 103 [HN] type species *Dicraeus bacillus* Marseul, 1876
akiyamai Masumoto, Akita & Lee, 2013c: 266 **A:** TAI
amplus Gebien, 1914b: 8 (*Dicraeosis*) **A:** TAI "India"
sauteri Gebien, 1914b: 11 (*Dicraeosis*)
apiciconcavus Yuan & Ren, 2014b: 69 **A:** SHA

bacillus bacillus Marseul, 1876c: 103 (*Dicraeus*) **A:** JA
bacillus hiroshii Akita & Masumoto, 2015b: 419 **A:** JA (Mikura Is.)
capucinus Gravely, 1915: 520 (*Dicraeosis*) **A:** NP SD
apterus Kaszab, 1978a: 174 (*Dicraeosis*)
carinatus Gebien, 1914b: 12 (*Dicraeosis*) **A:** JA NP TAI **ORR**
shirozui Nakane, 1956b: 164 (*Dicraeosis*)
cylindricus Gebien, 1914b: 9 (*Dicraeosis*) **A:** TAI
furcifer Shibata, 1980: 73 (*Dicraeosis*) **A:** TAI
hiraii Akita & Masumoto, 2015b: 417 **A:** JA (Ryukyus)
jinxiuensis Ren & Yang, 2004: 316 (*Dicraeosis*) **A:** GUX
leprosus Ren & Yang, 2004: 317 (*Dicraeosisi*) **A:** GUX
meshimanus Nakane, 1985: 163 (*Dicraeosis*) **A:** JA (Meshima Is., Danjo-Gunto Is.)
miyakoensis Nakane, 1985: 163 (*Dicraeosis*) **A:** JA (Ryukyus)
mysticus Masumoto, Akita & Lee, 2013c: 268 **A:** TAI
oshimanus Nakane, 1985: 162 (*Dicraeosis*) **A:** JA (Amami-Ôshima)
pokharicus Schawaller, 2006c: 477 **A:** NP
reticulatus Motschulsky, 1860a: 102 **A:** BT NP **ORR**
subparallelus Pic, 1926b: 21 (*Dicraeosis*) **A:** CH
unicornis Shibata, 1980: 69 (*Dicraeosis*) **A:** TAI
wakoi Masumoto & Akita, 2002: 147 **A:** JA (Ryukyus)
xinyicus Yuan & Ren, 2014b: 71 **A:** GUA
yonagunianus Nakane, 1985: 163 (*Dicraeosis*) **A:** JA (Ryukyus: Yonakuni Is.)

genus *Stenophanes* Solsky, 1876: 294 type species *Hedyphanes mesostena* Solsky, 1871
mesostena Solsky, 1871: 376 (*Hedyphanes*) **A:** FE JA JIL NC SC NE FE NMO
doii Kôno, 1935a: 89
rubripennis Marseul, 1876c: 137 (*Helops*)
strigipennis Marseul, 1876c: 138 (*Helops*)

genus *Taichius* Ando, 1996: 196 type species *Platycrepis hemiceroides* Blair, 1929
forticornis Pic, 1922a: 21 (*Steneucyrtus*) **A:** GUX ZHE **ORR**
frater Ando, 1998: 370 **A:** FUJ ZHE
fukudai Masumoto, 1981e: 81 (*Hemicera*) **A:** TAI

genus *Taiwanomenephilus* Masumoto, 1986b: 61 type species *Taiwanomenephilus chui* Masumoto, 1986
chui Masumoto, 1986b: 62 **A:** TAI

genus *Tearchus* Kraatz, 1880c: 110 type species *Tearchus annulipes* Kraatz, 1880
Heteromerotylus Pic, 1921e: 11 type species *Heteromerotylus bicoloripes* Pic, 1921
annulipes Kraatz, 1880c: 110 **A:** AP BT **ORR**

geniculatus Pic, 1925h: 76 (*Heteromerotylus*) **A**: GUX **ORR**
vitalisi Pic, 1922i: 209 (*Heteromerotylus*) **A**: YUN **ORR**

genus *Teles* Mulsant & Godart, 1876a: 181 type species *Teles eutymi* Mulsant & Godart, 1876
eutymi Mulsant & Godart, 1876a: 182 **A**: TR

genus *Tetragonomenes* Chevrolat, 1878c: clii type species *Tetragonomenes semiviridis* Chevrolat, 1878
Falsoaugolesthus Masumoto, 1993c: 38 type species *Eucyrtus pulcher* Pic, 1927
Obriomaia Gebien, 1927: 45 type species *Eucyrtus subcostatus* Fairmaire, 1893
Tetragonomecus Rye, 1880: 87 [UE]
boninensis Nakane, 1977: 159 (*Obriomaia*) **A**: JA (Bonin Is.)
hirasawai Masumoto, 1984: 20 **A**: TAI
japonicus Masumoto, 1985a: 23 **A**: JA (Ryukyus) TAI
komiyai Masumoto, 1981b: 31 (*Obriomaia*) **A**: JA (Ryukyus)
kurosawai Masumoto, 1986a: 251 (*Augolesthus*) **A**: TAI
palpalis Kaszab, 1941c: 64 (*Obriomaia*) **A**: TAI
palpaloides Nakane, 1963a: 28 (*Obriomaia*) **A**: JA
pseudorufiventris Masumoto, 1981c: 45 (*Obriomaia*) **A**: TAI
reoi Akita & Masumoto, 2019: 93 **A**: JA (Daitojima)
rufiventris Kaszab, 1964a: 4 (*Obriomaia*) **A**: JA (Ryukyus)
semiviolaceus Nakane, 1968b: 80 (*Obriomaia*) **A**: JA (Ryukyus)

genus *Tetraphyllus* Laporte & Brullé, 1831: 404 type species *Tetraphyllus latreillei* Laporte & Brullé, 1831
Addia Lewis, 1894: 465 type species *Addia scatebrae* Lewis, 1894
brunneipes brunneipes Kaszab, 1941c: 60 **A**: TAI
laevis M. Chûjô, 1977: 11 **A**: JA (Ryukyus)
latior Nakane, 1963a: 27 (*Addia*) **A**: JA (Ryukyus)
amamiensis Kaszab, 1964a: 2
masaakii Akita & Masumoto, 2015b: 415 **A**: JA (Iheya Is.)
monticola Nakakita, 1987: 63 (*Addia*) **A**: TAI
nakanei Masumoto, 1982c: 22 (*Addia*) **A**: TAI
paykullii Dalman, 1823: 60 (*Amarygmus*) **A**: JA
lunuliger Marseul, 1876c: 129 (*Artactes*)
punctatus yunnanus Kaszab, 1944b: 57 **A**: YUN
satoi Ando & Sakai, 1999: 157 **A**: TAI
scatebrae Lewis, 1894: 466 (*Addia*) **A**: JA
unicolorata Nakane, 1991: 4 (*Hemicera*)
shibatai Nakakita, 1987: 61 (*Addia*) **A**: TAI
tsaii Masumoto, 1986b: 66 (*Addia*) **A**: TAI

genus *Tonkinius* Fairmaire, 1903a: 13 type species *Tonkinius sculptilis* Fairmaire, 1903
sculptilis Fairmaire, 1903a: 13 **A:** FUJ YUN **ORR**
striatipennis Gravely, 1915: 527 **A:** AP
thibetanus Kaszab, 1943: 72 **A:** XIZ **ORR**

genus *Upis* Fabricius, 1792b: 515 type species *Attelabus ceramboides* Linnaeus, 1758
ceramboides Linnaeus, 1758: 388 (*Attelabus*) **E:** BY CT EN FI LT NR NT SV SZ UK **A:** CH ES FE HEI KZ MG TR WS **NAR**
reticulata Say, 1824a: 279 (*Zophosis*)
variolosa DeGeer, 1775: 32 (*Tenebrio*)

genus *Zabroideus* Fairmaire, 1894b: 219 type species *Zabroideus pinguis* Fairmaire, 1894
pinguis Fairmaire, 1894b: 220 **A:** XIZ

tribe Stenochiini Kirby, 1837

genus *Eucrossoscelis* Nakane, 1963a: 29 type species *Eucrossoscelis broscosomoides* Nakane, 1963
aokii Akita & Masumoto, 2017: 50 **A:** JA (Kuroshima Is.)
araneiformis Allard, 1876a: 67 (*Helops*) **A:** JA (Kyushu)
helopioides Lewis, 1894: 482 (*Strongylium*)
broscosomoides Nakane, 1963a: 29 **A:** JA (Amami-Ôshima)
hastatus Yuan & Ren, 2006: 851 **A:** GUI
michioi M. Chûjô, 1978a: 78 **A:** JA (Ryukyus)
maruyamai Masumoto, 1999a: 121

genus *Strongylium* Kirby, 1819: 417 [NP] type species *Strongylium chalconatum* Kirby, 1819
Anomoearthrum Mäklin, 1867: 482 type species *Anomoearthrum debile* Mäklin, 1867
Coelolophus Mäklin, 1867: 394 type species *Coelolophus bicarinatus* Mäklin, 1867
Crossoscelis Gebien, 1914b: 52 type species *Crossoscelis clauda* Gebien, 1914
Ebenolus Fairmaire, 1897j: 227 type species *Ebenolus vernicatus* Fairmaire, 1897
Eustrongylium Kolbe, 1894b: 366 type species *Strongylium episcopale* Kolbe, 1894
Falsolophocnemis Pic, 1917c: 13 type species *Falsolophocnemis sinuatipes* Pic, 1917
Gentinadis Laporte, 1840: 240 type species *Stenochia caerulea* Laporte, 1840 (= *Strongylium azureum* Germar, 1823)
Gibbostrongylium Pic, 1917c: 18 type species *Strongylium medanense* Pic, 1917
Messalia Pascoe, 1883: 442 type species *Messalia varians* Pascoe, 1883
Microstrongylium Pic, 1917c: 13 type species *Microstrongylium cyanicolle* Pic, 1917
Notostrongylium Carter, 1915: 523 type species *Notostrongylium rugosicolle* Carter, 1915
Pedostrongylium Pic, 1916f: 11 type species *Strongylium luteonotatum* Pic, 1916
Poecilesthostrongylium Pic, 1918b: 12 type species *Strongylium amazonicum* Pic, 1918
Reminius Casey, 1924: 321 type species *Reminius ocularis* Casey, 1924 (= *Tenebrio terminatus* Say, 1824)
Saerangodes Sturm, 1843: 163 type species *Helops interpunctatus* Germar, 1823

Stenochia Kirby, 1819: 423 type species *Stenochia rufipes* Kirby, 1819
Strongyliastrum Fairmaire, 1894a: 39 type species *Strongyliastrum braetii* Fairmaire, 1894 (= *Strongylium rufipenne* Redtenbacher, 1844)
Styrax Westwood, 1875: 227 type species *Styrax tricondyloides* Westwood, 1875
Xanthothopeia Mäklin, 1867: 223 type species *Xanthothopeia rufipennis* Mäklin, 1867
Xanthothopia Gemminger, 1870b: 2001 [UE]
Zuercheria Reitter, 1908f: 134 type species *Zuercheria matthiesseni* Reitter, 1908

albopilosum Gebien, 1914b: 44 **A:** TAI
alishanum Masumoto, 1981c: 15 **A:** TAI
amamianum Nomura, 1963b: 38 **A:** JA (Ryukyus)
andoi Masumoto, 1996b: 138 **A:** TAI
angusticolle Mäklin, 1864: 333 **A:** NP SD **ORR**
angustissimum Pic, 1922c: 27 **A:** NP SD UP YUN **ORR**
anhuiense Masumoto, 2000b: 168 **A:** ANH
anmashanum Masumoto, Akita & Lee, 2008: 23 **A:** TAI
antracinum Mäklin, 1864: 306 **A:** "China boreale"
arunense Masumoto & Schawaller, 2010: 37 **A:** NP
atritarse Pic, 1916e: 13 **A:** YUN
basifemoratum Mäklin, 1864: 326 **A:** JIA
beesoni Blair, 1937: 37 **A:** NP SD
bhutanicum Masumoto, 1999b: 337 **A:** BH SD
binhense Pic, 1922h: 101 **A:** YUN **ORR**
nodieri Pic, 1936b: 19
brancuccii Masumoto & Schawaller, 2010: 30 **A:** NP
brevicorne Lewis, 1894: 482 **A:** JA
brunneum Yuan & Ren, 2005: 399 **A:** GUX HAI
carbonarium Gebien, 1914b: 47 **A:** TAI
castanescens Fairmaire, 1896b: 112 **A:** SD
chihpenense Masumoto, 2005b: 205 **A:** TAI
chinense Fairmaire, 1891f: ccxiii **A:** HUB
chutungense Masumoto, 2005b: 202 **A:** TAI
claudum Gebien, 1914b: 53 (*Crossoscelis*) **A:** TAI
clavipes Mäklin, 1864: 331 **A:** JIA
crurale Fairmaire, 1893g: 320 **A:** YUN **ORR**
cultellatum araii M. Chûjô, 1980: 7 **A:** JA (Senkaku Is.)
costatum M. Chûjô, 1979: 54 [HN]
cultellatum cultellatum Mäklin, 1864: 345 **A:** HKG JA NP SC YUN Pacific (O'ahu Is.)
cultellatum taiwanum Nomura, 1963b: 40 **A:** CH TAI
cultellatum yuwanum Nomura, 1963b: 40 **A:** JA (Ryukyus)
curvicornis Gravely, 1915: 535 **A:** AP
dimidiatum Fairmaire, 1891f: cciv **A:** SCH SHN

dorsocupreum Fairmaire, 1903a: 19 **A:** GUX **ORR**
endoi Masumoto, 1982d: 55 **A:** TAI
erythrocephalum Fabricius, 1801a: 156 (*Helops*) **A:** HAI JA (Ryukyus) TAI **ORR**
flavilabre Fairmaire, 1891f: ccxii **A:** HUB HUN
foochouense Masumoto, 2004: 380 **A:** FUJ
formosanum Gebien, 1914b: 45 **A:** TAI
fugianense Masumoto, 2000b: 170 **A:** FUJ
fujitai Masumoto, 1981a: 46 **A:** TAI
gaoliense Masumoto, 2004: 378 **A:** YUN
gardneri Blair, 1930b: 180 **A:** BT NP SD
gibbosipenne Nakane, 1963a: 29 **A:** JA
gibbosulum Fairmaire, 1891f: ccxii **A:** GUA JIX
guizhouense Masumoto, 2000b: 166 **A:** GUI
habashanense habashanense Masumoto, 1999a: 114 **A:** YUN
habashanense lijiangense Masumoto, 1999a: 115 **A:** YUN
holcopterum Fairmaire, 1893g: 320 **A:** BT **ORR**
hsiaoi Masumoto, Akita & Lee, 2008: 25 **A:** TAI
hyacinthinum Kaszab, 1957c: 61 **A:** FUJ
impigrum Lewis, 1894: 481 **A:** JA
impressipenne Pic, 1916d: 12 **A:** FUJ
japanum Marseul, 1876d: 320 **A:** JA
  naraense Nakane, 1968b: 84
  tanegashimae Nomura, 1963b: 38
  yakushimae Nomura, 1963b: 37
jizushanense Masumoto, 1999a: 116 **A:** YUN
jucundum Mäklin, 1864: 353 **A:** HKG
katsumii Masumoto, 1999b: 347 **A:** TAI
kawadai Masumoto, 1981b: 29 **A:** JA (Ryukyus)
  akitai Masumoto, 1998a: 195
kentingense Masumoto, 1981a: 47 **A:** TAI
kimioi Utsunomiya, 2011: 313 **A:** GUI
klapperichi Kaszab, 1957c: 60 **A:** FUJ
kuantouense Masumoto, 2005b: 195 **A:** TAI
kubani Masumoto, 2004: 387 **A:** YUN
kulzeri Kaszab, 1954: 263 **A:** FUJ
lanhai Masumoto, Akita & Lee, 2013c: 264 **A:** TAI
laszlorum Masumoto, 2005b: 201 **A:** TAI
liangi Yuan & Ren, 2014a: 330 **A:** YUN
lini Masumoto, Akita & Lee, 2008: 17 **A:** TAI
lishanum Masumoto, 1981a: 43 **A:** TAI
longipenne Fairmaire, 1903c: 318 (*Coelolophus*) **A:** FUJ

longissimum Gebien, 1914b: 50 **A:** TAI
lutaoense Masumoto, 1992: 93 **A:** TAI
macrops Wiedemann, 1823: 41 (*Helops*) **A:** NP **ORR**
marseuli Lewis, 1894: 481 **A:** JA NC
apterum Nomura & Yamazaki, 1960: 14
nigripes Nomura & Yamazaki, 1960: 15
watanabei Nomura & Yamazaki, 1960: 15
martensi Masumoto & Schawaller, 2010: 29 **A:** NP
masatakai Masumoto, Lee & Akita, 2007: 156 **A:** TAI
masumotoi Yuan & Ren, 2005: 401 **A:** SCH
quadrimaculatum Yuan & Ren, 2005: 401
matthiesseni Reitter, 1908f: 135 (*Zuercheria*) **A:** IN
mediofoveatum Blair, 1931a: 200 **A:** NP SD **ORR**
medvedevi Yuan & Ren, 2019: 1062 **A:** FUJ
mesopotamicum Gebien, 1911: 598 [RN] **A:** IQ
cancellatum Abeille de Perrin, 1894a: 94 (*Stenochia*) [HN]
fossum Allard, 1896: 321
metallescens Dohrn, 1880: 377
miwai Masumoto, 1996b: 139 [RN] **A:** TAI
insolitum Miwa, 1939: 413 [HN]
multiimpressum Pic, 1917e: 23 **A:** CH
multipunctatum Pic, 1936b: 19 **A:** YUN
nakanei Masumoto, 1981a: 49 **A:** TAI
nanfangum Masumoto, 1982d: 53 **A:** TAI
nanrenense Masumoto, Akita & Lee, 2012b: 201 **A:** TAI
niponicum Lewis, 1894: 480 **A:** JA "Korea'
obscuratum Yuan & Ren, 2005: 402 **A:** CHQ
fuscum Yuan & Ren, 2005: 402
ochii Masumoto, 1992: 92 **A:** TAI
ochrolatum Ren, 1998: 111 **A:** ZHE
ohmomoi Masumoto, Akita & Lee, 2013a: 119 **A:** TAI
okumurai Masumoto, 1981a: 44 **A:** TAI
opacicolle Fairmaire, 1891f: ccxiv **A:** SCH
osawai Masumoto, 2005b: 196 **A:** TAI
oshimanum Fairmaire, 1897g: 167 **A:** JA (Ryukyus)
palingense Masumoto, 2005b: 200 **A:** TAI
phedongense Masumoto, 1998a: 174 **A:** BH SD
pilimarginum Yuan & Ren, 2005: 403 **A:** GUX
pilosulum Fairmaire, 1891f: ccviii **A:** SCH
pinfaense Masumoto, 2000c: 405 **A:** GUI
pseudogibbosipenne Masumoto, 1981a: 43 **A:** TAI

rufipenne L. Redtenbacher, 1844: 533 **A:** CHQ GUX HUB NP SCH SD UP XIZ YUN **ORR**
braetii Fairmaire, 1894a: 40 (*Strongyliastrum*)
semirufum Pic, 1917c: 15
rufitarse Pic, 1916d: 12 **A:** YUN
saracenum Reiche & Saulcy, 1857: 270 (*Stenochia*) **E:** GR **A:** IS TR
perdubium Gebien, 1911: 599 [RN]
sulcipenne Reitter, 1908f: 134 (*Zuercheria*) [HN]
schawallerianum Masumoto, 2004: 371 **A:** NP
schenklingi Gebien, 1914b: 48 **A:** TAI
shibatai Nomura, 1963b: 39 **A:** JA (Ryukyus)
shigeoi Masumoto, Akita & Lee, 2008: 20 **A:** TAI
simulator Dohrn, 1880: 373 **A:** SD **ORR**
interruptum Blair, 1930b: 179
sinuatipenne Miwa, 1939: 414 **A:** TAI
yasumatsui M. Chûjô, 1968a: 16
sobrinum Dohrn, 1880: 376 **A:** NP SD UP
stanislavium Masumoto, 2004: 381 **A:** YUN
stevensi Gravely, 1915: 534 **A:** AP
subaeneum Pic, 1917h: 12 **A:** YUN
sulcielytrum Yuan & Ren, 2005: 400 **A:** GUX
suridobhanum Masumoto & Schawaller, 2010: 33 **A:** NP
szentivanyi Kaszab, 1941c: 71 **A:** TAI
tabanai Masumoto, 1998a: 196 **A:** SCH
taliense Pic, 1940b: 13 **A:** YUN
taliopacium Masumoto, 2004: 374 **A:** YUN
tanakai Ando, 2003b: 79 **A:** JA
tanikadoi Masumoto, 1998a: 197 **A:** SCH
tehuashense Masumoto, 2005b: 204 **A:** TAI
thibetanum Pic, 1916d: 13 **A:** XIZ YUN
yunnanum Pic, 1936a: 36
uedai Nomura, 1964: 48 **A:** JA (Yakushima)
undulatum kuatunense Kaszab, 1957c: 60 **A:** FUJ
wadai Masumoto, 2005b: 193 **A:** TAI
westermanni Mäklin, 1864: 341 **A:** NP SD UP
cariosipenne Fairmaire, 1896b: 112
wittmeri Masumoto & Schawaller, 2010: 34 **A:** NP
wuyishanense Yuan & Ren, 2006: 852 **A:** FUJ
yakushimanum Nakane, 1963a: 29 **A:** JA
yasuhikoi Masumoto, 1996b: 137 **A:** TAI
yokoyamai Masumoto, 1981a: 48 **A:** TAI
yunnanatrum Masumoto, 2004: 373 **A:** YUN
yunnanicum Masumoto, 1999b: 338 **A:** YUN

yoroi Akita & Masumoto, 2009a: 255 **A**: BT
zhengi Yuan & Ren, 2007: 550 **A**: GUX
zoltani Masumoto, 1981c: 16 **A**: TAI

genus *Uenostrongylium* Masumoto, 1999: 123 type species *Cryptobates laosensis* Pic, 1928
becvari Masumoto, 2006: 70 **A**: GUI
hunanense Masumoto, 2006: 72 **A**: HUN
scaber Yuan, Li & Ren, 2018: 24 **A**: ZHE

family Tenebrionidae, nomina dubia
Diaperis cornigera Gistel, 1857: 12 **E**: HU
Erelus Mulsant & Rey, 1853f: 185
Erelus sulcipennis Mulsant & Rey, 1853f: 188 **E**: IT (Sicilia)
Heliopates rugipennis Faldermann, 1837: 56 **A**: IN
Helops lapidicola Gistel, 1857: 14 **E**: AU
Opatrum glabrum Gistel, 1857: 18 "Taurien"
Opatrum lutosum Gistel, 1857: 47 **E**: PT
Opatrum politum Gistel, 1857: 51 "Taurica chersonesus"
Pedinus sibiricus Gebler, 1833: 287 **A**: Altai
Pygidiphorus Mulsant, 1856f: 520 [= 1856a: unnumb. pp.] type species *Pygidiphorus caroli* Mulsant, 1856f
Pygidiphorus caroli Mulsant, 1856f: 521 [= 1856a: unnumb. pp.] **E**: FR (Bordeaux)

family PROSTOMIDAE C.G. Thomson, 1859

genus *Prostomis* Latreille, 1819a: 171 type species *Trogosita mandibularis* Fabricius, 1801
  Megagnathus Dejean, 1821: 103 [HN] type species *Trogosita mandibularis* Fabricius, 1801
beatae Schawaller, 1991a: 6 **A**: NP
edithae Schawaller, 1991a: 9 **A**: BT NP SCH YUN **ORR**
elburica Fleischer, 1919b: 211 **E**: AB AR **A**: IN RU (Dagestan)
grandis Fursov, 1935: 90 **E**: ST "Caucasus"
latoris Reitter, 1889i: 315 **A**: JA SC
mandibularis Fabricius, 1801a: 155 (*Trogosita*) **E**: AU BU BY CZ DE FR GR GG HU IT LA PL PT RO SK SP ST SV SZ UK YU **A**: TR
  brunneus Fursov, 1935: 90
  maxillosus P.W.J. Müller, 1805: 205 (*Cucujus*)
  strandi Roubal, 1936b: 58
mordax Reitter, 1887d: 287 **A**: FE JA SC
morsitans Pascoe, 1860b: 98 **A**: NP SD
okinawaensis Ito & Yoshitomi, 2017b: 171 **A**: JA (Amami-Ôshima, Okinawa)
subtilis Szallies, 1994: 259 **A**: TR
susannae Schawaller, 1991a: 9 **A**: NP **ORR**
taiwanensis Ito & Yoshitomi, 2017a: 113 **A**: TAI
yaeyamaensis Ito & Yoshitomi, 2017b: 172 **A**: JA (Okinawa)

family SYNCHROIDAE Lacordaire, 1859

genus *Synchroa* Newman, 1838: 378 type species *Synchroa punctata* Newman, 1838

Phaiona Haldeman, 1848: 99 type species *Phaiona murina* Haldeman, 1848

chinensis Nikitsky, 2000: 45 **A:** GAN SCH SHA

formosana Hsiao, 2015: 244 **A:** TAI

melanotoides Lewis, 1895a: 263 **A:** FE JA JIL "Korea"

genus *Thescelosynchroa* Hsiao, Konvička & Ko, 2018: 27 type species *Synchroa pangu* Hsiao, Li, Liu & Pang, 2016

pangu Hsiao, Li, Liu & Pang, 2016: 596 (*Synchroa*) **A:** SCH

family OEDEMERIDAE Latreille, 1810

subfamily **Calopodinae** A. Costa, 1852

genus *Calopus* Fabricius, 1775: 20 type species *Cerambyx serraticornis* Linnaeus, 1758

serraticornis Linnaeus, 1758: 395 (*Cerambyx*) **E:** AL AN AU BH BU BY CR CT CZ EN FI FR GE GG GR HU IT LA LS LT NR NT PL RO SK SP ST SV SZ UK **A:** ES TR WS

pretneri J. Müller, 1929: 63

genus *Sparedrus* Dejean, 1821: 72 type species *Calopus testaceus* Andersch, 1797

Ocularium Pic, 1922b: 18 type species *Ocularium rufum* Pic, 1922

Polacus Sasaji, 1985: 10 type species *Polacus longicornis* Sasaji, 1985 (= *Sparedrus sasajii* Švihla, 2007)

Saloninus Fairmaire, 1891e: cxxxiii type species *Saloninus nebulosus* Fairmaire, 1891

Sparedropsis Heyden, 1886a: 191 type species *Sparedropsis fuscus* Heyden, 1886

albipilis Švihla, 1983a: 108 (*Sparedropsis*) **A:** IN

affinis Švihla, 2006a: 20 **A:** PA

alesivani Švihla, 2006a: 6 **A:** NP

bulgaricus Švihla, 2006b: 124 **E:** BU

championi Švihla, 2006a: 19 **A:** HP UP

concolor Švihla, 2006a: 13 **A:** AF

davidis Fairmaire, 1888b: 131 **A:** HEB SHX

elegans Švihla, 2006a: 17 **A:** PA

fuscus Heyden, 1886a: 191 (*Sparedropsis*) **A:** KI TD UZ

holzschuhi Švihla, 2006a: 13 **A:** AF

hurkai Švihla, 2006a: 22 **A:** PA

kazakhstanicus Švihla, 2006a: 21 **A:** KZ

klapperichi Švihla, 2006a: 14 **A:** AF

kutiauensis Švihla, 2006a: 14 **A:** AF

lencinae Vázquez, 1988: 259 **E:** SP

lepidus Švihla, 2006a: 18 **A:** KA PA

longicollis Švihla, 2006a: 6 **A:** IN

longicornis Iablokoff-Khnzorian, 1970: 62 (*Sparedropsis*) **A:** TD

murinus Švihla, 2006a: 9 **A:** PA
nebulosus Fairmaire, 1891e: cxxxiii (*Salonmus*) **A:** KA
orsinii A. Costa, 1852: 5 **E:** IT
pakistanicus Švihla, 2006a: 16 **A:** PA
proximus Švihla, 2006a: 9 **A:** AF
sasajii Švihla, 2007: 160 [RN] **A:** TAI
longicornis Sasaji, 1985: 10 [HN]
sepiaceus Švihla, 2006a: 21 **A:** PA UP
similis Švihla, 2006a: 12 **A:** AF
subserratus Gressitt, 1939: 218 (*Sparedropsis*) **A:** HAI
testaceus Andersch, 1797: 166 (*Calopus*) **E:** AL AU BH BU CR CZ GR HU IT RO SK SL SZ YU **A:** TR
unicolor Švihla, 1982: 381 (*Sparedropsis*) **A:** KA
uzbekistanicus Švihla, 2006a: 11 **A:** UZ

subfamily **Oedemerinae** Latreille, 1810
tribe Asclerini Gistel, 1848

genus *Alloxantha* Seidlitz, 1899: 828 type species *Alloxantha lutea* Seidlitz, 1899
Alloxanthina Švihla, 1983a: 117 type species *Alloxanthina flava* Švihla, 1983
Ananconiomorpha Švihla, 1983a: 120 type species *Ananconiomorpha platytarsia* Švihla, 1983
albopubens Švihla, 1987: 23 **A:** IN
carinata Karsch, 1881: 50 (*Nacerdes*) **N:** AG EG LB **A:** SI
sinaita Pic, 1929g: 109 (*Colobostomus*)
fernandezi Švihla, 1996: 86 **N:** CI (Gran Canaria)
flava Švihla, 1983a: 117 (*Alloxanthina*) **A:** AE IN IQ OM
flavoides Švihla, 1995: 18 **A:** IN PA
fulva Wollaston, 1854: 523 (*Ditylus*) **N:** CI (Gran Canaria, Tenerife) SG (Pitao Grande)
fuscostrigosa Fairmaire, 1879c: 203 (*Nacerdes*) **N:** AG MO TU **AFR**
lutea Seidlitz, 1899: 830 **N:** CI (Tenerife, El Hierro, La Palma, Gomera, Gran Canaria)
novellae Vázquez, 2002: 24 **N:** CI (Gran Canaria)
obsoleta Fairmaire, 1894c: 312 (*Ananca*) **N:** AG TU **AFR**
ochracea Seidlitz, 1899: 829 **N:** CI (Tenerife, El Hierro)
platytarsia Švihla, 1983a: 120 (*Ananconiomorpha*) **A:** IN OM PA SA
sahariana Vázquez, 1991: 107 **N:** MO
seidlitzi Švihla, 1988b: 376 **N:** CI (Tenerife)
svihlai Vázquez, 1993: 127 **N:** AG
talhouki Švihla, 1984: 265 **A:** AE IN OM SA
wollastoni Švihla, 1988b: 375 **N:** CI (Tenerife)

genus *Anacerda* Champion, 1927b: 249 type species *Anacerda leptidioides* Champion, 1927
leptidioides Champion, 1927b: 249 **A:** AF HP UP

genus *Anisochroa* Semenov, 1900d: 652 type species *Anisochroa zarudnyi* Semenov, 1900
semenovi Švihla, 1983a: 111 **A:** IN
volkovitshi Švihla, 1995: 22 **A:** TD TM UZ
zarudnyi Semenov, 1900d: 653 **A:** IN

genus *Anisochrodes* Švihla, 1983a: 108 type species *Anisochrodes jelineki* Švihla, 1983
escalerai Švihla, 1999a: 81 **A:** IN
holzschuhi Švihla, 1983b: 337 **A:** AF
janavlada Švihla, 2012: 342 **A:** YE (Suqutra)
jelineki Švihla, 1983a: 110 **A:** IN PA

genus *Asclernacerdes* Švihla, 2009a: 545 type species *Asclernacerdes akiyamai* Švihla, 2009
akiyamai Švihla, 2009a: 545 **A:** GUA

genus *Ascleropsis* Seidlitz, 1899: 860 type species *Asclera maculicollis* Ganglbauer, 1889
jakowlewi Semenov, 1894c: 542 (*Asclera*) **A:** KI KZ
kolibaci Švihla, 1997b: 422 **A:** SCH
maculicollis Ganglbauer, 1889: 42 (*Asclera*) **A:** GAN SCH
maculicollis Pic, 1907j: 174 (*Probosca*) [HN]
miladae Švihla, 2005b: 74 **A:** SCH
oudai Švihla, 1999a: 82 **A:** SCH
similis Švihla, 1997b: 421 **A:** SHA SHX
volkovitshi Švihla, 1997b: 421 **A:** SCH

genus *Asessinia* Švihla, 1986a: 196 type species *Sessinia subopaca* Pic, 1924
flavomarginata Miyatake, 1985: 407 (*Anoxacis*) **A:** JA (Shikoku, Ryukyus) TAI
geniculata Kôno, 1937a: 10 (*Alloxacis*) **A:** JA (Ryukyus)
iriomotoensis Miyatake, 1985: 407 (*Anoxacis*) **A:** JA (Iriomote) TAI
ohmomoi Akiyama & Hori, 2016: 47 **A:** JA (Hokkaido, Honshu)
vittata Miyatake, 1985: 407 (*Anoxacis*) **A:** JA (Kyushu, Ryukyus)

genus *Chitona* W.L.E. Schmidt, 1844: 46 type species *Stenostoma variegatum* Germar, 1824 (= *Leptura connexa* Fabricius, 1798)
Dolichopyga Seidlitz, 1899: 832 type species *Probosca acuminata* Reitter, 1890 (= *Probosca cretica* Fairmaire, 1861)
baulnyi Fairmaire, 1863: 646 **N:** AG TU
callida Kocher, 1968: 109 **N:** MO
connexa Fabricius, 1798: 154 (*Leptura*) **E:** PT SP **N:** AG MO
variegata Germar, 1824b: 167 (*Stenostoma*)
cretica Fairmaire, 1861: 199 **E:** GR "Balkans" **A:** TR
acuminata Reitter, 1890d: 152 (*Probosca*)
balcanica Pic, 1913s: 169 (*Dolichopyga*)

fucata fucata Faldermann, 1837: 139 (*Nacerdes*) **E:** AB ST (Dagestan) **A:** IN TM

cinerea Motschulsky, 1849: 134 (*Probosca*)

fucata tristis Faldermann, 1837: 140 (*Nacerdes*) **E:** AB (Naxcivan) AR **A:** IN TR

sieversi Kiesenwetter, 1878: 257

ganglbaueri Reitter, 1889a: 38 **E:** AB (Naxcivan)

gracilicollis Fairmaire, 1879c: 205 **N:** AG

incana W.L.E. Schmidt, 1846: 132 (*Probosca*) **E:** GR (Kós) **A:** TR

innotaticollis Pic, 1913s: 170 **N:** TU

kocheri Pic, 1949c: 74 **N:** MO

macedonica Švihla, 2006c: 114 **E:** BU GR MC

semividua Fairmaire, 1876: 50 **N:** AG

suturalis G.-A. Olivier, 1811: 446 (*Oedemera*) **E:** SP **N:** CI (Fuerteventura) MO

ornata Küster, 1847c: 88

strigilata Suffrian, 1848: 170

genus *Chitonoidea* Švihla, 1983a: 116 type species *Chitonoidea persica* Švihla, 1983

persica Švihla, 1983a: 116 (*Chitonoidea*) **A:** IN IS

genus *Colobostomoides* Švihla, 1983a: 113 type species *Colobostomoides longepubens* Švihla, 1983

longepubens Švihla, 1983a: 115 **A:** IN OM PA

marshalli socotraensis Švihla, 2012: 338 **A:** YE (Suqutra)

genus *Colobostomus* Fairmaire, 1886a: 452 type species *Colobostomus griseovestitus* Fairmaire, 1886

Isoloxantha Semenov, 1902b: 353 [RN] type species *Ananconia handlirschi* Seidlitz, 1899

Xanthomima Semenov, 1900d: 645 [HN] type species *Ananconia handlirschi* Seidlitz, 1899

arabicus arabicus Švihla, 1984: 260 **A:** SA

arabicus persicus Švihla, 2006d: 3 **A:** IN

griseovestitus Fairmaire, 1886a: 453 **N:** EG **A:** SA YE **AFR**

handlirschi Seidlitz, 1899: 832 (*Ananconia*) **N:** AG EG

boisyi Ancey, 1900: 140 (*Ananca*)

obscuricolor Pic, 1925i: 228

nitidus Pic, 1919e: 14 **N:** EG

genus *Copidita* J.L. LeConte, 1866: 164 type species *Probosca quadrimaculata* Motschulsky, 1853

testaceopyga Pic, 1923c: 32 [DA] **A:** "China"

genus *Dainsclera* Švihla, 1997b: 423 type species *Asclera obsoleta* Ganglbauer, 1889

obscura Švihla, 1980: 51 (*Indasclera*) **A:** BT

obscuroides Švihla, 1987: 18 (*Ascleropsis*) **A:** NP SD XIZ

obscuroviridis Švihla, 1997b: 426 **A:** SCH YUN

obsoleta obsoleta Ganglbauer, 1889: 43 (*Asclera*) **A:** GAN SCH SHX
obsoleta sulcicollis Pic, 1929d: 6 (*Asclera*) **A:** FE HEI

genus *Dentostomus* Švihla, 1984: 261 type species *Ananca anceyi* Pic, 1920
Hypascleroides Švihla, 1986a: 197 type species *Hypascleroides socotrensis* Švihla, 1986
anceyi Pic, 1920c: 5 (*Ananca*) **A:** SA YE **AFR**
guichardi Švihla, 1987: 25 (*Paroxacis*) **A:** YE (Abdelkuri I., Suqutra)
socotrensis Švihla, 1986a: 198 (*Hypascleroides*) **A:** YE (Suqutra, Darsah Is.)

genus *Eobia* Semenov, 1894b: 455 type species *Asclera cinereipennis* Motschulsky, 1866
subgenus *Eobia* Semenov, 1894b: 455 type species *Asclera cinereipennis* Motschulsky, 1866
chinensis ambusta Lewis, 1895a: 438 **A:** JA TAI
chinensis chinensis Hope, 1843: 63 (*Nacerdes*) **A:** ZHE (Zhousan Is.)
chinensis kotoensis Kôno, 1937b: 138 **A:** TAI (Lanxu Is.)
cinereipennis cinereipennis Motschulsky, 1866: 173 (*Asclera*) **A:** JA SC
japonica Harold, 1878b: 133 (*Ananca*)
cinereipennis ishigakiana Kôno, 1937a: 9 **A:** JA (Ryukyus)
cinereipennis ogasawarensis Matsumura, 1915: 227 (*Xanthochroa*) **A:** JA (Bonin Is.)
curticornis Pic, 1911p: 188 **A:** TAI
diversithorax Pic, 1927i: 48
formosana Pic, 1923c: 31 **A:** TAI
fuscipennis Nakane, 1954: 178 **A:** JA (Honshu, Ryukyus)
magna M. Chûjô & Satô, 1972: 26 **A:** JA (Iriomote Is.)
matsumurai Kôno, 1937b: 138 **A:** JA (Bonin Is.)
nigriceps Pic, 1923c: 30 **A:** JA
pallidipennis Gressitt, 1939: 219 **A:** HAI
piceopyga Pic, 1923c: 31 **A:** "China"
sinensis Gemminger, 1870a: 124 (*Ananca*) [RN] **A:** HKG
chinensis Boheman, 1858: 111 (*Nacerdes*) [HN]
testaceiceps Pic, 1923c: 30 **A:** "China"
subgenus *Pareobia* Švihla, 1986a: 189 type species *Eobia florilega* Lewis, 1895
florilega Lewis, 1895a: 438 **A:** JA TAI "China"

genus *Ganglbaueria* Semenov, 1890: 378 type species *Ganglbaueria collaris* Semenov, 1890
subgenus *Ganglbaueria* Semenov, 1890: 378 type species *Ganglbaueria collaris* Semenov, 1890
collaris Semenov, 1890: 379 **A:** MG XIN
hedini Pic, 1933b: 8 (*Nacerdochroa*)
turkestanica Švihla, 1995: 10 **A:** KZ
subgenus *Pseudananconia* Švihla, 1983a: 121 type species *Ananconia mesopotamica* Reitter, 1908 (= *Ananca heluanensis* Pic, 1901)

heluanensis Pic, 1901c: 34 (*Ananca*) **N:** EG **A:** IN IQ IS JO SY
mesopotamica Reitter, 1908i: 246 (*Ananconia*)
kermanica Švihla, 1983a: 124 (*Pseudananconia*) **A:** IN
richteri Švihla, 1987: 24 (*Pseudananconia*) **A:** IN
vazquezi Švihla, 1995: 14 **A:** TD
subgenus *Zubkovia* Semenov, 1894b: 467 type species *Zubkovia turcomanica* Semenov, 1894
iranica Švihla, 1995: 12 **A:** IN
turcomanica Semenov, 1894b: 468 (*Zubkovia*) **A:** TD TM

genus *Gilotia* Pardo Alcaide, 1968: 42 type species *Gilotia perroti* Pardo Alcaide, 1968
perroti Pardo Alcaide, 1968: 43 **N:** MO

genus *Hyperopselaphus* Mizota, 1999b: 287 type species *Hyperopselaphus ikedai* Mizota, 1999
ikedai Mizota, 1999b: 288 **A:** JA (Ludao Is., Yaeyama Is.) TAI

genus *Indasclera* Švihla, 1980: 48 type species *Asclera indica* Fairmaire, 1894
binotata Pic, 1927c: 13 (*Asclera*) **A:** YUN
bocaki Švihla, 1997b: 443 **A:** SCH
brancuccii Švihla, 1997b: 440 **A:** BT NP SD
brevicollis Gressitt, 1939: 221 (*Asclera*) **A:** FUJ GUA GUI GUX
brunneipennis Lewis, 1895a: 440 (*Asclera*) **A:** JA
carinicollis Lewis, 1895a: 439 (*Oxacis*) **A:** FE (Kuriles: Kunashir Is.) JA (Hokkaido, Honshu)
chinensis Švihla, 1997b: 456 **A:** HUN
diluta Gressitt, 1939: 222 (*Asclera*) **A:** GUA
formosana Pic, 1910u: 95 (*Asclera*) **A:** TAI
haemorrhoidalis Pic, 1907j: 174 (*Proboscia*) **A:** YUN
himalaica Švihla, 1980: 50 **A:** NP SD
godawariensis Švihla, 1997b: 437
igai Nakane, 1954: 177 (*Asclera*) **A:** JA
incostata Pic, 1926c: 19 (*Asclera*) **A:** GAN SCH SHA XIZ **ORR**
indica Fairmaire, 1894a: 42 (*Asclera*) **A:** AP BT NP SD
excellens Fleischer, 1919a: 168 (*Ascleropsis*)
japonica amamiana Miyatake, 1985: 405 (*Asclera*) **A:** JA (Ryukyus)
japonica japonica Pic, 1910u: 95 (*Asclera*) **A:** JA (Honshu, Kyushu)
satana Nakane, 1954: 175 (*Asclera*)
loi Gressitt, 1939: 223 (*Asclera*) **A:** HAI
murzini Švihla, 2009b: 95 **A:** SCH
nakanei Mizota, 1999a: 82 **A:** JA (Honshu, Kyushu)
nepalensis lubosi Švihla, 2002: 336 **A:** SD **ORR**
nepalensis nepalensis Švihla, 1987: 17 (*Ascleropsis*) **A:** AP NP

peculiaris Pic, 1914b: 16 (*Asclera*) **A:** YUN **ORR**
puncticollis Švihla, 2006d: 5 **A:** AP
rolciki Švihla, 2002: 334 **A:** SD **ORR**
ruficollis Lewis, 1895a: 434 (*Ditylus*) **A:** JA
konoi Nakane, 1973a: 7 (*Asclera*)
rugosipennis Pic, 1910u: 95 (*Asclera*) **A:** TAI
similis Švihla, 2002: 335 **A:** YUN **ORR**
strangulata Fairmaire, 1889b: 50 (*Anoncodes*) **A:** SCH
subrugosa Kôno, 1937b: 137 (*Asclera*) **A:** JA (Kyushu, Ryukyus)
kyushuensis Nakane, 1954: 176 (*Asclera*)
takakuwai Akiyama, 2012: 190 **A:** TAI
thibetana Pic, 1915c: 5 (*Asclera*) **A:** "Tibet"
uenoi Švihla, 1997b: 434 **A:** TAI
unicostata Pic, 1914l: 68 (*Asclera*) **A:** "Tibet"
wittmeri Švihla, 1997b: 429 **A:** TAI

genus *Ischnomera* Stephens, 1832: 53 type species *Cantharis caerulea* Linnaeus, 1758
Asclera Dejean, 1834: 228 type species *Necydalis sanguinicollis* Fabricius, 1787
Ascleronia Seidlitz, 1899: 859 type species *Ischnomera semiflava* Reitter, 1891
Chromasclera Seidlitz, 1899: 859 type species *Asclera partitipennis* Fairmaire, 1892
abdominalis Heyden, 1887b: 304 (*Asclera*) **A:** BEI FE GAN JA (Honshu)
analis Fairmaire, 1888b: 131 (*Oedemera*) [HN]
flavipes Reitter, 1891a: 31
apicipennis Švihla, 1992: 40 **A:** KZ
auripennis Reitter, 1903b: 45 (*Asclera*) **E:** GR **A:** TR
bajsunensis Švihla, 2004: 64 **A:** UZ
besucheti Švihla, 1993a: 53 **A:** PA
bilyi Švihla, 1992: 41 **A:** TD UZ
caerulea Linnaeus, 1758: 403 (*Cantharis*) **E:** AB AL AR AU BH BU CR CZ FR GB GE GG GR HU IT MC NR PL RO SK SL SP ST SV SZ UK YU **A:** IN LE TM TR
carinithorax Roubal, 1936a: 259 (*Asclera*)
coerulescens Fabricius, 1775: 208 (*Necydalis*)
nigripes G.-A. Olivier, 1811: 446 (*Oedemera*)
obliterata Pic, 1926e: 6 (*Asclera*)
opacocoerulea Reitter, 1911a: 408
quadrinervosa Latreille, 1804b: 12 (*Oedemera*)
unicolor Melsheimer, 1846: 54
cinerascens bodemeyeri Ganglbauer, 1900: 161 **A:** TR
cinerascens cinerascens Pandellé, 1867: 182 (*Asclera*) **E:** AU BH CR CZ DE FR GB GE HU IT NR PL RO SK SL SP SV SZ YU
cinerascens kurdistana Švihla, 1988a: 219 **A:** IQ TR

cinerascens reitteri Ganglbauer, 1881: 106 **E:** AB AL BU GR **A:** TR

cyanea Fabricius, 1792b: 351 (*Necydalis*) **E:** AL AU BH BU CR CZ DE FR GB GE GR HU IT MC MD NL PL RO SK SL SP SV SZ TR UK YU **A:** TR

graeca Dahlgren, 1976: 22 (*Asclera*)

viridescens Pic, 1920c: 5 (*Asclera*)

danilevskiji Švihla, 1992: 39 **A:** TD

deyrollei Pic, 1895h: 243 (*Asclera*) **E:** GG ST (Caucasus) **A:** SY TR

milleri Pic, 1898c: 38 (*Asclera*)

dzhungarica Švihla, 1986b: 216 **A:** KZ MG XIN

erdemliensis Švihla, 1998c: 239 **A:** SY TR

fulvicollis Reitter, 1897a: 38 (*Asclera*) **E:** GR (Crete, Ikaria) **A:** CY TR

fuscipennis Švihla, 1988a: 209 **A:** TR

girardi Švihla, 1992: 37 **A:** YUN

haemorrhoidalis W.L.E. Schmidt, 1846: 43 (*Asclera*) **E:** GR TR **A:** TR

impressithorax Pic, 1897e: 42 (*Asclera*)

hauseri ferganica Švihla, 1992: 41 **A:** KI

hauseri hauseri L. Heyden, 1887c: 322 (*Asclera*) **A:** KZ TD UZ

holzschuhi Švihla, 1988a: 207 **A:** IN

jezeki Švihla, 1988a: 219 **A:** IN

klapperichi klapperichi Švihla, 1997a: 86 **A:** AF

klapperichi wittmeri Švihla, 1997a: 87 **A:** PA

kopetzi Švihla, 2004: 64 **A:** NP

lutescens Abeille de Perrin, 1892: 64 (*Asclera*) **A:** TR

maceki Švihla, 1988a: 208 **A:** IN IQ TR

nigricans Fairmaire, 1897j: 238 (*Asclera*) **A:** HP NP **ORR**

viator Champion, 1927b: 250 (*Asclera*)

nigrocyanea koreana Fleischer, 1919a: 168 **A:** FE "Korea"

nigrocyanea nigrocyanea Lewis, 1895a: 420 (*Asclera*) **A:** FE (Kuriles) JA

okushimai Švihla, 1997a: 85 **A:** JA (Okinawa)

partitipennis Fairmaire, 1892b: 158 (*Asclera*) **A:** IN IQ SY TR

iranica Zaitzev, 1941: 247 (*Asclera*)

persica Švihla, 1992: 43 **A:** IN

rosti Pic, 1914j: 57 (*Asclera*) **A:** KA PA

russowi Seidlitz, 1899: 851 (*Asclera*) **A:** KI KZ TD UZ

turkestanica Pic, 1912e: 26 (*Asclera*)

sanguinicollis Fabricius, 1787a: 170 (*Necydalis*) **E:** AL AU BE BH BU CR CZ DE FR GB GE GR HU IR IT LA LT MC NR PL RO SK SL SP SV SZ UK YU **A:** TR

flavicollis Panzer, 1794c: 18 (*Necydalis*)

semiflava Reitter, 1891a: 31 **A:** KZ UZ XIN

nigripennis Reitter, 1897a: 37 (*Asclera*)

similis Švihla, 1982: 384 (*Asclera*) **A:** IN TR

sveci Švihla, 2004: 62 **A:** TR
taiwana Švihla, 1992: 38 **A:** TAI
tenietensis Abeille de Perrin, 1896: 283 (*Asclera*) **N:** AG TU
testaceipes testaceipes Pic, 1914j: 57 (*Asclera*) **A:** KA
testaceipes viridipennis Švihla, 1986b: 215 **A:** PA
xanthoderes Mulsant, 1858a: 232 (*Asclera*) **E:** FR PT SP **N:** MO

genus *Nacerdochroa* Reitter, 1893a: 113 type species *Nacerdochroa plutschevskyi* Reitter, 1893 (= *Nacerdes caspia* Faldermann, 1836)
Holoxantha Semenov, 1894b: 470 type species *Ditylus concolor* Brullé, 1838
carinatopyga Švihla, 1983a: 113 **A:** AE IN OM SA YE
caspia Faldermann, 1836: 392 (*Nacerdes*) **E:** AB ST **A:** IN KZ TD TM UZ
glasunovi Semenov, 1894b: 466
plutschevskyi Reitter, 1893a: 114
concolor Brullé, 1839: 70 (*Ditylus*) **N:** CI SG
hierrensis Franz, 1985: 102 (*Alloxantha*)
pici Švihla, 2008a: 45 [RN] **N:** MO
concolor Pic, 1942c: 11 (*Sessinia*) [HN]

genus *Oschaninia* Semenov, 1898: 175 type species *Ganglbaueria wilkinsi* Heyden, 1894
subgenus *Homomorpha* Semenov, 1898: 177 type species *Homomorpha crucifera* Semenov, 1898
crucifera Semenov, 1898: 177 (*Homomorpha*) **A:** XIN
subgenus *Oschaninia* Semenov, 1898: 175 type species *Ganglbaueria wilkinsi* Heyden, 1894
semenovi Kuznetzova, 1935: 298 (*Homomorpha*) **A:** TM
wilkinsi Heyden, 1894a: 52 (*Ganglbaueria*) **A:** MG TM
zhelochovtzevi Kuznetzova, 1935: 300

genus *Parasessinia* Švihla, 1986a: 195 type species *Parasessinia nicobarica* Švihla, 1986
choui Akiyama, 2015a: 129 **A:** TAI

genus *Paroxacis* Arnett, 1951: 334 type species *Probosca lucana* LeConte, 1866
sahariana Švihla, 1995: 22 **N:** AG

genus *Probosca* W.L.E. Schmidt, 1846: 130 type species *Probosca viridana* W.L.E. Schmidt, 1846 (= *Callidium virens* Fabricius, 1792)
subgenus *Ananconia* Seidlitz, 1899: 831 type species *Chitona martini* Fairmaire, 1896
martini Fairmaire, 1896c: 224 (*Chitona*) **A:** SA YE **AFR**
spurcaticollis Fairmaire, 1886a: 452 (*Ananca*) **A:** SA YE **AFR**
subgenus *Asclerella* Semenov, 1900d: 648 type species *Asclerella tenera* Semenov, 1900
tenera Semenov, 1900d: 651 (*Asclerella*) **A:** IN PA
elongatipennis Pic, 1935e: 101 (*Ananconia*)

subgenus *Probosca* W.L.E. Schmidt, 1846: 130 type species *Probosca viridana* W.L.E. Schmidt, 1846 (= *Callidium virens* Fabricius, 1792)
infima Fairmaire, 1879c: 204 **N:** AG
letorneuxi Pic, 1894j: 93 **N:** EG **A:** CY
metallescens Fairmaire, 1866a: 57 (*Chitona*) **N:** MO
pallipes G.-A. Olivier, 1811: 446 (*Oedemera*) **N:** EG
unicolor Küster, 1847c: 87 (*Chitona*) **E:** SP **N:** MO
plumbea Suffrian, 1848: 169
virens Fabricius, 1792b: 338 (*Callidium*) **E:** IT (Sardinia, Sicilia) SP **N:** AG MO TU
boiteli Pic, 1936f: 148
hispanica Pic, 1920c: 5
nigrofemorata Pic, 1898h: 122
purpureomicans Pic, 1913s: 170
viridana W.L.E. Schmidt, 1846: 130
subgenus *Proboxantha* Švihla, 1995: 16 type species *Ananconia comata* Blair, 1923
afroarabica Švihla, 1984: 262 (*Ananconia*) **N:** EG **A:** IS YE **AFR**
batesi Švihla, 1984: 263 (*Ananconia*) **A:** IQ SA
comata Blair, 1923a: 125 (*Ananconia*) **A:** AE IN IQ JO SA
fuscipennis Blair, 1923a: 124 (*Isoloxantha*) **A:** AE IN IQ KU SA
persica Pic, 1929d: 6 (*Zubkovia*)
maindroni Pic, 1935e: 101 (*Ananconia*) **A:** AE PA YE (Suqutra)
marginata Walker, 1871: 17 **N:** EG LB **A:** SA YE **AFR**
notatithorax Pic, 1906c: 26
vanharteni Švihla, 2008c: 264 **A:** AE

genus *Pseudohyperasclera* Macnamara, 1965: 70 type species *Pseudohyperasclera michaeli* Macnamara, 1965
Pseudohypasclera Macnamara, 1971: 164 [UE]
satoi Akiyama, 2015b: 450 (*Pseudohyperasclera*) **A:** JA (Yaeyama Is.)

genus *Schellia* Reitter, 1914c: 49 type species *Schellia sitaroides* Reitter, 1914
sitaroides Reitter, 1914c: 50 **A:** KZ TM

genus *Schistopselaphus* Fairmaire, 1896b: 126 type species *Schistopselaphus apicatus* Fairmaire, 1896
sonani Kôno, 1937b: 140 **A:** TAI (Lanxu Is., Ludao Is.)

genus *Xanthochroina* Ganglbauer, 1881: 105 type species *Xanthochroa auberti* Abeille de Perrin, 1876
Anancosessinia Kôno, 1937b: 139 type species *Anancosessinia tarsalis* Kôno, 1937
Asclerostoma Fleischer, 1919a: 169 type species *Asclerostoma reitteri* Fleischer, 1919

auberti Abeille de Perrin, 1876: clxvi (*Xanthochroa*) **E:** CR FR GR HU IT SP **A:** CY IN IS LE TR

latetestacea Pic, 1920g: 18

reitteri Fleischer, 1919a: 170 (*Asclerostoma*)

tarsalis Kôno, 1937b: 140 (*Anancosessinia*) **A:** JA (Ryukyus) TAI

tribe Ditylini Mulsant, 1858

genus *Ascleranoncodes* Pic, 1915i: 10 type species *Ascleranoncodes distincticornis* Pic, 1915

Indanerces Pic, 1923c: 24 type species *Indanerces nitidicollis* Pic, 1923

suturalis haematodes Švihla, 2009b: 91 **A:** YUN **ORR**

suturalis suturalis Švihla, 1998b: 163 (*Pseudonerdanus*) **A:** NP SD

genus *Chrysanthia* W.L.E. Schmidt, 1844: 46 type species *Cantharis viridissima* Linnaeus, 1758

bhutanica Švihla, 1980: 54 **A:** BT NP

canaliculata Iablokoff-Khznorian, 1974: 190 **A:** TD

chalcochroa chalcochroa Fairmaire, 1892b: 159 **A:** TR

distinctithorax Pic, 1904e: 57

tauricola Pic, 1920c: 5

chalcochroa holzschuhi Pardo Alcaide, 1972: 214 **A:** TR (Prov. Antalya)

cyprica Pic, 1920c: 5 **A:** CY

flavipes Reitter, 1889f: 266 **E:** GR (mainland, Sporades) **A:** IS JO SY TR

oralis Fairmaire, 1892b: 158

semicuprea Pic, 1920c: 5

fuscimembris fuscimembris Fairmaire, 1891e: cxxxiv **A:** KA PA UP

cyanescens Pic, 1926c: 19

fuscimembris kaganensis Švihla, 2001a: 96 **A:** PA

geniculata chinensis Švihla, 1999a: 80 **A:** HEB HUB SHA SCH

geniculata geniculata W.L.E. Schmidt, 1846: 128 **E:** AN AU BE BH BU BY CR CT CZ DE EN FI FR GB GE GG GR HU IT LA LT LU MD NL NR NT PL RO SK SL SP SV SZ UK YU **A:** ES HEB TM WS

cupreomicans Westhoff, 1881: 195

geniculata L. Heyden, 1877: 293 [HN]

nigricornis Westhoff, 1881: 195

saturata Westhoff, 1881: 195

viridis W.L.E. Schmidt, 1846: 128 [HN]

geniculata integricollis Heyden, 1886b: 275 **A:** FE NC SC

geniculata viatica Lewis, 1895a: 443 **A:** FE (Sakhalin, Kuriles) JA (Hokkaido, Honshu, Tsushima)

krali Švihla, 1982: 382

hamata Vázquez, 1989: 130 **E:** SP

himalaica Champion, 1920a: 74 **A:** HP NP PA UP

maroccana Håk. Lindberg, 1932: 34 **N:** MO

reitteri Seidlitz, 1899: 866 **E:** PT SP
rugicollis Champion, 1920a: 75 **A:** HP **ORR**
superba Reitter, 1872: 182 **E:** PT SP **N:** AG LB MO TU
oranensis Seidlitz, 1899: 867
valens Champion, 1920a: 75 **A:** HP
varipes planiceps Kiesenwetter, 1878: 256 **E:** AR GG ST (Kavkaz) **A:** TR
turcica Pic, 1899a: 15
varipes varipes Kiesenwetter, 1861a: 252 **E:** AL BH BU CR GR MC TR YU
viridissima Linnaeus, 1758: 403 (*Cantharis*) **E:** AL AN AU BE BH BU BY CR CT CZ EN FI FR GE GG GR HU IT LA LS LT MC NR NT PL RO SK SL SP SV SZ YU **A:** ES WS
cuprina Rey, 1892: 138
diversipes Pic, 1932h: 30
korbi Reitter, 1894c: 104
thalassina Fabricius, 1792b: 350 (*Necydalis*)
viridis DeGeer, 1775: 15 (*Cantharis*)
wittmeri obscurimembris Švihla, 2001a: 96 **A:** KA PA
wittmeri wittmeri Švihla, 1980: 56 **A:** KA

genus *Diasclera* Reitter, 1913d: 663 type species *Diasclera viridescens* Reitter, 1913
Eumecomera Arnett, 1951: 297 type species *Asclera obscura* LeConte, 1854
Nacerdasclera Munster, 1921: 8 type species *Oedemera sibirica* Gebler, 1830
sibirica laevithorax Pic, 1915h: 9 (*Asclera*) **A:** FE HEI JIL NC
ussurica Pic, 1926e: 6 (*Asclera*)
sibirica sibirica Gebler, 1829a: 132 (*Oedemera*) **A:** ES MG WS "China"
viridescens oxiana Iablokoff-Khnzorian, 1974: 188 **A:** TD
viridescens viridescens Reitter, 1913d: 663 **E:** AB

genus *Diplectrus* Kirsch, 1866: 210 type species *Diplectrus ferrugineus* Kirsch, 1866
bistigmaeus Zhang, Ren & Ba, 2012: 407 **A:** XIZ
castaneicollis Švihla, 1999a: 75 **A:** "China" **ORR**
improvisus Švihla, 1999a: 77: 210 **A:** SCH
longipennis Fairmaire, 1896b: 126 (*Ananca*) **A:** SD
testaceipennis Pic, 1923c: 28 (*Patiala*)
piceipennis Pic, 1923c: 28 (*Patiala*) **A:** BT SD
variicollis Švihla, 1999a: 77 **A:** SCH

genus *Ditylus* Fischer von Waldheim, 1817: 469 type species *Ditylus helopioides* Fischer von Waldheim, 1817 (= *Helops laevis* Fabricius, 1787)
Mimetes Eschscholtz, 1818: 467 type species *Mimetes unicolor* Eschscholtz, 1818 (= *Helops laevis* Fabricius 1792)

catenulatus Švihla, 2008b: 59 **A:** SCH
laevis laevis Fabricius, 1787a: 213 (*Helops*) **E:** AU BY CZ EN FI GE KZ LA LT NT PL SK SV UK **A:** ES FE JA (Hokkaido, Honshu, Tsushima) NC NE
helopioides Fischer von Waldheim, 1817: pl. 5
unicolor Eschscholtz, 1818: 468 (*Mimetes*)
laevis oxianus Švihla, 1993a: 52 **A:** TM
praetermittus Švihla, 2005b: 72 **A:** SCH
sichuanus Švihla, 1996: 83 **A:** SCH
tibetanus Švihla, 2005b: 72 **A:** XIZ

tribe Nacerdini Mulsant, 1858

genus *Anogcodes* Dejean, 1834: 228 type species *Necydalis melanura* Fabricius, 1787
Anoncodes W.L.E. Schmidt, 1844: 46 type species *Necydalis melanura* Fabricius, 1787
Anoncodina Seidlitz, 1899: 786 type species *Necydalis ruficollis* Fabricius, 1781
Lethonymus Marseul, 1857: 130 type species *Lethonymus difformis* Marseul, 1857
Oedechira Motschulsky, 1872: 54 type species *Oedemera paradoxa* Faldermann, 1837 (= *Cantharis ustulatus* Scopoli, 1763)
Pachychirus L. Redtenbacher, 1845: 134 type species *Necydalis adusta* Panzer, 1795 (= *Cantharis ustulatus* Scopoli, 1763)
Peronocnemis Fairmaire, 1886c: 352 type species *Peronocnemis davidis* Fairmaire, 1886
caucasicus Pic, 1906b: 19 (*Nacerda*) **E:** "Caucasus"
latebasalis Pic, 1926e: 5 (*Nacerdes*)
coarctatus coarctatus Germar, 1823: 165 (*Oedemera*) **E:** BY CT **A:** ES KZ MG WS
nigriventris Motschulsky, 1859d: 493 (*Anoncodes*)
coarctatus croceiventris Motschulsky, 1859d: 453 (*Anoncodes*) **A:** FE JA (Hokkaido, Honshu) NC NE SC
manciuricus Magistretti, 1939: 159 (*Nacerda*)
sambuceus Lewis, 1895a: 439 (*Anoncodes*)
davidis Fairmaire, 1886c: 352 (*Peronocnemis*) **A:** GAN QIN SCH XIZ YUN
femoratus Ganglbauer, 1889: 38 (*Anoncodes*)
difformis Marseul, 1857: 130 (*Lethonymus*) **E:** AB AR GR (Rhodes) ST **A:** CY IN LE TR
bodemeyeri Pic, 1911m: 170 (*Anoncodes*)
signaticollis Pic, 1898a: 2 (*Anoncodes*)
tschitscherini Semenov, 1895b: 248 (*Nacerdes*)
versicolor Chevrolat, 1873: 205 (*Anoncodes*)
flaviceps Fairmaire, 1866a: 57 (*Anoncodes*) **N:** MO
fulvicollis Scopoli, 1763: 43 (*Cantharis*) **E:** AN AU BH BU CR CZ FR GE GR HU IT LS PL RO SK SL SP SZ UK YU **A:** TR
longicollis Roubal, 1936a: 257 (*Nacerda*)
medius Gredler, 1866: 295 (*Anoncodes*)
obscurus Gyllenhal, 1810: 630 (*Necydalis*)
therondi Méquignon, 1948: 78 (*Nacerda*)

geniculatus W.L.E. Schmidt, 1846: 99 (*Anoncodes*) **A:** TR
basalis W.L.E. Schmidt, 1846: 99 (*Anoncodes*)
melanurus Fabricius, 1787a: 170 (*Necydalis*) **E:** AU BH BU BY CR CT CZ EN FR GE GR HU IT LA LT MC NL PL RO SK SL SP ST SZ UK **A:** KZ TR WS XIN
fasciatus Villers, 1789: 283 (*Necydalis*)
rufus Fischer von Waldheim, 1820: pl. 5 (*Ditylus*)
scutellaris Waltl, 1839: 222 (*Anoncodes*)
ustulatus Fabricius, 1787a: 170 (*Necydalis*) [HN]
ruficollis Fabricius, 1781: 263 (*Necydalis*) **E:** AR AU BH BU CR CZ FR GE GG GR HU IT MC RO SK SP ST SZ TR UK YU **A:** IN IQ SY TR
coerulescens Rossi, 1790: 172 (*Necydalis*)
rufiventris Scopoli, 1763: 43 (*Cantharis*) **E:** AU BH BU BY CR CT CZ EN FR GE GR HU IT LA LT MC NT PL RO SK SL SP SV SZ UK YU **A:** TR
bipartitus Schrank von Paula, 1776: 68 (*Leptura*)
dorsalis G.-A. Olivier, 1811: 444 (*Oedemera*)
melanocephalus Fabricius, 1794: 453 (*Necydalis*)
schatzmayri H. Wagner, 1928: 117 (*Nacerdes*) **E:** FR SP
seladonius alpinus W.L.E. Schmidt, 1846: 121 (*Anoncodes*) **E:** AU FR GE IT SL SZ
azureus W.L.E. Schmidt, 1846: 123 (*Anoncodes*)
seladonius austriacus Ganglbauer, 1881: 103 (*Nacerdes*) **E:** AU CZ HU RO SK
seladonius seladonius Fabricius, 1792b: 352 (*Necydalis*) **E:** FR IT PT SP SZ **N:** AG TU
amoenus W.L.E. Schmidt, 1846: 118 (*Anoncodes*)
aurosus Fairmaire, 1863: 645 (*Nacerdes*)
dispar Dufour, 1841: 8 (*Oedemera*)
meridionalis A. Costa, 1852: 10 (*Anoncodes*)
viridipes W.L.E. Schmidt, 1846: 116 (*Anoncodes*)
seladonius turcicus W.L.E. Schmidt, 1846: 120 (*Anoncodes*) **E:** AL BH BU CR GR HU MC YU **A:** TR
stenoderus Fairmaire, 1876: 50 (*Anoncodes*) **N:** AG
ustulatus Scopoli, 1763: 43 (*Cantharis*) **E:** AB AL AR AU BH BU BY CR CT CZ EN FR GE GG HU IT LT MC PL RO SK SL SP ST SV SZ TR UK YU **A:** IN TR WS
adustus Panzer, 1795: 279 (*Necydalis*)
axillaris Ménétriés, 1832: 212 (*Oedemera*)
collaris Panzer, 1795: 278 (*Necydalis*)
ferrugineus Schrank von Paula, 1776: 68 (*Leptura*)
flavipennis Motschulsky, 1872: 55 (*Oedechira*)
flaviventris Faldermann, 1837: 147
hybridus Rossi, 1794: 56 (*Necydalis*)
inlateralis Pic, 1930a: 1 (*Anoncodes*)
melanocephalus Fischer von Waldheim, 1829b: 370 (*Ditylus*) [HN]
paradoxus Faldermann, 1837: 147 (*Oedemera*)
zimmermanni Gistel, 1857: 71 (*Anoncodes*)
wartmanni Pic, 1894m: 180 (*Anoncodes*) **E:** SP **N:** AG MO

genus *Nacerdes* Dejean, 1834: 228 type species *Necydalis notata* Fabricius, 1792 (= *Cantharis melanura* Linnaeus, 1758)

subgenus *Allagatha* Semenov & Ter-Minasjan, 1937: 227 type species *Allagatha decora* Semenov & Ter-Minasjan, 1937 (= *Xanthochroa crassifemur* Blair, 1935)

crassifemur Blair, 1935b: 302 (*Xanthochroa*) **A:** KA PA

decora Semenov & Ter-Minasjan, 1937: 228 (*Allagatha*)

subgenus *Nacerdes* Dejean, 1834: 228 type species *Necydalis notata* Fabricius, 1792 (= *Cantharis melanura* Linnaeus, 1758)

Nacerda Stephens, 1839: 337 type species *Cantharis melanura* Linnaeus, 1758

Nacerdoscuta Pic, 1915c: 6 type species *Nacerdoscuta semirufa* Pic, 1915

brancuccii Švihla, 1983b: 336 **A:** AF KA PA

melanura Linnaeus, 1758: 403 (*Cantharis*) **E:** AB AL AU BE BH BU CR CZ DE EN FI FR GB GE GG GR HU IR IT LA LT MA MR NL NR NT PL PT RO SK SP ST SV SZ UK YU **N:** AG CI MO MR TU **A:** AF IS JA NC SC TAI TR **AFR AUR NAR NTR**

acuta Marsham, 1802: 372 (*Cantharis*)

analis G.-A. Olivier, 1811: 443 (*Oedemera*)

apicalis Say, 1835: 188 (*Oedemera*) [HN]

erminea Germar, 1817: 225 (*Necydalis*)

italica Chevrolat, 1877b: 121 (*Xanthochroa*)

lepturoides Thunberg, 1784: 18 (*Cantharis*)

nigripes Fabricius, 1781: 260 (*Cantharis*)

notata Fabricius, 1792b: 353 (*Necydalis*)

rufa Brullé, 1839: 70 (*Ditylus*)

sardea W.L.E. Schmidt, 1846: 34

testacea Geoffroy, 1785: 155 (*Cantharis*)

zoufali Reitter, 1907d: 161

semirufa Pic, 1915c: 6 (*Nacerdoscuta*) **A:** KA PA

subgenus *Xanthochroa* W.L.E. Schmidt, 1844: 46 type species *Oedemera carniolica* Gistel, 1834

Asiochroa Švihla, 1998a: 56 type species *Xanthochroa metallipennis* Fairmaire, 1889 (= *Xanthochroa waterhousei* Harold, 1875)

Axanthochroa Švihla, 1986a: 174 type species *Xanthochroa brachyptera* Švihla, 1980

Patiala Lewis, 1895a: 434 type species *Patiala antennata* Lewis, 1895

Patialomorpha Nakane, 1954: 174 type species *Xanthochroa ainu* Lewis, 1895

ainu Lewis, 1895a: 436 (*Xanthochroa*) **A:** FE (Sakhalin, Kuriles) JA

akiyamai Švihla, 1998a: 68 **A:** TAI

antennata Lewis, 1895a: 434 (*Patiala*) **A:** JA

apicipennis Švihla, 1998a: 50 **A:** TAI

arcuata Tian, Ren & Li, 2014: 114 **A:** ANH

asahinai Nakane, 1958: 91 (*Xanthochroa*) **A:** JA (Honshu, Shikoku)

atriceps atriceps Lewis, 1895a: 436 (*Xanthochroa*) **A:** FE JA SC

atriceps shikokuana Iga & Nakane, 1954: 154 (*Xanthochroa*) **A:** JA
atripennis Pic, 1934b: 29 (*Xanthochroa*) **A:** SCH YUN
baibarana Kôno, 1932a: 139 (*Xanthochroa*) **A:** TAI
becvari Švihla, 1998a: 52 **A:** YUN
bicostata Lewis, 1895a: 436 (*Xanthochroa*) **A:** JA
brachyptera Švihla, 1980: 46 (*Xanthochroa*) **A:** BT
brendelli Švihla, 1987: 11 **A:** NP SD XIZ
carniolica atlantica Allemand, 1993: 8 **A:** FR SP
carniolica carniolica Gistel, 1834b: 15c (*Oedemera*) **E:** AL AU BH BU BY CR CZ FR GE GG GR HU IT LS MC PL RO SK SL SP SV SZ YU
blossevillei Guérin-Méneville, 1838: 40 (*Oedemera*)
lippichii Schmidt, 1834: 62 (*Oedemera*)
carniolica foveata Fairmaire, 1892b: 157 (*Xanthochroa*) **A:** TR
carniolica peloponesiaca Švihla, 1991: 3 **E:** GR (Pelopponisos, Crete)
caroli Švihla, 2001c: 226 **A:** BT
caudata Kôno, 1936d: 517 (*Xanthochroa*) **A:** JA
dedicata Švihla, 2001c: 228 **A:** AP **ORR**
deformis Lewis, 1895a: 435 (*Patiala*) **A:** JA
impressithorax Pic, 1937b: 24 (*Xanthochroa*)
fujiana Švihla, 1998a: 66 **A:** FUJ
fulvicrus Fairmaire, 1889b: 50 (*Xanthochroa*) **A:** GAN GUI HUB SHA SCH SHX YUN
semenowi Ganglbauer, 1889: 37 (*Xanthochroa*)
fuscipennis Champion, 1927b: 247 (*Xanthochroa*) **A:** NP UP
gracilis bellieri Reiche, 1861c: 209 (*Xanthochroa*) **E:** FR SZ
testacea Pic, 1920g: 17 *Xanthochroa*) [HN]
tournieri Pic, 1894h: 72 (*Xanthochroa*)
gracilis gracilis W.L.E. Schmidt, 1846: 37 (*Xanthochroa*) **E:** AU BH CR FR GG GR HU IT RO SP ST SZ UK
grandis Pic, 1923c: 23 (*Xanthochroa*) **A:** BT NP "India"
guizhouensis Švihla, 2001c: 228 **A:** GUI
hesperica Magistretti, 1941b: 227 (*Xanthochroa*) **E:** SP
hilleri hilleri Harold, 1878a: 81 (*Xanthochroa*) **A:** FE (Kuriles) JA SC
hilleri kurosai Nakane, 1988a: 3 (*Xanthochroa*) **A:** JA (Hokkaido, Honshu)
himalaica himalaica Blair, 1930b: 181 (*Xanthochroa*) **A:** NP SD
hiromichii Švihla, 2004: 72 [RN] **A:** TAI
apicalis Kôno, 1932a: 141 (*Xanthochroa*) [HN]
holzschuhi Švihla, 1987: 10 **A:** NP XIZ
kantneri Švihla, 2001c: 230 **A:** TAI
katoi Kôno, 1932a: 140 (*Xanthochroa*) **A:** JA
konoi Nakane, 1954: 172 (*Xanthochroa*) **A:** JA
kubani Švihla, 1998a: 60 **A:** SCH YUN
ludmilae Švihla, 1998a: 52 **A:** HUB

luteipennis Marseul, 1877: 484 (*Xanthochroa*) **A:** FE JA NC NE SC
martensi Švihla, 1993a: 53 **A:** NP
mimoncomeroides Švihla, 1998a: 60 **A:** XIZ **ORR**
ocularis Lewis, 1895a: 435 (*Patiala*) **A:** JA SC
osawai Nakane, 1954: 173 (*Xanthochroa*) **A:** JA
potanini Ganglbauer, 1889: 36 (*Xanthochroa*) **A:** GAN SCH
raymondi canyellesi Vázquez, 2006: 40 **E:** SP (Mallorca)
raymondi ciliciensis Švihla, 1991: 6 **E:** GR (Rhodes) **A:** TR
raymondi raymondi Mulsant & Godart, 1860: 158 (*Xanthochroa*) **E:** FR SP **N:** AG
barbara Peyerimhoff, 1918b: 143 (*Xanthochroa*)
schneideri Švihla, 2001c: 227 **A:** SCH
spinicoxis hirai Nakane, 1954: 175 (*Xanthochroa*) **A:** JA (Shikoku)
spinicoxis spinicoxis Nakane, 1954: 174 (*Xanthochroa*) **A:** JA (Honshu, Shikoku)
strandi Kôno, 1936d: 516 (*Xanthochroa*) **A:** FE (Kuriles) JA
subviolacea Pic, 1911f: 101 (*Xanthochroa*) **A:** YUN
yunnana Pic, 1923c: 23 (*Xanthochroa*)
taiwana Kôno, 1932a: 140 (*Xanthochroa*) **A:** TAI
transfretalis Švihla, 1998a: 66 **A:** TAI
umenoi okinawana M. Chûjô & Satô, 1972: 26 (*Xanthochroa*) **A:** JA (Okinawa)
umenoi umenoi Kôno, 1937a: 3 (*Xanthochroa*) **A:** JA (Amami-Ôshima, Okinawa)
violaceonotata Pic, 1922c: 22 (*Xanthochroa*) **A:** YUN **ORR**
wadai Nakane, 1954: 171 (*Xanthochroa*) **A:** JA
wardi Švihla, 1987: 5 **A:** NP XIZ
waterhousei Harold, 1875: 93 (*Xanthochroa*) **A:** FE (Sakhalin) GAN HUB JA SC SCH SHA
cyanipennis Marseul, 1877: 483 (*Xanthochroa*) [HN]
metallipennis Fairmaire, 1889b: 49 (*Xanthochroa*)

genus *Opsimea* Miller, 1881: 224 type species *Opsimea ventralis* Miller, 1881
Ezonacerda Kôno, 1934b: 28 type species *Oedemera nigripennis* Matsumura, 1911
nigripennis Matsumura, 1911: 128 (*Oedemera*) **A:** FE (Sakhalin, Kuriles) JA
ventralis Miller, 1881: 225 **E:** AL BH BU CR GG GR ST (Kavkaz) YU **A:** TR
parallela Fairmaire, 1892b: 158 (*Stenaxis*)
quadrinervosa Reiche, 1862b: 296 (*Oedemera*) [HN]

tribe Oedemerini Latreille, 1810

genus *Dryopomera* Fairmaire, 1897j: 238 type species *Dryops indica* Fairmaire, 1896
subgenus *Dryopomera* Fairmaire, 1897j: 238 type species *Dryops indica* Fairmaire, 1896
Paroncomera Nakane, 1954: 179 type species *Paroncomera yatoi* Nakane, 1954
havai Švihla, 2001b: 128 **A:** SHA
kurosai H. Akiyama, 1998: 167 **A:** JA (Ryukyus) TAI
longissima Švihla, 1994: 244 **A:** NP SD

magna Švihla, 1999a: 84 **A:** NP
nepalensis Švihla, 1994: 246 **A:** NP
taiwana Švihla, 1997a: 89 **A:** TAI
yatoi tokarensis Nakane, 1954: 180 (*Paroncomera*) **A:** JA (Tokara Is.)
yatoi yatoi Nakane, 1954: 179 (*Paroncomera*) **A:** JA (Honshu, Shikoku, Kyushu)

genus *Oedemera* G.-A. Olivier, 1789: 31 type species *Necydalis caerulea* Linnaeus, 1767 (= *Cantharis nobilis* Scopoli, 1763)
Oedemerastra Seidlitz, 1899: 923 type species *Necydalis podagrariae* Linnaeus, 1767
Oedemerella Seidlitz, 1899: 933 type species *Oedemera croceicollis* Gyllenhal, 1827
Oedemerina A. Costa, 1852: 31 type species *Necydalis lurida* Marsham, 1802
Oedemeronia Seidlitz, 1899: 919 type species *Necydalis flavipes* Fabricius, 1792
Stenolytra Dilwyn, 1829: 61 type species *Necydalis caerulea* Linnaeus, 1767 (= *Cantharis nobilis* Scopoli, 1763)
subgenus *Oedemera* G.-A. Olivier, 1789: 31 type species *Necydalis caerulea* Linnaeus, 1767 (= *Cantharis nobilis* Scopoli, 1763)
Oedemerastra Seidlitz, 1899: 923 type species *Necydalis podagrariae* Linnaeus, 1767
Oedemerella Seidlitz, 1899: 933 type species *Oedemera croceicollis* Gyllenhal, 1827
Oedemerina A. Costa, 1852: 31 type species *Necydalis lurida* Marsham, 1802
Oedemeronia Seidlitz, 1899: 919 type species *Necydalis flavipes* Fabricius, 1792
Stenolytra Dilwyn, 1829: 61 type species *Necydalis caerulea* Linnaeus, 1767 (= *Cantharis nobilis* Scopoli, 1763)
alaiensis Semenov, 1894c: 544 **A:** KI KZ UZ
algerica Pic, 1899h: 116 **E:** IT (Lampedusa Is.) **N:** AG LB MO TU
abdominalis Pic, 1897e: 42 [HN]
pici Schenkling, 1915: 58 [RN]
artvinensis Švihla, 1999b: 59 **A:** TR
atrata W.L.E. Schmidt, 1846: 77 **E:** AL CR FR GR IT SL SP YU **A:** TR
atriceps Abeille de Perrin, 1880: 261 **A:** IS JO
atrocyanea Švihla, 1999b: 12 **A:** KZ
barbara Fabricius, 1792b: 351 (*Necydalis*) **E:** CR FR GR IT MA PT SP TR **N:** AG MO SG TU **A:** TR
dalmatina Pic, 1898f: 73
subvirescens Pic, 1920c: 5
basalis Küster, 1849b: 70 **E:** SP **N:** TU
basipes Abeille de Perrin, 1892: 63 **A:** CY
brevipennis Ganglbauer, 1881: 108 **E:** GR (Sporades) TR **A:** SY TR
akbesiana Pic, 1901b: 11
pulchripes J.R. Sahlberg, 1908: 93
centrochinensis centrochinensis Švihla, 1999b: 15 **A:** SCH SHA
centrochinensis lixiana Švihla, 2003: 341 **A:** SCH
concolor Lewis, 1895a: 442 **A:** FE (Sakhalin, Kuriles) JA (Hokkaido, Honshu, Tsushima)

crassipes Ganglbauer, 1881: 114 **E:** AL BU CR FR GR IT MA PT SL SP TR **N:** AG LB MO TU TR **A:** CY IS JO LE SY TR
caudata Seidlitz, 1899: 919
cretica Švihla, 1999b: 33 **E:** GR (Crete)
croceicollis Gyllenhal, 1827: 523 (*Necydalis*) **E:** AB AR AU BE BU CR CT CZ DE EN FI FR GE GG GR HU IT LT NL PL RO SK SL ST SV SZ TR UK **A:** ES KZ
sarmatica F.F. Morawitz, 1861: 292
cuprata Reiche, 1864: 245 **N:** AG MO
atlasica Pic, 1932g: 28
danieli Magistretti, 1941b: 232 **E:** GG
femorata Scopoli, 1763: 45 (*Cantharis*) **E:** AB AL AN AR AU BE BU BY CR CT CZ DE EN FI FR GE GG GR HU IT LA LS LT MC MD NR NT PL RO SK SL SP ST SV SZ TR UK YU **A:** ES IN KZ TM TR WS
fallax Seidlitz, 1899: 913
flavescens Linnaeus, 1767: 642 (*Necydalis*)
praterrana Schrank von Paula, 1781: 168 (*Necydalis*)
similis W.L.E. Schmidt, 1846: 59
flavipennis W.L.E. Schmidt, 1846: 62 **E:** GR (Sporades) **A:** CY IS JO LE SY TR
coarcticollis Abeille de Perrin, 1880: 260
flavipes Fabricius, 1792b: 355 (*Necydalis*) **E:** AL AN AU BH BU CR CT CZ FR GE GR HU IT LA LS LT LU MC NL NR PL PT RO SK SL SP SV SZ TR UK YU **A:** IS JO LE SY TR
aenea G.-A. Olivier, 1795: 14
clavipes Fabricius, 1801b: 373 (*Necydalis*)
delagrangei Pic, 1923i: 13
pusilla A. Costa, 1852: 29
graeca Švihla, 1978: 38 **E:** AL GR MC
hispanica Pic, 1904h: 91 **E:** SP
laufferi Reitter, 1905c: 206
inapicalis Pic, 1910o: 65 **A:** CY
jucunda Heyden, 1890c: 363 **A:** AF TD UZ
kantneri Švihla, 2004: 71 **A:** SY
kocheri Pic, 1934d: 21 **N:** MO
nigra Kocher, 1953: 51
lateralis Gebler, 1829a: 131 **E:** AB BU CR FR GG HU IT RO SP ST SZ UK YU **A:** IN TM WS
caucasica Kolenati, 1847: 142
impressithorax Pic, 1920c: 6
korbi Seidlitz, 1899: 918
lucidicollis flaviventris Fairmaire, 1891f: ccxix **A:** FUJ GUI HEB HEI HEN HUB HUN JIX NC SCH SHA SHN ZHE
coreana Pic, 1926e: 6
sieversi Seidlitz, 1899: 919

lucidicollis lucidicollis Motschulsky, 1866: 174 **A:** FE (Kuriles) JA

montana Marseul, 1877: 485

vitticollis Motschulsky, 1866: 174 (*Oncomera*)

lurida lurida Marsham, 1802: 360 (*Necydalis*) **E:** AB AL AN AR AU BE BU BY CR CZ DE EN FI FR GB GE GG GR HU IR IT KZ LA LT LU MC NL NR NT PL PT RO RU SK SL SP ST SV SZ TR UK YU **A:** IN SY TM TR

viridula Seidlitz, 1899: 919

lurida sinica Švihla, 1999b: 41 **A:** SCH YUN

manicata Lewis, 1895a: 442 **A:** JA

masatakai Švihla, 2003: 341 **A:** SCH

melanopyga W.L.E. Schmidt, 1846: 66 **E:** IT

maculiventris A. Costa, 1852: 24

merkli Seidlitz, 1899: 919 **A:** TR

mogadorica Escalera, 1913: 56 **N:** MO

basalis Chevrolat, 1877b: 121 [HN]

rufina Escalera, 1913: 56

theryi Pic, 1936e: 30

monticola Švihla, 1978: 35 **E:** AU CZ FR GE IT PL RO SK SL SZ UK

nasalis Reitter, 1896k: 290 **A:** KI KZ UZ XIN

atrofemoralis Pic, 1915a: 2

rostrifera Reitter, 1896k: 290

nigripes Ganglbauer, 1889: 44 (*Stenaxis*) **A:** GAN SCH

nigroapicata Švihla, 1982: 384 **A:** IN

nobilis Scopoli, 1763: 45 (*Cantharis*) **E:** AL AU BE CR CZ DE FR GB GE GR IT LU MC NL PT SL SP SZ YU **N:** AG MO TU

caerulea Linnaeus, 1767: 642 (*Necydalis*)

ceramboides Forster, 1771: 47 (*Necydalis*)

gallica Pic, 1920c: 6

magica Dalla Torre, 1880: 23

sirgueyi Pic, 1927e: 5

subviolacea Pic, 1901b: 11

viridescens Pic, 1920c: 6

obscuripes Pic, 1912r: 220 **N:** MO

pallidicolor Pic, 1899h: 116 **N:** AG

pallidipes angustipennis Gressitt, 1939: 224 **A:** FUJ GUA

pallidipes pallidipes Pic, 1907j: 174 **A:** SCH

pallidipes shaanxiensis Švihla, 2003: 341 **A:** HUB SHA

parallelipennis Seidlitz, 1899: 919 **A:** FE NC

penicillata W.L.E. Schmidt, 1846: 54 **E:** BU GG GR MC RO TR **A:** TR

petroschmidti L. Heyden, 1894b: 265 **A:** KI KZ UZ

confusa Reitter, 1895d: 158

uniformis Seidlitz, 1899: 915

podagrariae acutipalpis Abeille de Perrin, 1880: 259 **A:** IS TR
podagrariae podagrariae Linnaeus, 1767: 642 (*Necydalis*) **E:** AB AN AR AU BE BH BU BY CR CZ FR GE GG GR HU IT LA LS LT MA MC MD NL PL PT RO SK SL SP ST SV SZ TR UK YU **A:** IN TM TR
discoidalis Pic, 1921a: 5 [HN]
distinctipes Pic, 1921e: 10
diversipes Pic, 1921a: 5
femoralis Seidlitz, 1899: 925 [HN]
flavicrus Seidlitz, 1899: 925
fulva Geoffroy, 1785: 155 (*Cantharis*)
incerta Mulsant, 1858a: 192
inlateralis Pic, 1921e: 10
limbaticollis Pic, 1921a: 5
melanocephala Panzer, 1795: 277 (*Necydalis*) [HN]
reducta Pic, 1921a: 5
sebastiani Pic, 1901b: 10
sericans Mulsant, 1858a: 190
simplex Donovan, 1801: 90 (*Necydalis*) [HN]
testacea Fabricius, 1792b: 355 (*Necydalis*)
podagrariae ventralis Ménétriés, 1832: 212 **E:** AB **A:** IN TM
immunda Reitter, 1885d: 386
menetriesii Faldermann, 1837: 144
obscura Ganglbauer, 1881: 109
schmidti Gemminger, 1870a: 124 [RN]
ventralis W.L.E. Schmidt, 1846: 55 [HN]
pthysica Scopoli, 1763: 44 (*Cantharis*) **E:** AB AN AR AU BE BU CR CZ DE FR GE GG GR HU IT LU MC NL PL PT RO SK SL SP ST SV SZ TR UK **A:** IN TR
angusticollis A. Costa, 1852: 20
discoidalis Latreille, 1804b: 13
femorata Panzer, 1795: 279 (*Necydalis*) [HN]
marginata Fabricius, 1798: 155 (*Necydalis*)
stenoptera Faldermann, 1837: 146
subulata G.-A. Olivier, 1795: 15
vittata J. Frivaldszky, 1877c: 325
qinlingensis Švihla, 1999b: 66 **A:** GAN HUB SCH SHA
reini L. Heyden, 1887d: 448 **N:** MO
robusta Lewis, 1895a: 443 **A:** FE HEI JA (Honshu) NC
laticollis Seidlitz, 1899: 919
rostralis deubeli Ganglbauer, 1897: 570 **E:** RO UK (Carpathes)
rostralis rostralis Reitter, 1885d: 386 **E:** AB AR GG ST (Kavkaz) UK **A:** IN TR

rufofemorata euboica Pic, 1902h: 64 **E:** GR **A:** TR
rufofemorata pruinosa Abeille de Perrin, 1880: 259 **A:** IS LE SY TR
rufofemorata rubricollis Reitter, 1885d: 386 **A:** TR
rufofemorata rufofemorata Germar, 1823: 166 **E:** AL AU BH BU CR GR MC SL YU
brevicornis W.L.E. Schmidt, 1846: 78 [HN]
cyanescens W.L.E. Schmidt, 1846: 72
femorata Brullé, 1832: 227 [HN]
satoi Švihla, 2003: 339 **A:** SHA
sauteri luteipes Švihla, 1999b: 18 **A:** TAI
sauteri sauteri Pic, 1927i: 48 **A:** TAI
schrammi Pic, 1913k: 137 **E:** SP
sexualis ryukyuana Švihla, 1999b: 19 **A:** JA (Ryukyus)
sexualis sexualis Marseul, 1877: 486 **A:** JA (Honshu, Shikoku, Kyushu, Tsushima, Goto Is.)
sichuana Švihla, 1999b: 65 **A:** SCH SHA
simplex Linnaeus, 1767: 643 (*Necydalis*) **E:** FR IT MA PT SP **N:** AG LB MO TU
basipennis Pic, 1936e: 30
basitestacea Pic, 1936e: 30
brevicollis W.L.E. Schmidt, 1846: 56
flavimana W.L.E. Schmidt, 1846: 57
tibialis P. Lucas, 1846: pl. 34
subrobusta Nakane, 1954: 180 (*Oedemerina*) **E:** AR CT CZ EN FI FR GG HU LA LT MC NR NT PL RO SK ST SV SZ **A:** ES FE GAN HUB IN JA (Honshu) KZ MG NC NMO QIN SCH SHA TM TR WS
testaceithorax okinawana Kôno, 1937a: 11 **A:** JA (Amami-Ôshima, Okinawa)
testaceithorax testaceithorax Pic, 1927i: 49 **A:** FUJ GUA GUI JA (Ishigaki-jima, Iriomote-jima, Yonaguni-jima) JIX SCH TAI **ORR**
tristis W.L.E. Schmidt, 1846: 75 **E:** AU BE CR CZ FR GE HU IT LS LU PL RO SK SL SZ
unicolor W.L.E. Schmidt, 1846: 74 **E:** PT SP **N:** MO
olcesei Pic, 1898h: 121
tangeriana Pic, 1920c: 6
vilis pontica Švihla, 1999b: 44 **A:** TR
vilis siberoaltaica Švihla, 1999b: 45 **A:** ES WS
vilis vilis Roubal, 1917a: 44 **E:** AB AR GG
virescens chalybea Faldermann, 1837: 149 **E:** AR ST (Kavkaz)
virescens virescens Linnaeus, 1767: 650 (*Cantharis*) **E:** AB AN AU BE BH BU BY CR CT CZ DE EN FI FR GB GE GR HU IT LA LT LU MC NL NR NT PL RO SK SL SP ST SV SZ UK YU **A:** ES FE HEI IN JA (Honshu) KZ MG SHA SHX WS
abdominalis G.-A. Olivier, 1795: 50: 14
cupreomicans Reitter, 1905a: 93
striata Herbst, 1784b: 109 (*Necydalis*)

subgenus *Oncomera* Stephens, 1829b: 251 type species *Dryops femorata* Fabricius, 1792 (= *Oedemera femoralis* G.-A. Olivier, 1803)

Oncomerella Reitter, 1911a: 406 type species *Oedemera marmorata* Erichson, 1841

Oncomerina Seidlitz, 1899: 885 type species *Oedemera murinipennis* Kiesenwetter, 1859

antoinei Pic, 1932h: 30 (*Oncomera*) **N:** MO

mimeuri Pic, 1949c: 74 (*Oncomera*)

bomfordi Fairmaire, 1897j: 237 (*Dryops*) **A:** KA NP UP

diversecostata afghana Švihla, 1993b: 197 **A:** AF PA

diversecostata diversecostata Pic, 1915c: 6 (*Oncomera*) **A:** KA PA **ORR**

grandis Pic, 1915c: 6 (*Oncomera*)

farsica demissa Švihla, 1999a: 83 **A:** IN

farsica farsica Švihla, 1983a: 124 (*Oncomera*) **A:** IN

femoralis femoralis G.-A. Olivier, 1803: 21 **E:** AN AU BE BH BU CR FR GB GE GR HU IT NL PT RO SL SP SZ UK **A:** TR

acutiventris Abeille de Perrin, 1896: 284 (*Oncomera*)

calopoides Germar, 1817: 225 (*Necydalis*)

femorata Fabricius, 1792b: 74 (*Dryops*) [HN]

femoralis purpureocoerulea Ganglbauer, 1882: 137 (*Oncomera*) **E:** BH CR GR YU

flavicans Fairmaire, 1860: 632 (*Dryops*) **E:** BU GR MC YU **A:** CY IS JO SY TR

parnassica Pic, 1920c: 5 (*Oncomera*)

stylifera Abeille de Perrin, 1896: 284 (*Oncomera*)

marmorata Erichson, 1841: 185 **E:** PT SP **N:** AG MO TU

murinipennis karpathosensis Švihla, 1999a: 83 **E:** GR (Karpathos)

murinipennis murinipennis Kiesenwetter, 1859: 192 **E:** GR (Crete)

paganettii Stolz, 1915: 91

natolica Reiche, 1862c: 544 (*Oncomera*) **E:** GR **N:** EG **A:** IS LE SY TR

apicipennis Fairmaire, 1892b: 157

libanica Pic, 1920c: 6

pallida Fleischer, 1919a: 170

rhodosica Fleischer, 1919a: 171

nepalensis Švihla, 1993b: 194 **A:** NP

reitteri Švihla, 1993b: 199 [RN] **E:** GR

atriceps Reitter, 1894e: 303 (*Oncomera*) [HN]

venosa Lewis, 1895a: 441 (*Oncomera*) **A:** FE JA

subgenus *Stenaxis* W.L.E. Schmidt, 1844: 46 type species *Oedemera annulata* Germar, 1824

amurensis L. Heyden, 1884a: 295 **A:** FE HEI NC SHX

obscurior Pic, 1904d: 27

annulata Germar, 1824b: 166 **E:** AL AU BH BU CR HU IT MC RO SL YU

tribe Stenostomatini Mulsant, 1858

genus *Stenostoma* Latreille, 1810: 217 type species *Leptura rostrata* Fabricius, 1787

cossyrense Bologna, 1995b: 516 **E:** IT (Pantelleria Is.)
lowei Wollaston, 1854: 524 (*Stenaxis*) **N:** MR
melitense Cameron, 1907: 226 **E:** MA
rostratum rostratum Fabricius, 1787a: 159 (*Leptura*) **E:** GR (Kós, Crete, Rhodes) PT SP **N:** AG MO TU **A:** CY IS
schatzmayri Magistretti, 1941a: 77
rostratum septentrionale Švihla, 2005a: 72 [RN] **E:** AL CR FR GR (Kerkyra, Peloponnisos) IT PT SP YU
caeruleum V. Petagna, 1787: 14 (*Rhinomacer*) [HN]
ceramboides Rossi, 1794: 99 (*Necydalis*) [HN]

family STENOTRACHELIDAE C.G. Thomson, 1859

subfamily **Stenotrachelinae** C.G. Thomson, 1859

genus *Scotodes* Eschscholtz, 1818: 454 type species *Scotodes annulatus* Eschscholtz, 1818
Pelmalopus Fischer von Waldheim, 1824: 163 type species *Pelmalopus hummeli* Fischer von Waldheim, 1824
annulatus Eschscholtz, 1818: 454 **E:** BY CT EN FI LA LI NT **A:** ES FE HEI JA MG SC WS
hummeli Fischer von Waldheim, 1824: 163 (*Pelmalopus*)
niponicus Lewis, 1895a: 275
uniformis Motschulsky, 1872: 46 **A:** FE (Kamchatka)

genus *Stenocephaloon* Pic, 1932a: 2 type species *Stenocephaloon metallicum* Pic, 1932
metallicum Pic, 1932a: 2 **A:** JA

genus *Stenotrachelus* Berthold, 1827: 371 type species *Dryops aeneus* Paykull, 1799
aeneaus Paykull, 1799: 152 (*Dryops*) **E:** FI NR NT SW **A:** ES FE **NAR**
arctatus Say, 1824b: 283 (*Helops*)
fennicus Weber & Mohr, 1804: 182 (*Dryops*)
obscurus Mannerheim, 1852: 347
rouillieri Motschulsky, 1845a: 84

subfamily **Nematoplinae** LeConte, 1862

genus *Nematoplus* LeConte, 1855: 275 type species *Nematoplus collaris* LeConte, 1855
konoi Nakane, 1963a: 30 (*Pedilocephaloon*) **A:** JA
semenovi Nikitsky, 1973a: 611 **A:** FE
yamato Nakane, 1987: 176 **A:** JA

subfamily **Cephaloinae** LeConte, 1862

genus *Cephaloon* Newman, 1838: 376 type species *Cephaloon lepturides* Newman, 1838
pallens Motschulsky, 1860c: 140 (*Cephalaon*) **A:** JIL FE JA NC SC

sakurai Lewis, 1895a: 444 **A:** JA
variabile Motschulsky, 1860c: 141 (*Cephalaon*) **A:** CH FE NC

subfamily **Stoliinae** Nikitsky, 1985

genus *Stolius* Lewis, 1895a: 433 type species *Stolius vagepictus* Lewis, 1895
Falsostolius Pic, 1932d: 27 type species *Falsostolius diversipes* Pic, 1932
diversipes Pic, 1932d: 27 (*Falsostolius*) **A:** JA
vagepictus Lewis, 1895a: 433 **A:** JA

family MELOIDAE Gyllenhal, 1810

subfamily **Eleticinae** Wellman, 1910

tribe Derideini Wellman, 1910

genus *Deridea* Westwood, 1875: 226 type species *Deridea curculionides* Westwood, 1875
notata Thomas, 1897a: 389 **A:** SA YE
minor Kaszab, 1960c: 153

tribe Eleticini Wellman, 1910

genus *Eletica* Dejean, 1834: 224 type species *Lytta rufa* Fabricius, 1801 (= *Lytta bipustulata* Fabricius, 1801)
Iletica Gemminger, 1870b: 2145 [UE]
subgenus *Meteletica* Kaszab, 1955a: 18 type species *Cantharis testacea* G.-A. Olivier, 1789
testacea G.-A. Olivier, 1789: 277 (*Cantharis*) **A:** PA SD **ORR**
bicolor Schoenherr, 1817: 23 (*Lytta*)
gigantea Dohrn, 1873: 70
rufocincta Pic, 1915d: 8

subfamily **Meloinae** Gyllenhal, 1810

tribe Cerocomini Leach, 1815

genus *Anisarthrocera* Semenov, 1896e: 517 type species *Rhampholyssa batesi* Marseul, 1872
batesi Marseul, 1872a: 416 (*Rhampholyssa*) **A:** IN IQ SA
villiersi Kaszab, 1968f: 750

genus *Cerocoma* Geoffroy, 1762: 357 type species *Meloe schaefferi* Linnaeus, 1758
subgenus *Cerocoma* Geoffroy, 1762: 357 type species *Meloe schaefferi* Linnaeus, 1758
bernhaueri Pardo Alcaide, 1977: 64 **E:** AB **A:** IN IS JO LE TR
dahli Kraatz, 1863: 112 **E:** AB AR AU BU GG GR RO TR **A:** IN IS JO LE SY TR
nigrivestis Muche, 1963a: 13
prochaskana Reitter, 1896j: 267 **A:** IN IQ IS JO SY TR
schaefferi Linnaeus, 1758: 420 (*Meloe*) **E:** AB AL AR AU BE BH BU BY CT CZ FR GE GR HU IT KZ LA MC MD NL PL PT RO SK SL SP ST SZ UK YU **A:** TR
affinis Baudi di Selve, 1878b: 358

obscuripes Motschulsky, 1872: 49
orensis Reitter, 1914f: 193
viridis Geoffroy, 1785: 163
viridula Reitter, 1885a: 14
simplicicornis Reitter, 1914f: 193 **E:** AB AR "Caucasus" **A:** IN TR
aeneipes Reitter, 1914f: 193
subgenus *Cerocomina* Kaszab, 1951d: 259 type species *Cerocoma vahli* Fabricius, 1787
vahli taraudanti Escalera, 1913: 55 **N:** MO
vahli vahli Fabricius, 1787a: 217 **N:** AG MO TU
chalybaeiventris Chevrolat, 1840b: 268
wagneri Küster, 1844: 32
subgenus *Meloides* Piller & Mitterpacher, 1783: 39 type species *Meloides adamovichiana* Piller & Mitterpacher, 1783
adamovichiana Piller & Mitterpacher, 1783: 39 (*Meloides*) **E:** AL AU BH BU CR CZ GG GR HU MC RO SK TR YU
cuprea Baudi, 1878b: 359
imitatrix Maran, 1944: 91
schraderi Kraatz, 1863: 111
albopilosa Dvořák, 1993a: 7 **A:** IN TR
azurea Reitter, 1914f: 192 (*Cerocoma*) **A:** CY IN IS JO LE SY
coeruleotincta Muche, 1963a: 13
humilis Muche, 1963a: 12
barthelemyi Baudi di Selve, 1878b: 358 **E:** AR GR **A:** IN IS SY TR
syriaca Abeille de Perrin, 1880: 235
haifensis Kaszab, 1951d: 268
bodemeyeri Reitter, 1909c: 103 **A:** IN IQ SY
obscuripes Reitter, 1885a: 14
confusa Turco & Bologna, 2011: 31 **E:** AB AL AR BU CT GG GR SK **A:** IN IS JO LE SY TR
gloriosa Mulsant, 1857a: 103 **E:** AB AR **A:** SY TR
pseudogloriosa Muche, 1963a: 14
vasvarii Kaszab, 1941f: 680
graeca Mařan, 1944c: 99 **E:** AL GR MC
kunzei I. Frivaldszky, 1835: 265 **E:** AL AR BH BU CR GG GR MC RO TR YU "Caucasus" **A:** IN IS JO SY TM TR
iranica Muche, 1963a: 11
palaestinensis Muche, 1963a: 11
patelligera Baudi di Selve, 1878b: 358
patkaii Kaszab, 1941: 679
longiseta Turco & Bologna, 2011: 33 **E:** BU **A:** TR
macedonica Mařan, 1944c: 98 **E:** BH BU GR MC TR YU **A:** TR
malatyensis Kaszab, 1951d: 266 **E:** "Caucasus" **A:** TR

marginiventris Reitter, 1889a: 34 **E:** AB AR **A:** IN TR

pictiventris Reitter, 1890e: 174

muehlfeldi Gyllenhal, 1817: 13 **E:** AB AL AR AU BH BU CR CT CZ GE GR HU MC MD PL RO SK UK YU **A:** IN IQ IS JO SY TM TR

faldermanni Laporte, 1840: 267

gonocera Motschulsky, 1872: 49

micans Ménétriés, 1832: 206

rapillyi Pardo Alcaide, 1977: 61 **A:** IN IQ

turcica Pardo Alcaide, 1977: 60 **E:** AR BU GG GR MC RO ST **A:** IN TR

subgenus *Mesocerocoma* Kaszab, 1951d: 259 type species *Cerocoma scovitzii* Faldermann, 1837

latreillei latreillei Baudi di Selve, 1878b: 359 **A:** IN IQ

collaris Reitter, 1885a: 12

latreillei sterbai Mařan, 1944c: 95 **A:** IN

schah Kaszab, 1968f: 751

scovitzi intermedia Mařan, 1944c: 94 **N:** EG **A:** IQ IN IS JO SA SY TR

jureceki Mařan, 1944c: 94

marani Kaszab, 1951d: 268

scovitzi scovitzi Faldermann, 1837: 117 **E:** AB AR "Caucasus" "Transcaucasia" **A:** IN

lateralis Reitter, 1890e: 174

mirabilis Dvořák, 1993a: 7

picticornis Motschulsky, 1872: 49

rufiventris Reitter, 1890e: 173

subgenus *Metacerocoma* Kaszab, 1951d: 259 type species *Cerocoma schreberi* Fabricius, 1781

ephesica Reitter, 1885a: 12 **E:** AB AR BU GR MC TR **A:** IN TR

festiva Faldermann, 1837: 118 **E:** AB AR **A:** IN IQ KZ TM TR UZ

martae Turco & Bologna, 2011: 48 **E:** BU MC **A:** TR

prevezaensis Dvořák, 1993a: 4 **E:** AL GR

schreberi Fabricius, 1781: 331 **E:** AB AL AU BH BU CR CT CZ FR GE GG GR HU IT KZ MC MD PL PT RO SK SL SP ST TR UK YU **A:** AF KY KZ MG TD TM TR UZ WS XIN

genus *Diaphorocera* Heyden, 1863: 126 type species *Diaphorocera hemprichi* Heyden, 1863

carinicollis Chobaut, 1921: 298 **N:** AG TU

chrysoprasis Fairmaire, 1863: 644 **N:** AG LB MO TU **AFR**

kerimii Fairmaire, 1875a: 530

hemprichi hemprichi Heyden, 1863: 127 **N:** AG EG MO TU **A:** IS JO

hemprichi saudita Kaszab, 1983a: 179 **A:** AE IN KU OM SA

johnsoni Kaszab, 1983a: 180 **A:** AE KU SA

neglecta Černý & Bologna, 2018: 59 **N:** MO

obscuritarsis Fairmaire, 1885a: xxxviii **N:** AG EG MO TU **A:** IS JO SI **AFR**

ruficornis Pic, 1923g: 10

peyerimhoffi Kocher, 1954: 284 **N:** MO

promelaena Fairmaire, 1876: 69 **N:** AG MO TU **A:** SI
sicardi Bedel, 1918: 364 **N:** AG LB MO TU

genus *Rampholyssa* Kraatz, 1863: 110 type species *Cerocoma stevenii* Fischer von Waldheim, 1823
antennata Reitter, 1906b: 36 **A:** TM UZ
stevenii Fischer von Waldheim, 1823: pl. 41 (*Cerocoma*) **E:** KZ ST **A:** AF IN KI KZ PA TD TM UZ XIN
intermedia Semenov, 1896e: 518
komarowi Reitter, 1889b: 119
pallida Semenov, 1896e: 519
plagicollis Reitter, 1909c: 102
varentzowi Semenov, 1896e: 518

genus *Rhampholyssodes* Kaszab, 1983a: 182 type species *Rhampholyssodes pitcheri* Kaszab, 1983
pichteri Kaszab, 1983a: 182 **A:** AE SA

tribe Epicautini Parker & Böving, 1924

genus *Cyaneolytta* Péringuey, 1909: 249 type species *Lytta signifrons* Fåhraeus, 1870
coerulea Pfaff, 1834: 209 (*Lytta*) **A:** PA SD UP **ORR**
acteon Laporte, 1840: 273 (*Lytta*)
fryi Wollaston, 1861a: 252 (*Cantharis*) **A:** SA **AFR**
holocyanea Fairmaire, 1896c: 225 (*Cantharis*)
saphirina Mäklin, 1875: 609 (*Cantharis*)
maculifrons Mäklin, 1875a: 608 (*Cantharis*) **A:** SA YE **AFR**
frontalis Kolbe, 1883: 24 (*Lytta*)
frontosa Wellman, 1912: 34 (*Lytta*)
violacea Brandt & Ratzenburg, 1833: 123 (*Lytta*) **A:** UP **ORR**

genus *Denierella* Kaszab, 1952a: 81 type species *Cantharis incompleta* Fairmaire, 1896
emmerichi Pic, 1934f: 86 (*Epicauta*) **A:** FUJ HAI HUB JIX SCH ZHE
apicipennis Tan, 1958: 159, 165 (*Epicauta*)
serrata Kaszab, 1952a: 88
minutiserra Tan, 1988: 291 **A:** GUX NP SD XIZ YUN **ORR**
promerotricha Dvořák, 1996: 37 (*Epicauta*) **A:** NP UP **ORR**
thailandica Dvořák, 1996: 40 (*Epicauta*) **A:** YUN **ORR**
striolata Yang & Ren, 2007: 54

genus *Epicauta* Dejean, 1834: 224 type species *Meloe erythrocephalus* Pallas, 1771
subgenus *Epicauta* Dejean, 1834: 224 type species *Meloe erythrocephalus* Pallas, 1771
Anomalonyx Denier, 1935: 161 type species *Lytta fumosa* Germar, 1824
Anomalonychus Saylor, 1940: 46 [RN]

Apterospasta LeConte, 1862: 272 type species *Lytta segmenta* Say, 1823

Causima Dejean, 1834: 226 type species *Lytta vidua* Klug, 1825

Gnathospasta Horn, 1875: 154 type species *Gnathospasta mimetica* Horn, 1875

Henous Haldeman, 1852: 377 type species *Henous techanus* Haldeman, 1852 (= *Meloe conferta* Say, 1824)

Isopentra Mulsant & Rey, 1858a: 180 type species *Lytta megalocephala* Gebler, 1817

Maculicauta Dillon, 1952: 416 type species *Epicauta stuarti* LeConte, 1868

Nomaspis LeConte, 1866: 156 type species *Meloe parvus* Haldeman, 1852: 337 (=*Meloe parvulus* Haldeman, 1854)

Pleuropompha LeConte, 1862: 273 type species *Lytta costata* LeConte, 1854

aethiops Latreille, 1827: 286 (*Cantharis*) **N:** EG **AFR**

ambusta Pallas, 1781: 102 (*Meloe*) **A:** ES MG NMO

aptera Kaszab, 1952e: 590 **A:** FUJ GUI GUX JIA SCH YUN ZHE

assamensis Waterhouse, 1871: 407 (*Cantharis*) **A:** FUJ NP **ORR**

atkinsoni Kraatz, 1880a: 16 **A:** AF IN PA UP "Himalaya" **ORR**

niveolineata Haag-Rutenberg, 1880: 82, 85 (*Lytta*)

badeni badeni Haag-Rutenberg, 1880: 82 (*Lytta*) **A:** TAI XIN

taipina Kôno, 1936b: 90

badeni sinica Kaszab, 1960d: 256 **A:** YUN

brevitibialis Kaszab, 1952e: 589 **A:** YUN

cheni Tan, 1958: 164 **A:** JIX **ORR**

cryptogramaca Yang & Ren, 2006: 272 **A:** YUN

curvispina Kaszab, 1952e: 587 **A:** FUJ GUI GUX SCH

dalihodi Dvořák, 1996: 39 **A:** TAI

erythrocephala Pallas, 1771: 466 (*Meloe*) **E:** AB AL AR BU GR RO SK ST TR UK **A:** AF GUA HAI IN KI KZ TD WS TM TR UZ XIN

albivittis Pallas, 1781: 101 (*Meloe*)

latelineolata Mulsant & Rey, 1858a: 177

lineata Thunberg, 1791: 109 (*Meloe*)

sonchi Gmelin, 1790: 1896 (*Meloe*)

fasciceps Walker, 1871: 16 **N:** EG **A:** YE **AFR**

nigronotata Haag-Rutenberg, 1880: 67 (*Lytta*)

flabellicornis Germar, 1817: 208 (*Lytta*) **E:** BH CR IT (Venezia Giulia) SL YU

depressicornis Laporte, 1840: 274 (*Lytta*)

sibirica Reitter, 1905b: 194 [HN]

gorhami Marseul, 1873a: 227 (*Cantharis*) **A:** AHN FUJ GUA GUI GUX HAI HEN HKG HUB HUN JA JIA JIX SCH SHX SHN TAI ZHE "Korea" **ORR**

taishoensis Lewis, 1879: 464

himalayica Kaszab, 1960g: 283 **A:** UP

hirticornis Haag-Rutenberg, 1880: 81 (*Lytta*) **A:** AP FUJ GUA GUI GUX HAI HEN JA SCH SD TAI XIZ YUN **ORR**

afra Kaszab, 1960d: 257

hirtipes Waterhouse, 1871: 406 (*Cantharis*) **A:** IN KI KZ NP PA SD UP
impressicornis Pic, 1913r: 163 (*Lytta*) **A:** GUI GUX YUN
indiana Kaszab, 1956e: 635 **A:** NP **ORR**
insularis Haag-Rutenberg, 1880: 82 (*Lytta*) **A:** SCH **ORR**
interrupta Fairmaire, 1889b: 48 (*Lytta*) **A:** SCH XIZ YUN
    desgodinsi J. Frivaldszky, 1892: 115
kwangsiensis Tan, 1958: 166 **A:** SCH
laticornis Haag-Rutenberg, 1880: 82 (*Lytta*) **A:** "South China" **ORR**
maekliniana Kaszab, 1958: 189 **A:** YUN
    maeklini Haag-Rutenberg, 1880: 88 (*Lytta*)
    mojiangensis Tan & Deng, 1993: 155
mandibularis Anand, 1977: 82 **A:** HP NP
mannerheimi Mäklin, 1875a: 623 (*Cantharis*) **A:** BT FUJ GUA GUX NP UP YUN **ORR**
megalocephala Gebler, 1817: 318 (*Lytta*) **E:** ST **A:** BEI ES FE HEB HEI HEN JIL LIA KZ MG NIN NMO SCH SHX YUN WS "Korea"
    albinae Reitter, 1905b: 196
    maura Faldermann, 1833: 61 (*Lytta*)
miroslavi Bologna, 2020 [see New Acts] **A:** YUN
    suturalis Dvorak, 2008: 5
montana Anand, 1977: 81 **A:** HP NP
nepalensis Hope, 1831: 32 (*Lytta*) **A:** AP NP PA SD **ORR**
obscurocephala Reitter, 1905b: 195 **A:** ANH HEB HEN HUB JIA JIL JIX NIN NMO SHA SHN TIA ZHE
probsti Dvořák, 1996: 38 **A:** NP
quadraticollis Fairmaire, 1891d: ci **A:** HP KA PA
reini Kiesenwetter, 1879: 307 (*Cantharis*) **A:** JA
rubriceps L. Redtenbacher, 1844: 535 (*Epicauta*) **A:** KA PA XIN **ORR**
    haagi Bates, 1879: 483
    reversa Gemminger, 1870a: 124 (*Cantharis*) [RN]
ruficeps Illiger, 1800: 140 (*Lytta*) **A:** ANH FUJ GUX HUB HUN JIX SCH YUN **ORR**
    plumicornis Laporte, 1840: 274 (*Lytta*)
rufidorsum Goeze, 1777: 704 (*Meloe*) **E:** AL AU BH BU CR CZ FR GR HU IT MC MD PL RO SK SL SZ UK YU
    algirica Sulzer, 1776: 66 (*Meloe*)
    dubia G.-A. Olivier, 1790: 279 (*Cantharis*)
    erytrocephala Rossi, 1790: 259 (*Lytta*) [HN]
    marginata Dorthes, 1787: 69 (*Lytta*)
    rufa Gmelin, 1790: 2016 (*Lytta*)
    verticalis Illiger, 1804: 172 (*Lytta*)
sanguiniceps Fairmaire, 1885a: xxxviii (*Cantharis*) **N:** AG
seriata Ren & Yang, 2007: 706 **A** YUN

sharpi Marseul, 1875a: xxvii **A:** IN IS JO OM SA YE **AFR**
suavis Haag-Rutenberg, 1880: 83 (*Lytta*)
sibirica Pallas, 1773: 720 (*Meloe*) **E:** ST **A:** AHN BEI ES FE GAN GUA HEB HEI HEN HUB HUN JA JIA JIL JIX KZ LIA MG NE NC NIN NMO NO QIN SC SCH SHN SHX TAI WS XIN XIZ YUN ZHE **ORR**
chinensis Laporte, 1840: 274 (*Lytta*)
dubia Fabricius, 1781: 329 (*Lytta*)
dubia Fischer von Waldheim, 1823: pl. 42 (*Cantharis*)
pectinata Goeze, 1777: 701 (*Meloe*)
sibirica LeConte, 1866: 159 [HN]
sparsicapilla Yang & Ren, 2007: 713 **A:** YUN
textilis Haag-Rutenberg, 1880: 82 (*Lytta*) **A:** IN IQ SA
tibialis Waterhouse, 1871: 406 (*Cantharis*) **A:** FUJ GUA GUI GUX HEN HGK HUN JIA JIX NP SCH TAI YUN **ORR**
waterhousei Haag-Rutenberg, 1880: 81 (*Lytta*) **A:** ANH FUJ GUA GUX HAI HEN HUB JA JIX TAI SHN YUN **ORR**
formosensis Wellman, 1912: 31
wellmani Kaszab, 1956e: 636 **A:** XIZ **ORR**
weixiensis Tan, 1992: 575 **A** YUN
wittmeri Kaszab, 1978c: 331 **A:** SD
xantusi Kaszab, 1952e: 592 **A:** GUX JIA JIX NIN SCH SHG SHX
yunnanensis Kaszab, 1960d: 256 **A:** SCH YUN

genus *Epicauta*, nomina dubia
punctata Pallas, 1781: 100 (*Meloe*) **E:** RU “Caucasus” **A:** “Siberia”
rishwani Makhan, 2013: 1 **A:** CH (Chongqing)

genus *Psalydolytta* Péringuey, 1909: 254 type species *Lytta lorigera* Gerstaecker, 1854
Gnathospastoides Fletcher, 1914: 302 type species *Lytta rouxii* Laporte, 1840
aegyptiaca Mäklin, 1875a: 612 (*Cantharis*) **N:** EG **AFR**
monardi Pic, 1951h: 219
antennata Saha, 1979: 34 **A:** UP **ORR**
atricollis Pic, 1920a: 18 (*Epicauta*) **A:** HP SD UP **ORR**
rouxi Laporte, 1840: 274 (*Lytta*) **A:** HP UP **ORR**
villipes Haag-Rutenberg, 1880: 87 (*Lytta*) **A:** KA PA **ORR**

tribe Lyttini Streubel, 1846

genus *Alosimus* Mulsant, 1857: 358 type species *Meloe syriacus* Linnaeus, 1758
Halosimus Gemminger & Harold, 1870b: 2157 [UE]
abeillei Escherich, 1896c: 215 (*Lydus*) **A:** TR
albovillosus Escalera, 1914: 391 (*Halosimus*) **N:** MO
araxis Reitter, 1892a: 63 (*Halosimus*) **E:** AB AR **A:** TR

armeniacus Faldermann, 1837: 135 (*Lytta*) **E:** AB AR BU GR RO **A:** IN IQ IS JO SY TR
babadjanidesi Kaszab, 1951b: 147
cupratus Abeille de Perrin, 1880: 251 (*Lydus*)
depilis Abeille de Perrin, 1880: 250 (*Lydus*)
elegantulus Mulsant & Rey, 1858a: 143
brevicornis Abeille de Perrin, 1880: 249 (*Lydus*) **A:** IQ IS JO LE SY
castaneus Escherich, 1896a: 27 (*Lydus*) **E:** AR "Caucasus" **A:** TR
chalybaeus Tauscher, 1812b: 153 (*Oenas*) **E:** AB AR BU CT GR KZ MC MD RO ST TR UK YU **A:** IQ IN SY TR
dichrous Fischer von Waldheim, 1824: 230 (*Lytta*)
obscurus Escherich, 1896c: 225 (*Lydus*)
cirtanus Lucas, 1846: pl. 34 (*Cantharis*) **N:** AG TU
purpureicollis Chobaut, 1897a: 39 (*Halosimus*)
crabrerai Escalera, 1923e: 299 (*Halosimus*) [DA] **N:** MO
crassicornis Escalera, 1914: 391 (*Halosimus*) [DA] **N:** MO
decolor Abeille de Perrin, 1880: 249 (*Lydus*) **E:** BU GR TR **A:** SY TR
filitarsis Escalera, 1925f: 502 (*Halosimus*) [DA] **N:** MO
gracilis Abeille de Perrin, 1880: 251 (*Lydus*) **A:** IN IS JO SY TR
latitarsis Kaszab, 1951b: 145 **A:** TR
longicollis Escherich, 1896c: 213 (*Lydus*) **A:** IN
luteus Waltl, 1838: 467 (*Lytta*) **A:** IN IQ IS JO SY TR
maculicollis Mulsant & Wachanru, 1853: 172 (*Lydus*) **E:** GR **A:** TR
cinctus Schaufuss, 1862b: 310 (*Alosimus*)
noticollis Mulsant & Rey, 1858a: 137
marginicollis Haag-Rutenberg, 1880: 72 (*Lytta*) **A:** TR
mendax Fairmaire, 1879b: 50 (*Cantharis*) **N** AG LB TU
pallidicollis Gyllenhal, 1806: 27 (*Lytta*) **A:** IQ IS JO SY TR
reitterianus Semenov, 1900b: 480 (*Lydus*) [RN] **E:** AB AR **A:** IN TR
armeniacus Reitter, 1889a: 36 (*Halosimus*) [HN]
robustus Escalera, 1914: 504 (*Halosimus*) **N:** MO
albopilosus J.R. Sahlberg, 1903e: 29
lateauratus Pic, 1949a: 9 (*Lydus*)
theryi Pic, 1949a: 9 (*Lydus*)
smyrnensis Mařan, 1942c: 91 (*Lydus*) **E:** GR MC **A:** IN TR
sulcicollis Abeille de Perrin, 1880: 248 (*Lydus*) **A:** IS JO SY
luteipennis Pic, 1899h: 115 (*Lydus*)
syriacus afghanicus Kaszab, 1973d: 272 **A:** AF
syriacus austriacus Schrank von Paula, 1781: 225 (*Meloe*) **E:** AL AU BU CR CT CZ GE GR HU IT KZ LA MC MD PL RO ST SK SZ TR UK YU **A:** KZ TR
basinotatus Mařan, 1942c: 94 (*Lydus*)
binotatus Mařan, 1942c: 94 (*Lydus*)
crambes Pallas, 1782: 95 (*Meloe*)

cyanipennis Mařan, 1942c: 94 (*Lydus*)
danubianus Mařan, 1942c: 94 (*Lydus*)
nigricollis Escherich, 1896c: 219 (*Lydus*)
immaculifrons Mařan, 1942c: 94 (*Lydus*)
syriacus rauterbergi Reitter, 1907i: 485 (*Halosimus*) **N:** EG LB **A:** IS SI
syriacus syriacus Linnaeus, 1758: 419 (*Meloe*) **A:** IN IQ IS JO LE SA SY TR
albopilosus Kaszab, 1951b: 140
azureipennis J.R. Sahlberg, 1913b: 197 (*Lydus*)
myagri Fischer von Waldheim, 1823: pl. 42 (*Lytta*)
ruficollis Herbst, 1786: 179 (*Lytta*)
tenuicornis Escalera, 1914: 392 (*Halosimus*) [DA] **N:** MO
kocheri Pic, 1949a: 9
tyrrhenicus Bologna, 1989a: 191 **E:** IT
viridissimus Lucas, 1846: pl. 34 (*Cantharis*) **N:** AG LB MO TU
bipartitus Pic, 1905d: 107
coeruleus Escherich, 1896c: 221 (*Lydus*)
intermedius Pic, 1897e: 42 (*Oelosimus*)
opacipennis Fairmaire, 1871a: 395 (*Cantharis*)
purpurascens Pic, 1939b: 17 (*Lydus*)

genus *Berberomeloe* Bologna, 1989b: 355 type species *Meloe majalis* Linnaeus, 1758
insignis Charpentier, 1818: 258 (*Meloe*) **E:** SP
majalis Linnaeus, 1758: 419 (*Meloe*) **E:** FR PT SP **N:** AG MO TU
fissicornis Gorriz y Muñoz, 1882: 38 (*Meloe*)
frontalis Baudi di Selve, 1878b: 355 (*Meloe*)
laevigatus A.G. Olivier, 1797: 650 (*Meloe*)
maculifrons Lucas, 1846: pl. 33 (*Meloe*)
maculicollis Reitter, 1872: 176 (*Meloe*)
sanguineus Escherich, 1889c: 334 (*Meloe*)

genus *Cabalia* Mulsant & Rey, 1858a: 154 type species *Lytta segetum* Fabricius, 1792
Paroenas Kolbe, 1894a: 190 type species *Paroenas limbata* Kolbe, 1894
arabica Kaszab, 1983a: 172 **A:** SA
longicollis Kaszab, 1948: 13 **N:** AG
rubriventris Fairmaire, 1860: 338 (*Cantharis*) **N:** AG LB MO TU
rufiventris Walker, 1871: 16 (*Cantharis*) **N:** AG LB MO TU **A:** SI
segetum Fabricius, 1792b: 84 (*Lytta*) **E:** IT (Sicilia) MA **N:** AG TU LB
bassii Laporte, 1840: 272 (*Cantharis*)
coerulea Escherich, 1895: 293
perroudi Mulsant & Rey, 1858a: 154 (*Cacalia*)

genus *Calydus* Reitter, 1896h: 192 [RN] type species *Caloenas pulcher* Reitter, 1889
Caloenas Reitter, 1889a: 34 [HM] type species *Caloenas pulcher* Reitter, 1889
ater Kaszab, 1960b: 129 **A:** IN IQ TR
escherichi Reitter, 1898f: 351 **A:** TR
anatolicus Muche, 1962a: 51 (*Lydus*)
ornaticollis Escherich, 1896b: 122 (*Caloenas*) **A:** IN SY
pulcher Reitter, 1889a: 35 (*Caloenas*) **E:** AB AR
semenovi Escherich, 1896b: 122 (*Caloenas*) **A:** IN SY TR
plagifrons Reitter, 1913d: 662 (*Zonabris*)
syriacus Kaszab, 1960b: 130 **A:** SY

genus *Eolydus* Denier, 1913: 231 type species *Eolydus atrocoeruleus* Denier 1913 (= *Alosimus atripes* Pic, 1905)
afghanicus Kaszab, 1958c: 247 **A:** AF
atripes Pic, 1905f: 130 (*Alosimus*) **A:** IN IQ
atrocoeruleus Denier, 1913: 231
conspicuus Waterhouse, 1889: 129 (*Cantharis*) **A:** AF IN PA "Turkestan"
binotatithorax Kaszab, 1956e: 635
testaceicornis Pic, 1910j: 34 (*Lydus*)

genus *Lagorina* Mulsant & Rey, 1858a: 74 type species *Lytta sericea* Waltl 1835
scutellata Laporte, 1840: 273 (*Lytta*) **N:** AG MO
janthina Fairmaire, 1860: 338 (*Cantharis*)
sericea Waltl, 1835: 76 (*Lytta*) **E:** SP **N:** AG MO

genus *Lydomorphus* Fairmaire, 1882a: 85 type species *Lydomorphus cinnamomeus* Fairmaire, 1882
subgenus *Lydomorphus* Fairmaire, 1882a 85 type species *Lydomorphus cinnamomeus* Fairmaire, 1882
Cylindrothorax Escherich, 1896a: 28 [RN] type species *Lytta angusticollis* Haag-Rutenberg, 1880: 62
Mimovesperus Pic, 1923g: 10 type species *Mimovesperus pilosus* Pic, 1923 (= *Lytta verrucicollis* Karsch, 1881)
Sagitta Escherich, 1895: 295 [HM] type species *Lytta angusticollis* Haag-Rutenberg, 1880
angusticollis fairmairei Kaszab, 1955b: 250 (*Cylindrothorax*) **A:** SA YE **AFR**
angusticollis maculithorax Kaszab, 1955b: 251 (*Cylindrothorax*) **A:** YE
angusticollis somaliensis Kaszab, 1955b: 251 **A:** AE **AFR**
angusticollis suturellus Haag-Rutenberg, 1880: 63 (*Lytta*) **A:** AE IN OM PA SA YE **ORR**
insuturella Pic, 1914p: 106 (*Lytta*)
varicolor Wellman, 1910a: 24 (*Lytta*) [RN]
beccarii Haag-Rutenberg, 1880: 71 (*Lytta*) **A:** SA YE
arabicus Pic, 1912g: 34 (*Lytta*)

braeti Fairmaire, 1894a: 41 (*Cantharis*) **A:** SD **ORR**
brittoni Kaszab, 1953b: 853 (*Cylindrothorax*) **A:** AE IQ OM SA
buettikeri Kaszab, 1983a: 177 (*Cylindrothorax*) **A:** AE KU OM SA
chanzyi Fairmaire, 1876: 38 (*Epicauta*) [DA] **N:** AG LB MO TU
femoralis Kocher, 1955a: 125 (*Cylindrothorax*) [DA] **N:** MO
janczyki Kaszab, 1959h: 437 (*Cylindrothorax*) **A:** IN IQ
kurdistanicus Pic, 1912g: 34 (*Lytta*) **A:** "Kurdistan"
leucophthalmus bipallidus Kaszab, 1953b: 861 (*Cylindrothorax*) **A:** SA
leucophthalmus leucophthalmus Fairmaire, 1893d: 153 (*Cantharis*) **A:** SA YE **AFR**
bimaculatithorax Pic, 1913v: 457 (*Lytta*)
citernii Pic, 1914p: 105 (*Lytta*)
palaestinus Kirsch, 1871: 390 (*Lagorina*) [DA] **N:** EG LB **A:** IS JO KU SA SI
picticollis picticollis Haag-Rutenberg, 1880: 70 (*Lytta*) **A:** SA YE **AFR**
breviceps Kaszab, 1955b: 247 (*Cylindrothorax*)
spurcaticollis Fairmaire, 1883c: 104 (*Cantharis*)
pictus Laporte de Castelnau, 1840: 275 (*Lytta*) **A:** PA **ORR**
reymondi Selander, 1988: 34 (*Cylindrothorax*) [DA] [RN] **N:** AG
pici Reymond, 1956a: 41 (*Cylindrothorax*) [HN]
saharanus Kaszab, 1961g: 348 (*Cylindrothorax*) [DA] **N:** AG LB
semitestaceus Fairmaire, 1896a: 51 (*Cantharis*) **A:** NP
tenuicollis Pallas, 1798: 102 (*Meloe*) **N:** EG **A:** IN PA SA UP YE **AFR ORR**
ornatus Laporte, 1840: 275 (*Lytta*)
ruficollis G.-A. Olivier, 1790: 278 (*Cantharis*) [HN]
verrucicollis Karsch, 1881: 49 (*Lytta*) [DA] **N:** AG EG LB MO TU
atriventris Pic, 1911d: 75 (*Sagitta*)
pilosus Pic, 1923g: 10 (*Mimovesperus*)
subgenus *Somalolydus* Bologna, 1992: 63 type species *Cantharis bifoveiceps* Fairmaire, 1897
szalaymarzsoi Kaszab, 1978c: 337 (*Cylindrothorax*) **A:** SA YE

genus *Lydulus* Semenov, 1893d: 265 type species *Lydulus albopilosus* Semenov, 1893
Lydoenas Reitter, 1894a: 51 type species *Lydus pygmaeus* Dokhtouroff, 1889
albopilosus Semenov, 1893d: 266 **A:** KZ TM UZ
tibialis Semenov, 1896b: 61
granulidorsis Semenov, 1896b: 59 **A:** KZ XIN
pygmaeus Dokhtouroff, 1889: 180 **A:** TD TM
discoloripes Semenov, 1896b: 58
semipurpureus Reitter, 1902c: 188 **A:** KZ TD UZ
uzbekistanus Kaszab, 1981a: 165 **A:** UZ

genus *Lydus* Dejean, 1821: 75 type species *Meloe algiricus* Linnaeus, 1758
algiricus Linnaeus, 1758: 420 (*Meloe*) **N:** AG MO TU

algericus Laporte, 1840: 271 (*Lytta*)
immaculatus Fabricius, 1775: 826 (*Mylabris*)
indicus Herbst, 1784c: 145 (*Lytta*)
europaeus Escherich, 1896c: 208 **E:** AL BH BU CR GR HU MC MD RO UK YU **A:** TR
gibbiger Escherich, 1896c: 208 **A:** SY TR
humeralis Gyllenhal, 1817: 16 (*Lytta*) **A:** IS JO SY TR
suturalis Abeille de Perrin, 1880: 250
marginatus Fabricius, 1792b: 88 (*Mylabris*) **N:** AG MO TU
codinai Pardo Alcaide, 1950b: 46
marginea Schoenherr, 1817: 27 (*Lytta*)
pilicollis Fairmaire, 1892b: 155 **A:** SY TR
praeustus L. Redtenbacher, 1850: 49 **A:** IN IQ SY
hauseri Mader, 1931: 136
stigmatifrons Marseul, 1872a: 419
quadrimaculatus Tauscher, 1812b: 141 (*Mylabris*) **E:** KZ ST UK **A:** KZ UK
sanguinipennis Chevrolat, 1840b: 279 **N:** AG MO
taraudanti Escalera, 1913: 54
marrakensis Escalera, 1909b: 243
tarsalis Abeille de Perrin, 1880: 247 **A:** CY IN IS JO LE SY TR
cerastes Abeille de Perrin, 1880: 247
tenuicollis Reitter, 1905d: 246 **A:** IN
tenuitarsis Abeille de Perrin, 1880: 247 **E:** "Caucasus" **A:** IN IS JO LE SI SY TR
impressicollis Fairmaire, 1892b: 156 **A:** TR
trimaculatus italicus Kaszab, 1952a: 98 **E:** IT
trimaculatus trimaculatus Fabricius, 1775: 261 (*Mylabris*) **E:** AL BH BU CR GR HU KZ MC MD RO TR UK YU **A:** TR
communimaculatus Escherich, 1896c: 206
vexator Escherich, 1896c: 206
turcicus Kaszab, 1952a: 98 **A:** IN SY TR
rufus Kaszab, 1952a: 97
vasvarii Kaszab, 1952a: 97
unicolor Reitter, 1887e: 525 **E:** AB AR ST **A:** IN IS LE JO SY TM TR
bitlisiensis Pic, 1901c: 34
glabricollis Escherich, 1896c: 211
halbhuberi Escherich, 1896c: 210
signatithorax Pic, 1902b: 9

genus *Lytta* Fabricius, 1775: 260 type species *Meloe vesicatorius* Linnaeus, 1758
subgenus *Asiolytta* Kaszab, 1962b: 295 type species *Lytta badakschanica* Kaszab, 1958
badakschanica Kaszab, 1958c: 259 **A:** AF UZ
clematidis Pallas, 1782: 95 (*Meloe*) **A:** AF ES KZ MG TD TM UZ XIN WS

bivitta Schoenherr, 1817: 28 (*Cantharis*)
bivittis Pallas, 1782: 93 (*Meloe*)
fischeri Gebler, 1817: 317
korbi Pic, 1906b: 18
discipennis Fairmaire, 1891d: c (*Cantharis*) **A:** KA **ORR**
fissicollis Fairmaire, 1886c: 350 (*Cantharis*) **A:** GAN GUI HEN SCH SHX XIZ YUN
limbata afghanica Kaszab, 1981a: 164 **A:** AF PA
limbata limbata L. Redtenbacher, 1844: 535 (*Epicauta*) **A:** AF KA NP PA UP
manicata J.R. Sahlberg, 1903a: 38 (*Lytta*) **A:** TD UZ
poeciloptera Semenov, 1893d: 270 **A:** TD
rubra Hope, 1831: 32 **A:** AF NP
satiata Escherich, 1904: 31 **A:** TD XIN XIZ
selanderi Saha, 1979: 98 [DA] **A:** UP XIZ
subgenus *Eolytta* Shapovalov, 2016: 100 type species *Cantharis luteovittata* Kraatz, 1882
kabakovi Kaszab, 1981a: 162 **A:** AF
flavoangulata Fairmaire, 1891e: cxxxii (*Cantharis*) **A:** AF HP KA PA
grumi Semenov, 1893d: 268 **A:** XIN
laeta Waterhouse, 1889: 129 (*Cantharis*) **A:** AF
luteovittata Kraatz, 1882d: 334 (*Cantharis*) **A:** KZ NW TD UZ WS
melancholica Wellman, 1910a: 24
tristis Escherich, 1895: 287 [HN]
rubrinotata Tan, 1981: 409, 414 **A:** XIN XIZ
skrylniki Shapovalov, 2016: 101 **A:** AF
subgenus *Indiolytta* Selander, 1960: 28 type species *Lytta apicalis* Haag-Rutenberg, 1880 (= *Lydus melanurus* Hope, 1831)
melanura Hope, 1831: 32 (*Lydus*) **A:** NP SD UP **ORR**
apicalis Haag-Rutenberg, 1880: 71
subgenus *Lytta* Fabricius, 1775: 260 type species *Meloe vesicatorius* Linnaeus, 1758
antennalis Marseul, 1873a: 230 (*Cantharis*) **A:** HP JA NP PA UP XIN
battonii Kaszab, 1962b: 291 **A:** QIN SCH
bieti Wellmana, 1912: 33 **A:** SCH WP XIZ
caraganae Pallas, 1781: 97 (*Meloe*) **A:** ANH ES FE GAN HEB HEI HEN HUB HUN JA JIA JIL LIA MG NIN NMO NC QIN SHA SHN SHX XIN
pallasi Gebler, 1829a: 141 (*Lytta*)
choui Wang, Wang & Ren, 2012: 376 **A:** HEB
flavipennis Motschulsky, 1860c: 145 (*Cantharis*) **A:** PA UP **ORR**
freudei Kaszab, 1958c: 257 **E:** AR GG **A:** TR
kashmirensis Bologna, 1983: 125 **A:** KA
kryzhanovskyi Kaszab, 1962b: 292 **A:** QIN
medvedevi Shapovalov, 2016: 104 **A:** KY TD UZ
menetriesi Faldermann, 1832: 210 **E:** AB AR **A:** IN KI KZ TD TM UZ
regiszahiri Kaszab, 1958c: 255 **A:** AF

roborowskyi Dokhtouroff, 1887a: 343 (*Cantharis*) **A:** QIN SCH XIN XIZ YUN
thibetana E. Olivier, 1888: lvi
sifanica Semenov, 1910a: 28 **A:** NIN XIN XIZ
kukunoorensis Wellman, 1910a: 24 [RN]
tibetana Escherich, 1904: 30 [HN]
suturella Motschulsky, 1860c: 144 (*Cantharis*) **A:** ES FE GUX HEB HEI JA JIL LIA NIN NMO QIN SHA SHX TD XIN "Korea"
togata Fischer von Waldheim, 1844: 135 **A:** AF ES KI KZ TD WS XIN
flavovittata Ballion, 1878: 346 (*Cantharis*)
vesicatoria heydeni Haag-Rutenberg, 1880: 73 **A:** TR
vesicatoria moreana Pic, 1941a: 1 **E:** GR
vittata Brullé, 1832: 232 (*Cantharis*) [HN]
vesicatoria vesicatoria Linnaeus, 1758: 419 (*Meloe*) **E:** AL AR AU BE BH BU BY CR CT CZ EN FR DE GB GE GG GR HU IR IT KZ LA LT LU MC MD NL PL PT RO SK SL SP ST SV SZ TR UK WS YU
armeniaca Reitter, 1886b: 99 [HN]
aurantiaca Escherich, 1895: 275
basipennis Pic, 1948b: 4
caeruleicollis Pic, 1948b: 3
costatella Reitter, 1916g: 294
crassicornis A. Costa, 1882: 3 (*Cantharis*)
cupreopurpurea Pic, 1948: 3
dibapha Reitter, 1890a: 15
fauconneti Pic, 1948b: 4
heyrovskyi Kaszab, 1956b: 1
lateopaca Pic, 1948b: 4
lebisi Pic, 1949b: 12
leodi Escherich, 1896a: 28
martialis Pic, 1948b: 4
paulocuprea Pic, 1948b: 3
purpureicolor Pic, 1948b: 2
semicoerulea Della Beffa, 1909: 200
semicuprea Pic, 1948b: 3
semipurpurea Pic, 1948: 3
theresae Pic, 1948b: 3
viridicuprea Fleischer, 1923: 89

subgenus *Mesolytta* Selander, 1960: 27 type species *Oenas coccineus* Ménétriés, 1849
coccinea Ménétriés, 1849: 247 (*Oenas*) **A:** AF IN KZ TD TM UZ
fuscicrus Brancsik, 1899: 102
glabricollis Waterhouse, 1889: 129 (*Cantharis*)
minor Dokhtouroff, 1889: 179 (*Cantharis*)

deserti Semenov, 1891a: 377 **A:** KZ TD TM UZ

hauseri Heyden, 1894a: 51

vrendenburgi Kaszab, 1962b: 295 **A:** PA UZ

subgenus *Pseudolytta* Selander, 1960: 27 type species *Lytta aeneiventris* Haag-Rutenberg, 1880

aeneiventris Haag-Rutenberg, 1880: 75 **A:** FUJ GUA HKG HUB HUN JIA JIX

impressithorax Pic, 1924a: 21

kwanhsiensis Mařan, 1941b: 182 **A:** SCH

spissicornis Fairmaire, 1891c: xxiii (*Cantharis*) **A:** JIA

taliana Pic, 1915e: 9 **A:** SCH XIZ YUN

genus *Lytta*, species incertae sedis

zubovi Bologna & Nikitsky, 2010: 71: UK (Krym)

genus *Lyttolydulus* Reitter, 1913d: 662 type species *Lyttolydulus suturalis* Reitter, 1913 (= *Lytta simplicicornis* Pic, 1899)

Lyttalydulus Reitter, 1913d: 662 [emended to *Lyttolydulus* by Reitter, 1915]

Lydopsis Escalera, 1914: 393 type species *Lydus susicus* Escalera, 1914

cinereovestitus Fairmaire, 1876: 49 (*Cantharis*) **N:** AG EG MO TU

mus Escherich, 1904: 32 (*Lagorina*)

deserticolus Kaszab, 1952b: 94 **N:** LB MO

giganteus Kaszab, 1952b: 93 **N:** TU

mozabita Chobaut, 1897c: 171 (*Lytta*) **N:** AG

nigronotatus Pic, 1907l: 178 (*Lagorina*) **N:** AG

nubeculosus Kaszab, 1952b: 92 **N:** AG EG MO

rufulus Fairmaire, 1863: 645 (*Lydus*) **N:** AG LB MO TU

djerbensis Escherich, 1895: 285 (*Lytta*)

simplicicornis Pic, 1899n: 279 (*Lytta*) **A:** IS JO

suturalis Reitter, 1913d: 662

susicus Escalera, 1914: 393 (*Lydus*) **N:** MO

thiebaulti arabicus Kaszab, 1983a: 174 **A:** KU SA

thiebaulti thiebaulti Fairmaire, 1876: 49 (*Cantharis*) **N:** AG EG MO TU **A:** AE SI

genus *Lyttonyx* Marseul, 1876b: 36 type species *Cantharis bilateralis* Marseul, 1876 (= *Epicauta bicolor* Walker, 1871)

bicolor Walker, 1871: 17 (*Epicauta*) **N:** AG EG LB MO TU **A:** AE IN IS KU OM SA SI YE

beluchistanica Kaszab, 1959b: 2

bilateralis Marseul, 1876b: 35 (*Cantharis*)

myrmido Fairmaire, 1876: 93 (*Cantharis*)

genus *Muzimes* Aksentjev, 1988: 577 type species *Litta* [sic !] *collaris* Fabricius, 1787

Micromerus Mulsant & Rey, 1858a: 147 [HN]

bitlisiensis Kaszab, 1958a: 15 (*Micromerus*) **A:** TR
brodskyi Dvořák, 1983: 448 (*Micromerus*) **E:** AR
caucasicus Mařan, 1940a: 82 (*Lydus*) **E:** AB AR **A:** IN TR
unicoloriceps Kaszab, 1958a: 18
collaris Fabricius, 1787a: 215 (*Lytta*) **E:** BU GR HU KZ MC MD ST TR UK YU **A:** TR
erythrocynea Pallas, 1782: 96 (*Meloe*)
dersimensis Kaszab, 1968d: 447 (*Micromerus*) **A:** TR
erivanicus Mařan, 1941a: 62 (*Lydus*) **E:** AR **A:** TR
luteiceps Kaszab, 1958a: 17
nigrilabris Kaszab, 1958a: 17
nigriceps Kaszab, 1958a: 17
seminigriceps Kaszab, 1958a: 17
iranicus Mařan, 1940a: 86 (*Lydus*) **A:** IN
jureckovae Mařan, 1940a: 80 (*Lydus*) **E:** AR GG
luristanicus Mařan, 1940a: 83 (*Lydus*) **A:** IN
maceki Dvořák, 1983: 446 (*Micromerus*) **A:** IQ
marani Kaszab, 1958a: 16 (*Micromerus*) **E:** GG
nigricornis Escherich, 1896c: 232 (*Lydus*) **E:** AB AR
obenbergeri Kaszab, 1958a: 14 (*Micromerus*) **A:** TR
ressli Dvořák, 1993b: 2 (*Micromerus*) **A:** TR
semiobscurus Pic, 1898f: 73 (*Alosimus*) **A:** TR
sterbai Mařan, 1940a: 88 (*Lydus*) **A:** SY TR
tauricus Mařan, 1941a: 60 (*Lydus*) **A:** TR

genus *Oenas* Latreille, 1802: 186 type species *Meloe afer* Linnaeus, 1767
afer Linnaeus, 1767: 680 (*Meloe*) **N:** AG MO TU
africanus Desmarest, 1825: 435
maculatus Normand, 1936a: 154
unicolor Laporte, 1840: 271 (*Aenas*)
brevicollis Abeille de Perrin, 1880: 244 **A:** IS JO SY
crassicornis Illiger, 1800: 142 (*Lytta*) **E:** AL AU BH BU CR CZ GR HU IT MC MD RO SK ST TR UK YU **A:** IN IQ SY TR
ruficollis A.G. Olivier, 1795: 47: 14 (*Mylabris*)
cribricollis Abeille de Perrin, 1880: 244 **A:** IS LE SY
fusicornis Abeille de Perrin, 1880: 242 **E:** PT SP
hispanus Abeille de Perrin, 1880: 243
graecus Bologna, 1991: 176 **E:** GR
laevicollis Abeille de Perrin, 1880: 243 **A:** IS JO
pseudafer Kaszab, 1951e: 278 **N:** EG
luteus Kaszab, 1951e: 276
rubriceps Dvořák, 2004: 103 **A:** IN

sericeus A.G. Olivier, 1795: 46: 18 **N:** AG MO TU
luctuosus Latreille, 1804a: 393
tarsensis Abeille de Perrin, 1880: 244 **E:** AR GG **A:** IN SY TR
tenuicornis Abeille de Perrin, 1880: 245
wilhelmsi Faldermann, 1832: 209 **E:** AB AR **A:** IN TR
armeniacus Kaszab, 1951e: 277

genus *Sybaris* Stephens, 1832: 70 type species *Sybaris immunis* Stephens, 1832
Prionotus L. Redtenbacher, 1844: 536 type species *Prionotus praeustus* Redtenbacher, 1844
flavus Thunberg, 1791: 108 (*Meloe*) **A:** "China"
garhwalensis Saha, 1979: 118 **A:** UP
praeustus L. Redtenbacher, 1844: 536 (*Prionotus*) **A:** KA UP
semivittatus L. Redtenbacher, 1844: 537 (*Prionotus*) **A:** HP KA
testaceus Fabricius, 1792b: 85 (*Lytta*) **A:** NP PA **ORR**
melophtalmos A.G. Olivier, 1795: 46: 10 (*Cantharis*)
tunicatus L. Redtenbacher, 1844: 537 (*Prionotus*) **A:** KA

genus *Syriolytta* Kaszab, 1962b: 295 type species *Lydus vulneratus* Fairmaire, 1892 (= *Lytta suturifera* Pic, 1899)
suturifera Pic, 1899n: 279 (*Lytta*) **A:** IQ IN SY TR
fairmairei Borchmann, 1917: 94 (*Lytta*)
vulnerata Fairmaire, 1892b: 155 (*Lydus*) [HN]

genus *Teratolytta* Semenov, 1894c: 533 type species *Cantharis dives* Brullé, 1832
capillata Serri & Bologna, 2018: 146 **A:** IN
carlae Bologna, 2006: 145 **A:** TR
dives Brullé, 1832: 232 (*Cantharis*) **E:** AB AL AR BU CR GR MC ME RO SB ST TR UK **A:** SY TR
excellens Escherich, 1895: 280 (*Lytta*)
phalerata I. Frivaldszky, 1837: 182 (*Lytta*)
vittata Kiesenwetter, 1861a: 250 (*Lytta*) [HN]
dvoraki Bologna, 2006: 154 **A:** TR
eylandti Semenov, 1894c: 534 **A:** IN TM UZ
holzshuhi Dvořák, 1983: 443
flavipes Mulsant & Rey, 1858a: 163 (*Cantharis*) **A:** LE SY TR
saphirina Abeille de Perrin, 1895b: xxiv (*Lytta*)
semividua Abeille de Perrin, 1895b: xxiv (*Lytta*)
gentilis J. Frivaldszky, 1877b: 136 (*Cantharis*) **E:** GR (Kos) **A:** TR
coeruleipennis Escherich, 1895: 280 (*Lytta*)
cooensis J. Müller, 1936: 91
violaceipennis Beauregard, 1890: 493 (*Cantharis*)
kaszabi Kryzhanovskij, 1959: 857 **A:** TD TM UZ
klapperichi Kaszab, 1958c: 249 **A:** AF

krejciki Bologna & Di Giulio, 2016: 88 **A:** TR
kulzeri Kaszab, 1958c: 252 **A:** TR
monticola Bologna, 2006: 155 **A:** TR
optabilis Faldermann, 1832: 209 (*Lytta*) **E:** AB AR ST **A:** IN
pilosella Solsky, 1881a: 259 (*Cantharis*) **A:** KI KZ TD UZ
  albovillosa Escherich, 1895: 282 (*Lytta*)
  tadzhika Kryzhanovskij, 1959: 858
regina Kaszab, 1958c: 250 **A:** AF
senilis Abeille de Perrin, 1895b: xxiv (*Lytta*) **A:** IS SY TR
  bytinskii Kaszab, 1957e: 230
  jodina Kaszab, 1957e: 231
taurica Bologna, 2006: 147 **A:** TR
tricolor Haag-Rutenberg, 1880: 76 (*Lytta*) **A:** IN UZ
vanensis Kaszab, 1968d: 448 **A:** TR

genus *Trichomeloe* Reitter, 1911a: 391 type species *Meloe chrysocomus* Miller, 1861
chrysocomus Miller, 1861: 206 (*Meloe*) **A:** CY IS JO SY TR
conicicollis Reitter, 1907g: 215 (*Meloe*) **E:** GR (Castelorizo Is.) **A:** TR
deflexus Reitter, 1889c: 106 (*Meloe*) **A:** IS JO SY TR
mesopotamicus Bologna, 2008: 241 **A:** IQ
ottomanus Pliginskij, 1914: 254 (*Meloe*) **A:** SY TR
  frivaldszkyi Kaszab, 1958c: 304 (*Meloe*)
ovatus Pripisnova, 1987a: 43 (*Meloe*) **A:** TD
sericellus Reiche, 1857: 273 (*Meloe*) **E:** IT (Sicilia) UK **A:** IS JO LE SY TR
syriacus Bologna, 2008: 244 **A:** SY

tribe Mylabrini Rafinesque, 1815

genus *Actenodia* Laporte, 1840: 268 type species *Actenodia guttata* Laporte, 1840
  Rusadiria Pardo Alcaide, 1952: 60 type species *Mylabris billbergi* Gyllenhal, 1817
billbergi Gyllenhal, 1817: 33 (*Mylabris*) **E:** FR IT (West Liguria) PT SP
  baudii Magistretti, 1943: 36 (*Mylabris*)
  circumfusa Baudi di Selve, 1878c: 1062 (*Coryna*)
  clavicornis Duméril, 1825: 13 (*Mylabris*)
  gallica Pic, 1916a: 7 (*Coryna*)
  parcepunctata Pardo Alcaide, 1950a: 81 (*Coryna*)
confluens Reiche, 1866: 629 (*Coryna*) **E:** BU TR **A:** SI SY TR
  caudanigra Abeille de Perrin, 1880: 237 (*Coryna*)
  mediodisjuncta Pic, 1930a: 1 (*Coryna*)
  mediopartita Mader, 1929: 109 (*Mylabris*)
  ottomana Escherich, 1899: 109 (*Coryna*)
  picta Baudi di Selve, 1878b: 361 (*Coryna*)
denticulata Marseul, 1872a: 418 (*Coryna*) **A:** AE SA

distincta Chevrolat, 1840b: 269 (*Hycleus*) **E:** IT (Lampedusa Is., Sicilia) SP **N:** AG MO TU

andalusica Pic, 1916a: 7 (*Coryna*)

anticefasciata Magistretti, 1943: 36 (*Mylabris*)

beccariana Pic, 1916a: 7 (*Coryna*)

gafsana Normand, 1950a: 81 (*Coryna*)

galloisi Pic, 1916a: 7 (*Coryna*)

hafidi Pic, 1916a: 7 (*Coryna*)

particularenotata Pic, 1949a: 10 (*Coryna*)

sexnotata Pic, 1949a: 10

sicula Baudi di Selve, 1878b: 361 (*Coryna*)

theresae Pic, 1949a: 10

lata Reiche, 1866: 628 (*Coryna*) **N:** EG **A:** IS SI

fasciata Baudi di Selve, 1878b: 360 (*Dices*)

perfuga Dvořák, 1993a: 3 (*Paractenodia*) [DA] **A:** SY

peyroni Reiche, 1866: 630 (*Coryna*) **N:** LB **A:** IS JO LE SI SY TR

septempunctata Baudi di Selve, 1878b: 361 (*Coryna*) **N:** AG MO TU

extera Dvořák, 1993a: 1

susica Escalera, 1913: 54 (*Hycleus*)

suturifera Pic, 1896d: 42 (*Coryna*) **N:** AG LB MO TU

alfredi Pic, 1913f: 112 (*Coryna*)

bleusei Chobaut, 1896b: 376 (*Coryna*)

indenticulata Pic, 1913f: 112 (*Coryna*)

vaulogeri Pic, 1913f: 112 (*Coryna*)

genus *Ammabris* Kuzin, 1954: 359 type species *Mylabris elegans* A.G. Olivier, 1811

Mesorbata Pardo Alcaide, 1954a: 66 type species *Mylabris elegans* A.G. Olivier, 1811

amoenula Ménétriés, 1849: 247 **A:** KZ TM

andresi Pic, 1911l: 162 (*Zonabris*) [DA] **N:** EG

avilai Ruiz & García-París, 2008: 115 **N:** MO

boghariensis Raffray, 1873: 380 **N:** AG MO

chudeaui Bedel, 1921: 199 (*Zonabris*) **N:** AG

elegans A.G. Olivier, 1811: 101 **N:** AG EG LB MO TU **A:** AE IS JO KU OM SA SY

nefischensis Pic, 1926k: 229

trijuncta Pic, 1899l: 208

elegantissima Zubkov, 1837: 70 **A:** AF IN KZ TD TM UZ XIN

kuzini Kryzhanovskij, 1956: 887 **A:** TM

laticollis Escherich, 1899: 99 (*Zonabris*) **E:** AB **A:** IN TM

alpha Sumakov, 1915: 50

myrmidon Marseul, 1870b: 154 **N:** AG MO TU **A:** IS SI

raml W. Schneider, 1991: 280 **A:** AE SA

sisymbrii Klug, 1845: nr. 12 **N:** EG MO

theryi Abeille de Perrin, 1894b: cxxxii (*Zonabris*) [DA] **N:** AG
zoltankaszabi Ruiz & García-París, 2007: 151 [RN] **A:** SA
 excellens Kaszab, 1983a: 159 [HN]

genus *Croscherichia* Pardo Alcaide, 1950a: 63 type species *Mylabris circumflexa* Chevrolat, 1840 (= *Mylabris paykulli* Billberg, 1813)
 Lybicisca Kuzin, 1954: 369 type species *Mylabris sanguinolenta* A.G. Olivier, 1811
albilaena Bedel, 1899: 82 (*Zonabris*) **N:** AG EG LB TU **A:** SA
armass Ruiz, François & García-París, 2019: 34 **N:** MO
bedeli Bleuse, 1899: 383 (*Zonabris*) **N:** AG LB MO
 ainensis Pic, 1919f: 17 (*Zonabris*)
 augusti Pic, 1919e: 13 (*Zonabris*)
 crosi Pic, 1919f: 17 (*Zonabris*)
delarouzei Reiche, 1866: 639 (*Mylabris*) **A:** IS JO LE
 sancta Pic, 1920e: 13 (*Zonabris*)
femorata Klug, 1845: nr. 8 (*Mylabris*) **A:** SA
fulgurita Reiche, 1866: 640 (*Mylabris*) **N:** AG EG LB MO **AFR**
 bisbiinterrupta Pic, 1919a: 18
 subundulata Pic, 1919f: 18
gilvipes Chevrolat, 1840b: 273 (*Mylabris*) **N:** AG EG LB MO TU **A:** IS JO SI
 angulata Klug, 1845: nr. 18 (*Mylabris*)
 diffinis Abeille de Perrin, 1880: 238 (*Mylabris*)
 lameyi Marseul, 1885: cxlvii (*Mylabris*)
 mediodestituita Pic, 1919f: 18 (*Zonabris*)
goryi Marseul, 1870b: 88 (*Mylabris*) **A:** AE AF IN IQ IS JO OM PA SA SY **AFR ORR**
litigiosa Chevrolat, 1840b: 271 (*Mylabris*) **N:** AG EG MO TU **A:** IN IS OM SA SI SY "Turkestan"
 alepensis Pic, 1913s: 185 (*Zonabris*)
 islamita Marseul, 1873b: 467 (*Mylabris*)
mozabita Pic, 1897a: 2 (*Zonabris*) **N:** AG LB MO TU **AFR**
 angulata Sumakov, 1915: 60 (*Mylabris*) [HN]
 postbifasciata Pic, 1921f: 21 (*Zonabris*)
paykulli Billberg, 1813: 63 (*Mylabris*) **N:** AG LB MO TU
 abbesensis Pic, 1921f: 22 (*Zonabris*)
 alegensis Pic, 1921f: 22 (*Zonabris*)
 antebifasciata Sumakov, 1934: 4 (*Mylabris*)
 antebinotata Pic, 1921f: 22 (*Zonabris*)
 antequadrinotata Pic, 1921f: 22 (*Zonabris*)
 antetrinotata Pic, 1921f: 22 (*Zonabris*)
 apicecincta Pic, 1921f: 22 (*Zonabris*)
 atrojuncta Pic, 1921f: 21 (*Zonabris*)
 aumalensis Pic, 1921f: 22 (*Zonabris*)

bisquinquenotata Pic, 1915k: 22 (*Zonabris*)
borealis Voigts, 1902: 178 (*Zonabris*)
circumflexa Chevrolat, 1840b: 273 (*Mylabris*)
comptei Pic, 1898b: 3 (*Zonabris*)
demoflyisi Battoni, 1961: 29
destituita Pic, 1921f: 23 (*Zonabris*)
discojuncta Pic, 1921f: 22 (*Zonabris*)
distincta Baudi di Selve, 1878b: 369 (*Mylabris*)
diversenotata Pic, 1921f: 22 (*Zonabris*)
edoughnensis Pic, 1921f: 21 (*Zonabris*)
fontainei Pic, 1921f: 23 (*Zonabris*)
goudoti Chevrolat, 1840b: 274 (*Mylabris*)
mediofasciata Pic, 1921f: 22 (*Zonabris*)
mediotrinotata Pic, 1921f: 22 (*Zonabris*)
octopunctata Sumakov, 1934: 4 (*Mylabris*)
oraniensis Pic, 1898b: 3 (*Zonabris*)
posticefasciata Pic, 1921f: 22 (*Zonabris*)
postrinotata Pic, 1921f: 22 (*Zonabris*)
quinquenotata Pic, 1921f: 22 (*Zonabris*)
reghaiensis Pic, 1921f: 22 (*Zonabris*)
rufoapicalis Pic, 1921f: 23 (*Zonabris*)
scapularis Chevrolat, 1840b: 278 (*Mylabris*)
septemnotata Pic, 1921f: 22 (*Zonabris*)
sexsignata Pic, 1921f: 22 (*Zonabris*)
subangulata Kocher, 1955b: 127 (*Mylabris*)
subapicalis Pic, 1921f: 21 (*Zonabris*)
subexternepunctata Pic, 1921b: 21 (*Zonabris*)
subfasciata Pic, 1921f: 21 (*Zonabris*)
tlemcensis Pic, 1915k: 22 (*Zonabris*)
trifasciata Chevrolat, 1840b: 275 (*Mylabris*)
unireducta Pic, 1921f: 21 (*Zonabris*)

quadrizonata Fairmaire, 1875a: 530 (*Mylabris*) **N:** AG LB TU
richteri Kaszab, 1957b: 50 **A:** AE IN OM SA YE
salavatiani Kaszab, 1968f: 756 **A:** AE AF IN OM PA SA YE
sanguinolenta arabica Bologna, 1991: 153 **A:** SA
sanguinolenta sanguinolenta A.G. Olivier, 1811: 95 (*Mylabris*) **N:** AG EG LB MO TU **A:** IN IS JO SI SY

andresi Pic, 1911d: 74 (*Zonabris*) [HN]
adolfi Pic, 1912o: 90 (*Zonabris*) [RN]
ghardaiensis Pic, 1897a: 2 (*Zonabris*)
incerta Klug, 1845: n. 17 (*Mylabris*)

latreillei Billberg, 1813: 36 (*Mylabris*)
margensis Pic, 1924: 30 (*Zonabris*)
nedjalensis Pic, 1920e: 13 (*Zonabris*)
notatipennis Pic, 1897l: 124 (*Zonabris*)
unijuncta Pic, 1897a: 2 (*Zonabris*)
usta Baudi di Selve, 1878c: 1143 (*Mylabris*)
sonyae Kaszab, 1983a: 165 (*Mylabris*) **A**: AE SA
tigrinipennis Latreille, 1827: 286 (*Mylabris*) **N**: AG EG LB MO TU **A**: IQ IS JO KU SA YE **AFR**
bisbiflexuosa Pic, 1911d: 73 (*Zonabris*)
diversithorax Pic, 1910e: 18 (*Zonabris*)
luxorensis Pic, 1911d: 74 (*Zonabris*)
menthae Klug, 1845: nr. 11 (*Mylabris*)
vigintipunctata A.G. Olivier, 1811: 97 (*Mylabris*) **N**: EG **A**: IN IS JO KU SI
wartmanni Pic, 1896f: 62 (*Zonabris*) **N**: AG MO
abrupta Wellman, 1910a: 22 (*Mylabris*) [RN]
interrupta Pic, 1896f: 62 [HN]

genus *Hycleus* Latreille, 1817: 317 type species *Mylabris argentata* Fabricius, 1792
Arithmema Chevrolat, 1834: pl. 35 type species *Meloe decemguttata* Thunberg, 1791
Androfoveata Pardo Alcaide, 1954a: 81 type species *Mylabris duodecimpunctata* A.G. Olivier, 1811
Ceroctis Marseul, 1870b: 168 type species *Mylabris corynoides* Reiche, 1866 (= *Mylabris trizonata* Reiche, 1866)
Coryna Billberg, 1813: 73 [HN] type species *Mylabris argentata* Fabricius, 1792
Decatoma Dejean, 1821: 74 [HN] type species *Meloe lunata* Pallas, 1782
Decapotoma Voigts, 1902: 177 [RN] type species *Meloe lunata* Pallas, 1782
Dices Dejean, 1821: 74 type species *Cerocoma ocellata* A.G. Olivier, 1791
Euzonabris Kuzin, 1954: 357 type species *Meloe cichorii* Linnaeus, 1758
Gorrizia Pardo Alcaide, 1954a: 61 type species *Mylabris duodecimpunctata* A.G. Olivier, 1811
Megabris Des Gozis, 1881: cxiii type species *Meloe cichorii* Linnaeus, 1758
Mesogorbata Pardo Alcaide, 1954a: 78 type species *Mylabris apicipennis* Reiche, 1866
Mesoscutata Pardo Alcaide, 1954a: 81 type species *Mylabris duodecimpunctata* A.G. Olivier, 1811
Mesotaeniata Pardo Alcaide, 1955a: 4 type species *Meloe lunata* Pallas, 1782
Paractenodia Péringuey, 1904: 285 type species *Paractenodia parva* Péringuey, 1904
Sphenabris Kuzin, 1954: 361 type species *Meloe balteata* Pallas, 1782
Tigrabris Kuzin, 1954: 364 type species *Meloe atrata* Pallas, 1773
Zonabris Harold, 1879: 134 type species *Meloe cichorii* Linnaeus, 1758
abiadensis Marseul, 1870b: 29 (*Mylabris*) **N**: EG **AFR**
aegyptiacus Marseul, 1870b: 142 (*Mylabris*) **N**: EG LB
chakouri Pic, 1907h: 161 (*Zonabris*)
conjugatus Baudi di Selve, 1878b: 369 (*Mylabris*)

aestuans Klug, 1845: nr. 31 (*Mylabris*) **A:** SA YE **AFR**
rugosissimus Pic, 1909j: 130 (*Zonabris*)
afrotropicus Kaszab, 1983a: 171 (*Actenodia*) **A:** YE
allardi Marseul, 1870b: 193 (*Coryna*) **N:** AG MO TU
punctofasciatus Fairmaire, 1875a: 531 (*Mylabris*)
sefrensis Pic, 1896b: 12 (*Coryna*)
apicipennis Reiche, 1866: 635 (*Mylabris*) **N:** EG **A:** IS JO SI
fourtuai Pic, 1921g: 45 (*Zonabris*)
innesi Pic, 1912c: 1 (*Zonabris*)
arabicus Pallas, 1781: 89 (*Meloe*) **A:** OM SA YE
argentatus Fabricius, 1792b: 90 (*Mylabris*) **N:** EG
caroli Pic, 1908b: 34 (*Coryna*)
externenotatus Pic, 1908b: 34 (*Coryna*)
argentifer argentifer Pic, 1895e: 81 (*Zonabris*) **N:** AG LB MO TU
auberti Chobaut, 1898a: 87 (*Decatoma*)
henoni Fairmaire, 1897c: 121 (*Mylabris*)
argentifer bytinskii Kaszab, 1969a: 283 (*Decapotoma*) **A:** IS SA
argentifer pallidissimus Kaszab, 1983a: 168 (*Decapotoma*) **A:** KU SA
atratus Pallas, 1773: 722 (*Meloe*) **E:** AB AR CT GG ST UK **A:** AF IN KI KZ MG UZ TM WS XIN
karelini Fischer von Waldheim, 1842: 25 (*Mylabris*)
minutus Fabricius, 1798: 121 (*Mylabris*)
mediomaculatus Sumakov, 1929a: 4
metatarsalis Marseul, 1870b: 81 (*Mylabris*)
balteatus Pallas, 1782: 88 (*Meloe*) **A:** PA **ORR**
belli Borchmann, 1940b: 616 (*Mylabris*) **A:** PA
biguttatus Gebler, 1841a: 374 (*Mylabris*) **A:** AF IN KI KZ TD TM UZ XIN
fasciculatus Dokhtouroff, 1889: 140 (*Zonabris*)
karateginensis Sumakov, 1915: 73 (*Mylabris*)
pullatus Heyden, 1883a: 65 (*Mylabris*)
sexmaculatus Dokhtouroff, 1889: 141 (*Zonabris*)
similis Voigts, 1902: 178 (*Zonabris*)
bipunctatus A.G. Olivier, 1811: 94 (*Mylabris*) **A:** IN IQ PA SA
bushirensis Mader, 1929: 109 (*Mylabris*)
reitterioides Mader, 1929: 108 (*Mylabris*)
birecurvus Marseul, 1870b: 184 (*Coryna*) **A:** IS JO SY TR
caudaluteus Pic, 1930a: 1 (*Coryna*)
biundulatus Pallas, 1782: 78 (*Meloe*) **A:** PA FUJ YUN **ORR**
cleroides Lichtenstein, 1796: 74 (*Mylabris*)
pustulatus Thunberg, 1791: 113 (*Meloe*)
undulatus Herbst, 1786: 179 (*Mylabris*)
borchmannianus Kaszab, 1983a: 164 (*Mylabris*) **A:** YE

brevetarsalis Kaszab, 1960d: 260 (*Mylabris*) **A:** FUJ GUX HAI HGK NP SD TAI YUN **ORR**

brevicollis Baudi di Selve, 1878b: 373 (*Mylabris*) **E:** SP (Almería) **N:** AG LB MO TU

beauprei Pic, 1909q: 185 (*Zonabris*)

mediobisbijunctus Pic, 1916a: 7 (*Zonabris*)

multibijunctus Pic, 1921a: 5 (*Zonabris*)

nigribasis Escalera, 1914: 399 (*Zonabris*)

ouiouanensis Pic, 1936e: 30 (*Zonabris*)

plagiella Escalera, 1914: 398 (*Zonabris*)

brunnipes Klug, 1845: nr. 32 (*Mylabris*) **N:** AG EG LB MO TU **A:** AE IS OM SA YE

adrarensis Pic, 1942d: 78 (*Zonabris*)

humeronotatus Pic, 1939e: 29 (*Zonabris*)

interruptus Pic, 1896f: 62 (*Zonabris*)

letourneuxi Pic, 1896f: 62 (*Zonabris*)

subruptus Wellman, 1910a: 23 (*Mylabris*)

chodschenticus Ballion, 1878: 337 (*Mylabris*) **E:** AB **A:** KI KZ TD TM UZ XIN

ataensis Pic, 1901h: 78 (*Zonabris*)

diabolicus Escherich, 1899: 108 (*Zonabris*)

mediojunctus Pic, 1901h: 78 (*Zonabris*)

mediopunctatus Pic, 1901h: 78 (*Zonabris*)

cichorii Linnaeus, 1758: 419 (*Meloe*) **A:** ANH AP FUJ GAN GUA GUI HAI HEB HEN HKG HP HUB HUN JA JIA JIX NIN NP PA SCH SD SHX TAI UP YUN ZHE **ORR**

formosensis Kaszab, 1959h: 438 (*Mylabris*)

postscutellomaculatus Kaszab, 1960d: 259 (*Mylabris*)

cingulatus Faldermann, 1837: 122 (*Mylabris*) **E:** “Transcaucasia” **A:** IN TR “Turkestan”

amrishi Makhan, 2012c: 1

flavipennis Dokhtouroff, 1889: 157 (*Zonabris*) [HN]

flavis Mader, 1927: 196 [RN]

succinctus Baudi di Selve, 1878b: 365 (*Mylabris*)

colligatus L. Redtenbacher, 1850: 49 (*Mylabris*) **A:** AF IN IQ PA TM TR “Turkestan”

javeti Marseul, 1870b: 63 (*Mylabris*)

umbilicatus Kaszab, 1958c: 291 (*Mylabris*)

concinnus Marseul, 1870b: 191 (*Mylabris*) **A:** IS JO SY

contaminatus Abeille de Perrin, 1880: 236 (*Coryna*)

impressithorax Pic, 1930a: 2 (*Zonabris*)

cruentatus Klug, 1845: nr. 1 (*Mylabris*) **A:** SA

curticornis Pic, 1919f: 19 (*Zonabris*) [DA] **E:** AR

damohensis Saha, 1979: 60 (*Mylabris*) [DA] **A:** UP **ORR**

designatus Reiche, 1850: 377 (*Mylabris*) **N:** EG **AFR**

hacolyssa Rochebrun, 1883: 182 (*Mylabris*)

tristis Reiche, 1850: 379 (*Mylabris*)

diversesignatus Pic, 1919f: 18 (*Zonabris*) **N:** AG TU

diversipilis Sumakov, 1915: 42 (*Mylabris*) [DA] **A:** IS SY
dohrni Marseul, 1873b: 506 (*Mylabris*) **A:** PA UP **ORR**
dolens Marseul, 1872a: 417 (*Coryna*) **A:** SY "Palestine"
dorsetiferus Pan, Ren & Wang, 2011: 179 **A:** FUJ GUX NP SCH XIN YUN ZHE **ORR**
dubiosus Marseul, 1870b: 35 (*Mylabris*) **N:** EG **AFR**
 antebiplagiatus Kaszab, 1960f: 278 (*Mylabris*)
 bipartitus Marseul, 1873b: 427 (*Mylabris*)
 cursorius Kaszab, 1960f: 278 (*Mylabris*)
 dabbenei Pic, 1914p: 110 (*Zonabris*)
 erythraeus Pic, 1908g: 60 (*Zonabris*)
 katesheanus Kaszab, 1960f: 278 (*Mylabris*)
 mimis Kaszab, 1960f: 278 (*Mylabris*)
 peregrinus Kaszab, 1960f: 278 (*Mylabris*)
 rollei Pic, 1907b: 119 (*Zonabris*)
 taenia Kaszab, 1960f: 278 (*Mylabris*)
dufourii Graells, 1849: 621 (*Mylabris*) **E:** PT SP
duodecimmaculatus A.G. Olivier, 1811: 98 (*Mylabris*) **N:** MO
duodecimpunctatus A.G. Olivier, 1811: 98 (*Mylabris*) **E:** FR PT SP
 fasciatus Fuente, 1910: 444 (*Zonabris*)
 massiliensis Pic, 1902b: 9 (*Zonabris*)
 postunipunctatus Mader, 1929: 109 (*Mylabris*)
dunalis Bologna & Turco, 2007: 21 **A:** AE
duplicatus Klug, 1845: nr. 4 (*Mylabris*) **A:** SA YE **AFR**
 zonatus Klug, 1845: nr. 5 (*Mylabris*)
esfandiarii Kaszab, 1969a: 279 (*Decapotoma*) **A:** IS
euphraticus Marseul, 1870b: 84 (*Mylabris*) **A:** IQ
fimbriatus Marseul, 1870b: 83 (*Mylabris*) **N:** EG **A:** SA YE
 bafrechianus Pic, 1955e: 1047
 bisbiseparatus Pic, 1955a: 1 (*Zonabris*)
 fisselensis Pic, 1944b: 7 (*Zonabris*)
 laurantianus Pic, 1955a: 1 (*Zonabris*)
 posticeluteonotatus Pic, 1955e: 1047
 villiersianus Pic, 1955e: 1047
flavohirtus Kaszab, 1958c: 298 (*Decapotoma*) **A:** AF
fuscus A.G. Olivier, 1811: 100 (*Mylabris*) **E:** AB AR CT ST **A:** IN IQ IS JO KZ SY TM TR
 alienus J.R. Sahlberg, 1913b: 196 (*Zonabris*)
 circascutellaris Pic, 1914o: 77 (*Zonabris*)
 deleta Marseul, 1870b: 92 (*Mylabris*)
 galatiensis Escherich, 1899: 39 (*Zonabris*)
fraudulentus fraudulentus Bologna, 2007: 26 **A:** SA

fraudulentus hyneki Batelka, 2009b: 195 **A:** AE
ghorfii Pardo Alcaide, 1961b: 215 (*Gorrizia*) **N:** MO TU
gratiosus Marseul, 1870b: 135 (*Mylabris*) **N:** EG **A:** AE OM SA
hanguensis Kaszab, 1958c: 290 (*Mylabris*) **A:** PA
hemprichi Klug, 1845: n. 3 **N:** AG EG LB MO TU **A:** IN IQ IS JO SI SY
  kocheri Pardo Alcaide, 1954a: 68
himalayaensis Saha, 1979: 66 (*Mylabris*) [DA] **A:** AP NP UP
hokumanensis Kôno, 1940: 58 (*Mylabris*) [DA] **A:** HEI
horai Saha, 1972: 20 (*Mylabris*) [DA] **A:** PA
hirtus Tan, 1992: 576 (*Mylabris*) **A:** FUJ HUB SCH YUN
humerosus Escherich, 1899: 97 (*Zonabris*) **E:** AB AR GG GR ST TR **A:** IN IQ JO KZ SY TM TR UZ
infasciculatus Pic, 1929e: 14 (*Zonabris*) **A:** SA
japonicus Sumakov, 1913b: 474 (*Mylabris*) [DA] **A:** JA
kirgisicus Aksentjev & Sadykova, 1990: 152 (*Mylabris*) **A:** KI
lacteus Marseul, 1870b: 140 (*Mylabris*) **N:** EG
ligatus Marseul, 1870b: 31 (*Mylabris*) **A:** SA **AFR**
lindbergi Kaszab, 1973d: 294 (*Mylabris*) **A:** AF
linnavuorii Pardo Alcaide, 1963: 295 (*Mylabris*) **A:** IS
maculiventris Klug, 1845: n. 2 (*Mylabris*) **N:** EG **A:** OM SA YE **AFR**
  rubronotatithorax Pic, 1911j: 146 (*Zonabris*)
mannheimsi Kaszab, 1961g: 349 (*Mylabris*) **A:** SCH YUN
marcipoli Pan & Bologna, 2014b: 13 **A:** GAN GUA TAI **ORR**
mediobipunctatus Pic, 1919f: 19 (*Zonabris*) [DA] **A:** IN
mediofasciatellus Pic, 1908b: 33 (*Zonabris*) **N:** EG **A:** SI
  heluanensis Pic, 1908b: 33 (*Zonabris*)
  massariensis Pic, 1908b: 33 (*Zonabris*)
  sesostrisi Pic, 1930a: 2 (*Zonabris*)
medioinsignatus Pic, 1909m: 166 (*Zonabris*) **A:** BEI FUJ GUI GUX HEB HEN HP HUB MG NP PA SCH SHA SHN TIA XIZ YUN **ORR**
  phaleratoides Kaszab, 1960d: 261
mediozigzagus Pic, 1924d: 25 (*Zonabris*) [DA] **A:** IN
mimosae A.G. Olivier, 1811: 99 **A:** SA SY
mylabroides Laporte, 1840: 268 **N:** EG **AFR**
  mediofasciatus Pic, 1914p: 114 (*Coryna*)
nigriplantis Klug, 1845: n. 9 (*Mylabris*) **A:** IS KU SA YE **AFR**
nigrohirtus Kaszab, 1983: 167 (*Decapotoma*) **A:** SA
novemdecimpunctatus A.G. Olivier, 1811: 98 (*Mylabris*) **N:** AG EG MO TU **A:** IS SA SI
  argenteopubescens Escalera, 1909b: 247 (*Zonabris*)
  bicolorepilosus Sumakov, 1913a: 473 (*Mylabris*)
  octodecimpunctatus Klug, 1845: no. 23 (*Mylabris*)
  semifasciatus Pic, 1895e: 81 (*Zonabris*)

ocellaris A.G. Olivier, 1795: 48: 6 (*Cerocoma*) **N:** EG LB **AFR**

multijunctus Pic, 1908b: 34 (*Coryna*)

octodecimmaculatus Marseul, 1870b: 147 (*Mylabris*) **N:** AG EG LB MO TU **A:** IS

melaini Escalera, 1909b: 249 (*Zonabris*)

ornatus Reiche, 1866: 630 (*Coryna*) **A:** IS JO LE SI SY

berytensis Pic, 1912q: 177 (*Coryna*)

diversesignatus Pic, 1939f: 147 (*Coryna*) [HN]

latebijunctus Pic, 1912q: 177 (*Coryna*)

pakistanus Bologna, 2918: 649 **A:** PA

parvulus J. Frivaldszky, 1892: 114 (*Zonabris*) **A:** "China" YUN **ORR**

lajoyei Pic, 1916i: 126 (*Zonabris*)

phaleratus Pallas, 1781: 78 (*Meloe*) **A:** AHN FUJ JIX GUA GUI GUX HAI HEN HP HUB KA NP PA UP XIZ YUN ZHE **ORR**

choaicus Saha, 1972: 79 (*Mylabris*) [DA]

irigator Kaszab, 1960d: 259 (*Mylabris*),

moquinianus L. Ferrer, 1859: 540 (*Mylabris*)

sidae Fabricius, 1798: 120 (*Mylabris*)

pierrei Kaszab, 1968f: 765 (*Mylabris*) **A:** IN

pintoi Bologna & Turco, 2007: 19 **A:** SA

pirata Bologna & Turco, 2007: 16 **A:** AE OM

pitcheri Kaszab, 1983a: 169 (*Coryna*) **A:** AE SA

polymorphus polymorphus Pallas, 1771: 465 (*Attelabus*) **E:** AB AN AR AU BH BU BY CR CT CZ FR GE GG GR HU IT KZ LS MC MD PL RO SK SL SP ST SZ TR UK YU **A:** IN FE KZ SY TD TM TR UZ WS XIN

agilis Baudi di Selve, 1878b: 367 (*Mylabris*)

alpestris Pic, 1925f: 13 (*Zonabris*)

australis Baudi di Selve, 1878b: 367 (*Mylabris*)

baregesius Pic, 1925f: 13 (*Zonabris*)

cichorii Schrank von Paula, 1781: 222 (*Meloe*) [HN]

fasciatus Fuessly, 1775: 20 (*Meloe*)

floralis Pallas, 1782: 82 (*Meloe*)

frivaldszkyi Kaszab, 1956b: 151 (*Mylabris*)

fuesslini Panzer, 1796a: 18 (*Mylabris*)

karateginensis Sumakov, 1915: 73

luteointerruptus Pic, 1919f: 18 (*Zonabris*)

monetierensis Pic, 1925f: 13 (*Zonabris*)

mulsanti Pic, 1925f: 13 (*Zonabris*)

multidisjunctus Pic, 1929a: 2 (*Zonabris*) [HN]

multijunctus Kaszab, 1958b: 190 (*Mylabris*) [RN]

nigritulus Kaszab, 1958b: 190 (*Mylabris*) [RN]

nigritus Baudi di Selve, 1878b: 368 (*Mylabris*) [HN]

octomaculatus Villers, 1789: 404 (*Meloe*)
pelissieri Pic, 1925f: 13 (*Zonabris*)
quadrifarius Marseul, 1870b: 79 (*Mylabris*)
spartii Germar, 1817: 210 (*Mylabris*)
subbjunctus Pic, 1929a: 2 (*Zonabris*)
variabilis Olivier, 1795: 47: 10 (*Mylabris*) [HN]
viturati Pic, 1925f: 13 (*Zonabris*)
polymorphus regeli Sumakov, 1929a: 4 (*Mylabris*) **A**: UZ
postbilunulatus Pic, 1919f: 21 (*Zonabris*) [DA] **A**: "Kurdistan"
kurdistanicus Pic, 1919f: 21 (*Zonabris*)
posticatus Fairmaire, 1892b: 154 (*Mylabris*) **A**: SY TR
praeustus Fabricius, 1792b: 88 (*Mylabris*) **N**: AG MO TU
apicalis Chevrolat, 1840b: 278 (*Mylabris*)
contextus Chevrolat, 1840b: 278 (*Mylabris*)
nigritus Marseul, 1870b: 57 (*Mylabris*)
semirufus Marseul, 1870b: 57 (*Mylabris*)
superfluus Marseul, 1870b: 57 (*Mylabris*)
pseudobrunnipes Kaszab, 1983a: 163 (*Mylabris*) **A**: AE OM SA
quatuordecimpunctatus Pallas, 1781: 80 (*Meloe*) **E**: AR CT ST UK **A**: ES IN KZ TR UZ WS XIN
conbustus Tauscher, 1812b: 143 (*Mylabris*)
confusus Fischer von Waldheim, 1827: 6 (*Mylabris*)
meliloti A.G. Olivier, 1811: 99 (*Mylabris*)
setiger Waltl, 1838: 466 (*Mylabris*)
quatuordecimsignatus Marseul, 1870b: 133 (*Mylabris*) **N**: AG EG MO **A**: IS JO SA SI
peyerimhoffi Pic, 1929h: 126 (*Mylabris*)
pharius Pic, 1916a: 7 (*Zonabris*)
rajah Marseul, 1872: 452 (*Mylabris*) **A**: PA **ORR**
raphael Marseul, 1876a: 29 (*Mylabris*) **N**: EG **A**: IN IS SI
rotroui Pic, 1930a: 1 (*Coryna*) **N**: AG LB MO
ringenbachi Bologna, 2009: 351
rouxii Laporte de Castelnau, 1840: 268 (*Mylabris*) **A**: PA **ORR**
rubricollis Marseul, 1875b: xxiv (*Coryna*) **A**: SA YE
ruficornis Fabricius, 1798: 121 (*Mylabris*) **N**: MO
deauratus Pardo Alcaide, 1954a: 84 (*Mylabris*)
igneopubescens Escalera, 1909b: 249 (*Zonabris*)
tangerianus Pic, 1903g: 114 (*Zonabris*)
rufipalpis Escalera, 1909b: 248 (*Zonabris*) **N**: CI MO
discrepans Pardo Alcaide, 1954a: 82 (*Mylabris*) [HN]
mamorensis Pardo Alcaide, 1954a: 82 (*Mylabris*)
pardoianus Kaszab, 1958b: 190 (*Mylabris*) [RN]
rungsi Peyerimhoff, 1935b: 20 (*Zonabris*) **N**: MO

saharicus Chobaut, 1901: 280 (*Mylabris*) **N:** AG MO TU
mayeti Chobaut, 1901: 2890 (*Mylabris*)
oasium Chobaut, 1901: 2890 (*Mylabris*)
vareillesi Chobaut, 1901: 2890 (*Mylabris*)
scabiosae djiroftanus Kaszab, 1959b: 3 (*Mylabris*) **A:** IN
scabiosae dressi Kaszab, 1958c: 296 (*Mylabris*) **A:** AF
scabiosae scabiosae A.G. Olivier, 1811: 99 (*Mylabris*) **E:** AB AR BU CT GG GR RO ST **A:** IQ IN IS JO KI KZ LE MG SY TD TM TR UZ XIN
alieniger Heyden, 1883a: 67 (*Mylabris*)
auliensis Pic, 1901l: 111 (*Zonabris*)
eichleri Mader, 1927: 194 (*Zonabris*)
ghorbandanus Kaszab, 1958c: 295 (*Mylabris*)
golnazae Makhan & Ezzatpanah, 2011c: 1
maracadensis Dokhtouroff, 1889: 130 (*Zonabris*)
marseuli Ballion, 1878: 336 (*Mylabris*)
sicheri Fleischer, 1921: 142 (*Zonabris*)
transcaspicensis Pic, 1928e: 5 (*Zonabris*)
scabratus Klug, 1845: nr. 33 (*Mylabris*) **A:** OM SA SI YE
scapularis Klug, 1845: nr. 6 (*Mylabris*) **A:** OM SA YE
schah schah Reiche, 1866: 632 (*Mylabris*) **A:** IN PA
postluteodisjunctus Kaszab, 1957b: 59 (*Mylabris*)
postluteoreductus Kaszab, 1957b: 59 (*Mylabris*)
schah tuxeni Kaszab, 1953a: 306 **A:** AF (*Mylabris*)
schauffelei Kaszab, 1957b: 57 (*Mylabris*) **A:** IN PA UZ
anticebiconjuncta Kaszab, 1957b: 58 (*Mylabris*)
bisbisconjuncta Kaszab, 1957b: 58 (*Mylabris*)
postmediojuncta Kaszab, 1957b: 58 (*Mylabris*)
schoenherri pretiosus Kaszab, 1961g: 350 (*Mylabris*) **A:** FUJ
schoenherri schoenherri Billberg, 1813: 14 (*Mylabris*) **A:** FUJ GUA GUX HAI JIX TAI **ORR**
scutellatus Rosenhauer, 1856: 231 (*Mylabris*) **E:** PT SP
arapilensis Redondo, 1915: 30
deficiens De la Fuente, 1910: 445
lichtensteini Gorriz y Muñoz, 1882: 124 (*Mylabris*)
salmanticensis Redondo, 1915: 31
sexpunctata De la Fuente, 1913: 476
sexmaculatus A.G. Olivier, 1811: 98 (*Mylabris*) **E:** AB AR GG ST UK **A:** IN IQ IS JO KZ SA SI SY TR "Turkestan"
bimaculatus Klug, 1845: nr. 32 (*Mylabris*) [HN]
ledereri Marseul, 1870b: 57 (*Mylabris*)
oneratus Escherich, 1899: 109 (*Zonabris*)
superbus Faldermann, 1837: 123 (*Mylabris*)
sialanus Pic, 1929g: 110 (*Zonabris*) [DA] **N:** EG

silbermanni Chevrolat, 1840b: 277 (*Mylabris*) **N:** AG MO
reducticornis Pic, 1939b: 22 (*Zonabris*)
saidensis Pic, 1897l: 124 (*Zonabris*)
vicinus Lucas, 1846: pl. 34 (*Mylabris*)
solanensis Saha, 1979: 83 (*Mylabris*) [DA] **A:** HP **ORR**
solonicus Pallas, 1782: 87 (*Meloe*) **A:** ES FE HEI KZ MG NE NW WS "Korea"
dianae J.R. Sahlberg, 1913b: 196 (*Zonabris*)
famelica Ménétriés, 1854: 36 (*Mylabris*)
subparallelus Pic, 1919f: 19 (*Zonabris*) [DA] **A:** IN
talhouki Kaszab, 1983a: 160 (*Mylabris*) **A:** KU SA
talyshensis Sumakov, 1929a: 3 (*Mylabris*) [DA] **E:** AB
tekkensis marani Kaszab, 1958c: 297 (*Mylabris*) **A:** AF
antebistrinotatus Kaszab, 1958c: 297
tekkensis tekkensis Heyden, 1883e: 360 (*Zonabris*) **A:** AF IN KZ TD TM UZ
adnexus Mader, 1929: 109 (*Mylabris*)
atavus Mader, 1929: 109 (*Mylabris*)
imitatus Mader, 1929: 109 (*Mylabris*)
latebifasciatus Pic, 1919f: 18 (*Zonabris*)
tenerus Germar, 1836b: 5 (*Mylabris*) **E:** AU BU CZ HU RO SK YU
tenuepictus Fairmaire, 1892b: 154 (*Mylabris*) **N:** EG **A:** IS JO SY TR
marsabensis Pic, 1898b: 3 (*Zonabris*)
terminatus Illiger, 1800: 143 (*Mylabris*) **N:** EG **AFR**
afzelii Billberg, 1813: 48 (*Mylabris*)
bertrandi Laporte, 1840: 270 (*Mylabris*)
ustulatus Reiche, 1866: 633 (*Mylabris*)
theryanus Pic, 1940a: 2 (*Zonabris*) **N:** MO
alluaudi Pic, 1936e: 30 (*Zonabris*) [HN]
lecerfi Pic, 1940a: 2 (*Zonabris*)
lehuensis Pic, 1940a: 2 (*Zonabris*)
tigrinus Klug, 1845: n. 24 (*Mylabris*) **N:** EG **A:** IS SA SI YE **AFR**
tigripennis Marseul, 1870b: 143 (*Mylabris*) **N:** LB EG **A:** IS
obscurior Pic, 1897l: 125 (*Zonabris*)
trianguliferus Heyden, 1883e: 359 (*Zonabris*) **A:** AF IN KZ TD TM UZ
trizonatus Reiche, 1866: 631 (*Mylabris*) **N:** AG EG LB MO TU **A:** IS JO SI
biinterrupta Pic, 1898l: 224
binotata Pic, 1897l: 126
breveinterrupta Pic, 1919a: 18 (*Zonabris*)
coronata Marseul, 1873b: 556 (*Mylabris*)
corynoides Reiche, 1866: 631 (*Mylabris*)
gamondana Pic, 1919a: 18 (*Zonabris*)
gafsana Pic, 1919a: 18 (*Zonabris*)
interrupta Pic, 1897l: 126

mediodivisa Pic, 1919a: 18 (*Zonabris*)
setafana Pic, 1919a: 18 (*Zonabris*)
subdivisa Pic, 1919a: 18 (*Zonabris*)
subhumeralis Pic, 1903g: 114
subhumeralis Normand, 1938b: 365 (*Mylabris*) [HN]
trinotata Pic, 1953d: 499
varius A.G. Olivier, 1811: 96 (*Mylabris*) **N:** EG **A:** IS
wagneri Chevrolat, 1840b: 274 (*Mylabris*) **N:** AG TU
brevis Wellman, 1910a: 23 (*Mylabris*)
curtus Chevrolat, 1840b: 277 (*Mylabris*)
ernesti Pic, 1905i: 170 (*Zonabris*)
insignipennis Pic, 1901h: 77 (*Zonabris*)
mauricei Wellman, 1910a: 23 (*Mylabris*)
wittmeri Kaszab, 1983a: 152 (*Ceroctis*) **A:** SA
yemenicus Kaszab, 1983a: 171 (*Actenodia*) **A:** YE
yerburyi Gahan, 1896b: 456 (*Mylabris*) **A:** SA YE
theresae Pic, 1908b: 33 (*Zonabris*)
zebraeus Marseul, 1870b: 79 (*Mylabris*) **E:** AB AL AR BH BU GG GR MC TR **A:** IN IQ IS JO SY TM TR UZ "Turkestan"
kervillei Pic, 1932e: 17 (*Zonabris*)
tenenbaumi Pic, 1928e: 5 (*Zonabris*)
zoufali Reitter, 1908j: 272 (*Zonabris*)

genus *Lydoceras* Marseul, 1870b: 12 type species *Mylabris fasciata* Fabricius, 1775
Denierus Pic, 1913t: 175 type species *Zonabris stanleyana* Duvivier, 1890
fasciatus Fabricius, 1775: 261 (*Mylabris*) **N:** EG **A:** JO SA YE **AFR**
unifasciatus A.G. Olivier, 1811: 92 (*Mylabris*)
lictor Gerstaecker, 1884: 57 (*Mylabris*) **A:** "Arabia" **AFR**
bilineatus Thomas, 1897b: 501 (*Mylabris*)

genus *Mylabris* Fabricius, 1775: 261 type species: see Löbl, 2008
subgenus *Argabris* Kuzin, 1954: 362 type species *Meloe ocellata* Pallas, 1773
audouini Marseul, 1870b: 141 **A:** IN KI KZ TD UZ XIN
bangi Heyden, 1886a: 181 (*Zonabris*)
impedita bogatshevi Kaszab, 1958c: 292 **A:** AF PA
impedita impedita Heyden, 1883a: 66 (*Zonabris*) **A:** AF KI KZ TD UZ
undulata J.R. Sahlberg, 1908: 90 [HN]
undulatella Kaszab, 1958b: 190 [RN]
impedita kryzhanovskii Pripisnova, 1987b: 563 **A:** TD
impedita lopatini Pripisnova, 1987b: 563 **A:** TD
klugi kabuliensis Kaszab, 1958c: 293 **A:** AF

klugi klugi L. Redtenbacher, 1850: 49 **A:** IN KZ PA TD TM TR XIN
graffi Pic, 1947b: 10 (*Zonabris*)
klugi lacustris Kaszab, 1958c: 294 **A:** AF
ocellata Pallas, 1773: 721 (*Meloe*) **A:** KZ TD TM UZ XIN
argus A.G. Olivier, 1811: 98
octonotata Fischer von Waldheim, 1842: 132
pupillata Schoenherr, 1817: 60
quatuordecimmaculata Dokhtouroff, 1889: 150 **A:** TD UZ
sedecimpunctata Gebler, 1825: 49 **A:** AF IN KI KZ MG TD TM UZ WS XIN
gilana Pic, 1924d: 26 (*Zonabris*)
quindecimpunctata Sumakov, 1909: 220
tenuepunctata Dokhtouroff, 1889: 165 (*Zonabris*)
wilkinsi Dokhtouroff, 1889: 155 (*Zonabris*)
subgenus *Calydabris* Kaszab, 1960b: 133 type species *Calydus allousei* Kaszab, 1960
allousei Kaszab, 1960b: 131 (*Calydus*) **A:** IN IQ
cinctus Kaszab, 1960b: 133 (*Calydus*)
barezensis Serri, Pan & Bologna, 2012: 82 **A:** IN
mirzayani Kaszab, 1968f: 769 (*Calydus*) **A:** IN PA
subgenus *Chalcabris* Kuzin, 1954: 346 type species *Meloe festiva* Pallas, 1773
bistillata Tan, 1981: 407, 409 **A:** SCH XIZ YUN
bivulnera Pallas, 1781: 94 (*Meloe*) **E:** ST **A:** ES KI KZ MG WS XIN
bimaculatus Pallas, 1771: 466 (*Attelabus*)
caspica Gmelin, 1790: 1896 (*Cantharis*)
luteosexmaculata Pic, 1919f: 21
coeruleomaculata L. Redtenbacher, 1843: 987 **A:** IN IS JO LE SY
jarrahiana Kaszab, 1959b: 4 (*Decapotoma*)
shadegana Kaszab, 1959b: 4 (*Decapotoma*)
subhumeralis Pic, 1908k: 74
viridiflua Marseul, 1873b: 654
coerulescens Gebler, 1841a: 374 **E:** ST **A:** AF IN KZ MG TD TM UZ XIN
cyaneovaria Reitter, 1889b: 121 (*Zonabris*) **A:** AF KZ MG TM
speciosoides Sumakov, 1915: 7
festiva Pallas, 1773: 721 (*Meloe*) **E:** AR BU CT GR MC RO ST TR UK **A:** ES IN KZ MG TR WS XIN
sericea Pallas, 1782: 85 (*Meloe*)
formosa Wellman, 1910a: 23 [RN] **E:** BU GR MC TR **A:** TR
decora I. Frivaldszky, 1835: 264 [HN]
koenigi Dokhtouroff, 1889: 143 (*Zonabris*) **A:** AF IN KZ TD TM UZ XIN
ledebouri fortunata Kaszab, 1969c: 319 **A:** MG
ledebouri ledebouri Gebler, 1829b: 166 [DA] **E:** ST **A:** ES KZ MG WS XIN
marginata Fischer von Waldheim, 1844: 136 **E:** AB AR **A:** ES IN XIN
mongolica Dokhtouroff, 1887b: 345 (*Zonabris*) **A:** HEN MG NMO XHA WS XIN

alpha Sumakov, 1915: 7

aurora Escherich, 1904: 34 (*Zonabris*)

beta Sumakov, 1915: 7

chinensis J. Frivaldszky, 1892: 114 (*Zonabris*)

monozona Wellman, 1910a: 23 [RN] **E:** CT ST **A:** ES KI KZ MG WS XIN

unifasciata Ballion, 1878: 343 [HN]

munda Escherich, 1897a: 38 (*Zonabris*) **A:** TR

pulchella Faldermann, 1833: 59 **A:** KI KZ MG TD UZ XIN

sairamensis Ballion, 1878: 342 [DA] **A:** XIN

semivittata Pic, 1947: 7 (*Zonabris*) **A:** "Central Asia"

speciosa Pallas, 1781: 84 (*Meloe*) **A:** AF ES FE GAN HEB HEI JIX LIA KZ MG NMO QIN SHA UZ WS

uhagonii Martinez y Sáez, 1873a: 66 [DA] **E:** SP

perezi Pic, 1916: 7

postinsignata Mader, 1929: 108

quaterna Mader, 1929: 108

terna Mader, 1929: 108

traizeti Pic, 1916: 7

subgenus *Chrysabris* Kuzin, 1954: 348 type species *Meloe trifascis* Pallas, 1773

doriai Marseul, 1870b: 164 **A:** IN IQ

boetcheri Pic, 1909n: 171

centropunctatella Kaszab, 1958b: 189 [RN]

centropunctata Bodemeyer, 1916: 112 [HN]

daligrandi Pic, 1947: 7 (*Zonabris*)

luristanica Pic, 1920d: 12

persicola Pic, 1909n: 171

magnificia Bodemeyer, 1916: 112

unifasciata Bodemeyer, 1916: 112 [HN]

viridibifasciata Pic, 1920e: 13 (*Zonabris*)

viridifasciata Pic, 1920e: 13 (*Zonabris*)

gemmula Dohrn, 1873: 73 **A:** IN TM

ghazniana Kaszab, 1973d: 278 **A:** AF

karakalensis Kryzhanovskij, 1956: 886 **A:** TM

kraatzi Heyden, 1881: 329 (*Decatoma*) **A:** IN

plurivulnera Dohrn, 1873: 73 **A:** IN TM

marseuli Kirsch, 1880: 77 [HN]

viridula Marseul, 1882c: 170 [RN]

smaragdina Gebler, 1841b: 597 **A:** KZ MG WS

trifascis Pallas, 1773: 721 (*Meloe*) **A:** AF IN KZ TM WS "Turkestan"

pallasi Gebler, 1829b: 170

vittata Kirsch, 1880: 77

subgenus *Dvorabris* Bologna, 2019: 257 type species *Croscherichia maceki* Dvořák, 1985

maceki Dvořák, 1985: 1 (*Croscherichia*) **A:** IN IQ

subgenus *Eumylabris* Kuzin, 1954: 355 type species *Mylabris decempunctata* Fabricius, 1792 (= *Mylabris fabricii* Sumakov, 1924)

Neabris Kaszab, 1958c: 284 type species *Mylabris klapperichi* Kaszab, 1958

Spinabris Pan, Wang & Ren, 2010: 192 type species *Mylabris spinungulata* Pan, Wang & Ren, 2010

aulica Ménétriés, 1832: 208 **E:** AB **A:** ES FE HEI KZ MG NMO NO WS

breveapicalis Pic, 1919f: 21 (*Zonabris*) **A:** TR

calida Pallas, 1782: 85 (*Meloe*) **E:** AB AR BU CT GG GR MC ST TR UK **N:** AG EG LB MO TU **A:** AF ES FE HEB HEI HUB IN IQ IS JIA JIL JO KI KU KZ LE LIA MG NC NE NMO SA SC SHA SHX SI SY TD TM TR UZ XIN WS YE

aschnae Makhan & Ezzatpanah, 2011a: 1

baicalica Pic, 1919f: 22 (*Zonabris*)

bijuncta Pic, 1897a: 2 (*Zonabris*)

bimaculata A.G. Olivier, 1811: 93

bimaculaticeps Pic, 1920e: 13 (*Zonabris*)

decora A.G. Olivier, 1811: 94 [HN]

interrupta Eichler, 1924: 66 (*Zonabris*) [HN]

latifasciata Pic, 1896f: 62 (*Zonabris*)

maculata A.G. Olivier, 1795: 47: 7

maculata Eichler, 1923: 66 (*Zonabris*) [HN]

maroccana Escherich, 1899: 107 (*Zonabris*)

maura Chevrolat, 1840b: 273

niligena Reiche, 1866: 638

tlemceni Pic, 1930a: 3 (*Zonabris*)

transcaspica Escherich, 1899: 107 (*Zonabris*)

cincta A.G. Olivier, 1811: 93 **E:** AB AR CT GG GR MC ST TR UK **N:** AG EG LB MO TU **A:** AF HP IN IQ IS JO KZ PA SA SI SY TR UP

alfierii Pic, 1921g: 45 (*Zonabris*)

algerica Pic, 1898b: 3 (*Zonabris*)

bicolor Waltl, 1838: 465

diversepunctata Pic, 1921a: 5 (*Zonabris*)

gafsaensis Pic, 1915a: 2 (*Zonabris*)

jugatoria Reiche, 1866: 633

latecincta Kocher, 1955b: 128

lateripuncta Kaszab, 1957b: 55

matthesi Faldermann, 1837: 120

mediocingulata Kaszab, 1957b: 55

minae Makhan & Ezzatpanah, 2011b: 1

pseudocincta Kaszab, 1957b: 55

rufobifasciata Pic, 1921a: 5 (*Zonabris*)

rufomaculata Pic, 1921a: 5 (*Zonabris*)
rufotrifasciata Pic, 1921a: 5 (*Zonabris*)
soubironi Pic, 1921a: 5 (*Zonabris*)
taeniata Waltl, 1838: 465
tiflensis Billberg, 1813: 13
tricingulata L. Redtenbacher, 1850: 49
cinctoides Kaszab, 1957b: 55 **A:** IN
crocata Pallas, 1781: 87 (*Meloe*) **E:** AB AR BU CT GE GG GR HU RO ST TR UK **A:** ES HEI IN IS JO KI KZ MG NO SY TD TR UZ WS XIN
daurica Baudi di Selve, 1878b: 372
duodecimpunctata Tauscher, 1812b: 139 [HN]
spectabilis Baudi di Selve, 1878b: 372
fabricii Sumakov, 1924: 27 **E:** AB AL AR BU CT GG GR IT MC MD ST SZ TR UK YU **A:** IN IS JO KI KZ SI SY TD TM TR UZ
decimpunctata V. Petagna, 1787: 27 [HN]
forti Mulsant, 1857a: 133
sexpunctata Fuente, 1913: 308 (*Zonabris*) [HN]
sexpunctulata Kaszab, 1958b: 190
groschkei Kaszab, 1957b: 53 **A:** AF IN
impressa hoggarensis Pic, 1929h: 127 (*Zonabris*) **N:** AG
impressa impressa Chevrolat, 1840b: 275 **N:** AG MO TU
aini Pic, 1921a: 4 (*Zonabris*)
antejuncta Pardo Alcaide, 1955b: 54
bisbijuncta Pardo Alcaide, 1955b: 55
externobliterata Pardo Alcaide, 1955b: 55
oranensis Pic, 1909q: 185 (*Zonabris*)
vaucheri Escalera, 1914: 400 (*Zonabris*)
impressa stillata Baudi di Selve, 1878b: 375 **E:** IT (Sicilia)
ragusai Pic, 1897o: 217 (*Zonabris*)
klapperichi hirtipedes Kaszab, 1958c: 286 **A:** AF PA
klapperichi klapperichi Kaszab, 1958c: 285 **A:** AF PA
magnoguttata iranica Kaszab, 1957b: 56 **A:** IN
magnoguttata magnoguttata Heyden, 1881: 329 (*Zonabris*) **E:** ST **A:** AF IN KI KZ PA TD TM UZ XIN
angelicae Reitter, 1889b: 121
aschabadensis Voigts, 1902: 178 (*Zonabris*)
disjuncteguttata Pic, 1919f: 19 (*Zonabris*)
fasciculata Escherich, 1899: 108 (*Zonabris*)
heydeniana Dokhtouroff, 1889: 145 (*Zonabris*)
margarithae Reitter, 1889b: 120 (*Zonabris*)
magnoguttata pardoi Kaszab, 1958c: 278 **A:** AF PA

omocrates Wellman, 1910a: 23 [RN] **A:** ES HEB MG

axillaris Motschulsky, 1872: 50 [HN]

posticalis Dokhtouroff, 1889: 136 (*Zonabris*) **A:** AF IN KI PA UZ XIN

pulchra Kaszab, 1973d: 287 **A:** AF PA

quinqueplagiata Kaszab, 1958c: 274 **A:** AF PA

schrenkii Gebler, 1841a: 374 **A:** AF IN KI KZ TD TM UZ XIN

reitteri Dokhtouroff, 1889: 157 (*Zonabris*)

sculptilis Kaszab, 1958c: 279 **A:** AF

spinungulata Pan, Wang & Ren, 2010: 193 **A** XIN

subbrevicornis Kaszab, 1958c: 276 **A:** AF

tiflensis Billberg, 1813: 13 **A:** PA **ORR**

adam Kaszab, 1958c: 287

antefracta Pardo Alcaide, 1958b: 2

subgenus *Lachnabris* Kuzin, 1953: 137 type species *Mylabris mannerheimii* Gebler, 1844

crux Escherich, 1899: 103 (*Zonabris*) **E:** ST **A:** TD TM UZ

opulenta Escherich, 1904: 33 (*Zonabris*)

mannerheimii Gebler, 1844: 103 **A:** AF KI KZ TD TM XIN

turkestanica Dokhtouroff, 1889: 168 (*Zonabris*) **A:** TD

undecimnotata Heyden, 1883a: 66 (*Zonabris*) **A:** IN KZ UZ

subgenus *Mauritabris* Pardo Alcaide, 1969: 375 type species *Mylabris tenebrosa* Laporte, 1840

Zitunabris Pardo Alcaide, 1969: 375 type species *Mylabris oleae* Chevrolat, 1840

abdelkaderi Escalera, 1909b: 246 (*Zonabris*) **N:** MO

baulnyi Marseul, 1870b: 49 **N:** AG MO

cyrenaica Bologna, 2009: 355 **N:** LB

damascena Reiche, 1866: 634 **A:** IN IQ IS JO LE SI SY

filicornis Marseul, 1870b: 64 **N:** EG **A:** IS JO SI

longipilis Pic, 1897l: 124 (*Zonabris*) **N:** EG

unifenestrata Pic, 1915a: 2 (*Zonabris*)

mateui Pardo Alcaide, 1954b: 338 **N:** MO

oleae Chevrolat, 1840b: 269 **N:** AG LB MO TU

bidivisa Pic, 1921f: 24 (*Zonabris*)

bireductenotata Pic, 1914a: 1 (*Zonabris*)

bisbireducta Pic, 1921f: 24 (*Zonabris*) [HN]

bisbireductella Kaszab, 1958b: 190 (*Mylabris*) [RN]

diversedivisa Pic, 1921f: 24 (*Zonabris*)

diversesuturalis Pic, 1933e: 13 (*Zonabris*)

drahamensis Pic, 1921f: 24 (*Zonabris*)

harterti Pic, 1921f: 90 (*Zonabris*)

hipponensis Pic, 1921f: 23 (*Zonabris*)

laghouatensis Pic, 1921f: 24 (*Zonabris*)

latenigra Pic, 1921f: 24 (*Zonabris*)

mediobliterata Pic, 1921f: 24 (*Zonabris*)
multinotata Pic, 1906c: 26 (*Zonabris*)
postdivisa Pic, 1921f: 24 (*Zonabris*)
rotroui Pic, 1933e: 13 (*Zonabris*) [HN]
separata Pic, 1897l: 123 (*Zonabris*)
tridivisa Pic, 1921f: 24 (*Zonabris*)
unidivisa Pic, 1921f: 24 (*Zonabris*)
rimosa Marseul, 1870b: 40 **N:** AG MO TU
tenebrosa Laporte, 1840: 270 **N:** AG EG LB MO TU **A:** SA SI
diverseluteonotata Pic, 1953d: 499 (*Zonabris*)
exuta Normand, 1950a: 81
mohtari Escalera, 1910a: 284 (*Zonabris*)
oberthuri Pic, 1953d: 498 (*Zonabris*)
pauloluteobifasciata Pic, 1953d: 499 (*Zonabris*)
postluteofasciata Pic, 1921a: 6 (*Zonabris*)
satanas Mader, 1929: 108
soufensis Pic, 1921a: 6 (*Zonabris*)
septemnotata Pic, 1953d: 498 (*Zonabris*)
triluteofasciata Pic, 1921a: 6 (*Zonabris*)
uniluteofasciata Pic, 1921a: 6 (*Zonabris*)
subgenus *Mesosulcata* Pardo Alcaide, 1950a: 65 type species *Mylabris hieracii* Graells, 1849
Glaucabris Kuzin, 1954: 350 type species *Mylabris hieracii* Graells, 1849
hieracii Graells, 1849: 621 **E:** FR PT SP
suspiciosa Rosenhauer, 1856: 229
hirtipennis Raffray, 1873: 379 **N:** AG MO
ifranensis Kocher, 1955b: 128
kabylica Escherich, 1899: 101 (*Zonabris*)
leprieuri Pic, 1896: 62
thamii Kocher, 1963a: 36 **N:** MO
subgenus *Micrabris* Kuzin, 1954: 351 type species *Mylabris geminata* Fabricius, 1798
Mesolaevigata Pardo Alcaide, 1954a: 68 type species *Mylabris geminata* Fabricius, 1798
afghanica Kaszab, 1953a: 308 **A:** AF PA
alpina Ménétriés, 1832: 208 **E:** AB AR ST
alpinoides Sumakov, 1929: 2 **A:** TD
apicefasciata Sumakov, 1929a: 3 [DA] **A:** TM
beauregardi Górriz Muñoz, 1884: 531 **E:** PT SP
pauper Escherich, 1899: 91 (*Zonabris*)
beckeri Escherich, 1890a: 24 **E:** ST
therondi Roubal, 1932a: 1
connata Rey, 1892: 138 (*Zonabris*) **E:** FR IT (Western Alps) SZ
apicalifrons Puel, 1907: 116
camarguensis Puel, 1907: 116

eryngii Puel, 1907: 116
gracilior Puel, 1907: 116
fenestrata Puel, 1907: 118 [HN]
meridionalis Puel, 1907: 116
novickii Puel, 1907: 116
pseudobsoleta Puel, 1907: 116
prolongata Puel, 1907: 116
pueli Mader, 1927: 196 [RN]
pueli Sumakov, 1930a: 30 [HN]
subobliterata Pic, 1908j: 72
dashidorzsi Kaszab, 1964e: 357 **A:** MG
deferreri Ruiz & García-París, 2004: 186 **E:** SP
dejeanii Gyllenhal, 1817: 35 **E:** PT SP
djebelina Pic, 1902i: 72 (*Zonabris*) **N:** MO
dokhtouroffi afghana Kaszab, 1958c: 269 **A:** AF
dokhtouroffi dokhtouroffi Escherich, 1899: 104 (*Zonabris*) **A:** IN TD UZ
coalescens Voigts, 1902: 178 (*Zonabris*)
excisofasciata Heyden, 1883a: 67 (*Zonabris*) **A:** AF IN KI TM UZ
externepunctata Faldermann, 1832: 208 **E:** AB AR ST **A:** IN TR
taurusiensis Pic, 1901h: 77
fenestrata Escherich, 1899: 108 (*Zonabris*) **A:** "Turkestan"
flexuosa A.G. Olivier, 1811: 101 **E:** AN FR IT SP SZ
abrutia Sumakov, 1915: 19 [RN]
aprutia Magistretti, 1943: 40 [HN]
cassetensis Pic, 1908j: 72
inflexuosa Pic, 1908j: 72
italica Magistretti, 1943: 34 [HN]
luteosubapicalis Pic, 1930a: 3 (*Zonabris*)
magistrettii Kaszab, 1958b: 190 [RN]
pyrenaea Pic, 1908j: 72
rarissima Pic, 1908j: 72
vittata Magistretti, 1943: 34 [HN]
vittatoides Kaszab, 1958b: 190 [RN]
frolovi aliensis Kuzin, 1927: 26 **A:** KZ
frolovi androssovi Sumakov, 1929a: 1 **A:** "Turkestan"
frolovi frolovi Fischer von Waldheim, 1823: pl. 40 **A:** AF ES IN KI KZ MG TD TM UZ WS XIN
confluens Fischer von Waldheim, 1823: 227
frolovi Germar, 1824b: 170 [HN]
obtecta Escherich, 1899: 108
sexguttata Pic, 1947: 7 (*Zonabris*)
sodalis Heyden, 1883a: 65 (*Zonabris*)

frolovi kopalensis Sumakov, 1930b: 13 [RN] **A:** KZ
schapinskii Sumakov, 1929a: 1 [HN]
frolovi kuziniana Kaszab, 1958c: 262 **A:** AF
frolovi shapinskyi Kuzin, 1927: 26 **A:** KZ
frolovi syrdarjensis Sumakov, 1929a: 2 **A:** KZ
furcimacula Sumakov, 1915: 20 **E:** AB AR ST **A:** IN TR
geminata Fabricius, 1798: 120 **E:** AB AR CT GG ST UK **A:** IN KI KZ TR WS XIN
kushakiewitschi Dokhtouroff, 1889: 167 (*Zonabris*)
hauseri Escherich, 1899: 101 (*Zonabris*) **A:** KI KZ XIN
inculta Escherich, 1899: 98 (*Zonabris*) **E:** AB AR "Caucasus" **A:** IN SY TR
humeromaculata Mader, 1929: 109
intermedia Fischer von Waldheim, 1844: 132 **A:** ES KI KZ TD XIN
karacaevica Roubal, 1914: 145 (*Zonabris*) **E:** ST **A:** TM
laevicollis Marseul, 1870b: 109 **E:** AB AL AR BH GG GR MC TR **A:** IN KZ TM TR
liliputana Escherich, 1899: 106 (*Zonabris*) **A:** KI KZ TD
lucens Escherich, 1904: 32 **A:** AF KI KZ TD TM UZ XIN
apicetenuimarginata Sumakov, 1913a: 304
macilenta Marseul, 1873b: 489 **A:** BT HP KA NP PA UP XIZ **ORR**
maculosopunctata Graells, 1858: 75 **E:** IT (Western Alps) PT SP
mendizabali Rosa, 2008: 80
rosinae Escherich, 1899: 104 (*Zonabris*)
unisticta Voigts, 1902: 178 (*Zonabris*)
marakensis badakschanica Kaszab, 1958c: 266 **A:** AF
marakensis marakensis Kaszab, 1953a: 311 **A:** AF TD
modesta elbursiensis Aksentjev, 1985: 247 **A:** IN
modesta modesta Escherich, 1899: 99 (*Zonabris*) **E:** AB AR **A:** IN TR
neglecta Escherich, 1899: 102 (*Zonabris*) **E:** GR (Peloponnese)
nevadensis Escalera, 1915: 493 (*Zonabris*) **E:** SP
nuristanica nuristanica Kaszab, 1958c: 269 **A:** AF PA
nuristanica subalpestris Kaszab, 1958c: 271 **A:** AF PA
obsoleta Nowicki, 1872: 6 **E:** IT
italica Sumakov, 1929a: 6
pannonica Kaszab, 1956c: 333 **E:** AU HU RO SK
pentheri Ganglbauer, 1905: 279 **E:** AB or AR "Araxes valley" **A:** TR
platai Pardo Alcaide, 1975: 45 **E:** SP
pusilla bosnica Reitter, 1903f: 230 (*Zonabris*) **E:** AL BH CR YU
hercegovinensis Mader, 1927: 106 [RN]
pusilla latialis Magistretti, 1943: 34 **E:** IT (Central Apennines)
pusilla pusilla A.G. Olivier, 1811: 101 **E:** AB AL AR BU CT GG GR MD ST TR YU UK **A:** KI KZ TR UZ WS XIN

quadrisignata Fischer von Waldheim, 1823: pl. 40 **A:** KI KZ MG UZ WS XIN
    antenigra Mader, 1929: 108
    longeundulata Pic, 1919f: 18 (*Zonabris*)
    voigtsi Escherich, 1904: 33
sedilethorax Sumakov, 1929a: 3 **E:** AR **A:** IN
serena Escherich, 1899: 97: 91 (*Zonabris*) **E:** AB AR **A:** TR
sexnotata L. Redtenbacher, 1843: 987 **E:** AR **A:** SY TR
sibirica Fischer von Waldheim, 1823: pl. 40 **E:** CT ST UK **A:** ES KI KZ MG WS XIN
    besseri Baudi di Selve, 1878b: 374
    kushakevitshi Dokhtouroff, 1889: 167
    luteobiinterrupta Pic, 1919f: 19
sinuata Klug, 1845: n. 19 **A:** CY LE SY TR
sobrina Graells, 1849: 621 **E:** PT SP
splendidula Pallas, 1781: 83 (*Meloe*) **A:** ES FE GAN KI KZ MG WS XIN
    apicetenuimarginata Sumakov, 1913a: 304
    staudingeri Heyden, 1881: 328 (*Zonabris*)
steppensis Dokhtouroff, 1889: 137 (*Zonabris*) **E:** GG **A:** KI KZ TD ZU XIN
tomentosa Escherich, 1899: 99 (*Zonabris*) E: AR **A:** IN TR
    externeundulata Pic, 1919f: 19 (*Zonabris*)
    kulpensis Pic, 1919f: 18 (*Zonabris*)
unicolor Faldermann, 1837: 127 **E:** AB AR **A:** IN TR
varians Gyllenhal, 1817: 34 **E:** PT SP
    bipunctata Fuente, 1909: 307 [HN]
    bipunctatofasciata Fuente, 1909: 308
    decimspilota Chevrolat, 1865: 393
    fuenteana Kaszab, 1958b: 190 [RN]
    fuentei Pic, 1909n: 171 (*Zonabris*)
    inconstans Chevrolat, 1865: 393
    luteipennis Marseul, 1870b: 106
    pozuelensis Pic, 1909n: 171 (*Zonabris*)
    quinquepunctata Fuente, 1909: 307
    sexpunctata Fuente, 1909: 307 [HN]
    sexpunctella Kaszab, 1958b: 190 [RN]
    tripunctatofasciata Fuente, 1909: 308
    unipunctata Fuente, 1909: 307
waziristanica Kaszab, 1958c: 299 (*Coryna*) **A:** AF PA
subgenus *Monabris* Kuzin, 1953: 137 type species *Mylabris undecimpunctata* Fischer von Waldheim, 1842
undecimpunctata Fischer von Waldheim, 1844: 17 **A:** AF IN KZ TD TM UZ XIN
    septempunctata Pic, 1901h: 78 (*Zonabris*)

subgenus *Mylabris* Fabricius, 1775: 261 type species: see Löbl, 2008

Mesopunctata Pardo Alcaide, 1954a: 74 type species *Meloe quadripunctata* Linnaeus, 1767

alpicola Pan & Bologna, 2014a: 15 **A:** IN

amorii Graells, 1858: 76 **E:** PT SP

korbi Escherich, 1891b: 53 (*Zonabris*)

apiceguttata Pan & Bologna, 2014a: 18 **A:** IN

apicenigra Sumakov, 1915: 24 **A:** IN SY TR

batnensis Marseul, 1870b: 121 **N:** AG MO

cernyi Pan & Bologna, 2014a: 22 **A:** IS JO LE PA TR

ciliciensis Escherich, 1899: 106 (*Zonabris*) **A:** TR

concolor Marseul, 1870b: 112 **E:** AB AR **A:** IN TM TR

desertica Bologna, 2007: 12 **A:** SA

emiliae Escherich, 1899: 105 (*Zonabris*) **E:** AR "Caucasus" **A:** SY TR

guerini Chevrolat, 1840b: 271 **N:** AG MO TU

lambaesis Pic, 1953d: 499 (*Zonabris*)

normandi Pic, 1936f: 155

pelletieri Baudi di Selve, 1878a: 364

rubripennis Chevrolat, 1840b: 270

kodymi Mařan, 1944a: 37 **E:** GR (Crete)

stepaneki Mařan, 1944a: 38

taborskyi Mařan, 1944a: 38

madani Escalera, 1909b: 247 (*Zonabris*) **N:** MO

tahari Escalera, 1909b: 246 (*Zonabris*)

transversa Kaszab, 1958b: 190 [RN]

transversalis Escalera, 1914: 396 (*Zonabris*) [HN]

madoni Marseul, 1883a: 178 **A:** CY

lividiana Pic, 1911n: 178

mediorientalis Pan & Bologna, 2014a: 32 **A:** IS JO IQ IN LE

olivieri Billberg, 1813: 71 **E:** AB AR CT GG GR MC RO ST TR UK **A:** IN KI KZ SY TD TM TR

caspia Ménétriés, 1832: 206

grisescens Reiche, 1866: 638 [HN]

impar Baudi di Selve, 1878b: 375 [HN]

impexa Escherich, 1899: 107 (*Zonabris*)

kopetensis Pic, 1930a: 2 (*Zonabris*)

parumpicta Heyden, 1883d: 353 (*Zonabris*) **A:** IN TM

poggii Bologna, 2009: 353 **N:** LB

pseudoemiliae Pan & Bologna, 2014a: 37 **A:** IN

quadripunctata Linnaeus, 1767: 680 (*Meloe*) **E:** AB AL AR BU CT FR GG GR IT MC PT RO SP ST TR UK YU **A:** AF ES IN IQ IS JO KI KZ LE MG SY TD TM TR UZ WS XIN

adamsii Fischer von Waldheim, 1824: 224

angustata Sumakov, 1915: 72 [DA]

antefasciata Sumakov, 1913a: 304 [RN]
armeniaca Faldermann, 1837: 125 [HN]
chobautiana Sumakov, 1930a: 37 [RN]
decimpunctata Fabricius, 1781: 86 (*Meloe*)
fasciatopunctata Adams, in Fischer von Waldheim, 1824: 224
husseini L. Redtenbacher, 1850: 49
maldinesi Chevrolat, 1865: 392
melanura Pallas, 1781: 86 (*Meloe*)
mesoasiatica Kuzin, 1953: 157 [DA]
octopunctata A.G. Olivier, 1811: 95
quadripunctata Billberg, 1813: 28 [HN]
restricta Motschulsky, 1849: 132 [DA]
rishwani Makhan, 2012b: 1
sexpunctata Nedelkov, 1905: 433
tetraspilota Wellman, 1910a: 23 [RN]
vareillesi Chobaut, 1921: 280 [HN]
schreibersi Reiche, 1866: 336 **E:** IT (Sicilia) **N:** AG MO TU
parumpunctata Magistretti, 1943: 35
terminata Chevrolat, 1840b: 276 [HN]
unifasciata Magistretti, 1943: 35 [HN]
unifasciatella Kaszab, 1958b: 190 [RN]
snizeki Černý & Vrabec, 2019: 147 **A:** JO
tauricola Marseul, 1870b: 73 **A:** TR
tricincta Chevrolat, 1840b: 270 **N:** AG MO
brevevittata Pic, 1935g: 159, [= 1935c: 4] (*Zonabris*)
nigrosuturata Escalera, 1914: 397 (*Zonabris*)
variabilis Pallas, 1781: 81 (*Meloe*) **E:** AB AL AR AU BH BU CR CT CZ FR GG GR HU IT MC MD PT RO SK SL SP ST SZ TR UK YU **A:** AF ES IN IS JO KZ TM TR UZ WS
apicenigroconjuncta Kaszab, 1956b: 152
disrupta Baudi di Selve, 1878c: 1160
erivanica Pic, 1901h: 77 (*Zonabris*)
guerini Magistretti, 1943: 34 [HN]
hypocrita Mulsant, 1857a: 125
italiana Kaszab, 1958b: 190 [RN]
lacera Fischer von Waldheim, 1827: 6
leonii Magistretti, 1943: 35
medioluteoabrupta Kaszab, 1956b: 151
medioexterneconjuncta Kaszab, 1956b: 152
medioluteobipunctata Kaszab, 1956b: 152
medioluteotripunctata Kaszab, 1956b: 153

mutabilis Marseul, 1873b: 492
mutans Guérin-Méneville, 1837: 551
similaris Mulsant, 1857a: 125
sturmi Baudi di Selve, 1878c: 1160
subgenus *Pardabris* Bologna, 2019: 259 type species *Mylabris syriaca* Klug, 1845
atrofasciata Pic, 1921a: 4 (*Zonabris*) **A:** IN
interrupta A.G. Olivier, 1801: 93 **N:** EG **A:** IN
excellens L. Redtenbacher, 1850: 49
mediounifasciata Pic, 1920e: 13 (*Zonabris*)
subunifasciata Pic, 1920e: 13 (*Zonabris*)
suturalis Pic, 1898b: 2 (*Zonabris*) **A:** IN SY TR
morgani Pic, 1920d: 12 (*Zonabris*)
panjaoensis Kaszab, 1953a: 306 **A:** AF PA
syriaca Klug, 1845: nr. 13 **E:** AB AR GR **A:** IN IQ IS JO LE PA SY SI TR
antedivisa Mader, 1929: 108
intersecta Reiche, 1857: 274
leseleuci Pic, 1898b: 2 (*Zonabris*)
smyrnensis Escherich, 1899: 109 (*Zonabris*)
subgenus *Pseudabris* Fairmaire, 1894b: 222 type species *Pseudabris tigriodera* Fairmaire, 1894
brevipilosa Pan & Bologna, 2013: 139 **A:** SCH
hingstoni Blair, 1927: 254 **A:** XIZ
waltoni Blair, 1927: 255
latimaculata Pan & Bologna, 2013: 142 **A:** QIN XIZ
longiventris Blair, 1927: 253 **A:** XIZ
przewalskyi Dokhtouroff, 1887a: 341 (*Zonabris*) **A:** QIN SCH SD XIZ YUN
goutelli Fairmaire, 1889b: 48
regularis Pan & Bologna, 2013: 143 **A:** QIN XIZ
tigriodera Fairmaire, 1894b: 222 (*Pseudabris*) **A:** SCH XIZ

genus *Mylabris*, species incertae sedis
aurociliata Escherich, 1899: 98 (*Zonabris*) distribution unknown
axillaris Billberg, 1813: 24 **N:** EG
brevicornis Motschulsky, 1872: 50 **E:** UK **A:** "Siberia"
caroli Pic, 1899l: 200 (*Zonabris*) **N:** EG
delagrangei Pic, 1899l: 208 (*Zonabris*) **A:** SY
flavipennis Motschulsky, 1872: 50 **A:** ES
isis Marseul, 1876a: 28 **N:** EG
paliji Kotschenov, 1970: 21 **A:** KI
quinquemaculata Okamoto, 1924: 1984 **A:** SC
sanguinosa Marseul, 1872a: 417 **A:** IN IQ

soumacovi Pic, 1930a: 2 (*Zonabris*) **A:** KZ TM
tarbagataiensis Pic, 1919f: 18 (*Zonabris*) **A:** KZ

genus *Mylabris*, nomina dubia
elongata Herbst, 1784c: 147
gebleri Faldermann, 1837: 124 **A:** IN
trivittis Pallas, 1782: 92 (*Meloe*) "South Russia"

genus *Semenovilia* Kuzin, 1954: 370 type species *Mylabris fischeri* Gebler, 1847
fischeri Gebler, 1847: 502 (*Mylabris*) **E:** ST **A:** AF ES IN KZ TM UZ WS XIN
tauscheri Fischer von Waldheim, 1844: 130 (*Mylabris*) [HN]
komarowi Reitter, 1889b: 121 (*Zonabris*) **A:** "Turkestan"
connecticornis Reitter, 1894a: 51 (*Zonabris*)
pilimana Reitter, 1894a: 51 (*Zonabris*)

genus *Xanthabris* Kaszab, 1956e: 637 type species *Xanthabris baluchistana* Kaszab, 1956
baluchistana Kaszab, 1956e: 638 **A:** PA

tribe Mylabrini, species incertae sedis
Mylabris deserticola Gistel, 1857: 17 **E:** UK (Krym)
Mylabris fuscicornis Drapiez, 1821: 277 **A:** "China"
Zonabris heteroclitus Pic, 1919f: 21 **E:** ST
Mylabris jacquemontii Blanchard, 1844: 26 **A:** KA
Zonabris kuschkensis Pic, 1919f: 21 **A:** IN TM UZ
Mylabris magica Gistel, 1857: 31 **N:** "Barbaria"
Mylabris pallidomaculatus L. Redtenbacher, 1850: 49 **N:** EG **A:** IN

tribe Meloini Gyllenhal, 1810

genus *Meloe* Linnaeus, 1758: 419 type species *Meloe proscarabaeus* Linnaeus, 1758
subgenus *Afromeloe* K. Schmidt, 1913: 332 type species *Meloe angulatus* Leach, 1815
Submeloegonius Pliginskij, 1935: 320 type species *Meloe adisabebae* Pliginskij, 1935
lefevrei Guérin-Méneville, 1849: 322 **A:** YE (Suqutra) **AFR**
sanaanus Borchmann, 1938b: 465 **A:** YE
trapeziderus Gahan, 1903: 282 **A:** YE (Suqutra Is.)
subgenus *Alveomeloe* Pripisnova, 1987a: 41 type species *Meloe tadzhikistanicus* Pripisnova, 1987
babatagicus Pripisnova, 1987a: 42 **A:** TD
tadzhikistanicus Pripisnova, 1987a: 41 **A:** TD
subgenus *Chiromeloe* Reitter, 1911a: 390 type species *Meloe nigropilosellus* Reitter, 1900
nigropilosellus Reitter, 1900b: 56 **A:** UZ
subgenus *Desertimeloe* Kaszab, 1964d: 521 type species *Meloe centripubens* Reitter, 1897

alashana Kaszab, 1964d: 322 **A**: GAN
centripubens Reitter, 1897e: 223 **A**: MG XIN
auropictus Denier, 1921: 112
citroeni Reymond, 1933: 14
subgenus *Eurymeloe* Reitter, 1911a: 391 type species *Meloe brevicollis* Panzer, 1793
Coelomeloe Reitter, 1911a: 389 type species *Meloe tuccius* Rossi, 1792
aeneus Tauscher, 1812b: 151 **E**: KZ ST **A**: KZ TD TM
affinis affinis Lucas, 1846: pl. 33 **N**: AG MO TU
setosus Escherich, 1890b: 90
affinis schatzmayri Bologna, 1988: 272 **N**: LB
apenninicus Bologna, 1988: 250 **E**: IT
apivorus Reitter, 1895a: 10 **N**: MO
simillimus Escalera, 1914: 383
asperatus Tan, 1981: 415 [DA] **A**: XIZ
austrinus Wollaston, 1854: 257 [DA] **N**: MR
baamarani Ruiz & García-París, 2015: 5 **N**: MO
baudii Leoni, 1907: 263 **E**: CR IT SL
scabriusculus Baudi di Selve, 1878b: 353 [HN]
baudueri Grenier, 1863: 92 **E**: FR IT (Sardinia) PT SP **N**: AG MO
brevicollis algiricus Escherich, 1890c: 283 **E**: IT (Sicilia) **N**: AG TU
brevicollis brevicollis Panzer, 1793d: 15 **E**: AB AL AR AU BE BH BU BY CR CT CZ DE EN FI FR GE GB GG GR HU IT KZ LA LT LU MC MO NL NR NT PL PT RO SK SL SP ST SV SZ TR UK **A**: AF ES FE HEB HEI IN JIX JO KI KZ MG NC NO TD TR WS
aestivus Baudi di Selve, 1878b: 353
cephalotes Curtis, 1829: 38
laticollis Baudi di Selve, 1878b: 353
pallidotarsalis Kaszab, 1956b: 158
puncticollis Motschulsky, 1872: 46
splendens Escherich, 1889a: 105
brevicollis curticornis Escalera, 1914: 384 **N**: MO
brevicollis mistaniensis Aksentjev, 1985: 248: **E**: AB **A**: IN
corvinus Marseul, 1877: 482 **A**: FE JA NE "Korea"
crosi Peyerimhoff, 1926a: 94 **N**: MO
curticollis Kraatz, 1882c: 117 **A**: UZ
escherichi Reitter, 1889b: 107 **A**: TR
fernandezi Pardo Alcaide, 1951: 252 **N**: CI
flavicomus Wollaston, 1854: 528 **N**: CI MR
frontalis Reitter, 1905d: 245 **A**: IN
gaberti Reitter, 1907g: 214 **A**: KZ
ganglbaueri Apfelbeck, 1905: 250 [= 1907: 643] **E**: AL BH BU CR IT FR (Corse) GR SP YU **A**: SY TR

glazunovi Pliginskij, 1910: 171 **E:** AR GR ST UK "Transcaucasia" **A:** AF IN KI KZ TD TM TR
rufotarsalis Pliginskij, 1911: 47
gomari Ruiz & García-París, 2009: 94 **N:** MO
heptapotamicus Pliginskij, 1910: 170 **A:** KZ
ibericus Reitter, 1895a: 7 **E** SP
kandaharicus Kaszab, 1958c: 302 **A:** AF IN
laevipennis Brandt & Erichson, 1832: 124 **A:** ES FE MG
lederi Reitter, 1895a: 7 **A:** MG NMO
lobicollis Fairmaire, 1891c: xxiii **A:** HUB HUN
luctuosus Brandt & Erichson, 1832: 122 **E:** IT (Sicilia) **N:** AG TU
mandli Borchmann, 1942b: 682 **A:** ES
mathiesseni Reitter, 1905d: 245 **A:** IN
mediterraneus J. Müller, 1925: 22 **E:** AL AR AU CR BH BU FR GB GR HU IT MA PT SK SP TR YU "Transcaucasia" **N:** AG CI EG LB MO MR TU **A:** IN IS TR
murinus Brandt & Erichson, 1832: 127 **E:** GR (Crete) FR (Corse) IT (Sicilia, Sardinia) SP **N:** AG LB MO TU
thalensis Pic, 1933e: 10
nanus Lucas, 1846: pl. 33 **E:** PT SP **N:** AG EG LB MO TU **A:** IS JO SI SY
murinus Reitter, 1895a: 12 [HN]
omanicus Kaszab, 1983a: 186 **A:** AE KU OM
pallidicolor Escalera, 1909b: 242 **N:** MO
primaeveris Kaszab, 1958c: 301 **A:** AF
primulus Semenov, 1903c: 356 **A:** IN TM
pubifer Heyden, 1887c: 315 **A:** AF KI KZ TD TM
punjabensis Kaszab, 1958: 303 **A:** PA
pusio Wellman, 1910a: 22 **A:** UZ
pygmaeus Kraatz, 1882d: 334 [HN]
rugosus Marsham, 1802: 483 **E:** AB AL AR AU BE BH BU CR CT CZ FR GB GE GG GR HU IT LU NL MC MD PL PT RO SK SL SP ST SZ UK YU **A:** AF IN KZ SY TD TR
abdominalis Escherich, 1890b: 89
autumnalis Leach, 1815b: 40 [HN]
saharensis Chobaut, 1898a: 86 **E** IT (Lampedusa Is.) SP **N:** AG CI EG MO TU **A:** IS SA
otini Peyerimhoff, 1949:b 275
marianii Kaszab, 1983a: 184
sarmaticus Shapovalov, 2012: 208 **E:** KZ ST **A:** KZ
scabriusculus Brandt & Erichson, 1832: 125 **E:** AL AU BH BU CR CT CZ FR GE GR HU IT MC MD PL RO SL SK ST SZ TR UK YU "Caucasus" **A:** ES IN KZ TR UZ
nigra Pliginskij, 1911: 46 [HN]
pliginskyi Kaszab, 1958b: 191 [RN]
pygmaeus Baudi di Selve, 1878b: 353 [HN]
semipunctatus Krynicki, 1832: 140
tarsalis Depoli, 1912: 101 [HN]

schmidi Kaszab, 1978c: 341 **A:** HP SD UP
scutellatus Reitter, 1895a: 11 **E:** AR
servulus Bates, 1879: 483 **A:** AF KA NP XIN XIZ
subsetosus Reitter, 1895a: 10 **A:** TM
transversicollis Fairmaire, 1891d: cii **A:** KA "Himalaya"
tuccia corrosus Brandt & Erichson, 1832: 122 **E:** IT
tuccia tuccia Rossi, 1792: 290 **E:** AB AL AR AU BH BU CR CZ FR GR HU IT MA PT RO SK SP ST UK YU **N:** AG CI EG LB MO MR TU **A:** AF CY IN IS JO KZ SI SY TD TM TR UZ XIN
ruficornis Balbi, 1897: 94
zolotarevi Pliginskij, 1914b: 264 **A:** KZ
subgenus *Lampromeloe* Reitter, 1911a: 388 type species *Meloe variegatus* Donovan, 1793
cavensis L. Petagna, 1819: 40 **E:** FR GR IT MA PT SP TR **N:** AG EG LB MO TU **A:** CY IN IQ IS JO LE SY TR **AFR**
aenea Laporte, 1840: 278
latreillei Reiche, 1867: 89
purpurascens Germar, 1836a: 12
sardous Gené, 1836b: iv [= 1836a: 198]
specularis Gredler, 1878: 518 [DA]
stellatus Pliginskij, 1923b: 143 [DA]
variegatus mandzhuricus Pliginskij, 1930: 110 **A:** HEI
variegatus variegatus Donovan, 1793: 81 **E:** AB AR AU BE BH BU BY CR CT CZ DE FR GB GE GG GR HU IT LT MD NL PL RO SK SL SP ST SV SZ UK YU **N:** AG MO TU **A:** AF ES HEI IN JO KA KI KZ LE TD TM TR WS
areolatus Reitter, 1895a: 8
cupreus Baudi di Selve, 1878b: 354
scabrosus Marsham, 1802: 483
scabrosus Laporte, 1840: 278 [HN]
subgenus *Lasiomeloe* Reitter, 1911a: 389 type species *Meloe olivieri* Chevrolat, 1833
olivieri Chevrolat, 1833d: pl. 57 **E:** AB AR TR **N:** EG **A:** IN IQ JO SY TM TR
subgenus *Listromeloe* Reitter, 1911a: 390 type species *Meloe hungarus* Schrank von Paula, 1776
hungarus Schrank von Paula, 1776: 71 **E:** AB AR AU BU CR CZ FR (Alsace) GE GR HU MC MD RO SK ST TR UK YU "Caucasus" "Transcaucasia" **A:** IN IQ IS JO SY TR
limbatus Fabricius, 1801b: 588
marginatus Tauscher, 1812b: 152
subgenus *Meloe* Linnaeus, 1758: 419 type species *Meloe proscarabaeus* Linnaeus, 1758
Cnestocera C.G. Thomson, 1859: 124 type species *Meloe proscarabaeus* Linnaeus, 1758
Melittophagus Kirby, 1819: 164 type species *Pediculus melittae* Kirby, 1802 (= *Meloe violaceus* Marsham, 1802)
Proscarabaeus Schrank von Paula, 1781: 225 type species *Meloe proscarabaeus* Linnaeus, 1758
Triungulinus Dufour, 1828: 63 type species *Triungulinus andrenatarum* Dufour, 1828

aegyptius aegyptius Brandt & Erichson, 1832: 119 **E** IT (Sicilia) **N:** AG EG LB MO TU **A:** IS SI
  plicatipennis Lucas, 1846: pl. 33
  siculus Baudi di Selve, 1878b: 352
aegyptius subcyaneus Wollaston, 1864: 514 [DA] **N:** CI MO
aruanachalae Saha, 1979: 128 **A:** AP BT NP UP
auriculatus Marseul, 1877: 480 **A:** JA NE NMO SC XIZ "Northen, Central East China"
bodemeyeri Ganglbauer, 1900: 160 **A:** TR
coarctatus Motschulsky, 1858a: 35 **A:** JA
formosensis Miwa, 1930: 12 **A:** TAI
gracilior Fairmaire, 1891c: xxii **A:** HUB
kulabensis Shapovalov, 2014: 1337 **A:** TD
lobatus Gebler, 1832: 57 **A:** BEI ES FE FUJ HEI JA MG NC SC WS YUN
  bellus Jakovlev, 1897: 248
  granulifer Motschulsky, 1872: 47
  patellicornis Fairmaire, 1887b: 325
medogensis Tan, 1988: 292 **A:** XIZ
menoko Kôno, 1936b: 92 **A:** JA
ovalicollis Reitter, 1908i: 246 **A:** KI
poteli Fairmaire, 1897a: 123 **A:** JIX
proscarabaeus afghanistanicus Kaszab, 1953a: 310 **A:** AF
proscarabaeus proscarabaeus Linnaeus, 1758: 419 **E:** AB AL AR AU BE EH BU BY CR CT CZ DE EN FI FR GB GE GG GR HU IR IT KZ LA LU LT MA MC MD NL NR NT PL PT RO SK SL SP ST SV SZ TR UK YU **N:** EG MO **A:** ANH ES FE GAN HEB HEI HUB IN IS JO KA KI KZ LE LIA MG NC NE NMO NO SC SCH SY TD TM TR UZ WS XIN XIZ YUN
  andrenatarum Dufour, 1828: 64 (*Triungulinus*)
  atrata Meyer, 1793: 15
  brunsvicensis Meyer, 1793: 25
  caerulans Brullé, 1832: 230
  crispatus Fairmaire, 1884a: 173
  cyanella Brullé, 1832: 229
  cyanescens Csiki, 1953: 117
  cyaneus Mulsant, 1857a: 47
  exaratus Faldermann, 1832: 210
  gallicus Baudi di Selve, 1878b: 351
  incerta Tauscher, 1812b: 149
  megacephala Fischer von Waldheim, 1842: 27
  pannonicus Baudi di Selve, 1878b: 352
  punctata Fabricius, 1792b: 518
  rugicollis Stephens, 1832: 66 (*Proscarabaeus*)
  rugipennis Mannerheim, 1825: 31

tauricus Baudi di Selve, 1878b: 351
tectus Panzer, 1793d: 14 (*Proscarabaeus*)
volgensis Tauscher, 1812b: 148
vulgaris Stephens, 1832: 66 (*Proscarabaeus*)
proscarabaeus sapporensis Kôno, 1936b: 91 **A:** JA
proscarabaeus sericeorugosus Aksentjev, 1987: 474 **A:** NP
rathjensi Borchmann, 1938b: 464 **A:** YE
scabrus Pan & Ren, 2018: 77 **A:** XIZ
semenowi Jakovlev, 1897: 251 **A:** CH ES FE JA "Korea"
semicoriaceus Fairmaire, 1891d: cii **A:** KA
simplicicornis Escherich, 1889a: 105 [DA] **E:** AR "Caucasus" **A:** TR
subcordicollis Fairmaire, 1887a: 129 **A:** GAN GUI HUN JIX NMO XIZ YUN
tarsalis Jakovlev, 1897: 249 **A:** ES
tenuipes Jakovlev, 1897: 252 **A:** FE MG
violaceus Marsham, 1802: 482 **E:** AB AL AN AR AU BE BH BU BY CR CT CZ DE EN FI FR GB GE GG GR HU IR IT KZ LA LT MC MD NL NR NT PL PT RO SK SL SP ST SV SZ UK YU **N:** AG MO **A:** ES FE IN KA KI KZ TD TR UZ WS
angusticollis Rey, 1892: 137 [HN]
aprilina Meyer, 1793: 21
melittae Kirby, 1802: 168 (*Pediculus*)
montanus Reitter, 1911: 388
montanuus Kaszab, 1958b: 190 [RN]
proscarabaeus Sulzer, 1761: 92 [HN]
rufipes Bremi-Wolf, 1856: 199
similis Marsham, 1802: 482
simplex Fleischer in Escherich, 1890a: 23
strigosus Motschulsky, 1872: 48
tenuicollinus Kaszab, 1958b: 191 [RN]
subgenus *Meloegonius* Reitter, 1911a: 389 type species *Meloe cicatricosus* Leach, 1815
cicatricosus Leach, 1815b: 39 **E:** AB AL AR AU BH BU BY CR CT CZ FR GB GE GG GR HU IT MC MO PL RO SL SK SP ST SZ TR UK YU **A:** AF IN KI KZ TD TM TR UZ WS "Palestine"
iluronensis Salvaña, 1870: 37
ineditus Salvaña, 1870: 38
punctatoradiatus Latreille, 1804: 391
variolosus Fischer von Waldheim, 1842: 27
rufiventris himalaycus Kaszab, 1978c: 339 **A:** KA
rufiventris rufiventris Germar, 1832: 6 **E:** AB AU CR CT CZ FR (Alsace) GE GR HU IT MD PL RO SK ST SZ UK **N:** AG LB **A:** AF IN IQ JO KI KZ TD TR UZ
coriarius Brandt & Erichson, 1832: 131
hoffmannseggi Germar, 1832: 6
reticulatus Brandt & Ratzenburg, 1833: 108

subgenus *Meloenellus* Reitter, 1911a: 390 type species *Meloe griseopuberulus* Reitter, 1890
griseopuberulus Reitter, 1890c: 152 **A:** "Turkestan"
sulciceps Reitter, 1911b: 54 **A:** KZ
turkestanicus Escherich, 1890b: 88 **A:** KI
vlasovi Semenov & Arnoldi, 1937: 222 **A:** AF TM
klapperichi Kaszab, 1958c: 300
subgenus *Mesomeloe* Reitter, 1911a: 392 type species *Meloe sulcicollis* Kraatz, 1882 (= *Meloe xanthomelas* Solsky, 1881)
bytinskii Kaszab, 1969a: 279 **A:** IN
cinereovariegatus Heyden, 1885a: 294 **A:** KI UZ
coelatus Reiche, 1857: 271 **N:** AG CI MO TU **A:** IN IS JO QA SA SI
aegyptius Reitter, 1895a: 5 [HN]
nudus Wollaston, 1864: 514
xanthomelas Solsky, 1881a: 257 **A:** TD TM UZ
sulcicollis Kraatz, 1882d: 334
xanthomelaena Semenov, 1903d: 392
subgenus *Micromeloe* Reitter, 1911a: 390 type species *Meloe uralensis* Pallas, 1773
Trapezimeloe Pliginskij, 1935: 322 type species *Meloe conradti* Heyden, 1888
cavicornis Reitter, 1898a: 16 **A:** UZ
conradti Heyden, 1888: 42 **A:** KI TD
trapezicollis Dokhtouroff, 1889: 129
decorus Brandt & Erichson, 1832: 137 **E:** AU BU CR CZ FR GE HU IT KZ MD PL RO SK ST SZ UK YU **A:** KZ TD TR WS
pygmaeus L. Redtenbacher, 1849: 619
intermedius Escherich, 1904: 30 **A:** KZ TD UZ
kabuliensis Kaszab, 1981a: 164 **A:** AF
lopatini Pripisnova, 1987a: 44 **A:** TD
pintoi Bologna, 2018: 651 **A:** KA
reitteri Escherich, 1889b: 112 **E:** KZ ST **A:** KI KZ
simulans Reitter, 1895a: 6 **A:** TM
terentjevi Kaszab, 1978c: 340 **A:** TD
uralensis Pallas, 1773: 25 **E:** AR AU CR CT CZ HU IT (South Tyrol) KZ MD RO SK ST UK **A:** ES KZ TM
marginalis Wellman, 1910a: 22
punctatus Meyer, 1793: 28
subgenus *Taphromeloe* Reitter, 1911a: 389 type species *Meloe erythrocnemus* Pallas, 1782
erythrocnemus Pallas, 1782: 76 **E:** AB AR BH BU CR CT GR IT KZ RO ST TR UK YU **N:** AG MO TU **A:** AF IN KI KZ TD TM TR UZ WS XIN XIZ
corallipes Gebler, 1847: 492
foveolatus Guérin-Méneville, 1842: 338 **E:** IT (Apulia) SP **N:** AG LB TU
roubali Mařan, 1942b: 56

subgenus *Treiodous* Dugès, 1869: 102 type species *Treiodous barranci* Dugès, 1869 (= *Meloe laevis* Leach, 1815)

Anchomeloe Iablokoff-Khnzorian, 1983: 88 type species *Meloe autumnalis* A.G. Olivier, 1797

autumnalis autumnalis A.G. Olivier, 1797: 650 **E:** AL AR AU BE BH BU CR CT CZ FR HU GE GB GG GR IT MC MO NL PL PT RO SK SL SP ST SZ TR UK YU **N:** AG LB MO TU **A:** CY IN IS JO SY TR

carnicus Katter, 1883: 41

cribripennis Baudi di Selve, 1878b: 352

cyaneus Fabricius, 1801b: 589

glabratus Leach, 1815b: 43

hiemalis Gredler, 1866: 289

impunctatus Wellman, 1910a: 22

laevis Gredler, 1866: 290

punctatus Marsham, 1802: 483

punctipennis Escherich, 1889a: 106

autumnalis heydeni Escherich, 1889a: 106 **E:** IT (Sicilia)

genus *Meloe*, species incertae sedis

elegantulus Semenov & Arnoldi, 1934: 215 **A:** XIZ

longipennis Fairmaire, 1891c: xxii **A:** HUB

modestus Fairmaire, 1887b: 128 **A:** ANH FIJ JIA SCH SHX XIZ YUN

genus *Meloe*, nomina dubia

barcinonella Cuní Martorell, 1876: 222

bilineatus Aragona, 1830: 20

cinereus Brandt & Ratzenburg, 1833: 111

fascicularis Aragona, 1830: 20

marginatus Tauscher, 1812b: 152 [HN]

ovilis Mulsant, 1857a: 82

prolifericornis Motschulsky, 1872: 48

rugulosus Brullé, 1832: 230

sulcicollis Latreille, 1804a: 391

viennensis Schrank von Paula, 1776: 71

genus *Oreomeloe* Tan, 1981: 411 type species *Oreomeloe spinulus* Tan, 1981

spinulus Tan, 1981: 415 [DA] **A:** XIZ

genus *Physomeloe* Reitter, 1911a: 388 type species *Meloe corallifer* Germar, 1818

corallifer Germar, 1818: 259 (*Meloe*) **E:** PT SP

evae Flach, 1907: 19

subfamily **Nemognathinae** Laporte, 1840

tribe Horiini Latreille, 1802

genus *Horia* Fabricius, 1787a: 164 type species *Horia fabriciana* Betrem, 1929

Hoplozonitis Blackburn, 1892: 228 type species *Hoplozonitis mira* Blackburn, 1892

fabriciana Betrem, 1929: xxvii [RN] **N:** EG **A:** AE PA SA YE **ORR**

africana Aurivillius, 1890: 203 (*Cissites*) [HN]

gahani Betrem, 1932: 95 **A:** SD

genus *Synhoria* Kolbe, 1897: 256 type species *Horia cephalotes* A.G. Olivier, 1793

maxillosa Fabricius, 1801b: 86 (*Horia*) **A:** HEN HKG JA SCH SD TAI **ORR**

anguliceps Fairmaire, 1885b: 111 (*Horia*)

sasakii Kôno, 1936b: 97 (*Cissites*)

tosana Kôno, 1936b: 96 (*Horia*)

tribe Nemognathini Laporte, 1840

genus *Allendesalazaria* Escalera, 1910b: 379 type species *Allendesalazaria nymphoides* Escalera, 1910

nymphoides Escalera, 1910b: 380 **N:** AG MO

genus *Apalus* Fabricius, 1775: 127 type species *Meloe bimaculatus* Linnaeus, 1760

Coriologiton Marseul, 1879b: 65 [DA] type species *Criolis hilaris* Marseul, 1879

Criolis Mulsant, 1858b: 240 type species *Criolis guerini* Mulsant, 1858

Deratus Motschulsky, 1872: 51 type species *Meloe necydalea* Pallas, 1773

Hapalus Illiger, 1801: 138 [UE]

asiaticus Escherich, 1897b: 119 **A:** AF "Turkestan"

bimaculatus Linnaeus, 1760: 288 (*Meloe*) **E:** AU BH BU CR CZ DE EN FI FR GE HU IT KZ LA LT NR NT PL RO SK ST SP SV SZ UK **N:** AG LB MO TU **A:** IN JA KZ SY TM TR "Siberia"

caruanae Prochazka, 1892: 265 (*Hapalus*)

flava Escherich, 1897b: 120

lecomptei Pic, 1896j: 140 (*Hapalus*)

binaevus Prochazka, 1892: 270 (*Hapalus*) **A:** KI

bipunctatus Megerle von Mühlfeld, 1812: 18 **E:** AU CR CZ GE GR HU IT KZ PL RO SK SP ST UK YU "Caucasus" **A:** IS KZ TD TR "Palestine"

nigripes Kaszab, 1942b: 43 (*Hapalus*)

nigritarsis Ragusa, 1898: 222 (*Hapalus*)

melanipes Kaszab, 1942b: 43 (*Hapalus*)

rufus Ragusa, 1906: 6

cinctus Pic, 1896j: 140 (*Hapalus*) **E:** SP **N:** AG EG TU

creticus J. Frivaldszky, 1877a: 83 (*Hapalus*) **E:** GR (Crete)

davidis Fairmaire, 1886c: 352 (*Hapalus*) **A:** HUB

fasciatus Faldermann, 1835: 416 **A:** ES MG NW
mongolicus Kaszab, 1965f: 558
tibialis Motschulsky, 1872: 51 (*Deratus*)
guerini Mulsant, 1858b: 240 (*Criolis*) **E:** FR IT SP **N:** LB MO
haemapterus Fairmaire, 1889b: 49 **A:** SCH SHN
montanus Escherich, 1897b: 121 **A:** TR
mulsanti Reiche, 1878: lxxiii (*Criolis*) **A:** TR
necydaleus Pallas, 1782: 92 (*Meloe*) **E:** GR HU KZ MC RU UK "Caucasus" **N:** LB **A:** IN KI KZ TR
inornatus Prochazka, 1892: 266 (*Hapalus*)
moreanus Pic, 1922d: 19
phoebus Prochazka, 1892: 267 (*Hapalus*)
rufipennis Gebler, 1829a: 142
spectabilis I. Frivaldszky, 1835: 265
spectabilis Schaum, 1859: 52 (*Hapalus*) [HN]
plagiatus Waterhouse, 1889: 130 **A:** AF
subfasciatus Pic, 1922d: 19 **A:** IN
sulcithorax Pic, 1901i: 79 (*Hapalus*) **A:** SY

genus *Cochliophorus* Escherich, 1891a: 16 type species *Cochliophorus reitteri* Escherich, 1891
reitteri Escherich, 1891a: 16 **E:** GR

genus *Ctenopus* Fischer von Waldheim, 1823: pl. 38 type species *Ctenopus melanogaster* Fischer von Waldheim, 1824
abdominalis Motschulsky, 1845a: 83 **E:** ST
aurantiacus Semenov, 1893a: 15 **A:** IN
carinifer Semenov, 1896d: 380 **A:** AF TM
eous Semenov, 1900b: 487 **A:** XIN
hauseri Reitter, 1894a: 50 **A:** KI TD TM
ahngeri Semenov, 1896d: 381
lama Escherich, 1904: 35 **A:** XIN
melanogaster Fischer von Waldheim, 1823: pl. 38 **E:** AB AR **A:** IN KI KZ MG TM
nudus Escherich, 1897b: 128 **A:** TD TM
oxianus Semenov, 1894c: 540 **A:** UZ
persicus Semenov, 1893a: 17 **A:** IN
pygmaeus Semenov, 1903c: 357 **A:** TM
reitteri Semenov, 1893a: 18 **E:** AR
roseni Escherich, 1897b: 128 **A:** TM
hauseri Semenov, 1896d: 388 [HN]
rufoscutellatus Reitter, 1889a: 38 **A:** TM
semenovi Reitter, 1895b: 157 **A:** XIN
sinuatipennis Fairmaire, 1892e: cliv (*Zonitis*) **A:** IN TM UZ XIN
testaceus Semenov, 1893a: 14

sogdianus Semenov, 1900b: 486 **A:** TM UZ
vitticollis Reitter, 1889b: 38 **E:** AB AR

genus *Euzonitis* Semenov, 1893d: 276 type species *Apalus sexmaculatus* A.G. Olivier, 1789
adustipennis Motschulsky, 1872: 53 (*Zonitis*) **E:** AR KZ SK ST "Caucasus"
inapicalis Pic, 1935d: 11
ahngeriana Semenov, 1896d: 379 (*Zonitis*) **A:** TM
alfierii Pic, 1929g: 110 (*Zonitis*) **N:** EG
fulvipennis Fabricius, 1792b: 49 (*Zonitis*) **E:** AB AR AU BU CR CT GG GR HU IT (Calabria) MC RO SK ST UK YU **A:** IN TR
funeraria Fairmaire, 1884g: cxlii (*Zonitis*)
nigra Tauscher, 1812b: 163 (*Zonitis*)
nigricollis Faldermann, 1832: 211 (*Zonitis*)
maculiceps Reitter, 1897c: 110 **A:** IN KI TD UZ
mzchetica Sumakov, 1929b: 113 (*Zonitis*) **E:** GG
paulinae Mulsant & Rey, 1858a: 193 (*Zonitis*) [DA] **A:** IN IS JO
quadrimaculata Pallas, 1773: 720 (*Meloe*) **E:** AB AR AU BH BU CR CT CZ FR GG HU IT KZ MC MD PT RO SK SP ST SZ UK YU **N:** AG CI MO MR TU **A:** CY IN IQ IS KZ MG SY TM TR WS
adanensis Pic, 1935d: 11
atra Schwarz, 1808: 340 (*Zonitis*)
auliensis Pic, 1913k: 138
bifasciata Schwarz, 1808: 310 (*Zonitis*)
bimaculata Tauscher, 1812b: 158 (*Zonitis*)
concolor Abeille de Perrin, 1880: 257 (*Zonitis*)
concolor Wellman, 1910a: 25 (*Zonitis*) [HN]
fasciata Tauscher, 1812b: 159 (*Zonitis*)
haroldi Heyden, 1870: 150 (*Zonitis*)
iconiensis Pic, 1901i: 79 (*Zonitis*)
immaculata Escherich, 1892: 247 (*Zonitis*) [HN]
imperialis Wollaston, 1861b: 106 (*Zonitis*)
korbi Escherich, 1890a: 24 (*Zonitis*)
latenotata Pic, 1907f: 154
obliquata Motschulsky, 1872: 53 (*Zonitis*)
palumboi Ragusa, 1898: 221 (*Zonitis*)
quadripunctata Fabricius, 1801b: 84 (*Mylabris*)
rufofasciata Fairmaire, 1884g: cxlii (*Zonitis*)
xanthoptera Fairmaire, 1876: 94 (*Zonitis*)
rubida Ménétriés, 1832: 211 (*Zonitis*) **E:** AB AR GR **A:** AF TM TR
pubescens Waltl, 1838: 467 (*Zonitis*)
sexmaculata G.-A. Olivier, 1789: 166 (*Apalus*) **E:** AB AR BH BU CR CZ FR GG GR HU IT KZ MD RO SP ST UK YU **A:** AE CY IN IQ IS JO KZ LE SY TD TM TR

bipunctata Ragusa, 1881: 42 (*Zonitis*) [HN]
discolor Escherich, 1892: 243 (*Zonitis*) [HN]
distigma Wellman, 1910a: 25 (*Zonitis*) [RN]
fenestrata Pallas, 1782: 90 (*Zonitis*)
lunaris Wellman, 1910a: 25 (*Zonitis*) [RN]
lunata Tauscher, 1812b: 159 (*Zonitis*) [HN]
medioconjuncta Kaszab, 1942b: 43
obliterata Pic, 1922d: 18
octomaculata Ragusa, 1923: 7
pedemontana Pic, 1901i: 79 (*Zonitis*)
subconjuncta Pic, 1901i: 79 (*Zonitis*)
suramensis Pic, 1922d: 18
sogdiana Semenov, 1893d: 275 (*Zonitis*) **A:** TD TM
spectabilis Kraatz, 1881c: 326 (*Zonitis*) **A:** AF TD TM "Turkestan"
atricornis Reichardt, 1934: 237
terminata Abeille de Perrin, 1880: 255 (*Zonitis*) **E:** GR HU IT RO TR **N:** EG **A:** AF IN IS JO SY TM TR
armeniaca Pic, 1901i: 79 (*Zonitis*)
auricoma Escherich, 1891b: 54 (*Zonitis*)
bombycina Escherich, 1896a: 27 (*Zonitis*)
pallens Escherich, 1897a: 47
semideleta Pic, 1903m: 163 (*Zonitis*)
suprema Escherich, 1897b: 110
unimaculata Pic, 1933e: 14 [DA] **N:** MO
varentzowi Semenov, 1894c: 537 (*Zonitis*) **A:** IN TM

genus *Euzonitis*, species incertae sedis
jacobsoni Reichardt, 1934: 223 **A:** TD

genus *Glasunovia* Semenov, 1896c: 357 type species *Glasunovia caspica* Semenov, 1896
afghanica Semenov, 1896c: 358 **A:** AF IN TM
caspica Semenov, 1896c: 360 **A:** TM
sillemi Borchmann, 1935: 308 **A:** AF XIZ
sinica Pic, 1938e: 16 **A:** CH

genus *Leptopalpus* Guérin-Méneville, 1834a: pl. 35 type species *Leptopalpus chevrolatii* Guérin-Méneville, 1834 (= *Zonitis rostrata* Fabricius, 1792)
quadrimaculatus Gahan, 1896b: 457 **A:** YE **AFR**
rostratus Fabricius, 1792b: 50 (*Zonitis*) **E:** IT (Sardinia, Sicilia) PT SP **N:** AG MO TU
chevrolatii Guérin-Méneville, 1834a: pl. 35

genus *Longizonitis* Pan & Bologna, 2018: 44 type species *Zonitis semiruber* Pic, 1911
semirubra Pic, 1911: 101 (*Zonitis*) **A:** FUJ UP XIZ YUN

genus *Megatrachelus* Motschulsky, 1845a: 83 type species *Zonitis polita* Gebler, 1832
Schroetteria Reitter, 1911a: 395 type species *Zonitis polita* Gebler, 1832
pallidipennis Motschulsky, 1845a: 83 (*Zonitis*) [DA] **A:** ES
politus Gebler, 1832: 58 (*Zonitis*) **A:** ES FE HEI JA MG NC
megacephalus Kôno, 1936a: 4
quadricollis Fairmaire, 1892e: cliii (*Zonitis*) [DA] **A:** "Turkestan"
sibiricus Tauscher, 1812b: 162 (*Zonitis*) **A:** KZ MG
quadrinotatus Wellman, 1910a: 25 (*Lydus*) [RN]
quadrisignatus Faldermann, 1835b: 415 (*Lydus*)
subpolitus Reitter, 1911a: 395 (*Schroetteria*) **A:** NE

genus *Nemognatha* Illiger, 1807: 333 type species *Zonitis vittata* Fabricius, 1801 (= *Zonitis piazata* Fabricius, 1794)
Gnathonemula Aksentjev, 1981: 862 type species *Gnathonemula gracilis* Aksentjev, 1981
Nematognatha Gemminger & Harold, 1870b: 2163 [UE]
chrysomelina Fabricius, 1775: 126 (*Zonitis*) **E:** AB AR FR GR IT (Sicilia, Sardinia) PT SP ST UK **N:** AG EG MO TU **A:** AE AF CY IN IQ IS JO KZ LE SA SY TD TM TR UZ
akbesiana Pic, 1897d: 35
asiatica Motschulsky, 1872: 54
atripennis Chobaut, 1898b: 143
flavipes Ménétriés, 1849: 284
iliensis Kolov, 2005: 41
latenigra Pic, 1941a: 5
maculicollis Fairmaire, 1875a: 532 (*Zonitis*)
nigripes Suffrian, 1853: 236
unipunctata Beauregard, 1890: 466
flavicornis Stierlin, 1876: 478 [DA] **E:** AB AR **A:** IN
gemina Suffrian, 1853: 235 **N:** EG
gracilis Aksentjev, 1981: 863 (*Gnathonemula*) **A:** KZ UZ
transcaspica Kaszab, 1981a: 166
quinquemaculata Suffrian, 1853: 235 **N:** EG MO **A:** IN IS SA
sexmaculata Pic, 1911d: 74
sibirica Gebler, 1833: 290 (*Zonitis*) [DA] **A:** "Siberia"

genus *Nyadatus* Aksentjev, 1981: 860 [DA] type species *Nyadatus telejus* Aksentjev, 1981
sphalerus Aksentjev, 1981: 862 **A:** TM
telejus Aksentjev, 1981: 861 **A:** KZ TD UZ

genus *Sitaris* Latreille, 1802: 187 type species *Necydalis humeralis* Fabricius, 1775 (= *Necydalis muralis* Förster, 1771)

subgenus *Filalia* Escalera, 1906a: 162 type species *Sitaris cerambycina* Escalera, 1906

cerambycinus Escalera, 1906a: 161 **N:** LB MO

subgenus *Sitaris* Latreille, 1802: 187 type species *Necydalis humeralis* Fabricius, 1775 (= *Necydalis muralis* Förster, 1771)

Sitaropsis Iablokoff-Khnzoryan, 1958: 1896 type species *Hapalus erevanensis* Iablokoff-Khnzoryan, 1958

balachowskyi Peyerimhoff, 1931b: 83 (*Apalus*) **N:** AG

emiliae Escherich, 1897b: 130 **E:** GR

erevanensis Iablokoff-Khnzorian, 1958: 1896 (*Hapalus*) [DA] **E:** AR

ferrantei Pic, 1911d: 74 **N:** AG EG MO **A:** "Palestine"

grassei Reymond, 1956a: 41 (*Ctenopus*)

homonimus Kaszab, 1958b: 191 [RN] **N:** EG

atricornis Pic, 1927e: 6 [HN]

incantatus Peyerimhoff, 1929a: 124 (*Apalus*) **N:** AG MO TU

mellicolor Biggs, 1945: 110 (*Apalus*) [DA] **A:** IN

minutus Pic, 1932e: 17 **A:** TR

muralis Förster, 1771: 48 (*Necydalis*) **E:** AU BE CR CZ FR GB GE GR HU IT MD NL PL PT RO SK SP ST SZ UK YU **N:** AG EG LB MO TU **A:** "Turkestan"

attenuatus Geoffroy, 1785: 154 (*Cantharis*)

flavus Hamm, 1909: 277

humeralis Fabricius, 1787a: 170 (*Necydalis*)

humuralis Fabricius, 1775: 209 (*Necydalis*)

lativentris Schaufuss, 1861: 49 (*Sitaris*)

mauritanicus Normand, 1950a: 82 (*Apalus*)

nitidicollis Abeille de Perrin, 1870: 42

splendidus Schaufuss, 1861: 49

pallens Escherich, 1904: 35 (*Hapalus*) **A:** "Turkestan"

pectoralis Bates, 1879: 484 (*Sitaris*) **A:** XIN

rufipennis Küster, 1849a: 83 **E:** SP

rufipes Gory, 1841: pl. 73 **N:** AG MO

ferdinandi Escalera, 1910b: 382

rufiventris Kraatz, 1884: 175 **E:** GR **A:** TR

solieri Pecchioli, 1839: 529 **E:** CR FR GR IT PT SP **N:** AG CI MO TU **A:** TR **AFR**

apicalis Escherich, 1897b: 130

cabrerai Escalera, 1921: 308

corzoi Pardo Alcaide, 1950b: 45

longicornis Kraatz, 1858: 387, 388 (*Sitaris*)

tibialis Ragusa, 1898: 222

tauricus Motschulsky, 1872: 54 **E:** GR UK

tenuicornis Schaufuss, 1870: 47 [DA] **N:** EG

genus *Sitaris*, nomina dubia
flavescens Betta, 1847: 29 **E:** "Ticino"
fuliginosa Betta, 1847: 30 **E:** "Ticino

genus *Sitarobrachys* Reitter, 1883b: 309 type species *Sitarobrachys brevipennis* Reitter, 1883 (= *Stenoria thoracica* Kraatz, 1862)
thoracicus Kraatz, 1862b: 126 (*Stenoria*) **E:** BU GR SP **N:** AG CI EG MO TU **A:** IS TR
alfierii Pic, 1913h: 122
brevipennis Reitter, 1883b: 309
buigasi Escalera, 1909b: 240
diversipes Pic, 1924f: 29
proximus Escalera, 1914: 385

genus *Sitaromorpha* Dokhtouroff, 1889: 190 type species *Sitaromorpha wilkinsi* Dokhtouroff, 1889
wilkinsi Dokhtouroff, 1889: 190 [DA] **A:** "Turkestan"

genus *Stenoria* Mulsant, 1857a: 186 type species *Sitaris apicalis* Latreille, 1804
subgenus *Gineremia* Pardo Alcaide, 1961a: 92 type species *Stenoria saharica* Pardo Alcaide, 1961
saharica Pardo Alcaide, 1961a: 92 **N:** MO
subgenus *Stenoria* Mulsant, 1857a: 186 type species *Sitaris apicalis* Latreille, 1804
analis Schaum, 1859: 51 **E:** BE CR CZ FR GE IT NL PL PT SK SP ST SZ **N:** LB MO **A:** TR
acutipennis Fairmaire, 1881d: xliv (*Sitaris*)
adusta Schaum, 1859: 51 (*Sitaris*)
colletis Mayet, 1874: cxcviii (*Sitaris*)
sareptana Pic, 1912c: 9
antoinei Pardo Alcaide, 1953: 289 **E:** SP **N:** AG MO
gutticollis Pardo Alcaide, 1953: 291
apicalis Latreille, 1804a: 403 (*Sitaris*) **E:** CR FR GR HU IT MD PL PT RO SP ST UK YU "Caucasus" **N:** MO **A:** IN LE TM TR
basicollis Kaszab, 1956b: 158
bipunctata Kaszab, 1956b: 158
catalonica Pardo Alcaide, 1958b: 18
communimacula Kaszab, 1956b: 159
iranica Kaszab, 1959h: 440
kaszabiana Pardo Alcaide, 1958b: 17
kraatzi Mulsant & Rey, 1861: 191
luteifrons Pardo Alcaide, 1958a: 18
melanurus Küster, 1849a: 84 (*Sitaris*)
nigroplagiata Kaszab, 1956b: 159
paucinigra Pardo Alcaide, 1958a: 17
picicollis Escherich, 1897b: 123 (*Apalus*)

sturmii Küster, 1846b: 72 (*Ctenopus*)
tristicula Kaszab, 1956b: 160
vitticollis Kaszab, 1956b: 158
brunneicollis Pic, 1914m: 74 **A:** LE
canariensis Pic, 1902k: 80 (*Sitaris*) **N:** CI
erythromelas Semenov, 1893d: 278 (*Hapalus*) **A:** UZ
falcipennis Semenov, 1893d: 277 (*Hapalus*) **A:** TM "Turkestan"
fasciata Faldermann, 1835b: 416 (*Hapalus*) **A:** HEI MG NMO
flaviventris Ballion, 1878: 346 (*Cantharis*) **E:** CT KZ **A:** ES KZ MG XIN
tristis Escherich, 1897b: 124 (*Apalus*)
grandiceps Semenov, 1893d: 280 (*Hapalus*) **A:** IN
hauseri Escherich, 1904: 34 (*Hapalus*) **A:** FE MG QIN XIN
steppensis Kaszab, 1966a: 259
intermedia Dokhtouroff, 1889: 185 (*Hapalus*) **E:** AB AR **A:** IN TD TM
immaculaticeps Pic, 1911e: 97 (*Hapalus*)
klapperichi Kaszab, 1958c: 310 **A:** AF
komarowi Dokhtouroff, 1889: 90 (*Hapalus*) **A:** TM "Turkestan"
immaculatus Prochazka, 1892: 270 (*Hapalus*)
reitteri Prochazka, 1892: 270 (*Hapalus*)
kraatziana Semenov, 1900b: 490 (*Hapalus*) **A:** UZ
apicalis Kraatz, 1882d: 335 [HN]
kraatzi Semenov, 1893d: 277 (*Hapalus*)
laterimaculata Reitter, 1898e: 124 (*Hapalus*) **A:** MG NMO TD
longipennis Pic, 1933b: 7 (*Apalus*) **A:** XIN
muiri Kaszab, 1956e: 632 **A:** YE
arabica Kaszab, 1956e: 633
nigricollis Semenov, 1893d: 278 (*Hapalus*) **E:** AB AR
oohatai Morimoto & Maeta, 2009: 475 **A:** JA
osellai Bologna, 2018: 639 **A:** PA
plagiata Escalera, 1914: 387 (*Sitaris*) **N:** MO
escalerai Pic, 1920d: 10 (*Apalus*)
richteri Kaszab, 1959b: 5 **A:** IN
sokolowi Semenov, 1900b: 491 (*Hapalus*) **A:** UZ
testaceicornis Pic, 1939f: 147 (*Apalus*) **N:** EG **A:** SI
thakkola Schawaller, 1996b: 148 **A:** NP
tibetana Escherich, 1904: 35 (*Hapalus*) **A:** QIN

genus *Zonitis* Fabricius, 1775: 126 type species *Zonitis flava* Fabricius, 1775
subgenus *Zonitis* Fabricius, 1775: 126 type species *Zonitis flava* Fabricius, 1775
afghanica Kaszab, 1958c: 307 **A:** AF PA

ballionis Escherich, 1892: 242 **A:** AF KZ TD TM UZ XIN
scutellaris Fairmaire, 1892e: cliv
tristicula Semenov, 1900b: 481
bellieri Reiche, 1860: 731 **E:** IT (Sicilia) **N:** AG LB MO TU **A** TR
turcica J. Frivaldszky, 1877b: 104
bomiensis Tan, 1988: 292 **A:** XIZ
escherichi Semenov, 1893d: 273 **A:** TD TM TR
ferganensis Pic, 1951c: 11 **A:** "Turkestan"
fernancastroi Pardo Alcaide, 1950b: 4c [DA] **E:** FR IT (Sicily) SP **N:** MO
flava Fabricius, 1775: 127 **E:** AB AL AR AU BU CR CZ FR GG GR HU IT KZ PL PT RO SK SP ST SZ UK YU **N:** AG LB MO TU **A:** CY IN IS JO KZ LE SY TM TR UZ XIN
afra Rossi, 1792: 292 (*Lytta*)
analis Abeille de Perrin, 1880: 256
ancoroides Escherich, 1892: 242
atripes Reitter, 1911: 397
atriventris Pic, 1951a: 12
flaviventris J. Müller, 1902: 116
gaditana Pardo Alcaide, 1956: 18
hipponensis Pic, 1900e: 38
impressicollis Motschulsky, 1872: 53
inscutellaris Pic, 1922d: 18
latenigra Pic, 1951c: 11
melanopus Wellman, 1910a: 25 [RN]
metasternalis Csiki, 1953: 120 [HN]
metasternaloides Kaszab, 1958b: 191 [RN]
moltonii Schatzmayr, 1941a: 81
nigriceps Pardo Alcaide, 1956: 18
nigripennis Fabricius, 1798: 103
nigripes Motschulsky, 1872: 52
nigripes J. Müller, 1902: 116 [HN]
nigrithorax Pic, 1904h: 90 [HN]
obscuriceps Pic, 1907l: 178
praeusta Fabricius, 1792b: 48
ramirezi Pardo Alcaide, 1956: 18
reitteri Kaszab, 1958b: 191 [RN]
scutellata Motschulsky, 1872: 52
semiobscura Schatzmayr, 1941a: 81
signatithorax Pic, 1900e: 38
sophiensis Nedelkov, 1905: 434
testacea Fabricius, 1781: 331 (*Mylabris*)

thoracica Laporte, 1840: 276
unicolor Ragusa, 1898: 220
fortuccii Fairmaire, 1887f: 326 **A:** BEI MG
testaceonotata Pic, 1914c: 3
fuscimembris Fairmaire, 1886c: 351 [DA] **A:** JIX YUN
glasunowi Semenov, 1893d: 271 **E:** KZ ST "Caucasus" **A:** KZ MG TD TM
atripennis Semenov, 1893d: 272 [HN]
semenovi Kaszab, 1958b: 191 [RN]
hauseri Escherich, 1897b: 127 **A:** TM UZ
immaculata G.-A. Olivier, 1789: 166 (*Apalus*) **E:** AL AU BH BU CR GR HU IT FR MC PT RO SP ST SZ UK YU **N:** AG MO TU **A:** IS SY TR ZU
biimpressa Chevrolat, 1882: 4
binotaticollis Pic, 1913k: 137
fulva Rossi, 1792: 94 (*Mylabris*)
gibbicollis Abeille de Perrin, 1880: 254
gracilenta Fairmaire, 1892b: 156
ligurica Schellenberg, 1802: 11 (*Lymexylon*)
mahia Hübner, 1789: 44
mutica L.G. Scriba, 1790: 23
scutellaris Escherich, 1897a: 46 [HN]
scutellifera Wellman, 1910a: 25 [RN]
seminigra Reitter, 1884b: 93
silvestrii Schatzmayr, 1941b: 362
iphigeniae Pliginskij, 1914c: 429 **E:** UK
rufipes Pliginskij, 1914c: 429
rufoscutellata Pliginskij, 1914c: 429
kozlowi Semenov, 1900b: 482 **A:** XIN
microcephala Escherich, 1897a: 46 **A:** TR
nana Ragusa, 1881: 43 **E:** AL BH BU CR CZ FR GR HU IT SK ST SZ YU **A:** IS JO SY SZ TR
apicalis Motschulsky, 1872: 53 [HN]
laticollis Escherich, 1892: 238
motschulskyi Kaszab, 1958b: 191 [RN]
nigripectus Fairmaire, 1891e: cxxxiii **A:** HP KA
nigritarsis Stierlin, 1876: 477 (*Nemognatha*) **E:** AR "Caucasus" **A:** IN
nigriventris Motschulsky, 1872: 53 **E:** AR GG GR "Caucasus" **A:** TR
novercalis Escherich, 1891b: 236 **N:** AG
pseudopraeusta Kaszab, 1958c: 306 **A:** AF
ruficollis J. Frivaldszky, 1877b: 107 **E:** GR (Crete) **N:** AG LB MO **A:** IS JO SY TR
abeillei Bedel, 1892: 248
rubricollis Abeille de Perrin, 1880: 256
thoracica Escherich, 1892: 237 [HN]
sikkimensis Pic, 1910t: 91 **A:** SD

sinensis Pic, 1935b: 24 **A:** SCH
turkestanica Semenov, 1900b: 480 **A:** TD UZ

genus *Zonitis*, species incertae sedis
scutellaris Fairmaire, 1892e: cliv **A:** UZ "Turkestan"
prescutellaris Pic, 1911i: 138

genus *Zonitomorpha* Péringuey, 1909: 272 type species *Zonitis sellata* Fåhraeus, 1870
davidis Fairmaire, 1886c: 351 (*Zonitis*) **A:** BEI

genus *Zonitoschema* Péringuey, 1909: 274 type species *Lytta coccinea* Fabricius, 1801
Stenoderistella Reitter, 1911a: 395 type species *Stenodera pallidissima* Reitter, 1908
Zonitoides Fairmaire, 1883a: 31[HM] type species *Zonitoides megalops* Fairmaire, 1883
Zonitopsis Wellman, 1910b: 395 [RN] type species *Zonitoides megalops* Fairmaire, 1883
angustithorax Pic, 1912: 31 (*Zonitomorpha*) **A:** SHN
arabicum Batelka & Bologna, 2014: 243 **A:** YE
chourribum Batelka & Bologna, 2014: 247 **N:** TU
cothurnatum Marseul, 1873a: 228 (*Zonitis*) **A:** JA NC TAI
elongatipenne Pic, 1915e: 16 (*Zonitis*) **A:** JIX
genicularis Wellman, 1910: 25 (*Zonitis*) **A:** SA YE **AFR**
gibdoanum Kaszab, 1956e: 631 (*Zonitopsis*) **A:** OM SA **AFR**
iranicum Kaszab, 1959b: 4 **A:** AE IN IS SA **AFR**
japonicum Pic, 1910t: 90 (*Zonitis*) **A:** JA NC SC TAI ZHE **ORR**
geniculatum Borchmann, 1912a: 12 (*Zonitis*) [HN]
pallidum Fabricius, 1794: 447 (*Zonitis*)
kaszabi Batelka & Bologna, 2014: 258 **A:** YE (Suqutra)
kimi Kôno, 1936b: 95 (*Zonitis*) **A:** "Korea"
klapperichi Borchmann, 1941: 23 **A:** FUJ
macroxanthum yunnanum Kaszab, 1960d: 262 **A:** YUN
melanarthrum Fairmaire, 1894a: 36 (*Zonitis*) **A:** PA **ORR**
miwai Kôno, 1936b: 94 (*Zonitis.*) **A:** TAI
oculatissimum Peyerimhoff, 1929a: 125 [= 1931b: 84] **N:** AG **AFR**
okinawense Miwa, 1928: 76 (*Zonitis*) **A:** JA
pallidissimum Reitter, 1908c: 50 (*Stenodera*) **N:** AG EG LB MO TU **A:** IS SA
rubricolor Pic, 1924g: 454 **A:** AE IS OM SA **AFR**
stramineum Fairmaire, 1894: 35 (*Zonitis*) **A:** PA **ORR**

tribe Palaestrini Bologna, Turco & Pinto, 2013

genus *Palaestra* Laporte de Castelnau, 1840: 251 type species *Palaestra rubripennis* Laporte de Castelnau, 1840
bytinskii Kaszab, 1957e: 231 (*Zonitis*) **A:** IS

tribe Stenoderini Selander, 1991

genus *Stenodera* Eschscholtz, 1818: 469 type species *Stenodera sexpunctata* Eschscholtz, 1818

subgenus *Stenodera* Eschscholtz, 1818: 469 type species *Stenodera sexpunctata* Eschscholtz, 1818 (= *Meloe caucasica* Pallas, 1782)

Zonitides Abeille de Perrin, 1880: 253 type species *Zonitides oculifer* Abeille de Perrin, 1880

caucasica Pallas, 1781: 94 (*Meloe*) **E:** AB AL AR BU CR GG GR MC MD RO ST TR UK YU **A:** IN IS JO LE SY TM TR

bisbinotata Pic, 1909j: 130

crocata Abeille de Perrin, 1880: 253

inornata Escherich, 1897b: 101

sexmaculata Fabricius, 1794: 120 (*Mylabris*)

sexpunctata Eschscholtz, 1818: 469

djakonovi Aksentjev, 1978: 124 **A:** FE

foveicollis Fairmaire, 1897b: 231 (*Megatrachelus*) **A:** FUJ GUX

oculifera Abeille de Perrin, 1880: 253 (*Zonitides*) **A:** IS LE SY

bipunctata Chevrolat, 1882: 5 (*Zonitis*)

chevrolati Ragusa, 1882: 251 (*Zonitis*)

impressicollis Escherich, 1891b: 54

subgenus *Stenoderina* Aksentjev, 1988: 579 type species *Zonitis puncticollis* Chevrolat, 1834

anatolica J. Frivaldszky, 1884: 1 (*Zonitis*) **A:** SY TR

coeruleiceps Fairmaire, 1892b: 157 (*Zonitis*) **A:** TR

palaestina Mařan, 1942a: 25 **A:** IS

puncticollis Chevrolat, 1834: pl. 35 (*Zonitis*) **A:** IS LE SY TR

family Meloidae, nomen dubium

Meloe lutea Pallas, 1773: 722

family MYCTERIDAE Oken, 1843

subfamily **Mycterinae** Oken, 1843

genus *Mycterus* Clairville, 1798: 124 type species *Rhinomacer curculioides* Fabricius, 1781

subgenus *Eutryptes* Gistel, 1856: 375 type species *Bruchus umbellatarum* Fabricius, 1792

Cyclopidius Seidlitz, 1890: 555 [RN] type species *Bruchus umbellatarum* Fabricius, 1792

Cyclops Mulsant, 1859: 18 [HN] type species *Bruchus umbellatarum* Fabricius, 1792

Triconatus Spilman, 1954: 92 [RN] type species *Bruchus umbellatarum* Fabricius, 1792

articulatus Reitter, 1911a: 419 **E:** AR

tibialis Küster, 1850a: 86 **E:** AU BH CR CZ GE GR HU IT SK ST SZ TR UK YU **A:** TR

ruficornis Mulsant & Godart, 1855d: 278

umbellatarum Fabricius, 1787a: 41 (*Bruchus*) **E:** FR (Corse) IT PT SP **N:** AG MO TU **A:** IS

algiricus Seidlitz, 1920: 1177

pulverulentus Chevrolat, 1834: pl. 33

pulverunculus Küster, 1850a: 85
siculus Baudi di Selve, 1883a: 6
subtruncatus Guillebeau, 1893a: 90
subgenus *Mycterus* Clairville, 1798: 124 type species *Rhinomacer curculioides* Fabricius, 1781
Rhinomacer Fabricius, 1781: 199 [HN] type species *Rhinomacer curculioides* Fabricius, 1781
curculioides Fabricius, 1781: 199 (*Rhinomacer*) **E:** AB AU CR CT CZ FR GB GE GR HU IT LA PL PT RO SK SP ST SZ UK **N:** MO TU **A:** CY IS LE TR
coriaceus Thunberg, 1791: 115 (*Curculio*) [URN]
fulvus Brahm, 1790: 110 (*Rhinomacer*)
griseus Clairville, 1798: 124
gracilior Fairmaire, 1874a: 388

subfamily **Eurypinae** Thomson, 1860

genus *Omineus* Lewis, 1895b: 119 type species *Omineus humeralis* Lewis, 1895
Phalysius Champion, 1916: 150 type species *Phalysius caeruleus* Champion, 1916
Pseudothisias Pic, 1930: 23 type species *Pseudothisias bicolor* Pic, 1930
chuangae Hsiao & Pollock, 2019: 538 **A:** TAI
humeralis Lewis, 1895b: 119 **A:** JA TAI
taiwanensis Hsiao & Pollock, 2019: 535 **A:** TAI
toyoshimai Young & Saitô, 2017: 417 **A:** JA (Honshu)

subfamily **Hemipeplinae** Lacordaire, 1854

genus *Hemipeplus* Berthold, 1827: 394 type species *Hemipeplus hemipterus* Lacordaire, 1854
Nemicelus LeConte, 1854: 79 type species *Nemicelus marginipennis* LeConte, 1854
Ochrosanis Pascoe, 1866: 444 type species *Ochrosanis dohrnii* Pascoe, 1866
miyamotoi H. Kamiya, 1961: 75 **A:** JA (Amami-Ôshima) TAI

family BORIDAE C.G. Thomson, 1859
subfamily **Borinae** C.G. Thomson, 1859

genus *Boros* Herbst, 1797: 318 type species *Boros elongatus* Herbst, 1797 (= *Helops schneideri* Panzer, 1796)
Borus Illiger, 1802a: 129 [UE]
Borus Agassiz, 1846b: 49 [UE]
schneideri Panzer, 1796b: 1 (*Helops*) **E:** BY CT CZ EN FI FR GE LA LT NT PL RO SK ST SV UK **A:** ES FE JA MG NC NE NMO SC WS
boros Fabricius, 1801b: 558 (*Hypophloeus*)
corticalis Paykull, 1798: 93 (*Trogosita*)
elongatus Herbst, 1797: 319
picipes Schoenherr, 1806: 155 (*Trogosita*)

family TRICTENOTOMIDAE Blanchard, 1845

genus *Autocrates* J. Thomson, 1860a: 28 type species *Trictenotoma aenea* Westwood, 1846
aeneus Westwood, 1846: 353 (*Trictenotoma*) **A:** AP BT HP NP SD UP XIZ YUN "Punjab" **ORR**
maqueti Drumont, 2006: 4 **A:** GUI GUX HUB SC QIN
oberthueri Vuillet, 1910: 347 **A:** FUJ XIZ YUN
diversus Pic, 1920f: 15
vitalisi Vuillet, 1912: 297 **A:** HAI GUA GUX SCH SHA XIZ YUN **ORR**
waldenfelsi De Lisle, 1973: 17

genus *Trictenotoma* Gray, 1832: 534 type species *Trictenotoma childreni* Gray, 1832
davidi H. Deyrolle, 1875a: 68 [= 1875b: lx)
A: ANH FUJ GUA GUI GUX HAI HUN JIX SCH SHA XIZ YUN ZHE **ORR**
formosana Kriesche, 1919: 150 **A:** TAI
mniszechi H. Deyrolle, 1875b: lx **A:** BT SD **ORR**

family PYTHIDAE Solier, 1834

genus *Pytho* Latreille, 1796: 23 type species *Cucujus coeruleus* Fabricius, 1787 (= *Tenebrio depressus* Linnaeus, 1767)
Enoptes Gistel, 1848: 10 [RN] type species *Cucujus coeruleus* Fabricius, 1792
Pytholus Rafinesque, 1815: 114 type species *Cucujus coeruleus* Fabricius, 1792
abieticola J.R. Sahlberg, 1875: 222 **E:** AU CZ FI GE NR NT PL SK SV **A:** ES FE
niger Reitter, 1911a: 414
depressus Linnaeus, 1767: 675 (*Tenebrio*) **E:** AU BH BY CT CZ DE EN FI FR GB GE GG HU IT KZ LA LI LT NL NR NT PL RO SK SP ST SV SZ UK **A:** ES FE WS
atriceps Pic, 1912j: 49
basipennis Pic, 1912k: 58
castaneus Fabricius, 1792b: 95 (*Cucujus*)
coeruleus Herbst, 1782: 6 (*Cucujus*)
coeruleus Fabricius, 1787a: 166 (*Cucujus*) [HN]
coeruleus Fabricius, 1792b: 94 (*Cucujus*) [HN]
festivus Fabricius, 1792b: 95 (*Cucujus*)
lignarius DeGeer, 1775: 38 (*Tenebrio*)
planus Herbst, 1799: 14 (*Cucujus*)
jezoensis Kôno, 1936c: 36 **A:** JA
kolwensis C.R. Sahlberg, 1833: 445 **E:** EN FI NT PL SV **A:** ES WS
sahlbergi Pic, 1912j: 49
nivalis Lewis, 1888: 109 **A:** FE JA

family Pythidae, genus incertae sedis
genus *Osphyoplesius* Winkler, 1915: 331 type species *Osphyoplesius anophthalmus* Winkler, 1915
anophthalmus Winkler, 1915: 333 **E:** UK (Krym)

loebli Español, 1975a: 107 **E:** GR (Epeiros)
colacurcioi Magrini & Uliana, 2015: 574 **E:** IT (Sicilia)

family PYROCHROIDAE Latreille, 1807
subfamily **Tydessinae** Nikitsky, 1986

genus *Tydessa* Peacock, 1982: 362 type species *Dasytes constrictus* Lewis (= *Dasytes lewisi* Pic, 1937)
lewisi Pic, 1937c: 74 (*Dasytes*) [RN] **A:** FE JA
constrictus Lewis, 1895b: 118 (*Dasytes*) [HN]

subfamily **Pedilinae** Lacordaire, 1859

genus *Pedilus* Fischer von Waldheim, 1820: pl. 5 type species *Pedilus fuscus* Fischer von Waldheim, 1820
Corphyra Say, 1835: 189 type species *Anthicus lugubris* Say, 1827
Tosadendroides Kôno, 1935c: 158 type species *Tosadendroides okamotoi* Kôno, 1935
balcanicus Apfelbeck, 1918: 90 **E:** SB/MC [border area]
canescens Semenov, 1902a: 258 **E:** ST "Caucasus"
constricticollis Semenov, 1893e: 327 **A:** TD
errans Faust, 1877b: 323 **E:** GG ST
fuscus Fischer von Waldheim, 1820: pl. 5 **A:** ES MG WS
reitteri Pic, 1903b: 82
laevicollis Reitter, 1901d: 115 **E:** CT (South Ural) **A:** WS
mongolicus Reitter, 1901d: 116 **A:** MG NMO "Siberia"
okamotoi Kôno, 1935c: 157 (*Tosadendroides*) **A:** JA
polyxanthus Semenov, 1900e: 687 **A:** KI TD UZ
rubricollis Motschulsky, 1858d: 190 **E:** UK
rufipes Motschulsky, 1858d: 190 [RN] **A:** KI
fulvipes Motschulsky, 1845a: 82 [HN]
serricornis Semenov, 1902a: 260 **E:** ST "Caucasus"
signatipennis Fleischer, 1918: 200 **A:** KI
tibialis Semenov, 1893e: 327 **A:** KI KZ XIN
pallidipes Pic, 1900a: 10
tristis Semenov, 1893e: 328 **A:** IN KI KZ XIN "Turkestan"
fulvipennis Heyden, 1886b: 181
lividipennis Reitter, 1901d: 114
pallidipennis Semenov, 1893e: 327
weberi Reitter, 1901d: 114 **E:** ST UK
willbergi Semenov, 1900e: 687 **A:** TD
zubowskii Reitter, 1901d: 114 [HN]
xanthopus Semenov, 1899b: 654 **A:** ES KZ TM WS "Turkestan"
zubowskii Semenov, 1900a: 319 **A:** KI UZ "Turkestan"

subfamily **Pyrochroinae** Latreille, 1807

genus *Dendroides* Latreille, 1810: 212 type species *Dendroides canadensis* Latreille, 1810
lesnei Blair, 1914a: 314 **A:** JA
    nakabusana Kôno, 1936b: 97
ussuriensis L.N. Medvedev, 1977: 48 **A:** FE NC NE SC

genus *Eupyrochroa* Blair, 1914a: 316 type species *Pyrochroa insignata* Fairmaire, 1894
insignata Fairmaire, 1894b: 220 (*Pyrochroa*) **A:** FUJ XIZ YUN
    rufoverticalis Pic, 1915c: 4
limbaticollis Pic, 1909r: 330 (*Pyrochroa*) **A:** JIX TAI
    immaculata Pic, 1923b: 19 (*Pseudopyrochroa*)

genus *Frontodendroidopsis* Young, 2004b: 208 type species *Dendroides ocularis* Lewis, 1887
ocularis Lewis, 1887a: 174 (*Dendroides*) **A:** JA
pennyi Young, 2017: 591 **A:** YUN

genus *Hemidendroides* Ferrari, 1869: 198 type species *Dendroides ledereri* Ferrari, 1869
daglariensis Young, 2004c: 933 (*Pyrochroa*) [RN] **A:** SY
    cyanipennis Pic, 1896j: 141 (*Dendroides*) [HN]
davidis Fairmaire, 1878a: 121 (*Pyrochhroa*[sic!]) **A:** SCH
ledereri Ferrari, 1869: 195 **A:** TR
peyroni Reiche, 1878: lxxiii **A:** TR

genus *Himalapyrochroa* Young, 2004a: 190 type species *Himalapyrochroa gibbosa* Young, 2004
gibbosa Young, 2004a: 191 **A:** NP SD YUN
nitidicollis Pic, 1955: 12 **A:** SD (Sikkim)

genus *Phyllocladus* Blair, 1914a: 315 type species *Dendroides magnificus* Blair, 1912
bilobatus Young, 2013: 58 **A:** HUB SCH
cinnabarus Young, 2013: 60 **A:** GUI
grandipennis Pic, 1906d: 29 (*Dendroides*) **A:** BT XIZ
kasantsevi Young, 2005b: 103 **A:** GAN SHA
magnificus Blair, 1912: 532 (*Dendroides*) **A:** YUN

genus *Pogonocerus* Fischer von Waldheim, 1812: 281 type species *Pogonocerus thoracicus* Fischer von Waldheim, 1812
thoracicus Fischer von Waldheim, 1812: 281 **E:** GG ST (Kavkaz)

genus *Pseudodendroides* Blair, 1914a: 314 type species *Dendroides niponensis* Lewis, 1887
amamiana Nakane, 1988a: 4 (*Pseudopyrochroa*) **A:** JA (Amami-Ôshima)
niponensis Lewis, 1887a: 174 (*Dendroides*) **A:** JA
    nitidicollis Pic, 1955b: 12

ocularis Lewis, 1887a: 174 (*Dendroides*) **A:** JA
sulcatithorax Pic, 1955: 13 **A:** XIZ
umenoi Kôno, 1936a: 5 **A:** JA
uraiana Kôno, 1935c: 158 **A:** TAI

genus *Pyrochroa* Geoffroy, 1762: 338 type species *Cantharis coccinea* Linnaeus, 1760
coccinea Linnaeus, 1760: 202 (*Cantharis*) **E:** AU AR BE BH BY CR CT CZ DE EN FI FR GB GE GG GR HU IT LA LI LT LU RO SK SL ST SV NL NR NT PL PT RO SK SP ST SZ **N:** AG
  coccinea Donovan, 1793: [51], pl. 56 [HN]
  kabyliana Pic, 1898g: 98
  purpurata Schrank von Paula, 1781: 173
  rubra DeGeer, 1775: 20
pubescens Pic, 1907g: 158 (*Pyrochroa*) **A:** SD
  notaticeps Pic, 1912d: 18
  rubricolor Pic, 1912d: 18
serraticornis kiesenwetteri Fairmaire, 1849: 424 **E:** IT (Sicilia, South Italy)
serraticornis serraticornis Scopoli, 1763: 42 (*Cantharis*) **E:** AU AR BE BH CZ DE EN FR HU GB GE IR IT LA LI LT LU NL PL SK SP ST SZ UK **A:** TR
  purpurata O.F. Müller, 1788: 80
  rubens Schaller, 1783: 301 (*Lampyris*)
  ruberrima Geoffroy, 1785: 153
  rutilans Voet, 1806: 72
  satrapa Schrank von Paula, 1781: 174
  tauricola Pic, 1912e: 26
subcostulata Fairmaire, 1891d: cii **A:** BT KA PA UT

genus *Pseudopyrochroa* Pic, 1906d: 28 type species *Pseudopyrochroa deplanata* Pic, 1906
  Pyrochomima Pic, 1955b: 13 type species *Pyrochromima dentaticollis* Pic, 1955 (=*Pseudopyrochroa melanocephala* Blair, 1912)
acuminata Young, 2000: 249 **A:** TAI
antennalis Blair, 1912: 528 (*Pyrochroa*) **A:** SD YUN
atripennis Lewis, 1887a: 171 (*Pyrochroa*) **A:** JA
brevitarsis Lewis, 1887a: 170 (*Pyrochroa*) **A:** JA
cardoni Fairmaire, 1894a: 30 (*Pyrochroa*) **A:** IN AP SD **ORR**
carinifrons Kôno, 1929b: 68 **A:** TAI
choui Young, 2000: 250 **A:** TAI
costatipennis Pic, 1908l: 75 **A:** YUN
depressa Pic, 1914d: 17 **A:** TAI
  taiwana Kôno, 1929b: 67
donckieri Pic, 1908l: 75 **A:** YUN
  lyciformis Pic, 1909r: 330 (*Pyrochroa*)
episcopalis Lewis, 1895a: 432 (*Pyrochroa*) **A:** JA

facialis Fairmaire, 1891c: xx (*Pyrochroa*) **A:** HUB
fainanensis Pic, 1911k: 158 (*Pyrochroa*) **A:** TAI
flavilabris Blair, 1914a: 321 **A:** JA
girardi Young, 2001: 274 **A:** SCH
harmandi Pic, 1912d: 18 (*Pyrochroa*) **A:** NP SD
nepalensis Paulus, 1972: 75
japonica Heyden, 1879: 88 (*Pyrochroa*) **A:** JA
gibbifrons Lewis, 1887a: 173 (*Schizotus*)
higoniae Lewis, 1895a: 432 (*Pyrochroa*)
kazuoi Ohbayashi, 1968: 35 **A:** JA
teruhisai Saitô, 1990: 105
klapperichi Pic, 1955c: 29 **A:** FUJ
lateraria Motschulsky, 1860c: 143 (*Pyrochroa*) **A:** FE NE "Korea"
latevittata Pic, 1939d: 135 **A:** ZHE
laticollis Lewis, 1887a: 169 (*Pyrochroa*) **A:** JA
aurita Lewis, 1887a: 173 (*Schizotus*)
lineaticollis Pic, 1914d: 17 **A:** SCH NP XIZ
lophicristata Young, 2015: 452 **A:** NP **ORR**
marginata Kôno, 1929b: 70 **A:** JA
melanocephala Blair, 1912: 527 **A:** SD
angustata Pic, 1955b: 12
dentaticollis Pic, 1955b: 13
obtusicristata Young, 2000: 252 **A:** FUJ TAI
peculiaris Lewis, 1887a: 170 (*Pyrochroa*) **A:** FE JA
punctifrons Young, 2000: 253 **A:** TAI
rufula Motschulsky, 1866: 173 (*Pyrochroa*) **A:** FE JA
aurita Lewis, 1887a: 173 (*Schizotus*)
etorofuensis Kôno, 1935a: 87
ogasawarai Kôno, 1929b: 69
rubricollis Lewis, 1887a: 172 (*Schizotus*)
satoi Young, 2003: 346 **A:** FUJ "Korea"
tumidifrons Young, 2000: 254 **A:** TAI
velutina Fairmaire, 1891c: xxi (*Pyrochroa*) **A:** HUB
vestiflua Lewis, 1887a: 169 (*Pyrochroa*) **A:** FE JA
yaeyamana Kai and Yoshitomi, 2013: 231 **A:** JA (Yaeyama Is.)

genus *Schizotus* Newman, 1838: 374 type species *Schizotus cervicalis* Newman, 1838
Pyrochroella Reitter, 1911a: 385 type species *Pyrochroa pectinicornis* Linnaeus, 1758
cardinalis Mannerheim, 1853a: 302 (*Pyrochroa*) **A:** ES FE
innotaticeps Pic, 1904d: 26
fuscicollis Mannerheim, 1853a: 301 (*Pyrochroa*) **A:** ES FE
punctus Motschulsky, 1860c: 143

pectinicornis Linnaeus, 1758: 403 (*Cantharis*) **E:** AU BE BH BY CT CZ DE EN FI FR GE HU IR IT LA LI LT LU NL NR NT PL RO SK SP SV SZ
nigra DeGeer, 1775: 22 (*Pyrochroa*)
yamaguchii Kôno, 1936a: 5 **A:** TAI

genus *Sinodendroides* Young, 2005a: 94 type species *Sinodendroides chinensis* Young, 2005
chinensis Young, 2005a: 96 **A:** GAN SHA
distinctipennis Pic, 1938e: 15 (*Pseudodendroides*) **A:** GAN

subfamily **Agnathinae** Lacordaire, 1859

genus *Agnathus* Germar, 1825: 4 type species *Notoxus decoratus* Germar, 1818
decoratus Germar, 1818: 229 (*Notoxus*) **E:** AU BH BU BY CR CZ FR GE GG HU IT LU PL RO SK SL ST SZ TR UK
secundus Jelínek & Kubáň, 2009: 258 **A:** YUN

family SALPINGIDAE Leach, 1815

subfamily **Othniinae** LeConte, 1861

genus *Elacatis* Pascoe, 1860a: 52 type species *Elacatis delusa* Pascoe, 1860
Othnius LeConte, 1861: 102 type species *Othnius umbrosus* LeConte, 1861
atrithorax Pic, 1920a: 9 (*Othnius*) **A:** JA TAI
bengalensis Sengupta, 1977: 537 **A:** AP SD
bhutanensis Sengupta, 1980: 28 **A:** BT
formosana Borchmann, 1916: 106 (*Othnius*) **A:** TAI
kraatzii Reitter, 1879: 226 **A:** FE JA SC
ocularis Lewis, 1891b: 248 (*Othnius*) **A:** JA SC
rugicollis Borchmann, 1916: 108 (*Othnius*) **A:** TAI
similis Borchmann, 1916: 107 (*Othnius*) **A:** TAI

subfamily **Prostominiinae** Grouvelle, 1914

genus *Ocholissa* Pascoe, 1863b: 85 type species *Ocholissa laeta* Pascoe, 1863
Sasajia Masumoto & Akita, 2007: 18 type species *Sasajia hiroyukii* Masumoto & Akita, 2007
harmandi Grouvelle, 1908b: 411 **A:** SD
hiranoi Masumoto & Akita, 2009: 301 **A:** JA
hiroyukii Masumoto & Akita, 2007: 19 (*Sasajia*) **A:** JA

genus *Prostominia* Reitter, 1889i: 315 type species *Prostominia lewisi* Reitter, 1889
lewisi Reitter, 1889i: 316 **A:** FE (Kunashir) JA

genus *Trogocryptoides* Champion, 1924a: 167 type species *Trogocryptoides brachyderus* Champion, 1924
brachyderus Champion, 1924a: 168 **A:** UP
shintaroi Sasaji, 1988a: 35 **A:** JA

subfamily **Agleninae** Horn, 1878

genus *Aglenus* Erichson, 1845: 285 type species *Hypophloeus brunneus* Gyllenhal, 1813

Monopis Dejean, 1835: 314 [HN]

brunneus Gyllenhal, 1813: 711 (*Hypophloeus*) **E:** AU AZ CR CZ DE FR GB GE GR HU IR IT NL PL PT SK SP ST SV SZ "Caucasus" **N:** EG CI LB MR **A:** IN **AFR NAR**

longior Pic, 1923g: 9

major Schaufuss, 1882a: 48

obsoletus Shuckard, 1839: 249

rosellae Ragusa, 1892a: 137 (*Anommatus*)

rugipennis Schaufuss, 1882a: 48

subfamily **Inopeplinae** Grouvelle, 1908

genus *Inopeplus* Smith, 1851: 4 [RN] type species *Ino picta* Laporte, 1835

Ino Laporte, 1835: 135 [HN] type species *Ino picta* Laporte, 1835

Euryplatus Motschulsky, 1860a: 95 type species *Euryplatus lateralis* Motschulsky, 1860

Pseudino Fairmaire, 1869b: 208 type species *Pseudino coquereli* Fairmaire, 1869

albonotatus Motschulsky, 1860a: 98 (*Euryplatus*) **A:** AP SD UP **ORR**

apatani Pal, 1992b: 347 **A:** AP

biocellatus Motschulsky, 1860a: 98 (*Euryplatus*) **A:** AP SD **ORR**

decisus Walker, 1858: 205 (*Prognatha*) **A:** AP SD UP **ORR**

distinctus Sengupta, Pal & Mukhopadhyay, 1977: 401 **A:** SD

jairajpurii Pal, 1992b: 344 **A:** AP

nitidus Sengupta, Pal & Mukhopadhyay, 1977: 400 **A:** AP SD

patkoicus Pal, 1992a: 238 **A:** AP

quadrinotatus Gorham, 1873: 258 (*Ino*) **A:** JA

sauteri Grouvelle, 1914: 58 **A:** TAI

syozoi Sasaji, 1986: 7 **A:** TAI

uenoi ohbayashii Hatta, 1988: 300 **A:** JA

uenoi uenoi Sasaji, 1984: 46 **A:** JA

genus *Uruminopeplus* Satô & Hatta, 1988: 85 type species *Uruminopeplus sakaii* Satô & Hatta, 1988

sakaii Satô & Hatta, 1988: 86 **A:** JA

monstrosus Sasaji, 1988: 41 (*Inopeplus*) **A:** JA

subfamily **Salpinginae** Leach, 1815

genus *Cariderus* Mulsant, 1859: 46 type species *Rhinosimus aeneus* G.-A. Olivier, 1807

aeneus G.-A. Olivier, 1807: 86: 455 (*Rhinosimus*) **E:** AU BY FR GE GR HU IT LA PL SZ **N:** AG

megricus Iablokoff-Khnzorian, 1956a: 135

numidicus Pic, 1904b: 10

ornithorrhynchus Abeille de Perrin, 1874b: 30

genus *Chilopeltis* Seidlitz, 1916: 424 type species *Chilopeltis insculpta* Seidlitz, 1916
Platylissodema Blair, 1919: 117 type species *Lanthanus rouyeri* Pic, 1914
laevipennis Marseul, 1876d: 331 (*Lissodema*) **A:** JA SC

genus *Colposis* Mulsant, 1859: 41 type species *Salpingus virescens* Mulsant, 1859
mutilatus Beck, 1817: 19 (*Salpingus*) **E:** AU BH CZ ES FR GE HU IT LA LT PL RO RU SK SZ YU **A:** FE "Siberia"
maritimus Pic, 1903k: 140 (*Salpingus*)
virescens Mulsant, 1859: 41 (*Salpingus*)

genus *Istrisia* Lewis, 1895a: 254 type species *Istrisia rufobrunnea* Lewis, 1895
rufobrunnea Lewis, 1895a: 254 **A:** FE JA

genus *Lissodema* Curtis, 1833: 187 type species *Lissodema heyana* Curtis, 1833 (= *Salpingus cursor* Gyllenhal, 1813)
Spinolissodema Pic, 1919b: 3 type species *Spinolissodema unifasciata* Pic, 1919
Stenolissodema Desbrochers des Loges, 1900a: 6 type species *Salpingus lituratus* A. Costa 1847
ainunua Lewis, 1895a: 255 **A:** JA
cursor Gyllenhal, 1813: 716 (*Salpingus*) **E:** AU BE CR CT CZ DE FI FR GB GE HU IT LT LU NL NR NT PL RO SK ST SV SZ "Caucasus" **A:** FE
heyana Curtis, 1833: 187
kirkae Donisthorpe, 1925: 106
dentata Lewis, 1895a: 257 **A:** FE JA
beatula Lewis, 1895a: 257
denticollis Gyllenhal, 1813: 715 (*Salpingus*) **E:** AR AU BE BH BU CR CZ DE FR GB GE GR HU IR IT LU NL PL RO RU SK SP ST SV SZ "Caucasus"
gallica Pic, 1933a: 2
humeralis A. Villa & G.B. Villa, 1833: 35 (*Salpingus*)
quadriguttata Le Peletier de Saint-Fargeau & Audinet-Serville, 1825: 288 (*Rhinosimus*)
quadrimaculata Mulsant, 1830: 302 (*Rhinosimus*)
quadripustulata Marsham, 1802: 297 (*Curculio*) [HN]
dolsandoensis Lee & Lim, 2019: 23 **A:** SC
japona Reitter, 1877g: 382 **A:** JA
liturata A. Costa, 1847: 158 (*Salpingus*) **E:** FR IT SP **N:** AG TU
cloueti Guillebeau, 1896: 245
minuta Lewis, 1895a: 258 **A:** JA
morimotoi Sasaji, 1988a: 30 **A:** JA
munaguro Sasaji, 1988a: 28 **A:** JA
myrmido Marseul, 1876d: 332 **A:** JA
pictipennis Lewis, 1895a: 256 **A:** JA
plagiata Lewis, 1895a: 255 **A:** FE JA

quadridentata Champion, 1923: 79 **A:** UP
rosti Reitter, 1897d: 125 **E:** ST "Circassia"
teruhisai Sasaji, 1988: 27 **A:** JA
tomaroides Lewis, 1895a: 258 **A:** JA
uenoi Sasaji, 1988: 32 **A:** JA
unifasciata Pic, 1919b: 3 **A:** JA
validicornis Lewis, 1895a: 258 **A:** JA

genus *Pseudorabocerus* Pic, 1903k: 140 type species *Salpingus lederi* Reitter, 1888
lederi Reitter, 1888b: 213 (*Salpingus*) **E:** ST "Circassia"

genus *Pseudosphaeriestes* Nikitsky, 1992c: 491 type species *Salpingus niponicus* Lewis, 1895
niponicus Lewis, 1895a: 255 (*Salpingus*) **A:** FE (Kunashir) JA

genus *Rabocerus* Mulsant, 1859: 43 type species *Salpingus foveolatus* Ljungh, 1823
foveolatus Ljungh, 1823: 269 (*Salpingus*) **E:** AU BE BY CT CZ DE EN ES FI FR GB GE HU IR IT LA LT NR NT PL RO SK SV SZ
  bishopi Sharp, 1909: 245
  championi Sharp, 1909: 246
  impressithorax Pic, 1903k: 140 (*Salpingus*)
  mutilatus Champion, 1886b: 160 (*Salpingus*)
gabrieli Gerhardt, 1901: 18 (*Salpingus*) **E:** AU BE BY CT CZ DE EN FI FR GB GE HU IR IT LA LT NL NR NT PL SK ST SV SZ

genus *Salpingus* Illiger, 1802a: 150 [= 1802b: 301] type species *Curculio roboris* Fabricius, 1787
  Anthribus Clairville, 1798: 124 [HN] type species *Curculio ruficollis* Linnaeus, 1760
  Rhinosimus Latreille, 1802: 192 type species *Curculio planirostris* Fabricius, 1787
caucasicus Reitter, 1905f: 312 **E:** AR ST "Caucasus"
depressifrons Nikitsky & Belov, 1983: 523 **A:** FE JA SC
  morishimai Sasaji, 1987: 30
planirostris Fabricius, 1787a: 119 (*Curculio*) **E:** AN BE BU BY CR CT CZ DE EN FI FR GB GE HU IR IT LA LT LU NL NR NT PL PT RO SP ST SV "Caucasus" **N:** MO TU
  fulvirostris Fabricius, 1787b: 381 (*Curculio*)
  luteonitens Fairmaire, 1879c: 206
  quadriimpressus Desbrochers des Loges, 1900a: 22
spinolae A. Costa, 1847: 109 (*Rhinosimus*)
ruficollis Linnaeus, 1760: 177 (*Curculio*) **E:** AU BY CR CT CZ DE EN FI FR GE HU IR IT LA LT LU NL PL NR NT RO SK SL SP SV SZ UK
  bicolor Stephens, 1831: 217
  coeruleocephalus Latreille, 1804b: 28 (*Rhinosimus*)

roboris Fabricius, 1787a: 100 (*Curculio*)
rostratus DeGeer, 1775: 252 (*Curculio*)
viridicollis Pic, 1892c: 32 (*Rhinosimus*)
tapirus Abeille de Perrin, 1874b: 29 (*Rhinosimus*) E: IT SP **N:** AG TU
angustirostris Pic, 1898d: 35
elephas Chobaut, 1903a: 143
tapiroides Reitter, 1912b: 196
taiwanus Masumoto, Hirano & Akita, 2016: 62 **A:** TAI
yaeyamensis Masumoto, Hirano & Akita, 2016: 60 **A:** JA (Ryukyus: Yaeyama Is.)

genus *Salpingus*, nomen dubium
striatus Geoffroy, 1785: 115 (*Rhinomacer*) **E:** FR

genus *Salpingoides* Nikitsky, 1987: 1851 type species *Salpingoides thoracicus* Nikitsky, 1987: 1852
thoracicus Nikitsky, 1987: 1852 **A:** FE

genus *Sphaeriestes* Stephens, 1831: 218 type species *Dermestes ater* Paykull, 1798 (= *Salpingus stockmanni* Biström, 1977)
Arabocerus Iablokoff-Khnzorian, 1985: 216 type species *Salpingus impressus* Wollaston, 1857
Salpingellus Reitter, 1911a: 415 type species *Sphaeriestes ater* Paykull, 1798
Salpingus Gyllenhal, 1810: 640 [RN, HN] type species *Dermestes ater* Paykull, 1798
Sphaeriesthes Schenkling, 1922: 194 [UE]
Trichocolposinus Seidlitz, 1916: 489 type species *Vincenzellus hirtus* Broun, 1886
aeratus Mulsant, 1859: 34 (*Salpingus*) **E:** AU BH CT CZ GE GR IT PL SK ST SZ **N:** AG TU
cedri Pic, 1903k: 141 (*Salpingus*)
nitidus Chevrolat, 1860: 77 (*Salpingus*)
jacqueti Pic, 1903k: 140 (*Salpingus*)
bimaculatus Gyllenhal, 1810: 644 (*Salpingus*) **E:** CT CZ EN FI FR GE IT LA LT NR NT PL RU SV UK
castaneus Panzer, 1796a: 16: (*Notoxus*) **E:** AU BE BH BY CR CT CZ DE EN FI FR GB GE HU IR IT LA LT NL NR PL RU SK SL SP ST SV SZ "Caucasus"
achilleae Bonelli, 1812: 175 (*Salpingus*)
brunnescens Pic, 1892c: 32 (*Salpingus*)
immaculatus Stephens, 1831: 218
piceae Germar, 1824: 9 (*Salpingus*)
exsanguis Abeille de Perrin, 1870: 89 (*Salpingus*) **E:** FR SP
impressus Wollaston, 1857a: 161 **N:** MR
reyi Abeille de Perrin, 1874b: 26 (*Salpingus*) **E:** DE FI FR GB GE GR (Crete) IR IT NL NR PL SP SV SZ **A:** CY SY
aeneus Stephens, 1835: 421

ater Stephens, 1831: 218 [HN]
fowleri Seidlitz, 1916: 467
palpalis Baudi di Selve, 1877a: 49
tetramerus Rey, 1892: 138
stockmanni Biström, 1977: 18 (*Salpingus*) E: AU BY CT CZ DE EN ES FI FR GB GE HU IR IT LA LT NL NR NT PL RU SK SP ST SV SZ UK **A:** ES FE WS **NAR**
ater Paykull, 1798: 298 (*Dermestes*) [HN]

genus *Vincenzellus* Reitter, 1911a: 418 type species *Rhinosimus viridipennis* Latreille, 1804 (= *Anthribus ruficollis* Panzer, 1794)
Colposinus Seidlitz, 1916: 390 type species *Rhinosimus viridipennis* Latreille, 1804
fascipennis Reitter, 1897d: 125 (*Rhinosimus*) E: ST "Caucasus"
ruficollis Panzer, 1794c: 19 (*Anthribus*) E: AU BE BH BU CR CZ DE FR GE HU IR IT LU NL PL RO SK SP SV SZ UK YU
genei A. Costa, 1847: 89 (*Rhinosimus*)
roboris Fabricius, 1798: 161 (*Anthribus*)
ruficeps Bose, 1858: 96 (*Rhinosimus*)
viridipennis Latreille, 1804b: 26 (*Rhinosimus*)
vaulogeri Abeille de Perrin, 1901b: 234 (*Salpingus*) **N:** AG

subfamily **Aegialitinae** LeConte, 1862

genus *Aegialites* Mannerheim, 1853b: 178 type species *Aegialites debilis* Mannerheim, 1853
Eurystethes Seidlitz, 1916: 127 [RN] type species *Aegialites debilis* Mannerheim, 1853
antsiferovaensis Zerche, 2004: 47 **A:** FE (Kuril Is.: Antsiferova)
beringensis Zerche, 2004: 40 **A:** FE (Bering Is.: Arii Kamen, Toporkov Is.)
chirinkotanensis Zerche, 2004: 75 **A:** FE (Kuril Is.: Chirinkotan)
ekarmaensis Zerche, 2004: 77 **A:** FE (Kuril Is.: Ekarma)
gracilis Zerche, 2004: 46 **A:** FE (Kuril Is.: Shumshu)
iturupensis Zerche, 2004: 88 **A:** FE (Kuril Is.: Iturup)
kharimkotanensis Zerche, 2004: 73 **A:** FE (Kuril Is.: Kharimkotan)
kunashirensis Zerche, 2004: 89 **A:** FE (Kuril Is.: Kunashir)
kusnetzovi Pütz, 2009: 29 **A:** JA
leleji Zerche, 2004: 71 **A:** FE (Kuril Is.: Onekotan)
matuaensis Zerche, 2004: 84 **A:** FE (Kuril Is.: Matua)
onekotanensis Zerche, 2004: 70 **A:** FE (Kuril Is.: Onekotan)
raikokensis Zerche, 2004: 82 **A:** FE (Kuril Is.: Raikoke)
shiashkotanensis Zerche, 2004: 80 **A:** FE (Kuril Is.: Shiashkotan)
shumshuensis Zerche, 2004: 44 **A:** FE (Kuril Is.: Shumshu)
stejnegeri Linell, 1898: 74 **A:** FE (Sachalin: Tjuleniy Ostrov)
sugiharai Kôno, 1938: 3 (*Aegialatis*) **A:** FE (Kuril Is.: Shiashkotan)
urupensis Zerche, 2004: 86 **A:** FE (Kuril Is.: Urup)

family ANTHICIDAE Latreille, 1819
subfamily **Anthicinae** Latreille, 1819
tribe Anthicini Latreille, 1819

genus *Amblyderus* LaFerté-Sénectère, 1847: 368 [= 1849n: 62] type species *Anthicus scabricollis* LaFerté-Sénectère, 1847
Inamblyderus Pic, 1911i: 134 type species *Amblyderus obesus* Casey, 1895
brunneus Pic, 1893a: 15 **E:** IT (Calabria Sicilia)
indicus Fairmaire, 1894a: 33 **A:** NP PA UP **ORR**
laxatus Bonadona, 1964: 228 **A:** SA **AFR**
longidentatus Pic, 1957: 436 **A:** YE
maculipennis Pic, 1898j: 181 **N:** EG **A:** SA YE **AFR**
sabulosus Pic, 1899j: 173 **N:** EG **A:** IS
scabricollis LaFerté-Sénectère, 1847: 368 (*Anthicus*) **E:** FR (Corse) IT PT SP **N:** AG MO TU
inhumeralis Pic, 1900j: 89
sulcithorax Pic, 1898e: 69 **A:** YE **AFR**
truncatus LaFerté-Sénectère, 1849e: 3 [= 1849n: 64] **N:** EG **AFR**
tuberculatus Krekich-Strassoldo, 1928a: 88 **A:** UP **ORR**

genus *Anthicomorphus* Lewis, 1895a: 428 type species *Anthicomorphus suturalis* Lewis, 1895
cruralis Lewis, 1895a: 429 **A:** JA JIL SC TAI
himalayanus Telnov, 2010b: 21 **A:** NP
mimicus Telnov, 2016: 45 **A:** SHA
niponicus amamiensis Nomura, 1962a: 3 **A:** JA (Ryukyu Islands)
niponicus niponicus Lewis, 1895a: 429 **A:** FE JA JIL SC
puberulus Marseul, 1877: 467 (*Anthicus*) **A:** JA
suturalis Lewis, 1895a: 428 **A:** JA

genus *Anthicus* Paykull, 1798: 253 type species *Meloe antherinus* Linnaeus, 1760
subgenus *Anthicus* Paykull, 1798: 253 type species *Meloe antherinus* Linnaeus, 1760
Birricollis J.R. Sahlberg, 1903f: 67 [SN] type species *Anthicus genei* LaFerté-Sénectère, 1849
Birricomus Pic, 1894e: 43 type species *Anthicus genei* LaFerté-Sénectère, 1849
Brevicollis J.R. Sahlberg, 1903f: 67 [SN] type species *Meloe antherinus* Linnaeus, 1760
Brevicomus Pic, 1894e: 46 [SN] type species *Meloe antherinus* Linnaeus, 1760
Cartolus Mulsant & Rey, 1866a: 104 type species *Notoxus sellatus* Panzer, 1796
Cartolus Mulsant & Rey, 1866b: 192 [HN]
Curticollis Pic, 1892f: 44 [SN] type species *Anthicus trotommidens* Pic, 1892
Curticomus Pic, 1894g: 76 [RN] type species *Anthicus trotommidens* Pic, 1892
Eonius Thomson, 1864: 366 type species *Notoxus bimaculatus* Illiger, 1801
Laticomus Pic, 1894g: 71 [SN] type species *Notoxus bimaculatus* Illiger, 1801
Nathicus Casey, 1895: 655 type species *Nathicus virginiae* Casey, 1895
Nodolinus Mulsant & Rey, 1866a: 83 type species *Anthicus schmidtii* Rosenhauer, 1847

Nodolinus Mulsant & Rey, 1866b: 171 [HN]
Platylorus Mulsant & Rey, 1866a: 84 type species *Notoxus bimaculatus* Illiger, 1801
Platylorus Mulsant & Rey, 1866b: 172 [HN]
Pubicollis J.R. Sahlberg, 1903c: 9 [SN] type species *Anthicus axillaris* Schmidt, 1842
Pubicomus Pic 1894e: 48 [RN] type species *Anthicus axillaris* Schmidt, 1842
abashirensis Nakane, 1988a: 5 **A:** JA
abyssinicus adenensis Pic, 1957: 442 **A:** YE
alaiensis Pic, 1906b: 19 **A:** AF TM
anceyi Pic, 1895c: 59 **A:** SA **AFR**
angustatus Curtis, 1838: 714 **E:** FR GB IR PT **N:** CI
antherinus antherinus Linnaeus, 1760: 228 (*Meloe*) **E:** AB AL AU AZ BE BH BU BY CR CT CZ DE EN FI FR GB GE GG GR HU IT KZ LA LT LU MC MD ME NL NR NT PL PT RO SB SK SL SP ST SV SZ TR UK **A:** AF CY ES IN IS KI KZ LE SA SY TM TR UZ WS
astrachanicus Csiki, 1901a: 113
cinctellus Rossi, 1792: 46 (*Notoxus*)
semitestaceus Pic, 1892e: 43
tripustulatus Fabricius, 1792a: 169 (*Cicindela*)
valens Pic, 1896m: 179
antherinus syriae Pic, 1892e: 43 **E:** FR **N:** "northern Africa" **A:** AF IN LE SY
anticemaculatus Pic, 1900i: 78 **A:** KZ TM
arabicus Pic, 1894s: cclxxvii [RN] **A:** IS SA TR
laticollis Marseul, 1879a: 186 [HN]
arisi Pic, 1905k: 186 **E:** ST **A:** IN KZ MG TM UZ
tedjensis Pic, 1913i: 129
armaticornis L.N. Medvedev, 1974: 152 **A:** MG
armatus Truqui, 1855: 359 **N:** EG **A:** AE CY IN IQ IS JO KU SA SI SY
heluanensis Pic, 1899j: 175
lessepsi Marseul, 1868a: 192
pseudoarmatus Telnov, 2008: 273
rufibasis Pic, 1903n: 170
armiger LaFerté-Sénectère, 1849i: 12 [= 1849n: 113] **A:** UP **ORR**
ater Thunberg, 1787: 45 (*Notoxus*) **E:** AR AU BU BY CT CZ DE EN FI FR GE GR HU IT KZ LA LT MD NR NT PL PT SK SP ST SV SZ TR UK **N:** AG **A:** ES FE IQ IS KI KZ MG WS
ater Panzer, 1796a: 15 (*Notoxus*) [HN]
simplex Marseul, 1879a: 156
axillaris W.L.E. Schmidt, 1842c: 186 **E:** AB AL AU BH BU BY CR CT CZ DE EN FA FI GE GG GR HU IT LA LT MC NR NT PL RO SB SK ST SV SZ UK YU **A:** TR WS
inflatus Kolenati, 1846: 36
varians Kolenati, 1846: 36
baicalicus Mulsant & Rey, 1866a: 114 [= 1866b: 202] **A:** BEI ES FE HEI JA JIL LIA NC MG WS
niponicus Lewis, 1895a: 431
usitatus Pic, 1896m: 179

bajulus LaFerté-Sénectère, 1849i: 13 [= 1849n: 114] **A:** SD **ORR**
balachanus Pic, 1906h: 49 **A:** AB AF TM TR UZ
rosinae Pic, 1913n: 146
basithorax Pic, 1941b: 3 **E:** GR (Crete)
bigoti Bonadona, 1988: 42 **N:** MO
biguttatus LaFerté-Sénectère, 1849k: 119 [= 1849n: 218] **E:** FR (Corse) GR IT PT SP
bilunulatus Pic, 1893j: 79 **A:** IS "Syria"
bimaculatus Illiger, 1801: 80 (*Notoxus*) **E:** AU BE BY CT CZ DE EN FI FR GB GE HU IT KZ LA LT MD NL NR NT PL RO SK ST SV SZ UK **A:** KZ TR WS
dauricus Motschulsky, 1845a: 83
fasciatus Schilsky, 1888: 190 [HN]
pallens Schilsky, 1888: 190 [HN]
pallescens Pic, 1894g: 78 [RN]
sagitta Krynicki, 1829: 196
schilskyi Pic, 1894d: 56 [RN]
bruleriei Desbrochers des Loges, 1875: 43 **N:** EG **AFR**
honestus Marseul, 1878b: 50
brunneipennis Pic, 1896j: 141 **E:** FR (Corse) IT (Sardinia)
brunneus LaFerté-Sénectère, 1842: 249 **E:** FR PT SP **N:** AG CI MO MR TU
bousaadae Pic, 1893i: 74
lucasianus Desbrochers des Loges, 1869c: 191 [RN]
quadrimaculatus Lucas, 1843: 146
buyssoni Pic, 1899g: 77 **A:** NP SD **ORR**
catalanus Bonadona, 1953: 264 **E:** FR IT
chitwanus Telnov, 2018b: 228 **A:** NP
coiffaiti Bonadona, 1952: 236 **A:** LE
collaciculus Marseul, 1879a: 113 **N:** AG MO
commaculatus Krekich-Strassoldo, 1931: 36 **A:** HP NP UP **ORR**
consimilis Krekich-Strassoldo, 1928a: 98 **A:** NP SD UP **ORR**
constellatus Krekich-Strassoldo, 1928c: 109 **E:** BH BU GR MC **A:** TR
crassicollis Pic, 1894c: 49 **N:** AG
cribripennis Desbrochers des Loges, 1875: 45 **E:** IT PT SP **N:** AG MO TU
maurus Marseul, 1879a: 160
crinitulus Pic, 1901k: 90 **A:** UZ
crinitus LaFerté-Sénectère, 1849k: 105 [= 1849n: 204] **E:** AB AL AR AZ GE IT MA PL ST TR **N:** AG CI EG LB MO MR TU **A:** AE AF AP CY FUJ GUA GUI GUX HP HUB IN IQ IS JA JO KA KU NP OM PA SA SD SI SY TAI TR UP UZ XIZ YE YUN **AFR AUS NARi NTRi ORR**
communimacula Fairmaire, 1896a: 47
externemaculatus Pic, 1901q: 44
flavisternus Marseul, 1879a: 149
floreus LaFerté-Sénectère, 1849k: 112 (*Anthicus*) [= 1849n: 211]
laevaticeps Marseul, 1879a: 245 [RN]

laeviceps Marseul, 1879a: 105 [HN]
longipennis Desbrochers des Loges, 1875: 45 [HN]
manillanus Pic, 1903r: 646
pyrrhonius LaFerté-Sénectère, 1849k: 114 [1849n: 213]
uninotatus Pic, 1903s: 350
curvitarsis Krekich-Strassoldo, 1928a: 101 **A:** UP **ORR**
cylindricus Pic, 1899b: 20 **E:** RO
czernshorskyi Pic, 1912g: 33 **Ei:** IT **A:** SHN
demaisoni biuninotatus Pic, 1939f: 146 **N:** EG
demaisoni demaisoni Pic, 1893c: 27 **N:** EG
dentaticornis Pic, 1898g: 99 **E:** AB **A:** IN TM UZ
desertus Marseul, 1879a: 134 **A:** ES
difformis Marseul, 1879a: 178 **E:** BU GR UK **A:** TR
dilatatus Krekich-Strassoldo, 1914b: 112 **A:** KI KZ UZ XIN
dimidiatipennis Desbrochers des Loges, 1875: 46 **N:** EG **A:** SA YE **AFR**
diversepunctatus Pic, 1913j: 132 **A:** TAI
diversus Marseul, 1879a: 174 **E:** AB AR ST **A:** IN KZ TR WS
dzhungaricus L.N. Medvedev & Micheetchev, 1979: 142 **A:** MG
episcopalis Pic, 1903n: 170 **A:** IN IQ
insignatus Pic, 1904e: 57
escorialensis Pic, 1893x: clxi [Pic, 1893q: 135] **E:** SP
euphraticus LaFerté-Sénectère, 1849j: 72 [= 1849n: 172] **A:** IN IQ IS JO SY
externus Pic, 1899g: 78 **A:** NP SD
fenestratus W.L.E. Schmidt, 1842c: 181 **E:** AL BU CR FR GR HU IT MA PT SP ST TR "Caucasus" **N:** AG CI EG LB MO TU **A:** CY IS LE SY TR
fauconneti Pic, 1928a: 2
nigricans Pic, 1898g: 98
submaculatus Pic, 1898g: 98
fihaka Kejval, 2015b: 207 **A:** TR
flavipes altaicus L.N. Medvedev, 1975: 92 **A:** FE MG
flavipes flavipes Panzer, 1796f: 22 (*Notoxus*) **E:** AL AU BE BH BU BY CR CT CZ DE EN FI FR GB GE GR HU IR IT LA LT MC MD NL NR NT PL RO SK SP ST SV SZ TR UK "Caucasus" **A:** ES TR WS **NAR**
flavescens Pic, 1899k: 193
nigriceps Mannerheim, 1843b: 97
obscurus Küster, 1848a: 73
rufipes Paykull, 1800: 444
scoticus Rye, 1872: 10
fortepunctatus fortepunctatus Pic, 1892b: 18 **N:** AG **A:** IS **AFR**
fortiterpunctatus Pic, 1892b: 54 [RN]

fumosus Lucas, 1843: 146 **E:** FR (Corse) IT (Sardinia, Sicilia) **N:** AG CI MO TU YE **A:** TR
bicolor Lucas, 1843: 146
lethierryensis Pic, 1955a: 1 [RN]
lethierryi Pic, 1916a: 5 [HN]
sardiniensis Pic, 1928a: 2
fuscicornis LaFerté-Sénectère, 1849k: 116 [= 1849n: 215] **E:** FR PT SP **A:** TR YE **N:** AG LB MO TU
barbarus Pic, 1893h: 64
gebleri Mäklin, 1881: 46 **A:** ES WS
genei LaFerté-Sénectère, 1849k: 120 [= 1849n: 219] **E:** AL AR CR FR GR IT ME PT RO SB SP TR **N:** AG CI EG LB TU **A:** AF CY TR
giganteus LaFerté-Sénectère, 1849j: 60 [= 1849n: 160] **E:** "Caucasus" **A:** AF IN IS IQ PA TD TM TR UZ
gobiensis Pic, 1917i: 256 **A:** MG NMO NW
gordeevae L.N. Medvedev, 1974: 154 **A:** KI MG
gratiosus Pic, 1896m: 179 **E:** GR ST TR "Caucasus" **A:** IN KZ TR TM UZ
auliensis Pic, 1900i: 78
gravidulus Krekich-Strassoldo, 1931: 22 **A:** UP **ORR**
guttifer Wollaston, 1864: 522 **N:** CI
haasi Pic, 1900j: 89 **A:** AF TM UZ
infasciata Pic, 1908f: 58
hamicornis Marseul, 1880: 418 **E:** IT (Sardinia) PT SP **N:** AG LB TU **A:** JO SA
bedeli Pic, 1892n: cxcvii
hissarensis Pic, 1912h: 41 **A:** TD UZ
hoggarensis Pic, 1929h: 126 **N:** AG
humeralis Gebler, 1841b: 596 **A:** KZ MG NC WS
impressicollis Chobaut, 1924a: 104 **N:** MO TU
inaequaliceps inaequaliceps Pic, 1911c: 61 **A:** OM YE **AFR**
incrassicornis Pic, 1913r: 164
inaequalis Marseul, 1879a: 173 **E:** AR GG RO ST UK **A:** KZ TD
inconspicuus alternans Krekich-Strassoldo, 1931: 21 **A:** UP
inconspicuus inconspicuus Krekich-Strassoldo, 1931: 21 **A:** AP BT HP NP SD UP YUN
inderiensis Marseul, 1879a: 78 **E:** ST **A:** KZ ES TM WS
grandiceps Pic, 1902b: 10
inhamicornis Pic, 1942c: 10 **N:** MO
inimicus Bonadona, 1964: 230 **A:** KI MG TM XIN
insolubilis Pic, 1906b: 19 **A:** AF UZ
insperatus Pic, 1906g: 46 **A:** KZ NW
invreai Koch, 1933: 151 **E:** AL BU CR FR GR IT SB SP YU **A:** LE TR
irregularis Pic, 1932e: 18 **N:** AG
jacobsoni Pic, 1903d: 89 **A:** IN KZ TM

joannis Pic, 1901k: 90 **A:** IQ
khorgas Kejval, 2015b: 208 **A:** KZ
korbi Pic, 1902b: 10 **E:** AB AR GG ST **A:** IN IQ SY TD TM TR UZ
laetipennis Krekich-Strassoldo, 1931: 20 **A:** NP SD UP
laetus L.N. Medvedev, 1974: 153 **A:** MG
laeviceps Baudi di Selve, 1877d: 688 **E:** BH BU CR FR GR HU IT MA ME PT SB SP SZ TR YU "Caucasus" **N:** AG CI MO TU **A:** JO SY TR
cruciferus Ragusa, 1898: 231
lucidipes Pic, 1892h: 102
lucidithorax Pic, 1896c: 39
rummeli Pic, 1893s: 161
lamprinulus Krekich-Strassoldo, 1931: 21 **A:** NP PA SD UP **ORR**
lapidosus Wollaston, 1864: 521 **N:** CI
latebasalis Pic, 1932f: 25 **A:** MG
latedecoratus Pic, 1892k: 140 **E:** AB **A:** IN TR
astrabadensis Pic, 1905e: 113
karsantiana Pic, 1908d: 43
latefasciatus Desbrochers des Loges, 1875: 46 **E:** ST
latens Bonadona, 1977b: 86 **N:** AG MO TU **A:** TR
lateriguttatus Marseul, 1879a: 107 **E:** KZ ST
inguttatus Pic, 1912o: 90 (*Anthicus*)
lepaumi Pic, 1893c: 28 **N:** EG **A:** "Syria"
lepcha Telnov, 2018: 230 **A:** SD
lepidus Marseul, 1876d: 462 **A:** "Inde boreale"
lesnei Pic, 1899c: 30 **A:** PA
leveillei Pic, 1893u: xxxiii **E:** FR (Corse) IT (Sardinia, Sicilia)
libanicus Desbrochers des Loges, 1875: 44 **A:** LE SY
lii Uhmann, 1997: 21 (*Pseudoleptaleus*) **A:** HEB
lobularis Kejval, 2015b: 209 **A:** IN PA
lubbockii Wollaston, 1857a: 166 **N:** CI MR
luteicornis W.L.E. Schmidt, 1842c: 187 **E:** AL AU CR CZ DE FR GE IT MC PL PT SB SK SL SP SZ UK
picicornis Rey, 1892: 101
luteipes Marseul, 1879a: 159 **A:** IN IS JO LE SY TR
luteobinotatus Pic, 1957: 442 **A:** YE
muehlei Uhmann, 1988a: 2
maderi Heberdey, 1938: 163 **A:** YUN **ORR**
magnus Pic, 1914f: 35 **A:** PA "China" **ORR**
maindroni Pic, 1899c: 30 **A:** PA
manas Kejval, 2015b: 212 **A:** KI
mariaeantoniae Chobaut, 1894c: ccxxviii **N:** AG
medius Pic, 1905m: 493 **A:** ES MG WS

mellyi Pic, 1894f: 65 **N:** EG **A:** OM
minutissimus Pic, 1894c: 43 **N:** MO
moka Marseul, 1878b: 51 **A:** SA
mongolicus Pic, 1896m: 181 **A:** MG SHA
monstrator Telnov, 2005: 173 **A:** SCH
monstrosiceps Pic, 1912h: 41 **A:** KI TM
monstrosicornis Marseul, 1877: 465 **A:** AF FUJ GUA HUB JA NP SD TAI **ORR**
inapicalis Pic, 1913j: 131
rubidus Pic, 1895d: 64
nankineus Pic, 1913a: 12 **A:** ES FE HEB HEI JIA KZ SHA
niger G.-A. Olivier, 1811: 397 (*Notoxus*) **E:** AL BH BU CR FR GR IT MC ME PT RO SB SP ST SZ TR UK **N:** AG TU **A:** CY IN IQ IS JO LE SA SY TR **AFR**
differens Pic, 1896m: 179
morio LaFerté-Sénectère, 1849k: 116 [= 1849n: 215]
paulobrunnescens Pic, 1951a: 10
romanus Pic, 1951a: 10
nigronotatus Pic, 1894n: 182 **A:** NP UP **ORR**
notabilis Krekich-Strassoldo, 1914c: 225 **A:** NP UP
perileptoides Lewis, 1895a: 431 **A:** JA LIA SC
pilosus Marseul, 1877: 466 **A:** FE JA
pindos Kejval, 2015b: 214 **E:** AL AR GR
piratus Telnov, 2008: 271 **A:** AE
plucheryi Pic, 1951c: 15 [RN] **N:** MO
testaceonitidus Pic, 1939b: 21 [HN]
posticatus Pic, 1895i: 667 **N:** AG EG LB **A:** IN OM SA YE **AFR**
protensus Marseul, 1877: 470 **A:** JA
proximus Marseul, 1879a: 179 **E:** BU GR IT TR **A:** CY IS OM TR
pseudoruficeps Nardi, 2004: 141 [RN] **A:** TM
uniformis Pic, 1938d: 13 [HN]
pullulus Bonadona, 1964a: 233 **A:** ES
punctiger Krekich-Strassoldo, 1931: 23 **A:** NP SD UP YUN
pygmoeus LaFerté-Sénectère, 1849j: 82 [= 1849n: 182] [RN] **A:** NP **ORR**
quadridecoratus Abeille de Perrin, 1885: 160 **E:** FR (Corse) IT (Sardinia)
quadrioculatus LaFerté-Sénectère, 1849k: 102 [= 1849n: 201] [RN] **E:** AL AU BH CR CT FR IT GR HU PT SK SP ST SZ YU **A:** KI KZ TR
pedemontanus Pic, 1901g: 66
quadriguttatus Waltl, 1835: 75 [HN]
quadrispilus Marseul, 1879a: 104 **A:** AF IN IQ SY TM UZ WS
rectus Bonadona, 1970: 379 **A:** IN
reymondi Pic, 1951d: 37 **N:** AG
robustus Pic, 1902m: 408 **A:** HKG **ORR**
nigrobifasciatus Pic, 1928b: 18

rubens Krekich-Strassoldo, 1931: 20 **A:** AP HP NP PA UP YUN
rubiginosus Krekich-Strassoldo, 1931: 21 **A:** NP SD UP YUN **ORR**
rubripes Pic, 1894o: 206 **A:** AF
ruficeps Pic, 1913n: 146 **A:** ST TM
rufivestis Marseul, 1879a: 108 **E:** FR (Corse) IT (Sardinia) **N:** AG
grenieri Pic, 1910a: 325 [= 1910n: 58]
rufoaeneus Pic, 1915c: 10 **A:** JIA
sabulosus Marseul, 1879a: 172 **E:** AB GG MD RO ST **A:** IN
scapularis LaFerté-Sénectère, 1849k: 125 [= 1849n: 224] **A:** ES FE HEI MG NC SC WS
schmidtii Rosenhauer, 1847: 35 **E:** AL AR AU BU CR CZ FR GG GR HU IT MD PL PT RO SK SP ST SZ TR UK **A:** AF KZ TM TR UZ
subfasciatus LaFerté-Sénectère, 1849j: 79 [= 1849n: 179]
subobliteratus Pic, 1899b: 19
unipunctatus LaFerté-Sénectère, 1849j: 80 [= 1849n: 180]
scurrula Truqui, 1855: 362 **E:** GR **A:** CY TR "Syria"
brenskei Pic, 1893z: cclxx
sellatus Panzer, 1796f: 20 (*Notoxus*) **E:** AU BU BY CR CT CZ DE EN FI FR GE GR HU IT LA LT MD NL NR NT PL PT RO SK SP SV SZ UK
semicupreus Pic, 1893o: 126 **E:** ST
sexmaculatus Gebler, 1829a: 137 **A:** TM WS
signatellus Krekich-Strassoldo, 1928c: 111 **E:** AB ST
sillemi Heberdey, 1935: 306 **A:** XIN
sinaiticus Schatzmayr & Koch, 1934a: 235 **A:** AE IS SI OM
sivaschensis Blinstein, 1988: 76 **E:** UK
solers LaFerté-Sénectère, 1849j: 80 [= 1849n: 180] **A:** IQ
sollicitus LaFerté-Sénectère, 1849k: 98 [= 1849n: 197] **E:** AB **A:** AF IN IQ KZ LE SY TD TR
spadiceus Bonadona, 1958: 287 **E:** SP **N:** MO
staudingeri Pic, 1893v: liv **A:** AF ST TM UZ
leonardi Pic, 1913n: 146
steppensis Marseul, 1879a: 151 **E:** ST **A:** AF IN KI KZ TM UZ
stygius stygius LaFerté-Sénectère, 1849j: 61 [= 1849n: 161] **A:** YE **AFR**
subarmatus Pic, 1899c: 30 **A:** AF IN KZ PA TM UZ
subcrucifer subcrucifer Pic, 1900i: 78 **A:** KZ UZ
subcrucifer kaszabi Bonadona, 1964: 231 **A:** AF IN
subnitidus Pic, 1902b: 9 **A:** KZ TM
suturadepressa Pic, 1892l: 219 **N:** AG
suturalis simillimus Krekich-Strassoldo, 1928a: 97 **A:** NP UP
suturalis suturalis LaFerté-Sénectère, 1849j: 81 [= 1849n: 181] **A:** NP PA UP **ORR**
theryi Pic, 1892i: 115 [= 1893f: 53] **E:** IT (Sardinia) **N:** AG EG MO TU **A:** JO
batnensis Pic, 1892l: 217
mayeti Chobaut, 1898a: 84

thyreocephalus Solsky, 1866: 93 **E**: AR GG ST **A**: AF IN IQ KZ MG TM
tristis schaumii Wollaston, 1857b: 5721 **E**: FR GB PT SP **N**: MO
ater Stephens, 1832: 74 [HN]
tristis tristis W.L.E. Schmidt, 1842c: 172 **E**: AB AR BH BU CR FR GE GG GR IT MA ME PT RO SK SP ST SV TR UK **N**: AG CI EG LB MO TU **A**: AF CY IN IQ IS JO KZ LE SA SY TD TM TR UZ YE
discrepans Krekich-Strassoldo, 1928c: 106
fenestratus LaFerté-Sénectère, 1847: 375 [HN]
nigroapicalis Pic, 1913n: 145
sericeus LaFerté-Sénectère, 1847: 375
subhumeralis Pic, 1913n: 146
tristiculus Reitter, 1884b: 92
tristissimus Chobaut, 1923b: 130
truquii Desbrochers des Loges, 1875: 44
trotommidens Pic, 1892d: 44 **N**: AG
turgidipes Krekich-Strassoldo, 1928a: 99 **A**: NP PA **ORR**
turkestanicus Pic, 1893r: 155 **E**: ST **A**: AF KZ MG PA TD TM UZ
apicenotatus Pic, 1913o: 154
cinctipennis Pic, 1913n: 146
umbrinus LaFerté-Sénectère, 1849k: 115 [= 1849n: 214] **E**: FA FI LA NT SV SZ **A**: ES HEB HEI FE IN JIL KZ MG NC SC TIA TM WS
setulosus Boheman, 1851: 220
urganensis Pic, 1896m: 181 **A**: AF BEI ES FE HEI KZ MG NMO TIA WS
dachuricus Csiki, 1901a: 114
shishmareffi Csiki, 1901a: 114
vagesignatus Pic, 1940a: 1 **A**: KI
velatus Marseul, 1879a: 115 **N**: AG EG LB MO **A**: IN OM PA SA YE **AFR ORR**
mauritanicus Pic, 1952c: 492 (*Anthicus*)
vicarius Telnov, 2005: 177 **A**: SCH YUN
vicinor Telnov, 2018: 232 **A**: NP
yunnanus Pic, 1915c: 9 **A**: SCH YUN
subgenus *Odacanthicus* Sakai & Ohbayashi, 1994: 556 type species *Anthicus watarasensis* Sakai & Ohbayashi, 1994
watarasensis nazarovi Lafer, 1996: 40 **A**: FE
watarasensis watarasensis Sakai & Ohbayashi, 1994: 558 **A**: JA

genus *Anthicus,* nomina dubia
karelinii Zubkov, 1833: 332 **A**: TM
subaeneus LaFerté-Sénectère, 1847: 378 **E**: SP
subtilis Gistel, 1831: 304 **E**: GR

genus *Cordicomus* Pic, 1894c: 45 type species *Anthicus instabilis* W.L.E. Schmidt, 1842

Cordicollis J.R. Sahlberg, 1903f: 67 [SN] type species *Anthicus instabilis* Schmidt, 1842

cyclodiniformis Nardi, 2004: 140 [RN] **N:** MO

subopaculus Pic, 1936d: 25 (*Anthicus*) [HN]

galloisi Pic, 1910f: 20 (*Anthicus*) **A:** JA

gracilior Abeille de Perrin, 1885: 159 (*Anthicus*) **E:** AB BU CR FR GG GR HU IT MC PT RO SP ST TR **N:** EG **A:** KZ TM

auliatanus Pic, 1940a: 1 (*Anthicus*)

gracilis Panzer, 1796f: 21 (*Notoxus*) **E:** AB AU BU CR CZ FR GE GR HU IT KZ LT NR PL RO SK SL ST SV SZ UK **A:** TR WS **N:** AG

atricollis Abeille de Perrin, 1885: 160 (*Anthicus*)

lateripunctatus J. Sturm, 1826: 70 (*Anthicus*)

sinuatefasciatus Pic, 1940a: 1 (*Anthicus*)

stevenii LaFerté-Sénectère, 1849j: 69 [= 1849n: 169] (*Anthicus*)

infitialis Pic, 1894u: lxxvii (*Anthicus*) **A:** HEB LIA MG SC TIA

infuscatus LaFerté-Sénectère, 1849j: 82 [= 1849n: 182] (*Anthicus*) **N:** EG

instabilis franzi Bonadona, 1954b: 274 (*Cordicomus*) **E:** SP

instabilis geminipilis Desbrochers des Loges, 1875: 44 (*Anthicus*) **E:** SP

instabilis instabilis W.L.E. Schmidt, 1842c: 184 (*Anthicus*) **E:** AL BH BU CR DE FR GB GE GR IT LU MA NL NR PL PT SP ST SV SZ TR UK **N:** AG AZi CI EG LB MO MR TU **A:** IN IS JO TR

agilis Küster, 1849a: 75 (*Anthicus*)

deslogesi Pic, 1892e: 43 (*Anthicus*)

instabilis LaFerté-Sénectère, 1842: 259 (*Anthicus*) [HN]

mauritanicus Lucas, 1843: 146 (*Anthicus*)

olcesei Pic, 1894f: 65 (*Anthicus*)

puberulus Rey, 1892: 90 (*Anthicus*) [HN]

quittardi Pic, 1892o: 140 (*Anthicus*)

reyi Pic, 1893y: ccxii(*Anthicus*) [RN]

semiruber Pic, 1894c: 45 (*Anthicus*)

stabilis Pic, 1892k: 102 (*Anthicus*)

talaris Schiödte, 1893: 309 (*Anthicus*)

tibialis Curtis, 1838: 714 (*Anthicus*) [HN]

kuthyi Csiki, 1901a: 113 (*Anthicus*) **A:** ES MG NMO

litoralis Wollaston, 1854: 535 (*Anthicus*) **N:** CI MR SG

margelanicus Pic, 1893r: 155 (*Anthicus*) **E:** AB AR ST UK **A:** AF IN IQ KI KZ MG SY TD TM TR UZ

aralensis Pic, 1901c: 34 (*Anthicus*)

caucasicus Pic, 1893t: 175 (*Anthicus*)

tournieri Pic, 1894f: 65 (*Anthicus*)

muelleri Nardi, 2004: 138 [RN] **E:** CR RO ST UK
rufescens J. Müller, 1908: 316 (*Anthicus*) [HN]
notoxoides Wollaston, 1864: 520 (*Anthicus*) **N:** CI
opaculus opaculus Wollaston, 1864: 519 (*Anthicus*) **E:** MA MR PT SP **N:** AG CI LB MO TU
mozabita Chobaut, 1898a: 84 (*Anthicus*)
semibrunneus Pic, 1892l: 216 (*Anthicus*)
opaculus wittmeri Schatzmayr & Koch, 1934a: 224 (*Anthicus*) **N:** EG LB **A:** IS
pexicollis Fairmaire, 1894a: 34 (*Anthicus*) **A:** SD UP **ORR**
trabeatus Fairmaire, 1896a: 47 (*Anthicus*)
posticus Motschulsky, 1849: 131 (*Anthicus*) **E:** PT SP
sabuleti LaFerté-Sénectère, 1849j: 64 [= 1849n: 164] (*Anthicus*) **N:** AG EG MO TU
harenosus Krekich-Strassoldo, 1913a: 223 (*Anthicus*)
saharensis Chobaut, 1896b: 375 (*Anthicus*) **N:** AG LB MO **AFR**
trigonocephalus LaFerté-Sénectère, 1849i: 45 [= 1849n: 146] (*Anthicus*) **A:** NP UP **ORR**
turca Marseul, 1879a: 99 (*Anthicus*) **E:** TR

genus *Cyclodinus* Mulsant & Rey, 1866a: 77 [= 1866b: 165] type species *Anthicus humilis* Germar, 1824
Lagenicollis J.R. Sahlberg, 1903f: 67 [SN] type species *Anthicus humilis* Germar, 1824
Thicanus Casey, 1895: 670 type species *Anthicus annectans* LeConte, 1851
angustulus Pic, 1892b: 18 (*Anthicus*) **E:** ST "Caucasus" **A:** AF
ataensis Pic, 1901c: 34 (*Anthicus*) **E:** ST **A:** KZ TM
basanicus Sahlberg, 1913a: 57 (*Anthicus*) **A:** SY
basanicus Krekich-Strassoldo, 1913a: 227 (*Anthicus*) [HN]
bicarinula Marseul, 1879a: 86 (*Anthicus*) **E:** GR TR **A:** CY IS JO TR
kraatzi Pic, 1893z: cclxviii (*Anthicus*)
biplicatulus Reitter, 1889b: 119 (*Anthicus*) **E:** AB AR ST **N:** EG **A:** IN MG SY TM UZ
latetestaceus Pic, 1917i: 256 (*Anthicus*)
biskrensis Pic, 1893b: 26 (*Anthicus*) **N:** AG TU
blandulus blandulus Baudi di Selve, 1877d: 677 (*Anthicus*) **E:** GR IT (Sicilia) MA SP **N:** EG MO TU **AFR**
blandulus zinrami Koch, 1934b: 122 (*Anthicus*) **E:** GR SP **N:** MO
bremei LaFerté-Sénectère, 1842: 252 (*Anthicus*) **E:** AL AR FR GG GR IT PT RO SB SP ST TR YU **N:** AG EG LB MO TU **A:** AF CY IN IQ IS JO KI KU LE PA SI SY TD TM TR UZ **AFR**
impressithorax Pic, 1903q: 304 (*Anthicus*) [RN]
inapicalis Pic, 1919g: 67 (*Anthicus*)
latenotatus Pic, 1893o: 128 (*Anthicus*)
nigrolineatus Pic, 1892b: 26 (*Anthicus*)
riparius LaFerté-Sénectère, 1842: 254 (*Anthicus*)
stricticollis Fairmaire, 1903b: 217 (*Anthicus*) [HN]

subconiceps Pic, 1955f: 1043 (*Anthicus*)
subustulatus Pic, 1894a: 16 (*Anthicus*)
variabilis LaFerté-Sénectère, 1842: 253 (*Anthicus*)
brivioi Bucciarelli, 1962: 115 **E:** IT (Sardinia)
brunnescens Pic, 1893o: 137 (*Anthicus*) **A:** IN IQ IS SY TR
capilliger Marseul, 1878b: 51 (*Anthicus*) **A:** SA
cerastes Truqui, 1855: 355 (*Anthicus*) **E:** AB GR ST TR **A:** AF CY IN IQ IS TM TR
corcyreus Pic, 1900i: 78 (*Anthicus*)
ciliatipes Pic, 1916a: 5 (*Anthicus*) **N:** TU
coniceps Marseul, 1879a: 83 (*Anthicus*) **E:** AL AU BH BU CR FR GR IT MA PT RO SL SP ST TR UK **N:** AG EG LB MO TU **A:** CY IS TR
femoralis Marseul, 1879a: 82 (*Anthicus*) [HN]
imitator Krekich-Strassoldo, 1919b: 61 (*Anthicus*)
lagunarum Krekich-Strassoldo, 1919b: 62 (*Anthicus*)
magistri Gozis, 1882: 203 (*Anthicus*) [RN]
marseulianus Desbrochers des Loges, 1899b: 171 (*Anthicus*) [RN]
parvulus Krekich-Strassoldo, 1919b: 61 (*Anthicus*)
schatzmayri Koch, 1931: 68 (*Anthicus*)
sedjoumiensis Koch, 1931: 67 (*Anthicus*)
subopacicollis Pic, 1915l: 37 (*Anthicus*)
constrictus constrictus Curtis, 1838: 714 (*Anthicus*) **E:** AB AL BH BU CR FR GB GE GR IR IT MA MC NL NR PT SL SP ST TR UK **N:** AG CI EG MO SI TU **A:** JO TR
albionis Krekich-Strassoldo, 1919b: 71 (*Anthicus*)
lameyi Marseul, 1879a: 88 (*Anthicus*)
marinus Krekich-Strassoldo, 1919b: 70 (*Anthicus*)
maroccanus Pic, 1893z: cclxix (*Anthicus*)
ruffoi Bucciarelli, 1959: 185
constrictus subconvexus Rey, 1892: 89 (*Anthicus*) **E:** FR
croissandeaui croissandeaui Pic, 1893z: cclxxi (*Anthicus*) **E:** CTi FR IT (Sicilia) MA **N:** AG MO TU **A:** TR **AFR**
vaucheri Chobaut, 1895a: 156 (*Anthicus*)
dagestanicus Pic, 1919d: 10 (*Anthicus*) **E:** ST (Dagestan)
debilis LaFerté-Sénectère, 1849i: 28 [= 1849n: 129] (*Anthicus*) **E:** AB IT (Pantelleria Is.) MA MC ST **N:** AG CI EG LB MO TU **A:** AE AF CY IN IQ IS JO LE KA KU KZ LB OM PA QA SA SHN SI SY TD TM TR UZ YE **AFR ORR**
coniger Krekich-Strassoldo, 1927: 116 (*Anthicus*)
dimidiatus Wollaston, 1864: 521 (*Anthicus*)
phoxus Marseul, 1879a: 87 (*Anthicus*)
wegeneri Pic, 1899j: 174 (*Anthicus*)
dentatus Pic, 1895b: 52 (*Anthicus*) **E:** AB GR HU KZ RO SK ST TR UK "Turkestan" **A:** AF IN KZ TM TR
densifer Pic, 1900i: 78 (*Anthicus*)

denticulatus Pic, 1900i: 78 (*Anthicus*)
transdanubianus Roubal, 1927: 116 (*Anthicus*)
deplanatus Desbrochers des Loges, 1875: 45 (*Anthicus*) **A:** SY
desbrochersi Pic, 1893m: 115 (*Anthicus*) [RN] **E:** PT SP **N:** AG MO
anguliceps Abeille de Perrin, 1885: 159 (*Anthicus*) [HN]
caroli Pic, 1893o: 128 (*Anthicus*)
doriae Pic, 1895j: 784 (*Anthicus*) **E:** AB **A:** IN IQ TM
drurei Pic, 1901d: 37 (*Anthicus*) **A:** IQ
erro Truqui, 1855: 348 (*Anthicus*) **E:** GR **N:** EG **A:** CY IS JO LE YE
lysholmi Pic, 1899j: 174 (*Anthicus*)
fatuus Truqui, 1855: 350 (*Anthicus*) **E:** BU GR ST UK **N:** EG **A:** AF CY IN IQ IS JO LE SY TM TR
dilutius Truqui, 1855: 350 (*Anthicus*)
forticornis Pic, 1893o: 128 (*Anthicus*) **E:** GR TR **A:** TR
hammami hammami Pic, 1892d: 43 (*Anthicus*) **N:** AG EG TU
scutelloniger Pic, 1892d: 43 (*Anthicus*)
hammami viberti Pic, 1906j: 284 (*Anthicus*) **N:** TU
humilis Germar, 1824a: 6 (*Anthicus*) **E:** AB AL AU AZ BU CR CZ DE FR GE GR HU IT MA MC MD NR PL PT RO SK ST SV SZ UK **N:** AZi EG MO TU **A:** CY IS KZ LE SE TM TR WS UZ
adriaticus Krekich-Strassoldo, 1919b: 68 (*Anthicus*)
beckeri Desbrochers des Loges, 1875: 43 (*Anthicus*)
depilis Rey, 1892: 89 (*Anthicus*)
detritus Rey, 1892: 89 (*Anthicus*)
fuscicrus Rey, 1892: 89 (*Anthicus*)
latinus Krekich-Strassoldo, 1919b: 69 (*Anthicus*)
leukoranus Pic, 1913n: 145 (*Anthicus*)
lucidulus LaFerté-Sénectère, 1849i: 29 [= 1849n: 130] (*Anthicus*)
nigrinus Zetterstedt, 1838: 159 (*Anthicus*)
nigrofasciatus Truqui, 1855: 354 (*Anthicus*)
orientalis Krekich-Strassoldo, 1919b: 69 (*Anthicus*)
peranxius Rey, 1892: 89 (*Anthicus*)
siciliae Pic, 1893z: cclxx (*Anthicus*)
incomptus Truqui, 1855: 350 (*Anthicus*) **E:** GR (Corfu) **N:** EG **A:** CY IN IS LE SY TR
italicus Pic, 1901n: 178 (*Anthicus*) **E:** IT
kryzhanovskii Blinstein, 1988: 75 **E:** UK
lajoyei Pic, 1915c: 5 (*Anthicus*) **A:** HEB
larvipennis Marseul, 1879a: 85 (*Anthicus*) **E:** FR GR IT MA TR **N:** AG EG LB TU **A:** CY IN IQ SI TM
graecus Pic, 1896m: 178 (*Anthicus*)
motschulskyi Pic, 1894s: cclxxvii (*Anthicus*)
punicus Krekich-Strassoldo, 1919b: 73 (*Anthicus*)
testaceipes Pic, 1892l: 214 (*Anthicus*)
latior Pic, 1895l: cxxix (*Anthicus*) **N:** AG TU

latithorax Pic, 1893i: 74 (*Anthicus*) **A:** TR
lindbergi lindbergi Bonadona, 1964: 235 **A:** AF TM
lindbergi litatus Bonadona, 1964: 236 **A:** AF
longipilis C. Brisout de Barneville, 1863: 89 (*Anthicus*) **E:** CR FR GR SP YU **N:** AG MO TU
punctatithorax Pic, 1913o: 153 (*Anthicus*)
lotus Marseul, 1879a: 98 (*Anthicus*) **E:** ST **A:** IN KZ UZ
lucidicollis Marseul, 1879a: 89 (*Anthicus*) **E:** IT (Sardinia) SP **N:** AG TU
maltzevi Blinstein, 1988: 74 **E:** UK
mediobrunneus Pic, 1893o: 128 (*Anthicus*) **E:** ST
minutus minutus LaFerté-Sénectère, 1842: 255 (*Anthicus*) **E:** BU CR FR GR IT MA PT SP "Caucasus" **N:** AG MO TU **A:** TR
algiriensis Pic, 1893o: 126 (*Anthicus*)
lateralis Küster, 1849a: 77 (*Anthicus*)
sardous W.L.E. Schmidt, 1842c: 175 (*Anthicus*)
minutus truncaticeps Koch, 1934b: 121 (*Anthicus*) **E:** GR SP **N:** MO
misoloughii Pic, 1893o: 127 (*Anthicus*) **E:** GR
moltonii Bucciarelli, 1961: 272 **E:** CR
mongolensis L.N. Medvedev, 1974: 153 (*Anthicus*) **A:** MG
montandoni Pic, 1909n: 171 (*Anthicus*) **E:** KZ MD RO ST UK
morawitzi Desbrochers des Loges, 1875: 43 (*Anthicus*) **E:** AB MD RO ST **A:** KZ MG TD UZ
detectus Pic, 1896m: 179 (*Anthicus*)
moricei Pic, 1899j: 174 (*Anthicus*) **A:** IS
nitidior Pic, 1893z: cclxix (*Anthicus*) **E:** AB **A:** AF IN IQ MG TM UZ WS
bagdatensis Pic, 1901k: 90 (*Anthicus*)
phragmiteticola Telnov, 2018b: 241 **E** ST
raffrayi Pic, 1894t: xi (*Anthicus*) **A:** NE NO
obscurior Pic, 1894t: xi (*Anthicus*)
reitteri Pic, 1892b: 18 (*Anthicus*) **E:** AB ST **A:** AF IN SY TM UZ WS
reductus Brancsik, 1899: 102 (*Anthicus*) [HN]
rhagis Marseul, 1879a: 98 (*Anthicus*) **A:** IQ SA SY
roberti Pic, 1892l: 213 (*Anthicus*) **N:** AG
robustior Pic, 1908f: 57 (*Anthicus*) **A:** TD TM
salinus carthageniensis Pic, 1892n: cxcvii (*Anthicus*) **E:** IT (Sardinia) SP
rosai Bucciarelli, 1978: 158
subopaculus Pic, 1919g: 68 (*Anthicus*)
salinus salinus Crotch, 1867b: 439 (*Anthicus*) **E:** FR GB IT PT SP
chobauti Puel, 1919: 52 (*Anthicus*) [HN]
crotchi Pic, 1893m: 115 (*Anthicus*) [RN]
doctoris Pic, 1919g: 66 (*Anthicus*) [RN]
sociabilis Krekich-Strassoldo, 1919b: 67 (*Anthicus*)
sareptanus Pic, 1893p: 126 (*Anthicus*) **E:** GR ST "Caucasus"

sauteri Pic, 1913j: 131 (*Anthicus*) **A:** TAI
semiopacus Reitter, 1887e: 524 (*Anthicus*) **E:** ST **A:** AF IN KZ TD TM TR UZ
cerastoides Reitter, 1891a: 30 (*Anthicus*)
sibiricus Pic, 1893o: 128 (*Anthicus*) **A:** WS
subbrunneus Pic, 1913n: 145 (*Anthicus*) **A:** IN
thessalius Marseul, 1879a: 97 (*Anthicus*) **E:** GR
ustulatus anatoliae Pic, 1893o: 126 (*Anthicus*) **A:** IN TR
persicus Pic, 1895j: 785 (*Anthicus*)
ustulatus ustulatus LaFerté-Sénectère, 1849i: 23 [= 1849n: 124] (*Anthicus*) **E:** CR GG GR RO TR **N:** EG **A:** CY IN IQ IS LE JO SI SY TR
lacustris Pic, 1901b: 11 (*Anthicus*)
verticalis Pic, 1893t: 175 (*Anthicus*) **E:** AB
villosulus Truqui, 1855: 351 (*Anthicus*) **A:** AF CY JO SY TR

genus *Endomia* LaPorte, 1840: 259 type species *Anthicus elongatissimus* LaPorte, 1840 (= *Notoxus tenuicollis* Rossi, 1792)
Ochthenomus W.L.E. Schmidt, 1842d: 196 type species *Ochthenomus sinuatus* W.L.E. Schmidt, 1842 (= *Anthicus unifasciatus* Bonelli, 1812)
arabica Pic, 1913q: 162 **A:** SA YE
longicornis Pic, 1957: 448
besucheti Bonadona, 1989b: 17 **A:** PA **ORR**
castelsi Pic, 1929b: 14 **A:** AE IN OM PA **ORR**
concolor Desbrochers des Loges, 1874: 176 (*Ochthenomus*) **A:** IN TR
coniceps Reitter, 1901c: 79 (*Ochthenomus*) **A:** AF IN KI TD TM TR UZ
euphratica Krekich-Strassoldo, 1928a: 105 **A:** IN IQ TR
gratiosa Krekich-Strassoldo, 1928a: 103 **A:** PA UP **ORR**
himalayana Telnov, 2016: 49 **A:** NP
lefebvrei LaFerté-Sénectère, 1849m: 9 [= 1849n: 289] (*Ochthenomus*) **N:** AG EG LB MO TU **A:** AE AF CY IN IQ IS JO OM PA QA SA SI TM TR YE **AFR**
bivittata Truqui, 1855: 369 (*Ochthenomus*)
leprieuri Pic, 1893d: 38 (*Ochthenomus*)
nigripennis Sahlberg, 1913b: 194
obscuriceps Pic, 1913q: 162
reducta Pic, 1919e: 13
susica Escalera, 1914: 380 (*Ochthenomus*)
lunulata Krekich-Strassoldo, 1928a: 104 **A:** NP UP YUN **ORR**
occipitalis Dufour, 1843: 71 (*Anthicus*) **E:** AL BH BU CR FR GR IT ME PT SP ST (Caucasus) SZ **N:** AG CI EG MO MR TU **A:** AE TR "Syria" **AFR**
pallidicolor Pic, 1913q: 161
punctata LaFerté-Sénectère, 1847: 380 (*Ochthenomus*)
quadrinotata Pic, 1913q: 161

quinquemaculata Uhmann, 1995b: 495 **A:** YUN **ORR**

senilis Wollaston, 1864: 525 (*Ochthenomus*) **N:** CI

tenuicollis tenuicollis Rossi, 1792: 47 (*Notoxus*) **E:** AB AL AR AU BH BU CR CZ FR GE GG GR HU IT MA MC ME PT RO SB SK SP ST SZ TR UK YU **N:** AG EG LB MO MR TU **A:** CY IN IQ IS LE SA SY TR **AFR**

angustata LaFerté-Sénectère, 1847: 381 (*Ochthenomus*)

elongatissima LaPorte, 1840: 259 (*Anthicus*)

melanocephala Küster, 1847b: 57 (*Ochthenomus*)

obscuripennis Pic, 1913q: 162

unifasciata unifasciata Bonelli, 1812: 174 (*Anthicus*) **E:** AB AL AR BH BU CR FR GG GR IT MC ME PT SB SP ST SZ TR YU **N:** AG CI EG LB MO TU **A:** AE AF CY IN IS LE OM PA SA SI TM TR UZ YE **AFR**

caucasica Baudi di Selve, 1878a: 20 (*Ochthenomus*)

maculata Pic, 1919d: 11

sinuata W.L.E. Schmidt, 1842d: 199 (*Ochthenomus*)

vanharteni Telnov, 2008: 284 **A:** AE OM

genus *Hirticomus* Pic, 1894g: 69 type species *Notoxus hispidus* Rossi, 1792

Hirticollis J.R. Sahlberg, 1903f: 67 [SN] type species *Notoxus quadriguttatus* Rossi, 1792

biplagiatus LaFerté-Sénectère, 1849k: 111 [= 1849n: 210] (*Anthicus*) **N:** EG **A:** SA YE (Suqutra) **AFR**

balteatus LaFerté-Sénectère, 1849k: 112 [= 1849n: 211] (*Anthicus*)

basalis LaFerté-Sénectère, 1849k: 113 (*Anthicus*)

basalis LaFerté-Sénectère, 1849n: 212 (*Anthicus*) [HN]

bisuniguttatus Pic, 1951c: 15 (*Anthicus*)

chembanus Pic, 1932c: 18 (*Anthicus*)

densatus Fairmaire, 1898c: 241 (*Anthicus*)

fuscobrunneus Bonadona, 1969: 267 (*Hirticomus*)

prebasalis Pic, 1955e: 130 (*Anthicus*)

hirsutus LaFerté-Sénectère, 1849k: 107 [= 1849n: 206] (*Anthicus*) **A:** HKG KA NP YUN **ORR**

hirsutulus Motschulsky, 1863: 492 (*Anthicus*)

subcruciatus Fairmaire, 1896a: 48 (*Anthicus*)

hispidus Rossi, 1792: 46 (*Notoxus*) **E:** AB AL AR AU BH BU CR CT CZ FR GB (Channel Islands) GE GG GR HU IT KZ LA MA MC MD ME NR PL PT RO SB SK SL SP ST SV SZ TR UK **N:** AG CI EG MO MR **A:** AF CY IN IQ IS JO KI KZ LE SA TD TM TR UZ WS YE XIZ **AFR AUR ORR**

bicolor G.-A. Olivier, 1795: 51: 5 (*Notoxus*) [HN]

hirtellus Creutzer, 1796b: 3 (*Notoxus*)

jacqueti Pic, 1893p: 126 (*Anthicus*)

longus Pic, 1896n: 422 (*Anthicus*) **N:** AG

ocellatus LaFerté-Sénectère, 1849k: 106 [= 1849n: 205] (*Anthicus*) **A:** HKG NP UP YUN **AUR ORR**

quadriguttatus Rossi, 1792: 48 (*Notoxus*) **E:** AB AL AU BH CR CZ FR GE GR HU IT MA ME PL PT RO SB SK SL SP SZ TR YU **N:** AG AZi CI LB MO TU **A:** TM TR **NTRi**

bifasciatus LaPorte, 1840: 259 (*Anthicus*)

guttatus LaFerté-Sénectère, 1842: 248 (*Anthicus*)

quadrinotatus Gyllenhal, 1810: 498 (*Anthicus*)

valettensis Pic, 1951b: 10 (*Anthicus*)

genus *Leptaleus* LaFerté-Sénectère, 1849i: 5 [= 1849n: 106] type species *Notoxus rodriguesi* Latreille, 1804

aemulus Krekich-Strassoldo, 1914a: 11 (*Anthicus*) **A:** FUJ SHA ZHE

arabs Marseul, 1879a: 72 (*Anthicus*) **N:** AG LB MO TU **A:** IS OM SA SY YE **AFR**

atricolor Pic, 1957: 438 **A:** YE

brittoni Pic, 1957: 431 **A:** YE

chaudoiri Kolenati, 1846: 34 (*Formicoma*) **E:** AB AL BU GG GR YU **A:** AF CY IN LE NP SD SY TM TR **AFR**

delicatulus LaFerté-Sénectère, 1849i: 9 [= 1849n: 110] (*Anthicus*) **A:** HKG NP UP YUN **ORR**

glabellus Truqui, 1855: 346 (*Anthicus*) **E:** GR **N:** AG EG MO **A:** IN IS JO LE SA SI SY YE **AFR**

glabricephalus Uhmann, 1983: 194 **A:** NP SD

klugii klugii LaFerté-Sénectère, 1849i: 6 [= 1849n: 107] (*Anthicus*) **N:** AG EG LB MO TU **AFR A:** AE AF IN IS OM SA SI YE

truncatulus Fairmaire, 1892a: 116

maximicollis Pic, 1893l: 111 **N:** AG EG MO **A:** IN IS OM SA SI YE **AFR**

mediaguttatus Uhmann, 1995a: 38 **A:** AE OM

punctatissimus Fairmaire, 1893e: 157 **N:** EG **A:** IN LE OM SA YE **AFR**

rodriguesi Latreille, 1804a: 357 (*Notoxus*) **E:** FR GE IT MA PT SP **N:** AG EG MO TU **A:** "Syria"

postreductus Pic, 1919d: 10

pulchellus LaFerté-Sénectère, 1842: 256 (*Anthicus*) [HN]

pulchellus W.L.E. Schmidt, 1842d: 195 (*Anthicus*)

rufescens Pic, 1891b: 117

subinfasciatus Pic, 1916a: 5

sasajii Sakai & Telnov, 2001: 328 **A:** JA

sennarensis Pic, 1907p: 322 (*Anthicus*) **A:** OM SA YE **AFR**

minutus Bonadona, 1969: 273

profanus Bonadona, 1969: 274

trisignatus Pic, 1955f: 1043

testaceonotatus Pic, 1922b: 20 **A:** NP **ORR**

triguttatus LaFerté-Sénectère, 1849m: 299 (*Anthicus*) **A:** IN IS LE SY TR

truncatipennis Desbrochers des Loges, 1875: 43

uralensis Pic, 1905m: 490 **E:** ST **A:** KZ MG TM

genus *Nitorus* Telnov, 2007a: 81 type species *Formicomus trigibber* Marseul, 1876
angustior Krekich-Strassoldo, 1931: 16 (*Pseudoleptaleus*) **A:** SD UP
apricans apricans Krekich-Strassoldo, 1931: 17 (*Pseudoleptaleus*) **A:** AF NP PA SD **ORR**
apricans lindbergi Bonadona, 1964: 238 (*Pseudoleptaleus*) **A:** AF
asiaticus Pic, 1907j: 173 (*Anthicus*) **A:** NP SD **ORR**
asperatus Krekich-Strassoldo, 1928a: 90 (*Pseudoleptaleus*) **A:** HP NP SD UP **ORR**
aspericollis Krekich-Strassoldo, 1928a: 92 (*Pseudoleptaleus*) **A:** NP SD UP **ORR**
bidentatus Pic, 1901p: 797 (*Anthicus*) **A:** HKG **ORR**
bigibbosus Pic, 1913l: 138 (*Leptaleus*) **A:** NP SD TAI YUN **ORR**
obscuriceps Pic, 1915c: 9 (*Pseudoleptaleus*)
brevitarsis Krekich-Strassoldo, 1931: 17 (*Pseudoleptaleus*) **A:** NP UP
carmichaeli Krekich-Strassoldo, 1931: 18 (*Pseudoleptaleus*) **A:** SD
coomani Pic, 1926b: 16 (*Anthicus*) **A:** FUJ **ORR**
lacosus Pic, 1926b: 16 (*Anthicus*)
fumatus Krekich-Strassoldo, 1931: 18 (*Pseudoleptaleus*) **A:** NP SD **ORR**
glabratus Uhmann, 1996: 30 (*Stricticomus*) **A:** NP SD **ORR**
guttulatus Krekich-Strassoldo, 1928a: 94 (*Pseudoleptaleus*) **A:** SD
laevipennis Marseul, 1877: 471 (*Anthicus*) **A:** JA LIA SC
lepidus Krekich-Strassoldo, 1913b: 137 (*Pseudoleptaleus*) **A:** HP NP SD
moestus moestus Krekich-Strassoldo, 1928a: 94 (*Pseudoleptaleus*) **A:** HP NP SCH UP **ORR**
germanus Krekich-Strassoldo, 1928a: 95 (*Pseudoleptaleus*)
obfuscatus Krekich-Strassoldo, 1928a: 88 (*Pseudoleptaleus*) **A:** HP NP SD UP
pulchripennis Krekich-Strassoldo, 1928a: 93 (*Pseudoleptaleus*) **A:** NP SD UP **ORR**
petri Pic, 1894o: 205 (*Anthicus*) **A:** KZ "Tian Shan"
sahlbergi Pic, 1898i: 124 (*Anthicus*)
sculptus Krekich-Strassoldo, 1928a: 93 (*Pseudoleptaleus*) **A:** NP SD UP **ORR**
sensitivus Krekich-Strassoldo, 1928a: 91 (*Pseudoleptaleus*) **A:** HP NP SD UP **ORR**
sichuanus Telnov, 2005: 169 (*Pseudoleptaleus*) **A:** SCH
stevensi Krekich-Strassoldo, 1928a: 90 (*Pseudoleptaleus*) **A:** UP
suasum Telnov, 2010a: 5 **A:** AF
subnudus Krekich-Strassoldo, 1928a: 90 (*Pseudoleptaleus*) **A:** HP NP UP
testaceicornis Uhmann, 1983: 197 (*Pseudoleptaleus*) **A:** SD **ORR**
trigibber Marseul, 1876d: 460 (*Formicomus*) **A:** FUJ JA NE NP SC SD SHN TAI TIA UP **ORR**
basalis Pic, 1915c: 9 (*Psedoleptaleus*)
mediomaculatus Pic, 1913j: 130 (*Anthicus*)
unifasciatus unifasciatus Desbrochers des Loges, 1875: 43 (*Leptaleus*) **N:** EG **A:** IN **AFR**
sublatus Pic, 1914q: 181 (*Leptaleus*)

genus *Omonadus* Mulsant & Rey, 1866a: 104 [= 1866b: 192] type species *Meloe floralis* Linnaeus, 1758
Hemantus Casey, 1895: 682 type species *Hemantus rixator* Casey, 1895
Trapezicomus Pic, 1894c: 45 [RN] type species *Meloe floralis* Linnaeus, 1758

Trapezicollis J.R. Sahlberg, 1903f: 67 [SN] type species *Meloe floralis* Linnaeus, 1758

Trapezonotus Sahlberg, 1913b: 191 type species *Anthicus phoenicius* Truqui, 1855

beloni Pic, 1892d: 44 (*Anthicus*) **N:** AG

bifasciatus Rossi, 1792: 48 (*Notoxus*) **E:** AB AL AR AU BE BH BU CR CT CZ DE FR GB GE GG GR HU IT MA MC ME PL PT RO SB SK SP ST SV SZ TR UK **N:** AG MO TU **A:** AF IN IS KI KZ LE TD TM TR WS UZ

kolenatii Kolenati, 1846: 35 (*Anthicus*)

brevicornis Pic, 1894i: 86 (*Anthicus*) [= 1894l: 142] **N:** AG LB

cinctomus Marseul, 1879a: 106 (*Anthicus*) **A:** AE IS LE SA SY

confucii addendus Krekich-Strassoldo, 1928a: 102 (*Anthicus*) **A:** AF KA NP UP YUN **ORR**

confucii confucii Marseul, 1876d: 464 (*Anthicus*) **A:** BEI ES FE FUJ GAN GUA HEB HEI HEN JA JIA JIL LIA NC NMO SC SD SHA SHG SHX TAI YUN ZHE **ORR**

obscuripennis Pic, 1913a: 11 (*Anthicus*)

quadraticollis Yawata, 1944: 1 (*Anthicus*)

confucii tokaraensis Nomura, 1962a: 5 (*Anthicus*) **A:** JA (Ryukyu Islands)

floralis Linnaeus, 1758: 420 (*Meloe*) **E:** AB AL AR AU BE BH BU BY CR CT DE EN FI FR GB GE GG GR HU IR IT LA LS LT LU MA MC MD ME NL NR NT PL PT RO SB SK SL SP ST SV SZ TR UK **N:** AG AZ CI EG LB MO MR TU **A:** AF BEI BT CY ES FE HEB HUN IN IQ IS JA JIL JO KI LIA LE MG NE NIN NP OM PA SA SC SCH SD SHN SI SY TM TR WS UP UZ XIN **COS**

basalis A. Villa & J.B. Villa, 1838: 63 (*Anthicus*)

basillaris Say, 1824a: 279 (*Anthicus*)

breviculus Philippi R.A., 1864: 353 (*Anthicus*)

calycinus Panzer, 1792b: 3 (*Notoxus*)

fallax Broun, 1893b: 1168 (*Anthicus*)

formicarius G.-A. Olivier, 1795: 51: 4 (*Notoxus*) [HN]

formicoides Geoffroy, 1785: 156 (*Cantharis*)

fuscus Marsham, 1802: 486 (*Lytta*)

massauensis Pic, 1900n: 575 (*Anthicus*)

myrmicocephalus Rossi, 1792: 46 (*Notoxus*)

pedicularius Schrank von Paula, 1781: 224 (*Meloe*)

reducteapicalis Pic, 1915k: 22 (*Anthicus*)

semirufus Fairmaire & Germain, 1860: 3 (*Anthicus*)

syriacus Baudi di Selve, 1881: 294 (*Anthicus*)

formicarius atropterus Nomura, 1962a: 5 (*Anthicus*) **A:** JA (Ryukyu Islands)

formicarius formicarius Goeze, 1777: 706 (*Meloe*) **E:** AB AL AR AU BE BH BU BY CR CT DE EN FI FR GB GE GG GR HU IR IT LA LS LT LU MA MC MD ME NL NR NT PL PT RO SB SK SL SP ST SV SZ TR UK **N:** AG AZ EG MO TU **A:** AE AF ANH AP BT CY ES FE FUJ GUA GUI GUX HAI HEB HEI HP HUB IN IQ IS JA JIA JO KA LE NIN NP OM PA QA SA SHA SCH SD SY TAI TM TR WS UP UZ XIN YE YUN ZHE **COS**

enodis Casey, 1895: 684 (*Hemantus*)

picianus Koch, 1931: 75 (*Anthicus*)

quisquilius C.G. Thomson, 1864: 360 (*Anthicus*)
rixator Casey, 1895: 686 (*Hemantus*)
scenicus Casey, 1895: 685 (*Hemantus*)
semirufus Fairmaire, 1896a: 49 (*Anthicus*) [HN]
hadjaziensis Pic, 1899n: 264 (*Anthicus*) **A:** SA
ineditus LaFerté-Sénectère, 1849j: 54 [= 1849n: 154] (*Anthicus*) **A:** AF SY TR
longemaculatus Pic, 1938a: 18 (*Anthicus*) **A:** FUJ HEB JIA NP PA XIN UP YUN **ORR**
bifenestratus Pic, 1938a: 19 (*Anthicus*)
bifenestratus Heberdey, 1938: 162 (*Anthicus*) [HN]
gardneri Heberdey, 1938: 163 (*Anthicus*)
patruelis LaFerté-Sénectère, 1849j: 54 [= 1849n: 154] (*Anthicus*) **A:** NP SD UP **ORR**
phoenicius Truqui, 1855: 356 (*Anthicus*) **E:** ST UK "Caucasus" **A:** AF CY IS KZ LE SA ST SY TR
dilutius Truqui, 1855: 357 (*Anthicus*)
obscurius Truqui, 1855: 357 (*Anthicus*)
semicalvus Pic, 1895j: 785 (*Anthicus*) **A:** TR

genus *Sapintus* Casey, 1895: 732 type species *Anthicus pubescens* LaFerté-Sénectère, 1849
andreaskopetzi Telnov, 2014: 321 **A:** NP **ORR**
anguliceps LaFerté-Sénectère, 1849i: 33 [= 1849n: 134] (*Anthicus*) **A:** NP YUN **ORR**
apicatus apicatus Fairmaire, 1896a: 48 (*Anthicus*)
apicatus birmanicus Pic, 1907r: 341 (*Anthicus*)
bataviensis Marseul, 1882a: 63 [= 1882b: 123] (*Anthicus*) **A:** NP **ORR**
breviceps LaFerté-Sénectère, 1849j: 70 [= 1849n: 170] (*Anthicus*) **A:** AF NP PA SD **ORR**
cohaeres Lewis, 1895a: 430 (*Anthicus*) **A:** FE JA
diversicornis Pic, 1907o: 195 (*Anthicus*) **A:** AF HP KA NP SD UP
extus Lewis, 1895a: 430 (*Anthicus*) **A:** JA
fallax Bonadona, 1964: 234 **A:** AF PA
flavomaculatus LaFerté-Sénectère, 1849j: 72 [= 1849n: 172] (*Anthicus*) **A:** NP **ORR**
formosanus Pic, 1911b: 11 (*Anthicus*) **A:** TAI
hamai Nomura, 1964: 55 (*Anthicus*) **A:** JA (Iriomote Is.)
hartmanni Telnov, 2014: 330 **A:** NP
immaturus LaFerté-Sénectère, 1849j: 76 [= 1849n: 176] (*Anthicus*) **A:** KA **ORR**
javanus Marseul, 1882a: 62 [= 1882b: 121] (*Anthicus*) **A:** JA SW TAI YUN **ORR**
sodalis Pic, 1895d: 64 (*Anthicus*)
luteonotatus Pic, 1913j: 132 (*Anthicus*)
kalash Kejval, 2015a: 51 **A:** PA
litorosus Lewis, 1895a: 429 (*Anthicus*) **A:** JA
longepilosus Pic, 1942a: 13 (*Anthicus*) **A:** CH
marseuli Pic, 1893y: ccxi (*Anthicus*) [RN] **A:** FUJ JA NP TAI **ORR**
scoticus Marseul, 1876d: 462 (*Anthicus*) [HN]
triparticornis Pic, 1926d: 16 (*Anthicus*)

minamiiwo Bonadona & Sakai, 1986: 247 **A:** JA (Iwo Jima Is.)
misellus LaFerté-Sénectère, 1849j: 74 [= 1849n: 174] (*Anthicus*) **A:** NP SCH **ORR**
nomurai Nardi, 2004: 136 [RN] **A:** JA (Ryukyu Islands)
irregularis Nomura, 1962a: 4 (*Anthicus*) [HN]
oceanicus LaFerté-Sénectère, 1849j: 70 [= 1849n: 170] (*Anthicus*) **A:** JA (Ryukyu Islands) **AFR AUR ORR**
francoisi Pic, 1902m: 40 (*Anthicus*)
guamensis Blair, 1942: 57 (*Anthicus*)
plectilis Pic, 1910p: 71 (*Anthicus*) **A:** SD SHG TAI **ORR**
rugulipennis rugulipennis Fairmaire, 1893e: 157 (*Anthicus*) **A:** YE **AFR**
diversicolor Pic, 1952d: 78 (*Anthicus*)
sakishimanus Nomura, 1964: 56 (*Anthicus*) **A:** JA (Ryukyu Islands)
serotinus Krekich-Strassoldo, 1931: 23 (*Anthicus*) **A:** PA **ORR**
siamensis Pic, 1914f: 35 (*Anthicus*) **A:** HKG PA **ORR**
takaosus Pic, 1913i: 130 (*Anthicus*) **A:** TAI **ORR**
testaceicolor Pic, 1913j: 133 (*Anthicus*) **A:** HKG TAI
subsuturalis Pic, 1913j: 133 (*Anthicus*)
vexator Werner, 1965: 264 (*Anthicus*) **A:** HKG **ORR**

genus *Stricticomus* Pic, 1894g: 70 type species *Anthicus transversalis* A. Villa & G.B. Villa, 1833
Stricticollis J.R. Sahlberg, 1903f: 67 [SN] type species *Anthicus longicollis* Schmidt, 1842
adleri Pic, 1911m: 170 (*Anthicus*) **E:** ST "Caucasus"
amaniensis amaniensis Pic, 1913r: 164 (*Anthicus*) **A:** SA YE **AFR**
ambulator LaFerté-Sénectère, 1849k: 91 [= 1849n: 190] (*Anthicus*) **A:** NP KA PA SD TD UP **ORR**
ampliatus Sahlberg, 1913a: 56 (*Anthicus*) **A:** SY
ampliatus Krekich-Strassoldo, 1913a: 228 (*Anthicus*) [HN]
araxicola Reitter, 1889a: 34 (*Anthicus*) **E:** AB AR IN
arcuaticeps Pic, 1900j: 90 (*Anthicus*) **A:** AF IN KI KZ TM UZ
argenteofasciatus Pic, 1894o: 205 (*Anthicus*) **A:** IN KI KZ TD TM
bistrimaculatus Pic, 1913g: 114 (*Anthicus*) **A:** TD TM
bodemeyeri Pic, 1913n: 146 (*Anthicus*) **A:** AF IN
coreanus Pic, 1938d: 13 (*Anthicus*) **A:** ES FE JA MG NC SC
nipponicus Nomura, 1960: 47 (*Anthicus*)
walteri Uhmann, 1998b: 153
desolatius Telnov, 2008: 277 (*Stricticollis*) **A:** AE
goebelii LaFerté-Sénectère, 1849k: 91 [= 1849n: 190] (*Anthicus*) **E:** AR GE MD TR **A:** AF ES IN IQ IS KU KZ PA TD TM TR UZ
buchariensis Pic, 1897g: 62 (*Anthicus*)
herzi Pic, 1905m: 493 (*Anthicus*) **A:** AF IN TM
incrinitus Pic, 1929h: 126 (*Anthicus*) **N:** AG

longicollis W.L.E. Schmidt, 1842b: 130 (*Anthicus*) **E:** AB AL AR AU BH BU CR FR GG GR HU IT MC ME PT SK SP ST SZ **N:** AG MO **A:** CY IN IQ IS JO LE SY TR YE

pallidipes Pic, 1896g: 87 (*Anthicus*)

tenenbaumi Pic, 1925a: 1 (*Anthicus*)

uniguttatus Pic, 1896m: 179 (*Anthicus*)

longulus Pic, 1905m: 491 (*Anthicus*) **A:** ES TD

modestus LaFerté-Sénectère, 1849k: 92 [= 1849n: 191] (*Anthicus*) **N:** AG EG LB MO TU **A:** AE IN IQ IS JO KZ OM QA SA SI SY TR YE **AFR**

ophthalmicus Rottenberg, 1871: 259 (*Anthicus*) **E:** IT (Sicilia) **N:** AG EG MO TU **A:** IN IQ IS JO OM SA **AFR**

dromioides Pic, 1894u: lxxvii (*Anthicus*)

megalops Marseul, 1879a: 127 (*Anthicus*)

palustris Pic, 1905m: 490 (*Anthicus*) **A:** AF KZ NE NO TM

peplifer Marseul, 1879a: 126 (*Anthicus*) **A:** IN OM PA SA YE

adensis Pic, 1911m: 170 (*Anthicus*)

peyerimhoffi Pic, 1902l: 187 (*Anthicus*) **N:** EG **A:** IN OM PA SA SI

pseudotobias Telnov, 2010c: 18 (*Stricticollis*) **A:** NP

rarepunctatus Marseul, 1879a: 126 (*Anthicus*) **A:** AF IS SA

rubricollis Pic, 1892j: 139 (*Anthicus*) **N:** MO

lecerfi Pic, 1936c: 23 (*Anthicus*)

obscuricollis Pic, 1892j: 139 (*Anthicus*)

rufithorax LaFerté-Sénectère, 1849k: 88 [= 1849n: 187] (*Anthicus*) **E:** GR **A:** IN LE TR "Syria"

subnotata Pic, 1910c: 10 (*Anthicus*)

sternbergsi Telnov, 2005: 171 **A:** YUN

tixieri Pic, 1908f: 57 (*Anthicus*) **A:** IN KZ TM UZ

tobias Marseul, 1879a: 125 (*Anthicus*) **E:** AB AL AR AU BE BH BU BY CR CT CZ DE FR GB GE GR HU IT LA LU MA MC MD NL NR PL PT RO SK SP ST SV SZ TR **N:** AG CI EG MO MR TU **A:** AF BEI CY IN IQ IS JA JO KA KU KZ LE MG NE NP OM PA SA SD SI SY TD TM TR UP UZ YE **COS**

binhanus Pic, 1927d: 30 (*Anthicus*)

corporaali Pic, 1923c: 20 (*Anthicus*)

mauritiensis Pic, 1898k: 182 (*Anthicus*)

mundulus Sharp, 1885: 168 (*Anthicus*)

parisiensis Saint-Albin, 1952: 21 (*Anthicus*)

postoculatus Fairmaire, 1896a: 49 (*Anthicus*)

tanakai Nomura, 1960: 48 (*Anthicus*)

turanicus Reitter, 1889b: 118 (*Anthicus*)

transcaspicus Pic, 1898i: 123 (*Anthicus*) **E:** AB **A:** IN KZ TM UZ

transversalis meridionalis Pic, 1896c: 40 (*Anthicus*) **E:** IT (Lampedusa Is.) MA **N:** AG LB MO TU

erichi Koch, 1931: 77 (*Anthicus*)

transversalis transversalis A. Villa & J.B. Villa, 1833: 35 (*Anthicus*) **E:** AB AL AR BU CR CZ FR GE GR IT MD ME PT SK SP ST (Dagestan) TR UK **N:** AG EG LB MO TU **A:** AF CY IN IQ IS KZ LE SA SY TM TR

amoenus W.L.E. Schmidt, 1842c: 176 (*Anthicus*)

elegans Steven, 1806: 161 (*Anthicus*) [= 1811: 124] [HN]

nigrovelutinus Pic, 1892h: 103 (*Anthicus*)

ornatus Truqui, 1855: 358 (*Anthicus*)

tenellus LaFerté-Sénectère, 1842: 251 (*Anthicus*)

truncatus Pic, 1895n: cclxxxiv (*Anthicus*) **E:** GR **A:** TR

decoloratus Pic, 1897j: 120 (*Anthicus*)

subcoloratus Pic, 1901h: 179 (*Anthicus*)

valgipes Marseul, 1875b: xxxviii (*Anthicus*) **A:** ES FE JA JIL NC SC SHA

zagrosanus Telnov, 2010c: 19 (*Stricticollis*) **A:** IN TR (East Kurdistan)

zeravshanus Telnov, 2010c: 21 (*Stricticollis*) **A:** TD

genus *Yunnanomonticola* Telnov, 2002b: 47 type species *Yunnanomonticola nanzhao* Telnov, 2002

latissima Zhao, Wang & Wang, 2019: 155 **A:** YUN

nanzhao Telnov, 2002b: 49 **A:** YUN

tenuipenis Zhao, Wang & Wang, 2019: 157 **A:** GUI

tribe Formicomini Bonadona, 1974

genus *Anthelephila* Hope, 1833: 63 [= 1834b: 101] type species *Anthicus cyaneus* Hope, 1833

Formicoma Motschulsky, 1845a: 83 type species *Carabus pedestris* Rossi, 1790

Formicomus LaFerté-Sénectère, 1849g: 1

Formicosoma Motschulsky, 1845c: [unnumbered page] [RN]

Myrmecosoma Mannerheim, 1846: 227 [RN]

Orthauchen Krekich-Strassoldo, 1925: 316 type species *Formicomus aestimabilis* Krekich-Strassoldo, 1925

achaemenis Kejval, 2018b: 207 **A:** IN

alfierii Pic, 1924c: 22 (*Formicomus*) **N:** EG

akela Kejval, 2017c: 6 **A:** YUN **ORR**

amaena LaFerté-Sénectère, 1849g: 12 [= 1849n: 81] (*Formicomus*) **N:** AG EG **AFR**

mellyi Pic, 1894f: 64 (*Formicomus*)

anastazei Pic, 1935c: 6 (*Formicomus*) [RN] **N:** EG

alfierii Pic, 1935c: 4 (*Formicomus*) [HN]

anceyi rathjensi Pic, 1957: 440 (*Formicomus*) **A:** SA YE **AFR**

ancoriferra Telnov, 2003a: 130 **A:** SD

angustata Bonadona, 1964: 244 (*Formicomus*) **A:** AF IN

angustiformis Fairmaire, 1893e: 155 (*Formicomus*) **N:** EG **A:** IS SA SI **AFR**

filicollis Pic, 1898d: 63 (*Formicomus*)

animata Pic, 1903h: 119 (*Formicomus*) [RN] **A:** YUN **ORR**

annamita Pic, 1903f: 107 (*Formicomus*) [HN]

anki Kejval, 2018b: 210 **A:** IQ
antiqua Krekich-Strassoldo, 1919a: 172 (*Formicomus*) **A:** HP KA NP PA SD UP **ORR**
iridipennis Krekich-Strassoldo, 1928a: 79 (*Formicomus*)
arabica Pic, 1899n: 264 (*Formicomus*) **A:** IN IQ SA SY YE
aratrix Kejval, 2007: 619 **A:** SD
arcana Kejval, 2017c: 11 **A:** YUN **ORR**
armata Boheman, 1858: 103 (*Formicomus*) **A:** JA SE UP **ORR**
arunachalensis Kejval, 2018a: 218 **A:** AP
arunvallis Kejval, 2000a: 86 (*Formicomus*) **A:** NP
atayalorum Kejval, 2018b: 212 **A:** TAI
badia Krekich-Strassoldo, 1931: 8 (*Formicomus*) **A:** HP NP SD UP **ORR**
baeri Pic, 1903r: 645 (*Formicomus*) **A:** TAI **ORR**
formosana Pic, 1912a: 7 (*Formicomus*)
longispina Pic, 1912a: 7 (*Formicomus*)
spinosa Pic, 1929b: 14 (*Formicomus*)
vitalisi Pic, 1914d: 19 (*Formicomus*)
barbipes Krekich-Strassoldo, 1931: 9 (*Formicomus*) **A:** NP PA UP **ORR**
basalis Krekich-Strassoldo, 1931: 10 (*Formicomus*) **A:** NP UP **ORR**
basirufa Krekich-Strassoldo, 1928a: 76 (*Formicomus*) **A:** AP BT HP NP SD UP **ORR**
bejceki Kejval, 2002a: 110 **A:** YE (Suqutra)
bengalensis Wiedemann, 1823: 70 (*Anthicus*) **A:** NP **ORR**
bheri Kejval, 2010: 195 **A:** NP
bifurcata Kejval, 2000a: 94 (*Formicomus*) **A:** BT
bimaculatipennis Pic, 1939f: 146 (*Formicomus*) **A:** IN IS JO SA SI
biskrensis Fairmaire, 1883j: clvii (*Formicomus*) **N:** AG TU **AFR**
bispilifasciata bispilifasciata Pic, 1897n: 134 (*Formicomus*) **A:** OM SA YE **AFR**
bonadonai Kejval, 2000a: 95 (*Formicomus*) **A:** BT
bramina bramina LaFerté-Sénectère, 1849g: 10 [= 1849n: 79] (*Formicomus*) **A:** HUB JIA JIL PA TAI YUN **ORR**
bramina candens Krekich-Strassoldo, 1931: 8 (*Formicomus*) **A:** JA SHA
bramina coiffaiti Bonadona, 1964: 243 (*Formicomus*) **A:** JA SC
bramina tokaraensis Nomura, 1962a: 3 (*Formicomus*) **A:** JA (Ryukyu Islands)
brisouti Pic, 1892c: 33 (*Formicomus*) **N:** AG MO
bukharensis Kejval, 2000b: 408 (*Formicomus*) **A:** KZ UZ
caeruleipennis LaFerté-Sénectère, 1847: 369 (*Anthicus*) **E:** BEi SP **N:** AG CI EG LB MO TU **A:** AE IN IQ IS JO LE OM SA SI SY TR YE **AFR**
cyanoptera LaFerté-Sénectère, 1849g: 5 [= 1849n: 74] (*Formicomus*)
canaliculata LaFerté-Sénectère, 1849g: 22 [= 1849n: 90] (*Formicomus*) **E:** IT (Sicilia) **N:** AG EG MO TU **AFR**
cinchonae Krekich-Strassoldo, 1931: 2 (*Formicomus*) **A:** NP SD
comes Kejval, 2010: 197 **A:** NP
coniceps Pic, 1913b: 13 (*Formicomus*) **A:** SD **ORR**

consul LaFerté-Sénectère, 1849g: 23 [= 1849n: 91] (*Formicomus*) **A:** GUA GUX HAI NP YUN **AUR ORR**
corrusca Krekich-Strassoldo, 1931: 4 (*Formicomus*) **A:** UP SD
davita Kejval, 2017c: 23 **A:** YUN **ORR**
degener Kejval, 2006: 177 **A:** HUB SCH YUN **ORR**
dhankuta Kejval, 2018a: 220 **A:** NP
dracaena Kejval, 2018a: 222 **A:** BT
fallax Kejval, 2017c: 30 **A:** YUN **ORR**
fossicollis Kejval, 2002b: 270 **A:** SCH XIZ YUN
fuscipes Krekich-Strassoldo, 1931: 9 (*Formicomus*) **A:** TAI
gandaki Kejval, 2000a: 92 (*Formicomus*) **A:** NP
gladia Telnov, 2003a: 133 **A:** NP SD
gloriosa Krekich-Strassoldo, 1919a: 168 (*Formicomus*) **A:** GUI SCH YUN
gorkha Kejval, 2000a: 90 (*Formicomus*) **A:** NP
hamata Kejval, 2018a: 224 **A:** AP
hauseri Pic, 1897g: 62 (*Formicomus*) **E:** AR ST **A:** AF IN TM UZ
  sterbae Reitter, 1905c: 205 (*Formicomus*)
himalayana Krekich-Strassoldo, 1914b: 109 (*Formicomus*) **A:** BT HP NP SD UP **ORR**
imperatrix LaFerté-Sénectère, 1849f: 2 **A:** FUJ GUA GUI GUX HUB JA NC NP PA SC SD SE UP TAI ZHE **AFR ORR**
  cribriceps Marseul, 1876d: 459 (*Formicomus*)
  formicarius Nietner, 1856a: 533 (*Anthicus*)
  formicarius Nietner, 1857b: 383 (*Anthicus*) [HN]
  imperator LaFerté-Sénectère, 1849n: 66 [HN]
  myrmecodes Gemminger, 1870a: 123 (*Anthicus*) [RN]
  punctaticeps Pic, 1916g: 13 (*Formicomus*)
  quisquiliarius Nietner, 1857a: 22 (*Anthicus*) [RN]
  ruficolor Pic, 1916a: 5 (*Formicomus*)
insperata Kejval, 2010: 201 **A:** OM YE
ionica LaFerté-Sénectère, 1849g: 12 [= 1849n: 81] (*Formicomus*) **E:** GR TR **N:** EG **A:** CY IN IS JO LE SA SI SY TR YE
jelineki Kejval, 2000b: 414 (*Formicomus*) **A:** IN
kadleci Kejval, 2012: 353 **A:** YE
kailasa Kejval, 2010: 206 **A:** UP
kaszabi Bonadona, 1964: 245 (*Formicomus*) **A:** AF PA UP
klapperichi Uhmann, 1988b: 399 (*Formicomus*) **A:** TAI
kresli Kejval, 2007: 638 **A:** NP SD
lacertosa Krekich-Strassoldo, 1931: 11 (*Formicomus*) **A:** NP SD UP **ORR**
lagenicollis Fairmaire, 1894a: 41 (*Formicomus*) **A:** BT NP SD
  aestimabilis Krekich-Strassoldo, 1919a: 168 (*Formicomus*)
  bhutanensis Pic, 1913u: 204 (*Formicomus*)
latipennis Pic, 1914i: 55 (*Formicomus*) **A:** NP SD

latro bicoloripes Pic, 1935c: 4 (*Formicomus*) **E:** SP

latro latro LaFerté-Sénectère, 1849g: 21 [= 1849n: 90] (*Formicomus*) **E:** CR IT (Sardinia, Sicilia) SP **N:** AG EG TU **A:** IN IS

lewisi Marseul, 1876d: 458 (*Formicomus*) **A:** JA TAI **ORR**

limaria Kejval, 2006: 181 **A:** NP UP

lobulicola Kejval, 2007: 641 **A:** NP

longiceps Pic, 1913u: 204 (*Formicomus*) **A:** BT HP NP SD UP YUN

championi Krekich-Strassoldo, 1928a: 75 (*Formicomus*)

macilenta Bonadona, 1962: 349 (*Formicomus*) **A:** SA OM YE **AFR**

maindroni Pic, 1903s: 349 (*Formicomus*) **A:** UP **ORR**

arguta Krekich-Strassoldo, 1928a: 78 (*Formicomus*)

mandarina Boheman, 1858: 104 (*Formicomus*) **A:** FUJ HKG SCH **ORR**

mansoori Kejval, 2018b: 214 **A:** UP

merkli Kejval, 2018b: 2016 **A:** TAI

mishmi Kejval, 2018a: 226 **A:** AP

moratrix Kejval, 2018b: 217 **A:** AP

muelleriana Pic, 1955b: 9 (*Formicomus*) **A:** YE **AFR**

mülleri Pic, 1953e: 97 (*Formicomus*) [HN]

multiformis Kejval, 2002a: 107 **A:** AE IN IS JO OM PA

mutillaria Saunders, 1834: 66 **A:** NP PA SD UP YUN **ORR**

fascifer Uhmann, 1983: 64 (*Formicomus*)

inapicalis Pic, 1899f: 106 (*Formicomus*)

innotatipennis Pic, 1913p: 155 (*Formicomus*)

tonkinea Pic, 1899f: 105 (*Formicomus*)

nadari Pic, 1893t: 175 (*Formicomus*) **N:** EG **A:** IN IS JO LE SI SY TR

sphinx Krekich-Strassoldo, 1919a: 171 (*Formicomus*)

nemrod LaFerté-Sénectère, 1849g: 6 [= 1849n: 75] (*Formicomus*) **A:** AF IN IQ TM UZ

phymopes Bonadona, 1964: 248 (*Formicomus*)

nepalensis Kejval, 2000a: 87 (*Formicomus*) **A:** NP SD

ninus LaFerté-Sénectère, 1849g: 11 [= 1849n: 79] (*Formicomus*) **E:** ST (Causacus) **A:** AF CY IN IQ IS LE SA SI SY TD TM TR UZ

nitescens Krekich-Strassoldo, 1931: 5 (*Formicomus*) **A:** HP KA SD UP

nuristanica Bonadona, 1964: 247 (*Formicomus*) **A:** AF KA PA

okinawana Nomura, 1962a: 2 (*Formicomus*) **A:** JA (Ryukyu Islands)

panelii Hille, 1977: 199 (*Formicomus*) **A:** SA YE **AFR**

paolii Krekich-Strassoldo, 1928b: 384 (*Formicomus*) **A:** YE **AFR**

parallela Marseul, 1879a: 50 (*Formicomus*) **A:** IN IQ

pedestris Rossi, 1790: 224 (*Carabus*) **E:** AB AL AR AU AZ BH BU CR CZ FR GE GG GR HU IT KZ LA MC MD ME ME PL PT RO SB SK SL SP ST SZ TR UK **N:** AG LB MO MR TU **A:** AF CY IN IS JO KI KZ SY TM TR UZ WS XIN

atratula Reitter, 1889a: 33 (*Formicomus*)

brevipilis Desbrochers des Loges, 1875: 42 (*Formicomus*)

cursor LaFerté-Sénectère, 1849g: 10 [= 1849n: 78] (*Formicomus*)
equestris Panzer, 1800: 8 (*Notoxus*)
fusca Geoffroy, 1799: 344 (*Cantharis*)
hispanica Motschulsky, 1849: 131 (*Formicoma*)
nobilis Faldermann, 1837: 107 (*Anthicus*)
rubida Reitter, 1878a: 249 (*Formicomus*)
sareptana Desbrochers des Loges, 1875: 42 (*Formicomus*)
thomsoni Pic, 1901j: 90 (*Formicomus*)
thoracica Panzer, 1794b: 6 (*Notoxus*)
tincta Reitter, 1889a: 33 (*Formicomus*)
pegasus Reitter, 1892a: 63 (*Formicomus*) **A:** IS KZ TD TM UZ
persica Kejval, 2000b: 415 (*Formicomus*) **A:** IN, IQ
pilitarsis Kejval, 2012: 357 **A:** YE (Suqutra)
pokharensis Kejval, 2000a: 91 (*Formicomus*) **A:** NP SD
praefecta Kejval, 2012: 361 **A:** YE
praeses LaFerté-Sénectère, 1849g: 17 [= 1849n: 85] (*Formicomus*) **A:** SD **ORR**
nigrivestis Krekich-Strassoldo, 1931: 13 (*Formicomus*)
probsti Kejval, 2000a: 88 (*Formicomus*) **A:** NP
pseudocorrusca Kejval, 2000a: 85 (*Formicomus*) **A:** NP
psiloptera Krekich-Strassoldo, 1931: 1 (*Formicomus*) **A:** NP SD
ruficollis Saunders, 1834: 65 **A:** AF BT GUA GUX JA NP SD SE SY **ORR**
annamita Pic, 1903t: 56 (*Formicomus*)
cyanochroa Nomura, 1962a: 2 (*Formicomus*)
inhumeralis Pic, 1903r: 564 (*Formicomus*)
uncinata Desbrochers, 1875: 42 (*Formicomus*)
rufobasalis Pic, 1913a: 9 (*Formicomus*) **A:** HP KA NP PA SD UP **ORR**
montana Krekich-Strassoldo, 1914b: 111 (*Formicomus*)
sausai Kejval, 2018a: 229 **A:** AP
sauteri Pic, 1912a: 6 (*Formicomus*) **A:** TAI
sculpta Kejval, 2007: 644 **A:** BT SD
serdanga Marseul, 1884: 163 (*Formicomus*) **A:** YUN **ORR**
sikkimensis Pic, 1913a: 9 (*Formicomus*) **A:** AP BT HP NP SD UP
montivaga Krekich-Strassoldo, 1928a: 80 (*Formicomus*)
simoni Pic, 1893u: xxxiii (*Formicomus*) **A:** IN OM SA YE
sinica Kejval, 2002b: 260 **A:** YUN
soror Kejval, 2002a: 95 **A:** IN PA
spiniventris Krekich-Strassoldo, 1914c: 226 (*Formicomus*) **A:** SD **ORR**
strandi Krekich-Strassoldo, 1931: 3 (*Formicomus*) **A:** NP PA SD UP
surata Baudi di Selve, 1879: 87 (*Formicomus*) **N:** MO
taliana Pic, 1913a: 9 (*Formicomus*) **A:** YUN **ORR**
kippenbergi Uhmann, 1978: 81 (*Formicomus*)
triungula Kejval, 2010: 224 **A:** AP

truncatipennis Pic, 1897g: 61 (*Formicomus*) **A:** TD
uhmanni Kejval, 2000a: 93 (*Formicomus*) **A:** NP
umbratilis Krekich-Strassoldo, 1928a: 73 (*Formicomus*) **A:** BT HP NP SD UP
viridipennis Krekich-Strassoldo, 1931: 7 (*Formicomus*) **A:** AE AF IN PA UP **ORR**
lindbergi Bonadona, 1960: 56 (*Formicomus*)
vishnumati Kejval, 2010: 229 **A:** NP
volkonskyi Pic, 1942d: 77 **N:** AG
walkeri Pic, 1895m: cxci (*Formicomus*) **A:** SA YE
subelongata Pic, 1899n: 264 (*Formicomus*)

genus *Stenidius* LaFerté-Sénectère, 1847: 371 [= 1849k: 93; = 1849n: 192] type species *Anthicus vittatus* Lucas, 1843
aristidis Pic, 1893e: 51 (*Anthicus*) **N:** EG **A:** SA
aulicus LaFerté-Sénectère, 1849g: 13 [= 1849n: 82] (*Formicomus*) **A:** NP SD **ORR**
coloratus Motschulsky, 1863: 492 (*Leptaleus*)
contractipennis Pic, 1911o: 185 (*Anthicus*) **E:** GR MC **N:** TR
curtulus Krekich-Strassoldo, 1928a: 84 (*Anthelephilus*) **A:** HP KA NP PA SD UP
dolosus Kejval, 2002c: 193 **A:** NP UP
femoralis Desbrochers des Loges, 1875: 44 (*Anthicus*) **E:** FRi **A:** IN IS JO LE SY TR
saulcyi Marseul, 1879a: 75 (*Anthicus*)
triangulum Desbrochers des Loges, 1875: 44 (*Anthicus*)
finicola Kejval, 2002c: 196 **A:** KA PA
glabratus Kejval, 2002c: 199 **A:** NP
lindbergi Bonadona, 1960a: 55 (*Clavicomus*) **A:** AF
luteonotatus Pic, 1909b: 38 (*Formicomus*) **A:** BT HP NP PA SD UP
obliquesetosus Telnov, 2018b: 262 **A:** NP
pygidialis Pic, 1895j: 784 (*Anthicus*) **A:** IN IQ
schuhi Uhmann, 1994a: 409 (*Anthelephilus*) **A:** UP
tenuipes LaFerté-Sénectère, 1849k: 95 [= 1849n: 194] (*Anthicus*) **E:** AB AR CR GR ST (Caucasus) **A:** AF CY IN IQ IS JO LE SA SY TM TR UZ
saturatus Pic, 1902l: 186 (*Anthicus*)
vittatus hartliebi Pic, 1899j: 174 (*Anthicus*) **N:** EG
vittatus vittatus Lucas, 1843: 145 (*Anthicus*) **E:** IT (Sicilia) GR **N:** AG AZ EG TU **A:** IN **AFR**

tribe Microhoriini Bonadona, 1974

genus *Aulacoderus* LaFerté-Sénectère, 1849l: 133 [= 1849n: 232] type species *Anthicus transversalis* LaFerté-Sénectère, 1849 (= *Anthicus mutatus* Gemminger, 1870)
adensis Kejval, 2017a: 111 **A:** YE
alitienensis Pic, 1897c: 27 (*Anthicus*) **A:** YE **AFR**
bouvieri Pic, 1898e: 71 (*Anthicus*) **N:** EG **A:** IN OM SA YE **AFR**
bucheti Pic, 1903a: 21 (*Anthicus*) **N:** CI

canariensis Wollaston, 1864: 523 (*Anthicus*) **N:** CI
coarcticollis Baudi di Selve, 1881: 295 (*Anthicus*) **A:** AF IN IQ IS JO LE SY TR
appli Pic, 1900i: 78 (*Anthicus*)
crenulifer Kejval, 2017a: 67 **N:** CI (La Gomera, Tenerife)
ferrantei Peyerimhoff, 1943: 16 (*Anthicus*) **A:** EG "Sahara"
friwaldszkyi LaFerté-Sénectère, 1849l: 168 [= 1849n: 267] (*Anthicus*) **E:** AL AR AU CR GR HU IT MA **A:** AF IN IQ IS JO LE SY TR
blechroides Baudi di Selve, 1883b: 150 (*Anthicus*)
distinctipes Pic, 1901n: 180 (*Anthicus*)
funebris Reitter, 1884e: 257 (*Tomoderus*)
melitensis Pic, 1903j: 139 (*Anthicus*)
nigrithorax Pic, 1897c: 27 (*Anthicus*)
pallidior Pic, 1941a: 5 (*Anthicus*)
sulcatus Pic, 1897c: 27 (*Anthicus*)
sulcicollis Pic, 1895j: 786 (*Anthicus*)
sulcithorax Desbrochers des Loges, 1875: 46 (*Anthicus*)
inoblitus Krekich-Strassoldo, 1913a: 225 **N:** AG **A:** SA **AFR**
inspinosus Kejval, 2017a: 102 **A:** AE OM YE
kamil Kejval, 2017a: 86 **A:** IN
klapperichi Kejval, 2017a: 114 **A:** IS JO
kormosi Kejval, 2017a: 70 **N:** CI (Hierro, Teneriffe)
maderae Bonadona, 1963: 105 **N:** MR
mecheriensis Chobaut, 1896b: 376 (*Anthicus*) **N:** AG MO
vaulogeri Pic, 1897r: 297 (*Anthicus*)
mithra Kejval, 2017a: 88 **A:** IN TR
mohammedis Marseul, 1879a: 232 (*Anthicus*) **N:** AG EG MO **A:** AE OM SA SI YE **AFR**
picheyrei Peyerimhoff, 1943: 16 (*Anthicus*)
manzer Bonadona, 1964: 242
peyroni Pic, 1900i: 78 (*Anthicus*) **A:** JO OM SA SI SY YE **AFR**
breveapicalis Pic, 1951g: 136
polygoni Kejval, 2017a: 71 **N:** CI (Tenerife)
postnotatus Pic, 1910l: 39 (*Anthicus*) **N:** EG **A:** SA SI
scydmaenoides Wollaston, 1864: 524 (*Anthicus*) **N:** CI (Tenerife)
sefrensis Pic, 1894u: lxxviii (*Anthicus*) **N:** AG MO **AFR**
zizyphi Pic, 1902h: 65 (*Anthicus*)
setifer Kejval, 2017a: 75 **N:** CI (Tenerife)
sibilans Kejval, 2017a: 76 **N:** CI (La Gomera, La Palma)
sulcifer Pic, 1893v: liv (*Anthicus*) **N:** AG
miniaticollis Chobaut, 1898a: 85 (*Anthicus*)
vulcanicus Kejval, 2017a: 79 **N:** CI (Tenerife)

genus *Clavicomus* Pic, 1894g: 70 type species *Anthicus longiceps* LaFerté-Sénectère, 1849

Clavicollis J.R. Sahlberg, 1903f: 67 [SN] type species *Anthicus longiceps* LaFerté-Sénectère, 1849

Pseudantichus Desbrochers des Loges, 1868: 80 type species *Formicomus oliverii* Desbrochers des Loges, 1868 [precedence suppressed]

absconditus Telnov, 2000: 175 **A:** NP

adustus Krekich-Strassoldo, 1931: 34 (*Anthicus*) **A:** NP SD UP **ORR**

afghanus Telnov, 2010a: 9 (*Clavicollis*) **A:** AF

almorae almorae Krekich-Strassoldo, 1931: 33 (*Anthicus*) **A:** HP SD UP

almorae paradoxus Krekich-Strassoldo, 1931: 33 (*Anthicus*) **A:** HP UP YUN

ambustus Krekich-Strassoldo, 1931: 33 (*Anthicus*) **A:** HP NP SD **ORR**

angulifer Pic, 1893z: cclxx (*Anthicus*) **A:** IS "Syria"

anomalus Telnov, 1998a: 97 **A:** SCH

apicordiger Bonadona, 1954b: 273 (*Anthicus*) **N:** MO

aquatilis Krekich-Strassoldo, 1931: 30 (*Anthicus*) **A:** NP SD

assamensis Pic, 1907q: 338 (*Anthicus*) **A:** NP **ORR**

assequens Krekich-Strassoldo, 1931: 35 (*Anthicus*) **A:** HUB JIA NC TAI **ORR**

atratus Krekich-Strassoldo, 1931: 32 (*Anthicus*) **A:** BT NP SD UP

austriacus Pic, 1901g: 66 (*Anthicus*) **E:** AU

bicarinifrons Pic, 1892c: 33 (*Anthicus*) **N:** AG

pygidiolongus Pic, 1892k: 140 (*Anthicus*)

biguttatus Bonadona, 1964: 239 **A:** AF IN KA PA

brevipilis Pic, 1893y: ccxi (*Anthicus*) [RN] **N:** AG

pilosus Pic, 1892e: 43 (*Anthicus*) [HN]

bruckii Kiesenwetter, 1870: 147 (*Anthicus*) **E:** SP **N:** AG

brunneipes brunneipes Krekich-Strassoldo, 1931: 28 (*Anthicus*) **A:** HP NP SD UP **ORR**

brunneipes comparabilis Krekich-Strassoldo, 1931: 29 (*Anthicus*) **A:** UP

callimus Baudi di Selve, 1877d: 700 [= 1878a: 13] (*Anthicus*) **E:** SP

comes Krekich-Strassoldo, 1931: 28 (*Anthicus*) **A:** BT SD UP **ORR**

cordatus Krekich-Strassoldo, 1931: 27 (*Anthicus*) **A:** NP SD UP

dichrous LaFerté-Sénectère, 1849l: 135 [= 1849n: 234] (*Anthicus*) **E:** IT (Sicilia) **N:** AG

lacunosus Pic, 1901g: 66 (*Anthicus*)

doderoi doderoi Pic, 1902c: 19 (*Anthicus*) **E:** IT (Sardinia)

doderoi melonii Degiovanni, 2014: 2 (*Clavicollis*) **E:** IT (Sardinia)

erythroderus Marseul, 1878c: 55 (*Anthicus*) **N:** AG TU

obscurans Pic, 1892c: 33 (*Anthicus*)

subcyaneus Pic, 1899b: 19 (*Anthicus*)

fugax LaFerté-Sénectère, 1849l: 136 [= 1849n: 235] (*Anthicus*) **A:** SD **ORR**

fugiens Marseul, 1876d: 463 (*Anthicus*) **A:** JA

cavipennis Uhmann, 2005: 205

garze Telnov, 2018: 236 **A:** SCH (Garzę Autonom. region)

gigas Pic, 1899h: 86 (*Anthicus*) **E:** GR TR **A:** CY TR
bergeri Bonadona, 1986: 44 (*Microhoria*)
rhodius Pic, 1899h: 86 (*Anthicus*)
gravidus Krekich-Strassoldo, 1931: 25 (*Anthicus*) **A:** NP SD UP
harmandi Pic, 1899g: 78 (*Anthicus*) **A:** NP SD
hauseri Pic, 1906h: 49 (*Anthicus*) **A:** MG QIN XIZ
henoni Pic, 1892h: 103 (*Anthicus*) **E:** SP **N:** AG MO
quinquenotatus Reitter, 1906a: 442 (*Anthicus*) [RN]
v-notatus Pic, 1899b: 20 (*Anthicus*)
heydeni Marseul, 1879a: 132 (*Anthicus*) **E:** PT SP
himalayanus Pic, 1909c: 57 (*Anthicus*) **A:** HP NP
hummeli Pic, 1933b: 10 (*Anthicus*) **A:** XIN
immaculipennis Krekich-Strassoldo, 1931: 29 (*Anthicus*) **A:** HP NP PA SD UP
inabsolutus Telnov, 2003b: 281 **A:** SD
indeprensus Telnov, 2000: 177 **A:** NP
kabylianus Pic, 1896l: 167 (*Anthicus*) **N:** AG TU
kham Telnov, 2018: 240 **A:** SCH (Garze Autonom. Region)
kocheri Pic, 1951f: 81 (*Anthicus*) **N:** MO
kuluensis Pic, 1914d: 20 (*Anthicus*) **A:** HP NP
lepidulus Marseul, 1876d: 461 (*Anthicus*) **A:** JA
longiceps LaFerté-Sénectère, 1849l: 134 [= 1849n: 233] (*Anthicus*) **E:** IT (Sicilia) SP **N:** AG MO TU
longicornis Uhmann, 1983: 203 **A:** SD **ORR**
manifestus Pic, 1907d: 135 (*Anthicus*) **A:** TAI YUN
martinezi Pic, 1932e: 18 (*Anthicus*) **E:** SP
mouzafferi Pic, 1910m: 42 (*Anthicus*) **A:** IN
muguensis Telnov, 2000: 178 **A:** NP
nigrocyanellus Marseul, 1877: 470 (*Anthicus*) **A:** FE FUJ GAN GUA JA LIA NMO SC XIN YUN ZHE **ORR**
nigrofuscus Telnov, 2000: 179 **A:** NP
nigroterminatus Pic, 1909d: 106 (*Anthicus*) **E:** SP
notatipennis Pic, 1909c: 56 (*Anthicus*) **A:** HP NP SD
obscuricolor Pic, 1909c: 57 (*Anthicus*)
oliverii Desbrochers des Loges, 1865: 79 (*Formicomus*) **N:** AG TU
optabilis LaFerté-Sénectère, 1849k: 88 [= 1849n: 187] (*Anthicus*) **E:** FR
paganettii Pic, 1909f: 114 (*Anthicus*) **E:** IT
piceus LaFerté-Sénectère, 1849l: 136 [= 1849n: 235] (*Anthicus*) **A:** NP SCH SD **ORR**
plagiostolus Bonadona, 1958: 288 (*Cordicomus*) **N:** MO
plicatipennis Pic, 1936e: 29 (*Anthicus*) **N:** MO
posthumus Krekich-Strassoldo, 1931: 32 (*Anthicus*) **A:** NP SD UP
postimpressus Pic, 1938b: 5 (*Anthicus*) **N:** MO

postluteofasciatus Pic, 1938b: 5 (*Anthicus*) **E:** SP
protervus Krekich-Strassoldo, 1931: 30 (*Anthicus*) **A:** SD UP
ragusae Pic, 1898m: 229 (*Anthicus*) **E:** IT (Sicilia)
schrammi Pic, 1913n: 146 (*Anthicus*) **E:** PT SP
semidepressus Pic, 1893h: 64 (*Anthicus*) **N:** AG
separatithorax Pic, 1914q: 184 (*Anthicus*) **A:** NP SD UP **ORR**
shibatai Nomura, 1962a: 3 (*Anthicus*) **A:** JA (Ryukyu Islands)
sikkimensis Pic, 1907o: 196 (*Anthicus*) **A:** BT PA SD
sinensis Pic, 1907d: 135 (*Anthicus*) **A:** SE TAI YUN
spinipennis Pic, 1898g: 99 (*Anthicus*) **N:** AG
spinosus Pic, 1912h: 42 (*Anthicus*) **A:** AF KI UZ
sporadicus Krekich-Strassoldo, 1931: 26 (*Anthicus*) **A:** NP SD UP
striaticollis Krekich-Strassoldo, 1931: 26 (*Anthicus*) **A:** NP SD UP
subpiceus Pic, 1914f: 36 (*Anthicus*) **A:** "Himalaya"
tersus Krekich-Strassoldo, 1931: 31 (*Anthicus*) **A:** HP NP SD UP
tonkinensis Krekich-Strassoldo, 1928: 101 (*Anthicus*) **A:** YUN **ORR**
turgidus Krekich-Strassoldo, 1928a: 100 (*Anthicus*) **A:** SD
uhagoni Pic, 1904g: 81 (*Anthicus*) **E:** SP
uniformis Krekich-Strassoldo, 1931: 27 (*Anthicus*) **A:** NP SD UP
variabilis Telnov, 2003b: 282 **A:** NP
versicolor Kiesenwetter, 1866: 383 (*Anthicus*) **E:** SP
weigeli Telnov, 2000: 180 **A:** NP
wuyishanensis Nardi, 2004: 137 [RN] **A:** FUJ
plicatipennis Pic, 1954b: 64 (*Anthicus*) [HN]

genus *Liparoderus* LaFerté-Sénectère, 1849k: 85 [= 1849n: 184] type species *Anthicus insignis* Lucas, 1843
argenteovestitus Pic, 1893e: 51 (*Anthicus*) **N:** AG EG
subincisipes Pic, 1924c: 21 (*Anthicus*)
galeatus Bonadona, 1988: 46 **N:** AG
insignis insignis Lucas, 1843: 145 (*Anthicus*) **E:** FR (Corse) PT SP **N:** AG LB MO TU
alutaceiceps Pic, 1916a: 5 (*Anthicus*)
insignior Pic, 1894g: 71 (*Anthicus*)
mutandus Desbrochers des Loges, 1899a: 157 (*Anthicus*)
noualhieri Pic, 1896i: 132 (*Anthicus*)
obscuripes Pic, 1896i: 132 (*Anthicus*)
rubrithorax Pic, 1897f: 50 (*Anthicus*)
tripolitanus Pic, 1900m: 165 (*Anthicus*)
insignis leseleuci Pic, 1893j: 78 (*Anthicus*) **N:** MO
luteobisignatus Pic, 1942c: 11 (*Anthicus*)
panousei Pic, 1948a: 2 (*Anthicus*)

undatus Bonadona, 1986: 36 (*Microhoria*) **N:** MO
venator Dufour, 1849: 229 (*Anthicus*) **E:** SP **N:** AG MO TU
rubromaculatus Pic, 1896i: 132 (*Anthicus*)

genus *Microhoria* Chevrolat, 1877c: 168 type species *Anthicus oedipus* Chevrolat, 1860
Bifossicollis J.R. Sahlberg, 1903e: 31 type species *Anthicus iscariotes* LaFerté-Senectère, 1849
Immicrohoria Pic, 1894g: 72 type species *Notoxus fasciatus* Chevrolat, 1834
Platyoria Bonadona, 1952: 234 type species *Anthicus terminatus* W.L.E. Schmidt, 1842
Submicrohoria Bonadona, 1952: 234 type species *Anthicus plumbeus* LaFerté-Senectère, 1842
abdeselami Escalera, 1914: 379 (*Anthicus*) **N:** MO
abeillei Pic, 1892b: 17 (*Anthicus*) **N:** AG
admirabilis Pic, 1894i: 87 (*Anthicus*) [= 1894l: 144] **N:** AG MO
advecta Krekich-Strassoldo, 1929: 173 (*Anthicus*) **N:** AG
adventicia Krekich-Strassoldo, 1929: 162 (*Anthicus*) **E:** CR
agtayi Bonadona, 1986: 42 **N:** MO
aguilari Bonadona, 1960a: 52 **E:** GR (Lesvos)
akbesiana Pic, 1896a: 5 (*Anthicus*) **A:** TR
albopilosa Krekich-Strassoldo, 1929: 179 (*Anthicus*) **E:** FR SP
amata Bonadona, 1958: 291 **N:** MO
amicitiae Dufour, 1849: 230 (*Anthicus*) **E:** SP
macularis Krekich-Strassoldo, 1929: 187 (*Anthicus*)
semicincta Desbrochers des Loges, 1875: 47 (*Anthicus*)
andalusiaca LaFerté-Sénectère, 1849l: 151 [= 1849n: 250] (*Anthicus*) **E:** SP
angulapex Koch, 1935c: 260 (*Anthicus*) **E:** GR **A:** TR
annulipes Pic, 1894i: 88 [= 1894l: 144] (*Anthicus*) **N:** AG
antoinei Chobaut, 1923a: 76 (*Anthicus*) **N:** MO
aphaenops Pic, 1902f: 41 (*Anthicus*) **A:** IN TM UZ
arcuatipes Krekich-Strassoldo, 1928a: 103 (*Anthicus*) **A:** HP NP SD UP
aspelia Truqui, 1855: 363 (*Anthicus*) **E:** GR **A:** CY
aspernata Chobaut, 1923b: 129 (*Anthicus*) **N:** MO
atlasica Pic, 1951f: 81 (*Anthicus*) **N:** MO
aubei LaFerté-Sénectère, 1849l: 152 [= 1849n: 251] (*Anthicus*) **E:** PT SP **N:** AG LB MO
calorificus Pic, 1900f: 56 (*Anthicus*)
balearica Pic, 1904g: 81 (*Anthicus*) **E:** SP (Baleares)
rubricula Krekich-Strassoldo, 1929: 182 (*Anthicus*)
barrosi Pic, 1938d: 13 (*Anthicus*) **E:** PT
baudii Pic, 1893w: lxxxvii (*Anthicus*) **N:** AG TU
obscurevestita Pic, 1894g: 73 (*Anthicus*)
baudueri Baudi di Selve, 1877d: 701 [= 1878a: 14] (*Anthicus*) **N:** AG
benigna Krekich-Strassoldo, 1929: 162 (*Anthicus*) **E:** PT SP
biargenteofasciata Pic, 1929d: 6 (*Anthicus*) **N:** MO

biauriculata Pic, 1920d: 10 (*Anthicus*) **N:** AG
bicoloriceps Pic, 1932b: 32 [= 1932g: 27] (*Anthicus*) **N:** MO
binotaticollis Pic, 1919d: 10 (*Anthicus*) **E:** FR (Corse)
bispilifasciata Marseul, 1878c: 55 (*Anthicus*) **E:** SP
bleusei Pic, 1892n: cxcviii (*Anthicus*) **E:** SP
boyadjeani Pic, 1904f: 73 (*Anthicus*) **A:** TR
bremondi Pic, 1936e: 29 (*Anthicus*) **N:** MO
breviuscula Desbrochers des Loges, 1875: 48 (*Anthicus*) **A:** LE
brisouti brisouti Desbrochers des Loges, 1875: 48 (*Anthicus*) **E:** FR PT SP
corrugata Krekich-Strassoldo, 1929: 188 (*Anthicus*)
incisa Baudi di Selve, 1877d: 697 (*Anthicus*)
nigrescens Krekich-Strassoldo, 1929: 188 (*Anthicus*)
brisouti peragalloi Pic, 1902h: 64 (*Anthicus*) **E:** SP
brisouti sternolobata Bonadona, 1954b: 280 **E:** PT SP
caliginosa caliginosa LaFerté-Sénectère, 1849l: 163 [= 1849n: 262] (*Anthicus*) **E:** AL BH BU CR GR IT SL TR YU
caliginosa epidaurosa Pic, 1936c: 22 (*Anthicus*) **E:** GR
caliginosa pannonica Kaszab, 1956c: 337 [= 1956b: 168] (*Anthicus*) **E:** HU
caliginosa sexualis Koch, 1933: 155 (*Anthicus*) **E:** CR
cantabrica cantabrica Marseul, 1879a: 212 (*Anthicus*) **E:** SP
manueli Pic, 1906a: 12 (*Anthicus*)
cantabrica relicta Krekich-Strassoldo, 1929: 172 (*Anthicus*) **E:** SP
capito LaFerté-Sénectère, 1849l: 160 (*Anthicus*) **E:** SP **A:** MO
capito LaFerté-Sénectère, 1849n: 259 (*Anthicus*) [HN]
caprai Bucciarelli, 1977: 179 **E:** IT
caspia Desbrochers des Loges, 1875: 45 (*Anthicus*) **E:** ST **A:** TM
cerrutii Bucciarelli, 1976: 139 **E:** GR (Crete)
chakouri Pic, 1909l: 143 (*Anthicus*) **N:** AG G
chardoni chardoni Pic, 1893s: 161 (*Anthicus*) **N:** AG
chardoni fezzarae Pic, 1893s: 162 (*Anthicus*) **N:** AG
chobauti Pic, 1892g: 75 (*Anthicus*) **E:** IT (Lampedusa Is.) SP **N:** AG TU
lucidior Pic, 1892g: 76 (*Anthicus*)
cinctuta Marseul, 1878c: 54 (*Anthicus*) **N:** AG LB TU
rossii Pic, 1931f: 66 (*Anthicus*)
codinai Pic, 1919d: 11 (*Anthicus*) **E:** SP
constricticollis Desbrochers des Loges, 1870: 124 (*Anthicus*) **N:** AG TU
corallicollis Reitter, 1889e: 258 (*Anthicus*) **E:** GR **A:** IS
curticollis Pic, 1894k: 140 (*Anthicus*) **E:** PT SP
cyrtopyga Bonadona, 1952: 236 **E:** AB ST **A:** TM
decora Krekich-Strassoldo, 1929: 174 (*Anthicus*) **E:** SP
degener Baudi di Selve, 1881: 294 (*Anthicus*) **A:** LE

dejeani corsica LaFerté-Sénectère, 1849l: 146 [= 1849n: 245] (*Anthicus*) **E:** FR (Corsica)
valdoniellensis Pic, 1919d: 10 (*Anthicus*)
dejeani dejeani LaFerté-Sénectère, 1849l: 145 [= 1849n: 244] (*Anthicus*) **E:** CR FR (Corsica) IT
dejeani stocki Krekich-Strassoldo, 1929: 171 (*Anthicus*) **E:** FR (Corsica)
delagrangei Pic, 1892e: 43 (*Anthicus*) **A:** SY/TR (Akbes region)
dentipalpis Bonadona, 1977a: 9 **N:** MO
digitalis Marseul, 1878c: 55 (*Anthicus*) **E:** SP **N:** AG MO
emaciata Pic, 1896m: 180 (*Anthicus*) **E:** GR (Crete) **A:** LE SY
espunana Pic, 1930d: 10 (*Anthicus*) **E:** SP
faceta Bonadona, 1960a: 54 **A:** AF
fairmairei C. Brisout de Barneville, 1863: 90 (*Anthicus*) **E:** FR PT SP **N:** AG
decoloratipes Chobaut, 1908: 162 (*Anthicus*)
fasciata fasciata Chevrolat, 1834: pl. 34 (*Notoxus*) **E:** AL CR FR GE GR IT ME SB SL SP SZ YU
affinis LaFerté-Sénectère, 1842: 248 (*Anthicus*)
antoniae LaFerté-Sénectère, 1842: 249 (*Anthicus*)
basidensata Pic, 1936c: 21 (*Anthicus*)
bicincta LaFerté-Sénectère, 1849l: 148 [= 1849n: 247] (*Anthicus*) [HN]
francogallicus Koch, 1933: 153 (*Anthicus*) [HN]
fulvicollis Pic, 1892d: 43 (*Anthicus*) [HN]
genistae Rosenhauer, 1847: 36 (*Anthicus*)
histrio LaFerté-Sénectère, 1849l: 148 [= 1849n: 247] (*Anthicus*) [HN]
lata Pic, 1892c: 33 (*Anthicus*)
monogramma W.L.E. Schmidt, 1842c: 174 (*Anthicus*)
perobscura Krekich-Strassoldo, 1929: 170 (*Anthicus*)
unifasciata Krekich-Strassoldo, 1929: 169 (*Anthicus*) [HN]
fasciata fortis Krekich-Strassoldo, 1929: 170 (*Anthicus*) **N:** MO
rutila Krekich-Strassoldo, 1929: 170 (*Anthicus*)
fasciata helvetica Krekich-Strassoldo, 1929: 169 (*Anthicus*) **E:** SZ
fasciata opaca Rey, 1892: 101 (*Anthicus*) **E:** FR
ferianensis Pic, 1900f: 56 (*Anthicus*) **N:** AG TU
feroni Bonadona, 1960a: 51 **E:** GR
fornicata Krekich-Strassoldo, 1931: 39 (*Anthicus*) **A:** HP
fortissima Pic, 1894g: 78 [= 1895g: 196] (*Anthicus*) **N:** AG
franzi Bonadona, 1958: 289 **E:** SP **N:** MO
fumeoalata Krekich-Strassoldo, 1931: 39 (*Anthicus*) **A:** NP UP
funeraria Marseul, 1879a: 225 (*Anthicus*) **E:** GR IT **A:** CY
fuscipes Marseul, 1879a: 199 (*Anthicus*) **E:** SP **N:** AG MO
ghilianiii LaFerté-Sénectère, 1849l: 150 [= 1849n: 249] (*Anthicus*) **E:** SP
globipennis Pic, 1897j: 120 (*Anthicus*) **A:** TR "Syria"
quercicola Krekich-Strassoldo, 1913a: 227 (*Anthicus*) [HN]
quercicola Sahlberg, 1913a: 59 (*Anthicus*)

gorgus Truqui, 1855: 366 (*Anthicus*) **E:** GR **A:** CY "Syria"
gourvesi Bonadona, 1986: 38 **N:** MO
hafidi Pic, 1923a: 3 (*Anthicus*) **N:** MO
hameti Escalera, 1914: 376 (*Anthicus*) **N:** MO
helenae Koch, 1933: 154 (*Anthicus*) **E:** IT
hindukushica Telnov, 2010a: 13 **A:** AF
hingstoni Blair, 1927: 251 (*Anthicus*) **A:** XIZ
hipponensis Pic, 1893d: 38 (*Anthicus*) **N:** AG
hispanica Pic, 1899b: 20 (*Anthicus*) **E:** SP
humerifer Pic, 1902f: 41 (*Anthicus*) **E:** GR TR
imbasicornis Pic, 1931c: 6 (*Anthicus*) **E:** SP
informipes Krekich-Strassoldo, 1919a: 173 (*Anthicus*) **A:** HP KA UP
inobscura Pic, 1908i: 66 (*Anthicus*) **E:** TR
ionica Pic, 1901e: 52 (*Anthicus*) **E:** BU GR TR **A:** TR
iscariotes LaFerté-Sénectère, 1849n: 301 (*Anthicus*) **E:** GR **N:** EG LB **A:** CY IS JO LE SI
kaifensis Pic, 1896m: 180 (*Anthicus*) **A:** IS JO LE SY TR
obscuriceps Pic, 1899j: 177 (*Anthicus*)
lafertei Truqui, 1855: 365 (*Anthicus*) **E:** BU GR **A:** CY IQ IS TR
lanata Krekich-Strassoldo, 1929: 177 (*Anthicus*) **E:** SP
latecincta Chobaut, 1923b: 128 (*Anthicus*) **N:** MO
latipennis Pic, 1892o: 313 (*Anthicus*) **A:** IS JO SI SY TR
lavocati Pic, 1951d: 37 (*Anthicus*) **N:** MO
lederi Marseul, 1879a: 252 (*Anthicus*) **E:** AR CR GR **A:** TM TR
leonhardi Krekich-Strassoldo, 1913a: 229 (*Anthicus*) **A:** IT
strassoldoi Pic, 1930d: 10 (*Anthicus*) [RN]
leprieuri Baudi di Selve, 1877d: 717 (*Anthicus*) **N:** AG TU
leptostemma Kolenati, 1846: 35 (*Anthicus*) **E:** AB AR GG KZ ST **A:** AF TM UZ
leuthneri Pic, 1897j: 120 (*Anthicus*) **E:** GR (Crete) **A:** IS LE SY
creticus Pic, 1903n: 170 (*Anthicus*)
dissimilis Pic, 1897j: 120 (*Anthicus*)
lindbergi albocrispa Bonadona, 1954b: 277 **E:** MO
lindbergi flavostola Bonadona, 1958: 293 **E:** SP
lindbergi lindbergi Pic, 1932b: 33 [= 1932g: 27] (*Anthicus*) **N:** MO
lividipes Desbrochers des Loges, 1875: 47 (*Anthicus*) **A:** LE SY
lobata Pic, 1905c: 99 (*Anthicus*) **N:** AG
loebli Uhmann, 1989: 250 **A:** PA
ludovici Pic, 1893l: 111 (*Anthicus*) **N:** AG MO
mactae Pic, 1894g: 73 (*Anthicus*) **N:** AG
testaceofasciatus Pic, 1894g: 73 (*Anthicus*)
maculicollis Pic, 1893r: 156 (*Anthicus*) **N:** AG TU
madoni Pic, 1931d: 9 (*Anthicus*) **N:** AG

magnifica magnifica Pic, 1936e: 29 (*Anthicus*) **N:** MO
magnifica roulleaui Bonadona, 1954b: 277 **N:** MO
major major Pic, 1896l: 168 (*Anthicus*) **N:** AG MO TU
major rugata Krekich-Strassoldo, 1929: 186 (*Anthicus*) **E:** SP
mateui Bonadona, 1954b: 277 **E:** SP
mateui Bonadona, 1954a: 100 [HN]
melanocephala Bonelli, 1812: 173 (*Anthicus*) **E:** IT
mesopotamica Pic, 1912c: 9 (*Anthicus*) **A:** IQ
miranda Krekich-Strassoldo, 1931: 37 (*Anthicus*) **A:** NP UP
mogadorica Escalera, 1914: 375 (*Anthicus*) **N:** MO
mollis Desbrochers des Loges, 1875: 46 (*Anthicus*) **E:** ST "Caucasus" **A:** IN
monodi Bonadona, 1977a: 5 **A:** IN
moroderi Pic, 1930d: 10 (*Anthicus*) **E:** SP
mylabrina Gené, 1839: 74 (*Anthicus*) **E:** FR (Corse) IT (Sardinia)
revelierei Pic, 1893b: 26 (*Anthicus*)
nectarina Panzer, 1794b: 7 (*Notoxus*) **E:** BU CR CZ GE GR HU PL RO SK UK ST **A:** AF IN KI KZ MG TD TM WS UZ XIN
atriceps Pic, 1904g: 82 (*Anthicus*)
balazuci Bonadona, 1986: 35 (*Clavicomus*)
bicincta Gebler, 1825: 49 (*Anthicus*)
kaszabi L.N. Medvedev, 1975: 92 [HN]
maltae Pic, 1892f: 66 (*Anthicus*) [RN]
malvae Pic, 1892e: 43 (*Anthicus*)
medvedevi Telnov, 2004: 137 [RN]
nepticula Bonadona, 1964: 240 **A:** AF PA TD TM
nicolasi Pic, 1919d: 10 (*Anthicus*) **A:** KZ/TM ("Transcaspia")
normandi Pic, 1915l: 37 (*Anthicus*) **N:** TU
oberthuri Baudi di Selve, 1877d: 709 [= 1878a: 17] (*Anthicus*) **N:** AG MO
brevinotata Pic, 1896j: 141 (*Anthicus*)
juncea Pic, 1899b: 20 (*Anthicus*)
obuncata Normand, 1950a: 79 (*Anthicus*) **N:** TU
oedipus Chevrolat, 1860: 210 (*Anthicus*) **N:** AG MO TU
scaura Fairmaire, 1875b: 495 (*Anthicus*)
subsinuatipes Pic, 1920d: 10 (*Anthicus*)
tortiscelis Marseul, 1875a: 38 (*Anthicus*)
oertzeni oertzeni Pic, 1901n: 181 (*Anthicus*) **E:** AL BU GR MC
sporadensis Pic, 1901n: 181 (*Anthicus*)
oertzeni sicula Bucciarelli, 1978: 162 **E:** IT (Sicilia)
opipara Bonadona, 1977a: 7 **E:** PT SP
ovata Marseul, 1879a: 229 (*Anthicus*) **E:** GR
palicari LaPorte, 1840: 258 (*Anthicus*) **E:** GR TR

pallidioritarsis Pic, 1936g: 213 (*Anthicus*) **N:** MO
pallidula Pic, 1892e: 43 (*Anthicus*) **E:** BU CR GR KZ MC PL RO ST YU **N:** MO **A:** IS KZ TM TR WS XIN XIZ
mutata Pic, 1892e: 43 (*Anthicus*) [HN]
notata Pic, 1901m: 131 (*Anthicus*) [RN]
terminata LaFerté-Sénectère, 1849l: 144 [= 1849n: 243] (*Anthicus*) [HN]
pardoi Bonadona, 1952: 234 **N:** MO
paykullii Gyllenhal, 1808: 55 (*Anthicus*) **E:** PT SP **N:** CI AG TU
bijubata Motschulsky, 1849: 129 (*Anthicus*)
petraea Pic, 1902l: 187 (*Anthicus*) **A:** IS JO SA
piceodiscoidalis Pic, 1936d: 25 (*Anthicus*) **N:** MO
piciceps Desbrochers des Loges, 1875: 47 (*Anthicus*) **E:** GR/MC ("Macedonia") ST **A:** IS TR WS
melanocephala Marseul, 1879a: 216 (*Anthicus*) [HN]
pinicola Reitter, 1889e: 258 (*Anthicus*) **E:** GR
feroni Bonadona, 1960a: 51
testacea Pic, 1941a: 5 (*Anthicus*)
plagifer Krekich-Strassoldo, 1931: 38 (*Anthicus*) **A:** HP NP SD UP
planiceps Desbrochers des Loges, 1875: 49 (*Anthicus*) **N:** AG
plumbea alpina Pic, 1951b: 10 (*Anthicus*) **E:** FR
plumbea plumbea LaFerté-Sénectère, 1842: 250 (*Anthicus*) **E:** CR FR IT PT SP
brevis W.L.E. Schmidt, 1842c: 180 (*Anthicus*)
callosus W.L.E. Schmidt, 1842c: 181 (*Anthicus*)
melanarius W.L.E. Schmidt, 1842c: 178 (*Anthicus*)
subrubricollis Pic, 1951b: 10 (*Anthicus*)
postluteomaculata Pic, 1938b: 5 (*Anthicus*) **N:** MO
quadraticeps Desbrochers des Loges, 1875: 47 (*Anthicus*) **A:** JO SY TR
rabinovitchi Koch, 1935b: 141 (*Anthicus*) **N:** EG
raveli Pic, 1899b: 20 (*Anthicus*) **E:** BU GR IT **A:** TR
rectipennis LaFerté-Sénectère, 1849l: 144 (*Anthicus*) **E:** AB GG
rectipennis LaFerté-Sénectère, 1849n: 243 (*Anthicus*) [HN]
robustioriceps Pic, 1938c: 11 (*Anthicus*) **N:** MO
roseicollis Pic, 1892h: 103 (*Anthicus*) **E:** PT **N:** AG
portalegrea Pic, 1909p: 178 (*Anthicus*)
semirosea Pic, 1896c: 41 (*Anthicus*)
rosti Pic, 1906i: 79 (*Anthicus*) **A:** HP KA NP PA SD
rubriceps Pic, 1896m: 180 (*Anthicus*) **E:** GR **A:** LE SY TR
rubronotata Pic, 1896m: 180 (*Anthicus*)
rubrofasciata Pic, 1894g: 76 (*Anthicus*) **N:** AG
rufescens Pic, 1893i: 74 (*Anthicus*) **A:** TR
saidi Escalera, 1914: 378 (*Anthicus*) **N:** MO
schmiedeknechti Pic, 1899j: 177 (*Anthicus*) **E:** GR **A:** TR
scrobicollis LaFerté-Sénectère, 1849l: 166 [= 1849n: 265] (*Anthicus*) **E:** PT SP **N:** AG

selvei bicolorithorax Pic, 1930d: 10 (*Anthicus*) **E:** SP
selvei consona Krekich-Strassoldo, 1929: 178 (*Anthicus*) **E:** FR SP
selvei pullata Krekich-Strassoldo, 1929: 179 (*Anthicus*) **E:** SP
selvei selvei Pic, 1895g: 195 (*Anthicus*) **E:** SP
separanda Krekich-Strassoldo, 1929: 180 (*Anthicus*) **E:** SP
sicardi Pic, 1893n: 123 (*Anthicus*) **N:** TU
thalensis Pic, 1920d: 10 (*Anthicus*)
sidonia Truqui, 1855: 365 (*Anthicus*) **E:** "Eastern Europe" **A:** IS LE SY TR
simplicipes Pic, 1936d: 26 (*Anthicus*) **N:** MO
stettini Pic, 1892e: 43 (*Anthicus*) [RN] **E:** FR
ruficollis W.L.E. Schmidt, 1842c: 172 (*Anthicus*) [HN]
subgracilis Krekich-Strassoldo, 1929: 179 (*Anthicus*) **E:** SP
subtruncata Pic, 1920d: 10 (*Anthicus*) **N:** AG
succincta reymondi Bonadona, 1954b: 275 **N:** MO
succincta succincta Chevrolat, 1877c: 169 (*Anthicus*) **N:** AG MO
jurjurensis Pic, 1902d: 24 (*Anthicus*)
tumidipes Marseul, 1879a: 195 (*Anthicus*)
superba pueli Chobaut, 1903b: 170 (*Anthicus*) **N:** AG
superba superba Pic, 1896c: 40 (*Anthicus*) **N:** AG TU
sydowi olympiae Pic, 1941a: 5 (*Anthicus*) **E:** GR
sydowi reductior Pic, 1936c: 21 (*Anthicus*) **E:** GR
sydowi sydowi Pic, 1936c: 21 (*Anthicus*) **E:** GR
syrensis Pic, 1902f: 41 (*Anthicus*) **E:** GR **A:** CY
taeniata Baudi di Selve, 1877d: 712 [= 1878a: 18] (*Anthicus*) **N:** AG
taurica Pic, 1904f: 74 (*Anthicus*) **A:** TR
tenebricosa Pic, 1896a: 5 (*Anthicus*) **A:** LE
terminata W.L.E. Schmidt, 1842b: 128 (*Anthicus*) **E:** AL BH BU CR CT FR GB GR IT MC ME PT SB SK SL SP ST TR UK YU **N:** AG TU **A:** IN TM TR
chiosicola Reitter, 1889e: 259 (*Anthicus*)
dalmatina Pic, 1901m: 130 (*Anthicus*)
emgei Pic, 1897r: 296 (*Anthicus*)
latebrunnea Pic, 1901m: 131 (*Anthicus*)
latebrunnea Pic, 1903o: 178 (*Anthicus*) [HN]
sanguinicollis LaFerté-Sénectère, 1849l: 140 [= 1849n: 239] (*Anthicus*)
torretassoi Pic, 1931f: 66 (*Anthicus*) **N:** LB
tunisica Pic, 1893n: 123 (*Anthicus*) **N:** TU
unicolor calliger Marseul, 1879a: 225 (*Anthicus*) **E:** FR GR
unicolor unicolor W.L.E. Schmidt, 1842c: 179 (*Anthicus*) **E:** AL AU BU CR CZ FR GR HU IT MC PL SK TR **A:** TR
patagiata Kiesenwetter, 1861a: 247 (*Anthicus*)
validicornis LaFerté-Sénectère, 1849l: 165 (*Anthicus*)
validicornis LaFerté-Sénectère, 1849n: 264 (*Anthicus*) [HN]

valga Fairmaire, 1875b: 496 (*Anthicus*) **N:** AG MO
varus Marseul, 1875a: 38 (*Anthicus*)
valida Pic, 1896m: 179 (*Anthicus*) **E:** SP
dispar Pic, 1899b: 21 (*Anthicus*)
velutina LaFerté-Sénectère, 1849l: 162 (*Anthicus*) **E:** FR SP
velutinus LaFerté-Sénectère, 1849n: 261 (*Anthicus*) [HN]
venusta A. Villa & J.B. Villa, 1833: 35 (*Anthicus*) **E:** AU FR IT SZ
atrithorax Pic, 1912o: 90 (*Anthicus*)
birubronotata Pic, 1912o: 90 (*Anthicus*)
nigerrima Pic, 1892h: 103 (*Anthicus*)
nigrissima Pic, 1892e: 43 (*Anthicus*)
unifasciata W.L.E. Schmidt, 1842c: 173 (*Anthicus*) [HN]
veris Pic, 1893e: 52 (*Anthicus*) **N:** AG
vespertina Rosenhauer, 1856: 224 (*Anthicus*) **E:** PT SP
villiersi Bonadona, 1984: 71 **E:** PT
violaris seminiger Pic, 1899r: 412 (*Anthicus*) **N:** AG
violaris violaris Marseul, 1878b: 51 (*Anthicus*) **N:** AG
volxemi Marseul, 1878a: xliii (*Anthicus*) **E:** PT SP
vosseleri melanoderus Chobaut, 1897a: 39 (*Anthicus*) **N:** AG
vosseleri vosseleri Pic, 1894i: 88 (*Anthicus*) **N:** AG MO
vosseleri Pic, 1894n: 145 (*Anthicus*) [HN]
winkleri Telnov, 2004: 137 [RN] **A:** "Syria"
longipygus Pic, 1919d: 10 (*Anthicus*) [HN]
yemenita Nardi, 2004: 138 [RN] **A:** YE
scotti Pic, 1957: 448 [HN]
zonata LaFerté-Sénectère, 1849l: 153 [= 1849n: 252] (*Anthicus*) **E:** IT (Sardinia)

genus *Tenuicomus* Pic, 1894g: 69 [= 1878a: 16] type species *Anthicus pumilus* Baudi di Selve, 1877d: 702
Tenuicollis J.R. Sahlberg, 1903f: 67 [SN] type species *Tenuicomus olivaceus* LaFerté-Sénectère, 1849
agriliformis Pic, 1893j: 79 (*Anthicus*) **N:** MO
alfierii Pic, 1923a: 1 (*Anthicus*) **N:** EG
alfierii Pic, 1923a: 1 (*Anthicus*)
angelinii Degiovanni, 2012: 98 (*Tenuicollis*) **A:** TR
arenarius Bonadona, 1956: 119 (*Microhoria*) **N:** AG
armeniacus Pic, 1899n: 264 (*Anthicus*) **E:** AR
barnevillei Pic, 1892b: 26 (*Anthicus*) **E:** SP
bonnairii Fairmaire, 1883e: xlv (*Anthicus*) **N:** AG LB TU
nubilicollis Pic, 1906j: 283 (*Anthicus*)
cyanipennis Grilat, 1886: 187 (*Anthicus*) **N:** AG TU
grilati Pic, 1898g: 99 (*Anthicus*)
depressus LaFerté-Sénectère, 1849l: 156 [= 1849n: 255] (*Anthicus*) **E:** AB ST ("Circassia") **A:** IN

dolichocephalus Baudi di Selve, 1877d: 702 [= 1878a: 15] (*Anthicus*) **N**: AG
duplex Nardi, 2004: 139 [RN] **E**: AR GG
subaeneus LaFerté-Sénectère, 1849l: 181 [= 1849n: 280] (*Anthicus*) [HN]
unicolor Kolenati, 1846: 37 (*Anthicus*) [HN]
edmondi Pic, 1893a: 16 (*Anthicus*) **A**: UZ
escalerai Pic, 1904g: 81 (*Anthicus*) **E**: SP
finalis Telnov, 2003b: 283 (*Tenuicomus*) **A**: AF IN PA
fuscomaculatus Pic, 1893v: liv (*Anthicus*) **N**: AG MO
tangerianus Pic, 1893j: 79 (*Anthicus*)
insignitus Pic, 1906j: 285 (*Anthicus*) **N**: TU
luristanicus Pic, 1911e: 97 (*Anthicus*) **A**: IN IQ
pietschmi Pic, 1938c: 11 (*Anthicus*)
meloiformis Reitter, 1890d: 152 (*Anthicus*) **N**: AG MO
ocreatus algecirensis Koch, 1940b: 76 (*Anthicus*) **E**: SP
ocreatus ocreatus LaFerté-Sénectère, 1847: 379 (*Anthicus*) **E**: IT PT SP **N**: AG MO TU
algerinus Pic, 1892b: 25 (*Anthicus*)
basicollis Pic, 1916a: 5 (*Anthicus*)
posticus LaFerté-Sénectère, 1849l: 157 [= 1849n: 256] (*Anthicus*)
olivaceus LaFerté-Sénectère, 1849l: 154 [= 1849n: 253] (*Anthicus*) **E**: IT (Sicilia) PT SP **N**: AG MO TU
ottomanus LaFerté-Sénectère, 1849l: 155 [= 1849n: 254] (*Anthicus*) **E**: GR (Lesvos) TR **A**: TR
merkli Pic, 1897r: 295 (*Anthicus*)
pallicrus Dufour, 1849: 230 (*Anthicus*) **E**: PT SP **N**: AG MO TU **A**: SA
paralleliceps Reitter, 1890d: 151 (*Anthicus*) **E**: AB "Arax valley"
pauperculus LaFerté-Sénectère, 1847: 378 (*Anthicus*) **E**: IT (Sicilia) SP **N**: AG MO TU
lucasi Marseul, 1879a: 163 (*Anthicus*)
pumilus Baudi di Selve, 1877d: 702 [= 1878a: 16] (*Anthicus*)
pierrei Bonadona, 1956: 118 (*Microhoria*) **N**: AG
platiai Degiovanni, 2000: 86 (*Tenuicomus*) **E**: IT (Calabria)
siccensis Normand, 1950b: 107 (*Anthicus*) **N**: TU
subaereus pardoi Bonadona, 1977a: 4 (*Tenuicomus*) **N**: MO
subaereus quaesitus Bonadona, 1977a: 3 (*Tenuicomus*) **N**: MO
subaereus ramirezi Bonadona, 1977a: 5 (*Tenuicomus*) **E**: SP
subaereus subaereus Reitter, 1890d: 151 (*Anthicus*) **N**: AG MO
sericeus Pic, 1896c: 39 (*Anthicus*) [HN]
subcaeruleus lampros Krekich-Strassoldo, 1931: 36 (*Anthicus*) **A**: UP
subcaeruleus subcaeruleus Pic, 1906i: 78 (*Anthicus*) **A**: AF HP KA NP PA SD UP
subsericeus Pic, 1898m: 231 (*Anthicus*) **E**: IT (Sicilia)
tarifanus Pic, 1904a: 4 (*Anthicus*) **E**: SP
tibialis Waltl, 1835: 75 (*Anthicus*) **E**: PT SP
velox bucciarellii Nardi, 1995: 161 (*Tenuicomus*) [RN] **E**: IT
relictus Bucciarelli, 1959: 188 (*Anthicus*) [HN]

velox velox LaFerté-Sénectère, 1849l: 162 [= 1849n: 261] (*Anthicus*) **E:** IT MA MC
viridipennis metallescens Pic, 1899e: 75 (*Anthicus*) **A:** UP
viridipennis viridipennis Pic, 1899e: 75 (*Anthicus*) **A:** UP **ORR**
viturati Pic, 1893s: 160 (*Anthicus*) **N:** AG
obscurithorax Pic, 1896l: 168 (*Anthicus*)

subfamily **Eurygeniinae** LeConte, 1862

tribe Eurygeniini LeConte, 1862

genus *Abdullahia* Abdullah, 1968: 193 type species *Eurygenius africanus* Kolbe, 1897
africana Kolbe, 1897: 263 (*Eurygenius*) **N:** AG MO **AFR**

genus *Neosteropalpus* Abdullah, 1967: 265 type species *Eurygenius niponicus* Lewis, 1895
imasakai Nakane, 1983a: 5 **A:** JA
kyushuensis Nakane, 1975: 168 **A:** JA
niponicus Lewis, 1895a: 423 (*Eurygenius*) **A:** JA
oitaensis kidoi Nakane, 1988a: 6 **A:** JA
oitaensis oitaensis Nakane, 1983a: 5 **A:** JA

genus *Pergetus* Casey, 1895: 636 type species *Eurygenius campanulatus* LeConte, 1874
eous Semenov, 1896c: 361 (*Hypsogenia*) **A:** XIN
mandschuricus Kaszab, 1957a: 49 (*Hypsogenia*) **A:** HEI

genus *Pseudostereopalpus* Abdullah, 1964: 74 type species *Stereopalpus angusticollis* Pic, 1899
angusticollis Pic, 1899g: 76 (*Stereopalpus*) **A:** AP BT NP SD UP XIZ YUN
fulvopictus Champion, 1925: 109 (*Eurygenius*)

genus *Stereopalpus* LaFerté-Sénectère, 1849c: 1 [= 1849n: 4] type species *Stereopalpus mellyi* LaFerté-Sénectère, 1849
Hypsogenia Semenov, 1893e: 332 type species *Hypsogenia centroasiatica* Semenov, 1893
Stereopselaphus Gemminger, 1870b: 2082 [UE]
afghanicus Kaszab, 1957a: 47 (*Hypsogenia*) **A:** AF
asiaticus Pic, 1942b: 12 (*Eurygenius*) **A:** SCH
centroasiaticus Semenov, 1893e: 333 (*Hypsogenia*) **A:** KI KZ TD TM UZ
obscuripes Pic, 1903e: 98 (*Hypsogenia*)
chinensis Abdullah, 1964: 38 **A:** FUJ
dentiventris Saitô & Young, 2016: 43 **A:** JA
femoralis Lewis, 1895a: 424 **A:** JA
gigas Marseul, 1876d: 447 (*Macratria*) **A:** FE JA
kentaroi Saitô, 2017: 184 **A:** JA
minutus Pic, 1942b: 12 **A:** SCH
nagayamai Nakane, 1983a: 6 **A:** JA
tokioensis Nakane, 1983a: 6 **A:** JA

genus *Steriphodon* Abeille de Perrin, 1895a: cclxxi type species *Stereopalpus chobauti* Abeille de Perrin, 1894
bedeli Abeille de Perrin, 1894a: 94 (*Stereopalpus*) **A:** SA YE (Périm) **AFR**
chobauti Abeille de Perrin, 1894a: 94 (*Stereopalpus*) **N:** AG MO TU
harenosus Telnov, 2016: 69 **A:** AE OM
indicum Pic, 1903h: 120 **A:** UP **ORR**

subfamily **Macratriinae** LeConte, 1862

genus *Macratria* Newman, 1838: 377 type species *Macratria linearis* Newman, 1838
Macrarthrius LaFerté-Sénectère, 1849d: 1 [= 1849n: 11] [UE]
albofasciata Champion, 1916c: 209 **A:** BT **ORR**
antennalis Lewis, 1895a: 425 **A:** JA
apicalis Lewis, 1895a: 424 **A:** JA
arabica Uhmann, 1998a: 104 **A:** SA
atrata Nomura, 1962a: 6 **A:** JA (Ryukyu Islands) TAI
basithorax Pic, 1905a: 48 **A:** SHG YUN
bicoloripes Pic, 1923b: 19 **A:** YUN
bipunctata Nomura, 1962a: 6 **A:** JA
chitwana Telnov, 2003b: 284 **A:** NP
cingulifera Marseul, 1876d: 449 **A:** JA
concolor LaFerté-Sénectère, 1849d: 8 [= 1849n: 18] (*Macrarthrius*) **A:** NP **ORR**
dichroa Champion, 1916c: 211 **A:** NP UP **ORR**
dotyali Telnov, 2018b: 242 **A:** NP
flavicornis Champion, 1916c: 214 **A:** AP BT YUN **ORR**
fluviatilis Lewis, 1895a: 425 **A:** JA
freyi Pic, 1938f: 71 **A:** JIA
fumosa Pascoe, 1860a: 56 **A:** UP **ORR**
rubroapicata Pic, 1903h: 120
griseosellata griseosellata Fairmaire, 1893b: 36 **A:** JA NP SCH TAI ZHE **ORR**
sauteri Pic, 1913j: 133
harmandi Pic, 1899g: 77 **A:** NP SD **ORR**
hungarica Hampe, 1873: 165 (*Steropes*) **E:** BH HU
infima Fairmaire, 1896a: 44 **A:** NP **ORR**
japonica Harold, 1877: 359 **A:** JA SC
kopetzi Telnov, 2018b: 244 **A:** NP
leprieuri gasconica Telnov, 2018: 246 **E:** FR
leprieuri leprieuri Reiche, 1864: 243 **N:** AG EG TU **AFR**
brunnipennis Pic, 1899b: 18
gracilis Pic, 1907p: 322
minima Motschulsky, 1858b: 67 (*Macrarthrius*) **A:** UP **ORR**
nankinea Pic, 1909e: 109 **A:** JIA **ORR**
nilgirica Champion, 1916c: 219 **A:** NP **ORR**

pelegrini Pic, 1923d: 28 **A:** JA
robustior Pic, 1923b: 19 **A:** NP SD **ORR**
rufescens Champion, 1916c: 218 **A:** NP SCH SD YUN
serialis Marseul, 1876d: 448 **A:** JA SC TAI
setigera Champion, 1916c: 225 **A:** NP SD **ORR**
sola Telnov, 2002a: 26 **A:** AF

subfamily **Notoxinae** Stephens, 1829

genus *Hypaspistes* Waterhouse, 1886: 39 type species *Hypaspistes armatus* Waterhouse, 1886
Pristocyphus Fairmaire, 1898b: 37 type species *Pristocyphus perrieri* Fairmaire, 1898
suturalis Pic, 1907a: 64 **N:** EG **A:** SA **AFR**
mateui Bonadona, 1959: 122

genus *Mecynotarsus* LaFerté-Sénectère, 1849b: 1 [= 1849n: 57] type species *Notoxus rhinoceros* Fabricius, 1798 (= *Notoxus serricornis* Panzer, 1796)
baeri Pic, 1903r: 644 **A:** TAI **ORR**
bicinctulus Marseul, 1879a: 42 **A:** AF IN IQ PA SY TR
bipustulatus Krekich-Strassoldo, 1928a: 70 **A:** HP NP PA UP **ORR**
bison bison G.-A. Olivier, 1811: 394 (*Notoxus*) **E:** GR (Lesvos) **N:** AG CI EG LB MO TU **A:** AE CY IN IQ IS KU LE OM PA SA SI TR YE **AFR**
algericus Desbrochers des Loges, 1881: 150
beccarii Pic, 1894r: 583
bimaculatus Desbrochers des Loges, 1898b: 4
bordonii Uhmann, 2004: 1
cornutus Pic, 1896m: 178
ferrantei Pic, 1910i: 26
latior Pic, 1897e: 6
macularis Baudi di Selve, 1877d: 661
mellyi Marseul, 1878b: 50
osiris Pic, 1893d: 38
sabulosus Pic, 1893s: 159
semicinctus Wollaston, 1865: 65 (appendix)
coronatus Chobaut, 1898a: 83 **N:** AG **AFR**
picheyrei Peyerimhoff, 1943: 15
fasciatus Motschulsky, 1863: 490 **A:** AF PA **ORR**
nigrobasalis Pic, 1942e: 16
notatus Pic, 1903s: 348
obliquemaculatus Marseul, 1882a: 59 [= 1882b: 117]
faustii Seidlitz, 1891: 591 **E:** AL BU FR (Corse) GG GR IT ME RO ST TR UK "Caucasus" **A:** KI TR UZ
flavipes Pic, 1913j: 129 **A:** TAI

forticornis Kejval, 2011a: 136 **A:** JA
antennalis Hashimoto & Sakai, 2011: 417
fragilis LaFerté-Sénectère, 1849b: 5 [= 1849n: 61] **A:** YUN **ORR**
jebali Kejval, 2017b: 304 **A:** AE
karakumensis Semenov, 1890: 375 **A:** SA TM UZ
ahngeri Semenov, 1900c: 598
minimus Marseul, 1876d: 461 **A:** FE JA NC NO TAI
laticornis Nomura, 1962a: 1
niponicus Lewis, 1895a: 427 **A:** JA
quadrimaculatus Pic, 1913j: 129 **A:** HKG NC NO TAI
sericellus Krekich-Strassoldo, 1931: 1
salam Kejval, 2017b: 306 **A:** SA
serricornis Panzer, 1796a: 17 (*Notoxus*) **E:** AL AU BH BU BY CR CZ FR GE GR HU IT ME PL PT RO SK SP ST SZ YU "Caucasus"
immaculatus Latreille, 1804a: 355 (*Notoxus*)
nigripennis Latreille, 1804a: 355 (*Notoxus*)
notatipennis Pic, 1915m: 41
rhinoceros Fabricius, 1798: 66 (*Notoxus*)
tenuipes Champion, 1891b: 189 **A:** ANH FE FUJ JA JIL LIA NC NO SC SHA SHN **ORR**
latenotatus Pic, 1928b: 18
obscurior Pic, 1943: 4
pallidior Pic, 1943: 4
sinensis Heberdey, 1942: 472
truquii Marseul, 1879a: 41 **N:** EG MO **A:** AE CY IN IS JO LE OM SA SY TR YE **AFR**
alatus Koch, 1935b: 134
lysholmi Pic, 1899j: 172
vagepictus Fairmaire, 1893b: 37 **A:** NP PA SD UP YUN **ORR**
abductus Nardi, 2008: 17
bimaculatus Pic, 1942e: 16 [HN]
vallicola Kejval, 2011a: 165 **A:** NP UP

genus *Notoxus* Geoffroy, 1762: 356 type species *Attelabus monoceros* Linnaeus, 1760
Ceratoderus Blanchard, 1845: 45 type species *Attelabus monoceros* Linnaeus, 1760
Monocerus A. Villa & J.B. Villa, 1833: 20 type species *Notoxus cornutus* Fabricius, 1792
abditus Nardi, 2004: 142 [RN] **A:** "Arabia"
unifasciatus Pic, 1947b: 15 [HN]
adygheicus Telnov, 2016: 58 **E:** AB AR ST (Caucasus) **A:** TR
anatolicus Heberdey, 1936b: 162 **A:** IN TR
angustulus Krekich-Strassoldo, 1919a: 166 **E:** AR GG GR **A:** IN TM TR
appendicinus Desbrochers des Loges, 1874: 165 **E:** AB AL AU BH BU CR CZ FR GE GG GR HU IT LU MC MD ME PL RO SB SK ST SZ TR UK **A:** AF CY IN IS KI KZ LE SY TM TR UZ

moldaviensis Pic, 1909n: 170
reductus Pic, 1911i: 138
separatus Pic, 1900h: 67
armina Kejval & Nardi, 2018: 42 **E:** AR **A:** TR
assamensis Krekich-Strassoldo, 1913b: 129 **A:** AP NP SD UP **ORR**
bangi Pic, 1902a: 1 **A:** KZ TM
quadrimaculatus Heyden, 1887c: 315 [HN]
bellus Krekich-Strassoldo, 1928a: 67 **A:** UP
binotatus Gebler, 1829a: 136 (*Anthicus*) **E:** CT (Chuvashia) **A:** ES FE KI KZ MG NC NMO TIA TM WS
suturalis LaFerté-Sénectère, 1849a: 41
brachycerus Faldermann, 1837: 106 (*Monocerus*) **E:** AB AL AU BH BU BY CR CZ FR GE GG GR HU IT MD PL PT RO SK SP ST SZ TR UK **N:** AG MO TU **A:** KI KZ LE TR
hipponensis Pic, 1894b: 22
major W.L.E. Schmidt, 1842a: 83
notaticollis Pic, 1915j: 13
obliteratus Pic, 1930c: 6
cechovskyi Kejval, 1999: 20 **A:** AF TM UZ
centralasiae Kejval, 1999: 16 **A:** UZ
confluens Krekich-Strassoldo, 1928a: 70 **A:** UP
donckieri Pic, 1908n: 274 **A:** YUN **ORR**
elongatus LaFerté-Sénectère, 1849a: 13 [= 1849n: 33] **E:** AB CT ST **A:** ES KI KZ MG WS
biconjunctus Pic, 1900k: 90
eurasicus Telnov, 2016: 60 **A:** WS
eurycerus Kiesenwetter, 1861a: 244 **E:** GG GR TR **A:** SY TR
histrix Pic, 1893t: 175
excisus Küster, 1848a: 68 **E:** FR PT SP **N:** AG MO TU
bicoronatus Bedel, 1869: 34
cavifrons LaFerté-Sénectère, 1849a: 11 [= 1849n: 31]
hispanicus Motschulsky, 1849: 128
maculatus Pic, 1916g: 13
meridionalis Pic, 1900h: 67
golestanus Kejval, 1999: 18 **A:** IN TM
haagi Marseul, 1879a: 28 **A:** FE JA NC
galloisi Pic, 1915n: 152
hokkaidensis Kôno, 1931: 2
henoni Bedel, 1895: 193 **N:** AG
disjunctus Pic, 1915j: 13
sefrensis Pic, 1915j: 13
hirsutulus Uhmann, 1980b: 163 **A:** KA PA
hirtus bolognai Kejval & Nardi, 2018: 44 **E:** SP **N:** MO

hirtus hirtus LaFerté-Sénectère, 1849a: 8 [= 1849n: 28] **E:** AB AR ST **A:** AF IN KI KZ MG TD TM TR UZ WS XIN XIZ

caucasicus Pic, 1900h: 65

nigrosignatus Pic, 1900h: 65

semipunctatus Reitter, 1888d: 431

inbasaliformis Kejval, 2011b: 646 **A:** AP NP **ORR**

lancifer afghanus Heberdey, 1936a: 143 **A:** AF

lancifer hoggarensis Heberdey, 1936a: 142 **N:** AG

lancifer lancifer G.-A. Olivier, 1811: 394 **A:** AF IN IQ IS JO SA SI UZ

chaldaeus LaFerté-Sénectère, 1849a: 29 [= 1849n: 49]

lancifer orientalis Heberdey, 1936a: 142 **N:** IN TM

lancifer sedilloti Pic, 1894g: 78 **N:** AG MO

bedeli Vauloger de Beaupré, 1895: 193

innotatus Pic, 1919e: 13

signatipennis Pic, 1919e: 13

latus Pic, 1928f: 9 **A:** UZ

lobicornis Reiche, 1864: 244 **E:** PT SP **N:** AG MO TU **A:** SY

impexus Kiesenwetter, 1870: 146

serridens Reitter, 1888d: 431

lonai Bucciarelli, 1973: 476 **E:** FR IT

martini Pic, 1903c: 83 **A:** YE **AFR**

mauritanicus latefasciatus Heberdey, 1936a: 153 **A:** IN TM

mauritanicus maroccanus Heberdey, 1936a: 153 **N:** MO

mauritanicus mauritanicus LaFerté-Sénectère, 1847: 365 **E:** GR PT SP **N:** AG TU

conjunctus Pic, 1900h: 67

lethierryi T. Pic, 1900: 32

semiobliteratus Pic, 1930c: 6

miles italicus Bucciarelli, 1978: 156 **E:** IT

miles miles W.L.E. Schmidt, 1842a: 86 **E:** AL AU BH BU CR GR HU RO SB TR YU

cinctus Pic, 1892a: 4

jacqueti Pic, 1892a: 4

miles rubrithorax Pic, 1900b: 160 **A:** TR

monoceros Linnaeus, 1760: 185 (*Attelabus*) **E:** AL AU BE BH BU BY CR CT CZ DE EN FI FR GB GE GR IT LA LT LU MC MD ME NL NR NT PL PT RO SB SK SL SP ST SV SZ UK TR **A:** ES KI KZ MG TM TR UK UZ WS **AFR**

biinterruptus Pic, 1900h: 66

cucullatus Geoffroy, 1785: 162

immaculatus Pic, 1899b: 19 [HN]

integer LaFerté-Sénectère, 1849a: 10 [= 1849n: 30]

latemaculatus Pic, 1892k: 139

obscuricollis Pic, 1916g: 13

platycerus LaFerté-Sénectère, 1849a: 12 [= 1849n: 32]
septemmaculata Roubal, 1933: 139
subjunctus Pic, 1911e: 97
unicornis Westhoff, 1882: 189
monpa Kejval, 2017d: 51 **A:** AP
montivagus Kejval, 1999: 20 **A:** UZ
noctivagus Krekich-Strassoldo, 1913b: 133 **A:** HP NP PA SD **ORR**
peregrinus Krekich-Strassoldo, 1914c: 215
numidicus Lucas, 1843: 145 (*Monocerus*) **E:** GB **N:** AG EG LB MO TU **A:** IS
aristidis Pic, 1893k: 88
cloueti Chobaut, 1895b: 347
mauritii T. Pic, 1900: 32
pretiosus Hille, 1975: 6 **A:** SA **AFR**
tiganii Uhmann, 1980a: 138
raddei Pic, 1903d: 88 **A:** ES FE MG
raffrayi brittoni Pic, 1957: 436 **A:** OM YE
raffrayi erythreus Pic, 1914q: 183 **A:** YE **AFR**
raffrayi raffrayi Pic, 1894r: 582 **A:** YE **AFR**
reuteri Telnov, 2018b: 258 **A:** TR (Kurdistan)
rubetorum Truqui, 1855: 343 **A:** CY IS LE TR
dilutius Truqui, 1855: 343
leuthneri Pic, 1900k: 90
obscurius Truqui, 1855: 343
sahlbergi Pic, 1898i: 122 **A:** AF KI TM UZ
sareptanus Heberdey, 1936a: 160 **E:** KZ ST **A:** KZ
senegalensis LaFerté-Sénectère, 1849a: 28 [= 1849a: 48] **N:** AG **AFR**
sergeyi Kejval & Nardi, 2018: 51 **A:** IN
siculus LaFerté-Sénectère, 1849a: 14 [= 1849n: 34] **E:** IT (Sicilia)
simulans coffini Bonadona, 1988: 40 [= 1989a: 11] **E:** AU FR IT SZ
simulans simulans Heberdey, 1935: 304 **E:** AB BH ST **A:** AF ES GAN IN KA KI KZ MG PA TD TM TR UZ WS XIN
sinensis Pic, 1907n: 190 **A:** YUN
spatulicornus Telnov, 2016: 62 **A:** NP
strejceki Kejval, 1999: 18 **A:** TD
suturalifer Pic, 1932j: 16 **A:** YUN **ORR**
syriacus LaFerté-Sénectère, 1849n: 298 **N:** EG LB **A:** IS LE SY
transcaspicus Pic, 1899b: 19 **E:** "Caucasus" **A:** AF KZ TM
tricoloratus Telnov, 2016: 63 **E:** AB
trifasciatus Rossi, 1792: 45 **E:** AB AL AR AU BH BU CR CT CZ DE FR GE GG GR HU IT KZ LT MC MD ME NL PL PT RO SB SK SL SP ST SZ TR UK **N:** AG LB MO TU **A:** CY IN IQ IS JO KI KZ LE SY TM TR UZ

armatus W.L.E. Schmidt, 1842a: 86
cornutus Fabricius, 1792a: 211 [HN]
discolor Pic, 1897g: 61
inhumeralis Pic, 1900h: 68
pesruchesi Chobaut, 1897a: 38
rosti Pic, 1905e: 113
sexmaculatus Pic, 1893a: 16
trinotatus Pic, 1894t: x **A:** BEI ES FE GAN HEB HEI JA JIL MG NC NMO SC SHN SHX TIA TM XIN
daimio Lewis, 1895a: 426
mongolicus Pic, 1914m: 74
subobliteratus Pic, 1903d: 88
turbidus Kejval & Nardi, 2018: 53 **A:** AF KA PA TD UZ

genus *Pseudonotoxus* Pic, 1899k: 189 type species *Notoxus testaceus* LaFerté-Sénectère, 1849
championi Krekich-Strassoldo, 1928a: 71 (*Mecynotarsus*) **A:** HP NP UP **ORR**
testaceus ringens Peyerimhoff, 1943: 14 **N:** AG
testaceus testaceus LaFerté-Sénectère, 1849a: 34 [= 1849n: 54] (*Notoxus*) **E:** IT (Sardinia) **N:** AG EG TU **A:** SA **AFR**
feminus Buck, 1965: 192 (*Mecynotarsus*)
photophilus Hille, 1961: 218
viberti Bonadona, 1964: 226 [= 1965: 862] **N:** AG

subfamily **Steropinae** Jacquelin du Val, 1863

genus *Steropes* Steven, 1806: 166 type species *Steropes caspius* Steven, 1806
Blastanus Illiger, 1807: 334 type species *Blastanus colon* Illiger, 1807
caspius Steven, 1806: 166 [= 1811: 129] **E:** AB CT MD ST UK **A:** AF KI KZ MG TM UZ WS
colon Illiger, 1807: 334 (*Blastanus*)
colasi Abdullah, 1966: 209 **A:** NMO
lateralis Kejval, 2012: 140 **A:** FUJ
latifrons Sumakov, 1908a: 133 **E:** ST (Caucasus) **A:** GAN KZ MG TD TIA TM UZ
obscurans Pic, 1894j: 93 **E:** AB GG RO ST UK **A:** AF KZ MG PA TM TR UZ XIZ
pici Abdullah, 1966: 213 **E:** BU GR YU
popei Abdullah, 1966: 213 **E:** GR TR
sinensis Kejval, 2019: 142 **A:** FUJ
vonhayekae Abdullah, 1966: 214 **A:** AE AF IN IQ SY

subfamily **Tomoderinae** Bonadona, 1961

genus *Macrotomoderus* Pic, 1901p: 741 type species *Macrotomoderus latipennis* Pic, 1901
Derarimus Bonadona, 1978: 655 type species *Derarimus carinatus* Bonadona, 1978
bukejsi Telnov, 2018a: 465 **A:** YUN

chingpo Telnov, 2018a: 465 **A:** YUN
clavipes Champion, 1890b: 325 (*Tomoderus*) **A:** JA
besucheti Uhmann, 1990a: 140
conus Telnov, 2018a: 467 **A:** YUN
darrenmanni Telnov, 2018a: 468 **A:** YUN
gracilis Telnov, 2018a: 469 **A:** YUN
jiuhuanus Telnov, 2007b: 92 **A:** ANH
kawa Telnov, 2018a: 470 **A:** YUN
kurbatovi Telnov, 1998a: 82 **A:** HUB (*Derarimus*)
makarovi Telnov, 2018a: 470 **A:** ZHE
microscopicus Telnov, 2018a: 472 **A:** YUN
mirabilis Telnov, 2018a: 472 **A:** YUN
monstratus Telnov, 2018a: 473 **A:** YUN
monstrificabilis Telnov, 2018a: 474 **A:** YUN
negator Telnov, 2007b: 94 **A:** SCH/SHA [border area]
nigripennis Uhmann, 1994b: 663 **A:** TAI
perforatus Telnov, 2018a: 475 **A:** YUN
periclitatus Telnov, 2018a: 476 **A:** YUN
satoi Saitô, 2003b: 321 (*Tomoderus*) **A:** JA (Ryukyu Islands)
schuelkei Telnov, 2018a: 477 **A:** YUN
sichuanus Telnov, 1998a: 84 **A:** SCH (*Derarimus*)
silvicolus Telnov, 2018a: 477 **A:** YUN
spurisi Telnov, 2018a: 478 **A:** YUN
uhmanni Telnov, 1998a: 85 **A:** HUB SCH (*Derarimus*)
wuliangshan Telnov, 2018a: 479 **A:** YUN
yunnanus Telnov, 1998a: 87 **A:** YUN (*Derarimus*)

genus *Pseudotomoderus* Pic, 1892m: 240 type species *Anthicus compressicollis* Motschulsky, 1839
compressicollis compressicollis Motschulsky, 1839: 59 (*Anthicus*) **E:** AB AL AU BH BU CR FR GG GR HU IT PT RO SK SP ST SZ UK **N:** AG MO TU **A:** AF IN IQ MG NP TR UZ WS **AFR ORR**
melanophthalmus LaFerté-Sénectère, 1842: 255 (*Anthicus*)
harwoodi Blair, 1923a: 124 (*Tomoderus*) **A:** IQ
schawalleri Uhmann, 1990b: 335 **A:** NP

genus *Rimaderus* Bonadona, 1978: 649 type species *Rimaderus mussardi* Bonadona, 1978
meaticollis Bonadona, 1978: 651 **A:** NP SD **ORR**
montivagus Telnov, 2003b: 280 **A:** NP

genus *Tomoderus* LaFerté-Sénectère, 1849h: 1 [= 1849n: 94] type species *Tomoderus cruciatus* LaFerté-Sénectère, 1849
Curtotomoderus Pic, 1900j: 89 type species *Tomoderus brevipennis* Pic, 1900
Sirambeus Pic, 1901p: 802 type species *Tomoderus depressipennis* Pic, 1901

bosnicus Pic, 1892m: 242 **E:** BH GR
dalmatinus Reitter, 1881a: 224 **E:** AL BH CR HU GR MC SB
ehlersi Heyden, 1882: 47 **E:** SP
fasciatonitidus Telnov, 2001: 119 **A:** NP SD
fortepunctatus Uhmann, 1988c: 692 **A:** SD
italicus Marseul, 1879a: 61 **E:** IT
lineatopunctatus Uhmann, 1994b: 666 **A:** TAI
major Pic, 1895f: 174 **E:** GG **A:** TR
martensi Uhmann, 1982: 233 **A:** NP SD
nitidus Pic, 1894a: 16 **N:** EG
piochardi Heyden, 1870: 147 **E:** PT SP
praemontanus Telnov, 2001: 121 **A:** SD
reitteri Heberdey, 1938: 161 **A:** XIN
scydmaenoides Reitter, 1878a: 249 **E:** AB GG ST (Caucasus) **A:** IN
smetanai Uhmann, 1989: 245 **A:** NP SD
schmidti Telnov, 2018b: 264 **A:** NP
sulcicollis LaFerté-Sénectère, 1849h: 3 [= 1849n: 96] **A:** NP SD UP **ORR**
syriacus Pic, 1907e: 138 **A:** IS SY TR
  innotatipennis Pic, 1919d: 10
tibialis Uhmann, 1988c: 696 **A:** SD
uhmanni Chandler, 2004: 142 [RN] **A:** TAI
  bimaculatus Uhmann, 1988c: 396 [HN]
unifasciatus Pic, 1907r: 341 **A:** SD UP YUN **ORR**
ventralis Marseul, 1878b: 50 **N:** AG TU

family ISCHALIIDAE Blair, 1920

genus *Ischalia* Pascoe, 1860a: 54 type species *Ischalia indigacea* Pascoe, 1860
subgenus *Ischalia* Pascoe, 1860a: 54 type species *Ischalia indigacea* Pascoe, 1860
akaishi Telnov & Barclay, 2019: 786 **A:** JA (Honshu)
anhuiensis Kazantsev & Young, 2010: 307 **A:** ANH
arakii Saitô, 2003: 55 **A:** JA (Hokkaido)
arisana Kôno, 1935c: 158 **A:** TAI
brachyptera Nikitsky, 1994a: 35 **A:** FE SC
chinensis Young, 1976: 213 **A:** JIX ZHE
horii Saitô & Young, 2015a: 177 **A:** JA (Hokkaido)
kitanoi Saitô, 2011: 310 **A:** JA (Shikoku)
kunashirica Nikitsky, 1994a: 33 **A:** FE (Kunashir)
lama Kazantsev & Young, 2010: 309 **A:** SCH
latemarginata Ohbayashi & Tôyama, 1994: 146 **A:** TAI
luteolineata Pic, 1912p: 142 **A:** JA (Honshu)
martensi Paulus, 1972: 78 **A:** AP NP
nepalensis Paulus, 1972: 78 **A:** NP

patagiata Lewis, 1879: 463 **A:** FUJ JA LIA SC YUN **ORR**
takane Saitô, 1994: 339 **A:** JA (Honshu)
toshikoae Saitô, 2011: 308 **A:** JA (Shikoku)
uenoi Satô, 1990: 102 **A:** TAI
subgenus *Eupleurida* LeConte, 1862: 267 type species *Eupleurida costata* LeConte, 1862
aptera Gusakov & Telnov, 2007: 40 **A:** SCH
sichuanensis Young, 2008: 326 **A:** GAN SCH

family ADERIDAE Csiki, 1909

genus *Aderus* Stephens, 1829b: 255 type species *Lytta boleti* Marsham, 1802 (= *Notoxus populneus* Creutzer, 1796)
Hylophilus Berthold, 1827: 375 [HN] type species *Notoxus populneus* Creutzer, 1796
Phomalus Casey, 1895: 785 type species *Xylophilus brunnipennis* LeConte, 1875
Xylophilus Latreille, 1829: 73 [HN] type species *Notoxus populneus* Creutzer, 1796
diversimembris Pic, 1954a: 55 (*Hylophilus*) **A:** FUJ JA
flavus Fairmaire, 1863: 644 (*Xylophilus*) [DA] **N:** AG
gibbulus Marseul, 1876d: 454 (*Xylophilus*) **A:** JA
grouvellei Pic, 1910q: 75 (*Hylophilus*) **A:** JA
laticeps Pic, 1913j: 134 (*Hylophilus*) **A:** JA TAI
populneus Creutzer, 1796: 4 (*Notoxus*) **E:** AB AL AN AR AU BH BY CR CT CY CZ DE EN GB GE GG GR FI FR IT HU LA LT LU MA NL NR PL PT RO SK SP ST SV SZ TR UK **N:** AG EG MO TU **A:** CY IN IS KZ LE SY TM TR UZ **NAR**i
biskrensis Pic, 1893s: 158 (*Xylophilus*)
boleti Marsham, 1802: 486 (*Lytta*)
rufulus Marseul, 1876d: 451 (*Xylophilus*) **A:** JA
yaeyamanus Nomura, 1964: 54 **A:** JA

genus *Anidorus* Mulsant & Rey, 1866a: 22 type species *Xylophilus nigrinus* Germar, 1842
Anidorus Mulsant & Rey, 1866b: 120 [HN]
caucasicus Pic, 1913g: 113 (*Hylophilus*) **E:** "Caucasus"
hipponensis Pic, 1893i: 75 (*Xylophilus*) **N:** AG
nigrinus Germar, 1842: 7 (*Xylophilus*) **E:** AN AU BE BY CT CZ DE EN FI FR GE HU IT LA LT LU NL NR NT PL SK SP ST SV SZ UK
alpestris Pic, 1899d: 51 (*Xylophilus*)
nadeshdae Semenov, 1899a: 613 (*Euglenes*)
patricius Abeille de Perrin, 1872: 37 (*Xylophilus*)
tirolensis Gredler, 1866: 282 (*Xylophilus*)
sanguinolentus Kiesenwetter, 1861a: 241 (*Xylophilus*) **E:** AU BE BH FR IT PT SP **N:** AG MO **A:** "Syria"
lateralis Gredler, 1866: 281 (*Xylophilus*)
tenietensis Pic, 1894e: 61 (*Xylophilus*)

sefrensis Pic, 1894i: 85 (*Xylophilus*) **N:** AG

sefrensis Pic, 1894l: 141 (*Xylophilus*) [HN]

testaceipes Pic, 1892k: 139 (*Xylophilus*) **N:** AG

genus *Ariotus* Casey, 1895: 788 type species *Xylophilus quercicola* Schwarz, 1878

Scanylus Casey, 1895: 800 type species *Scanylus pruinosus* Casey, 1895

takedai M. Saitô & Young, 2015b: 458 **A:** JP

genus *Cnopus* Champion, 1893: 460 type species *Cnopus flohri* Champion, 1893

minor Baudi di Selve, 1877d: 640 (*Xilophilus*) **E:** FR (Corse) IT

corsicus Pic, 1928a: 2 (*Hylophilus*)

minor Baudi di Selve, 1878a: 1 (*Xylophilus*) [HN]

pinicola Champion, 1919: 244 **A:** UP

genus *Cobososia* Collado & Alonso-Zarazaga, 1996: 98 type species *Hylophilus fulvicolor* Escalera, 1942

angulithorax Desbrochers des Loges, 1881: 146 (*Xylophilus*) **E:** IT (Sardinia) SP (Gibraltar) **N:** AG CI (La Palma, Tenerife) EG MO TU

benimorae Pic, 1893s: 157 (*Xylophilus*)

aristidis Pic, 1893i: 75 (*Xylophilus*) **N:** EG **A:** "Syria" **AFR**

pallescens Wollaston, 1854: 538 (*Xylophilus*) **E:** FR IT **N:** AG CI MO MR **A:** CY LE SY ("Kaifa") **AFR**

pallens Desbrochers des Loges, 1881: 147 (*Xylophilus*)

punctigera Mulsant & Rey, 1866a: 23 (*Xylophilus*)

punctigera Mulsant & Rey, 1866b: 111 (*Xylophilus*) [HN]

pruinosa pruinosa Kiesenwetter, 1861a: 241 (*Xylophilus*) **E:** AB AU BU CR CT CZ FR GE GG GR HU IT PT RO SK SP ST SZ UK "Caucasus" **N:** AG EG TU **A:** IN KI KZ TR UZ

montandoni Pic, 1925f: 13 (*Hylophilus*)

moreana Pic, 1925f: 13 (*Hylophilus*)

obscura Pic, 1892b: 27 (*Xylophilus*)

testacea Baudi di Selve, 1877c: 415 (*Xylophilus*) [HN]

pruinosa uniformis Pic, 1925f: 13 (*Hylophilus*) **E:** "Caucasus" **N:** EG

semiobscura Pic, 1893b: 26 (*Euglenes*) **E:** GG ST "Caucasus" **A:** "Turkestan"

symphoniaca Klinger, 2000: 277 **E:** AU BU GE HU

genus *Euglenes* Westwood, 1830: 59 type species *Anthicus oculatus* Paykull, 1798

Xylophilus Curtis, 1830: 299 [HN] type species *Anthicus oculatus* Paykull, 1798

kabylianus Pic, 1896c: 38 (*Xylophilus*) **N:** AG

klapperichi Pic, 1956: 6 **A:** AF

maronitus Pic, 1894q: 435 (*Xylophilus*) **A:** IS LE

nitidifrons C.G. Thomson, 1886: x (*Xylophilus*) **E:** AU CZ DE GE FI NR SV

oculatissimus Wollaston, 1864: 525 (*Xylophilus*) **N:** CI

oculatus Paykull, 1798: 256 (*Anthicus*) **E:** AU BE BY CT DE FI FR GE GB HU LA LT LU NL NR PL PT SP ST SV SZ
nigricollis Marsham, 1802: 487 (*Lytta*)
pygmaeus DeGeer, 1775: 80 (*Cerambyx*) **E:** AU BY CT CZ EN FI FR GE HU IT LA LT LU NR NT PL PT SK SP ST SV SZ **NARi**
fennicus Mannerheim, 1843b: 98 (*Euglenes*)
ferrugineus Paykull, 1798: 257 (*Anthicus*)
fulvus G.-A. Olivier, 1795: 51: 5 (*Notoxus*)
lokvenci Roubal, 1938: 40 (*Anidorus*)
melanocephalus Creutzer, 1796: 5 (*Notoxus*)
serricornis Reitter, 1885d: 384 **E:** RO
trisignatus Pic, 1956: 6 **A:** AF
wollastoni Israelson, 1971: 117 **N:** CI

genus *Gompelia* Alonso-Zarazaga, 2010 [RN] type species *Xylophilus neglectus* Jacquelin du Val, 1863.
Olotelus Mulsant & Rey, 1866a: 22 [HN] type species *Xylophilus neglectus* Jacquelin du Val, 1863
Olotelus Mulsant & Rey, 1866b: 110 [HN] type species *Xylophilus neglectus* Jacquelin du Val, 1863
atlasica Pic, 1932b: 28 (*Hylophilus*) [= 1932g: 28] **N:** MO
atomus A. Costa, 1884: 167 (*Euglenes*) **E:** IT (Sardinia)
brunneomaculata Pic, 1907m: 182 (*Hylophilus*) **A:** HP PA UP
diversithorax Pic, 1915n: 153 (*Hylophilus*) **A:** JA
drusus Pic, 1894q: 435 (*Xylophilus*) **A:** "Antiliban" "Syria"
flaveola Mulsant & Rey, 1866a: 26 (*Xylophilus*) [= 1866b: 114] **E:** CR FR GR IT MA SP SZ
galloisi Pic, 1910h: 23 (*Hylophilus*) **A:** JA
gratiosa alfierii Pic, 1924f: 30 (*Hylophilus*) **N:** EG
gratiosa gratiosa Pic, 1894q: 435 (*Xylophilus*) **N:** EG **A:** LE "Syria"
harmandi Pic, 1910h: 23 (*Hylophilus*) **A:** JA
longissima Pic, 1909b: 39 (*Hylophilus*) [DA] **A:** HP "Himalaya"
neglecta Jacquelin du Val, 1863: pl. 85 (*Xylophilus*) **E:** FR IT PT SP SZ **N:** AG EG MO TU
algirica Pic, 1897i: 81 (*Xylophilus*)
bicolor Schaufuss, 1882b: 536 (*Xylophilus*)
filicornis Schaufuss, 1882b: 536 (*Xylophilus*)
humeralis Favre, 1890: 249 (*Xylophilus*) [HN]
nigriceps Pic, 1899i: 172 (*Xylophilus*) **A:** LE TR "Syria"
obscurithorax Pic, 1899i: 172 (*Xylophilus*) **E:** GR (Rhodes)
reitteri Pic, 1897i: 81 (*Xylophilus*) **E:** AB (Araz Valley) "Caucasus"
atricornis Pic, 1900a: 12 (*Hylophilus*)
rosti Pic, 1909h: 118 (*Hylophilus*) **A:** HP "Himalaya"
ruficollis Rossi, 1794: 96 (*Notoxus*) **E:** CR GG GR IT "Caucasus" PT SP **N:** EG **A:** CY IS JO LE TR "Syria"

fulvipennis Pic, 1898l: 224 (*Xylophilus*) [RN]
nigripennis A. Villa & J.B. Villa, 1838: 63 (*Anthicus*)
nodicollis Reitter, 1885d: 384 (*Euglenes*)
subobscura Pic, 1910b: 1 (*Hylophilus*)
testaceipennis Pic, 1897i: 81 (*Xylophilus*) [HN]
testacea Kolenati, 1846: 38 (*Xylophilus*) **E:** RU "Transcaucasus"

genus *Mixaderus* Collado & Alonso-Zarazaga, 1996: 98 type species *Hylophilus simplicis* Escalera, 1942
Ginerosia Collado & Alonso-Zarazaga, 1996: 98 type species "*Mixaderus*" *infans* Baguena-Corella, 1948
Perisia Collado & Alonso-Zarazaga, 1996: 98 type species *Xylophilus clavicornis* Champion, 1917
cooperi Báguena Corella, 1948: 418 ("*Mixaderus* Báguena Corella") **N:** "EG (Zeltoum)"
tamaii Nomura, 1964: 54 ("*Mixaderus* Báguena Corella") **A:** JA

genus *Phytobaenus* R.F. Sahlberg, 1834: 276 type species *Phytobaenus amabilis* R.F. Sahlberg, 1834
amabilis amabilis R.F. Sahlberg, 1834: 277 **E:** BY CR CT CZ EN FI FR GE HU IT LA LT NT PL RO SK SV UK **A:** ES FE JA TAI "NE China"
bisbimaculatus Hampe, 1850: 356 (*Xylophilus*)
amabilis amamiensis Nomura, 1964: 51 **A:** JA (Ryukyus)
amabilis scapularis Marseul, 1876d: 453 (*Xylophilus*) **A:** JA

genus *Picemelinus* Nakane, 1987: 175 type species *Hylophilus flabellicornis* Pic, 1910
flabellicornis Pic, 1910h: 22 (*Hylophilus*) **A:** JA

genus *Pseudanidorus* Pic, 1893s: 158 type species *Xylophilus monstrosipes* Pic, 1893
Pseudeuglenes Pic, 1897i: 81 type species *Xylophilus pentatomus* C.G. Thomson, 1864
cyprius Baudi di Selve, 1877d: 641 (*Xilophilus* [sic!]) **A:** CY
cyprius Baudi di Selve, 1877c: 414 (*Xylophilus*) [HN]
laesicollis Fairmaire, 1883h: cxiii (*Xylophilus*) **E:** FR (Corse) **N:** AG
monstrosipes Pic, 1893s: 157 (*Xylophilus*)
pentatomus C.G. Thomson, 1864: 371 (*Xylophilus*) **E:** CT CZ DE EN FI FR GE IT NR NT PL SV
quinquetomus Seidlitz, 1891: 546 (*Euglenes*)
rossicus Semenov, 1899a: 614 (*Euglenes*)
peyroni Pic, 1900a: 19 (*Hylophilus*) **A:** LE "Syria"
rubrivestis Marseul, 1876d: 450 (*Xylophilus*) **A:** JA

genus *Pseudolotelus* Pic, 1901g: 67 type species *Euglenes punctatissimus* Reitter, 1885
cinctus Marseul, 1876d: 452 (*Xylophilus*) **A:** JA
curtipennis Pic, 1899i: 171 (*Xylophilus*) **A:** TR
distortus Champion, 1890a: 267 (*Xylophilus*) **A:** JA JIL
humeralis Nomura, 1964: 52 **A:** JA
japonicus Champion, 1890a: 268 (*Xylophilus*) **A:** JA

ogasawarensis Nakane, 1991: 8 **A:** JA
punctatissimus Reitter, 1885d: 384 (*Euglenes*) **E:** CR GR
kruperi Pic, 1901n: 178 (*Hylophilus*)
syriacus Pic, 1896e: 51 (*Xylophilus*) **A:** LE "Syria"

genus *Shigaderus* M. Saitô & Young, 2015b: 454 type species *Shigaderus nakagawayui* M. Saitô & Young, 2015
nakagawayui M. Saitô & Young, 2015b: 454 **A:** JA

genus *Syzeton* Blackburn, 1891: 337 type species *Syzeton laetus* Blackburn, 1891 (= *Anthicus abnormis* King, 1873)
brunnidorsis Marseul, 1876d: 450 (*Xylophilus*) **A:** JA
quadrimaculatus Marseul, 1876d: 452 (*Xylophilus*) **A:** JA TAI
shibatai Nomura, 1964: 53 **A:** JA

genus *Tokiophilus* Pic, 1921h: 415 type species *Tokiophilus applicatus* Pic, 1921
applicatus Pic, 1921h: 415 **A:** JA

genus *Vanonus* Casey, 1895: 791 type species *Vanonus wickhami* Casey, 1895
Tanilotes Casey, 1895: 798 type species *Tanilotes lacustris* Casey, 1895 (= *Vanonus wickhami* Casey, 1895)
brevicornis brevicornis Perris, 1869: 20 (*Xylophilus*) **E:** BE CZ DE FR GB GE IT NL PL SV
omissus Pic, 1948c: 7 (*Hylophilus*)
brevicornis rotundaticollis Israelson, 1971: 112 **E:** CI (La Palma)

family Aderidae, species incertae sedis
Xylophilus basifasciatus Champion, 1924b: 489 **A:** UP
Xylophilus bifossulatus Champion, 1924b: 491 **A:** UP
Xylophilus breviscopa Champion, 1924b: 486 **A:** UP
Xylophilus bulbifer Champion, 1919: 243 **A:** UP
Xylophilus chinensis Champion, 1916a: 43 **A:** CH
Hylophilus formosanus Pic, 1913j: 133 **A:** TAI
Xylophilus haldwaniae Champion, 1924b: 486 **A:** UP
Hylophilus hedini Pic, 1933b: 9 **A:** XIN "NW China"
Xylophilus himalaicus Champion, 1920b: 405 **A:** UP
Xylophilus holosericeus Champion, 1916a: 28 **A:** HP
Xylophilus kumaonae Champion, 1924b: 488 **A:** UP
Xylophilus latidens Champion, 1924b: 489 **A:** UP
Xylophilus luteitarsis Champion, 1924b: 484 **A:** SD UP
Xylophilus melanostictus Champion, 1924b: 487 **A:** UP
Xylophilus parvidens Champion, 1916a: 42 **A:** ZHE

Xylophilus pectinatus Champion, 1916a: 40 **A:** HKG
Hylophilus praescutellaris Pic, 1914s: 449 **A:** "Korea"
Hylophilus quadratipennis klapperichi Pic, 1954a: 55 **A:** FUJ
Xylophilus quadratipennis quadratipennis Champion, 1916a: 44 **A:** CH
Hylophilus sauteri Pic, 1913j: 134 **A:** TAI
Xylophilus seminiger Champion, 1924b: 485 **A:** UP **ORR**
Hylophilus singularicornis Pic, 1910h: 22 **A:** JA
Xylophilus spinimanus Champion, 1916a: 42 **A:** HKG
Hylophilus subnitidus Pic, 1924e: 27 [RN] **A:** SD
harmandi Pic, 1921h: 417 (*Hylophilus*) [HN]
Xylophilus tetrastictus Champion, 1924b: 487 **A:** UP
Xylophilus trapeziderus Champion, 1924b: 485 **A:** UP

family SCRAPTIIDAE Gistel, 1848
subfamily **Scraptiinae** Gistel, 1848
tribe Scraptiini Gistel, 1848

genus *Scraptia* Latreile, 1807: 199 type species *Melyris dubius* G.-A. Olivier, 1790
adenensis Franciscolo, 1956c: 453 **A:** YE
alutacea Reitter, 1889g: 268 **N:** MO
angustior Pic, 1944a: 9 **A:** JA
aureopubens Pic, 1903h: 119 **A:** SD
bifoveolata Küster, 1853: 90 **E:** GR IT **A:** TR
clairi Rey, 1892: 77
caucasicola Roubal, 1924: 182 **E:** ST
chinensis Pic, 1904c: 20 **A:** CH
cribriceps Champion, 1916d: 243 **A:** SE
curta Pic, 1944a: 9 **A:** JA
distincta Pic, 1944a: 10 **A:** "China"
dubia G.-A. Olivier, 1790: 21: 12 (*Melyris*) **E:** CR FR HU IT SK SL SP SZ
fusca Latreille, 1804a: 342 (*Melandrya*)
fuscula Gyllenhal, 1810: 525 (*Dircaea*)
sericea Gyllenhal, 1817: 49 (*Dircaea*)
subdubia Reitter, 1889g: 267
ferruginea Kiesenwetter, 1861b: 387 **E:** FR HU IT SK SL SZ
forticornis Champion, 1916d: 236 **A:** JA
fuscula P.W.J. Müller, 1821: 202 **E:** AU BY CZ DE EN FI FR GB HU IT LA LU NR NL PL SP ST SV UK **N:** AG MO
algirica Pic, 1898h: 122
innata Kangas, 1959: 188 (*Hallomenus*)

minuta Mulsant, 1856d: 142
nigricans Stephens, 1832: 39
nigriceps Heyden, 1870: 145
jakowleffi Reitter, 1889g: 268 **E:** ST
libanica Pic, 1917a: 2 **A:** IS LE
livens Marseul, 1876d: 456 **A:** JA
longicornis Kiesenwetter, 1861b: 243 **E:** GR
melina Champion, 1916d: 244 **A:** SD UP
nilgiriana Champion, 1916d: 246
mozabita Pic, 1898h: 123 (*Canifa*) **N:** AG
ophthalmica Mulsant, 1856d: 141 **E:** FR IT PT SP **N:** MO
pallescens Champion, 1916d: 245 **A:** SD
quadrisignata Champion, 1916d: 250 **A:** SD
roubali Winkler, 1927: 829 [RN] **E:** BU
forticornis Roubal, 1924: 181 [HN]
schotti Leblanc, 2012: 50 **E:** FR (Corse) **N:** AG MO TU
straminea Peyerimhoff, 1931b: 85 **N:** AG
testacea Allen, 1940: 56 **E:** GB HU FR SP SV **N:** AG
thoracica Baudi di Selve, 1877a: 83 **N:** MO
subfoveolata Reitter, 1889g: 267
trotommoides Pic, 1892a: 4 **A:** SY

genus *Trotomma* Kiesenwetter, 1851: 623 type species *Trotomma pubescens* Kiesenwetter, 1851
subgenus *Macrotrotomma* Pic, 1917a: 3 type species *Trotomma antennatum* Pic, 1902
antennatum Pic, 1902j: 80 **E:** IT
subgenus *Trotomma* Kiesenwetter, 1851: 623 type species *Trotomma pubescens* Kiesenwetter, 1851
baborensis Pic, 1902e: 32 **N:** AG MO
impressicolle Pic, 1917a: 3 **E:** GR
latipenne Pic, 1905f: 130 **N:** TU
longipenne Pic, 1910k: 36 **N:** AG
pubescens Kiesenwetter, 1851: 624 **E:** FR GR IT **N:** AG
tuniseum Pic, 1917g: 5
vaulogeri Pic, 1917a: 3 **N:** TU

genus *Trotommidea* Reitter, 1883b: 307 type species *Trotommidea salonae* Reitter, 1883
revelierei elongata Pic, 1896e: 51 **E:** GR **N:** AG TU **A:** LE
oertzeni Schilsky, 1900: 28 (*Scraptia*)
revelierei revelierei Abeille de Perrin, 1885: 157 (*Scraptia*) **E:** FR IT
baudii Pic, 1901g: 67
ruficeps Pic, 1932b: 27 **N:** MO

salonae Reitter, 1883b: 307 **E:** CR GR HU FR

corcyrea Pic, 1901e: 52

subfamily **Anaspidinae** Mulsant, 1856

tribe Anaspidini Mulsant, 1856

genus *Anaspis* Geoffroy, 1762: 315 type species *Mordella frontalis* Linnaeus, 1758

subgenus *Anaspis* Geoffroy, 1762: 315 type species *Mordella frontalis* Linnaeus, 1758

Plesianaspis A. Costa, 1854: 28 type species *Mordella thoracica* Linnaeus, 1758

acanthura Ermisch, 1962: 167 **A:** KI KZ UZ

apfelbecki Schilsky, 1894a: 98 **E:** BH SL YU

teratopyga Franciscolo, 1949c: 94

arctica Zetterstedt, 1828: 290 **E:** CT EN FI GE LA NR NT PL RO SK SV **A:** ES FE MG NE NC WS

imitata Schilsky, 1899: QQ

nigriceps Schilsky, 1895: 70

ruficeps Zetterstedt, 1840: 166

tenenbaumi Pic, 1928e: 5

atriventris Pic, 1932i: 32 **A:** GAN

auripes Ermisch, 1962: 165 **A:** KZ

basalis Emery, 1876: 118 **E:** AR

bohemica Schilsky, 1899: 88 **E:** AU CZ DE EN GB GE HU LA NL NR NT PL SV

forticornis Schilsky, 1899: 89

fortipes Reitter, 1911a: 379

norvegica Munster, 1924: 293

emarginata Schilsky, 1895: 64 **A:** MG

excellens Schilsky, 1908a: 100 **E:** HU

fasciata Forster, 1771: 65 (*Mordella*) **E:** AL BE CR DE FR GB GE IR IT LU NL PL PT SL SP SZ **N:** AG **A:** MG

bicolor Geoffroy, 1785: 142

biguttata Rossi, 1794: 103 (*Mordella*)

biinterrupta Pic, 1931c: 6

bipunctata Ragusa, 1898: 208 [HN]

bisbimaculata Pic, 1910b: 1 [RN]

cruciata A. Costa, 1854: 25

discicollis A. Costa, 1854: 25

fulvicollis Schilsky, 1895: 76

geoffroi P.W.J. Müller, 1821: 214

humeralis Fabricius, 1775: 264 (*Mordella*)

nigricollis Marsham, 1802: 492 (*Mordella*)

quadrimaculata A. Costa, 1854: 25 [HN]

quadrinotata Stephens, 1832: 46

quadripustulata Stephens, 1832: 46
scapularis Stephens, 1832: 47
sicula Pic, 1910b: 1 [RN]
subfasciata Stephens, 1832: 45
vulcanica A. Costa, 1854: 27

frontalis Linnaeus, 1758: 420 (*Mordella*) **E:** AU BE BH BU BY CT CZ DE EN FI FR GB GE GR HU IR IT LA LT LU MC NL NR NT PL RO SK SL SP ST SV SZ UK **A:** ES FE JA LE MG SC TR WS
abollata Gozis, 1882: 201
assimilis Snellen van Vollenhoven, 1854: 70
flavifrons Eschscholtz, 1818: 471 (*Mordella*)
flavoatra Letzner, 1857: 119
hopffgarteni Schilsky, 1895: 61
lateralis Gyllenhal, 1810: 616 (*Mordella*) [HN]
maculicollis Motschulsky, 1860c: 110
nigra Rossi, 1792: 55 (*Mordella*)
punctata Eschscholtz, 1818: 472 (*Mordella*)
verticalis Faldermann, 1837: 112

funagata Kôno, 1928b: 150 **A:** JA

garneysi Fowler, 1889: 333 **E:** DE FR GB GE IR LA NL SP SV SZ
bilaciniata Pic, 1918c: 14

graeca Schilsky, 1895: 62 **E:** BU GR

helvetica Pic, 1918c: 14 **E:** SZ

hispanica Schilsky, 1899: 90 **E:** SP

incognita Schilsky, 1895: 74 **E:** FR GR IT SP

inderiensis Motschulsky, 1860c: 142 **A:** KZ

infuscata Motschulsky, 1860c: 142 **A:** ES FE JA SC

kiesenwetteri Emery, 1876: 15 **E:** AL AU BH BU CR CZ GE HU PL RO SK SZ
melanaria Schilsky, 1899: RR
picea Schilsky, 1895: 70

koenigi Schilsky, 1899: 87 **A:** "Kurdistan"

lucana Franciscolo, 1943b: 76 **E:** IT

lurida Stephens, 1832: 44 **E:** AL AU BU CR CT FI FR GB GE GR HU IR IT LA LU NL SZ **A:** IS SY TR
analis A. Costa, 1854: 30 (*Plesianaspis*)
flava A. Costa, 1854: 29 (*Plesianaspis*)
fusca Schrank von Paula, 1781: 299 (*Mordella*)
subtestacea Stephens, 1832: 45
testacea Marsham, 1802: 493 (*Mordella*) [HN]
unicolor A. Costa, 1854: 30 (*Plesianaspis*)

maculata Geoffroy, 1785: 142 (*Mordella*) [NP] **E:** BE CR CZ DE FR GB GE IR IT LU NL PT SL SP SV SZ
bipunctata Bonelli, 1812: 173 (*Mordella*)
bimaculata Rossi, 1792: 245 (*Mordella*)
exoleta A. Costa, 1854: 27
florenceae Donisthorpe, 1928: 121
innotata Schilsky, 1892: 203
melanopa Forster, 1771: 64 (*Mordella*) [NO]
nigricollis Marsham, 1802: 492 (*Mordella*)
obscura Marsham, 1802: 492 (*Mordella*)
pallida Marsham, 1802: 492 (*Mordella*)
picta Hampe, 1871: 333 (*Silaria*)
testacea Snellen van Vollenhoven, 1854: 70
maroccana Fleischer, 1909: 241 **N:** MO
marseuli Csiki, 1915: 69 [RN] **A:** JA TAI
sericea Marseul, 1877: 477 [HN]
mongolica Ermisch, 1964b: 350 **A:** MG
nigripes C. Brisout de Barneville, 1866: 384 **E:** AB AR BU FR GG HU IT SP ST SZ
baudii Schilsky, 1894a: 97
curvicrus Schilsky, 1895: 65
nigritula Schilsky, 1894b: 331
orientalis Faldermann, 1837: 113 **E:** AB
oshimana Nomura, 1964: 42 **A:** JA
picimana Gebler, 1829a: 114 **A:** KZ XIN
poecila Faldermann, 1837: 114 **E:** AB
poggii Franciscolo, 1982: 198 **E:** IT
pulicaria A. Costa, 1854: 22 **E:** AB AL AU BH BU CR FR GB GE GG GR HU IT LU MC NL PT SL SP ST SZ YU **N:** AG MO TU **A:** IS
curtii Roubal, 1912c: 401
forcipata Mulsant, 1856f: 402
lateralis Stephens, 1832: 43
pyrenaea Fairmaire & C. Brisout de Barneville, 1859: 54 **E:** AL FR
regimbarti Schilsky, 1895: 69 **E:** BE DE FR GB GE IR IT NL PT SP SV SZ
fraudulenta Joy, 1912: 115
ruficollis Fabricius, 1792b: 115 (*Mordella*) **E:** AL AU BE BH BU DE FR GE IT LA LT NL NR PL SK SL SP ST SZ UK **A:** SY
alpicola Emery, 1876: 22
bickhardti Schilsky, 1906: 25
thoracica Herbst, 1784c: 149 (*Mordella*) [HN]
silvatica Gabriel, 1916: 183 **E:** PL

stussineri Fleischer, 1909: 242 **E:** AL IT MC
albanica Csiki, 1940a: 260
subsulcata Motschulsky, 1845b: 368 **A:** ES
thoracica Linnaeus, 1758: 420 (*Mordella*) **E:** AU BE BH BY CR CT CZ DE EN FI FR GB GE HU IR IT LA LT LU NL NR NT PL RO SK SL SP SV SZ UK YU **A:** ES TR
confusa Emery, 1876: 14
gerhardti Schilsky, 1892: 201
lateralis C.G. Thomson, 1864: 302 [HN]
latipalpis Schilsky, 1895: 73
marginicollis Håk. Lindberg, 1925: 22
martialis Pic, 1918c: 14
notatithorax Pic, 1918c: 14
obscuricolor Stchegoleva-Barovskaya, 1931: 61
occipitalis Stchegoleva-Barovskaya, 1931: 62
schilskyana Hellén, 1935: 22
scutellata Schilsky, 1899: OO
septentrionalis Champion, 1891a: 104
vaulogeri Chobaut, 1896a: 200 **N:** AG
versicolor Baudi di Selve, 1878b: 341 **E:** GR IT
ornata Schilsky, 1895: 98 (*Silaria*)
viennensis Schilsky, 1895: 66 **E:** AU BH BU CR CZ FR GR HU IT RO SK ST UK YU
subgenus *Larisia* Emery, 1876: 13 type species *Anaspis rufitarsis* Lucas, 1854
akaira Franciscolo, 1991: 173 **E:** IT
bernikovi bernikovi Apfelbeck, 1931: 37 **E:** BH
bernikovi levisternalis Roubal, 1935b: 107 **E:** RO
caucasica Schilsky, 1894a: 100 **E:** AR
lurida Schilsky, 1894a: 100
cavipalpis Apfelbeck, 1931: 37 **E:** MC
chevrolati Mulsant, 1856f: 434 **E:** GE IT PT SP **N:** AG MO **A:** IS
aemula Schilsky, 1895: 92
ferruginea Schilsky, 1895: 92
nigripennis Schilsky, 1899: EE, GG
tristis Schilsky, 1899: GG
dichroa Emery, 1876: 15 **E:** FR GR IT TR **A:** CY IS SY TR
fausti Schilsky, 1899: 93 **A:** KI TM
ganglbaueri Schilsky, 1899: 100 **E:** CR IT
heydeni Csiki, 1915: 58 **A:** TD UZ
flavipennis Heyden, 1893: 329 [HN]
horni Schilsky, 1899: 99 **E:** FR PT
discolor Schilsky, 1899: 99
luridipennis Ermisch, 1962: 168 **A:** TM UZ

mariae Roubal, 1910: 154 **E:** BU

meridionalis Franciscolo, 1956c: 465 **A:** YE

mulsanti C. Brisout de Barneville, 1859: 234 **E:** FR GE IT SP **N:** MO

eppelsheimi Schilsky, 1895: 91

nigriventris Schilsky, 1899: 94 **A:** TR

palpalis Gerhardt, 1876: 381 (*Silaria*) **E:** AL AU BH BU CR CT CZ FR GE GR HU IT MC PL ST SZ TR UK YU **A:** KZ TR

affinis Schilsky, 1895: 93

eucineticolor Roubal, 1935b: 106

impressa Guillebeau, 1893b: 328

rambouseki Roubal, 1910: 155 **E:** BU

reitteri Schilsky, 1895: 85 **A:** MG

revelieri Emery, 1876: 15 **E:** FR GR IT MC SP **A:** TR

rufitarsis Lucas, 1854: 35 **E:** GR MA **A:** CY IQ IS SY

salsolae cyrensis Bogatchev, 1939b: 132 **E:** AB

salsolae salsolae Bogatchev, 1939b: 131 **E:** AB

schneideri Emery, 1876: 119 **A:** ST

setulosa Schilsky, 1895: 90 **E:** GR

steppensis Motschulsky, 1860b: 142 **E:** BU ST

ochracea Stierlin, 1867: 223 (*Silaria*)

stierlini Emery, 1876: 15 **E:** BU GR ST **A:** LE SY TR

truquii Baudi di Selve, 1878b: 344 **E:** IT **A:** CY

turkmenica Schilsky, 1899: 92 **A:** TM

subgenus *Nassipa* Emery, 1876: 13 type species *Mordella flava* Linnaeus, 1758

adil Bogatchev, 1939b: 133 **E:** AB

costai Emery, 1876: 15 [NP] **E:** AU BE BH BU CR CZ DE FR GB GE HU IT LU MC NL PL SL SP SV SZ UK

fulvipes Costa, 1854: 29

fuscescens Stephens, 1832: 43 [NO]

skalitzkyi Schilsky, 1895: 80

thoracica A. Costa, 1854: 28 (*Plesianaspis*) [HN]

cypria Baudi di Selve, 1878b: 339 **E:** GR **A:** CY

flava Linnaeus, 1758: 420 (*Mordella*) **E:** AL AU BE BH BU BY CR CZ DE EN FI FR GE GR HU IT LA LT LU MC NL NR NT PL RO SK SL SP ST SV SZ UK YU **A:** LE

fauconneti Pic, 1918c: 13

pectoralis Schilsky, 1899: HH

unicolor Schilsky, 1899: KK

scutellaris Schilsky, 1899: II

himalayana Champion, 1927a: 54 **A:** UP

karafutona Kôno, 1933: 30 **A:** ES

melanostoma A. Costa, 1854: 23 **E:** AL AU BH CR CZ DE FR GB GE GR HU IT LT MC PL SL ST SV SZ UK **A:** CY TR

monilicornis Mulsant, 1856f: 392

occipitalis Emery, 1876: 118 **E:** ST

rufilabris Gyllenhal, 1827: 521 (*Mordella*) **E:** AU BE BH BU BY CR CT CZ DE EN FI FR GB GE GG HU IR IT LA LT LU MC NL NR NT PL RO SK SL SP ST SV SZ UK **A:** LE **AURi** (New Zealand)

atra Fabricius, 1792b: 115 (*Mordella*)

clavicornis Forster, 1771: 66 (*Mordella*)

hudsoni Donisthorpe, 1909: 800

nigra Rossi, 1792: 213 (*Mordella*)

obscuripes Schilsky, 1895: 81

testaceomarginata Roubal, 1935b: 107

takeii M. Chûjô, 1957a: 8 **A:** JA

subgenus *Pinassa* Ermisch, 1950: 73 type species *Anaspis angustata* Ermisch, 1950

eversi Ermisch, 1964a: 18 **N:** CI

imitator Ermisch, 1963d: 110 **N:** CI MR

lindbergi Franciscolo, 1956b: 48 **A:** CY

proteus Wollaston, 1854: 532 (*Silaria*) **E:** AZ **N:** CI MR

subgenus *Spanisa* Emery, 1876: 13 type species *Anaspis labiata* A. Costa, 1854

aharonii Reitter, 1911a: 379 **A:** IS

labiata A. Costa, 1854: 22 **E:** FR GR IT SP **N:** MO

subtilis Hampe, 1871: 333 **E:** AU BH CR HU IT RO RU SL SK ST YU

subgenus *Silaria* Mulsant, 1856f: 39 type species *Silaria brunnipes* Mulsant, 1856

algerica Ermisch, 1963b: 20 **N:** AG

almorana Champion, 1927a: 78 **A:** UP

alticola Champion, 1927a: 55 **A:** UP

balthasari Roubal, 1935b: 106 **E:** ST

bernhaueri Roubal, 1910: 156 **E:** BU

binotata Champion, 1927a: 56 **A:** NP SD UP

brunnipes Mulsant, 1856f: 427 (*Silaria*) **E:** AL AU BH BU BY CR CZ EN FI FR GE GR HU IT LA LT MC NL NT PL SK SV SZ TR UK YU **A:** KZ LE

brunnifrons Coucke, 1891: ccclxxxiv

fuscipes Mulsant, 1856f: 429

canariensis Ermisch, 1967c: 120 **E:** AZ **N:** CI

corcyrica Schilsky, 1899: 98 **E:** GR

distinguenda Ermisch, 1963b: 19 **E:** PT SP **N:** AG TU

diversipes Pic, 1933c: 28 **A:** YUN

major Pic, 1933c: 28

everestina Blair, 1927: 250 **A:** NP UP XIZ

hayashii M. Chûjô & Nakane, 1953: 23 **A:** JA
humerosa Csiki, 1915: 57 **E:** AL GR IT **A:** CY JO LE SY
  scapularis Emery, 1876: 46 [HN]
immaculata Pic, 1933c: 27 **A:** YUN
insularis Ermisch, 1963c: 67 **E:** GR **A:** CY
kochi Ermisch, 1944: 214 **E:** FR PT SP
kriegeri Ermisch, 1963a: 30 **E:** CZ GE IT FR MC RO SZ YU
latiuscula Mulsant, 1856f: 429 (*Silaria*) **E:** AU FR IT SZ
luteobrunnea Fleischer, 1909: 241 **E:** GR
luteola Marseul, 1877: 477 **A:** JA SC
mancinii Franciscolo, 1942b: 20 **E:** IT
minutula Champion, 1927a: 78 **A:** UP GAN
mitchyi M. Chûjô & Nakane, 1953: 22 **A:** JA
quadrimaculata Gyllenhal, 1817: 37 [NP] **E:** AU BE CR CZ FR GE IT PT SP SZ **N:** AG MO AFR
  fulvicollis Schilsky, 1895: 100
  fulvithorax Pic, 1910b: 1 [RN]
  lateobscura Pic, 1909n: 162
  bipustulata Bonelli, 1812: 172 (*Mordella*) [NO]
  quadripustulata P.W.J. Müller, 1821: 212 (*Silaria*)
  subobliterara Pic, 1909n: 162
saharaensis Ermisch, 1966b: 43 **N:** MO
serbica Apfelbeck, 1931: 38 **E:** CR GR IT MC TR YU
shibatai Nakane, 1961: 87 **A:** JA
sibirica Schilsky, 1899: 97 **A:** MG
  antennalis Sthegoleva-Barovskaya, 1930: 752
suturalis Emery, 1876: 16 **E:** FR IT
  limbata Schilsky, 1895: 97
tenebrica Champion, 1927a: 56 **A:** UP
thoracoxantha Franciscolo, 1943b: 79 **E:** IT
touzalini Pic, 1933c: 27 **A:** CH
trifasciata Chevrolat, 1860: 455 **E:** FR IT PT SP **N:** AG MO
  confluens Schilsky, 1895: 99
  interrupta Schilsky, 1895: 99
  quadripustulata Jacquelin du Val, 1863: 408
varians varians Mulsant, 1856f: 431 (*Silaria*) **E:** AL AU BE BH BU CZ EN FR GE GR HU IT LA NL PL PT RO SK SL SP ST SZ YU **A:** CY SY
  collaris Mulsant, 1856f: 433
  frontalis Costa, 1854: 21 (*Mordella*)
  juvenilis Schilsky, 1899: Z
varians longicornis Roubal, 1910: 156 **E:** BU

genus *Anaspis*, species incertae sedis
lindbergiana Roubal, 1934a: 6 **N:** MO (Atlas)

genus *Anaspis*, nomina dubia
testacea Forster, 1771: 63 (*Mordella*) **E:** GB
oxyacanthae Gmelin, 1790: 2024 (*Mordella*)

genus *Argyrabdera* J.R. Sahlberg, 1913a: 53 type species *Argyrabdera deserti* J.R. Sahlberg, 1913
deserti J.R. Sahlberg, 1913a: 53: **A:** TM

genus *Cryptanaspis* Franciscolo, 1956c: 463 type species *Cryptanaspis brittoni* Franciscolo, 1956
brittoni Franciscolo, 1956c: 464 **A:** YE

genus *Cyrtanaspis* Emery, 1876: 8 type species *Anaspis phalerata* Germar, 1847
phalerata Germar, 1847: 8 (*Anaspis*) **E:** AU BY CT CZ EN FI FR GE HU IT LA NL NT PL SL SP SV SZ
interrupta Schilsky, 1908c:606
obscura Schilsky, 1895: 60
sauteri Pic, 1911p: 189 **A:** TAI
yunnana Pic, 1933c: 27 **A:** GAN YUN
diversitarsis Pic, 1933c: 27

tribe Pentariini Franciscolo, 1954

genus *Anaspella* Schilsky, 1895: 59 type species *Anaspella clavicornis* Schilsky, 1895
antennata Franciscolo, 1953: 251 **N:** EG
clavicornis Schilsky, 1895: 59 **N:** AG MO TU
clavifera Marseul, 1876a: 25 (*Anaspis*) **N:** AG EG MO **A:** IS
desertifera Chobaut, 1898a: 86 (*Anaspis*) **N:** AG
inclavifera Pic, 1922e: 21 (*Anaspis*) **N:** EG
alfierii Pic, 1922e: 21 (*Anaspis*)
marioutensis Pic, 1922e: 21 (*Anaspis*)
lutea Marseul, 1876a: 26 (*Anaspis*) **N:** EG
subparallela Pankow, 1981: 273 **A:** SA

genus *Ectasiocnemis* Franciscolo, 1956a: 43 type species *Pentaria platycnema* Champion, 1920
Konaspis M. Chûjô, 1957b: 12 type species *Anthobates elongata* Kôno, 1929
Nipponanaspis M. Chûjô, 1957b: 11 type species *Nipponanaspis shirozui* M. Chûjô, 1957
anchoralis Nomura, 1961b: 87 **A:** JA
elongata Kôno, 1929c: 84 (*Anthobates*) **A:** JA
caneparii Franciscolo, 1989: 108 **A:** KA

chloroptera Champion, 1920a: 76 (*Pentaria*) **A:** UP
contorta Batten, 1983b: 487 **A:** AF
hazariana Franciscolo, 1956a: 57 **A:** PA
horaki Batten, 1983b: 485 **A:** AF
inermis Franciscolo, 1956a: 51 **A:** AF
laciniata Franciscolo, 1961: 53 **A:** AF
luteicollis Champion, 1921: 206 (*Pentaria*) **A:** AF UP
maxima Franciscolo, 1956a: 53 **A:** PA UP
ohkurai Nakane, 1955: 37 (*Pentaria*) **A:** JA
platycnema Champion, 1920a: 77 (*Pentaria*) **A:** UP
shirozui M. Chûjô, 1957b: 13 (*Nipponanaspis*) **A:** JA
straminea Champion, 1927a: 53 (*Pentaria*) **A:** UP
tibialis Schilsky, 1906: 24 (*Anaspis*) **A:** KA UP
kumaonensis Champion, 1920a: 76 (*Pentaria*)

genus *Pentaria* Mulsant, 1856d: 135 type species *Anaspis badia* Rosenhauer, 1847
Anthobates LeConte, 1850: 231 [HN]
Anthobatula E. Strand, 1929: 20 [RN]
abderoides Chobaut, 1893: 97 (*Silaria*) **N:** AG MO TU
arabica Pankow, 1981: 275 **A:** SA
badia Rosenhauer, 1847: 39 (*Anaspis*) **E:** AU FR GE HU IT SK SP ST SZ PT UK **N:** AG MO **A:** SY
sericaria Mulsant, 1856f: 439
bleusei Chobaut, 1897b: 133 **N:** AG
dimidiata Marseul, 1876a: 27 **A:** LE
gracilis Schilsky, 1894a: 99 (*Anaspis*) **E:** AB AR
reitteri Chobaut, 1894a: lxxv
mongolica Ermisch, 1969c: 176 **A:** MG
phoenix Franciscolo, 1956a: 459 **A:** YE
redemptus Peyerimhoff, 1931b: 86 **N:** AG
soderbomi Pic, 1933b: 10 (*Anthobates*) **A:** MG
spissicornis Peyerimhoff, 1946: 95 (*Anthobates*) **N:** MO

genus *Pseudopentaria* Franciscolo, 1956a: 460 type species *Anaspis defarguesi* Abeille de Perrin, 1885
defarguesi Abeille de Perrin, 1885: 161 (*Anaspis*) **E:** FR IT SP **N:** AG MO TU **A:** IS
immaculata Pic, 1895l: cxxvii (*Pentaria*)
oberthueri Champion, 1892: 109 (*Pentaria*)
pictipennis Reitter, 1891a: 31 (*Anaspis*)
unifasciata Chobaut, 1897: 135 (*Pentaria*) **N:** AG
mesopotamica Blair, 1923a: 126 (*Pentaria*) **A:** IQ
scotti Franciscolo, 1956a: 461 **A:** YE

References

Abdullah, M. 1966: A revision of the genus Steropes Steven, with a proposed new subfamily Steropinae of the family Anthicidae (Coleoptera). *The Entomologist's Monthly Magazine* **101**: 206–216, 1 pl.

Abdullah, M. 1967: A new genus (Neostereopalpus) of the Eurygeniini (Coleoptera: Anthicidae) from Japan. *Pakistan Journal of Scientific and Industrial Research* **10**: 265–267.

Abdullah, M. 1968: New Old World genera of the Eurygeniini (Coleoptera: Anthicidae). *Pakistan Journal of Scientific and Industrial Research* **11**: 190–198.

Abdurakhmanov, G.M. & Nabozhenko, M.V. 2011: *Opredelitel' i katalog zhukov-chernotelok (Coleoptera: Tenebrionidae s. str.) Kavkaza i yuga yevropeyskoy chasti Rossii*. Moscow: KMK Scientific Press Ltd, 361 pp. (in Russian).

Abeille de Perrin, E. 1869: [Coléoptères nouveaux]. *Petites Nouvelles Entomologiques* **1** (11) [1869–1875]: 42.

Abeille de Perrin, E. 1870: Salpingus exsanguis. Espèce nouvelle de la faune française. *L'Abeille, Mémoires d'Entomologie* **7**: 89–90.

Abeille de Perrin, E. 1872: *Études sur les coléoptères cavernicoles suivies de la description de 27 coléoptères nouveaux français*. Marseille: Marius Olive 40 + [1] pp.

Abeille de Perrin, E. 1874a: *Essai monographique sur les cisides européens et circaméditerranéens*. Marseille: E. Camoin, [1] + 99 + [1] pp.

Abeille de Perrin, E. 1874b: Étude sur les salpingiens européens. *Bulletin de la Société d'Histoire Naturelle de Toulouse* **8**: 24–31.

Abeille de Perrin, E. 1874c: [Diagnoses de cisides nouveaux]. *Bulletin de la Société Entomologique de France* **1874**: lii-liii.

Abeille de Perrin, E. 1876: Notes sur les cisides européens et circumméditerranéens. *Annales de la Société Entomologique de France* (5) **6**: 309–314.

Abeille de Perrin, E. 1876: [new taxa]. *Bulletin de la Société Entomologique de France* **1876**: clxvi.

Abeille de Perrin, E. 1880: Contribution à la faune coléoptérologique de la Méditerranée. *Bulletin de la Société d'Histoire Naturelle de Toulouse* **14**: 233–262.

Abeille de Perrin, E. 1885: Coléoptères rares ou nouveaux de France. *Revue d'Entomologie* **4**: 153–161.

Abeille de Perrin, E. 1892: Notices entomologiques. *Revue d'Entomologie* **11**: 62–67.

Abeille de Perrin, E. 1894a: Diagnoses de coléoptères reputés nouveaux. *L'Échange, Revue Linnéenne* **10**: 91–94.

Abeille de Perrin, E. 1894b: Description d'un coléoptère saharien nouveau. *Bulletin de la Société Entomologique de France* **1894**: cxxxii-cxxxiii.

Abeille de Perrin, E. 1895a: [Description de deux genres et d'une espèce française de coléoptères]. *Bulletin de la Société Entomologique de France* **1894**: cclxxi–cclxxii.

Abeille de Perrin, E. 1895b: [new taxa]. *Bulletin de la Société Entomologique de France* **1895**: xxiii-xxiv.

Abeille de Perrin, E. 1896: Descriptions d'un serropalpide et de trois oedémerides nouveaux (Col.). *Bulletin de la Société Entomologique de France* **1896**: 283–285.

Abeille de Perrin, E. 1899: Description d'une espèce nouvelle de colydien (Col.) des Alpes-Maritimes. *Bulletin de la Société Entomologique de France* **1899**: 408–409.

Abeille de Perrin, E. 1901a: Nouvelles espèces de coléoptères français. *L'Échange, Revue Linnéenne* **17**: 59–62.

Abeille de Perrin, E. 1901b: Descriptions de deux coléoptères nouveaux du nord de l'Afrique. *Bulletin de la Société Entomologique de France* **1901**: 234.

Abeille de Perrin, E. 1909: Étude d'un Rhipidius nouveau de Provence, R. Boissyi Abeille (Présentation de l'insecte). *Comptes-Rendus de Séances de la Société de Biologie* **67**: 854–858.

Adams, M. 1817: Descriptio insectorum novorum Imperii Rossici, inprimis Caucasi et Sibiriae. *Mémoires de la Société Impériale des Naturalistes de Moscou* **5**: 278–314, pl. x.

Agassiz, J.L.R. 1846a: Nomina systematica generum Coleopterorum, tam viventium quam fossilium, secundum ordinem alphabeticum disposita, adjectis auctoribus, libris in quibus reperiuntur, anno editionis, etymologia et familis ad quas pertinent. Pp. xi + [1] + 1–170. In: *Nomenclator zoologicus, continens nomina systematica generum animalium, tam viventium quam fossilium, secundum ordinem alphabeticum disposita, adjectis auctoribus, libris, in quibus reperiuntur, anno editionis, etymologia et familiis, ad quas pertinen, in singulis classibus*. [1842–1846]. Soloduri: Jent & Gassmann.

Agassiz, J.L.R. 1846b: *Nomenclatoris zoologici index universalis, continens nomina systematica classium, ordinum, familiarum et generum animalium omnium, tam viventium quam fossilium, secundum ordinem alphabeticum unicum disposita, adjectis homonymiis plantarum, nec non variis adnotationibus et emendationibus*. Soloduri: Jent & Gassmann, viii + 383 pp.

Ahrens, A. 1814: *Fauna Insectorum Europae, Fasciculus II*. Halae: Car. Aug. Kümmelii, [25] pp., 25 pls.

Akita, K. 2011: New or little-known tenebrionid species (Coleoptera) from Japan (10) Four new species and distribution record from Japan. Pp. 271–284. In: Kawai, S. (ed.): *Masumushi. Special Publication of the Japanese Society of Scarabaeidology No. 1*. Tokyo: Japanese Society of Scarabaeidology, 345 pp.

Akita, K. 2012: [new name]. In: Akita K. & Masumoto K.: New or Little-Known Tenebrionid Species (Coleoptera) from Japan (13) Three New Tenebrionid Species from Japan, and a Replacement of a Preoccupied Name. *Elytra, New Series* **2**: 207–216.

Akita, K. & Masumoto, K. 2003: [new taxon]. In: Masumoto K. & Akita K.: New or little-known tenebrionid species from Japan (Part 4) (Coleoptera). *Entomological Review of Japan* **58**: 173–182.

Akita, K. & Masumoto, K. 2008: New or Little-known Tenebrionid Species (Coleoptera) from Japan (7) Revisional Study of *Borboresthes* (Tenebrionidae: Alleculinae) Species from the Ryulyu Islands. *Entomological Review of Japan* **63**: 109–120.

Akita, K. & Masumoto, K. 2009a: A new Strongylium (Tenebrionidae: Stenochiini) species from Bhutan. *Entomological Review of Japan* **64**: 255–257.

Akita, K. & Masumoto, K. 2009b: New or little-known tenebrionid species (Coleoptera) from Japan (9) Two new species and a distribution record from Japan, and a replacement of a preoccupied name. *Elytra* **64**: 247–253.

Akita, K. & Masumoto, K. 2009c: New or little known tenebrionid beetles (Coleoptera, Tenebrionidae) from Japan (8) Two new species from Danjo-guntô and Ryukyu Islands. *Entomological Review of Japan* **64**: 77–82.

Akita, K. & Masumoto, K. 2012a: New or little known tenebrionid beetles (Coleoptera, Tenebrionidae) from Japan: (11) Description of seven new species, up-grading of a subspecies to the species rank, and a new distributional record of a species. *Elytra, New series* **1** [2011]: 275–294.

Akita, K. & Masumoto, K. 2012b: New or Little-Known Tenebrionid Species (Coleoptera) from Japan (13) Three New Tenebrionid Species from Japan, and a Replacement of a Preoccupied Name. *Elytra, New Series* 2: 207–216.

Akita, K. & Masumoto, K. 2013: New or little-known tenebrionid species (Coleoptera) from Japan (14) Six new species and three new subspecies from various areas in Japan. *Elytra, New Series* 3: 237–254.

Akita, K. & Masumoto, K. 2014: New or Little-known Tenebrionid Species (Coleoptera) from Japan. (15) A Revisional Study of the Genus *Mycetochara* BERTHOLD (Alleculinae, Alleculini). *Elytra, Tokyo, New Series* 4: 213–235.

Akita, K. & Masumoto, K. 2015a: New or little-known tenebrionid species (Coleoptera) from Japan (16) Descriptions of eight new species of Allecula and Borboresthes (Alleculinae, Alleculini), and redescription of the known species, with proposal for two new synonyms. *Elytra, New Series* 5: 357–377.

Akita, K. & Masumoto, K. 2015b: New or little-known tenebrionid species (Coleoptera) from Japan (17) Descriptions of new taxa, proposal for new taxonomical treatments and new occurrence records. *Elytra, New Series* 5: 409–428.

Akita, K. & Masumoto, K. 2015c: New or little-known tenebrionid species (Coleoptera) from Japan (18) Descriptions of new taxa and proposal for new treatments. *Elytra, New Series* 5: 429–448

Akita, K. & Masumoto, K. 2016: The Tenebrionid beetles of Japan. In: Fujita, H. (ed.): *Mushi-Sha's Iconographic Series of Insects 9*, Tokyo: 302 pp. (in Japanese with English title and descriptions of taxonomy).

Akita, K. & Masumoto, K. 2017: New or little-known tenebrionid species (Coleoptera) from Japan (19) Descriptions of three new species and a new occurring record. *Japanese Journal of Systematic Entomology* 17: 47–52.

Akita, K. & Masumoto, K. 2019: New or little-known tenebrionid species (Coleoptera) from Japan (20) Descriptions of two new species and new distributional records of Tetragonomenes palpaloides (Nakane, 1963). *Elytra, New Series* 9: 89–97.

Akiyama, H. 1998: A new species of the subgenus Dryopomera (Coleoptera, Oedemeridae) from the Yaeyama Islands of the Ryukyus. *Elytra* 26: 167–171.

Akiyama, H. 2012: A new species of the genus Indasclera (Coleoptera, Oedemeridae) from Taiwan. *Elytra, New Series* 2:189–191.

Akiyama, H. 2015a: A new species of the genus Parasessinia (Coleoptera, Oedemeridae) from Taiwan. *Elytra, New Series* 5:129–132.

Akiyama, H. 2015b: A new species of the genus Pseudohyperasclera (Coleoptera, Oedemeridae) from the Yaeyama Islands of the Ryukyus, Southwestern Japan. *Elytra, New Series* 5: 449–452.

Akiyama, H. & Hori, S. 2016: A new species of the genus Asessinia (Coleoptera, Oedemeridae) from central Honshu and southwestern Hokkaido Japan. *Elytra, New Series* 6: 47–50.

Aksentjev, S.I. 1978: Novyy vid naryvnikov roda Stenodera Esch. (Coleoptera, Meloidae) iz Primor'ya [A new species of meloid beetles of the genus Stenodera Esch. (Coleoptera, Meloidae) from Primorye Territory]. *Entomologicheskoe Obozrenie* 57: 124–125.

Aksentjev, S.I. 1981: Novye rody i vidy naryvnikov (Coleoptera, Meloidae) iz Sredney Azii [New genera and species of blister beetles (Coleoptera, Meloidae) from Middle Asia]. *Entomologicheskoe Obozrenie* 60: 860–864.

Aksentjev, S.I. 1985: Meloidae from Iran (Insecta: Coleoptera). *Senckenbergina Biologica* **65** [1984]: 245–250.

Aksentjev, S.I. 1987: Meloidae from the Nepal Himalayas (Insecta: Coleoptera). *Courier Forschungs-Institut Senckenberg* **93**: 471–476.

Aksentjev, S.I. 1988: Katalog taksonov rodovoy grupy zhestkokrylykh sem. Meloidae (Coleoptera) [A catalogue of the genus group taxa of the beetle family Meloidae (Coleoptera)]. *Entomologicheskoe Obozrenie* **67**: 569–582.

Aksentjev, S.I. & Sadykova Kh. M. 1990: A new species of blister beetles (Coleoptera, Meloidae) from Kirghiziya. *Zoologiskiy Zhurnal* **69**: 152–153.

Aliquiò, V. 1993: Dati nuovi e riassuntivi sui coleotteri tenebrionidi delle isole circum-siciliane (Coleoptera: Tenebrionidae). *Il Naturalista Siciliana* (Ser. IV) **17**: 111–125.

Aliquò, V., Leo, P. & Lo, Cascio P. 2006: I tenebrionidi dell'arcipelago Eoliano: nuovi dati faunistici e zoogeografici con descrizione di una nuova speciedi *Nalassus* Mulsant, 1854 (Coleoptera Tenebrionidae). *Naturalista siciliano, ser. IV* **30**: 293–314.

Allard, E. 1865a: Quatre nouvelles espèces du genre Erodius. *Annales de la Société Entomologique de France* (4) **4** [1864]: 387–388.

Allard, E. 1865b: Tableau synoptique des espèces du genre Erodius. *Annales de la Société Entomologique de France* (4) **4** [1864]: 389–398.

Allard, E. 1869a: Révision du genre Asida (Latr.). *L'Abeille, Journal d'Entomologie* **6**: 159–304.

Allard, E. 1869b: Description de quelques coléoptères nouveaux et notes diverses. *L'Abeille, Mémoires d'Entomologie* **5**: 465–478.

Allard, E. 1870: [Coléoptères nouveaux]. *Petites Nouvelles Entomologiques* **1** [1869–1875]: 49–50.

Allard, E. 1873: Monographie des espèces de coléoptères du genre Erodius, Fab. *Revue et Magasin de Zoologie Pure et Appliquée* (3) **1**: 122–234.

Allard, E. 1874: Mémoire sur les coléoptères ténébrionides formant les genres Sepidium, Fabr. & Vieta, Cast. *Revue et Magasin de Zoologie Pure et Appliquée* (3) **2**: 120–151.

Allard, E. 1876a: Révision des hélopines vrais de Lacordaire. *L'Abeille, Journal d'Entomologie* **14**: 1–80.

Allard, E. 1876b: Descriptions d'espèces nouvelles. *L'Abeille, Journal d'Entomologie* **14**: 21–25.

Allard, E. 1877a: Révision des helopides vrais. *Mitteilungen der Schweizerischen Entomologischen Gesellschaft* **5**: 13–268.

Allard, E. 1877b: Description de quelques hélopides nouveaux. *Petites Nouvelles Entomologique* **9**: 97–98.

Allard, E. 1880: Tableau des Blapsides. *Entomologische Monatsblätter* **15**: 71–74

Allard, E. 1881a: Essai de classification des blapsides de l'ancien monde. 1e partie. *Annales de la Société Entomologique de France* (5) **10** [1880]: 269–320.

Allard, E. 1881b: [new taxa]. *Bulletin de la Société Entomologique de France* **1881**: ciii-civ.

Allard, E. 1882a: Essai de classification des blapsides de l'ancien monde. 4e et dernière partie. *Annales de la Société Entomologique de France* (6) **2**: 77–140.

Allard, E. 1882b: [new taxa]. *Bulletin de la Société Entomologique de France* **1882**: lxxxvi-lxxxvii.

Allard, E. 1883: Mélanges Entomologiques. *Annales de la Société Entomologique de Belgique* **27**: 5–49.

Allard, E. 1884: Diagnoses de quatre nouveaux coléoptères. *Wiener Entomologische Zeitung* **3**: 248.

Allard, E. 1885a: Classification des adesmides et des mégagénides (Lacord.). *Annales de la Société Entomologique de France* (6) **5**: 154–208.

Allard, E. 1885b: Coleoptera nova. *Wiener Entomologische Zeitung* **4**: 180.

Allard, E. 1886: *Description de six nouvelles espèces de coléoptères hétéromères*. [7] pp.

Allard, E. 1888: [new taxon]. In: Allard, E. & Dodero, A.: Due nuovi coleotteri italiani raccolti in Sardegna da Umberto Lostia di S. Sofia. *Bullettino della Società Entomologica Italiana* **19** [1887]: 333–334.

Allard, E. 1894: Diagnoses de coléoptères nouveaux de l'Inde. *Le Naturaliste* (2) **8**: 104.

Allard, E. 1896: Hétèromères recueillis au Sikim par le DR. Harmand (Col.). *Bulletin de la Société Entomologique de France* **1896**: 318–321.

Allemand, R. 1993: Les Nacerdes ouest-paléarctiques du sous-genre Xanthochroa Schmidt, notes taxonomiques et commentaires sur la faune française (Coleoptera, Oedemeridae). *Bulletin de la Société Entomologique de France* **98**: 5–14.

Allen, A.A. 1940: Scraptia testacea nom. nov., and S. fuscula Mull. (Col., Scraptiidae). *The Entomologist's Monthly Magazine* **76**: 56–58.

Allen, A.A. 1995: An apparently new species of Mordellistena (Col.: Mordellidae) in Britain. *Entomologist's Record and Journal of Variation* **107**: 25–27.

Allen, A.A. 1999: Another new species of Mordellistena Costa (Col.: Mordellidae) in Britain. *Entomologist's Record and Journal of Variation* **111**: 205–206.

Allmann, F.R.S. 1864: On the construction and limitation of genera among the Hydroida. *The Annals and Magazine of Natural History* **13**: 345–380.

Alluaud, C. 1935: Les coléoptères des îles Salvages. *Revue Française d'Entomologie* **2**: 35–44.

Alonso-Zarazaga, M.A. 2010: Gompelia, a replacement name for Olotelus Mulsant & Rey, 1866 (Coleoptera, Aderidae). *Graellsia* **66**: 299–301.

Alonso-Zarazaga, M.A. 2013: Case 3624. A proposal for the rejection of 38 names in Anthicidae (Coleoptera). *Bulletin of Zoological Nomenclature* **70**: 171–184.

Anand, R.K. 1977: Two new species of the genus Epicauta Redtenbacher (Meloidae: Coleoptera) from Kulu Valley, India. *Entomon* **2**: 81–83.

Ancey, C.M.F. 1869: [Un coléoptère nouveau (Entoxylon Abeilei n. sp.)]. *Petites Nouvelles Entomologiques* **1** (12) [1869–1875]: 46.

Ancey, C.M.F. 1870: Description d'espèces nouvelles. *L'Abeille, Mémoires d'Entomologie* **7** [1869–1870]: 84–88.

Ancey, C.M.F. 1881a: Description de coléopères nouveaux d'Aden. *Le Naturaliste* **1**: 397.

Ancey, C.M.F. 1881b: Descriptions de coléoptères nouveaux. *Le Naturaliste* **1**: 412–413, 509.

Ancey, C.M.F. 1900: Description d'une nouvelle espèce d'oedéméride. *Miscellanea Entomologica* **8**: 140.

Andersch, J.D. 1797: Entomologische Bemerkungen. Pp. 152–166. In: Hoppe, D.H. (ed.) *Entomologisches Taschenbuch für die Anfänger und Liebhaber dieser Wissenschaft auf das Jahr 1797*. Regensburg: Montag und Weisse, 252 pp.

Ando, K. 1978: A new Ischnodactylus-species from Japan (Coleoptera, Tenebrionidae). *Entomological Review of Japan* **32**: 81–84.

Ando, K. 1984: A new tenebrionid beetle from Japan (Coleoptera, Tenebrionidae). *Entomological Review of Taiwan* **39**: 69–71.

Ando, K. 1989: Studies on the Tenebrionidae of Shibata collection mainly from S.E. Asia I. (Coleoptera). *Entomological Review of Japan* **44**: 105–11[illegible].

Ando, K. 1990: Studies on the Tenebrionidae of Shibata collection mainly from SE. Asia II. (Coleoptera). *Entomological Review of Japan* **44**: 103–112.

Ando, K. 1994: Two new species of the genus Hemicera Castelnau et Brullé (Coleoptera: Tenebrionidae). *Transactions of the Shikoku Entomological Society* **20**: 111–116.

Ando, K. 1996: Two new genera of the tribe Cnodalonini from Southeast Asia (Coleoptera, Tenebrionidae). *Japanese Journal of Systematic Entomology* **2**: 189–200.

Ando, K. 1997: Two new species of the genus Euhemicera Ando occuring in Taiwan (Coleoptera, Tenebrionidae). *Japanese Journal of Systematic Entomology* **3**: 105–112.

Ando, K. 1998: Revision of the genus Taichius Ando (Coleoptera, Tenebrionidae). *Japanese Journal of Systematic Entomology* **4**: 349–379.

Ando, K. 2000: A new subspecies of Oedemutes hirashimai M.T. Chûjô from Irabu Is., the Ryukyus, Japan (Coleoptera, Tenebrionidae). *Japanese Journal of Systematic Entomology* **6**: 309–311.

Ando, K. 2002: Notes on the genus Artactes (Coleoptera, Tenebrionidae), with description of a new species from Borneo. *Special Bulletin of the Japanese Society of Coleopterology* **5**: 363–369.

Ando, K. 2003a: A systematic revision of the genus Eucyrtus and its complex (Coleoptera, Tenebrionidae). *Japanese Journal of Systematic Entomology, Monographic Series* **1**, xi + 604 pp., 38 pls.

Ando, K. 2003b: A new Strongylium species (Coleoptera: Tenebrionidae) occurring in Japan. *Entomological Review of Japan* **58**: 79–81.

Ando, K. 2003c: A new genus of Tenebrionidae (Coleoptera) from Japan, with description of a new species. *Japanese Journal of Systematic Entomology* **5**: 135–141.

Ando, K. 2006: Contribution to the knowledge of Japanese Tenebrionidae (3) (Coleoptera). *Entomological Review of Japan* **61**: 29–32.

Ando, K. 2009: A new species of the genus Misolampidius from Japan (Coleoptera, Tenebrionidae). *Japanese Journal of Systematic Entomology* **15**: 187–190.

Ando, K. 2010. A remarkable new species of the genus Arthromacra from Japan (Coleoptera: Tenebrionidae: Lagriinae). *Entomological Review of Japan* **65**: 241–244.

Ando, K. 2013: A New Species of the Genus Neoplamius Masumoto from Taiwan (Coleoptera, Tenebrionidae). *Japanese Journal of Systematic Entomology* **19**: 187–190.

Ando, K. 2015: A proposition of new genus for Ischnodactylus loripes Lewis (Coleoptera, Tenebrionidae, Diaperini). *Elytra, New Series* **5**: 385–389.

Ando, K. 2017: Two new tenebrionid species (Coleoptera) from Japan. *Elytra, New Series* **7**: 297–304.

Ando, K. 2018: A new species of the genus Menimus Sharp (Coleoptera, Tenebrionidae) from Lan-yu Island, Taiwan. *Elytra, New Series* **8**: 129–133.

Ando, K. & Chûjô, M.T. 2005: Note on the genus Enanea (Coleoptera, Tenebrionidae) with description of a new species from Amami-Ôshima Is., Japan. *Elytra* **33**: 69–75.

Ando, K., Merkl, O., Jeng, M.-L., Chan, M.-L. & Hayashi, Y. 2016. Catalogue of Formosan Tenebrionidae (Insecta: Coleoptera). Japanese Journal of Systematic Entomology **21**, *Supplementa*: 1–88.

Ando, K. & Ren, G.-D. 2006: Contribution to the knowledge of Chinese Tenebrionidae I (Coleoptera). *Entomological Review of Japan* **61**: 81–94.

Ando, K. & Sakai, M. 1999: A new Tetraphyllus from Taiwan (Coleoptera, Tenebrionidae). *Japanese Journal of Systematic Entomology* **5**: 157–160.

Ando, K. & Schawaller, W. 2015: Two New Species of the Genus Phaedis (Coleoptera, Tenebrionidae) from Taiwan and China. *Elytra, New Series* 5: 403–408.

Ando, K. & Schawaller, W. 2018: Four New Species of the Genus Foochounus Pic from the Oriental Region (Coleoptera, Tenebrionidae). *Elytra, New Series* 8: 137–147.

Andres, A. 1920: I. Tenebrionidae (Coleoptera). Pp. 59–74. In: Zoologische Ergebnisse zweier in den Jahren 1902 und 1904 durch die Sinaihalbinsel unternommener botanischer Studienreisen. II. Teil. *Entomologische Blätter* **16**: 59–94.

Andres, A. 1922: Tenebrionidae. Pp. 26–27. In: Kneucker A.: Zoologische Ergebnisse zweier in den Jahren 1902 und 1904 durch die Sinaihalbinsel unternommener botanischer Studienreisen. II. Teil. *Entomologische Blätter* **18**: 20–28.

Andres, A. 1926: Notes et description d'un Tentyriinae cavernicole d'Egypte: Mesostenopa cavatica sp. nov. (Coléop.-ténébrionide). *Bulletin de la Société Royal Entomologique d'Egypte* 9 [1925]: 287–290.

Andres, A. 1928: Description du Zophosis dakhlensis n. sp. d'Egypte. *Bulletin de la Société Royal Entomologique d'Egypte* (N.S.) **12**: 48–50.

Andres, A. 1929: Note sur Pimelia angulata Fabr. et espèces voisines et description d'une variété nouvelle. *Bulletin de la Société Royal Entomologique d'Egypte* (N.S.) **13**: 10–25, pl. I.

Andres, A. 1931: A new tenebrionid from Egypt: Mitotagenia Priesneri. *Bulletin de la société Royale Entomologique d'Egypte* (N.S.) **15**: 56–57.

Anonymous 1828: Analyses d'ouvrages. 'Nouveaux Mémoires de la Société impériale des Naturalistes de Moscou'. *Bulletin du Nord* **1**(10): 125–148.

Antoine, M. 1926: Notes d'entomologie marocaine. *Bulletin de la Société des Sciences Naturelles du Maroc* 5: 248–259.

Antoine, M. 1931: Étude des espèces appartenant au genre Litoborus Mulsant et Rey (Col. Tenebrionidae). *Bulletin de la Société des Sciences Naturelles du Maroc* **10** [1930]: 175–209.

Antoine, M. 1932: Notes d'entomologie marocaine. *Bulletin de la Société des Sciences Naturelles du Maroc* 12: 173–188.

Antoine, M. 1933a: Notes d'entomologie marocaine. *Bulletin de la Société des Sciences Naturelles du Maroc* 13: 181–215.

Antoine, M. 1933b: Mission entomologique dans l'Atlas (Premier itinéraire. – Résultats partiels: Carabiques et ténébrionides). *Bulletin de la Société des Sciences Naturelles du Maroc* **13**: 215–230.

Antoine, M. 1934: Notes d'entomologie marocaine XX. Tenebrionides nouveaux. *Bulletin de la Société des Sciences Naturelles du Maroc* **14**: 104–124.

Antoine, M. 1936a: Diagnoses sommaires de Stenosis marocains (Col. Tenebrionidae). *Bulletin de la Société Entomologique de France* **41**: 13–16.

Antoine, M. 1936b: Notes d'entomologie marocaine XXIII. Mission entomologique dans le Grand Atlas (Deuxième itinéraire: le massif du Rhat). *Bulletin de la Société des Sciences Naturelles du Maroc* **16**: 65–76.

Antoine, M. 1936c: Notes d'entomologie marocaine XXII. – Description d'un Licinus (Col. Carabidae) et de deux Diplocyrthus (Col. Tenebrionidae) nouveaux du Maroc. *Bulletin de la Société Entomologique de France* **41**: 138–141.

Antoine, M. 1937a: Notes d'entomologie marocaine XXIV. Deuxième contribution à la faune du Djebel Rhat. *Bulletin de la Société des Sciences Naturelles du Maroc* **16** [1936]: 239–245.

Antoine, M. 1937b: Notes d'entomologie marocaine. XXVI. Le genre Sabularius Esc. *Bulletin de la Société des Sciences Naturelles du Maroc* **17**: 174–181.

Antoine, M. 1937c: Notes d'entomologie marocaine. XXVII. Alphasida nouvelles du Maroc (Coléopt. Ténébrionides). *Bulletin de la Société des Sciences Naturelles du Maroc* **17**: 182–196.

Antoine, M. 1939: Notes d'entomologie marocaine. XXX. – Coléoptères nouveaux du Moyen Atlas. *Bulletin de la Société Entomologique de France* **44**: 145–152.

Antoine, M. 1942a: Notes d'entomologie marocaine. XXXII [sic]. – Les Opatrum du Maroc (Col. Teneb.). *Bulletin de la Société d'Histoire Naturelle de l'Afrique du Nord* **33**: 19–30.

Antoine, M. 1942b: Notes d'entomologie marocaine. XXXII. Les Litoborinae du Maroc (Col. Teneb.). *Bulletin de la Société des Sciences Naturelles du Maroc* **21** [1941]: 19–52.

Antoine, M. 1942c: Notes d'entomologie marocaine XXXIV–XXXVI. *Bulletin de la Société des Sciences Naturelles du Maroc* **22**: 42–82.

Antoine, M. 1945: Notes d'entomologie marocaine XXXVIII. Les Crypticus du Maroc (Col. Tenebr.). *Eos, Revista Española de Entomología* **20** [1944]: 257–276.

Antoine, M. 1946a: Notes d'entomologie marocaine XLI. Sur les Arthrodeis du Maroc (Col. Tenebr.). *Bulletin de la Société des Sciences Naturelles du Maroc* **24**: 25–32.

Antoine, M. 1946b: Notes d'entomologie marocaine XLII. Les Thalpobia Fairm. (Col. Tenebr.). *Bulletin de la Société des Sciences Naturelles du Maroc* **24**: 33–39.

Antoine, M. 1946c: Notes d'entomologie marocaine. XLIII. – Notes complémentaires sur les Litoborinae et Opatrum du Maroc (Col. Tenebr.). *Bulletin de la Société d'Histoire Naturelle de l'Afrique du Nord* **37**: 51–59.

Antoine, M. 1947: Notes d'entomologie marocaine XLVI. – Trois ténébrionides nouveaux de la montagne marocaine. *Bulletin de la Société Entomologique de France* **52**: 147–149.

Antoine, M. 1948: Notes d'entomologie marocaine XLVII. Les Stenosis du groupe fulvipennis (Col. Tenebrionidae). *Revue Française d'Entomologie* **14** [1947]: 277–298.

Antoine, M. 1949a: Notes d'entomologie marocaine XLVIII. – Tableaux de détermination des Pimelia du Maroc (Col. Tenebrionidae). *Annales de la Société Entomologique de France* **116** [1947]: 17–58.

Antoine, M. 1949b: Notes d'entomologie marocaine XLIV. Matériaux pour l'étude des Helopinae du Maroc (Col. Tenebrionides). *Bulletin de la Société des Sciences Naturelles du Maroc* **25–27** [1945–1947]: 123–162.

Antoine, M. 1949c: Notes d'entomologie marocaine LI. – Les Stenosis du Maroc (suite et fin). *Revue Française d'Entomologie* **16**: 209–233.

Antoine, M. 1951a: Un ténébrionide nouveau du Sahara marocain. *Comptes Rendus de la Société des Sciences Naturelles du Maroc* **17**: 76–77.

Antoine, M. 1951b: Notes d'entomologie marocaine LII. – Coléoptères nouveaux ou intéressants. *Bulletin de la Société des Sciences Naturelles du Maroc* **29** [1949]: 311–321.

Antoine, M. 1951c: Notes d'entomologie marocaine LIV. – Captures et observations nouvelles (Coléoptères). *Bulletin de la Société des Sciences Naturelles du Maroc* **30** [1950]: 87–101.

Antoine, M. 1952: Notes d'entomologie marocaine LV. – Additions et remarques taxonomiques. *Bulletin de la Société des Sciences Naturelles du Maroc* **31** [1951]: 103–119.

Antoine, M. 1953a: Présentation de coléoptères troglobies. *Comptes Rendus de la Société des Sciences Naturelles du Maroc* **19**: 58–59.

Antoine, M. 1953b: Notes d'entomologie marocaine. LVI. – Mélanges coléoptérologiques. *Revue Française d'Entomologie* **20**: 202–223.

Antoine, M. 1954: Notes d'entomologie marocaine LX. – Sur quelques captures intéressantes (Coléopt. Carab. et Ténébr.). *Bulletin de la Société des Sciences Naturelles et Physiques du Maroc* **34**: 199–209.

Antoine, M. 1955: Notes d'entomologie marocaine LXII. – Trois ténebrionides nouveaux pour le Maroc. *Bulletin de la Société des Sciences Naturelles et Physiques du Maroc* **34** [1954]: 362–365.

Antoine, M. 1956: Notes d'entomologie marocaine LXIII. – Sur la systématique des Litoborini et révision des Melambiina marocains (Col. Téneb.). *Bulletin de la Société des Sciences Naturelles et Physiques du Maroc* **36**: 341–366.

Antoine, M. 1957: Notes d'entomologie marocaine LXIV. – Contribution à la connaissance des ténébrionides du Maroc. *Bulletin de la Société des Sciences Naturelles et Physiques du Maroc* **37**: 39–56.

Antoine, M. 1962: Notes d'entomologie marocaine LXIX. – Coléoptères nouveaux. En particulier du Rif et du Maroc oriental. *Bulletin de la Société des Sciences Naturelles et Physiques du Maroc* **42**: 47–56.

Aoki, J. 2008: A New Species of Colydiinae (Coleoptera, Zopheridae) from Tokunoshima Island, Southwestern Japan. *Elytra* **38**: 275–278.

Aoki, J. 2009: A Third Species of Neotrichus (Coleoptera, Zopheridae) from Japan. *Elytra* **37**: 143–147.

Aoki, J. 2011a: Revised Status of a Colydiid Species Known as 'Lasconotus niponius (Lewis)' (Coleoptera, Zopheridae). *Elytra, New Series* **1**: 97–102.

Aoki, J. 2011b: A new soil-inhabiting Zopherid beetle (Coleoptera: Zopheridae) from southwestern Japan. *Edaphologia* **89**: 13–17.

Aoki, J. 2012: A New Species of the Genus Microsicus from Ishigaki Island, South Japan (Coleoptera, Zopheridae, Colydiinae). *Elytra, New Series* **2**: 217–219.

Aoki, J. 2017: Japanese Species of the Genus Pycnomerus (Coleoptera, Zopheridae) with Description of Three New Species. *Elytra, New Series* **7**: 175–182.

Aoki, J. 2018: Japanese Species of the Genus Lasconotus (Coleoptera: Zopheridae: Colydiinae) with Description of a New Species from Ishigaki-jima, the Ryukyu Islands. *Species Diversity* **23**: 69–74.

Aoki, J. & Narukawa, N. 2016: Two Japanese Species of the Genus Namunaria (Coleoptera, Zopheridae). *Elytra, New Series* **6**: 101–109.

Aoki, J. & Narukawa, N. 2019: A new species of the genus Rhopalocerus (Coleoptera: Zopheridae: Colydiinae) from Taiwan. *Japanese Journal of Systematic Entomology* **25**: 165–167.

Aoki, J. & Okada, K. 2011: Species of the genus Glyphocryptus (Coleoptera, Zopheridae) from Japan and Taiwan. *Special Publications of the Japanese Society of Scarabaeidology* **11**: 243–251.

Apfelbeck, V. 1901: Bericht über eine entomologische Forschungsreise nach der Türkei und Griechenland im Jahre 1900. *Wissenschaftliche Mittheilungen aus Bosnien und der Herzegovina* (Wien) **8**: 447–469.

Apfelbeck, V. 1905: Paeninsulae balcanicae coleopterorum species novae. *Glasnik Zemaljskog Muzeja u Bosni i Hercegovini* **17**: 230–252.

Apfelbeck, V. 1907: Neue Koleopteren von der Balkanhalbinseln. *Wissenschaftliche Mitteilungen aus Bosnien und der Herzegowina* (Wien) **10**: 636–644.

Apfelbeck, V. 1907: Neu Koleopteren, gesammelt während einer im Jahre 1905 mit Subvention der Kaiserlichen Akademie der Wissenschaften in Wien durchgeführten Forschungsreise nach Albanien und Montenegro

(I. Serie). *Sitzungsberichte der Kaiserlichen Akademie der Wissenschaften in Wien, Mathematish-Naturwissenschaftiche Klasse* 115 (Abt. I): 1663–1674.

Apfelbeck, V. 1914: Revision der Palaearktischen Mordella-Arten aus der aculeata-Gruppe. *Annales Historico-Naturales Musei Nationalis Hungarici* 12: 605–622.

Apfelbeck, V. 1918: Diagnosen neuer Koleopteren von der Balkanhalbinsel. *Anzeiger der Kaiserlichen Akademie der Wissenschaften in Wien, Mathematisch-naturwissenschaftliche Klasse* 55: 88–90.

Apfelbeck, V. 1931: Neue Anaspis-Arten aus Jugoslavien. *Entomologischer Anzeiger* 11: 37–39.

Aragona, L.A. 1830: *De quisusquam insectis Italiae novis aut rarioribus.* Ticini Regii: Typographia Bizzomi, 31 pp.

Ardoin, P. 1958: Une nouvelle espèce d'Asida de la Faune de France (Coleopt. Tenebrionidae). *L'Entomologiste* 14: 37–41.

Ardoin, P. 1959: Contribution à l'étude des Helopinae de France (Col. Tenebrionidae). *Annales de la Société Entomologique de France* 127 [1958]: 9–49.

Ardoin, P. 1960a: Missions de MM.G. Colas et J. Mateu à l'archipel de Madère. Les ténébrionides (Col.). *Revue Française d'Entomologie* 27: 103–143.

Ardoin, P. 1960b: Une nouvelle espèce d'Uloma (Tenebrionidae) de l'Inde. *Entomologische Arbeiten aus dem Museum G. Frey* 11: 273–275.

Ardoin, P. 1964: Contribution à l'étude du genre Enicmosoma Geb. (Col. Tenebrionidae). *Entomologische Arbeiten aus dem Museum G. Frey* 15: 686–690.

Ardoin, P. 1965a: Contribution à l'étude des ténébrionides africains (Col.). *Bulletin de la Société Entomologique de France* 70: 73–85, pl. II.

Ardoin, P. 1965b: Note préliminaire sur les ténébrionides récoltés aux Monts Loma (Sierra-Leone). *Bulletin de l'Institut Français d'Afrique Noire* 27: 1315–1320.

Ardoin, P. 1969: Le Parc National du Niokolo-Koba, Fascicule III. XII. Coleoptera Tenebrionidae (note récapitulative) *Mémoires de l'Institut Fondamental d'Afrique Noire* 84: 247–261.

Ardoin, P. 1971a: Tenebrionidae (Coleoptera) récoltés par l'expédition Mares dans l'Archipel de la Galite, Tunisie. *Nouvelle Revue d'Entomologie* 1: 45–52.

Ardoin, P. 1971b: Contribution à l'étude des espèces africaines et malgaches du genre Anemia Laporte (Col. Tenebrionidae). *Annales de la Société Entomologique de France* (N.S.) 7: 357–422.

Ardoin, P. 1972: Deux nouvelles espèces du genre Isocerus Latr. (Col. Tenebrionidae). *Nouvelle Revue d'Entomologie* 2: 239–241.

Ardoin, P. 1973: Contribution à l'étude des Tenebrionidae (Coleoptera) de Sardaigne. *Annales de la Société Entomologique de France* (N.S.) 9: 257–307.

Ardoin, P. 1978a: Récoltes dans le Peloponnese oriental et description d'une sous-espèce nouvelle de Tenebrionidae (Coleoptera). *Entomops* 45: 133–140.

Ardoin, P. 1978b: Contribution à l'étude des Adesmia (Coleoptera: Tenebrionidae) du Sinaï, d'Israël et de Jordanie. *Nouvelle Revue d'Entomologie* 8: 295–313.

Ardoin, P. 1979: Une nouvelle espèce d'Akis (Coleoptera: Tenebrionidae) de l'île de Formentera, Baléares. *Entomologische Blätter* 74: 155–157.

Arnett, R.H. 1951: A revision of the Nearctic Oedemeridae (Coleoptera). *The American Midland Naturalist* 45: 257–391.

Aubé, C. 1842: Notes sur quelques coléoptères nouveaux. *Annales de la Société Entomologique de France* 11: 225–237.

Aubé, C. 1850: Description de quelques insectes coléoptères appartenant à l'Europe et à l'Algérie. *Annales de la Société Entomologique de France* (2) 8: 299–346.

Aubé, C. 1861: Description de six espèces nouvelles de coléoptères d'Europe dont deux appartenant a deux genres nouveaux et aveugles. *Annales de la Société Entomologique de France* (4) 1: 195–199.

Audinet-Serville, J.G.A. 1825: [new taxa]. In: *Encyclopédie méthodique ou par ordre de matières; par une Société de gens de lettres, de savants et d'artistes; Précédée d'un Vocabulaire universel, servant de Table pour tout l'Ouvrage, ornée des Portraits de MM. Diderot & d'Alembert, premiers Éditeurs de l'Encyclopédie. Histoire Naturelle. Entomologie, ou Histoire naturelle des Crustacés, des Arachnides et des Insectes. Tome dixième.* Paris: Agasse, (6) +832 + [1] pp.

Aurivillius, C. 1890: Neue Käfer aus Afrika. *Entomologisk Tidskrift* 11: 203–205.

Bach, M. 1856: *Käferfauna für Nord- und Mitteldeutschland mit besonderer Rücksicht auf die preussischen Rheinlande. Dritter Band.* Coblenz: J. Bölscher, 292 pp.

Báguena, Corella L. 1948: *Estudio sobre los Aderidae*. Instituto de Estudios Africanos, Madrid, xvi + 547 pp. + 1 p. [errata].

Báguena, Corella L. 1962: *Exploration du Parc National de la Garamba – Mission H. De Saeger. Fascicule 26 Aderidae (Coleoptera, Heteromeroidea)*. Bruxelles: Institut des Parcs Nationaux du Congo et du Ruanda-Urundi, 135 pp.

Ba, Y.-B. & Ren, G.-D. 2004: Taxonomic study on the genus Syachis Bates from China and its adjacent regions (Coleoptera, Tenebrionidae). *Acta Zootaxonomica Sinica* 29: 761–765.

Ba, Y.-B. & Ren, G.-D. 2009: Taxonomy of Promethis Pascoe (Coleoptera, Tenebrionidae) on Hainan Island, China, *Zootaxa* **2064**: 27–38.

Ba, Y.-B. & Ren, G.-D. 2013a: Two new species and one new combination of Stenosini (Coleoptera: Tenebrionidae) from Xizang, China. *Acta Entomologica Musei Nationalis Pragae* 53: 697–702.

Ba, Y.-B. & Ren, G.-D. 2013b: Taxonomy and distribution of Sternotrigon Skopin in China (Coleoptera, Tenebrionidae). *Zootaxa* **3693**: 568–578.

Bai, M. & Ren, G.-D. 2003: Four new species of the genus Cteniopinus Seidlitz, 1896 from China (Coleoptera, Tenebrionidae, Alleculinae). *Acta Zootaxonomica Sinica* 28: 682–687.

Bai, M. & Ren, G.-D. 2004: Coleoptera: Tenebrionidae: Alleculinae. Pp. 320–323. In: Yan, X.-K., Huang, F.-S. & Xue, D. (eds.): *Insects from Mt. Shiwandashan Area of Guangxi.* Beijing, China Forestry Publishing House, 668 pp. (in Chinese with English descriptions).

Bai, X.-L., Merkl, O. &, Ren, G.-D. 2019: Revision of the genus Bioramix Bates, 1879 from Nepal (Coleoptera, Tenebrionidae: Platyscelidini). *Entomological Review* **99**: 898–905.

Bai, X.-L. & Ren, G.-D. 2016: Two new species of the subgenus Cardiobioramix Kaszab from China (Coleoptera: Tenebrionidae: Bioramix). *Zoological Systematics* 41: 186–194.

Bai, X.-L. & Ren, G.-D. 2019a: Revision of the genus Platyscelis Latreille, 1818 from China (Coleoptera: Tenebrionidae: Platyscelidini). *Zootaxa* **4609**: 101–126.

Bai, X.-L. & Ren, G.-D. 2019b: [new taxa]. In: Bai X.-L., Li X.-M. & Ren G.-D.: Revision of the genus Oodescelis Motschulsky, 1845 from China (Coleoptera: Tenebrionidae: Platyscelidini). *Zootaxa* **4556**: 401–430.

Bakhuysen 1806: [new taxa]. In: Voet, J.E.: *Catalogus systematicus Coleopterorum – Catalogue systématique des coléoptères – Systematische naamlyst van dat geslacht van Insecten dat men Torren noemt. Volume* 2. La Haye: Bakhuysen, 82 + 85 + 87 pp.

Ballion, E.E. 1878: Verzeichniss der im Kreise von Kuldsha gesammelten Käfer. *Bulletin de la Société Impériale des Naturalistes de Moscou* **53** (1): 253–389.

Ballion, E.E. 1888a: Kurze Notizen ueber einige russische Blaps-Arten. II. Artickel. *Bulletin de la Société Impériale des Naturalistes de Moscou* (N.S.) **2**: 269–276.

Ballion, E.E. 1888b: Kurze Notizen ueber einige russische Blaps-Arten. III. Artickel. *Bulletin de la Société Impériale des Naturalistes de Moscou* (N.S.) **2**: 694–704.

Batelka, J. 2009a: A new species of the genus Ripiphorus Bosc, 1791 (Coleoptera: Ripiphoridae: Ripiphorinae) from the Arabian Peninsula. *Fauna of Arabia* **24**: 155–161.

Batelka, J. 2009b: [new taxon]. In: Batelka J. & Geisthardt M.: Order Coleoptera, family Meloidae. *Arthropod fauna of the UAE* **2**: 187–206.

Batelka, J. 2012: Ptilophorus purcharti sp. nov., the first ripiphorid from Socotra Island, with an account of the biogeography of the Ptilophorini (Coleoptera: Ripiphoridae). *Acta Entomologica Musei Nationalis Pragae* **52**, *Supplementum*: 269–285.

Batelka, J. & Hájek, J. 2009: A taxonomic review of the genus Eorhipidius (Coleoptera: Ripiphoridae: Ripidiinae), with descriptions of three new species from Asia. *Acta Entomologica Musei Nationalis Pragae* **49**: 769–782.

Bates, F. 1870: Descriptions of new genera and species of Heteromera. *The Entomologist's Monthly Magazine* **6** [1869–1870]: 268–275.

Bates, F. 1872: Notes on Heteromera, and description of new genera and species (No. 1). *The Entomologist's Monthly Magazine* **9** [1872–1873]: 97–99.

Bates, F. 1873a: Notes on Heteromera, and description of new genera and species (No. 8), (No. 9). *The Entomologist's Monthly Magazine* **9** [1873–1874]: 14–17, 45–52.

Bates, F. 1873b: Notes on Heteromera, and description of new genera and species. *The Entomologist's Monthly Magazine* **9** [1872–1873]: 201–204.

Bates, F. 1879: Characters of the new genera and species of Heteromera collected by Dr. Stoliczka during the Forsyth Expedition to Kashgar in 1873–74. *Cistula Entomologica* **2** [1875–1882]: 467–484.

Bates, F. 1890: Heteromera. Pp. 55–79. In: Ball, V. (ed.): *Scientific results of the second Yarkand Mission; based upon the collections and notes of the late Ferdinand Stoliczka, Ph. Dr.* Calcutta: Office of superintendent of Goverment Printing, India, 79 pp., 2 pls.

Batten, R. 1977: Two new Mordellidae (Coleoptera) from Southern Europe, and a key to the Mordellistena micans group. *Entomologische Berichten* **37**: 167–176.

Batten, R. 1978: New Mordellidae (Coleoptera) from Greece and Tunesia. *Entomologische Berichten* **38**: 78–80.

Batten, R. 1980: Notes on the Mordellistena gemellata group: two new species and a case of synonymy. *Entomologische Berichten* **40**: 41–46.

Batten, R. 1983a: A new Mordellistena from Asia Minor. *Entomologische Berichten* **43**: 88–89.

Batten, R. 1983b: Two new Ectasiocnemis (Anaspidinae, Coleoptera Mordellidae) from Afganistan and a case of synonymy. *Bonner Zoologische Beiträge* **34**: 485–488.

Batten, R. 1989: Two new Mordellidae (Coleoptera) from Kashmir and Sumatra. *Tijdschrift voor Entomologie* **132**: 105–107.

Baudi di Selve, F. 1870: Coleopterorum messis in insula Cypro et Asia minore ab Eugenio Truqui congregate recensitio: de Europaeis notis quibusdam additis. Pars tertia. *Berliner Entomologische Zeitschrift* **14**: 49–90.

Baudi di Selve, F. 1874a: Catalogo dei tenebrioniti della fauna europea e circummediterranea del Museo Civico di Genova. *Annali del Museo Civico di Storia Naturale di Genova* **6**: 89–115.

Baudi di Selve, F. 1874b: Coleotteri tenebrioniti delle collezioni italiane. Parte prima. *Bullettino della Società Entomologica Italiana* **6**: 183–200.

Baudi a Selve, F.1874c: Coleopterorum messis in insula Cypro et Asia Minore ab Eugenio Truqui congregatae recensitio: de Europaeis notis quibusdem additis. Pars quinta. *Berliner Entomologische Zeitschrift* **17** [1873]: 317–338.

Baudi di Selve, F. 1875a: Coleotteri tenebrioniti delle collezioni italiane. *Bullettino della Società Entomologica Italiana* **6** [1874]: 275–301.

Baudi di Selve, F. 1875b: Coleotteri tenebrioniti delle collezioni italiane. *Bullettino della Società Entomologica Italiana* **7**: 3–36.

Baudi di Selve, F. 1875c: Europaeae et circummediterraneae faunae Tenebrionidum specierum, quae Comes Dejean in suo Catalogo, editio 3ª consignavit, ex ejusdem collectione in R. Taurinensi Musaeo asservata, cum auctorum hodierne denominatione collatio. *Deutsche Entomologische Zeitschrift* **19**: 17–119.

Baudi di Selve, F. 1875d: Catalogo dei tenebrioniti della fauna europea e circummediterranea appartenenti alle collezioni del Museo Civico di Genova. Parte seconda. *Annali del Museo Civico di Storia Naturale di Genova* **7**: 684–703.

Baudi di Selve, F. 1876a: Europaeae et circummediterraneae faunae Tenebrionidum specierum, quae Comes Dejean in suo Catalogo, editio 3ᵉ consignavit, ex ejusdem collectione in R. Taurinensi Musaeo asservata, cum auctorum hodierne denominatione collatio. Pars altera. *Deutsche Entomologische Zeitschrift* **20**: 1–74.

Baudi di Selve, F. 1876b: Coleotteri tenebrioniti delle collezioni italiane. *Bullettino della Società Entomologica Italiana* **7** [1875]: 91–106, 137–165, 209–237.

Baudi di Selve, F. 1876c: Europaeae et circummediterraneae faunae Tenebrionidum specierum, quae Comes Dejean in suo Catalogo, editio 3ᵉ consignavit, ex ejusdem collectione in R. Taurinensi Musaeo asservata, cum auctorum hodierne denominatione collatio. Pars tertia. *Deutsche Entomologische Zeitschrift* **20**: 225–267.

Baudi di Selve, F. 1876d: Generis Helops specierum methodicae dispositionis tentamen. *Deutsche Entomologische Zeitschrift* **20**: 268–281.

Baudi di Selve, F. 1877a: *Coleotteri eteromeri esistenti nelle collezioni del R. Museo Zoologico di Torino ed in altre Italiane. Eteromeri delle famiglie susseguenti a quella dei tenebrioniti nei limiti della fauna europea e circum-mediterranea. Volume 1.* Torino: Stamperia Reale, 163 pp. [note: this is a separate issue from the *Atti della Reale Accademia delle Scienze di Torino* **12**: 571–729].

Baudi di Selve, F. 1877b: Coleotteri tenebrioniti delle collezioni italiane. *Bullettino della Società Entomologica Italiana* **9**: 25–54.

Baudi di Selve, F. 1877c: Europae et circummediterraneae faunae Heteromerum specierum, quae Comes Dejean in suo Catalogo, editio 3ª, consignavit, ex ejusdem collectione in R. Taurinensi Musaeo asservata, cum auctorum hodiernae denominatione collatio. Pars quarta. *Deutsche Entomologische Zeitschrift* **21**: 385–416.

Baudi di Selve, F. 1877d: Coleotteri eteromeri esistenti nelle collezioni del R. Museo Zoologico di Torino ed in altre Italiane. Eteromeri delle famiglie susseguenti a quella dei tenebrioniti nei limiti della fauna europea e circummeditteranea. *Atti della Reale Accademia delle Scienze di Torino* **12**: 571–729.

Baudi di Selve, F. 1878a: Europaeae et circummediterraneae faunae Heteromerum specierum, quae Comes Dejean in suo Catalogo, edito 3a, consignavit, ex ejusdem collectione in R. Taurinensi Musaeo asservata, cum auctorum hodierne recepta denominatione collatio. Pars quinta. *Deutsche Entomologische Zeitschrift* **22**: 1–20.

Baudi di Selve, F. 1878b: Europaeae et circummediterraneae faunae Heteromerum specierum, quae Comes Dejean in suo Catalogo, editio 3a, consignavit, ex ejusdem collectione in R. Taurinensi Musaeo asservata, cum auctorum hodierne recepta denominatione collatio. *Deutsche Entomologische Zeitschrift* **22**: 337–376.

Baudi di Selve, F. 1878c: Coleotteri eteromeri esistenti nelle collezioni del R. Museo zoologico di Torino ed in altre italiane. *Atti della Reale Accademia delle Scienze di Torino* **13**: 765–866, 1027–1183.

Baudi di Selve, F. 1879: [Une espèce nouvelle de *Formicomus*, de la collection du Musée royal.] *Annales de la Société Entomologique de Belgique* **22**: 87–88.

Baudi di Selve, F. 1881: Heteromerum species ex Aegypto, Syria et Arabia. *Deutsche Entomologische Zeitschrift* **25**: 273–296.

Baudi di Selve, F. 1883a: Note entomologiche. *Il Naturalista Siciliano* **3** [1883–1884]: 1–8.

Baudi di Selve, F. 1883b: Anthicus blechroides Baudi nov. spec. *Deutsche Entomologische Zeitschrift* **27**: 150.

Baudi di Selve, F. 1883c: Descrizione d'una nuova specie di Zophosis. *Annali del Museo Civico di Storia Naturale di Genova* **18**: 760–761.

Baudi di Selve, F. 1892: Di una nuova specie del genere Tritoma Geoffr. (Mycetophagus Hellw.). *Il Naturalista Siciliano* **11**: 153–154.

Baudi di Selve, F. 1894: Viaggio del Dr. E. Festa in Palestina, nel Libano e regioni vicine. *Bollettino dei Musei di Zoologia e Anatomia Comparata della Reale Università di Torino* **9** (173): 1–13.

Beauregard, H. 1890: *Les insectes vésicants*. Paris: Félix Alcan, xvi + 544 pp., 34 pls.

Beck, L. von. 1817: *Beiträge zur beierischen Insektenfaune, oder Beschreibung und Abbildung neuentdeckter Käfer, mit angehängtem Namensverzeichniss der Eleuteraten des Landgerichtbezirks Zusmeshausen*. Augsburg: J. Wolff, 45pp., 7 pls.

Bedel, E.M.L. 1869: [Une nouvelle espèce française du genre Notoxus.] *Bulletin de la Société Entomologique de France* **1869**: 34–35.

Bedel, E.M.L. 1891a: Description de deux coléoptères hétéromères du nord de l'Afrique. *Annali del Museo Civico di Storia Naturale di Genova* **30**: 794–796.

Bedel, E.M.L. 1891b: [Diagnose de trois coléoptères nouveaux du nord de l'Afrique.] *Bulletin de la Société Entomologique de France* **1890**: cxxxv-cxxxvi.

Bedel, E.M.L. 1892: Reinseignements sur les moeurs et le développement des Meloidae. *L'Abeille, Journal d'Entomologie* 27: 235–250.

Bedel, E.M.L. 1893: Diagnoses de coléoptères nouveaux d'Algérie. *L'Abeille, Journal d'Entomologie* 28: 102–104.

Bedel, E.M.L. 1894: Révision des Omophlus et Heliotaurus de Barbarie. *L'Abeille, Journal d'Entomologie* 28: 157–169.

Bedel, E.M.L. 1895: [new taxon]. In: Bedel L.E.M. & Chobaut A.: Étude sur les Macrosiagon Hentz (Emenadia Lap.-Cast.) de la région méditerranéenne. *L'Abeille, Journal d'Entomologie* 28: 181–192.

Bedel, E.M.L. 1895: [new taxon]. In: Vauloger de Beaupré M.: Description de deux Notoxus nouveaux du nord de l'Afrique. *L'Abeille, Journal d'Entomologie* 28: 193.

Bedel, E.M.L. 1899: Diagnose d'un nouveau mylabre saharien. *Bulletin de la Société Entomologique de France* **1899**: 382–383.

Bedel, E.M.L. 1903: Entomologie. Liste des principales espèces de coléoptères recueillis par le Mis de Segonzec, pp. 367–371. In: *Voyage au Maroc (1899–1901) par le Mis de Segonzec*. Paris: Armand Colin, xi + 408 + (1) pp., 11 pls, 1 map.

Bedel, E.M.L. 1904: Communications. Liste de coléoptères récoltés à la Ferté-Alais (Seine-et-Oiso). *Bulletin de la Société Entomologique de France* **1904**: 210–212.

Bedel, E.M.L. 1905: Communications. Description de deux coléoptères hétéromères nouveaux, de la côte occidentale du Maroc. *Bulletin de la Société Entomologique de France* **1905**: 272–273.

Bedel, E.M.L. 1906a: Description de deux coléoptères hétéromères nouveaux de la côte occidentale du Maroc. *Bulletin de la Société Entomologique de France* **1905**: 272–273.

Bedel, E.M.L. 1906b: Synonymies de coléoptères paléarctiques. *Bulletin de la Société Entomologique de France* **1906**: 91–93.

Bedel, E.M.L. 1906c: Observations divers. *Bulletin de la Société Entomologique de France* **1906**: 177–178.

Bedel, E.M.L. 1918: Diagnoses de coléoptères nouveaux du Maroc oriental. *Bulletin de la Société Entomologique de France* **1917**: 362–364.

Bedel, E.M.L. 1921: Description d'un Zonabris de l'Ahaggar (Sahara central) (Col. Meloidae). *Bulletin de la Société Entomologique de France* **1921**: 199.

Belon, R.P.M.J. 1882: [Des observations sur divers lathridiens]. *Bulletin de la Société Entomologique de France* **1882**: cxxiii-cxxv.

Berthold, A. 1827: *Latreille's natürliche Familien des Thierreichs. Aus dem Französischen mit Anmerkungen und Zusätzen von Dr. Arnold Adolph Berthold*. Weimar: Landes-Industrie-Comptoir, x + 602 pp.

Bertolini, S. 1868: Neue Käferarten des Trentinogebietes. *Verhandlungen der Kaiserlich-Königlichen Zoologisch-Botanischen Gesellschaft in Wien* 18: 119–120.

Bertolini, S. 1892: Contribuzione alla fauna Trentina dei Coleotteri. *Bullettino della Società Entomologica Italiana* **23**: 346–368.

Besser, W.S.J.G. von. 1832: Additamenta et observatiunculae in Cel. Steven Tentyrias et Opatra collectionis Stevenianae nunc musei Universitatis Mosquensis. *Nouvelles Mémoires de la Société Impériale des Naturalistes de Moscou* 2: 5–21.

Betrem, G. 1929: Systematiek der Horiini. *Tijdschrift voor Entomologie* 72: xxvi-xxix.

Betrem, G. 1932: Beiträge zur Kenntnis des Tribus Horiini der Familie der Meloidae (Col.). 1. Die Systematik der Horiini. *Treubia* **14**: 85–101, 1 pl.

Betta, V. 1847: *De quibusdam Coleopteris agri Ticinensis. Dissertatio*. Ticini regii: Fusi et Socii, 35

Biggs, H.E.J. 1945: Two new species of Coleoptera (Scarabaeidae and Meloidae). *The Entomologist's Monthly Magazine* **81**: 110–111.

Billberg, G.J. 1813: *Monographia Mylabridum*. Holmiae: Caroli Delén, 74 + 6 pp., 7 pls.

Billberg, G.J. 1815: Insecta ex ordine Coleopterorum descripta. *Nova Acta Upsalae* 7: 271–281, pl. IX.

Billberg, G.J. 1820: *Enumeratio insectorum in Museo Gust. Joh. Billberg*. Stockholm: Typis Gadelianis, [4] + 138 pp.

Binaghi, G. 1937: Studio sul genere Langelandia Aubé e descrizione di un nuovo genere di Coxelini (Col. Colydiidae). *Bollettino della Società Entomologica Italiana* **69**: 69–79.

Biström, O. 1977: Nomenclatoric notes on Coleoptera. *Notulae Entomologicae* **57**: 17–18.

Blackburn, T. 1888: Further notes on Australian Coleoptera, with description of new species. *Transactions of the Royal Society of South Australia* **10** [1886–1887]: 52–71, 177–287.

Blackburn, T. 1891: Further notes on Australian Coleoptera, with descriptions of new genera and species. X. *Transactions of the Royal Society of South Australia* **14**: 292–345.

Blackburn, T. 1907: Further notes on Australian Coleoptera, with description of new genera and species, xxxvii. *Transactions of the Royal Society of South Australia* **31**: 231–299.

Blair, K.G. 1912: Descriptions of new species of Pyrochroidae. *The Annals and Magazine of Natural History* (8) **9**: 527–533.

Blair, K.G. 1913: Some new species of Indian Tenebrionidae. *The Annals and Magazine of Natural History* (8) **12**: 56–58.

Blair, K.G. 1914a: A revision of the family Pyrochroidae (Coleoptera). *The Annals and Magazine of Natural History* (8) **13**: 310–326.

Blair, K.G. 1914b: On the Fabrician types of Tenebrionidae (Coleoptera) in the Banks collection. *The Annals and Magazine of Natural History* (8) **13**: 482–490.

Blair, K.G. 1914c: Notes on Coleoptera of the genus Azarelius Fairm. (Tenebrionidae), with descriptions of new species. *Annals and Magazin e of Natural History* (8) **14**: 315–316.

Blair, K.G. 1919: Notes on the Pythidae with descriptions of new species. *The Entomologist's Monthly Magazine* **55**: 112–124.

Blair, K.G. 1921: Notes on the Indian species of Hypophloeus, Fabr. with description of new species. *The Entomologist's Monthly Magazine* (3) 7 [1920]: 1–7.

Blair, K.G. 1922a: A new genus and some new species of Mordellidae. *The Entomologist´s Monthly Magazine* **58**: 221–226.

Blair, K.G. 1922b: The generic name Calycella Blair (Coleoptera). *The Entomologist´s Monthly Magazine* **58**: 250.

Blair, K.G. 1922c: Coleoptera of the Mt. Everest Expedition, 1921. *The Annals and Magazine of Natural History* (9) **9**: 558–562.

Blair, K.G. 1923a: New species of heteromerous Coleoptera from Mesopotamia. *The Entomologist's Monthly Magazine* **59**: 118–126.

Blair, K.G. 1923b: Coleoptera of the Second Mt. Everest Expedition, 1922. Part II. Heteromera. *The Annals and Magazine of Natural History* (9) **11**: 278–285.

Blair, K.G. 1927: Heteromera of the Third Mt.Everest Expedition, 1924. *The Annals and Magazine of Natural History* (9) **19**: 241–255.

Blair, K.G. 1928: Pythidae. Pars 99. In: Schenkling S. (ed.): *Coleopterorum Catalogus*. Berlin and s'Gravenhage: W. Junk, 41 pp.

Blair, K.G. 1929: Some new species of myrmecophilous Tenebrionidae (Col.). *Zoologischer Anzeiger* 28 (*Wasmann-Festband*): 238–247.

Blair, K.G. 1930a: The Indian species of Palorus Muls. (Coleoptera: Tenebrionidae) and some associated beetles. *Indian Forest Records, Entomological Series* (Calcutta) **14**: 133–152, 1 pl.

Blair, K.G. 1930b: Some new species of Indian Heteromera. *The Entomologist's Monthly Magazine* **66**: 177–181.

Blair, K.G. 1931a: Some new species of Indian Heteromera (Col.). *The Entomologist's Monthly Magazine* **67**: 199–205.

Blair, K.G. 1931b: A collection of Coleoptera from the Rub'Al Khali, southern Arabia with descriptions of some new species and varieties. *The Entomologist's Monthly Magazine* **67**: 269–271.

Blair, K.G. 1933: Tenebrionidae from Central Arabia collected by Major H. Sr. J.B. Philby. *The Entomologist's Monthly Magazine* **69**: 4–7.

Blair, K.G. 1935a: Some synonymic notes in the family Tenebrionidae (Col.). *The Entomologist's Monthly Magazine* **71**: 102–104.

Blair, K.G. 1935b: Oedemeridae. Pp. 302–303. In: Visser, C. & Visser-Hooft, J. (eds.): *Wissenschaftliche Ergebnisse der niederländischen Expeditionen in den Karakorum und die angrenzenden Gebiete in den Jahren 1922, 1925 und 1929/30*. Leipzig: F.A. Brockhaus, 501 + xviii pp., 8 pls.

Blair, K.G. 1937: Some new species of Indian Heteromera (Col.) (3). *The Entomologist Monthly Magazine* **73**: 35–37.

Blair, K.G. 1938: A new genus and species of tenebrionid beetles in bee-hives in India. *The Entomologist's Monthly Magazine* **74**: 222–223.

Blair, K.G. 1942: Coleoptera Heteromera from Guam, pp. 56–60. In: Insects of Guam 1. *Bulletin of the Bernice P. Bishop Museum* **172**: v + 218 pp., 7 pls.

Blanchard, É. 1843: Insectes de l'Amérique meridionale recueillis par Alcide d'Orbigny. In: Blanchard, É. & Brullé, A.: *Insectes du Voyage dans l'Amérique méridionale executé pendant les années 1828, 1829, 1830, 1831, 1832 et 1833 par Alcide d'Orbigny. Tome sixième 2e partie: Insectes.* [1837–1843]. Paris: P. Bertrand et Strasburg: V. Levrault, 222 pp., 32 pls.

Blanchard, É. 1844: Zoologie. Insectes recueillis à l'Himalaya par Victor Jacquemont. Pp. 13–31. In: Voyage dans l'Inde par Victor Jacquemont pendant les années 1828 à 1832. Publié sous les auspices de M. Guizot, Ministre de l'Instruction Publique. Descriptions des collections par MM. Isidore Geoffroy Saint-Hilaire. – Milne Edwards. – Émile Blanchard. Valenciennes. – Cambessèdes. – J. Decaisne. Tome quatrième. Paris: Typographie de Firmin Didot Frères [issued in parts].

Blanchard, É. 1845: *Histoire des insectes, traitant de leur moeurs et de leur métamorphoses en générale et comprenant une nouvelle classification fondée sur leur rapports naturelles*. In: *Traité Complet d'Histoire Naturelle. Histoires des insectes. Tome second.* Paris: Firmin Didot Frères, 524 pp., pls. 11–20.

Blanchard, É. 1853: Insectes. In: Hombrom, J.B. & Jacquinot, H. (eds): Atlas d'Histoire naturelle Zoologie. In: Voyage au Pôle Sud et dans l'Océanie sur les corvettes l'Astrolabe et la Zélée, executé par l'ordre du Roi pendant les année 1837–1838–1839–1840 sous le commandement de M.J. Dumont-d'Urville, Capitaine de vaisseau; Zoologie. Tome Quatrième. Deuxième partie. Coléoptères et autres ordres. Paris: J. Tastu, vii + 716 pp., 20 pls. [plates issued in 1847].

Blanco Villero, J.M. & Sáez Bolaño, J.A. 2011: Descripción de una nueva especie del género *Prionychus* Solier, 1835 de Marruecos y redescripción de *Prionychus anthracinus* (Mulsant, 1856) y *Prionychus lugens* (Küster, 1850) (Coleoptera, Tenebrionidae, Alleculinae). *Graellsia* **67**: 175–185.

Bland, J.H.B. 1863: Descriptions of a few supposed new species of North American Coleoptera. No. 2. *Proceedings of the Entomological Society of Philadelphia* **2**: 319.

Bleuse, L. 1899: Description d'un mylabre du Sud-Oranais (Col.). *Bulletin de la Société Entomologique de France* **1899**: 383.

Blinstein, S.J. 1988: Novye i maloizvestinye vidy bistryanok (Coleoptera, Anthicidae) s yuga Ukraini. Pp. 73–78. In: *Ekologiya i taksonomiya nasekomikh Ukrainskoi SSR. Sbornik Nauchnikh Trudov* [*Ecology and taxonomy of Ukrainian insects. Collection of Scientific Works*]. Kiev: Akademiya Nauk Ukrainskoy SSR.

Bodemeyer, B. von. 1916: Neuheiten der paläarktischen Coleopterenfauna. *Deutsche Entomologische Zeitschrift* **61**: 111–112.

Bodemeyer, B. von. 1927: *Ueber meine Entomologischen Reisen nach Kleinasien (1911), Ost-Sibirien, Schilka und Amur (1912), Tunis, Oasis Gafsa, Khroumerie (1913) und Iran, das Elbursgebirge (1914). Bd. IV. Iran, das Elbursgebirge*. Stuttgart: Albert Kerner, 96 pp., 1 pl.

Bogatchev, A.V. 1934: Materialy k poznaniyu fauny zhukov Apsheronskogo poluostrova. *Trudy Azerbaydzhanskogo Otdeleniya Zakavkazkogo Filiala Akademii Nauk SSSR, Sektsiya Zoologii* **7**: 14–71.

Bogatchev, A.V. 1939a: Novye vidy sem. Tenebrionidae. *Trudy Zoologicheskogo Instituta Akademii Nauk SSSR – Azerbaidschanskiy Filial* **10**: 123–130.

Bogatchev, A.V. 1939b: Novye formy Mordellidae (Coleoptera) iz AzSSR. *Trudy Zoologicheskogo Instituta Akademiya Nauk SSSR – Azerbaidschanskiy Filial* **10**: 131–134.

Bogatchev, A.V. 1939c: Fauna Binagadinskikh Kirovykh Plastov. Zhuki-Coleoptera. *Izvestiya Azerbaydzhanskogo Filiala Akademii Nauk SSSR* **2**: 137–160.

Bogatchev, A.V. 1940: Kharakternye pribavki k faune SSSR po otryadu zhestkokrylykh (Coleoptera). I. *Byulleten Moskovskova Obshchestva Ispitalelov Prirody, Otdel Biologii* **49**: 201–209.

Bogatchev, A.V. 1941a: Novye vidy roda Adesmia Fisch.-W. *Izvestiya Azerbaydzhanskogo Filiala Akademii Nauk SSSR* **3**: 96–100.

Bogatchev, A.V. 1941b: Novye vidy triby Blaptini (Coleoptera). *Izvestiya Azerbaydzhanskogo Filiala Akademii Nauk SSSR* **4**: 109–111.

Bogatchev, A.V. 1946a: Dve novykh Microdera Eschsch. (Coleoptera, Tenebrionidae). *Doklady Akademii Nauk AZSSR* (*Proceedings of the Academy of Sciences of the Azerbaijan SSR*) **2** (6): 210–211.

Bogatchev, A.V. 1946b: Novyy Cylindronotus iz doliny Kury. *Doklady Akademii Nauk AZSSR* (*Proceedings of the Academy of Sciences of the Azerbaijan SSR*) **2** (6): 245–247.

Bogatchev, A.V. 1946c: Novye chernotelki (Tenebrionidae) Palearctiki. *Doklady Akademii Nauk Azerbaydzhanskoy SSR* (*Proceedings of the Academy of Sciences of the Azerbaijan SSR*) **2** (9): 391–394.

Bogatchev, A.V. 1946d: Novye vidy triby Platyscelini (Coleoptera, Tenebrionidae). *Doklady Akademii Nauk Azerbaydzhanskoy SSR* **2** (10): 435–437.

Bogatchev, A.V. 1947a: Quelques espèces nouvelles des coléoptères d'Iran. *Entomologie et Phytopathologie Appliquées* (Teheran) **4**: 15–18.

Bogatchev, A.V. 1947b: Novyy vid roda Cyphostethe Mars. iz doliny Araksa (Coleoptera, Tenebrionidae). *Doklady Akademii Nauk Armyanskoy SSR* (*Proceedings of the Academy of Sciences of the Armenian SSR*) **7** (2): 79–81.

Bogatchev, A.V. 1947c: Novye vidy roda Platynoscelis Seidl. *Doklady Akademii Nauk Azerbaydzhanskoy SSR* (*Proceedings of the Academy of Sciences of the Azerbaijan SSR*) **3** (2): 84–88.

Bogatchev, A.V. 1947d: Novye vidy Tentyriini. *Doklady Akademii Nauk Arerbaydzhanskoy SSR* **3** (10): 460–463.

Bogatchev, A.V. 1947e: K systematike triby Blaptini (Tenebrionidae, Col.). *Doklady Akademii Nauk Arerbaydzhanskoy SSR* **3** (11): 513–515.

Bogatchev, A.V. 1949a: Novye vidy Tenebrionidae Azerbaydzhana i drugikh oblastey Palearktiki. *Doklady Akademii Nauk Azerbaydzhanskoy SSR* **5** (1): 38–41.

Bogatchev, A.V. 1949b: Novye palearcticheskie vidy Epitragini i Pimeliini (Tenebrionidae). *Doklady Akademii Nauk Azerbaydzhanskoy SSR* **5** (7): 277–280.

Bogatchev, A.V. 1950: Novye maloizvestnye vidy triby Erodiini (Coleoptera, Tenebrionidae). *Entomologicheskoe Obozrenie* **31**: 231–236.

Bogatchev, A.V. 1951: Novye formy roda Lobodera Muls. (Coleoptera, Tenebrionidae). *Trudi Krimskogo Filiala Academemii Nauk SSSR* **2**: 69–72.

Bogatchev, A.V. 1952a: Novyy rod i vid zhukov-chernotelok iz severnych Kara-Kumov. *Doklady Akademii Nauk UZSSR* **7**: 44–45.

Bogatchev, A.V. 1952b: Novye chernotelki (Tenebrionidae) iz SSSR i sopredelnykh stran. *Entomologicheskoe Obozrenie* **32**: 284–286.

Bogatchev, A.V. 1953: Novye formy roda Pimelia F. (Coleoptera, Tenebrionidae). *Entomologicheskoe Obozrenie* **33**: 302–306.

Bogatchev, A.V. 1957: Novye vidy chernotelok rodov Zophosis Latr., Erodius F. i Earophanta Sem. *Zoologicheskiy Zhurnal* **36** (4): 529–532.

Bogatchev, A.V. 1959: Novye vidy zhukov-chernotelok (Tenebrionidae) iz Tadzhikistana. In: Zapovednik 'Tigrovaya Balka'. *Trudy Akademii Nauk Tadzhikskoy SSR* **115**: 63–67.

Bogatchev, A.V. 1960a: Novyy rod zhukov-chernotelok iz Tadzhikistana Allotadzhikistania, gen. nov. (Tenebrionidae, Pimeliini). *Doklady Akademii Nauk Tadzhikskoy SSR* **3** (1): 43–46.

Bogatchev, A.V. 1960b: Novyy svoeobraznyy rod zhukov-chernotelok, naydennykh v Tadzhikistane – Tadzhikistania, gen. nov. (Coleoptera, Tenebrionidae). *Doklady Akademii Nauk Tadzhikskoy SSR* **3** (3): 35–37.

Bogatchev, A.V. 1960c: Novye vidy Prosodes (Coleoptera, Tenebrionidae) iz Tadzhikistana. *Izvestiya Otdeleniya Selskokhozyaystvennych i Biologicheskikh Nauk* **2**: 79–86.

Bogatchev, A.V. 1960d: [new taxa]. In: Bogatchev A.V. & Kryzhanovskiy O.L.: Novye i maloizvestnye vidy chernotelok (Coleoptera, Tenebrionidae) iz Zapadnoy Turkmenii. *Trudy Zoologicheskogo Instituta Akademii Nauk SSSR* **27**: 264–275.

Bogatchev, A.V. 1961a: Novye vidy zhukov-chernotelok (Tenebrionidae) iz Tadzhikskoy SSR, drugikh respublik Sredney Azii i sopredelnykh stran. *Trudy Instituta Zoologii i Parazitologii Akademii Nauk Tadzhikskoy SSR* **20**: 107–121.

Bogatchev, A.V. 1961b: Novyy vid roda Diesia Fisch.-W. (Coleoptera, Tenebrionidae). *Entomologicheskoe Obozrenie* **40** (2): 369–370.

Bogatchev, A.V. 1962: Novye vidy triby Akidini (Tenebrionidae) iz Srednej Azii. *Doklady Akademii Nauk Tadzhikskoy SSR* **5** (1): 44–46.

Bogatchev, A.V. 1963a: Novyy rod zhukov-chernotelok iz pustiny Kyzylkum (Col., Tenebrionidae, Tentyriini). *Doklady Akademii Nauk Uzbekskoy SSR* **6**: 57–59.

Bogatchev, A.V. 1963b: Novye vidy chernotelok (Tenebrionidae) s yuga Sredney Azii. *Trudy Instituta Zoologii i Parazitologii im. akad. E.N. Pavlovskogo Akademii Nauk Tadzhikskoy SSR* **24**: 94–109.

Bogatchev, A.V. 1964a: Dve novye formy roda Pimelia Fabr. (Coleoptera, Tenebrionidae). *Uzbekskiy Biologichesdkiy Zhurnal* **2**: 42–45.

Bogatchev, A.V. 1964b: Vnetrevidovye formy chernotelki Adesmia fischeri Fald. 1837 (Coleoptera, Tenebrionidae). *Entomologicheskoe Obozrenie* **43** (1): 175–179.

Bogatchev, A.V. 1965: [new taxon, p. 372]. In: Bey-Bienko G. Ya. (ed.): *Opredilitel nasekomykh evropeyskoy chasti SSSR v pyati tomakh. II. Zhestkokrylye i veerokrylye.* Nauka, Moskva-Leningrad.

Bogatchev, A.V. 1967a: Novyy rod i vid zhukov-chernotelok s Apsheronskogo poluostrova. *Trudy Instituta Zoologii Akademii Nauk Azerbaydzhanskoy SSR* **26**: 157–163.

Bogatchev, A.V. 1967b: Chernotelki roda Epitrichia Gebl. (Coleoptera, Tenebrionidae). *Entomologicheskoe Obozrenie* **46**: 348–352.

Bogatchev, A.V. 1967c: [new taxon]. In: Medvedev G.S.: Obzor chernotelok triby Leptodini (Coleoptera, Tenebrionidae). *Entomologicheskoe Obozrenie* **46**: 353–360.

Bogatchev, A.V. 1972: Novye vidy zhukov-chernotelok triby Opatrini (Coleoptera, Tenebrionidae). *Entomologicheskoe Obozrenie* **51**: 625–632.

Bogatchev, A.V. 1976: Novyy rod i vid zhukov-chernotelok (Coleoptera, Tenebrionidae) iz Tadzhikistana. *Entomologicheskoe Obozrenie* **55**: 98–100.

Bogatchev, A.V. & Kryzhanovskiy, O.L. 1955: Novyy vid chernotelok (Coleoptera, Tenebrionidae) iz zapadnoy Turkmenii. *Entomologicheskoe Obozrenie* **34**: 240–241.

Bogatchev, A.V. & Kryzhanovskiy, O.L. 1960: Novye i maloizvestnye vidy chernotelok (Coleoptera, Tenebrionidae) iz zapadnoy Turkmenii. *Trudy Zoologicheskogo Instituta Akademii Nauk SSSR* **27**: 264–275.

Bogatchev, A.V. & Medvedev, G.S. 1974: Novyy vid zhukov-chernotelok roda Blaps Fabr. (Coleoptera, Tenebrionidae) iz Turkmenii. *Entomologicheskoe Obozrenie* **53** (3): 645–646.

Bogatchev, A.V. & Nepesova, M.G. 1980: Zhuki-chernotelki roda Pachyscelis Sol. (Coleoptera, Tenebrionidae) fauny Turkmenii. *Entomologicheskoe Obozrenie* **59** (3): 607–611.

Bogdanov-Katjkov, N.N. 1915: De speciebus novis vel parum cognitis Tentyriinorum (Coleoptera, Tenebrionidae). Novye i maloizvestnye vidy podsemeystva Tentyriinae (Coleoptera, Tenebrionidae). *Russkoe Entomologicheskoe Obozrenie* **15** (1): 1–7.

Bogdanov-Katjkov, N.N. 1916a: Matériaux pour servir à l'étude des Gnathosia (Coleoptera, Tenebrionidae). *Revue Russe d'Entomologie* **16**: 68–71.

Bogdanov-Katjkov, N.N. 1916b: Espèces nouvelles ou peu connues des tentyriines (Coleoptera, Tenebrionidae). *Ezhegodnik Zoologicheskago Muzeya Imperatorskoy Akademii Nauk* **20** [1915]: xxix-xxx.

Boheman, C.H. 1851: Bidrag till Gottlands insekten-fauna. *Öfversigt af Kongliga Vetenskaps-Akademiens Förhandlingar* **1849**: 194–267.

Boheman, C.H. 1858: Coleoptera. Species nova descripsit. Pp. 1–112. In: Virgin, C.H. (ed.): *Kongliga Svenska Fregatten Eugenies Resa omkring jorden unter befäll af C.A. Virgin ären 1851–1853. Vetenskapliga Iakttagelser pa H.M. Konung Oscart den Försten befallning utgifna af K. Svenska Vetenskaps Akademien. II. Zoologi. 1. Insecta*. Uppsala et Stockholm: P.A. Norsteds & Söner, Almquist et Wiksells, 614 pp., 9 pls. [issued in parts, pp. 113–217 in 1859].

Böhm, R. 1908: Les Adesmides des déserts égyptiens. *Bulletin de la Société Entomologique d'Egypte* **1** [1908–1909]: 151–165.

Böhm, R. 1909: Les Zophosini et Erodiini de l'Egypte. *Bulletin de la Société Entomologique d'Egypte* **2**: 108–127.

Böhm, R. 1916a: Sur les caractères spécifiques des Tentyria et Tentyrina d'Egypte. *Bulletin de la Société Entomologique d'Egypte* **4** [1914–1918]: 49–51.

Böhm, R. 1916b: Sur les caractères spécifiques des Ocnera égyptiens. *Bulletin de la Société Entomologique d'Egypte* **4** [1914–1918]: 64–69.

Boieldieu, A. 1865: Quelques coléoptères nouveaux des îles d'Eubée et Baléares. *Annales de Société Entomologique de France* (4) **5**: 5–11.

Boisduval, J.B.A.D. de. 1835: [new taxa]. In: Hombrom, J.B. & Jacquinot, H. (eds): Atlas d'Histoire naturelle Zoologie. In: Voyage de découvertes de l'Astrolabe exécuté par l'ordre du Roi, pendant les années 1826–1827–1828–1829, de M. Dumont d'Urville. Faune Entomologique de l'Océan pacifique, avec l'illustration des insectes nouveaux recueillis pendant le voyage. Deuxième partie. Coléoptères et autres ordres. Paris: J. Tastu, vii + 716 pp., 20 pls. [plates issued in 1847].

Bolívar y Pieltain, C. 1914: Asida (Alphasida) Merceti nov. sp. (Col. Tenebrionidae). *Boletín de la Real Sociedad Española de Historia Natural* **14**: 237–238.

Bolívar y Pieltain, C. 1917: Descripción de una especie española del género Rhipidius (Col. Rhipiphoridae). *Boletín de la Real Sociedad Española de Historia Natural* **17**: 249–252.

Bologna, M.A. 1983: Una nuova specie di Lytta del Kashmir (Coleoptera, Meloidae). *Fragmenta Entomologica* **17**: 125–138.

Bologna, M.A. 1988: Note su Eurymeloe e revisione delle specie euromediterrenee del gr. *rugosus* (Coleoptera, Meloidae). *Fragmenta Entomologica* **20**: 233–301.

Bologna, M.A. 1989a: Un nuovo Alosimus del litorale tirrenico e note tassonomiche su altre specie congeneriche nord africane (Coleoptera, Meloidae). *Fragmenta Entomologica* **21**: 191–204.

Bologna, M.A. 1989b: Berberomeloe, a new west Mediterranean genus of Lyttini for Meloe majalis Linné (Coleoptera, Meloidae). Systematics and bionomics. *Bollettino di Zoologia* **55** [1988]: 359–366.

Bologna, M.A. 1991: Un nouvel Oenas de Grèce et note synonymique sur une autre espèce du Proche-Orient (Coleoptera, Meloidae). *Revue Française d'Entomologie* (N.S.) **13**: 175–179.

Bologna, M.A. 1992: [new taxon]. In: Bologna M.A. & Aloisi G.: Systematics of Lydomorphus Fairmaire 1882, with description of the first instar larva of L. dusalti (Dufour, 1821) (Coleoptera Meloidae). *Tropical Zoology* **5**: 55–71.

Bologna, M.A. 1995a: A new Eleticinae genus from Ethiopia and remarks on the tribe Morphozonitini (Coleoptera: Meloidae). *Annales de la Société Entomologique de France* (N.S.) **31**: 185–193.

Bologna, M.A. 1995b: Coleoptera Meloidae ed Oedemeridae di Lampedusa, Pantelleria e delle isole Maltesi. *Il Naturalista Siciliano*, Suppl. **19**: 505–525.

Bologna, M.A. 2006: [new taxa]. In: Bologna M.A. & Di Giulio A.: Revision of the genus Teratolytta (Coleoptera: Meloidae). *European Journal of Entomology* **103**: 137–161.

Bologna, M.A. 2008: [new taxa]. In: Bologna M.A. & Di Giulio A.: Revision of the genus Trichomeloe Reitter, with the description of new species and first instar larvae (Coleoptera: Meloidae). *Contributions to Zoology* 77: 227–248.

Bologna, M.A. 2009: The Meloidae (Coleoptera) of Libya: an annotated catalogue and description of three new species. *Annales de la Société Entomologique de France* **45**: 345–364.

Bologna, M.A. 2018: [new taxa]. In: Bologna M.A., Černý L. & Zubair A.: Meloidae (Coleoptera) of Pakistan and Kashmir with the description of three new species, new faunistic and taxonomic records and a zoogeographic analysis. *Turkish Journal of Zoology* **42**: 637–660.

Bologna, M.A. 2019: [new taxa]. In: Salvi D., Maura M., Pan Z. & Bologna M.A.: Phylogenetic systematics of Mylabris blister beetles (Coleoptera, Meloidae): a molecular assessment using species trees and total evidence. *Cladistics* **35**: 243–268.

Bologna, M.A. & Coco, E. 1991: Revisione del genere Croscherichia Pardo Alcaide, 1950 (Coleoptera Meloidae). *Memorie della Società Entomologica Italiana* **69** [1990]: 97–180.

Bologna, M.A. & Di Giulio, A. 2016: New species of Teratolytta from Turkey and a key to the females (Coleoptera Meloidae). *ZooKeys* **625**: 87–97.

Bologna, M.A. & Nikitsky, N.B. 2010: A new enigmatic Lytta from Crimea, eastern Ukraine (Coleoptera: Meloidae). *Zoosystematica Rossica* **19**: 70–76.

Bologna, M.A. & Pinto, J.D. 2001: Phylogenetic studies of Meloidae (Coleoptera), with emphasis on the evolution of phoresy. *Systematic Entomology* **26**: 33–72.

Bologna, M.A. & Turco, F. 2007: The Meloidae (Coleoptera) of the United Arab Emirates with an updating of the Arabian checklist. *Zootaxa* **1625**: 1–33.

Bonadona, P. 1952: Notes sur les Anthicidae paléarctiques. *Revue Française d'Entomologie* **19**: 233–237.

Bonadona, P. 1953: Les Anthicus français du groupe du tristis Schmidt. *Revue Française d'Entomologie* **20**: 253–264.

Bonadona, P. 1954a: In Mateu J.: Coleópteros de la Sierra Nevada. Familias: Dytiscidae, Hydrophilidae, Dryopidae, Catopidae, Silphidae, Pselaphidae, Histeridae, Cleridae, Cantharidae, Dasytidae, Aderidae, Anthicidae. *Archivos del Instituto de Aclimatación, Almería* **2**: 89–101, pl. 1.

Bonadona, P. 1954b: Notes sur les anthicides paléarctiques. *Revue Française d'Entomologie* **21**: 272–280.

Bonadona, P. 1956: Anthicides recueillis par. F. Pierre dans le Sahara algérien (Coleoptera). *Revue Française d'Entomologie* **23**: 116–120.

Bonadona, P. 1958: Notes sur les Anthicidae paléarctiques (Coleoptera). *Revue Française d'Entomologie* **25**: 287–297.

Bonadona, P. 1960a: Notes sur les Anthicidae paléarctiques (Coleoptera). *Revue Française d'Entomologie* **27**: 51–57.

Bonadona, P. 1960b: Les Endomia Castelnau d'Afrique et de Madagascar (Coléoptères Anthicidae). *Revue de Zoologie et de Botanique Africaines* **62**: 293–328.

Bonadona, P. 1962: Coléoptères Anthicidae d'Afrique Noire et de Madagascar. *Revue de Zoologie & de Botanique Africaines* **65**: 325–351.

Bonadona, P. 1963: XII. Anthicidae de l'Archipel de Madère. *Commentationes Biologicae* **25**: 105–107.

Bonadona, P. 1964: Note sur les Anthicidae paléarctiques. *Revue Française d'Entomologie* **31**: 225–249.

Bonadona, P. 1965: Anthicidae, pp. 862–865. In: Thérond J. and Hollande A.: Contribution à l'étude des coléoptères de la région de Béni-Abbès et de la Vallée de la Saoura. *Annales de la Société Entomologique de France* **1**: 851–877.

Bonadona, P. 1969: Le Parc National du Niokolo-Koba (Sénégal). XIII. Coleoptera Anthicidae. *Mémoires de l'Institut Fondamental d'Afrique Noire* **84**: 263–277.

Bonadona, P. 1970: Contribution a la faune de l'Iran. 18. Coléoptères Anthicidae. *Annales de la Société Entomologique de France* **6**: 379–383.

Bonadona, P. 1974: La classification des Anthicidae de la faune de France (Coleoptera). *L'Entomologiste* **30**: 101–111.

Bonadona, P. 1977a: Notes sur les Anthicidae paléarctiques (Col.). *L'Entomologiste* **33**: 2–11.

Bonadona, P. 1977b: Anthicus genei et formes affines (Col. Anthicidae). *L'Entomologiste* **33**: 84–87.

Bonadona, P. 1978: Les Tomoderini subendogés d'Afrique centrale et de l'Inde méridionale (Col. Anthicidae). *Revue Suisse de Zoologie* **85**: 645–656.

Bonadona, P. 1979: Une Isomira nouvelle de France méditerranéenne (Col. Alleculidae). *L'Entomologiste* **35**: 2–5.

Bonadona, P. 1984: Description d'une Microhoria nouvelle du Portugal (Coleoptera, Anthicidae). *Revue Française d'Entomologie* **6**: 71–72.

Bonadona, P. 1986: Notes sur les Anthicidae paléarctiques. (Insecta, Coleoptera). XVI. Descriptions d'un Clavicomus Pic et de quatre Microhoria Chevrolat. *Biocosme Mésogéen* **3**: 35–47.

Bonadona, P. 1988: Note sur les Anthicidae paléarctiques (suite). *Biocosme Mésogéen* **5**: 39–49.

Bonadona, P. 1989a: Les espèces françaises du genre Notoxus Geoffroy (Coleoptera Anthicidae). *L'Entomologiste* **45**: 9–13.

Bonadona, P. 1989b: Anthicidae (Coleoptera) nouveaux des collections du Muséum d'histoire naturelle de Genève. *Revue Suisse de Zoologie* **96**: 253–276.

Bonadona, P. & Sakai, M. 1986: [new taxon]. In: Sakai M.: Studies on Anthicidae of Japan (Coleoptera), 1. *Transactions of the Shikoku Entomological Society* **17**: 247–251.

Bonelli, F.A. 1812: Specimen faunae subalpinae sistens Insecta Pedemontii hucusque inedita, aut rariora, aut ea quae commodi damnive gratia quod inferunt, prudentis agricolae magis interest cognoscere. Per genera naturalia et species digesta. Fasciculus I. Coleoptera plerumque inedita comprehendens. *Memorie della R. Società di Agricultura di Torino* **9**: 149–183, pls. 1–6.

Bonometto, L. 1976: Le *Phaleria* di Sardegna. Note sistematiche ed ecologiche (Coleoptera Tenebrionidae). *Lavori, Socità Veneziana di Scienze Naturali* **1**: 28–31.

Borchmann, F. 1909a: Neue asiatische und australische Lagriiden hauptsächlich aus dem Museum in Genua. *Bollettino della Società Entomologica Italiana* **41** [1911]: 201–234.

Borchmann, F. 1909b: Systematische und synonymische Notizen über Lagriiden und Alleculiden. (Col.) *Deutsche Entomologische Zeitschrift* **53**: 712–714.

Borchmann, F. 1912a: H. Sauter's Formosa-Ausbeute. Lagriidae, Alleculidae, Cantharidae (Col.). *Supplementa Entomologica* **1**: 6–12.

Borchmann, F. 1912b: Neue afrikanische Lagriiden. (Material zur Monographie der afrikanischen Lagriiden.) *Coleopterologische Rundschau* **1**: 17–21.

Borchmann, F. 1915a: Die Lagriinae (Unterfamilie der Lagriidae.). *Archiv für Naturgeschichte* **A 81** [1916]: 46–188.

Borchmann, F. 1915b: Lagriidae und Alleculidae des 'Indian Museum'. *Records of the Indian Museum* **11**: 179–188.

Borchmann, F. 1915c: Eine neue Gattung der Statirinae (Col.). *Entomologische Mitteilungen* **4**: 296–299.

Borchmann, F. 1916: H. Sauter's Formosa-Ausbeute: Alleculidae und Othniidae (Col.). *Archiv für Naturgeschichte* **A 82** (5): 101–108.

Borchmann, F. 1917: Meloidae, Cephaloidae. Pars 69. In S. Schenkling (ed.): *Coleopterorum Catalogus*. Berlin: W. Junk, 208 pp.

Borchmann, F. 1925: Neue Heteromeren aus malayischen Gebiete. *Treubia* **6**: 329–354.

Borchmann, F. 1929: Ueber die von Herrn J.B. Corporaal in Ost-Sumatra gesammelten Lagriiden, Alleculiden, Meloiden und Othniiiden. *Tijdschrift voor Entomologie* **71**: 1–39.

Borchmann, F. 1930a: Eine neue Cistelina-Art aus der Ausbeute der Stötznerschen Szetschwan-Expedition. *Entomologische Blätter* **26**: 118–119.

Borchmann, F. 1930b: Die Gattung Cteniopinus Seidlitz. *Koleopterologische Rundschau* **15**: 143–164.

Borchmann, F. 1930c: Eine neue Alleculidenart aus Transbaikalien. *Koleopterologische Rundschau* **15**: 244–245.

Borchmann, F. 1930d: Die Lagriiden-Fauna der Philippinen. *Philippine Journal of Science* **41**: 403–533.

Borchmann, F. 1932: Fünf neue ostchinesische Heteromeren. (Aus der Kollektion Em. Reitter.) *Entomologisches Nachrichtenblatt* **6**: 91–96.

Borchmann, F. 1935: Meloidae (Col.). Pp. 308–309. In: Visser, C. & Visser-Hooft, J. (eds.): *Wissenschaftliche Ergebnisse der Niederländischen Expeditionen in den Karakorum und die angrenzenden Gebiete in den Jahren 1922, 1925 und 1929/1930. Vol. 1.* Leipzig: F.A. Brockhaus, xviii + 499 pp., 1 map, 8 pls.

Borchmann, F. 1936: Coleoptera Heteromera Fam. Lagriidae. In: Wytsman, P. (ed.): *Genera Insectorum. Fasc. 204*. Brussels: Louis Desmet-Verteneuil, 561 pp., 9 pls.

Borchmann, F. 1937: Neue Alleculiden aus dem Deutschen Entomologischen Institut, Berlin-Dahlem. (Coleoptera.). *Arbeiten über Morphologische und Taxonomische Entomologie aus Berlin Dahlem* **4**: 210–231.

Borchmann, F. 1938a: Neue Alleculiden aus dem Museum der Stadt Stettin (Col.). *Stettiner Entomologische Zeitung* **99**: 292–298.

Borchmann, F. 1938b: Über die von Herrn Dr. C. Rathjens im Jahre 1931 in Arabien gesammelten Meloidae. *Entomologische Blätter* **33**: 463–466.

Borchmann, F. 1940a: Alleculidae aus dem Deutschen Entomologischen Institut. (Coleoptera.). *Arbeiten über Morphologische und Taxonomische Entomologie aus Berlin Dahlem* **7**: 154–158.

Borchmann, F. 1940b: Neue Meloiden-Arten (Col.). *Mitteilungen der Münchener Entomologischen Gesellschaft* **30**: 592–621.

Borchmann, F. 1941: Über die von Herrn J. Klapperich in China gesammelten Heteromeren. *Entomologische Blätter* **37**: 22–29.

Borchmann, F. 1942a: Entomological results from the Swedish Expedition 1934 to Burma and British India. Coleoptera: Lagriidae und Alleculidae. Gesammelt von René Malaise. *Arkiv för Zoologi* **33** A (9): 1–32.

Borchmann, F. 1942b: Neue Meloiden-Arten (Col.) II. *Mitteilungen der Münchener Entomologischen Gesellschaft* **32**: 682–712.

Borges, P.A.V. 1991: Two new species of Tarphius Erichson, 1848 (Coleoptera, Colydiidae) from the Azores. *Bocagiana* **143**: 1–11.

Borges, P.A.V. & Serrano, A.R.M. 2017. In: Borges P.A.V., Amorim I.R., Terzopoulou S., Rigal F., Emerson B.C. & Serrano A.R.M.: Cryptic diversity in the Azorean beetle genus Tarphius Erichson, 1845 (Coleoptera: Zopheridae): An integrative taxonomic approach with description of four new species. *Zootaxa* **4236**: 401–449.

Bosc d'Antic, L.A.G. 1790: [new taxon]. In: Rossi, P. *Fauna Etrusca sistens insecta quae in provinciis Florentina et Pisana praesertim collegit Petrus Rossius. Tomus Primus.* Liburni: T. Masi & Sociorum, xxii + 272 pp., 10 pis.

Bosc d'Antic, L.A.G. 1791: Mémoire sur l'établissement d'un nouveau genre d'insectes coléoptères. Pp. 327–328. In: *La médecine éclairée par les sciences physiques, ou journal des découvertes relatives aux différentes parties de l'art de guérir. Vol. 1.*

Bosc d'Antic, L.A.G. 1792: [untitled]. *Journal d'Histoire Naturelle* **2**: 293–296, pl. 40

Bose, F.C. 1858: Rhinosimus ruficeps nov. spec. *Stettiner Entomologische Zeitung* **19**: 96.

Bouchard, P., Lawrence, J.F., Davies, A.E. & Newton, A.F. 2005: Synoptic classification of the world Tenebrionidae (Insecta: Coleoptera) with a review of family-group names. *Annales Zoologici* **55**: 499–530.

Bouchard, P. & Löbl, I. 2008: Tenebrionidae: Opatrini. New nomenclatural and taxonomic acts, and comments. P. 39. In: Löbl, I. & A. Smetana (eds.): *Catalogue of Palaearctic Coleoptera. Vol. 5. Tenebrionoidea.* Stenstrup: Apollo Books, 670 pp.

Bouchard, P., Löbl, I. & Merkl, O. 2007: Nomenclatural notes on tenebrionid beetles of the Palaearctic region (Insecta, Coleoptera). *Annales Zoologici* **57**: 385–394.

Bouskell, F. 1899: Notes on some aberrations of Osphya bipunctata Fab. *The Entomologist's Record and Journal of Variation* **11** (1): 265–267.

Bousquet, Y. & Bouchard, P. 2013: The genera in the second catalogue (1833–1836) of Dejean's Coleoptera collection. *ZooKeys* **282**: 1–219.

Bousquet, Y. & Bouchard, P. 2018: [new taxon]. In: Bousquet Y., Thomas D.B., Bouchard P., Smith A.D., Aalbu R.L., Johnston A.M., Steiner W.E. Jr: Catalogue of Tenebrionidae (Coleoptera) of North America. *ZooKeys* **728**: 1–455.

Bouyon, H. 2011: Heliotaurus (Atlasotaurus) casseti, nouveau sous-genre et nouvelle espèce du Maroc (Coleoptera, Tenebrionidae, Alleculinae). *Bulletin de la Société entomologique de France* **116**: 463–467.

Bouyon, H. 2014: Description de six nouvelles espèces du genre Podonta Solier (Coleoptera, Tenebrionidae, Alleculinae). *Bulletin de la Société entomologique de France* **119**: 249–255.

Bouyon, H. 2015a: Révision du genre Heliosthraema Reitter, 1890, endémique du Maroc (Coleoptera, Tenebrionidae, Alleculinae). *Bulletin de la Société entomologique de France* **120**: 313–322.

Bouyon, H. 2015b: Description d'une nouvelle espèce du genre Podonta Solier de Turquie (Coleoptera Tenebrionidae Alleculinae). *L'Entomologiste* 71: 321–322.

Brahm, N.J. 1790: *Insektenkalendar für Sammler und Oekonomen. Erster Theil.* Mainz: Kurfürstl. Privil. Universitätsbuchhandlung, lxvii + 248 pp.

Brancsik, C. 1874: Neue Coleopteren aus Ungarn. *Berliner Entomologische Zeitschrift* **18**: 227–231.

Brancsik, C. 1899: Aliquot Coleoptera novae Russiae asiaticae. *Évkönyve, A Trencsén Vármegyei Természettudományi Egylet [Jahresheft des Naturwissenschaftlichen Vereines des Trencsiner Comitates]* **21–22** [1898–1899]: 97–105 + [1], pl. 4.

Brancsik, C. 1914: Coleoptera nova. *A Trencsén Megyei Múzeumi Egyesület Értesítője* **1914**: 58–69.

Brandt, J.F. & Erichson, W.F. 1832: Monographia generis Meloe. *Nova Acta Academiae Leopoldina* **16** (1): 101–142, 1 pl.

Brandt, J.F. & Ratzenburg, J.T.C. 1833: *Medizinische Zoologie, oder getreue Darstellung und Beschreibung der Thiere die in Arzeneimittellehre in Betracht kommen, in systematischer Folge herausgegeben. Tome II.* [1830–1833]. Berlin: Hirschwald, 4 + 364 pp., 39 pls. [Tome I, 1827–1829: 6 + 198 pp., 24 pls].

Brenske, E. & Reitter, E. 1884: Neuer Beitrag zur Käferfauna Griechenlands. *Deutsche Entomologische Zeitschrift* **28**: 17–100.

Breit, J. 1908: Eine koleopterologische Sammelreise auf Mallorka (Balearen). *Verhandlungen der Kaiserlich-Königlichen Zoologisch-Botanischen Gesellschaft in Wien* **58**: 52–67.

Breit, J. 1914: Neue Coleopterenformen aus Süd-Europa. *Coleopterologische Rundschau* **3**: 51–62.

Brême, F.A.G. de. 1842: Note monographique sur le genre Misolampus. *Revue Zoologique* **1842**: 81–83.

Brême, F.A.G. de. 1846: *Essai monographique et iconographique de la tribu des Cossyphides. Deuxième Partie. (Cossyphus proprement dit).* Paris: Victor Masson, 31 pp., 3 pls.

Bremer, H.J. 1998: Revision der orientalischen Corticeus-Arten (Col., Tenebrionidae, Hypophloeini). I. Teil. *Acta Coleopterologica* **14**: 3–32.

Bremer, H.J. 1999: Revision der orientalischen Corticeus-Arten (Col., Tenebrionidae, Hypophloeini). II. Teil. *Acta Coleopterologica* **15**: 3–64.

Bremer, H.J. 2001: Revision der Gattung Amarygmus Dalman, 1823 und verwandten Gattungen. I. Allgemeine Bemerkungen, Status einiger Gattungen affine Amarygmus Dalman, 1823; neue Kombinationen von Arten der Gattung Amarygmus Dalman (Coleoptera: Tenebrionidae: Alleculinae: Amagygmini). *Coleoptera* **5**: 57–80.

Bremer, H.J. 2014a: Revision of Azarelius Fairmaire, Ziaelas Fairmaire and related Oriental termitophilous genera, with descriptions of two new genera and remarks on tribal placement (Coleoptera: Tenebrionidae: Amarygmini). *Stuttgarter Beiträge zur Naturkunde A, Neue Serie* **7**: 163–182.

Bremer, H.J. 2014b: Revision of the genus Amarygmus Dalman, 1823 and related genera. Part LXXII. The Amarygmini of Borneo (Coleoptera: Tenebrionidae), part IV. Species of the genus Plesiophthalmus and related genera (including species of Sumatra, Java and – partially – Peninsular Malaysia). *Mitteilungen der Münchener Entomologischen Gesellschaft* **104**: 31–83.

Bremer, H.J. & Grimm, R. 2019: A new Corticeus from China (Coleoptera, Tenebrionidae, Hypophlaeini). *Mitteilungen der Münchener Entomologischen Gesellschaft* **109**: 51–54.

Bremi-Wolf, J.J. 1856: *Catalog der Schweizerischen Coleopteren, als Vorläufer der Beiträge für schweizerische Entomologie.* Zürich: Schulthess, [iii]-vi +78 pp.

Brenske, E. 1884: [new taxa]. In: Brenske E. & Reitter E.: Neuer Beitrag zur Käferfauna Griechenlands. *Deutsche Entomologische Zeitschrift* **28**: 17–100, pls 1, 2.

Brisout de Barneville, C. 1859: Descriptions de quelques espèces nouvelles de coléoptères français. *Bulletin de la Société Entomologique de France* **1859**: 231–235.

Brisout de Barneville, C. 1862: Description de trois insectes coléoptères nouveaux qui se trouvent en France. *Revue et Magasin de Zoologie Pure et Appliquée* (2) **14**: 23–25.

Brisout de Barneville, C. 1863: [new taxa]. In: Grenier, A. (ed.): *Catalogue des coléoptères de France et matériaux pour servir à la faune des coléoptères français, par MM. E. Allard, Dr. Ch. Aubé, Ch. Brisout de Barneville, A. Chevrolat, L. Fairmaire, Al. Fauvel, Dr. A. Grenier, Dr. Kraatz, J. Linder, L. Reiche et Félicien de Saulcy*. Paris: L. Toinon, iv + 3–79 + 135 pp.

Brisout de Barneville, C. 1866: Coléoptères nouveaux trouvés en Espagne pendant l'excursion de la Société en 1865. *Annales de la Société Entomologique de France* (4) **6**: 355–426.

Brisout de Barneville, C. 1867: [new taxa]. In: Grenier, A. (ed.): *Matériaux pour servir à la faune des coléoptères de France, recueillis et publiés par le Dr. A. Grenier. 2e Cahier*. Paris: A. Grenier, [1] + iv + 131–194 pp.

Brisout de Barneville, C. 1888: [new taxa]. *Bulletin de la Société Entomologique de France* **1887**: cxcii.

Brisout de Barneville, C. & Brisout de Barneville, H. 1877: [new taxon]. *Bulletin de la Société Entomologique de France* **1877**: cvii.

Broun, T. 1880: *Manual of the New Zealand Coleoptera*. Wellington: J. Hughes, 652 pp.

Broun, T. 1893a: Descriptions of new Coleoptera from New Zealand. *The Annals and Magazine of Natural History* (6) **12**: 161–195.

Broun, T. 1893b: *Manual of the New Zealand Coleoptera. Parts V., VI., VII.* Wellington: Colonial Museum and Geological Survey Department, xvii + 975–1504 pp.

Brullé, G.A. 1832: IVe Classe. Insectes. Pp. 1–288. In: Bory de Saint Vincent, J.B.G.M.: *Expédition scientifique en Morée. Section des sciences physiques Tome III. – 1er partie. Zoologie. Deuxieme section. – Des animaux articulés*. Paris: F.G. Levraut, [1] + 400 + [2] pp.

Brullé, G.A. 1833: IVe Classe. Insectes. Pp. 289–400 + [2]. In: Bory de Saint Vincent, J.B.G.M.: *Expédition scientifique en Morée. Section des sciences physiques. Tome III. – 1er partie. Zoologie. Deuxième section. – Des animaux articulés*. Paris: F.G. Levraut, [1] + 400 + [2] pp.

Brullé, G.A. 1835: Coléoptères II. In: Audoin, J.V. (ed.): *Histoire naturelle des insectes. Vol. 5*. Paris: Pillot, 436 pp.

Brullé, G.A. 1839: Entomologie. Pp. 53–95. In: Barker-Webb, P. & Berthelot, S.: *Histoire naturelle des Iles Canaries. (Animaux articulés recueillis aux Iles Canaries.). Vol. II. Pars 2*. Paris: Mellier, 119 pp., 8 pls. [issued in parts, from 1836 to 1840].

Bucciarelli, I. 1959: Coleotteri anticidi dell'Aspromonte (Appenino calabrese). *Memorie del Museo Civico di Storia Naturale, Verona* 7: 183–189.

Bucciarelli, I. 1961: Una nuova specie dalmata di Cyclodinus Muls. e Rey. *Atti della Società Italiana di Scienze Naturali e del Museo Civico di Storia Naturale in Milano* **100**: 269–274.

Bucciarelli, I. 1962: Gli anticidi del Museo Entomologico del Pontificio Instituto Missioni estere di Monza. *Bollettino della Società Entomologica Italiana* **92**: 114–122.

Bucciarelli, I. 1973: Note sui Notoxus mediterranei. *Atti del Museo Civico di Storia Naturale Trieste* **28**: 471–484.

Bucciarelli, I. 1976: Su alcuni coleotteri anticidi raccolti da M.e.T. Cerruti nelle isole di Creta, Corfù e Thasos (Grecia), con descrizione di una nuova specie di Microhoria (Coleoptera, Anthicidae). *Fragmenta Entomologica* **12**: 133–142.

Bucciarelli, I. 1977: Microhoria (Immicrohoria) caprai n. sp. della Liguria. *Memorie della Società Entomologica Italiana* **55**: 179–182.

Bucciarelli, I. 1978: Anticidi nuovi, nuovi per l'Italia e sinonimie. *Bollettino della Società Entomologica Italiana* **110**: 154–163.

Buck, F.D. 1965: Some Anthicidae (Col.) from southern Rhodesia including a species new to science. *The Entomologist's Monthly Magazine* **101**: 192–194.

Buquet, J.B.L. 1840: Note sur quelques coléoptères nouveaux d'Algérie et particulièrement de Constantine. *Revue Zoologique, par la Société Cuvierienne* **1840**: 240–243.

Bujalance, J.L., Cárdenas, A.M., Ferrer, J. & Gallardo, P. 2016: The genus *Tentyria* (Latreille, 1802) in Doñana National Park (Southern Iberian Peninsula) with descriptions of two new species (Insecta: Coleoptera: Tenebrionidae). *Annales Zoologici* **66**: 345–355.

Bytinski-Salz, H. 1955: Prionotheca coronata Ol. freyi ssp. n. *Entomologische Arbeiten aus dem Museum G. Frey* **6**: 313–316.

Caillol, H. 1914: *Catalogue des coléoptères de Provence, d'après des documents recueillis et groupés. 3me Partie.* Marseille: Société Linnéenne de Provence, 594 pp.

Caillol, H. 1925: Une espèce présumée nouvelle d'Atritomus Reitter (Col. Mycetophagidae). *Bulletin de la Société Entomologique de France* **1925**: 101.

Cameron, M. 1907: Description of a new species of Stenostoma from Malta. *The Entomologist's Monthly Magazine* **43**: 226–227.

Campbell, J.M. 1980: Insects of Saudi Arabia Coleoptera: Fam. Alleculidae. *Fauna of Saudi Arabia* **2**: 133–136.

Candèze, E.C.A. 1861: Histoire de la métamorphose de quelques coléoptères exotiques. *Mémoires de la Société Royale des Scences de Liège* **16**: 325–410, 6 pls.

Canzoneri, S. 1959a: Note sugli Xanthomus Muls. (Coleoptera, Tenebrionidae). *Bollettino della Società Entomologica Italiana* **89**: 145–148.

Canzoneri, S. 1959b: Contributo ad una migliore conoscenza del genere Stenomax Allard (Col. Tenebrionidae). *Bollettino de la Società Entomologica Italiana* **89**: 148–149.

Canzoneri, S. 1960a: Diagnosi preliminare di una nuova entità del genere Phaleria. (III. Contributo allo studio dei tenebrionidi). *Bollettino della Società Entomologica Italiana* **90**: 43–45.

Canzoneri, S. 1960b: Note sistematiche e biogeografiche su alcuni tenebrionidi nuovi o poco noti. *Bollettino della Società Entomologica Italiana* **90**: 50–54.

Canzoneri, S. 1961a: Note sul genere Oochrotus Luc. VII Contributo allo studio dei tenebrionidi. *Bollettino del Museo Civico di Storia Naturale di Venezia* **14**: 7–27.

Canzoneri, S. 1961b: Sulla Phaleria bimaculata Lin. (IX Contributo allo studio dei tenebrionidi). *Bollettino della Società Entomologica Italiana* **91**: 24–27.

Canzoneri, S. 1961c: Sulla Phaleria provincialis Fauv. (X Contributo allo studia dei tenebrionidi). *Bollettino della Società Entomologica Italiana* **91**: 57–61.

Canzoneri, S. 1963a: Sulla Pimelia bipunctata Fabr. (XI Contributo allo studio di tenebrionidi). *Bollettino della Società Entomologica Italiana* **93**: 13–16.

Canzoneri, S. 1963b: Sulla Alphasida grossa Sol., con descrizione di una nuova sottospecie. (XII Contributo allo studia dei tenebrionidi). *Bollettino della Società Entomologica Italiana* **93**: 35–36.

Canzoneri, S. 1965: Una nuova specie del genere Falsaspila Koch (XV Contributo alla conoscenza dei Tenebrionidae). *Bollettino della Società Entomologica Italiana* **95**: 168–169.

Canzoneri, S. 1968: Materiali per una monografia delle Phaleria del sottogenere Phaleria Latr. 20. Contributo alla conoscenza dei Coleoptera, Tenebrionidae. *Memorie della Società Entomologica Italiana* **47**: 117–137.

Canzoneri, S. 1969: Blaps sulcata ssp. mercantii nov.: diagnosi preliminare (XIX Contributo allo studio dei Tenebrionidae). *Bollettino della Società Entomologica Italiana* **99–101**: 44.

Canzoneri, S. 1970: I Tenebrionidae delle Isole Egadi. *Memorie del Museo Civico di Storia Naturale di Verona* **18**: 55–89.

Canzoneri, S. 1971: Una nuova Akis di Persia (XXV Contrituto alla conoscenza dei Tenebrionidae). *Bollettino della Società Entomologica Italiana* **103**: 155.

Canzoneri, S. 1972a: Annotazioni sui Colpophorus italiani. *Bollettino del Museo Civico di Storia Naturale di Venezia* **22/23** [1969/1970]: 187–191.

Canzoneri, S. 1972b: Nuovi dati sui Tenebrionidae di 'Piccole Isole' Italiane, con descrizione di Alphasida tirelli moltonii n.ssp. *Atti della Societã Italiana di Scienza Naturali e del Museo Civico di Storia Naturale di Milano* **113**: 288–296.

Canzoneri, S. 1973: Una nuova specie del genere Phaleria Latr. *Bollettino della Società Entomologica Italiana* **105**: 41–42.

Canzoneri, S. 1976: I Tenebrionidae delle Isole Ponziane (Coleoptera). *Fragmenta Entomologica* **12**: 9–18.

Canzoneri, S. 1977: Contributo alla conoscenza de tenebrionidi appenninici. (XXXII Contributo allo studio dei Tenebrionidi). *Bollettino del Museo Civico di Storia Naturale Verona* **4**: 227–285.

Canzoneri, S. 1979: Primi appunti sui tenebrionidi dell'Arcipelago Maltese (XXXII Contributo allo studio dei Tenebrionidi). *Lavori – Società Veneziana di Sienza Naturali* **4**: 20–22.

Canzoneri, S. 1984: Una nuova specie di Pedinus di Sicilia. *Lavori – Società Veneziana di Sienza Naturali* **9**: 3–4.

Carl, M. 1989: Eine neue Tentyria-Art aus dem Irak. *Entomofauna* **10**: 25–30.

Carl, M. 1990a: Übersicht über die irakischen Adesmia-Arten und Neubeschreibung von Adesmia microgranulata sp. n. (Coleoptera, Tenebrionidae). *Mitteilungen der Münchener Entomologischen Gesellschaft* **80**: 71–83.

Carl, M. 1990b: Zwei neue Zophosis-Arten aus dem Vorderen Orient (Coleoptera: Tenebrionidae). *Nachrichtenblatt der Bayerischen Entomologen* **39**: 79–81.

Carl, M. 1990c: Eine neue Dichillus-Art aus dem Irak (Coleoptera, Tenebrionidae). *Entomofauna* **11**: 105–108.

Carl, M. 1991a: Die Gattungen Adelostomoides gen. n. und Adelostoma Dup. aus Mesopotamien (Coleoptera: Tenebrionidae). *Nachrichtenblatt der Bayerischen Entomologen* **40**: 23–27.

Carl, M. 1991b: Eine neue Tentyria-art aus Iraq (Coleoptera, Tenebrionidae) *Acta Zoologica Academiae Scientiarum Hungaricae* **37**: 171–175.

Carl, M. 1991c: Beschreibung von Micipsa monilifera sp. nov. aus Mesopotamien. Entomofauna12: 141–146.

Carl M. 1992a: A revision of the genus Microtelus Solier, 1838 (Coleoptera: Tenebrionidae). *Zoology of the Middle East* **6**: 69–87.

Carl, M. 1992b: Neue und bemerkenswerte Tenebrionidae (Coleoptera) aus dem Vorderen Orient. *Linzer Biologische Beiträge* **24**: 331–337.

Carl, M. 1994: The taxonomy, biogeography and phylogeny of the genus Prochoma in the Middle East and South-East Asia (Coleoptera: Tenebrionidae). *Fauna of Saudi Arabia* **14**: 77–96.

Carl, M. 2000: Beschreibung von Heinrichesia schaeferi gen. nov. et sp. nov. aus Sistan (Iran) (Coleoptera, Tenebrionidae). *Entomofauna* **21**: 257–261.

Carter, H.J. 1911: Revision of Pterohelaeus (continued) and of Sargus; with descriptions of new species of Australian Tenebrionidae. *Proceedings of the Linnean Society of New South Wales* **36**: 179–223, pl. VIII.

Carter, H.J. 1919: Notes on Australian Coleoptera, with description of new species. *Proceedings of the Linnean Society of New South Wales* **44**: 137–173, pls. IV, V.

Carter, H.J. 1920: Notes on some Australian Tenebrionidae, with description of new species; – also of a new genus and species of Buprestidae. *Proceedings of the Linnean Society of New South Wales* **45**: 222–249.

Carter, H.J. 1924a: Australian Coleoptera: notes and new species. No. III. *Proceedings of the Linnean Society of New South Wales* **49**: 19–45.

Carter, H.J. 1924b: Australian Coleoptera: notes and new species. No. IV. *Proceedings of the Linnean Society of New South Wales* **49**: 521–544.

Carter, H.J. 1928: Revision of the Australian species of the genera Curis, Neocuris and Trachys, together with notes and descriptions of new species of other Coleoptera. *Proceedings of the Linnean Society of New South Wales* **53**: 270–290.

Carter, H.J. & Zeck, E.H. 1937: A monograph of the Australian Colydiidae. *Proceedings of the Linnean Society of New South Wales* **62**: 181–208, pls. 8–9.

Casey, T.L. 1898: Studies in the Ptinidae, Cioidae and Sphindidae of America. *Journal of the New York Entomological Society* **6**: 61–93.

Casey, T.L. 1900: Review of the American Corylophidae, Cryptophagidae, Tritomidae and Dermestidae, with other studies. *Journal of the New York Entomological Society* **8**: 51–172.

Casey, T.L. 1891: Coleopterological Notices. III. *Annals of the New York Academy of Sciences* **6**: 9–214.

Casey, T.L. 1895: Coleopterological Notices. VI. *Annals of the New York Academy of Sciences* **8**: 435–838.

Casey, T.L. 1924: Additions to the knowledge of the Coleoptera of North America. *Memoirs on the Coleoptera* **11**: 1–347 pp.

Castro Tovar, A. 2013. Una nueva especie de Stenosis Herbst, 1799 (Coleoptera, Tenebrionidae) de la Península Ibérica. A new species of Stenosis Herbst, 1799 (Coleoptera, Tenebrionidae) from the Iberian Peninsula. *Zoologica Baetica* **24**: 141–154.

Castro Tovar A. 2014: Una nueva especie de Blaps Fabricius, 1775 del sureste de España (Coleoptera, Tenebrionidae). *Arquivos Entomolóxicos* **12**: 237–243.

Castro Tovar A. 2015: Una especie nueva de *Catomus* Allard, 1876 de España: *Catomus lopezi* n. sp. (Coleoptera, Tenebrionidae). *Boletín de la Sociedad Entomológica Aragonesa* **57**: 185–188.

Castro Tovar A. 2016: Aportación al conocimiento de los *Gunarus* Des Gozis, 1886 de España (Coleoptera: Tenebrionidae). *Boletín de la Sociedad Entomológica Aragonesa* **58**: 29–32.

Castro Tovar, A. & Ferrer, J. 2012: Nuevas especies y claves de las Pimelia franco-ibéricas (Coleoptera, Tenebrionidae) y notas sobre Pimelia (Amblyptera) de Marruecos. *Boletín de la Sociedad Entomológica Aragonesa* **50**: 219–253.

Černý, L. & Bologna, M.A. 2018: A new species of Diaphorocera from Morocco with unclear relationships and a key to the species (Coleoptera, Meloidae, Cerocomini). *Zookeys* **748**: 57–64.

Černý, L. & Vrabec, V. 2019: Mylabris (Mylabris) snizeki sp. n. from Jordan, with a key to the Jordanian specie of the nominotypical subgenus (Meloidae). *Zootaxa* **4555**: 146–150.

Champion, G.C. 1886a: Tenebrionidae. Pp. 137–264. In: *Biolologia Centrali-Americana. Insecta. Coleoptera Heteromera (part). Vol.* IV*, part 1.* [1884–1893]. London: R.H. Porter, 572 pp., 23 pls.

Champion, G.C. 1886b: Salpingus mutilatus, Beck, a British insect. *The Entomologist's Monthly Magazine* **23**: 160.

Champion, G.C. 1889: Melandryidae. Pp. 75–103, 5 pls. In: *Biologia Centrali-Americana. Insecta, Coleoptera Vol.* IV*, part 2.* [1889–1893]. London: R.H. Porter, x + 494 pp., 21 pls.

Champion, G.C. 1890a: Some remarks on the genus Xylophilus, with description of two species from Japan. *The Entomologist's Monthly Magazine* **26**: 264–269.

Champion, G.C. 1890b: On a new species of Tomoderus from Japan. *The Entomologist's Monthly Magazine* **26**: 325–326.

Champion, G.C. 1891a: Description of a new species of Anaspis from Scotland, with some remarks on the black species occurring in Britain. *The Entomologist's Monthly Magazine* **27** 104–105.

Champion, G.C. 1891b: On two new species of Anthicidae. *The Entomologist's Monthly Magazine* **27**: 188–189.

Champion, G.C. 1891c: Mordellidae. Pp. 257–350. In: *Biologia Centrali-Americana Insecta, Coleoptera. Vol.* IV*, part 2.* [1889–1893]. London: R.H. Porter, x + 494 pp., 21 pls.

Champion, G.C. 1891d: A list of the Coleoptera collected by Mr. J.J. Walker, R.N. F.L.S., in the region of the Straits of Gibraltar, with descriptions of four new species. *Transactions of the Entomological Society of London* **1891**: 375–401.

Champion, G.C. 1892: Note on Pentaria badia Rosenh. (= sericaria, Muls.), with description of a second species from the Pyrénées-Orientales. *The Entomologist's Monthly Magazine* **28**: 108–109.

Champion, G.C. 1893: Supplement (Fam. Xylophilidae). Pp. 455–461, pl. 21. In: *Biologia Centrali-Americana, Insecta, Coleoptera Heteromera (part). Vol.* IV*, part.2* [1889–1893]. London: R.H. Porter, x + 494 pp., 21 pls.

Champion, G.C. 1894a: On the Tenebrionidae collected in Australia and Tasmania by Mr. James J. Walker. R.N., F.L.S., during the voyage of H.M.S. 'Penguin' with descriptions of new genera and species. *Transactions of the Entomological Society of London* **1894**: 351–408, pl.viii.

Champion, G.C. 1894b: On two species of tenebrionid Coleoptera from Namoa Island *The Annals and Magazine of Natural History* (6) **14**: 448–449.

Champion, G.C. 1895: A list of Tenebrionidae supplementary to the 'Munich Catalogue. *Mémoires de la Société Entomologique de Belgique* **1** (3): 1–264 pp.

Champion, G.C. 1916a: On new or little-know Xylophilidae. *Transactions of the Entomological Society of London* **1916** (1): 1–64, I–II pls.

Champion, G.C. 1916b: Notes on Melandryidae (2–3). *The Entomologist's Monthly Magazine* **52**: 1–10, 32–40, 52–59, 75–83, 99–108, 144–156.

Champion, G.C. 1916c: On new and little-known Lagriidae and Pedilidae. *Transactions of the Entomological Society of London* **1916**: 181–242, pls. 63–64.

Champion, G.C. 1916d: On new exotic Scraptiina. *The Entomologist's Monthly Magazine* **52**: 233–256, 265–274, pl. 7.

Champion, G.C. 1917: Some Coleoptera from northen India. *The Entomologist's Monthly Magazine* **53**: 52–55.

Champion, G.C. 1919: Some Indian Coleoptera (1). *The Entomologist's Monthly Magazine* **55**: 236–246.

Champion, G.C. 1920a: Some Indian Coleoptera (2). *The Entomologist's Monthly Magazine* **56**: 68–77.

Champion, G.C. 1920b: On some Eastern Xylophilids (Coleoptera). *The Annals and Magazine of Natural History* (9) **5**: 393–406.

Champion, G.C. 1921: Some Indian Coleoptera (6). *The Entomologist's Monthly Magazine* **57**: 201–206.

Champion, G.C. 1922a: Some Indian Coleoptera (7). *The Entomologist's Monthly Magazine* **58**: 31–34, 68–76.

Champion, G.C. 1922b: Some Indian Coleoptera (8). *The Entomologist's Monthly Magazine* 58: 126–131, 145–152.

Champion, G.C. 1923: Some Indian Coleoptera (10). *The Entomologist's Monthly Magazine* **59**: 77–80.

Champion, G.C. 1924a: Some Indian Coleoptera (14). *The Entomologist's Monthly Magazine* **60**: 159–172.

Champion, G.C. 1924b: Additional species of Xylophilidae from the Philippines, India, Fiji and S. and W. Africa. *The Annals and Magazine of Natural History* (9) **12**: 481–498.

Champion, G.C. 1924c: Some Indian Coleoptera. *The Annals and Magazine of Natural History* (9) **13**: 249–264.

Champion, G.C. 1925: Some Indian (and Tibetan) Coleoptera (16). *The Entomologist's Monthly Magazine* (2) **36**: 101–112.

Champion, G.C. 1927a: Some Indian Coleoptera (22). *The Entomologist's Monthly Magazine* **63**: 49–57, 78–85.

Champion, G.C. 1927b: Some Indian Coleoptera (23). *The Entomologist's Monthly Magazine* **63**: 245–250.

Chandler, D.S. 2004: [new name]. In: Chandler D.S., Nardi G. & Telnov D.: Nomenclatural notes on the Palaearctic Anthicidae (Coleoptera). *Mitteilungen des Internationalen Entomologischen Vereins e. V.* (Frankfurt am Main) **29**: 109–173.

Chandler, D.S., Uhmann, G., Nardi, G. & Telnov, D. 2008: Family Anthicidae Latreille, 1807. In: Löbl, I. & Smetana, A. (eds): *Catalogue of Palaearctic Coleoptera. Volume 5 Tenebrionoidea*. Apollo Books, Stenstrup. 670 pp.

Charpentier, T. de. 1818: Meloe insignis, p. 258. In: Germar E.F.: VIII. Vermischte Bemerkungen über einige Käferarten. *Germar's Magazin der Entomologie* **3**: 228–260.

Charpentier, T. de. 1825: *Horae entomologicae, adjectis tabulis novem coloratis*. Wratislaviae: Apud A. Gosohorsky, xvi + 255 pp., 9 pls.

Chatanay, J. 1914a: Sur trois ténébrionides faisant partie de la collection de M. Leboeuf (Coléoptères). *Bulletin de la Société d'Étude des Sciences Naturelles de la Haute-Marne* **1914**: 19–24.

Chatanay, J. 1914b: Un nouveau Blaps nord-africain (Col. Tenebrionidae). *Bulletin de la Société Entomologique de France* **1914**: 75–77.

Chatanay, J. 1914c: Description de deux Tentyriinae nouveaux. *Annales de la Société Entomologique de France* **83**: 215–224.

Chatanay, J. 1914d: Diagnoses préliminaires des Zophosinae nouveaux (Col. Tenebrionidae). *Bulletin de la Société Entomologique de France* **1914**: 379–381.

Chatanay, J. 1915: Description d'un genre nouveau d'Épitragides de Madagascar (Coléoptères, Tenebrionidae). *Bulletin du Muséum National d'Histoire Naturelle, Paris* 21: 64–67.

Chatanay, J. 1917a: Matériaux pour servir à l'étude de la faune entomologique de l'Indo-Chine française reunis par M. Vitalis de Salvaza. *Bulletin du Muséum National d'Histoire Naturelle, Paris* 23: 229–255.

Chatanay, J. 1917b: Révision des Zophosis de l'est de l'Afrique. *Annales de la Société Entomologique de France* 85 [1916]: 503–624.

Chatzimanolis, S., Engel, M.S. & Trichas, A. 2002: Taxonomic changes for Aegean species of the Mediterranean darkling beetle genus Dendarus (Coleoptera: Tenebrionidae). *Journal of the Kansas Entomological Society* 75: 259–267.

Chen, B. 1997a: A study on the genus Donacolagria Pic (Coleoptera: Lagriidae) from China. *Entomologia Sinica* 4: 121–128.

Chen, B. 1997b: Two new genera and two new species of Lagriidae (Coleoptera) from China. *Entomologia Sinica* 4: 306–311.

Chen, B. 2002: Lagriidae. Pp. 170–180. In: Huang, B.-K. (ed.) *Fauna of Insects in Fujian Province of China* (vol. 6). *Fujian Science and Technology Press*. Fuzhou, 894 pp. (in Chinese).

Chen, B. & Chou, I. 1996: A new subgenus and two new species of the genus Cerogria Borchmann (Coleoptera: Lagriidae) from China. *Entomotaxonomia* 18: 265–269.

Chen, B. & Xia, X. 2001: The genus Schevodera Borchmann: phylogeny and historical biogeography with description of a new species (Coleoptera: Tenebrionidae: Lagriinae). *Oriental Insects* 35: 3–27.

Chen, B. & Yang, X. 1997: Three new species and key to species of the genus Arthromacra Kirby from China (Coleoptera: Lagriidae). *Entomologia Sinica* 4: 1–8.

Chen, B. & Yuan, F. 1996: A new genus and two new species of Statirinae (Coleoptera: Lagriidae) from China. *Entomotaxonomia* 18: 183–187.

Chevrolat, L.A.A. 1831: [new taxa, livraison 15]. In: Guérin-Méneville, F.E.: [1829–1838]: *Iconographie du règne animal de G. Cuvier, ou représentation d'après la nature de l'une des espèces les plus remarquables et souvent non encore figurées, de chaque genre d'animaux. Avec un texte descriptif mis au courant de la science. Ouvrage pouvant servir d'atlas à tous les traités de zoologie. Iconographie du règne animal de G. Cuvier. II. Planches des animaux invertébrés. Insectes*. Paris: J.B. Bailière, 110 pls.

Chevrolat, L.A.A. 1833a: [new taxa, livraison 28]. In: Guérin-Méneville, F.E.: [1829–1838]: *Iconographie du règne animal de G. Cuvier, ou représentation d'après la nature de l'une des espèces les plus remarquables et souvent non encore figurées, de chaque genre d'animaux. Avec un texte descriptif mis au courant de la science. Ouvrage pouvant servir d'atlas à tous les traités de zoologie. Iconographie du règne animal de G. Cuvier. II. Planches des animaux invertébrés. Insectes*. Paris: J.B. Bailière, 110 pls.

Chevrolat, L.A.A. 1833b: Mémoire sur un nouveau genre de coléoptères, de la famille de mélasomes. *Revue Entomologique* (G. Silbermann) 1: 25–26.

Chevrolat, L.A.A. 1833c: Notice sur un genre nouveau de la famille des diapérides. *Revue Entomologique* (G. Silbermann) 1: 30–32.

Chevrolat, L.A.A. 1833d: [new taxa]. *Magasin de Zoologie* 3, cl. IX, pl. 57.

Chevrolat, L.A.A. 1834: [new taxa, livraison 35]. In: Guérin-Méneville, F.E.: [1829–1838]: *Iconographie du règne animal de G. Cuvier, ou représentation d'après la nature de l'une des espèces les plus remarquables et souvent non encore figurées, de chaque genre d'animaux. Avec un texte descriptif mis au courant de la science.*

Ouvrage pouvant servir d'atlas à tous les traités de zoologie. Iconographie du règne animal de G. Cuvier. II. Planches des animaux invertébrés. Insectes. Paris: J.B. Bailière, 110 pls.

Chevrolat, L.A.A. 1835a: [new taxa, livraison 38]. In: Guérin-Méneville, F.E.: [1829–1838]: *Iconographie du règne animal de G. Cuvier, ou représentation d'après la nature de l'une des espèces les plus remarquables et souvent non encore figurées, de chaque genre d'animaux. Avec un texte descriptif mis au courant de la science. Ouvrage pouvant servir d'atlas à tous les traités de zoologie. Iconographie du règne animal de G. Cuvier. II. Planches des animaux invertébrés. Insectes.* Paris: J.B. Bailière, 110 pls.

Chevrolat, L.A.A. 1835b: Mémoire sur un coléoptère tétramère de la famille des xylophages et observations sur plusiers espèces de cet ordre, rencontrées dans diverses fourmillières. *Revue Entomologique* (G. Silbermann) **3**: 263–269, pl. 34.

Chevrolat, L.A.A. 1840a: Description de quelques coléoptères de la Galice et du Portugal provenant d'envoies de M. Deyrolles fils. *Revue Zoologique, par la Société Cuvierienne* **1840**: 8–18.

Chevrolat, L.A.A. 1840b: Description de mylabrides de Barbarie. *Revue Entomologique* (G. Silbermann) **5** [1837]: 266–279, 1 pl.

Chevrolat, L.A.A. 1844: [new taxa]. In: Guérin-Méneville, F.E.: [1829–1844]: *Iconographie du règne animal de G. Cuvier, ou représentation d'après la nature de l'une des espèces les plus remarquables et souvent non encore figurées, de chaque genre d'animaux. Avec un texte descriptif mis au courant de la science. Ouvrage pouvant servir d'atlas à tous les traités de zoologie. III. Texte explicatif. Insectes.* Paris: J.B. Baillière, 576 pp. [issued in parts].

Chevrolat, L.A.A. 1849: Trachyderme, p. 628. In: D'Orbigny, C.V.D. (ed.): *Dictionnaire universel d'histoire naturelle résumant et complétant tous les faits présentés par les encyclopédies, les anciens dictionnaires scientifiques, les œuvres complètes de Buffon, et les meilleurs traités spéciaux sur les diverses branches des sciences naturelles; donnant la description des êtres et des divers phénomènes de la nature, l'étymologie et la définition des noms scientifiques, les principales applications des corps organiques et inorganiques, à l'agriculture, à la médecine, aux arts industriels, etc.; dirigé par M. Charles d'Orbigny, et enrichi d'un magnifique atlas de 288 planches gravées sur acier. Tome douzième.* Paris: C. Renard, Martinet et Cie.

Chevrolat, L.A.A. 1860: Description de coléoptères nouveaux d'Algérie. *Revue et Magasin de Zoologie Pure et Appliquée* (2) **12**: 75–82, 208–212, 448–459.

Chevrolat, L.A.A. 1861: Description des coléoptères nouveaux d'Algérie. *Revue et Magasin de Zoologie Pure et Appliquée* (2) **13**: 205–208.

Chevrolat, L.A.A. 1865: Descriptions de coléoptères d'Espagne, nouveaux ou peu connus. *Revue et Magasin de Zoologie Pure et Appliquée* (2) **17**: 390–397.

Chevrolat, L.A.A. 1873: Description de quelques espèces nouvelles de coléoptères de Syrie. *Annales de la Société Entomologique de France* (5) **3**: 203–218.

Chevrolat, L.A.A. 1874: Coleopterorum species novarum descriptiones. *Anales de la Sociedad Española de Historia Natural* **3**: 157–159.

Chevrolat, L.A.A. 1877a: Coléoptères nouveaux de la famille des ténébrionides. *Petites Nouvelles Entomologiques* **2** [1876–1879]: 113–114, 117–118.

Chevrolat, L.A.A. 1877b: Coléoptères nouveaux. *Petites Nouvelles Entomologiques* **2** [1876–1879]: 121.

Chevrolat, L.A.A. 1877c: Description d'un sous-genre nouveau de coléoptères hétéromères de la famille des Anthicides et indication des espèces que s'y rapportent. *Annales de la Société Entomologique de France* (5) 7: 167–169.

Chevrolat, L.A.A. 1877d: Diagnoses de diapérides nouveaux. *Petites Nouvelles Entomologiques* 2 [1876–1879]: 170, 173, 177–178.

Chevrolat, L.A.A. 1878a: Diagnoses de diapérides. *Petites Nouvelles Entomologiques* 2 [1876–1879]: 221–222, 242–243.

Chevrolat, L.A.A. 1878b: Diagnoses de diaperides nouveaux. *Comptes-Rendus des Séances de la Société Entomologique de Belgique.* **1878**: clii-cliii.

Chevrolat, L.A.A. 1882: [Diagnoses de trois coléoptères nouveaux et des remarques sur une quatrième espèce]. *Bulletin de la Société Entomologique de France* **1882**: iv-v.

Chigray, I.A. 2019: A new genus and species of darkling beetles of the tribe Blaptini (Coleoptera: Tenebrionidae) from Afghanistan and taxonomic changes in the tribe. *Entomological Review* **99**: 914–923.

Chigray, I.A., Abdurakhmanov, G.M., Nabozhenko, M.V. & Shapovalov, A.M. 2016: On Blaps Fabricius, 1775 (Coleoptera: Tenebrionidae) from Western Kazakhstan with description of a new species from Tyuleniy Archipelago (Caspian Sea). *Zootaxa* **4173**: 1–17.

Chigray, I.A. & Nabozhenko, M. 2016: To the knowledge of the genus Blaps Fabricius, 1775 (Coleoptera: Tenebrionidae) from Iran and Transcaucasia. *Annales Zoologici* **66**: 267–275.

Chigray, I.A., Nabozhenko, M., Abdurakhmanov, G. & Keskin, B. 2019: A systematic review of the genus Dila Fischer von Waldheim, 1844 (= Caenoblaps König, 1906, syn.n.) (Coleoptera: Tenebrionidae) from the Caucasus, Turkey and boundary territories of Iran. *Insect Systematics & Evolution,* 1–30 (online published). DOI 10.1163/1876312X-00001006.

Chigray, I.A., Nabozhenko, M.V. & Keskin, B. 2015a: Novyy vid roda Blaps Fabricius, 1775 (Coleoptera: Tenebrionidae) iz Zapadnoy Turtsii. *Vestnik yuzhnogo nauchnogo tsentra* **11**: 63–65.

Chigray, I.A., Nabozhenko, M.V. & Keskin, B. 2015b: Obzor roda Gnaptor Brullé, 1832 (Coleoptera, Tenebrionidae) s opisaniyem novogo vida iz Turtsii. *Zoologicheskiy zhurnal* **94**: 1276–1281. [English version: A review of the genus Gnaptor Brullé 1832 (Coleoptera, Tenebrionidae) with description of a new species from Turkey. *Entomological Review* 2015, **95**: 1131–1136].

Chigray, I.A., Nabozhenko, M., Merkl, O. & Kovalev, A. 2018: A review of the genus Prosodes Eschscholtz, 1829 (Coleoptera: Tenebrionidae) of Iran. *Zootaxa* **4379**: 451–483.

Chigray, S.N. 2018: A revision of the genus Ascelosodis L. Redtenbacher, 1868 (Coleoptera: Tenebrionidae: Pimeliinae: Edrotini) of Pamir. *Caucasian Entomological Bulletin* **14**: 165–169.

Chigray, S.N. 2019a: [new taxa]. In: Chigray, S.N. & Abakumov, E.V. 2019: Taxonomic revision of the genus Calyptopsis Solier, 1835 (Coleoptera, Tenebrionidae: Pimeliinae: Tentyriini) from Iran. *Entomological Review* **99**: 209–244.

Chigray, S.N. 2019b: Revision of the Palaearctic Pimeliini (Coleoptera, Tenebrionidae): the genus Podhomala Solier, 1836. *Entomological Review* **99**: 924–948.

Chigray, S.N. & Nabozhenko, M.V. 2018: A review of the genus Calyptopsis Solier, 1835 (Coleoptera, Tenebrionidae: Pimeliinae: Tentyriini) from Kazakhstan and Middle Asia. *Entomological Review* **98**: 1124–1138.

Chigray, S.N., Nabozhenko, M.V., Keskin, B. & Abdurakhmanov, G. 2018. Taxonomic review of the genus Calyptopsis of Turkey and adjacent areas (Coleoptera: Tenebrionidae: Tentyriini). *Acta Entomologica Musei nationalis Pragae* **58**: 331–346.

Chobaut, A. 1891: Description d'une nouvelle espèce de Rhipidius avec quelques remarques sur ce genre. *Le Coléoptériste* **1**: 235–237, 1 pl.

Chobaut, A. 1893: Coléoptères nouveaux du Nord de l'Afrique. *L'Abeille, Journal d'Entomologie* **28** [1892–1896]: 97–101.

Chobaut, A. 1894a: [new taxa]. *Bulletin de la Société Entomologique de France* **1894**: lxxv-lxxvi.

Chobaut, A. 1894b: [new taxa]. *Bulletin de la Société Entomologique de France* **1894**: ci-ciii.

Chobaut, A. 1894c: [Description d'une espèce nouvelle d'Anthicus et des remarques sur Anthicus bedeli]. *Bulletin de la Société Entomologique de France* **1894**: ccxxviii-ccxxx.

Chobaut, A. 1895a: Notes sur quelques Danacaea et description d'une espèce nouvelle d'Anthicus (Col.). *Bulletin de la Société Entomologique de France* **1895**: 155–157.

Chobaut, A. 1895b: Description de deux variétés nouvelles de coléoptères circa-méditerranéens. *Bulletin de la Société Entomologique de France* **1895**: 346–347.

Chobaut, A. 1896a: Description d'une espèce nouvelle d'Anaspis d'Algérie. *Bulletin de la Société Entomologique de France* **1896**: 200.

Chobaut, A. 1896b: Descriptions de trois nouveaux coléoptères héteromères d'Algérie. *Bulletin de la Société Entomologique de France* **1896**: 375–377.

Chobaut, A. 1897a: Description de cinq variétés nouvelles de coléoptères de la faune circa-méditerranéenne. *Miscellanea Entomologica* **5**: 38–39.

Chobaut, A. 1897b: Notes sur le genre Pentaria et description d'une espèce nouvelle. *Bulletin de la Société Entomologique de France* **1897**: 132–135.

Chobaut, A. 1897c: Description de deux coléoptères nouveaux du Mzab (Sud-Algérien). *Bulletin de la Société Entomologique de France* **1897**: 170–171.

Chobaut, A. 1897d: Description de deux espèces nouvelles de coléoptères du Mzab (Sud-Algérien). *Bulletin de la Société Entomologique de France* **1897**: 204–206.

Chobaut, A. 1898a: Description de quelques espèces et variétés nouvelles de coléoptères algériens. *Revue d'Entomologie* **17**: 74–88.

Chobaut, A. 1898b: Voyage chez les Beni-Mzab (Sud de la Province d'Alger). Contribution à l'étude de la faune entomologique du Sahara algérien. *Memoires de l'Académie de la Vaucluse* **17**: 131–235, 11 pls., 1 map.

Chobaut, A. 1899: Description d'une espèces nouvelle de Zophosis (Col.) de la Tunisie mériodionale. *Bulletin de la Société Entomologique de France* **1899**: 409–411.

Chobaut, A. 1900a: Description d'un Opatride nouveau (Col. Tenebrion.) de la Tunisie méridionale. *Bulletin de la Société Entomologique de France* **1900**: 31–32.

Chobaut, A. 1900b: Description d'une Phaleria nouvelle d'Algérie. *Bulletin de la Société Entomologique de France* **1900**: 341–342.

Chobaut, A. 1901: Description d'un Mylabris nouveau du Sahara algérien (Col.). *Bulletin de la Société Entomologique de France* **1901**: 280–281.

Chobaut, A. 1902: Description d'un rhipiphoride nouveau (Col.) du nord de l'Afrique. *Bulletin de la Société Entomologique de France* **1902**: 89–90.

Chobaut, A. 1903a: Description d'un salpingide nouveau du nord de la Tunisie. *Bulletin de la Société Entomologique de France* **1903**: 143.

Chobaut, A. 1903b: Variabilité d'Anthicus superbus Pic (Col.). *Bulletin de la Société Entomologique de France* **1903**: 170–171.

Chobaut, A. 1904a: Description d'un Rhipidius nouveau de la France méridionale avec tableau dichotomique des Rhipidiini (Col.). *Bulletin de la Société Entomologique de France* **1904**: 228–232.

Chobaut, A. 1904b: Description de deux espèces nouvelles de coléoptères de l'Arabie. *Bulletin de la Société Entomologique de France* **1904**: 243–245.

Chobaut, A. 1906: Note sur le Myiodes subdipterus Bosc (Col.). *Bulletin de la Société Entomologique de France* **1906**: 223–224.

Chobaut, A. 1908: Variété nouvelle d'Anthicus fairmairei Bris. (Col.). *Bulletin de la Société Entomologique de France* **1908**: 162–163.

Chobaut, A. 1918: Description de trois ténébrionides nouveaux de la Tunisie méridionale. *Bulletin de la Société Entomologique de France* **1918**: 74–76.

Chobaut, A. 1919: Description des deux sexes, de l'oeuf et de la larve primaire d'un nouveau Rhipidius de Provence (Col. Rhipiphoridae). *Bulletin de la Société Entomologique de France* **1919**: 200–206.

Chobaut, A. 1921: Description d'un nouveau Diaphorocera (Col., Meloidae) de l'Algérie méridionale. *Bulletin de la Société Entomologique de France* **1921** 298–299.

Chobaut, A. 1922: Description d'un Rhipidius nouveau du Maroc (Coléoptère Rhipiphoridae). *Bulletin de la Société des Sciences Naturelles du Maroc* **2**: 127–128.

Chobaut, A. 1923a: Coléoptères nouveaux du Maroc. *Bulletin de la Société des Sciences Naturelles du Maroc* **3**: 75–77.

Chobaut, A. 1923b: Description de trois Anthicus nouveaux du Maroc (Coléoptères Anthicides). *Bulletin de la Société des Sciences Naturelles du Maroc* **3**: 128–130.

Chobaut, A. 1923c: Description d'un Oxycara nouveau (Col. Tenebrionidae) du Sud-Oranais. *Bulletin de la Société Entomologique de France* **1923** 165–167.

Chobaut, A. 1924a: Description d'un Anthicus nouveaux du Maroc (Coléoptères Anthicides). *Bulletin de la Société des Sciences Naturelles du Maroc* **4**: 104.

Chobaut, A. 1924b: Description d'un opatride (Col. Tenebrionidae) nouveau de l'Afrique du Nord. *Bulletin de la Société Entomologique de France* **1924**: 154–156.

Chobaut, A. 1924c: Description d'un Cyphostethe (Col. Tenebrionidae) nouveau de l'Algérie méridionale. *Bulletin de la Société Entomologique de France* **1924**: 166–167.

Chobaut, A. 1924d: Tableau synoptique des Stenalia paléarctiques à élytres maculés de jaune. *Bulletin de la Société Entomologique de France* **1924**: 179–183.

Chobaut, A. 1925: Description d'un Allophylax (Col. Tenebrionidae). *Bulletin de la Société Entomologique de France* **1924**: 228–229.

Chûjô, M. 1935a: Descriptions of three new mordellid-species from Formosa. *Sylvia, Taihoku* **6**: 72–75.

Chûjô, M. 1935b: A systematic catalogue of Formosan Rhipiphoridae (Coleoptera). *Sylvia, Taihoku* **6** 129–135.

Chûjô, M. 1939: On the Japanese Ciidae (Coleoptera). *Mushi* **12**: 1–10.

Chûjô, M. 1940a: Description of a new and a rare species of the Japanese ciid-beetles. *Insecta Matsumurana* **14**: 132–133.

Chûjô, M. 1940b: Descriptions of new ciid-beetles from Japan proper and Manchuria. *Insecta Matsumurana* **14**: 134–137.

Chûjô, M. 1941: Description of a new Ciid beetle from Japan. *Kontyû* **15** (2): 85–86.

Chûjô, M. 1956: Descriptions of two new species of the Japanese mordellid-beetles with notices on three known species. *Kontyû* (Tokyo) **24**: 174–178.

Chûjô, M. 1957a: Description of a new subgenus and five new species of the Japanese mordellid-beetles with notices on some known species. *Mushi* **30**: 1–10.

Chûjô, M. 1957b: Descriptions of a new genus and six new species of the Japanese mordellid-beetles with notices on some known species. *Mushi* **30**: 11–22.

Chûjô, M. 1959: Coleoptera of the Loo-choo Archipelago (I). *Memoirs of the Faculty of Liberal Arts & Education, Kagawa University, part II* **69**: 1–15.

Chûjô, M. 1960: Description of a new Mordellid-species from the Island Naka-No-Shima. *Niponnius* **1**: 1–3.

Chûjô, M. 1963a: A new species of the genus Basanus Lacordaire from Japan and Korea (Coleoptera, Tenebrionidae) *Niponius* **2**: 17–19.

Chûjô, M. 1963b: A new species of Gonocephalum Solier from Okinawa and Takara Is. (Coleoptera, Tenebrionidae). *Kontyû* (Tokyo) **31**: 149–153.

Chûjô, M. 1966: Taxonomic study on the Tenebrionidae (Coleoptera) of the Ryukyu Islands. *Journal of the Faculty of Agriculture, Kyushu University* **14**: 1–32, 2 pls.

Chûjô, M. 1967: A check list of the Formosan Tenebrionidae (Coleoptera) I. *Kontyû* (Tokyo) **35**: 375–381.

Chûjô, M. 1968a: A check list of the Formosan Tenebrionidae (Coleoptera) II. *Kontyû* (Tokyo) **36**: 14–17.

Chûjô, M. 1968b: A check list of the Formosan Tenebrionidae (Coleoptera) III. *Kontyû* (Tokyo) **36**: 393–397.

Chûjô, M. 1968c: Eine neue Basanus-Art aus Japan (Col., Tenebrionidae). *Kontyû* (Tokyo) **36**: 398–399.

Chûjô, M. 1972: Coleoptera of the Loo-choo Archipelago (IV). *Memoirs of the Faculty of Education, Kagawa University*, **2** (208): 1–18.

Chûjô, M. 1973: Five new species of the Tenebrionidae from Japan and her adjacent regions, with a new record of Gonocephalum coenosum Kaszab from the Senkaku island (Coleoptera). *Esakia* **8**: 13–23.

Chûjô, M. 1975: Tenebrionidae of the Nansei Islands I (Coleoptera). *Esakia* **9**: 15–24, 2 pls.

Chûjô, M. 1977: Tenebrionidae of the Nansei Islands II (Coleoptera). *Esakia* **10**: 1–19.

Chûjô, M. 1978a: Tenebrionidae of the Nansei Islands III (Coleoptera). *Esakia* **11**: 63–80.

Chûjô, M. 1978b: Tenebrionidae of the Nansei Islands IV (Coleoptera). *Esakia* **12**: 1–16.

Chûjô, M. 1979: Tenebrionidae of the Nansei Islands V (Coleoptera). *Esakia* **14**: 31–56.

Chûjô, M. 1980: Tenebrionidae of the Nansei Islands VI (Coleoptera). *Esakia* **15**: 1–10.

Chûjô, M. 1981: Four new species of the Formosan Setenis (Coleoptera, Tenebrionidae). *Esakia* **17**: 131–140.

Chûjô, M. 1983: Tenebrionidae of the Nansei Islands VIII (Coleoptera). *Esakia* **20**: 47–53.

Chûjô, M. 1984: Tenebrionidae of the Nansei Islands IX (Coleoptera). *Esakia* **22**: 1–4.

Chûjô, M. 1985: Notes on the Japanese Tenebrionidae (Coleoptera). *Esakia* **23**: 61–66.

Chûjô, M. 1985: Lagriidae of the Nansei Islands. *Mushi* **50**: 87–90.

Chûjô, M. & Imasaka, S. 1982: Five new species of Misolampidius Solsky from Japan and Korea (Coleoptera, Tenebrionidae). *Esakia* **19**: 123–134.

Chûjô, M. & Miyatake, M. 1961: [new taxa]. In: Miyatake M.: The tribe Leiochrini of Japan (Coleoptera: Tenebrionidae). *Transactions of the Shikoku Entomological Society* 7: 33–40.

Chûjô, M. et Nakane, T. 1953: Mordellidae of Shikoku, Japan. *Mushi* 25: 17–24.

Chûjô, M. & Nakane, T. 1955: Description of a new mordellid-beetle from Japan. *Akitu* 4: 27–29.

Chûjô, M. & Satô, M. 1972: 36. Family Oedemeridae. Coleoptera of the Loo-choo archipelago (IV). *Memoirs of the Faculty of Education Kagawa University* (2) 208: 1–28.

Chûjô, Mt. 1992: [new taxa]. In: Chûjô Mt. & Lee C.-E.: Tenebrionidae from Chejudo Island Korea (Insecta, Coleoptera). *Esakia* 32: 31–46.

Chûjô, Mt. 1995a: Gonocephalum ficifolium, a new species of Tenebrionidae from Japan (Coleoptera). *Esakia* 35: 113–116.

Chûjô, Mt. 1995b: Tenebrionid beetles of the genus Promethis Pascoe from the Nansei Islands (Coleoptera). *Esakia* 35: 117–122.

Chûjô, Mt. 1997a: Tenebrionidae of the Nansei Islands X (Coleoptera). *Esakia* 36: 135–141.

Chûjô, Mt. 1997b: Tenebrionidae from the Nansei Islands XI (Coleoptera). *Journal of the Faculty of Agriculture, Kyushu University* 41: 197–204.

Chûjô, Mt. & Lee, C.-E. 1993: Korean Tenebrionidae (Insecta, Coleoptera). *Esakia* 33: 109–122.

Clairville, J.P. de. 1798: [new taxa]. In: Clairville, J.P. de. & Schellenberg, J.R. 1798: *Helvetische Entomologie oder Verzeichniss der schweizerischen Insekten nach einer neuen Methode geordnet*. Erster Theil. Zürich: Orell, Füssli und Companie, 149 + [4] pp., 14 pls. [J.R. Schellenberg is author of the plates].

Clermont, J. 1925: Description d'une espèce nouvelle de Tenebrionidae (Col.). *Bulletin de la Société Entomologique de France* 1925: 183–184.

Cobos, A. 1950: Datos para el catálogo de los coleópteros de España. Especies de los alrededores de Málaga. *Boletín de la Real Sociedad Española de Historia Natural* 47 [1949]: 589–594.

Cobos, A. 1953: Especies nuevas de coleópteros de la provincia de Almería, I. *Archivos del Instituto de Aclimatación* (Almería) 1: 127–134.

Cobos, A. 1954: Una breve campaña entomológica por las Sierras de Tejada y Almijara (Provincia de Málaga): Ins. Coleópteros. *Archivos del Instituto de Aclimatación* (Almería) 3: 29–39.

Cobos, A. 1961: Exploratión entomológica del central del Tizi-Ifri (macizo del Iguelmalet, Rif central, Marruecos). Coleópteros. *Archivos del Instituto de Aclimatación* (Almería) 10: 63–98, 1 pl.

Cobos, A. 1962: Sobre las Alphasida del subgénero Betasida Rtt. y descripción de une nueva especie (Coleoptera, Tenebrionidae). *Archivos del Instituto de Aclimatación* (Almería) 11: 37–40, pl. 2.

Cobos, A. 1963a: Ensayo sobre las Asida Latr. de Marruecos (Col. Tenebr.). *Archivos del Instituto de Aclimatación* (Almería) 12: 9–37, 6 pls.

Cobos, A. 1963b: El género 'Coelometopus' Sol. (Col. Tenebrionidae). *Archivos del Instituto de Aclimatación* (Almería) 12: 49–53, 1 pl.

Cobos, A. 1965a: Adiciones y correcciones al 'Ensayo sobre las Asida Latr., de Marruecos' (Col. Tenebrionidae) *Graellsia* 21: 55–57.

Cobos, A. 1965b: Una nueva especie de Helioctamenus de Canarias (Col. Colydiidae). *Graellsia* 21: 59–63.

Cobos, A. 1968: Dos nuevas Alphasida (Col. Tenebrionidae) de Marruecos. *Annales de la Société Entomologique de France* (N.S.) 4: 911–913.

Cobos, A. 1988a: La Coleopterofauna endémica almeriense. *Graëllsia* **43**: 3–17.

Cobos, A. 1988b: Révision de las Alphasida Escalera 1905, del subgénero Betasida Reitter 1917 (Coleoptera, Tenebrionidae). *Eos, Revista Española de Entomología* **64**: 47–56.

Cockerell, T.D.A. 1910: The classification of the Meloidae. *Entomological News* **21**: 307.

Codina, A. 1918: Alguns coleópteros de les illes de Eivissa (Santa Eulalia i Illa Plana) i de Formentera (Saint Francesc Xavier). *Anuari Junta de Ciencies Naturals* (Barcelona) **1918**: 264–269.

Coiffait, H. 1947: Un Rhipiphoride nouveau de la faune de France. *Revue Française d'Entomologie* **14**: 15–19.

Collado, J. & Alonso-Zarazaga, M.A. 1996: Báguena-Corella's genera of Aderidae (Coleoptera): nomenclatural corrections and validations. *The Coleopterist's Bullettin* **50**: 96–98.

Comolli, A. 1837: *De Coleopteris novis ac rarioribus minusve cognitis provinciae Novocomi*. Ticini Regii: Fusi et Socii, 54 pp.

Compte, A. 1970: Los coléopteros de las Islas Columbretes. *Eos, Revista Española de Entomología* **45** [1969]: 97–137.

Compte, A. 1985: II. Congresso Ibérico de Entomologia: Mordellidae de las islas Baleares. *Boletim da Sociedade Portuguesa de Entomologia, Suplemento 1* **1985**: 57–66.

Costa, A. 1844: Osservazioni intorno ai Coleotteri Lamellicorni del Regno di Napoli. *Annali dell'Accademia degli Aspiranti Naturalisti di Napoli* **2**: 9–23.

Costa, A. 1847: Specie nuove e rare d'insetti delle montagne del Matese. *Annali dell'Accademia degli Aspiranti Naturalisti Napoli* (2) **1**: 89–131.

Costa, A. 1847: Descrizione di alcuni coleotteri del Regno di Napoli. *Annali dell'Accademia degli Aspiranti Naturalisti Napoli* (2) **1**: 134–162.

Costa, A. 1852: *Parte Ia. Coleotteri Eteromeri. Famiglia degli Edemeridei – Oedemeridea*. In: *Fauna del regno di Napoli ossia enumerazione de tutti gli animali che abitano le diverse regioni di questo regno e la acque che la bagnano contenente la descrizione de'nuovi o poco esattamente conosciuti noc figure ricavete da originali vivente e dipinte al naturale*. Napoli: Gaetano Sautto, 36 pp., pls. ix-xi.

Costa, A. 1853: *Parte Ia. Coleotteri, Eteromeri. Famiglia de'Mordellidei – Mordellidea*. In: *Fauna del regno di Napoli ossia enumerazione de tutti gli animali che abitano le diverse regioni di questo regno e la acque che la bagnano contenente la descrizione de'nuovi o poco esattamente conosciuti noc figure ricavete da originali vivente e dipinte al naturale*. Napoli: Gaetano Sautto, 1–8 pp.

Costa, A. 1854: *Parte Ia. Coleotteri, Eteromeri. Famiglia de'Mordellidei – Mordellidea*. In: *Fauna del regno di Napoli ossia enumerazione de tutti gli animali che abitano le diverse regioni di questo regno e la acque che la bagnano contenente la descrizione de'nuovi o poco esattamente conosciuti noc figure ricavete da originali vivente e dipinte al naturale*. Napoli: Gaetano Sautto, 9–32 pp.

Costa, A. 1859: *Parte IIa. Supplementi alle famiglie di coleotteri*. In: *Fauna del regno di Napoli ossia enumerazione de tutti gli animali che abitano le diverse regioni di questo regno e la acque che la bagnano contenente la descrizione de'nuovi o poco esattamente conosciuti noc figure ricavete da originali vivente e dipinte al naturale*. Naples: Gaetano Sautto, 24 pp.

Costa, A. 1878: Relazione di un viaggio per l'Egitto, la Palestina e le coste della Turchia asiatica per richerche zoologiche. *Atti della Realle Accademia delle Scienze Fisiche e Matematiche* 7 (2): 1–40.

Costa, A. 1882: Relazione di un viaggio nelle calabrie per ricerche zoologiche fatto nella state del 1876. *Atti della reale Accademia delle Scienze Fisiche e Matematiche di Napoli* 9: 1–62, 1 pl.

Costa, A. 1884: Notizie ed osservazioni sulla geo-fauna sarda. Memoria quarta. *Atti della Reale Accademia delle Scienze Fisiche e Matematiche di Napoli* (2) 1 (13): 1–31.

Costa, A. 1884: Notizie ed osservazioni sulla geo-fauna sarda; memoria quarta. *Rendiconto della Reale Accademia delle Scienze Fisiche e Matematiche* 23 (12): 167–174.

Coucke, L. 1891: Tables dichotomiques pour servir à la détermination des espèces Belges de coléoptères hètèromères. *Comptes-Rendus des Séances de la Société Entomologique de Belgique* 1891: ccclcccii-ccclxxxviii.

Creutzer, C. 1796: [new taxa]. In: Panzer, G.W.F. 1796b: *Faunae Insectorum Germanicae initia oder Deutschlands Insecten. Ed. 2. XXXV*. Nürnberg: Felsecker, 24 pp. + 24 pls.

Creutzer, C. 1796: [new taxa]. In: Panzer, G.W.F. 1796b: *Faunae Insectorum Germanicae initia oder Deutschlands Insecten. Ed. 2. XXXVI*. Nürnberg: Felsecker, 24 pp. + 24 pls.

Creutzer, C. 1799: *Entomologische Versuche*. Wien: Karl Schaumburg und Comp., 140 + [8] pp., 3 pls.

Crotch, G.R. 1867a: On the Coleoptera of the Azores. *Proceedings of the Scientific Meetings of the Zoological Society of London* 1867: 359–391, pl. 23.

Crotch, G.R. 1867b: [new taxa]. In: Crotch G.R. & Sharp D.: Additions to the Catalogue of British Coleoptera, with descriptions of new species. *Transactions of the Entomological Society of London* 5: 435–451.

Crotch, G.R. 1872: List of the Coleoptera found during the Progress of the Survey. Pp. 263–268. In: Wilson, C.W. & Palmer, H.S. (eds): *Ordonance Survey of the Peninsula of Sinaï*. South Hampton.

Crotch, G.R. 1873a: *Check list of the Coleoptera of America, north of Mexico*. Salem: Mass: Naturalists' Agency, 136 pp.

Crotch, G.R. 1873b: Synopsis of the Endomychidae of the United States. *Transactions of the American Entomological Society* 4: 359–363.

Csiki, E. 1901a: Bogarak. Coleopteren. Pp. 77–120. In: Horváth, G. (ed.): *Zichy Jenõ gróf harmadik ázsiai utazása. II. kötet. Zichy Jenõ gróf harmadik ázsiai utazásának állattani eredményei.* [*Dritte asiatische Forschungsreise des des Grafen Eugen Zichy. Band II. Zoologische Ergebnisse der dritten asiatischen Forschungsreise des Grafen Eugen Zichy*]. Budapest: Victor Hornyánszky and Leipzig: Karl w. Hiersemann, 470 pp., 28 pls.

Csiki, E. 1901b: Coleoptera nova ex Hungaria. *Természetrajzi Füzetek*. 24: 486–490.

Csiki, E. 1909: Újabb adatok Magyarország bogárfaunájához. *Rovartani Lapok* 16: 3–6.

Csiki, E. 1913: Rhipiphoridae. Pars 54. In: S. Schenkling (ed.): *Coleopterorum Catalogus*. Berlin: W. Junk, 29 pp.

Csiki, E. 1915: Mordellidae. Pars 63. In: S. Schenkling (ed.): *Coleopterorum Catalogus*. Berlin W. Junk, 84 pp.

Csiki, E. 1924: Serropalpidae. Pars 77. In: S. Schenkling (ed.): *Coleopterorum Catalogus*. Berlin: W. Junk 62 pp.

Csiki, E. 1940a: Bogarak. Coleopteren. In: Csiki E.: Allattani Kutatásai Albániában (Explorationes zoologicae ab E. Csiki in Albania paractae). *A Magyar Tudomanyos Academia Balkán-Kutatásainak Tudományos Eredményei, Budapest*, 1 (2): 208–288.

Csiki, E. 1940b: Magyarországi új bogarak. (Coleoptera nova ex Hungaria.). *Matematikai Természettudományi Értesito* 59: 916–920.

Csiki, E. 1949: Über eine neue Mordellistena aus der Krim. *Folia Entomologica Hungarica* (S.N.) 3: 45–46.

Csiki, E. 1953: Ueber neue und bekannte Coleoptera aus Ungarn und der angrenzendern Ländern. *Annales Historico-Naturales Musei Nationalis Hungarici* (N.S.) **3**: 115–135.

Cuni, Martorell M. 1876: [new taxon]. In: Cuni Martorell, M. & Martorell Peña, M.: *Catalogo metódico y razonado de los coleópteros observados en Cataluña*. Gorchs, Barcelona, 320 pp.

Curtis, J. 1826: *British Entomology; being illustrations and descriptions of the genera of insects found in Great Britain and Ireland: containing coloured figures from nature of the most rare and beautiful species, and in many instances of the plants upon which they are found. Vol. III*. London: J. Curtis, [8] pp + 99–146 pls. and sheets.

Curtis, J. 1827: *British Entomology; being illustrations and descriptions of the genera of insects found in Great Britain and Ireland: containing coloured figures from nature of the most rare and beautiful species, and in many instances of the plants upon which they are found. Vol. IV*. London: J. Curtis, [8] pp + 147–194 pls. and sheets.

Curtis, J. 1829: *A guide to an arrangement of British Insects*. London: Westely, vi + 248 pp.

Curtis, J. 1829: *British entomology; being illustrations and descriptions of the genera of insects found in Great Britain and Ireland: containing coloured figures from nature of the most rare and beautiful species, and in many instances of the plants upon which they are found. Vol. VI*. London: J. Curtis, [8] pp. + 242–289 pls and sheets.

Curtis, J. 1830: *British entomology; being illustrations and descriptions of the genera of insects found in Great Britain and Ireland: containing coloured figures from nature of the most rare and beautiful species, and in many instances of the plants upon which they are found. Vol. VII*. London: J. Curtis, [8] pp. + 290–337 pls and sheets.

Curtis, J. 1833: Characters of some undescribed genera and species indicated in the guide to an arrangement of British Insects. *The Entomological Magazine* **1**: 186–199.

Curtis, J. 1836: *British Entomology being illustrations and descriptions of the genera of insects found in Great britain and Ireland: containing coloured figures from nature of the most rare and beautiful species, and in many instances of the plants upon which they are found. Vol. XIII*. London: J. Curtis, 578–625 pls and sheets.

Curtis, J. 1838: *British Entomology; being illustrations and descriptions of the genera of insects found in Great Britain and Ireland: containing coloured figures from nature of the most rare and beautiful species, and in many instances of the plants upon which they are found. Vol. XV*. London: J. Curtis, [4] pp. + 674–721 pls and sheets.

Czetó, Z. 1990: New Mordellidae (Coleoptera) from the Mediterranean Region, and a key to the genus Mordellistenula Sthegolewa-Barowskaya. *Folia Entomologica Hungarica* (N.S.) **51**: 25–30.

Dadabaev, H.R. 1986: Novyy vid roda Prosodes Eschsch. (Coleoptera, Tenebrionidae) s Ghissarskogo Khrebta. *Doklady Akademii Nauk Tadzhikskoy SSR* **29**: 127–128.

Dadabaev, Kh. 1986: Dva novykh vida roda Prosodes Eschsch. (Coleoptera, Tenebrionidae) iz Tadzhikistana. *Vestnik Zoologii* **4**: 78–81.

Daffner, H. 1983: Eine neue Art der Gattung Langelandia Aubé von den Kanarischen Inseln (Coleoptera, Colydiidae). *Nachrichtenblatt der Bayerischen Entomologen* **32**: 20–22.

Daffner, H. 1985: Eine neue Art der Gattung Langelandia Aubé aus Griechenland. *Nachrichtenblatt der Bayerischen Entomologen* **34**: 36–38.

Dahlgren, G. 1976: Eine neue Asclera-Art aus Griechenland (Col., Oedemeridae). *Entomologisk Tidskrift* **97**: 21–22.

Dajoz, R. 1960: Les coléoptères mycetophiles de la forêt de la Massane (Pyrénées-Orientales). Note préliminaire. *Vie et Milieu* **11**: 195–208.

Dajoz, R. 1964: Contribution à l'étude des coléoptères Mycetophagidae I. Revision du genre Esarcus et description de trois espèces nouvelles d'Espagne. *L'Entomologiste* **20**: 7–15.

Dajoz, R. 1965: Catalogue des coléoptères de la forêt de la Massane. *Vie et Milieu, Supplement* **15** (4): 1–207 + [2] pp.

Dajoz, R. 1968: Révision des Colydiidae anophtalmes de la faune paléarctique (Col.) IV. Études sur les genres Anommatus et Langelandia. *Annales de la Société Entomologique de France* (N. S) **4**: 975–988.

Dajoz, R. 1969: Révision des Colydiidae anophtalmes de la faune paléarctique. 3. Étude morphologique et biometrique de Langelandia anophtalma Aubé et de Langelandia reitteri Bellon. *Bulletin du Muséum National d'Histoire Naturelle, Paris* **40**: 710–717.

Dajoz, R. 1971: Coléoptères Colydiidae nouveaux ou peu connus de la région paléarctique. *L'Entomologiste* **27**: 83–101.

Dajoz, R. 1973: Coléoptères Colydiidae paléarctiques nouveaux ou peu connus. *Cahiers des Naturalistes* **29**: 79–84.

Dajoz, R. 1975: Ergebnisse der Bhutan-Expedition 1972 des Naturhistorischen Museums in Basel. Coleoptera: Fam. Colydiidae & Cerylonidae. *Entomologica Basiliensia* **1**: 293–311.

Dajoz, R. 1977: Coléoptères Colydiidae et Anommatidae paléarctiques. *Faune de l'Europe et du Bassin Méditerranéen* 8. Paris: Masson, 275 pp.

Dajoz, R. 1983: Les peuplements de coléoptères terricoles de Tunisie. Études écologique et biogéographique des Tenebrionidae (incl. 2 taxa nouveaux) et Carabidae. *Cahiers des Naturalistes* **38** [1982]: 33–67.

Dajoz, R. 1984: Ténébrionides européens nouveaux ou peu connus (Coleoptera). *Cahiers des Naturalistes* (N.S.) **39** [1983]: 89–92.

Dajoz, R. 1985: Description de deux Tenebrionidae nouveaux de Tunisie (Coleoptera). *Nouvelle Revue d'Entomologie* (N.S.) **2**: 193–196.

Dalla Torre, K.W. von. 1879: Die Käferfauna von Oberösterreich. Systematisches Verzeichnis der in Oberösterreich bisher beobachteten Käfer. *Jahres-Bericht des Vereines für Naturkunde in Oesterreich ob der Enns zu Linz* **10**: 1–125.

Dalla Torre, K.W. von. 1880: Die Käferfauna von Oberösterreich. Systematisches Verzeichnis der in Oberösterreich bisher beobachteten Käfer. *Jahres-Bericht des Vereines für Naturkunde in Oesterreich ob der Enns zu Linz* **11**: 1–81.

Dalman, J.W. 1823: *Analecta Entomologica*. Holmiae: Typis Lindhianis, vii + 104 + [4] pp., 4 pls.

D'Amore-Fracassi, A. 1907: Gen. Colpotus Muls. et Rey. Tavola sinottica e descrizione delle specie appartenenti alla fauna europea, caucasica e circummediterranea. *Il Naturalista Siciliano* **19**: 177–199.

DeGeer, C. 1775: *Mémoires pour servir à l'histoire des Insectes. Tome cinquième.* Stockholm: Limprimerie Pierre Hesselberg, vii + 448 pp., 16 pls.

Degiovanni, A. 2000: Contributo alla conoscenza dei coleotteri anticidi italiani con descrizione di Tenuicomus platiai n. sp. (Insecta Coleoptera Anthicidae). *Quaderno di Studi e Notizie di Storia Naturale della Romagna* **13** (*Supplement*): 85–89.

Degiovanni, A. 2012: Una nuova specie della Turchia del genere Tenuicollis Marseul, 1879 (Insecta Coleoptera Anthicidae). *Quaderno di Studi Notizie di Storia Naturale della Romagna* **36**: 97–103.

Degiovanni, A. 2014: Clavicollis doderoi melonii n. ssp. della Sardegna (Coleoptera, Anthicidae). *DORIANA, Supplemento agli Annali del museo civico di storia naturale 'G. Doria'* **8**, No 389: 1–11.

Dejean, P.F.M.A. 1821: *Catalogue de la collection de coléoptères de M. le Baron Dejean*. Paris: Crevot, viii + 136 + 2 (errata) pp.

Dejean, P.F.M.A. 1834: *Catalogue des coléoptères de la collection de M. le Comte Dejean. Deuxième édition. 3e Livraison*. Paris: Méquignon-Marvis Pères et Fils, pp. 177–256.

Dejean, P.F.M.A. 1835: *Catalogue des coléoptères de la collection de M. le Comte Dejean. Deuxième édition. 4e Livraison*. Paris: Méquignon-Marvis Pères et Fils, pp. 257–360.

Dejean, P.F.M.A. 1836: *Catalogue des coléoptères de la collection de M. le Comte Dejean. Troisième édition revue, corrigée et augmentée. Livraisons 1–4*. Paris: Méquignon-Marvis Père et Fils, 1–384 pp.

De Lisle, M.O. 1973: Description d'un nouveau coléoptère Trictenotomidae de Malaisie. *Nouvelle Revue d'Entomologie* **3**: 17–18.

Della Beffa G. 1909: Note coleotterologiche. *Rivista Coleotterologica Italiana* 7: 192–201.

Della Beffa G. 1931: Himalaia cashmiriano. Appendici scientifiche. Insetti del Kashmir raccolti dal dott. Cesare Calciati nelle spedizione Mario Piacenza (1913). Milano: Rizzoli & C., 173–187.

Demaison, C. 1905: Description d'une variété et d'une espèce nouvelle d'Asie mineure (Col.). *Bulletin de la Société Entomologique de France* **1904**: 286–287.

Denier, P. 1913: Description d'un genre et d'une espèce nouvelle de Meloidae (Col.). *Bulletin de la Société Entomologique de France* **1913**: 231–232.

Denier, P. 1921: Description d'une espèce nouvelle de Meloe paléarctique. *Bulletin de la Société Entomologique de France* **1921**: 112–113.

Denier, P. 1935: Coleopterorum americanorum familiae Meloidarum enumeratio synonymica. *Revista de la Sociedad Entomólogica Argentina* 7: 139–176.

Denisova, L.A. 1940: Obzor palearkticheskikh vidov podsemeystva Tenebrioninae (Coleoptera, Tenebrionidae). *Trudy Zoologicheskogo Instituta Akademii Nauk* SSSR **6**: 222–251.

Desbrochers des Loges, J. 1868: Description de deux coléoptères nouveaux des environs de Bone. *Bulletin de l'Académie d'Hippone* **4**: 77–80.

Desbrochers des Loges, J. 1870: Descriptions de coléoptères nouveaux d'Europe et confins. *L'Abeille, Mémoires d'Entomologie* 7: 97–135.

Desbrochers des Loges, J. 1874: [new taxa]. In: Marseul S.-A.: Répertoire des coléoptères d'Europe décrits isolément depuis 1863. Troisième partie. *L'Abeille, Journal d'Entomologie* **12** [1874–1875]: 1–456.

Desbrochers des Loges, J. 1875: Anthicides nouveaux, pp. 42–49. In: *Opuscules Entomologiques* (*Coléoptères*). 1er Cahier [1874–1875]: 1–56. Paris: A. Gaudon.

Desbrochers des Loges, J. 1881: Insectes coléoptères du nord de l'Afrique nouveaux ou peu connus. 1er Mémoire. Ténébrionides. *Bulletin de l'Académie d'Hippone* **16**: 51–168.

Desbrochers des Loges, J. 1884: Diagnoses de coléoptères nouveaux algériens. *Bulletin de l'Academie d'Hippone* 19: 169.

Desbrochersdes Loges, J. 1890: [Diagnoses des coléoptères nouveaux d'Algérie]. *Bulletin de la Société Entomologique de France* 1894: clxxxv-clxxxvi.

Desbrochers des Loges, J. 1894: Révision monographique des ténébrionides appartenant à la tribu des cossyphides d'Europe et des contrées méditerranéennes en Afrique et en Asie. *Le Frelon* 4: 2–21.

Desbrochers des Loges, J. 1898a: Quelques matériaux pour la faune des coléoptères de Barbarie. *Le Frelon* 6: 65–82.

Desbrochers des Loges, J. 1898b: Quelques matériaux pour la faune des coléoptères de Barbarie. *Le Frelon* 7: 1–51.

Desbrochers des Loges, J. 1899a. Faunule des Coléoptères de la France et de la Corse. Anthicidae. *Le Frelon* 7 (10): 151–160.

Desbrochers des Loges, J. 1899b. Faunule des Coléoptères de la France et de la Corse. Anthicidae (Suite). *Le Frelon* 7 (11): 161–176.

Desbrochersdes Loges, J. 1899c. Faunule des Coléoptères de la France et de la Corse. Anthicidae (Suite). *Le Frelon* 7 (12): 177–192.

Desbrochers des Loges, J. 1900a. Desbrochers des Loges J. 1899c. Faunule des Coléoptères de la France et de la Corse. Salpingidae, Pyrochroidae. *Le Frelon* 8 (10–11): 1–32.

Desbrochers des Loges, J. 1900b. Faunule des Coléoptères de la France et de la Corse. Melandryidae. *Le Frelon* 9 (1): 1–16.

Des Gozis, M. 1886: *Recherche de l'espèce typique de quelques anciens genres. Rectifications synonymiques et notes diverses.* Motluçon: Imprimerie Herbin, 36 pp.

Desmarest, A.G. 1825: Oenas africana. P. 435. In: Cuvier, F.G. (ed.) *Dictionnaire de Sciences Naturelles*. Paris: Levrault, 1816–1830.

Deyrolle, A. 1864: [new taxa]. In: Wollaston, T.W.: *A catalogue of the coleopterous insects of the Canaries in the collection of the British Museum*. London: Trustees, xiii + 648 pp.

Deyrolle, A. 1867: Monographie de la tribu des zophosites. *Annales de la Société Entomologique de France* (4) 7: 73–248, 4 pls.

Deyrolle, H. 1875a: Tableau synoptique des espèces de Tricténotomides. *Bulletin des Séances de la Société entomologique de France* **43**: 67–69.

Deyrolle, H. 1875b: Tableau synoptique des espèces de Tricténotomides. Séance du 24 Mars 1875. *Annales de la Société entomologique de France, Cinquième série* 5: lix–lxi.

Dillon, L.S. 1952: The Meloidae (Coleoptera) of Texas. *The American Midland Naturalist* 48: 330–420.

Dilwyn, L.W. 1829: *Memoranda relating to Coleopterous insects, found in the neighbourhood of Swansea.* Swansea: W.C. Murray and D. Rees, Cambrian-office, 75 pp.

Dodero, A. 1916: Appunti Coleotterologici II. *Annali del Museo Civico di Storia Naturale Giacomo Doria* 47: 337–354.

Dohrn, C.A. 1873: Exotisches. *Entomologische Zeitung* (Stettin) **34**: 70–83.

Dohrn, C.A. 1880: Exotisches. *Entomologische Zeitung* (Stettin) **41**: 367–382.

Dohrn, C.A. 1882: Exotisches. *Entomologische Zeitung* (Stettin) **43**: 245–259, 457–470.

Dohrn, C.A. 1884: Exotisches. *Entomologische Zeitung* (Stettin) **45**: 39–46.

Dokhtouroff, W. 1887a: Insecta itinere Cl. N. Przewalskii in Asia centrali novissima lecta VIII. Cantharides nouveaux du Thibet. *Horae Societatis Entomologicae Rossicae* **21**: 341–344.

Dokhtouroff, W. 1887b: Description de deux coléoptères nouveaux de la faune aralo-caspienne. *Horae Societatis Entomologicae Rossicae* **21**: 344–345.

Dokhtouroff, W. 1889: Faune coléoptérologique aralo-caspienne. XIII partie. Meloïdes et cantharides (Mylabrides). *Horae Societatis Entomologicae Rossicae* **24** [1889–1890]: 121–192, 1 pl.

Donisthorpe, H.S.K. 1909: Anaspis hudsoni, Donisthorpe, fem. bred. *The Entomologist's Record and Journal of Variation* **23**: 800.

Donisthorpe, H.S.K. 1916: Cis dentatus, Mellie, not a British insect. *The Entomologist's Record and Journal of Variation* **28**: 155–156.

Donisthorpe, H.S.K. 1925: Lissodema kirkae n.sp. A species of Coleoptera new to science. *The Entomologist's Record and Journal of Variation* **37**: 106–107.

Donisthorpe, H.S.K. 1928: Anaspis florenceae nov. sp.; a species of Coleoptera new to science; with a few notes on some other species of the genus. *The Entomologist's Record and Journal of Variation* **40**: 121–122.

Donisthorpe, H.S.K. 1936: Cis savilli n. sp., a species of Coleoptera new to science. *The Entomologist's Monthly Magazine* **72**: 220.

Donovan, E. 1793: *The natural history of British Insects; explaining them in their several states with the period of their transformations, their food, oeconomy, &c. Together with the history of such minute insects as require investigations by the microscope. The whole illustrated by coloured figures designed and executed from living specimens. Vol. II.* London: printed for the author, and for F. and C. Rivington, 96 + [7] pp., pls. 37–72.

Donovan, E. 1801: *The natural history of British Insects; explaining them in their several states, with the period of their transformations, their food, oeconomy, &c. together with the history of such minute Insects as require investigation by the microscope, the whole illustrated by coloured figures, designed and executed from living specimens. Vol. X.* London: F.C. and J. Rivington, 95+vii pp., 35 pls.

Dorthes, J.A. 1787: Observations sur quelques insectes nuisibles aux blés et à la luzerne. *Mémoires de la Société Agricole de Paris* **1787** (trim. de printemps): 61–71.

Drapiez, P.A.J. 1819: Description de huit espèces d'insectes nouveaux. *Annales Générales des Sciences Physiques (Bruxelles)* **2**: 197–204, 1 pl.

Drapiez, P.A.J. 1820a: Description de six espèces d'insectes nouveaux. *Annales Générales des Sciences Physiques* (Bruxelles) **4**: 349–354.

Drapiez, P.A.J. 1820b: Description de huit espèces d'insectes nouveaux. *Annales Générales des Sciences Physiques* (Bruxelles) **5**: 117–123, pl. 82.

Drapiez, P.A.J. 1821: Description de cinq espèces d'insectes nouveaux. *Annales Générales des Sciences Physiques* (Bruxelles) **8**: 275–282, 1 pl.

Drogvalenko, A.N. 2002: Novye i redkie dlya fauny Ukrainy vidy zhestkokrylykh nasekomykh (Insecta: Coleoptera). Soobshcheniye 2 [New and rare species of beetles (Insecta: Coleoptera) for fauna of Ukraine. Information 2]. *Izvestiya Kharkovskogo Entomologicheskogo Obshchestva* **9**: 9–19.

Drogvalenko, A.N. 2005: New and rare species of beetles (Insecta: Coleoptera) for Fauna of Ukraine. Information 3. *The Kharkov Entomological Society Gazette* **12**: 86–92.

Drumont, A. 2006: Contribution à l'étude des Trictenotomidae avec la description d'une nouvelle espèce d'Autocrates Thomson, 1860: A. maqueti n. sp., originaire du sud de la Chine (Coleoptera, Trictenotomidae). *Les Cahiers Magellanes* **61**: 1–12

Dubrovin, N.N. 1992a: [new taxa]. In: Dubrovin N.N. & Kompantseva T.V.: Morpho-ecological review of larvae of Isomira (Coleoptera, Alleculidae) with the description of a new species from central Asia. *Zoologicheskiy Zhurnal* 71 (7): 14–22.

Dubrovin, N.N. 1992b: Sem. Alleculidae – pyl'ceedy. Pp. 510–517. In: Ler, P.A.: *Opredelitel' nasekomykh dalnego vostoka SSSR v shesti tomakh. Tom III, zhestkokrylye, ili zhuky, Chast' 2.* Sankt-Peterburg: Nauka, 704 pp.

Dubrovina, M.I. 1973a: Novyy vid roda Stenerophlina (Coleoptera, Alleculidae) iz Uzbekistana [A new species of the genus Stenerophlina (Coleoptera, Alleculidae) from Uzbekistan] *Zoologicheskiy Zhurnal* 52: 284–286.

Dubrovina, M.I. 1973b: Novyy podrod i novye vidy zhukov-pyl'ceedov roda Isomira Muls. (Coleoptera, Alleculidae) iz Stredney Azii [New subgenus and new species of pollen beetles of the genus Isomira Muls. (Coleoptera, Alleculidae) from Central Asia]. *Entomologichekoe Obozrenie* 52: 367–376.

Dubrovina, M.I. 1975: Novyy podrod i novye vidy zhukov-pyl'ceedov roda Hymenalia Muls. (Coleoptera, Alleculidae) iz Mongolii i sopredel'nykh rayanov Kitaya [A new subgenus and new species of pollen beetles of the genus Hymenalia Muls. (Coleoptera, Alleculidae) from Mongolia and from adjacent regions of China]. *Nasekomye Mongolii* 3: 165–172.

Dubrovina, M.I. 1976: Taksonomya podroda Pterna Siedl. roda Mycetochara Berth. (Coleoptera, Alleculidae). *Zoologicheskiy Zhurnal* 55: 1407–1409.

Dubrovina, M.I. 1978: Novye dannye o vidakh triby Alleculini (Coleoptera, Alleculidae) evropeyskoy chasti SSSR. *Nauchnye Doklady Vysshey Shkoly Biologicheskie Nauki* 7: 53–55.

Dubrovina, M.I. 1982: Obzor zhukov-pyl'ceedov roda Isomira Muls. (Coleoptera, Alleculidae) fauny SSSR [A review of pollen-beetles of the genus Isomira Muls. (Coleoptera, Alleculidae) of the USSR]. *Entomologichekoe Obozrenie* **61**: 131–144.

Dufour, L.J.M. 1820: Description de dix espèces nouvelles ou peu connues d'insectes recueillis en Espagne. *Annales Générales des Sciences Physiques* (Bruxelles) **6**: 307–317, pl. XCVI.

Dufour, L.J.M. 1828: Description d'un genre nouveau d'Insectes de l'ordre des Parasites. *Annales des Sciences Naturelles* (Paris) (1) **13**: 62–66, 1 pl.

Dufour, L.J.M. 1841: Histoire de metamorphoses d'une Oedémère. *Annales de la Société Entomologique de France* **10**: 5–10.

Dufour, L.J.M. 1843: Excursion entomologique dans les montagnes de la Vallée d'Ossau. *Bulletin des Sciences, Lettres et Arts de Pau* **3**: 1–118.

Dufour, L.J.M. 1849: Note sur trois espèces du genre Anthicus. *Annales des Sciences Naturelles. Zoologie* (3) **11**: 229–230, pl. 5.

Dufour, L.J.M. 1851: Des zones entomologiques dans nos Pyrénées occidentales, et designation des insectes qui les habitent (2). *Actes de la Société Linnéenne de Bordeaux* **17**: 304–364.

Duftschmid, C.E. 1812: *Fauna Austriaca oder Beschreibung der österreichischen Insekten für angehende Freunde der Entomologie. Zweiter Theil.* Linz und Leipzig: K.K. priv. akademischen Kunst-, Mussik- und Buchhandlung, viii + 311+ [4] pp.

Duftschmid, C.E. 1825: *Fauna Austriaca, oder Beschreibung der österreichischen Insekten, für angehende Freude der Entomologie. Dritter Theil.* Linz: Priv. akademischen Kunst-, Musik- und Buchhandlung, 289 pp.

Dugès, E. 1869: Descripción de algunos melóideos indigenas. *Naturaleza Mexicana* 1 [1869]: 100–113.

Dugès, E. 1870: Descripción de algunos melóideos indigenas. *Naturaleza Mexicana* 1 [1869]: 125–128, 157–168, 1 pl.

Duméril, A.M.C. 1825: Mylabre, Mylabris. Pp. 11–13. In: Cuvier, F.G. (ed.): *Dictionnaire des Sciences naturelles, dans lequel on traite méthodiquement des différents êtres de la nature, considérés soit en eux-mêmes, d'après l'état actuel de nos connaissances, soit relativement a l'utilité qu'en peuvent retirer la médecin, l'agriculture, le commerce et les arts. Suivi d'une biographie des plus célébres naturalistes. Tome 34.* Paris: F.G. Levrault, 560 pp.

Duponchel, P.-A.-J. 1827: Mémoires sur quelques insectes nouveaux ou peu connus. *Annales de la Société Linnéenne de Paris* **6**: 338–343, 2 pls.

Duponchel, P.-A.-J. 1845: [new taxa]. In: D'Orbigny, C.V.D. (ed.): *Dictionnaire universel d'histoire naturelle résumant et complétant tous les faits présentés par les encyclopédies, les anciens dictionnaires scientifiques, les œuvres complètes de Buffon, et les meilleurs traités spéciaux sur les divers branches des sciences natrurelle; donnant la description des êtres et des divers phénomènes de la nature, l'étymologie et la définition des noms scientifiques, les principales applications des corps organiques et inorganiques, à l'agriculture, à la médicine, aux arts industriels, etc.; gravé sur acier. Tome 6.* Paris: C. Renard, 792 pp.

Dury, C.A. 1902: A revised list of Coleoptera observed near Cincinnati, Ohio, with notes on localities, bibliographical references, and descriptions of new species. *Journal of the Cincinnati Society of Natural History* **20**: 107–196.

Dury, C.A. 1917: Synopsis of the coleopterous family Cisidae (Cioidae) of America north of Mexico. *The Journal of the Cincinnati Society of Natural History* **22** (2): 2–28.

Dury, C.A. 1919: A new ciside genus with new species from Manitoba. *Canadian Entomologist* **51**: 158.

Dvořák, M. 1983: Drei neue Arten und Bemerkungen zur Familie Meloidae (Coleoptera). *Acta Entomologica Bohemoslovaca* **80**: 441–450.

Dvořák, M. 1985: Croscherichia maceki sp. n. aus Irak (Coleoptera, Meloidae). *Annotationes Zoologicae et Botanicae* (Bratislava) **169**: 1–4.

Dvořák, M. 1990: Taxonomische Bemerkungen zur Gattung Cerocoma Geoffroy (Coleoptera, Meloidae, Cerocomini). *Annotationes Zoologicae et Botanicae* (Bratislava) **199**: 1–12.

Dvořák, M. 1993a: Einige neue paläarktische Meloiden-Arten (Coleoptera). *Annotationes Zoologicae et Botanicae* (Bratislava) **212**: 1–10.

Dvořák, M. 1993b: Revision der Gattung Micromerus Mulsant et Rey, 1858 (Coleoptera, Meloidae). *Annotationes Zoologicae et Botanicae* (Bratislava) **214**: 1–9.

Dvořák, M. 1996: Neue asiatische Epicauta-Arten (Coleoptera: Meloidae). *Folia Heyrovskiana* **4**: 35–42.

Dvořák, M. 2004: Neue Oenas Latreille, 1802 Art aus Iran (Coleoptera, Meloidae). *Acta Musei Moraviae, Scientiae Biologicae* **89**: 103–104.

Dvořák, M. 2008: Epicauta suturalis sp. n., eine neue chinesische Art (Coleoptera: Meloidae). *Studies and Reports of District Museum Prague-East, Taxonomical Series* **4**: 5–6.

Egorov, L.V. 1987: Novye vidy zhukov-chernotelok roda Oodescelis Motsch. (Coleoptera, Tenebrionidae). *Entomologicheskoe Obozrenie* **66**: 800–802.

Egorov, L.V. 1989a: Novye vidy i podvid roda Oodescelis Motsch. (Coleoptera, Tenebrionidae) fauny SSSR. *Trudy Zoologicheskogo Instituta Akademii Nauk SSSR* **208**: 102–106.

Egorov, L.V. 1989b: Obzor zhukov-chernotelok roda Platyscelis Latr. (Coleoptera, Tenebrionidae) fauny SSSR. *Entomologicheskoe Obozrenie* **68**: 336–352.

Egorov, L.V. 1990a: Novye vidy chernotelok roda Bioramix Bat. (Coleoptera, Tenebrionidae). *Trudy Zoologicheskogo Instituta Akademii Nauk SSSR* **211**: 117–134.

Egorov, L.V. 1990b: K sinonimike dvukh vidov roda Platyscelis Latr. i opisanie novogo vida roda Bioramix Bat. (Coleoptera, Tenebrionidae). *Entomologicheskoe Obozrenie* **69**: 890–892.

Egorov, L.V. 1991: Die Platyscelidini des Staatlichen Museums für Tierkunde in Dresden (Insecta, Coleoptera: Tenebrionidae). *Reichenbachia* **28**: 145–148.

Egorov, L.V. 1992: Novye vidy chernotelok triby Platyscelidini (Coleoptera, Tenebrionidae). *Entomologicheskoe Obozrenie* **71**: 796–799.

Egorov, L.V. 1998: Obzor zhukov-chernotelok roda Somocoelia Kraatz, 1882 (Coleoptera, Tenebrionidae). *Entomologicheskoe Obozrenie* **77**: 209–215.

Egorov, L.V. 2004: O sisteme zhukov-chernotelok Platyscelidini (Coleoptera, Tenebrionidae) mirovoy fauny. *Entomologicheskoe Obozrenie* **83**: 581–613.

Egorov, L.V. 2006: Novye vidy zhukov-chernotelok podroda Cardiobioramix Kasz. roda Bioramix Bat. (Coleoptera, Tenebrionidae, Platyscelidini) iz provintsiy Gan'su i Sychuan' Kitaya. *Entomologicheskoe Obozrenie* **85**: 789–797.

Egorov, L.V. 2009: Novyy vid roda Oodescelis Motschulsky, 1845 (Coleoptera: Tenebrionidae: Platyscelidini) is Zapadnogo Tjan-Shanja [New species of the tenebrionid beetles genus Oodescelis Motsch]. (Coleoptera, Tenebrionidae, Platyscelidini) from the West Tien Shan. *Caucasian Entomological Bulletin* **5**: 217–220. [in Russian].

Egorov, L.V. 2018: A new species of the genus Oodescelis Motsch. (Coleoptera, Tenebrionidae) from the China. *Caucasian Entomological Bulletin* **14**: 187–190.

Ehrenberg, C.G. 1861: Über den Tiefgrund des stillen Oceans zwischen Californien und den Sandwich-Inseln aus bis 15600' Tiefe nach Lieut. Brooke. *Monatsberichte der Königlichen Preussische Akademie des Wissenschaften zu Berlin* **1860**: 819–833.

Ehrenberg, C.G. 1862: Über die Tiefgrund-Verhältnisse des Oceans am Eingange der Davisstrasse und bei Island. *Monatsberichte der Königlichen Preussische Akademie des Wissenschaften zu Berlin* **1861**: 275–315.

Eichler, W. 1924: Nouvelles espèces et variétés de coléoptères du Caucase et de l'Asie mineure. *Polskie Pismo Entomologiczne* **3**: 61–68.

Emery, M.C. 1876: Essai monographique sur les mordellides de l'Europe et des contrées limitrophes. *L'Abeille, Journal d'Entomologie* **14** [1874–1876]: 1–128.

Emery, M.C. 1891: [new taxa]. *Bulletin de la Société Entomologique de France* **1891**: xxxix-xl.

Erichson, W.F. 1841: Über die Insecten von Algier mit besonderer Rücksicht auf ihre geographische Verbreitung. Pp. 140–194. In: Wagner, M.F. (ed.): *Reisen in der Regentschaft Algier 1836, 1837 und 1838 nebst einem naturhistorischen Anhang und einem Kupferatlas. Dritter Band.* Leipzig: L. Voss, xviii + 298 pp., 17 pls.

Erichson, W.F. 1842: Beitrag zur Insecten-Fauna von Vandiemensland, mit besonderer Berücksichtigung der geographischen Verbreitung der Insecten. *Archiv für Naturgeschichte* 8 (1): 83–287, pls 4, 5.

Erichson, W.F. 1843: Beitrag zur Insecten-Fauna von Angola in besonderer Beziehung zur geographischen Verbreitung der Insecten in Afrika. *Archiv für Naturgeschichte* **9** (1): 199–267.

Erichson, W.F. 1845: Lieferung II. Pp. 161–320. In: *Naturgeschichte der Insecten Deutschlands. Erste Abtheilung. Coleoptera. Dritter Band.* Berlin: Nicolaische Buchhandlung, 320 pp. [issued in parts, pp. 1–320 (Lieferung I, II) in 1845].

Erichson, W.F. 1846: Lieferung III. Pp. 321–480. In: *Naturgeschichte der Insecten Deutschlands. Erste Abtheilung. Coleoptera. Dritter Band.* Berlin: Nicolaischen Buchhandlung, vii + [2] + 968 pp., 1 pl. [issued in parts, pp. 321–480 in 1846, pp. 801–968 in 1848].

Ermisch, K. 1940: Revision der ostasiatischen Glipa-Arten. *Entomologische Blätter* **36**: 161–173.

Ermisch, K. 1941a: Stenomordella, eine neue Mordellidengattung aus der chinensischen Provinz Fukien. *Entomologische Blätter* **37**: 115–117.

Ermisch, K. 1941b: Mordellistenoda, eine neue Mordellidengattung aus der chinesischen Provinz Fukien. *Mitteilungen Münchener Entomologische Gesellschaft* **37**: 589–593.

Ermisch, K. 1941c: Tribus Mordellistenini (Col. Mordell.). *Mitteilungen der Münchener Entomologischen Gesellschaft* **31**: 710–726.

Ermisch, K. 1942a: Eine neue paläarktische Mordellide. *Arbeiten über Morphologische und Taxonomische Entomologie aus Berlin-Dahlem* **9**: 85–88.

Ermisch, K. 1942b: Tolidostena, ein neues Mordellidengenus aus der chinesischen Provinz Fukien. *Mitteilungen der Münchener Entomologischen Gesellschaft* **32**: 674–678.

Ermisch, K. 1943: Tolida Tournieri Em. und ihre Synonyme Schusteri Schilsky und similis Stchegoleva. *Entomologische Blätter* **39**: 18–21.

Ermisch, K. 1944: Mordellidae. In: G. Frey: Tagebuch meiner Spanienreise und deren wissenschaftliche Ergebnisse. I. Reisetagebuch. *Mitteilungen der Münchener Entomologischen Gesellschaft* **34**: 210–216.

Ermisch, K. 1948: Neue Arten der Gattung Tomoxia und verwandter Gattungen aus Afrika. *Entomologische Blätter* **41–44**: 98–110.

Ermisch, K. 1950a: Mordellidae. In: *Mission G.F. de Witte, Exploration du Parc National Albert 1933–1935.* Brussels: Institut des Parcs Nationaux du Congo Belge **71**: 1–95.

Ermisch, K. 1950b: Die Gattungen der Mordelliden der Welt. *Entomologische Blätter* **45–46**: 34–92.

Ermisch, K. 1952a: Mordellidae des Belgischen Congogebietes des Musée Royal du Congo Belge in Tervuren. *Annales du Musée Royal du Congo Belge, Tervuren* **22**: 7–105.

Ermisch, K. 1952b: Neue Mordelliden aus der chinesischen Provinz Fukien. *Entomologische Blätter* **47–48**: 143–157.

Ermisch, K. 1954a: Über Typen und neue Arten der afrikanischen Mordelliden. *Memorie della Società Entomologica Italiana* **33**: 167–200.

Ermisch, K. 1954b: Mordellistena-Studien I. Mordellistena micans Germar und die micans der Autoren. *Beiträge zur Entomologie* **4**: 173–180.

Ermisch, K. 1956: Mordellidae. Pp. 269–328. In: Horion, A.: *Faunistik der mitteleuropäischen Käfer. Band V: Heteromera.* Tutzing: A. Frey, 336 pp., 6 pls.

Ermisch, K. 1957: Revision der Gattung Mordellistenula Stshegol.-Bar. *Entomologische Blätter* **53**: 45–51.

Ermisch, K. 1962: Drei neue Anaspis-Arten aus Turkestan. *Senckenbergiana Biologica* **43**: 165–170.

Ermisch, K. 1963a: Neue Mordelliden (Heteromera, Mordellidae) aus Deutschland und Nachträge zur Faunistik der mitteleuropäischen Mordelliden. *Entomologische Blätter* **59**: 1–36.

Ermisch, K. 1963b: Beitrag zur Mordelliden-Fauna Portugals. *Notulae Entomologicae* **43**: 14–21.

Ermisch, K. 1963c: Die Mordelliden der Insel Cypern. *Notulae Entomologicae* **43**: 49–67.

Ermisch, K. 1963d: XIII. Anaspis (Pinassa) proteus Woll. und imitator n. sp. In: Lindberg Har.: A contribution to the study of beetles in the Madeira Islands. *Commentationes Biologicae* **25**: 108–111.

Ermisch, K. 1964a: Anaspis (Pinassa) eversi n. sp. von der Kanareninsel Fuerteventura. *Entomologische Blätter* **60**: 18–19.

Ermisch, K. 1964b: Ergebnisse der zoologischen Forschungen von Dr. Z. Kaszab in der Mongolei. *Annales Historico-Naturales Musei Nationalis Hungarici* **56**: 341–351.

Ermisch, K. 1965a: 49. Mordellidae II. Ergebnisse der zoologischen Forschungen von Dr. Z. Kaszab in der Mongolei. *Reichenbachia* **7**: 63–80.

Ermisch, K. 1965b: Mordellistena-Arten von den Kanarischen Inseln. *Entomologische Blätter* **61**: 67–73.

Ermisch, K. 1965c: Synonymische und nomenklatorische Feststellungen in der Familie Mordellidae. *Reichenbachia* **5**: 197–201.

Ermisch, K. 1965d: Neue Mordelliden von der Balkanhalbinsel. *Reichenbachia* **5**: 251–272.

Ermisch, K. 1966a: Neue westpaläarktische Mordellistena-Arten. *Entomologische Blätter* **62**: 30–39.

Ermisch, K. 1966b: Neue Mordelliden aus Nordafrika. *Entomologische Blätter* **62**: 40–43.

Ermisch, K. 1967a: 90. Mordellidae III. Ergebnisse der zoologischen Forschungen von Dr. Z. Kaszab in der Mongolei. *Reichenbachia* **9**: 45–52.

Ermisch, K. 1967b: Neue Mordellistena-Arten aus Mitteleuropa und der Balkanhalbinsel. *Entomologische Blätter* **63**: 110–119.

Ermisch, K. 1967c: Anaspis (Silaria) canariensis n. sp. von den Kanaren-Insel Gomera und Hierro. *Entomologische Blätter* **63**: 120.

Ermisch, K. 1967d: Ergebnisse der 1. Mongolisch-tschechoslowakischen entomologisch-botanischen Expedition in der Mongolei. *Acta Faunistica Entomologica Musei Nationalis Pragae* **12**: 125–129.

Ermisch, K. 1968a: Neue Mordellini aus der chinensischen Provinz Fukien. *Reichenbachia* **10**: 279–292.

Ermisch, K. 1968b: 132. Mordellidae IV. Ergebnisse der zoologischen Forschungen von Dr. Z. Kaszab in der Mongolei. *Reichenbachia* **11**: 17–24.

Ermisch, K. 1969a: Eine neue Hoshihananomia-Art aus Ussuri. *Entomologische Blätter* **65**: 102–103.

Ermisch, K. 1969b: Neue Mordelliden aus Europa, Nordafrika und Nahen Osten. *Entomologische Blätter* **65**: 104–115.

Ermisch, K. 1969c: 183. Mordellidae V. Ergebnisse der zoologischen Forschungen vin Dr. Z. Kaszab in der Mongolei. *Reichenbachia* **12**: 171–177.

Ermisch, K. 1970: 204. Mordellidae VI. Ergebnisse der zoologischen Forschungen von Dr. Z. Kaszab in der Mongolei. *Reichenbachia* **13**: 169–187.

Ermisch, K. 1977: Die Mordellistena-Arten Ungarns und benachbarter Gebiete sowie Beschreibung einer neuen Hoshihananomia-Art aus Siebenbürgen. *Folia Entomologica Hungarica* (N.S.) **30**: 151–177.

Escalera, M.M. de la. 1901: Materiales para una revisión del género Asida. *Boletín de la Real Sociedad Española de Historia Natural* **1**: 172–175.

Escalera, M.M. de la. 1903: Otra Asida nueva de España del grupo de las aterciopeladas. *Boletín de la Real Sociedad Española de Historia Natural* 3: 75–76.

Escalera, M.M. de la. 1905a: Sistema de las especies ibéricos del gen. Asida Latr. *Boletín de la Real Sociedad Española de Historia Natural* 5: 377–402, 430–451.

Escalera, M.M. de la. 1905b: Una nueva especie de «Eulipus» Woll. (Tentyriini) de Río de Oro. *Boletín de la Real Sociedad Española de Historia Natural* 5: 467–468.

Escalera, M.M. de la. 1906a: Especies nuevas de coleópteros de Marruecos. *Boletín de la Real Sociedad Española de Historia Natural* 6: 161–162.

Escalera, M.M. de la. 1906b: Adiciones al sistema de las especies ibéricas del género Asida. Una Globasida nueva de la provincia de Murcía. *Boletín de la Real Sociedad Española de Historia Natural* 6: 193.

Escalera, M.M. de la. 1906c: Una nuova Alphasida de Motril. *Boletín de la Real Sociedad Española de Historia Natural* **1906**: 235.

Escalera, M.M. de la. 1906d: Sistema de las especies ibéricas del gén. Asida Latr. *Boletín de la Real Sociedad Española de Historia Natural* 6: 306–317.

Escalera, M.M. de la. 1906e: Sobre la variabilidad de las especies de Alphasida de las ramas orientales y descrición de especies nuevas. *Boletín de la Real Sociedad Española de Historia Natural* **1906**: 381–385.

Escalera, M.M. de la. 1907: Especies nuevas de Marruecos. *Boletín de la Real Sociedad Española de Historia Natural* 6: 337–340.

Escalera, M.M. de la. 1909a: Especies nuevas de Marruecos. *Boletín de la Real Sociedad Española de Historia Natural* 9: 135–136.

Escalera, M.M. de la. 1909b: Especies nuevas de meloidos del SW. de Marruecos. *Boletín de la Real Sociedad Española de Historia Natural* 9: 240–249.

Escalera, M.M. de la. 1910a: Nuevos coleópteros de Marruecos. *Boletín de la Real Sociedad Española de Historia Natural* **1910**: 283–285.

Escalera, M.M. de la. 1910b: Coleópteros nuevos de Marruecos. *Boletín de la Real Sociedad Española de Historia Natural* 10: 379–382.

Escalera, M.M. de la. 1910c: Especies nuevas de Marruecos. *Boletín de la Real Sociedad Española de Historia Natural* 10: 409–416.

Escalera, M.M. de la. 1911: Coleópteros nuevos del S.W. de Marruecos. *Boletín de la Real Sociedad Española de Historia Natural* 11: 299–304.

Escalera, M.M. de la. 1912: Dos nuevas especies españolas del género Asida. *Boletín de la Real Sociedad Española de Historia Natural* 12: 166–168.

Escalera, M.M. de la. 1913: Coleópteros nuevos del Sus (Marruecos). *Trabajos del Museo de Ciencias Naturales, Serie Zoológica* (Madrid) 8: 3–56.

Escalera, M.M. de la. 1914: Los coleópteros de Marruecos. *Trabajos del Museo Nacional de Ciencias Naturales Serie Zoológica* (Madrid) 11: 1–553.

Escalera, M.M. de la. 1915: Una specie nueva de Zonabris de España. *Boletín de la Real Sociedad Española de Historia Natural* 15: 493–494.

Escalera, M.M. de la. 1921a: Descripciones de especies nuevas de Asida de la Península Ibérica. *Boletín de la Real Sociedad Española de Historia Natural* 21: 117–120.

Escalera, M.M. de la. 1921b: Especies nuevas de Asida de la Península Ibérica. *Boletín de la Real Sociedad Española de Historia Natural* 21: 350–363.

Escalera, M.M. de la. 1921c: Especies nuevas de Asida de la Península Ibérica. *Boletín de la Real Sociedad Española de Historia Natural* 21: 427–437.

Escalera, M.M. de la. 1922a: Especies ibéricos del género Asida (Col. Tenebriónidos). *Boletín de la Real Sociedad Española de Historia Natural* 22: 64–71.

Escalera, M.M. de la. 1922b: Especies nuevas de Asida de Marruecos. *Boletín de la Real Sociedad Española de Historia Natural* 22: 170–175.

Escalera, M.M. de la 1922c: Los Heliotaurus (Col. Cistelidae) de Marruecos de protorax rojo. *Boletín de la Real Sociedad Española de Historia Natural* 22: 358–362.

Escalera, M.M. de la. 1922d: Especies nuevas de Arthrodeis de Marruecos (Col. Tenebriónidos). *Boletín de la Real Sociedad Española de Historia Natural* 22: 373–377.

Escalera, M.M. de la. 1922e: Sistema de las especies del género Asida de la Península Ibérica. Subgén. Glabrasita Esc. 1. nota. *Boletín de la Real Sociedad Española de Historia Natural* 22: 388–391.

Escalera, M.M. de la. 1922f: Sistema de las especies españolas del género Asida Latr. *Boletín de la Real Sociedad Española de Historia Natural* 22: 431–451.

Escalera, M.M. de la. 1923a: Sistema de las especies del género Asida de la Península Ibérica. Subgén. Glabrasida Esc. 2.a nota (1). *Boletín de la Real Sociedad Española de Historia Natural* 22 [1922]: 463–473.

Escalera, M.M. de la. 1923b: Sistema de las especies del género Asida de la Península Ibérica. Subgén. Glabrasida Esc. 3.a nota. *Boletín de la Real Sociedad Española de Historia Natural* 23: 37–42.

Escalera, M.M. de la. 1923c: Sistema de las especies del género Asida de la Península Ibérica. Subgén. Glabrasida Esc. 4.a nota. *Boletín de la Real Sociedad Española de Historia Natural* 23: 79–87.

Escalera, M.M. de la. 1923d: Tenebriónidos nuevos de Marruecos y Rio de Oro. *Boletín de la Real Sociedad Española de Historia Natural* 23: 128–132.

Escalera, M.M. de la. 1923e: Una specie nueva de meloido de Marueccos. *Boletín de la Real Sociedad Española de Historia Natural* 23: 299–300.

Escalera, M.M. de la. 1925a: Varios tenebriónidos y un curculiónido nuevos de Marruecos. *Boletín de la Real Sociedad Española de Historia Natural* 24 [1924]: 324–337.

Escalera, M.M. de la 1925b: Una especie nueva de Glabrasida Esc. (Col. Tenebrionidae) de España. *Boletín de la Real Sociedad Española de Historia Natural* 25: 153–154.

Escalera, M.M. de la. 1925c: Especies nuevas de coleópteros del Mediano Atlas marroquí. *Boletín de la Real Sociedad Española de Historia Natural* 25: 311–316.

Escalera, M.M. de la. 1925d: Especies de Pachychila y otros tenebriónidos nuevos de Marruecos. *Boletín de la Real Sociedad Española de Historia Natural* 25: 372–385.

Escalera, M.M. de la. 1925e: Especies nuevas y variedades de Asida de Baleares (Col. Tenebr.). *Boletín de la Real Sociedad Española de Historia Natural* 25: 404–410.

Escalera, M.M. de la. 1925f: Tenebriónidos y meloidos del Mediano Atlas. *Boletín de la Real Sociedad Española de Historia Natural* 25: 498–505.

Escalera, M.M. de la 1926: Una especie más de Heliotaurus de Marruecos (Col. Cistelidae). *Boletín de la Real Sociedad Española de Historia Natural* **26**: 184–185.

Escalera, M.M. de la. 1927a: Las Machasida Esc. (Col. Tenebrionidae) de Marruecos. *Bulletin de la Société des Sciences Naturelles du Maroc* **7**: 135–149.

Escalera, M.M. de la 1927b: Una nueva especie española de Crypticus Latr. *Eos, Revista Española de Entomología* **3**: 501–504.

Escalera, M.M. de la. 1929: Las Glabrasida fortitertricosttae de Marruecos (Col. Tenebr.). *Memorias de la Real Sociedad Española de Historia Natural* **15**: 105–108.

Escalera, M.M. de la 1933: Una especie y tres variedades nuevas de Heliotaurus de protórax rojo de Marruecos (Col. Allecul.). *Boletín de la Real Sociedad Española de Historia Natural* **33**: 129–131.

Escalera, M.M. de la. 1934a: Especies nuevas de Crypticus Latr. de Sierras de Segura (Col. Tenebr.). *Boletín de la Real Sociedad Española de Historia Natural* **34**: 211–212.

Escalera, M.M. de la. 1934b: Especies nuevas de Akis de Marruecos y del Sahara occidental (Col. Ten.). *Eos, Revista Española de Entomología* **9**: 301–309, pl. XVII.

Escalera, M.M. de la. 1935a: Especies de Pachychila Esch. y otros tenebriónidos de Ifni y del Sáhara español. *Boletín de la Sociedad Española de Historia Natural* **35**: 295–309.

Escalera, M.M. de la. 1935b: Especies de Pachychila Esch. y otros tenebriónidos nuevos de Marruecos. *Boletín de la Sociedad Española de Historia Natural* **35**: 372–380.

Escalera, M.M. de la. 1940: Especies de Sepidium F. de Ifni (Col. Tenebrionidae). *Eos, Revista Española de Entomología* **13** [1937]: 5–11, 1 pl.

Escalera, M.M. de la. 1943a: Especies nuevas de Asida (Col. Tenebrionidae) del Reino de Valencia y zonas limitrofes existentes en las colecciones de D. Emilio Moroder y Dr. Baguena en 1936. *Boletín de la Real Sociedad Española de Historia Natural* **43**: 199–207.

Escalera, M.M. de la. 1943b: Datos sobre coléopteros de Ifni del Sahara español. 1. Una nueva especie de Eurycaulus Fairm. del litoral entre C. Juby y C. Bojador. *Eos, Revista Española de Entomología* **19**: 215–221.

Escalera, M.M. de la. 1944: De la disparidad especifica de Dendarus zariquieyi Esp. (Col. Tenebrionidae) de sus similares ibéricios y africanos. *Eos, Revista Española de Entomología* **20**: 83–92.

Escherich, K. 1889a: Beschreibung einer neuen deutschen Meloë-Art und mehrerer Varietäten. *Wiener Entomologische Zeitung* **8**: 105–106.

Escherich, K.L. 1889b: Meloë Reitteri, eine neue russische Meloë-Art. *Wiener Entomologische Zeitung* **8**: 112.

Escherich, K.L. 1889c: Nachträge und Berichtungen zum Catalogus Coleopterorum von Gemminger und Harold betreffend die Gattung Meloë. *Deutsche Entomologische Zeitschrift* **2**: 333–335.

Escherich, K.L. 1890a: Meloiden-Studien. *Wiener Entomologische Zeitung* **9**: 23–24.

Escherich, K.L. 1890b: Revision der behaarten Meloë-Arten der Alten Welt. *Wiener Entomologische Zeitung* **9**: 87–96.

Escherich, K.L. 1890c: Bemerkungen über Meloë algirus nov. spec. aus Algier. *Deutsche Entomologische Zeitschrift* **1890**: 283.

Escherich, K.L. 1891a: Cochliophorus nov. gen. Meloidarum. *Deutsche Entomologische Zeitschrift* **1891**: 16.

Escherich, K.L. 1891b: Meloiden-Studien. II. Theil. *Wiener Entomologische Zeitung* **10**: 53–55.

Escherich, K.L. 1892: Die palaärktischen Vertreter der Coleopteren-Gattung Zonitis Fab. Versuch einer Monographie. I. *Deutsche Entomologische Zeitschrift* 1891: 225–250, 1 pl.

Escherich, K.L. 1895: Beiträge zur Naturgeschichte der Meloidengattung Lytta Fab. *Verhandlungen der Kaiserlich-Königlichen Zoologisch-Botanischen Gesellschaft in Wien* 44 [1894]: 251–298, 4 pls.

Escherich, K.L. 1896a: Meloiden-Studien. IV. Theil. *Wiener Entomologische Zeitung* 15: 27–30.

Escherich, K.L. 1896b: Zur Kenntnis der Gattung Caloenas Reitt. *Wiener Entomologische Zeitung* 15: 121–123.

Escherich, K.L. 1896c: Revision der Meloiden-Gattung Lydus Latr. *Deutsche Entomologische Zeitschrift* 1896: 193–235.

Escherich, K.L. 1896d: Beitrag zur Fauna der tunisischen Insel Djerba. *Verhandlungen der Kaiserlich-Königlichen Zoologisch-Botanischen Gesellschaft in Wien* 46: 268–279.

Escherich, K.L. 1897a: Zoologische Ergebnisse einer von Dr. K. Escherich und L. Kathariner nach Central-Kleinasien unternommen Reise. IV. Theil. Coleopteren. Under the title of the paper the following text is read: 'Unter Mitwirkung von E. Brenske, L. Ganglbauer, Dr. L. von Heyden, E. Wasmann und J. Weise'. *Entomologische Zeitung* (Stettin) 58: 3–69.

Escherich, K.L. 1897b: Revision der palaearktischen Zonitiden, einer Unterfamilie der Meloiden. *Verhandlungen des Naturforschenden Vereines in Brünn* 35 [1896]: 96–133.

Escherich, K.L. 1899: Zur Kenntniss der Coleopteren-Gattung Zonabris Harold. *Wiener Entomologische Zeitung* 18: 84–92, 97–109.

Escherich, K.L. 1904: Neue paläarktische Meloiden aus der F. Hauser'schen Sammlung. *Münchener Entomologische Zeitschrift* 2: 30–36.

Eschscholtz, J.F. 1818: Decades tres Eleutheratorum novorum. *Mémoires de l'Académie Impériale des Sciences de St. Peterburg* 6: 451–484.

Eschscholtz, J.F. 1829: *Zoologischer Atlas, enthaltend Abbildungen und Beschreibungen neuer Tierarten, während des Flottscapitains v. Kotzebue zweiter Reise um die Welt, auf der Russisch-Kaiserlichen Kriegs-schlupp Predpriaetië in den Jahren 1823–1826 beobachtet. Drittes Heft*. Berlin: G. Reimer, 18 pp., pls xi–xv.

Eschscholtz, J.F. 1831: *Zoologischer Atlas, enthaltend Abbildungen und Beschreibungen neuer Tierarten, während des Flottscapitains V. Kotzebue zweiter Reise um die Welt, auf der Russisch-Kaiserlichen Kriegsschlupp Predpriaetië in den Jahren 1823–1826 beobachtet. Viertes Heft*. Berlin: G. Reimer, 19 pp., pls xvi-xx.

Español, F. 1937: Assaig monogràfic del subgènere Dendarus s.str. (Col. Tenebrionidae). *Treballs del Museu de Ciències Naturals de Barcelona* (*Sèrie Entomològica* 12) 11: 1–89, 13 pls.

Español, F. 1940: Contribución al conocimiento de los coleópteros de Ibiza y Formentera. *Las Ciencias* 5: 9.

Español, F. 1942: Los representantes catalanes de la familia 'Rhipiphoridae' (Col.). *Anales de la Escuela de Peritos Agrícolas y Superior de Agricultura y de los Servicios Técnicos de Agricultura* 2: 335–346.

Español, F. 1943a: Datos para el conocimiento de los Tenebrionidos del Mediterráneo occidental. *Graellsia* 1: 23–28.

Español, F. 1943b: Contribución al conocimiento de los coleópteros de Ibiza y Formentera y un nuevo Heliophilus ibérico (Col. Tenebrionidae). *Anales de la Asociación Española para el Progreso de las Ciencias* 8: 93–108.

Español, F. 1943c: Misión cientifica E. Morales Agacino, Ch. Rungs y B. Zolotarevsky a Ifni y Sáhara Español. Tenebrionidae (Col.). 1.a parte. *Eos, Revista Española de Entomología* 19: 119–148.

Español, F. 1944: Nuevos datos para el conocimiento de los tenebriónidos (Col.) del Sáhara Español. *Eos, Revista Española de Entomología* **20**: 7–30.

Español, F. 1945a: Nuevos comentarios sistematicos sobre la subfamilia Opatrinae Reitt. con la descripción de un nuevo representante del Sáhara Español (Col. Tenebrionidae). *Eos, Revista Española de Entomología* **20** [1944]: 213–232, pls XII–XVII.

Español, F. 1945b: Revisión de los Phylan ibéricos (Col. Tenebrionidae). *Eos, Revista Española de Entomología* **21**: 297–357, pl. xiii.

Español, F. 1946: Tenebriónidos (Col.) nuevos o interesantes recogidos por D. Eugenio Mordes y D. Joaquín Mateu en el Sáhara Español. *Eos, Revista Española de Entomología* **22**: 107–122, pls. VIII–X.

Español, F. 1947a: Revisión del género Micrositus (Col. Tenebrionidae). *Trabajos del Museo de Ciencias Naturales de Barcelona* (*Nueva Seria Zoológica*) **1**: 5–60.

Español, F. 1947b: Algunos Stenosinae y Opatrinae recogidos durante una breve campaña entomologica por la provincia de Castellon (Col. Tenebrionidae). *Eos, Revista Española de Entomología* **23**: 185–200.

Español, F. 1948a: Notas sombre coleópteros africanos. *Eos, Revista Española de Entomología* **24**: 311–318.

Español, F. 1948b: Contribución al estudio de los Crypticus paléarticos: El subgénero Lamprocrypticus. *Eos, Revista Española de Entomología* **24**: 475–491.

Español, F. 1949: Contribución al estudio de los Crypticus del grupo del pruinosus: el subgénero Pseudoseriscus Españ. (Col. Tenebrionidae). *Eos, Revista Española de Entomología* **25**: 199–239.

Español, F. 1950a: Los Crypticus s.str. de la Península Ibérica y del norte de Africa (Col. Tenebrionidae). *Eos, Revista Española de Entomología* **26**: 115–156.

Español, F. 1950b: Contribución al estudio de los Crypticus paléarticos (Col. Tenebrionidae). Pp. 124–129. In: *Transactions of the Eighth International Congress of Entomology, Stockholm 1948*. Stockholm: Ales R. Elfström, 1030 pp.

Español, F. 1951a: Tenebriónidos de Las Pitiusas (Baleares occidentales). *Eos, Revista Española de Entomología* **27**: 4–41.

Español, F. 1951b: Un nuevo Arthrodeis de la region del Draa: A. (Apentanes) bolivari n.sp. *Eos, Revista Española de Entomología* (Tomo extraordinario) [1950]: 363–366.

Español, F. 1952a: Mas Tenebriónidos del Sahara Español (Col.). *Eos, Revista Española de Entomología* **28**: 71–76.

Español, F. 1952b: Note sur les Crypticus appartenant au groupe de C. viaticus Fairm. (Col. Tenebrionidae). *Transactions of the IXth International Congress of Entomology, Amsterdam* **1**: 117–120.

Español, F. 1953a: Helopinae de la zona mediterránea de Marruecos (Col. Tenebrionidae). *Eos, Revista Española de Entomología* **29**: 71–83.

Español, F. 1953b: Los Colpotus del Mediterráneo Occidental (Col. Tenebrionidae). *Eos, Revista Española de Entomología* **29**: 123–142.

Español, F. 1953c: Misiones saharianas de l'Institut scientifique Chérifien (1950–1951) Col. Tenebrionidae. *Bulletin de la Société des Sciences Naturelles du Maroc* **31** [1951]: 287–312.

Español, F. 1954a: Los tenebriónidos (Col.) de Baleares. *Trabajos del Museo de Ciencias Naturales de Barcelona* (*Nueva Seria Zoológia*) **1**: 3–96.

Español, F. 1954b: Los Crypticus de Canarias (Col. Tenebrionidae). *Eos, Revista Española de Entomología* **30**: 99–117.

Español, F. 1954c: Coleópteros de Sierra Nevada 'Tenebrionidae'. *Archivos des Instituto de Aclimatación* **2**: 111–122, 1 pl.

Español, F. 1954d: Datos para el conocimiento de las tenebrionidos del mediterranea occidental. *Eos, Revista Española de Entomología* **30**: 161–162.

Español, F. 1954e: Los Opatrum andaluces del subgénero Colpophorus (Col. Tenebrionidae). *Eos, Revista Española de Entomología* **30**: 315–330.

Español, F. 1955a: Los Crypticini paléarticos (Col. Tenebrionidae). *Eos, Revista Española de Entomología* **31**: 7–38.

Español, F. 1955b: Nuevos datos sobre los Tenebrionidos de las Pitiusas (Baleares occidentales). *Boletin de la Sociedad de Historia Natural de Baleares* **1**: 25–29.

Español, F. 1955c: Soble algunos tenebriónidos gallegos recognidos por el profesor H. Franz. *Publicationes del Instituto de Biología Aplicada* **19**: 95–103.

Español, F. 1955d: Colydiidos endogeos del Levante español. *Publicaciones del Instituto de Biología Aplicada* **21**: 101–106.

Español, F. 1956a: Los Probaticus de España (Col. Tenebrionidae). *Eos, Revista Española de Entomología* **32**: 83–123.

Español, F. 1956b: Nuevos Opatrini de Marruecos y del Sahara atlántico (Col. Tenebrionidae). *Publicationes de la Facultad de Ciencias de Barcelona, Universidad de Barcelona* **1956**: 277–286.

Español, F. 1957a: Los Hegeter de las Canarias Orientales (Col. Tenebrionidae). *Trabajos del Museo Ciencas Naturales de Barcelona (Nueva Seria Zoológica)* **2**: 3–16.

Español, F. 1957b: Contribución al estudio de los Stenohelops Ibericos (Col. Tenebrionidae). *Eos, Revista Española de Entomología* **33**: 19–38.

Español, F. 1957c: Un nuevo Helioctamenus de Marruecos. *Publicaciones del Instituto de Biología Aplicada* **25**: 147–150.

Español, F. 1957d: Contributión al conocimiento de los Tentyriini de las Canarias orientales: sobre el pretendido gen. Pseudotalpophila Reitt. (Col. Tenebrionidae). *Eos, Revista Española de Entomología* **33**: 157–176.

Español, F. 1958a: Contribución al conocimiento de los artrópodos y moluscos terrestres de las Islas Columbretes. *Miscelánea Zoológica* **1**: 3–37.

Español, F. 1958b: Avance al estudio de los Stenosis ibéricos (Col. Tenebrionidae). *Eos, Revista Española de Entomología* **34**: 33–54.

Español, F. 1958c: Datos para el conocimiento de los tenebriónidos de Mediterráneo occidental (Coleoptera). *Graellsia* **96**: 107–116.

Español, F. 1959: Opatrinae del Rif, Marruecos (Col. Tenebrionidae). *Eos, Revista Española de Entomología* **35**: 243–255.

Español, F. 1960a: Datos para el conocimiento de los tenebriónidos de Mediterráneo occidental. *Miscellánea Zoológica* **1** (3): 109–118, 1 pl.

Español, F. 1960b: Un nuevo tipo de tenebriónido sabulicola de las costas del Perú. *Publicationes del Instituto de Biología Aplicada* **31**: 113–117.

Español, F. 1961a: Revisión de los Dendarus s.str. (Col. Tenebrionidae). *Eos, Revista Española de Entomología* **37**: 41–70.

Español, F. 1961b: El género Tarphius, nuevo par la fauna española (Col. Colydiidae). *Graellsia* **19**: 81–85.

Español, F. 1961c: Sobre alcunas Solskyia del Paquistán (Co. Tenebrionidae). *Miscellánea Zoológica* **1** (4): 123–131.

Español, F. 1961d: Los Cylindronotini de la Península Ibérica (Col. Tenebrionidae). *Eos, Revista Española de Entomología* **37**: 135–160.

Español, F. 1961e: Los Nesotes de España (Col. Tenebrionidae). *Eos, Revista Española de Entomología* **37**: 289–308.

Español, F. 1961f: Los Blaps de la Península Ibérica (Col. Tenebrionidae). *Eos, Revista Española de Entomología* **37**: 399–414.

Español, F. 1962a: Los Opatrinae de las Islas Canarias (Col. Tenebrionidae). *Eos, Revista Española de Entomología* **38**: 203–221.

Español, F. 1962b: Los Nesotes de Gran Canaria (Col. Tenebrionidae). *Eos, Revista Española de Entomología* **38**: 339–346.

Español, F. 1963a: El género 'Coelometopus' Sol. (Col. 'Tenebrionidae'). *Archivos del Instituto de Aclimatación* **12**: 49–53.

Español, F. 1963b: Nuevos tenebriónidos africanos. *Miscellánea Zoológica* **1** (5): 73–79.

Español, F. 1963c: Los Opatrum ibéricos (Col. Tenebrionidae). *Eos, Revista Española de Entomología* **38**: 471–483.

Español, F. 1963d: Datos para el conocimiento de los tenebriónidos del Mediterráneo occidental (Coleoptera). *Eos, Revista Española de Entomología* **39**: 185–209.

Español, F. 1963e: Datos para el conocimiento de los tenebriónidos del Mediterráneo occidental (Coleoptera). *Eos, Revista Española de Entomología* **39**: 379–388.

Español, F. 1963f: Más datos sobre los Helopinae de Marruecos (Col. Tenebrionidae). *Eos, Revista Española de Entomología* **39**: 389–396.

Español, F. 1966: Datos para el conocimiento de los tenebriónidos del Mediterráneo occidental (Coleoptera). *Graellsia* **22**: 167–175.

Español, F. 1967a: Misión entomológica Hakan Lindberg y M. Meinander a Marruecos. Col. Tenebrionidos. *Publicaciones del Instituto de Biología Aplicata* **42**: 17–47.

Español, F. 1967b: Blaps propheta fiorii subspecie nueva de la Tripolitana. *Studi Sassar* (*Sez. 3*) **14** [1966]: 197–198.

Español, F. 1969: Tenebrionidos de la región de Sax-Salinas, provincia de Alicante (Coleoptera). *Graellsia* **24**: [1968]: 79–86.

Español, F. 1974: Nuevas localizaciones de Nalassus ibéricos (Coleoptera, Tenebrionidae). *Boletín de la Real Sociedad Española de de Historia Natural* (*Biología*) **72**: 213–217.

Español, F. 1975a: Sobre un nuevo representante europeo del género Osphyoplesius Winkler (Col. Boridae). *Eos, Revista Española de Entomología* **49** [1973]: 105–111.

Español, F. 1975b: Un nuevo Litoborini de las Islas Canarias (Col. Tenebrionidae, Opatrinae). *Vieraea* **4** [1974]: 237–244.

Español, F. 1980: Sobre la presencia del género Probaticus Seidl., 1896, en las Islas Baleares (Col. Tenebrionidae, Helopinae). *Eos, Revista Española de Entomología* **56**: 107–110.

Español, F. 1981: Sobre un nuevo Stenosis ibérico (Col. Tenebrionidae). *Publicaciones del Departamento de Zoología, Barcelona* 7: 57–60.

Español, F. & Viñolas, A. 1981a: Eine neue Eurycaulus-Art aus der algerischen Sahara (Coleoptera, Tenebrionidae, Opatrinae). *Stuttgarter Beiträge zur Naturkunde Serie A (Biologie)* 352: 1–7.

Español, F. & Viñolas, A. 1981b: Una puesta al dia de los Phylan Steph. ibéricos (Col. Tenebrionidae). *Eos, Revista Española de Entomología* 57: 97–112

Español, F. & Viñolas, A. 1983a: Revisión de los Gonocephalum del grupo prolixum (Col., Opatrinae) *Eos, Revista Española de Entomología* 59: 31–39.

Español, F. & Viñolas, A. 1983b: Revisión del género Heliopathes Muls. 1854 (Col. Tenebrionidae). 1: Grupo del strigicollis. *Nouvelle Revue d'Entomologie* 13: 213–218.

Español, F. & Viñolas, A. 1987: Los Erodius ibéricos (Col. Tenebrionidae). *Eos, Revista Española de Entomología* 63: 21–29.

Esser, J. 2016: Revision der Gattung Typhaecla Ganglbauer, 1899 (Coleoptera: Mycetophagidae) – 1 Teil: Paläarktische Arten. *Entomologische Zeitschrift* 126: 169–175.

Esser, J., Varandi, H.B. & Farashian, M.E 2018: A new species of Eulagius Motschulsky, 1845 (Coleoptera: Mycetophagidae) from Iran. *Linzer biologische Beiträge* 50 : 239–243.

Evenhuis, N.L. 2012: Publication and Dating of the Exploration Scientifique de l'Algérie: Histoire Naturelle des Animaux Articulés (1846–1849) by Pierre Hippolyte Lucas. *Zootaxa* 3448: 1–61.

Eversmann, E. F. von. 1833: [new taxon]. In: Faldermann, F.: Species novae Coleopterorum Mongoliae et Sibiriae. *Bulletin de la Société des Naturalistes de Moscou* 6: 46–72.

Fabre, J.H. 1886: *Souvenirs entomologiques (troisième série). Études sur l'instinct et les moeurs des insectes. (troisième édition)*. Paris. Librairie Ch. Delagrave, 433 pp.

Fabricius, J.C. 1775: *Systema entomologicae, systens insectorum classes, ordines, genera, species, adiectis synonymis, locis, descriptionibus, observationibus*. Flensburgi et Lipsiae: Libraria Kortii, [32] + 832 pp.

Fabricius, J.C. 1777: *Genera insectorum eorumque characteres naturales secundum numerum, figuram, situm et proportionem omnium partium oris adiecta mantissa specierum nuper detectorum*. Chilonii: Mich. Friedr. Bartschii: 14 + 310 pp.

Fabricius, J.C. 1781: *Species insectorum, exhibens eorum differentias specificas, synonyma auctorum, loca natalia, metamorphosis, adiecitis observationibus, descriptionibus. Tom I*. Hamburgi et Kilonii: Carol Ernest Bohnii, viii +552 pp.

Fabricius, J.C. 1787a: *Mantissa insectorum sistens eorum species nuper detectas adiectis characteribus generis, differentiis specificis, emendationibus observationibus. Tom I*. Hafniae: Christ. Gottl. Proft, xx + 348 pp.

Fabricius, J.C. 1787b: *Mantissa insectorum sistens eorum species nuper detectas adiectis characteribus generis, differentiis specificis, emendationibus observationibus. Tom II*. Hafniae: Christ Gottl. Proft, 382 pp.

Fabricius, J.C. 1790: Nova insectorum genera. *Skrivter af Naturhistorie Selskabet* 1 213–228.

Fabricius, J.C. 1792a: *Entomologica systematica emendata et aucta. Secundum classes, ordines, genera, species adjectis synonimis, locis, observationibus, descriptionibus. Tom I. Pars I*. Hafniae: Christ. Gottl. Proft , xx + 330 pp.

Fabricius, J.C. 1792b: *Entomologica systematica emendata et aucta. Secundum classes, ordines, genera, species adjectis synonimis, locis, observationibus, descriptionibus. Tom I. Pars II*. Hafniae: Christ. Gottl. Proft., 538 pp.

Fabricius, J.C. 1794: *Entomologia systematica emendata et aucta. Sécundum classes, ordines, genera, species adjectis synonimis, locis, observationibus, descriptionibus. Tom. IV.* [Appendix specierum nuper detectarum: pp. 435–462]. Hafniae: C.G. Proft, Fil. et Soc., [6] + 472 + [5] pp.

Fabricius, J.C. 1798: *Supplementum entomologiae systematicae.* Hafniae: Proft et Storch, [2] + 572 pp.

Fabricius, J.C. 1801a: *Systema Eleutheratorum secundum ordines, genera, species adiectis synonymis, locis, observationibus, descriptionibus. Tomus I.* Kiliae: Bibliopolii Academici Novi, xxiv + 506 pp.

Fabricius, J.C. 1801b: *Systema Eleutheratorum secundum ordines, genera, species adiectis synonymis, locis, observationibus, descriptionibus. Tomus II.* Kiliae: Bibliopolii Academici Novi, 687 pp.

Fåhraeus, O.I. von. 1870: Coleoptera Caffrariae, annis 1838–1854 a J.A. Wahlberg collecta. Heteromera. *Öfversigt af Kongliga Svenska Vetenskaps-Akademiens Förhandlingar* 27: 234–358.

Failla-Tebaldi, L. 1887: Excursione entomologica all'Isola di Lampedusa. *Il Naturalista Siciliana* 6 [1886–1887]: 102–104.

Fairmaire, L. 1848: Description de quelques coléoptères nouveaux. *Annales de la Société Entomologique de France* (2) **6**: 167–176.

Fairmaire, L. 1849: Description de quelques coléoptères nouveaux d'Europe et de France. *Annales de la Société Entomologigue de France* (2) **7**: 419–427.

Fairmaire, L. 1849: Essai sur les coléoptères de la Polynésie (suite). *Revue et Magazin de Zoologie Pure et Appliquée* (2) **1**: 445–460.

Fairmaire, L. 1850a: Essai sur les coléoptères de la Polynésie. *Revue et Magasin de Zoologie* (2) **2**: 50–64, 115–122, 181–185.

Fairmaire, L. 1850b: Description de quelque coléoptères nouveaux d'Europe et de France. *Annales de la Société Entomologique de France* (2) 7 (1849): 419–427.

Fairmaire, L. 1851: Description de sept coléoptères nouveaux provenant du Midi de France et du Maroc. *Revue et Magazin de Zoologie Pure et Appliquée* (2) **3**: 527–528.

Fairmaire, L. 1852: Quelques coléoptères nouveaux du midi de l'Europe et du nord de l'Afrique. *Annales de la Société Entomologique de France* (2) **10**: 69–93.

Fairmaire, L. 1855: Rectifications et description de espèces nouvelles de coléoptères de la faune méditerranéenne. *Annales de la Société Entomologigue de France* (3) **3**: 306–322.

Fairmaire, L. 1856: Miscellanea entomologica. *Annales de la Société Entomologique de France* (3) **4**: 517–548, pl. 16.

Fairmaire, L. 1860: Diagnoses de nouvelles espèces de coléoptères. *Annales de la Société Entomologique de France* (3) **8**: 338, 629–632.

Fairmaire, L. 1861: [new taxon]. In: Jacquelin du Val, C.: *Manuel Entomologique. Genera des coléoptères d'Europe comprenant leur classification en familles naturelles, la description de tous les genres, des tableaux synoptiques destinés à faciliter l'étude, le Catalogue de toutes les espèces de nombreux dessins au trait de charactères et plus de treize cents types représentant un ou plusieurs insectes de chaque genre dessinés et peints d'apres nature avec le plus grand soin par M. Jules Migneaux. Tome troisième.* Paris: A. Deyrolle [1859–1863, 464 + 200 pp., 100 pls].

Fairmaire, L. 1862: Miscellanea entomologica. Quatrième partie. *Annales de la Société Entomologique de France* (4) **1** [1861]: 577–596.

Fairmaire, L. 1863: Coléoptères d'Algérie rapportés par M. Ogier de Baulny. *Annales de la Société Entomologique de France* (4) **3**: 637–648.

Fairmaire, L. 1864: Espèces nouvelles de coléoptères d'Algérie. *Annales de la Société Entomologique de France* (4) **4**: 233–246.

Fairmaire, L. 1866a: [new taxa]. In: Fairmaire L. & Coquerel C.: Essai sur les coléoptères de Barbarie. Quatrième partie. *Annales de la Société Entomologique de France* (4) **6**: 17–74.

Fairmaire, L. 1866b: Notice sur les coléoptères récoltés par M.J. Lédérer sur le Bosz-Dagh (Asie Mineure). *Annales de la Société Entomologique de France* (4) **6**: 249–280.

Fairmaire, L. 1867: Essai sur les coléoptères de Barbarie. Cinqième partie. *Annales de la Société Entomologique de France* (4) **7**: 387–416.

Fairmaire, L. 1868: Essai sur les coléoptères de Barbarie. Sixième partie. *Annales de la Société Entomologique de France* (4) **8**: 471–502.

Fairmaire, L. 1869a: Notes sur les coléoptères recueillis par Charles Coquerel à Madagascar et sur les côtes d'Afrique. *Annales de la Société Entomologique de France* (4) **8** [1868]: 753–820.

Fairmaire, L. 1869b: Notes sur les coléoptères recueillis par Charles Coquerel à Madagascar et sur les cotes d'afrique. *Annales de la Société Entomologique de France* (4) **9**: 179–260.

Fairmaire, L. 1871a: Essai sur les coléoptères de Barbarie. Septième partie. *Annales de la Société Entomologique de France* (4) **10** [1870]: 369–404.

Fairmaire, L. 1871b: Notes sur les coléoptères recueillis par Charles Coquerel à Madagascar et sur les côtes d'Afrique, 3e partie. *Annales de la Société Entomologique de France* (5) **1**: 29–78.

Fairmaire, L. 1873a: [new taxa]. In: Fairmaire L. & Raffray A.: Coléoptères du nord de l'Afrique. *Revue et Magasin de Zoologie Pure et Appliquée* (3) **1**: 331–385.

Fairmaire, L. 1873b: Diagnoses de divers coléoptères nouveaux découverts en Algérie. *Annales de la Société Entomologique de France* (5) **3**: 391–392.

Fairmaire, L. 1874a: Diagnoses de coléopteres nouveaux du nord de l'Afrique. *Petites Nouvelles Entomologiques* **1** [1869–1875]: 388–389.

Fairmaire, L. 1874b: Note sur quelques coléoptères des environs de Tuggurth. *Petites Nouvelles Entomologiques* **1** [1869–1875]: 407–408.

Fairmaire, L. 1875a: Coléoptères de la Tunisie récoltés par M.r Abdul Kerim. *Annali del Museo Civico di Storia Naturale di Genova* **7**: 475–540.

Fairmaire, L. 1875b: Diagnose de quelques coléoptères nouveaux d'Algérie. *Petites Nouvelles Entomologiques* **1** [1869–1875]: 495–496.

Fairmaire, L. 1875c: Descriptions de quelques coléoptères du Maroc (1). *Petites Nouvelles Entomologiques* **1** [1869–1875]: 543–544.

Fairmaire, L. 1875d: Diagnoses d'hétéromères de Madagascar. *Bulletin de la Société Entomologique de France* **1875**: xxxiii-xxxiv.

Fairmaire, L. 1876: Diagnoses de coléoptères du nord de l'Afrique. *Petites Nouvelles Entomologiques* **2**: 37–38, 49–50, 93–94.

Fairmaire, L. 1877: Coléoptères nouveaux du nord de l'Afrique. *Petites Nouvelles Entomologiques* **3**: 141.

Fairmaire, L. 1878a: [new taxa]. In: Deyrolle H. & Fairmaire L.: Descriptons de coléoptères recueillis par M. l'abbé David dans la Chine centrale. *Annales de la Société Entomologique de France* (5) **8**: 87–140.

Fairmaire, L. 1878b: [new taxa]. *Bulletin de la Société Entomologique de France* **1878**: lv-lvi.

Fairmaire, L. 1879a: Diagnoses de coléoptères du nord de l'Afrique. *Le Naturaliste* **1**: 3–4.

Fairmaire, L. 1879b: Descriptions de coléoptères nouveaux du nord de l'Afrique. *Annales de la Société Entomologique de France* (5) **9**: 155–172.

Fairmaire, L. 1879c: Coléoptères du nord de l'Afrique. *Revue et Magasin de Zoologie Pure et Appliquée* (3) **7**: 178–218.

Fairmaire, L. 1879d: Description de quelques coléoptères nouveaux recueillis en Espagne par D. Manuel Martorell y Peña. *Annales de la Société Entomologique de France* (5) **9**: 241–244.

Fairmaire, L. 1880a: Descriptions des coléoptères nouveaux du nord de l'Afrique. 3e partie. *Annales de la Société Entomologique de France* (5) **10**: 5–32.

Fairmaire, L. 1880b: Diagnoses de coléoptères du Maroc. *Le Naturaliste* **2**: 190.

Fairmaire, L. 1880c: Descriptions de coléoptères d'Espagne et de Turquie. *Annales de la Société Entomologique de France* (5) **10**: 237–244.

Fairmaire, L. 1880d: Descriptions des coléoptères nouveaux du nord de l'Afrique. 4e partie. *Annales de la Société Entomologique de France* (5) **10**: 245–254.

Fairmaire, L. 1881a: Essai sur les coléoptères des îles Viti (Fidgi). *Annales de la Société Entomologique de France* (6) **1**: 243–318.

Fairmaire, L. 1881b: [Crypticus pubens]. *Bulletin de la Société Entomologique de France* **1880**: cxxxii.

Fairmaire, L. 1881c: [new taxa]. *Bulletin de la Société Entomologique de France* **1881**: i-xii.

Fairmaire, L. 1881d: [Descriptions de deux espèces nouvelles de coléoptères trouvées en Catalogne: Abax, Sitaris]. *Bulletin de la Société Entomologique de France* 1881: xliii-xliv.

Fairmaire, L. 1882a: Coléoptères recueillis par M.G. Révoil chez les çomalis. In: Révoil, G.: *Faune et flore des pays çomalis*. Paris: Challames Ainé, iv + 104 pp., 1 pl.

Fairmaire, L. 1882b: Notes sur quelques coléoptères du Soudan et de l'Inde boréale. *Annales de la Société Entomologique de France* (6) **2**: 65–68.

Fairmaire, L. 1882c: Coléoptères hétéromères de Sumatra. *Notes from the Leyden Museum* **4**: 219–265.

Fairmaire, L. 1883a: Essai sur les coléoptères de l'archipel de la Nouvelle Bretagne. *Annales de la Société Entomologique du Belgique* **27**: 1–58.

Fairmaire, L. 1883b: Description de coléoptères hétéromères de l'île de Saleyer. *Notes from the Leyden Museum* **5**: 31–40.

Fairmaire, L. 1883c: Coléoptères d'Abyssinie. *Annales de la Société Entomologique de France* (6) **3**: 89–112.

Fairmaire, L. 1883d: Coléoptères de Tripoli. *Entomologische Zeitung* (Stettin) **44**: 459–460.

Fairmaire, L. 1883e: Description de coléoptères recueillis par M. le Baron Bonnaire en Algérie. *Comptes-Rendus des Séances de la Société Entomologique de Belgique* **1883**: xli-xlviii.

Fairmaire, L. 1883f: [Descriptions suivantes de deux coléoptères de l'Hindoustan]. *Bulletin de la Société Entomologique de France* **1883**: lv-lvi.

Fairmaire, L. 1883g: Description de quelques coléoptères marocains. *Comptes-Rendus des Séances de la Société Entomologique de Belgique* **1883**: cviii-cxi.

Fairmaire, L. 1883h: Description de coléoptères recueillis par M. le Baron Bonnaire en Algérie. *Comptes-Rendus des Séances de la Société Entomologique de Belgique* **1883**: cxi-cxiv.

Fairmaire, L. 1883i: [Communications]. *Bulletin de la Société Entomologique de France* 1883: cxxiv-cxxvi.

Fairmaire, L. 1883j: Descriptions de coléoptères recueillis par le Baron Bonnaire en Algérie. *Comptes-Rendus des Séances de la Société Entomologique de Belgique* 27: clvi-clix.

Fairmaire, L. 1883k: Descriptions de deux Helops nouveaux. *Bulletin de la Société Entomologique de France* 1882: clxix-clxx.

Fairmaire, L. 1883l: [new taxa]. *Bulletin de la Société Entomologique de France* 1882: clxxvii-clxxviii.

Fairmaire, L. 1884a: Liste des coléoptères reueillis par M. l'abbé David à Akbès (Asie-Mineure). *Annales de la Société Entomologique de France* (6) 4: 165–180.

Fairmaire, L. 1884b: Note sur les coléoptères recueillis par M. Ach. Raffray à Madagascar et descriptions des espèces nouvelles. 1. partie. *Annales de la Société Entomologique de France* (6) 4: 225–242.

Fairmaire, L. 1884c: Description de coléoptères marocains. *Le Naturaliste* 6: 446–447.

Fairmaire, L. 1884d: Descriptions de coléoptères recueillis par le Baron Bonnaire en Algérie. *Comptes-Rendus des Séances de la Société Entomologique de Belgique* 1884: lix-lxx.

Fairmaire, L. 1884e: Diagnoses de coléopteres del'Afrique orientale. *Comptes-Rendus des Séances de la Société Entomologique de Belgique* 1884: lxx-lxxviii.

Fairmaire, L. 1884f: Desciptions de coléoptères nouveaux du Maroc. *Comptes-Rendus des Séances de la Société Entomologique de Belgique* 1884: cxvii-cxxi.

Fairmaire, L. 1884g: [Description de trois coléoptères nouveaux, trouvés en Bulgarie par notre collègue M. Alléon]. *Bulletin de la Société Entomologique de France* 1883: cxli-cxliii.

Fairmaire, L. 1884h: Diagnoses de coléoptères del'Afrique orientale. *Comptes-Rendus des Séances de la Société Entomologique de Belgique* 1884: cxlii-cxlix.

Fairmaire, L. 1884i: Diagnoses de coléoptères del'Afrique orientale. *Comptes-Rendus des Séances de la Société Entomologique de Belgique* 1884: cccxlvi-ccexlvii.

Fairmaire, L. 1885a: [Descriptions de deux coléoptères algériens]. *Bulletin de la Société Entomologique de France* 1885: xxxviii-xxxix.

Fairmaire, L. 1885b: Malacodermes, lymexylonide et hétéromères nouveaux recueillis par M. Deby à Sumatra et à Bornéo. *Comptes-Rendus des Séances de la Société Entomologique de Belgique* 1885: cv-cxii.

Fairmaire, L. 1886a: Notes sur les coléoptères recueillis par M. Laligant à Obock. *Annales de la Société Entomologique de France* (6) 5 [1885]: 435–462.

Fairmaire, L. 1886b: Notes sur les coléoptères recueillis par M. Raffray, à Madagascar et descriptions des espèces nouvelles. 2e partie. *Annales de la Société Entomologique de France* (6) 6: 31–96, pl. 2.

Fairmaire, L. 1886c: Descriptions de coléoptères de l'intérieure de la Chine. *Annales de la Société Entomologique de France* (6) 6: 303–356.

Fairmaire, L. 1887a: Coléoptères des voyages de M.G. Revoil chez les Somâlis et dans l'intérieur du Zanguebar. *Annales de la Société Entomologique de France* (6) 7: 69–186.

Fairmaire, L. 1887b: Coléoptères de l'intérieur de la Chine. *Annales de la Société Entomologique de Belgique* 31: 87–136.

Fairmaire, L. 1887c: Coléoptères nouveaux ou peu connus du Musée de Leyden. *Notes from the Leyden Museum* 9: 145–162.

Fairmaire, L. 1887d: Description de cinq espèces nouvelles de la famille des Cantharides. *Notes from the Leyden Museum* 9: 193–196.

Fairmaire, L. 1887e: Coléoptères des voyages de M.G. Révoil chez les Somâlis et dans l'intérieur du Zanguebar. Heteromera (2e Parie). *Annales de la Société Entomologique de France* (6) 7: 277–368, pls 1–3.

Fairmaire, L. 1887f: Notes sur les coléoptères des environs de Pékin (1e partie). *Revue d'Entomologie* **6**: 312–336.

Fairmaire, L. 1887g: [new taxa]. *Bulletin de la Société Entomologique de France* **1886**: clxxviii-clxxix.

Fairmaire, L. 1888a: Coléoptères de l'intérieur de la Chine. *Annales de la Société Entomologique de Belgique* **32**: 7–46.

Fairmaire, L. 1888b: Notes sur les coléoptères des environs de Pékin (2e partie). *Revue d'Entomologie* 7: 111–160.

Fairmaire, L. 1888c: Énumération des coléoptères recuiellis par M. le Dr Hans Schinz dans le sud de l'Afrique. *Annales de la Société Entomologique de France* (6) **8**: 173–202.

Fairmaire, L. 1889a: Descriptions de coléoptères de l'Indo-Chine. *Annales de la Société Entomologique de France* (6) 8 [1888]: 333–378.

Fairmaire, L. 1889b: Coléoptères de l'intérieur de la Chine 5e partie. *Annales de la Société Entomologique de France* (6) **9**: 5–81.

Fairmaire, L. 1889c: [Diagnoses de quelques coléoptères recueillis à Pékin par M. Provost]. *Bulletin de la Société Entomologique de France* **1888**: cc-cci.

Fairmaire, L. 1891a: Note supplémentaire sur les coléoptères d'Obock. *Annales de la Société Entomologique de France* (6) **10** [1890]: 547–554.

Fairmaire, L. 1891b: Notes sur quelques coléoptères de l'Afrique intertropicale et descriptions d'espèces nouvelles. *Annales de la Société Entomologique de France* **60**: 231–274, pl. 5.

Fairmaire, L. 1891c: Description de coléoptères de l'intérieur de la Chine (Suite, 6e partie). *Comptes-Rendus des Séances de la Société Entomologique de Belgique* **1891**: vi-xxiv.

Fairmaire, L. 1891d: Descriptions de coléoptères des Montagnes de Kashmir. *Comptes-Rendus des Séances de la Société Entomologique de Belgique* **1891**: lxxxviii-ciii.

Fairmaire, L. 1891e: Descriptions de coléoptères des Montagnes de Kashmir (Suite). *Comptes-Rendus des Séances de la Société Entomologique de Belgique* **1891**: cxxi-cxxxiv.

Fairmaire, L. 1891f: Coléoptères de l'intérieur de la Chine (Suite, 7e partie). *Comptes-Rendus des Séances de la Société Entomologique de Belgique* **1891**: clxxxvii-ccxix.

Fairmaire, L. 1891g: Coléoptères de l'Afrique autrale. *Comptes-Rendus des Séances de la Société Entomologique de Belgique* **1891**: cckxxix-cccvii.

Fairmaire, L. 1892a: Coléoptères d'Obock. Troisième partie. *Revue d'Entomologie* **11**: 77–127.

Fairmaire, L. 1892b: Descriptions de coléoptères des environs d'Akbès (Syrie). *Annales de la Société Entomologique du Belgique* **36**: 144–159.

Fairmaire, L. 1892c: [new taxa]. *Bulletin de la Société Entomologique de France* **1892**: vii-viii.

Fairmaire, L. 1892d: [new taxa]. *Bulletin de la Société Entomologique de France* **1892**: cix-cxi.

Fairmaire, L. 1892e: Descriptions de quelques coléoptères nouveaux du Turkestan. *Bulletin de la Société Entomologique de France* **1892**: clii-clv.

Fairmaire, L. 1893a: Notes sur les coléoptères du Chua. *Annales de la Société Entomologique de Belgique* **37**: 9–50.

Fairmaire, L. 1893b: Contributions à la faune Indo-Chinoise. 11e Mémoire (1). Coléoptères hétéromères. *Annales de la Société Entomologique de France* 62: 19–38.

Fairmaire, L. 1893c: Coléoptères nouveaux des Indes Orientales, de la famille des Scarabaeidae, Rhipidoceridae, Tenebrionidae et Oedemeridae. *Notes from the Leyden Museum* 15: 17–64.

Fairmaire, L. 1893d: Note sur quelques coléoptères des pays Somalis. *Annales de la Société Entomologique de Belgique* 37: 144–156.

Fairmaire, L. 1893e: Matériaux pour la faune coléoptérique du Sénégal. *Annales de la Société Entomologique de France* 62: 147–158.

Fairmaire, L. 1893f: Notes sur quelques coléoptères des environs de Lang-Song. *Annales de la Société Entomologique de Belgique* 37: 287–302.

Fairmaire, L. 1893g: Coléoptères du Haute Tonkin. *Annales de la Société Entomologique de Belgique* 37: 303–325.

Fairmaire, L. 1893h: Coléoptères des îles Comores. *Annales de la Société Entomologique de Belgique* 37: 521–555.

Fairmaire, L. 1893i: Descriptions de quelques coléoptères de l'Afrique septentrionale. *Bulletin de la Société Entomologique de France* 1893: cxlvi-cxlix.

Fairmaire, L. 1894a: Hétèromères du Bengale. *Annales de la Société Entomologique de Belgique* 38: 16–43.

Fairmaire, L. 1894b: Quelques coléoptères du Thibet. *Annales de la Société Entomologique de Belgique* 38: 216–225.

Fairmaire, L. 1894c: Descriptions de coléoptères d'Algérie. *Annales de la Société Entomologique de Belgique* 38: 310–313.

Fairmaire, L. 1894d: Coléoptères de l'Afrique intertropicale et australe. Deuxième note. *Annales de la Société Entomologique de Belgique* 38: 314–335.

Fairmaire, L. 1894e: Coléoptères du Kilimandjaro et des environs. *Annales de la Société Entomologique de Belgique* 38: 386–395.

Fairmaire, L. 1894f: Coléoptères de l'Afrique intertropicale et australe. Troisième note. *Annales de la Société Entomologique de Belgique* 38: 651–679.

Fairmaire, L. 1895a: Description de coléoptères de Madagascar et îles voisines. *Annales de la Société Entomologique de Belgique* 39: 443–454.

Fairmaire, L. 1895b: Coléoptères nouveaux d´Akbès, des dernières chasses de M. Charles Delagrange. *Bulletin de la Société Entomologique de France* 66: cix-cxi.

Fairmaire, L. 1896a: Hétéromères de l´Inde recueillis par M. Andrewes. *Annales de la Société Entomologique de Belgique* 40: 6–62.

Fairmaire, L. 1896b: Coléopterès de l'Inde boréale, Chine et Malaisie. *Notes from the Leyden Museum* 18: 81–129.

Fairmaire, L. 1896c: Description de coléoptères recueillis pa le Dr. Ch. Matin à Obock, Djibouti et Aden. *Bulletin de la Société Entomologique de France* 1896: 223–225.

Fairmaire, L. 1896d: Coléopterès de l'Inde et de Malaisie. *Notes from the Leyden Museum* 18: 225–240.

Fairmaire, L. 1896e: Description de quatre coléoptères nouveaux de la région mediterranéenne. *Bulletin de la Société Entomologique de France* 1896: 344–345.

Fairmaire, L. 1897a: Pseudadrus, nouveau genre d'opatrides. *Revue d'Entomologie* 16: 39.

Fairmaire, L. 1897b: Description de six coléoptères de l'île Ishigaki-Sima. *Bulletin de la Société Entomologique de France* **1897**: 68–72.

Fairmaire, L. 1897c: Coléoptères de l'Inde boréale, Chine et Malaisie. *Notes from the Leyden Museum* **18**: 81–129.

Fairmaire, L. 1897d: Matériaux pour la faune coléoptérique de la région malgache. *Annales de la Société Entomologique de Belgique* **41**: 92–119.

Fairmaire, L. 1897e: Coléoptères nouveaux de l'Afrique intertropicale et australe. *Annales de la Société Entomologique de France* **66**: 109–155.

Fairmaire, L. 1897f: Description de trois coléoptères hétéromères d'Algérie et d'Egypte. *Bulletin de la Société Entomologique de France* **1897**: 120–121.

Fairmaire, L. 1897g: Descriptions de deux coléoptères nouveaux de Japon. *Bulletin de la Société Entomologique de France* **1897**: 166–167.

Fairmaire, L. 1897h: Description d'une espèce nouvelle du genre Hionthis (Col.). *Bulletin de la Société Entomologique de France* **1897**: 180–181.

Fairmaire, L. 1897i: Description de coléoptères nouveaux de la Malaisie, de l'Inde et de la Chine. *Notes from the Leyden Museum* **19**: 209–223.

Fairmaire, L. 1897j: Coléoptères de l'Inde et de la Malaisie. *Notes from the Leyden Museum* **19**: 225–240.

Fairmaire, L. 1897k: Coléoptères de Szé-Tschouen et de Koui-Tchéou (Chine). *Notes from the Leyden Museum* **19**: 241–255.

Fairmaire, L. 1898a: Description de trois coléoptères des îles Ishigaki (Japon méridional). *Bulletin de la Société Entomologique de France* **1898**: 14–15.

Fairmaire, L. 1898b: Descriptions de quatre coléoptères de Madagascar. *Bulletin de la Société Entomologique de France* **1898**: 37–39.

Fairmaire, L. 1898c: Matériaux pour la faune coléoptérique de la région malgache. 5e note. *Annales de la Société Entomologique de Belgique* **42**: 222–260.

Fairmaire, L. 1898d: Descriptions de coléoptères d'Asie et de Malaisie. *Annales de la Société Entomologique de France* **67**: 383–400.

Fairmaire, L. 1899: Description de coléoptères nouveaux recueillis en Chine par M. de Latouche. *Annales de la Société Entomologique de France* **68**: 616–643.

Fairmaire, L. 1901a: Matériaux pour la faune coléoptèrique de la région Malgache (11e Note). *Revue d'Entomologie* **20**: 101–248.

Fairmaire, L. 1901b: Description de coléoptères des montagnes de Sikkim. *Bulletin de la Société Entomologique de France* **1901**: 255–268.

Fairmaire, L. 1903a: Descriptions de quelques hétéromères recueillis par M. Fruhstorfer dans le Haut-Tonkin. *Annales de la Société Entomologique de Belgique* **47**: 13–20.

Fairmaire, L. 1903b: Matériaux pour la faune coléoptérologique de la région malgache. 16e note. *Annales de la Société Entomologique de France* **72**: 181–259.

Fairmaire, L. 1903c: Description de coléoptères recueillis en Chine par M. de Latouche. *Bulletin de la Société Entomologique de France* **1902**: 316–318.

Fairmaire, L. & Brisout de Barneville, C.N.F. 1859: [new taxon]. In: Fairmaire L.: Miscellanea entomologica. Troisième partie. *Annales de la Société Entomologique de France* (3) **7**: 21–64.

Fairmaire, L. & Coquerel, C. 1866: Essai sur les coléoptères de Babarie. Quatrième partie. *Annales de la Société Entomologique de France* (4) **6**: 17–74.

Fairmaire, L. & Germain, P. 1860: Coleoptera Chilensia. *Extrait des Annales de la Société Entomologique de France*. Paris: F. Malteste et Cie, 8 pp.

Fairmaire, L. & Raffray, A. 1873: Coléoptères du nord de l'Afrique. *Revue et Magazin de Zoologie Pure et Appliquée* **3**: 331–385.

Faldermann, F. 1832: [new taxa]. In: Ménétriés, E.: *Catalogue raisonné des objets de zoologie recueillis dans un voyage au Caucase et jusqu'aux frontières actuelles de la Perse entrepris par l'ordre de S.M. l'Empereur*. Académie des Sciences St.-Pétersbourg, xxxiii + 272 + iv + [1].

Faldermann, F. 1833: Species novae Coleopterorum Mongoliae et Sibiriae. *Bulletin de la Société des Naturalistes de Moscou* **6**: 46–72.

Faldermann, F. 1835a: Lettre sur deux nouveaux melasomes. *Bulletin de la Société des Naturalistes de Moscou* **8**: 165–170.

Faldermann, F. 1835b: Coleopterorum ab illustrissimo Bungio in China boreali, Mongolia, et Montibus Altaicis collectorum, nec non ab ill. Turczaninoffio et Stchukino e provincia Irkutsk missorum illustrationes. *Mémoires de l'Académie Impériale des Sciences de St. Pétersbourg. Sixième Série. Sciences Mathematiques, Physiques et Naturelles* **3** (1): 337–464, pls I–V.

Faldermann, F. 1836: Bereicherung zur Käfer-Kunde des Russischen Reiches. *Bulletin de la Société des Naturalistes de Moscou* **9**: 351–398, 3 pls.

Faldermann, F. 1837: Fauna entomologica trans-caucasica. Pars II. Coleoptera trans-caucasica. *Nouveaux Mémoires de la Société Impériale des Naturalistes de Moscou* **5**: 1–433, 15 pls.

Fan, X. 1995a: A new species and a new record species of genus Mordellistena. *Entomotaxonomia* **17**: 23–25. (in Chinese with English summary).

Fan, X. 1995b: Mordellidae. Pp.: 224–224. In: Wu, H. (ed.): *Insects of Baishanzu Mountain, eastern China*. Beijing: China Forestry Publishing House, xiii, + 586 pp. (in Chinese with English summary).

Fan, X. & Yang, J. 1993: Revision of the genus Glipa of China. *Memoirs of Beijing Natural History Museum* **53**: 45–68. (in Chinese with English summary).

Fan, X. & Yang, J. 1995: Mordellidae. Pp. 95–101. In: Zhu, T. (ed.): *Insects and macrofungi of Gutianshan, Zhejiang*. Hangzhou: Zhejiang Science and Technology Publishing House, 318 pp. (in Chinese with English summary).

Fan, Z.-T. & Huang, R.-X. 2005: [new taxon]. In: Wu W., Fan Z.-T. & Huang R.-X.: Three new species of Platyope (Coleoptera: Tenebrionidae) from Xinjiang, China. *Entomotaxonomia* **27**: 283–288. (in Chinese with English summary).

Fancello, L. & Leo, P. 1991: Lyreus septemstriatus, nuova specie di Colydiidae anoftalma dello Sardegna, appartenente ad un genere nuovo per la fauna italiana (Coleoptera). *Bollettino della Società Entomologica Italiana* **123**: 129–31.

Fancello, L. & Magrini, P. 2013: Descrizione di Langelandia (Normandella) montalbica, nuova specie di Sardegna (Coleoptera, Zopheridae). *Giornale Italiano di Entomologia* **13**: 283–288.

Faraj, A.M. 2018: A New species of comb-clawed beetles, Mycetochara Berthold, 1827 from Erbil-Iraq (Coleoptera: Alleculidae). *Tikrit Journal for Agricultural Sciences* **18**: 96–99.

Faust, J. 1875: Beiträge zur Kenntnis der Käfer des Europäischen und Asiatischen Russlands mit Einschluss der Küsten des Kaspischen Meeres. *Horae Societatis Entomologicae Rossicae* **11**: 163–252.

Faust, J. 1877: Aeltere und einige neue Käfer der russischen Fauna. *Bulletin de la Société Impériale des Naturalistes de Moscou* **52** (2): 34–45.

Faust, J. 1877: Beiträge zur Kenntnis der Käfer des europäischen und asiatischen Russlands mit Einschluss der Küsten des Kaspischen Meeres. *Horae Societatis Entomologicae Rossicae* **12** [1876–1877]: 300–330.

Fauvel, A. 1867: Catalogue des coléoptères de la Nouvelle Calédonie et dépendances, avec descriptions, notes et synonymies nouvelles. *Bulletin de la Société Linnéenne de Normandie* (2) **1** [1865–1866]: 172–215, pl. 1.

Fauvel, A. 1899: Tableau des Phaleria françaises. *Revue d'Entomologie* **18**: 77–79.

Fauvel, A. 1901: Phaleria nouvelle de France. *Revue d'Entomologie* **20**: 253–255.

Fauvel, A. 1904: Faune analytique des coléoptères de la Nouvelle Calédonie. 2e Partie. *Revue d'Entomologie* **23**: 113–208.

Favre, E. 1890: *Faune des coléoptères du Valais et des régions limitrophes*. Zürich: Zurcher & Furrer, xliv + 448 pp.

Ferrari, J.A. 1869: Drei neue westasiatische Käfer. *Verhandlungen der Kaiserlich-Königlichen Zoologisch-Botanischen Gesellschaft in Wien* **19**: 193–200.

Ferrer, J. 1983: Seven new species of Tenebrionidae from the Old World (Coleoptera). *Entomologica Scandinavica* **14**: 137–144.

Ferrer, J. 1992: Dos nuevas species de Tenebrionidae (Coleoptera). *Nouvelle Revue d'Entomologie* (N.S.) **9**: 83–89.

Ferrer, J. 1993a: Essai de révision des espèces africaines et européennes appartenant au genre Gonocephalum Solier (Coleoptera, Tenebrionidae). *Atti del Museo Civico di Storia Naturale di Trieste* **45**: 59–150.

Ferrer, J. 1993b: Description of new genus and species of Opatrini from the Canary Island (Coleoptera, Tenebrionidae). *Nouvelle Revue d'Entomologie* (N.S.) **10**: 121–125.

Ferrer, J. 1995: Essai de revision des espèces africaines et europèennes appartenant au genre Gonocephalum Solier (Coleoptera, Tenebrionidae) . Deuxième partie. *Atti del Museo Civico di Storia Naturale di Trieste* **46**: 1–75.

Ferrer, J. 1997a: Deux nouveaux Mesomorphus de la région orientale (Coleoptera, Tenebrionidae, Opatrini). *Entomofauna* **18**: 73–80.

Ferrer, J. 1997b: Nouvelle espèce de Gonocephalum (Solier) de l'Afrique et l'Oman. *Nouvelle Revue d'Entomologie* (N.S.) **14**: 211–216.

Ferrer, J. 2000a: Contributions to the knowledge of the genera Scleropatrum Reitter, 1890 and Polycoelogastridium Reichardt, 1936 (Coleoptera, Tenebrionidae, Opatrini). *Entomofauna* **21**: 33–48.

Ferrer, J. 2000b: Révision des espèces africaines et européennes appartennantes au genre Gonocephalum Solier (Coleoptera, Tenebrionidae, Opatriini). Troisième partie. *Atti del Museo Civico di Storia Naturale di Trieste* **48**: 69–153.

Ferrer, J. 2003: Les Caedius Mulsant et Cyptus Gerstaecker de la région afrotropicale (Coleoptera, Tenebrionidae, Opatrini). *Nouvelle Revue d'Entomologie* **19** [2002]: 295–349.

Ferrer, J. 2008: Contribución al conocimiento de los Asidini ibero-baleares. Secunda nota. Las Alphasida (Glabrasida) del grupo tricostatae Escalera 1922 (Coleoptera, Tenebrionidae, Pimeliinae). *Boletín de la Sociedad Entomológica Aragonesa* **43**: 61–73.

Ferrer, J. 2010a: Taxonomic notes on the genus Gonocephalum Solier, 1834, with description of new taxa (Coleoptera: Tenebrionidae). *Annales Zoologici* **60**: 231–238.

Ferrer, J. 2010b: Sinopsis del subgénero Lithoboriolus Español 1945, de Phylan Dejean, 1821, rehabilitación de Phylan reyi (Piochard de la Brûlerie, 1869) stat. rest. y descripción de nuevas especies (Coleoptera, Tenebrionidae). *Boletín de la Sociedad Entomológica Aragonesa* **46**: 127–141.

Ferrer, J. 2011: Contribución al conocimiento del género Phylan Dejean, 1821, y descripción de una especie nueva del género Heliopates Dejean, 1834 (Coleoptera, Tenebrionidae, Pedinini). *Boletín de la Sociedad Entomológica Aragonesa* **49**: 75–82.

Ferrer, J. 2013a: Análisis cladístico de la tribu Pedinini (Coleoptera, Tenebrionidae) de la Península Ibérica, descripción de un nuevo género y comentarios sobre el género Psammoardoinellus leo, 1981. *Boletín de la Sociedad Entomológica Aragonesa* **53**: 31–55.

Ferrer, J. 2013b: Sobre la identidad y distribución geográfica de Phylan gibbus (Fabricius, 1775) y sus presuntas sinonimias (Coleoptera, Tenebrionidae, Pedinini). *Boletín de la Sociedad Entomológica Aragonesa* **52**: 49–65.

Ferrer, J. 2018: Contribución al conocimiento del género Pachychila Eschscholtz, 1831 y descripción de dos especies nuevas de la Península Ibérica y una de Italia (Coleoptera, Tenebrionidae, Pimeliinae). *Boletín Sociedad Entomológica Aragonesa* **62**: 35–54.

Ferrer, J. & Beltran, F. 2009: La confusión taxonómica de 'Stenosis sardoa' Espanol, 1985 et auct. y descripción de un nuevo Stenosis franco-ibérico (Coleoptera: Tenebrionidae). *Boletín de la Sociedad entomológica Aragonesa,* **45**: 509–514.

Ferrer, J. & Castro Tovar A. 2010: Descripción de una nueva especie del género Melanimon Stevens, 1829 de la Península Ibérica (Coleoptera, Tenebrionidae). *Boletín de la Sociedad Entomológica Aragonesa* **47**: 39–44.

Ferrer, J. & Iwan, D. 2012: Revisión del género Heliopates Dejean, 1834 (Coleoptera, Tenebrionidae). 1: Las especies de la Península Ibérica, Baleares y Sicilia. *Boletín de la Sociedad Entomológica Aragonesa* **50**: 115–147.

Ferrer, J. & Martínez, J.C. 2012: Contribución al concimiento del género Trachynotus Latreille: un caso de homonimia en el género *Sepidium* Fabricius (Coleoptera: Tenebrionidae: Pimeliinae). *Boletin Sociedad Entomológica Aragonesa* **51**: 283–287.

Ferrer, J. & Picka, J. 1990: The Blaps species of Sweden, with review of the B. lethifera group (Coleoptera, Tenebrionidae). *Entomologisk Tidskrift* **111**: 25–32.

Ferrer, J. & Soldati, L. 1999: Contribution à l'étude des Tenebrionidae de Turquie (Insecta, Coleoptera). *Entomofauna* **20**: 53–92.

Ferrer, J. & Whitehead, P.F. 2002: The genus Xanthomus Mulsant, 1854 (Coleoptera: Tenebrionidae), its evolutionary history and conservation significance. *Annales Zoologici* **52**: 383–401.

Ferrer, J. & Yvinec, J.H. 2004: Révision de la tribu des Lychnogyini Reitter, 1904 sensu nov. et description d'un nouveau genre et d'une espèce nouvelle de Taklamakan, Chine (Coleoptera: Tenebrionidae: Pimeliinae). *Annales de la Société Entomologique de France* (N.S.) **40**: 41–49.

Ferrer, J. & Yvinec, J.H. 2005: Changement du nom homonymique d'un genre de coléoptères Tenebrionidae du désert froid de Taklamakan, Chine (Coleoptera). *Nouvelle Revue d'Entomologie* **22**: 121–122.

Ferrer, L. 1859: Note sur la Mylabris Moquinia n. sp. *Revue et Magasin de ZoologiePure et Appliquée* (2) 11: 539–540.

Fiori, A. 1903: Studio critico dei Tenebrio Lin. italiani. *Rivista Coleopterologica Italiana* 1: 221–227.

Fischer, J.B. 1827: *Tentamen conspectus Cantharidiarum. Dissertatio inauguralis quam pro summis in medicina et chirurgia honoribus legitime obtinendis eruditorum examini.* Monachi: Lindauer, 26 pp.

Fischer von Waldheim, G. 1809: Sur deux genres nouveaux du coléoptères. *Mémoires de la Société Impériale des Naturalistes de Moscou* 2: 293–304, pl. 18.

Fischer von Waldheim, G. 1812: Pogonocerus, novum genus insectorum Caucasi meridionalis. *Mémoires de la Société Impériale des Naturalistes de Moscou* 3: 281–283.

Fischer von Waldheim, G. 1817: Adversaria Zoologica. Fasciculus secundus. Insecta. De coleopteris quibusdam novis. *Mémoires de la Société Impériale des Naturalistes de Moscou* 5: 463–471.

Fischer von Waldheim, G. 1820: *Entomographie de la Russie* [*Entomographia Imperii Rossici*]. *Auctoritate Societatis Caesareae Mosquensis naturae scrutatorum collecta et in lucem edita. Volume I.* Mosquae: Auguste Semen, 17 pls [Coleoptera] + 9 pls [other insects].

Fischer von Waldheim, G. 1821: *Lettre adressée au nom de la Société Impériale des Naturalistes de Moscou, à l'un de ses membres M. le docteur Chrétien-Henri Pander, par Gotthelf Fischer de Waldheim, directeur de la Société; Contenant une notice sur un nouveau genre d'oiseau et sur plusiers nouveaux insectes.* Moscou: Auguste Semen, 15 pp.

Fischer von Waldheim, G. 1822: *Entomographie de la Russie* [*Entomographia Imperii Rossici*]. *Auctoritate Societatis Caesareae Mosquensis naturae scrutatorum collecta et in lucem edita. Volumen I.* Mosquae: Auguste Semen, viii + 210 pp.

Fischer von Waldheim, G. 1823: *Entomographie de la Russie, et genres des Insectes. Entomographia Imperii Rossici, sue Caesareae Majestati Alexandro I dicata. Volume II.* Plates XVIII–L [Coleoptera] + VI–XI [Lepidoptera].

Fischer von Waldheim, G. 1824: *Entomographie de la Russie, et genres des Insectes. Entomographia Imperii Rossici, sue Caesareae Majestati Alexandro I dicata. Volume II.* Mosquae: Augusti Semen Typographi, xx + 264 pp.

Fischer von Waldheim, G. 1829a: *Museum Historiae Naturalis Universitatis Caesareae Mosquensis. Pars 2. Insecta.* Mosquae: Typis Universitatis Caesareae, 147 pp.

Fischer von Waldheim, G. 1829b: Notice sur quelques insectes de la collection de C. Steven. *Bulletin de la Société Impériale des Naturalistes de Moscou* 1 (1–2): 368–370, 1 pl.

Fischer von Waldheim, G. 1830: Note sur quelques nouvelles espèces d'insectes; dans une lettre à Mr. de Steven. *Bulletin de la Société Impériale des Naturalistes de Moscou* 2: 183–188.

Fischer von Waldheim, G. 1832a: [new taxa]. In: Ménétriés, E.: *Catalogue raisonné des objets de zoologie recueillis dans un voyage au Caucase et jusqu'aux frontières actuelles de la Perse entrepris par ordre de S.M. l'Empereur.* St.-Pétersbourg: Imprimerie Impériale des Sciences, xxxiii + 271 + iv + (1) pp.

Fischer von Waldheim, G. 1832b: Analecta ad faunam insectorum rossicam. *Bulletin de la Société Impériale des Naturalistes de Moscou* 4: 423–440, pls. 5–6.

Fischer von Waldheim, G. 1835: Adesmiae genus speciebus quibusdam novis illustratum. *Bulletin de la Société Impériale des Naturalistes de Moscou* 8: 310–316, pl. VIII.

Fischer von Waldheim, G. 1836: Analecta ad faunam insectorum rossicam. *Bulletin de la Société Impériale des Naturalistes de Moscou* **9**: 154–166, pl. 5.

Fischer von Waldheim, G. 1837a: [new taxa]. In: Faldermann F.: Fauna entomologica trans-caucasica. Pars II. Coleoptera trans-caucasica. *Nouveaux Mémoires de la Société Impériale des Naturalistes de Moscou* **5**: 1–433, 15 pls.

Fischer von Waldheim, G. 1837b: Notice sur les mélasomes. *Bulletin de la Société Impériale des Naturalistes de Moscou* **10** (4): 3–18, pl. II.

Fischer von Waldheim, G. 1838: Nouvelle espèce d'Adesmia, genre de coléoptères. *Revue Zoologique, par la Société Cuvierienne* **1838**: 316.

Fischer von Waldheim, G. 1842: *Catalogus Coleopterorum in Siberia orientali a Cel. Gregorio Silide Karelin collectorum*. Moscou, 28 pp.

Fischer von Waldheim, G. 1844: Spicilegum Entomographiae Rossicae. II. Heteromera. *Bulletin de la Société Impériale des Naturalistes de Moscou* **17**: 3–144, pls 1–3.

Flach, C. 1882: Zwei neue deutsche Cis-Arten. *Deutsche Entomologische Zeitschrift* **26**: 249–251.

Flach, K. 1907: Beiträge zur Käferfauna der Iberischen Halbinsel. *Wiener Entomologiche Zeitung* **26**: 17–19.

Fleischer, A. 1900: Uebersichtstabelle der Arten der Coleopteren-Gattung Palorus Duv. *Wiener Entomologische Zeitung* **19**: 236–237.

Fleischer, A. 1909: Neue Mordellidae. *Wiener Entomologische Zeitung* **28**: 241–242.

Fleischer, A. 1916: Neue Varietät der Pelecotoma fennica Payk. aus Bosnien. *Wiener Entomologische Zeitung* **35**: 120.

Fleischer, A. 1918. Eine neue Pedilus art (Col. Pyrochroidae). *Wiener entomologische Zeitung* **37/38**: 200

Fleischer, A. 1919a: Neue palaärktische Oedemeriden. *Entomologische Blätter* **15**: 168–171.

Fleischer, A. 1919b: Eine neue Prostomis-Art aus dem Elbrusgebirge. *Entomologische Blätter* **15**: 211.

Fleischer, A. 1921: Bemerkenswerte Käfer-Aberrationen. *Wiener Entomologiche Zeitung* **38**: 142.

Fleischer, A. 1923: Lytta vesicatoria Linn var. dibapha Reitt. *Časopis České společnosti Entomologické* **20**: 89.

Fletcher, T.B. 1914: *Some South Indian Insects*. Madras: Superintendent Goverment Press, vii + 565 pp.

Foley, I.A. & Ivie, M.A. 2008: A revision of the genus Phellopsis LeConte (Coleoptera: Zopheridae). *Zootaxa* **1689**: 1–28.

Fonscolombe, E.L.J.H. de. 1845: Calendrier de faune et de flore pour les environs de Aix ou première apparition des principaux insectes et première floraison des végétaux qui s'y trouvent. *Mémoires de l'Accadémie des Sciences, Agriculture, Arts et Belles-Lettres* **5**: 357–682.

Forel, A. 1893: Sur la classification de la famille des formicides, avec remarques synonymiques. *Annales de la Société Entomologique de Belgique* **37**: 161–167.

Forskål, P. 1775: *Descriptiones animalium avius, amphibiorum, piscium, insectorum, vermium quae in itinere orientali observavit*. Hauniae: Mölleri, aulae Typographi, 164 pp., 1 pl.

Forster, J.R. 1771: *Novae species insectorum, Centuria I*. Londoni: T. Davies et B. White, viii + 100 pp.

Fouquè, R. 2008: Revision of the genus Pseudethas Fairmaire, with descriptions of four new species from Nepal and Thailand (Coleoptera: Tenebrionidae: Stenosini). *Stuttgarter Beiträge zur Naturkunde A, Neue Serie* **1**: 357–369.

Fouquè, R. 2013: Revision of the genus Herbertfranziella Kaszab and description of the new genus Nepalofranziella of Himalayan Dichillina (Coleoptera: Tenebrionidae: Stenosini). *Stuttgarter Beiträge zur Naturkunde A, Neue Serie* **6**: 183–197.

Fowler, M.A. 1889: On the British species of the genus Anaspis, Geoffroy, with description of a new species. *The Entomologist's Monthly Magazine* **25**: 331–336.

Franciscolo, M. 1942a: Mordellidi raccolti dal marchese G. Doria all'isola del Giglio negli anni 1900–1902. *Bollettino della Società Entomologica Italiana* **74**: 6–10.

Franciscolo, M. 1942b: Raccolte entomologiche nell'isola di Capraia fatte da C. Mancini e da F. Capra (1927–1931). *Bollettino della Società Entomologica Italiana* **74**: 18–23.

Franciscolo, M. 1942c: Diagnosi di nuovi mordellidi Italiani. *Bollettino della Società Entomologica Italiana* **74**: 76–80.

Franciscolo, M. 1942d: Mordellokoides grandii nuovo sottogenere e nuova specie di Mordellistena dell'Italia meridionale. *Bollettino dell'Instituto di Entomologia della R. Universita di Bologna* **13**: 133–136.

Franciscolo, M. 1943a: Machairophora, nuovo genere della tribú dei Mordellini ed alcune note sistematiche e sinonimiche sui Mordellidae. *Bollettino della Società Entomologica Italiana* **75**: 33–40.

Franciscolo, M. 1943b: Su due nuove Anaspis Geoffroy. *Bollettino della Società Entomologica Italiana* **75**: 76–80.

Franciscolo, M. 1949a: Un nuovo mordellino di Birmania. *Doriana* **1** (1): 1–4.

Franciscolo, M. 1949b: Note su alcuni mordellidi. *Bollettino della Società Entomologica Italiana* **79**: 52–58.

Franciscolo, M. 1949c: XIII° Contributo alla conoscenza dei mordellidi. *Memorie della Società Entomologica Italiana* **28**: 81–95.

Franciscolo, M. 1950: Ermischiella papuana n. ge. n. sp. di mordellide della Nuova Guinea e diagnosi preliminari di tre nuovi generi africani. *Doriana* **1** (5): 1–8.

Franciscolo, M. 1951: Una nuova Mordella dell'Albania. *Bollettino della Società Entomologica Italiana* **81**: 7–9.

Franciscolo, M. 1952: La Neoglipa e generi vicini della Nuova Guinea. *Annali del Museo Civico di Storia Naturale di Genova* **65**: 325–357.

Franciscolo, M. 1953: Results of the Armstrong College Expedition to Siwa Oasis (Lybian desert), 1935, under the leadership of Prof. J. Omer-Cooper. *Bulletin de la Société Fouad Ier d'Entomologie* **37**: 251–253.

Franciscolo, M. 1954: On two species of Anaspidinae (Col. Scraptiidae) taken on flowers of Protea abyssina in Natal. *Proceedings of the Royal Entomological Society of London* (B) **23** (3–4): 63–73.

Franciscolo, M. 1956a: Su un nuovo genere di Scraptiidae dell'India settentr. *Memorie della Società Entomologica Italiana* **35**: 43–70.

Franciscolo, M. 1956b: Su due interessanti specie del genere Anaspis L. del bacino del Mediterráneo. *Bollettino della Società Entomologica Italiana* **86**: 47–56.

Franciscolo, M. 1956c: 30. Coleoptera: Mordellidae and Scraptiidae. Pp. 451–469. In: *British Museum Expedition to South-West Arabia, 1937–38.* [1941–1957] *Volume 1*. London: Trustees of the British Museum, 504 pp., 1 pl., 1 map.

Franciscolo, M. 1961: On a new scraptiid beetle from Afganistan. *Proceedings of the Royal Entomological Society of London* **26**: 53–57.

Franciscolo, M. 1974: New and little-known Mordellistena Costa from Pakistan and India. *Oriental Insects* **8**: 71–84.

Franciscolo, M. 1982: Le Anaspis (s. str.) del gruppo maculata Fourcr., con descrizione di una nuova specie sarda. *Memorie della Società Entomologica Italiana* **60** [1981]: 197–207.

Franciscolo, M. 1989: About Ectasiocnemis Franciscolo, 1956. *Bollettino della Società Entomologica Italiana* **121**: 108–123.

Franciscolo, M. 1991: Su alcuni mordellidi e scraptiidi (Coleoptera-Heteromera) delle isole Pelagie. *Il Naturalista Siciliano* (4) **15**: 167–178.

Franz, H. 1967: Revision der Tarphius-Arten Europas, Nordwestafrikas und der Kanarischen Inseln (Coleopt. Colydiidae). *Eos, Revista Española de Entomología* **43**: 61–91.

Franz, H. 1970: Langelandia-Arten von Madeira-Gruppe und von den Kanaren (Coleoptera, Colydiidae). *Bocagiana* **26**: 1–5.

Franz, H. 1980: Neue und ungenügend bekannte Coleoptera (Scydmaenidae et Colydiidae) aus Mazedonien. *Zeitschrift der Arbeitsgemeinschaft Österreichischer Entomologen* **31** [1979]: 107–110.

Franz, H. 1982: Beiträg zur Kenntnis der Coleopterenfauna der Insel Galitia (mit einem Beitrag von Cl. Besuchet). *Sitzungsberichte der Österreichischen Akademie der Wissenschaften in Wien (Abteilung I)* **191**: 231–240.

Franz, H. 1984a: Beitrag zur Kenntnis der Coleopterenfauna Mazedoniens. *Zeitschrift der Arbeitsgemeinschaft Österreichischer Entomologen* **36**: 57–62.

Franz, H. 1984b: Weitere neue Coleopterenarten von den Kanarischen Inseln. *Vieraea* **13** [1983]: 75–81.

Franz, H. 1985: Revision der Genera Holoxantha Semenow und Alloxantha Seidlitz (Coleoptera, Oedemeridae). *Eos, Revista Española de Entomología* **61**: 95–105.

Franz, H. 1990a: Weitere Beiträge zur Kenntnis der Coleopterenfauna der Canarischen Inseln (Colydiidae und Ptinidae). *Zeitschrift der Arbeitsgemeinschaft Österreichischer Entomologen* **42**: 27–31.

Franz, H. 1990b: Stenosis canariensis n. sp. (Coleoptera, Tenebrionidae). *Zeitschrift der Arbeitsgemeinschaft der Österreichischer Entomologen* **42**: 93–94.

Franz, H. 1996: Die Ergebnisse meiner langjährigen Aufsammlungen der Coleopteenfauna [sic] auf der Insel hierro (Kanarische Inseln). *Sitzungsberichte Mathematisch-naturwissenschaftliche Klasse Abt. I* **202** [1995]: 71–138.

Freude, H. 1955: Die Monommiden der Welt. I. Teil: Die Monommiden der indo-australischen Region. *Entomologische Arbeiten aus dem Museum G. Frey* **6**: 1–73, pls 1–11.

Freude, H. 1958: Die Monommidae der Afrikanischen Region (Coleoptera). (IV. Teil der Monommiden der Welt mit Zusammenfassung der Ergebnisse). *Annales du Musée Royal du Congo Belge, Série in 8o, Sciences Zoologiques* **61**: 1–115.

Freude, H. 1978: Monommidae des Bishop Museums in Honolulu (Coleoptera). *Pacific Insects* **18**: 217–221.

Friedenreich, C.W. 1883: Pilzbewohnende Käfer in der Provinz Santa Catharina (Südbrasilien). *Entomologische Zeitung* (Stettin) **44**: 375–380.

Frivaldszky, J. 1865: Adatok a magyarhoni barlangok faunájához. *Mathematikai és Természettudományi Közlemények Vonatkozólag a Hazai Viszonyokra* **3**: 17–53.

Frivaldszky, J. 1877a: Coleoptera nova ex ins. Creta et Asia Minore. *Természetrajzi Füzetek* **1**: 83–87.

Frivaldszky, J. 1877b: Adnotationes coleopterologicae. *Természetrajzi Füzetek* **1**: 89–136.

Frivaldszky, J. 1877c: Adatók Temes és Krassó megyek faunájához. *Mathematikai és Természettudományi Közlemenyek* **13**: 285–378.

Frivaldszky, J. 1880: Coleoptera nova in Europa orientali et Asia Minor a D. Eduardo Merkl dedecta [sic.]. Ujtéhélyröpüek Európa keleti részében es Kis-Azsiában Merkl Ede urtól gyüjtve. *Természetrajzi Füzetek* **4**: 260–266.

Frivaldszky, J. 1884: Coleoptera nova ex Asia Minore. *Természetrajzi Füzetek* **8**: 1–8.

Frivaldszky, J. 1890: Coleoptera in expeditione D. Comitis Belae Széchenyi in China, praecipue borealis, a Dominis Gustavo Kreitner et Ludovico Lóczy anno 1879 collecta. *Természetrajzi Füzetek* **12** [1889]: 197–210.

Frivaldszky, J. 1892: Coleoptera in expeditione D. Comitis Bolae Széchenyi in China, praecipuae boreali, a Dominus Gustavo Kreitner et Ludovico Loczy, anno 1879 collecta. Pars seconda. *Természetrajzi Füzetek* **15**: 114–125.

Frivaldszky, J. 1894: Coleoptera nova. *Természetrajzi Füzetek* **16** [1893]: 85–86.

Frivaldszky, I. 1835: Közlések a' Balkány vidékén tett természettudományi utazásról. *A Magyar Tudós Társaság Évkönyvei* **2**: 235–276, 7 pls.

Frivaldszky, I. 1837: Balkány vidéki természettudományi utazás. Második közlés. *A Magyar Tudós Társaság Évkönyvei* **3**: 156–184, 8 pls.

Frivaldszky, I. 1845: Rövid áttekintése egy természetrajzi utazásnak, az európai Törökbirodalomban, egyszersmind nehány a közben újonnat fölfedezett állatnak leirása. *A Királyi Magyar Természettományi Társulat Évkönyvei* **1**: 163–187, 3 pls.

Frivaldszky, I. 1865: Jellemzö adatok Magyarország faunájához. *A Magyarország Tudományi Akadémia Évkönyvei* **11** (4). Pest: Emich Gustáv, 274 pp., xiii pls.

Frölich, J.A. von. 1792: Bemerkungen über einige seltene Käfer aus der Insectensammlung des Herrn Hofr. und Prof. Rudolph in Erlangen. *Naturforscher* **26**: 68–165.

Fuente, J.M. de la. 1907: Datos para la fauna de la provincia de Ciudad Real. XIX. Coleópteros. *Boletín de la Real Sociedad Española de Historia Natural* **7**: 317–323.

Fuente, J.M. de la. 1909: Datos para la fauna de la provincia de Cuiudad Real. *Boletín de la Real Sociedad Española de Historia Natural* **9**: 306–310.

Fuente, J.M. de la. 1910: Datos para la fauna de la provincia de Cuiudad Real. *Boletín de la Real Sociedad Española de Historia Natural* **10**: 442–446.

Fuente, J.M. de la. 1913: Descripción de algunos Coleópteros nuevos de España. *Boletín de la Real Sociedad Española de Historia Natural* **13**: 475–476.

Fuessly, G. 1775: *Verzeichniss der ihm bekannten Schweizerischen Inseckten mit einer ausgemahlten Kupfertafel nebst der Ankündigung eines neuen Insecten Werks.* Zürich und Winterthur: Heinrich Steiner und Compagnie, xii + 62 pp., 1 pl.

Fursov, N.I. 1935: Drei neue kaukasische Arten aus Familie Cucujidae nützlich in der Forstwirtschaft. *Bulletin de la Société des Naturalistes de Moscou, Section Biologique* (N.S.) **44**: 90–91.

Fursov, N.I. 1939: Eine neue Gattung und 6 neue Coleopterenarten von Mittelasien und Caucasus. *Bulletin de la Société des Naturalistes de Moscou, Section Biologique* (N.S.) **48**: 88–90.

Gabriel, J.J.C. 1916: Eine neue Anaspis-Art. *Entomologische Mitteilungen* **5**: 183–184.

Gahan, C.J. 1896a: On the Coleoptera obtained by Dr. Anderson's collector during Mr. T. Bent's expedition to the Hadramaut, South Arabia. *Journal of the Linnean Society* **25**: 285–291.

Gahan, C.J. 1896b: On Coleoptera from Aden and Somaliland. *The Annals and Magazine of Natural History* (6) **18**: 448–461.

Gahan, C.J. 1900: Description of new genera and species of Coleoptera collected by Mr. Ogilvie-Grant and Dr. Forbes in the Islands of Sokotra and Abd-el-Kuri. *Bulletin of the Liverpool Museum* 3: 8–13.

Gahan, C.J. 1903: Coleoptera. In: Forbes H.O.: Natural history of Sokotra and Abd-el-Kuri: Being the report upon the results of the conjoint expedition to these islands in 1898–9, by Mr. W.R. Ogilvie-Grant, of the British Museum, and Dr. H.O. Forbes, of the Liverpool Museums, together with information from other available sources. Forming a monogragh of the islands. *Bulletin of the Liverpool Museums* **1903**: 263–292, pl. 17.

Ganglbauer, L. 1881: Bestimmungs-Tabellen der europäischen Coleopteren. IVa. Oedemeridae. *Verhandlungen der Kaiserlich-Königlichen Zoologisch-Botanischen Gesellschaft in Wien* **31**: 97–116.

Ganglbauer, L. 1882: Coleopterologische Mittheilungen. *Wiener Entomologische Zeitung* **1**: 135–140.

Ganglbauer, L. 1889: Insecta a Cl.G.N. Potanin in China et in Mongolia novissime lecta. VII. Buprestidae, Oedemeridae, Cerambycidae. *Horae Societatis Entomologicae Rossicae* **24** [1889–1890]: 21–85.

Ganglbauer, L. 1891: Eine neue Art der Melandryiden-Gatung Zilora. *Wiener Entomologische Zeitung* **10**: 132–133.

Ganglbauer, L. 1897: Einige neue Coleopteren des mitteleuropäischen Faunengebietes. *Verhandlungen der Kaiserlich-Königlichen Zoologisch-Botanischen Gesellschaft in Wien* **47**: 565–573.

Ganglbauer, L. 1899: *Die Käfer von Mitteleuropa. Die Käfer der österreichisch-ungarischen Monarchie, Deutschlands, der Schweiz, sowie des französischen und italienischen Alpengebietes. Dritter Band. Familienreihe Staphylinoidea. II. Theil. Scydmaenidae, Silphidae, Clambidae, Leptinidae, Platypsyllidae, Corylophidae, Sphaeriidae, Trichopterygidae, Hydroscaphidae, Scaphidiidae, Histeridae. Familienreihe Clavicornia. Sphaeritidae, Ostomidae, Byturidae, Nitidulidae, Cucujidae, Erotylidae, Phalacridae, Thorictidae, Lathridiidae, Mycetophagidae, Colydiidae, Endomychidae, Coccinellidae*. Wien: Carl Gerold's Sohn, III +1046 pp.

Ganglbauer, L. 1900: [new taxa]. In: Bodemeyer E. von: *Quer durch Klein-Asien in den Bulghar-Dagh. Eine naturwissenschatliche Studien-Reise*. Emmendingen: Verlangs-Aktiengesellschaft vormals Dölter, v + 169 pp.

Ganglbauer, L. 1905: Coleoptera. Ergebnisse einer naturwissenschaftlichen Reise zum Erdschias-Dagh (Kleinasien). Ausgeführt von Dr. Arnold Penther und Dr. Emerich Zederbauer auf Kosten der 'Gesellschaft zur Förderung des naturhistorischen Erforschung des Orients in Wien im Jahre 1902'. *Annalen des K.K. Naturhistorischen Hofmuseums* **20**: 246–290.

Gao, Z.-H. & Ren, G.-D. 2009: Taxonomy of the genus Morphostenophanes Pic from China, with two new species (Coleoptera, Tenebrionidae). *Acta Zoologica Academiae Scientiarum Hungaricae* **55**: 307–319.

Garciá, R. & Oromí, P. 2001: Melansis reyesi n. sp., un nuevo Litoborini de las islas Canarias (Coleoptera, Tenebrionidae). *Vieraea* **28**: 51–59.

Gardini, G. 1975: Materiali per lo studio dei Tenebrionidi dell'arcipelago Toscano (Col. Heteromera). *Lavori della Società Italiana di Biogeografia* (Nuova Seria) **5** [1974]: 1–87.

Gardini, G. 1979: Nuovi dati su Tenebrionidi (Col.) dell'Archipelago Toscano. *Bollettino di Museo Civico di Storia Naturale Verona* **6**: 73–77.

Gardini, G. 2010: *Boromorphus italicus* n.sp. dell'Italia meridionale (Coleoptera, Tenebrionidae). *Doriana* 8: 1–12.

Gaubil, J. 1849: *Catalogue synonymique des coléoptères d'Europe et d'Algérie*. Paris: Maison, 296 pp.

Gebien, H. 1906a: Ueber die von Fabricius beschriebenen Typen von Tenebrioniden in den Museen von Kopenhagen und Kiel. *Deutsche Entomologische Zeitschrift* **1906**: 209–237.

Gebien, H. 1906b: Notiz über Asida convexicollis. *Wiener Entomologische Zeitung* **25**: 248.

Gebien, H. 1910a: Tenebrionidae I, II. Pars 15, 22. In: Schenkling, S. (ed.): *Coleopterorum Catalogus. Vol. 18.* Berlin: W. Junk, 1–354 pp.

Gebien, H. 1910b: 7. Coleoptera. 19. Tenebrionidae. Pp. 363–397. In: Sjöstedt Y. (ed.): *Wissenschaftliche Ergebnisse der Schwedischen Zoologischen Expedition nach dem Kilimandjaro, dem Meru und den umgebenden Massaisteppen Deutsch-Ostafrikas 1905–1906 unter der Leitung von Prof. Dr. Yngve Sjörstedt. 1. Band.* Stockholm: Königl. Schwedisches Akademie der Wissenschaften, 442 pp.

Gebien, H. 1911: Tenebrionidae III, IV. Pars 28, 37. In: Schenkling, S. (ed.): *Coleopterorum Catalogus. Vol. 18.* Berlin: W. Junk, 355–740 pp.

Gebien, H. 1913: Neue Tenebrioniden der palaearktischen Fauna (Col. Heter.) nebst einer Übersicht über die Gattung Iphthimus. *Coleopterologische Rundschau* **2**: 1–8.

Gebien, H. 1914a: Die Tenebrionidenfauna Borneos. Erster Teil. *Sarawak Museum Journal* **2** (5): 1–58, pl. 1.

Gebien, H. 1914b: H. Sauter's Formosa-Ausbeute. Tenebrionidae (Coleopt.). *Archiv für Naturgeschichte* A **79** (9) [1913]: 1–60, 1 pl.

Gebien, H. 1914c: Fauna Simularensis. Coleoptera, Fam. Tenebrionidae. *Notes from the Leyden Museum* **36**: 61–80.

Gebien, H. 1914d: Die Tenebrioniden der Philippinen. *The Philippine Journal of Science* 8 (D) [1913]: 373–433.

Gebien, H. 1918: Beitrag zur Kenntnis der Gattung Setenis (Col., Heterom.). *Entomologische Mitteilungen* **7**: 121–130, pl. 2.

Gebien, H. 1919a: Beitrag zur Kenntnis der Gattung Setenis (Col., Heterom.). *Entomologische Mitteilungen* **8**: 1–14.

Gebien, H. 1919b: Monographie der südamerikanischen Camarien (Coleopt. Heterom.) nebst einer Übersicht über die indischen Gattungen der Camariinen. *Archiv für Naturgeschichte* A **83** (3) [1917]: 25–167, 2 pls.

Gebien, H. 1920: *Käfer aus der Familie Tenebrionidae gesammelt auf der 'Hamburger deutsch-südwestafrikanischen Studienreise 1911'. Hamburgische Universität, Abhandlungen aus der Auslandskunde. Band 5. Reihe C Naturwissenschaften Band 2.* Hamburg: L. Friederichsen & Co., 168 pp., 2 pls.

Gebien, H. 1920: Coleoptera Tenebrionidae. Pp. 213–500, pls IX–XI. In: *Nova Guinea. Résultats de l'Expédition scientifique néerlandaise la Nouvelle-Guinée.* Vol. XIII (3), Zoologie. Leide: E.J. Brill.

Gebien, H. 1921: Die Tenebrioniden Westafrikas. *Archiv für Naturgeschichte* A **86** (6) [1920]: 1–256.

Gebien, H. 1922a: The Percy Sladen trust Expedition to the Indian Ocean in 1905. No. V. – Coleoptera, Heteromera: Tenebrionidae. *Transactions of the Linnean Society of London* (Series 2, Zoology) **18** [1922–1925]: 261–323, pl. 23.

Gebien, H. 1922b: Philippine Tenebrionidae, II. *The Philippine Journal of Science* **19** [1921]: 439–515, 2 pls.

Gebien, H. 1925a: Die Tenebrioniden (Coleoptera) des Indomalayischen Gebietes, unter Beruecksichtigung der benachbarten Faunen. IV. Die Gattungen Phloeopsidus, Dysantes, Basanus, und Diaperis. *The Philippine Journal of Science* 27 (1): 131–157, 257–288, 1 pl.

Gebien, H. 1925b: Die Tenebrioniden (Coleoptera) des Indomalayischen Gebietes, unter Beruecksichtigung der benachbarten Faunen. VI. Die Gattung Platydema Castenau und Brullé. *The Philippine Journal of Science* 27 (3): 423–453, 2 pls.

Gebien, H. 1925c: Die Tenebrioniden (Coleoptera) des Indomalayischen Gebietes, unter Beruecksichtigung der benachbarten Faunen. VII. Die Gattungen Ischnodactylus, Hoplocephala, und Martianus. *The Philippine Journal of Science* 27 (4): 539–595, 1 pl.

Gebien, H. 1925d: Die Tenebrioniden (Coleoptera) des Indomalayischen Gebietes, unter Beruecksichtigung der benachbarten Faunen. viii. *The Philippine Journal of Science* 28: 101–128, 1 pl.

Gebien, H. 1927: Fauna Sumatrensis (Beitrag Nr. 31) Tenebrionidae (Col.). *Supplementa Entomologica* 15: 22–58.

Gebien, H. 1935: Tenebrionidae. Pp. 53–77, 1 pl. In: Van Stralen, V. (ed.): Résultats Scientifiques du Voyage aux Indes Orientales Néerlandaises de LL. AA. RR. le Prince et la Princesse Léopold de Belgique. *Mémoires du Musée Royal d'Histoire Naturelle de Belgique. Hors Série*. Volume IV. Invertébrés: Hexapodes.

Gebien, H. 1937: Katalog der Tenebrioniden (Col. Heteromera) Teil I. *Pubblicazioni del Museo Entomologico 'Pietro Rossi'* 2: 505–883.

Gebien, H. 1938: [new taxa]. In: Schuster A: & Gebien H.: Tenebrioniden (Col.) aus Arabien. *Entomologische Blätter* 34: 49–62.

Gebien, H. 1939: Katalog der Tenebrioniden. *Mitteilungen der Münchener Entomologischen Gesellschaft* 29: 443–474.

Gebien, H. 1940a: Die von Dr. Trinkler in Tibet gesammelten Tenebrioniden (Col. Tenebr.). *Mitteilungen der Münchener Entomologischen Gesellschaft* 30: 1–3.

Gebien, H. 1940b: Katalog der Tenebrioniden, Teil II. *Mitteilungen der Münchener Entomologischen Gesellschaft* 30: 755–786.

Gebien, H. 1943: Katalog der Tenebrioniden, Teil III. *Mitteilungen der Münchener Entomologischen Gesellschaft* 33: 895–926.

Gebler, F.A. von. 1817: Insecta Siberia rariora, decas prima. *Mémoires de la Société Impériale des Naturalistes de Moscou* 5: 315–333.

Gebler, F.A. von. 1825: Coleopterorum Sibiriae species novae. Pp. 42–57. In: Hummel, A.D.: *Essais Entomologiques, Insectes de 1824. Novae species. Vol. 1, Nr. 4.* St. Petersbourg: Chancellerie privée du Ministère de l'Intérieur, 71 pp. + index.

Gebler, F.A. von. 1829a: Bemerkungen über die Insekten Sibiriens, vorzüglich des Altai. [Part 3]. Pp. 1–228. In: Lederbour C.F. von. (ed.): *Reise durch das Altai-Gebirge und die soongorische Kirgisen-Steppe. Auf Kosten der Kaiserlichen Universität Dorpat unternommen im Jahre 1826 in Begleitung der Herren D. Carl Anton Miecherund D. Alexander von Bunge K.K. Collegien Assessors. Zweiter Theil.* Berlin: G. Reimer, 427 pp.

Gebler, F.A. von. 1829b: Des mylabrides de la Sibérie occidentale et des confins de la Tartarie. *Nouvelles Mémoires de la Société Impériale des Naturalistes de Moscou* 1: 145–171.

Gebler, F.A. von. 1830: Bemerkungen über die Insekten Sibiriens, vorzüglich des Altai. [Part 3]. Pp. 1–228. In: Lederbour, C.F. von. (ed.): *Reise durch das Altai-Gebirge und die soongorische Kirgisen-Steppe. Auf Kosten der Kaiserlichen Universität Dorpat unternommen im Jahre 1826 in Begleitung der Herren D. Carl Anton Miecherund D. Alexander von Bunge K.K. Collegien Assessors. Zweiter Theil.* Berlin: G. Reimer, 427 pp. [Note: this seems to be a second edition or reprint of the 1829a work, dated from 1830].

Gebler, F.A. von. 1832: Notice sur les coléoptères qui se trouvent dans le district des mines de Nertschinsk, dans la Sibérie orientale, avec la description de quelques espèces nouvelles. *Nouveaux Mémoires de la Société des Naturalistes de Moscou* **2** (8): 23–78.

Gebler, F.A. von. 1833: Notae et addidamenta ad catalogum Coleopterorum Sibiriae occidentalis et confinis Tatariae operis, C.F. von Ledebours Reise in das Altaigebirge und die soongarische Kirgisensteppe (Zwyter Theil. Berlin 1830). *Bulletin de la Société Impériale des Naturalistes de Moscou* **6**: 262–309.

Gebler, F.A. von. 1836: Lettre de M. le Conseiller d'Etat et Chevalier de Gebler, contenant un rapport d'un voyage dans les hautes montagnes catouniennes jusqu'à la frontière de la Chine; et description de trois nouvelles espèces de coléoptères. *Bulletin de la Société Impériale des Naturalistes de Moscou* **9**: 329–345.

Gebler, F.A. von. 1841a: Characteristik mehrerer neuer sibirischer Coleopteren. *Bulletin Scientifique de l'Académie Impériale des Sciences de Staint-Pétersbourg* **8** (24): 369–376 [columns].

Gebler, F.A. von. 1841b: Notae et addimenta ad catalogum Coleopterorum Sibiriae occidentalis et confinis Tatariae operis: C.F. Ledebour's Reise in das Altaigebirge und die soongarische Kirgisensteppe. Zweyter Theil. Berlin 1830. Fasciculus secundus. *Bulletin de la Société Impériale des Naturalistes de Moscou* **14** (4): 577–625.

Gebler, F.A. von. 1843: Charakteristik der vom Hn. Dr. Schrenk im Jahre 1841 in den Steppen und Gebirgen der Songarei gefundenen neuen Coleopteren-Arten. *Bulletin de la Classe Physico-Mathématique de l'Académie Impériale des Sciences de St.-Pétersbourg* **1**: 36–42 [columns].

Gebler, F.A. von. 1844: Charakteristik der von Hn. Dr. Schrenk in den Jahren 1842 und 1843 in den Steppen der Dsungarei gefundenen neuen Coleopterenarten. *Bulletin de la Classe Physico-Mathématique de l'Académie Impériale des Sciences de St.-Pétersbourg* (2) **3** [1844–1845]: 97–106 [columns].

Gebler, F.A. von. 1847: Verzeichniss der im Kolywano-Woskresenskischen Hüttenbezirke Süd-West Sibiriens beobachteten Kaefer mit Bemerkungen und Beschreibungen. *Bulletin de la Société Impériale des Naturalistes de Moscou* **20** (3): 391–512.

Gebler, F.A. von. 1859: Verzeichniss der von Herrn Dr. Schrenk in den Kreisen Ajagus und Karakaly in der östlichen Kirkisiensteppe und in der Songarey in den Jahren 1840 bis 1843 gefundenen Kaeferarten. *Bulletin de la Société Impériale des Naturalistes de Moscou* **32** (1): 426–519, (4): 315–356.

Gebler, F.A. von. 1860: Coleopterorum species novae, a Dr. Schrenk in deserto kirgiso-songorico anno 1843 detectae. *Bulletin de la Société Impériale des Naturalistes de Moscou* **33** (2): 1–39.

Gemminger, M. 1851: *Systematische Uebersicht der Käfer um München.* Jena: Mauke, 65 pp., 1 pl.

Gemminger, M. 1870a: [new names]. In: Harold E. von.: Geänderte Namen. *Coleopterologische Hefte* **6**: 119–124.

Gemminger, M. 1870b: [new names]. In: Gemminger, M. & Harold, E. von.: *Catalogus coleopterorum hucusque descriptorum synonymicus et systematicus. Tom. VII. Tenebrionidae, Nilionidae, Pythidae, Melandryidae, Lagriidae, Pedilidae, Anthicidae, Pyrochroidae, Mordellidae, Rhipidophoridae, Cantharidae, Oedemeridae* Monachii: E.H. Gummi, pp. 1801–2180 + [9].

Gené, J. 1836a: De quisbusdam insectis Sardiniae novis aut minus cognitis. Fasciculus I. *Memoralia Regiae Scientiarum Academiae Taurinensis* **39**: 161–199, 1 pl.

Gené, J. 1836b: [Dix-sept coléoptères nouveaux de Sardeigne]. *Bulletin de la Société Entomologique de France* **1836**: ii-iv.

Gené, J. 1839: De quisbusdam insectis Sardiniae novis aut minus cognitis. Fasciculus II. *Memoralia Regiae Scientiarum Academiae Taurinensis* (2) **1**: 43–84, 2 pls.

Geoffroy, E.L. 1762: *Histoire abrégée des insectes qui se trouvent aux environs de Paris; dans laquelle ces animaux sout rangés suivant un ordre méthodique.* [1762–1763] *Tome première.* Paris: Durand, xxviii + 523 pp. + 22 pls.

Geoffroy, E.L. 1785: [new taxa]. In: Fourcroy, A.F. de.: *Entomologia parisiensis; sive Catalogus Insectorum quae in Agro Parisiensi reperiuntur; secundum methodum Geoffraeanam in sectiones, genera & species distributus: cui addita sunt nomina trivalia & fere trecentae novae species. Vol. 1.* Parisiis: Serpentineis, vii + 231 pp.

Geoffroy, E.L. 1799: *Histoire abrégée des insectes, dans laquelle ces animaux sont rangés suivant un ordre méthodique. Nouvelle édition, revue, corrigée et augmentée d'un supplément considérable. Vol. 1.* Paris: Delalain Fils, xxviii + 556 pp., 10 pls.

Gerhardt, J. 1876: Anaspis (Silaria) palpalis n. sp. *Deutsche Entomologische Zeitschrift* **1876**: 381–383.

Gerhardt, J. 1901: Eine neue Käferart. *Zeitschrift für Entomologie* (N.F.) (Breslau) **26**: 18–21.

Gerhardt, J. 1904: Eine verkannte deutsche Käferart. *Zeitschrift für Entomologie* (Breslau) **29**: 79–82.

Germar, E.F. 1813: Neue Insekten. *Magazin der Entomologie* **1** (1): 114–133.

Germar, E.F. 1814: [new taxon]. In: Ahrens, A.: *Fauna Insectorum Europae. Fasciculus 2.* Halae: Car. Aug. Kümmelii, [25] pp., 25 pls.

Germar, E.F. 1817: *Reise nach Dalmatien und in das Gebiet von Ragusa. Band 2.* Leipzig und Altenburg: F.A. Brockhaus, xii + 323 pp., 8 pls, 2 maps.

Germar, E.F. 1818: Vermischte Bemerkungen über einige Käferarten. *Magazin der Entomologie* **3**: 223–260, pl. 3.

Germar, E.F. 1821: *Fauna Insectorum Europae. Fasciculus 4.* Halae: Car. Aug. Kümmelii, [25] pp., 25 pls.

Germar, E.F. 1823: *Species insectorum novae aut minus cognitae, descriptionibus illustratae. Volumen Primum. Coleoptera.* Halae: J.C. Hendelii et filii, xxiv + 624 pp., 2 pls.

Germar, E.F. 1824: *Fauna Insectorum Europae. Fasciculus 10.* Halae: Car. Aug. Kümmelii, [25] pp., 25 pls.

Germar, E.F. 1827: *Fauna Insectorum Europae. Fasciculus 13.* Halae: Car. Aug. Kümmelii, [25] pp., 25 pls.

Germar, E.F. 1831: *Fauna Insectorum Europae. Fasciculus 14.* Halae: Car. Aug. Kümmelii, [25] pp., 25 pls.

Germar, E.F. 1832: *Fauna Insectorum Europae. Fasciculus 15.* Halae: Car. Aug. Kümmelii, [25] pp., 25 pls.

Germar, E.F. 1836a: *Fauna Insectorum Europae. Fasciculus 16.* Halae: Car. Aug. Kümmelii, [25] pp., 25 pls.

Germar, E.F. 1836b: *Fauna Insectorum Europae. Fasciculus 17.* Halae: Car. Aug. Kümmelii, [25] pp., 25 pls.

Germar, E.F. 1836c: *Fauna Insectorum Europae. Fasciculus 18.* Halae: Car. Aug. Kümmelii, [25] pp., 25 pls.

Germar, E.F. 1842: *Fauna Insectorum Europae. Fasciculus 22.* Halae: Car. Aug. Kümmelii, [25] pp., 25 pls.

Germar, E.F. 1847: *Fauna Insectorum Europae. Fasciculus 24.* Halae: [2 + 25] pp., 25 pls.

Gerstaecker, C.E.A. 1854: Diagnosen der von Peters in Mossambique gesammelten und von Dr. Gerstaecker bearbeiteten Kaefer aus der Familien der Vesicantia. *Monatsberichte der Königlichen Preussischen Akademie der Wissenschaften zu Berlin* **1854**: 694–695.

Gerstaecker, C.E.A. 1855a: *Rhipiphoridum Coleopterorum familiae dispositio systematica. Dissertatio Inauguralis Zoologica quam consensu et auctoritate amplissimi philosophorum ordinis in alma litterarum universitate Friderica Guilelma pro summis in philosophia honoribus rite capessendis die IV. M. Julii a MDCCCLV. Hora XII. Publice defendet auctor A. Gerstaecker, med. Dr. Berolinensis.* Berolini: Schlesinger, 36 pp.

Gerstaecker, C.E.A. 1855b: *Rhipiphoridum Coleopterorum familiae dispositio systematica.* Berolini: F. Nicolai, 36 pp., 1 pl.

Gerstaecker, C.E.A. 1871: Beitrag zur Insektenfauna von Zanzibar, nach dem während der Expedition des Baron v. d. Decken gesammelten Material zusammengestellt. *Archiv für Naturgeschichte* 37 (1): 42–86.

Gerstaecker, C.E.A. 1884: Bestimmung der von Herrn Dr. G.A. Fischer während seiner Reise nach dem Massai-Land gesammelten Coleopteren. *Jahrbuch der Hamburgischen Wissenschaflichen Anstalten* 1: 41–63.

Gestro, R. 1874: Diagnose d'un Coléoptère nouveau de la famille des Colydiens. *Petites Nouvelles Entomologiques* 1: 443.

Gestro, R. 1880: Appunti sull'entomologia tunisina. *Annali del Museo Civico di Storia Naturale di Genova* **15** [1879–1880]: 405–424.

Gestro, R. 1881: Aliquot Coleopterorum Musei Civici Januensis diagnoses. *Annali del Museo Civico di Storia Naturale di Genova* **16** [1880–1881]: 658–664.

Gestro, R. 1892: Di alcuni coleotteri raccolti nel paese dei Somali dall'Ing. L. Bricchetti Robecchi. *Annali di Museo Civico di Storia Naturale di Genova* **32**: 747–790.

Gestro, R. 1895: Esplorazione del Giuba e dei suoi affluenti compiuta dal cap. V. Bottego durante gli anni 1892–93 sotto gli auspicii della Società geografica italiana. XVI. Coleotteri. *Annali di Museo Civico di Storia Naturale di Genova* **35**: 247–478.

Gillerfors, G. 1985: Two new species of the genus Tarphius Erichson from the Azores and redescription of Tarphius wollastoni Crotch (Coleoptera: Colydiidae). *Bocagiana* **85**: 1–10.

Gillerfors, G. 1986a: Two new species of Tarphius Erichson (Coleoptera, Colydiidae) and a new species of Mniophlosoma Wollaston (Coleoptera, Chrysomelidae) from the Azores. *Bocagiana* **98**: 1–10.

Gillerfors, G. 1986b: Two new species of the genus Tarphius Erichson from La Gomera (Canary Islands) and desigantion of lectotypes of Tarphius humerosus Wollaston and T. gomerae Franz (Coleoptera, Colydiidae). *Vieraea* **16**: 3–10.

Gillerfors, G. 1990: Tarphius palmensis sp. n., a new species of the genus Tarphius Erichson from La Palma, Canary Islands (Col., Colydiidae). *Vieraea* **18**: 149–151.

Gillerfors, G. 1991: Three new species of Coleoptera from the Canary Islands. *Vieraea* **20**: 21–26.

Gillerfors, G. 1997: A new Tarphius (Coleoptera, Colydiidae) from Madeira Island. *Bocagiana* **187**: 1–4.

Gillerfors, G. & Oromi, P. 1991: Tarphius stagosus sp. n., a new species from La Gomera, Canary Islands (Coleoptera, Colydiidae). *Nouvelle Revue d'Entomologie* (N.S.) **8**: 85–87.

Giorna, M.S. 1804: Mémoire d'entomologie. *Mémoires de l'Académie des Sciences, Littérature et Beaux-Arts de Turin. Sciences Physiques et Mathématiques* 1: 215–223, pls. 1, 2.

Girard, C. 1965: Contribution à l'étude du genre Tentyria Latreille (Col. Tenebrionidae). *Bulletin de la Société Entomologique de France* **70**: 138–139.

Girard, C. 1966a: Contribution à l'étude du genre Tentyria Latreille (Col. Tenebrionidae). (2e note). *Bulletin de la Société Entomologique de France* **71**: 84–86.

Girard, C. 1966b: Note sur quelques Tentyriini du Moyen-Orient. *Bulletin de la Société Entomologique de France* 71: 203–206.

Girard, C. 1968: Contribution à la faune de l'Iran 9. Coléoptères Tenebrionidae, genre Tentyria. *Annales de la Société Entomologique de France* (N.S.) 4: 867–874.

Girard, C. & Pierre, F. 1965: Sur quelques Tentyriini du Sudan (Republic of Sudan) (Col. Tenebrionidae). *Bulletin de la Société Entomologique de France* 70: 134–137.

Gistel, J.N.F.X. 1831: Entomologische Fragmente. *Isis von Oken* 1831: 301–310.

Gistel, J.N.F.X. 1834a: *Die Insecten-Doubletten aus der Sammlung des Herrn Grafen Rudolph von Jenison Walworth zu Regensburg, welche sowohl im Kauf als im Tausche abgegeben werden. Nro. I. Käfer*. München: Jaquet, 36 pp.

Gistel, J.N.F.X. 1834b: XIV. Entomologische Notizen. *Faunus* 3: 128–151.

Gistel, J.N.F.X. 1848: *Naturgeschichte des Thierreichs, für höhere Schulen*. Stuttgart: R. Hoffmann'sche Verlags-Buchhandlung, xvi + 216 + [4] pp., 32 pls.

Gistel, J.N.F.X. 1856: *Die Mysterien der europäischen Insectenwelt. Ein geheimer Schlüssel für Sammler aller Insecten-Ordnungen und Stände, behufs des Fangs, des Aufenthalts-Orts, der Wohnung, Tag- und Jahreszeit u. s. w., oder autoptische Darstellung des Insectenstaats in seinem zusammenhange zum Bestehen des Naturhaushaltes überhaupt und insbesondere in seinem Einflusse auf die phanerogamische und cryptogamische Pflanzenbevölkerung Europa's. Zum ersten Male nach fünfundzwanzigjährigen eigenen Erfahrungen zusammengestellt und herausgegeben*. Kempten: Tobias Dannheimer. xii + 532 pp.

Gistel, J.N.F.X. 1857 *Achthundert und zwanzig neue oder unbeschriebene wirbellose Thiere*. Straubing: Verlag der Schorner'schen Buchhandlung, 92 pp. [note: separate issue from Vacuna].

Gmelin, J.F. 1790: *Systema Naturae per regna tria naturae, secundum classes, ordines, genera, species, cum characteribus, differentiis, synonymis, locis. Editio decima tertia, aucta, reformata. ToM.I. Pars IV. Classis V. Insecta*. Lipsiae: Georg Emanuel Beer, 1517–2224 pp.

Goeze, J.A.F. 1777: *Entomologische Bayträge zu des Ritter Linné zwölften Ausgabe des Natursystems. Erster Theil*. Leipzig: Weidemanns Erben und Reich, xvi +736 pp.

Gorham, H.S. 1873: Descriptions of a new genus and two new species of Coleoptera from Japan. *The Entomologist's Monthly Magazine* 9: 257–258.

Górriz y Muñoz, R.J. 1882: *Ensayo para la monografía de los coléopteros melóidos indígenas con aplicación a las ciencias médicas*. Zaragoza: Sanz y Navarro, 201 pp., 2 pls.

Gory, H.L. 1841: [new taxa, no. 71–74]. *Magasin de Zoologie* (2) 3: 9 pp., 4 pls.

Gozis, M. des. 1881: [Quelques rectifications synonymiques touchant différents genres et espèces de coléoptères français (1re partie)]. *Bulletin de la Société Entomologique de France* 1881: cxii-cxiii.

Gozis, M. des. 1882: Notes et remarques pour le futur catalogue des coleoptères gallo-rhénane. *Revue d'Entomologie* 1: 193–207.

Gozis, M. des. 1886: *Recherche de l'espèce typique de quelques anciens genres. Rectifications synonymiques et notes diverses*. Montluçon: Herbin, 36 pp

Gradl, J. 1882: Aus der Fauna des Egerlandes. *Entomologische Nachrichten* 8: 323–327.Graells, M. de la Paz 1849: [new taxa]. *Revue de Zoologie Pure et Appliquée* (2) 1: 620–622.Graells, M. de la Paz. 1851: Description de quelques coléoptères nouveaux de la faune centrale d'Espagne. *Annales de la Société Entomologique de France* (2) 9: 5–27, 1 pl.

Graells, M. de la Paz. 1858: Insectos nuevos de España. *Memorias de la Comision del Mapa Geológico de España, año de 1855. Parte zoológica.* Madrid: Imprenta Nacional, 111 + [1] pp., 7 pls.

Gravely, F.H. 1915: XLII. Coleoptera, IX: Tenebrionidae. In: Zoological results of the Abor Expedition, 1911–12. XLII. Coleoptera IX: Tenebrionidae. *Records of the Indian Museum* 8 [1912–1922]: 519–536, pls XLIII, XLIV.

Gray, G. 1832: *The class insecta arranged by the Baron Cuvier, with supplementary additions to each order by Edward Griffith, F.L.S., A.S., &c. and Edward Pidgeon, Esq. and notices of new genera and species by George Gray, Esq.* London: Whittaker, Treacher, and Co., 570 pp.

Gredler, I.V.M. 1854: Die Käfer von Passier. *Zeitschrift des Ferdinandeums für Tirol und Vorarlberg* **3**: 1–20. [issued also as *Die Käfer von Passier. 1.Heft.* Innsbruck: Wagner'sche Buchdruckerei, 20 pp.].

Gredler, I.V.M. 1866: *Die Käfer von Tirol nach ihrer horizontalen und vertikalen Verbreitung. ii. Hälfte: Dascillidae – Schluss.* Bozen: Druk u. Verlag von G. Ferrari, vormals Eberle, 235–491 pp.

Gredler, I.V.M. 1878: Zur Käfer-Fauna Central-Afrikas. *Verhandlungen der Kaiserlich-Königlichen Zoologisch-Botanischen Gesellschaft in Wien* **27**: 501–522.

Grenier, A. 1863: *Catalogue des coléoptères de France et matériaux pour servir à la faune des coléoptères français, par MM. E. Allard, Dr. Ch. Aubé, Ch. Brisout de Barneville, A. Chevrolat, L. Fairmaire, Al. Fauvel, Dr. A. Grenier, Dr. Kraatz, J. Linder, L. Reiche et Félicien de Saulcy.* Paris: L. Toinon, iv + 3–79 + 135 pp.

Gressitt, J.L. 1939: Some Oedemeridae from South China, Hainan I., Formosa and the Riukiu Islands (Coleoptera). *Lingnan Science Journal* **18**: 217–231.

Gressitt, J.L. 1941: Rhipiphoridae from south China (Coleoptera). *Annals of the Entomological Society of America* **34**: 525–536.

Gridelli, E. 1928: Appunti su alcuni Tenebrionidae dell'Africa settentrionale. *Bollettino della Società Entomologica Italiana* **60**: 15–18.

Gridelli, E. 1929a: Appunti su alcuni Tenebrionidae dell'Africa settentrionale. III. – Revisione del genere Doderoella. *Bollettino della Società Entomologica Italiana* **61**: 1–10.

Gridelli, E. 1929b: Appunti su alcuni Tenebrionidae dell'Africa settentrionale. IV. – Revisione del genere Tentyriina Reitt. *Bollettino della Società Entomologica Italiana* **61**: 49–63.

Gridelli, E. 1930: Risultati zoologici della Missione inviata dalla R. Società Geografica Italiana per l'esplorazione dell' oasi di Giarabub (1926–1927). Coleotteri. *Annali del Museo Civico di Storia Naturale di Genova* **54**: 1–485 + [1], 1 map.

Gridelli, E. 1931: Spedizione des Barone Raimondo Franchetti in Dancalia. Coleotteri – Tenebrionidae. *Annali del Museo Civico di Storia Naturale di Genova* **55**: 220–233.

Gridelli, E. 1933: Spedizione scientifica all'Oasi di Cufra (Marzo-Guglio 1931). Coleotteri. *Annali del Museo Civico di Storia Naturale di Genova* **56**: 155–258, pl.v.

Gridelli, E. 1934: Materiali zoologici raccolti dalla spedizione italiana al Karakoram (1929 – Anno VII). Coleoptera – Tenebrionidae. *Atti del Museo Civico di Storia Naturale Trieste* **12**: 37–68, pls. IX, X.

Gridelli, E. 1938: Coleotteri dell'Africa orientale italiana. 5° Contributo. Specie a me note appartenenti ai generi Mesostena e Mesostenopa auct. (Tenebrionidae). *Bollettino della Società Entomologica Italiana* **70**: 103–121.

Gridelli, E. 1939a: Coleotteri dell'Africa orientale italiana. 6e Contributo. *Bollettino della Società Entomologica Italiana* **71**: 75–79.

Gridelli, E. 1939b: Coleotteri dell'Africa orientale italiana 8. contributo. Revisione delle species riferite dagli autori ai generi Helopinus, Drosochrus, Emyon. (Coleopt. Tenebrionidae). *Atti del Museo Civico di Storia Naturale Trieste* **14**: 187–286.

Gridelli, E. 1939c: Coleotteri dell'Africa orientale italiana 10. contributo. Revisione delle species del genere Leichenum Blch. (Coleopt. Tenebrionidae). *Atti del Museo Civico di Storia Naturale Trieste* **14**: 207–242.

Gridelli, E. 1939d: Coleotteri del Fezzan e dei Tassili d'Agger (Missione Scortecci 1936). *Atti della Società Italiana di Scienze Naturali* **78**: 385–456, pl. XXIII.

Gridelli, E. 1940: Coleotteri dell'Africa orientale italiana 11° Contributo. Materiali per lo studio della fauna Eritrea raccolti nel 1901–03 dal Dott. Alfredo Andreini. *Memorie della Società Entomologica Italiana* 18 [1939]: 219–258.

Gridelli, E. 1945: Coleotteri dell'Africa orientale (XIV. contributo). Appunti per una monografia delle specie etiopiche del genere Gonocephalum Sol. (Coleopt. Tenebr.). *Atti del Museo Civico di Storia Naturale Trieste* **16**: 1–28.

Gridelli, E. 1948: Coleotteri dell'Africa tropicale XVII contributo. Ulteriori appunti per una monografia delle specie africaine del genere Gonocephalum Sol. (Coleopt. Tenebr.). *Atti del Museo Civico di Storia Naturale Trieste* **17**: 1–56.

Gridelli, E. 1949: [new taxon]. In: Porta, A.: *Fauna Coleopterorum italica. Supplementum ii.* Sanremo: Soc. An. G. Gandolfi, 386 pp.

Gridelli, E. 1950: Il problema della specie a diffusione transadritica, con particolare riguardo ai coleotteri. *Memorie di Biogeografia Adriatica* (Venezia) **1**: 1–233.

Gridelli, E. 1952: Contribution à l'étude de peuplement de la Mauritanie. Coléoptères ténébrionides. *Bulletin de l'Institut Français d'Afrique Noire* **14**: 60–96.

Gridelli, E. 1953: Catalogo regionato delle specie di coleotteri tenebrionidi dell'Arabia. *Atti del Museo Civico di Storia Naturale Trieste* **19**: 1–70, pl. 1.

Gridelli, E. 1954: Catalogo regionato delle species della famiglia Tenebrionidae a me nota dell'Afghanistan (Coleoptera). *Atti del Museo Civico di Storia Naturale Trieste* **19**: 169–292.

Gridelli, E. 1955: The 3rd Danish Expedition to Central Asia. Zoological Results 15. Tenebrionidae (Insecta) from Afghanistan. *Videnskabelige Meddelelser fra Dansk Naturhistorisk Forening i København* **117**: 23–74.

Gridelli, E. 1957: Una nuova specie del genera Syachis F. Bat. ed osservazioni su alcune specie note (Coleoptera Tenebrionidae). *Atti del Museo Civico di Storia Naturale Trieste* **21**: 49–61.

Gridelli, E. 1972: Materiali per lo studio dei tenebrionidi italiani. Seconda nota postuma. *Atti del Museo Civico di Storia Naturale Trieste* **27**: 187–256.

Grilat, R. 1886: [Une nouvelle espèce de coléoptère]. *Bulletin de la Société Entomologique de France* **1886**: 187–188.

Grimm, R. 1981: Die Fauna der Ägäis-Insel Santorin. Teil 2 Tenebrionidae (Coleoptera). *Stuttgarter Beiträge zur Naturkunde Serie A* (*Biologie*) **348**: 1–14.

Grimm, R. 1985: Zur Kenntnis der Tenebrioniden aus Süditalien (Insecta: Coleoptera). *Stuttgarter Beiträge zur Naturkunde Serie A* (*Biologie*) **379**: 1–32.

Grimm, R. 1987: Pachychila denticollis sp.n. aus Südspanien (Coleoptera, Tenebrionidae). *Nachrichtenblatt der Bayerischen Entomologen* **36**: 21–24.

Grimm, R. 1991: Tenebrioniden von der Insel Zypern (Insecta: Coleoptera). *Biocosme Mésogéen* **8**: 15–49.

Grimm, R. 2000: Die Gattung Eurycaulus Fairmaire, 1868, in Algerien und Tunesien. *Entomologische Zeitschrift* (Stuttgart) **110**: 89–91.

Grimm, R. 2003: Die Arten der Gattung Palorinus Blair, mit Beschreibungen von vier neuen Arten (Coleoptera: Tenebrionidae). *Stuttgarter Beiträge zur Naturkunde Serie A* (*Biologie*) **646**: 1–9.

Grimm, R. 2005: Taxonomic and faunistic notes on the genus Eurycaulus, with descriptions of two new species from the Arabian Peninsula (Coleoptera: Tenebrionidae). *Stuttgarter Beiträge zur Naturkunde Serie A* (*Biologie*) **672**: 1–11

Grimm, R. 2012: A new species of Ammobius Guérin-Méneville, 1844 from Oman (Coleoptera: Tenebrionidae: Tenebrioninae: Opatrini). *Mitteilungen der Münchener Entomologischen Gesellschaft* **102**: 37–39.

Grimm, R. 2013: A new species of Sclerum Dejean, 1834 from the United Arabic Emirates and Oman (Coleoptera: Tenebrionidae: Tenebrioninae: Opatrini). *Mitteilungen der Münchener Entomologischen Gesellschaft* **103**: 81–84.

Grimm, R. 2017: New taxonomic and faunistic data on the genus Cneocnemis Gebien, 1914 in Laos and Thailand, with description of two new species. *Mitteilungen der Münchner Entomologischen Gesellschaft* **107**: 19–22.

Grimm, R. 2018: Pedinus (Colpotus) skalei n. sp. (Insecta: Coleoptera: Tenebrionidae: Pedinini) from southern Turkey. *Vernate* **37**: 307–308.

Grimm, R. & Lillig, M. 2000: Erodius graniventris in der westlichen Sahara: eine Artengruppe (Coleoptera, Tenebrionidae). *Bulletin de la Société Entomologique de France* **105**: 31–35.

Grimm, R. & Merkl, O. 2018: A new species of *Thraustocolus* Kraatz from Iran (Insecta: Coleoptera: Tenebrionidae: Tentyriini). *Vernate* **37**: 317–319.

Grimm, R. & Schawaller, W. 2000: The genus Colpotus Mulsant & Rey (Coleoptera: Tenebrionidae) in the Eastern Mediterranean region with descriptions of two new species. *Stuttgarter Beiträge zur Naturkunde Serie A* (*Biologie*) **615**: 1–15.

Grouvelle, A. 1903: Clavicornes de l'Inde septentrionale récoltés par M. Harmand. Nitidulidae.-Colydiidae.-Cucujidae.-Monotomidae.-Dryopidae. *Annales de la Société Entomologique de France* **72**: 108–124.

Grouvelle, A. 1905: Nitidulides, Colydiides, Cucujides et Mycetophagides de le Guinée Espagnole. *Memorias de la Real Sociedad Española de Historia Natural* **1**: 241–259.

Grouvelle, A. 1908a: Etude sur les Colbicus vrais de l'Europe, de l'Asie et de l'Australie. *Notes from the Leyden Museum* **30**: 113–122.

Grouvelle, A. 1908b: Coléoptères de la région indienne. Rhysodidae, Trogositidae, Nitidulidae, Colydiidae, Cucujidae (1er mémoire). *Annales de la Société Entomologique de France* **77**: 315–495, pls 6–9.

Grouvelle, A. 1911: Descriptions of five new Indian species of clavicorn Coleoptera. *Records of the Indian Museum* **6**: 315–317.

Grouvelle, A. 1913: Zoological results of the Abor Expedition, 1912–1912. Coleoptera I. Rhysodidae, Nitidulidae, Colydiidae, Cucujidae, Passandridae, Discolomidae, Cryptophagidae, Mycetophagidae, Dryopidae. *Records of the Indian Museum* **8**: 99–117.

Grouvelle, A. 1914: H. Sauters Formosa-Ausbeute. Rhysodidae, Nitidulidae, Ostomidae, Colydiidae, Passandridae, Cucujidae, Cryptophagidae, Diphyllidae, Lathridiidae, Mycetophagidae, Dermestidae. *Archiv für Naturgeschichte* A 79 11 [1913]: 33–76.

Grouvelle, A. 1914: Coleoptera: Cucujidae, Cryptophagidae [of the Seychelles]. *Transactions of the Linnean Society of London (Zoology)* (2) 17: 141–156.

Grouvelle, A. 1918: Coleoptera of the families Ostomidae, Monotomidae, Colydiidae und Notiophygidae from the Seychelles and Aldabra Islands. *Transactions of the Entomological Society of London* 1918: 1–57, pls 1, 2.

Guérin-Méneville, F.E. 1825a: [new taxon]. In: Latreille, P.A., Le Peletier de Saint-Fargeau, A., Audinet-Serville, A.J.G. de. & Guérin-Méneville, F.: *Encyclopédie méthodique ou par ordre de matières; par une société de gens de lettres, de savants et d'artistes; Précédée d'un vocabulaire universel, servant de table pour tout l'ouvrage, ornée des portraits de MM. Diderot & d'Alembert, premiers Éditeurs de l'Encyclopédie. Histoire naturelle. Entomologie, ou histoire naturelle des crustacées, des arachnides et des insectes. Tome Dixième.* Paris: H. Agasse, [7] + 344 pp. [note: Pars 2, pp. 345–832, issued in 1828].

Guérin-Méneville, F.E. 1825b: Helops. Pp. 111–112. In: Bory de Saint-Vincent, J.B.G.M. (ed.): *Dictionnaire classique d'histoire naturelle, par Messieurs Audoin, Isid. Bordon, Ad. Brongniart, De Candolle, Daudebard de Férussac, A. Desmoulins, Drapiez, Edwards, Flourens, Geoffroy de Saint-Hilaire, A. de Jussieu, Kunth, G. de Lafosse, Lamouroux, Latreille, Lucas fils, Presle-Duplessis, C. Prévost, A Richard, Thiébaut de Berneaud, et Bory de Saint-Vincent. Ouvrage dirigé par ce dernier collaborateur, et dans lequel on a ajouté, pour le porter au niveau de la science, un grand nombre de mots qui n'avaient pu faire partie de la plupart des dictionnaires antérieurs. Tome huitième.* Paris: Ray et Gravier [et] Beaudoin Frères.

Guérin-Méneville, F.E. 1831 [livraison 12]: In: *Iconographie du règne animal de G. Cuvier, ou représentation d'après la nature de l'une des espèces les plus remarquables et souvent non encore figurées, de chaque genre d'animaux. Avec un texte descriptif mis au courant de la science. Ouvrage pouvant servir d'atlas à tous les traités de zoologie. II. Planches des animaux invertébrés. Insectes.* [1829–1838]. Paris: J.B. Baillière, 110 pls.

Guérin-Méneville, F.E. 1834a [livraison 35]: In: *Iconographie du règne animal de G. Cuvier, ou représentation d'après la nature de l'une des espèces les plus remarquables et souvent non encore figurées, de chaque genre d'animaux. Avec un texte descriptif mis au courant de la science. Ouvrage pouvant servir d'atlas à tous les traités de zoologie. II. Planches des animaux invertébrés. Insectes.* [1829–1838]. Paris: J.B. Baillière, 110 pls.

Guérin-Méneville, F.E. 1834b: Matériaux pour une classification des mélasomes. (Extraits d'une monographie de cette famille). *Magasin de Zoologie* 1: 1–39, pls. 101–118.

Guérin-Méneville, F.E. 1835: [new taxon]. In: Guérin-Méneville, F.E. & Percheron A.R.: *Genera des insectes ou exposition detaillée de tous les caractères propres à chacun des genres de cette classe d'animaux. Première livraison.* Paris: Mécquignon-Marvis, 10 pls + [unnumb. pp.]. (1) no. 2.

Guérin-Méneville, F.E. 1837: *Dictionnaire pittoresque d'histoire naturelle et des phénomènes de la nature par une société de naturalistes sous la direction de Guérin. Tome 5.* Paris, 640 pp., 1 pl.

Guérin-Méneville, F.E. 1838: Genre oedémère, Oedemera. *Revue Zoologique par la Société Cuvierienne* 1: 39–40.

Guérin-Méneville, F.E. 1842: Description abrégée de trois espèces nouvelles du genre Meloe. *Revue de Zoologique par la Société Cuvierienne* 5: 338–339.

Guérin-Méneville, F.E. 1843a: Description d'une nouvelle espèce de coléoptère du genre Myrmechixenus, découverte dans les serres aux ananas de M. Panckoucke; à sa campagne de Fleury, près Paris. *Annales de la Société Entomologique de France* (2) **1**: 69–71, pl. 2.

Guérin-Méneville, F.E. 1843b: [new taxa]. *Bulletin de la Société Entomologique de France* **1859**: clxxxvi-cxciii.

Guérin-Méneville, F.E. 1844: *Iconographie du Règne Animal de G. Cuvier, ou représentation d'après nature de l'une des espèces les plus remarquables, et souvent non encore figurées, de chaque genere d'animaux. Avec un texte descriptif mis au courant de la science. Ouvrage pouvant servir d'atlas à tous les traités de zoologie. III. Texte explicatif. Insectes*. Paris: J.B. Baillière, 576 pp.

Guérin-Méneville, F.E. 1849: Insectes. Pp. 239–390, 12 pls. In: Lefebvre, C.T. (ed.): *Voyage en Abyssinie exécuté pendant les années 1839–43 par une commission scientifique composée de Th. Lefebvre, A. Petit et R. Quartin-Dillon naturalistes, et Vignaud dessinateur. Histoire naturelle. Zoologie, par MM.O. des Murs, F. Prévost, Guicherot et Guérin de Méneville. Zoologie 6* (*4*). Paris: Bertrand.

Guérin-Méneville, F.E. 1857: Matériaux pour une monographie des coléoptères du group des eumorphides et plus spécialement du genre Eumorphus. Pp. 237–280. In: Thompson, J. (ed.): *Archives Entomologiques ou recueil contenant des illustrations d'insectes nouveaux ou rares. Tome premier.* Paris: Société entomologique de France, 512 + (1) pp., XXI pls.

Guérin-Méneville, F.E. 1858: Description de deux coléoptères du genre Sepidium, dont l'un est pentamère et paraît être le mâle, et l'autre est hétéromère. *Revue de Zoologie Pure et Appliquée* (2) **10**: 127–129, pl. IV.

Guérin-Méneville, F.E. 1859: [new taxa]. *Bulletin de la Société Entomologique de France* **1859**: clxxxvi-cxciii.

Guérin-Méneville, F.E. 1862: Description d'une nouvelle espèce de coléoptères du genre Melancrus et rectification relative à une note publiée dans le Bulletin entomologique de 1859. *Annales de la Société Entomologique de France* (4) **1** [1861]: 375–376.

Guillebeau, F. 1887: Révision du genre Orchesia Latreille. *L'Échange, Revue Linnéenne* **3** (29–35): 1–8.

Guillebeau, F. 1888: Révision du genre Orchesia Latreille. *L'Échange, Revue Linnéenne* **4** (37–41): 9–15.

Guillebeau, F. 1893a: Synopsis des espèces connues du genre Mycterus d'Europe et du Bassin de la Méditerranée. *L'Échange, Revue Linnéenne* **9**: 89–90.

Guillebeau, F. 1893b: Description d´une espèce inédite du genre Anaspis Geoffroy. *Mittheilungen der Schweizerischen Entomologischen Gesellschaft* **8**: 328.

Guillebeau, F. 1896: Description de quelques espèces de coléoptères inédites (suite). *Bulletin de la Société Entomologique de France* **1896**: 239–245.

Günther, K. 1937: Ueber einige von M. Beick am Tien-Shan und in Nordkansu gesammelte Käfer. *Koleopterologische Rundschau* **23**: 130–135.

Gusakov, A.A. 2005: New taxa of false darkling beetles (Coleoptera: Melandryidae) from China. *Russian Entomological Journal* **14**: 117–118.

Gusakov, A.A. 2009: Novye i maloizvestnye vidy zhukov-tenelubov (Coleoptera, Melandryidae) iz Vostochnoi Palearctiki. *Eversmania* **19–20**: 3–9.

Gusakov, A.A. & Telnov, D. 2007: Systematic changes and new species of Ischaliidae (Coleoptera). *Folia heyrovskyana, Series A,* **15**: 39–46.

Gyllenhal, L. 1806: [new taxa]. In Schoenherr, C.J.: *Synonymia Insectorum, oder: Versuch einer Synonymie Aller bisher bekannten Insecten; nach Fabricii Systema Eleutheratorum geordnet. Mit Berichtigungen und*

Anmerkungen, wie auch Beschreibungen neuer Arten und illuminierten Kupfern. Erster Band. Eleutherata oder Käfer. Erster Theil. Lethrus – Scolytes. Stockholm: A. Nordström, xxii + 294 pp., 3 pls.

Gyllenhal, L. 1808: [new taxa]. In Schoenherr, C.J.: *Synonymia Insectorum, oder: Versuch einer Synonymie aller bisher bekannten Insecten; nach Fabricii Systema Eleutheratorum &c. geordnet. Band 1. Eleutherata oder Käfer. Zweiter Theil. Spercheus – Cryptocephalus*. Stockholm: C.F. Marquard, ix – 1 + 424 pp., 4 pls.

Gyllenhal, L. 1810: *Insecta Suecica descripta. Classis I, Coleoptera sive Eleuterata. Tom I. Pars II*. Scaris: F.J. Leverentz, xx + 660 pp.

Gyllenhal, L. 1813: *Insecta Suecica descripta. Classis I. Coleoptera sive Eleutherata. Volume 1 Pars III*. Scaris: Leverentz, iv + 730 pp.

Gyllenhal, L. 1817: [new taxa]. In: Schoenherr C.J.: Synonymia Insectorum oder Versuch einer Synonymie aller bisher bekannten Insecten; nach Fabricii Systema Eleutheratorum etc. geordnet. Erster Band, Eleutherata oder Käfer, Dritter Theil, Hispa – Molorchus. Upsala, E. Bruzelius, xi + 506 pp.

Gyllenhal, L. 1827: *Insecta Svecica descripta. Classis I. Coleoptera sive Eleuterata. Tom I. Pars IV*. Lipsiae: Friedericum Fleischer, viii + [2] + 762 pp.

Haag-Rutenberg, G.J. 1870a: [new taxa]. In: Heyden L.F.J.D. von. 1870: Zweiter Theil. Beschreibungen der neuen Arten. Pp. 58–176. In: Heyden L.F.J.D von: Entomologische Reise nach dem südlichen Spanien, der Sierra Guadarrama und Sierra Morena, Portugal und den Cantabrischen Gebirgen. *Berliner Entomologische Zeitschrift* **14** *Beiheft*, 218 pp., 2 pls.

Haag-Rutenberg, G.J. 1870b: Beiträge zur Familie der Tenebrioniden. *Coleopterologische Hefte* **6**: 84–93.

Haag-Rutenberg, G.J. 1875a: *Beiträge zur näheren Kenntniss einiger Gruppen aus der Familie der Tenebrioniden*. Berlin: Berliner Entomologischer Verein, 56 pp.

Haag-Rutenberg, G.J. 1875b: Beiträge zur Familie der Tenebrioniden. (5. Stück). *Coleopterologische Hefte* **14**: 67–92.

Haag-Rutenberg, G.J. 1875c: Hidrosis, eine neue Adelostomiden-Gattung. *Deutsche Entomologische Zeitschrift* **19** (1–2): 120.

Haag-Rutenberg, G.J. 1875d: Monographie der Eurychoriden (Adelostomides Lacord.). *Berliner Entomologische Zeitschrift* **19** (4): 359–428. [Issued also as separat publication in Berlin: Nicolai'sche Verlags-Buchhandlung, 70 pp.

Haag-Rutenberg, G.J. 1876a: Eine neue Melasomen-Art aus Marocco. *Entomologische Monatsblätter* **1**: 10.

Haag-Rutenberg, G.J. 1876b: Neue Pimelien. *Entomologische Monatsblätter* **1**: 75–76.

Haag-Rutenberg, G.J. 1876c: Akis Heydeni n. sp. *Entomologische Monatsblätter* **1**: 76.

Haag-Rutenberg, G.J. 1877a: Revision der Gattung Himatismus Er. *Deutsche Entomologische Zeitschrift* **21**: 273–283.

Haag-Rutenberg, G.J. 1877b: Zwei neue Adesmiiden-Arten. *Deutsche Entomologische Zeitschrift* **21**: 283–284.

Haag-Rutenberg, G.J. 1878: Beschreibung neuer Arten von Heteromeren als Nachtrag zur Monographie. *Mittheilungen des Münchener Entomologischen Vereins* **2**: 77–95.

Haag-Rutenberg, G.J. 1879a: Fernere Nachträge zu den Heteromeren-Monographien der Moluriden, Eurychoriden und Adesmiiden. *Deutsche Entomologische Zeitschrift* **23**: 289–296.

Haag-Rutenberg, G.J. 1879b: Leptodopsis, eine neue Heteromeren-Gattung. *Deutsche Entomologische Zeitschrift* **24**: 408–410.

Haag-Rutenberg, G.J. 1880: Beiträge zur Kenntniss der Canthariden. *Deutsche Entomologische Zeitschrift* **24**: 17–90.

Haldeman, S.S. 1848: Description of North American Coleoptera, chiefly in the cabinet of J.L. Leconte, with references to described species. *Journal of the Academy of Natural Sciences of Philadelphia* (2) **1**: 95–100.

Haldeman, S.S. 1852: Appendix C, Insects. Pp. 366–378, 2 pls. In: Stansbury, H.: *Exploration and survey of the Valley of the Great Salt Lake of Utah, including a reconnaissance of a new route through the Rocky Mountains.* Philadelphia: Lippincott, Grambo & Co., 487 pp.

Hamm, A.H. 1909: Sitaris muralis, Forst., n. var. flava. *The Entomologist's Monthly Magazine* 45: 277.

Hampe, C. 1850: Beschreibung einiger neuen Käfer-Arten. *Entomologische Zeitung* (Stettin) **11**: 346–360.

Hampe, C. 1852: Verzeichnis der vom M. Wagner im Kaukasus, in Trans-Kaukasien, Armenien u. West-Persien gesammelten Koleopteren, mit Beschreibung von 16 neuen Arten. Pp. 302–315. In: Wagner, M.F.: *Reise nach Persien und dem Lande der Kurden. Band 2.* Leipzig: Arnoldi, 4 + 316 pp.

Hampe, C. 1871: Beschreibungen einiger neuen Käfer. *Berliner Entomologische Zeitschrift* **14** [1870]: 331–336.

Hampe, C. 1873: Zwei neue Anthicinen. *Verhandlungen der Kaiserlich-Königlichen Zoologisch-Botanischen Gesellschaft in Wien* **23**: 165–166, pl. 2.

Hanatsuka, M., Masumoto, K. & Kon, M. 2005: A new Isomira (Coleoptera, Tenebrionidae, Alleculinae) from Ishigaki Island of the Ryukyus. *Elytra* **33**: 413–416.

Hanatsuka, M., Masumoto, K. & Kon, M. 2006: Two new alleculine species (Coleoptera, Tenebrionidae, Alleculinae) from Japan. *Elytra* **34**: 179–183.

Harold, E. von. 1875a: V. Beschreibung eines neuen Oedemeriden aus Japan. *Coleopterologische Hefte* **14**: 93.

Harold, E. von. 1875b: Verzeichniss der von Herrn T. Lenz in Japan gesammelten Coleopteren. *Abhandlungen vom Naturwissenschaftlichen Vereine zu Bremen* **4**: 283–296.

Harold, E. von. 1877: Beiträge zur Käferfauna von Japan. (Zweites Stück.) Japanische Käfer des Berliner Königl. Museums. *Deutsche Entomologische Zeitschrift* **21**: 337–367.

Harold, E. von. 1878a: Beiträge zur Käferfauna von Japan. (Viertes Stück). *Deutsche Entomologische Zeitschrift* **22**: 65–88.

Harold, E. von. 1878b: Bericht über eine Sendung Coleopteren aus Hiogo. *Abhandlungen vom Naturwissenschaftlichen Vereine zu Bremen* **5** [1886]: 115–135.

Harold, E. von. 1878c: Zur Kenntniss der Gattung Ceropria (Coleoptera Heteromera). *Entomologische Zeitung* (Stettin) **39**: 345–355.

Harold, E. von. 1879: Bericht über die von den Herren A. von Homeyer und P. Pogge in Angola und im Luanda-Reiche gesammelten Coleopteren. *Coleopterologische Helfte* **16**: 1–224, 3 pls.

Harold, E. von. 1880: Einige neue Coleopteren. *Mittheilungen des Münchener Entomologischen Vereins* **4**: 148–171.

Hashimoto, K. & Sakai, M. 2011: A New Species of the Genus Mecynotarsus (Coleoptera, Anthicidae, Notoxinae) from Japan. *Japanese Journal of Systematic Entomology* **17**: 415–420.

Hatayama, T. 1985: Mordellidae. Pp. 376–397. In: Kurosawa, Y., Hisamatsu, S. & Sasaji, H.: *Colored Illustrations of the Coleoptera of Japan. Vol. 3.* Osaka: Hoikusha, x +500 pp., 72 pls. (in Japanese).

Hatch, M.H. 1965: *The beetles of the Pacific North-West. Part. IV. Macrodactyles, Palpicornes, and Heteromera with the collaboration of David V. Miller, David V. McCorkle, Floyd Werner and Dennis W. Boddy.* Seattle: University of Washington Press, 268 pp., 28 pls.

Hatta, K. 1988: The inopeplid beetles of Japan. *Kontyû* **56**: 299–302.

Hayashi, M. 1960: Studies on Melandryidae from Japan 1. (Col.). *Entomological Review of Japan* **11**: 42–46.

Hayashi, M. and Katô, A. 1956: A new species of Melandryidae from Japan (Col.). *Entomological Review of Japan* **7** (2): 37–38.

Heberdey, R.F. 1934: Entomological investigations on the spike disease of sandal (23). Anthicidae (Col.) *The Indian Forest Records* (*Entomology Series*) **20**: 1–14.

Heberdey, R.F. 1935: Anthicidae (Col.). Pp. 304–307. In: Visser, P.C. & Visser-Hooft, J. (eds.): *Wissenschaftliche Ergebnisse der niederländischen Expeditionen in den Karakorum und die angrenzenden Gebiete 1922, 1925 und 1929/30. Zoologie. Band 1*. Leipzig: F.A. Brockhaus, xviii + 499 pp., 8 pls.

Heberdey, R.F. 1936a: Revision der paläarktischen Arten der Gattung Notoxus Geoffr. I. Die Verwandten des Notoxus monoceros L. *Koleopterologische Rundschau* **22** (2/3): 125–160.

Heberdey, R.F. 1936b: Revision der paläarktischen Arten der Gattung Notoxus Geoffr. I. Die Verwandten des Notoxus monoceros L. *Koleopterologische Rundschau* **22** (4): 161–180.

Heberdey, R.F. 1938: Neue Anthiciden aus Ostasien. *Entomologisches Nachrichtenblatt* **12**: 161–164.

Heberdey, R.F. 1942: Revision der paläarktischen Arten der Gattung Mecynotarsus Laf. (Coleopt., Anthicidae). *Mitteilungen der Münchener Entomologischen Gesellschaft* **32**: 445–485.

Heineken, C. 1835: Description of a new species of Hemiptera, and of a species of Hegeter. *Zoological Journal* **5** [1832–1834]: 35–41.

Heinze, E. 1943b: Langelandia (Normandella) zacynthia sp. nov., eine neue Langelandia aus Griechenland. (Colydiidae). *Entomologische Blätter* **39**: 127–129.

Hellén, W. 1935: Koleopterologische Mitteilungen aus Finnland. XIII. *Notulae Entomologicae* **15**: 89–93.

Hellenius, C.N. 1787: Försök, til beskrifning paa et nytt genus blandinsecterna, som kunde kallas Serropalpus. *Kongliga Vetenskaps Academiens Nya Handlingar* **7**: 310–319, pl. VIII.

Heller, K.M. 1923: Die Coleopterenausbeute der Stötznerschen Sze-Tschwan-Expedition (1913–1915). *Entomologische Blätter* **19**: 61–79.

Hellwig, J.C.L. 1792: Dritte Nachricht von neuen Gattungen im entomologischen System. *Neuestes Magazin für die Liebhaber der Entomologie* **1** (3): 385–408.

Hentz, N.M. 1830: Remarks on the use of the maxillae in coleopterous insects, with an account of two species of the family Telephoridae, and of three of the family Mordellidae, which ought to be type of two distinct genera. *Transactions of the American Philosophical Society, Philadelphia* **2**: 458–463, pl. 15.

Herbst, J.F.W. 1782: Nähere Bestimmung des Käfergeschlechts. Platt= oder Rindkäfer. *Archiv der Insectengeschichte herausgegeben von J.C. Fuessly* **2**: 1–8, 1 pl.

Herbst, J.F.W. 1783: Kritisches Verzeichniss meiner Insektensammlung. Erste Klasse. Erste Abteilung. Scarabaeus-Lagria. *Archiv der Insectengeschichte, herausgegeben von J.C. Fuessly* **4**: 1–68, pls. 19–23.

Herbst, J.F.W. 1784a: Kurze Einleitung zur Kenntniss der Insecten für Ungeübte und Anfänger. In: Borowski, G.H.: *Gemeinzzige* [sic] *Naturgeschichte des Thierreichs, darin die merkwürdigsten und nüzlichsten Thiere in systematischer Ordnung beschrieben und die Geschlechter in Abbildungen nach der Natur vorgestellt werden. Band 6*. Berlin und Straslund: Lange, 278 pp., 48 pls.

Herbst, J.F.W. 1784b: Kritisches Verzeichniss meiner Insektensammlung. Fortsetzung. Curculio-Dytiscus. *Archiv der Insectengeschichte, herausgegeben von J.C. Fuessly* **5** (1): 73–128, pls. 24–28.

Herbst, J.F.W. 1784c: Kritisches Verzeichniss meiner Insektensammlung. Fortsetzung. Gyrinus-Paederus. *Archiv der Insectenschichte, herausgegeben von J.C. Fuessly* 5 (2): 129–151, pls. 29–30.

Herbst, J.F.W. 1786: Erste Mantisse zum Verzeichniss der ersten Klasse meiner Insektensammlung. *Archiv der Insectenschichte, herausgegeben von J.C. Fuessly* 7: 153–182, pls. 43–48.

Herbst, J.F.W. 1792: *Natursystem aller bekannten in- und ausländischen Insekten, als eine Fortsetzung der von Buffonschen Naturgeschichte. Der Käfer vierter Theil. Mit zwölf illuminierten Kupferstafeln.* Berlin: Joachim Pauli, viii + 197 pp., 12 pls.

Herbst, J.F.W. 1793: *Natursystem aller bekannten in- und ausländischen Insekten, als eine Fortsetzung der von Büffonschen Naturgeschichte. Der Käfer fünfter Theil.* Berlin: Pauli, 392 pp., pls. 44–59 + J-N.

Herbst, J.F.W. 1797: *Natursystem aller bekannten in- und ausländischen Insekten, als eine Fortsetzung der von Buffonschen Naturgeschichte. Der Käfer siebenter Theil.* Berlin: Geh. Commerzien-Raths Pauli, xi + 346 pp.

Herbst, J.F.W. 1799: *Natursystem aller bekannten in- und ausländischen Insekten, als eine Fortsetzung der von Buffonschen Naturgeschichte. Der Käfer achter Theil.* Berlin: Geh. Commerzien-Raths Pauli, xiv + 420 pp., 121–137 pls.

Heyden, L.F.J.D. von. 1863: Zwei neue Coleopteren-Gattungen aus dem Mittelmeergebiet. *Berliner Entomologische Zeitschrift* 7: 126–130.

Heyden, L.F.J.D. von. 1870: Entomologische Reise nach dem südlichen Spanien, der Sierra Guadarrama und Sierra Morena, Portugal und den Cantabrischen Gebirgen. *Berliner Entomologische Zeitschrift* **14** *Beiheft*, 218 pp., 2 pls.

Heyden, L.F.J.D. von. 1875: Bericht über die von Herrn Prof. Dr. Freiherr von Frisch und Dr. J.J. Ren auf den Canarischen Inseln gesammelten Käfer. *Bericht über die Senckenbergische Naturforschende Gesellschaft in Frankfurt am Main* **1874–1875**: 135–145.

Heyden, L.F.J.D. von. 1877: Die Käfer von Nassau und Frankfurt. *Jahrbücher des Nassauischen Vereins für Naturkunde* **29–30**: 55–412.

Heyden, L.F.J.D. von. 1879: Diagnoses Coleopterorum aliquot novorum in Japonia a Dom. Prof. J.J. Rein, Doct. phil., collectorum. *Bericht über die Senckenbergische Naturforschende Gesellschaft in Frankfurt am Main* **1878–1879**: 87–88.

Heyden, L.F.J.D. von. 1881: Beiträge zur Käferfauna von Turkestan. Zonabris-Arten von Margelan. *Deutsche Entomologische Zeitschrift* 25: 324–330.

Heyden, L.F.J.D. von. 1882: Beitrag zur Coleopteren-Fauna der Sierra de Cordoba. *Deutsche Entomologische Zeitschrift* **26**: 43–48.

Heyden, L.F.J.D. von. 1883a: Neue Zonabris (Mylabris Ol.) aus Turkestan. *Deutsche Entomologische Zeitschrift* **27**: 65–68.

Heyden, L.F.J.D. von. 1883b: Verzeichniss der von Herrn Dr. Med. W. Kobelt in Nordafrika und Spanien gesammelten Coleopteren. *Bericht über die Senckenbergische Naturforschende Gesellschaft* **1882–1883**: 217–237.

Heyden, L.F.J.D. von. 1883c: [new taxon]. In: Heyden L.F.J.D. von & Weise J.: Neue südost-europäische und klein-asiatische Käfer. *Deutsche Entomologische Zeitschrift* 27: 310–314.

Heyden, L.F.J.D. von. 1883d: Käfer aus Ochs (in Turkestan). *Deutsche Entomologische Zeitschrift* **27**: 337–353.

Heyden, L.F.J.D. von. 1883e: Ueber Käfer aus Tekke-Turkmenien. *Deutsche Entomologische Zeitschrift* **27**: 354–360.

Heyden, L.F.J.D. von. 1884a: Beitrag zur Coleopterenfauna der Insel Askold und andere Theile des Amurgebietes. *Deutsche Entomologische Zeitschrift* **28**: 273–300.

Heyden, L.F.J.D. von. 1884b: Coleopterologische Ausbeute einer Excursion nach der Sierra d'Espuñas. *Deutsche Entomologische Zeitschrift* **28**: 355–361.

Heyden, L.F.J.D. von. 1885a: [new taxon]. In: Heyden L.F.J.D. von. & Kraatz G.: Beiträge zur turkestanischen Coleopteren-Fauna. *Deutsche Entomologische Zeitschrift* **29**: 273–298.

Heyden, L.F.J.D. von. 1885b: Weitere Beiträge zur Coleopteren-Fauna des Amur-Gebietes. *Deutsche Entomologische Zeitschrift* **29**: 298–307, pl. iv.

Heyden, L.F.J.D. von. 1886a: [new taxa]. In: Heyden L.F.J.D. von. & Kraatz G.: Beiträge zur Coleopteren-fauna von Turkestan, namentlich des Alai-Gebirges, unter Beihilfe der Herren Dr. Candèze, Ganglbauer Dr. Stierlin, und Weise. *Deutsche Entomologische Zeitschrift* **30**: 177–194.

Heyden, L.F.J.D. von. 1886b: Die Coleopteren-Fauna des Suyush-Flusses (Amur). *Deutsche Entomologische Zeitschrift* **30**: 269–277.

Heyden, L.F.J.D. von. 1887a: Verzeichniss der von Herrn Otto Herz auf der chinesischen Halbinsel Korea gesammelten Coleopteren. *Horae Societatis Entomologicae Rossicae* **21**: 243–273.

Heyden, L.F.J.D. von. 1887b: Vierter Beitrag zur Kenntniss der Coleopteren-Fauna der Amurländer. *Deutsche Entomologische Zeitschrift* **31**: 297–304.

Heyden, L.F.J.D. von. 1887c: XI. Beitrag zur Coleopteren-Fauna von Turkestan. *Deutsche Entomologische Zeitschrift* **31**: 305–336.

Heyden, L.F.J.D. von. 1887d: Bericht über von den Herren Prof. Dr. Freiherrn von Fritsch und Prof. Dr. Rein in Marocco und dem Atlas-Gebirge gesammelten Käfer. *Deutsche Entomologische Zeitschrift* **31**: 433–448.

Heyden, L.F.J.D. von. 1888: Ueber die von L. Conradt aus den Alai-Gebirge mitgebrachten Coleopteren. – Beitrag XII. Mit Beiträgen von J. Faust. *Deutsche Entomologische Zeitschrift* **32**: 39–44.

Heyden, L.F.J.D. von. 1889a: Descriptiones Coleopterorum novorum Regionis Palaearcticae. *Deutsche Entomologische Zeitschrift* **1889**: 325–330.

Heyden, L.F.J.D. von. 1889b: Insecta, a Cl.G.A. Potanin in China et in Mongolia novissime lecta. XII. Scarabaeidae, Cantharidae, Cleridae, Lagriidae, Melandryidae, Pedilidae, Anthicidae. *Horae Societatis Entomologicae Rossicae* **23**: 654–677.

Heyden, L.F.J.D. von. 1890a: Aufzählung der Käfer-Arten aus Tunis und Tripolis aus Loosen von M. Quedenfeldt. *Deutsche Entomologische Zeitschrift* **1890**: 65–75.

Heyden, L.F.J.D. von. 1890b: Die Varietäten der Diaperis boleti L. *Deutsche Entomologische Zeitschrift* **1890**: 78.

Heyden, L.F.J.D. von. 1890c: XIII. Beitrag zur Coleopteren-Fauna von Turkestan. *Deutsche Entomologische Zeitschrift* **1890**: 353–367.

Heyden, L.F.J.D. von. 1890d: Beiträge zur Coleopteren-Fauna von Amasia und Samsoun in Nord-Kleinasien. *Deutsche Entomologische Zeitschrift* **1890**: 369–374.

Heyden, L.F.J.D. von. 1892a: XIV. Beitrag zur Coleopteren-Fauna von Turkestan. Unter Mitwirkung der Herren J. Faust, A. Kuwert und J. Weise. *Deutsche Entomologische Zeitschrift* **1892**: 99–104.

Heyden, L.F.J.D. von. 1892b: XV. Beitrag zur Coleopteren-Fauna von Turkestan, Turkmenien und Süd-West-Sibirien. Aus Sendungen des Herrn Dr. Staudinger. *Deutsche Entomologische Zeitschrift* **1892**: 105–110.

Heyden, L.F.J.D. von. 1894a: [new taxa]. In: Hauser F.: Beitrag zur Coleopteren-Fauna von Transcaspien und Turkestan. Zusammengestellt unter Mitwirkung der Herren Dr. Eppelsheim, Escherich, Faust, Dr. von Heyden, Kuwert, Reitter und Weise [u. Pastor Schmidt]. *Deutsche Entomologische Zeitschrift* **1894**: 17–74.

Heyden, L.F.J.D. von. 1894b: Beiträge zur Coleopteren-Fauna von Turkestan. *Horae Societatis Entomologicae Rossicae* **28**: 262–268.

Heyden, L.F.J.D. von. 1897: [new taxon]. In: Escherich K.: Zoologische Ergebnisse einer von Dr. K. Escherich und Dr. L. Kathariner nach Central-Kleinasien unternommenen Reise. IV. Theil. Coleopteren. Unter Mitwirkung von E. Brenske, L. Ganglbauer, Dr. L. von Heyden, E. Wasmann und J. Weise. *Entomologische Zeitung* (Stettin): **58**: 3–69.

Heyden, L.F.J.D. von. 1908: Über die Coleopteren-Gattungen Parablops. *Deutsche Entomologische Zeitschrift* **1908**: 132.

Heyden, L.F.J.D. von. 1909: Die kanarische Phaleria ornata Woll. und ihre Varietäten. *Deutsche Entomologische Zeitschrift* **1909**: 486.

Heyden, L.F.J.D. von. & Kraatz, G. 1882: Käfer um Samarkand, gesammelt von Haberhauer. *Deutsche Entomologische Zeitschrift* **26**: 297–338.

Heyden, L.F.J.D. von., Reitter, E. & Weise, J. 1891: *Catalogus Coleopterorum Europae, Caucasi et Armeniae rossicae*. Berlin: Friedländer & Sohn; Mödling: E. Reitter; Caen: Revue d'Entomologie, 812 pp.

Hille, J.C. van. 1961: Coleoptera: Anthicidae, pp. 217–258. In: *South African Animal Life, results of the Lund University Expedition in 1950–1951. Vol. VIII*. Uppsala: Almqvist & Wiksells Boktryckeri Ab., 557 pp.

Hille, J.C. van. 1975: Ten African species of Notoxus (Col.Anthicidae). *Koleopterologische Rundschau* **52**: 3–13.

Hille, J.C. van. 1977: Species of the genus Formicomus Laf. from North-East Africa and adjacent parts of the Arabian peninsula (Coleoptera: Anthicidae). *Entomologica Scandinavica* **9**: 197–204.

Hinton, H.E. 1948: A synopsis of the genus Tribolium Macleay, with some remarks on the evolution of its species-groups (Coleoptera, Tenebrionidae). *Bulletin of Entomological Research, London* **39**: 13–55.

Holdhaus, K. 1919: Ergebnisse der wissenschaflichen Expedition nach Mesopotanien, 1910. Koleopteren aus Mesopotamien. *Annalen des Naturhistorischen Museums in Wien* **33**: 39–58.

Hölzel, E. 1958: Die mitteleuropäischen Arten der Gattung Isomira Muls. (Col. Alleculidae) mit Beschreibung der Untergattung Heteromira subgen. nov. und Art moroi spec. nov. aus den Kärntner Karawanken. *Nachrichtenblatt der Bayerischen Entomologen* **7**: 17–25.

Hope, F.W. 1831: Synopsis of the new species of Nepaul insects in the collection of Major General Hardwicke. Pp. 21–31. In: Gray, J.E. (ed.): *The Zoological Miscellany. Vol. 1*. London: Treuttel, Wurtz and Co., 40 pp., 4 pls.

Hope, F.W. 1833: On the characters of several new genera and species of coleopterous insects. *Proceedings of the Zoological Society of London* **1**: 61–64.

Hope, F.W. 1834a: Descriptions of some hitherto uncharacterized exotic Coleoptera, chiefly from New Holland. *Transactions of the Entomological Society of London* **1**: 11–20, pls. 1, 2.

Hope, F.W. 1834b: Characters and descriptions of several new genera and species of coleopterous insects. *Transactions of the Zoological Society of London* **1**: 91–112, pls. 14–16.

Hope, F.W. 1841: *The Coleopterist's Manual, part the third, containing various families, genera, and species, of beetles, recorded by Linneus and Fabricius. Also, descriptions of newly discovered and unpublished insects*. London: J.G. Bridgewater and Bowdery & Keby, [6] + 191 pp., 3 pls.

Hope, F.W. 1842: Description of the coleopterous insects sent to England by Dr. Cantor from Chusan and Canton, with observation on the entomology of China. *Journal of Proceedings of the Entomological Society of London* **1842**: 49–52 [note: pp. 49 and 50 are lacking in original issue].

Hope, F.W. 1843: Description of the coleopterous insects sent to England by Dr. Cantor from Chusan and Canton, with observation on the entomology of China. *The Annals and Magazine of Natural History* (1) 11: 62–66.

Hope, F.W. 1845: Descriptions of the coleopterous insects sent to England by Dr. Cantor from Chusan and Canton, with observations on the Entomology of China. *Journal of the Proceedings of the Entomological Society of London*, pp. 49–52. In: *The Transactions of the Entomological Society of London* 4.

Horák, J. 1978: Stenalia bilyi sp.n. aus Tadschikistan. *Acta Entomologica Bohemoslovaca* 75: 400–403.

Horák, J. 1979: Mordella hofferi sp. n. von der Balkanhalbinsel. *Acta Entomologica Bohemoslovaca* 76: 104–107.

Horák, J. 1980: Drei neue Arten der Gattung Mordellistena aus Tadschikistan. *Acta Entomologica Bohemoslovaca* 77: 280–286.

Horák, J. 1981: Zwei neue Arten der Familie Mordellidae aus der paläarktischen Region. *Acta Entomologica Bohemoslovaca* 78: 335–339.

Horák, J. 1982: Mordellistenochroa gen.n. und Beschreibung vier neuer ostpaläarktischer Arten (Col. Mordellidae). *Acta Entomologica Bohemoslovaca* 79: 46–55.

Horák, J. 1983: Revision der Mordellistena-Arten aus der pentas-Gruppe. *Entomologische Abhandlungen* 47: 1–13.

Horák, J. 1985: Ergebnisse der tschechoslowakisch-iranischen entomologischen Expeditionen nach Iran 1970, 1973 und 1977. Coleoptera: Mordellidae 1 (Stenaliini, Mordellini). *Entomologische Abhandlungen* 49: 1–25.

Horák, J. 1986: Beitrag zur Systematik der Gattung Hoshihananomia KONO. *Reichenbachia* 23: 161–164.

Horák, J. 1989: Brodskyella gen.n. des Tribus Stenaliini aus der orientalischen Region (Col. Mordellidae). *Entomologische Abhandlungen* 53: 37–44.

Horák, J. 1990: Typenrevision einiger wenig bekannter Arten aus der Gattung Mordellistena Costa. *Entomologische Abhandlungen* 53: 125–142.

Horák, J. 1991: Zwei neue Arten der Gattung Mordellistena Costa aus Marokko. *Reichenbachia* 28: 135–140.

Horák, J. 1994: Description of Glipa (Macroglipa) thoracica sp.n. and revision of some species of the genus Glipa Leconte from the Oriental region. *Folia Heyrovskyana* 2: 101–110.

Horák, J. 1995a: The genus Stenalia from Oriental region. *Acta Societatis Zoologicae Bohemicae* 59: 89–99.

Horák, J. 1995b: Revision of species of the tribus Mordellistenini from Oriental region. Part 1. *Acta Societatis Zoologicae Bohemicae* 59: 165–176.

Horák, J. 1995c: Revision of some Oriental Mordellini with description of two new species. *Entomological Problems* 26: 165–184.

Horák, J. 1996: Revision of some little known species of genus Mordellistena with description of two new species. *Klapalekiana* 32: 171–184.

Horák, J. 2007: Four new genuses of tribe Mordellistenini from Southeastern Asia. *Studies and Reports of Distrikt Museum Prague-East, Taxonomical Series* 3: 59–81.

Horák, J. 2009a: Order Coleoptera, family Mordellidae. Pp. 207–215. In: Antonius van Harten (eds.): *Arthropod Fauna of the* UAE. *Volume* 2. Dar Al Ummah Printing, Distribution & Advertising. 786 pp.

Horák, J. 2009b: Revision of some Oriental Mordellini with description of four new species. Part 3. *Studies and Reports of Distrikt Museum Prague-East, Taxonomical Series* 5: 65–90.

Horák, J., Farkač, J. & Nakládal, O. 2012: Mordellidae (Coleoptera) from Socotra Island. *Acta Entomologica Musei Nationalis Pragae* 52, *supplementum*: 253–268.

Horn, G.H. 1868: New species of Coleoptera from the Pacific district of the United States. *Transactions of the American Entomological Society* **2**: 129–140.

Horn, G.H. 1870: Revision of the Tenebrionidae of America, north of Mexico. *Transactions of the Americal Philosophical Society, New Series* **14**: 253–404.

Horn, G.H. 1875: Synonymical notes and descriptions of new species of North American Coleoptera. *Transactions of the American Entomological Society* **5**: 125–156.

Horn, G.H. 1888: Miscellaneous coleopterous studies. *Transactions of the American Entomological Society* **15**: 26–48.

Houlbert, C. 1934: [new taxa]. In: Barthe E.: Tableaux analytiques des coléoptères de la faune franco-rhénane (France, Hollande, Belgique, région rhénane, Valais). Famille LXX Melandryidae. *Miscellanea Entomologica* **35** (12): 1–34.

Hozawa, S. 1914: Note on a new termitophilous Coleoptera found in Formosa (Ziaelas formosanus). *Annotationes Zoologicae Japonensis* **8**: 483–488, pl. VI.

Hsiao, Y. 2015: A new species of the genus Synchroa from Taiwan, with a key to the world fauna (Coleoptera: Synchroidae). *Acta Entomologica Musei Nationalis Pragae* **55**: 243–248.

Hsiao, Y., Konvička, O. & Ko, C.-C. 2018: The world fauna of Synchroidae Lacordaire, 1959 (Coleoptera, Tenebrionoidea, Synchroidae). *European Journal of Taxonomy* **407**: 1–33.

Hsiao, Y., Li, Y., Liu, Z. & Pang, H. 2016: A new species of Synchroa Newman from China (Coleoptera: Synchroidae). *Zootaxa* **4093**: 595–600.

Hsiao, Y. & Pollock, D.A. 2019: Contribution to the knowledge of the genus Omineus Lewis, 1895 in Taiwan, with description of two new species (Coleoptera, Mycteridae, Eurypinae). *Zootaxa* **4568**: 533–547.

Hsiao, Y, Pollock D.A. & Barclay, M. 2015: Two new species of Cyanopenthe Nikitsky from Taiwan (Coleoptera, Tetratomidae, Penthinae). *Zootaxa* **4058**: 578–588.

Hubenthal, W. 1905: Phryganophilus ruficollis F. var. nov. rosti aus Japan. *Deutsche Entomologische Zeitschrift* **1905**: 57–58.

Hubenthal, W. 1911: Beschreibungen neuer europäischer Coleopteren und Bemerkungen zu bekannten Arten. *Entomologische Blätter* **7**: 185–193.

Hübner, J. 1789: Beiträge zur Naturgeschichte der Insecten. *Der Naturforscher* **24**: 36–59, 1 pl.

Hummel, A.-D. 1827: *Essais Entomologiques. No. VI. Insectes de 1826. Novae species. Catalogus Insectorum quae-itinere Petropoli in Chersonesum Tauricum et Iberiam anno 1825 collegit Benedictus Jäger.* St. Pétersburg: Chancellerie privée du Ministère de l'Intérieur, 48 + (1) pp.

Iablokoff-Khnzorian, S.M. 1956a: Chetyre novykh vida zhestkokrylykh iz Armyanskoy SSR (Coleoptera, Insecta) [Four new species of Coleoptera from the Armenian SSR]. *Doklady Akademii Nauk Armyanskoy SSR* **22** (3): 135–139.

Iablokoff-Khnzorian, S.M. 1956b: Novyy vid chernotelok iz Armyanskoy SSR. *Doklady Akademii Nauk Armyanskoy SSR* **23**: 41–43.

Iablokoff-Khnzorian, S.M. 1957a: Novye vidy zhestkokrylykh iz Armyanskoy SSR i Nakh. ASSR. *Zoologicheskiy Sborkik Akademii Nauk Armyanskoy SSR* **3**: 153–183.

Iablokoff-Khnzorian, S.M. 1957b: Novyy veeronosets iz Armyanskoi SSR (Coleoptera, Rhipiphoridae). *Doklady Akademii Nauk Armyanskoi SSR* **24**: 231–233.

Iablokoff-Khnzorian, S.M. 1958: Dva novykh vida zhestkokrylykh – Heteromera iz Armenskoy SSR (Insecta, Coleoptera) [Deux nouveaux coléoptères hétéromères de l'Arménie de SSR (Insecta, Coleoptera)]. *Zoologiskiy Zhournal Akademii Nauk SSSR* **37**: 1896–1898.

Iablokoff-Khnzorian, S.M. 1959a: Novye vidy zhestkokrylykh iz Talysha. *Izvestiya Akademii Nauk Azerbaydzhanskoy SSR* **3**: 57–62.

Iablokoff-Khnzorian, S.M. 1959b: Novyy zhuk pylceed iz Armyanskoy SSR (Coleoptera, Alleculidae) Heliotaurus emmae Khnzorian, sp. nov. *Doklady Akademii Nauk Armyanskoy SSR* **28**: 191–192.

Iablokoff-Khnzorian, S.M. 1959c: Novyy pylceed iz doliny Araksa (Coleoptera, Alleculidae) Mycetocharina riabovi Khnzorian sp. nov. *Doklady Akademii Nauk Armyanskoy SSR* **29**: 141–143.

Iablokoff-Khnzorian, S.M. 1964: Bemerkungen über einige Reitter'sche Typen aus dem Ungarischen Naturwissenschaftlichen Museum (Coleoptera). *Folia Entomologica Hungarica* (N.S.) **17**: 293–315.

Iablokoff-Khnzorian, S.M. 1970: Novye vidy zhestkokrylykh iz Armenii i drugikh chastey SSSR (Insecta, Coleoptera). *Sbornik Zoologicheskogo Instituta Akademii Nauk Armyanskoy SSR* **15**: 49–80.

Iablokoff-Khnzorian, S.M. 1973: Dva novych vida veeronostsev iz SSSR (Coleoptera, Rhipiphoridae). *Doklady Akademii Nauk Armyanskoi SSR* **57**: 307–312.

Iablokoff-Khnzorian, S.M. 1974: Dva novykh vida zhestkokrylykh – uzkotelok iz Tadzhikistana. *Doklady Akademii Nauk Armyanskoy SSR* **59**: 188–192.

Iablokoff-Khnzorian, S.M. 1976: Les Isomira Muls. du Caucase (Coleoptera, Alleculidae). *Acta Zoologica Cracoviensia* **21**: 315–329.

Iablokoff-Khnzorian, S.M. 1978: Novyy vid zhestokrylykh-uzkotelok iz Primorya. *Doklady Akademii Nauk Armyanskoy SSR, Entomologia* **46**: 252–255.

Iablokoff-Khnzorian, S.M. 1983: *Fauna Armenia. Vol. 5. Coleoptera, Meloidae & Alleculidae.* Erevan: Akademia Nauk Armyanskoy SSR, 155 pp.

Iablokoff-Khnzorian, S.M. 1984: Novyy vid zhukov-chernotelok iz Armyanskoy SSR. *Doklady Akademii Nauk Armyanskoy SSR* **78**: 43–44.

Iablokoff-Khnzorian, S.M. 1985: Les Pythidae paléarctiques (Coleoptera). *Deutsche Entomologische Zeitschrift* (N.F.) **32**: 193–229.

Iablokoff-Khnzorian, S.M. 1986: Novyy rod i vid zhukov-veeronostsev iz Tadzhikistana (Coleoptera, Rhipiphoridae). *Doklady Akademii Nauk Armyanskoi SSR* **82**: 89–92.

ICZN 1999: *International Code of Zoological Nomenclature. Fourth Edition adopted by the International Union of Biological Sciences.* The International Trust for Zoological Nomenclature, London, xxix + 305 pp.

ICZN 2010: International Commission on Zoological Nomenclature. Opinion 2244 (Case 3443). *Tentyria* Latreille, 1802 (Coleoptera, Tenebrionidae): usage conserved. *Bulletin of Zoological Nomenclature* **67**: 104–105.

ICZN 2016: International Commission on Zoological Nomenclature. Opinion 2377 (Case 3624) A proposal for the rejection of 38 names in Anthicidae (Coleoptera): approved. *The Bulletin of Zoological Nomenclature* **73**: 65–69.

Iga, M. & Nakane, T. 1954: Oedemeridae Pp. 154–155. In: Nakane, T. (ed.): *Coloured illustrations of the Insects of Japan. Coleoptera.* Osaka: Hoikusha Co. Ltd., 196 pp., 64 pls. (in Japanese, with English descriptions).

Illiger, J.C.W. 1794: Beschreibung einiger neuen Kaferarten aus der Sammlung des Herrn Professore Hellwig in Braunschweig. *Neuestes Magazin für die Liebhaber der Entomologie* (5) **1**: 593–620.

Illiger, J.C.W. 1798: *Verzeichniss der Käfer Preussens. Entworfen von Johann Gottlieb Kugelann Apotheker in Osterode. Mit einer Vorrede des Professors und Pagenhofmeisters Hellwig in Braunschweig, und dem angehängten Versuche einer natürlichen Ordnungs- und Gattungsfolge der Insekten.* Halle: Johann Jacob Gebauer, xlii + [1] + 510 + [1] pp.

Illiger, J.C.W. 1800: Vierzig neue Insekten aus der Hellwigschen Sammlung in Braunschweig. *Archiv für Zoologie und Zootomie* (Wiedemann) 1 (2): 103–150.

Illiger, J.C.W. 1801: Nachtrag und Berichtigungen zum Verzeichnisse der Käfer-Preussens. *Magazin für Insektenkunde* 1: 1–94.

Illiger, J.C.W. 1802a: Namen der Insekten-Gattungen, ihr Genitiv, ihr grammatisches Geschlecht, ihr Silbenmass, ihre Herleitung; zugleich mit den Deutschen Benennungen. *Magazin für Insektenkunde* 1: 125–155.

Illiger, J.C.W. 1802b: Aufzählung der Käfergattungen nach der Zahl der Fussglieder. *Magazin für Insektenkunde* 1: 285–305.

Illiger, J.C.W. 1804: Bemerkungen zum zweiten Theile von Fabricius Systema Eleutheratorum. *Magazin für Insektenkunde* 3: 160–180.

Illiger, J.C.W. 1805: [new taxon]. In: Panzer, G.W.F.: *Kritische Revision der Insektenfauna Deutschlands nach dem System bearbeitet. Zum I–XCVI. Heft. I. Baendchen.* Nürnberg: Felseckerschen Buchhandlung, 12 + 144 pp.

Illiger, J.C.W. 1807: Vorschlag zur Aufnahme im Fabricischen Systeme fehlende Käfergattungen. *Magazin für Insektenkunde* 6: 318–349.

Illiger, J.C.W. 1807: Vermischte Nachrichten und Bemerkungen. Literatur. *Magazin für Insektenkunde* 6: 350–370.

Illiger, J.C.W. 1810: [new taxon]. In: Gyllenhal, L.: *Insecta Suecica descripta. Classis I, Coleoptera sive Eleuterata. Tom I. Pars II.* Scaris: F.J. Leverentz, xx + 660 pp.

Imasaka, S. 2005: Review of the genus Arthromacra (Coleoptera, Tenebrionidae, Lagriinae) in Japan. *Miscellaneous Reports of the Hiwa Museum for Natural History* 44: 61–163, pls 1–28.

Ishikawa, H. & Sakai, M. 2001: A new species of the genus *Lederina* (Coleoptera, Melandryidae) from Japan. *Japanese Journal of Systematic Entomology* 7: 333–336.

Ishikawa, H. & Sakai, M. 2004: The genus Microscapha (Coleoptera, Melandryidae) of Japan. *Japanese Journal of Systematic Entomology* 10: 127–132.

Ishikawa, H. & Sakai, M. 2007: A new species of the genus Lederina (Coleoptera, Melandryidae) from central Japan. *Japanese Journal of Systematic Entomology* 13: 55–58.

Ishikawa, H., Toyoshima, R. & Lee, C.-F. 2007: A new species of the genus Dircaeomorpha (Coleoptera, Melandryidae) from Taiwan. *Elytra, New Series* 35: 177–181.

Israelson, G. 1971: On the Aderidae of the Canary Islands (Coleoptera). *Eos, Revista Española de Entomología* 46 [1970]: 109–122 (in Japanese and English).

Israelson, G. 1980: Taxonomical and nomenclatoral notes on some Canarian Coleoptera. *Vieraea* 9 [1979]: 183–209.

Israelson, G. 1984: Coleoptera from the Azores. *Boletim do Museu Municipal do Funchal* 36: 142–161.

Israelson, G. 1985: A new Macaronesian genus of Cisidae (Coleoptera). *Entomologische Blätter* 81: 80–84.

Ito, T. & Yoshitomi, H. 2017a. Four New Species of the Genus Prostomis (Coleoptera: Prostomidae) from the Oriental Region. *Japanese Journal of Systematic Entomology* 23: 113–118.

Ito, T. & Yoshitomi, H. 2017b. Revision of the Genus Prostomis (Coleoptera: Prostomidae) of Japan. *Japanese Journal of Systematic Entomology* 23: 167–174.

Iwan, D. 1997: Revision of the Asian genera of the tribe Platynotini (Coleoptera, Tenebrionidae). *Annales Zoologici* 47: 243–272.

Iwan, D. 2001: Two new species of Pseudoblaps Guérin, 1834 (Coleoptera: Tenebrionidae: Platynotini) from Asia. *Annales Zoologici* 51: 147–152.

Iwan, D. 2003: A review of the genus Pseudoblaps Guérin, 1834 (Coleoptera: Tenebrionidae: Platynotini). *Annales Zoologici* 53: 489–507.

Iwan, D., Ferrer, J. & Raś, M. 2010: Catalogue of the world Gonocephalum Solier, 1834 (Coleoptera, Tenebrionidae, Opatrini). Part 1. List of the species and subspecies. *Annales Zoologici* 60: 245–304.

Iwan, D. & Kamiński, M.J. 2016: Toward a natural classification of opatrine darkling beetles: comparative study of female terminalia. *Zoomorphology* 135: 453–485.

Iwan, D. & Löbl, I. 2007: Nomenclatural notes on tenebrionid beetles of the Palaearctic region (Insecta: Coleoptera). *Annales Zoologici* 57: 733–739.

Jacquelin du Val, C. 1858: Pp. 129–232. In: *Manuel Entomologique. Genera des coléoptères d'Europe comprenant leur classification en familles naturelles, la description de tous les generes, des tableaux synoptiques destinés à faciliter l'étude, le catalogue de toutes les espèces de nombreux dessins au trait de charactères et plus de treize cents types représentant un ou plusieurs insectes de chaque genre dessinés et peints d'apres nature avec le plus grand soin par M. Jules Migneaux. Tome deuxième.* Paris: A. Deyrolle [1857–1859, 285 + [3] + 22 + [2] pp., 67 pls].

Jacquelin du Val, C. 1860: Pp. 137–272. In: *Manuel Entomologique. Genera des coléoptères d'Europe comprenant leur classification en familles naturelles, la description de tous les genres, des tableaux synoptiques destinés à faciliter l'étude, le Catalogue de toutes les espèces de nombreux dessins au trait de charactères et plus de treize cents types représentant un ou plusieurs insectes de chaque genre dessinés et peints d'après nature avec le plus grand soin par M. Jules Migneaux. Tome troisième.* Paris: A. Deyrolle Deyrolle [1859–1863, 464 + 200 pp., 100 pls].

Jacquelin du Val, C. 1861: Pp. 273–352. In: *Manuel Entomologique. Genera des coléoptères d'Europe comprenant leur classification en familles naturelles, la description de tous les genres, des tableaux synoptiques destinés à faciliter l'étude, le Catalogue de toutes les espèces de nombreux dessins au trait de charactères et plus de treize cents types représentant un ou plusieurs insectes de chaque genre dessinés et peints d'après nature avec le plus grand soin par M. Jules Migneaux. Tome troisième.* Paris: A. Deyrolle Deyrolle [1859–1863, 464 + 200 pp., 100 pls].

Jacquelin du Val, C. 1863: Pp. 353–464. In: *Manuel Entomologique. Genera des coléoptères d'Europe comprenant leur classification en familles naturelles, la description de tous les genres, des tableaux synoptiques destinés à faciliter l'étude, le Catalogue de toutes les espèces de nombreux dessins au trait de charactères et plus de treize cents types représentant un ou plusieurs insectes de chaque genre dessinés et peints d'après nature avec le plus grand soin par M. Jules Migneaux. Tome troisième.* Paris: A. Deyrolle [1859–1863, 464 pp., 100 pls].

Jakobson, G. 1896: Tria Coleoptera nova e Rossia europaea. *Horae Societatis Entomologicae Rossicae* 29 [1894–1895]: 520–524.

Jakowlew, B.E. 1897: Espèces nouvelles du genre Meloë (subg. Proscarabaeus Steph.) de la Sibérie orientale. *Horae Societatis Entomologicae Rossicae* 31: 248–252.

Jedlička, A. 1965: Neue Carabiden aus Nepal (Coleoptera). *Khumbu Himal* **1**: 98–107.

Jelínek, J. & Kubáň, V. 2009: A review of the genus *Agnathus* (Coleoptera: Pyrochroidae: Agnathinae), with description of Agnathus secundus sp. nov. from China. *Acta Entomologica Musei Nationalis Pragae* **49**: 253–281.

Johnson, C. 1989: Tenebrionidae (Coleoptera) collected in the Eastern Province of Saudi Arabia. *Fauna of Saudi Arabia* **10**: 123–133.

Jolivet, P. 1951: Contribution a l'Étude des Coléoptères Chrysomeloidae de la région Indo-Australienne (1re note). *Bulletin de l'Institut Royal de Sciences Naturelles du Belgique* **27** (52): 1–7.

Joy, N.H. 1912: A note on Anaspis ruficollis, Brit. Colls. *The Entomologist's Monthly Magazine* **48**: 115.

Kai, T. & Yoshitomi, H. 2013: A New Species of the Genus Pseudopyrochroa (Coleoptera, Pyrochroidae) from the Ryukyus, Japan. *Elytra, New Series* **3**: 229–235.

Kamiński, M.J. & Iwan, D. 2017: Revision of the subtribe Pedinina (Tenebrionidae: Pedinini). *Annales Zoologici* **67**: 585–607.

Kamiya, H. 1961: Discovery of the family Hemipeplidae in Japan (Coleoptera: Heteromera). *Mushi* **35**: 75–77.

Kamiya, H. 1963: On the systematic position of the genus Usechus Motschulsky, with a description of a new species from Japan (Coleoptera). *Mushi* **37**: 19–26.

Kangas, E. 1959: Eine neue Art der Gattung Hallomenus Panz. (Col., Serropalpidae) aus Finnland. *Annales Entomologici Fennici* **25**: 187–192.

Kangas, E. 1986: Eine neue Mordellistena-Art. *Annales Entomologici Fennici* **52**: 139–141.

Kangas, E. 1988: Mordellistena dalmatica Ermisch sensu Kangas (1976) (Coleoptera, Mordellidae), eine neue unbeschriebene Art. *Annales Entomologici Fennici* **54**: 89–92.

Karsch, F.A.F. 1881: Die Käfer der Rohlfs'schen Afrikanischen Expedition 1878–79. *Berliner Entomologische Zeitschrift* **25**: 41–50, pl. 2.

Kaszab, Z. 1938a: Species nova generis Platyscelis Latr. (Col., Tenebrionidae) Pl. székessyi spec. nov. *Folia Entomologica Hungarica* **4**: 1–3.

Kaszab, Z. 1938b: I. Gruppe des Subgenus Oodescelis Motsch. (Col. Tenebr.). *Stettiner Entomologische Zeitung* **99**: 49–56.

Kaszab, Z. 1938c: Die Arten der Gattung Cnemeplatia Costa und Psilachnopus Reitter (Col. Tenebr. Opatrinae). *Entomologisk Tidskrift* **59**: 77–83.

Kaszab, Z. 1938d: Zwei neue Platyscelis-Arten aus Turkestan (Col. Tenebr.). *Koleopterologische Rundschau* **24**: 91–94.

Kaszab, Z. 1938e: Neue Tenebrioniden-Arten aus Indien. *Entomologische Blätter* **34**: 226–231.

Kaszab, Z. 1939a: Neue indomalayische Tenebrioniden. *Arbeiten über Morphologische und Taxonomische Entomologie aus Berlin Dahlem* **6**: 95–111.

Kaszab, Z. 1939b: Tenebrioniden aus Neu-Guinea. *Nova Guinea (N.S.)* **3**: 185–267.

Kaszab, Z. 1940a: Beiträge zur Kenntnis der Tenebrioniden-Fauna Kleinasiens. *Folia Entomologica Hungarica* **5**: 2–7.

Kaszab, Z. 1940b: Revision der Tenebrioniden-Tribus Platyscelini (Col. Teneb.). *Mitteilungen der Münchener Entomologischen Gesellschaft* **30**: 119–235.

Kaszab, Z. 1940c: Neue Heterotarsinen (Coleopt. Tenebr.) aus der Sammlung des Ungarischen National-museums. *Annales Musei Nationalis Hungarici* **33**: 153–160.

Kaszab, Z. 1940d: Revision der Tenebrioniden-Tribus Platyscelini (Col. Teneb.). *Mitteilungen der Münchener Entomologischen Gesellschaft* **30**: 896–1003.

Kaszab, Z. 1941a: Die indomalayischen Misolampinen (Coleopt., Tenebr.). *Annales Musei Nationalis Hungarici* **34**: 1–44, pl. 1.

Kaszab, Z. 1941b: Die indischen Arten der Gattung Pachypterus Luc. *Folia Entomologica Hungarica* **6**: 16–20.

Kaszab, Z. 1941c: Tenebrioniden aus Formosa (Col.). *Stettiner Entomologische Zeitung* **102**: 51–72.

Kaszab, Z. 1941d: Neue orientalische Tenebrioniden (Coleoptera). *Arbeiten über Morphologische und Taxonomische Entomologie aus Berlin Dahlem* **8**: 118–127.

Kaszab, Z. 1941e: Zwei neue Pedinus-Arten (Coleopt., Tenebr.) von der Balkanhalbinsel. *Annales Musei Nationalis Hungarici* **34**: 161–164.

Kaszab, Z. 1941f: Zoologische Ergebnisse der ersten (VI.-X.1936.) und zweiten (V.-VIII.1937) Forschungsreisen Vasvári's in Kleinasien. IV. Meloidae. *Mathematischer und Naturwissenschaftlicher Anzeiger der Ungarischen Akademie der Wissenschaften* **60**: 674–681.

Kaszab, Z. 1942a: Beiträge zur Kenntnis der orientalischen Opatrinen (Col. Teneb.). *Mitteilungen der Münchener Entomologischen Gesellschaft* **32**: 1–43.

Kaszab, Z. 1942b: Die Meloiden Ungarns (Coleop.). *Fragmenta Faunistica Hungarica* **5**: 33–45.

Kaszab, Z. 1942c: Über die von J. Klapperich in China gesammelten Leiochrinen (Col., Tenebrionidae). *Entomologische Blätter* **38**: 214–217.

Kaszab, Z. 1943: Über die Gattungen Tearchus Kr. und Tonkinius Fairm. (Col., Tenebr.). *Folia Entomologica Hungarica* **8**: 71–74.

Kaszab, Z. 1944a: [new taxon]. In: Español F.: Nuevos datos para el conocimiento de los Tenebriónidos (Col.) del Sáhara Español. *Eos, Revista Española de Entomología* **20**: 7–30.

Kaszab, Z. 1944b: Über die Arten der Gattung Tetraphyllus Lap. & Brill. (Coleopt.). *Annales Musei Nationalis Hungarici* **37**: 25–71.

Kaszab, Z. 1946: *Monographie der Leiochrinen*. Budapest: Ungarisches Naturwissenschaftliches Museum, 221 pp., 1 pl.

Kaszab, Z. 1948: Die Arten der Gattung Cabalia Muls. (Col., Meloidae). *Folia Entomologica Hungarica* (S. N.) **3**: 12–14.

Kaszab, Z. 1951a: Eine neue Subspezies von Trigonoscelis fasciculitarsis Reitt. aus Turkmenien (Coleopt., Tenebr.). *Folia Entomologica Hungarica* (N.S.) **4**: 1–4, 1 pl.

Kaszab, Z. 1951b: Neue Revision der Gattung Alosimus Muls. (Col., Meloidae). *Annales Historico-Naturales Musei Nationalis Hungarici* **1**: 138–150

Kaszab, Z. 1951c: A new species of Gonocephalum (Col., Tenebrionidae) from the Philippine Islands. *The Annals and Magazine of Natural History* (12) **4**: 182–185.

Kaszab, Z. 1951d: Revision der Cerocominen. *Acta Biologica Academiae Scientiarum Hungaricae* **2**: 255–274.

Kaszab, Z. 1951e: Über die Arten der Gattung Oenas. *Acta Biologica Academiae Scientiarum Hungaricae* **2**: 275–279.

Kaszab, Z. 1952a: Neue Epicautinen (Col. Meloidae) aus der Orientalischen Region. *Entomologische Arbeiten aus dem Museum G. Frey* **3**: 79–89.

Kaszab, Z. 1952b: Über Lydus susicus Esc. und die neuen Arten der Gattung Lyttolydulus Reit. (Col. Meloidae) *Entomologische Arbeiten aus dem Museum G. Frey* **3**: 89–95.

Kaszab, Z. 1952c: Eine neue Lydus-Art aus Kleinasien, nebst einer Bestimmungstabelle der bisher bekannten Arten der Gattung Lydus sensu stricto (Col., Meloidae). *Annales Historico-Naturales Musei Nationalis Hungarici* (S.N.) **2**: 95–99.

Kaszab, Z. 1952d: Die indomalaischen und ostasiatischen Arten der Gattung Gonocephalum Solier (Coleoptera Tenebrionidae). *Entomologische Arbeiten aus dem Museum G. Frey* **3**: 416–688.

Kaszab, Z. 1952e: Die paläarktischen und orientalischen Arten der Meloiden-Gattung Epicauta Redtb. *Acta Biologica Academiae Scientiarum Hungaricae* **3**: 573–599.

Kaszab, Z. 1953a: The 3rd Danish expedition to central Asia. Zoological results 11. Meloiden (Insecta) aus Afghanistan. *Videnskabernas Meddelelser fra Dansk Naturhistorisk Forening i København* **115**: 305–311.

Kaszab, Z. 1953b: Einige neue Cylindrothorax-Arten (Col., Meloidae). *The Annals and Magazine of Natural History* (12) **6**: 857–861.

Kaszab, Z. 1954: Über die von Herrn J. Klapperich in der chinesischen Provinz Fukien gesammelten Tenebrioniden (Coleoptera). *Annales Historico-Naturales Musei Nationalis Hungarici* (S.N.) **5**: 248–264.

Kaszab, Z. 1955a: Revision der Meloiden-Gattung Eletica Lac. (Coleoptera-Heteromera). *Annales du Musée Royal du Congo Belge Tervuren, Sciences Zoologiques* **41**: 1–121, 11 pls.

Kaszab, Z. 1955b: Die Arten der Meloiden-Gattung Cylindrothorax Escher. (Coleoptera). *Annales Historico-Naturales Musei Nationalis Hungarici* (S.N.) **6**: 225–258.

Kaszab, Z. 1955c: Tenebrioniden der Fiji-Inseln. *Proceedings of the Hawaiian Entomological Society* **15**: 657–671.

Kaszab, Z. 1956a: Neue Tenebrioniden (Coleoptera) aus der papuanischern und indomalayischen Region. *Annales Historico-Naturales Musei Nationalis Hungarici* (S.N.) **7**: 93–108.

Kaszab, Z. 1956b: Neue Meloiden-, Alleculiden-, Serropalpiden-, Anthiciden- und Oedemeriden-Formen aus Ungarn und den angrenzenden Gebieten (Coleoptera). *Folia Entomologica Hungarica* (N.S.) **9**: 142–172.

Kaszab, Z. 1956c: Neue Heteromera (Coleoptera) aus Ungarn. *Acta Zoologica Academiae Scientiarum Hungaricae* **2**: 333–338.

Kaszab, Z. 1956d: Zwei neue Prosodes-Arten aus der Untergattung Oligoprosodes Reitt. (Coleoptera, Tenebrionidae). *The Annals and Magazine of Natural History* (12) **9**: 418–422.

Kaszab, Z. 1956e: Neue Meloiden aus der Sammlung des British Museum (Coleoptera). *The Annals and Magazine of Natural History* (12) **9**: 631–640.

Kaszab, Z. 1957a: Einige neue Heteromeren aus Asien (Coleoptera). *Opuscula Zoologica* (Budapest) **2**: 47–53.

Kaszab, Z. 1957b: Neue Meloiden aus Iran 1954 (Coleopt.) (Ergebnisse der Entomologischen Reisen Willi Richter, Stuttgart, in Iran 1954 und 1956 – Nr. 6). *Jahreshefte des Vereins für Vaterländische Naturkunde in Württemberg* **112**: 50–59.

Kaszab, Z. 1957c: Weitere neue Tenebrioniden (Coleoptera) aus Fukien. *Bonner Zoologische Beiträge* **8**: 56–63.

Kaszab, Z. 1957d: Neue Tenebrioniden aus Iran 1954 (Coleopt.) (Ergebnisse der Entomologischen Reisen Willi Richter, Stuttgart, im Iran 1954 und 1956 – Nr. 7). *Jahreshefte des Vereins für Vaterländische Naturkunde in Württemberg* **112**: 60–65.

Kaszab, Z. 1957e: New Meloidae (Coleoptera) from Israel. *Bulletin of the Research Council of Israel, Section Biology and Geology* **6 b**: 229–232.

Kaszab, Z. 1957f: Zehn neue Tenebrioniden aus Asien (Coleoptera). *Annales Historico-Naturales Musei Nationalis Hungarici* (S.N.) **8**: 289–299.

Kaszab, Z. 1958a: Die Arten der Meloiden-Gattung Micromerus Muls. und Rey (Coleoptera). *Acta Entomologica Musei Nationalis Pragae* **32** (488): 5–18.

Kaszab, Z. 1958b: Neue Meloiden-Namen (Coleoptera). *Annales Historico-Naturalis Musei Nationalis Hungarici* **50**: 189–191.

Kaszab, Z. 1958c: Die Meloiden Afghanistans (Coleoptera). *Acta Zoologica Academiae Scientiarum Hungaricae* **3**: 245–312.

Kaszab, Z. 1959a: Beiträge zur Kenntnis der Tenebrioniden-Fauna von Iran (Coleoptera). *Stuttgarter Beiträge zur Naturkunde* **19**: 1–3.

Kaszab, Z. 1959b: Neue Meloiden aus Iran 1954, 1956 (Coleoptera). *Stuttgarter Beiträge zur Naturkunde* **20**: 1–5.

Kaszab, Z. 1959c: Drei neue Blaps- und eine neue Laena-Art aus der Türkei und den angrenzenden Gebieten (Coleoptera Tenebrionidae). *Kungliga Fysiografiska Sällskapets i Lund Förhandlingar* **29**: 51–56.

Kaszab, Z. 1959d: Wissensschaftliche Ergebnisse der zoologischen Expedition des National-Museums in Prag nach der Türkei. 24. Coleoptera Tenebrionidae. *Acta Entomologica Musei Nationalis Pragae* **33**: 69–82.

Kaszab, Z. 1959e: The 3rd Danish Expedition to Central Asia. Zoological Results 24. Ergänzungen zur Tenebrioniden (Insecta). Aufarbeitung von Dr. E. Gridelli der 3. Dänischen Zentralasiatischen Expedition. *Videnskabelige Meddelelser fra Dansk Naturhistorisk Forening i København* **120** [1958]: 237–255.

Kaszab, Z. 1959f: Die Tenebrioniden Afghanistans, auf Grund der Ergebnisse der Sammelreise des Herrn J. Klapperich in den Jahren 1952/53 (Col.). *Entomologische Arbeiten aus dem Museum G. Frey* **10**: 321–404, 10 pls.

Kaszab, Z. 1959g: Neue Leptodes-Arten aus Asien, nebst einer Revision der Leptodini (Coleoptera, Tenebrionidae). *Acta Zoologica Academiae Scientiarum Hungaricae* **4**: 349–368.

Kaszab, Z. 1959h: Eine neue Meloiden-Art und einige neue Formen der Koleopteren-Sammlung des Naturhistorischen Museums in Wien. *Annalen des Naturhistorischen Museums in Wien* **63**: 436–441.

Kaszab, Z. 1960a: Die Tenebrioniden Afghanistans, auf Grund der Ergebnisse der Sammelreise des Herrn J. Klapperich in den Jahren 1952/53 (Col.). 1. Fortsetzung und Schluss. *Entomologische Arbeiten aus dem Museum G. Frey* **11**: 1–179.

Kaszab, Z. 1960b: Die Arten und systematische Stellung der Meloiden-Gattung Calydus Reitter (Coleoptera). *Acta Zoologica Academiae Scientiarum Hungaricae* **6**: 125–134.

Kaszab, Z. 1960c: Einige neue Meloiden (Coleoptera) aus der Sammlung des British Museum. *The Annals and Magazine of Natural History* (13) **3**: 153–159.

Kaszab, Z. 1960d: Wissenschaftliche Ergebnisse der chinesisch-sowjetischen zoologischen Expedition nach SW. China. Meloidae (Coleoptera). *Annales Historico-Naturalis Musei Nationalis Hungarici* **52**: 255–263.

Kaszab, Z. 1960e: Neue orientalische Misolampinen (Coleoptera, Tenebrionidae) *Annales Historico-Naturales Musei Nationalis Hungarici* **52**: 265–294.

Kaszab, Z. 1960f: Mission zoologique de l I.R.S.A.C. en Afrique orientale (P. Basilewsky et N. Leleup, 1957). XLV. Coleoptera Meloidae. *Annales du Musée Royal du Congo Belge Tervuren, Sciences Zoologiques* 88: 269–286.

Kaszab, Z. 1960g: Einige neue Meloiden (Coleoptera) des Museum G. Frey. *Entomologische Arbeiten aus dem Museum G. Frey* **11**: 283–290.

Kaszab, Z. 1961a: Wissenschaftliche Ergebnisse der von Dr. F. Schmid in Pakistan und Iran gesammelte Tenebrioniden (Coleoptera). *Miscelánea Zoológica* **1**: 115–121.

Kaszab, Z. 1961b: Revision der Tenebrioniden-Gattung Derispia Lewis (Coleoptera). *Acta Zoologica Academiae Scientiarum Hungarica* 7: 139–184.

Kaszab, Z. 1961c: Zwei neue Tenebrioniden (Coleoptera) aus Arabien und Kleinasien. *Abhandlungen und Berichte aus dem Staatlichen Museum für Tierkunde in Dresden* **26**: 169–175.

Kaszab, Z. 1961d: Neue Tenebrioniden (Coleoptera) aus der Zoologischen Staatssammlung in Münnchen. *Mitteilungen der Münchener Entomologischen Gesellschaft* **51**: 213–230.

Kaszab, Z. 1961e: Ergebnisse der Deutschen Afghanistan-Expedition 1956 der Landessamlungen für Naturkunde Karlsruhe. Tenebrionidae (Coleoptera). *Beiträge zur Naturkundlichen Forschung in Südwestdeutschland* **19**: 235–238.

Kaszab, Z. 1961f: Ergebnisse der österreichischen Karakorum-Expedition 1958 Coleoptera – Tenebrionidae. *Annales Historico-Naturales Musei Nationalis Hungarici* **53**: 341–355.

Kaszab, Z. 1961g: Neue Meloiden (Coleoptera) aus dem Museum A. Koenig. *Bonner Zoologische Beiträge* **12**: 343–353.

Kaszab, Z. 1961h: Beiträge zur Kenntnis der Tenebrioniden-Tribus Leiochrini (Coleoptera). *Annales Historico-Naturales Musei Nationalis Hungarici* **53**: 357–380.

Kaszab, Z. 1961i: Neue Arten der Gattung Leiochrodes Westwood (Coleoptera: Tenebrionidae). *Acta Zoologica Academiae Scientiarum Hungarica* 7: 433–466.

Kaszab, Z. 1962a: Beiträge zur Kenntnis der chinesischen Tenebrioniden-Fauna (Coleoptera). *Acta Zoologica Academiae Scientiarum Hungaricae* 8: 75–86.

Kaszab, Z. 1962b: Über das System der asiatischen Lytta-Arten, nebst Beschreibung drei neuer Arten (Coleoptera, Meloidae). *Annales Historico-Naturales Musei Nationalis Hungarici* **54**: 289–298.

Kaszab, Z. 1962c: Beiträge zur Kenntnis einiger asiatischen Tenebrioniden-Gattungen und Arten (Coleoptera). *Annales Historico-Naturales Musei Nationalis Hungarici* **54**: 299–317.

Kaszab, Z. 1963a: Ergebnisse der österreichischen Iran-Expedition 1949/50. X. Fünf neue Tenebrioniden aus Iran. *Anzeiger der Österreichischen Akademie der Wissenschaften Mathematisch-Naturwissenschaftliche Klasse* **100** (5): 69–77.

Kaszab, Z. 1963b: Die paläarktischen und orientalischen Arten der Gattung Mesomorphus Seidl. (Coleoptera, Tenebrionidae). *Acta Zoologica Academiae Scientiarum Hungaricae* 9: 333–354.

Kaszab, Z. 1963c: Angaben zur Kenntnis der Tenebrioniden des Tschadsee-Gebietes, nebst einer Revision der afrikanischen Mesomorphus-Arten (Coleoptera). *Revue de Zoologie et de Botanique Africaines* 67: 341–387.

Kaszab, Z. 1964a: Über die Tenebrioniden einiger japanischen Inseln (II). *Entomological Review of Japan* **17**: 1–10.

Kaszab, Z. 1964b: Tenebrionidae der mongolisch-deutschen biologischen Expedition 1962 (4. Coleoptera, Tenebrionidae). *Entomologische Abhandlungen* **32**: 1–26.

Kaszab, Z. 1964c: Über die Tenebrioniden einiger japanischen Inseln (I). *Entomological Review of Japan* **16**: 39–49.

Kaszab, Z. 1964d: Ergebnisse der mongolisch-deutschen biologischen Expedition 1962. Nr. 3. Meloidae (Coleoptera). *Folia Entomologica Hungarica* (N.S.) **17**: 317–324.

Kaszab, Z. 1964e: Ergebnisse der zoologischen Forschungen von Dr. Z. Kaszab in der Mongolei. 15. Meloidae (Coleoptera). *Folia Entomologica Hungarica* (N.S.) **17**: 345–363.

Kaszab, Z. 1964f: The Zoological Results of the Gy. Topál's collectings in South Argentina. *Annales Historico-Naturales Musei Nationalis Hungarici* 56: 353–387.

Kaszab, Z. 1964g: Beiträge zur Kenntnis der Tenebrioniden-Fauna des mittleren Teiles der Mongolischen Volksrepublik (Coleoptera). *Acta Zoologica Academiae Scientiarum Hungaricae* 10: 363–404.

Kaszab, Z. 1965a: Wissenschaftliche Ergebnisse der von Dr. F. Schmid in Indien gesammelten Tenebrioniden (Coleoptera). *Miscelánea Zoológica* 2: 107–120.

Kaszab, Z. 1965b: Meloiden und Tenebrioniden aus der Mongolei 1964 von H. Muche gesammelt (Coleoptera). *Reichenbachia* 5: 131–134.

Kaszab, Z. 1965c: Neue Tenebrioniden (Coleoptera) aus China. *Annales Historico-Naturales Musei Nationalis Hungarici* 57: 279–285.

Kaszab, Z. 1965d: Zoologische Ergebnisse der Forschungen von Dr. T. Pócs in der Volksrepublik Vietnam. Tenebrionidae (Coleoptera). *Annales Historico-Naturales Musei Nationalis Hungarici* 57: 287–296.

Kaszab, Z. 1965e: Angaben zur Kenntnis der Tenebrioniden-Fauna der Mongolischen Volksrepublik (Coleoptera). *Acta Zoologica Academiae Scientiarum Hungaricae* 11: 295–346.

Kaszab, Z. 1965f: Ergebnisse der zoologischen Forschungen von Dr. Z. Kaszab in der Mongolei. 38. Meloidae (Coleoptera). *Folia Entomologica Hungarica* (N.S.) 18: 551–561.

Kaszab, Z. 1965g: Ein neuer Käfer der Familie Tenebrionidae (Coleoptera) aus der Mongolei. *Annales Zoologici* 23: 1–4.

Kaszab, Z. 1966a: 74. Meloidae der III. Expedition. Ergebnisse der zoologischen Forschungen von Dr. Z. Kaszab in der Mongolei (Coleoptera). *Reichenbachia* 7: 259–264.

Kaszab, Z. 1966b: Revision der Tenebrioniden-Gattung Microdera Eschsch. (Coleoptera). *Acta Zoologica Academiae Scientarum Hungaricae* 12: 279–305.

Kaszab, Z. 1967a: Coleoptera: Tenebrionidae der mongolisch-deutschen biologischen Expedition 1964. *Mitteilungen aus dem Zoologischen Museum in Berlin* 43: 3–33.

Kaszab, Z. 1967b: Die Tenebrioniden der Westmongolei (Coleoptera). *Acta Zoologica Academiae Scientiarum Hungaricae* 13: 279–351.

Kaszab, Z. 1968a: Tenebrionidae und Meloidae (Coleoptera) aus Nord-Korea, gesammelt von M. Mroczkowski und A. Riedel im Jahre 1965. *Annales Zoologici* 26: 7–14.

Kaszab, Z. 1968b: Beiträge zur Kenntnis der Fauna Afghanistans (Sammelergebnisse von O. Jakeš 1963–64, D. Povolný & Fr. Tenora 1966, J. Šimek 1965–66, D. Povolný, J. Geiser, Z. Šebek & Fr. Tenora 1967) Tenebrionidae, Col. *Časopis Moravského Musea, Vědy přírodní* 53, *Supplementum*, 7–124 pp., 4 pls.

Kaszab, Z. 1968c: Ergebnisse der zoologischen Forschungen von Dr. Z. Kaszab in der Mongolei. 168. Tenebrionidae (Coleoptera). *Acta Zoologica Academiae Scientiarum Hungaricae* 14: 339–397.

Kaszab, Z. 1968d: Ergebnisse zoologischer Sammelreisen in der Türkey. Coleoptera: Meloidae. *Annalen des Naturhistorischen Museums in Wien* 72: 443–450.

Kaszab, Z. 1968e: Ergebnisse zoologischer Sammelreisen in der Türkei. Coleoptera: Tenebrionidae. *Annalen des Naturhistorischen Museums in Wien* 72: 451–463.

Kaszab, Z. 1968f: Contribution à la faune de l'Iran, 8. Coléoptères Meloidae. *Annales de la Société Entomologique de France* (N.S.) 4: 749–776.

Kaszab, Z. 1969a: Three new East Mediterranean meloids (Coleoptera, Meloidae). *Israel Journal of Entomology* 4: 279–284.

Kaszab, Z. 1969b: Ergebnisse der zologischen Forschungen von Dr. Z. Kaszab in der Mongolei. 168. Tenebrionidae (Coleoptera) der VI. Expedition. *Acta Zoologica Academiae Scientarum Hungaricae* **15**: 299–332.

Kaszab, Z. 1969c: 177. Meloidae der V. Expedition. Ergebnisse der zoologischen Forschungen von Dr. Z. Kaszab in der Mongolei (Coleoptera). *Faunistische Abhandlungen, Staatalisches Museum für Tierkunde in Dresden* **2**: 317–322.

Kaszab, Z. 1969d: Contribution à la faune de l'Iran 15. Coléoptères Tenebrionidae, genre Lobodera. *Annales de la Société Entomologique de France* (N.S.) **5**: 451–456.

Kaszab, Z. 1970a: Beiträge zur Kenntnis der Fauna Afghanistans (Sammelergebniss von O. Jakeš 1963–64, D. Povolný & Fr. Tenora 1966, J. Šimek 1965–66, D, Povolný, J. Geiser, Z. Šebek & Fr. Tenora 1967). Tenebrionidae, Col. *Časopis Moravského Musea, Vědy přírodní* **54**, *Supplementum*: 5–182 pp., 23 pls.

Kaszab, Z. 1970b: Fünf neue Tenebrioniden aus Asien (Coleoptera). *Entomologische Arbeiten aus dem Museum G. Frey* **21**: 112–122.

Kaszab, Z. 1970c: Beiträge zur Kenntnis der Tenebrioniden-Fauna von Nepal (Coleoptera). *Khumbu Himal* **3**: 422–434.

Kaszab, Z. 1972: Missione 1965 del Prof. Giuseppe Scortecci nello Yemen (Arabia meridionale). *Atti della Società Italiana di Scienze Naturali e del Museo Civico di Storia Naturale di Milano* **113** (4): 366–384, pl. 17.

Kaszab, Z. 1973a: Tenebrionidae (Coleoptera) aus Nepal. *Acta Zoologica Academiae Scientiarum Hungaricae* **19** (1972): 23–74, pls I–V.

Kaszab, Z. 1973b: Neue Beiträge zur Kenntnis der Tenebrioniden-Fauna van Afghanistan (Coleoptera). *Entomologia Scandinavica* **4**: 35–58.

Kaszab, Z. 1973c: [new taxon]. In: Medvedev G.S. & Kaszab Z. 1973: Ergebnisse der Mongolisch-sowietischen biologischen Expeditionen in der Mongolischen Volksrepublik Tenebrionidae, excl. Tentyriini (Coleoptera). *Folia Entomologica Hungarica* (N.S.) **26**: 79–111.

Kaszab, Z. 1973d: Beiträge zur Kenntnis der Fauna Afghanistans (Sammelergebnisse von O. Jakeš 1963–64, D. Povolný 1965, D. Povolný & Fr. Tenora 1966, J. Šimek 1965–66, D. Povolný, J. Gaisler, Z. Šebenek & Fr. Tenora 1967). Meloidae. Col. *Časopis Moravského Musea, Vědy přírodní* **56–57**: 267–308.

Kaszab, Z. 1974: Neue Angaben zur Kenntniss der Tenebrioniden (Coleoptera) von Afghanistan. *Annales Historico-Naturales Musei Nationalis Hungarici* **66**: 187–203.

Kaszab, Z. 1975a: Tenebrioniden aus Indien (Coleoptera). *Acta Zoologica Academiae Scientarum Hungaricae* **21**: 1–38, 7 pls.

Kaszab, Z. 1975b: Tenebrioniden (Coleoptera) aus Nepal, II. *Annales Historico-Naturales Musei Nationalis Hungarici* **67**: 119–126.

Kaszab, Z. 1975c: Die Arten der Gattung Scaphidema Redt. (Coleoptera, Tenebrionidae). *Annales Historico-Naturales Musei Nationalis Hungarici* **67**: 127–130.

Kaszab, Z. 1975d: Revision der asiatischen Platynotinen (Coleoptera: Tenebrionidae). *Acta Zoologica Academiae Scientiarum Hungaricae* **21**: 277–367, 7 pls.

Kaszab, Z. 1975e: Ergebnisse der Bhutan-Expedition 1972 des Naturhistorischen Museums in Basel. Coleoptera: Fam. Tenebrionidae. *Entomologia Basiliensia* **1**: 313–333.

Kaszab, Z. 1976a: Revision der Arten der Gattung Heterotarsus Latreille, 1829 (Coleoptera: Tenebrionidae). *Acta Zoologica Academiae Scientiarum Hungaricae* **22**: 33–63.

Kaszab, Z. 1976b: Zwei neue Arten der Gattung Epitrichia Mäklin, 1872 (Coleoptera, Tenebrionidae). *Annales Historico-Naturales Musei Nationalis Hungarici* 68: 99–104.

Kaszab, Z. 1976c: Ergebnisse der mongolisch-sowjetischen biologischen Expedition in der Mongolischen Volksrepublik seit 1967. Tenebrionidae: Tentyriini (Coleoptera), 11. *Annales Historico-Naturales Musei Nationalis Hungarici* 68: 105–115.

Kaszab, Z. 1976d: [new taxon]. In: Kaszab Z. & Knor I.: Zur Tenebrionidenfauna der Tuvinischen ASSR in Südsibirien (Coleoptera). *Acta Zoologica Academiae Scientiarum Hungaricae* 22: 293–297.

Kaszab, Z. 1977a: Neue Tenebrioniden (Coleoptera) von den Galapagos und Antillen, sowie aus Südamerika und Ostasien. *Annales Historico-Naturales Musei Nationalis Hungarici* 69: 117–131.

Kaszab, Z. 1977b: Tenebrionidae der Nepal-Expeditionen von Dr. J. Martens (1969–1974) (Insecta: Coleoptera). *Senckenbergiana Biologica* 57 [1976]: 241–283.

Kaszab, Z. 1977c: Ergänzungen zur Tenebrionidenfauna der Westmongolei (Coleoptera). Ergebnisse der Mongolisch-Deutschen biologischen Expeditionen seit 1962, Nr. 76. *Mitteilungen aus dem Zoologischen Museum in Berlin* 53: 277–279.

Kaszab, Z. 1978a: Vier neue Tenebrioniden aus Nordbengal (Coleoptera). *Folia Entomologica Hungarica* (N.S.) 31 (2): 173–176.

Kaszab, Z. 1978b: [new taxa]. In: Kaszab Z., Schawaller W. & Skopin N.G.: Systematik und Ökologie einiger Tenebrionidae aus Kashmir und Ladakh. *Senckenbergiana Biologica* 59: 215–234.

Kaszab, Z. 1978c: Neue Meloiden aus Asien, Afrika und Südamerika (Coleoptera). *Acta Zoologica Academiae Scientiarum Hungaricae* 24: 331–342.

Kaszab, Z. 1979a: Revision der asiatischen Arthrodosis-Artigen Erodiini (Coleoptera: Tenebrionidae). *Acta Zoologica Academiae Scientiarum Hungaricae* 25: 69–119, 1 pl.

Kaszab, Z. 1979b: Die Arten der Gattung Sivacrypticus Kaszab, 1964 (Coleoptera, Tenebrionidae). *Annales Historico-Naturales Musei Nationalis Hungarici* 71: 185–204.

Kaszab, Z. 1979c: Insects of Saudi Arabia. Coleoptera: Fam. Tenebrionidae. *Fauna of Saudi Arabia* 1: 257–288.

Kaszab, Z. 1980a: Neue Tenebrioniden (Coleoptera) aus Sri Lanka I. *Acta Zoologica Academiae Scientiarum Hungaricae* 26: 123–196.

Kaszab, Z. 1980b: Angaben zur Kenntnis der Tenebrioniden Nordvietnams (Coleoptera). *Annales Historico-Naturales Musei Nationalis Hungarici* 72: 169–221.

Kaszab, Z. 1981a: Faunistische und taxonomische Studien über Meloiden (Coleoptera). *Annales Historico-Naturales Musei Nationalis Hungarici* 73: 159–181.

Kaszab, Z. 1981b: Neue orientalische Stenosini (Coleoptera: Tenebrionidae). *Acta Zoologica Academiae Scientiarum Hungaricae* 27: 273–313.

Kaszab, Z. 1981c: Insects of Saudi Arabia. Coleoptera: Fam. Tenebrionidae (Part 2) Zusammenfassung der bis jetzt bekannt gewordenen Tenebrioniden (First part). *Fauna of Saudi Arabia* 3: 276–401.

Kaszab, Z. 1982a: Neue orientalische Tenebrioniden. *Acta Zoologica Academiae Scientiarum Hungaricae* 28 57–80.

Kaszab, Z. 1982b: Insects of Saudi Arabia. Coleoptera: Fam. Tenebrionidae (Part 2) Zusammenfassung der bis jetzt bekanntgewordenen Tenebrioniden (Second part). *Fauna of Saudi Arabia* 4: 124–243.

Kaszab, Z. 1983a: Insects of Saudi Arabia. Coleoptera: Fam. Meloidae. A synopsis of the Arabian Meloidae. *Fauna of Saudi Arabia* 5: 144–204.

Kaszab, Z. 1983b: Über die Gattung Leprocaulus Fairmaire, 1896 (Coleoptera, Tenebrionidae). *Annales Historico-Naturales Musei Nationalis Hungarici* 75: 177–183.

Kaszab, Z. 1985: Drei neue Tenebrioniden (Coleoptera) aus Asien. *Bollettino di Museo Civico di Storia Naturale di Verona* 12: 449–459.

Kaszab, Z. 1986: Beiträge zur Kenntnis der Tenebrioniden (Coleoptera) aus Arabien. *Annales Historico-Naturales Musei Nationalis Hungarici* 78: 147–150.

Kaszab, Z. 1988: Katalog und Bestimmungstabelle der Gattung Promethis Pascoe, 1869 (Coleoptera, Tenebrionidae). *Acta Zoologica Hungarica* 34: 67–170, pls I–XV.

Kaszab, Z. & Knor, I. 1976: Zur Tenebrionidenfauna der Tuvischen ASSR in Südsibirien (Coleoptera). *Acta Zoologica Academiae Scientiarum Hungaricae* 22: 293–297.

Kaszab, Z. & Medvedev, G.S. 1972: Ergebnisse der mongolisch-sowjetischen biologischen Expedition in der Mongolischen Volksrepublik seit 1967. Tenebrionidae: Tentyriini (Coleoptera). *Annales Historico-Naturales Musei Nationalis Hungarici* 64: 203–218.

Kaszab, Z. & Medvedev, G.S. 1976: Ergebnisse der mongolisch-sowjetischen biologischen Expeditionen in der Mongolischen Volksrepublik seit 1967. Tenebrionidae: Tentyriini (Coleoptera), II. *Annales Historico-Naturales Musei Nationalis Hungarici* 68: 105–115.

Kaszab, Z. & Medvedev, G.S. 1977: Tentyriini (Coleoptera, Tenebrionidae) aus der Mongolei, III. *Annales Historico-Naturales Musei Nationalis Hungarici* 69: 133–143.

Kaszab, Z. & Medvedev, G.S. 1984: Drei neue asiatische Tenebrioniden (Coleoptera). *Acta Zoologica Hungarica* 30: 79–85.

Kaszab, Z. & Pinheiro, M.F. 1972: Una nova espécie de Sepidium (Coleoptera, Tenebrionidae) em Portugal. *Estudos e Divulgaçao Téchnica (C) Entomologia Forestal* 1972: 5–17.

Katter, F. 1883: *Monographiae der europäischen Arten der Gattung Meloë mit besonderer Berücksichtigung der Biologie dieser Insekten.* Putbus: Dose, 61 pp.

Kawanabe, M. 1993: A new genus of the family Ciidae (Coleoptera) from Yakushima Island, Southwest Japan. *Japanese Journal of Entomology* 61: 763–76.

Kawanabe, M. 1994a: A new genus and species of the family Ciidae (Coleoptera) from Southwest Japan. *Japanese Journal of Entomology* 62: 186–192.

Kawanabe, M. 1994b: A new species of the genus Orthocis Casey from Japan (Coleoptera: Ciidae). *Transactions of the Shikoku Entomological Society* 20: 187–190.

Kawanabe, M. 1995: Taxonomic notes on the genus Nipponapterocis (Coleoptera, Ciidae), with descriptions of two new species from Japan. *Elytra* 23: 167–175.

Kawanabe, M. 1996a: A new species of the genus Ennearthron (Coleoptera, Ciidae) from Hokkaido, Japan. *Japanese Journal of Systematic Entomology* 2: 105–108.

Kawanabe, M. 1996b: A new genus of the family Ciidae (Coleoptera), with description of a new species from the Ryukyu Islands, southwest Japan. *Elytra* 24: 331–336.

Kawanabe, M. 1997a: A taxonomic study on the Japanese species of the genus Sulcacis (Coleoptera, Ciidae). *Elytra* 25: 67–78.

Kawanabe, M. 1997b: The Japanese species related to the group of Cis nitidus (Coleoptera, Ciidae). *Elytra* 25: 311–331.

Kawanabe, M. 2001: A new ciid beetle of the genus Cis from Hokkaido, Japan (Coleoptera: Ciidae). *Sukunahikona. Special Publication of the Japan Coleopterological Society* (Osaka) 1: 267–270.

Kawanabe, M. 2002: A revision of the genus Octotemnus (Coleoptera, Ciidae) from Japan. *Japanese Journal of Systematic Entomology* 8: 243–267.

Kawanabe, M. 2005: A new species of the genus Octotemnus(Coleoptera, Ciidae) from Nantou County, central Taiwan. *Elytra* 33: 47–50.

Kazantsev, S.V. & Young, D.K. 2010: Two new species of Ischalia Pascoe, 1860 from China, with observations on previously described Chinese species and notes on mimicry in the Palaearctic and Nearctic members of the genus (Coleoptera: Ischaliidae). *Russian Entomological Journal* 19: 307–312.

Kejval, Z. 1998: Review of oriental and some Palaearctic species of the genus Endomia (Coleoptera: Anthicidae). *European Journal of Entomology* 95: 99–131.

Kejval, Z. 1999: Five new species of Notoxus (Coleoptera: Anthicidae) from Palaearctic Asia with redescription of N. quadrimaculatus. *Klapalekiana* 35: 13–23.

Kejval, Z. 2000a: Revisional notes on Oriental Formicomus LaFerté-Sénectére (Coleoptera, Anthicidae) 2. Himalayan species related to F. lagenicollis Fairmaire, 1894. *Acta Musei Moraviae, Scientiae Biologicae* (Brno) 85: 73–111.

Kejval, Z. 2000b: Revisional notes on the Palaearctic Formicomus (Coleoptera: Anthicidae). *Acta Societatis Zoologicae Bohemicae* 64: 403–417.

Kejval, Z. 2002a: The species of Anthelephila (Coleoptera: Anthicidae) related to A. angustiformis and A. walkeri. *Folia Heyrovskyana* 10: 83–114.

Kejval, Z. 2002b: New species, synonymy and records of Anthelephila (Coleoptera: Anthicidae). *Acta Societatis Zoologicae Bohemicae* 66: 243–278.

Kejval, Z. 2002c: Contribution to the knowledge of the genus Stenidius (Coleoptera: Anthicidae). *Klapalekiana* 38: 185–212.

Kejval, Z. 2006: Studies of the genus Anthelephila Hope, 1833 (Coleoptera, Anthicidae). 9. Descriptions of five new species from the Oriental region. *Entomologica Basiliensia et Collectionis Frey* 28: 175–191.

Kejval, Z. 2007. Studies of the Genus Anthelephila Hope (Coleoptera: Anthicidae) – 10. Species related to A. imperatrix. *Revue Suisse de Zoologie* 114: 615–653.

Kejval, Z. 2010: Studies on the genus Anthelephila (Coleoptera: Anthicidae) 11. New species and records from India, Sri Lanka, Nepal and Oman. *Acta entomologica musei nationalis pragae* 50: 189–234.

Kejval, Z. 2011a: Revision of the Oriental, East Palaearctic and some Australasian species of Mecynotarsus (Coleoptera: Anthicidae). *Folia heyrovskyana, Series A*, 18: 111–179.

Kejval, Z. 2011b: Taxonomic revision of the Oriental species of Notoxus (Coleoptera: Anthicidae). *Acta entomological musei nationalis pragae* 51: 627–673.

Kejval, Z. 2012: Studies on the genus Anthelephila (Coleoptera: Anthicidae). 12 Review of the species from Yemen, including Socotra Island. *Acta entomologica musei nationalis pragae 52, supplementum*: 347–363.

Kejval, Z. 2015a: On some remarkable species of Sapintus Casey (Coleoptera: Anthicidae) from the Himalaya and Hindu Kush Mountain Ranges. *Klapalekiana* 51: 49–53.

Kejval, Z. 2015b: On the Anthicus giganteus LaFerté-Sénectère, 1849 species-group (Coleoptera: Anthicidae). *Klapalekiana* 51: 201–217.

Kejval, Z. 2017a: The Palaearctic species of Aulacoderus LaFerté-Sénectère (Coleoptera: Anthicidae) *Klapalekiana* 55: 55–133.

Kejval, Z. 2017b: New species, synonymy and records of Mecynotarsus (Coleoptera: Anthicidae). *Klapalekiana* **53**: 299–322.

Kejval, Z. 2017c: Studies of the genus Anthelephila Hope (Coleoptera: Anthicidae) 14. Twenty-four new species from Asia and new records of A. fossicollis Kejval. *Zootaxa* **4306**: 1–52.

Kejval, Z. 2017d: Description of Notoxus monpa sp. nov. and the first record of N. hilaris (Coleoptera: Anthicidae) from northeastern India. *Klapalekiana* **53**: 51–54.

Kejval, Z. 2018a: Studies of the genus Anthelephila Hope (Coleoptera: Anthicidae) 17. A. lagenicollis species-group. *Acta musei moraviae, scientiae biologicae* **103**, No 2: 207–248.

Kejval, Z. 2018b: Studies of the genus Anthelephila Hope (Coleoptera: Anthicidae) 18. New species and records from the Palaearctic region. *Klapalekiana* **54**: 207–220.

Kejval, Z. 2019: Two new species of Steropes Steven (Coleoptera: Anthicidae) from southern China. *Acta Musei Moraviae, Scientiae biologicae* **104**: 139–146.

Kejval, Z. & Nardi, G. 2018: On some Notoxus (Coleoptera: Anthicidae) of the N. monoceros species-group. *Klapalekiana* **54**: 41–60.

Kelejnikova, S.I. 1963: Novyy rod i vid triby Tentyriini (Coleoptera, Tenebrionidae) iz Kirgizii. *Zoologicheskiy Zhurnal* **42** (4): 622–623.

Kelejnikova, S.I. 1970: Novyy vid roda Blaps (Col., Tenebr.) iz Altaya. Pp. 27–29. In: Tscherepanov A.I. (ed.): *Novye i maloizvetsnye vidy sibirskoy fauny*. 3. Institut Biologii, Sibirskoye Otdelenie Akademiy Nauk SSSR. Novosibirsk: Nauka, 112 pp.

Kelejnikova, S.I. 1977: Novye chernotelki (Coleoptera, Tenebrionidae) iz Turkmenii. *Entomologicheskoe Obozrenie* **56** (3): 653–655.

Keskin, B. & Nabozhenko, M.V. 2010: A new species and new records of the genus Nalassus Mulsant, 1854 (Coleoptera: Tenebrionidae: Helopini) from Turkey. *Annales Zoologici* **60**: 23–28.

Keskin, B. & Nabozhenko, M.V. 2011: Review of the genus Odocnemis Allard, 1876: O. korbi species-group (Coleoptera: Tenebrionidae: Helopini). *Annales Zoologici* **61**: 339–354.

Keskin, B., Nabozhenko, M.V. 2012: Idahelops alpagutae (Coleoptera: Tenebrionidae: Helopini): a new genus and species from the Aegean region of Turkey. *Zootaxa* **3207**: 63–67.

Keskin, B. & Nabozhenko, M.V. 2015: The new genus Taurohelops (Coleoptera: Tenebrionidae) from Anatolia. *The Coleopterists Bulletin* **69** (m04): 83–92.

Keskin, B., Nabozhenko, M.V. & Alpagut, Keskin N. 2017a: Eustenomacidius egeuniversitatis sp.n. – the first record of the tenebrionid genus in Turkey. *Turkish Journal of Zoology* **41**: 237–240.

Keskin, B., Nabozhenko, M.V. & Alpagut-Keskin, N. 2017b: Taxonomic review of the genera Nalassus Mulsant, 1854 and Turkonalassus gen. nov. of Turkey (Coleoptera: Tenebrionidae). *Annales Zoologici* **67**: 725–747.

Kiesenwetter, E.A.H. von. 1851: Énumération des coléoptères. *Annales de la Société Entomologique France* (2) **9**: 577–656.

Kiesenwetter, E.A.H. von. 1859: Beitrag zur Käferfauna Griechenlands. Sechstes Stück: Malacodermata, Cleridae, Ptinidae, Anobiadae. *Berliner Entomologische Zeitschrift* **3**: 158–192.

Kiesenwetter, E.A.H. von. 1861a: Beitrag zur Käferfauna Griechenlands. Siebentes Stück: Tenebrionidae, Cistelidae, Lagriariae, Pedilidae, Anthicidae, Mordellonae, Meloidae, Oedemeridae. *Berliner Entomologische Zeitschrift* 5: 221–252.

Kiesenwetter, E.A.H. von. 1861b: Eine entomologische-Excursion in das Wallis und nach dem Monte Rosa. *Berliner Entomologische Zeitschrift* 5: 360–3[illegible]5.

Kiesenwetter, E.A.H. von. 1863: Beiträge zur Kenntniss der zur Untergattung Isomira Muls. gehörenden Cistela-Arten. *Berliner Entomologische Zeitschrift* 7: 423–427.

Kiesenwetter, E.A.H. von. 1866: Eine entomologische Excursion nach Spanien im Sommer 1865. *Berliner Entomologische Zeitschrift* 9 [1865]: 359–396.

Kiesenwetter, E.A.H. von. 1870: [new taxa]. In: Heyden L.F.J.D. von: Entomologische Reise nach dem südlichen Spanien, der Sierra Guadarrama und Sierra Morena, Portugal und den Cantabrischen Gebirgen. *Berliner Entomologische Zeitschrift* 14 *Beiheft*, 218 pp., 2 pls.

Kiesenwetter, E.A.H. von. 1873: Revision der Cisteliden-Gattung Podonta Muls. *Berliner Entomologische Zeitschrift* 17: 9–22.

Kiesenwetter, E.A.H. von. 1878: [new taxa]. In: Schneider O. & Leder H.: Beiträge zur Kenntniss der kaukasischen Käferfauna. *Verhandlungen des Naturforschenden Vereines in Brünn* 16 [1877]: 3–258, pls 1–4.

Kiesenwetter, E.A.H. von. 1879: Coleoptera Japoniae collecta a Domino Lewis et aliis. *Deutsche Entomologische Zeitschrift* 23: 305–320.

Kiesenwetter, E.A.H. von. 1880: Beschreibung neuer Pedinus-Arten. *Entomologische Monatsblätter* 2 (15): 65–70.

Kirby, W. 1802: *Monographia Apum Angliae*. Vol. 2. Ipswich: White, 1–388 pp., 4 pls.

Kirby, W. 1819: A century of insects, including several new genera described from his cabinet. *Transactions of the Linnean Society of London* 12: 375–453, pls 21–23.

Kirby, W. 1837: The insects. In: Richardson J. (ed.): *Fauna Boreali-Americana; or the zoology of the northern parts of British America: containing descriptions of the objects of natural history collected on the late northern land expedition, under the command of Captain Sir John Franklin, R.N. Part the fourth and last.* Norwich: Josian Flechter, xxxix + 325 + [2] pp., 8 pls.

Kirsch, T.F.W. 1866: Beiträge zur Käferfauna von Bogotà. *Berliner Entomologische Zeitschrift* 10: 173–216.

Kirsch, T.F.W. 1869: Beitrag zur Kenntniss der Gattung Omophlus Sol. *Berliner Entomologische Zeitschrift* 13: 97–128.

Kirsch, T.F.W. 1871: *Neue Käfer-Arten aus Egypten gesammelt von Dr. Schneider.* Berliner Entomologische Zeitschrift 14 *[1870]: 389–396.*

Kirsch, T.F.W. 1878: [new taxa]. In: Schneider O. & Leder H.: Beiträge zur Kenntniss der kaukasischen Käferfauna. *Verhandlungen des Naturforschenden Vereines in Brünn* 16 [1877]: 3–262.

Kirsch, T.F.W. 1880: Zwei neue Mylabris. *Entomologische Monatsblätter* 2: 77.

Kirchsberg, O. von. 1876: *Pimelia balcanica Kirchsberg var. nov.* Entomologische Monatsblätter 1: 30–3[illegible].

Kirchsberg, O. von. 1877a: Vieta Millingenii nov. spec. *Deutsche Entomologische Zeitschrift* 21: 203.

Kirchsberg, O. von. 1877b: Anthrodeis arabicus nov. spec. *Deutsche Entomologische Zeitschrift* 21: 204.

Kiyoyama, Y. 1975: A new species of genus Stenomordella Ermisch from Is. Amami-Ôshima with new record *The Entomological Review of Japan* 28: 46–48.

Kiyoyama, Y. 1987: Two new species of the genus Mordellochroa. *The Entomological Review of Japan* **42**: 31–36.

Kiyoyama, Y. 1991: A study on the genus Tolidostena Ermisch from Taiwan. *The Entomological Review of Japan* **46**: 53–61.

Kiyoyama, Y. 1993: A study on the genus Hoshohananomia Kôno from Japan and Taiwan. *Transaction of the Shikoku Entomological Society* **20**: 71–86.

Kiyoyama, Y. 1994: A new species of the genus Variimorda Méquignon from Taiwan. *Transaction of the Shikoku Entomological Society* **20**: 193–196.

Klinger, R. 2000: Revision des Otolelus pruinosus (Kiesenwetter) mit Beschreibung einer neuen Art (Col., Aderidae). *Entomologische Nachrichten und Berichte* **44**: 275–278.

Klug, J.C.F. 1830: *Symbolae Physicae, seu icones et descriptiones Insectorum quae ex itinere per Africam borealem et Asiam occidentalem Friderici Guilelmi Heimprich et Cristiani Godofredi Ehrenberg medicinae te chirurgiae doctorum studio novae aut illustratae redierunt. Decas secunda.* Berolini: Oficina Academica, [20] pp., pls xi–xx.

Klug, J.C.F. 1833: Bericht über eine auf Madagarscar veranstaltete Sammlung von Insecten aus der Ordnung Coleoptera. *Abhandlungen der Königlichen Akademie der Wissenschaften* (Berlin) **1832–1833**: 91–223, 5 pls.

Klug, J.C.F. 1845: *Symbolae physicae seu icones et dscriptiones Insectorum que ex itinere per Africam borealem et Asiam occidentalem Friderici Guilelmi Hemprich et Christiani Godofredi Ehrenberg medicinae et chirurgiae doctorum studio nova autillustrate redierunt. Insecta.* Berolini: Reumen, 1–24, pls. 31–32.

Knoch, A.W. 1801: *Neue Beyträge zur Insectenkunde. Erster Theil.* Leipzig: Schwickert 1: xii + 208 pp., pls i–ix.

Knor, I.B. 1970: [new taxon]. In: O faune chernotelok (Coleoptera, Tenebrionidae) altaysko-sayanskoy gornoy strany. Pp. 82–87: In: *Fauna Sibiri.* Novosibirsk: Nauka.

Knor, I.B. 1976: Novye dl'a fauny Sovietskogo Soyuza chernotelki (Coleoptera, Tenebrionidae) iz Tuvy. Pp. 102–109. In: Tsherepanov A.I. (ed.): *Novosti fauny Sibiri. Novye i maloizvestnye vidy fauny Sibiri. Vypusk 10.* Novosibirsk: Nauka, 208 pp.

Kobayashi, T. and Sota, T. 2019: Divergent host use among cryptic species in the fungivorous ciid beetle Octotemnus laminifrons (Motschulsky, 1860), with descriptions of three new species from Japan. *Systematic Entomology* **44**: 179–191.

Koch, C. 1931: Risultati scientifici delle spedizioni entomologiche di S.A.S. il Principe Alessandro della Torre e Tasso nell'Africa settentrionale e in Sicilia. II. Anthicus Payk. *Memorie della Società Entomologica Italiana* **10**: 66–79.

Koch, C. 1933: Risultati scientifici delle caccie entomologiche di S.A.S. il Principe Alessandro della Torre e Tasso in Italia. II Anthicus. *Bollettino della Società Entomologica Italiana* **65**: 149–159.

Koch, C. 1934a: Beitrag zur Kenntnis der Tenebrioniden vom Gebel Elba und Mersa Halaib (Coleoptera). *Bulletin de la Société Royale Entomologique d'Egypte* **18**: 92–103.

Koch, C. 1934b: Studienergebnisse der entomologischen Expeditionen Seiner Durchlaucht des Fürsten della Torre e Tasso in Nordafrika und Sizilien. IV. Anthicus (Col. Anthic.). *Deutsche Entomologische Zeitschrift* **1934**: 117–127.

Koch, C. 1935a: Wissenschaftliche Ergebnisse der entomologischen Expedition seiner Durchlaucht des Fürsten A. della Torre e Tasso nach Aegypten und auf der Halbinsel Sinai. VII. Tenebrionidae (Coleoptera). *Bulletin de la Société Royale Entomologique d'Egypte* **19**: 2–111.

Koch, C. 1935b: Wissenschaftliche Ergebnisse der entomologischen Expeditionen seiner Durchlaucht des Fürsten Alessandro C. della Torre e Tasso nach Aegypten und auf die Halbinsel Sinai. VIII. Anthicidae (Coleoptera). *Bulletin de la Société Royale Entomologique d'Égypte* **19**: 132–144.

Koch, C. 1935c: Risultati scientifici delle cacce entomologiche di S.A.S. il Principe della Torre e Tasso nelle Isole dell' Egeo. III. Anthicidae. *Bollettino del Laboratorio di Zoologia Generale e Agraria del R. Instituto Superiore Agrario in Portici* **28**: 257–281

Koch, C. 1935d: Wissenschaftliche Ergebnisse der entomologischen Expedition Seiner Durchlaucht des Fürsten A. Della Torre e Tasso in Palaestina. I. Tenebrionidae (Coleoptera). *Bulletin de la Société Royal Entomologique d'Egypte* **19**: 262–288.

Koch, C. 1935e: Risultati scientifici delle cacce entomologiche di S.A.S. il Principe Alessandro della Torre e Tasso nelle isole dell'Egeo. V. Tenebrionidae. *Bollettino del Laboratorio di Zoologia Generale e Agraria del R. Instituto Superiore Agrario di Portici* **28**: 309–320.

Koch, C. 1936: Neue Tenebrionidae aus dem östlichen Mediterrangebiet. *Pubblicazioni del Museo Entomologico 'Pietro Rossi' Duino* **1**: 291–301.

Koch, C. 1937: Wissenschaftliche Ergebnisse über die während der Expedition Seiner Durchlaucht des Fürsten Alessandro C. della Torre e Tasso in Lybien aufgefundenen Tenebrioniden. *Pubblicazioni del Museo Entomologico 'Pietro Rossi' Duino* **2**: 285–500.

Koch, C. 1939: Die Käfer der libyschen Ausbeute des Herrn Georg Frey. *Mitteilungen der Münchener Entomologischen Gesellschaft* **29**: 216–293, pls. viii, ix.

Koch, C. 1940a: Der saharo-sindhische Verbreitungs-Typus bei der ungeflügelten Tenebrioniden-Gattung Mesostena unter Berücksichtigung ähnlicher biogeographischer Verhältnisse der Tentyria-Stamm-Gruppe Tentyrina. *Rivista di Biologia Coloniale* **3**: 1–137.

Koch, C. 1940b: Resultados cientificos de los viajes entomológicos en España patrocinados por Su Alteza el Principe Alessandro C. della Torre e Tasso. Pp. 73–80. *VI Congreso Internacional de Entomología. Madrid, 6–12 de Septiembre de 1935. Vol. 1.* Madrid: Laboratorio de Entomología del Museo Nacional de Ciencias Naturales, 406 pp.

Koch, C. 1940c: Über einige neue und interessante, von Johannes Roth 1857 im Wadi Arabad gesammelte Tenebrioniden (Col.). *Mitteilungen der Münchener Entomologischen Gesellschaft* **30**: 84–107.

Koch, C. 1940d: Phylogenetische, biogeographische und systematische Studien über ungeflügelte Tenebrioniden (Col. Tenebr.). *Mitteilungen der Münchener Entomologischen Gesellschaft* **30**: 254–337.

Koch, C. 1940e: Phylogenetische, biogegraphische und systematische Studien über ungeflügelte Tenebrioniden (Col. Tenebr.). II. *Mitteilungen der Münchener Entomologischen Gesellschaft* **30**: 683–750, pls 18–20

Koch, C. 1941a: Die Verbreitung und Rassenbildung der marokkanischen Pimelien (Col. Tenebr.) (Eine biogeographische Studie). *Eos, Revista Española de Entomología* **16** [1940]: 7–123, pls 1–3.

Koch, C. 1941b: Contributo alla conoscenza del genere Mitotagenia Rtt. (Col. Tenebr.). *Memorie della Società Entomologica Italiana* **20**: 37–42.

Koch, C. 1941c: Phylogenetische, biogegraphische und systematische Studien über ungeflügelte Tenebrioniden (Col. Tenebr.). III. *Mitteilungen der Münchener Entomologischen Gesellschaft* **31**: 252–314, pls 12–13

Koch, C. 1943a: Phylogenetische, biogeographische und systematische Studien über ungeflügelte Tenebrioniden (Col. Tenebr.). IV. *Mitteilungen der Münchener Entomologischen Gesellschaft* **33**: 473–597.

Koch, C. 1943b: Revision der Tenebrionidengattungen Thalpophila und Rhytinota (Col. Tenebr.). *Mitteilungen der Münchener Entomologischen Gesellschaft* **33**: 759–889, pls 28–37.

Koch, C. 1944a: Die Adesmiini der tropischen und subtropischen Savannen Afrikas. *Revue de Zoologie et Botanique Africaine* **38**: 139–191.

Koch, C. 1944b: Tenebrionidae I (I. Beitrag zur Kenntnis der iberischen Fauna). *Mitteilungen der Münchener Entomologischen Gesellschaft* **34**: 216–254.

Koch, C. 1945: Über die Tenebrioniden der Gipfelfauna des Hohen und Mittleren Atlas Marokkos. Ergebnisse der Forschungsreise R. Paulian und A. Villiers nach dem Hohen Atlas Marokkos 1938. XIX. Beitrag. *Eos, Revista Española de Entomología* **20** [1944]: 387–433.

Koch, C. 1948a: Die Tenebrioniden Kretas (Col.). *Mitteilungen der Münchener Entomologischen Gesellschaft* **34** [1944]: 255–363, pls. XVIII–XXIX.

Koch, C. 1948b: Beitrag zur Kenntnis der Tribus Litoborini der Tenebrioniden-Unterfamilie der Opatrinae (Col. Ten.). *Eos, Revista Española de Entomología* **24**: 403–433.

Koch, C. 1949: Beitrag zur Kenntnis der asiatischen Tentyriini (Col., Tenebr.). *Entomologische Blätter* **41–44** [1945–48]: 76–81.

Koch, C. 1950a: The Tenebrionidae (Col.) of Southern AfricA.V. – Contribution to the knowledge of the Caedius-group of Opatrini and the Crypticini from south of the Sahara. *Memórias do Museu Dr. Álvaro de Castro* **1**: 35–90, 6 pls.

Koch, C. 1950b: Die Tenebrioniden des südlichen Afrikas. IV. – Über vier neue afrikanische Himatismus-Arten. *Entomologische Arbeiten aus dem Museum G. Frey* **1**: 186–197.

Koch, C. 1951: Beitrag zur Kenntnis der Gattung Adesmia (Col. Tenebrionid.). *Entomologische Blätter* **45–46** [1949–50]: 121–130.

Koch, C. 1952: The Tenebrionidae of Southern Africa. VIII. Materials for a monographic study on Euchychorini (Coleoptera). *Bulletin de la Société Fouad 1er d'Entomologie* **45**: 1–125, pls I–XIII.

Koch, C. 1953: Tenebrioniden aus südlichen Afrikas XIV. Über einige neue Molurini aus dem Ungarischen Naturwissenschaftlichen Museum zu Budapest (Vorarbeiten zu einer Monographie der Molurini 4.). *Annales Historico-Naturales Musei Nationalis Hungarici, Series nova* **3** [1952]: 137–181, pls. V, VI.

Koch, C. 1956: II. Tenebrionidae (Coleoptera Polyphaga), Opatrinae, First part: Platynotini, Litoborini and Loensini. *Exploration du Parc National de l'Upemba* **40**: 1–472, 35 pls.

Koch, C. 1959: Erster taxonomischer Beitrag zur Kenntnis der Tenebrioniden Somalis. *Entomologische Arbeiten aus dem Museum G. Frey* **10**: 568–596.

Koch, C. 1962: Vierter taxonomischer Beitrag zur Kenntnis der Tenebrioniden Somalias: Über die von ProF.G. Scortecci 1953 und 1957 in der Migiutinia-Provinz gesammelten Arten. *Atti della Società Italiana di Scienze Naturali e del Museo Civico di Storia Naturale in Milano* **101**: 237–271.

Koch, C. 1960: Zweiter taxonomischer Beitrag zur Kenntnis der Tenebrioniden Somalias. *Entomologische Arbeiten aus dem Museum G. Frey* **11**: 325–415, 2 pls.

Koch, C. 1965a: Missione 1962 del Prof. Giuseppe Scortecci nell'Arabie meridionale. Coleoptera Tenebrionidae. Includendo materiale di viaggi in Arabia del Sig. G. Popov (1962) e del Dr. G. Benardelli (1962–63). *Atti della Società Italiana di Scienze Naturali e del Museo Civico di Storia Naturale di Milano* **104**: 99–154, pls V–VII.

Koch, C. 1965b: Sur les types de Fairmaire des tribus Blaptini et Platyscelini conservés au Muséum de Paris. *Annales de la Société Entomologique de France* (N.S.) **1**: 125–135.

Koch, C. 1969: Storthocnemis kulzeri, ein bisher unbekanntes Bindeglied zwischen den Tribus Pimeliini/ Platyopini und Leucolaepini (Col., Tenebrionidae). *Entomologische Arbeiten aus dem Museum G. Frey* **20**: 484–493.

Koch, C. 1970: Die Tenebrioniden (Coleoptera) des Archipels von Socotra. *Monitore Zoologico Italiano* N.S. *Supplemento* **3**: 69–132.

Kocher, L. 1951a: Notes sur les Erodius marocains (Contribution à l'étude systématique du genre). *Bulletin de la Société des Sciences Naturelles et Physiques du Maroc* **30** [1950]: 195–230.

Kocher, L. 1951b: Sur le genre Cimipsa Peyerh. (Col. Tenebr.) avec description d'une espèce et d'une variété nouvelles. *Bulletin de la Société des Sciences Naturelles et Physiques du Maroc* **31**: 125–130.

Kocher, L. 1953: Localisations nouvelles ou intéressantes de coléoptères marocains. *Travaux de l'Institut Scientifique Chérifien* 7: 1–142 pp., 1 map.

Kocher, L. 1954: Prospection entomologique (Coléoptères) de la moyenne Moulouya. *Bulletin de la Société de Sciences Naturelles et Physiques du Maroc* **34**: 263–286.

Kocher, L. 1955a: Observations sur certains meloides marocains avec descriptions d'une nouvelle espèce. *Comptes Rendus des Séances Mensuelles de la Société des Sciences Naturelles et Physiques du Maroc* 7: 125–126.

Kocher, L. 1955b: Variation chez certains Mylabris marocains. *Comptes Rendus des Séances Mensuelles de la Société des Sciences Naturelles et Physiques du Maroc* 7: 127–128.

Kocher, L. 1956a: Contribution à l´étude des Heliomophlus de l´Atlas marocain. *Bulletin Société des Sciences Naturelles et Physiques du Maroc* **35** [1955]: 235–242.

Kocher, L. 1956b: Note complémentaire sur les Erodius marocains. *Bulletin de la Société des Sciences Naturelles et Physiques du Maroc* **35** [1955]: 243–249.

Kocher, L. 1957: Nouvelles espèces ou variétés marocaines de coléoptères. *Bulletin de la Société des Sciences Naturelles et Physiques du Maroc* **36** [1956]: 367–373.

Kocher, L. 1958a: Observations sur le genre Adelphinus Fairm. (Col. Ténébr.), avec description d'une nouvelle espèce du Maroc. *Comptes Rendus des Séances Mensuelles de la Société des Sciences Naturelles et Physiques du Maroc* **24**: 123–125.

Kocher, L. 1958b: Un nouveau Scaurus cavernicole du Maroc (Col. Ténébr.): Scaurus camelus sp. nov. *Comptes Rendus des Séances Mensuelles de la Société des Sciences Naturelles et Physiques du Maroc* **24**: 195–196.

Kocher, L. 1960: Nouvelles espèces de coléoptères du Maroc. *Bulletin de la Société des Sciences Naturelles et Physiques du Maroc* **40**: 233–241.

Kocher, L. 1962a: Coléoptères nouveaux ou intéressants du Maroc présaharien. *Bulletin de la Société des Sciences Naturelles et Physiques du Maroc* **42**: 195–214.

Kocher, L. 1962b: Contribution à l'étude du genre Diplocyrtus Quedf. (Col. Tenebr.). *Bulletin de la Société des Sciences Naturelles et Physiques du Maroc* **42**: 301–310.

Kocher, L. 1963a: Nouvelle recolte des coléoptères dans le Bas Drâ (Sahara occidental marocain). *Bulletin de la Société de Sciences Naturelles et Physiques du Maroc* **43**: 35–37.

Kocher, L. 1963b: Les Diplocyrtus de la collection Escalera (Col. Tenebrionidae). *Bulletin de la Société des Sciences Naturelles et Physiques du Maroc* **43**: 113–116.

Kocher, L. 1963c: Coléoptères nouveaux ou intéressants de la région rifaine. *Bulletin de la Société des Sciences Naturelles et Physiques du Maroc* **43**: 193–199.

Kocher, L. 1964: Présentation de coléoptères rares recueillis récemment au Maroc et de quelques espèces nouvelles pour la sciences. *Comptes Rendus des Séances Mensuelles de la Société des Sciences Naturelles et Physiques du Maroc* **30**: 43–45.

Kocher, L. 1965a: Contribution à l'étude des Stenosis du groupe elvirae Ant. (Col. Tenebr.). *Bulletin de la Société des Sciences Naturelles et Physiques du Maroc* **44** (3e et 4e trimestre) [1964]: 25–27.

Kocher, L. 1965b: Scaurus cavernicoles de la collection Antoine (Col. Ténébr.). *Bulletin de la Société des Sciences Naturelles et Physiques du Maroc* **44** (3e et 4e trimestre) [1964]: 33–36.

Kocher, L. 1966a: Alphasida nouvelles ou intéressants de la région rifaine (Col. Tenebr.). *Bulletin de la Société des Sciences Naturelles et Physiques du Maroc* **45** [1965]: 43–48.

Kocher, L. 1966b: Coléoptères marocains nouveaux ou intéressants. *Bulletin de la Société des Sciences Naturelles et Physiques du Maroc* **45** [1965]: 133–136.

Kocher, L. 1968: Description de nouveaux coléoptères du Maroc. *Bulletin de la Société des Sciences Naturelles et Physiques du Maroc* **48**: 107–113.

Kolbe, H.J. 1883: Neue Coleoptera von Westafrika. *Berliner Entomologische Zeitschrift* **27**: 15–36.

Kolbe, H.J. 1886: Beiträge zur Kenntnis der Coleopteren-Fauna Koreas, bearbeitet auf Grund der von Herrn C. Gottsche veranstalteten Sammlung; nebst Bemerkungen über die zoogeographische Verhältnisse dieses Faunengebietes und Untersuchungen über einen Sinnensapparat im Gaumen von Microlampidius morio. *Archiv für Naturgeschichte* **52** (1): 139–240, pls X, XI.

Kolbe, H.J. 1894a: Über Coleopteren aus Afrika. *Entomologische Zeitung* (Stettin) **55**: 178–190.

Kolbe, H.J. 1894b: Über Coleopteren aus Afrika. II. *Entomologische Zeitung* (Stettin) **55**: 361–397.

Kolbe, H.J. 1897: Coleopteren. Die Käfer Deutsch-Ost-Afrikas, 367 pp., 4 pls. In: Kolbe H.J.: Käfer und Netzflügler Ost-Africas. In: Möbius K. (ed.): *Deutsch-Ost-Afrika* **4** (4). Berlin: D. Reimer, 367 + 42 pp., 5 pls.

Kolbe, H.J. 1900: Ein vergessener Nyctobates. *Entomologische Nachrichten* **26**: 72–74.

Kolbe, H.J. 1901: Ein Schädling des Affenbrotbaumes, Adansonius fructum n. sp. aus der Familie der Curculioniden. *Allgemeine Zeitschrift für Entomologie* **6**: 341–343.

Kolbe, H.J. 1908: [new taxa]. In: Kolbe H.J., Obst P. & Weise J.: Coleoptera. Pp. 82–96, pl. III. In: *Expedition Filchner nach China und Tibet 1903–1905. Wissenschaftliche Ergebnisse X. Band – 1. Teil.* Berlin: Siegfried Mittler und Sohn.

Kolbe, H.J. 1915: Eine neue isolierte Tenebrionidengattung von den Philippineninseln. *Deutsche Entomologische Zeitschrift* **1915**: 261–264.

Kolbe, H. 1928: 2. Tiergeographie und Morphologie, neue Untersuchungen zur Entwicklungsgeschichte der Tiergattungen. *Zoologischer Anzeiger* **77**: 195–209.

Kolenati, F.A. 1846: *Meletemata Entomologica. Fasc. III. – Brachelytra Caucasi cum distributione geographica adnexis Pselaphinis, Scydmaenis, Notoxidibus et Xylophagis*. Petropoli: Typis Imperialis Academiae Scientiarum, 44 pp., pls 12–14.

Kolenati, F.A. 1847: Über einige russische Oedemeriden. *Bulletin de la Société Impériale des Naturalistes de Moscou* **20** (1–2): 132–147.

Kollar, V. & Redtenbacher, L. 1842: Aufzählung und Beschreibung der von Freiherr Carl v. Hügel auf seiner Reise durch Kaschmir und das Himalayagebirge gesammelten Insecten. Pp. 395–564. In: Hügel, C. von:

Kashmir und das Reich der Siek. Vierter Band. Zweiter Abtheilung. Stuttgart: Hallbergerische Verlag, [-] + 247–865 pp.

Kolov, S.V. 2005: Novyy dlya fauny Kazakhstana vid zhuka-naryvnika (Coleoptera, Meloidae) [New record of blister-beetle (Coleoptera, Meloidae) for the fauna of Kazakhstan]. *Euroasian Entomological Journal* 4: 41–42

Kompantsev, A.V. 1996: A new species of the genus Cis Latr. (Coleoptera, Ciidae) from Middle Asia and notes on Palaearctic species of the comptus group. *Entomologicheskoe Obozreniye* 75: 587–595.

König, E. 1906: Dritter Beitrag zur Coleopteren-Fauna des Kaukasus. *Wiener Entomologische Zeitung* 25: 23–27.

Kôno, H. 1927: Beitrag zu den Rhipiphoriden-Arten von Japan. *Insecta Matsumurana* 1: 177–182.

Kôno, H. 1928a: Die Mordelliden Japans. *Transaction of Sapporo Natural History Society* 10: 31–46.

Kôno, H. 1928b: Einige Mordelliden-Arten Japans. *Insecta Matsumurana* 2: 146–150.

Kôno, H. 1929a: Die Lagriiden Japans (Col.). *Insecta Matsumurana* 4: 25–35.

Kôno, H. 1929b: Die Pyrochroiden Japans. *Insecta Matsumurana* 3: 62–72.

Kôno, H. 1929c: Die Mordelliden Japans. *Insecta Matsumurana* 3: 83–86.

Kôno, H. 1930a: Cerogria miwai nova spec. (Col. Lagriidae). *Transactions of the Natural History Society of Formosa* 20: 18–19.

Kôno, H. 1930b: Einige Alleculiden-Arten aus Japan und Formosa (Col.). *Insecta Matsumurana* 4: 95–99.

Kôno, H. 1930c: Die Mordelliden Japans. *Insecta Matsumurana* 4: 104.

Kôno, H. 1930d: Die Serropalpiden aus Sachalin (Col.). *Insecta Matsumurana* 5: 48–51.

Kôno, H. 1931: Die Notoxus-Arten aus Japan (Col. Anthicidae). *Insecta Matsumurana* 6: 1–2.

Kôno, H. 1932a: Die Nacerdinen Japans (Col.). *Insecta Matsumurana* 6: 138–144.

Kôno, H. 1932b: Die Mordelliden Japans. *Transaction of Sapporo Natural History Society* 12: 152–160, 1 pl.

Kôno, H. 1933: Die Mordelliden Japans. *Transaction of Sapporo Natural History Society* 13: 29–31.

Kôno, H. 1934a: H. Sauter's Formosa Ausbeute: Mordellidae. *Insecta Matsumurana* 8: 116–118.

Kôno, H. 1934b: Die Heteromeren von Sachalin. *Insecta Matsumurana* 9: 27–40.

Kôno, H. 1935a: Die Heteromeren aus den Kurilen (Col.). (Erster Beitrag zur Kenntnis der Käferfauna der Kurilen). *Insecta Matsumurana* 9 [1934]: 85–89.

Kôno, H. 1935b: Die Mordelliden Japans. *Transaction of Sapporo Natural History Society* 14: 123–130.

Kôno, H. 1935c: Drei neue Heteromeren aus Japan und Formosa, mit einer Liste der geographischen Verbreitung der japanischen Pyrochroiden und Pediliden. *Insecta Matsumurana* 9: 157–161.

Kôno, H. 1936a: Über die Käfersammlung des Museums 'Umeno Konchû Kenkiujo' I. Pyrochroidae, Meloidae, Rhipiphoridae. *Bulletin of the Umeno Entomological Laboratory, Kurume* 3: 1–6.

Kôno, H. 1936b: Neue und wenig bekannte Käfer Japans. 1. *Insecta Matsumurana* 10: 87–98.

Kôno, H. 1936c: Beitrag zu den Pytho-Arten Japans (Col.). *Insecta Matsumurana* 11: 35–37.

Kôno, H. 1936d: Zwei neue Xanthochroa-Arten (Coleoptera, Oedemeridae). *Festschrift zum 60. Geburtstage von Professor Dr. Embrik Strand* (Riga) 1: 516–518.

Kôno, H. 1937a: Über die Käfersammlung des Museums 'Umeno Konchû Kenkiujo'. III Oedemeridae. Cephaloidae. *Bulletin of the Umeno Entomological Laboratory, Kurume* 5: 1–15. (in Japanese, German title and descriptions).

Kôno, H. 1937b: Neue und wenig bekannte Käfer Japans. II Oedemeridae. *Insecta Matsumurana* **11**: 135–146.

Kôno, H. 1938: An inter-tidal rock-dwelling beetle, *Aegialatis* [sic] *stejnegeri*, and its one new subspecies. *The Entomological World* **6** (46): 1–5 (in Japanese).

Kôno, H. 1939: Die Tenebrioniden in Tannen und Fichtenwald in Japan. *Insecta Matsumurana* **13**: 85–87.

Kôno, H. 1940: Die Meloiden von Mandschukuo. *Insecta Matsumurana* **15**: 57–62.

Kontkanen, P. 1956: Zur Kenntnis der paläarktischen Käferfauna. I. *Annales Entomologici Fennici* **22**: 55–63.

Konvička, O. 2014: Osphya lehnertae sp. nov. from Greece (Coleoptera: Melandryidae). *Klapalekiana* **50**: 161–166.

Konvička, O. 2015: Melandrya (Emmesa) jaromiri sp. nov. from China (Coleoptera: Melandryidae). *Acta Musei Silesiae, Scientiae Naturales* **64**: 97–99.

Konvička, O. 2016: Osphya brusteli sp. nov. from the Balkan Peninsula (Coleoptera: Melandryidae). *Acta Musei Silesiae, Scientiae Naturales* **65**: 271–277.

Kotschenov, V.G. 1970: Novyy vid naryvnika (Coleoptera, Meloidae) iz Kirghizii. Pp. 21–22. In: Novye i maloizvestnyye vidy bespozvonochnykh zhivotnykh. *Entomologicheskiya Issledovaniya v Kirgizii* [Vypusk 6]. Frunze: Ilim, 87 pp.

Korge, H. 1971: Beiträge zur Kenntnis der Koleopterenfauna Kleinasiens. *Annotationes Zoologicae et Botanicae* (Bratislava) **67**: 1–68.

Kraatz, G. 1853: Wissenschaftliche Mittheilungen. *Entomologische Zeitung (Stettin)* **14**: 255–257.

Kraatz, G. 1855: Synonymische Bemerkungen. *Entomologische Zeitung (Stettin)* **16**: 372–374.

Kraatz, G. 1858: Neue Literatur – Histoire naturelle des coléoptères de France par M.E. Mulsant. Vésicants. Paris, 1857, 8, 198 S., 1 Tafel. *Berliner Entomologische Zeitschrift* **2**: 387–392.

Kraatz, G. 1862a: Ueber die Arten der Gattungen Hyperops, Microtelus und Dichillus. *Berliner Entomologische Zeitschrift* **6**: 91–99.

Kraatz, G. 1862b: Beiträge zur Käferfauna Griechenlands.Von Dr. Krüper neu aufgefundene Arten. *Berliner Entomologische Zeitschrift* **6**: 121–126.

Kraatz, G. 1863: Revision der Cerocomiden-Gruppe. *Berliner Entomologische Zeitschrift* **7**: 109–115.

Kraatz, G. 1865a: *Revision der Tenebrinoiden der alten Welt aus Lacordaire's Gruppen der Erodiides, Tentyriides, Akisides, Piméliides, und der europäischen Zophosis-Arten*. Berlin, 393 pp.

Kraatz, G. 1865b: Synonymische Bemerkungen. *Berliner Entomologische Zeitschrift* **9**: 414.

Kraatz, G. 1868a: Beiträge zur Kenntniss der deutschen Käferfauna. *Berliner Entomologische Zeitschrift* **12**: 283–304.

Kraatz, G. 1868b: Ueber deutsche Käferarten. *Berliner Entomologische Zeitschrift* **12**: 331–338.

Kraatz, G. 1870: [new taxa]. Heyden L.F.J.D. von. 1870: Zweiter Theil. Beschreibungen der neuen Arten. Pp. 58–176. In: Heyden L.F.J.D. von: Entomologische Reise nach dem südlichen Spanien, der Sierra Guadarrama und Sierra Morena, Portugal und den Cantabrischen Gebirgen. *Berliner Entomologische Zeitschrift* **14** *Beiheft*, 218 pp., 2 pls.

Kraatz, G. 1872: [new taxa]. In: Reitter E.: Neue Käferarten von Oran gesammelt von Hans Leder. (Erstes Stück). *Berliner Entomologische Zeitschrift* **16**: 167–186.

Kraatz, G. 1873: Ueber Uloma Perroudi Muls. aus Deutschland und eine neue Art von Cypern. *Berliner Entomologische Zeitschrift* **17**: 196–197.

Kraatz, G. 1874: Asida pusillima, eine neue Art aus der Sierra Nevada. *Berliner Entomologische Zeitschrift* **18**: 112–114.

Kraatz, G. 1877: Apolites graecus Kraatz n. sp. *Deutsche Entomologische Zeitschrift* **21**: 304.

Kraatz, G. 1880a: Uebersicht der in den folgenden 'Beiträgen zur Kenntniss der Canthariden von Dr. Haag-Rutenberg' beschriebenen und in der Uebersichtstabellen erwähnten Lytta-Arten. *Deutsche Entomologische Zeitschrift* **24**: 15–16.

Kraatz, G. 1880b: Ueber die Heteromeren-Gattung Rhytinota Eschsch. und Verwandte. *Deutsche Entomologische Zeitschrift* **24**: 91–95.

Kraatz, G. 1880c: Beitrag zur Kenntnis der asiatischen Cnodaloniden (Tenebrioniden Coh. II. Trib. XL. Lacord.). *Deutsche Entomologische Zeitschrift* **24**: 97–120.

Kraatz, G. 1880d: [new taxa]. In: Heyden L.F.J.D. von. Verzeichnis von Coleopteren aus Asturien mit Beschreibungen neuer Arten von Candèze, v. Heyden, Kirsch, Kraatz, Stierling. *Deutsche Entomologische Zeitschrift* **24**: 281–303.

Kraatz, G. 1881a: Bemerkungen über typische Exemplare von Arten der Heteromeren-Gattung Blaps aus Fischer von Waldheim's Sammlung. *Deutsche Entomologische Zeitschrift* **25**: 53–60.

Kraatz, G. 1881b: Ueber Veränderlichkeit der Prosodes-Arten. *Deutsche Entomologische Zeitschrift* **25**: 61–62.

Kraatz, G. 1881c: Zonitis spectabilis nov. sp. *Deutsche Entomologische Zeitschrift* **25**: 326.

Kraatz, G. 1881d: Einige neue Pimeliinen von Margelan. *Deutsche Entomologische Zeitschrift* **25**: 331–333.

Kraatz, G. 1882a: In: Heyden L.F.J.D. von.: Beitrag zur Coleopteren-Fauna der Sierra de Cordoba. *Deutsche Entomologische Zeitschrift* **26**: 42–48.

Kraatz, G. 1882b: Beiträge zur Käferfauna von Turkestan. II. Neue Tenebrioniden von Margelan. *Deutsche Entomologische Zeitschrift* **26**: 81–95.

Kraatz, G. 1882c: [new taxa]. In: Heyden L.F.J.D. von. & Kraatz G.: Käfer um Margelan gesammelt von Haberhauer. *Deutsche Entomologische Zeitschrift* **26**: 99–118.

Kraatz, G. 1882d: [new taxa]. In: Heyden L.F.J.D. von. & Kraatz G.: Käfer um Samarkand, gesammelt von Haberhauer. *Deutsche Entomologische Zeitschrift* **26**: 297–338.

Kraatz, G. 1883a: [new taxa]. In: Heyden L.F.J.D. von. & Kraatz G.: Käfer aus Osch (in Turkestan). *Deutsche Entomologische Zeitschrift* **27**: 337–353.

Kraatz, G. 1883b: [new taxa]. In: Heyden L.F.J.D. von. & Kraatz G.: Käfer aus Tekke-Turcmenien. *Deutsche Entomologische Zeitschrift* **27**: 354–360.

Kraatz, G. 1883c: Zwei neue Nalassus- (Helops-) Arten. *Deutsche Entomologische Zeitschrift* **27**: 395.

Kraatz, G. 1884: [new taxa]. In: Heyden L.F.J.D. von. & Kraatz G.: Neue Käfer-Arten aus Osch (Turkestan). *Deutsche Entomologische Zeitschrift* **28**: 217–228.

Kraatz, G. 1885: [new taxa]. In: Heyden L.F.J.D. von. & Kraatz G.: Beiträge zur turkestanischen Coleopteren-Fauna. *Deutsche Entomologische Zeitschrift* **29**: 273–298.

Kraatz, G. 1886: [new taxa]: In: Heyden L.F.J.D. von. & Kraatz G.: Beiträge zur Coleopteren-fauna von Turkestan, namentlich des Alai-Gebirges unter Beihilfe der Herren Dr. Candèze, Ganglbauer, Dr. Stierlin, und Weise. *Deutsche Entomologische Zeitschrift* **30**: 177–194.

Kraatz, G. 1887: Pimelia laevisulcata Kraatz aus Morocco, eine auffällige Varietät der P. Fairmairei Kraatz. *Deutsche Entomologische Zeitschrift* **31**: 361.

Kraatz, G. 1888a: Zwei neue Blaps von Alai (Turkestan). *Deutsche Entomologische Zeitschrift* **32**: 195–197.

Kraatz, G. 1888b: Ueber die Tenebrioniden-Gattungen Microdera und Dordanea. *Deutsche Entomologische Zeitschrift* **32**: 294.

Kraatz, G. 1891: Rhipidius apicipennis nov. spec. aus Thüringen. *Deutsche Entomologische Zeitschrift* **35**: 358–360, pl. 5.

Kraatz, G. 1896: [new taxa]. In: Ragusa E.: Catalogo ragionato dei coleotteri di Sicilia. *Naturalista Siciliano* (N.S.) **1**: 69–106.

Krauss, H. 1911: Typhaea fumata var. n. obscura. *Rivista Coleopterologica Italiana* **9**: 109.

Krekich-Strassoldo, H. von. 1913a: Neue Anthiciden und Mitteilungen über die Verbreitung bekannter Anthiciden. *Wiener Entomologische Zeitung* **32**: 223–232.

Krekich-Strassoldo, H. von. 1913b: Beschreibungen neuer Anthiciden. *Verhandlungen der Kaiserlich-Königlichen Zoologisch-Botanischen Gesellschaft in Wien, Bericht der Sektion für Koleopterologie* **63**: 129–140.

Krekich-Strassoldo, H. von. 1914a: Neue Anthiciden und Mitteilungen über die Verbreitung bekannter Anthiciden. *Wiener Entomologische Zeitung* **33**: 1–14.

Krekich-Strassoldo, H. von. 1914b: Neue Anthiciden. *Verhandlungen der Kaiserlich-Königlichen Zoologisch-Botanischen Gesellschaft in Wien, Bericht der Sektion für Koleopterologie* **64**: 108–114.

Krekich-Strassoldo, H. von. 1914c: Beiträge zur Kenntnis der Anthiciden. *Verhandlungen der Kaiserlich-Königlichen Zoologisch-Botanischen Gesellschaft in Wien, Bericht der Sektion für Koleopterologie* **64**: 215–228.

Krekich-Strassoldo, H. von. 1919a: Beiträge zur Kenntnis der Anthiciden und Hylophiliden (Col.). *Entomologische Mitteilungen* **8**: 166–175.

Krekich-Strassoldo, H. von. 1919b: Über Anthicus humilis Germ. und verwandte Arten. *Koleopterologische Rundschau* **8**: 60–76.

Krekich-Strassoldo, H. von. 1925: Anthicidae of the Philippines I. *The Philippine Journal of Science* **27**: 515–535, pls. 1–3.

Krekich-Strassoldo, H. von. 1927: Ein neuer Anthicus von China. *Entomologische Blätter* **22** [1926]: 116–117.

Krekich-Strassoldo, H. von. 1928a: Beiträge zur Kenntnis indischer Anthiciden. *Archiv für Naturgeschichte (Abteilung A)* **5** [1926]: 67–105.

Krekich-Strassoldo, H. von. 1928b: Il Formicomus canaliculatus LaF. e specie affini d'Africa. *Annali del Museo Civico di Storia Naturale Giacomo Doria, Genova* **52** [1925–1928]: 383–386.

Krekich-Strassoldo, H. von. 1928c: Ueber Anthicus tristis Schm. und einige verwandte Arten. *Coleopterologisches Centralblatt* **3**: 104–111.

Krekich-Strassoldo, H. von. 1928d: Beiträge zur Kenntnis malayischer Anthiciden. *Tijdschrift voor Entomologie* **71**: 1–10.

Krekich-Strassoldo, H. von. 1929: Revision der paläarktischen Anthicus-Arten der Gruppe 'Bifossicolles' (Gruppe XVIII in Marseuls Monographie). *Koleopterologische Rundschau* **15**: 151–189.

Krekich-Strassoldo, H. von. 1931: Beiträge zur Kenntnis indischer Anthiciden II. *Folia Zoologia et Hydrobiologia* **3**: 1–41, pls. 1–3.

Kriechbaumer, J. 1848: Wissenschaftliche Mittheilungen. Osphya ?aeneipennis. *Entomologische Zeitung* (Stettin) **9**: 163–165.

Kriesche, R. 1917: Bemerkungen über einige Trictenotomiden. (Coleopt.). *Archiv für Naturgeschichte* **83A**(11): 150.

Królik, R. 2002: Cis tauriensis n. sp. from Turkey (Coleoptera: Ciidae). *Genus* **13**: 197–202.

Królik, R. 2016: Two new species and nomenclatural changes in the Cis castaneus species group (Coleoptera, Ciidae). *Zootaxa* **4114**: 492–500.

Krynicki, I. 1829: [Corréspondence 'Literae Cel. Professoris J. Krynicki, Directori datae, Charkoviae 28 Julii. 1829']. *Bulletin de la Société Impériale des Naturalistes de Moscou* **1**: 187–199.

Krynicki, I. 1832: Enumeratio Coleopterorum Rossiae meridionalis et praecipue in Universitatis Caesareae Charkoviensis circulo obvenientium, quae annorum 1827–1831 spatio observavit. *Bulletin de la Société Impériale des Naturalistes de Moscou* **5**: 65–179, 2 pls.

Krynicki, I. 1834: Adenda et nonnulla synonyma Rosiae meridionalis Coleopterorum. *Bulletin de la Société Impériale des Naturalistes de Moscou* **7**: 166–173.

Kryzhanovskij, O.L. 1956: Two new species of genus Mylabris (Col. Meloidae) from Turkmenian SSR. *Entomologicheskoe Obozrenie* **35**: 886–889.

Kryzhanovskij, O.L. 1959: Obzor vidov roda Teratolytta (Coleoptera, Meloidae) vstrechayushchikhsya v SSSR, s opisaniem dvukh novykh form [The key of the species Teratolytta (Coleoptera, Meloidae), inhabiting the U.S.S.R., with the description of two new forms from Tadjikistan]. *Entomologicheskoe Obozrenie* **38**: 855–859.

Kugelann, J.G. 1794: Verzeichniss der in einigen Gegenden Preussens bis jetzt entdeckten Käferarten, nebst kurzen Nachrichten von denselben. *Neuestes Magazin für die Liebhaber der Entomologie* **1** (5): 513–582.

Kühnelt, W. 1951: Zwei bemerkenswerte Tenebrioniden aus Iran. *Verhandlungen der Naturforschenden Gesellschaft in Basel* **62**: 245–247.

Kühnelt, W. 1957: Ergebnisse der österreichischen Iran-Expedition 1949/50. Die Tenebrioniden Irans. *Sitzungsberichte der Österreichischen Akademie de Wissenschaften, Mathematisch-Naturwissenschaftliche Klasse* **166**, 2: 65–102.

Kulzer, H. 1950: Drei neue Arten der Gattung Toxicum und Cryphaeus aus dem palaearktischen Faunengebiet (Tenebr.). *Entomologische Arbeiten aus dem Museum G. Frey* **1**: 37–42.

Kulzer, H. 1952: Siebenter Beitrag zur Kenntnis der Tenebrioniden (Col.). Einige neue Gattungen und Arten der Tribus Cnodalonini aus dem Nachlass von H. Gebien in col. G. Frey. *Entomologische Arbeiten aus dem Museum G. Frey* **3**: 19–764, 2 pls.

Kulzer, H. 1954: Achter Beitrag zur Kenntnis der Tenebrioniden (Col.). *Entomologische Arbeiten aus dem Museum G. Frey* **5**: 20–73, pls V–VII.

Kulzer, H. 1956: Bemerkenswerte Tenebrioniden aus der Thar-Wüste. 14. Beitrag zur Kenntnis der Tenebrioniden (Col.). *Entomologische Arbeiten aus dem Museum G. Frey* **7**: 635–653.

Kulzer, H. 1957: Insect of Micronesia Coleoptera: Tenebrionidae. *Insect of Micronesia* **17**: 185–256.

Kulzer, H. 1959: Neue Tenebrioniden aus Süd-America (Col.). (18. Beitrag zur Kenntnis der Tenebrioniden). *Entomologische Arbeiten aus dem Museum G. Frey* **10**: 523–567.

Kulzer, H. 1960: Einige neue Tenebrioniden (Col.) (20. Beitrag zur Kenntnis der Tenebrioniden). *Entomologische Arbeiten aus dem Museum G. Frey* **11**: 304–317.

Kulzer, H. 1964: Über neue Tenebrionidenarten. *Entomologische Arbeiten aus dem Museum G. Frey* 15: 221–276.

Küster, H.C. 1844: *Die Käfer Europa's. Nach der Natur Beschrieben.* Nürnberg: Bauer & Raspe. Heft 1: [4] + 100 sheets, 2 pls.

Küster, H.C. 1845a: *Die Käfer Europa's. Nach der Natur Beschrieben.* Nürnberg: Bauer & Raspe. Heft 2: [4] + 100 sheets, 2 pls.

Küster, H.C. 1845b: *Die Käfer Europa's. Nach der Natur Beschrieben.* Nürnberg: Bauer & Raspe. Heft 3: [4] + 100 sheets, 2 pls.

Küster, H.C. 1846a: *Die Käfer Europa's. Nach der Natur Beschrieben.* Nürnberg: Bauer & Raspe. Heft 4: [6] + 100 sheets., 3 pls.

Küster, H.C. 1846b: *Die Käfer Europa's. Nach der Natur Beschrieben.* Nürnberg: Bauer & Raspe. Heft 5: [4] + 100 sheets, 3 pls.

Küster, H.C. 1846c: *Die Käfer Europa's. Nach der Natur Beschrieben.* Nürnberg: Bauer & Raspe. Heft 7: [4] + 100 sheets, 2 pls.

Küster, H.C. 1847a: *Die Käfer Europa´s. Nach der Natur beschrieben.* Nürnberg: Bauer & Raspe. Heft 8: [4] + 100 sheets, 2 pls.

Küster, H.C. 1847b: *Die Käfer Europa's. Nach der Natur Beschrieben.* Nürnberg: Bauer & Raspe. Heft 9: [4] + 100 sheets, 3 pls.

Küster, H.C. 1847c: *Die Käfer Europa's. Nach der Natur Beschrieben.* Nürnberg: Bauer & Raspe. Heft 12: [4] + 100 sheets, 2 pls.

Küster, H.C. 1848a: *Die Käfer Europa's. Nach der Natur Beschrieben.* Nürnberg: Bauer & Raspe. Heft 13: [4] + 100 sheets, 3 pls.

Küster, H.C. 1848b: *Die Käfer Europa's. Nach der Natur Beschrieben.* Nürnberg: Bauer & Raspe. Heft 14: [4] + 100 sheets, 2 pls.

Küster, H.C. 1849a: *Die Käfer Europa's. Nach der Natur Beschrieben.* Nürnberg: Bauer & Raspe. Heft 16: [4] + 100 sheets, 3 pls.

Küster, H.C. 1849b: *Die Käfer Europa´s. Nach der Natur beschrieben.* Nürnberg: Bauer & Raspe. Heft 18: [4] + 100 sheets pls.

Küster, H.C. 1849c: *Die Käfer Europa's. Nach der Natur Beschrieben.* Nürnberg: Bauer & Raspe. Heft 19: [4] + 100 sheets, 3 pls.

Küster, H.C. 1850a: *Die Käfer Europa's. Nach der Natur Beschrieben.* Nürnberg: Bauer & Raspe. Heft 20: [4] + 100 sheets, 2 pls.

Küster, H.C. 1850b: *Die Käfer Europa's. Nach der Natur Beschrieben.* Nürnberg: Bauer & Raspe. Heft 21: [4] + 100 sheets, 2 pls.

Küster, H.C. 1851: *Die Käfer Europa's. Nach der Natur Beschrieben.* Nürnberg: Bauer & Raspe. Heft 22: [4] + 100 sheets, 3 pls.

Küster, H.C. 1852a: *Die Käfer Europa's. Nach der Natur Beschrieben.* Nürnberg: Bauer & Raspe. Heft 24: [6] + 100 sheets, 2 pls.

Küster, H.C. 1852b: *Die Käfer Europa's. Nach der Natur Beschrieben.* Nürnberg: Bauer & Raspe. Heft 25: [4] + 100 sheets, 2 pls.

Küster, H.C. 1853: *Die Käfer Europa's. Nach der Natur Beschrieben*. Nürnberg: Bauer & Raspe. Heft **25**: [8] + 100 sheets, 2 pls.

Kuzin, V.S. 1927: O geographytcheskoy izmentchivosti v rode Mylabris Fabr. (Coleoptera Meloidae) [Zur geographischen Variabilität in der Gattung Mylabris Fabr. (Coleoptera Meloidae)]. *Revue Russe d'Entomologie* **21**: 17–26.

Kuzin, V.S. 1934: Novyy vid roda Psammocryptus (Coleoptera, Tenebrionidae) iz Srednei Azii. *Sbornik Trudov Gosudarstvenogo Zoologicheskogo Muzeya (pri Moskovskom Gosudarstvennom Universitete)* **1**: 160.

Kuzin, V.S. 1953: Naryvniki Kazakhstana [The blister beetles of Kazakhstan]. *Trudy Respublikanskoy Stantcii Zashchity Rasteniy Kazakhstanskiy Filial Vaskhnil (Alma-Ata)* **1**: 72–154.

Kuzin, V.S. 1954: K poznanyyu systemy narybnikov (Coleoptera, Meloidae, Mylabrini). *Trudy Vsesoyznogo Entomologicheskogo Obshchestva* **44**: 336–379.

Kuznetzova, E.A. 1935: Zwei neue zentralasiatische Oedemeriden (Coleoptera). *Entomologicheskoe Obozrenie* **25**: 298–303.

Kwieton, E. 1976: Une espèce nouvelle du genre Pimelia F. (Coleoptera, Tenebrionidae). *Annotationes Zoologicae et Botanicae* (Bratislava) **113**: 1–3.

Kwieton, E. 1977a: Révision phylogénétique du groupe de Pimelia obsoleta (Col., Tenebrionidae). *Bulletin de la Société Entomologique de Mulhouse* **1977**: 17–24.

Kwieton, E. 1977b: Esquisse phylogénétique du genre Pimelia F. *Acta Entomologica Musei Nationalis Pragae* **39**: 559–589.

Kwieton, E. 1978a: Espèces nouvelles des genres Adesmia Fisch., Pimelia Sol. et Vieta (Col. Tenebrionidae). *Bulletin de la Société Entomologique de Mulhouse* **1978**: 8–12.

Kwieton, E. 1978b: Sur la division de l'ancien genre Pachyscelis Sol. (Col., Tenebrionidae). *Bulletin de la Société Entomologique de Mulhouse* **1978**: 28–34.

Kwieton, E. 1980a: Synopsis des espèces du genre Sepidium F. d'Algérie et de Tunisie (Col. Tenebrionidae). *Annotationes Zoologicae et Botanicae* (Bratislava) **138**: 1–19.

Kwieton, E. 1980b: Contribution à la connaissance des genres Erodius Fab. et Morica Sol. (Col. Tenebrionidae) *Bulletin de la Société Entomologique de Mulhouse* **1980**: 25–28.

Kwieton, E. 1981a: Coléoptères Tenebrionidae constatés par l'expédition de Tis au Pakistan. *Annotationes Zoologicae et Botanicae* (Bratislava) **143**: 1–20.

Kwieton, E. 1981b: Insects of Saudi Arabia Coleoptera: Fam. Tenebrionidae, Tribe Pimeliini. *Fauna of Saudi Arabia* **3**: 402–407.

Kwieton, E. 1982a: Contributions ultérieures à la connaissance du genre Pimelia F. (Col., Tenebrionidae). *Annotationes Zoologicae et Botanicae* (Bratislava) **145**: 1–38.

Kwieton, E. 1982b: Notes sur les Tenebrionidae Pimeliinae du Pakistan. *Bulletin de la Société Entomologique de Mulhouse* **1982**: 21–26.

Kwieton, E. 1986a: Synopsis des Asidini d'Algérie et de Tunisie I (Coleoptera, Tenebrionidae). *Annotationes Zoologicae et Botanicae* (Bratislava) **175**: 1–29.

Kwieton, E. 1986b: Synopsis des Asidini d'Algérie et de Tunisie II (Coleoptera, Tenebrionidae). *Annotationes Zoologicae et Botanicae* (Bratislava) **176**: 1–14.

Kwieton, E. 1987: Mise au point sur la phylogénie et la paléogéographie du g. Pimelia F. (Col. Tenebrionidae). *Bulletin de la Société Entomologique de Mulhouse* **1987**: 9–15.

Labrique, H. 1992: Description d'une nouvelle espèce de Scaurus du Maroc (Coleoptera, Tenebrionidae). *Revue Française d'Entomologie* (N.S.) **14**: 123–125.

Labrique, H. 1995: Les Scaurus du groupe tingitanus Peyerimhoff (Coleoptera, Tenebrionidae). *Bulletin de la Société Entomologique de France* **100**: 119–126.

Labrique, H. & Chavanon, G. 2005: Une nouvelle sous-espèce de Scaurus F. (Coleoptera Tenebrionidae). *Bulletin Mensuel de la Société Linnéenne de Lyon* **74**: 41–43.

Labrique, H., Chavanon, G. & Janati-Idrissi, A. 2011: Les Phylan Dejean, 1821 du sousgenre Meladeras Mulsant et Rey, 1854 (Coleoptera; Tenebrionidae; Pedinini; Dendarina) au Maroc. *Les cahiers du Musée des Confluences – Études scientifiques* **2**: 75–82.

Lacordaire, J.T. 1859: *Histoire naturelle des insectes. Genera des coléoptères ou exposé méthodique et critique de tous les genres proposés jusqu'ici dans ce ordre d'Insectes. Tome Cinquième.* Paris: Librairie Encyclopédique de Roret. Première partie, 1–400 pp., seconde partie, 401–750 pp.

Lacordaire, J.T. 1866: *Histoire naturelle des insectes. Genera des coléoptères ou exposé méthodique et critique de tous les genres proposés jusqu'ici dans ce ordre d'Insectes. Tome Septième.* Paris: Librairie Encyclopédeique de Roret, 620 pp.

Lacordaire, J.T. 1876: *Histoire naturelle des insectes. Genera des coléoptères ou exposé méthodique et critique de tous les genres proposés jusqu'ici dans ce ordre d'Insectes. Atlas.* Paris: Librairie Encyclopédique de Roret, 47 pp., 134 pls.

Lafer, L.S. 1996: 100a, 101. Sem. Anthicidae (Pedilidae) – bistryanki. Pp. 26–45. In: Ler P.A. (ed.): *Opredelitel' nasekomykh Dal'nego Vostoka Rossii* [*Key to the insects of the Russian Far East*]. *Tom III. Zhestkokrylye, ili zhuki. Chast' 1.* Vladivostok: Nauka, 555 pp.

LaFerté-Sénectère, F.T. de. 1842: Sur les Anthicus recueillis aux environs de Perpignan, et description de quelques espèces nouvelles. *Annales de la Société Entomologique de France* **11**: 247–260, pl. 10.

LaFerté-Sénectère, F.T. de. 1847: [new taxa of Anthicidae, pp. 365–381, pl. 32]. In: Lucas P.H.: *Exploration scientifique de l'Algérie pendant les années 1840, 1841, 1842 publiée par ordre du gouvernement et avec le concours d'une Commission Académique. Sciences physiques Zoologie. Vol. II. Histoire naturelle des animaux articulés. Cinquième classe. Insectes. Premier ordre. Les coléoptères.* Paris: Imprimerie Nationale [1849], 590 pp., 47 pls.

LaFerté-Sénectère, F.T. de. 1849: [new taxa]. In: Guérin-Méneville F.-E. (ed.): *Species et iconographie générique des animaux articulés ou representation des genres, avec leur description et celle de toutes les espèces de cette grande division du règne animal. Première Partie: Insectes Coléoptères (1846–1847).* Paris: de Fain et Thunot.

LaFerté-Sénectère, F.T. de. 1849a. Notoxus. Livraison 4, No. 21, 36 pp., 1 pl.

LaFerté-Sénectère, F.T. de. 1849b. Mecynotarsus, Livraison 4, No. 22, 5 pp., 1 pl.

LaFerté-Sénectère, F.T. de. 1849c. Stereopalpus, Livraison 5, No. 18, 3 pp., 1 pl.

LaFerté-Sénectère, F.T. de. 1849d. Macrarthrius, Livraison 5, No. 20, 10 pp., 1 pl.

LaFerté-Sénectère, F.T. de. 1849e. Amblyderus, Livraison 6, No. 23, 3 pp., 1 pl.

LaFerté-Sénectère, F.T. de. 1849f. Anthelephilus, Livraison 6, No. 24, 5 pp., 1 pl.

LaFerté-Sénectère, F.T. de. 1849g. Formicomus, Livraison 7, No. 25, 25 pp., erratum p., 1 pl.

LaFerté-Sénectère, F.T. de. 1849h. Tomoderus, Livraison 7, No. 26, 8 pp., 1 pl.

LaFerté-Sénectère, F.T. de. 1849i. Anthicus (Première division), Livraison 7, No. 27, pp. 1–45, 1 pl.

LaFerté-Sénectère, F.T. de. 1849j. Anthicus (Seconde Division), Livraison 8, No. 28, pp. 47–83, 1 pl.

LaFerté-Sénectère, F.T. de. 1849k. Anthicus (Troisième Division), Livraison 8, No. 29, pp. 85–132, 1 pl.

LaFerté-Sénectère, F.T. de. 1849l. Anthicus (Quatrième Division), Livraison 9, No. 30, pp. 133–181, 1 pl.

LaFerté-Sénectère, F.T. de. 1849m. Ochthenomus, Livraison 9, No. 31, 9 pp., 1 pl.

LaFerté-Sénectére, F.T. de. 1849n: *Monographie des Anthicus et genres voisins, coléoptères hétéromères de la tribu des Trachélides (1848)*. Paris: De Sapia, xxii + 340 pp., pls. 17–32.

Lagar, A. & Fresneda, J. 2014: Stenohelops barcinensis sp. n. de Barcelona (España) (Coleoptera, Tenebrionidae, Helopini). *Boletín de la Sociedad Entomológica Aragonesa* 55: 93–99.

Lang, J.-T. & Ren, G.-D. 2009: A review of the genus Catapiestus Perty, 1831 (Coleoptera: Tenebrionidae: Cnodalonini) with descriptions of two new species from China. *Caucasian Entomological Bulletin* 5: 195–199.

Laporte F.L.N. de Caumont de Castelnau. 1832: Mémoire sur cinquante espèces nouvelles ou peu connues d'insectes. *Annales de la Société Entomologique de France* 1: 386–415.

Laporte F.L.N. de Caumont de Castelnau. 1833: Coléoptères et hémiptères nouveaux. *Revue Entomologique* (G. Silbermann) 1: 32–36.

Laporte F.L.N. de Caumont de Castelnau. 1835: *Études entomologiques, ou description d'insectes nouveaux, et observations sur la synonymie.* Paris: Méquignon-Marvis, 159 pp.

Laporte F.L.N. de Caumont de Castelnau. 1840: *Histoire naturelle des insectes coléoptères; avec une introduction renfermant l'anatomie et la physiologie des animaux articulés, par M. Brullé. Tome deuxième.* Paris: P. Duménil, 563 + [1] pp., pls 20–37.

Laporte F.L.N. de Caumont de Castelnau & Brullé G.A. 1831: Monographie du genre Diaperis. *Annales des Sciences Naturalles* 23: 325–410, pl. 10.

Latreille, P.A. 1796: *Précis des caractères génériques des insectes, disposés dans un ordre naturel.* Paris: F. Bourdeaux, xiii+201 + [7] pp.

Latreille, P.A. 1802: *Histoire naturelle, générale et particulière des crustacés et des insectes. Ouvrage faisant suite à l'histoire naturelle générale et particulière, composée par Leclerc de Buffon, et rédigée par C.S. Sonnini, membre de plusieurs sociétés savantes. Familles naturelles des genres. Tome troixième.* Paris: F. Dufart, xii +13–467 + [1] pp.

Latreille, P.A. 1804a: *Histoire naturelle, genérale et particuliere, des crustacés et des insectes. Ouvrage faisant suite aux oeuvres de Leclerc de Buffon, et partie du cours complet d'histoire naturelle rédigé par C.S. Sonnini, membre de plusieurs Sociétés savantes. Tome dixième.* Paris: F. Dufart, 445 pp., pls 81–90.

Latreille, P.A. 1804b: *Histoire naturelle, genérale et particuliere, des crustacés et des insectes. Ouvrage faisant suite aux oeuvres de Leclerc de Buffon, et partie du cours complet d'histoire naturelle rédigé par C.S. Sonnini, membre de plusieurs Sociétés savantes. Tome onzième.* Paris: F. Dufart, 422 pp.

Latreille, P.A. 1807: *Genera crustaceorum et insectorum secundum ordinem naturalem in familias disposita, iconibus exemplisque plurimis explicita. Tomus secundus.* Parisiis et Argentorati: Amand Koenig, 280 pp.

Latreille, P.A. 1809: *Genera crustaceorum et insectorum secundum ordinem naturalem in familias disposita, iconibus exemplisque plurimis explicita. Tomus quartus et ultimus.* Parisiis et Argentorati: Amand Koenig, 399 pp.

Latreille, P.A. 1810: *Considérations générales sur l'ordre naturel des animaux composant les classes des crustacés, des arachnides, et des insectes; avec un tableau méthodique de leurs genres, disposés en familles.* Paris: F. Schoell, 444 pp.

Latreille, P.A. 1816: [new taxa]. In: *Nouveau Dictionnaire d'Histoire Naturelle, appliquée aux arts, à l'agriculture, à l'économie rurale et domestique, à la médecine, etc. Par une société de naturalistes et d'agriculteurs. Nouvelle édition presque entièrement refondue et considérablement augmenté; avec des figures tirées des trois règnes de la nature. Tome I*. Paris: Deterville, [4] + lxxix + 552 pp.

Latreille, P.A. 1817: [new taxa]. In: Cuvier G.: *Le règne animal distribué d'après son organisation, pour servir de base à l'histoire naturelle des animaux et d'introduction à l'anatomie comparée. Tom III, contenant les crustacés, les arachnides et les insectes*. Paris: Deterville, xxiv + 660 pp., 2 pls.

Latreille, P.A. 1818a: [new taxa]. In: *Nouveau Dictionnaire d'Histoire Naturelle, appliquée aux arts, à l'agriculture, à l'économie rurale et domestique, à la Médecine, etc. Nouvelle édition presque entièrement refondue et considérablement augmentée, avec des figures tirées des trois règnes de la nature. Med-Min. Tome XX*. Paris: Deterville, [2] + 586 pp.

Latreille, P.A. 1818b: [new taxa]. In: *Nouveau Dictionnaire d'Histoire Naturelle, appliquée aux arts, à l'agriculture, à l'économie rurale et domestique, à la Médecine, etc. Nouvelle édition presque entièrement refondue et considérablement augmentée, avec des figures tirées des trois règnes de la nature. Mu-Nil. Tome XXII*. Paris: Deterville, [2] + 618 pp.

Latreille, P.A. 1818c: [new taxa]. In: *Nouveau Dictionnaire d'Histoire Naturelle, appliquée aux arts, à l'agriculture, à l'économie rurale et domestique, à la médecine, etc. Nouvelle édition presque entièrement refondue et considérablement augmentée, avec des figures tirées des trois règnes de la nature. Pas-Pho. Tome XXV*. Paris: Deterville, [4] + 610 pp., 7 pls.

Latreille, P.A. 1818d: [new taxa]. In: *Nouveau Dictionnaire d'Histoire Naturelle, appliquée aux arts, à l'agriculture, à l'économie rurale et domestique, à la médecine, etc. Nouvelle édition presque entièrement refondue et considérablement augmentée, avec des figures tirées des trois règnes de la nature. Pla-Por. Tome XXVII*. Paris: Deterville, [4] + 586 pp., 4 pls.

Latreille, P.A. 1818e: [new taxa]. In: *Nouveau Dictionnaire d'Histoire Naturelle, appliquée aux arts, à l'agriculture, à l'économie rurale et domestique, à la médecine, etc. Nouvelle édition presque entièrement refondue et considérablement augmentée, avec des figures tirées des trois règnes de la nature. Cor-Cun. Tome VIII*. [1817] Paris: Deterville, [4] + 602 pp., 10 pls.

Latreille, P.A. 1819a: [new taxa]. In: *Nouveau Dictionnaire d'Histoire Naturelle appliquée aux arts, à l'agriculture, à l'économie rurale et domestique, à la médecine, etc. par une société de naturalistes et d'agriculteurs. Nouvelle édition presqu'entièrement refondue et considérablement augmentée; avec des figures des trois règnes de la nature. Por-Ral. Tome XXVIII*. Paris: Deterville, [4] + 570 pp.

Latreille, P.A. 1819b: [new taxa]. In: *Nouveau Dictionnaire d'Histoire Naturelle appliquée aux arts, à l'agriculture, à l'économie rurale et domestique, à la médecine, etc. par une société de naturalistes et d'agriculteurs. Nouvelle édition presqu'entièrement refondue et considérablement augmentée; avec des figures des trois règnes de la nature. Ram-Ryz. Tome XXIX*. Paris: Deterville, [4] + 576 pp.

Latreille, P.A. 1827: Description d'insectes d'Afrique recueillis par M. Caillaud. Pp. 271–292. In: Cailliaud, F.: *Voyage à Méroé, au fleuve Blanc, au-delà de Fazoql dans le midi du royaume de Sennâr, à Syouah et dans cinq autres oasis; fait dans les années 1819, 1820, 1821 et 1822. Tome IV*. Paris: Imprimerie Royale, 416 pp.

Latreille, P.A. 1828: Piméliaires. Pp. 573–583. In: Bory de Saint-Vincent, J.B.G.M. (ed.): *Dictionnaire classique d'histoire naturelle, par Messieurs Audoin, Isid. Bordon, Ad. Brongniart, De Candolle, Daudebard de Férussac, A. Desmoulins, Drapiez, Edwards, Flourens, Geoffroy de Saint-Hilaire, A. de Jussieu, Kunth, G. de Lafosse, Lamouroux, Latreille, Lucas fils, Presle-Duplessis, C. Prévost, A. Richard, Thiébaut de Berneaud, et Bory de*

Saint-Vincent. Ouvrage dirigé par ce dernier collaborateur, et dans lequel on a ajouté, pour le porter au niveau de la science, un grand nombre de mots qui n'avaient pu faire partie de la plupart des dictionnaires antérieurs. Paris: Ray et Gravier [et] Beaudoin Frères.

Latreille, P.A. 1829: Suite et fin des insectes. In: Cuvier G.: *Le règne animal distribué d'après son organisation, pour servir de base à l'histoire naturelle des animaux et d'introduction à l'anatomie comparée. Nouvelle édition, revue et augmentée. Tome v.* Paris: Déterville, xxii + 556 pp.

Lauffer, G. 1905: Neue Arten und Varietäten von Coleopteren der pyrenäischen Halbinsel. *Boletín de la Real Sociedad Española de Historia Natural* 5: 403–407.

Lawrence, J.F. 1980: A new genus of Indo-Australian Gempylodini with notes on the constitution of the Colydiidae (Coleoptera). *Journal of the Australian Entomological Society* 19: 293–310.

Lawrence, J.F. 1991: Three new Asiatic Ciidae (Coleoptera: Tenebrionoidea) associated with commercial, dried fungi. *The Coleopterists Bulletin* 45: 286–292.

Lazorko, W. 1974: Zwei neue Stenalia-Arten aus der Ukraine. *Reichenbachia* 15: 109–115.

Leach, W.E. 1815a: Entomology. Pp. 57–172. In: Brewster, D. (ed.): *The Edinburg Encyclopedia* 9 (1). Edinburg: Baldwin, 384 pp.

Leach, W.E. 1815b: III. An essay on the British species of the genus Meloë, with descriptions of two exotic species. *Transactions of the Linnean Society of London* 11 [1810]: 35–49, 2 pls.

Lebedev, A.G. 1931: Zwei neue Coleopteren aus Central-Asien. *Entomologisches Nachrichtenblatt* 5: 78–79.

Lebedev, A.G. 1932: Caediexis arenicola gen. et sp. n. (Tenebrionidae). *Entomologische Blätter* 28: 125–126.

Leblanc, P. 2002: Description de deux nouvelles espèces ouest-européennes du genre Mediimorda Méquignon, 1946. *Bulletin Mensuel de la Société Linnéenne de Lyon* 71: 317–324.

Leblanc, P. 2007: Révision des espéces ouest-paléarctiques du gebre Mediimorda Méquignon, 1946 et description de deux nouvelles espéces d'Afrique du Nord. *Bulletin menssuel de la Soccíété linnéenne de Lyon* 76 (9–10): 235–250.

Leblanc, P. 2012: Miscellaneous notes on western palaearctic Scraptia and description of a new species (Coleoptera Scraptiidae). *Bulletin Mensuel de la Société Linnéenne de Lyon* 81: 47–53.

LeConte, J.L. 1850: *General remarks upon the Coleoptera of Lake Superior.* Pp. 201–242. In: Agassiz J.L.R. & Cabot J.E. (eds): *Lake Superior, its physical character, vegetation and animals, compared with those of other and similar regions. With a narrative of the tour, by J. Eliott Cabot. And contributions by other scientific gentlemen.* Boston: Gould, Kendall and Lincoln, x + 428 pp., 8 pls.

LeConte, J.L. 1851: Descriptions of new species of Coleoptera, from California. *Annals of the Lyceum of Natural History, New York* 5: 125–216.

LeConte, J.L. 1854: Notice of some Coleopterous Insects, from the Collections of the Mexican Boundary Commission. *Proceedings of the Academy of Natural Sciences of Philadelphia* 7: 79–85.

LeConte, J.L. 1855: Synopsis of the Pyrochroidae of the United States. *Proceedings of the Academy of Natural Sciences of Philadelphia* 7: 270–277.

LeConte, J.L. 1856: Synopsis of the Mycetophagidae of the United States. *Proceedings of the Academy of Natural Sciences of Philadelphia* 8: 12–15

LeConte, J.L. 1857: The Coleoptera of Kansas and eastern New Mexico. *Smithsonian Contribution (to Zoology)* 11: 1–58.

LeConte, J.L. 1861: Classification of the Coleoptera of North America. Part 1. Prepared for the Smithsonian Institution. *Smithsonian Miscellaneous Collections* 3: i-xxv + 1–208.

LeConte, J.L. 1862: Classification of the Coleoptera of North America. Part 1. (cont.). Prepared for the Smithonian Institution. *Smithsonian Miscellaneous Collections* **3**: 209–286.

LeConte, J.L. 1863: New species of North American Coleoptera. *Smithsonian Miscellaneous Collections* **167** [1863–1866]: 1–92.

LeConte, J.L. 1866: New species of North American Coleoptera. *Smithsonian Miscellaneous Collections* **6** (167): 87–168 [index 169–177].

LeConte, J.L. 1874a: Descriptions of new Coleoptera, chiefly from the Pacific Slope of North America. *Transactions of the American Entomological Society* **5**: 43–72.

LeConte, J.L. 1874b: [new taxa]. In: Austin E.P.: Catalogue of the Coleoptera of Mt. Washington, N.H. by E.P. Austin; with descriptions of new species, by J.L. LeConte, M.D. *Proceedings of the Boston Society of Natural History* **16**: 265–276.

LeConte, J.L. 1879: New Coleoptera. *The North American Entomologist* **1** (1): 1–5, 1 pl.

Lee, S.-G. & Lim, J. 2019: A new record of the genus Lissodema Curtis 1833 (Coleoptera: Salpingidae) with description of a new species in Korea. *Entomological Research* **49**: 21–25.

Lemaire, J.-M. & Soldati, F. 2009: Contribution to the Natural History of Cyprus: Stenosis austini, a new species from the Troodos Range (Coleoptera, Tenebrionidae, Pimeliinae). *Biocosme Mésogéen* **26**: 127–132.

Leo, P. 1980a: Psammoardoinellus, nuovo genere di Opatrinae della Sardegna (Coleoptera Tenebrionidae). *Memorie della Società Entomologica Italiana* **59**: 34–36.

Leo, P. 1980b: Un nuovo Stenohelops Reitt. della Sardegna (Coleoptera Tenebrionidae Helopinae). *Bollettino dela Società Entomologica Italiana* **112**: 36–39.

Leo, P. 1980c: Asida argentierae n.sp. di Sardegna (Coleoptera Tenebrionidae). *Bollettino della Società Sarda di Scienzie Naturale* **19**: 189–192.

Leo, P. 1982a: Nalassus genei melonii ssp. nov., dell'isola di San Pietro (Sardegna sud-occidentale) (Coleoptera Tenebrionidae). *Doriana* **5** (246): 1–4.

Leo, P. 1982b: Gli Pseudoseriscius italiani del gruppo helvolus (Küster) (Coleoptera Tenebrionidae). *Natura* **73**: 97–107.

Leo, P.1984a: Belopus (Centorus) proceroides n. sp. della Sardegna (Coleoptera Tenebrionidae). *Bollettino della Società Entomologica Italiana* **116**: 165–167.

Leo, P. 1984b: Due nuove Asida Latr. di Sardegna (Coleoptera Tenebrionidae). *Annali del Museo Civico di Storia Naturale «Giacomo Doria», Genova* **85**: 73–77.

Leo, P. 1985: Le popolazioni italiane di Nalassus aemulus (Küster) e descrizione di una nuova razza della Sardegna (Coleoptera Tenebrionidae Helopinae). *Bollettino del Museo Civico di Storia Naturale di Verona* **12**: 141–148.

Leo, P. 1998: Nuovi dati sui tenebrionidi delle isole Toscane e descrizione di Asida (s. str.) gestroi Leoni lanzai n. ssp. (Coleoptera, Heteromera). *Atti del Museo di Storia Naturale della Maremma* **17**: 73–77.

Leo, P. 2008: Osservazioni su Dichillus corsicus e descrizione di tre nuove specie del Mediterraneo occidentale (Coleoptera, Tenebrionidae). *Annali del Museo Civico di Storia Naturale* **99**: 603–627.

Leo, P. 2009: Observations on some Tenebrionidae (Coleoptera) from Sardinia, with description of three new Asida. *Zootaxa* **2318**: 400–420.

Leo, P. 2012: Tre nuove specie di Asida della Sardegna (Coleoptera, Tenebrionidae). *Annali del Museo Civico di Storia Naturale 'G. Doria'* **104**: 97–113.

Leo, P. 2017: Short nomenclatural notes on Pachychila and Dendarus (Coleoptera: Tenebrionidae Tentyriini and Pedinini). *Revista gaditana de Entomología* 8: 173–175.

Leo, P. 2018: Cyclocnera azarovi, un nuevo género y especie de Pimeliini de Omán (Coleoptera, Tenebrionidae, Pimeliinae). *Revista gaditana de Entomología* 9: 49–57.

Leo, P. & Liberto, A. 2003: Un nuovo genere di Helopini della Grecia (Coleoptera, Tenebrionidae). *Fragmenta Entomologica* **34** [2002]: 299–309.

Leo, P. & Liberto, A. 2011: Un nuovo genere e una nuova specie di Melambiina dell'Oman (Coleoptera, Tenebrionidae). *Fragmenta entomologica* **43**: 157–166.

Leo, P. & Pisano, P. 1978: Tassonomia, geonemia ed ecologia delle razze di Pimelia angusticollis Sol. e descrizione di una nuova sottospecie (Coleoptera Tenebrionidae). *Atti XI Congresso Nazionale Italiano di Entomologia Portici-Sorrento, 10–15 maggio 1976*: 79–83.

Leoni, G. 1909: Le Asida italiane. *Rivista Coleopterologica Italiana* 7: 142–189, 211–226.

Leoni, G. 1910: Le Asida italiane. *Rivista Coleopterologica Italiana* 8: 9–16, 40–55, 69–95.

Leoni, G. 1911: Specie e varietã nuove o rare di coleotteri italiani. *Rivista Coleopterologica Italiana* 9: 1–10.

Leoni, G. 1929: Apunti sulle Asida (Coleoptera-Tenebrionidae) italiane e descrizione di nuove forme. *Bollettino dal Laboratorio di Entomologia del R. Instituto Superiore Agraria di Bologna* 2: 148–154.

Le Peletier de Saint-Fargeau, A.L.M. & Audinet-Serville, A.J.G. 1825: [new taxa]. In: Latreille P.A., Le Peletier de Saint-Fargeau A., Audinet-Serville A.J.G. de. & Guérin-Méneville F.: *Encyclopédie méthodique. Histoire naturelle. Entomologie, ou histoire naturelle des crustacés, des arachnides et des insectes. Tome dixième.* Paris: Agasse, [4] + 832 + [1] pp.

Lepesme, P. 1943: Un Tribolium inédit du Sénegal (Col. Tenebrionidae). *Revue Française d'Entomologie* **10**: 45–46.

Leske, N.G. 1785: *Reise durch Sachsen in Rücksicht der Naturgeschichte und Őkonomie unternommen und beschrieben. Erstes Heft.* Leipzig: J.G. Müllersche Buchhandlung, xxx + 549 pp., 38 pls.

Lesne, P. 1902a: Diagnose d'un Rhipidius nouveau des environs de Paris (Col.). *Bulletin de la Société Entomologique de France* **1902**: 181–182.

Lesne, P. 1902b: Notes sur deux espèces françaises de coléoptères du genre Rhipidius. *Bulletin du Muséum d'Histoire Naturelle* 8: 420–423.

Lesne, P. 1915: Les érodiens de l'Afrique orientale (coléoptères ténébrionides). *Bulletin du Muséum National d'Histoire Naturelle, Paris* **21**: 225–240.

Lesne, P. 1917: Notes sur divers cisides (Col. Ciidae). *Bulletin de la Société Entomologique de France* **1917**: 190–192.

Lesne, P. 1920: Notes sur divers zophosites. *Bulletin du Muséum National d'Histoire Naturelle, Paris* **26**: 14–19.

Lesne, P. 1922: Bostrychides, clérides, sphindides et ténébrionides. Pp. 649–704. In: *Voyage M. le Baron Maurice de Rothschild en Éthiopie et en Afrique orientale anglaise (1904–1905). Résultats scientifiques. Animaux articulés. Deuxième partie.* Paris, Imprimerie Nationale, 1041 pp., 3 maps, 100 pls.

Letzner, K. 1857: Ueber Anaspis flavo-atra, Mordella pumila, die Puppe von Opilus domesticus, die früheren Zustgände von Chrysom. cuprea F. u. 20-punctata Scop., litura F., Helodes phellandrii L., beccabungae Ill., Puppe u. Larvensack von Cryptocephalus sericeus L., Crioceris asparagi L., Rhagium bifasciatum F., Puppe von Ampedus nigrinus Payk. *Arbeiten und Veränderungen der Schlesischen Gesellschaft für Vaterländische Kultur* (Breslau) **1857**: 119–138.

Levrat, J.N.B.G. 1853: Description d'une espèce nouvelle du genre Pimelia. *Annales de la Société Lynnéenne* (2) **1**: 1, 1 pl.

Levrat, J.N.B.G. 1859a: Description de quelques coléoptères nouveaux des environs de Tunis. *Études Entomologiques* (Lyon) **1**: 33–36.

Levrat, J.N.B.G. 1859b: Description d'une nouvelle espèce du genre Pimelia. *Études Entomologiques* (Lyon) **1**: 41–42.

Lewis, G. 1879: On certain new species of Coleoptera from Japan. *The Annals and Magazine of Natural History* (5) **4**: 459–467.

Lewis, G. 1887a: On the Pyrochroidae of Japan. *The Annals and Magazine of Natural History* (5) **20**: 165–175.

Lewis, G. 1887b: On a new species of Diaperis from Japan. *The Entomologist* **20**: 217–218.

Lewis, G. 1887c: On a new species of Phellopsis found in Japan and Siberia. *The Entomologist* **20**: 218–220.

Lewis, G. 1888: On the capture of a new species of Pytho in Japan. *The Entomologist* **21**: 108–109.

Lewis, G. 1891a: On two new species of Heteromera from Japan. *The Entomologist's Monthly Magazine* **27**: 70–71.

Lewis, G. 1891b: On two species of Elacatis (Othnius) found in Japan. *The Entomologist's Monthly Magazine* **27**: 247–248.

Lewis, G. 1894: On the Tenebrionidae of Japan. *The Annals and Magazine of Natural History* (6) **13**: 377–400, 465–485, pl. XIII.

Lewis, G. 1895a: On the Cistelidae and other heteromerous species of Japan. *The Annals and Magazine of Natural History* (6) **15**: 250–278, 422–448, pl. 8.

Lewis, G. 1895b: On the Dascillidae and malacoderm Coleoptera of Japan. *The Annals and Magazine of Natural History* (6) **16**: 98–122, pl. 6.

Lewis, G. 1896: On new species of Coleoptera from Japan, and notice of others. *The Annals and Magazine of Natural History* (6) **17**: 329–343.

Li, B.-Y. & Jiang R.-X. 2019: Discovery of the genus Micropeneta Pic in China, with description of a new species (Coleoptera: Tenebrionidae: Gnathidiini). *Zootaxa* **4608**: 196–200.

Li, J.-K. 1992: *The Coleoptera Fauna of Northeast China*. Jilin: Education Publishing House, [8] + 205 pp. (in Chinese with English title).

Li, J.-K. & Chen, P. 1993: *Studies on fauna and ecogeography of soil animal*. Chanchun: Northeast Normal University Press, [8] + 265 pp. (in Chinese with English title).

Li, J.-K. & Gao, M.-X. 2015: [new taxa]. In: Li J.-K., Zhang L.-M. & Zhang X.-P. 2015: [*Color illustrated book of soil beetles in Northeast China, Staphylinoidea and Tenebrionoidea*]. Harpin Map publisher, Harbin: 199 pp. (in Chinese with English description).

Li, J.-K. & Lin, L. 2015: [new taxa]. In: Li J.-K., Zhang L.-M. & Zhang X.-P.: [*Color illustrated book of soil beetles in Northeast China, Staphylinoidea and Tenebrionoidea*]. Harpin Map publisher, Harbin: 199 pp. (in Chinese with English description).

Li, X.-M., Bai, X.-L. & Ren, G.-D. 2018: A new species of the genus Blaptogonia from the Himalayas with four DNA markers (Coleoptera, Tenebrionidae, Blaptini). *ZooKeys* **773**: 69–78.

Li, X.-M., Bai, X.-L. & Ren, G.-D. 2019: Two new species of the genus Gnaptorina Reitter from the Hengduan Mountains, China (Coleoptera: Tenebrionidae: Blaptini). *Zootaxa* **4695**: 83–89.

Li, Y.-C., Egorov, L.V. 2013: [new taxa]. In: Li Y.-C., Egorov L.V. & Shi A.-M.: Two new species of the genus Bioramix Bates, 1879 (Coleoptera: Tenebrionidae: Platyscelidini), from the Chinese Provinces Sichuan and Tibet. *Caucasian Entomological Bulletin* **9**: 39–94.

Li, Y.-C., Egorov, L.V. 2016: [new taxa]. In: Li Y.-C., Egorov L.V. & Shi A.M.: Three new species of the subgenus Leipopleura Seidlitz from Tibet, China (Coleoptera, Tenebrionidae, Bioramix Bates). *ZooKeys* **609**: 29–41.

Li, Y.-C., Egorov, L.V. & Shi, A.M. 2016: Two new species of the subgenus Cardiobioramix Kaszab (Coleoptera: Tenebrionidae: Bioramix Bates) from the Sichuan Province, China. *Zootaxa* **4111**: 584–590.

Li, Y.-C. & Shi, A.-M. 2019: A new species of Agnaptoria Reitter (Coleoptera: Tenebrionidae: Blaptini) from Dhaulagiri Zone in Nepal. *Oriental Insects*: 1–12.

Li, Z. & Ren, G.-D. 2004: A systematic study on Gnaptorina Reitter (Coleoptera: Tenebrionidae) from China. *Oriental Insects* **38**: 251–275.

Li, Z. & Ren, G. 2019: Two new species of comb-clawed beetles (Coleoptera: Tenebrionidae: Alleculinae) from Yunnan, China. *Zootaxa* **4576**: 162–170.

Liang, J. & Ren, G.-D. 2013: Description of a new species of the genus Colobicus Latreille (Coleoptera, Zopheridae) from China. *Acta Zootaxonomica Sinica* **38**: 109–113.

Liberto, A. & Leo, P. 2002: Descrizione di una nuova specie di Melansis Wollaston, 1864 di Gran Canaria, Isole Canarie (Coleoptera, Tenebrionidae). *Vieraea* **30**: 19–29.

Lichtenstein, A.A. 1796: *Catalogus musei Zoologici ditissimi Hamburgi, d III. Februar 1796. Auctionis lege distrahendi. Sectio Tertia Continens Insecta. Verzeichniss von höchstseltenen, aus allen Welttheilen mit vieler Mühe und Kosten zusammen gebrachten, auch aus unterschiedlichen Cabinettern, Sammlungen und Auctionen ausgehobenen Naturalien welche von einem Liebhaber, als Mitglied der Batavischen und verschiedener anderer Naturforschenden Gesellschaften gesammlet worden. Dritter Abschnitt, bestehend in wohlerhaltenen, mehrentheils ausländischen und höchstseltenen Insecten, die theils einzeln, theils mehrere zusammen in Schachteln festgesteckt sind, und welche am Mittwochen, den 3ten Februar 1796 und den folgenden Tagen auf dem Eimbeckschen Hause öffentlich verkauft werden sollen durch den Mackler Peter Hinrich Pachischefsky*. Hamburg: Schniebes, 13 + 222+ 2 pp.

Liljeblad, E. 1945: *Monograph of the family Mordellidae (Coleoptera) of North America, north of Mexico.* Miscellaneous publications Museum of Zoology, University of Michigan, no. 62. 226 pp., 7 pls.

Lillig, M. 1994: The subgenus Eodirosis Kwieton of the genus Erodius Fabricius, with two new species and a key to the species (Coleoptera: Tenebrionidae: Pimeliinae: Erodiini). *Israel Journal of Entomology* 28: 151–158.

Lillig, M. 1997: A new species of the genus Erodius from Israel and Egypt (Coleoptera: Tenebrionidae: Pimeliinae). *Israel Journal of Entomology* 31: 55–58.

Lillig, M. 2001: Revision der Untergattung *Symphoxycara* Koch, 1943 der Gattung *Oxycara* Solier, 1835 (Coleoptera: Tenebrionidae, Tentyriini) *Coleoptera* 5: 363–387.

Lillig, M. 2006: Cryphaeus laticeps sp. n. from the Middle East (Coleoptera, Tenebrionidae, Toxidini). *Mittleilungen der Münchener Entomologischen Gesellschaft* **95**: 49–53.

Lillig, M. 2009a: Two new species of the genus Erodius Fabricius, 1775 from Oman and Syria (Coleoptera Tenebrionidae: Pimeliinae). *Zoology in the Middle East* **48**: 75–80.

Lillig, M. 2009b: New species of the genus Oxycara Solier, 1835 from Oman (Coleoptera: Tenebrionidae). *Caucasian entomological Bulletin* **5**: 227–230.

Lillig, M. 2010: A new species of the genus Phaleria Latreille, 1802 from Dhofar in Oman (Coleoptera: Tenebrionidae: Diaperinae). *Zoology in the Middle East* **51**: 89–93.

Lillig, M. 2014: The genera Stegastopsis Kraatz, 1865 and Orostegastopsis Koch, 1962 with description of O. planioculata sp. n. from Oman (Coleoptera: Tenebrionidae: Tentyriini). *Zoology in the Middle East* **60**: 345–352.

Lillig, M. 2015: A new species of the genus Arthrodosis Reitter, 1900 (Coleoptera: Tenebrionidae: Erodiini) from Iran. *Annales Zoologici* **65**: 167–169.

Lillig, M. & Pavlíček, T. 2002: Astorthocnemis becvarorum, a new genus and new species from Middle East (Coleoptera: Tenebrionidae: Pimeliinae: Platyopini). *Mitteilungen des International Entomologischen Vereins* **27**: 97–104.

Lindberg Håk. 1925: Bidrag till kännedomen om åländska ögruppens skalbaggsfauna. *Meddelanden af Societas pro Fauna et Flora Fennica* **48**: 29–60.

Lindberg Håk. 1932: [new taxa]. In: Pic M. & Lindberg Håk.: Inventa entomologica, itineris Hispanici et Maroccani, quod a. 1926 fecerunt Harald et Håkan Lindberg. XII. Anobiidae, Cleridae, Malacodermata, Heteromera (ex parte). *Commentationes Biologicae* **3** (18): 1–37.

Lindberg Håk. 1962: Coleoptera Insularum Canariensium. III. Tenebrionidae. *Commentationes Biologicae* **25** (1): 1–85, 11 plates.

Lindberg Har. 1950: Beitrag zur Kenntnis der Käferfauna der Kanarischen Inseln. *Commentationes Biologicae* **10** (18) 1–20.

Lindberg Har. 1953: Zweiter Beitrag zur Kenntnis der Käferfauna der Kanarischen Inseln. *Commentationes Biologicae* **13** (12): 1–18.

Linder, M.J. 1864: Description de trois coléoptères européens nouveaux. *Annales de la Société Entomologique de France* (4) **4**: 250–252.

Linell, M.L. 1898: A new species of Aegialites. *The Canadian Entomologist* **30**: 74–75.

Linnaeus, C. 1758: *Systema Naturae per regna tria naturae, secundum classes, ordines, genera, species, cum characteribus, differentiis, synonymis, locis. Tomus I. Ed. Decima, Reformata.* Holmiae: Laurentii Salvii, [5] + 6–823 + [1] pp.

Linnaeus, C. 1760: *Fauna Suecica sistens animalia Sveciae regni: Mammalia, Aves, Amphibia, Pisces, Insecta, Vermes, distributa per classes & ordines, genera & species, cum differentiis specierum, synonymis auctorum, locis natalium, descriptionibus insectorum. Editio altera, auctiora.* '1761'. Stockholmiae: Laurentii Salvii, [48] + 578 pp., 2 pls.

Linnaeus, C. 1764: *Museum S:ae R:ae Ludovicae Ulricae Reginae. Svecorum, Gothorum, Vandalorumque &c. &c. &c. In quo Animalis rariora, exotica, imprimis Insecta & conchilia describuntur & determinantur. Prodromus instar editum.* Holmiae: L. Salvii, vi + 720 + (2).

Linnaeus, C. 1767: *Systema naturae per regna tria naturae, secundum classes, ordines, genera, species, cum characteribus, differentiis, synonymis, locis. Tom I. Pars II. Editio duodecima reformata.* Holmiae: Laurentii Salvii, 533–1327 + [37] pp.

Linnaeus, C. 1771: *Mantissa plantarum altera generum editionis VI & specierum editionis II.* Holmiae: Impensis Direct. Laurentii Salvii, [4] + 588 pp.

Liu, H.-Y. & Ren, G.-D. 2009a: The Chinese species of the genus Itagonia Reitter (Coleoptera Tenebrionidae: Blaptini) with description of one new species. *Acta Zootaxonomica Sinica* **34**: 446–450.

Liu, H.-Y. & Ren, G.-D. 2009b: Review of the genus Pseudognaptorina Kaszab from China (Coleoptera, Tenebrionidae). *Zootaxa* **2174**: 45–50.

Liu, H.-Y. & Ren, G.-D. 2010: A new species of the genus Catomus Allard from China (Coleoptera, Tenebrionidae). *Acta Zootaxonomica Sinica* **35**: 71–73.

Liu, S.-S. & Ren, G.-D. 2013: Taxonomy of the genus Blindus Mulsant et Rey from China, with description of two new species (Coleoptera, Tenebrionidae, Pedinini). *Acta Zoologica Academiae Scientiarum Hungaricae* **59**: 197–211.

Ljungh, S.I. 1823: Nya insecter, ur egan samling. *Svenska Vetenskaps Akademien Nya Handlingar* **1823** (2): 265–273.

Löbl, I. 2005: Laenagenius apterus gen. et sp. n. from China (Coleoptera: Pterogeniidae). *Entomological Problems* **35**: 153–155.

Löbl, I. 2008: Comments, p. 45. In: Löbl I. & Smetana A. (eds): *Catalogue of Palaearctic Coleoptera. Volume 5 Tenebrionoidea.* Apollo Books, Stenstrup. 670 pp.

Löbl, I., Bouchard, P. & Merkl, O. [new taxon] 2008. In: Löbl I., Bouchard P., Merkl O. & Iwan D.: Tenebrionidae. New nomenclatural and taxonomic acts, and comments. Pp. 40–45. In: Löbl I. & A. Smetana (eds.): *Catalogue of Palaearctic Coleoptera. Vol. 5. Tenebrionoidea.* Stenstrup: Apollo Books, 670 pp.

Löbl, I., Bouchard, P., Merkl, O. & Iwan, D. 2008: Tenebrionidae. New nomenclatural and taxonomic acts, and comments. Pp. 40–45. In: Löbl I. & A. Smetana (eds.): *Catalogue of Palaearctic Coleoptera. Vol. 5. Tenebrionoidea.* Stenstrup: Apollo Books, 670 pp.

Löbl, I. & Merkl, O. 2003: On the type species of several tenebrionid genera and subgenera (Coleoptera, Tenebrionidae). *Acta Zoologica Academiae Scientorum Hungaricae* **49**: 243–253.

Löbl, I. & Schawaller, W. 2008: Tenebrionidae: Laenini. New nomenclatural and taxonomic acts, and comments. P. 30. In: Löbl I. & A. Smetana (eds.): *Catalogue of Palaearctic Coleoptera. Vol. 5. Tenebrionoidea.* Stenstrup: Apollo Books, 670 pp.

Löbl, I. & Smetana, A. 2010: Errata for Volume 5, pp. 20–36. In: Löbl I. & Smetana A. (eds): *Catalogue of Palaearctic Coleoptera. Volume 6 Chrysomeloidea.* Apollo Books, Stenstrup. 924 pp.

Löbl, I. & Smetana, A. 2011: Errata for Volume 5, pp. 32–35. In: Löbl I. & Smetana A. (eds): *Catalogue of Palaearctic Coleoptera. Volume 7 Curculionoidea I.* Apollo Books, Stenstrup. 373 pp.

Löbl, I. & Smetana, A. 2013: Errata for Volume 5, pp. 36–40. In: Löbl I. & Smetana A. (eds): *Catalogue of Palaearctic Coleoptera. Volume 8 Curculionoidea II.* Apollo Books, Stenstrup. 700 pp.

Lo, Cascio P. 2000: [new taxon]. In: Lo Cascio P., Bartolozzi L., Cecchi B., Dapporto L. & Sforzi A.: Contributi alla conoszenza dell'arthropodofauna dell'Isola di Pianoso (Arcipelago Toscano). 3. Coleoptera Tenebrionidae. *Bollettino della Società Entomologica Italiano* **132**: 157–174.

Lo Cascio & F. Garita: A new Trachyscelis from the Socotra Archipelago (Yemen) (Coleoptera Tenebrionidae). Tenebrionidae. *Bollettino della Società Entomologica Italiano* **145**: 85–92.

Lohse, G.A. 1964: Die in Mitteleuropa vertretene Gattungen der Cisidae. (1. Beitrag zur Kenntnis der mitteleuropäischen Cisidae). *Entomologische Blätter* **60**: 116–122.

Lohse, G.A. 1965: Über einige umstrittene Cis-Arten (2. Beitrag zur Kenntnis europäischer Cisidae). *Entomologische Blätter* **60** [1964]: 174–181.

Lohse, G.A. 1966: Cisiden-Studien (3. Beitrag zur Kenntnis der Mitteleuropäischen Cisiden). *Entomologische Blätter* **62**: 25–29.

Lohse, G.A. 1969: Cisiden Studien IV. *Entomologische Blätter* **65**: 48–52.

Lohse, G.A. 1989: Typhaea stercorea (L.), T. crenata (Melsheimer) und T. decipiens sp. n. *Entomologische Blätter* **85**: 144–146.

Lokay, E. 1907: Nový druh myrmekofilního rodu Oochotus Lucas. *Časopis České Společnosti Entomologické* **4**: 91–92.

Lopes-Andrade, C. 2002: Two cases of homonymy in the genus Cis Latreille (Coleoptera: Tenebrionoidea: Ciidae). *Dugesiana* **9**: 54.

Lopes-Andrade, C. 2008a: The first record of Cis chinensis Lawrence from Brazil, with the delimitation of the Cis multidentatus species-group (Coleoptera: Ciidae). *Zootaxa* **1755**: 35–46.

Lopes-Andrade, C. 2008b: An essay on the tribe Xylographellini (Coleoptera: Tenebrionoidea: Ciidae). *Zootaxa* **1832**: 1–110.

Lopes-Andrade, C. & Grebennikov, V.V. 2015: First record and five new species of Xylographellini (Coleoptera: Ciidae) from China, with online DNA barcode library of the family. *Zootaxa* **4006**: 463–480.

López-Pérez, J.J. 2010: Descripción de una nueva especie de Akis Herbst, 1799 de la Isla de Saltés, provincia de Huelva (Coleoptera, Tenebrionidae, Akidini). *Boletín de la Asociación española de Entomología* **33** [2009]: 503–515.

Lucas, P.H. 1843: Note sur quelques nouvelles espèces d'insectes de la famille des Trachélides qui habitent les possessions françaises du nord de l'Afrique. *Revue Zoologique, par la Société Cuvierienne* (2) **6**: 145–147.

Lucas, P.H. 1844: Note sur quelques nouvelles espèces d'insectes qui habitent les possessions françaises du nord de l'Afrique. *Revue Zoologique, par la Société Cuvierienne* (2) **7**: 262–267.

Lucas, P.H. 1846: Pp. 1–360. In: *Exploration scientifique de l'Algérie pendant les années 1840, 1841, 1842 publiée par ordre du gouvernement et avec le concours d'une Commission Académique. Sciences physiques Zoologie. Vol. II. Histoire naturelle des animaux articulés. Cinquième classe. Insectes. Premier ordre. Les coléoptères.* Paris: Imprimerie Nationale [1849], 590 pp., 47 pls.

Lucas, P.H. 1847: Pp. 361–448. In: *Exploration scientifique de l'Algérie pendant les années 1840, 1841, 1842 publiée par ordre du gouvernement et avec le concours d'une Commission Académique. Sciences physiques Zoologie. Vol. II. Histoire naturelle des animaux articulés. Cinquième classe. Insectes. Premier ordre. Les coléoptères.* Paris: Imprimerie Nationale [1849], 590 pp., 47 pls.

Lucas, P.H. 1849: Pp. 449–590. In: *Exploration scientifique de l'Algérie pendant les années 1840, 1841, 1842 publiée par ordre du gouvernement et avec le concours d'une Commission Académique. Sciences physiques Zoologie. Vol. II. Histoire naturelle des animaux articulés. Cinquième classe. Insectes. Premier ordre. Les coléoptères.* Paris: Imprimerie Nationale, 590 pp., 47 pls.

Lucas, P.H. 1850a: [new taxa]. *Bulletin de la Société Entomologique de France* **1850**: iii-iv.

Lucas, P.H. 1850b: [new taxa]. *Bulletin de la Société Entomologique de France* **1850**: vi-vii.

Lucas, P.H. 1850c: [new taxa]. *Bulletin de la Société Entomologique de France* **1850**: xi-xiii.

Lucas, P.H. 1852: [new taxa]. *Bulletin de la Société Entomologique de France* **1852**: xxviii-xxix.

Lucas, P.H. 1853: Essai sur les animaux articulés qui habitent l'île de Crête. *Revue et Magasin de Zoologie Pure et Appliquée* (2) **5**: 565–576.

Lucas, P.H. 1854: Essai sur les animaux articulés qui habitent l'île de Crête. *Revue et Magasin de Zoologie Pure et Appliquée* (2) 6: 28–44.

Lucas, P.H. 1855a: Note sur quelques nouvelles espèces de coléoptères qui habitent les possessions français de nord de l'Afrique. *Revue et Magasin de Zoologie Pure et Appliquée* (2) 7: 290–292.

Lucas, P.H. 1855b: [new taxa]. *Bulletin de la Société Entomologique de France* 1855: xxxiv.

Lucas, P.H. 1856: [new taxa]. *Bulletin de la Société Entomologique de France* 1856: lv-lvii.

Lucas, P.H. 1857: [new taxa]. *Bulletin de la Société Entomologique de France* 1857: lv-lvii.

Lucas, P.H. 1858a: [new taxa]. *Bulletin de la Société Entomologique de France* 1858: clxxvii-clxxxi.

Lucas, P.H. 1858b: [new taxon]. *Bulletin de la Société Entomologique de France* 1858: clxxvii-clxxxi.

Lucas, P.H. 1858c: [new taxa]. *Bulletin de la Société Entomologique de France* 1858: ccxx-ccxxiii.

Lucas, P.H. 1859a: [new taxa]. *Bulletin de la Société Entomologique de France* 1859: xxii-xxiv.

Lucas, P.H. 1859b: [new taxa]. *Bulletin de la Société Entomologique de France* 1859: cxiii-cxv.

Luigioni, P. 1921: Una varietà nuova dell'Akis italica Sol. (Coleoptera – Fam. Tenebrionidae). *Atti della Pontificia Academia Romana dei Nuovi Lincei* 74: 51.

Machado, A. 1979: Consideraciones sobre el género Pimelia (Col. Tenebrionidae) en las Islas Canarias y descriptión de una nueva especie. *Boletín de la Asociación Española de Entomología* 3: 119–127.

Machado, A. 2012: Two new Tarphius species from Macaronesia (Coleoptera, Zopheridae). *Journal of Natural History* 46: 637–643.

MacLeay, W.J. 1873a: Notes on a collection of insects from Gayndah. *Transactions of the Entomological Society of New South Wales* 2 [1871]: 79–205.

MacLeay, W.J. 1873b: Notes on a collection of insects from Gayndah. *Transactions of the Entomological Society of New South Wales* 2 [1872]: 239–318.

MacLeay, W.J. 1887: The insects of the Cairns District, Northern Queensland. Part II. *Proceedings of the Linnean Society of New South Wales* 2: 307–328.

MacLeay, W.S. 1825: *Annulosa Javanica, or an attempt to illustrate the natural affinities and analogies of the insects collected in Java by Thomas Horsfield, M.D., F.L. & G.S. and deposited by him in the honourable East-India Company.* London: Kingsbury, Parbury, and Allen, xii + 9–50 pp., 1 pl.

Macnamara, J.P. 1965: *The origin of Pacific fauna of Oedemeridae: the Oedemeridae of Micronesia, 65–13,006.* University Microfilms, Inc., Ann Arbor, Michigan. 114 pp.

Macnamara, J.P. 1971: Emendation of taxa of Micronesian Oedemeridae (Coleoptera). *Entomological News* 82: 164.

Mader, L. 1927: Coleopterologische Notizen. *Entomologischer Anzeiger* 7: 193–197.

Mader, L. 1929: Zur Variabilität der Meloiden und Coccinelliden. *Entomologischer Anzeiger* 9: 108–114.

Mader, L. 1931: Über paläarktische Coleopteren II. *Entomologischer Anzeiger* 11: 134–136.

Mader, L. 1943: Beitrag zur Kenntnis der Koleopteren. *Entomologische Blätter* 39: 41–46.

Maeda, M. & Nakane, T. 1988: New or little-known Coleoptera from Japan and its adjacent regions, XL. Family Alleculidae. *Review of Miyazaki Sangyo-Keiei University* 1: 1–10.

Maeda, M. & Nakane, T. 1991: [new taxa]. In: Nakane, T.: Notes on some little known beetles (Coleoptera) in Japan. 7. *Kita-Kyūshū no Konchū* 38: 1–9, pl. 1 (in Japanese, with title and descriptions in English).

Magistretti, M. 1939: Contributo alla conoscenza degli Oedemeridi paleartici II. *Bollettino della Società Entomologica Italiana* 71: 159–160.

Magistretti, M. 1941a: Contributo alla conoscenza degli Oedemeridi paleartici IV. *Bollettino della Società Entomologica Italiana* **73**: 76–77.

Magistretti, M. 1941b: Contributo alla conoscenza degli Oedemeridi paleartici III. *Mitteilungen der Münchener Entomologischen Gesselschaft* **31**: 227–234.

Magistretti, M. 1943: Le specie italiane del genere Mylabris (Col. Meloidae). *Annali del Museo Civico di Storia Naturale di Genova* **62**: 30–54.

Magrini, P. & Uliana, M. 2015: A new species of the enigmatic genus Osphyoplesius from Sicily (Coleoptera: Tenebrionoidea: Pythidae). *Zootaxa* **3925**: 573–580.

Makhan, D. 2012a: Prosodes rishwani sp. nov., a new darkling beetle from Jangal-e Abr, Semnan province, Iran (Coleoptera, Tenebrionidae: Blaptini). *Calodema* **205**: 1–5.

Makhan, D. 2012b: Hycleus amrishi sp. nov. (Coleoptera: Meloidae), an additional new blister beetle from Ahvanu, Damghan, Semnan Province, Iran. *Calodema* **229**: 1–3.

Makhan, D. 2012c: Mylabris rishwani sp. nov. (Coleoptera: Meloidae), an additional new blister beetle from Ahvanu, Damghan, Semnan Province, Iran. *Calodema* **232**: 1–4.

Makhan, D. 2013: Epicauta rishwani sp. nov. (Coleoptera: Meloidae), a new blister beetle from Mt. Jinyun, Beibei, Chongqing, China. *Calodema* **292**: 1–3.

Makhan, D. 2018: Laena amrishi sp. nov. (Coleoptera: Tenebrionidae, Lagriinae) from the Jinyun mountain in Beibei, Chongqing, China. *Calodema* **649**: 1–3.

Makhan, D. & Ezzatpanah, S. 2011a: Mylabris aschnae sp. nov., a new blister beetle from Mahallat, Markazi Province, Iran (Coleoptera: Meloidae). *Calodema* **192**: 1–6

Makhan, D. & Ezzatpanah, S. 2011b: Mylabris minae sp. nov., an additional new blister beetle from Mahallat, Markazi Province, Iran (Coleoptera: Meloidae). *Calodema* **193**: 1–6.

Makhan, D. & Ezzatpanah, S. 2011c: Hycleus golnazae sp. nov., a new blister beetle from Mahallat, Markazi Province, Iran (Coleoptera: Meloidae). *Calodema* **194**: 1–5.

Makita, S. 2017a: Description of a new species of Trachypholis Erichson, 1845 (Coleoptera, Tenebrionoidea, Zopheridae) from Lanyu Island, Taiwan. *Journal of Entomological Research* **41**: 369–371.

Makita, S. 2017b: A new species of the genus Hyberis Pascoe, 1860 (Coleoptera: Zopheridae) from Taiwan. *Far Eastern Entomologist* **333**: 10–12.

Mäklin, F.G. 1846. Coleoptera myrmecophila Fennica. *Bulletin de la Société Impériale des Naturalistes de Moscou* **19**: 157–187.

Mäklin, F.W. 1863: Die Gattung Praogena und deren Repräsentanten. *Acta Societatis Scientiarum Fennicae* **7**: 555–583.

Mäklin, F.W. 1864: *Monographie der Gattung Strongylium Kirby, Lacordaire und der damit zunächst verwandten Formen*. 410 pp., 3 pls.

Mäklin, F.W. 1872: Synonymische und systematische Bemerkungen. *Entomologische Zeitschrift* (Stettin) **33**: 242–249.

Mäklin, F.W. 1875a: Neue Canthariden. *Acta Societatis Scientificae Fennicae* **10**: 597–632.

Mäklin, F.W. 1875b: Neue Cisteliden. *Acta Societatis Scientiarum Fennica* **10**: 661–682.

Mäklin, F.W. 1881: Coleoptera insamlade under den Nordenskiöld'ska Expeditionen 1875 på några öar vid Norges Nordvestkust. på Novaja Semlja och ön Waigatsch samt vid Jenissej i Sibirien. *Kongliga Svenska Vetenskaps-Akademiens Handlingar* **18** (4): 1–48.

Mal, N. 1984: Les espèces de Sepidium affines à S. bidentatum Solier, et description d'une espèce nouvelle du Portugal (Col. Tenebrionidae). *L'Entomologiste* **40**: 193–204.

Mal, N. 2019: Remarques sur les Sepidiina Eschsholtz, 1829 de la péninsule arabique et description de deux taxa inédits (Coleoptera, Tenebrionidae, Pimeliinae, Sepidiini). *Entomologia Africana* **24**: 2–6.

Mamaev, B.M. 1977: Redkie vidy zhestkokrylykh iz semeystva Colydiidae (Coleoptera) v faune sredney polosy evropeyskoy chasti SSSR. Pp. 27–30. In: Skarlato, O. (ed.): *Novye i maloizvestnye vidy nasekomykh evropeyskoy chasti SSSR*. Leningrad: Nauka, 116 pp.

Mamaev, B.M. 1978: [new taxon]. In: Mamaev B.M. & Kompantsev A.V.: Novye dannye o mycetophilnykh tenelubakh, vkluchaya rod Euryzilora Lew. (Coleoptera, Melandryidae). *Nauchnye Doklady Vyshey Shkoly. Biologicheskye Nauki* **3**: 46–49.

Mannerheim, C.G. von. 1825: Novae coleopterorum species Imperii Rossici incolae descriptae. Pp. 19–4[illegible]. In: Hummel, A.D.: *Essais entomologiques. Bd. 1. Nr. 4*. St. Pétersbourg: Chancellerie privée du Ministère de l'Intérieur, 72 pp.

Mannerheim, C.G. von. 1843a: Beitrag zur Kaefer-fauna der Aleutischen Inseln, Insel Sitkha und Neu-Californiens. *Bulletin de la Société des Naturalistes de Moscou* **16** (2): 175–314.

Mannerheim, C.G. von. 1843b: Description de quelques autres nouvelles espèces de coléoptères de Finlande. *Bulletin de la Société Impériale des Naturalistes de Moscou* **16** (3): 88–99.

Mannerheim, C.G. von. 1844a: Description de quelques nouvelles espèces de coléoptères de Finlande. *Bulletin de la Société Impériale des Naturalistes de Moscou* **17** (1): 189–202.

Mannerheim, C.G. von. 1844b: Lettre à S.E. Mr. Fischer de Waldheim ou relation d'un voyage fait en 184[illegible], en Suède, en Danemarck et dans le Nord de l'Allemagne. *Bulletin de la Société Impériale des Naturalistes de Moscou* **17**: 844–872.

Mannerheim, C.G. von. 1846: Revue critique de quelques ouvrages récents de M. Victor de Motschulsky. *Bulletin de la Société Impériale des Naturalistes de Moscou* **19** (1): 194–254.

Mannerheim, C.G. von. 1849: Insectes coléoptères de la Sibérie orientale nouveaux ou peu connus. No. 1. *Bulletin de la Société Impériale des Naturalistes de Moscou* **22** (1): 220–249.

Mannerheim, C.G. von. 1852: Zweiter Nachtrag zur Kaefer-Fauna der nord-amerikanischen Laender des russischen Reiches. *Bulletin de la Société Impériale des Naturalistes de Mouscou* **25** (2) [1852–1853]: 283–372.

Mannerheim, C.G. von. 1853a: Insectes coléoptères de la Sibérie orientale, nouveaux ou peu connus. *Bulletin de la Société Impériale des Naturalistes de Moscou* **25** (4) [1852–1853]: 273–309.

Mannerheim, C.G. von. 1853b: Dritter Nachtrag zur Käferfauna der Nord-Amerikanischen Länder des russischen Reiches. *Bulletin de la Société Impériale des Naturalistes de Moscou* **26** (3) [1853–1854]: 95–273.

Mařan, J. 1933: Beitrag zur Kenntnis der Gattung Marolia Muls. *Entomologisches Nachrichtenblatt* **7**: 110–113.

Mařan, J. 1935a: De novis Alleculidarum speciebus formisque regionis palaearcticae. *Sborník Entomologického Oddělení Národního Musea v Praze* **13** (118): 141–146.

Mařan, J. 1935b: Specierum subgeneris Dichillocerus Rttr. revisio (Col. Tenebrionidae Gen. Dichillus Duv.). *Sborník Entomologického Oddělení Národního Musea v Praze* **13**: 147–148.

Mařan, J. 1936: De novis Alleculidarum speciebus formisque regionis palaearcticae Part II. *Sborník Entomologického Oddělení Národního Musea v Praze* **14** (137): 205–207.

Mařan, J. 1940a: Conspectus specierum generis Lydus Latr. (subg. Alosimus Muls.) ex affinitate speciei Lydus collaris Fabr. *Věstník Československé Zoologické Společnosti v Praze* **8**: 73–90.

Mařan, J. 1940b: Nový druh rodu Penthe Newm. z Číny. Novae speciei generis Penthe Newm. descriptio (Coleoptera: Melandryidae). *Časopis Československé Společnosti Entomologické* **37**: 87–88.

Mařan, J. 1940c: Generis Allecula Fabr. regionis palaearcticae orientalis species novae. *Sborník Entomologického Oddělení Národního Musea v Praze* **18** (191): 168–171.

Mařan, J. 1941a: Duae novae species Lydus Latr. (subg. Alosimus) ex affinitate speciei Lydus collaris Fabr. *Sborník Entomologického Oddělení Národního Musea v Praze* **19** (202): 60–63.

Mařan, J. 1941b: De novis parumque cognitis chinensibus generis Lytta F. speciebus ex affinitate L. aeneiventris Haag. Coleoptera Meloidae. *Sborník Entomologického Oddělení Národního Musea v Praze* **19** (223): 180–185.

Mařan, J. 1942a: Revise rodu Stenodera Eschschz. [Generis Stenodera Eschschz. revisio]. Coleoptera. Meloidae. *Sborník Entomologického Oddělení Národního Musea v Praze* **20**: 14–29.

Mařan, J. 1942b: Revise podrodu Taphromeloë Rttr. [Subgeneris Taphromeloë Rttr. revisio]. Coleoptera Meloidae. Gen. Meloë L. *Sborník Entomologického Oddělení Národního Musea v Praze* **20**: 50–58.

Mařan, J. 1942c: Přehled druhů rodu Lydus (subg. Alosimus Muls.) ze skupiny Lydus syriacus L. [Specierum generis Lydus (subg. Alosimus Muls.) ex affinitate speciei Lydus syriacus L. revisio]. Coleoptera Meloidae. *Sborník Entomologického Oddělení Národního Musea v Praze* **20** (234): 78–98.

Mařan, J. 1944a: Novae speciei generis Mylabris F. ex insula Creta descriptio. Coleoptera. Meloidae. *Časopis České Společnosti Entomologické* **41**: 35–39.

Mařan, J. 1944b: De novis Alleculidarum speciebus chinensibus. *Sborník Entomologického Oddělení Národního Musea v Praze* **21–22**: 76–77.

Mařan, J. 1944c: Vorarbeiten für eine monographische Bearbeitung der Gattung Cerocoma Geoffr. *Věstník Československé Zoologické Společnosti v Praze* **9** [1941–1944]: 78–101.

Mařan, J. 1944d: Generis Gonodera Muls. revisio. Coleoptera Alleculidae. *Sborník Entomologického Oddělení Národního Musea v Praze* **21–22**: 184–196.

Mařan, J. 1954a: Nový palearktický druh rodu Bitoma Herbst z entomologických sbírek Národního musea v Praze. Bitoma iranica m. nov. spec. (Col., Colydiidae). *Časopis Československé Společnosti Entomologické* **50** [1953]: 132–133.

Mařan, J. 1954b: Descriptions de deux espèces nouvelles du genre Mycetochara Berth. (Coleoptera, Alleculidae). *Sborník Entomologického Oddělení Národního Musea v Praze* **29**: 169–170.

Marquet, C. 1897: Catalogue des coléoptères du Languedoc. Espèces observées dans quelques régions de cette province, notamment à Toulouse, Bériers, Cette, etc. *Bulletin de la Société d'Histoire Naturelle de Touluse* **31**: 5–240.

Marcuzzi, G. 2001: Description of a new species of tenebrionid [sic] from the Isle of Djerba (Gulf of Gabès), (Tunisia) (Coleoptera, Heteromera). *Doriana* 7 (328): 1–3.

Marcuzzi, G. & Turchetto, M. 1982: Eutagenia elvirae, nuova species di coleotteri tenebrionide dell'Italia meridionale (Heteromera). *Memorie dela Società Entomologica Italiana* **60** [1981]: 235–238.

Marseul, S.A. de. 1857: *Catalogue des coléoptères d'Europe*. Paris: S.-A. de Marseul, xvi + 200 pp.

Marseul, S.A. de. 1867: Descriptions d'espèces nouvelles. *L'Abeille, Mémoires d'Entomologie* 4: xxxiii-lx.

Marseul, S.A. de. 1868a: Descriptions des espèces nouvelles. *L'Abeille, Journal d'Entomologie* 5: 171–218.

Marseul, S.A. de. 1868b: Archéologie entomologique. *L'Abeille, Mémoires d'Entomologie* 6 [1867–1869]: 1–52.

Marseul, S.A. de. 1869: Genre Iphthimus par E. Truqui. *L'Abeille, Journal d'Entomologie* 5: 271–275.

Marseul, S.A. de. 1870a: Descriptions de coléoptères nouveaux. *L'Abeille, Mémoires d'Entomologie* 6 [1867–1869]: 368–386.

Marseul, S.A. de. 1870b: Monographie des mylabrides d'Europe et des contrées limitrophes en Afrique et en Asie. *L'Abeille, Mémoires d'Entomologie* 7 (2): 1–204, 2 pls.

Marseul, S.A. de. 1872a: Description d'espèces nouvelles. *L'Abeille, Mémoires d'Entomologie* 8 [1871–1872]: 413–420.

Marseul, S.A. de. 1872b: Pp. 1–160. In: Repertoire des coléoptères d'Europe décrits isolément depuis 1863 (2e partie). *L'Abeille, Mémoire d'Entomologie* 9: 1–448. [pp. 161–448 issued in 1873].

Marseul, S.A. de. 1873a: Coléoptères du Japon recueillis par M. Georges Lewis. Énumération des histerides et des hétéromères avec la description des espèces nouvelles. *Annales de la Société Entomologique de France* (5) 3: 219–230.

Marseul, S.A. de. 1873b: Monographie des mylabrides. *Mémoires de la Société Royale des Sciences de Liège* (2) 3: 363–662, 6 pls.

Marseul, S.A. de. 1875a: Nouvelles et faits divers de l'Abeille No. 7: *L'Abeille, Journal d'Entomologie* 14 [1874–1876]: xxv-xxviii.

Marseul, S.A. de. 1875b: Nouvelles et faits divers de L'Abeille No. 10. Mélanges (suite). *L'Abeille, Journal d'Entomologie* 14 [1874–1876]: xxxviii-xxxix.

Marseul, S.A. de. 1876a: Descriptions d'espèces nouvelles. *L'Abeille, Journal d'Entomologie* 14 [1874–1876]: 25–30.

Marseul, S.A. de. 1876b: Nouvelles et faits divers de l'Abeille. Mélanges (suite). *L'Abeille, Journal d'Entomologie* 16 [1876–1878]: 35–36, 38–40.

Marseul, S.A. de. 1876c: Coléoptères du Japon recueillis par M. Georges Lewis. Énumération des Hétéromères avec la description des espèces nouvelles. *Annales de la Société Entomologique de France* (5) 6: 93–142.

Marseul, S.A. de. 1876d: Coléoptères du Japon recueillis par M. Georges Lewis. 2e Mémoire. Énumération des Hétéromères avec la description des espèces nouvelles. 2e Partie. *Annales de la Société Entomologique de France* (5) 6: 315–349, 447–464.

Marseul, S.A. de. 1877: [Coléoptères du Japon recueillis par M. Georges Lewis. 2e Mémoire. Énumération des Hétéromères avec la description des espèces nouvelles. 3e et dernière partie.] *Annales de la Société Entomologique de France* (5) 6 [1876]: 465–486.

Marseul, S.A. de. 1878a: Anthicides recueillis par Camille van Volxem dans son voyage en Portugal, Andalousie et partie boréale du Maroc, en 1871. *Bulletin de la Société Entomologique de Belgique* 21: xlii-xliii.

Marseul, S.A. de. 1878b: Nouvelles et faits divers de l'Abeille. Deuxième Série, No. 13. Mélanges. *L'Abeille, Journal d'Entomologie* 16 [1876–1878]: 50–51.

Marseul, S.A. de. 1878c: Nouvelles et faits divers de l'Abeille, Deuxième Série. – Nos. 14 et 15. Mélanges (suite). *L'Abeille, Journal d'Entomologie* 16 [1876–1878]: 54–56.

Marseul, S.A. de. 1879a: Monographie des Anthicides de l'Ancien-Monde. *L'Abeille, Journal d'Entomologie* 17: 1–268, pls. 1–2.

Marseul, S.A. de. 1879b: Études sur les insectes d'Angola qui se trouvent au Muséum National de Lisbonne. *Journal de Sciences Mathématiques, Physiques et Naturelles* **25**: 43–67.

Marseul, S.A. de. 1879c: Nouvelles et faits divers de l'Abeille. Mélanges (suite). *L'Abeille, Journal d'Entomologie* **17**: 74–76, 99–100.

Marseul, S.A. de. 1880: [new taxa]. In: Gestro, R.: Appunti sull'entomologia Tunisina. *Annali del Museo Civico di Storia Naturale di Genova* **15** [1879–1880]: 405–424.

Marseul, S.A. de. 1882a: Espèces nouvelles de coléoptères de la famille des Pédilides et Anthicides du Musée Royal d'hist. nat. à Leyde. *Tijdschrift voor Entomologie* **25** [1881–1882]: 54–64.

Marseul, S.A. de. 1882b: New species of Coleoptera belonging to the families Pedilidae and Anthicidae. *Notes from the Leyden Museum* **4**: 112–124.

Marseul, S.A. de. 1882c: Nouvelles et faits divers de l'Abeille. Deuxième série – No 43. *L'Abeille, Journal d'Entomologie* **20**: 169–172.

Marseul, S.A. de. 1883a: Nouvelles et faits divers de l'Abeille. Deuxième série – No 45. *L'Abeille, Journal d'Entomologie* **21**: 178.

Marseul, S.A. de. 1883b: Nouvelles et faits divers de l'Abeille. Deuxième série – No 46. *L'Abeille, Journal d'Entomologie* **21**: 181–184.

Marseul, S.A. de. 1885: Nouvelles et faits divers de l'Abeille. Première sàrie No. 36. Mélanges. *L'Abeille, Journal d'Entomologie* **22** [1886–1887]: cxlvii.

Marseul, S.A. de. 1887: Catalogue des coléoptères de l'Ancien Monde Europe et contrées limitrophe en Afrique et en Asie. *L'Abeille, Journal d'Entomologie* **24** [1886–1887]: 193–360.

Marsham, T. 1802: *Entomologia britannica, sistens Insecta Britanniae indigena, secundum methodum Linnaeanum disposita. Tomus 1. Coleoptera*. London: Wilks & Tailor, xxxi + 547 + [1] pp.

Martinez y Sáez, A. 1873a: Datos sobre algunos coleópteros de los alredadores de Cuenca. *Anales de la Sociedad Española de Historia Natural* **2**: 53–75, 1 pl.

Martinez y Sáez F. 1873b: Descripciones de coleópteros de España. *Anales de la Sociedad Española de Historia Natural* **2**: 407–417.

Martínez Fernández, J.C. 2010: Un nuevo representante del género Blaps Fabricius, 1775 de la Península Ibérica: Blaps tichyi n. sp. (Coleoptera, Tenebrionidae). *Boletín de la Sociedad Entomológica Aragonesa* **47**: 181–185.

Martínez Fernández, J.C. & Soldati, F. 2014: Contribución a la revisión del sugénero Alphasida Escalera, 1905 (s. str.) (Coleoptera: Tenebrionidae): el grupo de Alphasida lorcana (Pérez Arcas, 1865). *Boletín de la Sociedad Entomológica Aragonesa* **54**: 35–50.

Martínez Fernández, J.C. & Soldati, F. 2017: Contribución a la revisión del sugénero Alphasida Escalera, 1905 (s. str.) (Coleoptera: Tenebrionidae): 2ª parte, el grupo de Alphasida holosericea (Germar, 1824). *Boletín de la Sociedad Entomológica Aragonesa* **60**: 115–131.

Mas-Peinado, P., Buckley, D., Ruiz, J.L. & García París, M. 2018: Recurrent diversification patterns and taxonomic complexity in morphologically conservative ancient lineages of Pimelia (Coleoptera: Tenebrionidae). *Systematic Entomology* **43**: 522–548.

Masumoto, K. 1981a: Tenebrionidae of Formosa (1). *Elytra* **8**: 37–53.

Masumoto, K. 1981b: Notes and descriptions of Japanese Tenebrionidae (I). *Entomological Review of Japan* **35**: 29–32.

Masumoto, K. 1981c: Tenebrionidae of Formosa (2). *Elytra* **9**: 15–52.

Masumoto, K. 1981d: New or little known Tenebrionidae from Formosa (I). *Entomological Review of Japan* **36**: 15–26.

Masumoto, K. 1981e: Tenebrionidae of Formosa (3). *Elytra* **9**: 79–99.

Masumoto, K. 1982a: New or little known Tenebrionidae from Formosa (II). *Entomological Review of Japan* **36**: 143–152.

Masumoto, K. 1982b: New or little known Tenebrionidae from Formosa (III). *Entomological Review of Japan* **37**: 49–53.

Masumoto, K. 1982c: Tenebrionidae of Formosa (4). *Elytra* **10**: 17–32.

Masumoto, K. 1982d: Tenebrionidae of Formosa (5). *Elytra* **10**: 53–72.

Masumoto, K. 1983a: Notes and descriptions of Japanese Tenebrionidae (II). *Entomological Review of Japan* **38**: 81–92.

Masumoto, K. 1983b: Notes and descriptions of Japanese Tenebrionidae (III). *Entomological Review of Japan* **38**: 173–176.

Masumoto, K. 1984: Tenebrionidae of Formosa (6). *Elytra* **11**: 16–24.

Masumoto, K. 1985a: Notes and descriptions of Japanese Tenebrionidae (IV). *Entomological Review of Japan* **40**: 21–27.

Masumoto, K. 1985b: The Japanese species of Gonocephalum (Coleoptera, Tenebrionidae). *Elytra* **12**: 27–37.

Masumoto, K. 1985c: Two new species of tenebrionid beetles from China. *Entomological Review of Japan* **40**: 121–124.

Masumoto, K. 1986a: Two new Augolesthus (Coleoptera, Tenebrionidae) from East Asia, with notes on the known species of the genus, 250–256 pp. Entomological papers presented to Yoshihiko Kurosawa on the Occasion of his retirement. Tokyo: Coleopterist's Association of Japan.

Masumoto, K. 1986b: Tenebrionidae of East Asia (III). A new genus and three new species from Taiwan. *Elytra* **14**: 61–68.

Masumoto, K. 1987a: A study of the Japanese Lagriidae. *Entomological Review of Japan* **42**: 37–60.

Masumoto, K. 1987b: Notes and descriptions of Japanese Tenebrionidae (V). *Entomological Review of Japan* **42**: 87–94.

Masumoto, K. 1988: A study of the Taiwanese Lagriidae. *Entomological Review of Japan* **43**: 33–52 + 6 pls.

Masumoto, K. 1989a: Plesiophthalmus and its allied genera (Coleoptera, Tenebrionidae, Amarygnini) (Part 3). *Japanese Journal of Entomology* **57**: 96–121.

Masumoto, K. 1989b: Plesiophthalmus and its allied genera (Coleoptera, Tenebrionidae, Amarygnini) (Part 4). *Japanese Journal of Entomology* **57**: 295–317.

Masumoto, K. 1989c: Plesiophthalmus and its allied genera (Coleoptera, Tenebrionidae, Amarygnini) (Part 5). *Japanese Journal of Entomology* **57**: 536–564.

Masumoto, K. 1989d: Plesiophthalmus and its allied genera (Coleoptera, Tenebrionidae, Amarygnini) (Part 6). *Japanese Journal of Entomology* **57**: 742–757.

Masumoto, K. 1989e: Four new species of flat large-sized camariine genera from Southeast Asia (Coleoptera, Tenebrionidae, Cnodalonini). *Kanagawa-Chûjô* (Yokohama) **90**: 229–236.

Masumoto, K. 1990a: Plesiophthalmus and its allied genera (Coleoptera, Tenebrionidae, Amarygnini) (Part 7). *Japanese Journal of Entomology* **58**: 35–64.

Masumoto, K. 1990b: A new Phellopsis species from South Korea (Zopheridae, Coleoptera). *Entomological Review of Japan* **45**: 87–91.

Masumoto, K. 1990c: Tenebrionidae of East Asia (VI) On the genus Falsocamaria Pic (Cnodalonini). *Elytra* **18**: 93–100.

Masumoto, K. 1990d: Plesiophthalmus and its allied genera (Coleoptera, Tenebrionidae, Amarygnini) (Part 8). *Japanese Journal of Entomology* **58**: 243–274.

Masumoto, K. 1990e: Plesiophthalmus and its allied genera (Coleoptera, Tenebrionidae, Amarygnini) (Part 9). *Japanese Journal of Entomology* **58**: 475–505.

Masumoto, K. 1990f: New Himalayan species of Laena (Coleoptera, Tenebrionidae) preserved in the collection of the National Science Museum, Tokyo. *Bulletin of the National Science Museum, Tokyo* (Ser. A) **16**: 175–196.

Masumoto, K. 1990g: Plesiophthalmus and its allied genera (Coleoptera, Tenebrionidae, Amarygnini) (Part 10). *Japanese Journal of Entomology* **58**: 693–724.

Masumoto, K. 1991a: Plesiophthalmus and its allied genera (Coleoptera, Tenebrionidae, Amarygnini) (Part 11). *Japanese Journal of Entomology* **59**: 1–36.

Masumoto, K. 1991b: Two new tenebrionid species (Coleoptera, Tenebrionidae) from Japan. *Elytra* **19**: 25–29.

Masumoto, K. 1992: Tenebrionidae of East Asia (VIII). Three new species of Campsiomorpha and Strongylium from Taiwan. *Elytra* **20**: 89–94.

Masumoto, K. 1993a: Larger flattened species of Camariine genera from Asia (Part 1). *Japanese Journal of Entomology* **61**: 137–148.

Masumoto, K. 1993b: Larger flattened species of Camariine genera from Asia (Part 2). *Japanese Journal of Entomology* **61**: 217–234.

Masumoto, K. 1993c: A study of the genus Augolesthus Motschulsky, 1872 (Coleoptera, Tenebrionidae, Cnodalini). *Entomological Review of Japan* **48**: 37–43.

Masumoto, K. 1993d: Notes and descriptions of Japanese Tenebrionidae, VI. *Entomological Review of Japan* **48**: 127–136.

Masumoto, K. 1994a: Notes and descriptions of Japanese Tenebrionidae, VII. *Entomological Review of Japan* **49**: 41–46.

Masumoto, K. 1994b: A study of the Asian species of the genus Ceropria (Coleoptera, Tenebrionidae, Diaperini) (part 1). *Japanese Journal of Entomology* **62**: 763–774.

Masumoto, K. 1995a: A study of the Asian species of the genus Ceropria (Coleoptera, Tenebrionidae, Diaperini) (Part 2). *Japanese Journal of Entomology* **63**: 1–13.

Masumoto, K. 1995b: A study of the Asian species of the genus Ceropria (Coleoptera, Tenebrionidae, Diaperini) (Part 3). *Japanese Journal of Entomology* **63**: 723–734.

Masumoto, K. 1996a: Two new tenebrionid species (Coleoptera) from the Ryukyu Islands. *Japanese Journal of Entomology* **64**: 211–214.

Masumoto, K. 1996b: Study of Asian Strongyliini (Coleoptera, Tenebrionidae) I. Six new species from Thailand, Laos and Taiwan, together with a new replacement name. *Elytra* **24**: 131–140.

Masumoto, K. 1996c: A new toxicine genus from Taiwan (Tenebrionidae, Coleoptera). *Entomological Review of Japan* **51**: 67–69.

Masumoto, K. 1996d: Fourteen new Laena (Coleoptera, Tenebrionidae) from China, Vietnam and Thailand. *Bulletin of the National Science Museum, Tokyo* (Ser. A) 22: 165–187.

Masumoto, K. 1998a: Study on Asian Strongyliini (Coleoptera, Tenebrionidae) V. Twenty new species of the genus Strongylium from East Asia. *Elytra* **26**: 173–200.

Masumoto, K. 1998b: New tenebrionid beetles from East Asia (Coleoptera, Tenebrionidae). *Japanese Journal of Systematic Entomology* **4**: 305–319.

Masumoto, K. 1999a: Study on Asian Strongyliini (Coleoptera, Tenebrionidae) VII. Brachypterous Strongyliines. *Elytra* 27: 113–125.

Masumoto, K. 1999b: Study on Asian Strongyliini (Coleoptera, Tenebrionidae) VIII. Ten new species of the genus Strongylium from East Asia. *Elytra* 27: 335–352.

Masumoto, K. 1999c: Additions to Plesiophthalmus and its allied genera (Coleoptera, Tenebrionidae, Amarygmini) from East Asia, Part 1. *Elytra* 27: 353–370.

Masumoto, K. 2000a: Additions to Plesiophthalmus and its allied genera (Coleoptera, Tenebrionidae, Amarygmini) from East Asia, Part 2. *Elytra* 28: 145–161.

Masumoto, K. 2000b: Study of Asian Strongyliini (Coleoptera, Tenebrionidae) IX Hairy Strongylium species from Southeast Asia. (Part 1. Species group of Strongylium gibbosulum). *Elytra* 28: 163–172.

Masumoto, K. 2000c: Study of Asian Strongyliini (Coleoptera, Tenebrionidae) X. Ten new species of the genus Strongylium from Southeast Asia. *Elytra* **28**: 391–407.

Matsumoto, K. 2001a: A new genus and species of the tribe Adeliini (Coleoptera, Tenebrionidae) from the upper hypogean zone of Southwest China. *Journal of the Speleological Society of Japan* 26: 44–49.

Masumoto, K. 2001b: Additions to Plesiophthalmus and its allied genera (Coleoptera, Tenebrionidae, Amarygmini) from East Asia, Part 3. *Elytra* 29: 57–73.

Masumoto, K. 2002: Two new species of the genus Allopezus (Coleoptera, Tenebrionidae, Cnodalonini). *Special Bulletin of the Japanese Coleopterological Society, Tokyo* **5**: 371–376.

Masumoto, K. 2004: Study on Asian Strongyliini (Coleoptera, Tenebrionidae) . XV. Ten new Strongylium species from Southeast Asia. *Elytra* **32**: 371–388.

Masumoto, K. 2005a: A new Plesiophthalmus species (Coleoptera, Tenebrionidae) from Taiwan. *Elytra* 33: 76–78.

Masumoto, K. 2005b: Study on Asian Strongyliini (Coleoptera, Tenebrionidae) . XVI. Eight new Strongylium species from Taiwan. *Elytra* 33: 193–209.

Masumoto, K. & Akita, K. 2001a: New or little-known tenebrionid species from Japan (Part 1). *Entomological Review of Japan* **56**: 15–22.

Masumoto, K. & Akita, K. 2001b: New or little-known tenebrionid species from Japan (Part 2). *Entomological Review of Japan* **56**: 75–79.

Masumoto, K. & Akita, K. 2001c: Two new tenebrionid species from Japan. *Special Publication of the Japan Coleopterological Society* (Osaka) **1**: 247–250.

Masumoto, K. & Akita, K. 2002: New or little-known tenebrionid species from Japan (Part 3). *Entomological Review of Japan* **57**: 147–153.

Masumoto, K. & Akita, K. 2003: New or little-known tenebrionid species from Japan (Part 4) (Coleoptera) *Entomological Review of Japan* **58**: 173–182.

Masumoto, K. & Akita, K. 2007: A new genus and species of the subfamily Colydiinae (Zopheridae: Coleoptera) from the Ogasawara Islands. *Entomological Review of Japan* **62**: 17–20.

Masumoto, K. & Akita, K. 2008: New or little-known tenebrionid species (Coleoptera) from Japan (Part 6) (Coleoptera). Two new species from Honshu and Ishigaki-jima Island. *Entomological Review of Japan* **63**: 35–41.

Masumoto, K. & Akita, K. 2009: The true affinity of the genus Sasajia Masumoto et Akita (Coleoptera, Tenebrionoidea), with description of a new Ocholissa from the Ryukyu Islands. *Elytra* **37**: 297–304.

Masumoto, K., Akita, K. & Lee, C.-F. 2005: New tenebrionid beetles from Taiwan (1). *Entomological Review of Japan* **60**: 247–254.

Masumoto, K., Akita, K. & Lee, C.-F. 2008: New tenebrionid beetles from Taiwan (4). *Entomological Review of Japan* **63**: 15–26.

Masumoto, K., Akita, K. & Lee, C.-F. 2012a: New tenebrionid beetles (Coleoptera) from Taiwan (5) Descriptions of a species belonging to a new genus and three new species of three different Tribes, and records of six species in new occurrence. *Elytra, New Series* **2**: 25–37.

Masumoto, K., Akita, K. & Lee, C.-F. 2012b: New tenebrionid beetles (Coleoptera) from Taiwan (6) Descriptions of four new species, and a comment concerning a male of the previously described species with figures of the male genitalia. *Elytra, New Series* **2**: 193–205.

Masumoto, K., Akita, K. & Lee, C.-F. 2013a: Tenebrionid beetles (Coleoptera) from the Kenting area in Taiwan. Three New Species and Three New Records of Occurrence. *Elytra, New Series* **3**: 113–122.

Masumoto, K., Akita, K. & Lee, C.-F. 2013b: New tenebrionid beetles (Coleoptera) from Taiwan (7) Descriptions of six new species and a new record of the named species of the genus Derispia (Diaperinae, Leiochrinini). *Elytra, New Series* **3**: 123–138.

Masumoto, K., Akita, K. & Lee, C.-F. 2013c: New tenebrionid beetles (Coleoptera) from Taiwan (8) Descriptions of six new species and new occurrence records of four species from Taiwan. *Elytra, New Series* **3**: 255–273.

Masumoto, K., Akita, K. & Lee, C.-F. 2014a: New tenebrionid (Coleoptera) beetles from Taiwan (9) Descriptions of eight new Cnodalonini species and an additional account of the previously described species. *Elytra, New Series* **4**: 73–93.

Masumoto, K., Akita, K. & Lee, C.-F. 2014b: New tenebrionid (Coleoptera) beetles from Taiwan (10) Descriptions of four new Tenebrionidae and a revival of synonymized species. *Elytra, New Series* **4**: 237–247.

Masumoto, K., Akita, K. & Lee, C.-F. 2015a: New tenebrionid beetles (Coleoptera) from Taiwan (11) Descriptions of six new species and proposing three new treatments. *Elytra, New Series* **5**: 161–176.

Masumoto, K., Akita, K. & Lee, C.-F. 2015b: New tenebrionid (Coleoptera) beetles from Taiwan (12) Descriptions of five new species and reports of two new occurring Records. *Japanese Journal of Systematic Entomology* **21**: 305–312.

Masumoto, K., Akita, K. & Lee, C.-F. 2017: New or little-known tenebrionid beetles (Coleoptera) from Taiwan (14) Descriptions of five new Tarpela species (Tenebrioninae, Helopini) with a diagnostic key to all the Taiwanese species. *Elytra, New Series* **7**: 1–12.

Masumoto, K., Akita, K. & Lee, C.-F. 2019a: New or little-known tenebrionid beetles (Coleoptera) from Taiwan. (16) Descriptions of new species, proposals of new systematic position and reports of distributions. *Japanese Journal of Systematic Entomology* **25**: 123–128.

Masumoto, K., Akita, K. & Lee, C.-F. 2019b: Three new taiwanese taxa from the darkling beetle subtribe Gnathidiina (Coleoptera: Tenebrionidae: Diaperinae). *Annales Zoologici* **69**: 159–164

Masumoto, K., Akita, K. & Utsunomiya, Y. 2003: A new Misolampidius (Coleoptera, Tenebrionidae) from Sichuan, Southwest China. *Special Bulletin of the Japanese Society of Coleopterology* 6: 327–330.

Masumoto, K., Ando, K. & Akita, K. 2006: New or little known tenebrionid species from Japan (Part 5). A new species belonging to a new genus (Coleoptera: Tenebrionidae: Helopini). *Entomological Revue of Japan* 61: 33–38.

Masumoto, K. & Grimm, R. 2004: A new genus and a species of the Palorinae (Coleoptera: Tenebrionidae) from Japan. *Entomological Review of Japan* 59: 127–130.

Masumoto, K., Hirano, Y. & Akita, K. 2016: Revision of the Salpingus Species (Coleoptera, Salpingidae) from Asia, with Descriptions of Two New Species from the Ryukyus and Taiwan. *Elytra, New Series* 6: 57–64.

Masumoto, K. & Makihara, H. 1997: Study on the tenebrionid beetles in South Sumatra. *Bulletin of the Forestry and Forest Products Research Institute Ibaraki* 374: 115–153.

Masumoto, K. & Merkl, O. 2003: A new Spiloscapha and a new Basanus from Taiwan (Coleoptera: Tenebrionidae: Scaphidemini). *Entomological Review of Japan* 58: 165–171.

Masumoto, K. & Nishikawa, N. 1986: A revisional study of the species of the genus Uloma from Japan, Korea and Taiwan (Tenebrionidae, Coleoptera). *Insecta Matsumurana* (N.S.) 35: 17–43.

Masumoto, K., Novák, V., Lee, C.-F. & Akita, K. 2017: A Revisional Study of the Subfamily Alleculinae (Coleoptera: Tenebrionidae) of Taiwan. *Miscellaneous Reports of the Hiwa Museum for Natural History* 58: 1–46 + 2 plates.

Masumoto, K., Novák, V., Akita, K. & Lee, C.-F. 2018: A Revisional Study of the Subfamily Alleculinae (Coleoptera: Tenebrionidae) of Taiwan (Part 2). *Miscellaneous Reports of the Hiwa Museum for Natural History* 59: 75–121 + 2 plates.

Masumoto, K., Novák, V., Akita, K. & Lee, C.-F. 2019a: A Revisional Study of the Subfamily Alleculinae (Coleoptera: Tenebrionidae) from Taiwan (Part 3). *Miscellaneous Reports of the Hiwa Museum for Natural History* 60: 89–118 + 2 plates.

Masumoto, K., Novák, V., Akita, K. & Lee, Ch.-F. 2019b: A Revisional Study of the Subfamily Alleculinae (Coleoptera: Tenebrionidae) from Taiwan (Part 4). Ten New Species of Genera Allecula, Borboconalia, Borboresthes, Pseudohymenalia and Upinella. *Folia Heyrovskyana, Series A* 27: 20–51.

Masumoto, K. & Schawaller, W. 2010: Strongylium species (Tenebrionidae: Coleoptera) from Nepal and its neighbouring areas. *Entomological Review of Japan* 65: 21–40.

Masumoto, K. & Yin Weng-y. 1993: Two new Yunnanese beetles of the genus Laena (Tenebrionidae) and Sivacrypticus (Archeocrypticidae) (Coleoptera). *Elytra* 21: 239–243.

Masumoto, K. & Yin, Weng-Y. 1994: Four new Lena species (Coleoptera, Tenebrionidae) from Yunnan, Southwest China. *Elytra* 22: 165–170.

Matsumura, S. 1911: Erster Beitrag zur Insekten-Fauna von Sachalin. *The Journal of the College of Agriculture, Tohoku Imperial University, Sapporo, Japan* 4: 1–145.

Matsumura, S. 1906: *Thousand Insects of Japan* 3. Tokyo: Keisesha Shoten, p. 1–8, no. 450–737, pl. 36–55.

Matsumura, M. 1915: *Taxonomy of Insects* 2. Tokyo: Keiseisha-shotem, 316 + 10 – 10 pp., 5 pls (in Japanese).

Mayet, V. 1874: [Note sur Sitaris colletis]. *Bulletin de la Société Entomologique de France* 1873: cxcviii-cxcix.

Mayet, V. 1890: [Description d'un Opatrum nouveau de Tunisie]. *Bulletin de la Société Entomologique de France* 1890: civ-cv.

Méhely, L. 1890: Ujabb adatok Erdély s különösen a Barczaság bogárvilágának ismeretéhez [Supplementa ad faunam Coleopterorum Transsylvaniae]. *Orvos-természettudományi Értesítő* **2**: 257–360.

Medvedev, G.S. 1961: Novye vidy roda Sarathropus (Coleoptera, Tenebrionidae) iz Sredney Azii. *Entomologicheskoe Obozrenie* **40**: 365–368.

Medvedev, G.S. 1962: Novoe podsemeystvo chernotelok (Coleoptera, Tenebrionidae) iz Turkmenii. *Zoologicheskiy Zhurnal* **51**: 1184–1189.

Medvedev, G.S. 1964a: Zoogeograficheskaya kharakteristika chernotelok (Coleoptera, Tenebrionidae) Kugitanga i opisanie novykh form. *Zoologicheskiy Zhurnal* **43**: 54–64.

Medvedev, G.S. 1964b: Novye formy zhukov-chernotelok (Coleoptera, Tenebrionidae) iz Kazakhstana. *Trudy Zoologicheskogo Instituta* **34**: 159–163.

Medvedev, G.S. 1964c: Novye formy chernotelok (Coleoptera, Tenebrionidae) iz Stredney Azii. *Entomologicheskoe Obozrenie* **43**: 651–661.

Medvedev, G.S. 1966: Novyy rod i vid zhukov-chernotelok (Coleoptera, Tenebrionidae) iz Turkmenii. *Trudy Zoologicheskogo Instituta* **37**: 98–99.

Medvedev, G.S. 1967: Obzor chernotelok triby Leptodini (Coleoptera, Tenebrionidae). *Entomologicheskoe Obozrenie* **46**: 353–360.

Medvedev, G.S. 1968a: *Zhuki-chernotelki (Tenebrionidae) podsemeystvo Opatrinae Triby Platynotini, Dendarini, Pedinini, Dissonomini, Pachypterinmi, Opatrini (chast) i Heterotarsini.* Fauna SSSR Zhestkokrylye Tom XIX vypusk 2. Leningrad: Nauka, 285 pp

Medvedev, G.S. 1968b: Novyy vid roda Pseudoblaps Guérin (Coleoptera, Tenebrionidae) iz Irana. *Entomologicheskoe Obozrenie* **47**: 179–180.

Medvedev, G.S. 1968c: Novye chernotelki triby Akidini (Coleoptera, Tenebrionidae). *Entomologicheskoe Obozrenie* **47**: 892–898.

Medvedev, G.S. 1970: Novye vidy zhukov-chernotelok (Coleoptera, Tenebrionidae) iz aridnych rayonov SSSR i Afganistana. *Entomologicheskoe Obozrenie* **49**: 393–408.

Medvedev, G.S. 1972a: Novyy vid roda Leptodes Sol. (Coleoptera, Tenebrionidae) iz Tadzhikistana. *Trudy Zoologicheskogo Instituta Akademia Nauk SSSR* **52**: 309–310.

Medvedev, G.S. 1972b: Novyy vid roda Dilamus Jacquelin du Val (Coleoptera, Tenebrionidae) iz Mongolskoy Narodnoy Respubliki. *Nasekomye Mongolii* **1**: 491–494.

Medvedev, G.S. 1973: Polozhenie rodov Leichenum Dej. i Idisia Pask. (Coleoptera, Tenebrionidae) v sisteme i opisanie novogo roda iz severnykh Karakumov. *Entomologicheskoe Obozrenie* **52**: 644–650, 2 pls.

Medvedev, G.S. 1974: Novyy vid roda Eurycaulus (Coleoptera, Tenebrionidae) v sredney Azii i Afganistane. *Zoologicheskiy Zhurnal* **53** (6): 946–948.

Medvedev, G.S. 1975: Obzor zhukov-chernotelok roda Dichillus Jacquelin du Val (Coleoptera, Tenebrionidae) fauny SSSR. *Entomologicheskoe Obozrenie* **54**: 591–604, 2 pls.

Medvedev, G.S. 1976: Novye vidy zhukov-chernotelok (Coleoptera, Tenebrionidae) iz Irana. *Entomologicheskoe Obozrenie* **55**: 889–902.

Medvedev, G.S. 1978a: Novye vidy zhukov-chernotelok roda Dilamus Jacquelin du Val (Coleoptera, Tenebrionidae) iz Sredney Azii. *Trudy Zoologicheskogo Institute Akademia Nauk SSSR* **59**: 147–152.

Medvedev, G.S. 1978b: Novye vidy zhukov-chernotelok triby Helopini (Coleoptera, Tenebrionidae) iz stredney Azii. *Trudy Zoologicheskogo Instituta Akademia Nauk SSSR* **71** [1977]: 50–52.

Medvedev, G.S. 1978c: Dva novye vida zhukov-chernotelok (Coleoptera, Tenebrionidae) iz Irana. *Entomologicheskoe Obozrenie* 57: 855–857.

Medvedev, G.S. 1979a: Novyy vid zhukov-chernotelok roda Sternoplax Friv. (Coleoptera, Tenebrionidae) iz Kugitanga. *Trudy Zoologicheskogo Instituta Akademia Nauk SSSR* 82: 40–42.

Medvedev, G.S. 1979b: Zhuky-chernotelki roda Philhammus Farim. (Coleoptera, Tenebrionidae) fauny SSSR. *Trudy Zoologicheskogo Instituta Akademii Nauk SSSR* 83: 42–45.

Medvedev, G.S. 1979c: Novye vidy roda Leptodes Sol. (Coleoptera, Tenebrionidae) iz Yugo-Zapadnogo Tadzhikistana. *Trudy Zoologicheskogo Instituta Akademia Nauk SSSR* 88: 69–73.

Medvedev, G.S. 1979d: Novyy vid zhukov-chernotelok roda Gnathosia Fischer-Waldheim (Coleoptera, Tenebrionidae) iz Tadzhikistana. *Trudy Vsesoyuznogo Entomologicheskogo Obshchestva* 61: 78–79.

Medvedev, G.S. 1981: In: Angelov P.A. & Medvedev G.S.: Zhuki chernotelki (Coleoptera, Tenebrionidae) Bolgarii. *Entomologicheskoe Obozrenie* 60: 302–315.

Medvedev, G.S. 1982: Novyy vid zhukov-chernotelok roda Cnemeplatia Costa (Coleoptera, Tenebrionidae) iz Karakalpakskoy ASSR. *Trudy Zoologicheskogo Instituta Akademii Nauk SSSR* 110: 40–41.

Medvedev, G.S. 1984: Novyy vid zhukov-chernotelok roda Penthicus Faldermann (Coleoptera, Tenebrionidae) iz Tuvy. *Nasekomye Mongolii* 9: 298–300.

Medvedev, G.S. 1987a: Zhuki-chernotelki roda Zophohelops Rtt. i blizkikh rodov (Coleoptera, Tenebrionidae) sredney Azii i Kazakhstana. *Trudy Zoologicheskogo Instituta Akademii Nauk SSSR* 164: 95–129.

Medvedev, G.S. 1987b: Obzor zhukov-chernotelkok roda Cylindrinotus Fald. (Coleoptera, Tenebrionidae) Kazakhastana i Sredney Azii. *Trudy Zoologicheskogo Instituta Akademii Nauk SSSR* 170: 99–104.

Medvedev, G.S. 1988a: Obzor chernotelok roda Penthicinus Rtt. (Coleoptera, Tenebrionidae). Pp. 55–65. In: Tobias, V.I. (ed.): Sistematika nasekomykh is kleschey. *Trudy Vsesoyuznogo Entomologicheskogo Obschestva* 70. Leningrad: Nauka, 26 pp.

Medvedev, G.S. 1988b: Obzor zhukov-chernotelok roda Alcinoeta Strand (Coleoptera, Tenebrionidae). *Entomologicheskoe Obozrenie* 67 (4): 798–802.

Medvedev, G.S. 1989a: Novyy vid zhukov-chernotelok roda Remipedella Sem. i polozhenie triby Remipedellini (Coleoptera, Tenebrionidae) v sisteme. *Trudy Zoologicheskogo Instituta Akademii Nauk SSSR* 208: 90–97.

Medvedev, G.S. 1989b: K nomeklature i systematike zhukov-chernotelok (Coleoptera, Tenebrionidae) Mongolii. *Nasekomye Mongolii* 10: 371–388.

Medvedev, G.S. 1990a: Novye vidy zhukov-chernotelok roda Leptodes Solier (Coleoptera, Tenebrionidae). *Trudy Zoologicheskogo Instituta Akademii Nauk SSSR* 211: 104–116.

Medvedev, G.S. 1990b: Novye vidy zhukov-chernotelok (Coleoptera, Tenebrionidae) iz Mongolii i Kitaya, pp. 132–138. In: *Nasekomye Mongolii.* Vol. 11. Leningrad, Nauka, 624 pp.

Medvedev, G.S. 1991: Novye chernotelki trib Stenosini i Cnemeplatiini (Coleoptera, Tenebrionidae) mirovoy fauny. *Entomologicheskoe Obozrenie* 70: 557–570.

Medvedev, G.S. 1992: Novyy vid chernotelok roda Pachyscelis Sol. (Coleoptera, Tenebrionidae) iz yuzhnogo Kyrgyzstana. *Entomologicheskoe Obozrenie* 71: 793–795.

Medvedev, G.S. 1993: K poznaniyu zhukov-chernotelok roda Lasiostola Sol. (Coleoptera, Tenebrionidae). *Entomologicheskoe Obozrenie* 73: 106–118.

Medvedev, G.S. 1994a: Tenebrionid beetles of the tribe Stenosini from China (Coleoptera: Tenebrionidae). *Zoosystematica Rossica* 2 (2) [1993]: 307–309.

Medvedev, G.S. 1994b: Novye dannye po sistematike zhukov-chernotelok triby Stenosini (Coleoptera, Tenebrionidae). *Entomologicheskoe Obozrenie* 73: 844–867.

Medvedev, G.S. 1995: Novye chernotelki roda Prosodes Eschsch. (Coleoptera, Tenebrionidae) iz Sredney Azii. I. Podrody Prosodes s. str., Mesoprosodes subgeN.N., Peltarium F.-W. i Uroprosodes Rtt. *Entomologicheskoe Obozrenie* 74: 811–854.

Medvedev, G.S. 1996: Novye chernotelki roda Prosodes Eschsch. (Cvoleoptera, Tenebrionidae) iz Sredney Azii. II. Podrody Mesopersina Rtt., Dilopersina Rtt., Iranosodes subgeN.N., Prosodura Rtt., Megaprosodes Rtt. i Prosodella Rtt. *Entomologicheskoe Obozrenie* 75: 596–636.

Medvedev, G.S. 1997a: Novye vidy zhukov-chernotelok roda Leptodes Sol. (Coleoptera, Tenebrionidae) iz gornykh sistem Tyan-Shanya i Paropamiza. *Entomologicheskoe Obozrenie* 76: 176–182.

Medvedev, G.S. 1997b: Novye chernotelki roda Prosodes Eschsch. (Coleoptera, Tenebrionidae) iz Sredney Azii. III. Podrody Prosodinia Rtt., Diprosodes Rtt. i Ferganoprosodes subgen. n. *Entomologicheskoe Obozrenie* 76: 563–605.

Medvedev, G.S. 1998a: A new species of tenebrionid beetle of the genus Cylindronotus from Tukmenistan (Coleoptera: Tenebrionidae). *Zoosystema Rossica* 7 (1): 138.

Medvedev, G.S. 1998b: K poznaniyu zhukov-chernotelok triby Blaptini (Coleoptera, Tenebrionidae) vostochnoy chasti Tibetskogo nagorya. *Entomologicheskoe Obozrenie* 77: 171–208.

Medvedev, G.S. 1998c: Novye vidy zhukov-chernotelok triby Blaptini (Coleoptera, Tenebrionidae) iz Gissaro-Darvaza i s Tibetskogo Nagorya. *Entomologicheskoe Obozrenie* 77: 555–586.

Medvedev, G.S. 1999a: Taksonomicheskoe znachenie stroenia polovykh putey samok zhukov-chernotelok triby Blaptini (Coleoptera, Tenebrionidae) i opisanie novogo podroda i novogo vida. *Entomologicheskoe Obozrenie* 78: 391–406.

Medvedev, G.S. 1999b: Sistema zhukov-chernotelok roda Prosodes Eschsch. (Coleoptera, Tenebrionidae). I. Podrody Oliprosodes Rtt. i Prosodina Rtt. *Entomologicheskoe Obozrenie* 78: 849–886.

Medvedev, G.S. 2001: Novye vidy zhukov-chernotelok rodov Asidoblaps Fairm. i Prosodes Eschsch. (Coleoptera, Tenebrionidae) iz Kitaya i Mesopotamii. *Entomologicheskoe Obozrenie* 80: 81–89.

Medvedev, G.S. 2002: Obzor zhukov-chernotelok roda Asidoblaps Fairm. (Coleoptera, Tenebrionidae). *Entomologicheskoe Obozrenie* 81: 64–110.

Medvedev, G.S. 2003: Sistema zhukov-chernotelok roda Prosodes Eschsch, (Coleoptera, Tenebrionidae) . II. Podrod Prosodinia Rtt. (Dopolnenie) *Entomologicheskoe Obozrenie* 82: 362–415.

Medvedev, G.S. 2004a: Novye vidy zhukov-chernotelok triby Blaptini (Coleoptera, Tenebrionidae) iz Indii, Nepala i Kitaya. *Entomologicheskoe Obozrenie* 83: 163–189.

Medvedev, G.S. 2004b: Novye vidy zhukov-chernotelok (Coleoptera, Tenebrionidae) iz pustyn Paleraktiki. *Entomologicheskoe Obozrenie* 83: 553–580.

Medvedev, G.S. 2005a: Sistema zhukov-chernotelok roda Prosodes Eschsch. (Coleoptera, Tenebrionidae). IV. Podrody Prosodestes Rtt. i Gebleria Motsch. *Entomologicheskoe Obozrenie* 84: 62–107.

Medvedev, G.S. 2005b: On collections of the sand desert faunas of Tenebrionidae (Coleoptera) of Middle Asia, Iran, and Afghanistan. Pp. 299–314. In: *Contribution to Systematics and Biology of Beetles. Papers Celebrating the 80th Birstday of Igor Konstantinovich Lopatin.* Sofia, Moscow: Pensoft.

Medvedev, G.S. 2005c: Novye vidy zhukov-chernotelok roda Asidoblaps Fairm. (Coleoptera, Tenebrionidae) iz provinciy Gansu i Sychuanu Kitaya. *Entomologicheskoe Obozrenie* 84: 531–568.

Medvedev, G.S. 2006: New data on the tenebrionid genus Agnaptorina Rtt. (Coleoptera, Tenebrionidae) of south-western China. *Entomologicheskoe Obozrenie* 85: 176–205, 261.

Medvedev, G.S. 2007a: New species of the Tenebrionid Genus Menimus Sharp, 1876 (Coleoptera, Tenebrionidae) from Southern Palaearctic. *Entomological Review* 87: 865–879.

Medvedev, G.S. 2007b: K sistematike i morfologii zhukov-chernotelok triby Blaptini (Coleoptera, Tenebrionidae). *Entomologicheskoe obozrenie* 86: 132–170. [A contribution to the taxonomy and morphology of the tribe Blaptini (Coleoptera, Tenebrionidae). *Entomological Review* 2007, 87: 181–214].

Medvedev, G.S. 2008: Novyye vidy zhukov-chernotelok rodov Tagonoides Fairm., Gnaptorina Rtt. i Agnaptoria Rtt. (Coleoptera, Tenebrionidae) iz Kitaya. *Entomologicheskoe obozrenie* 87: 808–830. [New species of the tenebrionid-beetle genera Tagonoides Fairm., Gnaptorina Rtt. and Agnaptoria Rtt. (Coleoptera, Tenebrionidae) from China. *Entomological Review* 2008, 88: 1142–1160].

Medvedev, G.S. 2009a: Sostav rodov Gnaptorina Reitter i Pseudognaptorina Kaszab zhukov-chernotelok triby Blaptini (Coleoptera, Tenebrionidae). *Entomologicheskoe obozrenie* 88: 416–429 [Composition of the genera Gnaptorina Reitter and Pseudognaptorina Kaszab of the tribe Blaptini (Coleoptera, Tenebrionidae). *Entomological Review* 2009, 89: 451–461].

Medvedev, G.S. 2009b: Desyat' novykh vidov zhukov-chernotelok roda Asidoblaps Fairmaire (Coleoptera, Tenebrionidae) iz Yugo-Zapadnogo Kitaya. *Entomologicheskoe obozrenie* 88: 819–851. [Ten new species of the tenebrionid genus Asidoblaps Fairmaire (Coleoptera, Tenebrionidae) from southwestern China. *Entomological Review* 2009, 89: 1104–1130].

Medvedev, G.S. & Abdurakhmanov, G.M. 1984: Novyy vid zhukov-chernotelok roda Dila (Coleoptera, Tenebrionidae) iz Dagestana. *Zoologicheskiy Zhurnal* 63 (2): 291–293.

Medvedev, G.S. & Iljina, V. 2007. A new subgenus and a new species of the tenebrionid genus Leptodes Dejean, 1834 (Coleoptera, Tenebrionidae). *Entomologicheskoe Obozrenie* 86: 880–882 (in Russian, English translation in *Entomological Review* 87: 1035–1037).

Medvedev, G.S. & Iwan, D. 2006: Notes on the taxonomy of tenebrionid beetles (Coleoptera: Tenebrionidae). *Annales Zoologici* 56: 613–621.

Medvedev, G.S. & Kabakov 1994 [new taxon]. In: Medvedev G.S.: Novye dannye po sistematike zhukov-chernotelok triby Stenosini (Coleoptera, Tenebrionidae). *Entomologicheskoe Obozrenie* 73: 844–867.

Medvedev, G.S. & Kabakov 1996 [new taxon]. In: Medvedev G.S.: Novye chernotelki roda Prosodes Eschsch. (Cvoleoptera, Tenebrionidae) iz Sredney Azii. 11. Podrody Mesopersina Rtt., Dilopersina Rtt., Iranosodes subgen. n., Prosodura Rtt., Megaprosodes Rtt. i Prosodella Rtt. *Entomologicheskoe Obozrenie* 75: 596–636.

Medvedev, G.S. & Kaltaev, T. 1979: Novye vidy zhukov-chernotelok (Coleoptera, Tenebrionidae) iz Kirgizii. *Izvestya Akademii Nauk Kirgizskoy SSR* 4: 57–60.

Medvedev, G.S. & Kaszab, Z. 1973: Ergebnisse der Mongolisch-sowietischen biologischen Expeditionen in der Mongolischen Volksrepublik. Tenebrionidae, excl. Tentyriini (Coleoptera). *Folia Entomologica Hungarica* (N.S.) 26: 79–111.

Medvedev, G.S. & Kompantseva, T.V. 1989: Novye vostochnopalearkticheskie vidy zhukov-chernotelok (Coleoptera, Tenebrionidae). *Trudy Zoologicheskogo Instituta Akademii Nauk SSR* 208: 98–101.

Medvedev, G.S. & Merkl, O. 2001: Novye vidy zhukov-chernotelok triby Blaptini (Coleoptera, Tenebrionidae) iz yugo-zapadnogo Kitaya. *Entomologicheskoe Obozrenie* **80**: 620–626.

Medvedev, G.S. & Merkl, O. 2005: Two new species of Prosodes Eschscholtz, 1829 (Coleoptera, Tenebrionidae: Blaptini) from Iran. *Acta Zoologica Academiae Scientarum Hungaricae* **51**: 171–180.

Medvedev, G.S. & Mihkailov, V.A. 1978: Novyy vid roda Prosodes Eschsch. (Coleoptera, Tenebrionidae) iz Darvazskogo Khrebta. *Doklady Akademii Nauk Tadzikskoy* SSR **21**: 61–63.

Medvedev, G.S. & Mordkovich, V.G. 1970: Novyy vid zhukov chernotelok (Coleoptera, Tenebrionidae) iz Tuvy. Pp. 30–32. In: Tsherepanov, A.I. (ed.): *Novye i maloizvestnye vidy sibirskoy fauny*. 3. Institut Biologii, Sibirskoye Otdelenie Akademiy Nauk SSSR. Novosibirsk: Nauka, 112 pp.

Medvedev, G.S. & Nepesova, M.G. 1985a: *Opredelitel zhukov-chernotelok Tukmenistana*. Ashkhabad: Ylym, 177 pp.

Medvedev, G.S. & Nepesova, M.G. 1985b: Novye vidy zhukov-chernotelok (Coleoptera, Tenebrionidae) iz Turkmenii i Karakalpakii. *Entomologicheskoe Obozrenie* **64**: 358–362.

Medvedev, G.S. & Pirnazarov, B.P. 1972: Novyy vid zhuka-chernotelki roda Leptodes Sol. (Coleoptera, Tenebrionidae) s poberezhya Aralskogo Morya. *Entomologicheskoe Obozrenie* **51**: 125–126.

Medvedev, L.N. 1974: On the fauna of Anthicidae (Coleoptera) of Mongolia. Pp. 149–155. In: *Nasyekomiye Mongolii [Insects of Mongolia]. Vol. 4, Nr. 2*. Leningrad: Nauka.

Medvedev, L.N. 1975: Anthicidae (Coleoptera) from Mongolia. *Acta Zoologica Academiae Scientiarum Hungaricae* **21**: 87–95.

Medvedev, L.N. & Micheetchev, A.N. 1979: New data of Anthicidae from Mongolia (Coleoptera). *Faunistische Abhandlungen Staatliches Museum für Tierkunde in Dresden* **7**: 141–143.

Megerle von Mühlfeld, J.K. 1812: *Bemerkungen, Berichtigungen und Zusaetze zu Illiger's Zusaetzen, Berichtigungen und Bemerkungen zu Fabricii Systema Eleytheratorum*. Linz: Akad. Buchhandlung, 44 pp.

Melichar, L. 1912: Neue palaearktische Coleopteren. *Coleopterologische Rundschau* **1**: 33–36.

Mellié, J. 1847: [Mélanges et nouvelles]. *Revue Zoologique, par la Société Cuvierienne* **10**: 108–110.

Mellié, J. 1849a: Monographie de l'ancien genre Cis des auteurs. *Annales de la Société Entomologique de France* (2) **6**: 205–274, 313–396, pls 9–12. Mellié J. 1849a: Monographie de l'ancien genre Cis des auteurs. Annales de la Société Entomologique de France (2) 6 [1848]: 205–274 [No. 3], 313–396 + pl. 9–12 [No. 4].

Mellié, J.1849b: [Communication]. *Bulletin de la Société Entomologique de France* (2) **7**: lxxxvi-lxxxvii.

Melsheimer, F.E. 1844: Descriptions of new species of Coleoptera of the United States. *Proceedings of the Academy of Natural Sciences of Philadelphia* **2** [1844–1845]: 98–118.

Melsheimer, F.E. 1846: Descriptions of new species of Coleoptera of the United States. *Proceedings of the Academy of Natural Sciences of Philadelphia* **3**: 53–66.

Melsheimer, F.E. 1847: Descriptions of new species of Coleoptera of the United States. *Proceedings of the Academy of Natural Sciences of Philadelphia* **3**: 158–181.

Ménétriés, E. 1832: *Catalogue raisonné des objets de zoologie recueillis dans un voyage au Caucase et jusqu'aux frontières actuelles de la Perse entrepris par l'ordre de S.M. l'Empereur*. St. Petersburg: Académie des Sciences, xxxiii + 272 + iv + [1].

Ménétriés, E. 1836a: Insectes nouveaux de la Turquie. *Bulletin Scientifique publié par l'Académie Impériale des Sciences de Saint-Pétersbourg* **1**: 149–151.

Ménétriés, E. 1836b: Sur quelques insectes de la Russie. *Bulletin Scientifique publié par l'Académie Impériale des Sciences de Saint-Pétersbourg* 1: 180–181.

Ménétriés, E. 1840: Catalogue des insectes recueillis entre Constantinople et le Balkan. *Mémoires de l'Académie Impériale des Sciences de St.-Pétersbourg* (6) 5: 1–52, 2 pls.

Ménétriés, E. 1848: Descriptions des insectes recueillis par feu M. Lehmann. *Mémoires de l'Académie Impériale des Sciences de St.-Pétersbourg* 8 (2) (6: *Sciences Mathématiques, Physiques et Naturelles* 6 livr. 1/2): 17–56.

Ménétriés, E. 1849: Catalogue des insectes recueillis par feu M. Lehmann avec les descriptions des nouvelles espèces. (Seconde et dernière partie) Coléoptères hétéromères. *Mémoires de l'Académie Impériale des Sciences de St.-Pètersbourg* 8 (2) (6: *Sciences Mathématiques, Physiques et Naturelles* 6 livr. 4): 217–328, pls 3–5.

Ménétriés, E. 1854: Coléoptères recueillis dans la Mongolie chinoise et aux environs de Pékin. *Études Entomologiques* 3: 26–41.

Meng, L. & Ren, G.-D. 2005: A systematic study of the genus Myatis Bates from China and adjacent areas (Coleoptera, Tenebrionidae). *Acta Zootaxonomica Sinica* 30: 104–110.

Méquignon, A.E. 1937: Note synonymique. *Bulletin de la Société Entomologique de France* 42: 194.

Méquignon, A.E. 1941: Description d'un Mordellidae (Col.) nouveau paléarctique. *Bulletin de la Société Entomologique de France* 46: 147.

Méquignon, A.E. 1946: Contribution à l'étude des mordellides paléarctiques. *Revue Française d'Entomologie* 13: 52–76.

Méquignon, A.E. 1947: Notes diverses sur lers coléoptères de France (5ième note). *Bulletin de la Société Entomologique de France* 52: 58–61.

Méquignon, A.E. 1948: Notes diverses sur coléoptères de France. *Bulletin de la Société Entomologique de France* 53: 75–80.

Merkl, O. 1990: A review of Bothynogria Borchmann (Coleoptera, Tenebrionidae: Lagriini). *Acta Zoologica Hungarica* 36: 279–294.

Merkl, O. 1991a: Lagriini of the Nepal-Himalayas (Coleoptera: Tenebrionidae). *Stuttgarter Beiträge zur Naturkunde, Series A (Biologie)* 470: 1–18.

Merkl, O. 1991b: Epitragini from Iran (Coleoptera, Tenebrionidae). *Acta Zoologica Hungarica* 37: 59–66.

Merkl, O. 2004: On taxonomy, nomenclature, and distribution of some Palaearctic Lagriini, with description of a new species from Taiwan (Coleoptera: Tenebrionidae). *Acta Zoologica Academiae Scientiarum Hungaricae* 50: 283–305.

Merkl, O. 2007: Notes on Asian Lagriini, with description of Cerogria gozmanyi sp. n. (Coleoptera: Tenebrionidae). *Acta Zoologica Academiae Scientiarum Hungaricae* 53 (Suppl. 1): 255–272.

Merkl, O. 2011a: Arthromacra Kirby, 1837 in the Himalayas (Coleoptera, Tenebrionidae, Lagriini). Pp. 301–312. In: Kawai, S. (ed.): *Masumushi. Special Publication of the Japanese Society of Scarabaeidology No. 1.* Tokyo: Japanese Society of Scarabaeidology, 345 pp.

Merkl, O. 2011b: Donaciolagria malgorzatae sp. nov. from Indochina, and new records of the genus (Coleoptera, Tenebrionidae: Lagriini). *Annales Zoologici* 61: 361–366.

Merkl, O. 2019a: Lagriini from Bhutan (Coleoptera, Tenebrionidae: Lagriinae). *Annales Zoologici* 69: 65–81.

Merkl, O. 2019b: Two New Species of Donaciolagria Pic, 1914 (Coleoptera, Tenebrionidae: Lagriinae) from Myanmar and China. *Entomological Review* 99: 1014–1020.

Merkl, O. & Chen, B. 1997: A review of Mimoborchmania Pic (Coleoptera, Tenebrionidae: Lagriini). *Acta Zoologica Academiae Scientiarum Hungaricae* **43**: 111–119.

Merkl, O. & Egorov, L.V. 2015: Somocoelia triplehorni Merkl and Egorov (Coleoptera: Tenebrionidae), the first species of Platyscelidini in Iran. *The Coleopterists Bulletin* **69**: 73–77.

Merkl, O. & Masumoto, K. 2007: A review of Taiwanese Paramisolampidius (Coleoptera, Tenebrionidae: Cnodalonini). *Acta Zoologica Academiae Scientiarum Hungaricae* **53**: 127–137.

Meyer, F.A.A. 1793: *Tentamen monographiae generis Meloës*. Gottingae: Vanderhoek et Ruprecht, 32 pp.

Miedel, J. 1880: Observations sur les Opatrum. *Deutsche Entomologische Zeitschrift* **24**: 136–140.

Mikhaylov, G.S. & Chikatunov, M.G. 1987: *Zhetskokrylye yugo-vostochnoy chasti Pamiro-Alaya (v predelach Gorno-Badakhshanckoy avtonomnoy oblasti Tadzhikistoy SSR)*. Dushambe, 59 pp.

Miller, L. 1851: [new taxa, in:Versammlung am 3. Dezember 1851]. *Verhandlungen des Zoologisch-Botanischen Vereins in Wien* **2**: 83–113.

Miller, L. 1858: Zwei neue Erodiinen-Gattungen. *Wiener Entomologische Monatschrift* **2**: 115–124.

Miller, L. 1861: Neue Käfer aus Kindermann's Vorräthen. *Wiener Entomologische Monatsschrift* **5**: 169–182, 201–209, 2 pls.

Miller, L. 1881: [new taxa]. In: Reitter, E.: Coleopterologische Ergebnisse einer Reise nach Croatien, Dalmatien und der Herzegowina im Jahre 1879. *Verhandlungen der Kaiserlich-Königlichen Zoologisch-Botanischen Gesellschaft in Wien* **30** [1880]: 201–228.

Miller, L. 1883: Neue Coleopteren aus Griechenland, gesammelt von E. v. Oertzen. *Verhandlungen der Kaiserlich-Königlichen Zoologisch-Botanischen Gesselschaft in Wien* **33**: 263–266.

Miwa, Y. 1928: A study on the species of Meloidae in the Japanese Empire. *Transactions of the Sapporo Natural History Society* **10**: 63–78.

Miwa, Y. 1930: A new Formosan blister-beetle of the genus Meloe. *Transactions of Natural History Society of Formosa* **20** (106): 12–13.

Miwa, Y. 1931: A systematic catalogue of Formosan Coleoptera. *Reports of the Department of Agriculture of the Governement Research Institute, Taihoku* **55**: 1–359.

Miwa, Y. 1933: An enumeration of Coleoptera from the Island Irimote in Loochoo, with descriptions of new species. *Transaction of the Natural History Society of Formosa* **23**: 4–11.

Miwa, Y. 1935: Coleoptera from Amami-Islands in Loo-Choo Archipelago. *Transactions of the Kansai Entomological Society* **6**: 11–30, 2 pls. [in Japanese, English title]

Miwa, Y. 1939: Descriptions of four new species belonging to the family Tenebrionidae from Formosa. *Zoological Magazine* (Tokyo) **51**: 412–415.

Miyatake, M. 1954: Studies on the Japanese Ciidae, I. *Scientific Reports of the Matsuyama Agricultural College* no.**14**: 40–67, pls. 1–11.

Miyatake, M. 1955a: Descriptions of a new species of the Tetratomidae from Japan (Coleoptera). *Transactions of the Shikoku Entomological Society* **4**: 93–94.

Miyatake, M. 1955b: Notes on some Japanese Ciidae (Coleoptera). *Transactions of the Shikoku Entomological Society* **4**: 94–95.

Miyatake, M. 1957: Two genera of Mycetophagidae Litargus Erichson and Litargops Reitter of Japan (Coleoptera). *Transactions of the Shikoku Entomological Society* **5**: 33–45, 1 pl.

Miyatake, M. 1959a: A revision of the Japanese species related to the genera Triphyllus and Pseudotriphyllus (Coleoptera, Mycetophagidae). *Entomological Review of Japan* 5: 24–30.

Miyatake, M. 1959b: The Ciidae of Amami-Oshima. (Studies on Japanese Ciidae, II). *Memoirs of the Ehime University* (6) **4**: 119–124.

Miyatake, M. 1960: The genus Pisenus Casey and some notes on the family Tetratomidae (Coleoptera). *Transactions of the Shikoku Entomological Society* **6**: 121–135.

Miyatake, M. 1961: The tribe Leiochrini of Japan (Coleoptera: Tenebrionidae). *Transactions of the Shikoku Entomological Society* 7: 33–40.

Miyatake, M. 1964: Notes on the tribe Bolitophagini of Japan, with the descriptions of four new genera and two new species Japan (Coleoptera: Tenebrionidae). *Transactions of the Shikoku Entomological Society*. **8**: 59–84.

Miyatake, M. 1968: Two new species of Mycetophagidae from Japan (Coleoptera). *Entomological Review of Japan* **20**: 29–32.

Miyatake, M. 1970: A revision of the genus Byrsax Pascoe of Japan, with some notes on the Japanese Bolitophagini (Coleoptera: Tenebrionidae). *Transactions of the Shikoku Entomological Society* **10**: 116–124, pls 4, 5.

Miyatake, M. 1974: A new species of the genus Pisenus Casey from Formosa (Coleoptera, Tetratomidae). *Transactions of the Shikoku Entomological Society* **12**: 60–62.

Miyatake, M. 1982a: A new genus and two new species of the tribe Hypulini from Japan (Coleoptera, Melandryidae). *Transactions of the Shikoku Entomological Society* **16**: 29–35.

Miyatake, M. 1982b: Three new species of the Ciidae from Taiwan (Coleoptera). Pp. 73–79. In: Satô, M., Hori, Y., Arita, Y. & Okadome, T. (eds): *Special Issue to the Memory of Retirement of Emeritus Professor Michio Chûjô*. Nagoya: Biological Laboratory, Nagoya Women's University, [2] + 185 pp. 1 pl.

Miyatake, M. 1985: Ciidae, pp. 278–285. In: Kurosawa, Y., Hisamatsu, S. & Sasaji, H. (eds): *The Coleoptera of Japan in Color, Vol. III*. Osaka: Hoikusha Publishing Co., x + 500 pp., 192 pls. (in Japanese).

Mizota, K. 1999a: A revision of the genus Indasclera in Japan (Coleoptera, Oedemeridae). *Insecta Matsumurana* **56**: 69–94.

Mizota, K. 1999b: A new genus and species of Asclerini (Coleoptera: Oedemeridae) from the Yaeyama Islands (Japan) and Taiwan. *The Coleopterist's Bulletin* **53**: 287–292.

Mollandin de Boissy, R. 1940: Un Rhipidius nouveau de Provence. *Revue Française d'Entomologie* 7: 94–95

Montrouzier, X. 1860: Essai sur la faune entomologique de la Nouvelle-Calédonie (Balade). *Annales de la Société Entomologique de France* (3) **8**: 229–308.

Morawitz, A.F. 1865: Ueber die in Russland und den angrenzenden Ländern vorkommenden Akis-Arten. *Horae Societatis Entomologicae Rossicae* **3** [1865–1866]: 3–48.

Morawitz, F.F. 1861: Einige für die russisch-europäische Fauna neue Käfer. *Bulletin de la Société Impériale des Naturalistes de Moscou* **34** (1) [1861–1862]: 284–294.

Morimoto, K. & Maeta, Y. 2009: A new species of Stenoria (Coleoptera, Meloidae) from Japan, with description of the first instar larva. *Japanese Journal of Systematic Entomology* **15**: 475–480.

Morishima, N. 1988: A new genus and new species of the tribe Melandryini (Coleoptera, Melandryidae). *Elytra* **16**: 41–44.

Motschulsky, V. de. 1839: Insectes du Caucase et des provinces transcaucasiennes. *Bulletin de la Société Impériale des Naturalistes de Moscou* **12**: 44–93, pls I, II, V, VI.

Motschulsky, V. de. 1845a: Remarques sur la collection de coléoptères russes de Victor de Motschulsky. *Bulletin de la Société Impériale des Naturalistes de Moscou* **18** (1): 3–127, pls. 1–3.

Motschulsky, V. de. 1845b: Observations sur le Musée entomologique de l'Université Impériale de Moscou. *Bulletin de la Société des Naturalistes de Moscou* **18** (4): 332–388.

Motschulsky, V. de. 1845c: Corrections à faire dans l'article intitulé: 'Remarques sur la Collection de coléoptères russes de Victor de Motchoulsky'. Bulletin N° 1. 1845. *Bulletin de la Société Impériale des Naturalistes de Moscou* **18** (4): [unnumbered page at end of issue].

Motschulsky, V. de. 1846: Remarques sur la collection de coléoptères Russes de Victor de Motschulsky, 3me Article. *Bulletin de la Société Impériale des Naturalistes de Moscou* **19** (1): 372–418.

Motschulsky, V. de. 1849: Coléoptères reçus d'un voyage de M. Handschuh dans le midi de l'Espagne, énumerés et suivis de notes. *Bulletin de la Société Impériale des Naturalistes de Moscou* **22** (3–4): 52–163.

Motschulsky, V. de. 1851a: Lettres adressées à Monsieur le premier secrétaire de la Société impériale des naturalistes de Moscou. *Bulletin de la Société Impériale des Naturalistes de Moscou* **24** (2): 592–616.

Motschulsky, V. de. 1851b: Lettre à Monsieur le Dr. Renard, sécretaire de la Société impériale des naturalistes de Moscou. *Bulletin de la Société Impériale des Naturalistes de Moscou* **24** (2): 649–657.

Motschulsky, V. de. 1853: Nouveautés. *Études Entomologiques* **1** [1852–1853]: 77–80.

Motschulsky, V. de. 1854: Diagnoses de coléoptères nouveaux, trouvés par MM. Tatarinoff et Gaschkéwitsch aux environs de Pékin. *Études Entomologiques* **2**: 44–51.

Motschulsky, V. de. 1856: Voyages. Lettres de M. de Motschulsky à M. Ménétriés. N. 4. St. Pétersburg le 25 oct. 1855. *Études Entomologiques* **5**: 3–38, 1 pl.

Motschulsky, V. de. 1858a: Insectes du Japon. *Études Entomologiques* **6** [1857]: 25–41.

Motschulsky, V. de. 1858b: Insectes des Indes orientales. 1:ière Série. *Études Entomologiques* **7**: 20–122, 1 pl.

Motschulsky, V. de. 1858c: Sur les collections coléoptèrologiques de Linné et de Fabricius. *Études Entomologiques* **7**: 123–152.

Motschulsky, V. de. 1858d: Nouveautés. *Études Entomologiques* **7**: 187–190.

Motschulsky, V. de. 1859a: Coléoptères rapportés de la Songarie par M. Séménof. *Mélanges Biologiques tirés du Bulletin de l'Académie Impériale des Sciences de St.-Pétersbourg* **3**: 290–309.

Motschulsky, V. de. 1859b: Coléoptères rapportés de la Songarie par M. Séménov. *Bulletin de l'Académie Impériale des Sciences de Staint-Pétersburg* **1**: 301–314 [columns].

Motschulsky, V. de. 1859c: [new taxa]. In: Gebler, F.A.: Verzeichniss der von Herrn Schrenk in den Kreisen Ajagur und Kakaraly in den östlichen Kirgisensteppen und in der Songarey in den Jahren 1840 bis 1843 gefundeten Kaeferarten. *Bulletin de la Société Inpériale des Naturalistes de Moscou* **32** (1): 426–519.

Motschulsky, V. de. 1859d: Catalogue des Insectes rapportés de environs du fl. Amour, dépuis la Schilka jusqu'à Nikolaëvsk. *Bulletin de la Société Impériale des Naturalistes de Moscou* **32** (3–4): 487–507.

Motschulsky, V. de. 1860a: Insectes des Indes orientales, et de contrées analogue. *Études Entomologiques* 8 [1859]: 25–118.

Motschulsky, V. de. 1860b: Insects nouveaux ou peu connus des bassins de la Méditerranée et de la mer Noire jusqu'à la mer Caspienne. *Études Entomologiques* 8 [1859]: 119–144.

Motschulsky, V. de. 1860c: Coléoptères rapportes de la Sibérie orientale et notamment des pays situés sur les bords du fleuve Amour par MM. Schrenck, Maack, Ditmar, Voznessenski etc. Pp. 77–257 + [1] pp., pls 5–11, 1 map. In: Schrenck, P.L.: *Reisen und Forschungen im Amur-Lande in den Jahren 1854–1856 im Auftrage der Kaiserl. Akademie der Wissenschaften zu St. Peterburg ausgeführt und in Verbindung mit mehreren Gelehrten herausgegeben von Dr.Leopold Schrenck. Band II. Zweite Lieferung. Coleopteren.* St. Peterburg: Kaiserliche Akademie der Wissenschaften, 976 pp.

Motschulsky, V. de. 1860d: Coléoptères rapportés en 1859 par M. Sévertsef des steppes méridionales des Kirghises, et énumérés. *Mélanges Biologiques tirés du Bulletin de l'Académie Impériale des Sciences de St.-Pétersbourg* **3**: 408–452.

Motschulsky, V. de. 1860e: Coléoptères rapportés en 1859 par M. Sévertsef des steppes méridionales des Kirghises, et énumérés. *Bulletin de l'Académie Impériale des Sciences de St.-Pétersbourg* **2**: 513–544 [columns].

Motschulsky, V. de. 1861: Insectes du Japon. *Études Entomologiques* **9** [1860]: 4–39.

Motschulsky, V. de. 1862: Insectes du Japon. *Études Entomologiques* **10** [1861]: 3–24.

Motschulsky, V. de. 1863: Essai d'un catalogue des insectes de l'île de Ceylan. *Bulletin de la Société Impériale des Naturalistes de Moscou* **36** (2): 421–532.

Motschulsky, V. de. 1866: Catalogue des insectes reçus du Japon. *Bulletin de la Société Impériale des Naturalistes de Moscou* **39** (1–2): 163–200.

Motschulsky, V. 1868: Énumération des nouvelles espèces de coléoptères rapportés de ces voyages. 6-ième Article. *Bulletin de la Société Impériale des Naturalistes de Moscou* **41** (2) [1868–1869]: 170–201, pl. VIII.

Motschulsky, V. de. 1870: Énumération des nouvelles espèces de coléoptères rapportés de ses voyages *Bulletin de la Société Impériale des Naturalistes de Moscou* **43** (1): 379–407, pls III, IV.

Motschulsky, V. de. 1872: Énumération des nouvelles espèces de coléoptères rapportés de ses voyages. *Bulletin de la Société Impériale des Naturalistes de Moscou* **45** (3–4): 23–55.

Motschulsky, V. de. 1873: Énumération des nouvelles espèces de coléoptères rapportés de ses voyages. *Bulletin de la Société Impériale des Naturalistes de Moscou* **46** (1): 466–482.

Muche, W.H. 1962a: Lydus anatolicus spec. nov. (Col., Meloidae). *Reichenbachia* **1**: 51–53.

Muche, W.H. 1962b: Die Podonta meiner Anatolienausbeute (Coll., Alleculidae, Omophlini) *Reichenbachia* **1**: 145–147.

Muche, W.H. 1963a: Beitrag zur Kenntnis der Genus Cerocoma Geoffr. (Coleoptera, Meloidae, Cerocomini). *Reichenbachia* **2**: 11–15.

Muche, W.H. 1963b: Über einige von Herrn Dr. Eckerlein in Kleinasien gesammelte Omophlini (Coleoptera, Alleculidae). *Reichenbachia* **2**: 61–65.

Muche, W.H. 1964a: Neue Omophlus-Arten (Coleoptera, Alleculidae, Omophlinae). *Reichenbachia* **4**: 235–240.

Muche, W.H. 1964b: Revision des Genus Omophlus Sol. (Coleoptera, Alleculidae, Omophlinae). *Entomologische Abhandlungen* **29**: 591–624.

Muche, W.H. 1965a: Eine neue Unterart von Cteniopinus altaicus Gebl. (Alleculidae, Omophlinae). *Reichenbachia* **5**: 207–210.

Muche, W.H. 1965b: Neue Omophlus des Naturhistorischen Museums, Budapest (Coll. Alleculidae, Omophlini). *Entomologische Blätter* **61**: 85–99.

Muche, W.H. 1965c: Neue Podonta-Arten (Coleoptera, Alleculidae, Omophlinae). *Reichenbachia* **6**: 83–87.

Muche, W.H. 1967: Revision des Genus Heliotaurus Mulsant (Col., Alleculidae, Omophlinae). *Entomologische Zeitschrift* 77: 17–26.

Muche, W.H. 1972a: 2. Beitrag zur Alleculidenfauna der Kaukasusländer. *Entomologische Nachrichten* **10**: 127–131.

Muche, W.H. 1972b: Ergebnisse der zoologischen Forschungen von Dr. Z. Kaszab in der Mongolei 288. Coleoptera, Alleculidae. *Annales Historico-Naturales Musei Nationalis Hungarici* **64**: 223–225.

Muche, W.H. 1973: Die Omophlinen meiner Mittelasienausbeute 1972 (Col. Alleculidae). *Mitteilungen der Münchener Entomologischen Gesellschaft* **63**: 36–41.

Muche, W.H. 1974: Eine neue Allecula aus Anatolien (Coleoptera, Alleculidae). *Fragmenta Entomologica* **10**: 219–222.

Muche, W.H. 1978: Beitrag zur Kenntnis der Insektenfauna Mittelasiens (Hymenoptera, Tenthredinidae; Coleoptera, Alleculidae et Rhipiphoridae). *Entomologische Abhandlungen* **42**: 355–361.

Muche, W.H. 1979a: Zwei neue Omophlus-Arten aus dem Naturhistorischen Museum Basel. (Coleoptera, Alleculidae). *Reichenbachia* **17**: 171–174.

Muche, W.H. 1979b: Eine neue Allecula aus dem Zoologischen Museum Berlin (Coleoptera, Alleculinae). *Mitteilungen aus dem Zoologischen Museum in Berlin* **55**: 269–270.

Muche, W.H. 1980: Omophlina muminovi sp. n. aus Tadshikistan, mit einem Beitrag zur Verbreitung tadshikischer Omophlina-Arten. *Faunistische Abhandlungen* 7: 291–293.

Muche, W.H. 1981: Eine neue Isomira-Art (Untergattung Asiomira) aus China (Coleoptera, Alleculidae). *Reichenbachia* **19**: 157–158.

Muche, W.H. 1982: Insects of Saudi Arabia Coleoptera: Fam. Alleculidae. *Fauna of Saudi Arabia* **4**: 116–123.

Muche, W.H. 1984: Eine neue Isomira-Art von Mallorca (Spanien). (Coleoptera, Alleculidae). *Reichenbachia* **22**: 207–208.

Muche, W.H. 1986: Eine neue Mycetocharina-Art aus Afghanistan (Coleoptera, Alleculidae). *Reichenbachia* **23** (32): 181–183.

Muche, W.H. 1987: Eine neue Art der Gattung Megischia Solier aus Griechenland (Coleoptera, Alleculidae, Omophlinae). *Reichenbachia* **24**: 135–137.

Müller, C. 1887: Vierzehn neue Heteromeren, von Bradshaw im Zambesi-Gebiete aufgefunden und im Museum der Königlichen Zoologischen Gesellschaft „Natura Artis Magistra" zu Amsterdam befindlich. *Tijdschrift voor Entomologie* **30**: 297–308.

Müller, J. [G.] 1902: Coleopterologische Notizen. III. *Wiener Entomologische Zeitung* **21**: 115–116.

Müller, J. [G.] 1904: Beschreibung neuer dalmatinischer Koleopteren. *Münchener Koleopterologische Zeitschrift* **2** [1904–1906]: 208–210.

Müller, J. 1908: Beschreibungen neuer dalmatinischer Koleopteren. *Münchner Koleopterologische Zeitschrift* **3**: 314–316.

Müller, J. [G.] 1912*: Ueber zwei neue Dunkelkäfer (Tenebrionidae) von der Insel Pelagosa.* Entomologische Blätter **8***:* 289–291.

Müller, J. [G.] 1916: *Coleopterologische Beiträge zur Fauna der österreichischen Karstprovinzen und ihrer Grenzgebiete.* Entomologische Blätter 12: 73–109.

Müller, J. [G.] 1917: Die ostadriatischen Asida-Arten. *Wiener Entomologische Zeitung* 36: 1–17.

Müller, J. [G.] 1925: Sulle Meloë europee del gruppo rugosus. *Studi Entomologici* 1: 21–25.

Müller, J. [G.] 1929: Un nuovo Calopus (Coleopt. Oedemeridae). *Bollettino della Società Entomologica Italiana* 61: 63–64.

Müller, J. [G.] 1936: Sopra alcuni Meloidi dell'Isola di Coo (Egeo). *Bollettino della Società Entomologica Italiana* 68: 91–92.

Müller, J. [G.] 1946: Nuovi coleotteri della regione balcanica occidentale. *Redia* 31: 107–122.

Müller, J. [G.] 1950: Appunti sulle Asida di Pelagosa, Pomo ed altri scogli dell'Adriatico meridionale (Coleoptera, Tenebrionidae). *Bollettino della Società Entomologica Italiana* 80: 12–13.

Müller, O.F. 1764: *Fauna insectorum Fridrichsdalina sive methodica descriptio insectorum agri fridrichsdalensis cum characteribus genericis et specificis, nominibus trivialibus, locis natalibus, iconibus allegatis, novisve pluribus speciebus additis.* Hafniae et Lipsiae: Officina libraria Io. Frid. Gleditschii, xxiv + 96 pp.

Müller, O.F. 1788: *Zoologiae Danicae prodromus, seu animalium Daniae et Norvegicae indigenarum characteres, nomina, et synonyma imprimis popularium.* Hafniae: Hallager, 32 + 282 pp.

Müller, P.W.J. 1805: Bemerkungen über die Fussgliederzahl einiger Käfergattungen, in Beziehung auf Illiger's Abhandlungen über diesen Gegenstanden in dem Magazin für Insektenk. 1.B. p. 285 und folg. *Magazin der Insektenkunde* 4: 197–219.

Müller, P.W.J. 1821: Neue Insecten. *Magazin der Entomologie* (Halle) 4: 184–230.

Mulsant, E. 1830: *Lettres à Julie sur l'entomologie (en prose et en vers), suivies d'une description méthodique de la plus grande partie des insectes de France. Volume* 2. Lyon: Babeuf, 402 pp.

Mulsant, E. 1851: Description d'un genre et d'un coléoptère nouveaux dans la famille des cistéliens. *Mémoires del'Academie Société de Lyon* 1: 201–204.

Mulsant, E. 1852: Description d'un genre et d'un coléoptère nouveaux dans la famille des cistéliens. *Opuscules Entomologiques* 1: 68–71.

Mulsant, E. 1854: *Histoire naturelle des coléoptères de France. Latigènes.* Paris: L. Maison, x + 396 + 2 pp.

Mulsant, E. 1856a: *Histoire naturelle des coléoptères de France. Latipennes.* Paris: L. Maison, [4] + 44 + [5] pp.

Mulsant, E. 1856b: *Histoire naturelle des coléoptères de France. Pectinipèdes.* Paris: L. Maison, [6] + 96 pp.

Mulsant, E. 1856c: *Histoire naturelle des coléoptères de France. Barbipalpes.* Paris: L. Maison, [8] + 116 pp.

Mulsant, E. 1856d: *Histoire naturelle des coléoptères de France. Longipèdes.* Paris: L. Maison, [8] + 172 pp., 2 pls.

Mulsant, E. 1856e: Notes relatives à quelques insectes coléoptères de la tribu des pectinipèdes. *Opuscules Entomologiques* 7: 17–59.

Mulsant, E. 1856f: Histoire des coléoptères de France. Barbipalpes, longipèdes, latipennes. *Annales de la Société Linnéenne de Lyon* (N.S.) 3: 193–544.

Mulsant, E. 1857a: *Histoire naturelle de coléoptères de France. Vésicants.* Paris: Magnin & Blanchard et Cie., 201 pp., 1 pl. + [9 pp. suppl.].

Mulsant, E. 1857b: [new taxon]. In: Mulsant E. & Rey C.: Description de quelques coléoptères de la tribu des longipèdes. *Annales de la Société Linnéenne de Lyon* (N.S.) 4: 410–413.

Mulsant, E. 1858a: Tableau méthodique des coléoptères angustipennes de France. *Annales de la Société Linnéenne de Lyon* (N.S.) **5**: 65–238.

Mulsant, E. 1858b: Description d'un coléoptère nouveau de la tribu des vésicants. *Annales de la Société Linnéenne de Lyon* (N.S.) **8**: 239–242.

Mulsant, E. 1859: *Histoire naturelle des coléoptères de France. Rostrifères*. Paris: Magnin et Blanchard et Cie, [8] + 56 pp.

Mulsant, E. & Godart, A. 1855a: Description d'une espèce nouvelle d'Helops. *Opuscules Entomologiques* **6**: 83–86.

Mulsant, E. & Godart, A. 1855b: Description d'une espèce nouvelle d'Helops. *Annales de la Société Linnéenne de Lyon* (N.S.) **2**: 153–156.

Mulsant, E. & Godart, A. 1855c: Description de quelques coléoptères nouveaux ou peu connus. *Opuscules Entomologiques* **6**: 161–183.

Mulsant, E. & Godart, A. 1855d: Description de quelques coléoptères nouveaux ou peu connus. *Annales de la Société Linnéenne de Lyon* (N.S.) **2**: 261–283.

Mulsant, E. & Godart, A. 1856: Espèce nouvelle de coléoptère du genre Orchesia. *Annales de la la Société Linnéenne de Lyon* (N.S.) **3**: 108–110.

Mulsant, E. & Godart, A. 1859a: Description de trois espèces nouvelles de coléoptères. *Opuscules Entomologiques* **9**: 177–183.

Mulsant, E. & Godart, A. 1859b: Description de trois espèces nouvelles de coléoptères. *Annales des Sciences Physiques et Naturelles, d'Agriculture et d'Industrie publiées par la Société Impériale d'Agriculture, etc. de Lyon* (3) **3**: 241–252.

Mulsant, E. & Godart, A. 1860: Description d'une espèce nouvelle de coléoptère angustipenne. *Annales de la Société Linnéenne de Lyon* (N.S.) **7**: 158–161.

Mulsant, E. & Godart, A. 1866: Description de quelques coléoptères nouveaux ou peu connus. *Annales de la Société Linnéenne de Lyon* (N.S.) **12**: 447–456.

Mulsant, E. & Godart, A. 1868: Description de deux espèces nouvelles d'Alphitobius (Coléoptères de la tribu des latigènes, famille des ulomiens). *Annales de la Société Linnéenne de Lyon* (N.S.) **16**: 288–291.

Mulsant, E. & Godart, A. 1870: Description de trois espèces nouvelles de coléoptères. *Opuscules Entomologiques* **14**: 49–53.

Mulsant, E. & Godart, A. 1876a: Description d'une espèce nouvelle de coléoptères latigènes servant à former un genre nouveau. *Annales de la Société Linnéenne de Lyon* (N.S.) **22** [1875]: 181–183.

Mulsant, E. & Guillebeau, F. 1855a: Notes pour servir à l'histoire des lagries (Insectes coléoptères hétéromères). *Annales de la Société Linnéenne de Lyon* (N.S.) **2**: 65–74.

Mulsant, E. & Guillebeau, F. 1855b: Description d'une espèce nouvelle d'Uloma (Coléoptère de la tribu des latigènes). *Annales de la Société Linnéenne de Lyon* (N.S.) **2**: 421–424.

Mulsant, E. & Guillebeau, F. 1857: Description d'une espèces nouvelle du genre Orchesia de la tribu des coléoptères brevipalpes. *Annales de la la Société Linnéenne de Lyon* (N.S.) **4**: 414–416.

Mulsant, E. & Mulsant, V. 1855: Description d'une espèce nouvelle de genre mycétochare (Coléoptère de la tribu des pectinipèdes). *Opuscules Entomologiques* **6**: 155–157.

Mulsant, E. & Revelière, E. 1860: Description d'un coléoptère nouveau constituant un genre nouveau dans la tribu des opatates. *Annales de la Société Linnéenne de Lyon* (N.S.) 7: 153–157.

Mulsant, E. & Rey, C. 1853a: Description de quelques coléoptères nouveaux ou peu connus. *Annales de la Société Linnéenne de Lyon* (N.S.) 1 [1852–1853]: 2–9.

Mulsant, E. & Rey, C. 1853b: Description de quelques coléoptères nouveaux ou peu connus. *Opuscules Entomologiques* 2: 4–11.

Mulsant, E. & Rey, C. 1853c: Essai d'une division des derniers mélasomes. *Opuscules Entomologiques* 4: 1–235 + [4] pp., pls I–IV.

Mulsant, E. & Rey, C. 1853d: Essai d'une division des derniers mélasomes. *Mémoires de l'Académie des Sciences, Belles-Lettres et Arts de Lyon, Classe des Sciences* (2) 2 [1852]: 226–329.

Mulsant, E. & Rey, C. 1853e: Essai d'une division des derniers mélasomes. *Mémoires de l'Académie des Sciences, Belles-Lettres et Arts de Lyon, Classe des Sciences* (2) 3: 20–151 + [4], pls I–IV.

Mulsant, E. & Rey, C. 1853f: Description d'un coléoptère constituant un genre nouveau parmi les taxicornes. *Opuscules Entomologiques* 2: 185–188.

Mulsant, E. & Rey, C. 1854a: Essai d'une division des derniers mélasomes. *Opuscules Entomologiques* 5: 9–255.

Mulsant, E. & Rey, C. 1854b: Essai d'une division des derniers mélasomes. *Mémoires de l'Académie Impériale des Sciences, Belles-Lettres et Arts de Lyon. Classe des Sciences* (N.S.) 4: 155–332.

Mulsant, E. & Rey, C. 1855a: Description des omocrates et des héliopathes derniers insectes de la tribu des pandarites. *Annales de la Société Linnéenne de Lyon* (N.S.) 2: 77–142.

Mulsant, E. & Rey, C. 1855b: Description d'une nouvelle espèce de Pandarinus. *Opuscules Entomologiques* 6: 94–96.

Mulsant, E. & Rey, C. 1855c: Description d'une nouvelle espèce de Pandarinus (Coléoptère de la tribu des pandarites). *Annales de la Société Linnéenne de Lyon* (N.S.) 2: 194–197.

Mulsant, E. & Rey, C. 1856: Description d'une espèce nouvelle de coléoptères de la tribu des latigènes. *Opuscules Entomologiques* 7: 15–16.

Mulsant, E. & Rey, C. 1858a: Coup d'oeil sur les insectes de la famille des cantharidiens accompagné de la description de diverses espèces nouvelles ou peu connues. *Mémoires de l'Académie Impériale des Sciences, Belles-Lettres et Arts de Lyon. Classe des Sciences* (N.S.) 8: 122–220, 3 pls.

Mulsant, E. & Rey, C. 1858b: Description d'une espèce nouvelle de Phaleria (Coléoptères latigène). *Annales des Sciences Physiques et Naturelles, d'Agriculture et d'Industrie publiées par la Société Impériale d'Agriculture, etc. de Lyon* (3) 2: 310–312.

Mulsant, E. & Rey, C. 1858c: Description d'une espèce constituant un genre nouveau dans la famille des mordelliens. *Annales des Sciences Physiques et Naturelles, d'Agriculture et d'Industrie publiées par la Société Impériale d'Agriculture, etc. de Lyon* (3) 2: 313–315.

Mulsant, E. & Rey, C. 1859a: Description d'une nouvelle espèce de Phaleria (Coléoptères latigènes). *Opuscules Entomologiques* 9: 52–54.

Mulsant, E. & Rey, C. 1859b: Essai d'une division des derniers mélasomes. *Mémoires de l'Académie des Sciences, Belles-Lettres et Arts de Lyon. Classe des Sciences* (N.S.) 9: 62–155.

Mulsant, E. & Rey, C. 1859c: Essai d'une division des derniers mélasomes. *Opuscules Entomologiques* **9**: 65–137.

Mulsant, E. & Rey, C. 1859d: Description de quelques coléoptères nouveaux. *Opuscules Entomologiques* **9**: 159–176.

Mulsant, E. & Rey, C. 1859e: Essai d'une division des derniers mélasomes. Familles des parvilabres quatrième tribu opatrites. *Opuscules Entomologiques* **10**: 1–160.

Mulsant, E. & Rey, C. 1859f: Description de quelques coléoptères nouveaux. *Annales des Sciences Physiques et Naturelles, d'Agriculture et d'Industrie, publiées par du Société Impériale d'Agriculture, etc., de Lyon* (3) **3**: 222–240.

Mulsant, E. & Rey, C. 1860: Essai d'une division des derniers mélasomes. Famille des parvilabres quatrième tribu opatrites. *Mémoires de l'Académie des Sciences, Belles-Lettres et Arts de Lyon. Classe des Sciences* (N.S.) **10**: 1–68.

Mulsant, E. & Rey, C. 1861: Description de deux coléoptères nouveaux ou peu connus. *Opuscules Entomologiques* **12**: 189–192.

Mulsant, E. & Rey, C. 1866a: *Histoire naturelle des coléoptères de France. Colligères. Volume 17*. Paris: F. Savy, 4 + 188 pp., 3 pls.

Mulsant, E. & Rey, C. 1866b: Histoire naturelle des coléoptères de France. Tribu des colligères. *Annales de la Société Linnéenne de Lyon* (N.S.) **3**: 89–282.

Mulsant, E. & Wachanru, A. 1853: Première série de coléoptères nouveaux ou peu connus. *Mémoires de l'Académie des Sciences, Belles-Lettres et Arts de Lyon, Classe des Sciences* **2** [1852]: 1–17.

Munster, T.G. 1921: Insecta, ex Sibiria meridionali et Mongolia, in itinere Ørjan Olsen 1914 collecta. A. Coleoptera, a Fritz Jensen lecta. I. Oedemeridae. *Norsk Entomologisk Tidsskrift* **1**: 8–10.

Munster, T.G. 1924: Nova etc. ex Norvegia. *Norsk Entomologisk Tidsskrift* **1**: 288–294.

Munster, T.G. 1926: Tillaeg og bemaerkninger til norges koleopterfauna. *Norsk Entomologisk Tidsskrift* **2** (3): 158–168 [cont. **2** (4) (1927): 169–200, **2**(5) (1928): 262–298].

Munster, T.G. 1928: Tillaeg og bemaerkningen til norges koleopterfauna. *Norsk Entomologisk Tidsskrift* **2**: 262–298.

Nabeel, A.M., Gazang, T.O. & Hazim, S.A. 2016: A New species of comb-clawed beetles Cteniopus Solier, 1835 (Coleoptera: Alleculidae) from Erbil Governorate Kurdistan Region-Iraq. *Kirkuk University Journal, Scientific Studies* **11**: 214–221.

Nabozhenko, M.V. 2000: Novye vidy zhukov-chernotelok roda Cylindronotus Fald. (Coleoptera, Tenebrionidae) s Severnogo Kavkaza. *Entomologicheskoe Obozrenie* **79**: 107–111.

Nabozhenko, M.V. 2001a: Taxonomic notes on the genus Zophohelops Reitter, 1901with description of a new species from Tadzhikistan and a new genus Pseudoprobaticus gen. nov. *Annales Zoologici* **51**: 511–515.

Nabozhenko, M.V. 2001b: O sisteme zhukov-chernotelok tribi Helopini i obzor rodov Nalassus Mulsant i Odocnemis Allard (Coleoptera, Tenebrionidae) evropeyskoy chasti SNG i Kavkaza. *Entomologicheskoe Obozrenie* **80**: 627–668.

Nabozhenko, M.V. 2002a: Novyy rod zhukov-chernotelok triby Helopini (Coleoptera, Tenebrionidae). *Vestnik Zoologii* **36**: 41–46.

Nabozhenko, M.V. 2002b: Zhuki-chernotelki rodov Hedyphanes Fischer i Entomogonus Solier (Coleoptera, Tenebrionidae, Helopini) Kavkaza. *Entomologicheskoe Obozrenie* 81: 684–692.

Nabozhenko, M.V. 2005: New synonymies and new species of the genus Hedyphanes Fischer von Waldheim, 1822 (Coleoptera, Tenebrionidae, Helopini). *Acta Zoologica Academiae Scientarum Hungaricae* 51: 349–355.

Nabozhenko, M.V. 2006a: Obzor iranskikh vidov podroda Helopocerodes Reitter, 1922 roda Nalassus Mulsant, 1854 (Coleoptera: Tenebrionidae). *Trudy Russkogo Entomologicheskogo Obshestva* 77: 245–249.

Nabozhenko, M.V. 2006b: Revisiya roda Catomus Allard, 1876 i sblizhayemykh s nim rodov (Coleoptera, Tenebrionidae) fauny Kavkaza, Sredney Azii i Kitaya. *Entomologicheskoe Obozrenie* 85: 798–857.

Nabozhenko, M.V. 2007: Taxonomic notes on the genus Catomus Allard, 1876 (Coleoptera Tenebrionidae: Helopini) with description of a new species from Afghanistan. *Caucasian Entomological Bulletin* 3: 41–45.

Nabozhenko, M.V. 2008: New species of the genus Zophohelops Reitter, 1902 (Coleoptera: Tenebrionidae) with review of the species of Northern Tien-Shan. *Caucasian Entomological Bulletin* 4: 295–301.

Nabozhenko, M.V. 2008b. Review of the genus Pseudoprobaticus Nabozhenko, 2001 (Coleoptera: Tenebrionidae). *Annales Zoologici* 58: 721–724.

Nabozhenko, M.V. 2008c: Tenebrionidae: Blaptini. New nomenclatural and taxonomic acts and comments. Pp. 35–36. In: Löbl, I. & A. Smetana (eds.): *Catalogue of Palaearctic Coleoptera. Vol. 5. Tenebrionoidea.* Stenstrup: Apollo Books, 670 pp.

Nabozhenko, M.V. 2008d: Tenebrionidae: Helopini. New nomenclatural and taxonomic acts, and comments. Pp. 36–38. In: Löbl, I. & A. Smetana (eds.): *Catalogue of Palaearctic Coleoptera. Vol. 5 Tenebrionoidea.* Stenstrup: Apollo Books, 670 pp.

Nabozhenko, M.V. 2009: A new species and new records of darkling beetles of the tribe Helopini (Coleoptera, Tenebrionidae) from Iran and Armenia. *Caucasian Entomological Bulletin* 5: 185–187.

Nabozhenko, M.V. 2010: Contribution to the knowledge of the subgenus Helopondrus Reitter, 1922 of the genus *Nalassus* Mulsant, 1854 (Coleoptera: Tenebrionidae) of Iran. *Caucasian Entomological Bulletin* 6: 51–55.

Nabozhenko, M.V. 2011a: New species of the genus Armenohelops Nabozhenko, 2002 (Coleoptera: Tenebrionidae: Helopini) from Turkey. *Caucasian Entomological Bulletin* 7: 135–138.

Nabozhenko, M.V. 2011b: Two new species of the genus Nalassus Mulsant, subgenus Helopondrus Reitter (Coleoptera: Tenebrionidae) from Turkey. *Stuttgarter Beiträge zur Naturkunde A, Neue Serie* 4: 263–267.

Nabozhenko, M.V. 2011c: Two new species of the tribe Helopini (Coleoptera: Tenebrionidae) from Artvin Province, Turkey. *Annales Zoologici* 61: 335–338.

Nabozhenko, M.V. 2013a: New taxa of the genus Nalassus Mulsant, 1854 from Georgia. *Caucasian Entomological Bulletin* 9: 261–264.

Nabozhenko, M.V. 2013b: Taxonomic notes on the genera Hedyphanes Fischer von Waldheim, 1820 and Entomogonus Solier, 1848 (Coleoptera: Tenebrionidae) of Turkey. *Journal of Insect Biodiversity* 1: 1–3.

Nabozhenko, M.V. 2014: New darkling beetles of the tribe Helopini (Coleoptera: Tenebrionidae) from Iran. *Caucasian Entomological Bulletin* 10: 237–241.

Nabozhenko, M.V. 2015a: Review of the genus Cylindrinotus Faldermann, 1837 (Coleoptera: Tenebrionidae: Helopini). *The Coleopterists Bulletin* **69** (mo4): 101–114.

Nabozhenko, M.V. 2015b: To the knowledge of Helopini (Coleoptera: Tenebrionidae) of Morocco. *Caucasian Entomological Bulletin* **11**: 33–37.

Nabozhenko, M.V. 2015c: New species of the genus Adelphinus Fairmaire et Coquerel, 1866 (Coleoptera: Tenebrionidae) from Iran. *Caucasian Entomological Bulletin* **11**: 363–365.

Nabozhenko, M.V. 2018a: Review of the genus Hedyphanes Fischer von Waldheim, 1822 (Coleoptera: Tenebrionidae: Helopini) of Kazakhstan, Middle Asia, Iran and Afghanistan. *Entomological Review* **98**: 594–628.

Nabozhenko, M.V. 2018b: [new taxon]. In: Nabozhenko M.V., Özgen I. & Ivanushenko Yu. 2018: A new species of the genus Entomogonus Solier, 1848 (Coleoptera: Tenebrionidae) from Eastern Anatolia. *Zootaxa* **4441**: 549–554.

Nabozhenko, M.V. 2018c: [new taxa]. In: Nabozhenko, M.V. & Chigray, I.A.: New and little known species of Alleculini (Coleoptera: Tenebrionidae: Alleculinae): extinct from Eocene Baltic Amber and extant from Lebanon. *Caucasian Entomological Bulletin* **14**: 171–176.

Nabozhenko, M.V. 2019a: [new taxa]. In: Nabozhenko, M. V & Purchart, L.: Socotraphanes – a new endemic genus of darkling beetles (Coleoptera: Tenebrionidae) from Socotra Island. *Annales Zoologici* **69**: 149–157.

Nabozhenko, M.V. 2019b: New and little known species of lichenophagous Tenebrionidae (Coleoptera) of the tribe Helopini from Iran. *Entomological Review* **99**: 1026–1034.

Nabozhenko, M.V. & Abdurakhmanov, G.M. 2007: Review of the genus Nalassus Mulsant, 1854 (Coleoptera, Tenebrionidae) in Daghestan. *Caucasian Entomological Bulletin* **3**: 187–191

Nabozhenko, M.V. & Abdurakhmanov, M. 2009: Review of tenebrionid beetles of the genus Calyptopsis Solier, 1835 (Coleoptera: Tenebrionidae) of the Northern Caucasus. *Yug Rossii: Ecologiya, Razvitiye* **1**: 79–84.

Nabozhenko, M.V. & Ando, K. 2018: Subtribal, generic and subgeneric composition of darkling beetles of the tribe Helopini (Coleoptera: Tenebrionidae) in the Eastern Palaearctic. *Acta zoologica Academia Scientiarum Hungaricae* **64**: 277–327.

Nabozhenko, M.V. & Dzambazishvili, M. Ya. 2001: Novye vidy zhukov-chernotelok roda Nalassus Muls. (Coleoptera, Tenebrionidae) iz Gruzii. *Entomologicheskoe Obozrenie* **80**: 90–95.

Nabozhenko, M.V., Ferrer, J., Kalashian, M. & Abdurakhmanov, G. 2016: Contribution to the knowledge of darkling beetles of the tribe Ceratanisini (Coleoptera: Tenebrionidae) from the Caucasus and Anatolia. *Annales Zoologici* **66**: 607–620.

Nabozhenko, M.V. & Grimm, R. 2018: New and little known species of the tribe Helopini (Insecta: Coleoptera: Tenebrionidae) from Iran. *Veröffentlichungen des Naturkundemuseums Erfurt* **37**: 309–315.

Nabozhenko, M.V. & Grimm, R. 2019: New species and new records of darkling beetles of the tribe Helopini (Coleoptera: Tenebrionidae) from the Western Palaearctic. *Caucasian Entomological Bulletin* **15**: 107–116.

Nabozhenko, M.V. & Gurgenidze, L.N. 2006: Description of the larva of Cylindrinotus gibbicollis Faldermann, 1837 and notes on the classification of the subtribe Cylindrinotina Español, 1956 (Coleoptera, Tenebrionidae: Helopini). *Kavkazskiy Entomologichesky Bulleten* **2**: 79–82.

Nabozhenko, M.V. & Ivanov, S.N. 2015: A new Nalassus Mulsant, 1854 (Coleoptera: Tenebrionidae: Helopini), the first representative of the genus from the Russian Far East. *Zootaxa* **3955**: 137–141.

Nabozhenko, M. V. & Ivanov, S. N. 2018: Novy vid i podrod roda Toxicum Latreille, 1802 (Coleoptera, Tenebrionidae, Toxicini) s Dal'nego Vostoka Rossii. *Zoologicheskii Zhurnal* **97**: 545–551. [English version: A new species and subgenus of the genus Toxicum Latreille, 1802 (Coleoptera, Tenebrionidae, Toxicini) from the Russian Far East. *Entomological Review* **98**: 892–898].

Nabozhenko, M.V. & Keskin, B. 2009: Two new species of the genus Gunarus Des Gozis, 1886 (Coleoptera: Tenebrionidae: Helopini) from Southern Turkey. *Zootaxa* **2170**: 53–60.

Nabozhenko, M.V. & Keskin, B. 2010: Novyy rod i vid zhukov-chernotelok triby Helopini (Coleoptera: Tenebrionidae) iz Turtsii. *Entomologicheskoe obozrenie* **89**: 840–844. [English version: A new genus and species of darkling beetles of the tribe Helopini (Coleoptera, Tenebrionidae) from Turkey. *Entomological Review* **90**: 1215–1218].

Nabozhenko, M.V. & Keskin, B. 2014: New data about „nalassoid" genera from south-eastern Anatolia with description of a new species of Zophohelops (Coleoptera: Tenebrionidae). *Acta Entomologica Musei Nationalis Pragae* **54**: 243–249.

Nabozhenko, M.V. & Keskin, B. 2016: Revision of the genus Odocnemis Allard, 1876 (Coleoptera: Tenebrionidae: Helopini) from Turkey, the Caucasus and Iran with observations on feeding habits. *Zootaxa* **4202**: 1–97.

Nabozhenko, M.V. & Keskin, B. 2017: Taxonomic review of the genus Helops Fabricius, 1775 (Coleoptera: Tenebrionidae) of Turkey. *Caucasian Entomological Bulletin* **13**: 41–49.

Nabozhenko, M.V. & Lillig, M. 2013: A new subgenus and species of the genus Hedyphanes Fischer von Waldheim, 1820 (Coleoptera: Tenebrionidae: Helopini) from Israel and Egypt. *Zootaxa* **3641**: 188–192.

Nabozhenko, M.V., Keskin, B. & Alpagut Keskin N. 2016: Taxonomic review of the genus Armenohelops Nabozhenko, 2002 (Coleoptera: Tenebrionidae) with additional support of the mitochondrial COI gene sequences. *Kavkazskiy Entomologichesky Bulleten* **12**: 255–268.

Nabozhenko, M.V. & Medvedev, G.S. 2006: [new taxon]. In: Nabozhenko M.V.: Revisiya roda Catomus Allard, 1876 i sblizhayemykh s nim rodov (Coleoptera, Tenebrionidae) fauny Kavkaza, Sredney Azii i Kitaya. *Entomologicheskoe Obozrenie* **85**: 798–857.

Nabozhenko, M.V. & Nabozhenko, S.V. 2015: A new species of Odocnemis Allard, 1876 (Coleoptera: Tenebrionidae) from Greece. *Caucasian Entomological Bulletin* **11**: 367–368.

Nabozhenko, M.V., Nikitsky, N.B. & Keskin, B. 2017: Taxonomic review of the genus Euboeus s. str. Boieldieu, 1865 (= Probaticus s. str. Seidlitz, 1896, syn. n.) (Coleoptera, Tenebrionidae). *Zootaxa* **4358**: 494–506.

Nabozhenko, M.V. & Sadeghi, S. 2017: Foranotum perforatum gen. et sp. nov. – a new troglobitic darkling beetle (Coleoptera: Tenebrionidae: Kuhitangiinae: Foranotini trib. nov.) from a cave in Southern Zagros, Iran. *Zootaxa* **4338**: 163–172.

Nabozhenko, M.V. & Tichý, V. 2006: A new species of the genus Odocnemis Allard, 1876 (Coleoptera, Tenebrionidae) from Turkey. *Kavkazskiy Entomologichesky Bulleten* **2**: 183–185.

Nabozhenko, M.V. & Tichý, V. 2011: Two new species of the genus Entomogonus Solier, 1848 (Coleoptera: Tenebrionidae) from Turkey. *Caucasian Entomological Bulletin* **7**: 45–49.

Nabozhenko, M.V. & Tichý, V. 2019: Darkling beetles of the tribe Helopini (Coleoptera: Tenebrionidae) of Jordan. *Far Eastern Entomologist* **384**: 1–9.

Nabozhenko, M. & Ren, G.-D. 2018: [new taxa]. In: Yuan, C.-X., Nabozhenko, M. & Ren, G.-D.: Contribution to the knowledge of the genus Ainu Lewis, 1894 (Coleoptera: Tenebrionidae: Stenochiinae). *Journal of Asia-Pacific Entomology* **21**: 1292–1298.

Naezen, D.E. 1794: Beskrifning pa nagrsa vid Umea funde Insekter dels okande, dels forut otydeligen bemarkte, och i Fauna Suecica ej uptagne. *Kongliga Vetenskaps Academiens nya Handlingar* **15**: 256–274, pl. 1.

Nakakita, T. 1987: Two new species of the genus Addia from Taiwan (Coleoptera, Tenebrionidae). *Entomological Review of Japan* **42** *Supplement*: 61–64.

Nakane, T. 1949: Descriptions of two new species of mordellid-beetles from Japan. *Mushi* **20**: 39–41.

Nakane, T. 1950: New or little known Coleoptera from Japan and its adjacent regions V. *Transactions of Kansai Entomological Society* **15**: 20–24.

Nakane, T. 1954: New or little-known Coleoptera from Japan and its adjacent regions, XI. Oedemeridae. *The Scientific Reports of the Saikyo University* (A) **1**: 171–188.

Nakane, T. 1955: New or little-known Coleoptera from Japan and its adjacent regions. XII. *The Scientific Reports of the Saikyo University* (A) **2**: 24–42.

Nakane, T. 1956a: Entomological results from the scientific survey of the Tokara islands. *Publication of the Osaka Municipal Museum of Natural History* **9**: 1–10.

Nakane, T. 1956b: New and little-known Coleoptera from Japan and its adjacent regions. XIII. *The Scientific Reports of the Saikyo University* (A) **2** (3): 164–174.

Nakane, T. 1957: Einige Mordelliden aus Japan. *Insecta Matsumurana* **21**: 48–52.

Nakane, T. 1958: On the Coleoptera of Shimokita Peninsula, northern end of Honshu, Japan (Insecta) I. *Miscellaneous Reports of the Research Institute for Natural Resources* **46–47**: 83–92 (in Japanese, with English title and summary).

Nakane, T. 1960: Einige Bemerkungen über die japanischen Mordelliden. *Entomological Review of Japan* **12**: 17–19.

Nakane, T. 1963a: New or little known Coleoptera from Japan and its adjacent regions, XIX. *Fragmenta Coleopterologica* **6–7**: 26–30.

Nakane, T. 1963b: New or little-known Coleoptera from Japan and its adjacent regions. *The Scientific Reports of Kyoto prefectural University* (*Natural Science and Living Science*) **3** (5): A 221–A 226.

Nakane, T. 1966: Coleoptera from Afghanistan. *Results of the Kyoto University Scientific Expedition to the Karakorum and Hindukush, 1955*. Vol. **8**: 231–248.

Nakane, T. 1968a: New or little known Coleoptera from Japan and its adjacent regions, XXVI. *Fragmenta Coleopterologica* **19**: 75–76.

Nakane, T. 1968b: New or little known Coleoptera from Japan and its adjacent regions, XXVII. *Fragmenta Coleopterologica* **19–21**: 76–85.

Nakane, T. 1968c: New or little-known Coleoptera from Japan and its adjacent regions. XXVIII. *Fragmenta Coleopterologica* **21**: 85–86.

Nakane, T. 1973a: Notes on the synonymy and on some types of Japanese Coleoptera in certain European collections II: Heteromera (1) (Insecta). *Bulletin of the National Science Museum* (Tokyo) **16**: 1–8.

Nakane, T. 1973b: Notes on some species of Tenebrionidae from the Yaeyama Islands (Insecta: Coleoptera). *Memoirs of the National Science Museum* (Tokyo) **6**: 103–108, pl. 11.

Nakane, T. 1975: On the heteromerous Coleoptera occurring in Kyushu including Yakushima and Tanegashima. *Memoirs National Science Museum* (Tokyo) **8**: 161–172, pl. 16.

Nakane, T. 1977: Notes on some species of Coleoptera (Insecta) occurring in the Ogasawara and the Volcano Islands, with descriptions of new species. *Memoirs of the National Science Museum* (Tokyo) **10**: 147–162 (in Japanese, with title and descriptions in English).

Nakane, T. 1978: On some interesting species of Coleoptera from Izu and Agasawara Islands, with descriptions of new species. *Memoirs of the National Science Museum* (Tokyo) **11**: 155–161 (in Japanese, with title and descriptions in English).

Nakane, T. 1979: New or little-known Coleoptera from Japan and its adjacent regions. XXXI. *Fragmenta Coleopterologica* **25/28**: 112–114.

Nakane, T. 1983a: On the Eurygeniini in Japan (Anthicidae). *Kita-Kyūshū no Konchū* **30** (1): 1–6, pl.

Nakane, T. 1983b: New or little-known Coleoptera from Japan and its adjacent regions. XXXI. *Fragmenta Coleopterologica* **35/37**: 139–150.

Nakane, T. 1984: On the Coleoptera occurring in Yaku-shima Island. Pp. 587–631. Conservation Reports of the Yaku-shima Wilderness Area, Kyushu, Japan.

Nakane, T. 1987: Notes on some little known beetles (Coleoptera) in Japan. *Kita-Kyūshū no Konchū* **34**: 171–176, 1 pl.

Nakane, T. 1988a: Notes on some little-known beetles (Coleoptera) in Japan, 2. *Kita-Kyūshū no Konchū* **35**: 1–6, 1 pl. (in Japanese, with English title and descriptions).

Nakane, T. 1988b: Notes on some little-known beetles (Coleoptera) in Japan. *Kita-Kyūshū no Konchū* **35**: 77–83, pl. 9 (in Japanese, with English title and descriptions).

Nakane, T. 1988c: New or little-known Coleoptera from Japan and its adjacent regions. XXXVIII. *Fragmenta Coleopterologica* **38–40**: 153–164.

Nakane, T. 1989: Notes on some little-known beetles (Coleoptera) in Japan. *Kita-Kyūshū no Konchū* **36**: 1–10.

Nakane, T. 1990: Notes on some little-known beetles (Coleoptera) in Japan. 6. *Kita-Kyūshū no Konchū* **37**: 61–68, 1 pl.

Nakane, T. 1991: Notes on some little known bettles (Coleoptera) in Japan. 7. *Kita-Kyūshū no Konchū* **38**: 1–9, pl. 1 (in Japanese, with English title and descriptions).

Nakane, T. 1994a: Notes on some little-known beetles (Coleoptera) in Japan. 12. *Kita-Kyūshū no Konchū* **41**: 81–86 (in Japanese, with English title and descriptions).

Nakane, T. 1994b: On the species of the genus Lagria Fabricius occurring in the main islands of Japan (Coleoptera, Lagriidae). *Essa-Konchū dokokai Tokubetsu hokoku* (2): 241–246 (in Japanese, with English title and descriptions).

Nakane, T. 1997: Notes on some little-known beetles (Coleoptera) in Japan. 14. *Kita-Kyūshū no Konchū* **44**: 1–6 (in Japanese, with English title and descriptions).

Nakane, T. & Hayashi, N. 1955: Melandryidae. Pp. 77–82. In: *Coloured illustrations of the insects of Japan. Coleoptera*. The Kinki Coleopterologigal Society, [xii] + 196 + [19] pp., 64 pls.

Nakane, T. & Masumoto, K. 1979: [new taxon]. In: New or little-known Coleoptera from Japan and its adjacent regions. XXXI. *Fragmenta Coleopterologica* **25/28**: 112–114.

Nakane, T. & Nobuchi, A. 1955: On a new genus and six new species of ciid-beetles from Japan (Ciidae Coleoptera). *Scientific Report of Saikyo University. Natural Science and Living Science* (A) **2** (1): 47–52.

Nakane, T. & Nobuchi, A. 1956: Entomological results from the scientific survey of the Tokara Islands. II. Coleoptera: Ciidae. *Scientific Report of Saikyo University. Mathematics and Natural Science*(A) **2** (2): 107–108.

Nakane, T. & Nomura, S. 1950: New or little known Coleoptera from Japan and its adjacent regions. I. *Transactions of the Kansai Entomological Society* **15**: 1–17.

Nakane, T. & Nomura, S. 1957: Eine neue Hoshihananomia-Art aus Japan. *Entomological Review of Japan* **8**: 3–4.

Nardi, G. 1995: Tenuicomus velox bucciarellii nomen novum per Tenuicomus velox relictus (Coleoptera, Anthicidae). *Bollettino dell'Associazione Romana di Entomologia* **49** [1994]: 161–163.

Nardi, G. 2004: [new names]. In: Chandler D.S., Nardi G. & Telnov D.: Nomenclatural notes on the Palaearctic Anthicidae (Coleoptera). *Mitteilungen des Internationalen Entomologischen Vereins e. V.* **29**: 109–173.

Nardi, G. 2008: Miscellaneous notes on World Anthicidae. *Zootaxa* **1779**: 1–32.

Nedelkov, N. 1905: Prinos kam entomologichnata fauna na Bulgariya [A contribution to the entomological fauna of Bulgaria]. *Priodichesko Spisanije ma Bulgaskoto Knizhovno Druzhestvo v Sofia* (17) **66**: 404–439.

Neervoort van de Poll, J.R.H. 1886: A new species of the Heteromerous genus Leiochrinus, Westwood. *Notes from the Leyden Museum* **8**: 34.

Nevinson, B.G. 1889: On a new Pimelia brought by Mr. Joseph Thomson from Morocco. *The Entomologist's Monthly Magazine* **25**: 255.

Newman, E. 1838: Entomological notes. *The Entomological Magazine* **5**: 168–181, 202–205, 372–402, 483–500.

Nicolas, J.-P. 1977: Description de Marolia leseigneuri n. sp. (Col. Melandryidae). *Bulletin de la Société Linnéenne de Lyon* **46**: 281–283.

Nietner, J. 1856: Entomological papers – being descriptions of new Ceylon Coleoptera, with such observations on their habits as appear any way interesting. *Journal of the Asiatic Society of Bengal* **25**: 381–394, 523–539.

Nietner, J. 1857a: Entomological papers; being chiefly descriptions of new Ceylon Coleoptera, with such observations on their habits, &c., as appear in any way interesting. *Journal of the Ceylon Branch of the Royal Asiatic Society* **3** (9): 1–75.

Nietner, J. 1857b: Descriptions of new Ceylon Coleoptera. *The Annals and Magazine of Natural History* (2) **9**: 374–388.

Nikitsky, N.B. 1972: Novye i maloizvestnye vidy zhukov-tenelubov (Coleoptera, Melandryidae) iz yuzhnogo Primorya razvivauschsiesia v drevesine. *Entomologicheskoe Obozrenie* **51**: 203–207.

Nikitsky, N.B. 1973a: Novyy vid zhukov-golovachey roda Nematoplus LeC. (Coleoptera, Cephaloidae) iz yuzhnogo Primorya [A new species of the genus Nematoplus LeC. (Coleoptera, Cephaloidae) from the South Maritim Territory]. *Zoologicheskiy Zhurnal* **52**: 609–612.

Nikitsky, N.B. 1973b: Novyy vid zhukov-tenelubov roda Enchodes (Coleoptera, Melandryidae) iz yuzhnogo Primorya. *Zoologicheskiy Zhurnal* **52**: 1728–1730.

Nikitsky, N.B. 1974: Melandrya fasciata n. sp. du sud de l'Extrême-Orient (Coleoptera, Melandryidae). *Bulletin de la Société d'Entomologique Mulhouse* **1974**: 61–62.

Nikitsky, N.B. 1983: Un nouveau genre de la famille de Melandryidae (Coleoptera) de Chine. *Bulletin de la Société d'Entomologique Mulhouse* **1983**: 37–40.

Nikitsky, N.B. 1984: Vidy rodov Wanachia i Abdera (Coleoptera, Melandryidae) vostochnoy Palearktiki. *Zoologicheskiy Zhurnal* **63** (5): 677–681.

Nikitsky, N.B. 1985a: Zhuki semeyst Tetratomidae i Cephaloidae vostochnoy palearktiki. *Sbornik Trudov Zoologicheskogo Muzea MGU* 23: 26–37.

Nikitsky, N.B. 1985b: Vidy roda Dircaea F. (Coleoptera, Melandryidae) vostochnoy Palearktiki. *Byulleten' Moskovskogo Obshchestva Ispytateley Prirody, Otdel Biologicheskiy* (N.S.) 90 (4): 53–64.

Nikitsky, N.B. 1985c: The genera Mycetoma Mulsant, 1856 and Orchesia Latreille, 1807 in the Eastern Palaearctic (Insecta: Coleoptera: Tetratomidae, Melandryidae). *Senckenbergiana Biologica* 65: 265–277.

Nikitsky, N.B. 1985d: Zhuki-teneluby roda Melandrya F. (Coleoptera, Melandryidae) vostochnoy Palearktiki. *Entomologicheskoe Obozrenie* 64: 748–759.

Nikitsky, N.B. 1986: Semeystvo Pilipalpidae stat. nov. (Coleoptera, Heteromera), ego sostav i sistematicheskoie Svyazi [The family Pilipalpidae stat. n. (Coleoptera, Heteromera), its composition and taxonomic relationships]. *Zoologicheskiy Zhurnal* 65: 1178–1189.

Nikitsky, N.B. 1988a: Vidy roda Phryganophilus (Coleoptera, Melandryidae) Palearktiki. *Zoologicheskiy Zhurnal* 67: 1426–1430.

Nikitsky, N.B. 1988b: Novye taksony zhukov-griboedov (Col. Mycetophagidae) i Salpingidae Dalnego Vostoka SSSR [New taxa of Coleoptera of the families Mycetophagidae and Salpingidae from the Far East of the USSR]. *Zoologicheskiy Zhurnal* 67: 1845–1853.

Nikitsky, N.B. 1989a: Zhestkokrylye semeystv Tetratomidae i Melandryidae Dalnego Vostoka SSSR. Pp. 3–87. In: Shatalkin, A.I. (ed.): *Nasekomye v ecosistemakh Sibiri i Dalnego Vostoka*. Sbornik trudov Zoologicheskogo muzeja MGU. Vol. 27. Izdatelstvo Moskovskogo Universiteta.

Nikitsky, N.B. 1989b: Novye taksony zhukov-griboedov (Col. Mycetophagidae) i Tetratomidae Dalnego Vostoka SSSR. *Zoologicheskiy Zhurnal* 68: 56–64.

Nikitsky, N.B. 1992a: A new Microscapha from the Nepal Himalayas (Coleoptera, Melandryidae). *Mitteilungen der Münchener Entomologischen Gesellschaft* 82: 83–85.

Nikitsky, N.B. 1992b: Sem. Mycetophagidae. Pp. 406–423. Sem. Tetratomidae. Pp. 426–435. In: Ler, P.A. (ed.): *Opredelitel nasekomykh Dalnego Vostoka SSSR. Tom III. Zhestkokrylye, ili zhuki. Chast 2 [Keys to the identification of insects of the Soviet Far East. 3 – Coleoptera, or beetles. Part 2.]*. St. Petersburg: Nauka 704 pp.

Nikitsky, N.B. 1992c: Sem. Salpingidae. Pp. 482–493. In: Ler, P.A. (ed.): *Opredelitel' nasekomykh Dal'nego Vostoka SSSR. Tom 3 – Zhestkokrylye, ili zhuki. Chast 2 [Keys to the identification of insects of the Soviet Far East. 3 – Coleoptera, or beetles. Part 2.]*. St. Petersburg: Nauka, 704 pp.

Nikitsky, N.B. 1992d: Sem. Melandryidae. P. 659. In: Ler, P.A. (ed.): *Opredelitel nasekomykh Dal'nego Vostoka SSSR. Tom 3 – Zhestkokrylye, ili zhuki. Chast 2 [Keys to the identification of insects of the Soviet Far East. 3 – Coleoptera, or beetles. Part 2.]*. St. Petersburg: Nauka, 704 pp.

Nikitsky, N.B. 1994a: Vidy roda Ischalia (Coleoptera, Ischaliidae) dal'nego Vostoka Rossii. *Zoologicheskiy Zhurnal* 73: 33–38.

Nikitsky, N.B. 1994b: New species of the genus Lederina Nikitsky et Belov, 1982, stat. nov. from the Nepal Himalayas (Coleoptera, Melandryidae). *Entomofauna* 15: 345–352.

Nikitsky, N.B. 1996a: Novye zhestkokrylye (Coleoptera) iz Kitaya. *Zoologicheskiy Zhurnal* 75: 1366–1375.

Nikitsky, N.B. 1996b: Novyy vid roda Langelandia (Coleoptera, Colydiidae) s severo-zapadnogo Kavkaza. *Zoologicheskiy Zhurnal* 75: 1434–1435.

Nikitsky, N.B. 1998a: *Generic classification of the beetle family Tetratomidae (Coleoptera, Tenebrionoidea) of the world, with description of new taxa*. Sofia-Moscow: Pensoft, Series Faunistika (9), 80 pp.

Nikitsky, N.B. 1998b: [new taxa]. In: Nikitsky, N.B., Semenov, V.B. & Dolgin, M.M. (eds): *Zhestkokrylye – xylobionty, mycetobionty i plastinchatoyusye Prioksko-Terrasnogo Biosphernogo zapovednika (s obzorom fauny etikh grupp Moskovskoy oblasti). Dopolnenie 1 (s zamechaniami po nomenklature i systematike nekotorykh zhukov Melandryidae mirovoy fauny)*. Sbornik trudov Zoologicheskogo muzeja MGU. Vol. 36 (1). Moskva: Izdatelstvo Moskovskogo Universiteta, 60 pp.

Nikitsky, N.B. 2000: K poznaniyu zhestkokrylykh semeystva Synchroidae (Coleoptera, Tenebrionoidea) mirovoy fauny. *Zoologicheskiy Zhurnal* 78 (1): 42–48.

Nikitsky, N.B. 2001: Novye vidy zhukov-tenelubov (Coleoptera, Melandryidae). Pp. 19–25. In: *Ekologia, monitoring i ratsionalnoye prirodopolzovanie*. Vol. 307 (1). Moskva: MGUL.

Nikitsky, N.B. 2002: Lichinki zhukov-tenelubov (Coleoptera, Melandryidae) Rossii, s zametkami po taksonomii. *Byulleten' Moskovskogo Obshchestva Ispytateley Prirody, Otdel Biologicheskiy* (N.S.) **107** (1): 8–30.

Nikitsky, N.B. 2003: Novyy vid roda Pseudotryphillus Reitter, 1880 (Coleoptera, Mycetophagidae) iz Nepala. *Byulleten' Moskovskogo Obshchestva Ispytateley Prirody, Otdel Biologicheskiy* (N.S.) **108** (3): 49–50.

Nikitsky, N.B. 2005: Zhestkokrylye podsemeystva Penthinae Lacordaire, 1859 (Coleoptera, Tenebrionoidea, Tetratomidae) mirovoi fauny. *Byulleten' Moskovskogo Obshchestva Ispytateley Prirody, Otdel Biologicheskiy* (N.S.) **110** (5): 16–26.

Nikitsky, N.B. 2007: Vidy roda Holostrophus Horn, 1888 (Coleoptera, Tetratomidae, Eustrophinae, Holostrophini) mirovoy fauny. *Byulleten' Moskovskogo Obshchestva Ispytateley Prirody, Otdel Biologicheskiy* (N.S.) **112**: 13–30.

Nikitsky, N.B. 2007: Novye taksony zhukov-gryboedov i zhukov-tenelubov (Coleoptera, Mycetophagidae, Melandryidae). *Bulleten' Moscovskogo Obschestva Ispytateley Prirody, Otdel Biologicheskiy* (N.S.) **112**: 58–62.

Nikitsky, N.B.M.V. 2008a: Mycetophagidae. New nomenclatural and taxonomic acts, and comments. P. 28. In: Löbl, I. & A. Smetana (eds.): *Catalogue of Palaearctic Coleoptera. Vol. 5. Tenebrionoidea*. Stenstrup: Apollo Books, 670 pp.

Nikitsky, N.B.M.V. 2008b: Melandryidae. New nomenclatural and taxonomic acts, and comments. Pp. 28–29. In: Löbl, I. & A. Smetana (eds.): *Catalogue of Palaearctic Coleoptera. Vol. 5. Tenebrionoidea*. Stenstrup: Apollo Books, 670 pp.

Nikitsky, N.B. 2016: A new species of the genus Tetratoma Fabricius (Coleoptera, Tetratomidae) from China. *Zootaxa* **4154**: 346–350.

Nikitsky, N.B. & Belov, V.V. 1982: The false darkling beetle genus Lederia Rtt. (Coleoptera, Melandryidae). *Folia Entomologica Hungarica* (N.S.) **43**: 111–123.

Nikitsky, N.B. & Belov, V.V. 1983: Palearkticheskie vidy roda Salpingus Ill. (Col. Salpingidae) [Palaearctic species of the genus Salpingus (Coleoptera, Salpingidae)]. *Zoologicheskiy Zhurnal* **62** (4): 516–529.

Nikitsky, N.B. & Saitô, M. 2014: Description of a new species of the genus Zilora (Coleoptera, Melandryidae) from Central Honshu, Japan, with a provisional key to species of the world. *Elytra, New Series* **4**: 205–211.

Nobuchi, A. 1955a: Studies on the ciid-beetles from Japan (I) With the descriptions of a new genus and some new species. *Entomological Review of Japan* **6**: 53–58.

Nobuchi, A. 1955b: Studies on the ciid-beetles from Japan (II). *Akitu* **4**: 105–108.

Nobuchi, A. 1959: Two new species of ciid-beetles. *Entomological Review of Japan* **10**: 51–52.

Nobuchi, A. 1960a: Ciidae from Hokkaido (Coleoptera). *Entomological Review of Japan* **11**: 37–41, pl. 5.

Nobuchi, A. 1960b: On a new species of Cis Latreille from Japan (Coleoptera, Ciidae). *Akitu* **9** (3–4): 65–66.

Nobuchi, A. & Wada, Y. 1955a: [new taxon]. In: Nobuchi A. 1955a: Studies on the ciid-beetles from Japan (I) with the descriptions of a new genus and some new species. *Entomological Review of Japan* **6**: 52–58.

Nobuchi, A. & Wada, Y. 1955b: [Cis mikagensis sp. n.]. P. 106. In: Nobuchi A.: Studies on the ciid-beetles from Japan (II). *Akitu* **4**: 105–108.

Nobuchi, A. & Wada, Y. 1956: A new species of Japanese Xylographus (Ciidae Coleoptera). *Entomological Review of Japan* **7**: 35–36.

Nobuchi, A. & Wada, Y. 1959: [new taxon]. In: Nobuchi A.: Two new species of ciid-beetles. *Entomological Review of Japan* **10**: 51–52.

Nomura, S. 1951: Zur Kenntnis der Mordellistenini (Col. Mordellidae) aus Japan, Korea und Formosa. *Tôhô-Gakuhô* **1**: 41–70.

Nomura, S. 1957: Mordellid- and elmid-beetles of Yakushima. *Entomological Review of Japan* **8**: 40–44.

Nomura, S. 1958: Zur Kenntnis der Gattung Mordella aus Japan und dessen Umgebung. *Tôhô-Gakuhô* **7**: 35–58.

Nomura, S. 1959: Notes on the Japanese Melandryidae (Coleoptera) . I. *Entomological Review of Japan* **10**: 43–45.

Nomura, S. 1960: Two new species of the genus Anthicus from Japan (Coleoptera: Anthicidae). *The Entomological Review of Japan* **11**: 47–48.

Nomura, S. 1961a: Two new Cteniopinus-species from Japan and Formosa (Alleculidae). *Entomological Review of Japan* **12**: 38–40 + pl. 6.

Nomura, S. 1961b: Some new species of Coleoptera from Japan. *Tôhô-Gakuhô* **11**: 70–89.

Nomura, S. 1962a: The anthicid-beetles from the Loochoo Islands. *Entomological Review of Japan* **14**: 1–7, pl. 1.

Nomura, S. 1962b: Some new and remarkable species of Coleoptera from Japan and its adjacent regions. *Tôhô-Gakuhô* **12**: 35–51.

Nomura, S. 1963a: [new taxa]. In: Nakane, T. Obhayashi, K., Nomura, S. & Kurosawa, Y. (eds): *Iconographia Insectorum Japonicum. Vol. 2 (Coleoptera)* Tokyo: Hokuryukan, 443 pp., 192 pls (in Japanese).

Nomura, S. 1963b: Some new species of the Coleoptera from Loochoo Is. and its adjacent regions. *Entomological Review of Japan* **15**: 37–42, pl. 5.

Nomura, S. 1964: Some new species of the Coleoptera from Loochoo Is. and its adjacent regions, II. *Entomological Review of Japan* **17**: 47–57, pl. 3.

Nomura, S. 1966: Mordellid-Fauna of the Loochoo Islands, with descriptions of some new forms. *Entomological Review of Japan* **18**: 41–53.

Nomura, S. 1967a: The Mordellidae from Formosa. *Entomological Review of Japan* **19**: 5–34.

Nomura, S. 1967b: Some new and remarkable species of the Coleoptera from Japan and its adjacent regions, II. *Entomological Review of Japan* **19**: 52–62.

Nomura, S. 1975: Mordellidae of the Bonin Islands. *Entomological Review of Japan* **28**: 29–45.

Nomura, S. & Katô, A. 1957: On the subgenus Pseudomordellistena Ermisch, from Japan, Formosa and south China. *Transactions of the Shikoku Entomological Society* **5**: 77–90.

Nomura, S. & Katô, A. 1958: A new subgenus and two new species of the family Melandryidae from Japan (Coleoptera). *Entomological Review of Japan* **9**: 1–3.

Nomura, S. & Katô, A. 1959: Descriptions of some new mordellid- and melandriid-species and notes on others. *Entomological Review of Japan* **10**: 5–9.

Nomura, S. & Katô, A. 1968: A new species of the genus Melandrya from Japan (Coleoptera, Melandryidae). *Entomological Review of Japan* **21**: 1–2

Nomura, S. & Nakane, T. 1959: A new species of the genus Pelecotomoides from Japan (Coleoptera, Rhipiphoridae). *Insecta Matsumurana* **22**: 115–118.

Nomura, S. & Yamazaki, H. 1960: A new species and two new subspecies of the genus Strongylium from Japan (Tenebrionidae). *Entomological Review of Japan* **12**: 14–16.

Normand, H. 1916: Nouveaux coléoptères de la faune tunisienne. *Bulletin de la Société Entomologique de France* **1916**: 283–287.

Normand, H. 1936a: Contribution au catalogue des coléoptères de la Tunisie (9e fascicule). *Bulletin de la Société d'Histoire Naturelle de l'Afrique du Nord* **27**: 81–100, 144–164.

Normand, H. 1936b: Contribution au catalogue des coléoptères de la Tunisie, 10e fascicule. *Bulletin de la Société d'Histoire Naturelle de l'Afrique du Nord* **27**: 355–383.

Normand, H. 1938a: Contribution au catalogue des coléoptères de la Tunisie. Premier supplément. *Bulletin de la Société d'Histoire Naturelle de l'Afrique du Nord* **29**: 274–308.

Normand, H. 1938b: Contribution au catalogue des coléoptères de la Tunisie. 13e fascicule. *Bulletin de la Société d'Histoire Naturelle de l'Afrique du Nord* **29**: 337–371.

Normand, H. 1939: Contribution au catalogue des coléoptères de la Tunisie, 14e fascicule. Deuxième supplément. *Bulletin de la Société d'Histoire Naturelle de l'Afrique du Nord* **29** [1938]: 481–515, pl. 20.

Normand, H. 1950a: Contribution au catalogue des coléoptères de la Tunisie. Troisième Supplement (Fascicule 4). *Bulletin de la Société des Sciences Naturelles de Tunisie* **2** [1949]: 79–104, pls. 6–7.

Normand, H. 1950b: Nouveaux coléoptères du Nord de l'Afrique. *Revue Française d'Entomologie* **17**: 105–108.

Normand, H. 1955: Heliotaurus nouveaux de Tunisie. *Bulletin de la Société des Sciences Naturelles de Tunisie* **8**: 85–88.

Novák, V. 2005: Three new species of the genus Borboresthes Fairmaire, 1897 (Coleoptera: Tenebrionidae: Alleculinae). *Studies and Reports of District Museum Prague-East, Taxonomical Series* (1–2) **1**: 121–132.

Novák, V. 2006a: New species of the genus Mycetocharina Seidlitz, 1891 (Coleoptera: Tenebrionidae: Alleculinae) from Iran. *Studies and reports of District Museum Prague-East, Taxonomical Series* **2**: 85–98.

Novák, V. 2006b: Megischina bozdaglariensis sp. n. (Coleoptera: Tenebrionidae: Alleculinae) from Turkey. *Studies and reports of District Museum Prague-East, Taxonomical Series* **2**: 99–104.

Novák, V. 2006c: Hymenophorus evae sp. nov. and H. gerdae sp. nov. (Coleoptera: Tenebrionidae: Alleculinae) from Iran. *Acta Societatis Zoologicae Bohemicae* **69**: 317–325.

Novák, V. 2007a: New species of the genus Hymenalia Mulsant, 1856 (Coleoptera: Tenebrionidae: Alleculinae) from Palaearctic region. *Studies and reports of District Museum Prague-East, Taxonomical series* **3**: 149–170.

Novák, V. 2007b: New Alleculinae (Coleoptera: Tenebrionidae) species from Nepal. *Veröffentlichungen Naturkundesmuseum Erfurt* **26**: 233–240.

Novák, V. 2007c: A new genus and three new species of Alleculinae (Coleoptera: Tenebrionidae) from Socotra Island, Yemen. *Fauna of Arabia* 23: 319–334.

Novák, V. 2008a: New Alleculinae from China (Coleoptera: Tenebrionidae). *Vernate* 27: 207–220.

Novák, V. 2008b: Mycetocharina (Alleculopsis) bahukalatensis sp. nov. (Coleoptera: Tenebrionidae: Alleculinae) from Iran. *Acta Entomologica Musei Nationalis Pragae* 48: 73–78.

Novák, V. 2008c: *Order Coleoptera, family Tenebrionidae*. Pp. 257–263. In: Van Harten, A. (ed.): *Arthropod fauna of the United Arab Emirates, Volume 1*. Abu Dhabi: Dar Al Ummah Printing, Publishing, Distribution & Advertising. 754 pp.

Novák, V. 2008d: Tenebrionidae: Alleculinae. New nomenclatural and taxonomic acts, and comments. Pp. 38–39. In: Löbl, I. & A. Smetana (eds.): *Catalogue of Palaearctic Coleoptera. Vol. 5. Tenebrionoidea*. Stenstrup: Apollo Books, 670 pp.

Novák, V. 2009a: New species of *Isomira* from Nepal and China (Insecta: Coleoptera: Tenebrionidae: Alleculinae). *Vernate* 28: 363–376.

Novák, V. 2009b: Cteniopinus (Lechinius) holzschuhi sp. n. from Bhutan (Coleoptera: Tenebrionidae: Alleculinae: Cteniopodini). Studies *and reports of District Museum Prague-East, Taxonomical series* 5: 249–256.

Novák, V. 2010a: Review of Hymenalia species (Coleoptera: Tenebrionidae: Alleculinae) from China. *Studies and Reports, Taxonomical Series* 6: 190–231

Novák, V. 2011a: [new taxa]. In: Novák, V., Jansson, N., Avci, M., Sarikaya, O., Coskun, M., Atay, E. & Gurkan, T.: New Allecula species (Coleoptera: Tenebrionidae: Alleculinae) from Turkey. *Studies and Reports, Taxonomical Series* 7: 335–346.

Novák, V. 2011b: New genera of Alleculinae (Coleoptera: Tenebrionidae) from Oriental region Part III. Microsthes gen. nov. *Studies and Reports, Taxonomical Series* 7: 319–334.

Novák, V. 2011c: Gonodera gilanica n. sp. (Coleoptera: Tenebrionidae: Alleculinae: Gonocerini) from Iran. *Stuttgarter Beiträge zur Naturkunde A, Neue Serie* 4: 185–190.

Novák, V. 2011d: Revision of the genus Paracistela Borchmann, 1941 (Coleoptera: Tenebrionidae: Alleculinae). *Studies and Reports, Taxonomical Series* 7: 347–382.

Novák, V. 2012a: [new taxa]. In: Novák, V., Jansson, N., Chiari, S., Zauli, A., Audisio, P. & Carpaneto, G.M.: A new species of Allecula (Coleoptera: Tenebrionidae: Alleculinae) from cork oak stands of Italy. *Zootaxa* 3483: 29–38.

Novák, V. 2012b: New „yellow" Borboresthes (Coleoptera: Tenebrionidae: Alleculinae) species from China and Oriental Region. *Studies and Reports, Taxonomical Series* 8: 227–267.

Novák, V. 2012c: New genera of Alleculinae (Coleoptera: Tenebrionidae: Alleculinae) from Palaearctic and Oriental Regions. *Studies and Reports, Taxonomical Series* 8: 269–293.

Novák, V. 2013a: New genera of Alleculinae (Coleoptera: Tenebrionidae: Alleculinae: Gonoderina) from Palaearctic and Oriental Regions. Part II – Kralia gen. nov. *Studies and Reports, Taxonomical Series* 9: 499–508.

Novák, V. 2013b: Review of the West Palaearctic Pseudocistela with description of P. hajeki sp. nov. from Iran (Coleoptera: Tenebrionidae: Alleculinae). *Acta Entomologica Musei Nationalis Pragae* 53: 293–301.

Novák, V. 2013c: [new taxa]. In: Novák, V., Avci, M., Jansson, N., Sarikaya, O., Atay, E., Kayis, T., Coskun, M. & Aitar, F.: A New Mycetochara Species (Coleoptera: Tenebrionidae: Alleculinae) from Turkey. *Journal of the Entomological Research Society* 15: 51–58.

Novák, V. 2013d: Megischina renei sp. nov. (Coleoptera: Tenebrionidae: Alleculinae) from Rhodes, Greece. *Acta Societatis Zoologica Bohemica* 77: 265–270.

Novák, V. 2014a: New genera of Alleculinae (Coleoptera: Tenebrionidae) from the Palaearctic Region. Part I – Borbonalia gen. nov. *Studies and Reports, Taxonomical Series* **10**: 135–159.

Novák, V. 2014b: Revision of the genus Liodocistela Pic, 1930 stat. nov. (Coleoptera: Tenebrionidae: Alleculinae). *Studies and Reports, Taxonomical Series* **10**: 527–552.

Novák, V. 2014c: New species of the genus Microsthes Novák, 2011 (Coleoptera: Tenebrionidae: Alleculinae) from Palaearctic and Oriental Regions. *Studies and Reports, Taxonomical Series* **10**: 553–576.

Novák, V. 2014d: New species of Isomira Mulsant, 1856 (Coleoptera: Tenebrionidae: Alleculinae) from Nepal and China – Part II. *Stuttgarter Beiträge zur Naturkunde A, Neue Serie* 7: 153–161.

Novák, V. 2015a: New genera of Alleculinae (Coleoptera: Tenebrionidae) from Palaearctic and Oriental Regions. Part III – Bobina gen. nov. *Studies and Reports, Taxonomical Series* **11**: 123–141.

Novák, V. 2015b: New Borboresthes species (Coleoptera: Tenebrionidae: Alleculinae) from Palaearctic and Oriental Regions. *Folia Heyrovskyana, Series A* **22** [2014]: 74–98.

Novák, V. 2015c: New genera of Alleulinae (Coleoptera: Tenebrionidae) from Palaearctic Region. Part II. Chitwania gen. nov. *Folia Heyrovskyana, Series A* **23**: 90–95.

Novák, V. 2015d: New genera of Alleculinae (Coleoptera: Tenebrionidae) from Palaearctic and Oriental Regions. Part IV – Gerdacula gen. nov. *Studies and Reports, Taxonomical Series* **11**: 143–158.

Novák, V. 2015e: New Hymenalia species (Coleoptera: Tenebrionidae: Alleculinae) from China and Oriental Region. *Studies and Reports, Taxonomical Series* **11**: 371–389.

Novák, V. 2015f: Contribution to the knowledge of Hymenalia badia species group from the Palaearctic Region (Coleoptera: Tenebrionidae: Alleculinae). *Folia Heyrovskyana, Series A* **23**: 71–88.

Novák, V. 2015g: New genera of Alleulinae (Coleoptera: Tenebrionidae: Alleculinae) from Palaearctic and Oriental Regions. Part V – Mycetocula gen. nov. *Folia Heyrovskyana, Series A* **23**: 77–89.

Novák, V. 2015h: Review of the genus Upinella Mulsant, 1856 stat. nov. (Coleoptera: Tenebrionidae: Alleculinae). *Folia Heyrovskyana, Series A* **23**: 89–111.

Novák, V. 2015i: [new taxa]. In: Novák, V., Abacigil, O., Varli, S.V. & Jansson, N.: Mycetochara kazdagiica sp. nov. from Turkey (Coleoptera: Tenebrionidae: Alleculinae: Mycetocharini) *Folia Heyrovskyana, Series A* **22** [2014]: 134–141.

Novák, V. 2016a: Allecula olexai sp. nov. (Coleoptera: Tenebrionidae: Alleculinae) from Abkhazia. *Acta Societatis Zoologica Bohemica* **80**: 123–126.

Novák, V. 2016b: New Borboresthes species (Coleoptera: Tenebrionidae: Alleculinae) from China and Oriental Region. *Folia Heyrovskyana, Series A* **24**: 25–43.

Novák, V. 2016c: New genera of Alleculinae (Coleoptera: Tenebrionidae) from Palaearctic and Oriental Regions VI – Loricula gen. nov. *Folia Heyrovskyana, Series A* **24**: 44–56.

Novák, V. 2016d: Mycetocharina (Alleculopsis) gafsaica sp. nov. (Coleoptera: Tenebrionidae:Alleculinae) from Tunisia. *Acta Societatis Zoologica Bohemica* **80**: 117–121.

Novák, V. 2016e: Review of the genus Asiomira (Dubrovina, 1973) stat. nov. (Coleoptera: Tenebrionidae: Alleculinae: Gonoderini). *Studies and Reports, Taxonomical Series* **12**: 177–191.

Novák, V. 2016f: New species of Pseudohymenalia Novák, 2008 (Coleoptera: Tenebrionidae: Alleculinae: Gonoderina). *Studies and Reports, Taxonomical Series* **12**: 193–218.

Novák, V. 2016g: New Podonta Solier species (Coleoptera: Tenebrionidae: Alleculinae) from the Palaearctic Region. *Studies and Reports, Taxonomical Series* 12: 449–470.

Novák, V. 2017a: New species of Allecula Fabricius, 1801 (Coleoptera: Tenebrionidae: Alleculinae) from Palaearctic and Oriental Regions. *Folia Heyrovskyana, Series A* 25: 20–40.

Novák, V. 2017b: Alogista kadleci sp. nov. and genus Synallecula Kolbe stat. nov. with S. borchmanni sp. nov. from the Palaearctic Region (Coleoptera: Tenebrionidae: Alleculinae: Alleculini). *Folia Heyrovskyana, Series A* 25: 123–134.

Novák, V. 2017c: New Anthracula Fairmaire (Coleoptera: Tenebrionidae: Alleculinae) species from Palaearctic Region. *Folia Heyrovskyana, Series A* 25: 41–55.

Novák, V. 2017d: New species and nomenclatory acts in Alleculini (Coleoptera: Tenebrionidae: Alleculinae) from the Palaearctic Region. *Studies and Reports, Taxonomical Series* 13: 429–446.

Novák, V. 2017e: Review of the subgenus Rhinobarus Reitter, 1906 (Coleoptera: Tenebrionidae: Alleculinae: Cteniopodini: Cteniopus) with descriptions of two new species from the Palaearctic Region. *Folia Heyrovskyana, Series A* 25: 135–146.

Novák, V. 2018a: New species, new combination and distributional data in Alleculini Laporte, 1840 (Coleoptera: Tenebrionidae: Alleculinae) from the Palaearctic Region. *Folia Heyrovskyana, Series A* 26: 57–70.

Novák, V. 2018b: New Borboresthes (Coleoptera: Tenebrionidae: Alleculinae: Alleculini) species from Southeast of China. *Folia Heyrovskyana, Series A* 26: 16–56.

Novák, V. 2018c: New genera of Alleculinae (Coleoptera: Tenebrionidae) from Palaearctic and Oriental Regions VII – Fifina gen. nov. *Studies and Reports, Taxonomical Series* 14: 469–477.

Novák, V. 2018d: A Review of Cistelomorpha L. Redtenbacher, 1868 species with bicolour dorsal surface (Coleoptera: Tenebrionidae: Alleculinae: Cteniopodini) with a new species from Nepal and Oriental Region. *Folia Heyrovskyana, Series A* 26: 35–80.

Novák, V. 2018e: New species, new combination and new distributional data in Cteniopodini Solier 1835 (Coleoptera: Tenebrionidae: Alleculinae) from the Palaearctic Region. *Folia Heyrovskyana, Series A* 26: 81–94.

Novák, V. 2019a: New species of the genus Cteniopinus Seidlitz, 1896 (Coleoptera: Tenebrionidae: Alleculinae: Cteniopodini). *Folia Heyrovskyana, Series A* 27: 105–127.

Novák, V. 2019b: Two new species of Cistelopsis Fairmaire, 1896 (Coleoptera: Tenebrionidae: Alleculinae: Alleculini) from the Palaearctic Region. *Folia Heyrovskyana, Series A* 27: 63–69.

Novák, V. 2019c: New genera of Alleculinae (Coleoptera: Tenebrionidae) from Palaearctic and Oriental Regions. Part XI – Oracula gen. nov. *Folia Heyrovskyana, Series A* 27: 52–86.

Novák, V. 2019d: New genera of Alleculinae (Coleoptera: Tenebrionidae) from Palaearctic and Oriental Regions IX – Psis gen. nov. *Folia Heyrovskyana, Series A* 27: 70–80.

Novák, V. 2019e: New genera of Alleculinae (Coleoptera: Tenebrionidae) from Palaearctic and Oriental Regions. Part X – Spinecula gen. nov. *Studies and Reports, Taxonomical Series* 15: 435–457.

Novák, V. 2019f: Contribution to the knowledge of the genus Upinella Mulsant (Coleoptera: Tenebrionidae Alleculinae: Alleculini) with description of a new species and a Thornella subgen. nov. and Tibinella subgen. nov. *Folia Heyrovskyana, Series A* 27: 81–104.

Novák, V. 2019g: New genera of Alleculinae (Coleoptera: Tenebrionidae: Alleculinae: Alleculini) from Palaearctic and Oriental Regions VIII – Zizu gen. nov. *Studies and Reports, Taxonomical Series* 15: 185–203

Novák, V. 2019h: Four new Isomira (Coleoptera: Tenebrionidae: Alleculinae: Alleculini) species from Iran, Jordan, and Malta, along with notes on another species of the genus. *Folia Heyrovskyana, Series A* **27**: 87–104.

Novák, V. 2019i: New species and new distributional data on Cteniopodini from Palaearctic Region (Coleoptera: Tenebrionidae: Alleculinae). *Studies and Reports, Taxonomical Series* **15**: 459–473.

Novák, V. 2019j: New Borbonalia species (Coleoptera: Tenebrionidae: Alleculinae) from China. *Acta Societatis Zoologicae Bohemicae* **82** [2018]: 165–176

Novák, V. & Purchart, L. 2012: New species of the genus Socotralia and first record of the genus Alogista from Socotra Island (Coleoptera: Tenebrionidae: Alleculinae). *Acta Entomologica Musei Nationalis Pragae* **52** *Supplementum*: 323–336.

Novák, V. & Nabozhenko, M. 2019: A new species of *Cornucistela* Campbell, 1980 (Coleoptera: Tenebrionidae: Alleculinae: Gonoderina) from Iran. *Zootaxa* **4612**: 267–274.

Nowicki, M. 1873: *Beschreibung neuer Käferarten*. Krakau: Jagellonische Universitäts Buchdrückerei, 6 pp.

Nyholm, T. 1953: Bemerkungen über einige schwedische Cis-Arten (Col., Ciidae). *Entomologisk Tidskrift* **74**: 187–196.

Obenberger, J. 1914: Beitrag zur Kenntnis der palaearktischen Käferfauna. (Fortsetzung und Schluss). *Coleopterologische Rundschau* **3**: 97–115, 129–144.

Obenberger, J. 1917: II. Beitrag zur Kenntnis der palaearktischen Käferfauna. *Archiv für Naturgeschichte* A **82** (4) [1916]: 9–45, 2 pls.

Obenberger, J. 1918: Vier neue palearktische Coleopteren. *Entomologische Blätter* **14**: 58–61.

Obenberger, J. 1922: II. Beitrag zur Kenntnis der palaearktischen Käferfauna. *Archiv für Naturgeschichte* A **82** (4) [1917]: 9–45, pls. 2.

Oberthür, R. 1903: Une nouvelle Asida appartenant au groupe des espèces espagnoles aterciopeladas. *Boletín de la Sociedad Española de Historia Natural* **1903**: 74–75.

Obregón, R. & Verdugo, A. 2012: Alphasida (Alphasida) subbaetica, nueva especie de Asidini de las Sierras Subbéticas de Córdoba, España (Coleoptera: Tenebrionidae). *Boletín de la Sociedad Andaluza de Entomología* **20**: 50–75.

Odnosum, V.K. 1990: Novyy vid zhuka-gorbatki roda Mordellistena. *Novosti faunistiki i systematiky* **1990**: 54–55.

Odnosum, V.K. 2000: Mordellid beetles of the genus Stenalia (Coleoptera, Mordellidae) of Central and Eastern Palaearctics. Communication 1. *Vestnik Zoologii* **34**: 37–50.

Odnosum, V.K. 2001: Mordellid beetles of the genus Stenalia (Coleoptera, Mordellidae) of Central and Eastern Palaearctics. Communication 2. *Vestnik Zoologii* **35**: 21–26.

Odnosum, V.K. 2002: Mordellid beetles (Coleoptera, Mordellidae) in the fauna of Kazakhstan and Middle Asia. Communication 1. *Vestnik Zoologii* **36**: 13–23.

Odnosum, V.K. 2003: Mordellid beetles (Coleoptera, Mordellidae) in the fauna of Kazakhstan and Middle Asia. Communication 2. *Vestnik Zoologii* **37**: 33–49.

Odnosum, V.K. 2005: Mordellistena dolini sp. n., a new mordellid beetle of the gemellata group. *Vestnik Zoologii* **39**: 53–54.

Ogloblin, D.A. 1950: [new taxa]. In: Ogloblin, D.A. & Znojko, D.V.: *Fauna SSSR, Zhestkokrylye, Tom 18, vyp. 8. Pylceedy (Cem. Alleculidae), ch. 2, podsem. Omophlinae*. Moskva, Leningrad: Akademiya nauk SSSR, 133 pp.

Ohbayashi, N. & Tôyama, M. 1994: Two new pyrochroid species (Coleoptera, Pyrochroidae) from Taiwan and Thailand. *Japanese Journal of Entomology* 62: 146–150.

Okada, K. 2005: Occurrence of the genus Colobicones Grouvelle (Coleoptera, Zopheridae, Colydiinae) in Japan, with descriptions of two new species. *Elytra* 33: 425–431.

Okamoto, H. 1924: The insect-fauna of Quelpart Island (Saishuto). *Bulletin of the Agricultural Experimental Station Chosen, Korea* 1: 47–233, 4 pls., 1 map.

Olexa, A. 1955: Dircaea australis Fairm. var. imitata var. nov. *Bulletin de la Société d'Entomologique Mulhouse* 1955: 38–39.

Olivier, G.-A. 1789: *Encyclopédie méthodique, ou par ordre de matières; par une société de gens de lettres, de savans et d'artistes; précédée d'un vocabulaire universel, servant de table pour tout l'ouvrage, ornée des portraits de Mm. Diderot & d'Alembert, premiers éditeurs de l'Encyclopédie. Histoire naturelle. Insectes. Tome quatrième.* Paris: Panckoucke, 1–331 pp.

Olivier, G.-A. 1790: *Entomologie, ou histoire naturelle des insectes, avec leurs caractères génériques et spécifiques, leur description, leur synonymie, et leur figure enluminée. Coléoptères. Tome second.* Paris: de Baudoin, 485 + xix + [3] pp., 63 pls. [Note: genera 9–34, each genus with separate pagination].

Olivier, G.-A. 1791: *Encyclopédie méthodique, ou par ordre de matières; par une société de gens de lettres, de savans et d'artistes; précédée d'un vocabulaire universel, servant de table pour tout l'ouvrage, ornée des portraits de Mm. Diderot & d'Alembert, premiers éditeurs de l'Encyclopédie. Histoire naturelle. Insectes. Tome sixième. Pars I.* Paris: C.J. Panckoucke, 1–368 pp.

Olivier, G.-A. 1792: *Encyclopédie méthodique, ou par ordre de matières; par une société de gens de lettres, de savans et d'artistes; précédée d'un vocabulaire universel, servant de table pour tout l'ouvrage, ornée des portraits de Mm. Diderot & d'Alembert, premiers éditeurs de l'Encyclopédie. Histoire naturelle. Insectes. Tome sixième. Pars II.* Paris: C.J. Panckoucke, 369–704 pp.

Olivier, G.-A. 1793: *Encyclopédie méthodique, ou par ordre de matières; par une société de gens de lettres, de savans et d'artistes; précédée d'un vocabulaire universel, servant de table pour tout l'ouvrage, ornée des portraits de Mm. Diderot & d'Alembert, premiers éditeurs de l'Encyclopédie. Histoire naturelle. Insectes. Tome septième.* (1). Paris: Panckoucke, 1–368 pp.

Olivier, G.-A. 1795: *Entomologie, ou histoire naturelle des insectes, avec leurs caractères génériques et spécifiques, leur description, leur synonymie, et leur figure enluminée. Coléoptères. Tome troisième.* Paris: de Lanneau, 557 + xxviii pp., 65 pls. [Note: genera 35–65, each genus with separate pagination].

Olivier, G.-A. 1797: *Encyclopédie méthodique, ou par ordre de matières; par une société de gens de lettres, de savans et d'artistes; précédée d'un vocabulaire universel, servant de table pour tout l'ouvrage, ornée des portraits de Mm. Diderot & d'Alembert, premiers éditeurs de l'Encyclopédie. Histoire naturelle. Insectes. Tome septième.* (2). Paris: Panckoucke, 369–827 pp.

Olivier, G.-A. 1807: *Entomologie, ou histoire naturelle des insectes, avec leurs caractères génériques et spécifiques, leur description, leur synonymie et leur figure enluminée. Coléoptères. Tome cinqième.* Paris: Baudonin, 612 pp. [Note: each genus with separate pagination].

Olivier, G.-A. 1803: L´Oedemère fémorale (Oedemera) femoralis. P. 21. In: *Nouveau dictionaire d´histoire naturelle, appliqée aux arts, principalement à l´agriculture et à l´economie rurale et domestique. Tome XVI.* Paris: Société de naturalistes et d´agriculteurs, 575 pp.

Olivier, G.-A. 1811: *Encyclopédie méthodique, ou par ordre de matières; par une société de gens de lettres, de savans et d'artistes; précédée d'un vocubulaire universel, servant de table pour tout l'ouvrage, ornée des portraits de Mm. Diderot & d'Alembert, premiers éditeurs de l'Encyclopédie. Histoire naturelle. Insectes. Tome huitième*. Paris: H. Agasse, 722 pp.

Olivier, E. 1888: Description d'une nouvelle espèce de vésicants. *Bulletin de la Société Entomologique de France* **1888**: lxi-lxii.

Olliff, A.S. 1883a: Description of two larvae and new genera and species of clavicorn Coleoptera, and a synopsis of Helota, MacLeay. *Cistula Entomologica* **3**: 49–61.

Olliff, A.S. 1883b: Remarks on a small collection of Clavicorn Coleoptera from Borneo, with descriptions of new species. *Transactions of the Entomological Society of London* **1883**: 173–186.

Oromí, P. 1977: Los Arthrodeis Sol. de las Islas Canarias (Col. Tenebrionidae, Erodiini). *Vieraea* **7**: 3–22.

Oromí, P. 1978: Sobre los Crypticus (Col. Tenebrionidae) (Coleoptera) de Canarias: Descripción de C. gomerensis sp. n. *Boletín de la Asociación Española de Entomología* 7 (*Suppl. A*): 215–231.

Oromí, P. 1982: Los Tenebrioniodos de las Islas Canarias. – Inst. Estud. Canarias, 50 Aniversario 1932–1982: 267–292.

Oromí, P. 1984a: Helioctamenus occidentalis, un nuevo Colydiidae de Canarias (Coleoptera). *Vieraea* **13** [1983]: 225–232.

Oromí, P. 1984b: Nuevos aportiacones al conocimiento de la distribución de los coléopteros de Canarias. *Vieraea* **13** [1983]: 283–240.

Oromí, P. 1990: Una nueva especie del género Pimelia de la Isla de Grand Canaria (Coleoptera, Tenebrionidae). *Vieraea* **19**: 245–249.

Oromí, P. & Garcia, R. 1986: Canaricerus aeonii, un nuevo Orthocerini de Tenerife, Islas Canarias (Coleoptera, Colydiidae). *Nouvelle Revue d'Entomologie* (N.S.) **3**: 303–307.

Osculati, G. 1844: *Note d'un viaggio nella Persia e nelle India orientale negli anni 1841, 1842*. Monza: Luca Corbetta, 73 pp., 1 pl.

Pal, T.K. 1984: Two new species of Pseudendestes Lawrence (Coleoptera: Colydiidae) from India. *Bulletin of the Zoological Survey of India* **6**: 31–35.

Pal, T.K. 1992a: A new inopeplid species (Coleoptera: Cucujoidea) from Patkoi Hill Range of Arunachal Pradesh. *Journal of the Bombay Natural History Society* **89**: 238–241.

Pal, T.K. 1992b: Inopeplidae and elacatidae (Coleoptera: Heteromera) from Arunachal Pradesh, India. *Records of the Zoological Survey of India* **91**: 343–351.

Pal, T.K. 2003: Insecta: Coleoptera: Colydiidae (sensu lato). Pp. 163–176. In: *State Fauna Series* **9**, *Fauna of Sikkim* (*Part 3*).

Pal, T.K. & Ślipiński, S.A. 1984: Notes on the Nematidium Erichson (Coleoptera, Colydiidae) with description of new species. *Polskie Pismo Entomologiczne* **53**: 531–43.

Palisot de Beauvois, A.M.F.J. 1805–1821: *Insectes recueillis en Afrique et en Amérique, dans le royaumes d'Oware et de Benin, à Saint-Domingue et dans les États-Unis, pendant les années 1786–1797*. Paris: Levrault, Schoell et Cie, xvi + 224 pp., 34 + 20 + 9 +23 +3 +7+ 14 pls.

Pallas, P.S. 1771: *Reise durch verschiedene Provinzen des russischen Reichs. Erster Theil.* St. Petersburg: Kayserliche Akademie der Wissenschaften, [10] + 3–504 pp., 11 pls.

Pallas, P.S. 1773: *Reise durch verschiedene Provinzen des russischen Reichs. Zweyter Theil. Zweytes Buch vom Jahr 1771*. St. Petersburg: Kayserliche Akademie der Wissenschaften, pp. 371–744.

Pallas, P.S. 1781: *Icones Insectorum praesertim Rossiae Sibiriaeque peculiarum quae collegit et descritionibus illustravit. Fasciculus primus*. Erlangae: W. Walther, 1–56 pp.

Pallas, P.S. 1782: *Icones Insectorum praesertim Rossiae Sibiriaeque peculiarum quae collegit et descritionibus illustravit. Fasciculus secundus.* Erlangae: W. Waltheri, 57–96 pp., pls A-F.

Pallas, P.S. 1798: *Icones Insectorum praesertim Rossiae Sibiriaeque peculiarum quae collegit et descritionibus illustravit. Fasciculus tertius.* Erlangae: W. Waltheri, 97–104 pp., pls. G-H.

Palm, T. 1976: Eine neue Mordellistena-Art aus Tenerife. *Entomologisk Tidskrift* **97**: 31–33.

Pan, Z. & Bologna, M.A. 2013: [new taxa]. In: Pan, Z., Ren, G.D., Wang, X.P. & Bologna, M.A.: Revision of the genus Pseudabris Fairmaire (Coleoptera, Meloidae), an endemic to the Tibetan Plateau, with biogeographic comments. *Systematic Entomology* **38**: 134–150.

Pan, Z. & Bologna, M.A. 2014a: Taxonomy, Bionomics and Faunistics of the Nominate Subgenus of Mylabris Fabricius, 1775, with the description of five new species (Coleoptera: Meloidae Mylabrini). *Zootaxa* **3806**: 1–78.

Pan, Z. & Bologna, M.A. 2014b: [new taxa]. In: Pan, Z., Carosi, M. & Bologna, M.A.: A new Eastern Asian Hycleus and key to the Chinese species of the phaleratus group (Coleoptera, Meloidae, Mylabrini). *Zookeys* **463**: 11–19.

Pan, Z. & Ren, G.-D. 2018: Taxonomic revision of the subfamily Meloinae (Coleoptera: Meloidae) from Xizang, China, with description of a new species. *Zoological Systematics* **43**: 66–88.

Pan, P., Ren, G.-D. & Bologna, M.A. 2018. Longizonitis, a new nemognathine genus from the Himalayas (Coleoptera: Meloidae). *Zookeys* **765**: 43–59.

Pan, Z., Ren, G.-D. & Wang, X.-P. 2011. Taxonomy of the genus Hycleus Latreille (Coleoptera, Meloidae) from China, with description of one new species. *Acta Zootaxonomica Sinica* **36**: 179–197 (in Chinese with English summary).

Pan, Z., Wang, X.-P. & Ren, G.-D. 2010: A new subgenus and species of Mylabris Fabricius (Coleoptera: Meloidae) from China. *Entomological Review of Japan* **65**: 191–197.

Pandellé, L. 1867: [new taxon]. In: Grenier, A. (ed.): *Matériaux pour servir à la faune des coléoptères de France, recueillis et publiés par le Dr. A. Grenier. 2e Cahier*. Paris: A. Grenier, [1] + iv + 131–194 pp.

Panin, S. 1943: Supplément à la connaissance des coléoptères roumains (2-ème supplément). *Bulletin de la Section Scientifique de l'Académie Roumaine* **24** [1942]: 178–182.

Pankow, W. 1981: Insect of Saudi Arabia. Coleoptera: Fam. Mordellidae and Scraptiidae, Subfam. Anaspidinae. *Fauna of Saudi Arabia* **3**: 273–275.

Panzer, G.W.F. 1792a: *Faunae insectorum germanicae initia oder Deutschlands Insecten*. Heft 2. Norimbergae: Felsecker, 24 pp. + 24 pls.

Panzer, G.W.F. 1792b: *Faunae insectorum germanicae initia oder Deutschlands Insecten*. Heft 3. Norimbergae Felsecker, 24 pp. + 24 pls.

Panzer, G.W.F. 1793a: *Faunae insectorum germanicae initia oder Deutschlands Insecten*. Heft 6. Norimbergae: Felsecker, 24 pp. + 24 pls.

Panzer, G.W.F. 1793b: *Faunae insectorum germanicae initia oder Deutschlands Insecten*. Heft 8. Norimbergae: Felsecker, 24 pp. + 24 pls.

Panzer, G.W.F. 1793c: *Faunae insectorum germanicae initia oder Deutschlands Insecten*. Heft 9. Norinbergae: Felsecker, 24 pp. + 24 pls.

Panzer, G.W.F. 1793d: *Faunae insectorum germanicae initia oder Deutschlands Insecten*. Heft 10. Norinbergae: Felsecker, 24 pp. + 24 pls.

Panzer, G.W.F. 1793e: *Faunae insectorum germanicae initia oder Deutschlands Insecten*. Heft 16. Norinbergae: Felsecker, 24 pp. + 24 pls.

Panzer, G.W.F. 1794a: *Faunae insectorum germanicae initia oder Deutschlands Insecten*. Heft 22. Norinbergae: Felsecker, 24 pp. + 24 pls.

Panzer, G.W.F. 1794b: *Faunae insectorum germanicae initia oder Deutschlands Insecten*. Heft 23. Norinbergae: Felsecker, 24 pp. + 24 pls.

Panzer, G.W.F. 1794c: *Faunae insectorum germanicae initia oder Deutschlands Insecten*. Heft 24. Norinbergae: Felsecker, 24 pp. + 24 pls.

Panzer, G.W.F. 1795: *Deutschlands Insectenfaune oder entomologisches Taschenbuch für das Iahr 1795* [*Entomologia Germanica exhibens insecta per Germania indigena*]. Nürnberg: Felsecker, iv + 372 pp., 12 pls.

Panzer, G.W.F. 1796a: *Faunae insectorum germanicae initia oder Deutschlands Insecten*. Heft 31. Norinbergae: Felsecker, 24 pp. + 24 pls.

Panzer, G.W.F. 1796b: *Faunae insectorum germanicae initia oder Deutschlands Insecten*. Heft 34. Norinbergae: Felsecker, 24 pp. + 24 pls.

Panzer, G.W.F. 1796c: *Faunae insectorum germanicae initia oder Deutschlands Insecten*. Heft 35. Norinbergae: Felsecker, 24 pp. + 24 pls.

Panzer, G.W.F. 1796d: *Faunae insectorum germanicae initia oder Deutschlands Insecten*. Heft 36. Norinbergae: Felsecker, 24 pp. + 24 pls.

Panzer, G.W.F. 1796e: *Faunae insectorum germanicae initia oder Deutschlands Insecten*. Heft 37. Norinbergae: Felsecker, 24 pp. + 24 pls.

Panzer, G.W.F. 1796f: *Fauna insectorum germanicae initia oder Deutschlands Insecten*. Heft 38. Norinbergae: Felsecker, 24 pp. + 24 pls.

Panzer, G.W.F. 1797: *Faunae insectorum germanicae initia oder Deutschlands Insecten*. Heft 45. Norinbergae: Felsecker, 24 pp. + 24 pls.

Panzer, G.W.F. 1798: *Faunae insectorum germanicae initia oder Deutschlands Insecten*. Heft 57. Norinbergae: Felsecker, 24 pp. + 24 pls.

Panzer, G.W.F. 1799a: *Faunae insectorum germanicae initia oder Deutschlands Insecten*. Heft 61. Norinbergae: Felsecker, 24 pp. + 24 pls.

Panzer, G.W.F. 1799b: *Faunea insectorum germanicae initia oder Deutschlands Insecten*. Heft 67. Norinbergae: Felsecker, 24 pp. + 24 pls.

Panzer, G.W.F. 1799c: *Faunae insectorum germanicae initia oder Deutschlands Insecten*. Heft 68. Norinbergae: Felsecker, 24 pp. + 24 pls.

Panzer, G.W.F. 1800: *Faunae insectorum germanicae initia oder Deutschlands Insecten*. Heft 74. Norinbergae: Felsecker, 24 pp. + 24 pls.

Panzer, G.W.F. 1803a: *Faunae insectorum germanicae initia oder Deutschlands Insecten*. Heft 90. Norinbergae: Felsecker, 24 pp. + 24 pls.

Panzer, G.W.F. 1803b: *Faunae insectorum germanicae initia oder Deutschlands Insecten*. Heft 94. Norinbergae: Felsecker, 24 pp. + 24 pls.

Panzer, G.W.F. 1805a: *Kritische Revision der Insektenfaune Deutschlands nach dem System bearbeitet. Zum I–XCVI. Heft. I. Baendchen*. Nürnberg: Felfseckerschen Buchhandlung, 12 + 144 pp.

Panzer, G.W.F. 1805b: *Faunae insectorum germanicae initia oder Deutschlands Insecten*. Heft 98. Norinbergae: Felsecker, 24 pp. + 24 pls.

Papp, C.H. 1946: Verzeichnis der bei Madarasz-See (Ungarn) gesammelten Käferarten im Jahre 1939–1943. (II. Stück Hygrobiidae bis Ptinidae). *Additamenta Faunistica Coleopterorum* **6** (2): 34–38.

Pardo, Alcaide A. 1948: Estudios sobre Meloidae. I. Acerca de la validez especifica de Mylabris maculosopunctata Grll., rosinae Escher. y pauper Escher. *Eos, Revista Española de Entomología* **24**: 493–502.

Pardo, Alcaide A. 1950a: Estudios sobre Meloidae. II. Los Mylabrini de la Península Ibérica. *Boletín de Patólogia Vegetal y de Entomología Agricola* **17**: 61–82.

Pardo, Alcaide A. 1950b: Contribución al conocimiento de la fauna entomológica marroquí III. *Instituto General Franco de Estudios e investigación Hispano-Arabe. Publicaciones fuera de series*, 70 pp.

Pardo, Alcaide A. 1951: Estudios sobre Meloidae III. Una nueva especie de Meloe de la isla de Tenerife y comentarios sobre algunos meloideos de la citada isla. *Eos, Revista Española de Entomología* **27**: 249–255.

Pardo, Alcaide A. 1952: Los generos de Meloidae de la fauna hespérica, *Graellsia* **8** [1950]: 39–79.

Pardo, Alcaide A. 1953: Contribución al conocimiento de la fauna entomólogica marroquí V. Una nueva especie de Stenoria Muls. Del sahara marroquí (col. Meloidae). *Tamuda* **1**: 288–293.

Pardo, Alcaide A. 1954a: Études sur les Meloidae. V. Les Mylabrini du Maroc et du Sahara occidental espagnol (Col. Meloidae). *Bulletin de la Société de Sciences Naturelles et Physiques du Maroc* **34**: 55–88.

Pardo, Alcaide A. 1954b: Estudios sobre Meloidae IV. Sobre algunos Mylabris Fabr. Del Norteafricano. *Eos, Revista Española de Entomología* **30**: 331–344.

Pardo, Alcaide A. 1955a: Estudios sobre Meloidae Vi. Sobre algunos Mylabris (s. lat.) de la región etiópica. *Bulletin de l'Institut Royal de Sciences Naturelles du Belgique* **31** (49): 1–32.

Pardo, Alcaide A. 1955b: Contribución al conocimiento de la fauna entomólogica marroquí VI. Coleopteros del valle inferior del Uad Muluya. *Tamuda* **3**: 39–74.

Pardo, Alcaide A. 1956: Analectas entomologicas V. Los Zonitis de la fauna ibérica (Col. Meloidae). *Graellsia* **14**: 13–19.

Pardo, Alcaide A. 1958a: Analectas entomologicas. VII. Los Sitarini de la Peninsula Ibérica (Col. Meloidae). *Graellsia* **16**: 13–22.

Pardo, Alcaide A. 1958b: Études sur les Meloidae (Col.) IX. Observations sur quelques Mylabris des régions éthiopienne et orientale. *Bulletin de l'Institut Royal de Sciences Naturelles de Belgique* **34** (13): 1–39.

Pardo, Alcaide A. 1961a: Estudios sobre 'Meloidae'. XIII. Meloideos del Sahara occidental español y de la región du Draa (Marueccos) (Coleoptera). *Eos, Revista Española de Entomología* **37**: 91–111.

Pardo, Alcaide A. 1961b: Études sur les Meloidae. XV. Sur deux Mylabris du Maroc méridional saharien. *Bulletin de la Société de Sciences Naturelles et Physiques du Maroc* **42**: 215–220.

Pardo, Alcaide A. 1963: Études sur les Meloidae. XIV. Trois nouvelles espèces de Mylabris Fabricius (Coleoptera). *Eos, Revista Española de Entomología* **39**: 295–306.

Pardo, Alcaide A. 1968: Contribution à la connaisance de la faune entomologique Marocaine X. Un nouveau type d'Oedemeridae (Col.) du Maroc Saharien: Gilotia perroti nov. gen. et nov. sp. *Arquivos do Museu Bocage* **2**: 41–49.

Pardo, Alcaide A. 1969: Études sur les Meloidae. XXI. Matériaux pour une révision des mylabrides de l'Afrique du Nord et du Moyen orient (2e partie). *Eos, Revista Española de Entomología* **54**: 367–376.

Pardo, Alcaide A. 1972: Études sur Oedemeridae II. Une nouvelle et singulière espèce de Chrysanthia Schmidt d'Asie Mineure (Coleoptera). *Fragmenta Entomologica* **8**: 213–221.

Pardo, Alcaide A. 1975: Estudios sobre Meloidae (Coleoptera) XXIV. Una nueva especie de Mylabris Fabricius (Mylabris platai n.sp.) de Sierra nevada. *Quadernos Cientificas de Biologia* **4**: 45–48.

Pardo, Alcaide A. 1977: Études sur les Meloidae (Col.). XXIII. Trois nouvelles taxa de Cerocoma Geoffroy d'Iran et Turquie d'Asie. *Nouvelle Revue d'Entomologie* **7**: 59–66.

Pardo, Alcaide A. 1978: Un nuevo Alleculidae de Marruecos: Hymenorus candeli nov. sp. (Ins. Col.). *Butlleti de la Institució Catalana d'Historie Natural* (*Sec. Zool.* 2) **42**: 89–91.

Pascoe, F.P. 1860a: Notices of new or little-known genera and species of Coleoptera. *Journal of Entomology* **1**: 36–64, pl. 2, 3.

Pascoe, F.P. 1860b: Notices of new or little-known genera and species of Coleoptera. *Journal of Entomology* **1**: 98–132, pls 5–8.

Pascoe, F.P. 1862: Notices of new or little-known genera and species of Coleoptera. Part III. *Journal of Entomology* **1**: 319–370, pls 16, 17.

Pascoe, F.P. 1863a: Notices of new or little-known genera and species of Coleoptera. *Journal of Entomology* **2**: 26–56.

Pascoe, F.P. 1863b: List of Colydiidae collected in the Amazons Valley by H.W. Bates, Esq., and descriptions of new species. *Journal of Entomology* **2**: 79–99.

Pascoe, F.P. 1863c: List of Colydiidae collected in the Indian Islands by Alfred R. Wallace, Esq., and descriptions of new species. *Journal of Entomology* **2**: 121–143, pl. 8.

Pascoe, F.P. 1866: Notices of new or little-known genera and species of Coleoptera. *Journal of Entomology* **2**: 443–493, pls. 17, 19.

Pascoe, F.P. 1868: Contribution to the knowledge of the Coleoptera. *Transactions of the Entomological Society of London* **1868**: xi-xiii.

Pascoe, F.P. 1869: Descriptions of new genera and species of Tenebrionidae from Australia and Tasmania. *The Annals and Magazine of Natural History* (4) **3**: 132–153, pl. 11, 277–296, pl. 12.

Pascoe, F.P. 1870: Addtitions to the Tenebrionidae of Australia &c. *The Annals and Magazine of Natural History* (4) **5**: 94–107.

Pascoe, F.P. 1871: Notes on Coleoptera, with descriptions of new genera and species. Part I. *The Annals and Magazine of Natural History* (4) **8**: 345–361, pl. 14.

Pascoe, F.P. 1882: Notes on Coleoptera, with descriptions of new genera and species. Part IV. *The Annals and Magazine of Natural History* (5) **9**: 25–37.

Pascoe, F.P. 1887: Notes on Coleoptera, with descriptions of new genera and species. Part VI. *The Annals and Magazine of Natural History* (5) **20**: 8–20, pl. 1.

Patti, M.Z. 1840: Descrizione di un nuovo coleoptero. *Giornale di Scienze, Lettere ed Arti Sicilia* **72**: 292–294.

Paulus, H.F. 1972: Neue Pyrochroidae aus Nepal (Coleoptera, Heteromera), mit einer Diskussion der verwandtschaftlichen Verhältnisse der Familie, 1. *Zeitschrift der Arbeitsgemeinschaft Österreichischer Entomologen* **23** [1971]: 75–85.

Paykull, G. von. 1798: *Fauna Suecica. Insecta. Tomus I*. Uppsala: Joh. F. Edman, [8] + 358 + [2] pp.

Paykull, G. von. 1799: *Fauna Suecica. Insecta. Tomus II*. Upsaliae: Joh. F. Edman, 234 pp.

Paykull, G. von. 1800: *Fauna Suecica. Insecta. Tomus III*. Upsaliae: Joh. F. Edman, 459 pp.

Peacock, E.R. 1982: Tydessa n. gen. from Japan (Coleoptera: Pyrochroidae). *Entomologica Scandinavica* **13**: 361–365.

Pecchioli, V. 1839: Description d'une nouvelle espèce de Sitaris. *Annales de la Société Entomologique de France* **8**: 527–530, 1 pl.

Penecke, K.A. 1912: Zwei neue Käferarten aus Bulgarien. *Wiener Entomologische Zeitung* **31**: 240–244.

Penrith, M.-L. 1977: The Zophosini (Coleoptera: Tenebrionidae) of Western Southern Africa. *Cimbebasia, Memoir* **3**: 1–291.

Penrith, M.-L. 1982a: Revision of the Zophosini (Coleoptera: Tenebrionidae). Part 5. A derived subgenus from Northern Africa. *Cimbebasia* (A) **6**: 165–226.

Penrith, M.-L. 1982b: Revision of the Zophosini (Coleoptera: Tenebrionidae). Part 6. The subgenus Hologenosis Deyrolle. *Cimbebasia* (A) **6**: 227–289.

Penrith, M.-L. 1983a: Revision of the Zophosini (Coleoptera: Tenebrionidae). Part 7. The African species of the subgenus Oculosis Penrith. *Cimbebasia* (A) **6**: 291–367.

Penrith, M.-L. 1983b: Revision of the Zophosini (Coleoptera: Tenebrionidae). Part 8. The palaearctic species of the subgenus Oculosis Penrith, the subgenus Cheirosis Deyrolle, and a monotypical subgenus from Socotra. *Cimbebasia* (A) **6**: 369–384.

Penrith, M.-L. 1984: Revision of the Zophosini (Coleoptera: Tenebrionidae). Part 9. The remaining species of the subgenus Septentriophosis Penrith. *Cimbebasia* (A) **6**: 385–416.

Penrith, M.-L. 1986: Revision of the Zophosini (Coleoptera: Tenebrionidae). Part 10. Key to the subgenera, supplement, evolution and biogegraphy of the tribe, and catalogue. *Cimbebasia* (A) **6**: 417–502.

Perez, Arcas L. 1865a: Insectos nuevos ó poco conocidos de la fauna española. Primera parte. *Revista de los Progresos de las Ciencias Exactas, Fisicas y Naturales* **15**: 166–186 [issued also separately, pp. 1–23].

Perez, Arcas L. 1865b: Insectos nuevos ó poco conocidos de la fauna española. Secunda parte. *Revista de los Progresos de las Ciencias Exactas, Fisicas y Naturales* **15**: 413–444 [issued also separately, pp. 25–56].

Perez, Arcas L. 1868: Insectos nuevos ó poco conocidos de la fauna española. Tercera parte. *Revista de los Progresos de las Ciencias Exactas, Fisicas y Naturales* **18**: 490–520 [issued also separately, pp. 57–86].

Perez, Arcas L. 1872: Especies nuevas ó críticas de la fauna española. *Anales de la Sociedad Española de Historia Natural* **1**: 89–137, pls. I–II.

Perez, Arcas L. 1874: Especies nuevas ó criticas de la fauna española. Tercera parte. *Anales de la Sociedad Española de Historia Natural* **3**: 111–155, pls I–III.

Pérez-Vera, F. & Ávila, J.M. 2012a: Contribución al conocimiento de los Asidini marroquíes II. Descripción de una nueva especie de *Alphasida* Escalera, 1905 del subgénero *Machlasida* Escalera, 1907 (Coleoptera, Tenebrionidae). *Boletín de la Asociación española de Entomología* **36**: 75–85.

Pérez-Vera, F. & Ávila, J.M. 2012b: Contribución al conocimiento de los Asidini marroquíes IV. Una nueva subespecie de Asida ruficornis Sol.: Asida (Planasida) ruficornis escalerae n. ssp. (Coleoptera, Tenebrionidae). *Boletín de la Asociación española de Entomología* **36**: 185–191.

Pérez-Vera, F. & Ávila, J.M. 2012c: Asidini Marroquíes VI. Descripción de tres nuevas especies pertenecientes al género Alphasida Escalera, 1905 y de una subespecie nueva del género Asida Latreille, 1802 (Coleoptera, Tenebrionidae). *Graellsia* **68**: 17–29.

Pérez-Vera, F. & Ávila, J.M. 2012d: Asidini Marroquíes VIII. Descripción de dos nuevas subespecies del género Alphasida Escalera, 1905, y Asida Latreille, 1802 (Coleoptera, Tenebrionidae). *Bulletin de la Société entomologique de France* **117**: 393–400.

Pérez-Vera, F., Ruiz, J.L. & Ávila, J.M. 2012: Descripción de una nueva especie de Asida Latreille, 1802 del subgénero Planasida Escalera, 1907, del norte de África (Coleoptera, Tenebrionidae). *Boletín de la Asociación española de Entomología* **36**: 381–400.

Pérez-Vera, F. & Ávila, J.M. 2014: Deux nouveaux taxons d'Alphasida (Glabrasida) Escalera, 1910, d'Andalousie centrale (Coleoptera, Tenebrionidae, Pimeliinae). *Bulletin de la Société entomologique de France* **119**: 329–337.

Pérez-Vera, F. & Ávila, J.M. 2015a: Dos nuevas especies del género Asida Latreille, 1802 del noroeste de España (Coleoptera: Tenebrionidae). *Boletín de la Asociación española de Entomología* **39**: 101–113.

Pérez-Vera, F. & Ávila, J.M. 2015b: Révision du complexe d'Alphasida (Machlasida) olcesei (Fairmaire, 1867), avec la description d'une nouvelle espèce (Coleoptera, Tenebrionidae). *Bulletin de la Société entomologique de France* **120**: 213–226.

Pérez-Vera, F. & Ávila, J.M. 2015c: Révision du complexe supra-spécifique Alphasida (Glabrasida) goudoti (Solier, 1836) (Coleoptera, Tenebrionidae). *Boletín de la Asociación española de Entomología* **39**: 21–49.

Pérez-Vera, F. & Ávila, J.M. 2015d: Révision des Alphasida (Glabrasida) Escalera, 1910, groupe I (section incostulatae Escalera, 1922) (Coleoptera, Tenebrionidae, Asidini). *Bulletin de la Société entomologique de France* **120**: 485–532.

Pérez-Vera, F. & Ávila, J.M. 2016: Revisión del complejo supra-específico Alphasida (Glabrasida) castellana (Graells, 1858) (Coleoptera, Tenebrionidae). *Graellsia* **72**: 1–18.

Pérez-Vera, F., Ávila, J.M. & Martínez Fernández, J.C. 2017: Nueva ordenación taxonómica del subgénero Elongasida Escalera, 1906 del género Alphasida Escalera, 1905 (Coleoptera, Tenebrionidae). *Graellsia* **73**: 1–26.

Péringuey, L. 1904: Sixth Contribution to the South African Coleopterous Fauna. Descriptions of new species of Coleoptera in the Collections of the South African Museum. *Annals of the South African Museum* **3**: 167–300, 1 pl.

Péringuey, L. 1909: Descriptive Catalogue of the Coleoptera of South Africa. Family Meloidae. *Transactions of the Royal Society of South Africa* **1**: 165–297, 4 pls.

Perris, E. 1852: Seconde excursion dans les Grandes-Landes. *Annales de la Société Linnéenne de Bordeaux* **1852**: 1–72.

Perris, E. 1864: Description de quelques espèces nouvelles de coléoptères et notes diverses. *Annales de la Société Entomologique de France* (4) **4**: 275–310.

Perris, E. 1865: Descriptions de quelques nouvelles espèces de coléoptères, rectifications et notes. *Annales de la Société Entomologique de France* (4) **5**: 505–512.

Perris, E. 1869: Descriptions de quelques coléoptères nouveaux. Rectifications et notes. *L'Abeille, Mémoires d'Entomologie* 7 [1869–1870]: 3–33.

Perris, E. 1874: [new taxon]. In: Abeille de Perrin, E. 1874a: [Diagnoses de Cisides nouveaux]. *Bulletin de la Société Entomologique de France* **1874**: lii-liii

Perris, E. 1875: Descriptions de quelques insectes jugés nouveaux. *L'Abeille, Journal d'Entomologie* **13**: 1–1.

Perroud, B.-P. & Montrouzier, X. 1864: La Faune entomologique de Kanala (Nouvelle-Calédonie) et description de quelques espèces nouvelles ou peu connues. *Annales de la Société Linnéenne de Lyon* **11**: 44–256, pl. 50.

Perroud, B.-P. & Mulsant, E. 1856: Description de deux nouvelles espèces de coléoptères constituant un genre nouveau dans la famille des ulomiens. *Opuscules Entomologiques* 7: 160–165.

Perty, J.A.M. 1831: *Observationes nonnullae in Coleoptera Indiae orientalis. Dissertatio philosophico-entomologica, quam unacum praemissis thesibus auctoritate et consensus illustris philosophorum ordinis in Academia Ludovico-Maximilianea facultatem legendi rite adepturus.* Monachii: Mich. Lindauer, xxxxiv pp., 1 pl.

Petagna, L. 1819: Memoria su di alcuni insetti del Regno di Napoli. *Memorie della Accademia delle Scienze. Classe de Fisica e di Storia Naturale* (Napoli) **1**: 19–40, 4 pls.

Petagna, V. 1786: *Specimen Insectorum ulterioris Calabriae.* Neapoli: Typis Petri Perger, vii + 46 pp., 1 pl.

Petagna, V. 1787: *Specimen Insectorum ulterioris Calabriae.* Francofurti et Moguntiae: apud Varrentrapp et Wenner, vi +46 pp., 1pl.

Petri, K. 1912: *Siebenbürgens Käferfauna auf Grund ihrer Erforschung bis zum Jahre 1911.* Hermannstadt: Jos. Drotleff, xi + 374 columns.

Petri, K. 1926: Ergänzungen und Berichtigungen zur Käferfauna Siebenbürgens, 1912. Zusammengestellt unter Mitwirkung der Herren Friedrich Deubel, Kronstadt, Oberstabsarzt Dr. Eugen Worell und Dr. Arnold Müller, Hermannstadt. *Verhandlungen und Mitteilungen des Siebenbürgischen Vereins für Naturwissenschaften zu Hemannstadt* **75–76**: 165–206.

Peyerimhoff, P. de. 1907a: Liste de coléoptères du Sinaï. *L'Abeille, Journal d'Entomologie* **31**: 1–48.

Peyerimhoff, P. de. 1907b: Recherches sur la synonymie des coléoptères du Nord-Africain (première note). *L'Abeille, Journal d'Entomologie* **31**: 185–187

Peyerimhoff, P. de. 1911: Nouveaux coléoptères du Nord-Africain (treizième note) (Tenebrionidae) récoltes du Dr. Edm. Sergent dans l'Extrême-Sud-Oranais. *Bulletin de la Société Entomologique de France* **1911**: 346–349.

Peyerimhoff, P. de. 1913a: Nouveaux coléoptères du Nord-Africain (seizième note, faune du Djurdjura). *Bulletin de la Société Entomologique de France* **1913**: 253–255.

Peyerimhoff, P. de. 1913b: Nouveaux coléoptères du Nord-Africain (quizième note: faune marocaine). *Annales de la Société Entomologique de France* **81**: 515–521.

Peyerimhoff, P. de. 1915: Notes sur la biologie de quelques coléoptères phytophages du Nord-Africain (deuxième série). *Annales de la Société Entomologique de France* **84**: 19–61.

Peyerimhoff, P. de. 1916a: Nouveaux coléoptères du Nord-Africain (vingt-deuxième note): Récoltes de M. le Dr. H. Foley dans l'extrême Sud-Oranais. *Bulletin de la Société Entomologique de France* **1916**: 71–76.

Peyerimhoff, P. de. 1916b: Nouveaux coléoptères du Nord-Africain (vingt-troisieme note): Récoltes de M. le Lt Nicloux dans le Sud-Ouest Marocain. *Bulletin de la Société Entomologique de France* **1916**: 82–84.

Peyerimhoff, P. de. 1917a: Nouveaux coléoptères du Nord-African. Vingt-cinquième note avec des renseignements synonymique sur les Silphidae et les Histeridae. *Annales de la Société Entomologique de France* **86**: 117–148.

Peyerimhoff, P. de. 1917b: Nouveaux coléoptères du Nord-Africain. Vingt-quatrième note: Faune de Djurdjura. *Bulletin de la Société Entomologique de France* **1916**: 318–319.

Peyerimhoff, P. de. 1917c: Nouveaux coléoptères du Nord-African. (Vingt-sixième note: Faune du cèdre et du sapin de Numidie). *Bulletin de la Société Entomologique de France* **18**: 329–332.

Peyerimhoff, P. de. 1918a: Nouveaux coléoptères du Nord-African. (Vingt-septième note: Faune du massif de Mouzaïa). *Bulletin de la Société Entomologique de France* **18** [1917]: 371–374.

Peyerimhoff, P. de. 1918b: Nouveaux coléoptères du Nord-Africain (Vingt-huitième note: Faune du Pin d'Alep). *Bulletin de la Société Entomologique de France* **1918**: 141–144.

Peyerimhoff, P. de. 1918c: Nouveaux coléoptères du Nord-Africain: (Vingt-neuvième note: Faune de Djurdjura). *Bulletin de la Société Entomologique de France* **1918**: 257–260.

Peyerimhoff, P. de. 1919a: Notes sur la biologie de quelques coléoptères phytophages du Nord-Africain (troisième série). *Annales de la Société Entomologique de France* **88**: 169–258.

Peyerimhoff, P. de. 1919b: Nouveaux coléoptères du Nord-Africain. Trentième note: Faune de l'Aurès. *Bulletin de la Société Entomologique de France* **1918**: 276–278.

Peyerimhoff, P. de. 1920a: Nouveaux coléoptères du Nord-Africain. Trente-deuxième note: Faune saharienne. *Bulletin de la Société Entomologique de France* **1919**: 325–328.

Peyerimhoff, P. de. 1920b: Nouveaux coléoptères du Nord-Africain. Trente-troisième note: Récoltes de M. le capitaine Surcouf dans le Grand Erg occidental. *Bulletin de la Société Entomologique de France* **1920**: 21–23.

Peyerimhoff, P. de. 1920c: Nouveaux coléoptères du Nord-Africain. Trente-cinquième note: Faune del'Aurès. *Bulletin de la Société Entomologique de France* **1920**: 243–246.

Peyerimhoff, P. de. 1921a: Sur quelques Tenebrionidae d'Orient (Col.). Description d'un Pimelia nouveau. *Bulletin de la Société Entomologique de France* **1920**: 327–329.

Peyerimhoff, P. de. 1921b: Nouveaux coléoptères du Nord Africain. Trente-huitième note: Tenebrionidae. *Bulletin de la Société Entomologique de France* **1921**: 62–66.

Peyerimhoff, P. de. 1921c: Nouveaux coléoptères du Nord Africain. Trente-neuvième note: Faune saharienne. *Bulletin de la Société Entomologique de France* **1921**: 231–236.

Peyerimhoff, P. de. 1923: Nouveaux coléoptères du Nord-Africain. Quarante-troisième note: Récoltes de M. le Dr Heimsch dans l'Erg Iguidi (Sahara occidental). *Bulletin de la Société Entomologique de France* **1922**: 318–321.

Peyerimhoff, P. de. 1924: Nouveaux coléoptères du Nord-Africain. Quarante-septième note: Récoltes de M.F. Peltier dans le Grand-erg occidental, le Gourara et le Touat. *Bulletin de la Société Entomologique de France* **1924**: 74–78.

Peyerimhoff, P. de. 1925: Nouveaux coléoptères du Nord-Africain. Cinquantième note (avec table). *Annales de la Société Entomologique de France* **94**: 1–29.

Peyerimhoff, P. de. 1926a: [new taxa]. In: Sainte Claire Deville J. & Peyerimhoff P. de: Nouveaux coléoptères du Nord Africain. Cinquante-cinquième note. Faune du Grand-Atlas marocain (suite). *Bulletin de la Société Entomologique de France* **1926**: 93–96.

Peyerimhoff, P. de. 1926b: Nouveaux coléoptères du Nord-Africain. Cinquante-septième note. Faune de la Zousfana et du Guir. *Bulletin de la Société Entomologique de France* **1926**: 193–196.

Peyerimhoff, P. de. 1927a: Études sur la systématique des coléoptères du Nord-Africain. I. – Les Pachychila (Tenebrionidae). *L'Abeille, Journal d'Entomologie* **34**: 1–57.

Peyerimhoff, P. de. 1927b: Nouveaux coléoptères du Nord-Africain. Cinquante-huitième note. Récoltes de M. le Dr. A. Sicard dans le Maroc oriental. *Bulletin de la Société Entomologique de France* **1926**: 218–221.

Peyerimhoff, O. de. 1928: Nouveaux coléoptères du Nord-African soixante-deuxième note. Un type nouveau de Pimeliinae (Tenebrionidae). *Bulletin de la Société Entomologique de France* **1928**: 61–63.

Peyerimhoff, P. de. 1929a: Nouveaux coléoptères du Nord Africain. Soixante-septième note. Faune du Hoggar et des massifs voisins. *Bulletin de la Société Entomologique de France* **1929**: 123–126.

Peyerimhoff, P. de. 1929b: Nouveaux coléoptères du Nord-Africain. Soixante-neuvième note. Faune du Hoggar et des massifs voisins. *Bulletin de la Société Entomologique de France* **1929**: 153–156.

Peyerimhoff, P. de. 1929c: Nouveaux coléoptères du Nord-Africain. Soixante-treizième note. Faune du Hoggar et des massifs voisins. *Bulletin de la Société Entomologique de France* **1929**: 282–286.

Peyerimhoff, P. de. 1931a: Nouveaux coléoptères du Nord-Africain. Soixante-quizième note (avec table et résumé). *Annales de la Société Entomologique de France* **100**: 1–20.

Peyerimhoff, P. de. 1931b: Mission scientifique du Hoggar. Coléoptères. *Mémoires de la Société d'Histoire Naturelle de l'Afrique du Nord* **2**: 5–173, 3 pls, 2 maps.

Peyerimhoff, P. de. 1933: Description de quatre Tenebrionidae (Col.) du Sahara central. *Bulletin de la Société d'Histoire Naturelle de l'Afrique du Nord* **24**: 344–356.

Peyerimhoff, P. de. 1934: Mission transsaharienne de S.A.R. le Prince Sixte de Bourbon-Parme (1932). Insectes coléoptères. *Revue Française d'Entomologie* **1**: 104–108.

Peyerimhoff, P. de. 1935a: Coléoptères (avec la description d'un genre et de deux espèces nouvelles). Mission au Tibesti (1930–1931) dirigée par M. Dalloni professeur à la faculté des Sciences d'Alger. *Mémoires de l'Académie des Sciences de l'Institut de France* **62**: 1–7, pl. 1.

Peyerimhoff, P. de. 1935b: Coléoptères marocains récoltés par M. Ch. Rungs. *Bulletin de la Société des Sciences Narutelles du Maroc* **15**: 19–25.

Peyerimhoff, P. de. 1935c: Coléoptères nouveaux ou mal connus de Berbérie. I. Le genre Storthocnemis Karsch (Tenebrionidae). *Bulletin de la Société Entomologique de France* **40**: 89–91.

Peyerimhoff, P. de. 1936: Coléoptères (avec la description d'un nouveau genre et de deux espèces nouvelles). In: Dalloni M.M.: Mission au Tibesti (1930–31). *Mémoires de l'Academie des Sciences de l'Institut de France* **62**: 74–81.

Peyerimhoff, P. de. 1938: Coléoptères nouveaux ou mal connus de Berbérie. XI. – Le genre Euthriptera Reitt. (Tenebrionidae). *Bulletin de la Société Entomologique de France* **43**: 178–180.

Peyerimhoff, P. de. 1939a: Coléoptères nouveaux ou mal connus de Berbérie XII – Le genre Belopus Gebien (Tenebrionidae). *Bulletin de la Société Entomologique de France* **44**: 25–28.

Peyerimhoff, P. de. 1939b: Description d'un nouvel Heliotaurus Algérien (Col. Alleculidae). *Annales de la Société Entomologique de France* **108**: 120.

Peyerimhoff, P. de. 1942a: Description d'un nouveau Crypticus (Col. Tenebrionidae) des Canaries. *Bulletin de la Société Entomologique de France* **47**: 11–12.

Peyerimhoff, P. de. 1942b: Récoltes entomologiques faites par L. Berland à Villa-Cisneros (Rio de Oro). Coléoptères. *Bulletin du Muséum National d'Histore Naturelle, Paris* (2) **14**: 215–222.

Peyerimhoff, P. de. 1942c: Sur un coléoptère nouveau du Rio de Oro. *Bulletin du Muséum National d'Histore Naturelle, Paris* (2) **14**: 276–277.

Peyerimhoff, P. de. 1943: Matériaux pour un catalogue des coléoptères sahariens. II. – Descriptions d'espèces nouvelles. *Bulletin de la Société d'Histoire Naturelle de l'Afrique du Nord* **34**: 7–35.

Peyerimhoff, P. de. 1945: Genres de coléoptères figurant à tort dans la faune euro-méditerranéenne. *Revue Française d'Entomologie* **12**: 1–5.

Peyerimhoff, P. de. 1946: Coléopterès du Sahara marocain et du Sahara occidental. *Bulletin de la Société des Sciences Naturelles du Maroc* **24** [1944]: 90–110.

Peyerimhoff, P. de. 1948a: Insectes coléoptères. Mission scientifique du Fezzan (1944–1945). *Mission Scientifique du Fezzan*: 5–84.

Peyerimhoff, P. de. 1948b: Études sur la systematique des coléoptères du Nord Africain. II. Les Scaurus (Tenebrionidae). *Revue Française d'Entomologie* **14** [1947]: 157–193.

Peyerimhoff, P. de. 1949a: Coléoptères nouveaux ou mal connus de Berbérie XXI. – Découverte dans le Grand Erg occidental du genre tauranien Sphenaria Mén. (Col. Tenebrionidae). *Bulletin de la Société Entomologie de France* **54**: 70–71.

Peyerimhoff, P. de. 1949b: Études et descriptions de coléoptères marocains ii. *Bulletin de la Société des Sciences Naturelles et Physiques du Maroc* **25–27** [1945–47]: 248–308.

Peyron, E. 1854: Description d'une nouvelle espèces du genre Cossyphus. *Revue et Magasin de Zoologie Pure et Appliquée* (2) **6**: 224–225.

Pfaff, C.H. 1824: Scharfer Stoff der Canthariden. In*: System der Materia medica nach chenmischen Principien mit Rücksicht auf die sinnlichen Merkmale und die Heilverhältnisse der Arzneymittel.* 7 (Zweyter Supplemenband). Leipzig: Vogel, xiv + 389 pp.

Philippi, R.A. 1864: [new taxa]. In: Philippi, R.A. & Philippi, F.: Beschreibung einiger neuen chilenischen Käfer. *Entomologische Zeitung* **25**: 313–406.

Pic, M. 1891a: Sur Crioceris tibialis Villa et Allecula morio Fabr. *L'Échange, Revue Linnéenne* **7**: 51.

Pic, M. 1891b: Notes coléoptérologiques. *L'Échange, Revue Linnéenne* **7**: 117–118.

Pic, M. 1892a: Descriptions et corrections. *L'Échange, Revue Linnéenne* **8**: 4.

Pic, M. 1892b: Descriptions. – Beschreibungen. *Miscellanea Entomologica* **1**: 17–19, 25–27, 53–54.

Pic, M. 1892c: Coléoptérologie descriptive. *L'Échange, Revue Linnéenne* **8**: 32–33.

Pic, M. 1892d: Anthicus algériens. *Miscellanea Entomologica* **1**: 43–44.

Pic, M. 1892e: Quelques mots sur les anthicides. *L'Échange, Revue Linnéenne* **8**: 43–44.

Pic, M. 1892f: A corriger. *L'Échange, Revue Linnéenne* **8**: 66.

Pic, M. 1892g: [new taxa]. *Bulletin de la Société Entomologiqe de France* **1892**: 75–77.

Pic, M. 1892h: Contribution à l'étude des anthicides d'Algérie. *L'Échange, Revue Linnéenne* **8**: 102–103.

Pic, M. 1892i: Notes entomologiques. *L'Échange, Revue Linnéenne* **8**: 114–115.

Pic, M. 1892j: Contribution à l'étude des coléoptères. *L'Échange, Revue Linnéenne* **8**: 139.

Pic, M. 1892k: Variétés. *L'Échange, Revue Linnéenne* **8**: 139–140.

Pic, M. 1892l: Anthicides recueillis en Algérie en avril et mai 1892. *Revue Scientifique du Bourbonnais et du Centre de la France* 5: 212–220.

Pic, M. 1892m: Sur les Tomoderus Laf. Coléoptères de la famille des anthicides. *Revue Scientifique du Bourbonnais et du Centre de la France* 5: 240–242.

Pic, M. 1892n: [new taxa]. *Bulletin de la Société Entomologique de France* 1892: cxcvi-cxcviii.

Pic, M. 1892o: Deux coléoptères nouveaux. *Revue d'Entomologie* 11: 313–314.

Pic, M. 1893a: Anthicides nouveaux. *L'Échange, Revue Linnéenne* 9: 15–16.

Pic, M. 1893b: Descriptions de coléoptères. *L'Échange, Revue Linnéenne* 9: 26–27.

Pic, M. 1893c: Anthicides nouveaux. *L'Échange, Revue Linnéenne* 9: 27–28.

Pic, M. 1893d: Examen des anthicides de la collection Leprieur. *L'Échange, Revue Linnéenne* 9: 38.

Pic, M. 1893e: Étude sur les anthicides de la collection Leprieur. *L'Échange, Revue Linnéenne* 9: 51–52.

Pic, M. 1893f: Descriptions – Beschreibungen. *Miscellanea Entomologica* 1: 53–54.

Pic, M. 1893g: Rectifications. *Miscellanea Entomologica* 1: 54.

Pic, M. 1893h: Examen des anthicides de la collection Leprieur (Suite). *L'Échange, Revue Linnéenne* 9: 64–65.

Pic, M. 1893i: Examen des anthicides de la collection Leprieur (Suite et fin.). *L'Échange, Revue Linnéenne* 9: 74–75.

Pic, M. 1893j: Description d'espèces nouvelles de la faune européenne et circa – Anthicus nouveaux (Coléoptères). *Feuille des Jeunes Naturalistes* (3) 23: 78–79.

Pic, M. 1893k: Rectification. *L'Échange, Revue Linnéenne* 9: 88.

Pic, M. 1893l: Coléoptères nouveaux de la faune circa-européenne. *Feuille des Jeunes Naturalistes* (3) 23: 111.

Pic, M. 1893m: Essai d'une étude sur les Anthicus (lagenicolles) Cyclodinus Muls. *L'Échange, Revue Linnéenne* 9: 113–116.

Pic, M. 1893n: Anthicides tunisiens. *L'Échange, Revue Linnéenne* 9: 122–123.

Pic, M. 1893o: [Essai d'une étude sur les Anthicus (lagenicolles) Cyclodinus Muls, continuation]. *L'Échange, Revue Linnéenne* 9: 125–128.

Pic, M. 1893p: Anthicides nouveaux. *Miscellanea Entomologica* 1: 126–127.

Pic, M. 1893q: Description de deux coléoptères nouveaux. *L'Échange, Revue Linnéenne* 9: 135.

Pic, M. 1893r: Anthicides nouveaux. *Revue d'Entomologie* 12: 155–156.

Pic, M. 1893s: Xylophilides et anthicides recueillis en Algérie en Avril et Mai, 1893. *Revue Scientifique du Bourbonnais et du Centre de la France* 6: 155–162.

Pic, M. 1893t: Anthicides nouveaux de l'Ancien monde. *Feuille des Jeunes Naturalistes* (3) 23: 175–176.

Pic, M. 1893u: [Une espèce et une variété nouvelles de coléoptères]. *Bulletin de la Société Entomologique de France* 1893: xxxiii-xxxiv.

Pic, M. 1893v: [Description de trois Anthicus nouveaux (Coléoptères)]. *Bulletin de la Société Entomologique de France* 1893: liii-lv.

Pic, M. 1893w: Descriptions d'espèces nouvelles de coléoptères et notes synonymiques. *Bulletin de la Société Entomologique de France* 1893: lxxxv-lxxxviii.

Pic, M. 1893x: [Deux coléoptères nouveaux]. *Bulletin de la Société Entomologique de France* 1893: clxi.

Pic, M. 1893y: [Quelques notes synonymiques sur certains Anthicus (Coléoptères)]. *Bulletin de la Société Entomologique de France* 1892: ccxi-ccxii.

Pic, M. 1893z: [Descriptions suivantes de nouveaux Anthicus (Coléoptères)]. *Bulletin de la Société Entomologique de France* **1892**: cclviii-cclxxi.

Pic, M. 1894a: Descriptions de coléoptères hétéromères africains. *Bulletin de la Société Zoologique de France* **1894**: 15–17.

Pic, M. 1894b: Catalogue géographique des anthicides de France, Corse, Algérie et Tunisie (Suite). *Revue Scientifique du Bourbonnais et du Centre du France* **7**: 19–25.

Pic, M. 1894c: Catalogue géographique des anthicides de France, Corse, Algérie et Tunisie (Suite). *Revue Scientifique du Bourbonnais et du Centre du France* **7**: 40–49.

Pic, M. 1894d: Liste des anthicides décrits postérieurement au Catalogus de MM. Gemminger et Harold (1870–1893). *Annales de la Société Entomologique de Belgique* **38**: 43–58.

Pic, M. 1894e: Description de coléoptères de la faune circa-européenne. *Feuille des Jeunes Naturalistes* (3) **24**: 60–61.

Pic, M. 1894f: Examen des anthicides de la collection H. Tournier. *L'Échange, Revue Linnéenne* **10**: 64–65.

Pic, M. 1894g: Catalogue geographique des anthicides de France, Corse, Algérie et Tunisie (Suite). *Revue Scientifique du Bourbonnais et du Centre du France* **7**: 69–79.

Pic, M. 1894h: Quelques notes en passant sur plusieurs coléoptères de la collection Henri Tournier. *L'Échange, Revue Linnéenne* **10**: 71–72.

Pic, M. 1894i: Extrait de la Revue Scientifique du Bourbonnais et du Centre de la France. Xylophilides & Anthicides recueillis en Algérie en Mai et Juin 1894. *L'Échange, Revue Linnéenne* **10**: 84–88.

Pic, M. 1894j: Descriptions de trois coléoptères hétéromères nouveaux. *Le Naturaliste* (2) **16**: 93.

Pic, M. 1894k: Deux Coléoptères nouveaux. *Feuille des Jeunes Naturalistes* (3) **24**: 140.

Pic, M. 1894l: Xylophilides et anthicides recueillis en Algérie en Mai et Juin 1894. *Revue Scientifique du Bourbonnais et du Centre du France* **7**: 140–146.

Pic, M. 1894m: Contribution à la faune coléoptérologique d'Algérie. *Le Naturaliste* (2) **16**: 180.

Pic, M. 1894n: Anthicides du Bengale. *Annales de la Société Entomologique de Belgique* **38**: 181–184.

Pic, M. 1894o: Descriptions de coléoptères. *Mémoires de la Société Zoologique de France* **7**: 203–207.

Pic, M. 1894p: Contribution à la faune coléoptérique d'Algérie. Description d'espèces nouvelles. *Le Naturaliste* **16**: 247–248.

Pic, M. 1894q: Liste complète des xylophilides décrits jusqu'en 1894, avec descriptions d'espèces nouvelles. *Mémoires de la Société Zoologique de France* **7**: 427–436.

Pic, M. 1894r: Nouvelles espèces d'anthicides appartenant au Musée Civique de Gênes. *Annali del Museo Civico di Storia Naturale di Genova* **34** [(2) **14**] [1894–1895]: 582–587.

Pic, M. 1894s: [new taxa]. *Bulletin de la Société Entomologique de France* **1893**: cclxxvi-cclxxix.

Pic, M. 1894t: [Descriptions de deux coléoptères nouveaux de la Chine orientale]. *Bulletin de la Société Entomologique de France* **1894**: x-xi.

Pic, M. 1894u: [Trois Anthicus nouveaux et observations sur quelques anthicides]. *Bulletin de la Société Entomologique de France* **1894**: lxxvii-lxxix.

Pic, M. 1895a: Examen des anthicides de la collection Lethierry. *L'Échange, Revue Linnéenne* **11**: 18–20.

Pic, M. 1895b: Observations et renseignements divers. *L'Échange, Revue Linnéenne* **11**: 51–53.

Pic, M. 1895c: Description de coléoptères nouveaux de la famille des anthicides. *Le Naturaliste* (2) **17**: 59–60.

Pic, M. 1895d: Anthicides récoltés dans les tabacs. *Bulletin de la Société Zoologique de France* **20**: 61–64.

Pic, M. 1895e: Descriptions de coléoptères d'Algérie. *L'Échange, Revue Linnéenne* **11**: 78–82.

Pic, M. 1895f: Un nouveau Tomoderus de la Russie d'Asie (Col.). *Bulletin de la Société Entomologique de France* **1895**: 174.

Pic, M. 1895g: Description de trois Anthicus méditerranéens. *L'Abeille, Journal d'Entomologie* **28** [1892–1896]: 195–196.

Pic, M. 1895h: Descriptions de coléoptères. *Le Naturaliste* (2) **9**: 243.

Pic, M. 1895i: Anthicides d'Afrique de la collection de M. Charles Alluaud. *Annales de la Société Entomologique de France* **63** [1894]: 665–668.

Pic, M. 1895j: Anthicus nouveaux du Musée Civique de Gênes. *Annali del Museo Civico di Storia Naturale di Genova* **34**: 784–786.

Pic, M. 1895l: Notes sur des coléoptères rares ou nouveaux d'Algérie. *Bulletin de la Société Entomologique de France* **1895**: cxxvi-cxxx.

Pic, M. 1895m: [Description de deux coléoptères nouveaux et une note sur Anthicus dromioides Pic]. *Bulletin de la Société Entomologique de France* 1894: cxci-cxcii.

Pic, M. 1895n: [Diagnoses de plusiers coléoptères rapportés de Syrie par M.C. Delagrange]. *Bulletin de la Société Entomologique de France* 1894: cclxxxiv-cclxxxv.

Pic, M. 1896a: Descriptions de deux Anthicus syriens. *Miscellanea Entomologica* **4**: 5–6.

Pic, M. 1896b: Excursion entomologique dans la province d'Oran (Algérie). *Revue Scientifique du Bourbonnais et du Centre de France* **8**: 10–13.

Pic, M. 1896c: Xylophilides et Anthicides recueillis en Algérie Avril-Juin 1895. *Revue Scientifique du Bourbonnais et du Centre de France* **9**: 37–41.

Pic, M. 1896d: Descriptions et notes sur divers coléoptères. *Miscellanea Entomologica* **4**: 41–43.

Pic, M. 1896e: Notes et renseignements sur les xylophilides (Coléoptères hétéromères). *Bulletin de la Société Zoologique de France* **21**: 49–53.

Pic, M. 1896f: Descriptions et notes diverses. *L'Échange, Revue Linnéenne* **12**: 61–62.

Pic, M. 1896g: Notes et diagnoses. *L'Échange, Revue Linnéenne* **12**: 87–88.

Pic, M. 1896h: Coléoptères du Nord de L'Afrique. *Miscellanea Entomologica* **4**: 93–94.

Pic, M. 1896i: Notes et renseignements descriptifs sur Anthicus insignis, Luc et races voisines. *L'Échange, Revue Linnéenne* **12**: 131–132.

Pic, M. 1896j: Descriptions de coléoptères d'Algérie et de Syrie. *Miscellanea Entomologica* **4**: 140–142.

Pic, M. 1896l: Xylophilides et anthicides recueillis en Algérie (Mai-Juin 1896). *Revue Scientifique du Bourbonnais et du Centre de France* **9**: 162–169.

Pic, M. 1896m: Examen des anthicides de la collection Reitter. *La Feuille des Jeunes Naturalistes* (3) **26**: 178–181.

Pic, M. 1896n: Description d'un Anthicus d'Algérie (Col.). *Bulletin de la Société Entomologique de France* **1896**: 422–423.

Pic, M. 1896o: Observations, remarques et renseignements entomologiques divers. *Bulletin de la Société Entomologique de France* **1896**: 72–73.

Pic, M. 1897a: Excursion entomologique dans le sud de l'Algérie. *Miscellanea Entomologica* **5**: 2–3, 35–38.

Pic, M. 1897b: *Variétés II*. Lyon: L. Jacquet.

Pic, M. 1897c: Coléoptères nouveaux. *Miscellanea Entomologica* **5**: 26–29.

Pic, M. 1897d: Description de coléoptères nouveaux, d'Europe et circa. *Le Frelon* **6** [1897–1898]: 35–36.

Pic, M. 1897e: Descriptions de coléoptères. *Miscellanea Entomologica* **5**: 5–6, 41–43.

Pic, M. 1897f: Sur le groupe Liparoderus, Laf. dans le genre Anthicus, Payk. *L'Échange, Revue Linnéenne* **13**: 50–51.

Pic, M. 1897g: Descriptions de coléoptères. *Miscellanea Entomologica* **5**: 61–62.

Pic, M. 1897h: Description d'un Corticus nouveau d'Algérie (Col.). *Bulletin de la Société Entomologique de France* **1897**: 78–79.

Pic, M. 1897i: Note sur les xylophilides (coléoptères hétéromères). *Bulletin de la Société Zoologique de France* **22**: 80–82.

Pic, M. 1897j: Descriptions de coléoptères. *La Feuille des Jeunes Naturalistes* (3) **27**: 119–120.

Pic, M. 1897l: Notes sur quelques Zonabris Har. (Mylabris F.). *Revue Scientifique du Bourbonnais et du Centre de France* **10**: 123–126.

Pic, M. 1897m: En route pour la Mzab.- Bou Saada (fin). *Miscellanea Entomologica* **10**: 124–127.

Pic, M. 1897n: Anthicides exotiques nouveaux. *Le Naturaliste* (2) **19**: 134.

Pic, M. 1897o: [new taxon]. In: Ragusa, E.: Catalogo ragionato dei coleotteri di Sicilia. Meloidae. *Il Naturalista Siciliano* (N.S.) **2**: 212–222.

Pic, M. 1897r: Descriptions de coléoptères. *Comptes Rendus des Séances et des Excursions de la Société d'Histoire Naturelle d'Autun* **10**: 295–300.

Pic, M. 1898a: Descriptions de coléoptères d´Anatolie & de Syrie. *Revue Scientifique du Bourbonais et du Centre de la France* **11**: 1–4.

Pic, M. 1898b: Variétés et nouvelles espèces de coléoptères. *Miscellanea Entomologica* **6**: 2–4.

Pic, M. 1898c: Description de coléoptères nouveaux, d'Europe et circa. *Le Frelon* **6**: 35–38.

Pic, M. 1898d: Description de coléoptères nouveaux. *Le Naturaliste* (2) **20**: 63.

Pic, M. 1898e: Anthicides (Col. hétéromères) africains nouveaux des collections du Muséum de Paris. *Bulletin de la Museum d'Histoire Naturelle, Paris* **4**: 67–72.

Pic, M. 1898f: Notes descriptives sur plusieurs coléoptères et sur un Ichneumon (Hyménoptère). *Miscellanea Entomologica* **6**: 73–75.

Pic, M. 1898g: Descriptions d'espèces ou variétés de coléoptères de la faune d'Eupope [sic] et circa. *Miscellanea Entomologica* **6**: 97–99.

Pic, M. 1898h: Description de coléoptères. *Bulletin de la Société d'Histoire Naturelle d'Autun* **11**: 116–125.

Pic, M. 1898i: Description de trois Anthicidae du Turkestan. *Revue d'Entomologie* **17**: 122–124.

Pic, M. 1898j: Un Amblyderus (Col. Hétéromères) nouveau d'Abyssinie. *Bulletin du Museum d'Histoire Naturelle, Paris* **1898**: 181.

Pic, M. 1898k: Description de trois coléoptères nouveaux de Madagascar et de l'île Maurice. *Bulletin de la Société Entomologique de France* **1898**: 182–183.

Pic, M. 1898l: Répertoire des publications zoologiques (1889–1897). *Annales de la Société Entomologique de France* **67**: 191–224.

Pic, M. 1898m: [new taxa]. In: Ragusa, E.:Catalogo ragionato dei coleotteri di Sicilia. *Il Naturalista Siciliano* **2** [1897]: 197–237.

Pic, M. 1899a: Contribution à l´étude du genre Chrysanthia Schm. *Feuille des Jeunes Naturalistes* **30**: 14–16.

Pic, M. 1899b: Notes sur les anthicides d'Europe avec diagnoses. *Miscellanea Entomologica* **7**: 18–21.

Pic, M. 1899c: Diagnoses de ptinides et anthicides (Col.) des collections du Muséum de Paris. *Bulletin du Muséum d'Histoire Naturelle, Paris* 5: 28–32.

Pic, M. 1899d: Liste de coléoptères de mes chasses en 1898 dans les Alpes. *Miscellanea Entomologica* 1: 50–51.

Pic, M. 1899e: Descriptions d'Anthicidae. *Revue de Entomologie* 18: 73–76.

Pic, M. 1899f: Anthicide et pedilide (Coléoptères hétéromères) recueillis au Sikkim par M. Harmand, et offerts par lui au Muséum d'Histoire Naturelle. *Bulletin du Muséum d'Histoire Naturelle, Paris* 5: 76–79.

Pic, M. 1899g: Diagnoses de coléoptères communiqués par M. L. Bleuse. *Bulletin de la Société Entomologique de France* **1899**: 85–86.

Pic, M. 1899h: Sur divers coléoptères de la faune paléarctique. *Miscellanea Entomologica* 7: 113–116.

Pic, M. 1899i: Xylophides et anthicides recueillis en Orient, en mars, avril et mai 1899. *Revue Scientifique du Bourbonnais et du Centre de la France* **12**: 170–178.

Pic, M. 1899j: Description de coléoptères. *Le Naturaliste* (2) **21**: 189.

Pic, M. 1899k: Les coléoptères anthicides du centre de la France. *Bulletin Trimestriel de la Société d'Histoire Naturelle de Macon* **1**: 189–196.

Pic, M. 1899l: Diagnoses de coléoptères d'Orient. *Bulletin de la Société d'Histoire Naturelle d'Autun* **12** (2): 204–210.

Pic, M. 1899m: Description de coléoptères nouveaux. *Le Naturaliste* (2) **21**: 264.

Pic, M. 1899n: Diagnoses de deux coléoptères nouveaux de Palestine. *Bulletin de la Société Entomologique de France* **1899**: 278–279.

Pic, M. 1899o: Description d'une variété nouvelle d'Alleculidae (Col.) de l'Espagne centrale. *Bulletin de la Société Entomologique de France* **1899**: 300.

Pic, M. 1899p: Diagnoses préliminaires d'Helopidae d'Asie Mineure (Col.). *Bulletin de la Société Entomologique de France* **1899**: 411.

Pic, M. 1899q: Notes sur trois Anthicus de la région méditerranéenne. *Bulletin de la Société Entomologique de France* **1899**: 412.

Pic, M. 1899r: Notes sur trois Anthicus de la région méditerranéenne. *Bulletin de la Société Entomologique de France* **1899**: 412.

Pic, M. 1900a: *Bestimmungs-Tabellen der europäischen Coleopteren.* XL. *Heft. Hylophilidae (früher Euglenini und Xylophilini)*. Paskau: Edm. Reitter, 21 pp.

Pic, M. 1900b: [new taxa]. In: Bodemeyer, E. von.: *Quer durch Klein-Asien in der Bulghar-Dagh. Eine naturwissenschaftliche Studien-Reise*. Emmendingen: Verlangs-Aktiengesellschaft vormals Dölter, v + 169 pp.

Pic, M. 1900c: Neue Coleopteren des Hamburger Museums. *Mitteilungen des Zoologischen Staatsinstitut und Zoologischen Museum in Hamburg* 17 (2): 9–10.

Pic, M. 1900d: Diagnoses de coléoptères d'Orient récoltés en 1899. *Bulletin de la Socété Naturelle d'Autun* **1900**: 24–31.

Pic, M. 1900e: Notes sur divers coléoptères. *L'Échange, Revue Linnéenne* **16**: 37–38.

Pic, M. 1900f: Descriptions de coléoptères algériens et tunisiens. *L'Èchange, Revue Linnéenne* **16**: 53–56.

Pic, M. 1900g: Notes et diagnoses. *L'Échange, Revue Linnéenne* **16**: 61–63.

Pic, M. 1900h: Contribution à l'étude des Notoxus (Coléoptères) d'Europe et de régions avoisinantes. *Feuille des Jeunes Naturalistes* **30**: 64–68.

Pic, M. 1900i: Notes et diagnoses. 2e article. *L'Èchange, Revue Linnéenne* **16**: 77–80.

Pic, M. 1900j: Diagnoses de coléoptères du globe. *L'Èchange, Revue Linnéenne* **16**: 89–91.

Pic, M. 1900k: Contribution a l'étude des Notoxus (Coléoptères) d'Europe et des régions avoisinantes. *Feuille des Jeunes Naturalistes* **30**: 89–91.

Pic, M. 1900l: Descriptions et habitats nouveaux de divers coléoptères d'Algérie et d'Orient. *Bulletin de la Société Entomologique de France* **1900**: 123–125.

Pic, M. 1900m: Contribution à l'étude des coléoptères de la Tripolitaine et de la Tunisie. *Bulletin de la Société Entomologique de France* **1900**: 164–166.

Pic, M. 1900n: Anthicidae de l'Erythrée. *Annali del Museo Civico di Storia Naturale di Genova* (2) **20**: 575–576.

Pic, M. 1900o: Diagnoses d'Anthicidae de la Nouvelle Guinée. *Annali del Museo Civico di Storia Naturale di Genova* **20**: 602–608.

Pic, M. 1901a: Notes diverses et diagnoses. *L'Échange, Revue Linnéenne* **17**: 2–4.

Pic, M. 1901b: Notes diverses et diagnoses (2e article). *L'Échange, Revue Linnéenne* **17**: 9–12.

Pic, M. 1901c: Notes diverses et diagnoses (5e article). *L'Échange, Revue Linnéenne* **17**: 33–35.

Pic, M. 1901d: Coléoptères d'Orient. *Comptes Rendus des Séances et des Excursions de la Société d'Histoire Naturelle d'Autun* **14**: 33–39.

Pic, M. 1901e: Descriptions abrégées de coléoptères provenant de Grèce. *L'Échange, Revue Linnéenne* **17**: 49–52.

Pic, M. 1901f: Notes diverses et diagnoses (6e Article). *L'Échange, Revue Linnéenne* **17**: 57–59.

Pic, M. 1901g: Notes diverses et diagnoses (7e article). *L'Échange, Revue Linnéenne* **17**: 65–68.

Pic, M. 1901h: Notes diverses sur le genre Zonabris Harold. *Bulletin de la Société Zoologique de France* **26**: 77–78.

Pic, M. 1901i: Communications diverses. 1. Espèces et variétés nouvelles de coléoptères paléarctiques. *L'Échange, Revue Linnéenne* **17**: 79–80.

Pic, M. 1901j: Diagnoses d'Anthicidae exotiques. *Annales de la Société Entomologique de Belgique* **45**: 89–90.

Pic, M. 1901k: Coléoptères nouveaux d'Orient, Arménie et Tunisie. *L'Échange, Revue Linnéenne* **17**: 89–91.

Pic, M. 1901l: Sur quelques variétés de Zonabris, du Turkestan (Col.). *Bulletin de la Société Entomologique de France* **1901**: 110–111.

Pic, M. 1901m: Note sur les Anthicus terminatus Schmt. et terminatus Laf. *Miscellanea Entomologica* **9**: 129–131.

Pic, M. 1901n: Xylophilides et anthicides capturés en Italie et Grèce, du 17 avril au 14 juin 1901. *Revue Scientifique du Bourbonnais et du Centre de la France* **14**: 173–182.

Pic, M. 1901o: Liste de quelques coléoptères recueillis en Grèce en 1901. *Bulletin de la Société Entomologique de France* **1901**: 253–255.

Pic, M. 1901p: Diagnoses de Macratria et d'Anthicidae de la Malaysie. *Annali del Museo Civico di Storia Naturale di Genova* (2) **20** [1900]: 740–742, 791–803.

Pic, M. 1901q: Description de Coléoptères nouveaux. *Le Naturaliste* **23**: 44.

Pic, M. 1902a: Coléoptères asiatiques nouveaux. *L'Échange, Revue Linnéenne* [18 sic] **17**: 1–3.

Pic, M. 1902b: Descriptions et notes diverses. *L'Échange, Revue Linnéenne* [18 sic] **17**: 9–10.

Pic, M. 1902c: Descriptions et notes diverses (2e article). *L'Échange, Revue Linnéenne* [18 sic] **17**: 17–19.

Pic, M. 1902d: Diagnoses de coléoptères de l'ancien et du nouveau monde. *L'Échange, Revue Linnéenne* **17**: 23–26.

Pic, M. 1902e: Nouvelles espèces et variétés de coléoptères paléarctiques. *L'Échange, Revue Linnéenne* **17**: 31–38.

Pic, M. 1902f: Deux notes entomologiques avec synopsis. *L'Échange, Revue Linnéenne* [18 sic] **17**: 41–42.

Pic, M. 1902g: Descriptions et notes diverses (4e article). *L'Échange, Revue Linnéenne* [18 sic] **17**: 55–57.

Pic, M. 1902h: Coléoptères présumés nouveaux de la faune circaméditerranéenne. *L'Échange, Revue Linnéenne* [18 sic] **17**: 63–65.

Pic, M. 1902i: Descriptions et notes diverses. *L'Échange, Revue Linnéenne* [18 sic] **17**: 72.

Pic, M. 1902j: Nouvelles espèces et variétés de coléoptères paléarctiques. *L'Échange, Revue Linnéenne* [18 sic] **17**: 79–80.

Pic, M. 1902k: Coléoptères nouveaux de l'Ile de Ténériffe. *L'Échange, Revue Linnéenne* [18 sic] **17**: 80–81.

Pic, M. 1902l: Anthicides recueillis en Orient par M. Paul de Peyerimhoff (Col.). *Bulletin de la Société Entomologique de France* **1902**: 185–187.

Pic, M. 1902m: Anthicidae exotiques nouveaux du Musee national Hongrois. *Természetrajzi füzetek* **25**: 405–410.

Pic, M. 1903a: Anthicides nouveaux des collections du Muséum de Paris (Col.). *Bulletin du Muséum d'Histoire Naturelle, Paris* **9**: 20–21.

Pic, M. 1903b: Notes entomologiques. *Bulletin de la Société Entomologique de France* **1903**: 79–83.

Pic, M. 1903c: Notes entomologiques. vi. Description d'un Notoxus de Djibouti. *Bulletin de la Société Entomologique de France* **1903**: 83.

Pic, M. 1903d: Nouveaux anthicides des collections du Musée Zoologique de l'Académie Impériale des Sciences à St. Pétersbourg. (Coleoptera, Anthicidae). *Annuaire du Musée Zoologique de l'Académie Impériale des Sciences de St. Pétersbourg* **8**: 88–89.

Pic, M. 1903e: Descriptions et notes complémentaires sur divers coléoptères paléarctiques. *L'Échange, Revue Linnéenne* **19**: 97–98.

Pic, M. 1903f: Coléoptères exotiques nouveaux (2e article). *L'Échange, Revue Linnéenne* **19**: 105–107.

Pic, M. 1903g: Descriptions et habitats de divers coléoptères du Nord de l'Afrique. *L'Échange, Revue Linnéenne* **19**: 113–115.

Pic, M. 1903h: Coléoptères nouveaux d'Asie et d'Afrique. *L'Échange, Revue Linnéenne* **19**: 119–120.

Pic, M. 1903i: Notes entomologiques et descriptions. *L'Échange, Revue Linnéenne* **19**: 121–125

Pic, M. 1903j: Coléoptères européens et Clytus syrien nouveaux. *L'Échange, Revue Linnéenne* **19**: 138–139.

Pic, M. 1903k: Contribution à l'étude du genre «Salpingus» Gyllh. *L'Échange, Revue Linnéenne* **19**: 140–141.

Pic, M. 1903l: Deux captures intéressantes. Diagnoses de divers coléoptères. *L'Échange, Revue Linnéenne* **19**: 145–147.

Pic, M. 1903m: Espèces et variétés nouvelles de coléoptères. *L'Échange, Revue Linnéenne* **19**: 161–163.

Pic, M. 1903n: Espèces et variétés nouvelles de coléoptères (2me article). *L'Échange, Revue Linnéenne* **19**: 169–171.

Pic, M. 1903o: Espèces et variétés nouvelles de coléoptères (3me article). *L'Échange, Revue Linnéenne* **19**: 177–178.

Pic, M. 1903p: Coléoptères rares ou nouveaux pour le département, recueilles en 1903. *Bulletin de la Société Histoire Naturelle d'Autun* **16**: 187–192.

Pic, M. 1903q: Note synonymique (Col.). *Bulletin de la Société Entomologique de France* **1903**: 304.

Pic, M. 1903r: Anthicides de Manille. *Annales de la Société Entomologique de France* **71** [1902]: 643–648.

Pic, M. 1903s: Voyage de M. Maurice Maindron dans l'Inde méridionale (Mai à Novembre 1901). 3e mémoire. *Annales de la Société Entomologique de France* **72**: 348–352.

Pic, M. 1903t: Anthicides exotiques nouveaux. *Le Naturaliste* **25**: 56.

Pic, M. 1904a: Notes et descriptions. *L'Échange, Revue Linnéenne* **20**: 2–5.

Pic, M. 1904b: Notes et descriptions (Suite). *L'Échange, Revue Linnéenne* **20**: 9–11.

Pic, M. 1904c: Diagnoses de seize coléoptères exotiques appartenant à diverses familles. *L'Échange, Revue Linnéenne* **20**: 18–20.

Pic, M. 1904d: Diagnoses de coléoptères asiatiques provenant surtout de Sibéria. *L'Échange, Revue Linnéenne* **20**: 25–27.

Pic, M. 1904e: Diagnoses de divers coléoptères d'Europe et Turquie d'Asie. *L'Échange, Revue Linnéenne* **20**: 57–58.

Pic, M. 1904f: Coléoptères nouveaux de la Turquie d'Asie. *L'Échange, Revue Linnéenne* **20**: 73–74.

Pic, M. 1904g: Nouvelles espèces et variétés de coléoptères paléarctiques. *L'Échange, Revue Linnéenne* **20**: 81–82.

Pic, M. 1904h: Diagnoses ou descriptions abrégées de coléoptères paléarctiques. *L'Échange, Revue Linnéenne* **20**: 89–94.

Pic, M. 1904i: Espèces et variétés nouvelles de coléoptères paléarctiques. *L'Échange, Revue Linnéenne* **21**: 97–100.

Pic, M. 1905a: Notes entomologiques diverses (Col.). *Bulletin de la Société Entomologique de France* **1905**: 47–48.

Pic, M. 1905b: Espèces et variétés nouvelles de coléoptères paléarctiques. *L'Échange, Revue Linnéenne* **21**: 97–110.

Pic, M. 1905c: Description de deux Anthicus nouveaux (Col.). *Bulletin de la Société Entomologique de France* **1905**: 99–100.

Pic, M. 1905d: Descriptions abrégées et notes diverses. *L'Échange, Revue Linnéenne* **21**: 105–107.

Pic, M. 1905e: Descriptions abrégées et notes diverses (2e article). *L'Échange, Revue Linnéenne* **21**: 113–115.

Pic, M. 1905f: Descriptions abrégées et notes diverses. *L'Échange, Revue Linnéenne* **21**: 129–131.

Pic, M. 1905g: Coléoptères nouveaux provenant de France, Grèce, Algérie et Turquie d'Asie. *L'Échange, Revue Linnéenne* **21**: 161–163.

Pic, M. 1905h: Petite contribution à l'étude des Myiodes ou Myodites Latr. paléarctiques. *L'Échange, Revue Linnéenne* **21**: 165–166.

Pic, M. 1905i: Sur les Zonabris Wagneri et curta Chevrolat. *Revue Scientifique du Bourbonnais et du Centre de France* **1905**: 167–170.

Pic, M. 1905j: Espèces et variétés nouvelles de coléoptères de France et Turquie d'Asie. *L'Échange, Revue Linnéenne* **21**: 177–179.

Pic, M. 1905k: Notes entomologiques et descriptions. *L'Échange, Revue Linnéenne* **21**: 185–187.

Pic, M. 1905l: Coléoptères asiatiques nouveaux. *Bulletin de la Société Entomologique de France* 1904: 287–289.

Pic, M. 1905m: Notes sur diverses espèces d'anthicides de Motschulsky et descriptions de plusiers nouveautés du même groupe. *Annuaire du Musée Zoologique de l'Académie Impériale des Sciences de St.Pétersbourg* 9 [1904]: 490–494.

Pic, M. 1906a: Deux nouveaux coléoptères d'Espagne. *L'Échange, Revue Linnéenne* 22: 12–13.

Pic, M. 1906b: Coléoptères européens, algériens et asiatiques nouveaux. *L'Échange, Revue Linnéenne* 22: 17–20.

Pic, M. 1906c: Habitats et descriptions de divers coléoptères paléarctiques. *L'Échange, Revue Linnéenne* 22: 25–27.

Pic, M. 1906d: Contribution à l'étude des pyrochroides. *L'Échange, Revue Linnéenne* 22: 28–30.

Pic, M. 1906e: Habitats et descriptions de divers coléoptères paléarctiques. *L'Échange, Revue Linnéenne* 22: 33–35.

Pic, M. 1906f: Coléoptères exotiques nouveaux ou peu connus. *L'Échange, Revue Linnéenne* 22: 43–45.

Pic, M. 1906g: Un nouvel Anthicus du Turkestan (Col.). *Bulletin de la Société Entomologique de France* 1906: 46–47.

Pic, M. 1906h: Coléoptères en partie nouveaux provenant de la Perse et des régions avoisinantes. *L'Échange, Revue Linnéenne* 22: 49–51.

Pic, M. 1906i: Deux nouveaux Anthicus du Kashmir (Col.). *Bulletin de la Société Entomologique de France* 1906: 78–79.

Pic, M. 1906j: Anthicides et ptinides d'Algérie et de Tunisie (Col.) récoltés par M. le Capitaine Vibert. *Bulletin de la Société Entomologique de France* 1906: 282–285.

Pic, M. 1907a: Anthicides recueillis en Erythrée. *Annali del Museo Civico di Storia Naturale, Genova* 43: 63–65.

Pic, M. 1907b: Coléoptères exotiques nouveaux ou peu connus (Suite). *L'Échange, Revue Linnéenne* 23: 119–120.

Pic, M. 1907c: Notes entomologiques diverses. *L'Échange, Revue Linnéenne* 23: 121–123.

Pic, M. 1907d: Coléoptères exotiques nouveaux ou peu connus (Suite). *L'Échange, Revue Linnéenne* 23: 134–135.

Pic, M. 1907e: Notes entomologiques diverses (Suite). *L'Échange, Revue Linnéenne* 23: 137–139.

Pic, M. 1907f: Notes entomologiques diverses. *L'Échange, Revue Linnéenne* 23: 153–154.

Pic, M. 1907g: Coléoptères exotiques nouveaux ou peu connus (Suite). *L'Échange Revue Linnéenne* 23: 158.

Pic, M. 1907h: Notes entomologiques diverses. *L'Échange, Revue Linnéenne* 23: 161–162.

Pic, M. 1907i: Coléoptères exotiques nouveaux ou peu connus (Suite). *L'Échange, Revue Linnéenne* 23: 165–166.

Pic, M. 1907j: Diagnoses de coléoptères asiatiques nouveaux. *L'Échange, Revue Linnéenne* 23: 171–174.

Pic, M. 1907k: Coléoptères exotiques nouveaux ou peu connus (Suite). *L'Échange, Revue Linnéenne* 23: 175–176.

Pic, M. 1907l: Notes entomologiques diverses. *L'Échange, Revue Linnéenne* 23: 177–179.

Pic, M. 1907m: Coléoptères exotiques nouveaux ou peu connus. (Suite.) *L'Échange, Revue Linnéenne* 23: 181–183.

Pic, M. 1907n: Coléoptères exotiques nouveaux ou peu connus (Suite). *L'Échange, Revue Linnéenne* **23**: 190–191.

Pic, M. 1907o: Un Silis et deux anthicides nouveaux provenant des Indes (Col.). *Bulletin de la Société Entomologique de France* **1907**: 195–196.

Pic, M. 1907p: Anthicides recueillis par M. Ch. Alluaud dans le Soudan Égyptien oriental (Col.) (Novembre 1905-Mai 1906). *Bulletin de la Société Entomologique de France* **1907**: 321–323.

Pic, M. 1907q: Ueber neue oder wenig bekannte Käfer aus Asien, Amerika und Deutsch-Ostafrika. *Stettiner Entomologische Zeitung* **68**: 337–342.

Pic, M. 1907r: Macratria et Anthicidae recueillis par feu L. Fea en Birmanie et regions avoisinantes. *Annali del Museo Civico di Storia Naturale di Genova* **43** [1907–1908]: 339–342.

Pic, M. 1908a: Contribution à l'étude des Osphya paléarctiques. *L'Échange, Revue Linnéenne* **24**: 31–32.

Pic, M. 1908b: Descriptions ou diagnoses et notes diverses (Suite). *L'Échange, Revue Linnéenne* **24**: 33–35.

Pic, M. 1908c: Étude synoptique sur divers „Cteniopini" (Col. hétéromères). *L'Échange, Revue Linnéenne* **24**: 38–40.

Pic, M. 1908d: Descriptions ou diagnoses et notes diverses (Suite). *L'Échange, Revue Linnéenne* **24**: 41–46.

Pic, M. 1908e: Étude synoptique sur divers „Cteniopini" (Col. hétéromères). *L'Échange, Revue Linnéenne* **24**: 47–48.

Pic, M. 1908f: Descriptions ou diagnoses et notes diverses (Suite). *L'Échange, Revue Linnéenne* **24**: 57–58.

Pic, M. 1908g: Coléoptères exotiques nouveaux ou peu connus. *L'Échange, Revue Linnéenne* **24**: 59–61.

Pic, M. 1908h: Étude synoptique sur divers „Cteniopini" (Col. hétéromères). *L'Échange, Revue Linnéenne* **24**: 61–62.

Pic, M. 1908i: Descriptions ou diagnoses et notes diverses (Suite). *L'Échange, Revue Linnéenne* **24**: 65–66.

Pic, M. 1908j: Captures intéressantes. Moeurs et nouvelles variétés de coléoptères français. *L'Échange, Revue Linnéenne* **24**: 71–72.

Pic, M. 1908k: Descriptions ou diagnoses et notes diverses (Suite). *L'Échange, Revue Linnéenne* **24**: 73–75.

Pic, M. 1908l: Deux nouveaux Pyrochroa Geoff. de Chine. *Bulletin de la Société Entomologique de France* **1908**: 75.

Pic, M. 1908m: Description d'un Anthicus nouveau d'Égypte. *Bulletin de la Société Entomologique d'Egypte* **1** [1908–1909]: 143–144.

Pic, M. 1908n: Deux nouveaux hétéromères exotiques (Col.). *Bulletin de la Société Entomologique de France* **1908**: 274–275.

Pic, M. 1909a: Divers coléoptères exotiques mouveaux. *Le Naturaliste* **31**: 19.

Pic, M. 1909b: Nouveaux coléoptères (malacodermes et hétéromères) des Indes. *Bulletin de la Société Entomologique de France* **1909**: 37–39.

Pic, M. 1909c: Deux nouveaux Anthicus de l'Himalaya (Col.). *Bulletin de la Société Entomologique de France* **1909**: 56–57.

Pic, M. 1909d: Descriptions ou diagnoses et notes diverses (Suite). *L'Échange, Revue Linnéenne* **25**: 105–106.

Pic, M. 1909e: Coléoptères exotiques nouveaux ou peu connus (Suite). *L'Échange, Revue Linnéenne* **25**: 108–110.

Pic, M. 1909f: Descriptions ou diagnoses et notes diverses (Suite). *L'Échange, Revue Linnéenne* **25**: 113–115.

Pic, M. 1909g: Coléoptère hétéromère pedilide nouveau du Bengale. *Bulletin du Muséum Nationale d'Histoire Naturelle, Paris* **15**: 109–110.

Pic, M. 1909h: Coléoptères exotiques nouveaux ou peu connus (Suite). *L'Échange, Revue Linnéenne* **25**: 117–118.

Pic, M. 1909i: Descriptions ou diagnoses et notes diverses (Suite). *L'Échange, Revue Linnéenne* **25**: 121–122.

Pic, M. 1909j: Descriptions ou diagnoses et notes diverses (Suite). *L'Échange, Revue Linnéenne* **25**: 129–131.

Pic, M. 1909k: Coléoptères exotiques nouveaux ou peu connus (Suite). *L'Échange, Revue Linnéenne* **25**: 133–134.

Pic, M. 1909l: Coléoptères exotiques nouveaux ou peu connus (Suite). *L'Échange, Revue Linnéenne* **25**: 142–143.

Pic, M. 1909m: Coléoptères exotiques nouveaux ou peu connus (Suite). *L'Échange, Revue Linnéenne* **25**: 146–148, 160, 166–167.

Pic, M. 1909n: Descriptions ou diagnoses et notes diverses (Suite). *L'Échange, Revue Linnéenne* **25**: 161–162, 169–171.

Pic, M. 1909o: Coléoptères exotiques nouveaux ou peu connus (Suite). *L'Échange, Revue Linnéenne* **25**: 173–175.

Pic, M. 1909p: Descriptions ou diagnoses et notes diverses (Suite). *L'Échange, Revue Linnéenne* **25**: 177–179.

Pic, M. 1909q: Descriptions ou diagnoses et notes diverses (Suite). *L'Échange, Revue Linnéenne* **25**: 185–186.

Pic, M. 1909r: Pyrochroidae nouveaux. *Bulletin de la Société Entomologique de France* **1909**: 330–331.

Pic, M. 1910a: [new taxa]. In: Sainte-Claire Deville J.: Catalogue critique des coléoptères de la Corse. *Revue d'Entomologie* (Caen), *Supplément* **25–28** [1906–1910]: 1–400.

Pic, M. 1910b: Description ou diagnoses et notes diverses (Suite). *L'Échange, Revue Linnéenne* **26**: 1.

Pic, M. 1910c: Descriptions ou diagnoses et notes diverses (Suite). *L'Échange, Revue Linnéenne* **26**: 9–11.

Pic, M. 1910d: Descriptions ou diagnoses et notes diverses. *L'Échange, Revue Linnéenne* **26**: 12–14.

Pic, M. 1910e: Descriptions, diagnoses et notes diverses (Suite). *L'Échange, Revue Linnéenne* **26**: 17–18.

Pic, M. 1910f: Coléoptères: pédilides et anthicides recueillis au Japon, par M.M.J. Harmand et E. Gallois. *Bulletin du Muséum National d'Histoire Naturelle, Paris* **16**: 19–21.

Pic, M. 1910g: Coléoptères exotiques nouveaux ou peu connus (Suite.). *L'Échange, Revue Linnéenne* **26**: 20–22.

Pic, M. 1910h: Coléoptères: hylophilides nouveaux originaires d'Abyssinie et du Japon (Col.). *Bulletin du Museum d'Histoire Naturelle, Paris* **16**: 21–24.

Pic, M. 1910i: Description d'un nouveau Mecynotarsus d'Egypte. *Bulletin de la Société Entomologique d'Egypte* **3**: 26–27.

Pic, M. 1910j: Descriptions, diagnoses et notes diverses. (Suite). *L'Échange, Revue Linnéenne* **26**: 33–34.

Pic, M. 1910k: Description d'un nouveau Trotomma d'Algérie. *Bulletin de la Société Entomologique de France* **1910**: 36.

Pic, M. 1910l: Coléoptères malacodermes et hétéromères en partie nouveaux. *Bulletin de la Société d'Histoire Naturelle d'Autun* **23**: 37–39.

Pic, M. 1910m: Descriptions ou diagnoses et notes diverses. (Suite). *L'Échange, Revue Linnéenne* **26**: 41–43.

Pic, M. 1910n: Descriptions ou diagnoses et notes diverses (Suite). *L'Échange, Revue Linnéenne* **26**: 57–58.

Pic, M. 1910o: Descriptions ou diagnoses et notes diverses (Suite). *L´Échange, Revue Linnéenne* **26**: 65–66.

Pic, M. 1910p: Coléoptères exotiques nouveaux ou peu connus (Suite). *L'Échange, Revue Linnéenne* **26**: 69–71.

Pic, M. 1910q: Coléoptères exotiques nouveaux ou peu connus (Suite). *L'Échange, Revue Linnéenne* **26**: 74–78.

Pic, M. 1910r: Description d'un Hylophilus du Japon (Col. Heteromera). *Bulletin de la Société Entomologique de France* **1910**: 75.

Pic, M. 1910s: Coléoptères exotiques nouveaux ou peu connus (Suite.) *L'Échange, Revue Linnéenne* **26**: 86–87.

Pic, M. 1910t: Hétéromères nouveaux du group des Zonitini. *Bulletin de la Société Entomologique de France* **1910**: 90–91.

Pic, M. 1910u: Coléoptères exotiques nouveaux ou peu connus (Suite). *L´Échange, Revue Linnéenne* **26**: 94–95.

Pic, M. 1910v: Malacodermes et hétéromères nouveaux d'Afrique et d'Asie. *Le Naturaliste* **32**: 259.

Pic, M. 1910w: Malacodermes et hétéromères nouveaux d´Afrique et d´Asie. *Le Naturaliste* **32**: 271–272.

Pic, M. 1911a: Anthicidae. Pars 36. In: Schenkling, S. (ed.): *Coleopterorum Catalogus*. Berlin: W. Junk, 102 pp.

Pic, M. 1911b: Descriptions bréves de vingt-huit hétéromères exotiques. *Mélanges Exotico-Entomologiques* **1**: 5–13.

Pic, M. 1911c: Mission Niger-Tchad dirigée par le Capitaine Tilho. Collections recueillies par le Dr Gaillard. Coléoptères: Anthicides nouveaux (Hétéromères). *Bulletin du Muséum National d'Histoire Naturelle, Paris* **17**: 58–63.

Pic, M. 1911d: Coléoptères égyptiens nouveaux ou plus ou moins rares. *Bulletin de la Société Entomologique d'Egypte* **4**: 71–76.

Pic, M. 1911e: Descriptions ou diagnoses et notes diverses (Suite). *L'Échange, Revue Linnéenne* **27**: 97–98.

Pic, M. 1911f: Coléoptères exotiques nouveaux ou peu connus (Suite). *L'Échange, Revue Linnéenne* **27**: 99–101.

Pic, M. 1911g: Coléoptères exotiques nouveaux ou peu connus (Suite). *L'Échange, Revue Linnéenne* **27**: 117–118.

Pic, M. 1911h: [new taxon]. In: Andres A.: Note sur quelques espèces de Zonabris de la région de Mariout (environs d'Alexandrie). *Bulletin de la Société Entomologique d'Egypte* **4**: 129–135.

Pic, M. 1911i: Coléoptères exotiques nouveaux ou peu connus (Suite). *L'Échange, Revue Linnéenne* **27**: 132–134, 138.

Pic, M. 1911j: Descriptions ou diagnoses et notes diverses (Suite). *L'Échange, Revue Linnéenne* **27**: 145–146.

Pic, M. 1911k: Coléoptères exotiques nouveaux ou peu connus (Suite). *L'Échange, Revue Linnéenne* **27**: 155–159.

Pic, M. 1911l: Notes pour aider à l'étude de divers coléoptères paléarctiques principalement gallo-rhénans. *L'Échange, Revue Linnéenne* **27**: 161–162.

Pic, M. 1911m: Descriptions ou diagnoses et notes diverses (Suite). *L´Échange, Revue Linnéenne* **27**: 169–170.

Pic, M. 1911n: Descriptions ou diagnoses et notes diverses. *L'Échange, Revue Linnéenne* **27**: 177–178.

Pic, M. 1911o: Descriptions ou diagnoses et notes diverses (Suite). *L'Échange, Revue Linnéenne* **27**: 185–186.

Pic, M. 1911p: H. Sauter´s Formosa-Ausbeute. Cantharidae, Lampyridae, Mordellidae. (Col.). *Deutsche Entomologische National-Bibliothek* **24**: 188–189.

Pic, M. 1911q: Coléoptères exotiques nouveaux ou peu connus (Suite.) *L'Échange, Revue Linnéenne* **27**: 190–191.

Pic, M. 1912a: Descriptions d'espèces & variétés appartenant a diverses familles. *Mélanges Exotico-Entomologiques* 2: 2–11.

Pic, M. 1912b: Descriptions d'espèces et variétés diverses. *Mélanges Exotico-Entomologiques* 3: 6–20.

Pic, M. 1912c: Descriptions ou diagnoses et notes diverses (Suite). *L'Échange, Revue Linnéenne* 28: 1–3, 9–10.

Pic, M. 1912d: Synopsis pour aider à l'étude des Pyrochroa Geof. des Indes. *L'Échange, Revue Linnéenne* 28: 18.

Pic, M. 1912e: Descriptions ou diagnoses et notes diverses (Suite). *L'Échange, Revue Linnéenne* 28: 25–26

Pic, M. 1912f: Coléoptères exotiques nouveaux ou peu connus. (Suite). *L'Échange, Revue Linnéenne* 28: 29–32.

Pic, M. 1912g: Descriptions ou diagnoses et notes diverses (Suite). *L'Échange, Revue Linnéenne* 28: 33–35.

Pic, M. 1912h: Descriptions ou diagnoses et notes diverses (Suite). *L'Échange, Revue Linnéenne*, 28: 41–43.

Pic, M. 1912i: Coléoptères exotiques nouveaux ou peu connus (Suite). *L'Échange, Revue Linnéenne* 28: 46–47.

Pic, M. 1912j: Descriptions ou diagnoses et notes diverses (Suite). *L'Échange, Revue Linnéenne* 28: 49–51.

Pic, M. 1912k: Descriptions ou diagnoses et notes diverses (Suite). *L'Échange, Revue Linnéenne* 28: 57–58.

Pic, M. 1912l: Coléoptères exotiques nouveaux ou peu connus (Suite). *L'Échange, Revue Linnéenne* 28: 63–64.

Pic, M. 1912m: Coléoptères exotiques nouveaux ou peu connus (Suite). *L'Échange, Revue Linnéenne* 28: 68–69.

Pic, M. 1912n: Descriptions ou diagnoses et notes diverses. *L'Échange, Revue Linnéenne* 28: 73–74.

Pic, M. 1912o: Nouveaux coléoptères paléarctiques. *L'Échange, Revue Linnéenne* 28: 89–91.

Pic, M. 1912p: Sur les Pyrochroidae du Muséum de Paris (Hétéromères). *Bulletin du Muséum National d'Histoire Naturelle, Paris* 18: 142–144.

Pic, M. 1912q: Notes sur quelques vésicants coléoptères. *Bulletin de la Société Entomologique de France* 1912: 176–177.

Pic, M. 1912r: Coléoptères du Maroc. *Bulletin du Muséum National d'Histoire Naturelle, Paris* 18: 219–221.

Pic, M. 1912s: Anthicides et hylophilides provenant des chasses de M.G.E. Bryant (iles Ceylan et Trinité. – Australie). *Annales de la Société Entomologique de France* 81: 279–288.

Pic, M. 1913a: Descriptions de 29 espèces et de plusiers variétés. *Mélanges Exotico-Entomologiques* 5: 7–20.

Pic, M. 1913b: Espèces et variétés nouvelles appartenant à diverses familles. *Mélanges Exotico-Entomologiques* 6: 8–16.

Pic, M. 1913c: Nouvelles espèces de 'Toxicum' Latreille. *Mélanges Exotico-Entomologiques* 6: 17–20.

Pic, M. 1913d: Espèces et variétés nouvelles, ou peu connues, rentrant dans diverses familles. *Mélanges Exotico-Entomologiques* 7: 10–20.

Pic, M. 1913e: Coléoptères divers du Tonkin et de l'Indo-Chine. *Mélanges Exotico-Entomologiques* 9: 2–20.

Pic, M. 1913f: 'Coryna denticulata' Mars. et 'C. suturifera' Pic (Col. vésicants, hétéromères). *Revue Scientifique du Bourbonnais* 26: 110–112.

Pic, M. 1913g: Note diverses, descriptions et diagnoses (Suite.). *L'Échange, Revue Linnéenne* 29: 113–115.

Pic, M. 1913h: Notes diverses, descriptions et diagnoses(Suite.). *L'Échange, Revue Linnéenne* 29: 121–122.

Pic, M. 1913i: Notes diverses, descriptions et diagnoses (Suite). *L'Échange, Revue Linnéenne* 29: 129–130.

Pic, M. 1913j: H. Sauter's Formosa-Ausbeute. Anthicidae, Pedilidae und Hylophilidae. *Archiv für Naturgeschichte* (A) 79: 129–135.

Pic, M. 1913k: Notes diverses, descriptions et diagnoses (Suite). *L'Échange, Revue Linnéenne* 29: 137–138.

Pic, M. 1913l: Nouveaux coléoptères anthicides. *Bulletin de la Société Zoologique de France* 38: 137–139.

Pic, M. 1913m: Coléoptères exotiques en partie nouveaux (Suite.). *L'Échange, Revue Linnéenne* **29**: 142.

Pic, M. 1913n: Notes diverses, descriptions et diagnoses (Suite). *L'Échange, Revue Linnéenne* **29**: 145–147.

Pic, M. 1913o: Notes diverses, descriptions et diagnoses (Suite). *L'Échange, Revue Linnéenne* **29**: 153–154.

Pic, M. 1913p: Nouveaux Anthicides exotiques (Deuxième article). *L'Échange, Revue Linnéenne* **29**: 155–158.

Pic, M. 1913q: Notes diverses, descriptions et diagnoses (Suite). *L'Échange, Revue Linnéenne* **29**: 161–162.

Pic, M. 1913r: Coléoptères exotiques en partie nouveau (Suite). *L'Échange, Revue Linnéenne* **29**: 163–166.

Pic, M. 1913s: Notes diverses, descriptions et diagnoses (Suite.). *L'Échange, Revue Linnéenne* **29**: 169–171, 185–187.

Pic, M. 1913t: Coléoptères exotiques en partie nouveaux (Suite). *L'Échange, Revue Linnéenne* **29**: 175–176, 185–186.

Pic, M. 1913u: Deux nouveaux Formicomus Laf., de l'Inde (Col. Heteromera). *Bulletin de la Société Entomologique de France* **1913**: 204–205.

Pic, M. 1913v: Deux vésicants africains nouveaux (Col. Heteromera). *Bulletin de la Société Entomologique de France* **1913**: 456–457.

Pic, M. 1914a: Notes diverses, descriptions et diagnoses (Suite). *L'Échange, Revue Linnéenne* **30**: 1–2.

Pic, M. 1914b: Coléoptères divers du Tonkin et de l'Indo-Chine. *Mélanges Exotico-Entomologiques* **9**: 2–20.

Pic, M. 1914c: Nouveau genre, espèces et variétés nouvelles. *Mélanges Exotico-Entomologiques* **11**: 2–8.

Pic, M. 1914d: Descriptions abrégées de malacodermes et hétéromères. *Mélanges Exotico-Entomologiques* **11**: 8–20.

Pic, M. 1914e: Notes diverses, descriptions et diagnoses (Suite). *L'Échange, Revue Linnéenne* **30**: 10–13.

Pic, M. 1914f: Descriptions abrégées d'anthicides exotiques. *L'Échange, Revue Linnéenne* **30**: 34–36.

Pic, M. 1914g: Notes diverses, descriptions et diagnoses (Suite). *L'Échange, Revue Linnéenne* **30**: 41–42.

Pic, M. 1914h: Coléoptères exotiques en partie nouveaux (Suite). *L'Échange, Revue Linnéenne* **30**: 44.

Pic, M. 1914i: Coléoptères exotiques en partie nouveaux (Suite). *L'Échange, Revue Linnéenne* **30**: 54–56.

Pic, M. 1914j: Deux nouvelles espèces d'Oedemerides du Kashmir (Col. Heteromera). *Bulletin de la Société Entomologique de France* **1914**: 57–58.

Pic, M. 1914k: Coléoptères exotiques en partie nouveaux (Suite). *L'Échange, Revue Linnéenne* **30**: 61–62.

Pic, M. 1914l: Coléoptères exotiques en partie nouveaux (Suite). *L'Échange, Revue Linnéenne* **30**: 67–68.

Pic, M. 1914m: Notes diverses, descriptions et diagnoses (Suite). *L'Échange, Revue Linnéenne* **30**: 73–75.

Pic, M. 1914n: Coléoptères exotiques en partie nouveaux (Suite). *L'Échange, Revue Linnéenne* **30**: 75–76.

Pic, M. 1914o: Notes diverses, descriptions et diagnoses (Suite). *L'Échange, Revue Linnéenne* **30**: 77–79.

Pic, M. 1914p: Notes sur divers vésicants des collections du Musée Civique de Gênes avec descriptions des nouveautés. *Annali del Museo Civico di Storia Naturale di Genova* (3) **6**: 103–115.

Pic, M. 1914q: Anthicides exotiques nouveaux ou peu connus (Col.). *Bulletin de la Société Zoologique de France* **39**: 181–184.

Pic, M. 1914r: Nouveaux Nemostira Fairm. asiatiques. *Bulletin de la Société Entomologique de France* **1914**: 304–305.

Pic, M. 1914s: Description de trois nouveaux Hylophilus Berthold asiatiques (Col. Hétéromères). *Bulletin de la Société Entomologique de France* **1914**: 448–450.

Pic, M. 1915a: Notes diverses, descriptions et diagnoses (Suite.). *L'Échange, Revue Linnéenne* **31**: 1–3.

Pic, M. 1915b: Descriptions abrégées diverses. *Mélanges Exotico-Entomologiques* **12**: 1–20.

Pic, M. 1915c: Nouveautés de diverses familles. *Mélanges Exotico-Entomologiques* **13**: 2–13.

Pic, M. 1915d: Nouveautés rentrant dans diverses familles. *Mélanges Exotico-Entomologiques* **14**: 2–20.

Pic, M. 1915e: Nouvelles espèces de diverses familles. *Mélanges Exotico-Entomologiques* **15**: 2–24.

Pic, M. 1915f: Nouveau genre, espèces et variétés nouvelles. *Mélanges Exotico-Entomologiques* **16**: 2–13.

Pic, M. 1915g: Diagnoses d'hétéromères. *Mélanges Exotico-Entomologiques* **16**: 14–24.

Pic, M. 1915h: Notes diverses, descriptions et diagnoses (Suite). *L'Échange, Revue Linnéenne* **31**: 9–10.

Pic, M. 1915i: Coléoptères exotiques en partie nouveaux (Suite). *L'Échange, Revue Linnéenne* **31**: 10–11.

Pic, M. 1915j: Notes diverses, descriptions et diagnoses (Suite.). *L'Échange, Revue Linnéenne* **31**: 13–14.

Pic, M. 1915k: Notes diverses, descriptions et diagnoses (Suite.). *L'Échange, Revue Linnéenne* **31**: 21–22.

Pic, M. 1915l: Notes diverses, descriptions et diagnoses (Suite). *L'Échange, Revue Linnéenne* **31**: 37–38.

Pic, M. 1915m: Notes diverses, descriptions et diagnoses (Suite). *L'Échange, Revue Linnéenne* **31**: 41–42.

Pic, M. 1915n: Anthicides et Hylophilides des chasses de E. Gallois au Japon. *Bulletin du Muséum National d'Histoire Naturelle, Paris* **21**: 152–153.

Pic, M. 1916a: Notes diverses, descriptions et diagnoses (Suite). *L'Échange, Revue Linnéenne* **32**: 5–7.

Pic, M. 1916b: Coléoptères exotiques en partie nouveaux. *L'Échange, Revue Linnéenne* **32**: 7–8.

Pic, M. 1916c: Diagnoses spécifiques. *Mélanges Exotico-Entomologiques* **17**: 8–20.

Pic, M. 1916d: Espèces et variétés nouvelles. *Mélanges Exotico-Entomologiques* **19**: 6–20.

Pic, M. 1916e: Descriptions abrégées diverses. *Mélanges Exotico-Entomologiques* **20**: 1–20.

Pic, M. 1916f: Descriptions abrégées diverses. *Mélanges Exotico-Entomologiques* **21**: 2–20.

Pic, M. 1916g: Notes diverses, descriptions et diagnoses (Suite). *L'Échange, Revue Linnéenne* **32**: 13–14.

Pic, M. 1916h: Notes diverses, descriptions et diagnoses (Suite). *L'Échange, Revue Linnéenne* **32**: 21–23.

Pic, M. 1916i: Nouveaux Zonabris de l'Inde et de Cochinchine (Col. Lyttidae). *Bulletin de la Société Entomologique de France* **1916**: 125–126.

Pic, M. 1917a: Notes diverses, descriptions et diagnoses. *L'Échange, Revue Linnéenne* **33**: 1–3.

Pic, M. 1917b: Descriptions abrégées diverses. *Mélanges Exotico-Entomologiques* **22**: 1–20.

Pic, M. 1917c: Descriptions abrégées diverses. *Mélanges Exotico-Entomologiques* **23**: 2–20.

Pic, M. 1917d: Descriptions abrégées diverses. *Mélanges Exotico-Entomologiques* **24**: 2–24.

Pic, M. 1917e: Descriptions abrégées diverses. *Mélanges Exotico-Entomologiques* **25**: 2–24.

Pic, M. 1917f: Descriptions abrégées diverses. *Mélanges Exotico-Entomologiques* **26**: 2–24.

Pic, M. 1917g: Notes diverses, descriptions et diagnoses (Suite). *L'Échange, Revue Linnéenne* **33**: 5–6.

Pic, M. 1917h: Coléoptères exotiques en partie nouveaux (Suite). *L'Échange, Revue Linnéenne* **33**: 11–12.

Pic, M. 1917i: Anthicides de Mongolie (Col. hétéromères). *Bulletin du Muséum National d'Histoire Naturelle, Paris* **23**: 256–257.

Pic, M. 1918a: Courtes descriptions diverses. *Mélanges Exotico-Entomologiques* **27**: 1–24.

Pic, M. 1918b: Courtes descriptions diverses. *Mélanges Exotico-Entomologiques* **28**: 1–24.

Pic, M. 1918c: Notes diverses, descriptions et diagnoses (Suite). *L'Échange, Revue Linnéenne* **34**: 13–15.

Pic, M. 1918d: Coléoptères exotiques en partie nouveaux (Suite). *L'Échange, Revue Linnéenne* **34**: 15–16.

Pic, M. 1919a: Étude détaillé de diverses insectes. *L'Échange, Revue Linnéenne* **35** [hors texte]: 1–20.

Pic, M. 1919b: Nouveautés diverses. *Mélanges Exotico-Entomologiques* **30**: 1–24.

Pic, M. 1919c: Notes diverses, descriptions et diagnoses (Suite). *L'Échange, Revue Linnéenne* **35**: 5–6.

Pic, M. 1919d: Notes diverses, descriptions et diagnoses (Suite). *L'Échange, Revue Linnéenne* **35**: 9–11.

Pic, M. 1919e: Notes diverses, descriptions et diagnoses (Suite). *L'Échange, Revue Linnéenne* **35**: 13–14.

Pic, M. 1919f: Notes diverses, descriptions et diagnoses (Suite). *L'Échange, Revue Linnéenne* **35**: 17–20, 21–22.

Pic, M. 1919g: Note sur divers coléoptères. *Miscellanea Entomologica* **24**: 66–69.

Pic, M. 1920a: Nouveautés diverses. *Mélanges Exotico-Entomologiques* **32**: 1–32.

Pic, M. 1920b: Coléoptères exotiques en partie nouveaux (Suite). *L'Échange, Revue Linnéenne* **36**: 2–3.

Pic, M. 1920c: Notes diverses, descriptions et diagnoses (Suite). *L'Échange, Revue Linnéenne* **36**: 5–8.

Pic, M. 1920d: Notes diverses, descriptions et diagnoses (Suite). *L'Échange, Revue Linnéenne* **36**: 9–12.

Pic, M. 1920e: Notes diverses, descriptions et diagnoses (Suite). *L'Échange, Revue Linnéenne* **36**: 13.

Pic, M. 1920f: Diagnoses de coléoptères exotiques. *L'Échange, Revue Linnéenne* **36**: 15–16.

Pic, M. 1920g: Notes diverses, descriptions et diagnoses (Suite). *L'Échange, Revue Linnéenne* **36**: 17–18.

Pic, M. 1921a: Notes diverses, descriptions et diagnoses (Suite). *L'Échange, Revue Linnéenne* **37**: 1–4, 5–6.

Pic, M. 1921b: Nouveautés diverses. *Mélanges Exotico-Entomologiques* **33**: 1–32.

Pic, M. 1921c: Nouveautés diverses. *Mélanges Exotico-Entomologiques* **34**: 1–33.

Pic, M. 1921d: Notes diverses, descriptions et diagnoses (Suite). *L'Échange, Revue Linnéenne* **37**: 9–10.

Pic, M. 1921e: Diagnoses de coléoptères exotiques (Suite). *L'Échange, Revue Linnéenne* **37**: 10–12, 13–15.

Pic, M. 1921f: Étude détaillée de diverses insectes. *L'Échange, Revue Linnéenne* **37** [hors texte] (403): 21–24.

Pic, M. 1921g: [new taxa]. In: Alfieri A.: Contributions à la faune entomologique du nord-est du Sinai. *Bulletin de la Société Entomologique d'Egypte* **6** [1919–1920]: 40–53.

Pic, M. 1921h: Coléoptères nouveaux de la famille des hylophilides. *Bulletin du Muséum National d'Histoire Naturelle, Paris* **27**: 415–418.

Pic, M. 1922a: Nouveautés diverses. *Mélanges Exotico-Entomologiques* **35**: 1–32.

Pic, M. 1922b: Nouveautés diverses. *Mélanges Exotico-Entomologiques* **36**: 1–36.

Pic, M. 1922c: Nouveautés diverses. *Mélanges Exotico-Entomologiques* **37**: 1–32.

Pic, M. 1922d: Notes diverses, descriptions et diagnoses (Suite). *L'Échange, Revue Linnéenne* **38**: 17–19.

Pic, M. 1922e: Notes diverses, descriptions et diagnoses (Suite). *L'Échange, Revue Linnéenne* **38**: 21–23.

Pic, M. 1922f: Coléoptères exotiques en partie nouveaux. *L'Échange, Revue Linnéenne* **38**: 23–24.

Pic, M. 1922g: Notes diverses, descriptions et diagnoses (Suite). *L'Échange, Revue Linnéenne* **38**: 29–30.

Pic, M. 1922h: Coléoptères hétéromères exotiques nouveaux. *Bulletin de la Société Zoologique de France* **47**: 100–103.

Pic, M. 1922i: Diagnoses d'hétéromères (Col.) de l'Indo-Chine. *Bulletin de la Société Entomologique de France* **1922**: 208–210.

Pic, M. 1922j: Sur les hétéromères Amarygminae (Col.). *Bulletin de la Société Zoologique de France* **47**: 303–306.

Pic, M. 1923a: Notes diverses, descriptions et diagnoses (Suite). *L'Échange, Revue Linnéenne* **39**: 1–3.

Pic, M. 1923b: Nouveautés diverses. *Mélanges Exotico-Entomologiques* **38**: 1–32.

Pic, M. 1923c: Nouveautés diverses. *Mélanges Exotico-Entomologiques* **39**: 3–32.

Pic, M. 1923d: Nouveautés diverses. *Mélanges Exotico-Entomologiques* **40**: 3–32.

Pic, M. 1923e: Coléoptères exotiques en partie nouveaux (Suite). *L'Échange, Revue Linnéenne* **39**: 4.

Pic, M. 1923f: Notes diverses, descriptions et diagnoses (Suite). *L'Échange, Revue Linnéenne* **39**: 5–7.

Pic, M. 1923g: Notes diverses, descriptions et diagnoses (Suite). *L'Échange, Revue Linnéenne* **39**: 9–11.

Pic, M. 1923h: Coléoptères exotiques en partie nouveaux (Suite). *L'Échange, Revue Linnéenne* **39**: 11–12.

Pic, M. 1923i: Mutations, synonymies, nouveautés. *L'Échange, Revue Linnéenne* **39**: 13–14.

Pic, M. 1923j: Nouveaux coléoptères exotiques. *Annales de la Société Linnéenne de Lyon* (N.S.) **69** [1922]: 73–76.

Pic, M. 1923k: Nouveaux coléoptères du Tonkin. *Bulletin de la Société Zoologique de France* **48** 269–271.

Pic, M. 1924a: Nouveautés diverses. *Mélanges Exotico-Entomologiques* **41**: 1–32.

Pic, M. 1924b: Nouveautés diverses. *Mélanges Exotico-Entomologiques* **42**: 1–32.

Pic, M. 1924c: Notes diverses, descriptions et diagnoses (Suite). *L'Échange, Revue Linnéenne* **39**: 21–23.

Pic, M. 1924d: Notes diverses, descriptions et diagnoses (Suite). *L'Échange, Revue Linnéenne* **39**: 25–27.

Pic, M. 1924e: Coléoptères exotiques en partie nouveaux (Suite.). *L'Échange, Revue Linnéenne* **39**: 27–28.

Pic, M. 1924f: Notes diverses, descriptions et diagnoses (Suite.). *L'Échange, Revue Linnéenne* **39**: 29–30.

Pic, M. 1924g: Vésicants en partie originaires du Congo. *Revue de Zoologie et Botanique Africaine* **12**: 450–454.

Pic, M. 1925a: Notes diverses, descriptions et diagnoses (Suite). *L'Échange, Revue Linnéenne* **41**: 1–3.

Pic, M. 1925b: Nouveautés diverses. *Mélanges Exotico-Entomologiques* **44**: 1–32.

Pic, M. 1925c: Notes diverses, descriptions et diagnoses (Suite). *L'Échange, Revue Linnéenne* **41**: 5–6.

Pic, M. 1925d: Notes diverses, descriptions et diagnoses (Suite). *L'Échange, Revue Linnéenne* **41**: 9–11.

Pic, M. 1925ea: Coléoptères exotiques en partie nouveaux (Suite). *L'Échange, Revue Linnéenne* **41**: 11–12.

Pic, M. 1925f: Notes diverses, descriptions et diagnoses (Suite). *L'Échange, Revue Linnéenne* **41**: 13–14.

Pic, M. 1925g: Coléoptères exotiques en partie nouveaux (Suite). *L'Échange, Revue Linnéenne* **41** 15–16.

Pic, M. 1925h: Coléoptères exotiques nouveaux ou peu connus. *Annales de la Société Linnéenne de Lyon* (N.S.) **72**: 73–77.

Pic, M. 1925i: Coléoptères nouveaux. *Bulletin de la Société Royale Entomologique d'Egypte* **18** 227–231.

Pic, M. 1925j: Contribution à l'étude des ténébrionides du genre Basides Mots. *Bulletin du Muséum National d'Histoire Naturelle, Paris* **32**: 431–438.

Pic, M. 1926a: Entomologische Ergebnisse der Schwedischen Kamtchatka-Expedition 1920–1922. 7. Coleoptera. *Arkiv för Zoologi* **18 B** (3): 1–5.

Pic, M. 1926b: Nouveautés diverses. *Mélanges Exotico-Entomologiques* **45**: 1–32.

Pic, M. 1926c: Nouveautés diverses. *Mélanges Exotico-Entomologiques* **46**: 1–32.

Pic, M. 1926d: Nouveautés diverses. *Mélanges Exotico-Entomologiques* **47**: 1–32.

Pic, M. 1926e: Notes diverses, descriptions et diagnoses (Suite). *L'Échange, Revue Linnéenne* **42**: 5–6.

Pic, M. 1926f: Notes diverses, descriptions et diagnoses. *L'Échange, Revue Linnéenne* **42**: 9–10.

Pic, M. 1926g: Coléoptères exotiques en partie nouveaux. *L'Échange, Revue Linnéenne* **42**: 15–16.

Pic, M. 1926h: Nouveaux coléoptères du Tonkin (2e article). *Bulletin de la Société Zoologique de France* **51**: 45–48.

Pic, M. 1926i: H. Sauter's Formosa Ausbeute. *Entomologische Mitteillungen* **15**: 67–69.

Pic, M. 1926j: Coléoptères exotiques nouveaux ou peu connus. *Annales de la Société Linnéenne de Lyon* (N.S.) **72** [1925]: 73–77.

Pic, M. 1926k: Coléoptères nouveaux. *Bulletin de la Société Entomologique d'Egypte* **9** [1925–1926]: 227–231.

Pic, M. 1926l: Coléoptères asiatiques nouveaux. *Bulletin de la Société Entomologique de France* **1925**: 301–303.

Pic, M. 1926m: Nouveaux coléoptères exotiques. *Bulletin du Muséum National d'Histoire Naturelle, Paris* **32** 354–359.

Pic, M. 1927a: Entomologische Ergebnisse der Schwedischen Kamtchatka expedition 1920–1922. 14. Coleoptera 2. *Arkiv för Zoology* **19 B** (3): 1–3.

Pic, M. 1927b: Nouveautés diverses. *Mélanges Exotico-Entomologiques* **48**: 1–32.

Pic, M. 1927c: Coléoptères de l'Indochine. *Mélanges Exotico-Entomologiques* **49**: 1–36.

Pic, M. 1927d: Coléoptères du globe. *Mélanges Exotico-Entomologiques* **50**: 1–36.

Pic, M. 1927e: Notes diverses, descriptions et diagnoses (Suite). *L'Échange, Revue Linnéenne* 43: 5–7.

Pic, M. 1927f: Diagnoses de coléoptères exotiques. *Bulletin Bi-Mensuel de la Société Linnéenne de Lyon* **6**: 5–6.

Pic, M. 1927g: Coléoptères exotiques en partie nouveaux (Suite). *L'Échange, Revue Linnéenne* **43**: 7–8.

Pic, M. 1927h: Notes diverses, descriptions et diagnoses (Suite). *L'Échange, Revue Linnéenne* **43**: 9–10.

Pic, M. 1927i: H. Sauter's Formosa-Ausbeute: Heteromera [ex parte] (Col.). *Entomologische Mitteilungen* **16**: 48–49.

Pic, M. 1927k: Coléoptères asiatiques nouveaux. *Bulletin Bi-Mensuel de la Société Linnéenne de Lyon* **6**: 132–133.

Pic, M. 1927j: Coléoptéres nouveaux de Chine et du Japon. *Bulletin de la Société Entomologique de France* **1927**: 152–154.

Pic, M. 1927l: Nouveaux coléoptères exotiques. *Bulletin de la Société Entomologique de France* **1927**: 183–184.

Pic, M. 1927m: Nouveau coléoptères du Globe (IV). *Bulletin de la Société Zoologique de France* **52**: 185–189.

Pic, M. 1928a: Notes diverses, descriptions et diagnoses (Suite). *L'Échange, Revue Linnéenne* **44**: 1–2.

Pic, M. 1928b: Notes et descriptions. *Mélanges Exotico-Entomologiques* **51**: 1–36.

Pic, M. 1928c: Nouveautés diverses. *Mélanges Exotico-Entomologiques* **52**: 1–32.

Pic, M. 1928d: Coléoptères exotiques en partie nouveaux (Suite). *L'Échange, Revue Linnéenne* **44**: 4.

Pic, M. 1928e: Notes diverses, descriptions et diagnoses (Suite). *L'Échange, Revue Linnéenne* **44**: 5–6.

Pic, M. 1928f: Notes diverses, descriptions et diagnoses (Suite). *L'Échange, Revue Linnéenne* **44**: 9–10.

Pic, M. 1928g: Coléoptères asiatiques nouveaux. *Bulletin de la Société Entomologique de France* **1928**: 247–248.

Pic, M. 1928h: Nouveaux coléoptères exotiques. *Bulletin du Muséum National d'histoire Naturelle, Paris* **34**: 257–258.

Pic, M. 1928i: Nouveaux coléoptères surtout tonkinois. *Bulletin de la Société Zoologique de France* **53**: 377–380.

Pic, M. 1929a: Notes diverses, nouveautés. *L'Échange, Revue Linnéenne* **45**: 1–2.

Pic, M. 1929b: Nouveautés diverses. *Mélanges Exotico-Entomologiques* **53**: 1–36.

Pic, M. 1929c: Nouveautés diverses. *Mélanges Exotico-Entomologiques* **54**: 1–36.

Pic, M. 1929d: Notes diverses, nouveautés. *L'Échange, Revue Linnéenne* **45**: 5–7.

Pic, M. 1929e: Notes diverses, nouveautés. *L'Échange, Revue Linnéenne* **45**: 13–14.

Pic, M. 1929f: Coléoptères exotiques nouveaux ou peu connus. *Bulletin de la Société Zoologique de France* **54**: 43–46.

Pic, M. 1929g: Coléoptères d'Egypte et Sinaï. *Bulletin de la Société Royal Entomologique d'Egypte* (N.S.) **13**: 108–145.

Pic, M. 1929h: Coléoptères nouveaux du Hoggar. *Bulletin de la Société Entomologique de France* **1929**: 126–128.

Pic, M. 1930a: Notes diverses, nouveautés. *L'Échange, Revue Linnéenne* **46**: 1–3.

Pic, M. 1930b: Nouveautés diverses. *Mélanges Exotico-Entomologiques* **56**: 1–36.

Pic, M. 1930c: Notes diverses, nouveautés. *L'Échange, Revue Linnéenne* **46**: 5–7.

Pic, M. 1930d: Notes diverses, nouveautés. *L'Échange, Revue Linnéenne* **46**: 9–11.

Pic, M. 1930e: Trois nouveaux coléoptéres. *Bulletin de la Société Entomologique de France* **45**: 76–77.

Pic, M. 1931a: Nouveautés diverses. *Mélanges Exotico-Entomologiques* **57**: 1–36.

Pic, M. 1931b: [new taxa]. In: Miwa Y.: A systematic catalogue of Formosan Coleoptera *Reports of the Department of Agriculture of the Governement Research Institute, Taihoku* **55**: 1–359.

Pic, M. 1931c: Notes diverses, nouveautés. *L'Échange, Revue Linnéenne* **47**: 5–6.

Pic, M. 1931d: Notes diverses, nouveautés (Suite). *L'Échange, Revue Linnéenne* **47**: 9–10.

Pic, M. 1931e: Notes diverses, nouveautés (Suite). *L'Échange, Revue Linnéenne* **47**: 13–14.

Pic, M. 1931f: [new taxa]. In: Koch C.: Risultati scientifici delle spedizioni entomologiche di S.A.S. il Principe Alessandro Della Torre e Tasso nell'Africa settentrionale e in Sicilia. *Memorie della Società Entomologica Italiana* **10**: 66–79.

Pic, M. 1932a: Diagnoses préliminaires. *Mélanges Exotico-Entomologiques* **59**: 1–9.

Pic, M. 1932b: [new taxa]. In: Pic M. & Lindberg Håk.: Inventa entomologica, itineris Hispanici et Maroccani, quod a. 1926 fecerunt Harald et Håkan Lindberg. XII. Anobiidae, Cleridae, Malacodermata, Heteromera (ex parte). *Commentationes Biologicae* **3** (18): 1–37.

Pic, M. 1932c: Contributions à l'étude de la faune du Mozambique. Voyage de M.P. Lesne (1928–1929). 9.e note. – Coléoptères, Anthicidae, Hylophilidae, Pedilidae, Scraptiidae. *Memórias e Estudos do Museu Zoológico da Universidade de Coimbra* (1) **60**: 5–26.

Pic, M. 1932d: Nouveautés diverses. *Mélanges Exotico-Entomologiques* **59**: 10–36.

Pic, M. 1932e: Notes diverses, nouveautés (Suite.). *L'Échange, Revue Linnéenne* **47**: 17–18.

Pic, M. 1932f: Notes diverses, nouveautés (Suite). *L'Échange, Revue Linnéenne* **47**: 25–26.

Pic, M. 1932g: Diagnoses préliminaires. *L'Échange, Revue Linnéenne* **47**: 26–28.

Pic, M. 1932h: Notes diverses, nouveautés (Suite). *L'Échange, Revue Linnéenne* **48**: 29–31.

Pic, M. 1932i: Coléoptères exotiques en partie nouveaux (Suite). *L'Échange, Revue Linnéenne* **48**: 32.

Pic, M. 1932j: Nouveautés diverses. *Mélanges Exotico-Entomologiques* **60**: 1–36.

Pic, M. 1933a: Notes diverses, nouveautés. *L'Échange, Revue Linnéenne* **49**: 1–2.

Pic, M. 1933b: Schwedisch-chinesische wissenschaftliche Expedition nach den nordwestlichen Provinzien Chinas unter Leitung von Dr. Sven Hedin und Prof. Sü Ping-chang. Insekten gesammelt vom schwedischen Arzt der Expedition Dr. David Hummel 1927–1930. 16. Coleoptera. 2. Helmidae, Dermestidae, Anobiidae, Cleridae, Malacodermata, Dascillidae, Heteromera (ex p.), Bruchidae, Cerambycidae, Phytophaga (ex p.). *Arkiv för Zoologi* **27 A** (2) [1935]: 1–14.

Pic, M. 1933c: Nouveautés diverses. *Mélanges Exotico-Entomologiques* **61**: 3–36.

Pic, M. 1933d: Nouveautés diverses. *Mélanges Exotico-Entomologiques* **62**: 1–36.

Pic, M. 1933e: Notes diverses, nouveautés (Suite). *L'Échange, Revue Linnéenne* **49**: 9–11, 13–15.

Pic, M. 1933f: Coléoptères du Japon recueillis par M.E. Gallois. *Bulletin du Muséum National d'Histoire Naturelle, Paris* (2) 5: 318–321.

Pic, M. 1934a: Nouveautés diverses. *Mélanges Exotico-Entomologiques* **63**: 1–36.

Pic, M. 1934b: Nouveautés diverses. *Mélanges Exotico-Entomologiques* **64**: 1–36.

Pic, M. 1934c: Coléoptères exotiques en partie nouveaux. *L'Échange, Revue Linnéenne* **50**: 22–23.

Pic, M. 1934d: Notes diverses, nouveautés (Suite). *L'Échange, Revue Linnéenne* **50**: 21–22, 25–27.

Pic, M. 1934e: Coléoptères exotiques en partie nouveaux. *L'Échange, Revue Linnéenne* **50**: 28, 31–32.

Pic, M. 1934f: Nouveaux coléoptères de Chine. *Entomologisches Nachrichtenblatt* **8**: 84–87.

Pic, M. 1935a: Nouveautés diverses. *Mélanges Exotico-Entomologiques* **65**: 1–36.

Pic, M. 1935b: Nouveautés diverses. *Mélanges Exotico-Entomologiques* **66**: 1–36.

Pic, M. 1935c: Notes diverses, nouveautés (Suite). *L'Échange, Revue Linnéenne* **51**: 3–4, 5–8.

Pic, M. 1935d: Notes diverses, nouveautés (Suite). *L'Échange, Revue Linnéenne* **51**: 11–12.

Pic, M. 1935e: Coléoptères oedémérides nouveaux. *Revue Française d'Entomologie* **2**: 98–102.

Pic, M. 1935f: Sur diverses insectes paléarctiques. *Bulletin de la Société Entomologique de France* **40**: 108–111.

Pic, M. 1935g: Descriptions de coléoptères paléarctiques. *Bulletin de la Société Entomologique de France* **40**: 158–160.

Pic, M. 1936a: Nouveautés diverses. *Mélanges Exotico-Entomologiques* **67**: 1–36.

Pic, M. 1936b: Nouveautés diverses. *Mélanges Exotico-Entomologiques* **68**: 10–36.

Pic, M. 1936c: Notes diverses, nouveautés (Suite). *L'Échange, Revue Linnéenne* **52**: 21–23.

Pic, M. 1936d: Notes diverses, nouveautés (Suite). *L'Échange, Revue Linnéenne* **52**: 25–27.

Pic, M. 1936e: Notes diverses, nouveautés (Suite). *L'Échange, Revue Linnéenne* **52**: 29–30.

Pic, M. 1936f: [new taxon]. In: Normand H.: Contribution au catalogue des coléoptères de la Tunisie (9e fascicule). *Bulletin de la Société d'Histoire Naturelle de l'Afrique du Nord* **27**: 144–164.

Pic, M. 1936g: Descriptions de coléoptères. *Bulletin de la Société Entomologique de France* **41**: 213–216.

Pic, M. 1936h: Nouveau coléoptères exotiques. *Bulletin de la Société Zoologique de France* **61**: 298–300.

Pic, M. 1937a: Notes diverses, nouveautés (Suite). *L'Échange, Revue Linnéenne* **53**: 1.

Pic, M. 1937b: Nouveautés diverses. *Mélanges Exotico-Entomologiques* **67**: 1–36.

Pic, M. 1937c: Dasytidae: Dasytinae. Pars 155. In: Schenkling S. (ed.): *Coleopterorum Catalogus. Vol. 19*. Berlin and s'Gravenhage: W. Junk, 130 pp.

Pic, M. 1937d: Notes diverses, nouveautés (Suite). *L'Échange, Revue Linnéenne* **53**: 1. 9–10.

Pic, M. 1937e: Nouveautés diverses. *Mélanges Exotico-Entomologiques* **68**: 10–36.

Pic, M. 1937f: Nouveautés diverses. *Mélanges Exotico-Entomologiques* **69**: 1–36.

Pic, M. 1937g: Cis Latr. et Anobiides nouveaux des Indes (Coléoptères). *Indian Forest Records, Entomology* **3**: 123–126.

Pic, M. 1937h: Nouveaux coléoptères d'Afrique et de Chine. *Revue Française d'Entomologique* **4**: 148–149.

Pic, M. 1937i: Coléoptères nouveaux de Chine. *Notes d'Entomologie Chinoise, Musée Heude* (Shanghai) **4**: 161–176.

Pic, M. 1938a: Nouveautés diverses, Mutations. *Mélanges Exotico-Entomologiques* **70**: 1–36.

Pic, M. 1938b: Notes diverses, nouveautés (Suite). *L'Échange, Revue Linnéenne* **54**: 5–6.

Pic, M. 1938c: Notes diverses, nouveautés (Suite). *L'Échange, Revue Linnéenne* **54**: 9–11.

Pic, M. 1938d: Notes diverses, nouveautés. (Suite). *L'Échange, Revue Linnéenne* **54**: 13–14.

Pic, M. 1938e: Mutations et coléoptères nouveaux. *L'Échange, Revue Linnéenne* **54**: 14–16.

Pic, M. 1938f: Nouveaux coléoptères. *Bulletin de la Société Entomologique de France* **43**: 71–72.

Pic, M. 1938g: Coléoptères asiatiques nouveaux. *Bulletin de la Société Entomologique de France* **43**: 220–221.

Pic, M. 1939a: Mutations et nouveautés diverses. *Mélanges Exotico-Entomologiques* **71**: 1–36.

Pic, M. 1939b: Notes diverses, nouveautés (Suite). *L'Échange, Revue Linnéenne* **55**: 17–19, 21–22.

Pic, M. 1939c: Coléoptères exotiques en partie nouveaux (Suite). *L'Échange, Revue Linnéenne* 55: 24.

Pic, M. 1939d: Deux nouveaux coléoptères de Chine. *Notes d'Entomologie Chinoise. Musée Heude* (Shanghai) 6: 135–136.

Pic, M. 1939e: Tenthredid (Hymén.) et coléoptères nouveaux. *L'Échange, Revue Linnéenne* 55: 29–30.

Pic, M. 1939f: Coléoptères nouveaux d'Egypte et du Sinaï. *Bulletin de la Société Fouad 1er d'Entomologie* 23: 143–149.

Pic, M. 1939g: Nouveautés et observations sur divers coléoptères. *VII. Internationaler Kongress für Entomologie, Verhandlungen* 1: 342–344.

Pic, M. 1939h: Nouveaux coléoptères de la presqu'île Malaise (VI.). *Journal of the Federal Malay State Museums* 18: 372–376.

Pic, M. 1940a: Tenthredides (Hymén.) et coléoptères nouveaux. *L'Échange, Revue Linnéenne* 56: 1–2.

Pic, M. 1940b: Diagnoses de coléoptères exotiques. *L'Échange, Revue Linnéenne* 56: 13–16.

Pic, M. 1941a: Notes diverses, nouveautés. *L'Échange, Revue Linnéenne* 57: 1, 5.

Pic, M. 1941b: Opuscula martialis. *L'Échange, Revue Linnéenne. Numéro Spécial* 2: 1–16.

Pic, M. 1941c: Opuscula martialis. *L'Échange, Revue Linnéenne. Numéro Spécial* 3: 1–16.

Pic, M. 1941d: Opuscula martialis. *L'Échange, Revue Linnéenne. Numéro Spécial* 5: 1–16.

Pic, M. 1942a: Opuscula martialis. *L'Échange, Revue Linnéenne. Numéro Spécial* 7: 1–16.

Pic, M. 1942b: Opuscula martialis. *L'Échange, Revue Linnéenne. Numéro Spécial* 8: 1–16.

Pic, M. 1942c: Coléoptères du globe (Suite). *L'Échange, Revue Linnéenne* 58: 3–12.

Pic, M. 1942d: Coléoptères nouveaux de l'Adrar du Iforas. *Bulletin de la Société Entomologique de France* 47: 77–79.

Pic, M. 1942e: Coléoptères du globe (suite). *L'Échange, Revue Linnéenne* 58: 15–16.

Pic, M. 1943: Opuscula martialis. *L'Échange, Revue Linnéenne. Numéro Spécial* 9: 1–16.

Pic, M. 1944a: Opuscula martialis. *L'Échange, Revue Linnéenne. Numéro Spécial* 12: 1–16.

Pic, M. 1944b: Coléoptères du globe (Suite). *L'Échange, Revue Linnéenne* 60: 5–8.

Pic, M. 1945: Coléoptères du globe (Suite). *L'Échange, Revue Linnéenne* 61: 1–4, 7–8.

Pic, M. 1946: Coléoptères du globe (Suite). *L'Échange, Revue Linnéenne* 62: 1–4.

Pic, M. 1947a: Diversités entomologiques. *Diversités Entomologiques* 1: 1–16.

Pic, M. 1947b: Coléoptères du globe (Suite.). *L'Échange, Revue Linnéenne* 63: 9–16.

Pic, M. 1948a: Coléoptères du globe (Suite.). *L'Échange, Revue Linnéenne* 64: 1–4.

Pic, M. 1948b: Étude variétiste de Lytta vesicatoria L. *Diversités Entomologiques* 4: 1–4.

Pic, M. 1948c: Coléoptères du globe (Suite). *L'Échange, Revue Linnéenne* 64: 6–8.

Pic, M. 1949a: Coléoptères du globe. *L'Échange, Revue Linnéenne* 65: 1–4, 9–12.

Pic, M. 1949b: Notes et descriptions. *Diversités Entomologiques* 5: 9–16.

Pic, M. 1949c: Nouveaux coléoptères du Maroc. *Bulletin de la Société des Sciences Naturelles du Maroc* 29: 73–75.

Pic, M. 1950a: Descriptions et notes variées. *Diversités Entomologiques* 7: 1–16.

Pic, M. 1950b: Coléoptères du globe (Suite). *L'Échange, Revue Linnéenne* 66: 5–8.

Pic, M. 1951a: Coléoptères du globe (Suite). *L'Échange, Revue Linnéenne* 67: 1–4, 10–18.

Pic, M. 1951b: Descriptions diverses et notes. *Diversités Entomologiques* 8: 7–16.

Pic, M. 1951c: Notes et descriptions. *Diversités Entomologiques* 9: 4–16.

Pic, M. 1951d: Deux nouveaux Anthicus Payk. du Sahara marocain (Col. Hétéromères). *Comptes Rendus de la Société des Sciences Naturelles du Maroc* **17**: 37–38.

Pic, M. 1951e: Un Alleculide nouveau du Maroc (Col. hétéromère). *Comptes Rendus de la Société des Sciences Naturalles du Maroc* **17**: 75.

Pic, M. 1951f: Description de deux Anthicus nouveaux du Maroc. *Bulletin de la Société des Sciences Naturelles du Maroc* **29** [1949]: 81–82.

Pic, M. 1951g: Entomological expeditions to Abyssinia, 1926–27 and 1948–49. Coleoptera. Anthicidae. *The Annals and Magazine of Natural History* (12) **4**: 125–136.

Pic, M. 1951h: Résultats de la Mission Zoologique Suisse au Cameroun. Malacodermes, hétéromères vésicants et phytophages (p. p.). *Mémoires de l'Institut Français d'Afrique Noire, Centre du Cameroun, Duala, Sciences Naturelles* **1**: 217–222.

Pic, M. 1952a: Coléoptères du globe (Suite). *L'Échange, Revue Linnéenne* **68**: 1–4.

Pic, M. 1952b: Coléoptères du globe (Suite.). *L'Échange, Revue Linnéenne* **68**: 5–8.

Pic, M. 1952c: Contribution à l'étude du peuplement de la Mauritanie. Coléoptères divers (deuxième note). *Bulletin de l'Institut Française d'Afrique Noire* **14**: 489–494.

Pic, M. 1952d: Hétéromères africains nouveaux (Coléoptères). *Revue de Zoologie et de Botanique Africaines* **46** (1/2): 60–80.

Pic, M. 1953a: Notes et descriptions. *Diversités Entomologiques* **12**: 5–16.

Pic, M. 1953b: Coléoptères nouveaux de la région marocaine. *Comptes Rendus de la Société des Sciences Naturalles du Maroc* **19**: 61–63.

Pic, M. 1953c: Melandryidae (Col. Hétéromères) nouveaux. *The Annals and Magazine of Natural History* (12) **6**: 333–336.

Pic, M. 1953d: Nouveaux coléoptères Africains. *Bulletin de l'Institut Français d'Afrique Noire* **15**: 495–506.

Pic, M. 1953e: Nouveaux Coléoptères africains. *Atti del Museo Civico di Storia Naturale di Trieste* **19** [1953–1954]: 93–109.

Pic, M. 1954a: Coléoptères nouveaux de Chine. *Bulletin de la Société Entomologique de Mulhouse* **1954**: 53–59.

Pic, M. 1954b: Coléoptères nouveaux de Chine (Suite). *Bulletin de la Société Entomologique de Mulhouse* **1954**: 60–64.

Pic, M. 1954c: Lagriidae et Alleculidae nouveaux du Musée royal du Congo belge (Coléoptères). *Revue de Zoologie et de Botanique Africaines* **49**: 225–264.

Pic, M. 1955a: Coléoptères du globe (Suite). *L'Échange, Revue Linnéenne* **71**: 1–3.

Pic, M. 1955b: Descriptions et notes. *Diversités Entomologiques* **15**: 9–16.

Pic, M. 1955c: Coléoptères nouveaux de Chine. *Bulletin de la Société Entomologique de Mulhouse* **1955**: 29–32.

Pic, M. 1955d: Contributions à l'étude de la faune entomologique du Ruanda-Urundi (Mission P. Basilewsky 1953). XII. Coleoptera Rhipiphoridae, Oedemeridae, Scraptiidae, Pedilidae, Aderidae, Anthicidae, Dermestidae, Byrrhidae, Anobiidae, Ptinidae, Cleridae, Lymexylonidae et Helodidae. *Annales du Musée Royal du Congo Belge, Tervuren* (*Série 8° Sciences Zoologiques*) **34**: 125–135.

Pic, M. 1955e: Contribution à l'étude du peuplement de la Mauritanie (3e note). *Bulletin de l'Institut Français d'Afrique Noire, Série A: Sciences Naturelles* **67**: 1041–1050.

Pic, M. 1955f: Contribution à l'étude du peuplement de la Mauritanie (3e note). *Bulletin de l'Institut Français d'Afrique Noire, Série A: Sciences Naturelles* **67**: 1041–1050.

Pic, M. 1956: Anobiidae, Ptinidae et Aderidae (Coleoptera) de l'expédition J. Klapperich en Afghanistan (1952–1953). *Bulletin de la Société Entomologique de Mulhouse* **1956**: 5–7.

Pic, M. 1957: [new taxa]. In: Pic M. & Hawkins C.N.: 29. Coleoptera: Anthicidae, pp. 435–450. In: *British Museum (Natural History) Expedition to South-West Arabia 1937–8*. [1941–1957] *Volume 1*. London: Trustees of the British Museum, 504 pp., 1 pl., 1 map.

Pic, T. 1900: Deux variétés de Notoxus d'Algérie. *L'Échange, Revue Linnéenne* **16**: 52.

Picka, J. 1979: Eine neue mitteleuropäische Art der Gattung Blaps Fabricius, 1775 (Coleoptera, Tenebrionidae). *Reichenbachia* **17**: 99–103.

Picka, J. 1983a: Eine neue Scaphidema-Art aus dem Kaukasus: Scaphidema rousi sp. n. (Coleoptera, Tenebrionidae). *Acta Entomologica Musei Nationalis Pragae* **41**: 261–263.

Picka, J. 1983b: Eine neue Melanimon-Art aus Bulgarien: Melanimon inermus sp.n. (Coleoptera, Tenebrionidae). *Acta Entomologica Musei Nationalis Pragae* **41**: 265–268.

Picka, J. 1984: Zur Faunistik und Taxonomie der Tenebrionidae (Coleoptera) der Insel Kreta. *Türkiye Bitki Koruma Dergisi* **7–8**: 17–31.

Picka, J. 1987: Aphaleria brodskyi new species from Soviet Middle Asia (Coleoptera, Tenebrionidae). *Elytron* **1**: 33–35.

Picka, J. 1988: Revision of the genus Brachyestes Fairmaire from North Africa (Coleoptera, Tenebrionidae). *Elytron* **2**: 107–112.

Picka, J. 1990: Revision of the Plamius quadrinotatus species group (Coleoptera, Tenebrionidae). *Annales Historico-Naturales Musei Nationalis Hungarici* **81**: 109–114.

Pierre, F. 1954: Le genre Leptonychus Chevrolat (Col. Tenebrionidae). *Annales de la Société Entomologique de France* **123**: 17–33.

Pierre, F. 1958: Coléoptères ténébrionides du Tasili n'Ajjer (Mission F. Bernard). In: Bernard F. (ed.): *Mission scientifique au Tassili n'Ajjer* **3**: 133–145. [Publication de l'Institut de Recherches Sahariennes de l'Université d'Alger].

Pierre, F. 1961: Expédition française au Jannu (Népal oriental) (Coleoptera, Tenebrionidae). *Bulletin de la Société Entomologique de France* **66**: 212–214.

Pierre, F. 1964a: Notes sur quelque Blaps de Syrie et de Palestine (Tenebrionidae). *Livre Jubilaire dédié à George Petit, Supplément 17 Vie et Milieu* **1964**: 407–409.

Pierre, F. 1964b: Storthocnemis nouveaux de la zone sahélienne et du Sahara. Remarques concernant les Leucolaephini, trib. nov. (Col. Tenebrionidae). *Bulletin de l'Institut Français d'Afrique Noire* (A) **26**: 866–874.

Pierre, F. 1964c: Contribution à la connaissance de la faune du Moyen-Orient (Missions G. Remaudière 1955 et 1959). 11. Coléoptères tenebrionides. *Vie et Milieu* **15** (4): 1043–1055.

Pierre, F. 1968: Contribution à la faune de l'Iran 12. Étude écologique et biogéographique des ténébrionides Pimeliinae s. nov. des régions désertiques et semi-désertiques de l'Iran. *Annales de la Société Entomologique de France (N.S.)* **4**: 997–1036.

Pierre, F. 1974: Contribution à la faune de l'Iran. Insectes coléoptères du désert du Lut. *Annales de la Société Entomologique de France* (N.S.) **10**: 229–235.

Pierre, F. 1976: Une nouvelle espèce de Pimelia du Sahara nord-occidental et remarques sur les possibilités d'hybridation de quelques formes (Col. Tenebrionidae). *Bulletin de la Société Entomologique de France* **81** 16–21.

Pierre, F. 1978: Note sur la distribution et la variation géographique des Pimelia du Maroc méridional, avec descriptions de nouvelles sous-espèces (Col. Tenebrionidae). *Bulletin de la Société Entomologigue de France* **83**: 197–206.

Pierre, F. 1982: Variation spécifique des Pimelia du groupe d'interstitialis dans le Nord de l'Afrique et au Sahara, avec description de nouveaux taxa (Col. Tenebrionidae). *Bulletin de la Société Entomologigue de France* **87**: 2–13.

Pierre, F. 1985: Sur quelques Pimelia recueillis par le Pr. A. Hollande dans le Haut Atlas Marocain (Coleoptera, Tenebrionidae). *Nouvelle Revue d'Entomologie* (N.S.) **1** [1984]: 287–297.

Piller, M. & Mitterpacher, L. 1783: *Iter per Poseganam Slavoniae provincian mensibus Junio, et Julio MDCCLXXXII susceptum*. Budae: Regiae Universitatis, 147 pp., 16 pls.

Piño, M. 1985: Introduction al conocimiento de los Mordellidae (Coleoptera) de Chile, clave para generos y lista especies. *Revista Chilena de Entomologia* **12**: 101–103.

Piochard de la Brûlerie C.J. 1869: Descriptions de nouvelles espèces espagnoles du groupe des Pandarites. *Annales de la Société Entomologique de France* (4) **9**: 31–38.

Plavilshchikov, N.N. 1927: Pseudocistela pamirica n. sp. (Col., Allecul.). *Entomologische Mitteilungen* **16**: 133–134.

Plaza, E. 1977: Mordellidae de las Islas Canarias. *Nouvelle Revue d´Entomologie* **7**: 171–181.

Plaza, E. 1986a: Contributión al conocimiento del género Mediimorda Meq., 1946; Mediimorda batteni nov. sp. de la fauna española. *Eos, Revista Española de Entomología* **61** [1985]: 265–273.

Plaza, E. 1986b: Las especies españolas de Mordellistena Costa del grupo episternalis. *Eos, Revista Española de Entomología* **61** [1985]: 265–273.

Plaza, E. & Compte, A. 1980: Contributión al conocimiento de los Mordellidae de España. *Eos, Revista Española de Entomología* **56**: 195–214.

Pliginskij [= Pliginsky] V.G. 1910: Deux espèces nouvelles du genre Meloë Linn. (Coleoptera, Meloidae). *Revue Russe d'Entomologie* **10**: 170–172.

Pliginskij, V.G. 1911: Notice sur un caractère chez les representant du genre Meloë. *Revue Russe d'Entomologie* **11**: 44–47.

Pliginskij, V.G. 1914a: Les meloïnes (Coleoptera, Meloidae) de la collection de v. Motschulsky. *Revue Russe d'Entomologie* **14**: 254–261.

Pliginskij, V.G. 1914b: Eine neue Meloë-Art (Coleoptera, Meloidae). *Revue Russe d'Entomologie* **14**: 264–265.

Pliginskij, V.G. 1914c: Eine neue Zonitis-Art aus der Krim (Coleoptera, Meloidae). *Revue Russe d'Entomologie* **14**: 429–430.

Pliginskij, V.G. 1923a: Novyy vid roda Mycetophagus Hellw. iz Kryma (Coleoptera, Mycetophagidae) – Mycetophagus tauricus sp. n. *Bulletin de la Société Entomologique de Moscou* **2** (2): 78–79.

Pliginskij, V.G. 1923b: Espèces nouvelles des coléoptères paléarctiques. *Revue Russe d'Entomologie* **18**: 141–143.

Pliginskij, V.G. 1930: Zwei neue Meloiden-Formen (Coleoptera). *Revue Russe d'Entomologie* **24**: 109–110.

Pliginskij, V.G. 1935: Notizen über Meloiden (Coleoptera, Meloidae), III. *Revue Russe d'Entomologie* **25**: 320–323.

Pomel, A. 1871: Le Sahara (Géologie, Géographie et Biologie). *Bulletin de la Société Algérienne de Climatologie Science Physique et Naturelle* **8**: 133–265.

Pool, C.J.C. 1917: The Coleoptera of the family Cassidae found in Britain, with descriptions of two new species. *Proceedings of the Zoological Society of London* **1917**: 83–93.

Pope, R.D. 1961: Colydiidae (Coleoptera Clavicornia). *Exploration du Parc National de la Garamba – Mission H. de Saeger* **25**: 1–115.

Poppius, B. 1904: Neue paläarktische Coleopteren. *Öfversigt af Finska Vetenskaps Forhandlingar* **46** (16)[1903–1904]: 1–14.

Pripisnova, M.G. 1987a: Novye podrod i vid roda Meloe L. (Coleoptera, Meloidae) iz Tadzhikskoy S.S.R [New subgenus and species of the genus Meloe L. (Coleoptera, Meloidae) from the Tadzhik S.S.R.]. *Izvestia Akademii Nauk Respubliki Tadzikistan, Otdelenie Biologiskikh Nauk* **3**: 41–45.

Pripisnova, M.G. 1987b: Fauna i ekologia zhukov-naryvnikov (Coleoptera, Meloidae) Tadzhikistana [Fauna and ecology of the meloid beetles (Coleoptera, Meloidae) of Tazhik SSR. *Entomologicheskoe Obozrenie* **66**: 555–571.

Prochazka, J. 1892: Revision der Arten der Coleopteren-Gattung Hapalus Fabr. Aus der palaärctischen Fauna. *Wiener Entomologische Zeitung* **11**: 263–270.

Puel, L. 1907: Tableau sur les variétés du 'Zonabris geminata' F. *L'Échange, Revue Linnéenne* **23**: 115–119.

Puel, L. 1919: Sur les Anthicus français du groupe de l'humilis Germ. *Miscellanea Entomologica* **24**: 49–6[illegible].

Purchart, L. 2009: Review of the genus Adelostoma Duponchel, 1827 (Coleoptea: Tenebrionidae: Adelostomini) from the Socotra Archipelago, with description of a new species. *African Entomology* **17**: 23–27.

Purchart, L. 2012a: Biodiversity research of darkling beetles on Socotra Island. Part 1. The genus Deretus Gahan, 1900 (Coleoptera: Tenebrionidae). *Zootaxa* **3153**: 57–68.

Purchart, L. 2012b: Revision of the genus Adelostoma (Coleoptera: Tenebrionidae). Part 1: subgenus Zarudnionymus Semenov & Bogatchev, 1947. *Zootaxa* **3164**: 17–31.

Purchart, L. 2013: A new species of the genus Deretus Gahan, 1900 (Coleoptera: Tenebrionidae) from the island of Socotra. *Annales Zoologici* **63**: 79–83.

Purchart, L. 2014a: Revision of the genus Histeromorphus (Coleoptera: Tenebrionidae) from the Socotra Archipelago with descriptions of three new species. *Acta Entomologica Musei Nationalis Pragae* **54**, *Supplementum*: 211–230.

Purchart, L. 2014b: Two new species of the genera Zophosis and Oxycara, and a new record of the genus Freyula from the Island of Socotra (Coleoptera: Tenebrionidae). *Acta Entomologica Musei Nationalis Pragae* **54**, *Supplementum*: 231–240.

Purchart, L. 2017: Revision of the genus Adelostoma (Coleoptera: Tenebrionidae). Part 2: A new subgenus and species from Oman. *Zootaxa* **4258**: 281–286.

Purchart, L. & Kamiński, M.J. 2017: A taxonomic review of the genus Clitobius with description of a new species from Oman (Coleoptera: Tenebrionidae). *Acta Entomologica Musei Nationalis Pragae* **57**, *Supplementum*: 139–163.

Purchart, L. & Schawaller, W. 2012: A new species of the genus Corticeus (Coleoptera: Tenebrionidae) from Socotra Island. *Acta Entomologica Musei Nationalis Pragae* **52**, *Supplementum*: 315–322.

Pütz, A. 2009: Eine neue Art der Gattung Aegialites Mannerheim, 1853 aus Japan (Coleoptera, Salpingidae, Aegialitinae). *Entomologische Nachrichten und Berichte* **53**: 29–31.

Quedenfeldt, G. 1887: Ein neues Helopiden-Genus von Marokko. *Entomologische Nachrichten* **13**: 257–259.

Quedenfeldt, G. 1889: Zwei neue afrikanische Tenebrioniden. *Entomologische Nachrichten* **15**: 353–35[illegible].

Quedenfeldt, G. 1890: Drei neue Tenebrioniden aus Tripolitanien. *Berliner Entomologische Zeitschrift* 33 [1899]: 395–400.

Quedenfeldt, G. 1891: Brachycryptus n. gen. Cistelidarum, prope Omophlus. *Entomologische Nachrichten* 17: 129–130.

Quedenfeldt, M. 1890: Akis Schweinfurthi n. sp. *Berliner Entomologische Zeitschrift* 35: 139.

Quensel, K. 1790: *Dissertatio historico-naturalis, ignotas insectorum species continents. 4.* Lundae: Resp. C.P. Lundgard, 20 pp.

Quensel, K. 1806: In: Schoenherr, C.J.: *Synonymia Insectorum, oder: Versuch einer Synonymie aller bisher bekannten Insecten; nach Fabricii Systema Eleutheratorum geordnet. Erster Band. Eleutherata oder Käfer.* Stockholm: Heinr. A. Nordström, xxii + 293 pp., 3 pls.

Raffray, A. 1873: [new taxa]. In: Fairmaire, L. & Raffray, A.: Coléoptères du Nord de l'Afrique. *Revue et Magasin de Zoologie Pure et Appliquée* (3) 1: 331–385.

Rafinesque, C.S. 1815: *Analyse de la nature ou tableau de l'univers et des corps organisés.* Palermo, 224 pp.

Ragusa, E. 1875: Gita entomologica all'isola di Pantelleria. *Bollettino della Società Entomologica Italiana* 7: 238–256.

Ragusa, E. 1881: Coleotteri nuovi o poco conosciuti della Sicilia. *Il Naturalista Siciliano* 1 [1881–1882]: 42–44.

Ragusa, E. 1882: Coleotteri nuovi o poco conosciuti della Sicilia. *Il Naturalista Siciliano* 1 [1881–1882]: 251–252.

Ragusa, A. 1892a: Catalogo ragionale dei coleotteri di Sicilia. *Il Naturalista Siciliano* 11 [1892–1893]: 122–143.

Ragusa, E. 1892b: Coleotteri nuovi o poco conosciuti della Sicilia. *Il Naturalista Siciliano* 12 [1892–1893]: 26–31.

Ragusa, E. 1898: Catalogo ragionato dei coleotteri di Sicilia. *Il Naturalista Siciliano* (N.S.) 2 [1897–1897]: 105–130, 197–237

Ragusa, E. 1906: Coleotteri nuovi o poco conosciuti della Sicilia. *Il Naturalista Siciliano* (N.S.) 19 [1906–1907]: 4–8.

Ragusa, E. 1923: Coleotteri nuovi o poco conosciuti nella Sicilia. *Atti della Reale Accademia delle Scienze e Lettere e Belle Arti di Palermo* (3) 12 [1920–1922]: 1–18.

Rallo, G. 1974: Annotazioni sugli Oochrotus italiani (Coleoptera, Heteromera: IV contributo). *Bollettino del Museo Civico di Storia Naturale di Venezia* 25: 61–63.

Rambur, J.P. 1838: Pls. XIX, XX. In: *Faune Entomologique de l'Andalousie. Volume 1.* [1837–1838]. Paris: Arthus Bertrand, 144 pp., pls. i, ii, ixx, xx. [Note: livraison 1 published in 1837, livraison 2 in 1838; the plates XIX and XX issued presumedly in 1838].

Razzauti, A. 1919: *Contributi alla conoscenza faunistica delle isole toscane II. Forme insulare tirreniche dell'Opatrum sculpturatum Fairmaire. (Coleoptera-Tenebrionidae).* Pisa: Succ. FF. Nistri, 7 pp.

Redondo, A. 1915. Coleópteros de Salamanca. *Brotéria (Série Zoologica)* 13: 14–48.

Redtenbacher, L. 1843: Illustrationes et descriptiones Coleopterorum novorum Syriae. Pp. 971–990. In: Russegger J. von: *Reisen in Europa, Asien und Afrika, mit besonderer Rücksicht auf die naturwissenschaftliche Verhältnisse der betreffenden Länder unternommen in den Jahren 1835 bis 1841. Band 1 (Teil 2).* Stuttgart: Schweizerbart'sche Verlagshandlung.

Redtenbacher, L. 1844: [new taxa]. In: Kollar, V. & Redtenbacher, L.: Aufzählung und Beschreibung der von Freiherr Carl von Huegel auf seiner Reise durch Kaschmir u. das Himalayagebirge gesammelten Insecten. Pp. 393–564. In: Hügel K.F. von. (ed.): *Kaschmir und das Reich der Siek. Vierter Band. Zweite Abtheilung.* Stuttgart: Hallbergerischer Verlag, pp. 244–586 pp. [1844]; 587–865 + [6] pp., 31 pls, 1 map [1848].

Redtenbacher, L. 1845: *Die Gattungen der deutschen Käfer-Fauna nach der analytischen Methode bearbeitet, nebst einem kurzgefassten Leitfaden zum Studium dieses Zweiges der Entomologie.* Wien: Carl Überreuter, [14] +177 + [1] pp., 2 pls.

Redtenbacher, L. 1847: [Part 3, pp. 321–480]. In: *Fauna Austriaca. Die Käfer. Nach der analytischen Methode bearbeitet* [1849]. Wien: C. Gerold, xxvii + 883 pp. (issued in parts).

Redtenbacher, L. 1849: *Fauna Austriaca. Die Käfer. Nach der analytischen Methode bearbeitet.* Wien: C. Gerold, xxvii + 883 pp. (issued in parts).

Redtenbacher, L. 1850: [new taxa]. In: Kollar, V. & Redtenbacher, L.: Ueber den Character der Insecten-Fauna von Südpersien. *Denkschriften der Kaiserlichen Akademie der Wissenschaften. Mathematisch-Naturwissenschaftliche Classe* 1: 42–53.

Redtenbacher, L. 1857: [Pp. 129–976]. In: *Fauna Austriaca. Die Käfer. Nach der analytischen Methode bearbeitet. Zweite, gänzlich umgearbeitete, mit mehreren Hunderten von Arten und mit der Charakteristik sämmtlicher europäischen Käfergattungen vermehrte Auflage* [1858]. Wien: Carl Gerold's Sohn, cxxxvi +1017 pp., 2 pls. (issued in parts).

Redtenbacher, L. 1868: Zoologischer Theil. Zweiter Band. I. Abtheilung A. 2. Coleopteren. In: *Reise der Oesterreichischen Fregatte Novara um die Erde in den Jahren 1857, 1858, 1859 unter den Befehlen des Commodore B. von Wüllerstorf-Urbair.* [1867]. Wien: Karl Gerold's Sohn, iv + 249 pp., 5 pls.

Redtenbacher, L. 1874: *Fauna Austriaca. Die Käfer. Nach der analytischen Methode bearbeitet. Dritte, gänzlich umgearbeitete und bedeutend vermehrte Auflage. Zweiter Band.* Wien: Carl Gerold's Sohn, 571 + viii pp.

Redtenbacher, W. 1842: *Quaedam genera et species Coleopterorum Archiducatus Austriae nondum descriptiorum.* Vindobonae: Carolus Ueberreuter, 31 pp.

Reichardt, A.N. 1934: Opredelitel i spisok zhukov naryvnikov v Turkmenii. *Trudy Sovieta po Izutcheniu Proizvoditelnykh Sil. Seriya Turkmenskaya* 6: 205–239.

Reichardt, A.N. 1935: Novye formy chernotelok triby Opatrini iz Tadzhikistana. *Trudy Tadzhikistanskoy Bazy ANSSR* 5: 251–256.

Reichardt, A.N. 1936a: Zhuki-chernotelki triby Opatrini palearkticheskoy oblasti. Révision des opatrines (Coleoptera Tenebrionidae) de la région paléarctique. *Tableaux analytiques de la Faune de l'URSS* 19, Moskva, Leningrad: Nauka, 1–224 pp.

Reichardt, A.N. 1936b: Obzor vidov roda Pedinus Latr. SSSR i Dalnego Vostoka (Coleoptera Tenebrionidae). *Trudy Zoologicheskogo Instituta Akademii Nauk SSSR* 3: 669–692.

Reiche, L.J. 1850: Entomologie. Pp. 257–532. In: Ferret, P.V.A. & Galinier J.G. (eds): *Voyage en Abyssinie dans les provinces du Tigre, du Samen et de l'Amhara. Tome Troisième.* [1847–1848]. Paris: Paulin, 536 pp.

Reiche, L.J. 1860: Coléoptères de Sicile recueillis par M.E. Bellier de la Chavignerie et déterminés par M.L. Reiche et description de dix espèces nouvelles. *Annales de la Société Entomologique de France* (3) 8: 717–737.

Reiche, L.J. 1860: Description d'une nouvelle espèce de coléoptères du genre Microtelus. *Annales de la Société Entomologique de France* (3) **8**: 334.

Reiche, L.J. 1861a: Species novae Coleopterorum descriptae, quae in Syria invenit Dom. Kindermann. *Wiener Entomologische Monatschrift* **5**: 1–8.

Reiche, L.J. 1861b: Sur quelques espèces de coléoptères du Nord de l'Afrique. *Annales de la Société Entomologique de France* (4) **1**: 87–92.

Reiche, L.J. 1861c: Coléoptères nouveaux recueillis en Corse par M.E. Bellier de la Chavignerie. *Annales de la Société Entomologique de France* (4) **1**: 201–210.

Reiche, L.J. 1861d: Notes synonymiques. *Annales de la Société Entomologique de France* (4) **1**: 211–212.

Reiche, L.J. 1862a: Espèces nouvelles de coléoptères appartenant à la faune circa-méditerranéenne. *Annales de la Société Entomologique de France* (4) **1** [1861]: 361–374.

Reiche, L.J. 1862b: Espèces nouvelles de coléoptères découvertes en Corse par M.E. Bellier de la Chavignerie, en 1861. *Annales de la Société Entomologique de France* (4) **2**: 293–300.

Reiche, L.J. 1862c: Espèces nouvelles de coléoptères appartenant à la faune circa-méditerranéenne. (Suite) (1). *Annales de la Société Entomologique de France* (4) **2**: 539–546.

Reiche, L.J. 1864: Espèces nouvelles de coléoptères d'Algérie. *Annales de la Société Entomologique de France* (4) **4**: 233–246.

Reiche, L.J. 1866: Étude sur les epèces de mylabrides de la collection de L. Reiche, suivie d'une note sur le genre Trigonurus Mulsant et description d'une espèce nouvelle. *Annales de la Société Entomologique de France* (4) **5** [1865]: 627–642.

Reiche, L.J. 1867: [new taxon]. In: Marseul, S.-A. de: Catalogue des coléoptères d'Europe et des pays limitrophes. *L'Abeille, Mémoires d'Entomologie* **4**: 1–131.

Reiche, L.J. 1878: [new taxon]. *Bulletin de la Société Entomologique de France* **1878**: lxxiii–lxxiv.

Reiche, L.J. & Saulcy, F. 1857: Espèces nouvelles ou peu connues de coléoptères, recueillis par M.F. de Saulcy, membre de l'Institut, dans son voyage en Orient. *Annales de la Société Entomologique de France* (3) **5**: 169–276, 1 pl.

Reinig, W.F. 1931: Entomologische Ergebnisse der Deutsch-Russischen Alai-Pamir-Expedition 1928 (II). 5. Coleoptera II. Tenebrionidae. *Mitteilungen aus dem Zoologischen Museum Berlin* **16** [1930]: 865–912.

Reinig, W.F. 1934: 6. Coleoptera VII. (Tenebrionidae, Nachtrag). P. 162. In: Entomologische Ergebnisse der Deutsch-Russischen Alai-Pamir-Expedition 1928 (II). *Deutsche Entomologische Zeitschrift* **1933**: 129–176.

Reitter, E. 1872: Neue Käferarten von Oran gesammelt von Hans Leder. *Berliner Entomologische Zeitschrift* **16**: 167–186.

Reitter, E. 1874: Beschreibungen neuer Käfer-Arten nebst synonymischen Notizen. *Verhandlungen der Kaiserlich-Königlichen Zoologisch-Botanischen Gesellschaft in Wien* **24**: 509–528.

Reitter, E. 1875: [new taxa]. In: Weise, J.: Neue Käferarten aus Ungarn. *Deutsche Entomologische Zeitschrift* **19** [1930]: 355–364.

Reitter, E. 1876a: Neue Gattungen und Arten aus der Familie der Cucujidae. *Coleopterologische Hefte* **15**: 37–64.

Reitter, E. 1876b: Neue transcaucasische Coleopteren, gesammelt von H. Leder. *Berliner Entomologische Zeitschrift* **20**: 289–294.

Reitter, E. 1877a: Coleopterologische Ergebnisse einer Reise nach Südungarn und die Transsylvanischen Alpen. *Verhandlungen des Naturforschenden Vereines in Brünn* 15 [1876]: 3–30, pl. 1.

Reitter, E. 1877b: Neue Arten aus der Gattung Sitophagus Muls. *Mittheilungen des Münchener Entomologischen Vereins* 1: 8–11.

Reitter, E. 1877c: Neue Arten aus den Familien der Cucujidae, Nitidulidae, Colydiidae und Cryptophagidae. *Mittheilungen des Münchener Entomologischen Vereins* 1: 22–28.

Reitter, E. 1877d: Neuer Mycetophagus aus Japan. *Deutsche Entomologische Zeitschrift* 21: 116.

Reitter, E. 1877e: Neue kaukasische Coleopteren, gesammelt von Hans Leder (Zweites Stück). *Deutsche Entomologische Zeitschrift* 21: 289–296.

Reitter, E. 1877f: Beitrag zur Kenntniss der Colydier. *Entomologische Zeitung* (Stettin) 38: 323–355.

Reitter, E. 1877g: Beiträge zur Käferfauna von Japan. (Drittes Stück). *Deutsche Entomologische Zeitschrift* 21: 369–384.

Reitter, E. 1877h: Atritomus nov. gen. Tritomidarum. *Deutsche Entomologische Zeitschrift* 21: 384.

Reitter, E. 1877i: Coleopterorum species novae. *Verhandlungen der Kaiserlich-Königlichen Zoologisch-Botanischen Gesellschaft in Wien* 27: 165–194.

Reitter, E. 1878a: [new taxa]. In: Schneider, O. & Leder, H.: Beiträge zur Kenntnis der kaukasischen Käferfauna. *Verhandlungen des Naturforschen Vereines in Brünn* 16 [1887]: 3–259, 3 pls.

Reitter, E. 1878b: Nachträge und Ergänzungen zur Bearbeitung der Cioiden von H. v. Kiesenwetter. *Deutsche Entomologische Zeitschrift* 22: 21–30.

Reitter, E. 1878c: Neue Cioidae. *Mittheilungen des Münchener Entomologischen Vereins* 2: 32–37.

Reitter, E. 1878d: Beiträge zur Coleopteren-Fauna der Carpathen. *Deutsche Entomologische Zeitschrift* 22: 33–64.

Reitter, E. 1878e: Neue Colydiidae des Berliner Museums. *Deutsche Entomologische Zeitschrift* 22: 113–125.

Reitter, E. 1879: Verzeichniss der von H. Christoph in Ost-Sibirien gesammelten Clavicornier etc. *Deutsche Entomologische Zeitschrift* 23: 208–226.

Reitter, E. 1880a: Bestimmungs-Tabellen der europäischen Coleopteren. I. Enthaltend die Familien: Cucujidae, Telmatophilidae, Tritomidae, Mycetaeidae, Endomychidae, Lyctidae und Sphindidae. *Verhandlungen der Kaiserlich-Königlichen Zoologisch-Botanischen Gesellschaft in Wien* 29 [1879]: 71–100.

Reitter, E. 1880b: Neue Lagria-Arten aus Africa und Ostindien. *Deutsche Entomologische Zeitschrift* 24: 253–257.

Reitter, E. 1880c: [new taxa]. In: Leder, H.: Beitrag zur kaukasischen Käfer-Fauna. Unter Mitwirkung hervorragender Fachgenossen. *Verhandlungen der Kaiserlich-Königlichen Zoologisch-Botanischen Gesellschaft in Wien* 29 [1879]: 451–488.

Reitter, E. 1880d: Neue Coleopteren aus dem südöstlichen Russland aufgefunden von Herrn J. Faust aus Helsingfors. *Verhandlungen der Kaiserlich-Königlichen Zoologisch-Botanischen Gesellschaft in Wien* 29 [1879]: 543–546.

Reitter, E. 1881a: Neue und seltene Coleopteren, im Jahre 1880 in Süddalmatien und Montenegro gesammelt und beschrieben. *Deutsche Entomologische Zeitschrift* 25: 177–229, pls. 6–7.

Reitter, E. 1881b: Coleopterologische Ergebnisse einer Reise nach Croatien, Dalmatien und der Herzegowina im Jahre 1879. *Verhandlungen der Kaiserlich-Königlichen Zoologisch-Botanischen Gesellschaft in Wien* **30** [1880]: 201–228.

Reitter, E. 1881c: Neue Käferarten aus Asturien. *Deutsche Entomologische Zeitschrift* **25**: 231–232.

Reitter, E. 1882a: Bestimmungs-Tabellen der europäischen Coleopteren. VI. Enhaltend die Familien: Colydiidae, Rhysodidae, Trogositidae. *Verhandlungen des Naturforschenden Vereines in Brünn* **22** [1881]: 113–149.

Reitter, E. 1882b: Neuere Literatur. Histoire naturelle des coléoptères de France. (Par E. Mulsant.) Famille des Lathridiens, par le R.P. Fr. Marie-Joseph Belon. Lyon 1881. *Deutsche Entomologische Zeitschrift* **26**: 161–165.

Reitter, E. 1883a: Neue Coleopteren aus Russland und Bemerkungen über bekannte Arten. *Revue Mensuele d'Entomologie* **1**: 111–118.

Reitter, E. 1883b: Zwei neue Heteromeren-Genera aus Europa. *Wiener Entomologische Zeitung* **2**: 307–310, pl. 4.

Reitter, E. 1884a: Diagnosen neuer Coleopteren aus Lenkoran. *Verhandlungen des Naturforschenden Vereines in Brünn* **22** [1883]: 3–10.

Reitter, E. 1884b: [new taxa]. In: Brenske, E. & Reitter, E.: Neuer Beitrag zur Käferfauna Griechenlands. *Deutsche Entomologische Zeitschrift* **28**: 17–100, pls 1, 2.

Reitter, E. 1884c: Resultate einer coleopterologischen Sammel-Campagne während den Monaten Februar bis April 1883 auf den jonischen Inseln. *Deutsche Entomologische Zeitschrift* **28**: 101–122.

Reitter, E. 1884d: Revision der europäischen Mycetochares-Arten. *Deutsche Entomologische Zeitschrift* **28**: 241–250.

Reitter, E. 1884e: Einige neue Coleopteren aus Süd-Europa. *Deutsche Entomologische Zeitschrift* **28**: 255–259.

Reitter, E. 1884f: Ueber die syrischen Arten der Gattung Anemia Lap. *Deutsche Entomologische Zeitschrift* **28**: 259–260.

Reitter, E. 1885a: Uebersicht der Arten der Gattung Cerocoma Geoff. *Deutsche Entomologische Zeitschrift* **29**: 1–14.

Reitter, E. 1885b: Coleopterologische Ergebnisse einer Excursion nach Bosnien im Mai 1884. *Deutsche Entomologische Zeitschrift* **29**: 193–216.

Reitter, E. 1885c: Coleopterologische Notizen. XIV. *Wiener Entomologische Zeitung* **4**: 313–317.

Reitter, E. 1885d: Neue Coleopteren aus Europa und den angrenzenden Ländern, mit Bemerkungen über bekannte Arten. *Deutsche Entomologische Zeitschrift* **29**: 353–397.

Reitter, E. 1885e: Uebersicht der bekannten Laena-Arten. *Deutsche Entomologische Zeitschrift* **29**: 398–402.

Reitter, E. 1886a: Revision der mit Stenosis verwandten Coleopteren der alten Welt. *Deutsche Entomologische Zeitschrift* **30**: 97–144.

Reitter, E. 1886b: Coleopterologische Notizen. XV. *Wiener Entomologische Zeitung.* **5**: 99–100.

Reitter, E. 1886c: Ueber eine neue europäische Allecula. *Wiener Entomologische Zeitung* **5**: 140.

Reitter, E. 1886d: Coleopterologische Notizen. XVI. *Wiener Entomologische Zeitung* **5**: 151–152.

Reitter, E. 1886e: Coleopterologische Notizen. XIX. *Wiener Entomologische Zeitung* **5**: 347–352.

Reitter, E. 1887a: Coleopterologische Notizen. XXI. *Wiener Entomologische Zeitung* **6**: 76–77.

Reitter, E. 1887b: Sulle specie del genere di Coleotteri Tetratoma Fabricius. *Il Naturalista Siciliano* **6** [1886–1887]: 84–85.

Reitter, E. 1887c: Neue Coleopteren aus Europa, den angrenzenden Ländern und Sibirien, mit Bemerkungen über bekannte Arten. *Deutsche Entomologische Zeitschrift* **31**: 241–288.

Reitter, E. 1887d: Insecta in itinere Cl. N. Przewalskii in Asia Centrali novissime lecta. IX. Tenebrionidae. *Horae Societatis Entomologicae Rossicae* **21**: 355–389.

Reitter, E. 1887e: Neue Coleopteren aus Europa, den angrenzenden Ländern und Sibirien, mit Bemerkungen über bekannte Arten. *Deutsche Entomologische Zeitschrift* **31**: 497–528.

Reitter, E. 1887f: Coleopterologische Notizen. XX. *Wiener Entomologische Zeitung* **6**: 28–29.

Reitter, E. 1888a: Coleopteren aus Circassien, gesammelt von Hans Leder in Jahr 1887. IV. Theil. *Wiener Entomologische Zeitung* **7**: 169–180.

Reitter, E. 1888b: Coleopteren aus Circassien, gesammelt von Hans Leder im Jahre 1887. V. Theil. *Wiener Entomologische Zeitung* **7**: 207–214.

Reitter, E. 1888c: Ueber einige mit Pimelia nahe verwandte Coleopteren-Gattungen. *Deutsche Entomologische Zeitschrift* **1888**: 329–331.

Reitter, E. 1888d: Neue Coleopteren aus Europa, den angrenzenden Ländern und Sibirien, mit Bemerkungen über andere Arten. Fünfter Theil. *Deutsche Entomologische Zeitschrift* **1888**: 417–432.

Reitter, E. 1889a: Neue Coleopteren aus Europa, den angrenzenden Ländern und Sibirien, mit Bemerkungen über bekannte Arten. Sechster Theil. *Deutsche Entomologische Zeitschrift* **1889**: 17–44.

Reitter, E. 1889b: *Coleopterologische Ergebnisse der im Jahre 1886 und 1887 in Transcaspien von Dr. G. Radde, Dr. A. Walter und A. Konchin ausgeführten Expedition.* Verhandlungen des Naturforschenden Vereines in Brünn 27 [1888]: 95–133.

Reitter, E. 1889c: Zwei neue Meloë-Arten aus Syrien und Armenien. *Wiener Entomologische Zeitung* 8: 106–107.

Reitter, E. 1889d: Beschreibungen der bekannten Tritomiden Japans mit Berücksichtigung der neuen Sammelergebnisse der Herrn George Lewis in den Jahren 1880 und 1881. *Wiener Entomologische Zeitung* 8: 245–249.

Reitter, E. 1889e: Berichte über die von E. v. Oertzen in Jahre 1887 in Griechenland u. Klein-Asien gesammelten Coleopteren. IX. Neue Arten aus verschiedenen Familien. *Deutsche Entomologische Zeitschrift* **1889**: 251–259.

Reitter, E. 1889f: Uebersicht der palaearctischen Chrysanthia-Arten. *Deutsche Entomologische Zeitschrift* **1889**: 266.

Reitter, E. 1889g: Uebersicht der bekannten Scraptia-Arten der palaearctischen Fauna. *Deutsche Entomologische Zeitschrift* **1889**: 267–268.

Reitter, E. 1889h: Neue Coleopteren aus Europa, den angrenzenden Ländern und Sibirien, mit Bemerkungen über bekannte Arten. *Deutsche Entomologische Zeitschrift* **1889**: 273–288.

Reitter, E. 1889i: Verzeichniss der Cucujiden Japans mit Beschreibungen neuer Arten. *Wiener Entomologische Zeitung* **8**: 313–320.

Reitter, E. 1889j: Neue Coleopteren aus Europa, den angrenzenden Ländern und Sibirien, mit Bemerkungen über bekannte Arten. *Deutsche Entomologische Zeitschrift* **1889**: 369–376.

Reitter, E. 1889k: Insecta, a cl. G.N. Potanin in China et in Mongolia novissime lecta. XIII. Tenebrionidae. *Horae Societatis Entomologicae Rossicae* **23**: 678–710.

Reitter, E. 1890a: Coleopterologische Notizen. *Wiener Entomologische Zeitung* **9**: 13–15.

Reitter, E. 1890b: Neue analytische Uebersicht der bekannten Arten der Coleopteren-Gattung Omophlus. *Deutsche Entomologische Zeitschrift* **1890**: 33–52.

Reitter, E. 1890c: Eine neue behaarte Meloë-Art aus Turkestan. *Wiener Entomologische Zeitung* **9**: 152.

Reitter, E. 1890d: Neue Coleopteren aus Europa, den angrenzenden Ländern und Sibirien, mit Bemerkungen über bekannte Arten. Neuenter Theil. *Deutsche Entomologische Zeitschrift* **1890**: 145–164.

Reitter, E. 1890e: Neue Coleopteren aus Europa, den angrenzenden Ländern und Sibirien, mit Bemerkungen über bekannte Arten. Zehnter Theil. *Deutsche Entomologische Zeitschrift* **1890**: 165–176.

Reitter, E. 1890f: Coleopterologische Notizen. XXXVIII. *Wiener Entomologische Zeitung* **9**: 210–213.

Reitter, E. 1890g: Uebersicht der Cteniopus-Arten aus Europa und den angrenzenden Ländern, soweit mir dieselben in Natura bekannt sind. *Wiener Entomologische Zeitung* **9**: 256–258.

Reitter, E. 1890h: [new taxa]. In: Heyden L.F.J.D. von.: XIII. Beitrag zur Coleopteren-Fauna von Turkestan. Unter Mitwirkung der Herren Reitter und Weise. *Deutsche Entomologische Zeitschrift* **1890**: 353–367.

Reitter, E. 1890i: In: Heyden L.F.J.D. von: Beiträge zur Coleopteren-Fauna von Amasia und Samsoun in Nord-Kleinasien. *Deutsche Entomologische Zeitschrift* **1890**: 369–374.

Reitter, E. 1890j: Neue Coleopteren aus Europa, den angrenzenden Ländern und Sibirien, mit Bemerkungen über bekannte Arten. Elfter Theil. *Deutsche Entomologische Zeitschrift* **1890**: 383–396.

Reitter, E. 1891a: Neue Coleopteren aus Europa, den angrenzenden Ländern und Sibirien, mit Bemerkungen über bekannte Arten. *Deutsche Entomologische Zeitschrift* **1891**: 17–36.

Reitter, E. 1891b: Drei neue Coleopteren. *Wiener Entomologische Zeitung* **10**: 33–34.

Reitter, E. 1891c: Zweiter Beitrag zur Coleopteren-Fauna des russischen Reiches. *Wiener Entomologische Zeitung* **10**: 195–199.

Reitter, E. 1891d: Dritter Beitrag zur Coleopterenfauna des russischen Reiches. *Wiener Entomologische Zeitung* **10**: 221–224.

Reitter, E. 1891e: Coleopterologische Notizen. XLI. *Wiener Entomologische Zeitung* **10**: 226–228.

Reitter, E. 1891f: Erster Beitrag zur Coleopteren-Fauna von Europa und den angrenzenden Ländern. *Wiener Entomologische Zeitung* **10**: 246–249.

Reitter, E. 1891g: Zweiter Beitrag zur Coleopteren-Fauna von Europa und den angrenzenden Ländern. *Wiener Entomologische Zeitung* **10**: 259–262.

Reitter, E. 1892a: Fünfter Beitrag zur Coleopterenfauna des russischen Reiches. *Wiener Entomologische Zeitung* **11**: 59–68.

Reitter, E. 1892b: Ueber die Arten der Coleopteren-Gattung Leptodes Sol. *Wiener Entomologische Zeitung* **11**: 93–96.

Reitter, E. 1892c: Sechster Beitrag zur Coleopteren-Fauna des russischen Reiches. *Wiener Entomologische Zeitung* **11**: 133–136.

Reitter, E. 1893a: Achter Beitrag zur Coleopteren-Fauna des russischen Reiches. *Wiener Entomologische Zeitung* **12**: 109–114.

Reitter, E. 1893b: Bestimmungs-Tabelle der unechten Pimeliden aus der palaearctischen Fauna. *Verhandlungen des Naturforschenden Vereines in Brünn* **31** [1892]: 201–250.

Reitter, E. 1893c: Revision der Coleopteren-Gattung Prosodes Esch. *Deutsche Entomologische Zeitschrift* **1893**: 261–312.

Reitter, E. 1893d: Ueber die Genus-Charaktere der Gattungen Blaps Fr., Prosodes Eschsch. und Verwandte. *Deutsche Entomologische Zeitschrift* **1893**: 313–316.

Reitter, E. 1893e: Uebersicht der mir bekannten Ammobius-Arten. *Wiener Entomologische Zeitung* **12**: 315.

Reitter, E. 1894a: [new taxa]. In: Hauser F. Beitrag zur Coleopteren-Fauna von Transcaspien und Turkestan. *Deutsche Entomologische Zeitschrift* **1894**: 17–74.

Reitter, E. 1894b: Einige neue Coleopteren von der Insel Malta. *Wiener Entomologische Zeitung* **13**: 81–83.

Reitter, E. 1894c: Vierter Beitrag zur Coleopteren-Fauna von Europa und den angrenzenden Ländern. *Wiener Entomologische Zeitung* **13**: 101–107.

Reitter, E. 1894d: Elfter Beitrag zur Coleopteren-Fauna des russischen Reiches. *Wiener Entomologische Zeitung* **13**: 238–248.

Reitter, E. 1894e: Fünfter Beitrag zur Coleopteren-Fauna von Europa und den angrenzenden Ländern. *Wiener Entomologische Zeitung* **13**: 299–306.

Reitter, E. 1895a: *Bestimmungs-Tabellen der europäischen Coleopteren Meloidae. 1 Theil: Meloini. 32.* Paskau: Verfasser, 13 pp.

Reitter, E. 1895b: Beschreibung neuer oder wenig gekannter Coleopteren aus der Umgebung von Akbes in Syrien. *Wiener Entomologische Zeitung* **14**: 79–88.

Reitter, E. 1895c: Uebersicht der mir bekannten Arten der Coleopteren-Gattung Phthora Germ. (Cataphronetis Lucas). *Entomologische Nachrichten* **21**: 147–149.

Reitter, E. 1895d: Zwölfter Beitrag zur Coleopteren-Fauna des russischen Reiches. *Wiener Entomologische Zeitung* **14**: 149–162.

Reitter, E. 1895e: Sechster Beitrag zur Coleopteren-Fauna von Europa und den angrenzenden Ländern. *Wiener Entomologische Zeitung* **14**: 199–204.

Reitter, E. 1895f: Beschreibungen mit Abbildungen neuer Coleopteren, gesammelt von Herrn Hans Leder bei Urga in der nördlichen Mongolei. *Wiener Entomlogische Zeitung* **14**: 280–286.

Reitter, E. 1896a: Beschreibungen neuer Coleopteren aus dem russischen Reich. *Deutsche Entomologische Zeitschrift* **1896**: 33–48.

Reitter, E. 1896b: Dreizehnter Beitrag zur Coleopteren-Fauna des russischen Reiches. *Wiener Entomologische Zeitung* **15**: 64–77.

Reitter, E. 1896c: Dichotomische Uebersicht der bekannten Arten der Coleopteren Gattung Capnisa Lac. = Gnathosia Fisch. *Entomologische Nachrichten* **22**: 129–135.

Reitter, E. 1896d: Uebersicht der mir bekannten, palaearctischen Arten der Coleopteren-Gattung Crypticus Latreille. *Entomologische Nachrichten* **22**: 145–151.

Reitter, E. 1896e: Uebersicht der Arten der Coleopteren-Gattung Heterophylus Muls. *Deutsche Entomologische Zeitschrift* **1896**: 159–160.

Reitter, E. 1896f: Uebersicht der mir bekannten, mit Penthicus Fald. verwandten Coleopteren-Gattungen und Arten aus der paläarktischen Fauna. *Deutsche Entomologische Zeitschrift* **1896**: 161–172.

Reitter, E. 1896g: Beitrag zur Kenntnis der Platysceliden. *Deutsche Entomologische Zeitschrift* **1896**: 173–176.

Reitter, E. 1896h: Namensänderung. *Wiener Entomologische Zeitung* **15**: 191–192.

Reitter, E. 1896i: Abbildungen mit Beschreibungen zehn neuer Coleopteren aus der palaearctischen Fauna. *Wiener Entomologische Zeitung* **15**: 233–240.

Reitter, E. 1896j: Achter Beitrag zur Coleopteren-Fauna von Europa und den angrenzenden Ländern. *Wiener Entomologische Zeitung* **15**: 265–272.

Reitter, E. 1896k: Vierzehnter Beitrag zur Coleopteren-Fauna des russisches Reiches. *Wiener Entomologische Zeitung* **15**: 285–291.

Reitter, E. 1897a: Uebersicht der Arten der Coleopteren-Gattung Asclera Schm. In: *Fest-Schrift zur Feier des Fünfzigjährigen Bestehens des Vereins für Schlesische Insektenkunde in Breslau 1847–1897*. Breslau: Maruschke & Berendt, 141 pp.

Reitter, E. 1897b: Ueber die Arten der Coleopteren-Gattung Anemia Cast., aus der palaearctischen Fauna. In: *Fest-Schrift zur Feier des Fünfzigjährigen Bestehens des Vereins für Schlesische Insektenkunde in Breslau 1847–1897*. Breslau: Maruschke & Berendt. 141 pp.

Reitter, E. 1897c: [new taxon]. In: Escherich, K.: Revision der palaearktischen Zonitiden, einer Unterfamilie der Meloiden. *Verhandlungen des Naturforschenden Vereines in Brünn* **35** [1896]: 96–133.

Reitter, E. 1897d: Fünfzehnter Beitrag zur Coleopteren-fauna des russischen Reiches. *Wiener Entomologische Zeitung* **16**: 121–127.

Reitter, E. 1897e: Dreissig neue Coleopteren aus russisch Asien und der Mongolei. *Deutsche Entomologische Zeitschrift* **1897**: 209–228.

Reitter, E. 1897f: Analytische Revision der Coleopteren-Gattung Microdera Esch. *Deutsche Entomologische Zeitschrift* **1897**: 229–235.

Reitter, E. 1897g: Dircaea subtilis n. sp. *Deutsche Entomologische Zeitschrift* **23**: 246.

Reitter, E. 1897h: Dichotomische Uebersicht der mir bekannten Gattungen aus der Tenebrioniden-Abtheilung: Tentyrini. *Deutsche Entomologische Zeitschrift* **1896**: 297–303.

Reitter, E. 1897i: Uebersicht der Arten der Coleopteren-Gattung Calyptopsis Sol. *Deutsche Entomologische Zeitschrift* **1896**: 305–311.

Reitter, E. 1897j: Psammocryptus rugiceps Reitter n. sp. aus Turkestan. *Deutsche Entomologische Zeitschrift* **1896**: 311.

Reitter, E. 1897k: Uebersicht der mir bekannten Arten der Coleopteren-Gattung: Dailognatha Stev. *Deutsche Entomologische Zeitschrift* **1896**: 313–317.

Reitter, E. 1898a: Eine Decade neuer Coleopteren aus der Buchara. *Wiener Entomologische Zeitung* **17**: 10–16.

Reitter, E. 1898b: Uebersicht der bekannten Arten der Coleopterengattung Scleropatrum Seidl. aus der palaearctischen Fauna. *Wiener Entomologische Zeitung* **17**: 36–39.

Reitter, E. 1898c: Drei neue Mycetocharina-Arten. (Coleoptera, Alleculidae). *Entomologische Nachrichten* **24**: 65–66.

Reitter, E. 1898d: Über die Gattungen: Apolites Duv. und Anisocerus Fald. (Coleopt. Tenebrionidae). *Wiener Entomologische Zeitung* **17**: 105–106.

Reitter, E. 1898e: Siebzehnter Beitrag zur Coleopteren-Fauna des russischen Reiches. *Wiener Entomologische Zeitung* **17**: 109–127.

Reitter, E. 1898f: Neue Coleopteren aus Europa und den angrenzenden Ländern. *Deutsche Entomologische Zeitschrift* **1898**: 337–360.

Reitter, E. 1899a: Uebersicht der mir bekannten, palaearctischen Arten der Coleopteren-Gattung Leichenum Blanch. *Entomologische Nachrichten* **25**: 83–86.

Reitter, E. 1899b: Zur weiteren Kenntniss der Coleopteren-Gattung Mycetochara Berthold. *Entomologische Nachrichten* **25**: 155–159.

Reitter, E. 1899c: Elfter Beitrag zur Coleopteren-Fauna von Europa und den angrenzenden Ländern. *Wiener Entomologische Zeitung* **18**: 155–161.

Reitter, E. 1899d: Beitrag zur Coleopteren-Fauna des russischen Reiches und der angrenzenden Ländern. *Deutsche Entomologische Zeitschrift* **1899**: 193–209.

Reitter, E. 1899e: Ueber einige Coleopteren aus der palaearctischen Fauna und aus Japan. *Entomologische Nachrichten* **25**: 216–220.

Reitter, E. 1900a: [new taxa]. In: Bodemeyer, E. von.: *Quer durch Klein-Asien in den Bulghar-Dagh. Eine naturwissenschaftliche Studien-Reise.* Emmendingen: Verlangs-Aktiengesellschaft vormals Dölter, v + 169 pp.

Reitter, E. 1900b: Beitrag zur Coleopteren-Fauna des russischen Reiches. *Deutsche Entomologische Zeitschrift* **1900**: 49–59.

Reitter, E. 1900c: Beitrag zur Coleopteren-Fauna von Europa und den angrenzenden Ländern. *Deutsche Entomologische Zeitschrift* **1900**: 81–88

Reitter, E. 1900d: *Bestimmungs-Tabelle der Tenebrioniden-Abtheilungen: Tentyrini und Adelostomini aus Europa und den angrenzenden Ländern.* Paskau: Edmund Reitter, [1] + 82–197. [note: same issued in 1901 in *Verhandlungen des Naturforschendes Vereines in Brünn* **39** [1900]: 82–197.

Reitter, E. 1900e: Coleoptera, gesammelt im Jahre 1898 in chinesisch Central-Asien von Dr. Holderer in Lahr. *Wiener Entomologische Zeitung* **19**: 152–166.

Reitter, E. 1900f: Neue, von Herrn Dr. John Sahlberg auf seinen Reisen in Corfu, Palästina und Centralasien gesammelte Coleopteren. *Wiener Entomologische Zeitung* **19**: 217–220.

Reitter, E. 1900g: Neue Coleopteren aus Europa, den angrenzenden Ländern, russisch Asien und der Mongolei. *Deutsche Entomologische Zeitschrift* **1899**: 273–281.

Reitter, E. 1900h: Weitere Beiträge zur Kenntniss der Coleopteren-Gattung Laena Latr. *Deutsche Entomologische Zeitschrift* **1899**: 282–286.

Reitter, E. 1900i: Uebersicht der mir bekannten Arten der Coleopteren-Gattung Dila Fisch. *Entomologische Nachrichten* **26**: 295–296.

Reitter, E. 1900j: Uebersicht der mit Erodius verwandten palaearktischen Coleopteren-Genera. *Entomologische Nachrichten* **26**: 298–300.

Reitter, E. 1900k: Uebersicht der bekannten Arten der Coleopteren-Gattung Ammozoum Sem. *Entomologische Nachrichten* **26**: 300–301.

Reitter, E. 1900l: Uebersicht der bekannten zwei Arten der neuen Coleopteren-Gattung Diaphanidius Reitt. *Entomologische Nachrichten* **26**: 301–302.

Reitter, E. 1901a: *Byrrhidae (Anobiidae) und Cioidae. Bestimmungs-Tabelle der europäischen Coleopteren. XLVII. (Heft 47).* Brünn: Edm. Reitter, 64 pp. [separate issue of Reitter E. 1902. Analytische Uebersicht der palaearctischen Gattungen und Arten der Coleopteren-Familien: Byrrhidae (Anobiidae) und Cioidae. *Verhandlungen des Naturforschenden Vereines in Brünn* **40** [1901]: 3–64].

Reitter, E. 1901b: Ueber die turkestanische Arten der Coleopteren-Gattung Laena Latr. *Wiener Entomologische Zeitung* **20**: 61–63.

Reitter, E. 1901c: Weitere Beiträge zur Coleopteren-Fauna des russischen Reiches. *Deutsche Entomologische Zeitschrift* **1901**: 65–84.

Reitter, E. 1901d: Uebersicht der Coleopteren-Gattung Pedilus Fisch. der palaearctischen Fauna. *Wiener Entomologische Zeitung* **20**: 114–116.

Reitter, E. 1901e: Abbildungen und Beschreibungen neuer Coleopteren aus der palaearctischen Fauna. *Wiener Entomologische Zeitung* **20**: 157–164.

Reitter, E. 1902a: Neue und seltene Coleopteren, gesammelt im Jahre 1901, in der Herzegowina, in Dalmatien und Bosnien. *Wiener Entomologische Zeitung* **21**: 1–9, pl. 1.

Reitter, E. 1902b: Eine Serie neuer Coleopteren aus dem russischen Reiche. *Deutsche Entomologische Zeitschrift* **1901**: 177–186.

Reitter, E. 1902c: Beschreibung neuer Coleopteren aus dem russischen Reiche. *Wiener Entomologische Zeitung* **21**: 185–190.

Reitter, E. 1902d: Dichotomische Übersicht der Coleopteren-Gattung Pterocoma Sol. *Deutsche Entomologische Zeitschrift* **1901**: 189–192.

Reitter, E. 1902e: Fünfzehnter Beitrag zur Coleopteren-Fauna von Europa und den angrenzenden Ländern. *Wiener Entomologische Zeitung* **21**: 191–196.

Reitter, E. 1902f: Coleopterologische Studien. *Wiener Entomologische Zeitung* **21**: 203–217.

Reitter, E. 1902g: Verschiedenes über die Coleopteren der Tenebrioniden-Abtheilung Helopina. *Deutsche Entomologische Zeitschrift* **1901**: 209–224.

Reitter, E. 1902h: Coleopterologische Notizen. *Wiener Entomologische Zeitung* **21**: 221–222.

Reitter, E. 1903a: Uebersicht der Arten der Coleopteren-Gattung Entomogonus Sol. *Wiener Entomologische Zeitung* **22**: 18–20.

Reitter, E. 1903b: Sechzehnter Beitrag zur Coleopteren-Fauna von Europa und den angrenzenden Ländern. *Wiener Entomologische Zeitung* **22**: 43–46.

Reitter, E. 1903c: Die Arten der Gattung Cephalostenus Sol. *Wiener Entomologische Zeitung* **22**: 132.

Reitter, E. 1903d: Siebzehnter Beitrag zur Coleopterenfauna von Europa und den angrenzenden Ländern. *Wiener Entomologische Zeitung* **22**: 173–178.

Reitter, E. 1903e: Neue, von Hauptmann E. von Bodemeyer im Jahre 1903 in Kleinasien gesammelte Coleopteren, vorzüglich Pselaphiden und Scydmaeniden. *Wiener Entomologische Zeitung* **22**: 221–224.

Reitter, E. 1903f: Übersicht der mit Zonabris variabilis Pall. zunächst verwandten Coleopteren-Arten. *Wiener Entomologische Zeitung* **22**: 230.

Reitter, E. 1903g: Übersicht der palaearctischen Arten der Coleopteren-Gattung Evaniocera Guer. *Wiener Entomologische Zeitung* **22**: 237–238.

Reitter, E. 1903h: Zwei neue Cteniopus-Arten aus Kleiasien. *Wiener Entomologische Zeitung* **22**: 257–258.

Reitter, E. 1904a: Bestimmungs-Tabelle der Tenebrioniden-Unterfamilien: Lachnogyini, Akidini, Pedinini, Opatrini und Trachyscelini aus Europa und den angrenzenden Ländern. *Verhandlungen des Naturforschenden Vereines in Brünn* **42** [1903]: 25–189.

Reitter, E. 1904b: Una nuova varietà della Akis spinosa L. *Il Naturalista Siciliano* **17**: 97.

Reitter, E. 1904c: Übersicht über die mit M. piceus F. zunächst verwandten Mycetophagus-Arten aus Europa und den angrenzenden Ländern. *Wiener Entomologische Zeitung* **23**: 165–166.

Reitter, E. 1904d: Sechs neue Coleopteren aus der palaearktischen Region. *Wiener Entomologische Zeitung* **23**: 255–258.

Reitter, E. 1905a: Acht neue Coleopteren aus der palaearktischen Fauna. *Wiener Entomologische Zeitung* **24**: 90–94.

Reitter, E. 1905b: Übersicht der mir bekannten Arten der Coleopteren-Gattung Epicauta Redtb. aus der palaearktischen Fauna. *Wiener Entomologische Zeitung* **24**: 194–196.

Reitter, E. 1905c: Neun neue Coleopteren aus der palaearktischen Fauna. *Wiener Entomologische Zeitung* **24**: 201–206.

Reitter, E. 1905d: Sechzehn neue Coleopteren aus der palaearktischen Fauna. *Wiener Entomologische Zeitung* **24**: 241–251.

Reitter, E. 1905e: [new taxa]. In: Ganglbauer, L.: Coleoptera. Ergebnisse einer naturwissenschaftlichen Reise zum Erdschias-Dagh (Kleinasien). Ausgeführt von Dr. Arnold Penther und Dr. Emerich Zederbauer auf Kosten der 'Gesellschaft zur Förderung des naturhistotrischen Erforschung des Orients in Wien' im Jahre 1902. *Annalen des k. k. Naturhistorischen Hofmuseums* **20**: 246–290.

Reitter, E. 1905f: Drei neue Coleopteren aus der palaearktischen Fauna. *Wiener Entomologische Zeitung* **24**: 311–312.

Reitter, E. 1906a: [new names]. In: Heyden, L F.J.D., Reitter, E. & Weise, J: *Catalogus Coleopterorum Europae, Caucasi et Armeniae Rossicae. Editio secunda*. Berlin, Paskau, Caen, 4 pp. + 774 columns.

Reitter, E. 1906b: Neue Coleopteren aus der paläarktischen Fauna. *Wiener Entomologische Zeitung* **25**: 31–37.

Reitter, E. 1906c: Drei neue im Quellgebiet des Indus von Professor Dr. Koken gesammlte Coleopteren. *Wiener Entomologische Zeitung* **25**: 40–42.

Reitter, E. 1906d: Uebersicht der Coleopteren-Unterfamilie: Omophlini der Alleculidae aus Europa und den angrenzenden Ländern. *Verhandlungen des Naturforschenden Vereins in Brünn* **44**: 115–175.

Reitter, E. 1906e: Vier neue von Herrn Paganetti-Hummler in Calabrien gesammelte Coleopteren. *Wiener Entomologische Zeitung* **25**: 129–130.

Reitter, E. 1906f: Dreizehn neue Coleopteren aus der palaearktischen Fauna. *Wiener Entomologische Zeitung* **25**: 237–244.

Reitter, E. 1906g: Neue Coleopteren aus Spanien. *Boletín de la Real Sociedad Española de Historia Natural* **6**: 375–380.

Reitter, E. 1906h: Neue Uebersicht der mir bekannten Arten der Coleopterengattung Laena Latr. aus Turkestan und Kaschmir, nebst Beschreibung von 2 neuen Arten aus Osteuropa. *Deutsche Entomologische Zeitschrift* **1906**: 442–448.

Reitter, E. 1907a: Coleopterologische Notizen. *Wiener Entomologische Zeitung* **26**: 27–30.

Reitter, E. 1907b: Nachträge zur Bestimmungtabelle der unechten Pimeliden aus der palaearktischen Fauna. *Wiener Entomologische Zeitung* **26**: 8[illegible]–92

Reitter, E. 1907c: Ein neuer Microtelus (Col.) aus Aegypten. *Wiener Entomologische Zeitung* **26**: 115.

Reitter, E. 1907d: Nacerdes melanura var. nov. Zoufali. *Societatis Entomologica* **2[illegible]**: 161.

Reitter, E. 1907e: Parablops subchalybaeus n. sp. *Wiener Entomologische Zeitung* **26**: 204.

Reitter, E. 1907f: Sechs neue Coleopteren aus Turkestan. *Wiener Entomologische Zeitung* **26**: 205–208.

Reitter, E. 1907g: Zwei neue Meloë aus der palaearktischen Fauna. *Wiener Entomologische Zeitung* **26**: 214–215.

Reitter, E. 1907h: Übersicht der Arten der Pimeliden-Gattung Podhomala Sol. (Col.). *Deutsche Entomologische Zeitschrift* **1907**: 412–414.

Reitter, E. 1907i: Einige neue paläarktische Coleopteren. *Deutsche Entomologische Zeitschift* **51**: 484–487.

Reitter, E. 1908a: Siebzehn Coleopteren-Neuheiten aus der palaearktischen Fauna. *Wiener Entomologische Zeitung* **27**: 17–24.

Reitter, E. 1908b: Übersicht der Entomogonus-Aten aus der Verwandschaft des E. Barthelemyi Sol. *Wiener Entomologische Zeitung* **27**: 30.

Reitter, E. 1908c: Beschreibung einiger neuen Käferarten von Egypten. *Bulletin de la Société Entomologique d'Egypte* **1** [1908–1909]: 39–40, 41–56.

Reitter, E. 1908d: Omophlus subcostatus (♀) n. sp. *Wiener Entomologische Zeitung* **27**: 56.

Reitter, E. 1908e: Neue paläarktische Käfer. *Entomologische Blätter* **4**: 129.

Reitter, E. 1908f: Zwei neue Coleopteren aus Adana in Kleinasien. *Wiener Entomologische Zeitung* **27**: 133–136.

Reitter, E. 1908g: Sieben neue Coleopteren aus Europa und den angrenzenden Ländern. *Wiener Entomologische Zeitung* **27**: 213–216.

Reitter, E. 1908h: Übersicht über die Arten der Gattung Boromorphus Woll. mit Ausschluss der Spezies von Madeira. *Wiener Entomologische Zeitung* **27**: 217.

Reitter, E. 1908i: Acht neue Coleopteren aus Europa, Kleinasien und Turkestan. *Wiener Entomologische Zeitung* **27**: 243–247.

Reitter, E. 1908j: Eine auffällige Zonabris-Art aus der Herzegowina. *Wiener Entomologische Zeitung* **27**: 272.

Reitter, E. 1908k: Die Arten der Gattung Laena Latr. (Tenebrionidae) aus dem westlichen Himalaya, gesammelt auf seinen Reisen von Herrn Karl Rost (Col.). *Deutsche Entomologische Zeitschrift* **1908**: 275–280.

Reitter, E. 1909a: Conspectus specierum genere Tentyrina Reitt. (Tenebrionidae). *Bulletin de la Société Entomologique d'Egypte* **2**: 28–29.

Reitter, E. 1909b: Übersicht der Arten der Coleopterengattung Hypulus Payk. *Entomologische Blätter* **5**: 51.

Reitter, E. 1909c: Neun neue Coleopterenarten und Varietäten aus der paläarktischen Fauna. *Wiener Entomologische Zeitung* **27**: 99–103.

Reitter, E. 1909d: Neue Revision der Arten der Coleopterengattung Prosodes Esch. *Wiener Entomologische Zeitung* **28**: 113–168.

Reitter, E. 1909e: Dila leptoscelis n. sp. *Entomologische Blätter* **5**: 239.

Reitter, E. 1909f: Solskia concavidorsis n. sp. *Entomologische Blätter* **5**: 296.

Reitter, E. 1909g: Neue Coleopteren aus der paläarktischen Fauna. *Wiener Entomologische Zeitung* **28**: 303–312.

Reitter, E. 1910a: Eine neue Gattung der Coleopterenfamilie der Tenebrionidae, zugleich ein Vertreter einer neuen Tribus bei den Lachnogiini. *Entomologische Blätter* 6: 20–21.

Reitter, E. 1910b: Due nuovi coleotteri di Sardegna. *Rivista Coleotterologica Italiana* 8: 115–116.

Reitter, E. 1910c: Siagonadia, nov. subgen. von Macrosiagon. *Wiener Entomologische Zeitung* **29**: 131.

Reitter, E. 1910d: Psilachnopus Aharonii. *Wiener Entomologische Zeitung* **29**: 197.

Reitter, E. 1911a: *Fauna Germanica. Die Käfer des Deutschen Reiches. Nach der analytischen Methode bearbeitet. III. Band*. Stuttgart: K.G. Lutz, 436 pp., pls. 81–128.

Reitter, E. 1911b: Paläarktische Coleopterennovitäten. *Wiener Entomologische Zeitung* **30**: 47–55.

Reitter, E. 1911c: Vier neue Koleopteren-Arten aus der palaearctischen Fauna. *Wiener Entomologische Zeitung* **30**: 234–236.

Reitter, E. 1911d: Übersicht der palaearctischen Alphitophagus-Arten (Col., Tenebrionidae). *Wiener Entomologische Zeitung* **30**: 268.

Reitter, E. 1911e: Zwei neue Coleopteren-Varietäten. *Wiener Entomologische Zeitung* **30**: 291.

Reitter, E. 1912a: Über die Myiodes-Arten aus der paläarktischen Fauna. *Wiener Entomologische Zeitung* **31**: 43.

Reitter, E. 1912b: Rhinosimus (Cariderus) sapiroides n. sp. *Entomologische Blätter* **8**: 196–197.

Reitter, E. 1912c: Übersicht der bekannten Arten der Coleopteren-Gattung Langelandia Aubé. *Wiener Entomologische Zeitung* **31**: 279–285.

Reitter, E. 1912d: Tenebrio molitor L. subsp. nov. Codinae. *Wiener Entomologische Zeitung* **31**: 310.

Reitter, E. 1913a: Zwei neue Coleopteren von der Balkan-Halbinsel. *Entomologische Blätter* **9**: 26–27.

Reitter, E. 1913b: Coleopterologische Novitäten der palaearktischen Fauna. *Coleopterologische Rundschau* **2**: 121–125.

Reitter, E. 1913c: Neue paläarktische Coleopteren. *Entomologische Blätter* **9**: 289–292.

Reitter, E. 1913d: Eine Serie neuer Coleopteren aus der palaearktischen Fauna. *Deutsche Entomologische Zeitschrift* **1913**: 649–666.

Reitter, E. 1914a: Die Microtelus-Arten aus der Verwandtschaft des M. careniceps Rche. (Col. Tenebrionidae). *Entomologische Blätter* **10**: 13–14.

Reitter, E. 1914b: Bestimmungs-Tabelle für die Unterfamilie: Erodini der Tenebrionidae aus Europa und den angrenzenden Ländern. *Deutsche Entomologische Zeitschrift* **1914**: 43–85.

Reitter, E. 1914c: Schellia nov. gen. Oedemeridarum. *Entomologische Blätter* **10**: 49–50.

Reitter, E. 1914d: Neue Übersicht der palaearktischen Alphitobius-Arten. *Coleopterologische Rundschau* **3**: 81–82.

Reitter, E. 1914e: Sechs neue Arten der Coleopteren-Gattung Hedyphanes Fischer. *Berliner Entomologische Zeitschrift* **58** [1913]: 184–187.

Reitter, E. 1914f: Zweite Übersicht der Arten der Gattung Cerocoma Geoffr. (Col. Meloidae.). *Berliner Entomologische Zeitschrift* **58** [1913]: 190–193.

Reitter, E. 1914g: Bestimmungs-Tabelle der Tenebrioniden-Abteilung der Scaurini. *Deutsche Entomologische Zeitschrift* **1914**: 369–380.

Reitter, E. 1914h: Bestimmungs-Tabelle der Tenebrioniden-Abteilung der Sepidiini. *Deutsche Entomologische Zeitschrift* **1914**: 381–392.

Reitter, E. 1915a: Bestimmungs-Tabelle der echten Pimeliiden aus der paläarktischen Fauna. *Wiener Entomologische Zeitung* **34**: 1–63.

Reitter, E. 1915b: Über die Arten der Coleopterengattung Gnathosia Firch. (Capnisa Lac.) mit ganz gerandeter Basis der Flügeldecken. *Coleopterologische Rundschau* **4**: 60–62.

Reitter, E. 1915c: Cis (Eridaulus) Mell. v. Matchanus nov. *Wiener Entomologische Zeitung* **34**: 66.

Reitter, E. 1915d: Dichotomische Übersicht der Tenebrinoden-Gattung Scythis Schaum. *Entomologische Blätter* **11**: 66–72.

Reitter, E. 1915e: Eine Serie neuer Coleopteren aus der Familie der Tenebrionidae der paläarktischen Fauna. *Wiener Entomologische Zeitung* **34**: 83–93.

Reitter, E. 1915f: Neue Arten aus der Tenebrioniden-Gattung Microdera Eschsch. *Entomologische Blätter* **11**: 92–96.

Reitter, E. 1915g: Coleopterologische Notizen. *Wiener Entomologische Zeitung* **34**: 124.

Reitter, E. 1915h: Die Arten der Tenebrioniden-Gattung Kokeniella Reitt. *Wiener Entomologische Zeitung* **34**: 127–128.

Reitter, E. 1915i: Über Arthrodosis-Arten aus der Buchara. *Wiener Entomologische Zeitung* **34**: 130–131.

Reitter, E. 1916a: Bestimmungstabelle der Arten der Gattung Adesmia Fisch. aus der paläarktischen Fauna (Col. Tenebrionidae). *Wiener Entomologische Zeitung* **35**: 1–31.

Reitter, E. 1916b: Bestimmungstabelle der Tenebrioniden-Gruppe der Phaleriini, aus der palaearktischen Fauna. *Entomologische Blätter* **12**: 3–10.

Reitter, E. 1916c: Bestimmungstabelle der Tenebrioniden-Unterfamilie Zophosini aus der paläarktischen Fauna. *Wiener Entomologische Zeitung* **35**: 81–99.

Reitter, E. 1916d: Bestimmungstabelle der Tenebrioniden, enthaltend die Zopherini, Elenophorini, Leptodini, Stenosini und Lachnogyini aus der paläarktischen Fauna. *Wiener Entomologische Zeitung* **35**: 129–171.

Reitter, E. 1916e: Bestimmungs-Tabelle der Tenebrioniden-Abteilung der palaearctischen Epitragini. *Entomologische Blätter* **12**: 139–149.

Reitter, E. 1916f: Lederia seidlitzi n. sp. *Wiener Entomologische Zeitung* **35**: 290.

Reitter, E. 1916g: Coleopterologische Notizen. *Wiener Entomologische Zeitung* **35**: 294.

Reitter, E. 1917a: Bestimmungs-Tabelle der palaearctischen Arten der Tenebrioniden-Abteilung Asidini. *Verhandlungen des Naturforschenden Vereins in Brünn* **55**: 1–74.

Reitter, E. 1917b: Bestimmungstabelle der Cossyphini und Misolampini (Tribus der Tenebrionidae). *Wiener Entomologische Zeitung* **36**: 129–150.

Reitter, E. 1918: Vier neue Coleopteren aus Albanien. *Wiener Entomologische Zeitung* **37**: 153–155.

Reitter, E. 1920a: *Bestimmungstabelle der Unterfamilien: Belopinae, Borinae, Tenebrioninae und Coelometopinae der Tenebrioniden (Col.). Bestimmungs-Tabellen der europäischen Coleopteren 87*. Paskau: E. Reitter, 24 pp.

Reitter, E. 1920b: Coleopterologische Notizen. *Entomologische Blätter* **15** [1919]: 220–221.

Reitter, E. 1922a: *Bestimmungs-Tabellen der europaeischen Coleopteren. VI. Heft. Enhaltend die Familien: Colydiidae, Rhysodidae, Ostomidae. Zweite, gänzlich umgearbeitete und auf die palaearktische Fauna ausgedehnte Auflage*. Troppau: E. Reitter, 73 pp.

Reitter, E. 1922b: Bestimmungstabelle der palaearktischen Helopinae (Col. Tenebrionidae). *Wiener Entomologische Zeitung* **39**: 1–44, 113–171.

Ren, G.-D. 1992: Two new species of genus Melanesthes Lacord. from desert of N.W. China. *Zoological Research* **13**: 229–332 (in Chinese with English summary).

Ren, G.-D. 1993a: [new taxa]. In: Ren, G.-D., Yu, Y., Ma, F. & Jiang, H.: Study on the genus Melanesthes Lacord. and description of its three new species and five new records species [sic] (Coleoptera: Tenebrionidae) in China. *Journal of the Ningxia Agricultural College* **14** (Supplement): 24–33 (in Chinese with English summary).

Ren, G.-D. 1993b: Three new species of the genus Melanesthes Lacord. from China (Coleoptera: Tenebrionidae). *Acta Entomologica Sinica* **36**: 486–489 (in Chinese with English summary).

Ren, G.-D. 1995a: Notes on Chinese species of Caedius Mulsant et Rey (Coleoptera: Tenebrionidae) with description of a new species. *Entomotaxonomia* 17 (Supplement): 22–24.

Ren, G.-D. 1995b: Coleoptera: Tenebrionidae. Pp. 233–236. In: Wu, H. (ed.): *Insects of Baishanzu Mountain, Eastern China*. China Forestry Publishing House, xiii + 586 pp. (in Chinese with English summary).

Ren, G.-D. 1998: Coleoptera: Tenebrionidae. Pp. 108–114. In: Wu, H. (ed.): *Insects of Longwangshan Nature Reserve*. Beijing: China Forestry Publishing house, ix + 404 pp.

Ren, G.-D. 1999: [new taxa]. In: Ren, G.-D. & Yu, Y.: *The darkling beetles from deserts and semideserts of China*. Hebei University Publishing House, [3] + 395 pp. (in Chinese with English summary).

Ren, G.-D. 2004 [new taxa]. In: Ren, G.-D. & Yin, H.: Coleoptera: Tenebrionidae. Pp. 67–72. In: Yang X.-K. (ed.): *Insects of the Great Yarlug Zangbo Caryon of Xizang, China*. Beijing: China Science and Technology Press, 339 pp., 12 pls. (in Chinese with English summary).

Ren, G.-D. 2016: [new taxa]. In: Ren, G.-D. Ba, Y., Liu, H., Niu, Y., Zhu, X., Li Zh. & Shi, A.: *Fauna Sinica: Insecta. Volume 63: Coleoptera: Tenebrionidae (I)*. Beijing: Science Press, 532 pp., XLIX pls. (in Chinese with English summary).

Ren, G.-D. & Ba, Y.-B. 2009: Revision of the genus Ocnera Fischer von Waldheim, 1822 (Coleoptera: Tenebrionidae: Pimeliini) from China. *Caucasian entomological Bulletin* 5: 22–225.

Ren, G.-D. & Ba, Y.-B. 2010: *Fauna of soil darkling beetles in China. Vol. 2. Tentyriformis (Coleoptera: Tenebrionidae). Epitragini, Akidini, Leptodini, Tentyriini*. Beijing: Science Press, 225 pp. + XIII colour plates.

Ren, G.-D. & Bai, M. 2005: Coleoptera: Tenbrionidae. Pp. 379–389. In: Yang, X.-K. (ed.): *Insects Fauna of Middle-West Qinling Range and South Mountain of Gansu Province*. 2005. Beijing: Science Press, 1055 pp. (in Chinese with English summary).

Ren, G.-D. & Dong, S. 2001: The beetles of the tribe Platyopini in China and description of two new species (Coleoptera: Tenebrionidae). *Journal of Hebei University (Natural Science Edition)* 21: 296–302.

Ren, G.-D. & Hua, H.-R. 2006: Tenebrionidae. Pp. 265–275. In: *Fanjingshan jingguan kunchong*. Guiyang: Guizhou Science and Technology Publishing House (in Chinese with English summary).

Ren, G.-D. & Li, Z. 2001: A new genus and two new species of the tribe Blaptini in China (Coleoptera: Tenebrionidae). *Zoological Research* 4: 310–314.

Ren, G.-D. & Liu, H.-Y. 2009: A new species of the genus Agnaptoria Reitter from Xizang, China (Coleoptera: Tenebrionidae: Blaptini). *Zootaxa* 2169: 55–62.

Ren, G.-D. & Liu, S.-S. 2004: Six new species of the genus Uloma from Gaoligong Mountain in China (Coleoptera, Tenebrionidae). *Acta Zootaxonomica Sinica* 29: 294–304.

Ren, G.-D. & Luo, Z. 1995: One new record genus [sic] and species of Blaptini (Coleoptera: Tenebrionidae) from China. *Entomotaxonomia* 17: 27–30 (in Chinese with English summary)

Ren, G.-D. & Meng, L. 1995: Fauna of the Tenebrionidae (Coleoptera) from Xingjiang: I. *Acta Agriculturae Boreali-Occidentalis Sinica* 4 (3): 52–57 (in Chinese with English summary).

Ren, G.-D., Ning, J. & Jia, L. 2011: A new species of the genus Cheirodes Gené from Alxa Plateau (Coleoptera, Tenebrionidae, Melanimini). *Acta Zootaxonomica Sinica* 36: 564–567.

Ren, G.D., Niu, Y., Li, Zh. & Shi, A. 2016: [new taxon]. In: Ren, G.-D., Ba, Y., Liu, H. Niu, Y., Zhu, X., Li Zh & Shi, A.: Fauna Sinica: Insecta. Volume 63: *Coleoptera: Tenebrionidae (I)*. Beijing: Science Press, 532 pp. XLIX pls. (in Chinese with English summary)

Ren, G.-D. & Shi, A.-M. 2006a: A taxonomic review of the Chinese species of the genus Ascelosodes Redtenbacher (Coleoptera: Tenebrionidae), with descriptions of three new species. *Zootaxa* **1228**: 1–24.

Ren, G.-D. & Shi, A.-M. 2006b: Two new record [sic] genera of Stenosini (Coleoptera, Tenebrionidae) from China, with descriptions of two new species. *Acta Zootaxonomica Sinica* **31**: 180–184.

Ren, G.-D. & Shi, A. 2016: [new taxon]. In: Ren, G.-D., Ba, Y., Liu, H., Niu, Y., Zhu, X., Li Zh. & Shi, A.: Fauna Sinica: Insecta. Volume 63: *Coleoptera: Tenebrionidae (1)*. Beijing: Science Press, 532 pp., XLIX pls. (in Chinese with English summary)

Ren, G.-D. & Wang, X.-P. 1993a: Three new species of the genus Adesmia Fisch. from China (Coleoptera, Tenebrionidae, Adesmini). *Journal of the Ningxia Agricultural College* **14** (*Supplement*): 1–8 (in Chinese with English summary).

Ren, G.-D. & Wang, X.-P. 1993b: In: Ren, G.-D., Wang, X.-P. & Ma, F.: Two new species and one new record of Platyope F.-W. from China. (Coleoptera: Tenebrionidae). *Journal of the Ningxia Agricultural College* **14** (Supplement): 44–49 (in Chinese with English summary).

Ren, G.-D. & Wang, X.-P. 2001: Eight new species of the genus Blaps Fabricius (Coleoptera: Tenebrionidae: Blaptini) of China. *Entomotaxonomia* **23**: 15–27 (in Chinese with English summary).

Ren, G.-D., Wang, X.-P. & Yu, Y.-Z. 2000: Nine records and one new species of the tribe Blaptini from China (Coleoptera: Tenebrionidae). *Journal of Hebei University* (*Natural Science Edition*) **20**: Supplement: 26–33.

Ren, G.-D. & Wu 1994: A new species of Belopus Gebin [sic] from China (Coleoptera: Tenebrionidae). *Acta Zootaxonomica Sinica* **19**: 351–353 (in Chinese with English summary).

Ren, G.-D. & Xu, J.-S. 2011: The genus Hexarhopalus Fairmaire, 1891 in China, with description of three new species (Coleoptera, Tenebrionidae: Cnodalonini). *Acta Zoologica Academiae Scientiarum Hungaricae* **57**: 23–34.

Ren, G.-D. & Yang, X.-J. 2004: Coleoptera: Tenebrionidae. Pp. 311–319. In Yan, X.-K., Huang, F.-S. & Xue, D. (eds.): *Insects from Mt. Shiwandashan Area of Guangxi*. Beijing, China Forestry Publishing House, 668 pp. (In Chinese with English summary).

Ren, G.D. & Yang, Y.-X. 2007. A new species of the genus Epicauta Dejan from China (Coleoptera, Meloidae). *Acta Zootaxonomica Sinica* **32**: 705–710.

Ren, G.-D. & Yu, Y.-Z. 2000: On the darkling beetle tribe Akidini from China (Coleoptera: Tenebrionidae). *Acta Zootaxonomica Sinica* **25**: 320–329 (in Chinese with English summary).

Ren, G.-D. & Yuan, C.-X. 2005: Taxonomic study on the arboreal darkling beetles of the genus Ainu Lewis, 1894 from China (Coleoptera, Tenebrionidae). *Acta Zootaxonomica Sinica* **30**: 98–103.

Ren, G.-D. & Yuan, C.-X. 2015: Taxonomy of the genus Ainu Lewis, 1894 (Coleoptera: Tenebrionidae) from China, with description of a new species. *Entomotaxonomia* **37**: 129–133.

Ren, G.-D. & Zhang, C.-L. 2010: The Chinese Blindus Mulsant et Rey (Coleoptera, Tenebrionidae), with description of a new species. *Acta Zootaxonomica Sinica* **35**: 63–70.

Ren, G.-D. & Zheng, Z. 1993a: Six new species of the genus Crypticus Latr. from northwest of China (Coleoptera: Tenebrionidae). *Journal of the Ningxia Agricultural College* **14** (Supplement): 9–23 (in Chinese with English summary).

Ren, G.-D. & Zheng, Z. 1993b: Six new species of the genus Scytosoma Fisch. 1895 (Coleoptera, Tenebrionidae, Tentyriini). *Journal of the Ningxia Agricultural College* **14** (Supplement): 34–43 (in Chinese with English summary).

Ren, G.-D. & Zheng, Z. 1993c: Four new species of the tribe Epitragini from desert in Northwest of China (Coleoptera, Tenebrionidae). *Journal of the Ningxia Agricultural College* **14** (Supplement): 50–58 (in Chinese with English summary).

Ren, G.-D. & Zheng, Z. 1993d: One new species of genus Belopus GB. from Inner Mongolia (Coleoptera: Tenebrionidae: Belopini). *Journal Inner Mongolia Normal University* (*Natural Science Edition*) **4**: 52–54. (in Chinese with English summary).

Ren, G.-D. & Zheng, Z. 1993e: One new record tribe [sic] and genus from China with one new species of Phthora (Coleoptera: Tenebrionidae). *Journal of Ningxia University* (*Natural Science Edition*) **14** (4): 77–80 (in Chinese with English summary).

Ren, G.-D. & Zhu, X. 2016: [new taxon]. In: Ren, G.D., Ba, Y., Liu, H., Niu, Y., Zhu, X., Li Zh. & Shi, A.: Fauna Sinica: Insecta. Volume 63: *Coleoptera: Tenebrionidae (1)*. Beijing: Science Press, 532 pp., XLIX pls. (in Chinese with English summary).

Rey, C. 1857: [new taxon]. In: Mulsant E. & Rey C.: Description de quelques coléoptères de la tribu des longipèdes. *Annales de la Société Linnéenne de Lyon* (2) **4**: 410–413.

Rey, C. 1878: Insecta. In: *The Zoological Record for 1876; being the volume thirteenth of the Record of Zoological Literature*. London: John van Voost, 240 pp.

Rey, C. 1889: Remarques en passant. *L'Échange, Revue Linnéenne* **5**: 23, 55.

Rey, C. 1890: Note sur le genre Phaleria. *Revue d'Entomologie* **9**: 327–329.

Rey, C. 1891a: Nouvelle note sur le genre Phaleria. *Revue d'Entomologie* **10**: 83–86.

Rey, C. 1891b: Troixième note sur le genre Phaleria et description d'une espèce nouvelle ou peu connue. *Revue d'Entomologie* **10**: 236–239.

Rey, C. 1892: Remarques en passant. *L'Échange, Revue Linnéenne* **8**: 2–3, 17, 30–31, 41–42, 53–54, 65, 77, 89–90, 101, 113–114, 137–138.

Reymond, A. 1933: Description d'une espèce nouvelle de Meloë paléarctique. *Bulletin de la Société Entomologique de France* **38**: 14–16.

Reymond, A. 1937: Un nouvel Earophanta d'Asie Centrale (Col. Tenebrionidae). *Bulletin de la Société Entomologique de France* **42**: 247–248.

Reymond, A. 1938: Résultats scientifiques d'un voyage en Asie centrale (Mission Haardt – Audoin-Dubreuil) 1931. Sur le peuplement entomologique de l'Asie centrale. *Revue de Géographie Physique* **10**: 1–285, 4 maps, 20 pls.

Reymond, A. 1956a: Présentation d'insectes (Coléoptères) nouveaux de la faune marocaine. *Comptes Rendus de la Société des Sciences Naturelles et Physiques du Maroc* **22**: 40–42.

Reymond, A. 1956b: Description d'un nouveau coléoptère troglophile du Moyen-Atlas septentrional: Scaurus harbuloti nov. sp. (Col. Tenebrionidae). *Bulletin de la Société des Sciences Naturelles et Physiques du Maroc* **36**: 265–269.

Richter, A.A. 1945: Novyy vid Scaurus iz doliny Araksa (Coleoptera, Tenebrionidae). *Doklady Akademii Nauk Armyanskoy SSR* **3**: 103–106.

Richter, A.A. 1949: Novyy paraziticheskyj zhuk iz Zakavkazia (Coleoptera, Rhipiphoridae). *Entomologicheskoe Obozrenie* **30**: 359–361.

Říha, P. 1958: Beitrag zur Kenntnis der Colydiiden von Adsharistan nebst einer Monographie der Gattung Dechomus Jacq. du Val (Coleoptera, Colydiidae). *Acta Entomologica Musei Nationalis Pragae* **32**: 465–69.

Rochebrune, A.T. de. 1883: [new taxon]. In: *Bulletin de la Société des Philomatique de Paris* (7) **7**: 182–188, pl. 3.

Roemer, J.J. 1789: *Genera insectorum Linnaei et Fabricii iconibus illustrata a Joanne Jacobo Roemer*. Vitoduri Helvetorum: apud Henrjc. Steiner et socios, viii + 86 + [3] pp., 37 pls.

Rosa, J.J. de la 2008: Descripción de Mylabris (Micrabris) maculosopunctata mendizabali ssp. n. de los Alpes occidentales (Coleoptera, Meloidae). *Boletín de la Sociedad Entomológica Aragonesa* **42**: 79–84.

Rosenhauer, W.G. 1847: *Beiträge zur Insekten-Fauna Europas. Erstes Bändchen; enthält die Beschreibung von sechzig neuen Käfern aus Bayern, Tyrol, Ungarn, etc., so wie die Käfer Tyrols, nach dem Ergebnisse von vier Reisen*. Erlangen: Theodor Blaesing, x + 160 pp., 1 pl.

Rosenhauer, W.G. 1856: *Thiere Andalusiens nach dem Resultate einer Reise zusammengestellt, nebst den Beschreibungen von 249 neuen oder bis jetzt unbeschriebenen Gattungen und Arten*. Erlangen: Theodor Blaesing, viii + 429 pp., 3 pls.

Rossi, P. 1790: *Fauna Etrusca sistens insecta quae in provinciis Florentina et Pisana praesertim collegit Petrus Rossius. Tomus Primus*. Liburni: T. Masi & Sociorum, xxii + 272 pp., 10 pls.

Rossi, P. 1792: *Mantissa insectorum exhibens species nuper in Etruria collectas a Petro Rossio adiectis faunae Etruscae illustrationibus ac emendationibus*. [Tomus primus]. Pisis: Polloni, 148 pp.

Rossi, P. 1794: *Mantissa insectorum exhibens species nuper in Etruria collectas a Petro Rossio adiectis faunae Etruscae illustrationibus ac emendationibus. Tomus secundus*. Pisis: Polloni, 154 pp., 8 pls.

Rotrou, M. 1938: Première contribution à l'étude des Planasida du groupe lanceocollis Esc. *Bulletin de la Société des Sciences Naturelles du Maroc* **18**: 119.

Rotrou, M. 1943: Description d'une nouvelle espèce de Sepidium de l'Algérie. *Bulletin de la Société d'Histoire Naturelle de l'Afrique du Nord* **34**: 235–236.

Rottenberg, A. von. 1871: Beiträge zur Coleopteren-Fauna von Sicilien. *Berliner Entomologische Zeitschrift* **14** [1870]: 235–260, pl. 2.

Rottenberg, A. von. 1873: Synonymische Bemerkungen 9. *Berliner Entomologische Zeitschrift* **17**: 213–217.

Roubal, J. 1910: Beitrag zur Kenntnis der Mordelliden von Bulgarien. *Entomologische Blätter* **6**: 153–157.

Roubal, J. 1911a: Neuheiten der palaearktischen Koleopterenfauna. *Wiener Entomologische Zeitung* **30**: 133–134.

Roubal, J. 1911b: Neue Käfer der palaearktischen Fauna. *Wiener Entomologische Zeitung* **30**: 272.

Roubal, J. 1912a: Koleopterologické výsledky mé cesty na Kavkaz v červenci 1910. Quid novi de Coleopterorum Caucasi ad occidentem vergentis fauna in meo itinere Julio mense anni 1910 suscepto cognoverim. v. *Časopis České Společnosti Entomologické* **9**: 25–28.

Roubal, J. 1912b: Drei palaearktische Nova. *Coleopterologische Rundschau* **1**: 70–72.

Roubal, J. 1912c: Deux nouveaux coléoptères de la région européenne. *Bulletin de la Société Entomologique de France* **1911**: 401–402.

Roubal, J. 1914: Eine neue Zonabris aus dem Kaukasus. *Coleopterologische Rundschau* **3**: 145–146.

Roubal, J. 1916: Cis ragusai sp. n. *Časopis České Společnosti Entomologické.* **13**: 89–90.

Roubal, J. 1917a: Eine neue Oedemera. *Neue Beiträge zur Systematischen Insektenkunde* **1**: 44.

Roubal, J. 1917b: Cteniopus expulsus nova species. *Archiv für Naturgeschichte* A 82 (4): 47–48.

Roubal, J. 1919: Drei coleopterologische Nova. *Deutsche Entomologische Zeitschrift* 1919: 63–64.

Roubal, J. 1921: Description d'un Mordella nouveau de Carniole. *Bulletin de la Société Entomologique de France* 1921: 248–249.

Roubal, J. 1924: Zwei neue Scraptia-Arten aus der paläarktischen Region. *Wiener Entomologische Zeitung* 41: 181–182.

Roubal, J. 1926: Sieben neue Coleopteren aus der paläarktischen Region. *Coleopterologisches Centralblatt* 1: 244–249.

Roubal, J. 1927: Ein Beitrag zur Kenntnis der Coleopteren-Fauna der westrumänischen Ebene. *Coleopterologisches Centralblatt* 2: 112–118.

Roubal, J. 1929a: Vier neue Coleopteren aus SSSR. *Entomologische Blätter* 25: 46–48.

Roubal, J. 1929b: Coleoptera nova asiatica. *Bollettino della Società Entomologica Italiana* 59: 96–98.

Roubal, J. 1929c: Descriptiones duorum Coleopterorum novorum regionis palaearcticae. *Entomologische Blätter* 25: 192–193.

Roubal, J. 1930: Coleoptera nova ex Europa descripta. *Časopis Československé Společnosti Entomologické* 27: 75–76.

Roubal, J. 1931a: Beschreibungen von vier neuen Coleopteren. *Časopis Československé Společnosti Entomologické* 28: 131–132.

Roubal, J. 1931b: Coleoptera nova pro Bohemia. *Časopis Československé Společnosti Entomologické* 28: 132.

Roubal, J. 1932a: Descriptions de quelques nouveaux coléoptères paléarctiques. *Miscellanea Entomologica* 34: 1.

Roubal, J. 1932b: Fragmente zur Koleopterenfaunistik des balkanischen Festlandes. *Entomologischer Anzeiger* 12: 129–130.

Roubal, J. 1933a: Beschreibungen von zwei neuen Cis-Arten aus Ostasien. *Stylops* 2: 72.

Roubal, J. 1933b: Beschreibungen einiger Coleopteren-Neuheiten. *Entomologisches Nachrichtenblatt* 7: 77–78.

Roubal, J. 1933c: Descriptions de deux mélandryides (Col.) d'Europe centrale. *Bulletin de la Société Entomologique de France* 38: 128.

Roubal, J. 1933d: Beschreibung einiger neuen Coleopterenformen aus der Slovakei. *Časopis Československé Společnosti Entomologické* 30: 139.

Roubal, J. 1934a: Inventa entomologica itineris Hispanici et Maroccani, quod a 1926 fecerunt Harald et Håkan Lindberg. XIX. Scydmaenidae, Pselaphidae, Mordellidae. *Commentationes Biologicae* 4 (11): 1–8.

Roubal, J. 1934b: Eine neue Mycetochara Berth. aus Mitteleuropa: M. melandryina n. *Entomologische Blätter* 30: 35–36.

Roubal, J. 1935a: Descriptio novae speciei generis Mycetochara Berth. (Alleculidae) et adnotationes diversae de speciebus aliis. *Časopis Československé Společnosti Entomologické* 32: 24–25.

Roubal, J. 1935b: Quelques notes sur les mordellides paléarctiques avec la description de cinq formes nouvelles. *Bulletin de la Société Entomologique de France* 40: 106–108.

Roubal, J. 1935c: Beschreibung vier neuer Coleopteren-Formen. *Entomologischer Anzeiger* 15: 10.

Roubal, J. 1936a: *Katalog coleopter (brouků) Slovenska a Podkarpatské Rusi na základě bionomickém a zoogeografickém a spolu systematický doplněk Ganglbauerových 'Die Käfer von Mitteleuropa' a Reitterovy 'Fauna germanica' Díl I*. Bratislava: Učená společnost Šafaříkova v Bratislavě, 434 pp.

Roubal, J. 1936b: Beschreibung einer neuen Untergattung, einer Varietät und einer Form der Coleopteren. *Festschrift zum 60. Geburtstage von Professor Dr. Embrik Strand* (Riga) **1**: 53–58.

Roubal, J. 1937a: Koleopterologische Notizen. *Entomologisches Nachrichtenblatt* **11**: 38–40.

Roubal, J. 1937b: Dvě nové broučí variety z ČSR. Zwei neue Käfervarietäten aus ČSR. *Časopis Československé Společnosti Entomologické* **34**: 67–68.

Roubal, J. 1938: Anidorus Lokvenci, n. sp. aderide (Col.) nouveau de l'Europe centrale. *Bulletin de la Société Entomologique de France* **43**: 40–41.

Ruiz, J.L, François, A. & García-París, M. 2019: A new singular species of Croscherichia Pardo Alcaide, 1950 (Coleoptera, Meloidae, Mylabrini) from arid zones of eastern Morocco. *ZooKeys* **885**: 27–50.

Ruiz, J.L. & García-París, M. 2004: Una specie nueva de Mylabris Fabricius, 1775 (Coleoptera, Meloidae) del sureste de la Península Ibérica. *Graellsia* **60**: 185–196.

Ruiz, J.L. & García-París, M. 2007: Nombre de reemplazo en el género Mylabris Fabricius, 1775 (Coleoptera, Meloidae). *Graellsia* **63**: 151–152.

Ruiz, J.L. & García-París, M. 2009: Descripción de una especie nueva de Meloe Linnaeus, 1758 del subgenero Eurymeloe Reitter, 1911 (Coleoptera, Meloidae) del Norte de Marueccos. *Graellsia* **65**: 91–109.

Ruiz, J.L. & García-París, M. 2015: Una nueva especie de Meloe Linnaeus, 1758 del suroeste de Marruecos incluida en el grupo de M. (Eurymeloe) rugosus Marsham, 1802 (Coleoptera: Meloidae). *Graellsia* **71**: e018.

Ruzzier, E. 2014: A new species of Ainu Lewis, 1894 (Coleoptera: Tenebrionidae) from Southeast Asia. *The Coleopterists Bulletin* **68**: 659–662.

Ruzzier, E. & Kovalev, A. 2016: First record of Calycina Blair, 1922 (Coleoptera, Mordellidae) in the Russian Far East with description of a new species. *Zootaxa* **4103**: 75–78.

Rye, E.C. 1872: Additions, &c., to the list of British Coleoptera, with descriptions of three new species. *The Entomologist's Monthly Magazine* **9**: 5–11.

Rye, E.C. 1873: Insecta. Coleoptera. Pp. 222–329. In: Rye, E.C. (eds.) *The Zoological Record for 1871; being the volume eighth of the Record of Zoological literature*. London: John Van Voost.

Rye, E.C. 1878: Insecta. 240pp. In: Rye, E.C. (eds.) *The Zoological Record for 1876; being the volume thirteenth of the Record of Zoological literature*. London: John Van Voost.

Saha, G.N. 1972: Studies on blister beetles (Coleoptera: Meloidae) from India. I. Two new species of the genus Eletica Lac. *Zoologischer Anzeiger* **189**: 98–101.

Saha, G.N. 1979: Revision of Indian blister beetles (Coleoptera: Meloidae: Meloinae). *Records of the Zoological Survey of India* **74**: 1–146.

Saha, G.N. 1988: Discovery of genus Platolenes (Coleoptera: Tenebrionidae) from India with description of two new species. *Records of the Zoological Survey of India* **85**: 433–437.

Sahlberg, C.R. 1823: *Periculi Entomographici, species insectorum nondum descriptas proposituri, Fasciculus*. Aboae: Typis Frenckelliorum, 82 + (1), 4 pls.

Sahlberg, C.R. 1833: Pars XXIX. Pp. 441–456. In: *Insecta Fennica enumerans, dissertationibus academicis. A. 1817–1834 editis. Tomus I*. Aboae: J.C. Frenckel [note: all parts issued as book in 1834,with viii + 520 pp.].

Sahlberg, C.R. 1836: *Dissertatio entomologica insecta Fennica enumerans, cujus particulam decimam. Tomus II*. Aboae: J.C. Frenckel, 145–160 pp.

Sahlberg, C.R. 1837: *Dissertatio entomologica Insecta Fennica enumerans, cujus particulam undecimam. Tomus II*. Aboae: J.C. Frenckel, 161–176 pp.

Sahlberg, J.R. 1871: Othismopteryx ett nytt genus bland Finlands Coleoptera af familijen Colydiidae. *Notiser ur Sällskapets pro Fauna et Flora Fennica Förhandlingar* 11: 441–444, pl. 1.

Sahlberg, J.R. 1875: Ueber die finnischen Arten der Käfer-Gattung Pytho und deren Larven. *Deutsche Entomologische Zeitschrift* 19: 219–224.

Sahlberg, J.R. 1881: En ny art af Coleopter-slägtet Zilora Muls. af melandryidernas familij funne i Finland och beschrifven. *Meddelanden af Societas pro Fauna et Flora Fennica* 7: 132–134.

Sahlberg, J.R. 1901: Coleoptera nova vel minus cognita faunae Fennicae. *Acta Societatis pro Fauna et Flora Fennica* 19 (3):1–23.

Sahlberg, J.R. 1903a: Coleoptera mediterranea et rosso-asiatica nova et minus cognita. II. *Öfversigt af Finska Vetenskaps-Societetens Förhandlingar* 15 (10): 1–40.

Sahlberg, J.R. 1903b: Coleoptera mediterranea et rosso-asiatica nova et minus cognita, maxima ex parte itineribus annis 1895–1896 et 1898–1899 collecta. *Öfversigt af Finska Vetenskaps-Societetens Förhandlingar* 45 (10): 1–40.

Sahlberg, J.R. 1903c: Ad cognitionem faunae Coleopterorum graecae fragmenta. *Öfversigt af Finska Vetenskaps-Societetens Förhandlingar* 45 (12): 1–9.

Sahlberg, J.R. 1903d: Ad cognitionem faunae Coleopterorum italicae fragmenta. *Öfversigt af Finska Vetenskaps-Societetens Förhandlingar* 45 (13): 1–14.

Sahlberg, J.R. 1903e: Coleoptera Levantina mensibus februario et martio 1896 in Palaestina et Aegypto inferiore collecta. *Öfversigt af Finska Vetenskaps-Societetens Förhandlingar* [1902–1903] 45(18): 1–36.

Sahlberg, J.R. 1903f: Coleoptera numido-punica mensibus martio, aprili et majo 1899 in Tunisia et Algeria orientali collecta. *Öfversigt af Finska Vetenskaps-Societetens Förhandlingar* 45 (19): 1–70.

Sahlberg, J.R. 1908: Coleoptera mediterranea et rosso-asiatica nova et minus cognita, maxima ex parte itineribus annis 1895–1896, 1898–1899 et 1903–1904 collecta. *Öfversigt af Finska Vetenskaps-Societetens Förhandlingar* 50 (7) [1907–1908]: 1–94.

Sahlberg, J.R. 1913a: Coleoptera mediterranea et rosso-asiatica nova et minus cognita maxima ex parte itineribus annis 1895–1896, 1898–1899 et 1903–1904 collecta. *Öfversigt af Finska Vetenskaps-Societetens Förhandlingar* 55 [1912–1913] A (8): 1–88.

Sahlberg, J.R. 1913b: Coleoptera mediterranea orientalis quae in Aegypto, Palestina, Syria, Caramania collegerunt John Sahlberg et Unio Saalas. *Öfversigt af Finska Vetenskaps-Societetens Förhandlingar* 55 [1912–1913] A (19): 1–282.

Sahlberg, R.F. 1834: Novae Coleopterorum fennicorum species. *Bulletin de la Société Impériale des Naturalistes de Moscou* 7: 267–280.

Saint-Albin, E. de. 1952: Note sur les anthicides (Col. Hétéromères). *Bulletin de la Société Entomologique de France* 57: 20–23.

Saitô, M. 1990: A new subspecies of pyrochroid beetle (Coleoptera, Pyrochroidae) from Okinawa-hontô, southwest Japan. *Elytra* 18: 105–107.

Saitô, M. 1995: Notes on Taiwanese species of the genus Arthromacra (Coleoptera: Tenebrionidae). *Elytra* 23: 177–185.

Saitô, M. 1999: Notes on the Japanese species of the genus Usechus (Coleoptera, Zopheridae). *Elytra* 27 103–111.

Saitô, M. 2001: A new species of the genus Hymenalia (Coleoptera: Tenebrionidae: Alleculinae) from the Yaeyama Islands, Japan with notes of Japanese species. *Special Publication of the Japan Coleopterological Society of Osaka* 1: 347–350.

Saitô, M. 2003a: A new species of the genus Ischalia (Coleoptera, Anthicidae, Ischaliinae) from Hokkaido, Japan. *Elytra* **31**: 55–60.

Saitô, M. 2003b: A new species of the genus Tomoderus (Coleoptera: Anthicidae) from the Ryukyu Islands, Southwest Japan. *Elytra* **31**: 321–323.

Saitô, M. 2006: A new species of the genus Mycetophagus from Japan (Coleoptera: Mycetophagidae). *Entomological Review of Japan* **61**: 201–204.

Saitô, M. 2011: Descriptions of two new species of the genus Ischalia (Coleoptera, Ischaliidae) from the island of Shikoku, Japan. *Elytra, New Series,* 1: 307–314.

Saitô, M. 2013: Two new species of the genus Mycetophagus (Coleoptera, Mycetophagidae) from the Yaeyama Islands, Southwest Japan. *Elytra, New Series* 3: 97–100.

Saitô, M. 2015: A new species and a newly recorded species of the genus Mycetochara (Coleoptera, Tenebrionidae, Alleculinae) from Japan, with notes on the subgeneric position of three Mycetochara species. *Elytra, New Series* 5: 379–384.

Saitô, M. 2017: A new Stereopalpus (Coleoptera, Anthicidae, Eurygeniinae), second representative having the character of tuberculate abdomen, from Honshu, Japan. *Elytra, New Series* 7: 183–186.

Saitô, M. & Ando, K. 2016: Description of a mew species of Arthromacra (Coleoptera, Tenebrionidae, Lagriinae) from Taiwan. *Elytra, New Series* **6**: 97–100.

Saitô, M. & Konvička, O. 2017: A new species of Holostrophus (Paraholostrophus) (Coleoptera: Tetratomidae). *Acta Musei Silesiae, Scientiae Naturales* **66**: 1–5.

Saitô, M. & Young, D.K. 2015a: A new Ischalia (Coleoptera, Ischaliidae) from Hokkaido, with a key to the Japanese Ischaliidae. *Elytra, New Series* 5: 177–181.

Saitô, M. & Young, D.K. 2015b: Descriptions of a new species of Ariotus (Coleoptera, Aderidae) and a new genus and species of Aderidae from Honshu, Central Japan, with a key to the genera of Japanese Aderidae. *Elytra, Tokyo, New Series* 5: 453–462.

Saitô, M. & Young, D.K. 2016. A new Stereopalpus (Coleoptera, Anthicidae, Eurygeniinae) from Shikoku, Japan. *Elytra* **6**: 43–46.

Sakai, M. & Ohbayashi, N. 1994: A new anthicid beetle (Coleoptera, Anthicidae) associated with ditch reeds in marshy places of Japan. *Japanese Journal of Entomology* **62**: 555–561.

Sakai, M. & Telnov, D. 2001: Leptaleus sasajii, a new species and the first representative of the genus Leptaleus from Japan (Coleoptera, Anthicidae). *Sukunahikona, Special Publication of the Japan Coleopterological Society, Osaka* 1: 327–332.

Salvaña, J.M. 1870: *Apuntes para la geografia y fauna entomologica de Mataró*. Madrid: Junte, 44 pp.

Sandoval-Gómez, V.E. & Lopes-Andrade, C. 2015: Aliocis, a replacement name for the preoccupied ciid genus Anoplocis Kawanabe, 1996 (Coleoptera: Tenebrionoidea: Ciidae). *Zootaxa* **3946**: 599–600.

Sasaji, H. 1971: Description of a new Cicones-species in Japan (Coleoptera: Colydiidae). *The Life Study (Fukui)* **15**: 43–45.

Sasaji, H. 1974: Notes on the Japanese Melandryidae, with descriptions of new species (Coleoptera). *Memoirs of the Faculty of Education, Fukui University Series II (Natural Science)* 24: 1–14.

Sasaji, H. 1983: Contribution to the taxonomy of the superfamily Cucujoidea (Coleoptera) of Japan and her adjacent districts, I. *Memoirs of the Faculty of Education, Fukui University Series II (Natural Science)* 33: 17–52.

Sasaji, H. 1984: Contribution to the taxonomy of the superfamily Cucujoidea (Coleoptera) of Japan and her adjacent districts, II. *Memoirs of the Faculty of Education, Fukui University Series II (Natural Science)* 34: 21–63.

Sasaji, H. 1985: Discovery of the subfamily Calopodinae (Coleoptera: Oedemeridae) in Formosa, with descriptions of a new genus and a new species. *Gekkan-Mushi* 177: 7–12. (in Japanese, English title and descriptions).

Sasaji, H. 1986: The Cucujoidea (Insecta: Coleoptera) collected by the Nagoya University Scientific Expedition to Formosa in 1984. *The Memoirs of the Faculty of Education, Fukui University, Series II (Natural Science)* 36: 1–14.

Sasaji, H. 1986: Notes on the Colydiidae (Coleoptera) of Japan and Formosa. Pp. 243–249. *Entomological papers presented to Yoshihiko Kurosawa on the occasion of his retirement.* Tokyo: The Coleopterists' Association of Japan, v + 342 pp.

Sasaji, H. 1987: Contribution to the taxonomy of the superfamily Cucujoidea (Coleoptera) of Japan and her adjacent districts, III. *Memoirs of the Faculty of Education, Fukui University Series II (Natural Science)* 37: 23–55.

Sasaji, H. 1988: Contribution to the taxonomy of the superfamily Cucujoidea (Coleoptera) of Japan and her adjacent districts, IV. *Memoirs of the Faculty of Education, Fukui University Series II (Natural Science)* 38: 13–48.

Sasaji, H. 1995: On the adaptive characteristics of the genus Lederia (Coleoptera, Melandryidae), with description of a new species from Japan. *Special Bulletin of the Japanese Society of Coleopterology* 4: 425–431.

Satô, M. 1990: The genus Ischalia (Coleoptera, Pyrochroidae) from Taiwan. *Elytra* 18: 101–104.

Satô, M. & Hatta, K. 1988: A new inopeplid beetle (Coleoptera, Inopeplidae) from the Ryukyu Islands. *Elytra* 16: 85–88.

Say, T. 1824a: Descriptions of coleopterous insects collected in the late expeditions to the Rocky Mountains, performed by order of Mr. Calhoun, Secretary of War, under the command of Major Long. *Journal of the Academy of Natural Sciences of Philadelphia* 3: 238–282.

Say, T. 1824b: Appendix. Part 1. – Natural History. 1. Zoology: E. Class Insecta. Coleoptera. Pp. 268–348. In: *Narrative of an expedition of the source of St. Peter's river, lake Winnepeek, lake of the Woods, etc. etc. performed in the year 1823, by order of the Hon. J.C. Calhoun, secretary of war, under the command of Stephen H. Long, Major U.S.T.E. compiled from the notes of major Long, Messrs. Say, Keating, and Calhoun by William H. Keating, A.M. etc. Vol. 2.* Philadelphia

Say, T. 1827: Descriptions of new species of coleopterous insects inhabiting the United States. *Journal of the Philadelphia Academy of Natural Sciences* 5: 150–204, 237–284, 293–304.

Say, T. 1835: Descriptions of new North American coleopterous insects, and observations on some already described. *Boston Journal of Natural History* 1 (2): 151–203.

Saylor, L.W. 1940: Two new generic names for South American Coleoptera. *Proceedings of the Entomological Society of Washington* **42**: 46.

Schaller, J.G. 1783: Neue Insecten. *Schriften der Naturforschenden Gesellschaft zu Halle* **1**: 217–328.

Schatzmayr, A. 1926: Bestimmungstabelle der Gattung Nacerda nebst Bemerkungen über die bisher bekannten europäischen Arten. *Coleopterologisches Centralblatt* **1**: 51–65.

Schatzmayr, A. 1938: Coleotteri raccolti dal Dott. Edgardo Moltoni durante le sue missioni ornitologiche in Libia. *Atti della Società Italiana di Scienze Naturali e del Museo Civico di Storia Naturale in Milano* **77**: 159–190.

Schatzmayr, A. 1941a: Coleotteri raccolti nella Grande Sila del Dott. E. Moltoni. *Atti della Società Italiana di Scienze Naturali e del Museo Civico di Storia Naturale in Milano* **80**: 49–102.

Schatzmayr, A. 1941b: Nuovi contributi alla conoscenza della fauna delle isole Italiane dell'Egeo. XIV: Cacce coleotterologiche nel Dodecaneso. *Bollettino del Laboratorio di Zoologia Generale ed Agraria della Facoltà di Agraria di Portici* **31**: 342–368.

Schatzmayr, A. & Koch, C. 1934a: Wissenschaftliche Ergebnisse der entomologischen Expeditionen seiner Durchlaucht des Fürsten Alexander C. della Torre e Tasso nach Aegypten und auf die Halbinsel Sinai. *Bulletin de la Société Royale Entomologique d'Egypte* **17** [1933]: 204–242, pl. 12.

Schatzmayr, A. & Koch, C. 1934b: Wissenschaftliche Ergebnisse der entomologischen Expedition Seiner Durchlaucht des Fürsten Alexander C. Della Torre e Tasso nach Aegypten und auf der Halbinsel Sinai. VII. Tenebrionidae. *Bulletin de la Société Royal Entomologique d'Egypte* **18**: 17–27.

Schaufuss, L.W. 1861: Ueber 22 zum Theil neue Käferarten. *Sitzungsberichte der Naturwissenschaftlichen Gesellschaft Isis zu Dresden* **1821**: 47–51.

Schaufuss, L.W. 1862a: [new taxa]. *Sitzungsberichte der Naturwissenschaftlichen Gesellschaft Isis zu Dresden* **1861**: 91–92.

Schaufuss, L.W. 1862b: Diagnoses de coléoptères nouveaux. *Annales de la Société Entomologique de France* (4) **2**: 309–312.

Schaufuss, L.W. 1869: [new taxa]. In: Ludwig Salvador & L.W. Schaufuss: *Beitrag zur Kenntniss der Coleopteren-Fauna der Balearen*. Prag: Selbstverlag, 31 pp.

Schaufuss, L.W. 1870: Coleopterologisches. *Nunquam Otiosus* **1** [1870–1872]: 29–49.

Schaufuss, L.W. 1872: Beschreibung einiger Käfer. *Nunquam Otiosus* **2** [1872–1877]: 275–278.

Schaufuss, L.W. 1879: Zoologische Ergebnisse von Excursionen auf den Balearen. III. (Addenda und Fortsetzung). *Nunqaam Otiosus* **3** [1879–1890]: 527–552.

Schaufuss, L.W. 1882a: Coléoptères aveugles de la famille des Colydiidae. *Annales de la Société Entomologique de France* (6) **2**: 46–48.

Schaufuss, L.W. 1882b: Zoologische Ergebnisse von Excursionen auf den Balearen. III. (Addenda und Fortsetzung). *Nunquam Otiosus* **3** [1879–1890]: 527–552.

Schaufuss, L.W. 1882c: Zoologische Ergebnisse von Excursionen auf den Balearen. *Verhandlungen der Kaiserlich-Königlichen Zoologisch-Botanischen Gesellschaft in Wien* **31** [1881]: 619–624.

Schaufuss, L.W. 1887: Beitrag zur Fauna der Niederländischen Besitzungen auf den Sunda-Inseln. II. *Horae Societatis Entomologicae Rossicae* **21**: 103–147.

Schaum, H.R. 1859: Beiträge zur europäischen Käferfauna. *Berliner Entomologische Zeitschrift* **3**: 42–59.

Schaum, H.R. 1864: Discussion critique sur la synonymie de plusieurs espèces de coléoptères. Notes sur des observations synonymiques de M. Reiche. *Annales de la Société Entomologique de France* (4) **3** [1863]: 669–672.

Schaum, H.R. 1865: [new taxon]. In: Kraatz, G.: *Revision der Tenebrioniden der alten Welt.* Berlin: Nicolaische Verlagsbuchhandlung, 393 pp.

Schawaller, W. 1987: Revision westpaläarktischer Tenebrionidae (Coleoptera). Teil 1. Die Arten der Gattung Akis Herbst. *Stuttgarter Beiträge zur Naturkunde, Serie A* (*Biologie*) **403**: 1–21.

Schawaller, W. 1990a: Revision of the Western Palaearctic Tenebrionidae (Coleoptera). Part 2. The genus Ammogiton in Arabia with description of a new species. *Fauna of Saudi Arabia* **11**: 40–48.

Schawaller, W. 1990b: Revision of the Western Palaearctic Tenebrionidae (Coleoptera). Part 3. A new genus of Erodiini from Arabia. *Fauna of Saudi Arabia* **11**: 49–54.

Schawaller, W. 1991a: Prostomidae (Coleoptera) aus dem Himalaya mit einem Beitrag zur Larvalmorphologie. *Stuttgarter Beiträge zur Naturkunde Serie A* (*Biologie*) **461**: 1–17.

Schawaller, W. 1991b: The genus Diphyrrhynchus in Arabia, with description of a new species (Coleoptera, Tenebrionidae). *Fauna of Saudi Arabia* **12**: 289–294.

Schawaller, W. 1992: Leiochrini (Coleoptera: Tenebrionidae) aus dem Nepal-Himalaya. *Stuttgarter Beiträge zur Naturkunde Serie A* (*Biologie*) **482**: 1–15.

Schawaller, W. 1993a: New and little known Tenebrionidae (Coleoptera) from the Arabian Peninsula. *Fauna of Saudi Arabia* **13**: 102–109.

Schawaller, W. 1993b: Leiochrini (Coleoptera: Tenebrionidae) aus dem kontinentalen Südost-Asien. *Entomologia Basiliensia* **16**: 349–358.

Schawaller, W. 1995a: Revision der Laena Mittelasiens (Insecta, Coleoptera, Tenebrionidae). *Spixiana* **18**: 65–73.

Schawaller, W. 1995b: The genus Basanus Lacordaire (Coleoptera: Tenebrionidae) in continental south-eastern Asia. *Acta Zoologica Academiae Scientiarum Hungaricae* **41**: 327–334.

Schawaller, W. 1996a: The genus Uloma Dejean (Coleoptera: Tenebrionidae) in the Himalayas. *Acta Zoologica Academiae Scietiarum Hungariae* **42**: 111–125.

Schawaller, W. 1996b: Stenoria thakkola sp. n., the first species of the tribe Sitarini from the Himalayas (Coleoptera: Meloidae). *Elytron* **10**: 147–152.

Schawaller, W. 1996c: Spinolyprops pakistanicus sp. n. (Coleoptera, Tenebrionidae), an Oriental element in the fauna of Pakistan. *Revue Suisse de Zoologie* **103**: 691–695.

Schawaller, W. 1997: Luprops kaszabi n.sp. aus dem Himalaya, mit einer Bestimmungstabelle der Luprops-Arten Nepals (Coleoptera: Tenebrionidae). *Entomologische Zeitschrift* **107**: 295–298.

Schawaller, W. 1998: New taxonomic and faunistic data on the genus Cneocnemis Gebien (Coleoptera: Tenebrionidae) with description of two new species from Nepal. *Linzer Biologische Beiträge* **30**: 463–468.

Schawaller, W. 1999: Notes on Palaearctic and Oriental Phrenapatini (Coleoptera: Tenebrionidae), with description of four new species. *Revue Suisse de Zoologie* **106**: 419–428.

Schawaller, W. 2000a: The genus Mesomorphus Seidlitz in Nepal with description of a new species (Insecta: Coleoptera: Tenebrionidae). *Faunistische Abhandlungen* **22**: 39–48.

Schawaller, W. 2000b: The genera Chariophenus Blair, 1929, and Foochounus Pic, 1921, in the Himalayas, with description of a new species from Nepal (Coleoptera: Tenebrionidae). *Entomologische Zeitschrift* (Stuttgart) **100**: 354–357.

Schawaller, W. 2001a: The genus Laena Latreille (Coleoptera: Tenebrionidae) in China, with descriptions of 47 new species. *Stuttgarter Beiträge zur Naturkunde Serie A (Biologie)* **632**: 1–62.

Schawaller, W. 2001b: Probaticus kalavriticus n. sp., der Untergattung Helopotrichus Reitter, 1922, aus Griechenland (Coleoptera, Tenebrionidae). *Entomologische Blätter* **97**: 43–48.

Schawaller, W. 2001c: Pseudethas jaegeri sp. n. from Tibet (Insecta: Coleoptera: Tenebrionidae: Stenosini). *Reichenbachia* **34**: 143–146.

Schawaller, W. 2001d: Nepalolaena kira gen. nov., sp. nov. from the Nepal Himalayas (Coleoptera, Tenebrionidae). *Entomologia Basiliensia* **23**: 277–282.

Schawaller, W. 2002: The genus Laena Latreille in Nepal, with taxonomic and faunistic notes on species from the adjacent Himalayas (Coleoptera: Tenebrionidae). *Stuttgarter Beiträge zur Naturkunde Serie A (Biologie)* **641**: 1–69.

Schawaller, W. 2003a: Two new species of Scaphidema Redtenbacher (Coleoptera: Tenebrionidae) from Nepal and China. *Stuttgarter Beiträge zur Naturkunde Serie A (Biologie)* **653**: 1–7.

Schawaller, W. 2003b: The genus Platydema Laporte & Brullé in the Himalayas and adjacent regions. With descriptions of five new species. Pp. 269–277. In: Hartmann, M. & Baumbach, H. (eds.): *Biodiversität und Naturausstattung im Himalaya – Biodiversity and Natural Heritage of the Himalaya*. Erfurt: Verein der Freunde & Forscher des Naturkundemuseum Erfurt e. v., 390 pp., XIV pls.

Schawaller, W. 2004a: New species and records of the genus Spiloscapha Bates (Coleoptera: Tenebrionidae) from the Oriental and Papuan regions. *Stuttgarter Beiträge zur Naturkunde Serie A (Biologie)* **661**: 1–10.

Schawaller, W. 2004b: The Oriental species of Platydema Laporte & Brullé, with descriptions of 16 new species (Coleoptera: Tenebrionidae). *Stuttgarter Beiträge zur Naturkunde Serie A (Biologie)* **671**: 1–49.

Schawaller, W. 2004c: New species and records of Tenebrionidae (Coleoptera) from the Socotra Archipelago. *Fauna of Arabia* **20**: 439–458.

Schawaller, W. 2005: New species and records of Leiochrini (Coleoptera, Tenebrionidae) from continental south-eastern Asia. *Entomologia Basiliensia et Collectionis Frey* **27**: 209–226.

Schawaller, W. 2006a: Revision of the Oriental species of the genus Bradymerus Perroud with descriptions of 29 new species. *Stuttgarter Beiträge zur Naturkunde Serie A (Biologie)* **694**: 1–64.

Schawaller, W. 2006b: First record of the subfamily Cossyphodinae (Coleoptera, Tenebrionidae) from the Socotra Archipelago. *Fauna of Arabia* **21**: 247–250.

Schawaller, W. 2006c: New species, new records and new synonyms of Tenebrionidae (Insecta: Coleptera) from Nepal. *Veröffentlichungen des Naturkundemuseums Erfurt* **25**: 475–482, pls 9, 10.

Schawaller, W. 2007: A new species of Heterotarsus Latreille and other records of arboreal Tenebrionidae (Coleoptera) from Yemen. *Carolinea* **65**: 179–181.

Schawaller, W. 2008a: The genus Laena Latreille (Coleoptera: Tenebrionidae) in China (part 2), with descriptions of 30 new species and a new identification key. *Stuttgarter Beiträge zur Naturkunde A (NS)* **1**: 387–411.

Schawaller, W. 2008b: Three new species of Scaphidema Redtenbacher (Coleoptera: Tenebrionidae) from China. *Stuttgarter Beiträge zur Naturkunde A (NS)* **1**: 381–385.

Schawaller, W. 2009a: Falsammidium medvedevi sp. n. (Coleoptera: Tenebrionidae: Opatrini) from Masirah Island, Oman. Caucasian entomological Bulletin 5: 201–202.

Schawaller, W. 2009b: A new species of Rhopalobates Fairmaire (Coleoptera: Tenebrionidae) from Burma and Bhutan. *Annales Zoologici* 59: 309–312.

Schawaller, W. 2009c: Two new epigean species of the genus Menimus Sharp, 1873 from Yunnan (China) (Insecta: Coleoptera: Tenebrionidae). In: Hartmann, M. & Weipert, J. (eds.): *Biodiversität und Naturausstattung im Himalaya III*. Erfurt (Verein der Freunde und Förderer des Naturkundemuseums), 363–365 pp., XIV pls.

Schawaller, W. 2009d: The genus Laena Dejean, 1821 in Iran (Elburz Mountains) with description of a new species (Coleoptera: Tenebrionidae). *Entomologica Basiliensia* 31: 121–126.

Schawaller, W. 2009e: The genus Laena Latreille (Coleoptera: Tenebrionidae) in the Indian Himalayas of Uttar Anchal and Himachal Pradesh, with descriptions of four new species. *Stuttgarter Beiträge zur Naturkunde A (NS)* 2: 347–362.

Schawaller, W. 2010: The genus Laena Dejean, 1821 (Coleoptera: Tenebrionidae) in Afghanistan, with descriptions of three new species. *Entomologica Basiliensia* 32: 129–135.

Schawaller, W. 2011a: A new species and new records of the genus Crypsis Waterhouse, 1877 (Coleoptera: Tenebrionidae). *Annales Zoologici* 61: 355–360.

Schawaller, W. 2011b [new taxon]. In: Schawaller W., Sharaf M.R. & Aldawood A.S.: The tribe Cossyphodini (Coleoptera: Tenebrionidae: Pimeliinae) of the Arabian Peninsula, with notes on biology and ecology. *Annales Zoologici* 61: 439–444.

Schawaller, W. 2012a: Two new species and new records of the genus Spinolyprops Pic, 1917 from the Oriental Region (Coleoptera, Tenebrionidae, Lupropini). *ZooKeys* 243: 83–94.

Schawaller, W. 2012b: The genus Laena Dejean (Coleoptera: Tenebrionidae) in the eastern Himalayas (Darjeeling, Sikkim, Bhutan, Assam, Arunachal Pradesh), with descriptions of seven new species. *Stuttgarter Beiträge zur Naturkunde A (NS)* 5: 257–269.

Schawaller, W. 2014a: Order Coleoptera, family Tenebrionidae: new records and the description of a new species. *Arthropod Fauna of the UAE* 5: 272–279.

Schawaller, W. 2014b: The genus Laena Dejean (Coleoptera: Tenebrionidae) in Pakistan and adjacent Kashmir and Ladakh, with descriptions of ten new species. *Stuttgarter Beiträge zur Naturkunde A (NS)* 7: 199–217.

Schawaller, W. 2015: The genus Laena Dejean, 1821, in Nepal (part 2), with descriptions of twelve new species and new records of known species (Insecta: Coleoptera: Tenebrionidae). In: Hartmann, M. & Weipert, J (eds.): *Biodiversität und Naturausstattung im Himalaya V.* Erfurt (Verein der Freunde und Förderer des Naturkundemuseums), 457–467 pp.

Schawaller, W. 2016a: The genus Menimus Sharp (Coleoptera: Tenebrionidae: Gnathidiini) in India, with description of two new species. *Stuttgarter Beiträge zur Naturkunde A (NS)* 9: 191–195.

Schawaller, W. 2016b: Taxonomy and distribution of Palaearctic Metaclisa Jaquelin du Val (Coleoptera: Tenebrionidae). *Annales Zoologici* 66: 551–556.

Schawaller, W. 2016c: Leiochrinini (Coleoptera: Tenebrionidae: Diaperinae) from north-eastern India and China, with descriptions of six new species. *Stuttgarter Beiträge zur Naturkunde A (NS)* 9: 197–205.

Schawaller, W. 2018: New species of the genus Laena Dejean (Coleoptera: Tenebrionidae) from southeastern Asia (3). *Integrative Systematics* **1**: 71–78.

Schawaller, W. 2019: A new species and new records of Foochounus Pic (Coleoptera, Tenebrionidae: Cnodalonini) from China. *Entomological Review* **99**: 1042–1045.

Schawaller, W., Al, Dhafer H. & Fadl, H. 2013: Boromorphus saudicus sp. n., a new record of the genus from the Arabian Peninsula (Tenebrionidae: Pimeliinae: Boromorphini). *Zoology in the Middle East* **59**: 148–151.

Schawaller, W. & Ando, K. 2009: Revision of the Genus Foochounus Pic (Coleoptera, Tenebrionidae) from the Oriental Region. *Entomological Review of Japan* **64**: 259–286.

Schawaller, W. & Aston, P. 2017: Two new species of the genus Laena Dejean (Coleoptera: Tenebrionidae: Lagriinae) from Hong Kong. *Zootaxa* **4344**: 169–173.

Schawaller, W., El Torkey, A. & Al, Dhafer H. 2013: The genera Cneocnemis Gebien and Cenoscelis Wollaston in Saudi Arabia (Coleoptera: Tenebrionidae: Ulomini). *Zoology in the Middle East* **59**: 342–346.

Schawaller, W. & Merkl, O. 2012: A new species of Pentaphyllus Dejean, 1821 (Tenebrionidae: Diaperinae) from Cyprus. *Annales Zoologici* **62**: 721–724.

Schellenberg, J.R. 1802: *Entomologische Beyträge*. Winterthur: Steiner, 24 pp., 10 pls.

Schenkling, S. 1915: Oedemeridae. Pars 65. In: Schenkling, S. (ed.): *Coleopterorum Catalogus*. Berlin: W. Junk, 82 pp.

Schenkling, S. 1922: *Nomenclator Coleopterologicus. Eine etymologische Erklärung sämtlicher Gattungs- und Artnamen der Käfer der Deutschen Fauna sowie der Angrenzenden Gebiete. Zweite Auflage*. Jena: Verlag von Gustav Fischer, iv + 249 pp.

Schilder, F.A. 1923a: Rhipiphoriden-Studien. IV. Bemerkungen zur Literatur und Nomenklatur. *Entomologische Blätter* **19**: 167–169.

Schilder, F.A. 1923b: Sauter's Formosa-Ausbeute: Rhipiphoridae (Col.). (Zugleich: Rhipiphoriden-Studien III.). *Entomologische Mitteilungen* **12**: 202–204.

Schilder, F.A. 1924: Rhipiphoriden-Studien. VI. Revision des Genus Metoecus Dej. (1833). *Deutsche Entomologische Zeitschrift* **66**: 237–244.

Schilsky, J. 1888: Beitrag zur Kenntniss der deutschen Käferfauna. *Deutsche Entomologische Zeitschrift* **32**: 177–190.

Schilsky, J. 1889: Synonymische und andere Bemerkungen zu Dr. Carl W.V. Dalla Torre's ‚Synopsis der Inseckten Oberösterreichs' und „Die Käferfauna von Oberösterreich". *Deutsche Entomologische Zeitschrift* **1889**: 345–356.

Schilsky: 1892: VII. Beitrag zur Kenntnits der deutschen Käferfauna. *Deutsche Entomologische Zeitschrift* **1892**: 193–208.

Schilsky, J. 1894a: *Die Käfer Europa's. Nach der Natur Beschrieben von Dr. H. C. Küster und Dr. G. Kraatz Fortgesetzt von J. Schilsky*. Nürnberg: Bauer und Raspe (Emil Küster) **30**: viii + no. taxa 1–100.

Schilsky, J. 1894b: IX. Beitrag zur deutschen Käferfauna. *Deutsche Entomologische Zeitschrift* **1894**: 329–332.

Schilsky, J. 1895: *Die Käfer Europa's. Nach der Natur Beschrieben von Dr. H.C. Küster und Dr. G. Kraatz Fortgesetzt von J. Schilsky*. Nürnberg: Bauer und Raspe (Emil Küster) **31**: viii + no. taxa 1–100 + [1] pp.

Schilsky, J. 1899: *Die Käfer Europa´s. Nach der Natur Beschrieben von Dr. H.C. Küster und Dr. G. Kraatz Fortgesetzt von J. Schilsky.* Nürnberg: Bauer und Raspe (Emil Küster) **35**: viii + no. taxa 1–100 + 41 pp from A to SS [= 268] pp.

Schilsky, J. 1900: *Die Käfer Europa's. Nach der Natur Beschrieben von Dr. H.C. Küster und Dr. G. Kraatz Fortgesetzt von J. Schilsky.* Nürnberg: Bauer und Raspe (Emil Küster) **37**: [vi] + no. taxa 1–100 + A-W pp.

Schilsky, J. 1903: *Die Käfer Europa´s. Nach der Natur Beschrieben von Dr. H.C. Küster und Dr. G. Kraatz Fortgesetzt von J. Schilsky.* Nürnberg: von Bauer und Raspe (Emil Küster) **40**: viii no. taxa 1–100 + A-PP pp.

Schilsky, J. 1906: *Die Käfer Europa´s. Nach der Natur Beschrieben von Dr. H.C. Küster und Dr. G. Kraatz Fortgesetzt von J. Schilsky.* Nürnberg: von Bauer und Raspe (Emil Küster) **43**: viii + i-cxix pp. + no. taxa 1–100 + [1] pp.

Schilsky, J. 1908a: *Die Käfer Europa´s. Nach der Natur Beschrieben von Dr. H.C. Küster und Dr. G. Kraatz Fortgesetzt von J. Schilsky.* Nürnberg: von Bauer und Raspe (Emil Küster) **45**: iv + no. taxa 1–100 + [2] pp.

Schilsky, J. 1908b: Zoologische und Anthropologische Ergebnisse einer Forschungsreise in westlichen und zentralen Südafrika, Bd. 1, Lfg. 1, Mordellidae. *Denkschriften der Medicinisch-Naturwissenschaftlichen Gesellschaft* (Jena) **13**: 137–138.

Schilsky, J. 1908c: XII. Beitrag zur deutschen Käferfauna. *Deutsche Entomologische Zeitschrift* **1908**: 604–608.

Schilsky, J. 1910: Mordellistena engelharti. Eine neue Mordellide aus Dänemark. *Entomologiske Meddelelser* **3**: 255–256.

Schiödte, J.C. 1893: [new taxa]. In: Bergroth, E.: Bemerkungen zum 'Catalogus Coleopterorum Europae, Caucasi et Armeniae rossicae. Ed. E. Reitter'. *Entomologische Nachrichten* **19**: 305–311.

Schmidt, K. 1913: Zur Kenntis der äthiopisch-afrikanischen Meloeformen (Coleopt.). *Stettiner Entomologische Zeitung* **74**: 327–339.

Schmidt, W. 1834: [new taxon]. In: Lippich, F.W.: *Topographie der k.k. Provinzialhauptstadt Laibach, in Bezug auf Natur- und Heilkunde, Medicinalordnung und Biostatik.* Joseph Blasnik, Laibach [= Ljubljana], xv + 403 + [1 (Zusäge und Verbesserungen)] pp. [+ 5 tables].

Schmidt, W.L.E. 1842a: Die europäischen Arten der Gattung Anthicus Fbr. *Entomologische Zeitung* (Stettin) **3**: 74–88.

Schmidt, W.L.E. 1842b: Die europäischen Arten der Gattung Anthicus Fbr. *Entomologische Zeitung* (Stettin) **3**: 122–135.

Schmidt, W.L.E. 1842c: Die europäischen Arten der Gattung Anthicus Fbr. *Entomologische Zeitung* (Stettin) **3**: 170–189.

Schmidt, W.L.E. 1842d: Die europäischen Arten der Gattung Anthicus Fbr. *Entomologische Zeitung* (Stettin) **3**: 193–200.

Schmidt, W.L.E. 1844: [new taxa]. In: Schmidt, O.G.C.F. 1844: *Catalogus Coleopterorum Europae.* Stettin: 76 + vi pp.

Schmidt, W.L.E. 1846: Revision der europäischen Oedemeriden. *Linnaea Entomologica* **1**: 1–146.

Schneider, D.H. 1792: Verzeichniss und Beschreibung der in der Sammlung des Herausgebers befindlichen zur Gattung Coccinella gehörigen europäischen Käfer (Sonnen-Käfer oder Blattlaus-Käfer). *Neuestes Magazin für die Liebhaber der Entomologie* **2**: 128–256.

Schneider, O. 1902: Phaleria ligurica n.sp. nebst Bemerkungen über Ph. Reyi und Ph. Cava. *Wiener Entomologische Zeitung* **21**: 10–12.

Schneider, W. 1991: New records of Meloidae (Insecta: Coleoptera) from Arabia with description of a new species. *Fauna of Saudi Arabia* **12**: 273–288.

Schoenherr, C.J. 1806: *Synonymia Insectorum, oder: Versuch einer Synonymie aller bisher bekannten Insecten; nach Fabricii Systema Eleutheratorum geordnet. Erster Band. Eleutherata oder Käfer. Erster Theil. Lethrus – Scolytes.* Stockholm: Heinr. A. Nordström, xxii + 293 pp., 3 pls.

Schoenherr, C.J. 1817: *Synonymia Insectorum, oder: Versuch einer Synonymie aller bisher bekannten Insecten; nach Fabricii Systema Eleutheratorum geordnet. Erster Band. Eleutherata oder Käfer. Dritter Theil. Hispa – Molorchus.* Uppsala: Em. Bruzelius, xi + 506 pp.

Schrank von Paula, F. 1776: *Beyträge zur Naturgeschichte. Mit sieben von Verfasser gezeichneten und in Kupfer gestochenen Tabellen.* Augsburg: Veith Bros, [6] + 137 + [3] pp., 7 pls.

Schrank von Paula, F. 1781: *Enumeratio insectorum Austriae indigenorum.* Augustae Vindelicorum: Eberhardi Klett et Franck, [xxiv] + 548 + [4] pp., 4 pls.

Schrank von Paula, F. 1786: Erstes Brief. Pp. 1–23. In: *Baierische Reise.* München: Baptist Strobl, [14] + 418 + [10] pp., 2 pls.

Schrank von Paula, F. 1789: Entomologische Beobachtungen. *Naturforscher* **24**: 60–90.

Schrank von Paula, F. 1798: *Fauna Boica. Durchgedachte Geschichte der in Baiern einheimischen und zahmen Thiere. Erster Band.* Nürnberg: Stein'schen Buchhandlung, xii + 720 pp.

Schuh, R. 1998a: Franzorphius franzi gen. et sp. n. from China (Coleoptera: Zopheridae: Colydiinae). *Koleopterologische Rundschau* **68**: 227–232.

Schuh, R. 1998b: Revision of the Synchita variegata species group (Coleoptera: Zopheridae, Colydiinae). *Annales Zoologici* **48**: 313–324.

Schuh, R. 1999: Revision of the Oriental species of the genus Namunaria Reitter, 1882 (Coleoptera: Zopheridae, Colydiinae). *Coleopterologische Rundschau* **69**: 137–152.

Schuh, R. 2012: Colydiinae (Coleoptera: Zopheridae) of Socotra Island. *Acta Entomologica Musei Nationalis Pragae* **52**, *Supplementum*: 287–294.

Schuh, R. & Hlaváč, P. 2015: Endogean and cavernicolous Coleoptera of the Balkans XIII. Langelandia (Paganettia) ozimeci sp. n. from Southern Dalmatia (Coleoptera: Zopheridae: Colydiinae). *Koleopterologische Rundschau* **85**: 243–248.

Schuh, R. & Mifsud, D. 2000: The cylindrical bark beetles of Malta (Insecta: Coleoptera: Zopheridae, Colydiinae). *Annalen des Naturhistorischen Museums Wien* **102 B**: 259–267.

Schulze, P. 1912: Wanachia n. n. pro Cardinia Seidlitz (non M.-E.). *Internationale Entomologische Zeitschrift* **6** (2): 11.

Schuster, A. 1914a: Itagonia Ganglbaueri nov. spec. (Col., Tenebr.). *Entomologische Mitteilungen* **3**: 58–59.

Schuster, A. 1914b: Blaps kolbei nov. spec. (Col., Tenebr.). *Entomologische Blätter* **10**: 142–143.

Schuster, A. 1914c: Dila Kuntzeni nov. spec. (Col., Tenebr.). *Entomologische Blätter* **10**: 182–183.

Schuster, A. 1915a: Bemerkungen über die Tenebrionidenausbeute Paganettis auf Creta. *Entomologische Blätter* **11**: 1–6.

Schuster, A. 1915b: Neue paläarktische Tenebrionidae (Col.) I. *Entomologische Blätter* **11**: 86–92.

Schuster, A. 1915c: Neue Laena-Arten. *Wiener Entomologische Zeitung* **34**: 347–361.

Schuster, A. 1916: Monographie der Coleopterengattung Laena Latreille. *Verhandlungen der Zoologisch-Botanischen Gesellschaft in Wien* **66**: 495–629.

Schuster, A. 1919a: Neue paläarktische Tenebrioniden (Col.) II. *Entomologische Blätter* **15**: 27–33.

Schuster, A. 1919b: Revision der Tenebrioniden-Gattungen Pachychilina Reitter und Pachychile Eschscholz. *Koleopterologische Rundschau* 8 [1919–1920]: 27–43.

Schuster, A. 1920a: Neue paläarktische Tenebrioniden (Col.) III. *Entomologische Blätter* **16**: 199–201.

Schuster, A. 1920b: Zwei neue Hionthis-Arten (Col., Tenebr.). *Entomologische Blätter* **16**: 234–236.

Schuster, A. 1921a: Revision der Tenebrioniden-Gattungen Pachychilina Reitter und Pachychile Eschscholtz (Schluss). *Koleopterologische Rundschau* 9: 1–27.

Schuster, A. 1921b: Die paläarktischen Tenebrioniden des Deutschen Entomologischen Institutes I. (Col.). *Entomologische Mitteilungen* **10**: 41–47.

Schuster, A. 1922a: [new taxon]. In: Dodero A.: Missione zoologica del Dr. E. Festa in Cirenaica. *Bollettino di Musei di Zoologica ed Anatomia Comparata della R. Università di Torino* 37 N. 743: 1–7.

Schuster, A. 1922b: Tenebrioniden aus Cyrenaica (Col.). *Memorie della Società Entomologica Italiana* **1**: 14–23.

Schuster, A. 1922c: Die paläarktischen Tenebrioniden des Deutschen Entomologischen Institutes. No. II. (Col.). *Entomologische Mitteilungen* **11**: 47–50.

Schuster, A. 1922d: Insecta, ex Sibiria meridionali et Mongolia in itinere Ørjan Olsen 1914 collecta. A. Coleoptera, a Fritz Jensen lecta. V. Tenebrionidae. *Norsk Entomologisk Tidsskrift* 1 [1922–1923]: 163–164.

Schuster, A. 1922e: Die paläarktischen Tenebrioniden des Deutschen Entomologischen Institutes III. (Col.). *Entomologische Mitteilungen* **11**: 212–213.

Schuster, A. 1923a: Neue chinesische Blaps (Col. Tenebrionidae). *Wiener Entomologische Zeitung* **40**: 30–32.

Schuster, A. 1923b: Neue paläarktische Tenebrioniden (Coleopt.). *Wiener Entomologische Zeitung* **40**: 156–162.

Schuster, A. 1925a: Neue Tenebrioniden aus der Cyrenaica. *Bollettino della Società Entomologica Italiana* 57: 25–31.

Schuster, A. 1925b: Zwei neue Helopini aus Libien (Coleopt.). *Bollettino della Società Entomologica Italiana* 57: 88–90.

Schuster, A. 1926a: Bestimmungstabelle der Laena-Arten aus dem Himalaya und den angrenzenden Gebieten. Mit Beschreibungen neuer Arten. *Koleopterologische Rundschau* **12**: 31–54.

Schuster, A. 1926b: Neue Tenebrioniden aus der Cyrenaica II. (Col.). *Bollettino della Società Entomologica Italiana* **58**: 130–135.

Schuster, A. 1926c: Heliophilus (s. str.) Liebei n. sp. (Col., Tenebr.). *Coleopterologisches Centralblatt* **1**: 243–244.

Schuster, A. 1928a: Eine neue Pimelia (Col., Tenebr.) aus Tripolis. *Bollettino della Società Entomologica Italiana* **60**: 26–28.

Schuster, A. 1928b: Neue Tenebrioniden aus Cyrenaica (IV). *Bollettino della Società Entomologica Italiana* **60**: 122–124.

Schuster, A. 1928c: Die Gattung Caenoblaps König. (Col. Tenebr.). *Koleopterologische Rundschau* **14**: 122–125.

Schuster, A. 1928d: Beitrag zur Kenntnis der Verbreitung der Tenebrioniden in Libien. *Bollettino della Società Entomologica Italiana* **60**: 149–154.

Schuster, A. 1928e: Neue Tenebrioniden aus der Cyrenaica. III. *Annali del Museo Civico di Storia Naturale Giacomo Doria* **52**: 373–378.

Schuster, A. 1928f: Tenebrionidae. Pp. 914–1024. In: Winkler, A. (ed.): *Catalogus Coleopterorum regionis palaearcticae*. Vol. II [1927–1932]. Wien: Albert Winkler, 817–1698 pp.

Schuster, A. 1929: Neue Pimelien (Col., Tenebr.) aus Marokko. *Koleopterologische Rundschau* **15** [1929–1930]: 38–41.

Schuster, A. 1930a: Zwei neue Tenebrioniden aus dem Mittelmeergebiet. *Bollettino della Società Entomologica Italiana* **62**: 65–67.

Schuster, A. 1930b: Neue Tenebrioniden (Coleopt.) aus Belutschistan. *Koleopterologische Rundschau* **15** [1929–1930]: 235–239.

Schuster, A. 1931: [new taxa]. In: Reinig, W.F. 1931: Entomologische Ergebnisse der Deutsch-Russischen Alai-Pamir-Expedition 1928 (II). 5. Coleoptera II. Tenebrionidae. *Mitteilungen aus dem Zoologischen Museum Berlin* **16**: 865–912.

Schuster, A. 1933a: [new taxon]. In: Lindberg Håk.: Inventa entomologica Hispanici et Maroccani, quod a. 1926 fererunt Harald et Håkan Lindberg. XVI. Tenebrionidae. *Commentationes Biologicae* **4** (5): 1–31, 1 pl.

Schuster, A. 1933b: Die Gattung Myladina Rtt. (Coleoptera, Tenebrionidae). *Sborník Entomologického Oddělení Národního Musea v Praze* **11**: 96–98.

Schuster, A. 1934a: Zur Nomenklatur des Subgenus Aulonoscelis Rtt. (Col., Tenebr.). *Koleopterologische Rundschau* **20**: 75.

Schuster, A. 1934b: Tenebrioniden vom Toten Meer. *Entomologische Blätter* **30**: 180–182.

Schuster, A. 1935a: Neue Tenebrioniden (Col.) aus der Cyrenaica. v. *Bollettino de la Società Italiana di Entomologica* **67**: 13–21.

Schuster, A. 1935b: Die Tenebrioniden-Ausbeute Dr. Gabriels in Persien. *Koleopterologische Rundschau* **21**: 20–29.

Schuster, A. 1935c: Die Gattung Asidoblaps,Fairm. (Col. Tenebr.). *Stylops* **4**: 161–165.

Schuster, A. 1935d: Neue Laena-Arten aus dem Himalaya (Col., Fam. Tenebrionidae). *The Annals and Magazine of Natural History* (10) **14**: 437–466.

Schuster, A. 1936a: Schwedish-chinesische wissenschaftliche Expedition nach den nordwestlichen Provinzen Chinas, unter Leitung von Dr. Hedin Svin und Prof. Sü Ping-chang. Insekten gesammelt vom schwedischen Arzt der Expedition Dr. David Hummel 1927–1930. 44. Coleoptera. 8. Tenebrionidae. *Arkiv für Zoologie* **27 A** (24): 1–6.

Schuster, A. 1936b: Die Tenebrioniden (Col.)-Ausbeute von Prof. Dr. Franz Werner und Kustos Dr. Otto Wettstein von den Griechischen, Ägäischen Inseln und dem italienischen Dodekanes in den Jahren 1934 und 1935. *Sitzberichte der Akademie der Wissenschaften in Wien, Mathematisch-naturwissenschaftliche Klasse, Abteilung I* **145**: 65–80.

Schuster, A. 1936c: Tenebrionidae. Entomologische Sammelergebnisse der Deutschen Hindukusch-Expedition 1935 der Deutschen Forschungsgemeinschaft. *Arbeiten über Morphologische und Taxonomische Entomologie aus Berlin-Dahlem* **3**: 192–201.

Schuster, A. 1937a: Eine Tenebrionides-Ausbeute von Mesopotamien, Palästina, Syrien und Aegypten. *Koleopterologische Rundschau* **23**: 44–52.

Schuster, A. 1937b: Zwei neue palaearktische Tenebrioniden (Col.) und Bemerkungen zu einer ungedeuteten Art. *Entomologische Blätter* **33**: 79–83.

Schuster, A. 1938a: [new taxa]. In: Schuster A: & Gebien H.: Tenebrioniden (Col.) aus Arabien. *Entomologische Blätter* **34**: 49–62.

Schuster, A. 1938b: Neue Tenebrioniden (Col.) von Persich-Belutschistan. *Koleopterologische Rundschau* **24**: 77–90.

Schuster, A. 1940a: Die Tenebrioniden (Col.) des Museums Hoang ho-Pei ho in Tientsin. *Koleopterologische Rundschau* **26**: 15–24.

Schuster, A. 1940b: Über drei neue asiatische Tenebrioniden (Col. Tenebr.). *Mitteilungen der Münchener Entomologischen Gesellschaft* **30**: 861–866.

Schuster, A. & Reymond, A. 1937: Quatre nouveaux ténébrionides provenant de la Mission Citroën-Centre-Asie (Col.). *Bulletin de la Société Entomologique de France* **42**: 234–238.

Schwarz, C. 1808: [new taxa]. In: Schoenherr, C.J.: *Synonymia Insectorum, oder: Versuch einer Synonymie aller bisher bekannten Insecten; nach Fabricii Systema Eleutheratorum geordnet. Erster Band. Eleutherata oder Käfer. Zweiter Theil. Spercheus – Cryptocephalus*. Stockholm: Carl Friedr. Merquardsix + [1] + 424 pp., pl. 4.

Scopoli, J.A. 1763: *Entomologia Carniolica exhibens insecta Carnioliae indigena et distributa in ordines, genera, species, varietates, methodo linnaeana*. Vindobonae: Ioannis Thomae Trattner, xxxvi + 420 pp., 3 pls.

Scriba, L.G. 1790: *Journal für die Liebhaber der Entomologie*. Frankfurt: Varrentrapp und Werner, vol. **1**: 1–96.

Scriba, W. 1865: Die Käfer im Grossherzogthum Hessen und seiner nächsten Umgebung. *Bericht der Obershessischen Gesellschaft für Natur- und Heilkunde* **11**: 1–59.

Scupola, A. 1984: Una nuova Idastrandiella Strand di Grecia (Coleoptera Tenebrionidae). *Bollettino di Museo Civico di Storia Naturale di Verona* **11**: 69–75.

Scupola, A. 1994: Una nuova specie di Mesostena Esch. dello Yemen (Col. Tenebrionidae). (III. Contributo alla conoscenza dei Tenebrionidi). *Frustula Entomologica N.S.* **17**: 25–28.

Scupola, A. 2000: Revisione della tribu Cossyphini Latreille, 1802. Parte I. Introduzione e genere Cossyphus Olivier, 1791 (Coleoptera Tenebrionidae). *Atti del Museo Civico di Storia Naturale di Trieste* **48**: 185–249.

Scupola, A. 2002: A proposito di Ulomina carinata Baudi di Selve, 1876 (Coleoptera, Tenebrionidae). *Bollettino del Museo Regionale di Scienze Naturali di Torino* **19**: 185–189.

Scupola, A. & Mifsud, D. 2001: Descrizione di una nuova sottospecie di Heliopathes avarus Mulsant & Rey, 1854 dell'Isola di Gozo (Arcipelago Maltese) (Coleoptera, Tenebrionidae). *Doriana* 7 (**324**): 1–7.

Seidlitz, G.C.M. von. 1889: Ueber die Coleopteren-Gattung Esarcus Reiche. *Deutsche Entomologische Zeitschrift* **1889**: 147–149.

Seidlitz, G.C.M. von. 1890: *Fauna Baltica. Die Kaefer (Coleoptera) der deutschen Ostseeprovinzen Russlands. Zweite neu bearbeitete Auflage*. Königsberg: Hartungsche Verlagsdruckerei, [10] + lvi + 192 + 818 pp., 1 pl. [issued in parts: pp. i-xl, Gattungen 1–16, Arten 1–96 in 1887; pp. xli-xlviii, Gattungen 17–80, Arten 97–336 in 1888; pp. Gattungen 81–128, Arten 337–512 in 1889; pp. Gattungen 129–160, Arten 513–608 in 1890; pp. xlix-lvi, Gattungen 161–192, Arten 609–818 in 1891].

Seidlitz, G.C.M. von. 1891: Gattungen pp. 129–160, Arten pp. 549–914. In: *Fauna Transsylvanica. Die Kaefer (Coleoptera) Siebenbürgens*. Königsberg: Hartungsche Verlagsdruckerei, [12] + lvi + 192 + 914 pp., 1 pl. [issued in parts: pp. i-xl, Gattungen 1–48, Arten 1–240 in 1888; pp. xli-lvi, Gattungen 49–128, Arten 241–544 in 1889; pp. xlix-lvi, Gattungen 129–192, Arten 545–914 in 1891].

Seidlitz, G.C.M. von. 1893: Tenebrionidae. Pp. 201–400. In: Kiesenwetter, H. von. & Seidlitz, G.C.M. von.: *Naturgeschichte der Insecten Deutschlands. Begonnen von Dr. W. F. Erichson, fortgesetzt von Prof. Dr. H. Schaum, Dr. G. Kraatz, H. v. Kiesenwetter, Julius Weise, Edm. Reitter und Dr. G. Seidlitz. Erste Abteilung Coleoptera. Fünfter Band. Erste Hälfte*. Berlin, Nicolaische Verlags-Buchhandlung, xxviii + 877 + [1] pp.

[issued in parts: pp. 201–400 in March 1893, pp. 401–608 in May 1894, pp. 609–800 in September 1896, pp. i-xxviiii + 801–877 in September 1898].

Seidlitz, G.C.M. von. 1894: Tenebrionidae. Pp. 401–608. In: Kiesenwetter, H. von. & Seidlitz, G.C.M. von.: *Naturgeschichte der Insecten Deutschlands. Begonnen von Dr. W. F. Erichson, fortgesetzt von Prof. Dr. H. Schaum, Dr. G. Kraatz, H. v. Kiesenwetter, Julius Weise, Edm. Reitter und Dr. G. Seidlitz. Erste Abteilung Coleoptera. Fünfter Band. Erste Hälfte.* Berlin, Nicolaische Verlags-Buchhandlung, xxviii + 877 + [1] pp. [issued in parts: pp. 201–400 in 1893, pp. 401–608 in 1894, pp. 609–800 in 1896, pp. i-xxviiii + 801–877 1898].

Seidlitz, G.C.M. von. 1895: Tenebrionidae. Pp. 609–800. In: Kiesenwetter, H. von. & Seidlitz, G.C.M. von.: *Naturgeschichte der Insecten Deutschlands. Begonnen von Dr. W. F. Erichson, fortgesetzt von Prof. Dr. H. Schaum, Dr. G. Kraatz, H. v. Kiesenwetter, Julius Weise, Edm. Reitter und Dr. G. Seidlitz. Erste Abteilung Coleoptera. Fünfter Band. Erste Hälfte.* Berlin, Nicolaische Verlags-Buchhandlung, xxviii + 877 + [1] pp. [issued in parts: pp. 201–400 in 1893, pp. 401–608 in 1894, pp. 609–800 in 1896, pp. i-xxviiii + 801–877 in September 1898].

Seidlitz, G.C.M. von. 1896: Alleculidae. Pp. 1–305. In: *Naturgeschichte der Insecten Deutschlands. Begonnen von Dr. W.F. Erichson, fortgesetzt von Prof. Dr. H. Schaum, Dr. G. Kraatz, H. v. Kiesenwetter, Julius Weise, Edm. Reitter und Dr. G. Seidlitz. Erste Abtheilung Coleoptera. Fünfter Band. Zweite Hälfte. Lieferungen 1–3.* Berlin: Nicolaische Verlags-Buchhandlung, 968 pp.

Seidlitz, G.C.M. von. 1898a: Tenebrionidae. Pp. 801–877. In: Kiesenwetter, H. von. & Seidlitz, G.C.M. von.: *Naturgeschichte der Insecten Deutschlands. Begonnen von Dr. W.F. Erichson, fortgesetzt von Prof. Dr. H. Schaum, Dr. G. Kraatz, H. v. Kiesenwetter, Julius Weise, Edm. Reitter und Dr. G. Seidlitz. Erste Abteilung Coleoptera. Fünfter Band. Erste Hälfte.* Berlin, Nicolaische Verlags-Buchhandlung, xxviii + 877 + [1] pp. [issued in parts: pp. 201–400 in 1893, pp. 401–608 in 1894, pp. 609–800 in 1896, pp. i-xxviiii + 801–877 in 1898].

Seidlitz, G.C.M. von. 1898b: Lagriidae, pp. 306–364; Melandryidae, pp. 365–680. In: *Naturgeschichte der Insecten Deutschlands. Begonnen von Dr. W.F. Erichson, fortgesetzt von Prof. Dr. H. Schaum, Dr. G. Kraatz, H. v. Kiesenwetter, Julius Weise, Edm. Reitter und Dr. G. Seidlitz. Erste Abtheilung Coleoptera. Fünfter Band. Zweite Hälfte. Lieferungen 1–3.* Berlin: Nicolaische Verlags-Buchhandlung, 968 pp.

Seidlitz, G.C.M. von. 1899: Oedemeridae. Pp. 681–968. In: *Naturgeschichte der Insecten Deutschlands. Begonnen von Dr. W.F. Erichson, fortgesetzt von Prof. Dr. H. Schaum, Dr. G. Kraatz, H. v. Kiesenwetter, Julius Weise, Edm. Reitter und Dr. G. Seidlitz. Erste Abtheilung Coleoptera. Fünfter Band. Zweite Hälfte. Lieferungen 1–3.* Berlin: Nicolaische Verlags-Buchhandlung, 968 pp.

Seidlitz, G.C.M. von. 1916: Die letzten Familien der Heteromeren. *Deutsche Entomologische Zeitschrift* **1916**: 113–128, 397–498.

Seidlitz, G.C.M. von. 1917: Die letzten Familien der Heteromeren. *Deutsche Entomologische Zeitschrift* **1917**: 65–116.

Seidlitz, G.C.M. von. 1920: *Naturgeschichte der Insecten Deutschlands. Begonnen von Dr. W.F. Erichson, fortgesetzt von Prof. Dr. H. Schaum, Dr. G. Kraatz, H. v. Kiesenwetter, Julius Weise, Edm. Reitter und Dr. G. Seidlitz. Erste Abtheilung. Coleoptera. Fünfter Band. Zweite Hälfte. Vierte Lieferung und Register.* Leipzig: K.F. Koehlers Antiquarium, 969–1206 pp.

Selander, R.B. 1960: Bionomics, systematics, and phylogeny of Lytta, a genus of blister beetles (Coleoptera, Meloidae). *Illinois Biological Monographs* **28**: i–vi, 1–295.

Selander, R.B. 1988: An annotated catalog and summary of bionomics of blister beetles of the genus Cylindrothorax (Coleoptera, Meloidae). *Transactions of the American Entomological Society* 114: 15–[illegible].

Selander, R.B. 1991: On the nomenclature and classification of the Meloidae (Coleoptera). *Insecta Mundi* 5: 65–94.

Semenov, A.P. 1889a: Diagnoses Coleopterorum novorum ex Asia centrali et orientali. *Horae Societatis Entomologicae Rossicae* 24 [1889–1890]: 193–226.

Semenov, A.P. 1889b: Ueber einige Arten der Gattung Sphenaria Mén. und Asphena Sem. *Wiener Entomologische Zeitung* 8: 309–310.

Semenov, A.P. 1890: Diagnoses Coleopterorum novorum ex Asia centrali et orientali. *Horae Societatis Entomologicae Rossicae* 25 [1890–91]: 262–382.

Semenov, A.P. 1891: Ueber die Arten der Coleopteren-Gattung Leptodes Sol. *Wiener Entomologische Zeitung* 10: 268–272.

Semenov, A.P. 1893a: Revisio synoptica meloidarum generis Ctenopus Fisch. *Bulletin de l'Académie de St. Petersburg* 13: 11–20.

Semenov, A.P. 1893b: Revisio synoptica Tenebrionidarum generis Leptodes Sol. *Horae Societatis Entomologicae Rossicae* 27 [1892–1893]: 225–235.

Semenov, A.P. 1893c: Symbolae ad cognitionem pimeliidarum. I–III. *Horae Societatis Entomologicae Rossicae* 27 [1892–1893]: 249–264.

Semenov, A.P. 1893d: Meloidarum species novae. *Horae Societatis Entomologicae Rossicae* 27 [1892–1893]: 265–281.

Semenov, A.P. 1893e: Symbolae ad cognitionem pedilidarum. *Horae Societatis Entomologicae Rossicae* 27 [1892–1893]: 326–334.

Semenov, A.P. 1893f: De coleopterorum familia nova. *Mélanges Biologiques tirés du Bulletin Physico-Mathématique de l'Académie Impériale des Sciences de St.-Pétersbourg* 13: 359–366.

Semenov, A.P. 1893g: Coleoptera asiatica nova. *Horae Societatis Entomologicae Rossicae* 27 [1892–1893]: 493–507.

Semenov, A.P. 1894a: Supplementum ad Edm. Reitteri «Revisionem» Tenebrionidarum generis Prosodes Eschsch. *Horae Societatis Entomologicae Rossicae* 28 [1893–1894]: 179–223.

Semenov, A.P. 1894b: Symbolae ad cognitionem oedemeridarum. *Horae Societatis Entomologicae Rossicae* 28 [1893–1894]: 449–474.

Semenov, A.P. 1894c: Coleoptera asiatica nova. II. *Horae Societatis Entomologicae Rossicae* 28 [1893–1894]: 526–547.

Semenov, A.P. 1894d: De coleopterorum familia nova. *Bulletin de l'Académie des Sciences de St.-Pétersbourg.* (N.S.) (35) 3: 607–614.

Semenov, A.P. 1895a: Coleoptera asiatica nova. III. *Horae Societatis Entomologicae Rossicae* 29 [1894–1895]: 125–144.

Semenov, A.P. 1895b: Coleoptera nova rossiae europae caucasique I. *Horae Societatis Entomologicae Rossicae* 29 [1894–1895]: 242–250.

Semenov, A.P. 1896a: Revisio specierum generis Petria Sem. *Annuaire du Musée Zoologique de l'Académie Impériale des Sciences de St.-Pétersbourg* 1: 26–30.

Semenov, A.P. 1896b: Coleopterorum genera faunae turanicae endemica vel praecipue peculiaria eorumque species. *Annuaire du Musée Zoologique de l'Académie Impériale des Sciences de St.-Pétersburg* 1: 42–61.

Semenov, A.P. 1896c: Coleoptera asiatica nova. VI. *Horae Societatis Entomologicae Rossicae* **29** [1894–1895]: 336–362.

Semenov, A.P. 1896d: Insectorum quondam novorum faunae transcaspiae, diagnoses. *Annuaire du Muséum Zoologique de l'Académie Impériale des Sciences de St.-Pétersburg* **1**: 378–386.

Semenov, A.P. 1896e: De genere Rhampholyssa Krtz. *Horae Societatis Entomologicae Rossicae* **29** [1894–1895]: 515–519.

Semenov, A.P. 1898a: Zametki o zhestkokrylich europeyskoy Rossii i Kaukaza. *Bulletin de la Société Impériales des Naturalistes de Moscou* **71**: 68–115.

Semenov, A.P. 1898b: Dua nova oedemeridarum genera. *Wiener Entomologische Zeitung* **17**: 175–178.

Semenov, A.P. 1898c: Coleoptera nova Rossiae europaeae caucasique. III. *Horae Societatis Entomologicae Rossicae* **31** [1896–1897]: 543–554.

Semenov, A. 1898d: Coleoptera nova Rossiae europaeae caucasique. *Horae Societatis Entomologicae Rossicae* **32** [1898–1899]: 280–290.

Semenov, A.P. 1898e: Specierum palaearcticarum ad subgenus Zimioma Gozis generis Ostoma Laichart. spectancium synopsis. *Horae Societatis Entomologicae Rossicae* **32** [1898–1899]: 286–290.

Semenov, A.P. 1899a: Coleoptera nova Rossiae europaeae caucasique. VI. *Horae Societatis Entomologicae Rossicae* **32** [1898–1899]: 604–615.

Semenov, A.P. 1899b: Coleoptera asiatica nova. VI. *Horae Societatis Entomologicae Rossicae* **32** [1898–1899]: 632–656.

Semenov, A.P. 1900a: Coleoptera asiatica nova. IX. *Horae Societatis Entomologicae Rossicae* **34**: 303–334.

Semenov, A.P. 1900b: Commentarii de meloidis. *Horae Societatis Entomologicae Rossicae* **34**: 479–497.

Semenov, A.P. 1900c: Coleoptera asiatica nova. X. *Horae Societatis Entomologicae Rossicae* **34**: 577–605.

Semenov, A.P. 1900d: De nonnullis oedemeridarum generibus. *Horae Societatis Entomologicae Rossicae* **34**: 643–655.

Semenov, A.P. 1900e: Coleoptera asiatica nova. XI. *Horae Societatis Entomologicae Rossicae* **34**: 676–688.

Semenov, A.P. 1901a: K litoralnoy fauny Kryma: I. Phaleria pontica, sp. n. (Coleoptera, Tenebrionidae). *Revue Russe d'Entomologie* **1**: 90–97.

Semenov, A.P. 1901b: Pervyy predstavitel′ v Rossii roda Hymenorus Muls. (Coleoptera, Alleculidae) i zoogeograficheskoe znachenie etoy pribavki k russkoy faune. *Revue Russe d'Entomologie* **4–5**: 167–172.

Semenov, A. 1902a: Coleoptera nova Rossiae europeae Caucasique. VIII. *Horae Societatis Entomologicae Rossicae* **35** [1901–1902]: 253–265.

Semenov, A.P. 1902b: Generica quaedam nomina mutanda vel emendanda. *Revue Russe d'Entomologie* **2**: 353.

Semenov, A.P. 1903a: Coleoptera novae faunae turanicae. *Revue Russe d'Entomologie* **3**: 8–13.

Semenov, A.P. 1903b: Analecta coleopterologica. III. *Revue Russe d'Entomologie* **3**: 169–173.

Semenov, A.P. 1903c: Coleoptera novae faunae turanicae. II. *Revue Russe d'Entomologie* **3**: 353–357.

Semenov, A.P. 1903d: Analecta coleopterologica. *Revue Russe d'Entomologie* **6**: 390–392.

Semenov, A.P. 1905: K litoralnoy faune Kryma: II. Novyy predstavitel roda Ammobius Guér. = Ammophthorus Lac. (Coleoptera, Tenebrionidae). *Revue Russe d'Entomologie* **5**: 123–125.

Semenov, A.P. 1907a: Synopsis generum tribus Platyopinorum (Coleoptera, Tenebrionidae Pimeliini). *Horae Societatis Entomologicae Rossicae* **38** [1907–1908]: 175–184.

Semenov, A.P. 1907b: De novo Pimeliinorum genere, quod tribum peculiarem representat (Coleoptera, Tenebrionidae). *Revue Russe d'Entomologie* 6 [1906]: 257–260.

Semenov, A.P. 1907c: Coleoptera nova heptapotamica. II. *Revue Russe d'Entomologie* 6 [1906]: 261–265.

Semenov, A.P. 1908: Coleoptera nova faunae dzhungaro-tianshanicae. I. *Revue Russe d'Entomologie* 8: 115–125.

Semenov, A.P. 1909: Coleoptera nova faunae turanicae. III. *Revue Russe d'Entomologie* 8 [1908]: 295–30[illegible].

Semenov, A.P. 1910a: Analecta coleopterologica. XV. *Revue Russe d'Entomologie* 9: 24–35.

Semenov, A.P. 1910b: Analecta coleopterologica. *Revue Russe d'Entomologie* 9: 433–435.

Semenov, A.P. 1910c: Symbolae ad faunam desertorum mesasiaticoruM.I. Synopsis specierum generis Argyrophana Sem. 1889 (Coleoptera, Tenebrionidae). *Revue Russe d'Entomologie* 10: 42–44.

Semenov, A.P. 1910d: Coleoptera nova faunae kirgisicae. I. *Revue Russe d'Entomologie* 10: 150–153.

Semenov, A.P. 1914: Ob odnom novom vide roda Sympiezocnemis (Coleoptera, Tenebrionidae) i o nekotorykh ego osobennostyakh. *Revue Russe d'Entomologie* 13 [1913]: 506–509.

Semenov, A.P. & Arnoldi, C. 1937: Trois formes nouvelles de coléoptères, caractéristiques pour la faune hyémale des dèserts de la Transcaspienne. *Bulletin de la Société Entomologique de France* 42: 221–224.

Semenov, A.P. & Bogatchev, A.V. 1936: Supplément à la révision du genre Blaps F. (Coleoptera, Tenebrionidae) de G. Seidlitz, 1893. *Festschrift zum 60. Geburtstage von Professor Dr. Embrik Strand* (Riga) 1: 553–568.

Semenov, A.P. & Bogatchev, A.V. 1947: In Bogatchev A.V. & Semenov A.P.: Novyy rod tribu Eurychorini (Col., Tenebrionidae). *Doklady Akademii Nauk Azerbaydzhanskoy SSR* 3: 175–177.

Semenov, A.P. & Skopin, N.G. 1951: Novyy predstavitel roda Stalagmoptera Sols. (Coleoptera, Tenebrionidae) iz pustiny Betpak-Dala. *Uchenye Zapiski Kazaskhskogo Gos. Universiteta* (Alma Ata) 13 (1): 127–128.

Semenov, A.P. & Ter-Minasjan, M.E. 1937: Sur un nouveau genre paléarctique de la tribu des Nacerdini (Col. Oedemeridae). *Bulletin de la Société Entomologique de France* 1937: 227–229.

Sénac, H. 1880: Description de quatre nouvelles espèces appartenant au genre Pimelia. *Annales de la Société Entomologique de France* (5) 10: 261–268.

Sénac, H. 1881: [new taxa]. *Bulletin de la Société Entomologique de France* 1881: xix-xx.

Sénac, H. 1882a: [new taxa]. *Bulletin de la Société Entomologique de France* 1882: xxx–xxxi.

Sénac, H. 1882b: [new taxa]. *Bulletin de la Société Entomologique de France* 1882: lvi–lviii.

Sénac, H. 1884a: *Essai monographique sur le genre Pimelia (Fabricius). Première partie. Espèces à tarses postérieurs et intermédiaires comprimés.* Paris: Jacques Chevalier, xix + 106 pp.

Sénac, H. 1884b: [new taxa]. *Bulletin de la Société Entomologique de France* 1884: x–xi.

Sénac, H. 1884c: [new taxa]. *Bulletin de la Société Entomologique de France* 1884: xxiv–xxv.

Sénac, H. 1885: [new taxa]. *Bulletin de la Société Entomologique de France* 1885: lxxxi–lxxxii.

Sénac, H. 1886: [new taxa]. *Bulletin de la Société Entomologique de France* 1886: xlv–xlvi.

Sénac, H. 1887a: *Essai monographique sur le genre Pimelia (Fabricius). Deuxième et dernière partie. Espèces à tarses postérieurs et intermédiaires non comprimés (2e Division de Solier).* Paris: Jacques Lechevalier, xv + 160 pp.

Sénac, H. 1887b: Sous-genre Pachyscelodes (subdivision nouvelle du genre Pachyscelis Sol.). *Annales de la Société Entomologique de France* (6) 7: 187–194.

Sénac, H. 1892: Voyage de M. Ch. Alluaud aux îles Canaries (novembre 1889 – juin 1890). *Annales de la Société Entomologique de France* 61: 103–108.

Sengupta, T. 1977: Discovery of the family Elacatidae (=Othniidae), Coleoptera from India with description of a new species. *Oriental Insect* **11**: 537–539.

Sengupta, T. 1980: Ergebnisse der Bhutan-Expedition 1972 des Naturhistorischen Museums in Basel. Coleoptera: Fam. Elacatidae (= Othniidae). *Entomologica Basiliensia* **5**: 28–30.

Sengupta, T., Pal, T.K. & Mukhopadhyay, P. 1977: On the family Inopeplidae (Coleoptera) from India. *Oriental Insects* **11**: 395–407.

Serrano, A.R.M., Amorim, I.R. & Borges, P.A.V. 2013: A new species of Tarphius Erichson, 1845 (Coleoptera: Zopheridae) from North Africa and notes on an Iberian species. *Zootaxa* **3613**: 493–500.

Serri, S. & Bologna, M.A. 2018: A new species of Teratolytta from Iran, with an updated identification key to the males of this genus (Coleoptera: Meloidae). *Fragmenta Entomologica* **50**: 145–148.

Serri, S., Pan, Z. & Bologna, M.A. 2012: A new Mylabris species from south-eastern Iran and a key to the Iranian species of the nominate subgenus (Coleoptera, Meloidae). *Zookeys* **219**: 81–86.

Shapovalov, A.M. 2012: A new species of the subgenus Eurymeloe Reitter, 1911 (Coleoptera, Meloidae: Meloe) from the steppe zone of Russia and Kazakhstan. *Caucasian Entomological Bulletin* **8**: 205–210.

Shapovalov, A.M. 2014: A new species of the blister-beetle genus Meloe L. (Coleoptera, Meloidae) from Tajikistan. *Entomological Review* **94**: 1337–1341.

Shapovalov, A.M. 2016: Remarks on taxonomy of the genus Lytta Fabricius, 1775 (Coleoptera: Meloidae) with description of a new subgenus and two new species from Asia. *Caucasian Entomological Bulletin* **12**: 99–108.

Sharp, D.S. 1876: Descriptions of some new genera and species of new Zealand Coleoptera. *The Entomologist's Monthly Magazine* **13** [1876–1877]: 70–79.

Sharp, D.S. 1879: On some Coleoptera from Hawaiian Islands. *Transaction of the Entomological Society of London* **1879**: 77–105.

Sharp, D.S. 1885a: On the Colydiidae collected by Mr. G. Lewis in Japan. *The Journal of the Linnaean Society. Zoology* **19**: 58–84, pl. III.

Sharp, D.S. 1885b: On some Colydiidae obtained by Mr. Lewis in Ceylon. *The Journal of the Linnaean Society. Zoology* **19**: 117–131, pl. VI.

Sharp, D.S. 1885: [new taxa]. In: Blackburn, T. & Sharp, D.S.: Memoirs on the Coleoptera of the Hawaiian Islands. *The Scientific Transactions of the Royal Dublin Society* (2) **3**: 119–300, pls. 4–5.

Sharp, D.S. 1891: Descriptions of two new species and a new genus of Japanese Coleoptera. *The Entomologist* **24** (*Supplement*): 6–7.

Sharp, D.S. 1894: Fam. Colydiidae. Pp. 441–488. In: Godman, F. & Salvin, O. (eds): *Biologia-Centrali-Americana. Insecta. Coleoptera. Vol. II. Part I.* London: Dulau & Co., 717 pp., 19 pls.

Sharp, D.S. 1902: Mycetophagidae. Pp. 637–642. [1887–1905]. In: Godman, F. & Salvin, O. (eds): *Biologia Centrali-Americana. Insecta. Coleoptera. Vol. II. Part. I.* [1887–1905]. London: Dulau & Co. 717 pp., 19 pls.

Sharp, D.S. 1909: Some critical remarks on the genus Rabocerus, Mulsant with descriptions of two new species. *The Entomologist's Monthly Magazine* (2) **20**: 245–247.

Shestoperov, E.L. 1930: Materialy po Coleoptera ekspeditsii po obsledovaniyu saranchyovykh nasekomykh v Persii. Pp. 99–107. In: *Otchet o deyatelnosti 1926–1927, 1927–1928 i 1928–1929 operatsionnye goda*. Askhabad: Izdanie Narkomzema Turkmenskoy SSR, 107 + (10) pp.

Shi, A.-M. 2010: [new taxa]. In: Shi, A.-M., Yuan, S.-B. & Wu, C.-L.: Three new species of Itagonia Reitter, 1887 (Coleoptera, Tenebrionidae: Blaptini) from Sichuan, China. *Zootaxa* **2573**: 62–67.

Shi, A.-M. 2013a: Three new species of Gnaptorina Reitter (Coleoptera, Tenebrionidae: Blaptini) from Tibet, China. *Zootaxa* 3637: 462–471.

Shi, A.-M. 2013b: Two new species of Itagonia Reitter (Coleoptera, Tenebrionidae, Blaptini) from China. *Zookeys* 309: 1–12.

Shi, A.-M. & Ren, G.-D. 2005: Study on the genus Agnaptoria Reitter (Coleoptera, Tenebrionidae) from China, with the description of a new species. *Entomotaxonomia* 27: 282. (in Chinese with English summary).

Shi, A.-M. & Ren, G.-D. 2007a: Two new species of Agnaptoria Reitter, 1887 (Coleoptera: Tenebrionidae) from Xizang, China. *Entomotaxonomia* 29: 281–286.

Shi A,-M. & Ren G.-D. 2007b: Two new species of Itagonia Reitter (Coleoptera: Tenebrionidae: Blaptini). *Zootaxa* 1483: 33–39.

Shi, A.-M. & Ren, G.-D. 2007c: A new species of Itagonia Reitter, 1887 (Coleoptera: Tenebrionidae) from Xizang, China. *Journal of China West Normal University* (*N.S.*) 28: 176–178.

Shi, A.-M., Ren, G.-D. & Merkl, O. 2005: Two new species of Pseudognaptorina Kaszab, 1977 (Coleoptera, Tenebrionidae: Blaptini) from the Tibet Plateau. *Acta Zoologica Academiae Scientarum Hungaricae* 51: 163–170.

Shi, A.-M., Ren, G.-D. & Merkl, O. 2007: Six new species of Gnaptorina Reitter, 1887 (Coleoptera, Tenebrionidae: Blaptini) from the Tibet Plateau. *Acta zoologica Academiae Scientiarum Hungaricae* 53: 219–238.

Shibata, T. 1978: Notes on Tenebrionidae from Taiwan and Japan, I. (Coleoptera). *Entomological Review of Japan* 32: 19–27.

Shibata, T. 1979: Notes on Tenebrionidae from Taiwan and Japan, II. (Coleoptera). *Entomological Review of Japan* 33: 67–73.

Shibata, T. 1980: Notes on Tenebrionidae from Taiwan and Japan, III. (Coleoptera). *Entomological Review of Japan* 34: 63–74.

Shiyake, S. 1996: Redescription of Falsomordellistena auromaculata (Kôno) with description of an allied new species from Japan (Coleoptera: Mordellidae). *Bulletin of the Osaka Museum of Natural History* 50: 9–15.

Shiyake, S. 2000: Mordellidae from East Asia with description of a new species. *Bulletin of the Osaka Museum of Natural History* 54: 25–30.

Shiyake, S. 2001: A new species of Glipostenoda from Shikoku, Japan. *Special Publication of the Japan Coleopterologal Society, Osaka* 1: 333–336.

Shuckard, W.E. 1838: Description of some new genera of Coleoptera in the author's collection. *The Entomological Magazine* 5 (5): 505–513 pp., pl. xviii.

Shuckard, W.E. 1839: *Elements of British Entomology. Containing a general introduction to the science, a systematic description of all the genera, and a list of all the species of British insects, with a history of their transformation, habits, economy, and distribution, with outline figures of the families and their larvae and pupae, an explanation of the technical terms, and full directions. Vol. 1.* London: Hippolyte Baillière. 240 pp.

Shuckard, W.E. 1840: *The British Coleoptera delineated, consisting of figures of all the genera of British Beetles, drawn in outline by W. Spry, M.E.S.* London, 83 pp., 94 pls.

Silfverberg, H. 1984: The Coleopteran genera of Dejean 1821. II. Polyphaga 1. *Annales Entomologici Fennici* 50: 58–60.

Silfverberg, H. 1991: Nomenclatorial corrections in North European Coleoptera. *Entomologica Fennica* 2: 21–22.

Silvestri, F. 1906: Descrizione di uno nuovo genere di Rhipiphoridae. *Redia* 3: 315–324, pl. 20.

Skopin, N.G. 1951: Novye i maloizvetnye vidy saranchovykh i zhestkokrylykh iz pustini Betpak-Dala. *Uchenye Zapiski Kazaskhskogo Gos. Universiteta* (Alma Ata) **13** (1): 121–126.

Skopin, N.G. 1958: Novyy vid zhukov-chernotelok triby Platyscelini iz yugo-vostochnogo Kazakhastana. *Trudy Instituta Zoologii Akademii Nauk Kazakhskoy SSR* **8**: 146–150.

Skopin, N.G. 1960a: Neue Tenebrioniden aus Zentralasien. I. *Annales Historico-Naturales Musei Nationalis Hungarici* **52**: 295–311.

Skopin, N.G. 1960b. Materialy po morfologii i ekologii lichinok triby Blaptini (Coleoptera, Tenebrionidae). *Trudy Instituta zoologii AN Kazakhskoy SSR,* **11**: 36–71.

Skopin, N.G. 1961a: Materialy po faune i ekologii chernotelok (Coleoptera, Tenebrionidae) yugo-vostochnogo Kazakhstana. *Trudy Nauchno-Isledovatelskogo Instituta Zashchtity Rasteniy Kazakhstanskoy Akademii Selskokhozyastvennykh Nauk* **6**: 172–208.

Skopin, N.G. 1961b: Neue Tenebrioniden aus Zentralasien II, nebst einigen systematischen sowie synonymischen Bemerkungen. *Annales Historico-Naturales Musei Nationalis Hungarici* **53**: 381–406.

Skopin, N.G. 1962: Lichinki podsemeystva Pimeliinae (Coleoptera, Tenebrionidae). *Trudy Nauchno-Issledovatelskogo Instituta Zashchity Rastenii Kazakhstanskoy Akademii Selskokhozyastvennykh Nauk* **7**: 191–298.

Skopin, N.G. 1964a: Materialy po faune i ekologii chernotelok (Coleoptera, Tenebrionidae) Mangyshlaka i severo-zapadnogo Ustyurta. *Trudy Nauchno-Issledovatelskogo Institute Zashchity Rastenii Kazakhstanskoy Akademii Selskokhozyastvennykh Nauk* **8**: 277–297.

Skopin, N.G. 1964b: Novye vidy chernotelok (Coleoptera, Tenebrionidae) iz smezhnykh s Kazakhtanom rayonov Centralnoy Azii. *Trudy Nauchno-Issledovatelskogo Institute Zashchity Rastenii Kazakhstanskoy Akademii Selskokhozyastvennykh Nauk* **8**: 371–388.

Skopin, N.G. 1964c: Neue Tenebrioniden aus Zentralasien. III, nebst einigen systematischen sowie synonymischen Bemerkungen. *Annales Historico-Naturales Musei Nationalis* **56**: 389–412.

Skopin, N.G. 1965: Zhuki-chernotelki (Coleoptera, Tenebrionidae) triby Platyscilini Kazakhstana.*Trudy Kazakhskogo Nauchno-Issledovatelskogo Instituta Zashchtity Rasteniy* **9**: 51–74.

Skopin, N.G. 1966a: Die Arten der Gattung Tentyria Latreille aus dem westlichen Zentralasien (Coleoptera, Tenebrionidae). *Entomologische Arbeiten aus dem Museum G. Frey* **17**: 1–6.

Skopin, N.G. 1966b: Neue Tenebrioniden aus Zentralasien IV., nebst einigen systematischen so wie synonymischen Bemerkungen. *Annales Historico-Naturales Musei Nationalis Hungarici* **58**: 325–342.

Skopin, N.G. 1966c: Eine neue Microdera-Art aus dem Balchashseegebiet (Coleoptera, Tenebrionidae). *Reichenbachia* **8**: 69–70.

Skopin, N.G. 1967: K poznaniyu zhukov-chernotelok roda Melanesthes Lac. (Coleoptera, Tenebrionidae) Centralno-Kazakhstanskoy Fauny. *Entomologicheskoe Obozrenie* **46** (1): 206–211.

Skopin, N.G. 1968a: Eine neue Tenebrionidengattung und – Art aus Zentralasien (Coleoptera-Tenebrionidae-Pimeliini). *Entomologische Arbeiten aus dem Museum G. Frey* **19**: 300–304.

Skopin, N.G. 1968b: Zhuki-chernotelki (Coleoptera, Tenebrionidae) Yuzhnogo Kazakhstana i ikh khozyaystvennoe znachenie. *Trudy Kazakhckogo Nauchno-Issledovatelskogo Institute Zashchity Rastenii* **10**: 73–114.

Skopin, N.G. 1969: Dopolnenie k faune zhukov-chernotelok (Coleoptera, Tenebrionidae) khrebeta Kutitang-Tay v Turkmenii. *Entomologicheskoe Obozrenie* **48** (4): 881–887.

Skopin, N.G. 1970: Revision der Tenebrioniden-Gattung Tagona Fisch.-W., 1822 (Coleoptera, Tenebrionidae). *Reichenbachia* **13**: 157–168.

Skopin, N.G. 1971a: Über einige Prosodes-Arten aus Mittelasien (Coleoptera: Tenebrionidae). *Beiträge zur Entomologie* **21**: 495–501.

Skopin, N.G. 1971b: Monographische Übersicht der Arten der Gattungen Diesia Fisch.-W. und Platyesia gen. nov. *Entomologische Abhandlungen des Staatlichen Museum für Tierkunde in Dresden* **38**: 325–349.

Skopin, N.G. 1972a: Zhuki-chernotelki (Coleoptera, Tenebrionidae) rodov Diesia Fisch.-W. i Platyesia, gen. n. *Entomologicheskoe Obozrenie* **51** (1): 114–124.

Skopin, N.G. 1972b: Revision der Tenebrioniden-Gattung Stalagmoptera Solsky, 1875 (Coleoptera). *Entomologische Arbeiten aus dem Museum G. Frey* **23**: 298–313.

Skopin, N.G. 1973a: Zhuki-chernotelki (Coleoptera, Tenebrionidae) kollekcii Balliona iz Kuldzhi i blizkikh rayonov Kazakhstana i Kirgizii. [Tenebrionid beetles (Coleoptera, Tenebrionidae) from Ballion's collection collected in Kuldzha and adjacent regions of Kazakhstan and Kirgizia]. *Entomologicheskoe Obozrenie* **52** (4): 866–875.

Skopin, N.G. 1973b: Revision der Tenebrioniden-Gattungsgruppe Trigonoscelis-Sternoplax (Coleoptera). *Entomologische Arbeiten aus dem Museum G. Frey* **24**: 104–185.

Skopin, N.G. 1974a: Zur Revision der eurasiatischen Arten der Gattung Belopus Gb. *Entomologische Abhandlungen des Staatlichen Museum für Tierkunde in Dresden* **40**: 65–103

Skopin, N.G. 1974b: Revision der Gattung Pterocoma Dejean; Solier, 1836 (Coleoptera Tenebrionidae). *Entomologische Abhandlungen des Staatlichen Museum für Tierkunde in Dresden* **40**: 127–164.

Skopin, N.G. 1974c: Nachtrag zur 'Revision der Tenebrioniden-Gattungsgruppe Trigonoscelis-Sternoplax (Coleoptera)'. *Entomologische Arbeiten aus dem Museum G. Frey* **25**: 259–260.

Skopin, N.G. 1977: Über Blaps transversinsulcata Ball. und ihre nähere Verwandten (Coleoptera, Tenebrionidae). *Annales Historico-Naturales Musei Nationalis Hungarici* **69**: 149–151.

Skopin, N.G. 1978: In: Kaszab Z., Schawaller W. & Skopin N.G.: Systematik und Ökologie einiger Tenebrioniden aus Kashmir und Ladakh. *Senckenbergiana Biologica* **59**: 215–234.

Skopin, N.G. 1979: Systematische Stellung der Gattung Scythis Schaum, 1865, sowie Revision der Arten (Coleoptera, Tenebrionidae). *Annales Historico-Naturales Musei Nationalis Hungarici* **71**: 169–183.

Skopin, N.G. & Kaszab, Z. 1978: Über die Arten der Gattung Blaps F. (Coleoptera, Tenebrionidae), gesammelt von Herrn Dr. W. Wittmer im Jahre 1976 in Kashmir. *Folia Entomologica Hungarica* (N.S.) **31**: 207–212.

Ślipiński, S.A. 1981a: Notes on the Oriental Colydiidae (Coleoptera). I. *Entomologica Basiliensia* **6**: 422–426.

Ślipiński, S.A. 1984: Insects of Saudi Arabia: Fam. Colydiidae. *Fauna of Saudi Arabia* **6**: 251–254.

Ślipiński, S.A. 1985: New and little known species of Colydiidae (Coleoptera) from Asia Madagascar and Comoro Islands. *Annales Zoologici* (Warszawa) **39**: 181–195.

Smith, B. 1851: *Nomenclature of coleopterous insects in the British Museum. Volume 5. Cucujidae.* London, iv + 25 pp.

Snellen van Vollenhoven, S.C. 1854: [new taxa]. In: Herklots, J.A.: *Bouwstoffen voor eene Fauna fon Niederland, twende deel, stück 1: 88 pp. Korte beschrijving der nieuwe soorten van Schildvleugelige Insecten in dezenaamlijst opgenoem.* Leiden: E.J. Brill, 70 p.

Soldati, F. 2007: A new species of the genus Asida Latreille, 1802 from French Southern Alpes (Insecta: Coleoptera: Tenebrionidae). *Annales Zoologici* **57**: 271–274.

Soldati, F. 2012a: A new species of the genus Probaticus Seidlitz, 1896 from Greece (Insecta: Coleoptera: Tenebrionidae). *Annales Zoologici* **62**: 221–225.

Soldati, F. 2012b: Une nouvelle espèce du genre Dendarus Dejean, 1821 décrite de Grèce (Coleoptera, Tenebrionidae). *Revue de l'Association Roussillonnaise d'Entomologie* (*RARE*) **21**: 52–59.

Soldati, F. 2016a: A new species of Asida Latreille, 1802 from the Northeastern Corsican Mountains (Insecta: Coleoptera: Tenebrionidae). *Annales Zoologici* **66**: 571–576.

Soldati, F. 2016b: Opatrum iwani sp. nov., espèce nouvelle du sud de la Corse (Coleoptera, Tenebrionidae) *Revue de l'Association Roussillonnaise d'Entomologie* (*RARE*) **25**: 1–10.

Soldati, F. 2019: [new taxon]. In: Soldati, F., Jansson, N., Avci, M., Atay, E., Coskun, M., Kayis, T. & Aytar, F.: A new species of Corticeus Piller & Mitterpacher from Turkey, with an updated key to turkish species belonging to this genus (Coleoptera: Tenebrionidae). *Annales Zoologici* **69**: 165–172.

Soldati, F. & François, A. 2015: A new species of Alphasida Escalera, 1905 from the Northeastern part of the Moroccan Middle Atlas (Insecta: Coleoptera: Tenebrionidae). *Annales Zoologici* **65**: 523–529.

Soldati, F. & Lemaire, J.M. 2010: A new species of the genus *Isomira* Mulsant, 1856 from Cyprus (Insecta: Coleoptera: Tenebrionidae). *Annales Zoologici* **60**: 305–307.

Soldati, F. & Leo, P. 2005: Revision of French and Corsican species of the genus Asida Latreille, 1802 (Insecta: Coleoptera: Tenebrionidae). *Annales Zoologici* **55**: 335–373.

Soldati, F. & Martínez Fernández, J.C. 2019: A new species of Alphasida Escalera, 1905 from South-Eastern Spain (Insecta: Coleoptera: Tenebrionidae). Annales Zoologici **69**: 641–644.

Soldati, F. & Rougon, D. 2017: Présence de Pimelia muricata Olivier, 1795, en Corse, description d'une nouvelle sous-espèce et découverte d'hybrides naturels avec P. angusticollis Solier, 1836 (Coleoptera, Tenebrionidae, Pimeliinae). *Bulletin de la Société entomologique de France* **122**: 401–408.

Soldati, F. & Soldati, L. 2000: A new species of Dendarus Latreille of the subgenus Pandarus Mulsant (Coleoptera: Tenebrionidae: Dendarini). *Bulletin de la Société Linnéenne de Bordeaux* **28**: 67–71.

Soldati, F. & Soldati, L. 2001a: Asida (Polasida) poneli n. sp., nouvelle espèce de la Riviera, et remarques taxonomiques sur les Polasida de France et d'Italie (Coleoptera, Tenebrionidae, Asidini). *Bulletin de la Société Linnéenne de Bordeaux* **29**: 103–108.

Soldati, F. & Soldati, L. 2001b: Une nouvelle Asida de Corse, Asida ascoense n. sp. (Coleoptera: Tenebrionidae: Asidini). *Bulletin de la Société Linnéenne de Bordeaux* **29**: 179–184.

Soldati, F. & Soldati, L. 2001c: Le genre Asida en Grèce (Coleoptera, Tenebrionidae, Asidini). *L'Entomologiste* **57**: 185–198.

Soldati, F. & Soldati, L. 2002a: Un nouveau Stenohelops du sud-est de l'Espagne (Coleoptera, Tenebrionidae, Helopini). (31. Contribution à la connaissance des Tenebrionidae). *Bulletin de la Société Linnéenne de Bordeaux* **30**: 137–142.

Soldati, F. & Soldati, L. 2002b: Compte-rendu d'une expedition scientifique en Grèce et description de nouveaux taxons (Coleoptera: Tenebrionidae). *Biocosme Mésogéen* **18** [2001]: 167–185.

Soldati, F. & Soldati, L. 2003a: Une espèce nouvelle de Phthora appartenant à un sous-genre inédit (Coleoptera, Tenebrionidae, Phalerini). (33.Contribution à la connaissance des Tenebrionidae). *Bulletin de la Société Linnéenne de Bordeaux* **31**: 1–8.

Soldati, F. & Soldati, L. 2003b: Une nouvelle espèce du genre Asida de l'île de Skyros, Grèce (Coleoptera, Tenebrionidae, Asidini). *Revue de l'Association Roussillonnaise d'Entomologie (RARE)* **12**: 43–45.

Soldati, F. & Soldati, L. 2014: Description de Corticeus vanmeeri n. sp., espèce nouvelle de Pyrénées occidentales françaises et clé de détermination des espèces du genre Corticeus Piller & Mitterpacher, 1783 en France. *Revue de l'Association Roussillonnaise d'Entomologie* **23**: 114–122.

Soldati, L. 2017 [new taxa]. In: Soldati L., Condamine F.L., Clamens A.-L. & Kergoat G.J.: Documenting tenebrionid diversity: progress on Blaps Fabricius (Coleoptera, Tenebrionidae, Tenebrioninae, Blaptini) systematics, with the description of five new species. *European Journal of Taxonomy* **282**: 1–29.

Soldati, L. & Soldati, F. 1999: A new Blaps species from Turkey (Coleoptera: Tenebrionidae). *Bulletin de la Société Linnéenne de Bordeaux* **27**: 29–32.

Soldati, L. 2009. *The Darkling Beetles (Coleoptera: Tenebrionidae) of Qatar.* Natura optima dux Foundation, Warszawa, 101 pp. + 17 plates.

Solier, A.J.J. 1834: Essai d'une division des coléoptères hétéromères, et d'une monographie de la famille des collaptèrides. *Annales de la Société Entomologique de France* **3**: 479–636, pls XII–XVI.

Solier, M. 1835a: Prodrome de la famille des xystropides. *Annales de la Société Entomologiqué de France* **4**: 229–248.

Solier, A.J.J. 1835b: Essai sur les collaptèrides (suite). 2e tribu. Tentyrites. *Annales de la Société Entomologique de France* **4**: 249–419, pls VI–IX.

Solier, A.J.J. 1835c: Essai sur les collaptèrides (suite). 3e Tribu. Macropodites. *Annales de la Société Entomologique de France* **4**: 509–574, pls XIV, XV.

Solier, A.J.J. 1836a: Essai sur les collaptèrides (suite). 4e Tribu. Pimélites. *Annales de la Société Entomologique de France* **5** [1835–1836]: 5–200, pls I–IV.

Solier, A.J.J. 1836b: Essai sur les collaptèrides (suite). 5e Tribu. Nyctélites. *Annales de la Société Entomologique de France* **5** [1835–1836]: 303–355, pls VI, VII.

Solier, A.J.J. 1836c: Essai sur les collaptèrides (suite). 6e Tribu. Asidites. *Annales de la Société Entomologique de France* **5** [1835–1836]: 403–512, pls XI–XIII.

Solier, A.J.J. 1836d: Essai sur les collaptèrides (suite). 7e Tribu. Akisites. *Annales de la Société Entomologique de France* **5** [1835–1836]: 635–684, pl. XXIII.

Solier, A.J.J. 1837: Essai sur les collaptèrides (suite). 8e Tribu. – Adélostomites. *Annales de la Société Entomologique de France* **6** [1836–1837]: 151–172, pl. VII.

Solier, A.J.J. 1838a: Essai sur les collaptèrides (suite). Deuxième division. *Annales de la Société Entomologique de France* **7**: 5–73, pls I–III.

Solier, A.J.J. 1838b: Essai sur les collaptèrides (suite). 10e Tribu. Scaurites. *Annales de la Société Entomologique de France* **8**: 159–199, pls VII, VIII.

Solier, A.J.J. 1838c: [new taxa]. In: Laporte F.L.N. de Caumont de Castelnau: *Histoire naturelle des insectes coléoptères; avec une introduction renfermant l'anatomie et la physiologie des animaux articulés, par M. Brullé. Tome deuxième.* Paris: P. Duménil, 563 + [1] pp., pls 20–37.

Solier, A.J.J. 1843: Essai sur les collaptèrides de la tribu des Molurites. *Memorie della Reale Accademia delle Scienze di Torino, Serie II*, **6**: 213–339, 4 pls.

Solier, A.J.J. 1848: Essai sur les collaptérides. 14e Tribu. Blapsites. Pp. 149–370, pls iv-xv. In: Baudi di Selve: *Studi Entomologici. Publ. Per di Flamino Baudi e di Eugenio Truqui. Tom. 1.* Torino: Stamporia Degli Artisti Tipografi, 376 pp.

Solier, A.J.J. 1851: Insectos. Coleopteros. Pp. 1–285. In: Gay, C.: *Historia Fisica y Politica de Chile. Fauna Chilena. Zoologie Tomo. 5.* Paris: Maulde et Renou, 563 pp.

Solsky [= Solskij] S.M. 1866: Matériaux pour servir à l'étude des insectes de la Russie. I. Notes sur quelques coléoptères nouveaux ou peu connus. *Horae Societatis Entomologicae Rossicae* **4** [1866–1867]: 79–96.

Solsky [= Solskij], S.M. 1871: Coléoptères de la Sibirie orientale. *Horae Societatis Entomologicae Rossicae* **7** [1870–1871]: 334–406.

Solsky [= Solskij], S.M. 1876: Matériaux pour l'entomographie des provinces asiatiques de la Russie. *Horae Societatis Entomologicae Rossicae* **11** [1875–1876]: 253–272, 273–299.

Solsky [= Solskij], S.M. 1881: Novyya ili maloizve'stnyya zhestkokrylyya okrain Rossiyskoy Imperii i prilegayushchikh k ney stran. *Trudy Russkago Entomologicheskago Obshchestva* **12** [1879–1881]: 231–265.

Solsky [= Solskij], S.M. 1882: Novyya ili maloizve'stnyya zhestkokrylyya okrain Rossiyskoy Imperii i prilegayushchikh k ney stran. *Trudy Russkago Entomologicheskago Obshchestva* **13** [1881–1882]: 31–84.

Sparacio, I. 2007: Nuovi coleotteri di Sicilia (Coleoptera Carabidae e Tenebrionidae). *Il Naturalista siciliano* **31**: 249–259.

Spilman, T.J. 1954: Generic names of the Salpingidae and their type species (Coleoptera). *Journal of the Washington Academy of Sciences* **44** (3): 85–94.

Spilman, T.J. 1973: Nomenclatural problems in six genera of Tenebrionidae (Coleoptera). *Proceedings of the Entomological Society of Washington* **75**: 39–44.

Starck, A. 1889: Coleoptera nova Imperii Rossici. *Wiener Entomologische Zeitung* **8**: 53–55.

Stchegoleva-Barovskaya, T. 1927: De mordellidis (Coleoptera) palaearcticis tribus novis. *Annuaire du Musée Zoologique de l'Académie des Sciences de l'URSS* **28**: 158–161.

Stchegoleva-Barovskaya, T. 1930a: De duabus novis mordellidarum speciebus e ponticis. *Revue Russe d'Entomologie* **24**: 56–58.

Stchegoleva-Barovskaya, T. 1930b: De mordellidis (Coleoptera) novis in collectione Musei Zoologici Academiae Scientiarum. *Comptes Rendus de l'Académie des Sciences de l'URSS* **1930**: 750–752.

Stchegoleva-Barovskaya, T. 1931: Aperçu des coléoptères de la fam. Mordellidae de la région nord-ouest de l'URSS (Ingrie). *Annuaire du Musée Zoologique de l'Académie des Sciences de l'URSS* **32**: 51–65.

Stchegoleva-Barovskaya, T. 1932: Nouvelles espèces asiatiques de la fam. Mordellidae. *Comptes Rendus de l'Académie des Sciences de l'URSS* **1932**: 219–222.

Stephens, J.F. 1829a: *The nomenclature of British insects; being a compendious list of such species as are contained in the systematic catalogue of the British insects, and forming a guide to their classification, &c. &c.* London: Baldwin & Craddock, (2) + 68 pp.

Stephens, J.F. 1829b: *A systematic catalogue of British insects: being an attempt to arrange all the hitherto discovered indigenous insects in accordance with their natural affinities. Containing also the references to every English writer on entomology, and to the principal foreign authors. With all the published British genera to the present time.* London: Baldwin & Cradock, xxxiv + [2]+ 416 + 388 pp.

Stephens, J.F. 1830: *Illustrations of British Entomology; or, a synopsis of indigenous insects: containing their generic and specific distinctions; with an account of their metamorphoses, times of appearance, localities, food, and economy, as far as practicable. Mandibulata. Volume III.* London: Baldwin and Craddock, 447 pp., pls. XVI–XIX.

Stephens, J.F. 1831: *Illustrations of British Entomology; or, a synopsis of indigenous insects. containing their generic and specific distinctions; with an account of their metamorphoses, times of appearance, localities, food, and economy, as far as practicable. Mandibulata. Volume IV.* London: Baldwin & Cradock, 413 + [1] pp., pls XX–XXIII. [issued in parts, pp. 1–366 in 1831; pp. 367–413 in 1832].

Stephens, J.F. 1832: Pp. 1–240, pls. XXIV–XXVI. In: *Illustrations of British Entomology; or, a synopsis of indigenous insects: containing their generic and specific distinctions; with an account of their metamorphoses, times of appearance, localities, food, and economy, as far as practicable. Mandibulata* [1832–1835] *Volume V.* London: Baldwin and Cradock, 447 pp. + list of illustrations, pls. XXIV–XXVII. [issued in parts: pp. 1–240, 1832; pp. 241–304, 1833; pp. 305–368, 1934; pp. 369–448, 1835].

Stephens, J.F. 1835: Pp. 369–448. In: *Illustrations of British Entomology; or, a synopsis of indigenous insects containing their generic and specific distinctions; with an account of their metamorphoses, times of appearance, localities, food, and economy, as far as practicable. Mandibulata. Volume V.* London: Baldwin and Cradock, 447 pp. + list of illustrations, pls. XXIV–XXVII. [issued in parts: pp. 1–240, 1832; pp. 241–304, 1833; pp. 305–368, 1934; pp. 369–448, 1835].

Stephens, J.F. 1839: *Manual of British Coleoptera, or beetles; containing a brief description of all the species of beetles hitherto ascertained to inhabit Great Britain and Ireland; together with a notice of their chief localities, times and places of appearances etc.* London: Longman, Orme Brown, Green and Longmans, xii + 443 pp.

Steven, C. von. 1806: Déscription de dix insectes de la Russie méridionale. *Mémoires de la Société des Naturalistes de l'Université Impériale de Moscou* 1: 155–167, pl. 10.

Steven, C. von. 1811: Description de dix insectes de la Russie méridionale. *Mémoires de la Société des Naturalistes de l'Université Impériale de Moscou* 1: 118–130, pl. 10.

Steven, C. von. 1828: Tentyriae et Opatra collectionis Stevenianae nunc Musei Universitatis Mosquensis. *Nouveaux Mémoires de la Société Impériale des Naturalistes de Moscou* 1: 83–100.

Stierlin, G. 1867: Beschreibung einiger neuen Käfer-Arten. *Mitteilungen der Schweizerischen Entomologischen Gesellschaft* 2: 218–228.

Stierlin, G. 1876: Beschreibung einiger neuer Käferarten. *Mitteilungen der Schweizerischen Entomologischen Gesellschaft* 4: 473–493.

Stierlin, G. 1902: Beschreibung einiger neuen Species von Coleopteren. *Mitteilungen der Schweizerischen Entomologischen Gesellschaft* **10**: 425–427.

Stolz, H. 1915: Eine neue Oedemera von der Insel Kreta. *Coleopterologische Rundschau* 4: 91–93.

Strand, A. 1965: Über die nordischen Arten der Gattung Cis Latr., Untergattung Eridaulus Thoms., mit Beschreibung einer neuen Art, Cis hanseni n. sp. (Col., Cisidae). *Norsk Entomologisk Tidsskrift* 13: 61–65.

Strand, E. 1917: Übersicht der in Gistel's 'Achthundert und zwanzig neue oder unbeschriebene wirbellose Thiere' (1857) behandelten Insekten. *Archiv für Naturgeschichte* (A) 82 (5) [1916]: 75–101.

Strand, E. 1929: Zoological and palaeontological nomenclatorial notes. *Acta Universitatis Latviensis* **20**: 1–29.

Strand, E. 1935: Revision von Gattungsnamen palaearktischer Coleoptera. *Folia Zoologica et Hydrobiologica* 7: 282–299.

Strassen, R. zur. 1957: Zur Kenntnis der Arten-Gruppe Akis spinosa Linnaeus, genei Solier und granulifera Sahlberg (Ins. Col. Tenebr.). *Senckenbergiana Biologica* **38**: 41–59.

Sturm, J. 1807: *Deutschlands Fauna in Abbildungen nach der Natur mit Beschreibungen. V. Abtheilung. Die Insecten. Zweites Bändchen. Käfer.* Nürnberg: J. Sturm, 279 pp., pls. 21–52.

Sturm, J. 1826: *Catalog meiner Insecten-Sammlung. Erster Theil. Käfer.* Nürnberg: Verfasser, viii +207pp., 4 pls.

Sturm, J. 1843: *Catalog der Käfersammlung von J. Sturm.* Nürnberg: Verfasser, 12 + 386 pp.

Sturm, J.H.C.F. 1849: *Dr. Jacob Sturm's Deutschlands Fauna in Abbildungen nach der Natur mit Beschreibungen. V Abtheilung. Die Insekten. Zwanzigstes Bändchen. Käfer.* Nürnberg: Author, vi + 103 pp. pls. 361–376.

Suffrian, C.W.L.E. 1848: Entomologische Bemerkungen. *Entomologische Zeitung* (Stettin) **9**: 98–108, 165–171.

Suffrian, C.W.L.E. 1853: Synonymische Miscellaneen. *Entomologische Zeitschrift* (Stettin) **13**: 232–237.

Sulzer, J.H. 1761: *Die Kennzeichen der Insecten.* Zürich: Heiddeger, 28 + 204 + 67 pp., 24 pls.

Sulzer, J.H. 1776: *Abgekürzte Geschichte der Insecten nach dem Linneischen System.* Winterthur: H. Steiner & Co., Teil 1: 27 + 274 pp.; Teil 2: 71 pp., 32 pls.

Sumakov, G.G. 1902: Materialy dlya fauny zhestokrylykh Zakaspitskoy oblasti. Beiträge zur Fauna der Coleopteren des Transkaspischen Gebietes. *Protokoly Obshchestva Estestvoispytateley pri Imperatskom Yuryevskom Universitete* **13** (1) [1901]: 21–48.

Sumakov, G.G. 1908a: Novyy vid roda Steropes Stev. (Coleoptera, Anthicidae). *Revue Russe d'Entomologie* **8**: 133–134.

Sumakov, G.G. 1908b: Novye sredneaziatskie zhestkokrylye (Col. Tenebrionidae i Carabidae). *Russkoe Entomologicheskoe Obozrenie* **8**: 245–247.

Sumakov, G.G. 1909: Beiträge zur Fauna der Coleopteren von Turkestan und dem transkaspischen Gebiet. *Sitzungsberichte der Naturforschenden Gesellschaft, Universität Dorpat* **17**: 209–224.

Sumakov, G.G. 1913a: Nouvelles variations de dessin sur les élytres des espèces asiatiques de Mylabris F. (Coleoptera, Meloidae). *Revue Russe d'Entomologie* **13**: 304–307.

Sumakov, G.G. 1913b: Deux nouvelles espèces du genre Mylabris F. (Coleoptera Meloidae). *Revue Russe d'Entomologie* **13**: 473–475.

Sumakov, G.G. 1915: Les espèces paléarctiques du genre Mylabris Fabr. (Coleoptera, Meloidae). *Horae Societatis Entomologicae Rossicae* **42**: 1–71.

Sumakov, G.G. 1924: Corrections synonymiques sur quelques espèces du genre Mylabris F. *Sitzungsberichte der Naturforschenden Gesellschaft, Universität Dorpat* **31**: 80–90.

Sumakov, G.G. 1929a: Nouvelles espèces, sous-espèces et morphes paléarctiques du genre Mylabris F. et quelques corrections synonymiques et systématiques. *L'Échange, Revue Linnéenne* **45** [hors texte] (435): 1–7.

Sumakov, G.G. 1929b: Diagnosis einer neuen Art der Gattung Zonitis aus dem Kaukasus. *Entomologisches Nachrichtenblatt* **3**: 113–114.

Sumakov, G.G. 1930a: Catalogue des espèces paléarctiques de tribu Mylabrina (Coleoptera, Meloidae). *Acta Instituti et Musei Zoologici Universitatis Tartuensis* **37**: 1–114.

Sumakov, G.G. 1930b: [new name]. In: Pic, M.: Annonces et avis divers. *L'Échange, Revue Linnéenne* 46: [9].

Sumakov, G.G. 1934: Inventa entomologica itineris Hispanici et Maroccani quod a 1926 fecerunt Harald et Håkan Lindberg. XXI Méloides du basin Méditérranéen et du Maroc. *Commentationes Biologicae* **4** (13): 1–6.

Sundevall, J.C. 1831: Beschreibung einer neuen Coleopteren-Gattung, Symbius Blattarum. *Isis von Oken* 11: 1222–1228, pl. VIII.

Švihla, V. 1978: Two new Oedemera species from Europe (Coleoptera, Oedemeridae). *Acta Entomologica Bohemoslovaca* 78: 35–41.

Švihla, V. 1980: Ergebnisse der Bhutan-Expedition des Naturhistorischen Museums in Basel. Coleoptera Fam. Oedemeridae. *Entomologica Basiliensia* 5: 45–58.

Švihla, V. 1982: New species of the family Oedemeridae (Coleoptera) from the Palaearctic Region. *Entomologica Basiliensia* 7: 381–386.

Švihla, V. 1983a: Results of the Czechoslovak-Iranian entomological expeditions to Iran 1970, 1973 and 1977. Coleoptera, Oedemeridae. *Acta Entomologica Musei Nationalis Pragae* 41: 107–132.

Švihla, V. 1983b: Contribution to the knowledge of the palaearctic Oedemeridae (Coleoptera). *Entomologica Basiliensia* 8: 334–341.

Švihla, V. 1984: Insects of Saudi Arabia. Coleoptera: Fam. Oedemeridae (Part 2). *Fauna of Saudi Arabia* 6: 260–266.

Švihla, V. 1986a: Revision of the generic classification of the Old World Oedemeridae (Coleoptera). *Sborník Národního Muzea v Praze* 41 B: 141–238.

Švihla, V. 1986b: Revision of Ischnomera species from Central Asia and the western Himalayas (Coleoptera, Oedemeridae). *Acta Entomologica Bohemoslovaca* 83: 213–220.

Švihla, V. 1987: Contribution to the knowledge of the Old World Oedemeridae (Coleoptera). *Annotationes Zoologicae et Botanicae* (Bratislava) 181: 1–27.

Švihla, V. 1988a: Revision of Ischnomera species from the western Palaearctic (Coleoptera, Oedemeridae). *Acta Entomologica Bohemoslovaca* 85: 205–222.

Švihla, V. 1988b: Review of the family Oedemeridae (Coleoptera) from Macaronesia. *Acta Entomologica Bohemoslovaca* 85: 373–379.

Švihla, V. 1991: Contribution to the knowledge of the Old World Oedemeridae (Coleoptera). *Annotationes Zoologicae et Botanicae* (Bratislava) 202: 1–14.

Švihla, V. 1992: Revision of Ischnomera species from the eastern Palaearctic (Coleoptera, Oedemeridae). *Acta Entomologica Bohemoslovaca* 89: 35–46.

Švihla, V. 1993a: Contribution to the knowledge of the Palaearctic Oedemeridae (Coleoptera). *Folia Heyrovskyana* 1: 52–55.

Švihla, V. 1993b: Revision of the genus Oedemera subgenus Oncomera (Coleoptera: Oedemeridae). *European Journal of Entomology* 90: 189–208.

Švihla, V. 1994: Revision of the subgenus Dryopomera s. str. (Coleoptera: Oedemeridae). *European Journal of Entomology* 91: 237–254.

Švihla, V. 1995: Contribution to the knowledge of the family Oedemeridae (Coleoptera) of arid regions of the Palaearctic. *Folia Heyrovskyana* 3: 9–23.

Švihla, V. 1996: Contribution to the knowledge of the Old World Oedemeridae (Coleoptera). *Acta Musei Nationalis Pragae, Series B, Historia Naturalis* 51: 83–88.

Švihla, V. 1997a: Contribution to the knowledge of the Palaearctic and Oriental Oedemeridae (Coleoptera). *Folia Heyrovskyana* 5: 83–96.

Švihla, V. 1997b: Revision of the genus Ascleropsis Seidlitz and related genera (Coleoptera, Oedemeridae). *Entomologica Basiliensia* **20**: 417–466.

Švihla, V. 1998a: Revision of Nacerdes subgenera Xanthochroa and Asiochroa subgen. n. (Coleoptera: Oedemeridae) from western and southern China and adjacent regions. *Folia Heyrovskyana* **6**: 49–71.

Švihla, V. 1998b: Revision of the genera Ascleranoncodes, Pseudonerdanus and Falsonerdanus (Coleoptera: Oedemeridae). *Folia Heyrovskyana* **6**: 147–166.

Švihla, V. 1998c: Notes on some species of the genus Ischnomera (Coleoptera: Oedemeridae) from Turkey. *Klapalekiana* **34**: 239–241.

Švihla, V. 1999a: Revision of the Old World Diplectrus with notes on the other genera (Coleoptera: Oedemeridae). *Folia Heyrovskyana* **7**: 73–86.

Švihla, V. 1999b: Revision of the subgenera Stenaxis and Oedemera s. str. of the genus Oedemera (Coleoptera: Oedemeridae). *Folia Heyrovskyana*, Supplementum **4**: 1–117.

Švihla, V. 2001a: New taxa of the Old World Ditylini (Coleoptera: Oedemeridae). *Folia Heyrovskyana* **9**: 89–98.

Švihla, V. 2001b: Contribution to the knowledge of the genus Dryopomera (Coleoptera: Oedemeridae). *Acta Societatis Zoologicae Bohemicae* **65**: 127–134.

Švihla, V. 2001c: Revision of the Nacerdes subgenus Xanthochroa (Coleoptera: Oedemeridae) from Malaysia and Indonesia with notes on species from other regions. *Folia Heyrovskyana* **9**: 217–234.

Švihla, V. 2002: New species of the genus Indasclera Švihla, 1980 (Insecta: Coleoptera: Oedemeridae). *Reichenbachia* **34**: 333–339.

Švihla, V. 2003: New taxa of the genus Oedemera (Coleoptera, Oedemeridae) from China. *Satonius, Special Bulletin of the Japanese Society of Coleopterology, Tokyo* **6**: 339–344.

Švihla, V. 2004: Contribution to the knowledge of the Old World Oedemeridae (Coleoptera). *Acta Societatis Zoologicae Bohemiae* **68**: 61–78.

Švihla, V. 2005a: On the taxonomy and nomenclature of Stenostoma rostratum (Coleoptera: Oedemeridae). *Časopis Národního Muzea, Řada Přírodovědná* **174**: 71–73.

Švihla, V. 2005b: New species and synonymies in Oedemeridae (Coleoptera) from the Palaearctic and Oriental regions. *Folia Heyrovskyana* **13**: 71–80.

Švihla, V. 2006a: A revision of the geus Sparedrus (Coleoptera: Oedemeridae) from central part of the Palaearctic Region and from western part of the Oriental Region. *Folia Heyrovskyana, Series A* **14**: 1–35.

Švihla, V. 2006b: New species of the genus Sparedrus (Coleoptera: Oedemeridae) from Bulgaria. *Acta Entomologica Musei Nationalis Pragae* **46**: 123–125.

Švihla, V. 2006c: Revision of Chitona species (Coleoptera: Oedemeridae) from eastern Mediterranean. *Acta Entomologica Musei Nationalis Pragae* **46**: 107–121.

Švihla, V. 2006d: On the taxonomy of the Palaearctic Oedemeridae (Coleoptera). *Časopis Národního Muzea v Praze, Řada Přírodovědná* **175**: 1–12.

Švihla, V. 2007: Preliminary revision of the genus Sparedrus (Coleoptera: Oedemeridae) from eastern and southeastern Asia. *Acta Entomologica Musei Nationalis Pragae* **47**: 153–168.

Švihla, V. 2008a: Oedemeridae. New nomenclatural and taxonomic acts, and comments. P. 45. In: Löbl, I. & A. Smetana (eds.): *Catalogue of Palaearctic Coleoptera. Vol. 5. Tenebrionoidea*. Stenstrup: Apollo Books, 670 pp.

Švihla, V. 2008b: Three new species of the Old World Oedemeridae (Coleoptera). *Zootaxa* **1912**: 59–65.

Švihla, V. 2008c: Order Coleoptera, family Oedemeridae. Pp. 264–269. In: Van Harten, A. (ed.): Arthropod fauna of the United Arab Emirates. Volume 1. Dar Al Ummah Printing, Publishing, Distribution & Advertising, Abu Dhabi, 754 pp.

Švihla, V. 2009a: A new genus of the tribe Asclerini (Coleoptera: Oedemeridae) from southeastern China. *Annales Zoologici* **59**: 545–548.

Švihla, V. 2009b: [new taxa]. In: Švihla, V. & Akiyama, H.: New species of the family Oedemeridae (Coleoptera) from south-eastern Asia. *Entomologica Basiliensia et Collectionis Frey* **31**: 89–99.

Švihla, V. 2012: A review of the family Oedemeridae (Coleoptera) of the Socotra Archipelago *Acta Entomologica Musei Nationalis Pragae* **52**, *Supplementum*: 337–346.

Szallies, A. 1994: Drei neue Coleopteren-Arten aus der Türkei (Coleoptera: Prostomidae, Cerambycidae). *Entomologische Zeitschrift* **104**: 259–263.

Tahami, M.S., Merkl, O. & Sadeghi, S. 2016: Leptodes of Iran, with description of six new cavernicolous species (Coleoptera: Tenebrionidae: Pimeliinae: Leptodini). *Annales Zoologici* **66**: 589–606.

Takakuwa, M. 1977: A new species of the genus Glipa LeConte from the Ryukyu Islands. *Elytra* **5**: 9–11.

Takakuwa, M. 1978: Occurence of a new species of the genus Yukahananomia Kôno (Mordellidae) in Central Honshu, Japan. *Elytra* **6**: 5–8.

Takakuwa, M. 1985: Notes on the tribe Mordellini from the Ogasawara Islands with descriptions of three new species. *Gekkan-Mushi* **176**: 4–11.

Takakuwa, M. 1985a: Two new species of the mordellid beetles from the Ryukyus. *Elytra* **13**: 71–74.

Takakuwa, M. 1986: The group of Glipa formosana (Coleoptera, Mordellidae) from Amami-Oshima Island of the Ryukyus. *Entomological Papers presented to Yoshihiko Kurosawa on the Occasion of his Retirement* **1986**: 257–263.

Takakuwa, M. 1986a: The group of Hoshihananomia hananomi (Coleoptera, Mordellidae) in Japan. *Papers on Entomology presented to Prof. Takehiko Nakane in Commemoration of his Retirement* **1986**: 237–244.

Takakuwa, M. 1993: A peculiar new species of the mordellid genus Glipa (Coleoptera) from Yakushima Island, Southwest Japan. *Bulletin of the Kanagawa Prefectural Museum* (*Nat. Sci.*) **22**: 91–94.

Takakuwa, M. 1995: A new species of the genus Yukahananomia (Coleoptera, Mordellidae) from Taiwan. *Special Bulletin of the Japannese Society of Coleopterology, Tokyo* **4**: 433–436.

Takakuwa, M. 1999: A new mordellid species allied to Glipa oshimana (Coleoptera, Mordellidae) from the Ryukyu Islands. *Elytra* **27**: 65–69.

Takakuwa, M. 2000: A taxonomic study of the mordellid subgenus Stenoglipa. *Bulletin Kanagawa Prefectural Museum* (*Nat. Sci.*) **29**: 53–105.

Takakuwa, M. 2001: A remarkable new species of the genus Mordella (Coleoptera: Mordellidae) from the Central Ryukyus. *Special Publication of the Japan Coleopterologal Society, Osaka* **1**: 293–296.

Takakuwa, M. 2001a: A beautiful new species of the genus Variimorda (Coleoptera, Mordellidae) from the Southern Ryukyus. *Elytra* **29**: 310–314.

Takakuwa, M. 2002: Glipa watanabeorum (Coleoptera, Mordellidae) a new mordellid beetle from Southern Ryukyus. *Special Bulletin of the Japannese Society of Coleopterology, Tokyo* **5**: 357–361.

Tan, J.J. 1958: On the genus Epicauta Redtb. of China (Meloidae, Coleoptera). *Acta Entomologica Sinica* **8**: 152–167 (in Chinese with English summary).

Tan, J.J. 1981: Coleoptera Meloidae. In: Zhonggue & al. (ed.): *The series of the comprehensive scientific expedition to the Qhinghai-Xizang Plateau. Insects of Xizang. Vol. 1.* Peking: Sciences Press, xi + 600 pp. (in Chinese with English summary).

Tan, J.J. 1988: [Meloidae]. Pp. 289–292. In: Climbing Expedition of Chinese Academy of Sciences (ed.): *Insects of Mt. Narjabarwa Region of Xizang.* Beijing: Science Press.

Tan, J.J. 1992. Meloidae. Pp. 575–579. In: Comprehensive Science Expedition of Chinese Academy of Sciences (ed.). *Insects of the Hengduan Mountains Region.* Vol. 1. , Bejing: Science Press.

Tan, J.J. & Deng, Z.X. 1993: On a new species of the genus Epicauta Dejean from the Yunnan Province, China (Coeleoptera: Meloidae). *Sinecologia* **10**: 155–156.

Tauscher, A.M. 1812a: Tentyriae ruthenicae descriptionibus iconibusque illustratae. *Mémoires de la Société Impériale des Naturalistes de Moscou* **3**: 22–42, 313–316, pls 2, 3.

Tauscher, A.M. 1812b: Enumeratio et descriptio insectorum e familia Cantharidarum quae in Russia observavit. *Memories de la Société Impériale des Naturalistes de Moscou* **3**: 129–164, 2 pls.

Telnov, D. 1998a: Anthicidae (Coleoptera) der Sammlung Sergej Kurbatov, mit Beschreibung von sechs neuen Arten aus der Orientalis. *Bulletin et Annales Société Royale Belge Entomologique* **134**: 81–100.

Telnov, D. 1998b: Anthicidae aus Nordvietnam, 2 (Insecta: Coleoptera). *Mitteilungen des Internationalen Entomologischen Vereins e.V.* **23**: 85–93

Telnov, D. 2000: Neue Anthicidae (Insecta: Coleoptera) aus der Sammlung des Naturkundemuseums Erfurt. *Veröffentlichungen des Naturkundemuseums Erfurt* **19**: 173–184.

Telnov, D. 2001: Anthicidae (Coleoptera) aus der Sammlung des Naturhistorischen Museums in Basel, Teil 1: Einleitung, Steropinae, Tomoderinae, Ischaliinae. *Entomologische Zeitschrift, Stuttgart* **111**: 117–123.

Telnov, D. 2002a: Vorläufige Auflistung der Anthicidae (Coleoptera) von Afghanistan mit Beschreibung einer neuen Art. *Latvijas Entomologs* **39**: 20–29.

Telnov, D. 2002b: Yunnanomonticola nanzhao – eine neue Gattung und Art der orientalischen Anthicini (Coleoptera: Anthicidae, Anthicini). *Entomologische Zeitschrift, Stuttgart* **112**: 47–50.

Telnov, D. 2003a: Neue Arten der Gattung Anthelephila Hope, 1833 (Coleoptera: Anthicidae). *Entomologische Zeitschrift* **113**: 130–137.

Telnov, D. 2003b: Trictenotomidae und Anthicidae (Insecta: Coleoptera) des Himalayas und angrenzender Regionen, Teil 1. Systematik, Faunistik, Zoogeographie, pp. 279–303. In: Hartmann, M. & Baumbach, H. (eds.): *Biodiversität und Naturausstattung im Himalaya.* Erfurt: Verein der Freunde und Förderer des Naturkundemuseums Erfurt e. v., 389 pp. + 16 pls.

Telnov, D. 2004: [new names]. In: Chandler, D.S., Nardi, G., & Telnov, D.: Nomenclatural notes on the Palaearctic Anthicidae (Coleoptera). *Mitteilungen der Internationalen Entomologischen Vereins e. v.* (Frankfurt am Main) **29**: 109–173.

Telnov, D. 2005: Anthicidae (Coleoptera) aus der Sammlung des Naturhistorischen Museums in Basel. Teil II: Bemerkenswerte und neue Notoxini, Endomiini und Anthicini (Anthicinae) aus der paläarktischen und orientalischen Region. *Entomologica Basiliensia et Collectionis Frey* **27**: 161–180.

Telnov, D. 2007a: Redefinition of Pseudoleptaleus Pic, 1900 (Coleoptera: Anthicidae, Anthicinae). *Entomologische Zeitschrift* (Stuttgart) **117**: 71–82.

Telnov, D. 2007b. Zur Kenntnis asiatischer Anthicidae, IV (Insecta: Coleoptera). *Mitteilungen des Internationalen Entomologischen Vereins e.V.* **32**: 89–105

Telnov, D. 2008: Order Coleoptera, family Anthicidae. Pp. 270–292. In: van Harten, A. (ed.) *Arthropod Fauna of the UAE, volume 1*. Dar Al Ummah Printing, Publishing, Distribution & Advertising, Abu Dhabi: 1–754.

Telnov, D. 2010a: Anthicidae von Afghanistan II (Insecta: Coleoptera). *Mitteilungen des Internationalen entomologischen Vereins* 35: 1–18.

Telnov, D. 2010b: Erster Vertreter der Gattung Anthicomorphus Lewis, 1895 (Coleoptera: Anthicidae) aus der Himalaya. *Mitteilungen des Internationalen Entomologischen Vereins* 35: 19–26.

Telnov, D. 2010c: Paläarktische und orientalische Stricticollis-Arten (Coleoptera: Anthicidae). *Latvijas Entomologs* 48: 17–32.

Telnov, D. 2014: Taxonomic revision of the genus Sapintus Casey, 1895 (Coleoptera: Anthicidae: Anthicinae) from the Indo-Australian transition zone, with remarks on some Oriental and Australian taxa. Pp. 253–344, pls 44–63. In: Telnov, D. (ed.) *Biodiversity, biogeography and nature conservation in Wallacea and New Guinea. Volume II*. Rīga: the Entomological Society of Latvia: 458 pp, 126 pls.

Telnov, D. 2016: Nomenclatural notes on Anthicidae and Pyrochroidae (Coleoptera). 5. *Latvijas Entomologs* 53: 41–88.

Telnov, D. 2018a: Revision of the Tomoderinae (Coleoptera: Anthicidae). Part II. Macrotomoderus Pic, 1901 species from China and Taiwan. *Annales Zoologici* 68: 463–492.

Telnov, D. 2018b: Nomenclatural notes on Anthicidae and Pyrochroidae (Coleoptera). 6. *Baltic Journal of Coleopterology* 18: 219–278.

Telnov, D. & Barlcay, M.V.L. 2019. Ischaliidae (Insecta: Coleoptera) in the collection of the Natural History Museum, London. *Annales Zoologici* 69: 785–804.

Tenenbaum, S. 1915: Neue Käferarten von den Balearen. *Bulletin de l'Académie des Sciences de Cracovie* 1914: 837–843, pls 63, 64.

Théry, A. 1892: [new taxon]. *Bulletin de la Société Entomologique de France* 1892: clxxx-clxxxi.

Théry, A. 1921: Les Stenosini du Maroc (Coléoptères Tenebrionidae). *Bulletin de la Société des Sciences Naturelles du Maroc* 1: 51–54.

Théry, A. 1924: Description d'un Stenosis nouveau du Maroc (Coléoptère Tenebrionidae). *Bulletin de la Société des Sciences Naturelles du Maroc* 4: 149–150.

Théry, A. 1931a: Notes d'entomologie marocaine et nord-africaine (Huitième note). *Bulletin de la Société des Sciences Naturelles du Maroc* 11: 26, pl. 1.

Théry, A. 1931b: Notes d'entomologie marocaine et nord-africaine (Dixième note). *Bulletin de la Société des Sciences Naturelles du Maroc* 12: 119–127.

Théry, A. 1932a: Notes d'entomologie marocaine et nord-africaine (Onzième note). *Bulletin de la Société des Sciences Naturelles du Maroc* 12 [1931]: 189–198.

Théry, A. 1932b: Notes d'entomologie marocaine et nord-africaine (Douxième note). *Bulletin de la Société des Sciences Naturelles du Maroc* 13: 129–138.

Théry, A. 1934: Notes d'entomologie marocaine et nord-africaine (Treizième note). *Bulletin de la Société des Sciences Naturelles du Maroc* 14: 47.

Thomas, M.K. 1897a: Note on Deridea Westwood (Lyttidae) with the description of a new species. *The Annals and Magazine of Natural History* (6) 19: 389–390.

Thomas, M.K. 1897b: Description of five new species of Mylabrina (Lyttidae) in the collection of the British Museum. *The Annals and Magazine of Natural History* (6) 19: 501–503.

Thomson, C.G. 1859: *Skandinaviens Coleoptera, synoptiskt bearbetade. 1. Tom.* Lund: Berlinska Boktryckeriet, 290 pp.

Thomson, C.G. 1863: *Skandinaviens Coleoptera, synoptiskt bearbetade. 5. Tom.* Lund: Lundbergska Boktryckeriet, 340 pp.

Thomson, C.G. 1864: *Skandinaviens Coleoptera, synoptiskt bearbetate. 6. Tom.* Lund: Lundbergska Boktryckeriet, 385 pp.

Thomson, C.G. 1868: *Skandinaviens Coleoptera, synoptiskt bearbetade. 10. Tom.* Lund: Lundbergska Boktryckeriet, 420 pp.

Thomson, C.G. 1886: [new taxa]. *Bulletin de la Société Entomologique de France* **1886**: ix-xi.

Thomson, J. 1858: Insectes. I. Ordre des Coléoptères. Pp. 30–343. In: Voyage au Gabon. Histoire naturelle des insectes et des arachnides recueillis pendant un voyage fait au Gabon en 1856 et en 1857 par M. Henry C. Deyrolle sous les auspices de MM. le Comte de Mniszech et Jamen Thomson, précédée de l'hisotire du voyage par J. Thomson; Arachnides par H. Lucas. *Archives Entomologiques* **2**: 1–469 + 14 pp., XIV pls.

Thomson, J. 1860a: *Musée scientifique ou recueil d'histoire naturelle.* [1860–1868]. Paris: J. Thomson, 96 pp., 9 pls.

Thomson, J. 1860a: *Musée scientifique ou recueil d'histoire naturelle.* [1860–1868]. Paris: J. Thomson, 96 pp., 9 pls.

Thomson, J. 1860b: Monographie de la famille des Monommides. *Annales de la Société Entomologique de France* (3) **8**: 5–38, pls 1–3.

Thunberg, C.P. 1784: Novae insectorum species descriptae. *Nova Acta Regiae Societatis Scientarum Upsaliensis* **4**: 1–28, pl. 1.

Thunberg, C.P. 1784: *Dissertatio entomologica novas insectorum species. Pars III.* Upsaliae: Johan Edman, 53–69 pp., 1 pl.

Thunberg, C.P. 1787: [new species]. In: Donation. Thunbergianae 1785. Continuat. II, pp. 43–58, 1 pl. In: Publico examini subjicit, Petrus A. Bjerkén. Cujus partem quartam. *D.D. Museum Naturalium Academiae Upsaliensis, parts 1–8.* Upsaliae: Joh. Edman, 106 pp., 1 pl.

Thunberg, C.P. 1791: *Dissertatio entomologica novas insectorum species. Pars VI. ResP.A.J. Lagus.* Upsaliae: Johan Edman, [4] + 108–130 pp., 1 pl.

Thunberg, C.P. 1806: Tvänne nya insect-slägten, Ptyocerus och Ripidius. *Kongliga Vetenskaps Academiens nya Handlingar* **27**: 1–6, 1 pl.

Thunberg, C.P. 1814: Beskifning på tvänne nya insect-slågten, Gnathocerus och Taumacera. *Kongliga Vetenskaps Academiens Handlingar* **1814**: 46–50, pl. IV.

Thunberg, C.P. 1821: *Dissertatio resp. A. Loffman. Opatrum Insecti genus,* Uppsala. **4**: 27–34.

Tian, Y., Ren, G.-D. & Li, Q. 2014: Taxonomy of the Nacerdes (Xanthochroa) carniolica species-group from China (Coleoptera, Oedemeridae, Nacerdini). *ZooKeys* **426**: 111–118.

Tokeji, M. 1953: Some notes on the mordelline-beetles from Japan and Saipan Island. *The Chûjô* **8**: 1–21.

Tournier, H. 1868: Coléoptères nouveaux d'Europe, d'Algérie et d'Egypte. *L'Abeille, Mémoires d'Entomologie* **5**: 141–148.

Tournier, H. 1874: Diagnoses de quelques coléoptères européens et circumeuropéens. *Comptes-Rendus des Séances de la Société Entomologique de Belgique* 1874: lxxxvi-lxxxix.

Tôyama, M. 1986: A new rhipiphorid Beetle (Coleoptera, Rhipiphoridae) from Malaysia. *Elytra* 14: 43–45.

Tôyama, M. 1987: A new rhipiphorid beetle (Coleoptera, Rhipiphoridae) from Japan. *Elytra* 15: 33–35.

Tôyama, M. & Hatayama, T. 1985: Occurence of the Rhipiphorid subfamily Hemirhipidiinae in Japan and Taiwan (Coleoptera, Rhipiphoridae). *Gekkan-Mushi* 176: 18–24.

Toyoshima, R. 1999: A new species of the genus Phloeotrya (Coleoptera, Melandryidae) from Central Japan. *Elytra* 27: 97–101.

Toyoshima, R. & Ishikawa, Y. 2000: A new Taiwanese species of the Tetratomidae and three new Japanese species of the Melandryidae (Coleoptera, Heteromera). *Elytra* 28: 173–191.

Trichas, A. 2008: The genus Dendarus Latreille, 1829 (Coleoptera, Tenebrionidae: Dendarini) in Greece (A systematic account of the genus with description of a new species and four new systematic combinations). *Advances in Arachnology and Developmental Biology*: 12: 419–464.

Trost, P. 1801: *Kleiner Beitrag zur Entomologie in einem Verzeichniss der Eichstettischen bekannten und neu entdecken Insecten mit Anmerkungen fuer Kenner und Liebhaber*. Erlangen: Palm, 71 pp.

Truqui, E. 1855: Anthicini Insulae Cypri et Syriae. *Memorie della Reale Accademia delle Scienze di Torino* (2) 16: 339–371, pl. 1.

Truqui, E. 1857: Generis Iphthimi characteres. *Entomologische Zeitung* (Stettin) 18: 92–94.

Tshernyshev, S.E. & Mordkovich, V.G. 2002: Novye nakhodki chernotelok roda Blaps Fabricius, 1775 (Coleoptera: Tenebrionidae) v Sibiri. *Russian Entomological Journal* 11: 383–385.

Tsuru, T. 2002: A new species of the genus Mordellina (Coleoptera, Mordellidae) from the southern Ryukyus. *Special Bulletin of the Japanese Society of Coleopterology* 5: 351–355.

Tsuru, T. 2004: A taxonomic study of the genus Mordellistenoda (Coleoptera, Mordellidae) from the Ryukyu Islands, southwest Japan, with description of a new species. *Elytra* 32: 405–413.

Tsuru, T. & Takakuwa, M. 2005: Two new subspecies of Variimorda miyarabi Nomura (Coleoptera, Mordellidae) from Yonaguni-jima Island of the southwesternmost Ryukyus and Taiwan. *Elytra* 33: 79–85.

Turco, F. & Bologna, M.A. 2011: Systematic revision of the genus Cerocoma Geoffroy, 1762 (Coleoptera: Meloidae: Cerocomini). *Zootaxa* 2853: 1–71.

Uhmann, G. 1978: Ein neuer Formicomus aus China (Coleopt., Anthicidae). *Entomologische Blätter* 74: 81–83.

Uhmann, G. 1980a: Eine bisher unbekannte Notoxus-Art aus dem Sudan, *Notoxus tiganii* nov. spec. (Coleoptera, Anthicidae). *Entomologische Blätter* 75: 138–140.

Uhmann, G. 1980b: Anthicidae aus Nepal, Kashmir und Ladakh (Insecta: Coleoptera) *Senckenbergiana Biologica* 60 [1979]: 163–169.

Uhmann, G. 1982: Anthicidae aus Nepal (Insecta: Coleoptera) *Senckenbergiana Biologica* 62 [1981]: 233–236.

Uhmann, G. 1983: Anthiciden der orientalischen Region (Coleoptera, Anthicidae). *Annales Historico-Naturales Musei Nationalis Hungarici* 75: 185–206.

Uhmann, G. 1988a: Die von Hans Mühle 1987 im Yemen gesammelten Anthicidae (Coleoptera). *Acta Coleopterologica* 4: 1–4.

Uhmann, G. 1988b: Die Anthicidae der Sammlung Johann Klapperich (Coleoptera). *Bonner Zoologische Beiträge* **39**: 395–407.

Uhmann, G. 1988c: Anthiciden der orientalischen Region aus dem Museum in Genf (Coleoptera, Anthicidae). *Revue Suisse de Zoologie* **94** [1987]: 687–701.

Uhmann, G. 1989: Anthicidae (Coleoptera) der orientalischen Region aus dem Naturhistorischen Museum in Genf, II. *Revue Suisse de Zoologie* **96**: 243–252.

Uhmann, G. 1990a: Weitere Anthiciden verschiedener Regionen aus dem Museum in Genf (Coleoptera, Anthicidae). *Revue Suisse de Zoologie* **97**: 139–151.

Uhmann, G. 1990b: Anthiciden aus Nepal III (Coleoptera). *Entomofauna* **11**: 333–347.

Uhmann, G. 1994a: Die von Rudolf Schuh in Asien gefundenen Anthiciden (Insecta, Coleoptera, Anthicidae). *Entomofauna* **15**: 405–413.

Uhmann, G. 1994b: Südostasiatische Anthiciden aus dem Naturhistorischen Museum in Genf, 4 (Coleoptera, Anthicidae). *Revue Suisse de Zoologie* **101**: 655–676.

Uhmann, G. 1995a: Anthicidae aus dem Sultanat von Oman (Coleoptera, Anthicidae). *Acta Coleopterologica* **11**: 38–40.

Uhmann, G. 1995b: Neue Anthicidae aus Irian Jaya und Thailand (Coleoptera). *Entomologische Zeitschrift* **24**: 492–496.

Uhmann, G. 1996: Anthiciden aus dem Naturhistorischen Museum in Wien (Coleoptera, Anthicidae). *Entomologische Blätter* **92**: 19–36.

Uhmann, G. 1997: Anthiciden aus China (Coleoptera, Anthicidae). *Acta Coleopterologica* **13**: 21–22.

Uhmann, G. 1998a: Anthicidae (Insecta: Coleoptera) from Saudi Arabia with the description of a new species. *Fauna of Arabia* **17**: 93–105.

Uhmann, G. 1998b: Eine neue Anthiciden-Art aus der Mongolei (Col., Anthicidae). *Entomologische Nachrichten und Berichte* **42**: 153–154.

Uhmann, G. 2004: A new Mecynotarsus from Pakistan (Coleoptera, Anthicidae). *Bonner Zoologische Beiträge* **52**: 1–2.

Uhmann, G. 2005: Eine eigenartige, neue Clavicomus-Art aus Japan (Coleoptera, Anthicidae). *Entomologische Blätter* **100** [2004]: 205–208.

Utsunomiya, Y. 2011: A new beautiful Strongylium from Guizhou, China (Tenebrionidae, Stenochiinae, Stenochiini). Pp. 313–316. In: Kawai, S. (ed.): *Masumushi. Special Publication of the Japanese Society of Scarabaeidology No. 1*. Tokyo: Japanese Society of Scarabaeidology, 345 pp.

Uyttenboogaart, D.L. 1929: Contributions to the knowledge of the fauna of the Canary Islands. IX. Description of new Tenebrionidae (Col.). *Tijdschrift voor Entomologie* **72**: 341–350.

Uyttenboogaart, D.L. 1930: Contribution to the knowledge of the fauna of the Canary-Island. Synopsis of the results of the collecting-excursions 1925–1927. *Tijdschrift voor Entomologie* **73**: 211–235.

Uyttenboogaart, D.L. 1933: [in Verslag]. *Tijdschrift voor Entomologie* **76**: xli-xlii.

Uyttenboogaart, D.L. 1934: Revision des Genus Tribolium (Col. Ten.). *Entomologische Blätter* **30**: 20–31.

Uyttenboogaart, D.L. 1937: Contributions to the knowledge of the fauna of the Canary Islands xix. *Tijdschrift voor Entomologie* **40**: 75–118.

Vauloger de Beaupré, M. 1895: Description de deux Notoxus nouveaux du nord de l'Afrique. *L'Abeille, Journal d'Entomologie* 28: 193.

Vauloger de Beaupré, M. 1897: Descriptions de deux erodiides nouveaux (Col.). *Bulletin de la Société Entomologique de France* 1897: 294.

Vauloger de Beaupré, M. 1897: Description de trois Erodius nouveaux du nord de l'Afrique (Col.). *Bulletin de la Société Entomologique de France* 1897: 307–308.

Vauloger de Beaupré, M. 1900: Contribution au catalogue des coléoptères du Nord de l'Afrique. *Annales de la Société Entomologique de France* 68 [1899]: 669–722.

Vázquez, X.A. 1988: A new species of Oedemeridae from the Iberian Peninsula. *Nouvelle Revue d'Entomologie* 5: 259–261.

Vázquez, X.A. 1989: El género Chrysanthia Schmidt en la Península Ibérica (Coleoptera, Oedemeridae). *Elytron* 3: 125–136.

Vázquez, X.A. 1991: Contribución al conociemiento de los Oedemeridae de Marruecos, con descripción de una nueva especie (Coleoptera). *Elytron* 5: 107–114.

Vázquez, X.A. 1993: New data on northern African Alloxantha Seidlitz with description of a new species (Coleoptera, Oedemeridae). *Nouvelle Revue d'Entomologie* 10: 127–132.

Vázquez, X.A. 2002: Una nueva especie de Alloxantha de las Canarias (Coleoptera: Oedemeridae). *Boletín de la Asociación Española de Entomología* 26: 23–29.

Vázquez, X.A. 2006: Una nueva subespecie de Nacerdes (Xanthochroa) raymondi (Mulsant & Godart 1860) de la isla de Mallorca (Coleoptera: Oedemeridae). *Bolletin de la Societat d'Història Natural de les Balears* 49: 39–44.

Viedma, M.G. de. 1959: Revisión del género Proctenius Reitt. (Coleoptera). *Eos, Revista Española de Entomología* 35: 257–266.

Viedma, M.G. de. 1962: El género Rhipidius Thunberg en España (Coleoptera). *Eos, Revista Española de Entomología* 38: 47–49.

Villa, A. & Villa, J.B. 1833: *Coleoptera Europae dupleta in Collectione Villa quae pro mutua commutatione offerri possunt*. Mediolani: pp. 1–36.

Villa, A. & Villa, J.B. 1835: *Coleopterorum species novae in catalogo dupletorum extantes diagnosibus, adumbrationibus atque observationibus illustratae*. Mediolani, pp. 32–36.

Villa, A. & Villa, J.B. 1838: *Alterum supplementum Coleopterorum Europae sive additio ad Catalogum et Supplementum I. Dupletorum collectionis Villa continens species alias, nunc pro mutua commutatione adhunc offerrendas; nec non aliquarum specierum emendationes et synonima quae in catalogo anni 1833 et supplemento anni 1835 oblatae fuerunt*. Mediolani, pp. 51–66.

Villers, C.J. de. 1789: *Caroli Linnaei entomologia, faunae Sueciae descriptionis aucta; DD. Scopoli, Geoffroy, de Geer, Fabricii, Schrank, &c. speciebus vel in systemate non enumeratis, vel nuperrime detectis, vel speciebus Galliae australis locupletana, generum specierumque rariorum iconibus ornata, curante & augente Carolo de Villiers, Acad. Lugd. Massil. Villa-Fr. Rhotom. necnom geometriae regio professore. Tomus primus*. Lugduni: Piestre et Delamolliere, xiv + [2] + 4 + 765 + [1] pp., 3 pls.

Viñolas, A. 1987: Révision de la Asida (Granulasida) ricoi (Martinez) (Col. Tenebrionidae). *Sessió Conjunta d'Entomologia Instució Catalana d'Historia Natural – Sociedad Catalana de Lepidopterologia* 5: 31–34.

Viñolas, A. 1990: Nueva ordenación de los géneros Phylan Stephens, 1832, y Micrositus Mulsant & Rey, 1854, de la tribu Dendarini (Coleoptera: Tenebrionidae). *VI. Sessió Conjunta d'Entomología Instució Catalana d'Historia Natural – Societat Catalana de Lepidopterologia* 6 [1989]: 53–68.

Viñolas, A. 1994: El género Pimelia Fabricius, 1775 en la Península Ibérica y Baleares, con nota sistemática sobre una especie de Canarias (Coleoptera, Tenebrionidae, Pimeliinae). *Sessió d'Entomología Instució Catalana d'Historia Natural* 8 [1993]: 125–140.

Viñolas, A. 1997a: Phylan (Litoboriolus) companyae sp. n. de Zamora, Península Ibérica (Coleoptera: Tenebrionidae, Dendarini). *Zapateri Revista Aragonesa de Entomología* 6 [1996]: 97–102.

Viñolas, A. 1997b: Pseudoseriscius munyozi sp. n. d'Almería, Península Ibèrica (Coleoptera: Tenebrionidae, Crypticini). *Butlletí de la Institució Catalana d'Historia Natural* 65: 97–104.

Viñolas, A. & Cartagena, M.C. 2003: *Revisión del género Phylan Stephens, 1837 (Coleoptera: Tenebrionidae: Dendarini)*. Entomological monographs of Arganio editio 1, Barcelona: 93 pp.

Viñolas, A. & Cartagena, M.C. 2005: *Fauna de Tenebrionidae de la Península Ibérica y Baleares*. Barcelona: Agrania editio, 428 pp.

Viñolas, A. & Muñoz-Batet, J. 2015: Una nueva especie del género Cis Latreille, 1796 de la comarca de Osona, Cataluña, España (Coleoptera: Tenebrionoidea: Ciidae). *Arquivos Entomolóxicos* 13 [2014]: 123–130.

Vít, S. 1986: Révision du genre Prostheca Wollaston (Coleoptera, Colydiidae). *Mittheilungen der Schweizerischen Entomologischen Gesellschaft* 59: 447–457.

Voet, J.E. 1806: *Catalogus systematicus Coleopterorum – Catalogue systématique des coléoptères – Systematische Naamlyst van dat geslacht van Insecten dat men Torren noemt. Volume 2*. La Haye: Bakhuysen, 82 + 85 + 87 pp.

Voigts, H. 1902: Zur Synonymie der Meloiden-Gattung Zonabris. *Wiener Entomologische Zeitung* 21: 177–178.

Vuillet, A. 1910: Description d'une nouvelle espèce du genre Autocrates Thoms. (Col. Trictenotomidae). *Bulletin de la Société Entomologique de France* 1910: 347–348.

Vuillet, A. 1912: Description d'une nouvelle espèce du genre Autocrates Thoms. *Insecta* 2: 297–299.

Wagner, G. 2003: Zwei neue Arten der Gattung Mesostena Eschscholtz, 1831 aus Jemen mit einer Bestimmungstabelle der Untergattung Mesostenopa Kraatz, 1865 für die arabische Halbinsel (Coleoptera: Tenebrionidae: Tentyriini). *Mitteilungen des Internationalen Entomologischen Vereins* 28: 111–125.

Wagner, G. 2005: Zwei neue Arten der Gattung Capnisiceps Chatanay, 1914 aus dem Oman (Coleoptera: Tenebrionidae: Tentyriini). *Mitteilungen des Internationalen Entomologischen Vereins* 30: 177–183.

Wagner, H. 1917: Ueber Isomira semiflava Küst. und icteropa Küst. (Col.). *Entomologische Mitteilungen* 6: 341–346.

Wagner, H. 1928: Beschreibung neuer Coleopteren der europäischen Fauna nebst kritischen Bemerkungen zu bekannten Arten. *Coleopterologisches Centralblatt* 3: 111–125.

Walker, F. 1837: Characters of two undescribed British Coleoptera. *The Entomologists Monthly Magazine* 4: 83.

Walker, F. 1858: Characters of some apparently undescribed Ceylon insects. *The Annals and Magazine of Natural History* (3) 2: 202–209, 280–286.

Walker, F. 1859a: Characters of some apparently undescribed Ceylon insects. *The Annals and Magazine of Natural History* (3) 3: 258–265.

Walker, F. 1859b: Characters of some apparently undescribed Ceylon insects. *The Annals and Magazine of Natural History* (3) 4: 217–224.

Walker, F. 1871: *List of Coleoptera collected by J.K. Lord, Esq. in Egypt, Arabia and near the African shore of the Red Sea. With characters of the undescribed species*. London: E.W. Janson, 19 pp.

Waltl, J. 1835: *Reise durch Tyrol, Oberitalien und Piemont nach dem südlichen Spanien*. Passau: Pustet'sche Buchhandlung, [8] + 247 pp. (Erster Theil) + 120 pp. (Zweiter Theil: Ueber die Thiere Andalusien).

Waltl, J. 1838: Beiträge zur Kenntniss der Coleopteren der Türken. *Isis Encyclopädische Zeitschrift vorzüglich für Naturgeschichte, vergleichende Anatomie und Physiologie, von Oken* [31] 1838 (6): 449–472.

Waltl, J. 1839: Käfer um Passau (Fortsetzung von Heft IV S. 273. 1838). *Isis Encyclopädische Zeitschrift vorzüglich für Naturgeschichte, vergleichende Anatomie und Physiologie, von Oken* [32] 1839 (3): 221–227.

Wang, H., Wang, X. & Ren, G.D. 2012: A new species in the genus Lytta Fabricius (Coleoptera: Meloidae) from China. *Entomotaxonomia* **2**: 375–377.

Wang, L.-F., Ren, G.-D. & Liu, C. 2012: Review of the Chinese species of Falsocamaria Pic with descriptions of two new species (Coleoptera: Tenebrionidae: Stenochiinae). *Acta Zoologica Academiae Scientiarum Hungaricae* **58**: 305–324.

Wańkowicz, J. 1867: Descriptions de trois coléoptères nouveaux trouvés en Lithuanie. *Annales de la Société Entomologique de France* (4) **7**: 249–254.

Wańkowicz, J. 1869: Notices entomologiques VI. Descriptions de nouvelles espèces de coléoptères. *Annales de la Société Entomologique de France* (4) **9**: 418–422.

Wasmann, E. 1896a: Neue Termitophilen und Termiten aus Indien. *Annali del Museo Civico di Storia Naturale di Genova* **36**: 613–630, pl. II.

Wasmann, E. 1896b: Neue Termitophilen und Termiten aus Indien. Nachtrag. *Annali del Museo Civico di Storia Naturale di Genova* **37** (2) [1896–1897]: 149–152.

Wasmann, E. 1897: Zoologische Ergebnisse einer von Dr. K. Escherich und Dr. L. Kathariner nach Central-Kleinasien unternommenen Reise. – Myrmecophilen. *Deutsche Entomlogische Zeitschrift* **1896**: 234–241.

Wasmann, E. 1899: Neue Termitophilen und Myrmecophilen aus Indien. *Deutsche Entomlogische Zeitschrift* **1899**: 145–169.

Waterhouse, C.O. 1871: On some black species of Cantharis with red heads and filiform antennae. *Transactions of the Entomological Society of London* (4) **3**: 405–408.

Waterhouse, C.O. 1874: Description of a new species of Heteromerous Coleoptera belonging to the genus Toxicum. *The Entomologist's Monthly Magazine* **11** [1874–5]: 126.

Waterhouse, C.O. 1875: Descriptions of some new coleopterous insects belonging to the genera Myodites and Rhipidius (Rhipiphoridae). *Cistula Entomologica* **1**: 369–371.

Waterhouse, O.C. 1877: Descriptions of new Coleoptera from various localities *The Entomologist's Monthly Magazine* **14** [1877–78]: 72–75.

Waterhouse, C.O. 1880: Description of new genus and species of Heteromerous Coleoptera. *The Annals and Magazine of Natural History* (5) **5**: 147–148.

Waterhouse, C.O. 1881: On the coleopterous insects collected by Prof. I. Bailey Balfour in the island of Socotra. *Proceedings of the Zoological Society of London* **1881**: 469–478, pl. XLIII.

Waterhouse, C.O. 1882: [new taxa]. *Proceedings of the Entomological Society of London* **1882**: iv-v.

Waterhouse, C.O. 1883: Descriptions of new species of Coleoptera belonging to the Rhipiphoridae. *Annals and Magazine of Natural History* (5) **11**: 279–281.

Waterhouse, C.O. 1884. On the Coleopterous insects collected by Mr. H.O. Forbes in the Timor-Laut Island. *Proceedings of the Zoological Society of London* **1884**: 213–219.

Waterhouse, C.O. 1886: A new genus of heteromerous Coleoptera allied to Notoxus. *The Annals and Magazine of Natural History* (5) **17**: 39.

Waterhouse, C.O. 1889: Coleoptera. Pp. 123–131, pl. 14. In: Aitchison, J.E.T.: On the Zoology of the Afghan Delimination Commission. *Transactions of the Linnean Society* (*Zoology*) (2) **5** [1888–1894]: 53–142, pls 6–14.

Waterhouse, C.O. 1894: Coleoptera (partim). Pp. 64–71. In: Walker, J.: A visit to Damma Island, East Indian Archipelago. *The Annals and Magazine of Natural History* (6) **14**: 49–71.

Weber, F. 1801: *Observationes entomologicae, continentes novorum quae condidit generum characteres, et nuper detectarum specierum descriptiones*. Kiliae: Bibliopolii Academici, xii + 116 +[1] pp.

Weber, F. & Obst, M.H. 1804: *Naturhistorische Reise durch einen Theil Schwedens.* Göttingen: Heinrich Dieterich, xii + 13–207 + [1] pp., 3 pls.

Wei, Z.-H. & Ren, G.-D. 2019a: Two new species of the genus Laena Dejean, 1821 (Coleoptera, Tenebrionidae: Laenini) from Jiangsu in China. *Entomological Review* **99**: 1056–1060.

Wei, Z.-H. & Ren, G.-D. 2019b: Eight new species of the genus Laena Dejean from China (Coleoptera: Tenebrionidae: Lagriinae). *Oriental Insects* **53**: 510–539.

Weise, E. 1974: Die Isomira-Arten (Col., Alleculidae) Mitteleuropas und des Mittelmeer-Raums. *Entomologische Blätter* **70**: 65–128.

Weise, J. 1877: [new taxa]. In Reitter, E.: Coleopterologische Ergebnisse einer Reise nach Südungarn und die Transsylvanischen Alpen. *Verhandlungen des Naturforschendes Vereines in Brünn* **15** [1876]: 3–30, pl. I.

Weise, J. 1878a: [new taxa]. In: Schneider, O. & Leder, H.: Beiträge zur Kenntniss der kaukasischen Käferfauna. *Verhandlungen des Naturforschendes Vereines in Brünn* **16** [1877]: 3–262.

Weise, J. 1878b: [Bemerkungen zur: Catalogi Coleopterorum Europae editio secunda, auctoribus Dr. J.P.E. Frdr. Stein et Jul. Weise. Berol. 1877]. *Deutsche Entomologische Zeitschrift* **22**: 189–195.

Weise, J. 1880a: Laena Heydeni Weise n. sp. *Entomologische Monatsblätter* **2**: 102–103.

Weise, J. 1880b: [new taxa]. In: Leder, H.: Beitrag zur kaukasischen Käfer-Fauna. *Verhandlungen der Kaiserlich-Königlichen Zoologisch-Botanischen Gesellschaft in Wien* **29** [1879]: 451–488.

Weise, J. 1883: [new taxa]. In: Heyden, L.F.J.D. von. & Weise, J.: Neue südost-europäische und klein-asiatische Käfer. *Deutsche Entomologische Zeitschrift* **27**: 310–314.

Wellman, F.C. 1910a: On the synonymy of the Meloidae (Coleoptera). *Deutsche Entomologische Zeitschrift* **54**: 22–26.

Wellman, F.C. 1910b: The generic and subgeneric types of the Lyttidae (Meloidae s. Cantharidae auct.), (Col.). *Canadian Entomologist* **41**: 389–396.

Wellman, F.C. 1912: New species of Lyttidae, with notes on described species (Col.). *Entomological News* **23**: 29–38.

Werner, F.G. 1965. Anthicidae. In: *Insects of Micronesia. Coleoptera* **16**: 255–269.

Westhoff, F. 1881: *Die Käfer Westfalens.* Bonn: Max Cohen & Sohn, xxiii + 323 pp.

Westhoff, F. 1882: Die Käfer Westfalens, II. Abtheilung. Pp. 141–323. *Verhandlungen des Naturhistorischen Vereins der Preussischen Rheinlande und Westfalens, Supplement* **38** [1881–1882]: xxviii + 323 pp.

Westwood, J.O. 1830: Observations upon the Notoxidae, a family of coleopterous insects, with characters of two new British genera separated therein. *Zoological Journal* **5** [1829]: 57–61, pl. 41.

Westwood, J.O. 1838: *An introduction to the modern classification of insects; founded on the natural habits and corresponding organisation of the different families. Vol. 1.* London: Longman, Orme, Brown, Green and Longmans: xii + 462 pp.

Westwood, J.O. 1839: Descriptions of insects figured in plates 9 and 10. In: Royle, J.F.: *Illustrations of the botany and other branches of the natural history of the Himalayan mountains, and of the flora of Cashmere. Volume 1.* London: W.H. Allen and Co., xxxviii–lvii pp., pls 9–10.

Westwood, J.O. 1841: Descriptions of some new genera of Australian heteromerous beetles in the collection of the Rev. F.W. Hope. *Arcana Entomologica* 1: 43–44.

Westwood, J.O. 1846: [new taxon]. In: Entomological society. *The Annals and Magazine of Natural History* 18: 353.

Westwood, J.O. 1851: Descriptions of three new genera of exotic Coleoptera. *Transactions of the Entomological Society of London, New Series* 1: 167–172, pl. XIII.

Westwood, J.O. 1883: Descriptions of some new exotic Coleoptera. *Tijdschrift voor Entomologie* 26 [1882–83]: 61–78, pl. III.

White, A. 1846: Insects. Pp. 1–24, pls. i-vi. In: J. Richardson and E. Gray (eds): *Zoology of the voyage of H.M.S. Erebus & Terror, under the command of Captain Sir James Clark Ross, R.N., F.R.S, during the years 1839 to 1843. Part XI. – April. Insects of New Zealand.* London: Longman, Brown, Green, and Longmans, 24 pp., 6 pls.

Wiedemann, C.R.W. 1819: Neue Käfer aus Bengalen und Java. *Zoologisches Magazin* 1 (3): 157–183.

Wiedemann, C.R.W. 1821: In: Wiedemann, C.R.W. & Germar, E.F.: Neue exotische Käfer. *Magazin der Entomologie* 4: 107–183.

Wiedemann, C.R.W. 1823: Zweihundert neue Käfer von Java, Bengalen und dem Vorgebirge der Guten Hoffnung. *Zoologisches Magazin* 2 (1): 3–133.

Wilke, S. 1921: Beiträge zur Systematik und geographischen Verbreitung ungeflügelten Tenebrioniden (Unterfam. Asidinae). *Archiv für Naturgeschichte* A 87: 248–312, pls I–IV.

Winkler, A. 1915: Ein neuer blinder Tenebrionide aus der Krim. *Wiener Entomologische Zeitung* 24: 331–335.

Winkler, A. 1927: Pp. 753–880. In: *Catalogus coleopterorum regionis palaearcticae.* Wien [1927–1932]: Albert Winkler, 1698 pp. [single column/page edition].

Wissmann, O.L. 1848: Entomologische Notizen. *Entomologische Zeitung* (Stettin) 9: 76–80.

Wittmer, W. 1936: Neue Koleopteren aus Aegypten und dem Sinaï aus der Sammlung des Museums 'Pietro Rossi', Duino. *Bulletin de la Société Royale Entomologique d'Egypte* 20: 16–18.

Wollaston, T.V. 1854: *Insecta Maderensia; being an account of the insects of the islands of the Madeiran group.* London: John van Voorst, xliii + 634 pp., 13 pls.

Wollaston, T.V. 1857a: *Catalogue of the coleopterous insects of Madeira in the Collection of the British Museum.* London: Taylor and Francis, xvi + 234 pp.

Wollaston, T.V. 1857b: Description of an uncharacterized British Anthicus. *The Zoologist: a Popular Miscellany of Natural History* 15: 5721–5722.

Wollaston, T.V. 1860a: On additions to the Madeiran Coleoptera. *The Annals and Magazine of Natural History* (3) 5: 252–267, 358–365.

Wollaston, T.V. 1860b: On additions to the Madeiran Coleoptera. *The Annals and Magazine of Natural History* (3) 6: 48–54.

Wollaston, T.V. 1861a: On certain Coleoptera from the Island of St. Vincent. *The Annals and Magazine of Natural History* (3) **7**: 197–206, 246–253.

Wollaston, T.V. 1861b: An additions to the Madeiran Coleoptera. *The Annals and Magazine of Natural History* (3) **8**: 99–111.

Wollaston, T.V. 1862a: On the Euphorbia-infesting Coleoptera of the Canary Islands. *Transactions of the Entomological Society of London* (3) **1**: 136–189.

Wollaston, T.V. 1862b: Notes on Tarphii; with the description of an allied genus. *Journal of Entomology* **1**: 371–387, pls. XVIII, XIX.

Wollaston, T.V. 1862c: Brief diagnostic characters of new Canarian Coleoptera. *The Annals and Magazine of Natural History* (3) **9**: 437–442.

Wollaston, T.V. 1862d: On additions to the Madeiran Coleoptera. *The Annals and Magazine of Natural History* (3) **10**: 287–293, 331–342.

Wollaston, T.V. 1864: *Catalogue of the Coleopterous insects of the Canaries in the Collection of the British Museum.* London: Taylor and Francis, xiii + 648 pp.

Wollaston, T.V. 1865: *Coleoptera Atlantidum, being an enumeration of the Coleopterous Insects of the Madeiras, Salvages, and Canaries.* London: John van Voorst, xlvii + 526 + 140 (Appendix) pp., 1 map.

Wollaston, T.V. 1867: *Coleoptera Hesperidum, being an enumeration of the Coleopterous insects of the Cape Verde Archipelago.* London: John van Voorst, xxxix + 285 pp.

Wollaston, T.V. 1871: On additions to the Atlantic Coleoptera. *Transactions of the Entomological Society of London* **1871**: 203–314.

Wollaston, T.V. 1873: On a new genus of Colydiidae, from Japan. *Transactions of the Entomological Society of London* **1873**: 1–4.

Wollaston, T.V. 1873: On new Coleoptera from Japan. *Entomologists Monthly Magazine* **10** [1873–1874]: 167–172.

Wollaston, T.V. 1874: On new Coleoptera from Japan. *Entomologists Monthly Magazine* **10** [1873–1874]: 200–203.

Wollaston, T.V. 1875: [new taxa]. In: Bates, H.W. & Wollaston, T.V.: Descriptions of new species referred to in Mr. Blackmore's notes on Coleoptera from South Morocco. *The Entomologist's Monthly Magazine* **11** [1874–1875]: 217–220.

Wood, T. 1884: A new species of Cis. *The Entomologist's Monthly Magazine* **21** [1884–1885]: 130–131, 147 [footnote].

Wu, W. & Huang, R.-X. 2005: [new taxa]. In: Wu, W., Fan, Z.-t. & Huang, R.-x.: Three new species of Platyope (Coleoptera: Tenebrionidae) from Xinjiang, China. *Entomotaxonomia* **27**: 283–288. (in Chinese with English summary).

Xu, J.-S. & Ren, G.-D. 2012: First record of the subgenus Augolesthus s. str. Motschulsky from China, with description of a new species (Coleoptera, Tenebrionidae, Cnodalonini). *Acta Zootaxonomica Sinica* **37**: 773–776.

Yamazaki, H. 1964: Plesiophthalmus nigrocyaneus Motschulsky and its allied species from Japan. *Entomological Review of Japan* **17**: 17–27.

Yamazaki, H. 1968: A new species of the genus Plesiophthalmus from the Ryukyu Isls (Coleoptera: Tenebrionidae). *Tohoku Konchu Kenkyu* **3**: 27–28.

Yamazaki, H. 2002: A new species of Plesiophthalmus (Coleoptera, Tenebrionidae) from Japan. *Special Bulletin of the Japanese Society of Coleopterology* 5: 377–379.

Yang, C. & Fan, X. 1988: A new species of Glipa from Fanjingshan mountain. *Entomological investigations Fanxingshan. Institute of Biology, Guizhou Academy of Science* 1988: 109–110 (in Chinese with English summary).

Yang, S.-L. & Guo, J. 2018: A new species of Catapiestus Perty, 1831 from China (Coleoptera, Tenebrionidae, Cnodalonini). *ZooKeys* 809: 49–54.

Yang, X.-J. & Ren, G.-D. 2004: A new species and twelve new records of the tribe Opatrini in China (Coleoptera, Tenebrionidae). *Acta Zootaxonomica Sinica* 29: 305–309.

Yang, X.-J. & Ren, G.-D. 2010: Three new species of Cteniopinus Seidlitz (Coleoptera, Tenebrionidae, Alleculinae) from Xizang, China. *Acta Zootaxonomica Sinica* 35: 57–62.

Yang, X.-J. & Ren, G.-D. 2011: A new species and a new name of the genus Cteniopinus Seidlitz from China (Coleoptera, Tenebrionidae, Alleculinae). *Acta Zootaxonomica Sinica* 36: 574–576.

Yang, Y.-X. & Ren, G.-D. 2006: A new species of the genus Epicauta from Yunnan (Coleoptera: Meloidae). *Entomotaxonomia* 28: 271–274 (in Chinese with English summary).

Yang, Y.-X & Ren, G.-D. 2007: A taxonomic study on Epicauta tibialis Group from China (Coleoptera, Meloidae). *Acta Zootaxonomia Sinica* 32: 711–715.

Yawata, H. 1944: Description d'une espèce nouvelle appartenant au genre Anthicus du Japon (Col. Anthicidae). *Transactions of the Kansai Entomological Society* 14: 1–2.

Young, D.K. 1996: Status of Pseudopyrochroa types in the Museum National d'Histoire Naturelle Paris (Coleoptera: Pyrochroidae: Pyrochroinae). *Oriental Insects* 30: 213–235.

Young, D.K. 2000: Five new species of Pseudopyrochroa (Coleoptera: Pyrochroidae: Pyrochroinae from Taiwan. *Oriental Insects* 34: 245–262.

Young, D.K. 2001: A new species of Pseudopyrochroa (Coleoptera: Pyrochroidae: Pyrochroinae) from China. *Oriental Insects* 35: 273–278.

Young, D.K. 2003: A new species of Pseudopyrochroa (Coleoptera: Pyrochroidae: Pyrochroinae) from China and Korea. *Special Bulletin of the Japanese Society of Coleopterology* (Tokyo) 6: 345–351.

Young, D.K. 2004a: Himalapyrochroa gibbosa, a new genus and species of fire-colored beetle (Coleoptera: Pyrochroidae: Pyrochroinae) from Darjeeling, India. *Oriental Insects* 38: 189–196.

Young, D.K. 2004b: Dendroidopsis and Frontodendroidopsis: two new Asian genera of fire-colored beetles (Coleoptera: Pyrochroidae: Pyrochroinae), with descriptions of two new species of Dendroidopsis, and lectotype designations for Dendroides assamensis Blair and Pseudopyrochroa binhana Pic. *Oriental Insects* 38: 197–217.

Young, D.K. 2004c: Pyrochroa daglariensis: replacement name for Dendroides cyanipennis Pic 1896, primary junior homonym of Dendroides cyanipennis Latreille 1817 (Coleoptera: Pyrochroidae: Pyrochroinae) *Proceedings of the Washington Entomological Society* 106(4): 933–934.

Young, D.K. 2005a: Sinodendroides chinensis, a new genus and species of fire-colored beetle (Coleoptera: Pyrochroidae: Pyrochroinae) from China. *Oriental Insects* 39: 93–100.

Young, D.K. 2005b: A new species of Phyllocladus (Coleoptera: Pyrochroidae: Pyrochroinae) from China, with a key to males of the two known Chinese species. *Oriental Insects* 39: 101–107.

Young, D.K. 2008: Three new Asian species of Ischalia Pascoe, 1860 (Coleoptera: Ischaliidae), with a world checklist of subgenera and species. *The Pan-Pacific Entomologist* 83: 321–331.

Young, D.K. 2013: Two new species of Phyllocladus (Coleoptera: Pyrochroidae: Pyrochroinae) from China, with a key to males of the Chinese species. *Zootaxa* **3669**: 56–64.

Young, D.K. 2015: A new species of Pseudopyrochroa (Insecta: Coleoptera: Pyrochroidae: Pyrochroinae) from Nepal and Thailand with the proposal of a new synonym in Pseudopyrochroa. Pp. 451–455. In: Hartmann, M. & Weipert, J. (eds.): *Biodiversität und Naturausstattung im Himalaya V*. Verein der Freunde und Förderer des Naturkundemuseums Erfurt e.V., 580 pp.

Young, D.K. 2017: A new species of Frontodendroidopsis (Coleoptera: Pyrochroidae: Pyrochroinae) from China, with a key to males of the three species. *Zootaxa* **4319**: 590–594.

Young, D.K. & Pollock, D.A. 2002: 99. Tetratomidae Billberg, 1820. Pp. 413–416. In: Arnett, R.H. Jr, Thomas, M.C., Skelley, P.E. & Frank, J.H. (Eds): *American beetles. Volume 2. Polyphaga: Scarabaeoidea through Curculionoidea*. CRC Press, Boca Raton, xiv, 861pp.

Young, D.K. & Saitô, M. 2017: A New Omineus (Coleoptera, Mycteridae; Eurypinae) from Honshu, Japan. *Elytra, New Series* 7: 417–420.

Yu, Y. & Ren, G.-D. 1997: Five new species of the genus Cteniopinus from China (Tenebrionidae: Alleculinae). *Sichuan Journal of Zoology* **16** (1): 8–12 (in Chinese with English summary).

Yuan, C.-X., Li, P. & Ren, G.-D. 2018: One new species of the genus Uenostrongylium (Coleoptera: Tenebrionidae) from China. *Entomotaxonomia* **40**: 23–26.

Yuan, C.-X. & Ren, G.-D. 2005: Study of the arboreal darkling beetles of the genus Strongylium from China (Coleoptera, Tenebrionidae). *Acta Zootaxonomica Sinica* **30**: 399–406.

Yuan, C.-X. & Ren, G.-D. 2006: A new record genus of Strongyliini (Coleoptera, Tenebrionidae) from China, with two new species and two new names. *Acta Zootaxonomica Sinica* **31**: 851–854.

Yuan, C.-X. & Ren, G.-D. 2007: A new species of the genus Strongylium Kirby (Coleoptera, Tenebrionidae) from China. *Acta Zootaxonomica Sinica* **32**: 550–552.

Yuan, C.-X. & Ren, G.-D. 2014a: Note on brachypterous Stenochiini from China (Coleoptera, Tenebrionidae) with description of a new species. *ZooKeys* **415**: 329–336.

Yuan, C.-X. & Ren, G.-D. 2014b: Two new species of the Stenochinus amplus species-group from China (Coleoptera, Tenebrionidae, Stenochiini). *ZooKeys* **416**: 67–76.

Yuan, C.-X. & Ren, G.-D. 2019 [new taxa]. In: Yuan, C.-X., Wang, W.-Q. & Ren, G.-D.: Taxonomy of Strongylium dorsocupreum Species-Group from China, with Description of a New Species (Coleoptera, Tenebrionidae: Stenochiinae). *Entomological Review* **99**: 1061–1067.

Zaitzev, P.A. 1941: Rasprostranenie vidov sem. uzkotelok v Zakavkaz"e (Coleoptera, Oedemeridae). *Trudy Zoologicheskogo Sektora* **3**: 229–243.

Zerche, L. 2004: Revision der Gattung Aegialites Mannerheim (Coleoptera: Salpingidae: Aegialitinae). *Stuttgarter Beiträge zu Naturkunde Serie A* (*Biologie*) **666**: 1–116 pp.

Zetterstedt, J.W. 1828: *Fauna insectorum Lapponica. Pars 1*. Hammone: Libraria Schulziana, xx + 563 pp.

Zetterstedt, J.W. 1838: Coleoptera. Columns 7–868. In: *Insecta Lapponica descripta. Sectio prima. Coleoptera, Orthoptera et Hemiptera*. Lipsiae: Leopoldi Voss, vi pp. + 1140 columns [issued in parts: 1838–1840].

Zhan, Ch.-L. & Ren, G.-D. 2009: Chinese species of the genus Centorus Mulsant, 1854 (s. str.) (Coleoptera: Tenebrionidae: Belopini) with description of two new species. *Caucasian Entomological Bulletin* **5**: 211–216.

Zhang, Z.-X., Ren, G.-D. & Ba, Y.-B. 2012: Taxonomy of the genus Diplectrus Kirsch, 1866 from China (Coleoptera, Oedemeridae). *Journal of Hebei University (Natural Science Edition)* **32**: 406–409.

Zhao, X.-L. & Ren, G.-D. 2011: A new species of the genus Laena Dejean from Xizang, China (Coleoptera, Tenebrionidae, Lagriinae). *Acta Zootaxonomia Sinica* **36**: 568–570.

Zhao, X.-L. & Ren, G.-D. 2012a: Six new species of the genus Laena Dejean from China (Coleoptera, Tenebrionidae, Lagriinae): *ZooKeys* **177**: 15–36.

Zhao, X.-L. & Ren, G.-D. 2012b: Three new species of the genus Laena Dejean from China (Coleoptera, Tenebrionidae, Lagriinae). *Zootaxa* **3246**: 43–50.

Zhao, Y.-C., Wang, Z.-X., Wang, X.-P. 2019: Two new species of Yunnanomorticola Telnov (Coleoptera, Anthicidae) from China. *ZooKeys* **842**: 153–161.

Zhou, D.-Y, & Song, X.-B. 2019: [new taxon]. In: Song, X.-B., Zhou, D.-Y., Chen, Z.-B., Lin, M.-Z. & Zhu, J.-C. First record of tribe Laenini (Coleoptera, Tenebrionidae, Lagriinae) from Shanghai, with description of a new species of the genus Laena Dejean. *Zootaxa* **4712**: 90–100.

Znojko, D.V. 1936: Ueber die systematische Stellung der Familie Petriidae (Coleoptera). *Travaux de l'Institut Zoologique de l'Académie des Sciences de l'URSS* **1936** (3): 67–77.

Znojko, D.V. 1950: [new taxa]. In: Oglobin, D.A. & Znojko, D.V.: *Fauna SSSR, Zhestkokrylye, Tom 18, vyp. 8. Pylceedy (Cem. Alleculidae), ch. 2, podsem. Omophlinae*. Moskva, Leningrad: Akademiya Nauk SSSR, 133 pp.

Zubkov [= Zoubkoff], B. 1829: Sur un nouveau genre et quelques nouvelles espèces de coléoptères. *Bulletin de la Société Impériale des Naturalistes de Moscou* **1**: 147–170, pls. 4, 5.

Zubkov [= Zoubkoff] B. 1833: Nouveaux coléoptères recueillis en Turcménie. *Bulletin de la Société Impériale des Naturalistes de Moscou* **6**: 310–340.

Zubkov [= Zoubkoff], B. 1837: Description de quelques coléoptères nouveaux. *Bulletin de la Société Impériale des Naturalistes de Moscou* **10** (6): 59–72, pls 3, 4.

Zoufal, V. 1892: Uebersicht der mir bekannten Arten der Coleopteren-Gattung Tenebrio L. *Wiener Entomologische Zeitung* **12**: 283–286.

Zoufal, V. 1893: Revision der Gattungen Centorus und Calcar aus Europa und den angrenzenden Ländern. *Wiener Entomologische Zeitung* **12**: 115–119.

Zürcher, C. 1911: Colpotus Reitteri n. sp. (Col. Tenebr.). *Wiener Entomologische Zeitung* **30**: 292.

Supporting References

Tenebrionidae by P. Bouchard & Y. Bousquet. In: New Nomenclatural and Taxonomic Acts, and Comments

Belopus Gebien, 1911 ***nomen protectum*** over *Hemeralopius* Gistel, 1848 ***nomen oblitum***

Al-Houty, W. 1984: A preliminary check list of Tenebrionidae (Coleoptera) from Kuwait. *Journal of the University of Kuwait (Sciences)* **11**: 261–268.

Al-Houty, W. 1997: Checklist of the insect fauna of Kuwait. *Kuwait Journal of Science & Engineering* **24**: 145–162.

Aliquò, V. & Soldati, F. 2014: Updating the CD-rom on Coleoptera Tenebrionidae of Italy and the check-list of the same family. *Biodiversity Journal* **5**: 429–442.

Allsopp, P.G. 1980: The biology of false wireworms and their adults (soil-inhabiting Tenebrionidae) (Coleoptera): a review. *Bulletin of Entomological Research* **70**: 343–379.

Bujalance, J.L. & Ferreras, M. 1987: Distribución y fenología de los Tenebriónidos (Col.) de la Sierra Subbética (Córdoba). *Miscellània Zoològica* **11**: 155–163.

Chikatunov, V., Lillig, M., Pavlíček T., Blaustein L. & Nevo E. 1997: Biodiversity of insects at a microsite, 'Evolution Canyon', Nahal Oren, Mt. Carmel, Israel. Coleoptera: Tenebrionidae. *Journal of Arid Environments* **37**: 367–377.

Dalmon, J. 1993: Présence de Cyphaeus cornutus (Fisch.) en France méridionale (Coleoptera Tenebrionidae). *Bulletin Mensuel de la Société Linnéenne de Lyon* **62**: 199–204.

Doyen, J.T. 1988: Descriptions of some phylogenetically important larvae of Tenebrionidae (Coleoptera). *The Coleopterists' Bulletin* **42**: 285–301.

Doyen, J.T. & Tschinkel, W.R. 1982: Phenetic and cladistic relationships among tenebrionid beetles (Coleoptera). *Systematic Entomology* 7: 127–183.

Erber, D. 1990: New or little known Coleoptera from Madeira. Results of excursions to Madeira in the years 1986–1990. *Boletim do Museu Municipal do Funchal* **42**: 147–181.

Iwan, D., Kubisz, D. & Tykarski, P. 2012: *Coleoptera Poloniae: Tenebrionoidea (Tenebrionidae, Boridae). Critical checklist, distribution in Poland and meta-analysis.* University of Warsaw – Faculty of Biology, Natura optima dux Foundation, Warszawa, 480 pp.

Juan, C. & Petitpierre, E. 1989: New chromosomal findings on the Spanish Tenebrionidae (Coleoptera). *Caryologia* **42**: 259–266.

Labrique, H. & Chavanon, G. 2002 : Résultats de la deuxième campagne de prospection entomologique dans le Maroc oriental: Tenebrionidae (Coleoptera). *Cahiers Scientifiques – Muséum d'Histoire Naturelle*, Lyon **2**: 65–81.

Leo, P. 1984: Belopus (Centorus) proceroides n. sp. della Sardegna (Coleoptera Tenebrionidae). *Bolletino della Società Entomologica Italiana* **116**: 165–167.

Löbl, I., Merkl, O., Ando, K., Bouchard, P., Lillig, M., Masumoto, K. & Schawaller, W. 2008: *Tenebrionidae. Catalogue of Palaearctic Coleoptera (Löbl I., Smetana A. eds.). Vol. 5. Tenebrionoidea.* Stenstrup: Apollo Books. 670 p.

Mordkovich, V.G. 2006: Darkling beetles (Coleoptera: Tenebrionidae) in the steppe catenae [in Russian]. *Proceedings of the Russian Entomological Society* 77: 235–240.

Mordkovich, V.G. 2012: How many forest and steppe ground and darkling beetles (Coleoptera: Carabidae, Tenebrionidae) are there in West Siberian faunal assemblages? *Russian Entomological Journal* **21**: 169–178.

Pavlícek, T., Chikatunov, V. & Nevo, E. 2005: Arthropods in the mounds of mole rats, Spalax ehrenbergi superspecies, in Israel. *Ecologia Mediterranea* **31**: 5–13.

Ren, G.-D. & Yu, Y.-Z. 1994: A new species of Belopus Gebien from China (Coleoptera, Tenebrionidae). *Acta Zootaxonomica Sinica* **19**(33): 351–353.

Ren, G.-D. & Zeng, Z.-M. 1993: One new species of genus Belopus Gb. from Inner Mongolia (Coleoptera: Tenebrionidae: Belopini) [in Chinese with English summary]. *Journal Inner Mongolia Normal University (Natural Science Edition)* **4**: 52–54.

Romero, R.O. 2015: La influencia del manejo del suelo en las comunidades de coleópteros edáficos (Coleoptera) en la ZEPA-LIC La Serena y Sierras Periféricas (Badajoz, España). *Boletín de la Sociedad Entomológica Aragonesa (S.E.A.)* **56**: 291–299.

Skopin, N.G. 1974: Zur Revision der Eurasiatischen Arten der Gattung Belopus Gb. *Entomologische Abhandlungen Staaliches Museum für Tierkunde in Dresden* **40**(2): 65–103.

Soldati, F. 1995: Les Tenebrionidae de l'Hérault: inventaire et écologie (Coleoptera). *Bulletin de la Société Entomologique de France* **100**: 261–275.

Soldati, F. & Soldati, L. 2003: Une espèce nouvelle de Phtora appartenant à un sous-genre inédit (Coleoptera, Tenebrionidae, Phaleriini) (33. Contribution à la connaissance des Tenebrionidae). *Bulletin de la Société Linnéenne de Bordeaux* **31**: 1–8.

Tschinkel, W.R. & Doyen, J.T. 1980: Comparative anatomy of the defensive glands, ovipositors and female genital tubes of tenebrionid beetles (Coleoptera). International *Journal of Insect Morphology and Embryology* **9**: 321–368.

Watt, J.C. 1974: A revised subfamily classification of Tenebrionidae (Coleoptera). *New Zealand Journal of Zoology* **1**: 381–452.

Zinovyev, E. 2011: Sub-fossil beetle assemblages associated with the 'mammoth fauna' in the Late Pleistocene localities of the Ural Mountains and West Siberia. *ZooKeys* **100**: 149–169.

Praeugena Laporte, 1840 ***nomen protectum*** over *Adelphus* Dejean, 1834 ***nomen oblitum***

Abdel-Dayem, M.S., Fad, H.H., El-Torkey, A.M., Elgharbawy, A.A., Aldryhim, Y.N., Kondratieff, B.C., Al Ansi, A.N. & Aldhafer, H.M. 2017: The beetle fauna (Insecta, Coleoptera) of the Rawdhat Khorim National Park, Central Saudi Arabia. *ZooKeys* **653**: 1–78.

Bouchard, P., Bousquet, Y., Davies, A.E., Alonso-Zarazaga, M.A., Lawrence, J.F., Lyal, C.H.C., Newton, A.F., Reid, C.A.M., Schmitt, M., Ślipiński, S.A. & Smith, A.B.T. 2011: Family-group names in Coleoptera (Insecta). *ZooKeys* **88**: 1–972.

Bouchard, P., Lawrence, J.F., Davies, A.E. & Newton, A.F. 2005: Synoptic classification of the world Tenebrionidae (Insecta: Coleoptera) with a review of family-group names. *Annales Zoologici* **55**: 499–530.

Datinon, B.D., Glitho, A.I., Tamò, M., Amevoin, K., Goergen, G. & Douro Kpindou, O.K. 2013: Inventory of major insects of *Jatropha curcas* L. (Euphorbiaceae) and their natural enemies in southern Benin. *ARPN Journal of Agricultural and Biological Science* **8**: 711–718.

Doyen, T.J. & Tschinkel, W.R. 1982: Phenetic and cladistic relationship of Tenebrionid beetles (Coleoptera). *Systematic Entomology* **7**: 127–183.

Ferrer, J. 2001: Contribución al conocimiento de los tenebriònidos de Tanzania y descripción de una nueva especie de Praeugena Mäklin (Insecta, Coleoptera). *Rivista Piemontese di Storia Naturale* **22**: 225–237.

Ferrer, J. 2004: Tenebrionidae (Col.) de Namibia avec descriptions de 12 espèces nouvelles. *Mitteilungen der Museum fürr Naturkunde Berlin, Zoolochische Reiche* **80**: 81–250.

Ferrer, J. 2006: Tenebrionidos nuevos o interesantes de Mozambique, colectados por Gianfranco Curletti (1996, 1999) (Coleoptera, Tenebrionidae). *Rivista Piemontese di Storia Naturale* **27**: 275–308.

Ferrer, J. 2013: Contribution to the knowledge of the genus Paoligena Pic 1927, with systematic comments, description of two new species and the creation of a new tribe Paoligenini tribus nova (Coleoptera, Tenebrionidae, Tenebrioninae). *Boletín de la Sociedad Entomológica Aragonesa (S.E.A.)* **52**: 105–112.

Ferrer, J. 2016: Pioneros de la Entomología Gustaf de Vylder, naturalista sueco, explorador del sur de África. *Boletín de la Sociedad Entomológica Aragonesa (S.E.A.)* **58**: 253–257.

Ferrer, J., Sakalian, V. & Georgiev, G. 2016: Darkling and ironclad beetles (Coleoptera: Tenebrionoidea: Tenebrionidae and Zopheridae) from Kenya, with descriptions of two new species. *Acta Zoologica Bulgarica* **68**: 159–170.

Gillon, Y. & Gillon, D. 1974: Recherches écologiques sur une savane sahéienne du Ferlo septentrional, Sénégal: données quantitatives sur les ténébrionides. *La Terre et la Vie* **28**: 296–306.

Grimm, R. 2002: Tenebrionidae von Gambia (Insecta, Coleoptera). *Entomofauna* **23**: 353–380.

Kaszab, Z. 1982: Insects of Saudi Arabia. Coleoptera: Fam. Tenebrionidae (Part 2). *Fauna of Saudi Arabia* **4**: 124–243.

Kenya, A.R. & Njoroge, L.N. 2006: An Entomological survey of Sabaki River Mouth, Kenya: A Conservation Research Report. *A Rocha Kenya Occasional Research Report* 10, 24 pp.

Kouassi, P., Yapi, A., Tano, Y. & Akpesse, A. 2005: Diversité et écologie des macro-invertébrés du sol (coléoptères, diplopodes) dans les écosystemes de Côte-d'Ivoire. *BIOTERRE, Revue Internationale des Sciences de la Vie et de la Terre* **5**: 111–138.

Löbl, I., Merkl, O., Ando, K., Bouchard, P., Lillig, M., Masumoto, K. & Schawaller, W. 2008: *Tenebrionidae. Catalogue of Palaearctic Coleoptera (Löbl I., Smetana A. eds.). Vol. 5. Tenebrionoidea.* Stenstrup: Apollo Book, 670 pp.

MacFadyen, D.N. & Compion, S. 2009: *A landscape of insects and other invertebrates.* Jacana Books, Johannesburg, 278 pp.

Merkl, O., Grabant, A. & Soltész, Z. 2015: *A Magyar Természettudományi Múzeum gyászbogártípusainak (Tenebrionidae) katalógusa. (Type catalogue of darkling beetles (Tenebrionidae) preserved in the Hungarian Natural History Museum).* Magyar Természettudományi Múzeum, Budapest, 735 pp.

Monks, J., Ross, S., Geiser, M., De Prins, J., Sharaf, M., Wyatt, N., Al, Rijeibi S. & Polaszek, A. 2019: A preliminary survey of the insect fauna of the Hajar Mountain Range, Oman. *Journal of Natural History* **53**: 939–963.

Moor, P.P. de 1970: Monograph of the Praeugenina (Coleoptera, Tenebrionidae, Strongyliini). *Transvaal Museum Memoir* **17** (1), 177 pp.

Penny, D. 2009: *Field guide to wildlife of The Gambia: An introduction to common flowers and animals.* Siri Scientific Press, Manchester.

Robiche, G. 2001: Description d'un nouveau genre de Tenebrionidae et de son espèce récolté en Afrique du Sud. *Lambillionea* **101**: 191–195.

Robiche, G., Le Gall, P. & Goergen, G. 2002: Contribution à l'étude de la biodiversité des Coléoptères Tenebrionidae de la République du Bénin : premier inventaire. *Lambillionea* **102** (4 suppl.): 381–431.

Schawaller, W. 2010: Check-list of the Tenebrionidae (sensu stricto) from Madagascar (Coleoptera). *Stuttgarter Beiträge zur Naturkunde A (Neue Serie)* **3**: 277–289.

Derosphaerus J. Thomson, 1858 ***nomen protectum*** over *Euphron* Dejean, 1834 ***nomen oblitum***.

Ando, K. 2003: A systematic revision of the genus Eucyrtus and its complex (Coleoptera, Tenebrionidae). *Japanese Journal of Systematic Entomology Monographic Series* **1**: i–xi, 1–604.

Bouchard, P. 2002: Phylogenetic revision of the flightless Australian genus Apterotheca Gebien (Coleoptera: Tenebrionidae: Coelometopinae). *Invertebrate Systematics* **16**: 449–554.

Bouchard, P. & Steiner, W.E. Jr. 2004: First descriptions of Coelometopini pupae (Coleoptera Tenebrionidae) from Australia, Southeast Asia and the Pacific Region with comments on phylogenetic relationships and anti-predator adaptations. *Systematic Entomology* **29**: 101–114.

Bouchard, P. & Yeates, D.K. 2001: Phylogenetic relationships of the Australasian Coelometopini (Coleoptera: Tenebrionidae); a quantitative cladistic analysis with a review of biology. *Organisms, Diversity and Evolution* **1**: 17–43.

Ferrer, J. 1995: Contribution to the knowledge of the Tenebrionidae of Somalia (Coleoptera). *Frustula Entomologica* **18**: 1–76.

Ferrer, J. 1996: Contributions to the knowledge of the Tenebrionidae of Kenya (Insecta: Coleoptera). *Frustula Entomologica* **19**: 51–109.

Grimm, R. 2014: New and little known species of Tenebrionidae (Coleoptera) from Borneo (4). *Stuttgarter Beiträge zur Naturkunde A, Neue Serie* **7**: 183–197.

Grimm, R. 2015: New and little known species of Tenebrionidae (Coleoptera) from Borneo (5). *Stuttgarter Beiträge zur Naturkunde A, Neue Serie* **8**: 215–225.

Hegde, V.D. 2019: Additions to the knowledge of darkling beetles (Coleoptera: Tenebrionidae) from the Indo-Burma Biodiversity Hotspot, Meghalaya, India. *Journal of Threateneed Taxa* **11**: 15074–15078.

Hegde, V.D. & Vasanthakumar, D. 2018a: Updated check-list of darkling beetles (Coleoptera: Tenebrionidae) of West Bengal, India. *Records of the Zoological Survey of India* **118**: 281–286.

Hegde, V.D. & Vasanthakumar, D. 2018b: New records of darkling beetles (Tenebrionidae: Coleoptera) from Maharashtra, India. *Journal of Entomology and Zoology Studies* **6**: 84–86.

Hisamatsu, S. 1970: Derosphaerus foveolatus (Marseul) captured at Nagasaki (Coleoptera: Tenebrionidae). *Ageha* **4**: 20.

Kaszab, Z. 1987: Die papuanisch-australischen Arten der Gattung Derosphaerus Thomson, 1858 (Coleoptera, Tenebrionidae). *Acta Zoologica Academiae Scientiarum Hungaricae* **33**: 41–85.

Löbl, I., Merkl, O., Ando, K., Bouchard, P., Lillig, M., Masumoto, K. & Schawaller, W. 2008: *Tenebrionidae. Catalogue of Palaearctic Coleoptera (Löbl I., Smetana A. eds.). Vol. 5. Tenebrionoidea*. Stenstrup: Apollo Books. 670 p.

Masumoto, K. & Makihara, H. 1997: Study on the Tenebrionid Beetles in South Sumatra. *Bulletin of the Forestry and Forest Products Research Institute Ibaraki* **374**: 115–153.

Matthews, E.G. 2019. 36. Tenebrionidae Latreille, 1802: In: Ślipiński SA, Lawrence JF (Eds) *Australian Beetles. Volume 2. Archostemata, Myxophaga, Adephaga, Polyphaga (part)*. Pp. 582–652.

Matthews, E.G. & Bouchard, P. 2008: *Tenebrionid Beetles of Australia: Descriptions of tribes, keys to genera, catalogue of species*. Australian Biological Resources Study, Canberra. viii + 398 pages.

Merkl, O. 1992: Tenebrionidae (Coleoptera) from Laos and Vietnam, with reclassification of Old World "Doliema". *Acta Zoologica Hungarica* **38**: 261–280.

Poussereau, J. & Ferrer, J. 2007: Quelques ténébrionides nouveaux pour les îles de la Réunion et des Comores (Coleoptera). *Bulletin de la Société Entomologique de France* **112**: 89–91.

Robiche, G. 2005: Description d'une nouvelle espèce de la région afrotropicale appartenant au genre *Derosphaerus* Thomson, 1858 (Coleoptera, Tenebrionidae). *Entomologia Africana* **10**: 31–34.

Robiche, G. 2010. Nouvelles espèces du Mozambique appartenant au genre Derosphaerus Thomson, 1858 (Coleoptera, Tenebrionidae). *Bulletin de la Société Entomologique de France* **115**: 501–504.

Schawaller, W. 2005: Three new tenebrionids (Coleoptera: Tenebrionidae) from Sumatra - New species or new genera? *Annales Zoologici* **55**: 565–569.

Schawaller, W. 2006: Revision of the Oriental species of the genus Bradymerus Perroud, with descriptions of 29 new species (Coleoptera: Tenebrionidae). *Stuttgarter Beiträge zur Naturkunde A* (*Biologie*) **694**: 1–64.

Schawaller, W. 2007: The genus Derosphaerus Thomson, 1858 in the Nepal Himalayas (Coleoptera, Tenebrionidae). *Entomologica Basiliensia* **29**: 171–175.

Schawaller, W. 2011: The genus Derosphaerus Thomson (Coleoptera: Tenebrionidae: Cnodalonini) in Borneo, with description of a new species. *Stuttgarter Beiträge zur Naturkunde A, Neue Serie* **4**: 289–296.

Index to Family-group Names

Index to Genus-group Names

genus-group synonyms are in *Italics*